Phosphor Handl

Phosphor Handbook

Novel Phosphor, Synthesis, and Applications

Edited by

Ru-Shi Liu
Xiao-Jun Wang

Third Edition

CRC Press
Taylor & Francis Group
Boca Raton London New York

CRC Press is an imprint of the
Taylor & Francis Group, an **informa** business

Third edition published 2022
by CRC Press
6000 Broken Sound Parkway NW, Suite 300, Boca Raton, FL 33487-2742

and by CRC Press
2 Park Square, Milton Park, Abingdon, Oxon, OX14 4RN

© 2022 Taylor & Francis Group, LLC

First edition published by CRC Press 1998

CRC Press is an imprint of Taylor & Francis Group, LLC

ISBN: 978-0-367-55514-6 (hbk)
ISBN: 978-1-032-15968-3 (pbk)
ISBN: 978-1-003-09867-6 (ebk)

DOI: 10.1201/9781003098676

Typeset in Times
by KnowledgeWorks Global Ltd.

In Memoriam the Early Editors of the Handbook

Shigeo Shionoya
Formerly of The
   University of Tokyo
Tokyo, Japan

Hajime Yamamoto
Formerly of Tokyo
   University of Technology
Tokyo, Japan

William M. Yen
Formerly of The
   University of Georgia
Athens, GA, USA

Contents

Foreword to the Third Edition of the *Phosphor Handbook*

The field of luminescence and phosphors has a long history, starting from early observations of light in the dark from afterglow materials. Centuries of extensive research followed aimed at providing insight into optical phenomena, now resulting in an increasing role of phosphors in our daily lives. Applications of luminescence grow more diverse and include, for example, phosphors in the color displays that our eyes seem to be glued to, energy-efficient LED lighting, data communication, luminescent probes in medical imaging and sensing, gadgets relying on afterglow phosphors and even luminescent lanthanides in our banknotes. It is interesting to note the central role that Asia has played in the discovery and development of new luminescent materials. Early applications involved afterglow paints in China, creating alternative images in the dark. While fundamental luminescence research was carried out in the 20th century at all continents, there has been a remarkably strong role of Japan and China in research, development and discovery of new luminescence processes and phosphors. It is, therefore, not surprising that the first edition of the *Phosphor Handbook* (*Keikotai Handobukku*) was initiated by the Phosphor Research Society in Japan in the 1980s.

The first *Phosphor Handbook* was a great book but with an impact limited to those speaking Japanese. Fortunately, about ten years later, the book was translated into English and edited by two giants in the field of luminescence: Shigeo Shionoya and William Yen. It is this version of the book that I acquired soon after it was released, and it has been a source of information ever since. All aspects of luminescent materials were covered: phosphor synthesis, optical measuring techniques, fundamentals of luminescence processes, operation principles of light emitting devices, light and color perception and of course an almost complete overview of all luminescent materials known, indexed by host material and activator ion. I cannot count how often I consulted this book, to quickly look up the optical properties of an ion-host combination, find a suitable material with specific luminescence characteristics, understand the operation principles of phosphors in various applications and learn about careful measurements and analysis of phosphor properties. The authors, except for one, were Japanese, underpinning the central role of Japan in phosphor research.

As the field of luminescence continued to evolve and expand, it became clear that a second edition of the *Phosphor Handbook* was needed. Sadly Shigeo Shionoya has passed away, and in 2006, William Yen together with Hajime Yamamoto edited the second edition of the *Phosphor Handbook*. The new edition was updated mostly by asking the original authors to adapt the various chapters to include recent developments. The *Phosphor Handbook* continued to play a prominent role in the luminescence community as a source of information on any topic related to phosphors. Almost 15 years later, it was again time to adapt the Handbook to cover important new developments in the rapidly changing phosphor field where new applications and new materials emerge, and also measuring techniques have changed with the introduction of for example cheap (pulsed) diode lasers, fiber optics and compact CCD-based spectrometers. Our great colleagues William Yen and Hajime Yamamoto are unfortunately no longer with us. and also many of the authors of the various chapters of the first and second edition of the *Phosphor Handbook* have passed away. This made it far from trivial to realize a third edition. We can be extremely grateful that Ru-Shi Liu and Xiao-Jun Wang have taken the initiative to edit and write this third edition of the *Phosphor Handbook*. It is very appropriate that the book is dedicated to the three founders, Shigeo Shionoya, William Yen and Hajime Yamamoto. At the same time, it is appropriate to sincerely thank Ru-Shi Liu and Xiao-Jun Wang for their strong commitment and time invested to organize, write and edit this third edition.

The third edition of the *Phosphor Handbook* is in some aspects different from the two previous editions. The authors are not the same, and it is wonderful to see that so many highly respected colleagues in the field have taken the time to contribute their expertise and knowledge to this third edition. Interestingly, again almost all of the authors of this third edition are Asian (with well over 100 contributing authors, you can count the non-Asian authors on the fingers of one hand). This illustrates the continued strong position of Asia in phosphor research. Just as in the previous editions, all aspects and the broad scope of phosphor research are covered, which makes this Handbook a worthy successor of the previous editions. It will serve as a comprehensive resource describing a wide variety of topics that were also included in the previous editions. It will educate newcomers and help everyone in the field to quickly access all relevant knowledge in the exciting field of phosphor research. In addition to the "classic" topics that continue to be relevant (but sometimes forgotten), many new topics are included, in Theory (e.g., first principle calculations), Materials (e.g., recent developments in quantum dots and upconversion nanocrystals) and Applications (e.g., LED phosphors for NIR sensing and agriculture). All this information no longer fits in a single volume and this third edition is, therefore, divided into three volumes.

At the time of writing this foreword, I have not read the new edition of the Handbook but did receive an overview of all the chapters and contributing authors. Based on this information, it is clear that the full phosphor community, from students to professors, can benefit from this new comprehensive source of everything you always wanted to know about phosphors – and more. The third edition of the *Phosphor Handbook* will be a classic and continue to promote progress and development of phosphors, in the spirit of the first edition. I look forward to reading it and hope that you as a reader will enjoy exploring this great book and be inspired by it in your research on luminescent materials.

Andries Meijerink
Utrecht, June 2021

Preface to the Third Edition

The last version of the *Phosphor Handbook* was well received by the phosphor research community since its publication in 2007. However, in last 14 years, many notable advances have occurred. The success of the blue LED (Nobel Prize in Physics, 2014) and its phosphor-converted solid illumination greatly advanced the traditional phosphor research. New phosphorescent materials such as quantum dots, nanoparticles and efficient upconversion, quantum cutting phosphors and infrared broadband emission phosphors have been quickly developed to find themselves in ever-broader applications, from phototherapy to bioimaging, optics in agriculture to solar cell coating. These applications have all expanded beyond the traditional use in lighting and display. All of these developments should be included in the popular Handbook, making it necessary to publish a new version that reflects the most recent developments in phosphor research. Unfortunately, all the three well-respected editors of the previous version have passed away. As their former students and colleagues, we, the editors, feel a strong sense of responsibility to carry on the legacy of the Handbook and to update accordingly to continue serving the phosphor community. The aim of the third edition of the Handbook is to continue to provide an initial and comprehensive source of knowledge for researchers interested in synthesis, characterization, properties and applications of phosphor materials.

The third edition of the Handbook consists of three separate volumes. Volume 1 covers the theoretical background and fundamental properties of luminescence as applied to solid-state phosphor materials. New sections include the rapid developments in principal phosphors in nitrides, perovskite and silicon carbide. Volume 2 provides the descriptions of synthesis and optical properties of phosphors used in different applications, including the novel phosphors for some newly developed applications. New sections include, Chapters 5 – Smart Phosphors, 6 – Quantum Dots for Display Applications, 7 – Colloidal Quantum Dots and Their Applications, 8 – Lanthanide-Doped Upconversion Nanoparticles for Super-Resolution Imaging, 9 – Upconversion Nanophosphors for Photonic Application, 16 – Single-Crystal Phosphors, 19 – Phosphors-Converting LEDs for Agriculture, 20 – AC-Driven LED Phosphors and 21 – Phosphors for Solar Cells. Volume 3 addresses the experimental methods for phosphor evaluation and characterization, and the contents are widely expanded from the second edition, including the theoretical and experimental designs for new phosphors as well as the phosphor analysis through high pressure and synchrotron studies. Almost all the chapters in the third edition, except for some sections in the Fundamentals of Luminescence, have been prepared by the new faces who are actively and productively working in phosphor research and applications.

We commemorate the memory of the three mentors and editors of the previous editions – Professors Shigeo Shionoya, Hajime Yamamoto and William M. Yen. It was their efforts that completed the original Handbook that guided and inspired numerous graduate students and researchers in phosphor studies and applications. We wish to dedicate this new edition to them.

As the editors, we sincerely appreciate all the contributors from across the world who overcame various difficulties through such an unprecedented pandemic year to finish their chapters on time.

We are grateful to Professor Andries Meijerink of Utrecht University for writing the foreword to the Handbook. We also highly appreciate the help from Nora Konopka, Prachi Mishra, and Jennifer Stair of CRC Press/Taylor & Francis Group and perfect editing work done by Garima Poddar of

KGL. Finally, we hope that this third edition continues the legacy of the Handbook to serve as a robust reference for current and future researchers in this field.

Co-editors:
Ru-Shi Liu
Taipei, Taiwan

Xiao-Jun Wang
Statesboro, GA, USA
May, 2021

Preface to the Second Edition

We, the editors as well as the contributors, have been gratefully pleased by the reception accorded to the *Phosphor Handbook* by the technical community since its publication in 1998. This has resulted in the decision to reissue an updated version of the Handbook. As we had predicted, the development and the deployment of phosphor materials in an ever increasing range of applications in lighting and display have continued its explosive growth in the past decade. It is our hope that an updated version of the Handbook will continue to serve as the initial and preferred reference source for all those interested in the properties and applications of phosphor materials.

For this new edition, we have asked all the authors we could contact to provide corrections and updates to their original contributions. The majority of them responded, and their revisions have been properly incorporated in the present volume. It is fortunate that the great majority of the material appearing in the first edition, particularly those sections summarizing the fundamentals of luminescence and describing the principal classes of lightemitting solids, maintains its currency and, hence, its utility as a reference source.

Several notable advances have occurred in the past decade, which necessitated their inclusion in the second edition. For example, the wide dissemination of nitride-based LEDs opens the possibility of white light solid-state lighting sources that have economic advantages. New phosphors showing the property of "quantum cutting" have been intensively investigated in the past decade and the properties of nanophosphors have also attracted considerable attention. We have made an effort, in this new edition, to incorporate tutorial reviews in all of these emerging areas of phosphor development.

As noted in the preface of the first edition, the Handbook traces its origin to one first compiled by the Phosphor Research Society (Japan). The society membership supported the idea of translating the contents and provided considerable assistance in bringing the first edition to fruition. We continue to enjoy the cooperation of the Phosphor Research Society and value the advice and counsel of the membership in seeking improvements in this second edition.

We have been, however, permanently saddened by the demise of one of the principals of the society and the driving force behind the Handbook itself. Professor Shigeo Shionoya was a teacher, a mentor and a valued colleague who will be sorely missed. We wish to dedicate this edition to his memory as a small and inadequate expression of our joint appreciation.

We also wish to express our thanks and appreciation of the editorial work carried out flawlessly by Helena Redshaw of Taylor & Francis.

William M. Yen
Athens, GA, USA

Hajime Yamamoto
Tokyo, Japan
December, 2006

Preface to the First Edition

This volume is the English version of a revised edition of the *Phosphor Handbook* (*Keikotai Handobukku*) that was first published in Japanese in December, 1987. The original Handbook was organized and edited under the auspices of the Phosphor Research Society (in Japan) and issued to celebrate the 200th Scientific Meeting of the Society which occurred in April, 1984.

The Phosphor Research Society is an organization of scientists and engineers engaged in the research and development of phosphors in Japan which was established in 1941. For more than half a century, the Society has promoted interaction between those interested in phosphor research and has served as a forum for discussion of the most recent developments. The Society sponsors five annual meetings; in each meeting, four or five papers are presented, reflecting new cutting-edge developments in phosphor research in Japan and elsewhere. A technical digest with extended abstracts of the presentations is distributed during these meetings and serve as a record of the proceedings of these meetings.

This Handbook is designed to serve as a general reference for all those who might have an interest in the properties and/or applications of phosphors. This volume begins with a concise summary of the fundamentals of luminescence and then summarizes the principal classes of phosphors and their light emitting properties. Detailed descriptions of the procedures for synthesis and manufacture of practical phosphors appear in later chapters and in the manner in which these materials are used in technical applications. The majority of the authors of the various chapters are important members of the Phosphor Research Society, and they have all made significant contributions to the advancement of the phosphor field. Many of the contributors have played central roles in the evolution and remarkable development of lighting and display industries of Japan. The contributors to the original Japanese version of the Handbook have provided English translations of their articles; in addition, they have all updated their contributions by including the newest developments in their respective fields. A number of new sections have been added in this volume to reflect the most recent advances in phosphor technology.

As we approach the new millennium and the dawning of a radical new era of display and information exchange, we believe that the need for more efficient and targeted phosphors will continue to increase and that these materials will continue to play a central role in technological developments. We, the co-editors, are pleased to have engaged in this effort. It is our earnest hope that this Handbook becomes a useful tool to all scientists and engineers engaged in research in phosphors and related fields and that the community will use this volume as a daily and routine reference, so that the aims of the Phosphor Research Society in promoting progress and development in phosphors are fully attained.

Co-Editors:
Shigeo Shionoya
Tokyo, Japan

William M. Yen
Athens, GA, USA
May, 1998

About the Authors

Ru-Shi Liu

Professor Ru-Shi Liu received his bachelor's degree in chemistry from Soochow University (Taiwan) in 1981. He got his master's degree in nuclear science from the National Tsing Hua University (Taiwan) in 1983. He obtained two PhD degrees in chemistry from National Tsing Hua University in 1990 and from the University of Cambridge in 1992. He joined Materials Research Laboratories at Industrial Technology Research Institute as an Associate Researcher, Research Scientist, Senior Research Scientist and Research Manager from 1983 to 1995. Then he became an Associate Professor at the Department of Chemistry of the National Taiwan University from 1995 to 1999. Then he was promoted to a Professor in 1999. In July 2016, he became the Distinguished Professor.

He got the Excellent Young Person Prize in 1989, Excellent Inventor Award (Argentine Medal) in 1995 and Excellent Young Chemist Award in 1998. He got the 9th Y. Z. Hsu scientific paper award due to the excellent energy-saving research in 2011. He received the Ministry of Science and Technology awards for distinguished research in 2013 and 2018. In 2015, he received the distinguished award for Novel and Synthesis by IUPAC and NMS. In 2017, he got the Chung-Shang Academic paper award. He got "Highly Cited Researchers" by Clarivate Analytics in 2018 and 2019. He got Hou Chin-Tui Award in 2018 due to the excellent research on basic science. He got the 17th Y. Z. Hsu Chair Professor award for the contribution to the excellent research on "Green Science & Technology" in 2019. He then got the 26th TECO award for the contribution to make the combination of the academic and practical application of materials chemistry in 2019. He got the Academic Award of the Ministry of Education and the Academic Achievement Award of the Chemical Society Located in Taipei in 2020.

His research is concerning with materials chemistry. He is the author and co-author of more than 600 publications in international scientific journals. He has also granted more than 200 patents.

Xiao-Jun Wang

Professor Xiao-Jun Wang obtained his BS degree in physics from the Jilin University in 1982, his MS degrees in physics from the Chinese Academy of Sciences in 1985 and the Florida Institute of Technology in 1987 and his PhD in physics from The University of Georgia in 1992 (supervisors: William M. Yen and William M. Dennis). He served as a Research Associate from 1992 to 1993 at the University Laser Center of Oklahoma State University and then received a fellowship from the National Institutes of Health (NIH) as a Postdoctoral Fellow from 1993 to 1995 at the Beckman Laser Institute of the University of California, Irvine. In 1995, he received an NIH training grant and joined Georgia Southern University as Assistant Professor. He was promoted to Full Professor in 2004 and continues to teach there today.

List of Contributors

Sheng Cao
Guangxi University
Nanning, China

Yaxin Cao
Lanzhou University
Lanzhou, China

Ho Chang
National Taipei University of Technology
Taipei, Taiwan

Baojiu Chen
Dalian Maritime University
Dalian, China

Chaohao Chen
University of Technology Sydney
Sydney, Australia

Weibin Chen
South China Agricultural University
Guangzhou, China

Shuo Ding
Ningbo Institute of Materials
    Technology and Engineering, CAS
Ningbo, China

Zuoling Fu
Jilin University
Changchun, China

Jianhua Hao
The Hong Kong Polytechnic University
Hong Kong, China

Lin Huang
Sun Yat-sen University
Guangzhou, China

Mochen Jia
Jilin University
Changchun, China

Dayong Jin
University of Technology Sydney
Sydney, Australia

Bingfu Lei
South China Agricultural University
Guangzhou, China

Jiayan Liao
University of Technology Sydney
Sydney, Australia

Hang Lin
Fujian Institute of Research on
    the Structure of Matter, CAS
Fuzhou, China

Jun Lin
Changchun Institute of Applied
    Chemistry, CAS
Changchun, China

Bo-Mei Liu
Sun Yat-sen University
Guangzhou, China

Feng Liu
Northeast Normal University
Changchun, China

Ru-Shi Liu
National Taiwan University
Taipei, Taiwan

Yuxue Liu
Northeast Normal University
Changchun, China

Lei Qian
Ningbo Institute of Materials
    Technology and Engineering, CAS
Ningbo, China

Veeramani Rajendran
National Taipei University
    of Technology
Taipei, Taiwan

Hongwei Song
Jilin University
Changchun, China

Nathan Evan Stott
Ningbo Institute of Materials
    Technology and Engineering, CAS
Ningbo, China

Zhaobing Tang
Ningbo Institute of Materials
    Technology and Engineering, CAS
Ningbo, China

Feng Wang
City University of Hong Kong
Hong Kong, China

Jing Wang
Sun Yat-sen University
Guangzhou, China

Mingwei Wang
Northeast Normal University
Changchun, China

Xiao-Jun Wang
Georgia Southern University
Statesboro, GA, USA

Yuansheng Wang
Fujian Institute of Research on the
    Structure of Matter, CAS
Fuzhou, China

Yuhua Wang
Lanzhou University
Lanzhou, China

Zhengliang Wang
Yunnan Minzu University
Kunming, China

Zhiying Wang
Jilin University
Changchun, China

Man-Chung Wong
The Hong Kong Polytechnic University
Hong Kong, China

Wai-Yeung Wong
The Hong Kong Polytechnic University
Hong Kong, China

Chunyan Wu
Ningbo Institute of Materials
    Technology and Engineering, CAS
Ningbo, China

Chaoyu Xiang
Ningbo Institute of Materials
    Technology and Engineering, CAS
Ningbo, China

Rong-Jun Xie
Xiamen University
Xiamen, China

Jiating Xu
Northeast Forestry University
Harbin, China

Tong-Tong Xuan
Xiamen University
Xiamen, China

Shihai You
Xiamen University
Xiamen, China

Dechao Yu
South China University
    of Technology
Guangzhou, China

Jialong Zhao
Guangxi University
Nanning, China

Hongyang Zhang
The Hong Kong Polytechnic University
Hong Kong, China

Qinyuan Zhang
South China University
    of Technology
Guangzhou, China

Xizhen Zhang
Dalian Maritime University
Dalian, China

Xuanyu Zhang
Ningbo Institute of Materials
    Technology and Engineering, CAS
Ningbo, China

Yixi Zhuang
Xiamen University
Xiamen, China

Yang Zhang
Nankai University
Tianjin, China

Donglei Zhou
Jilin University
Changchun, China

Qi Zhu
City University of Hong Kong
Hong Kong, China

1 Phosphor-Converting LED for Lighting

Bo-Mei Liu, Lin Huang, Tong-Tong Xuan, and Jing Wang

CONTENTS

1.1 INTRODUCTION

Inorganic light-emitting diode, well-known as LED, is the semiconductor device containing a PN junction that can generate single-color electroluminescence via the migration of electrons and subsequent recombination with holes under certain current. Since the first feasible red-light-emitting LEDs based on GaAsP were reported in the 1960s,[1] significant developments of LED technology had been gradually achieved, such as rich and tunable color emission, enhanced light output, and low cost. Especially around the 1990s, a series of high-efficiency ultraviolet (UV)/blue LEDs based on GaN and InGaN were successfully fabricated by Akasaki, Amano and Nakamura, et al.[2–4] The successful manufacture of efficient LEDs with high photon energies (>2.5 eV) makes it possible to obtain white light for general solid-state lighting source, which leads to the 2014 Nobel Prize in Physics.

Compared to traditional light sources, for instance, fluorescent lamps and tungsten halogen lamps, the white LEDs (WLEDs) have superior properties, such as high brightness, low energy consumption, high reliability, and long working lifetime. Therefore, they have played considerable roles in our daily life, from general illumination to display back-light sources for televisions and smartphones, etc.

Nowadays, the realization of commercially available WLEDs is combining UV/blue LED chips and luminescence materials (which are usually so-called phosphors that can convert high-energy photons into low-energy photons). These kinds of WLEDs are called phosphors-converting WLEDs, i.e., pc-WLEDs. The available options are "UV LED + red/green/blue (R/G/B) tri-phosphors" and "blue LED + yellow phosphor or blue LED + green/red phosphors". In typical pc-WLEDs, part

DOI: 10.1201/9781003098676-1

of UV or blue lights from LED chips is firstly converted into visible light by phosphors. Then the remaining lights from LED chips and visible lights converted by phosphors mix together into an ideal white light. Obviously, the white-light properties (luminous efficiency, color coordinate, color-rendering index, correlated color temperature [CCT], color gamut, etc.) of pc-WLEDs are mainly determined by the properties of LED chips and phosphors, respectively.

In principle, phosphors for pc-WLEDs should meet the following criteria:

1. Phosphors should have broadband excitation spectra that should match well with the emission spectra of UV or blue LED chips and give broadband visible emission.
2. Phosphors should have high photoluminescence efficiency, i.e., high photoluminescence quantum yield (PLQY).
3. Phosphors should have excellent chemical and thermal stability at working temperatures (up to approximately 150–200°C).
4. Phosphors should be nonhazardous or nontoxic and have a suitable particle-size distribution (0.5–3 um).
5. Synthetic processes of phosphors should be environment-friendly and simple and the raw materials for synthesizing phosphors should be easily obtained and cheap.

Hereafter, we will summarize the developments of phosphors and the related pc-WLEDs for lighting sources. In the first section, phosphors, including rare-earth ion and transition-metal ion–activated different compounds for general lighting are systematically summarized according to its emitting color and luminescence properties of commercially available or representative yellow-emitting phosphors, red-emitting phosphors, green-emitting phosphors, blue-emitting phosphors, and single-phase white-emitting phosphors, are mainly introduced. In the second section, narrow green- and red-emitting phosphors, especially Mn^{4+} ion–activated red-emitting phosphors, for back-lighting source in traditional LED display are mainly introduced. In the third section, phosphors, especially quantum dots (QDs), for back-lighting source in ultrahigh-definition micro-LED displays are systematically summarized and luminescence properties of QDs are described. In the fourth section, progresses on phosphor-converting LEDs (pc-LEDs) for general lighting, traditional LED displays and ultrahigh-definition micro-LED displays are mainly introduced. Finally, the challenges of phosphors for pc-LEDs in lighting source are presented.

1.2 PHOSPHORS FOR GENERAL LIGHTING

For illumination application, the emission spectrum of pc-WLED should be as close as possible to the solar spectrum that provides high-radiation flux, high color-rendering index (CRI), suitable CCT, etc. Consequently, phosphors with different emitting colors are needed to fabricate pc-WLEDs based on UV or blue LED chips. In general, phosphors consist of a suitable crystalline host material and one or more activators that act as an emitting center in the host. To date, many new and promising phosphors for general lighting have been invented, modified, or developed. According to the types of the host, phosphors include many compound systems, oxides (such as silicates, aluminates, borates, and phosphates), sulfides/oxysulfides, nitrides/oxynitrides, and fluorides/oxyfluorides. According to the types of the activator, phosphors are classified into transition-metal ion–activated ones or rare-earth ion–activated ones. Hereafter, we will describe these phosphors according to their emitting colors.

1.2.1 Yellow-Emitting Phosphors

It is no doubt that $Y_3Al_5O_{12}:Ce^{3+}$ (YAG:Ce) phosphor is the first reported and the most commercially successful yellow-emitting phosphors for general lighting. Figure 1.1(a) shows the photoluminescence excitation (PLE) and photoluminescence emission (PL) spectra of YAG:Ce.[5] The PLE

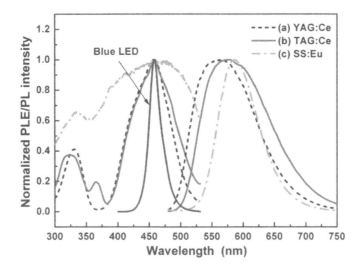

FIGURE 1.1 Normalized PLE and PL spectra of various yellow-emitting phosphors: (a) YAG:Ce ($Y_3Al_5O_{12}$:Ce^{3+}), (b) TAG:Ce ($Tb_3Al_5O_{12}$:Ce^{3+}), and (c) SS:Eu (Sr_3SiO_5:Eu^{2+}), respectively. (From Jang HS, Won YH, Jeon DY *Appl. Phys. B* 95, 715, 2009. With permission.)

spectrum of YAG:Ce is mainly composed of two broadbands that are centered at ~330 and 450 nm, which are assigned to the 4f–5d electronic transitions of Ce^{3+}. The strong and broad excitation bands of YAG:Ce match very well with the emission of UV and blue LED chips. The PL spectrum shows a very broad yellow emission band with a peak at about 558 nm and a full width at a half maximum (FWHM) of ~130 nm. YAG:Ce^{3+} shows the PLQY of about 97%. By coating a yellow-emitting YAG:Ce on a blue LED chip, the CRI and CCT of the fabricated WLEDs are about 70 and > 6000K.

Unfortunately, the PLQY of YAG:Ce declines more than 20% when the working temperature reaches above 200°C. In order to achieve more excellent thermal stability and modify emission spectrum of YAG:Ce, different synthesis methods had been adopted and the structural modifications had been made. Many other garnet-type aluminate/gallate/silicate phosphors, such as $Tb_3Al_5O_{12}$:Ce^{3+} (λ_{em} = 554 nm, λ_{ex} = 460 nm, as shown in Figure 1.1(b)), and $Lu_2CaMg_2Si_3O_{12}$:Ce^{3+}, have been reported recently and their chemical compositions are derived from cation/anion substitutions based on the same garnet model, as listed in Table 1.1. Also, the family of silicate phosphors, especially, M_2SiO_4 (M = Ca, Sr and Ba) and Sr_3SiO_5 (λ_{em} = 580 nm, λ_{ex} = 475 nm, as shown in Figure 1.1(c)), have received great interests.

In the meantime, several new yellow-emitting (oxy) nitride phosphors with excellent luminescence properties and chemical stability have emerged as suitable candidates for pc-WLEDs, as shown in Table 1.1. For instance, $La_3Si_6N_{11}$:Ce^{3+} exhibits broad excitation bands with peaks at around ~450 nm, which can be effectively excited by blue LED chip (Figure 1.2(a)). In addition, the optimized $La_3Si_6N_{11}$:Ce^{3+} exhibits excellent luminescence performance at high temperature. It remains 98.2% of the initial emission intensity when heated to 200°C in air. Its PLQY just decreases from 78.2% to 76.9% when the heating temperature increases from room temperature to 200°C in the air (Figure 1.2(b)) and it recovers to 77.82% upon cooling to room temperature in air.[6]

The excellent thermal stability of $La_3Si_6N_{11}$:Ce^{3+} may be due to the crystal structure of the high dense SiN_4 tetrahedra. The basic frame of nitride phosphor is constructed with high dense Si–N or Al–N tetrahedra, which forms a rigid skeleton structure and nitrogen-rich environments.[8] Based on the principle of structure determining the properties, nitride phosphors show high thermal and chemical stabilities and the center of gravity for Ce^{3+}/Eu^{2+} ions; 5d energy levels were shifted to lower values leading to yellow-emitting.

TABLE 1.1

Summary of Some Yellow-Emitting Phosphors Applicable for pc-LEDs

Group	Component	Excitation Range and Peak Position (nm)	Emission Rang, Peak and FWHM (nm)	Ref.
Oxide	$Y_3Al_5O_{12}:Ce^{3+}$	457	460–700, 558	[9]
	$Tb_3Al_5O_{12}:Ce^{3+},Gd^{3+}$	200–350, 375, 457	500–700, 554 → 610	[10]
	$Sr_3SiO_5:Eu^{2+}$	250–550, 412	500–700, 559 → 570	[11]
	$Li_2SrSiO_4:Eu^{2+}$	300–530, 308, ~420	500–700, 562	[12]
	$Lu_2CaMg_2Si_3O_{12}:Ce^{3+}$	380–550, 470	475–750, 605	[13]
	$Sr_8MgGd(PO_4)_7:Eu^{2+},Mn^{2+}$	220–500, 350	400–800, 510, 616	[14]
Oxyhalide	$Sr_3(Al_2O_5)Cl_2:Eu^{2+}$	250–500, 400	475–775, 610 (FWHM 160)	[15]
Nitride	$La_3Si_6N_{11}:Ce^{3+}$	350–550, 378, 450	400–700, 425, 578	[16]
	$(La,Ba)_3Si_6N_{11}:Ce^{3+}$	350–550, 378, 475	500–700, 595	[16]
	$Ba_2Si_5N_8:Eu^{2+}$	250–500, 260,	550–750, 572 → 650	[17]
Oxynitride	$CaSi_2O_2N_2:Eu^{2+}$	370–460, 259, 395	460–700, 560	[18]
	$SrSi_2O_2N_{8/3}:Eu^{2+}$	370–460, 387, 440	480–740, 530 → 570	[18]

1.2.2 RED-EMITTING PHOSPHORS

As described above, the white light of pc-WLED by combining a blue chip with YAG:Ce suffers high CCT and poor CRI values, which is mainly due to the lack of the red emission component of YAG:Ce.[19] In order to sufficiently cover the red spectral region (the maximum emission at about 610–620 nm) to achieve a high CRI and minimize the efficacy loss caused by low eye sensitivity at long wavelengths, new red phosphors emitting near the optimal wavelength are required.

Table 1.2 shows a summary of the performance of some red-emitting phosphors which could be excited by blue LED chips. These red-emitting phosphors can be further classified into rare-earth ions–activated red-emitting phosphors and Mn^{4+} ion–activated red-emitting phosphors. For rare-earth ions–activated red-emitting phosphors, two kinds of rare-earth ions with 4f–4f transitions

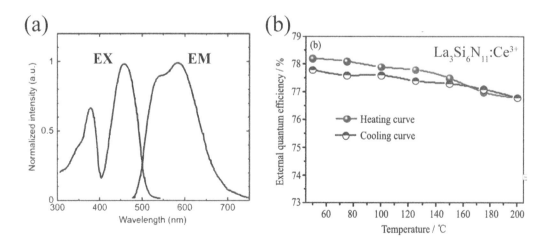

FIGURE 1.2   (a) Emission spectrum and excitation spectrum of $La_3Si_6N_{11}:Ce^{3+}$.[7] (From Kijima N, Seto T, Hirosaki N *ECS Trans.* 25, 247, 2009. With permission.) (b) Temperature dependence of the external quantum efficiency for $La_3Si_6N_{11}:Ce^{3+}$ sample. (From Du F, Zhuang W, Liu R, Liu Y, Gao W, Zhang X, Xue Y, Hao H *J. Rare Earths* 35, 1059, 2017. With permission.)

TABLE 1.2

Summary of Some Red-Emitting Phosphors Applicable for pc-LEDs

Group	Component	Excitation Range and Peak Position (nm)	Emission Range, Peak and FWHM (nm)	Ref.
Oxide	$Y_2O_3:Eu^{3+}$	360–420, sharp lines	610	[20]
	$Na_3Sc_2(PO_4)_3:Eu^{3+}$	250–300, sharp lines	621	[21]
	$Ca_2ZnWO_6:Sm^{3+}$	230–350, sharp lines	604	[22]
	$MgO:Sm^{3+}$	325–500, sharp lines	603	[23]
	$Gd_2(MoO_4)_3:Sm^{3+}$	250–310, sharp lines	601	[24]
	$LiSrBO_3:Eu^{2+}$	240–350, 470	618	[25]
	$YAG:Mn^{4+}$	250–530, 345, 480	673	[26]
Oxysulfide	$Y_2O_2S:Eu^{3+}$	220–300, sharp lines	614	[27]
Nitride	$CaAlSiN_3:Eu^{2+}$	270–570, 335, 450	660	[28]
	$Ca_2Si_5N_8:Eu^{2+}$	250–550, 350, 400, 450	600 (FWHM 80)	[29]
	$Sr[LiAl_3N_4]:Eu^{2+}$	375–600, 466	654 (FWHM 85)	[30]
Oxynitride	$Sr[Li_2Al_2O_2N_2]:Eu^{2+}$	—	614	[30]
Fluoride	$K_2SiF_6:Mn^{4+}$	300–510, 360, 460	631	[31]
	$K_2TiF_6:Mn^{4+}$	300–550, 362, 462	630	[32]
Oxyhalide	$3.5MgO \cdot 0.5MgF_2 \cdot GeO_2:Mn^{4+}$	275–500, 300, 420	632	[33]

and 4f–5d transitions are involved as red-emitting centers. Rare-earth ions with 4f–4f transitions for red-emitting phosphor are mainly Eu^{3+} ions.

Eu^{3+}-activated phosphors are famous red-emitting phosphors, e.g. $Y_2O_3:Eu^{3+}$,[20] for traditional fluorescent lamps, which have a suitable red emission around 612 nm, due to the $^5D_0 \rightarrow {}^7F_2$ transition of Eu^{3+}. In the last decades, Eu^{3+}/Sm^{3+}-activated oxide phosphors have been extensively researched again for potential application in pc-WLEDs. Unfortunately, they commonly have characteristic sharp and weak excitation peaks due to forbidden f–f transitions, which cannot efficiently absorb and convert the emission from UV or blue LED chips. Therefore, the performance of these phosphors is not good enough for potential application in pc-WLEDs, especially based on blue LED chip.

Rare-earth ions with 4f–5d transitions for red-emitting phosphor are mainly Eu^{2+} ions. Among them, Eu^{2+}-doped (oxy) nitride materials have attracted more attentions. Theoretically, the 4f–5d transitions of Eu^{2+} ion are sensitive to the crystal field and covalence, which makes it possible for Eu^{2+} in (oxy)nitride compounds to have a strong absorption in the UV to the visible spectral region and exhibit broad red emission bands. For example, Eu^{2+}-doped commercially available red-emitting phosphors are $(Ba,Sr)_2Si_5N_8:Eu^{2+}$ (λ_{max} ~590–625 nm, FWHM ~71–101 nm),$CaSiAlN_3:Eu^{2+}$ (λ_{max} ~630 nm, FWHM ~86 nm).[28, 34] Uheda et al.[35] first reported the PL properties of $CaSiAlN_3:Eu^{2+}$, which has a broad excitation band from the UV region to 590 nm and gives a broadband emission with a maximum between 630 and 680 nm (Figure 1.3(a)). In particular, at 200°C, the emission intensity remains ~90% of the initial intensity at 25°C (Figure 1.3(b)).[36] This phosphor exhibits high external PLQY (~60%), small thermal quenching, and high chemical stability.

However, these red-emitting nitride phosphors mentioned above have much broad emission bands, especially covering deep-red region above 700 nm, outside the sensitivity range of the human eye, which greatly limits the maximum achievable luminous efficacies of high-quality warm pc-WLEDs (CRI > 90). Recently several narrow red-emitting phosphors were reported, for instance, $Sr[LiAl_3N_4]:Eu^{2+}$ (λ_{max} = ~650 nm, FWHM = ~50 nm, internal PLQY = ~76%),[30] $Sr[Li_2Al_2O_2N_2]:Eu^{2+}$ (λ_{max} = ~614 nm, FWHM = ~48 nm, internal PLQY = ~80%),[38] as shown in Figure 1.4, which will make their way into commercially available red-emitting phosphors for pc-WLEDs. Even so, the chemical stability of these new narrow red-emitting nitride phosphors is poor,

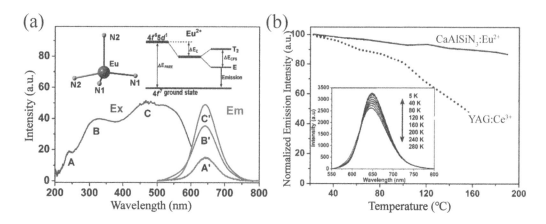

FIGURE 1.3 (a) Typical photoluminescence excitation (λ_{em} = 649 nm) and emission (λ_{ex} = 460 nm) spectra of $Ca_{0.98}Eu_{0.02}AlSiN_3$ and the schematic picture of the influence of the environment of an Eu^{2+} on the positions of electronic states. (b) The temperature-dependent emission intensity of $Ca_{1-x}Eu_xAlSiN_3$ samples with x = 0.02, and the inset shows the low-temperature emission of $Ca_{0.98}Eu_{0.02}AlSiN_3$.[37] (From Piao X, Machida KI, Horikawa T, Hanzawa H, Shimomura Y, Kijima N *Chem. Mater.* 19, 4592, 2007. With permission.)

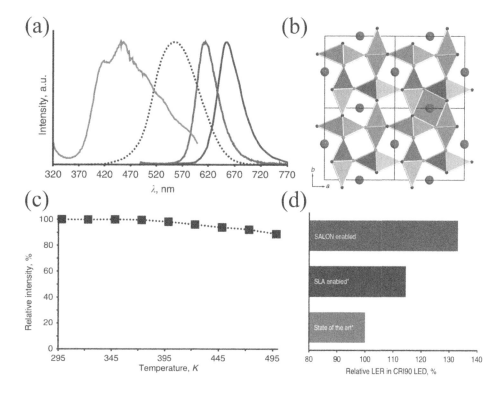

FIGURE 1.4 (a) Normalized excitation (gray, for the emission at λ_{max} equals 614 nm) and emission spectrum (red, excited with λ_{exc} equals 460 nm) of $Sr[Li_2Al_2O_2N_2]:Eu^{2+}$ in comparison to $Sr[LiAl_3N_4]:Eu^{2+}$, reference (purple) and the human-eye sensitivity curve (black dotted). (b) Structural overview of $Sr[Li_2Al_2O_2N_2]:Eu^{2+}$. Red spheres represent strontium, blue spheres oxygen, and green spheres nitrogen atoms. (c) Relative photoluminescence intensity of SALON measured from 298 to 500K (d) Relative luminous efficacy of radiation (LER) values for different warm-white pc-LED solutions (data normalized to the state of the art). (From Hoerder GJ, Seibald M, Baumann D, Schroder T, Peschke S, Schmid PC, Tyborski T, Pust P, Stoll I, Bergler M, Patzig C, Reissaus S, Krause M, Berthold L, Hoche T, Johrendt D, Huppertz H *Nat. Commun.* 10, 1824, 2009. With permission.)

which greatly limits their actual application. Consequently, from the viewpoint of industrial applications, new synthesis routes or coating methods for the high-performance narrow red-emitting nitride phosphors are still urgently required.

Besides rare-earth ions–activated red-emitting phosphors, Mn^{4+} ion–activated fluorides are also promising red-emitting phosphors for pc-WLEDs. Among them, $K_2SiF_6:Mn^{4+}$ (λ_{max} = ~654 nm, FWHM = ~50 nm) is one of the commercially available red-emitting phosphors,[39] which will be later reviewed in detail.

1.2.3 GREEN-EMITTING PHOSPHORS

Table 1.3 summarizes the green-emitting phosphors. Almost all of them are involved with Ce^{3+} or Eu^{2+} ions as emitting center. The host compound includes oxides (aluminate, silicate, etc.), sulfide, oxyhalide, nitride and oxynitride. Among them, $Lu_3Al_5O_{12}:Ce^{3+}$ (LuAG:Ce^{3+}) and β-$Si_{6-z}Al_zO_zN_{8-z}:Eu^{2+}$ (β-SiAlON:Eu) are well known as commercially available green-emitting phosphors. LuAG:Ce^{3+} is one of garnet-isostructural solid solutions, $(Y,Gd,Lu)_3(Al,Ga)_5O_{12}:Ce^{3+}$. As shown in Figure 1.5(a), LuAG:Ce^{3+} has broad PLE bands at 345 and 460 nm matching well with the emission of UV and blue LED chips and gives broad green emission band at 513 nm, due to 4f–5d transitions of Ce^{3+}. Also it has high PLQY of ~97% and excellent thermal stability. As shown in Figure 1.5(b), the PL intensity of LuAG:Ce^{3+} almost remains unchanged with increasing temperature under blue light excitation. Green β-SiAlON:Eu phosphor has an emission peak at ~530 nm with a FWHM of ~55 nm, as shown in Figure 1.6(a), and shows excellent thermal quenching behavior (~10% emission loss at 150°C), as shown in Figure 1.6(b), and high quantum efficiency (external/internal PLQY: 65/82%).[40]

Besides, there are many other green-emitting Eu^{2+}-doped silicates, for instance, $Ca_{2-x}Sr_xSiO_4$: Eu^{2+},[43] $SrBaSiO_4:Eu^{2+}$,[44] $Ca_{15}(PO_4)_2(SiO_4)_6:Eu^{2+}$,[45] $M_2SiO_4:Eu^{2+}$ (M = Ca, Sr, Ba),[46] Ba_2Ca $Zn_2Si_6O_{17}:Eu^{2+}$,[47] $Sr_{3.5}Mg_{0.5}Si_3O_8Cl_4:Eu^{2+}$.[48] Especially, $Ca_3SiO_4Cl_2:Eu^{2+}$ shows green luminescence under 400 nm excitation.[49] Its thermal stability was found to be higher than the commercially used green-emitting phosphor $Ba_2SiO_4:Eu^{2+}$.[50]

On the other hand, Tb^{3+} ion or Mn^{2+} ion–activated green-emitting phosphors have been extensively researched for potential application in pc-WLEDs. Unfortunately, both of them commonly have characteristic sharp and weak excitation peaks due to forbidden 4f–4f or 3d–3d transitions, which cannot efficiently absorb and convert the emission from UV or blue LED chips. Therefore,

TABLE 1.3
Summary of Some Green-Emitting Phosphors Applicable for pc-LEDs

Group	Component	Excitation Range and Peak Position (nm)	Emission Range, Peak and FWHM (nm)	Ref.
Oxide	$Lu_3Al_5O_{12}:Ce^{3+}$	325	450–750, 540	[51]
	$Tb_3Al_5O_{12}:Ce^{3+}$	250–520, 275, 325, 463	475–750, 545 → 555	[52]
	$CaSc_2O_4:Ce^{3+}$	400–500, 450	480–700, 520 (FWHM 105)	[53]
	$Sr_3SiO_5:Ce^{3+},Li^+$	200–500, 415	465–700, ~520	[54]
	$LaSr_2AlO_5:Ce^{3+}$	300–530, 330, 450	480–750, 556	[55]
	$RbLi(Li_3SiO_4)_2:Eu^{2+}$	300–500, ~390, 460	480–610, 530 (FWHM 42)	[56]
Sulfide	$Ca_2SiS_4:Eu^{2+}$	250–450, 425	400–700, 475, 505	[57]
	$SrGa_2S_4:Eu^{2+}$	250–500, 254, ~400	475–625, 534 (FWHM 55)	[58]
Oxyhalide	$Ba_4Gd_3Na_3(PO_4)_6F_2:Eu^{2+}$	230–430, 350	400–650, 481, 550	[59]
Nitride	$CaSiAlN_3:Ce^{3+}@CeB_6$	350–500, 370, 460	500–700, ~550	[60]
Oxynitride	$Si_{6-z}Al_zO_zN_{8-z}:Eu^{2+}$ (β-Sialon:Eu)	250–500, 304, ~406	528–550, 536 (FWHM 63)	[61]
	$SrSi_2O_2N_{8/3}:Eu^{2+}$	370–460, 341, 387, 440	480–740, 530 → 570	[18]

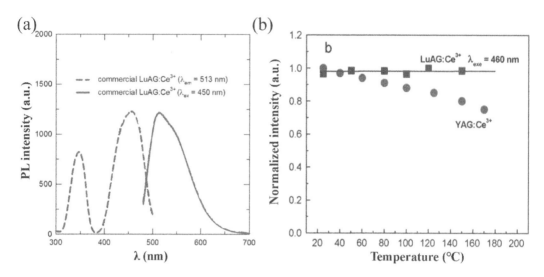

FIGURE 1.5 (a) Photoluminescence excitation spectra and emission spectra of commercial LuAG:Ce^{3+} phosphor at room temperature.[41] (b) PL spectra at different temperatures under 460 nm excitation and temperature-dependent normalized integrated PL intensities of LuAG:Ce^{3+} and YAG:Ce^{3+} garnet.[42] (From Praveena R, Shi L, Jang KH, Venkatramu V, Jayasankar CK, Seo HJ *J. Alloys Compd.* 509, 859, 2011. With permission.)

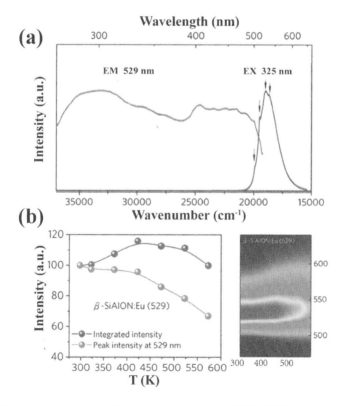

FIGURE 1.6 (a) Photoluminescence excitation and emission spectra of β-SiAlON:Eu green phosphors (dark arrows show the visible vibronic structure). (b) Thermal quenching behavior of photoluminescence for β-SiAlON:Eu; temperature-dependent photoluminescence spectra are shown on the right-hand side. (From Zhang X, Fang MH, Tsai YT, Lazarowska A, Mahlik S, Lesniewski T, Grinberg M, Pang WK, Pan F, Liang C, Zhou W, Wang J, Lee JF, Cheng BM, Hung TL, Chen YY, Liu RS *Chem. Mater.* 29, 6781, 2017. With permission.)

TABLE 1.4

Summary of Some Blue-Emitting Phosphors Applicable for pc-LEDs

Group	Component	Excitation Range and Peak Position (nm)	Emission Range, Peak and FWHM (nm)	Ref.
Oxide	$BaMgAl_{10}O_{17}:Eu^{2+}$	250–410, 305	400–550, 460 (FWHM 51)	[72, 73]
	$RbNa_3(Li_3SiO_4)_4:Eu^{2+}$	320–440, 395	450–510, 471 (FWHM 22.4)	[74]
	$Na_{0.5}K_{0.5}Li_3SiO_4:Eu^{2+}$	300–500, 400	450–600, 486 (FWHM 20.7)	[75]
	$KMg_4(PO_4)_3:Eu^{2+}$	250–420, 330, 388	375–525, 442 (FWHM 50)	[76]
	$SrZnP_2O_7:Eu^{2+}$	200–400, 330	390–460, 420 (FWHM 30)	[70]
	$RbBaPO_4:Eu^{2+}$	220–420, 250, 330	380–520, 430	[77]
	$SrCaP_2O_7:Eu^{2+}$	300–380, 330	380–510, 427	[78]
	$NaSc_2(PO_4)_3:Eu^{2+}$	280–400, 370	400–530, 453 (FWHM 44)	[79]
	$NaCaBO_3:Ce^{3+}$	250–400, 347	350–600, ~420	[80]
	$Li_4SrCa(SiO_4)_2:Ce^{3+}$	300–370, 288, 360	370–550, 420	[81]
Oxyhalide	$Ba_5SiO_4Cl_6:Eu^{2+}$	200–420, ~365	420–530, 440 (FWHM 29)	[73]
Nitride	$AlN:Eu^{2+}$	270–400, 330	400–570, 475	[82]
Oxynitride	$BaSi_2O_2N_2:Eu^{2+}$	370–460, 327, ~460	480–600, 499 (FWHM ~35)	[18]
	$Sr[Be_6ON_4]:Eu^{2+}$	400–500, ~420,	460–560, 495 (FWHM 35)	[83]

the performance of these phosphors is not good enough for potential application in pc-WLEDs, especially based on blue LED chip.

1.2.4 BLUE-EMITTING PHOSPHORS

Table 1.4 summarizes blue-emitting phosphors for pc-WLEDs based on the combination of UV LED + R/G/B phosphors. One of the most commercially used blue phosphors is $BaMgAl_{10}O_{17}:Eu^{2+}$ (known as BAM phosphor as shown in Figure 1.7) phosphor.[63, 64] Besides, a vast number of

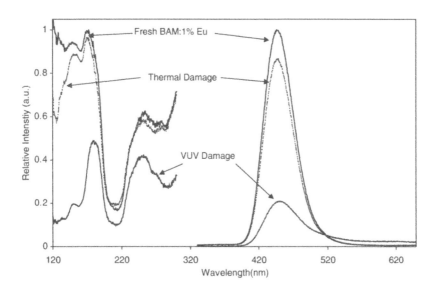

FIGURE 1.7 Excitation ($\lambda_{em} = 450$ nm) and emission ($\lambda_{ex} = 147$ nm) spectra of BAM doped at 1% Eu^{2+}, after thermal and VUV damage.[62] (From Howe B, Diaz A. L *J. Lumin.* 109, 51, 2004. With permission.)

Ce^{3+}/Eu^{2+}-doped phosphors have shown excellent blue-emitting performance and stability under near-UV excitation. For Ce^{3+} or Eu^{2+} ion, due to the exposure of its 5d electrons to a crystal environment, the centroid shift of the 5d levels of Ce^{3+} or Eu^{2+} is easily tuned to fit the UV excitation and they finally give blue emissions in oxide-based phosphors, for instance, $Na_4CaSi_3O_9:Ce^{3+}$,[65] $Gd_5Si_3O_{12}N:Ce^{3+}$,[66] $Gd_5Si_3O_{12}N:Ce^{3+}$,[66] $(CaCl_2/SiO_2):Eu^{2+}$,[67] $Sr_3MgSi_2O_8:Eu^{2+}$,[68] $Ba_2Si_3O_8:Eu^{2+}$,[69] $SrZnP_2O_7:Eu^{2+}$, and [70] $LiSrPO_4:Eu^{2+}$,[71].

1.2.5 SINGLE-PHASE WHITE-EMITTING PHOSPHORS

In addition to yellow-, red-, and green-emitting phosphors, single-phase white-emitting phosphors have attracted more attentions. Usually, the pc-WLEDs are fabricated by UV/blue LED chip and the blend of two or more different emitting color phosphors. Unfortunately, these phosphors will degrade in a different manner. Consequently, the overall white-light qualities of pc-WLEDs, including the CCT and CRI, will change as the working time increases. In principle, it is possible to combine a UV LED chip and a single-phase white-emitting phosphor that fully determine the white-light qualities of pc-WLEDs. Till date, a huge number of single-phase white-light-emitting phosphors have been reported. Table 1.5 summarizes the luminescence properties of single-phase white-emitting phosphors. The strategy of producing white-emitting light in single-phase compounds can be divided into, single luminescent center with white emitting at single crystalline site, single luminescent center with white emitting at multi-crystalline sites, and multi-luminescent centers with white emitting at single or multi-crystalline sites.

Doping with a Dy^{3+} ion in a host is the most common way to obtain white-light emission since Dy^{3+} ion has its characteristic line shape emissions in the blue (470–500 nm) and yellow (570–590 nm)

TABLE 1.5
Summary of Some Single-Phase White-Emitting Phosphors for pc-LEDs

Group	Component	Excitation Range and Peak Position (nm)	Emission Range, Peak and FWHM (nm)	Ref.
Oxide	$NaLa(PO_3)_4:Dy^{3+}$	275–475, sharp lines	440–490, 485, 550–580, 571	[84]
	$Ca_8MgBi(PO_4)_7:Dy^{3+}$	320–450, sharp lines	435–500, 555–595	[85]
	$Ba_5Zn_4Y_8O_{21}:Eu^{3+}$	220–300, sharp lines	400–750, sharp peaks	[86]
	$BaSrMg(PO_4)_2:Eu^{2+}$	250–440, 350	400–700, 447, 536	[87]
	$Ca_{9-x}Sr_xMgK(PO_4)_7:Eu^{2+}$	250–450, 360	400–720, ~428, ~470, ~545	[88]
	$Y_2Mg_2Al_2Si_2O_{12}:Eu^{2+},Mn^{2+}$	250–425, 380	400–725, 475, 620	[89]
	$Sr_3Bi(PO_4)_3:Eu^{2+},Mn^{2+}$	250–450, 297, 361	450–750, 520, 667	[90]
	$Li_4SrCa(SiO_4)_2:Eu^{2+},Mn^{2+}$	250–375, 280	375–650, 430, 590	[91]
	$Sr_3Gd_2(Si_3O_9)_2:Eu^{2+},Mn^{2+}$	250–370, 274, 313	350–650, 418, 468, 554	[92]
	$K_2BaCa(PO_4)_2:Eu^{2+},Mn^{2+}$	200–450, 270, 330	400–700, 470, 560	[93]
	$KCaY(PO_4)_2:Eu^{2+},Mn^{2+}$	270–500, 367	400–750, 480, 652	[94]
	$Ba_3CaK(PO_4)_3:Eu^{2+},Mn^{2+}$	240–450, 365	400–700, 460, 590	[95]
	$Sr_4Al_{14}O_{25}:Eu^{2+},Mn^{4+},Eu^{3+}$	200–450, 325	375–775, 496, 619, 659	[96]
	$Ca_{20}Al_{26}Mg_3Si_3O_{68}:Eu^{2+},Eu^{3+}$	275–400, 324	400–710, 449, 490, 615	[97]
	$\beta–Ca_2SiO_4:Eu^{2+},Eu^{3+}$	225–450, 275, 320	475–750, 508, 602	[98]
	$BaY_2Si_3O_{10}:Ce^{3+},Tb^{3+},Eu^{3+}$	250–420, sharp lines	400–700, 436, 544, 612	[99]
	$Ba_9Lu_2Si_6O_{24}:Eu^{2+},Ce^{3+},Mn^{2+}$	240–450, 330, 400	460, 490, 610	[100]
	$Ca_9MgNa(PO_4)_7:Eu^{2+},Mn^{2+}$	250–450, 365	400–800, 475, 648	[101]
Fluoride	$\beta-NaYF_4:Eu^{3+}$	Sharp lines	464–693, sharp lines	[102]
Oxyhalide	$Ba_3GdNa(PO_4)_3F:Eu^{2+}$	250–450, 342	400–675, 472, 608	[103]

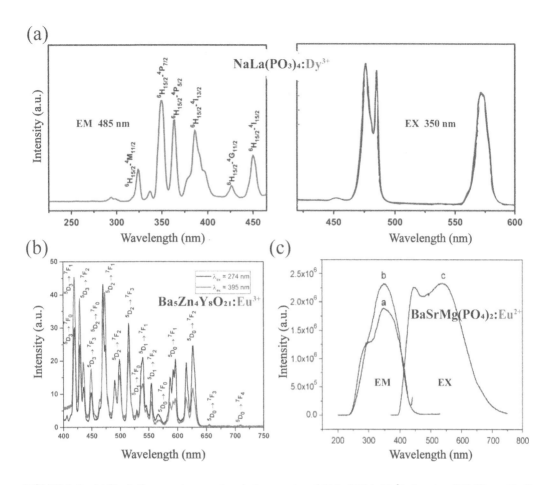

FIGURE 1.8 (a) Excitation spectrum and emission spectra of NaLa(PO$_3$)$_4$:Dy^{3+} phosphor.[84] (From Liu F, Liu Q, Fang Y, Zhang N, Yang B, Zhao G *Ceram. Int.* 41, 1917, 2015. With permission.) (b) Dual emission profile of Ba$_5$Zn$_4$Y$_{7.92}$Eu$_{0.08}$O$_{21}$ nanophosphor.[86] (From Dalal M, Taxak VB, Dalal J, Khatkar A, Chahar S, Devi R, Khatkar SP *J. Alloys Compd.* 698, 662, 2017. With permission.) (c) Excitation spectra (a: λ_{em} = 447, b: λ_{em} = 536 nm) and emission spectra (c: λ_{ex} = 350 nm) of BaSrMg(PO$_4$)$_2$:Eu^{2+} powder.[87] (From Wu ZC, Liu J, Hou WG, Xu J, Gong J *J. Alloys Compd.* 498, 139, 2010. With permission.)

regions, due to the transitions of $^4F_{9/2} \rightarrow {}^6H_{15/2}$ and $^4F_{9/2} \rightarrow {}^6H_{13/2}$, respectively. This strategy is so-called single luminescent center with white emitting at single crystalline site. For instance, NaLa(PO$_3$)$_4$:Dy^{3+},[84] Ca$_8$MgBi(PO$_4$)$_7$:Dy^{3+},[85] CaZr$_4$(PO$_4$)$_6$:Dy^{3+},[104] Ca$_3$B$_2$O$_6$:Dy^{3+},[105] Sr$_3$Gd(PO$_4$)$_3$:Dy^{3+}, and [106] Ca$_3$Mg$_3$(PO$_4$)$_4$:Dy^{3+}[107] were reported. In the case of NaLa(PO$_3$)$_4$:Dy^{3+} as shown in Figure 1.8(a), strong excitation peaks are found around 350–400 nm, which matches well with those of the near-UV LED chips. Under 350 nm excitation, the NaLa(PO$_3$)$_4$:Dy^{3+} exhibits intense white emission by combining the two emission peaks at 485 and 571 nm, attributed to the characteristics $^4F_{9/2} \rightarrow {}^6H_{15/2}$ and $^4F_{9/2} \rightarrow {}^6H_{13/2}$ transitions of Dy^{3+}.[84]

Apart from Dy^{3+}, it is also feasible to produce white light by singly doping Eu^{2+} ions in a host, such as BaSrMg(PO$_4$)$_2$:Eu^{2+}. BaSrMg(PO$_4$)$_2$:Eu^{2+} shows two main emission bands peaking at 447 and 536 nm under UV light (Figure 1.8(c)), and its CIE chromaticity coordinates are calculated (x = 0.291, y = 0.349), which are located in the white-light region in CIE-1931 chromaticity diagram.[87] It was found that the two emission bands result from the 4f^{6}5d^1–4f^7 transitions of Eu^{2+} ions occupying Sr^{2+} and Ba^{2+} sites in host lattice, respectively. This strategy is so-called single luminescent center with white emitting at multi-crystalline sites.

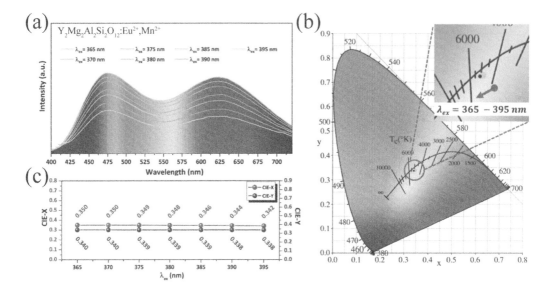

FIGURE 1.9 (a) PL spectra of $Y_2Mg_2Al_2Si_2O_{12}$:Eu^{2+},Mn^{2+} dependence on the excitation wavelength (λ_{ex} = 365–395 nm). (b) The variation of the CIE color coordinates for $Y_2Mg_2Al_2Si_2O_{12}$:Eu^{2+},Mn^{2+} with the excitation wavelength. (c) CIE chromaticity coordinates of $Y_2Mg_2Al_2Si_2O_{12}$:$0.03Eu^{2+}$,$0.25Mn^{2+}$ at various excitation wavelengths. (From Zhang X, Zhang D, Zheng Z, Zheng B, Song Y, Zheng K, Sheng Y, Shi Z, Zou H *Dalton Trans.* 49, 17796, 2020. With permission.)

In addition, a number of phosphors have been reported to give white luminescence with the combination of Eu^{2+}–Mn^{2+} (or Eu^{2+}–Eu^{3+}, Ce^{3+}–Eu^{2+}) pair. This strategy is so-called multi-luminescent centers with white emitting at single or multi-crystalline sites. For instance, Figure 1.9(a) shows the PL spectra of $Y_2Mg_2Al_2Si_2O_{12}$:Eu^{2+},Mn^{2+} monitored at different excitation wavelengths. Benefitted from the energy transfer process between Eu^{2+} and Mn^{2+}, the color point can be tuned from cyan to cold white as shown in Figure 1.9(b). The PLQY of $Y_2Mg_2Al_2Si_2O_{12}$:Eu^{2+},Mn^{2+} varies from 23% to 29% with color change. Also, the emission intensity declines about ~20% when the temperature reaches 150°C.[89]

1.3 PHOSPHORS FOR BACK-LIGHTING SOURCE FOR HIGH-DEFINITION DISPLAY

Nowadays, pc-WLEDs are also used as back-lighting source of the thin film transistor liquid crystal display (TFT-LCD, or LCD for short). In LCD device, bandpass color filters are used to obtain RGB primary monochromatic colors from pc-WLEDs as back-lighting source. More specifically, RGB primary monochromatic colors are proposed to be at 630 nm for red, 532 nm for green, and 467 nm for blue according to the International Telecommunication Union for Recommendation ITU-R BT-2020 (Rec. 2020) of ultrahigh-definition television. Therefore, not all the phosphors for general lighting can meet the requirements of the LCD back-light application. Phosphors for back-lighting source in LCD display should at least have emissions with narrow emission band and suitable emission peak in order to match well with the RGB bandpass color filters.

Among all the currently available phosphors for pc-WLEDs based on blue LED chip, β-SiAlON:Eu and K_2MF_6:Mn^{4+} (M = Si and Ge) are the most promising and commercially available green- and red-emitting phosphors, respectively. As mentioned above, β-SiAlON:Eu has an emission peak at ~530 nm with a FWHM of ~55 nm, which matches well with the green bandpass color filter. Additionally, it has an excellent thermal quenching behavior (~10% emission loss at

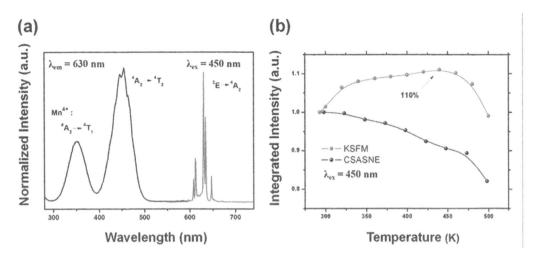

FIGURE 1.10 (a) PLE and PL spectra of K_2SiF_6:Mn^{4+}. (b) Thermal quenching behaviors of K_2SiF_6:Mn^{4+} and commercial (Ca,Sr)AlSiN$_3$:Eu^{2+} (CSASNE) at T = RT–500K. (From Huang L, Zhu YW, Zhang XJ, Zou R, Pan F, Wang J, Wu MM *Chem. Mater.* 28, 1495, 2016. With permission.)

150°C) and external/internal PLQY of 65%/82%.[40] Hereafter, we will mainly focus on research progresses of K_2SiF_6:Mn^{4+}.

Figure 1.10(a) shows typical PL and PLE spectra of K_2SiF_6:Mn^{4+}. Monitored at 630 nm, there are two broad excitation bands in the UV and blue region between 300 and 500 nm, attributed to the $^4A_{2g} \rightarrow {}^4T_{1,2g}$ transitions of Mn^{4+} ion. Under excitation at 450 nm, K_2SiF_6:Mn^{4+} gives sharp red emission predominating at 630 nm, due to the $^2E_g \rightarrow {}^4A_{2g}$ transitions of Mn^{4+}. The FWHM is only ~7 nm, which matches well with the red bandpass color filter. The PLQY of commercial K_2SiF_6:Mn^{4+} under 450 nm excitation is high up to ~90%. Moreover, K_2SiF_6:Mn^{4+} has an excellent thermal quenching behavior. At 340K (67°C), 420K (147°C), 440K (167°C), and 500K (227°C), it is about 108%, 110%, 110%, and 98% of the initial intensity at 298K, respectively. As a reference, the commercial red LED-used phosphor (Ca,Sr)AlSiN$_3$:Eu^{2+} only maintains 81% of its initial intensity at 500K.

Although K_2SiF_6:Mn^{4+} exhibits an excellent luminescence performance, there are still two main challenges, one is nongreen synthesis method and the other is its poor water resistance. In the last decade, many synthesis methods have been applied to synthesize K_2SiF_6:Mn^{4+}, including traditional solid-state and liquid-phase syntheses.[109] The latter is the most used one and can be divided into wet chemical etching, hydrothermal synthesis, cation exchange, and coprecipitation. All these liquid-phase synthesis methods inevitably use hydrofluoric acid (HF), which is well known as a highly corrosive and toxic reagent. Therefore, nongreen synthesis becomes one of the main challenges for K_2SiF_6:Mn^{4+}, especially in mass-scale synthesis. In 2016, Wang et al. reported a hydrothermal method and successfully synthesized K_2SiF_6:Mn^{4+} with the assistance of low-toxic H_3PO_4/KHF_2 liquid instead of high-toxic HF solution.[108] Thereafter, Wickleder et al. reported a microwave-assisted synthesis with the assistance of ionic liquids [Bmim]PF$_6$ as the solvent and fluorine source and successfully obtained undoped K_2SiF_6 nanoparticles.[110] And McKittrick et al. synthesized Na_2SiF_6 powders were by using an ionic liquid [Bmim]BF$_4$ and successfully prepared red-emitting $Na_2Si_{1-x}Mn_xF_6$ phosphors by dissolving $NaMnO_4$ in a low concentration of HF (6 M) and using the Na_2SiF_6 powders as the host lattice.[111] Unfortunately, the PLQY of the as-obtained K_2SiF_6:Mn^{4+} is not good enough for actual application in pc-WLEDs. Even so, these "HF-free" syntheses provide new thoughts for green synthesis of K_2SiF_6:Mn^{4+} in the future.

The poor water resistance is the other challenge for K_2SiF_6:Mn^{4+}. Due to the nature of ionic compound, K_2SiF_6 is extremely sensitive to humidity and the surface $[MnF_6]^{2-}$ dopant would easily be hydrolyzed into mixed-valence Mn oxides and hydroxides in the humid environment without HF, which greatly reduces the absorption and luminescence efficiency of K_2SiF_6:Mn^{4+}. For example, the luminescence intensity of K_2SiF_6:Mn^{4+} will sharply reduce by more than 70% after being immersed in water.[112] Therefore, how to maintain the luminescence performance of K_2SiF_6:Mn^{4+} in humid environment becomes a hot issue.

Coating is strongly expected to be a general approach to prevent the $[MnF_6]^{2-}$ structural units from hydrolyzing and enhance the water resistance of K_2SiF_6:Mn^{4+}. Till now, inorganic protective layers, organic polymers, etc. have been considered to be excellent moisture-resistant coating layers. [32,113–117] For organic polymers, Liu et al. reported a facile approach for K_2SiF_6:Mn^{4+} by coating with a moisture-resistant alkyl phosphate layer with a thickness of 50–100 nm (Figure 1.11(a)). The as-coated phosphor particles exhibit a high water tolerance and retain approximately 87% of their initial external PLQY after aging under high-humidity (85%) and high-temperature (85°C) conditions for one month.[115] For inorganic protective layers, General Electric Company developed a homogenous coating approach and successfully synthesized high humidity-stable K_2SiF_6:Mn^{4+}@K_2SiF_6 by treating original K_2SiF_6:Mn^{4+} phosphor with a mix solution of K_2SiF_6/HF/H_2SiF_6 (Figure 1.11(b)).[118] Further, Wang et al. developed a surface deactivation approach for K_2SiF_6:Mn^{4+} phosphor by treating it with H_2O_2, instead of using toxic HF acid as posttreatment regent and also successfully achieved the homogeneous K_2SiF_6:Mn^{4+}@K_2SiF_6 composite core-shell structure (Figure 1.11(c)).[112]

All the coating methods mentioned above are based on the idea of isolating the $[MnF_6]^{2-}$ clusters from water with the assistance of protective layers. However, if the protective layer is peeled off, hydrolysis of $[MnF_6]^{2-}$ would happen anytime and consequently decrease the emission intensity of K_2SiF_6:Mn^{4+}. Accordingly, Wang et al. further demonstrated a new strategy, a reductive DL-mandelic acid loading process, to prepare excellent moisture-resistant red-emitting Mn^{4+}-doped fluoride phosphors (Figure 1.11(d)–(g)).[119] For instance, the as-prepared DL-mandelic acid–loaded K_2GeF_6:Mn^{4+} not only maintains the same high luminescence properties as the original commercial K_2GeF_6:Mn^{4+} but also exhibits supreme moisture-resistant performance. It retains 98% of its initial emission intensity even after 168 hours of water erosion. The loaded reducing substance can *in situ* decompose the dark-brown hydrolysate generated from hydrolysis of K_2GeF_6:Mn^{4+}. This reductive DL-mandelic acid–loading strategy is novel and applicable to other commercial systems such as K_2SiF_6:Mn^{4+} and K_2TiF_6:Mn^{4+}. It opens up opportunities for excellent anti-moisture narrow red-emitting Mn^{4+}-doped fluoride phosphors.

1.4 PHOSPHORS FOR BACK-LIGHTING SOURCE FOR ULTRAHIGH-DEFINITION DISPLAY

Micro-LED has advantages of excellent optical properties (such as high brightness, high contrast, wide color gamut), fast response time, high power efficiency, high resolution, and high stability. Therefore, micro-LED has been strongly expected to show great potential to replace organic LED (OLED) and LCD.[120,121] Nowadays, "Blue Micro-LED chips + multi-color phosphors" is expected to be a promising strategy to realize full-color micro-LED with advantages of large-scale, low production cost and high transfer yield.

Unfortunately, the traditionally available phosphors usually have a large particle size (5–10 μm) and FWHM (> 50 nm), which cannot meet the requirements of the fabrication process and wide color gamut of device and finally greatly limits their actual application in ultrahigh resolution micro-LED displays.[121] Therefore, submicro- or nano-scale luminescent materials with high emission performance (narrow emission and suitable emission peak, matching well with bandpass color filters) are in great needs. Colloidal QDs usually show small particle size (< 25 nm), wide tunable bandgaps, narrow band emission (FWHM < 35 nm), high PLQY (near unity), short PL lifetime, etc.[122–124] All these features make QDs the best choice as nano-scale phosphors for micro-LED displays.

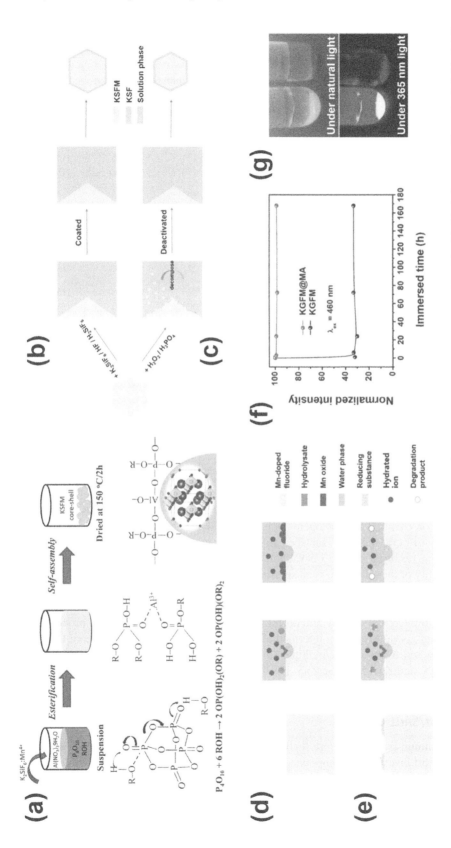

FIGURE 1.11 (a) Formation of the alkyl phosphate layer on the K_2SiF_6:Mn^{4+} phosphor surface. (From Nguyen HD, Lin CC, Liu RS *Angew. Chem. Int. Ed.* 54, 10862, 2015. With permission.) Designing strategies of K_2SiF_6:Mn^{4+}@K_2SiF_6 composite: (b) coating and (c) surface deactivation. (From Huang L, Liu Y, Yu J, Zhu Y, Pan F, Xuan T, Brik MG, Wang C, Wang J *ACS Appl. Mater. Interfaces* 10, 18082, 2018. With permission.) Different water erosive situations of (d) Mn^{4+}-doped fluoride and (e) reducing substance-loaded Mn^{4+}-doped fluoride. (f) The emission intensity of K_2GeF_6:Mn^{4+} and DL-mandelic acid–loaded K_2GeF_6:Mn^{4+} after immersion in water for t hours (t = 0–168; normalized at t = 0), and (g) photographs of K_2GeF_6:Mn^{4+} (right) and DL-mandelic acid–loaded K_2GeF_6:Mn^{4+} (left) after immersion in water for 1 hour, taken under natural light and 365 nm UV light, respectively. (From Huang L, Liu Y, Si S, Brik MG, Wang C, Wang J *Chem. Commun.* 54, 11857, 2018. With permission.)

TABLE 1.6

Summarized the Optical Performances of the QDs

Color	QDs	PL Peak (nm)	FWHM (nm)	PLQY (%)	Ref.
Blue	ZnTeSe/ZnSe/ZnS	457	36	100	[130]
	ZnSe/ZnS	455	20	83	[150]
	$CdZnS/Cd_xZn_{1-x}/ZnS$	441	15	100	[151]
	CdSe/ZnSe	475	NA	73	[152]
	$CsPbBr_3$	460	12	96	[153]
	$CsPb(Br_xCl_{1-x})_3$	471	17	100	[154]
Green	CdSe/ZnSe	525	NA	90	[152]
	$CdSe/Cd_{1-x}Zn_xZnSe_{1-y}S_y/ZnS$	533	NA	75	[155]
	InP/ZnSe/ZnS	528	36	95	[156]
	InP/GaP/ZnS	527	58	70	[138]
	$Cs_4PbBr_6/CsPbBr_3$	519	23.8	95	[157]
	$CsPbBr_3$	515	22	100	[158]
Red	CdSe/CdS	NA	NA	> 90	[159]
	$CdSe/Zn_{1-x}Cd_xS$	623	30	80	[132]
	InP/ZnSe/ZnS	630	35	100	[137]
	InP/ZnSe/ZnS	618	42	93	[139]
	$CsPb(I_{0.6}Br_{0.4})_3$	625	30	70	[160]
	$CsPbBr_{0.71}I_{2.23}$	649	40	80	[161]

Till now, there are three kinds of QDs, i.e., the II–VI core/shell, III–V core/shell, and perovskite QDs, which have a great potential application for micro-LEDs. The optical properties of the three types of QDs are summarized in Table 1.6 and will be described below in detail.

1.4.1 II–VI CORE/SHELL QDs

The II–VI QDs, such as CdSe, typically exhibit surface-related trap states, which leads to fast non-radiative de-excitation channels for photogenerated charge carries and thereby reduces the PLQY. [125] To enhance the PLQY, the QDs' surface can be passivated through overgrowth with a shell with broad bandgaps, resulting in type I core/shell QDs (Figure 1.12).[126–128] Interestingly, the efficiency and stability of the QDs have been significantly improved. For example, Hines et al. first overcoated CdSe QDs (3 nm) with ZnS layers and the as-obtained CdSe/ZnS core/shell show an enhanced PLQY of about 50%.[129] Unfortunately, the lattice mismatch between the core and shell material still leads to interface vacancy and defect, both of which deteriorate the PL properties of QDs. Then the use of a strain-reducing intermediate shell sandwiched between the core QDs and shell has been proposed to build core/multiple shells, such as CdSe/ZnSe/ZnS, CdSe/CdS/ZnS, $CdS/Zn_{0.5}Cd_{0.5}S/ZnS$, and ZnTeSe/ZnSe/ZnS, [130] which exhibits high PLQY of 75%–95% with narrow band emission (FWHM < 35 nm) in the spectral region of 400–750 nm.[131–133]

1.4.2 III–V CORE/SHELL QDs

Because of the cadmium-free, III–V core/shell QDs have attracted more attention. Indium phosphide (InP) is the most investigated compound, and the emission can be tuned from visible to near infrared by controlling the size.[135] However, the InP QDs show rather poor optical properties as compared to CdSe due to the defects in the deep in gap states of InP QDs. Although the ZnSe/ZnS shells (Figure 1.13) have been coated on InP core QDs, the PLQY and FWHM are still poor (PLQY: 20%–60%, FWHM: 46–63 nm). Oxidative defects are suspected to be the major cause of the poor optical properties of InP. Thus, many efforts have been made to reduce oxygen sources.[136] One of the most effective methods is in situ etching of the oxide surface of the InP QDs using HF during the growth stage of ZnSe shell.[137] The ZnS shell was then coated onto the InP/ZnSe QDs through adding Zn and S precursors sequentially. As-prepared InP/ZnSe/ZnS QDs have a near-unity PLQY (98%) and narrow band red emission centered at 630 nm with a FWHM of 35 nm. The PL peak of the InP-based core/shell QDs (e.g. InP/ZnSe/ZnS, InP/GaP/ZnSe/ZnS) can be tuned from 450 to 700 nm by changing the size of InP core with high PLQY of 65%–100% and narrow emission with the FWHM of 35–65 nm.[138–141]

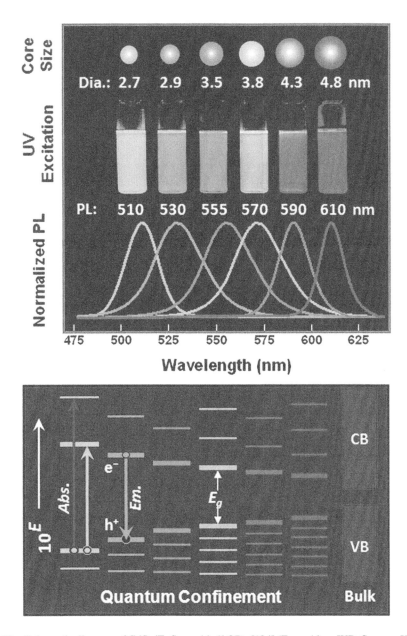

FIGURE 1.12 Schematic diagram of CdSe/ZnS core/shell QDs.[134] (From Algar WR, Susumu K, Delehanty JB, Medintz IL *Anal. Chem.* 83, 8826, 2011. With permission.)

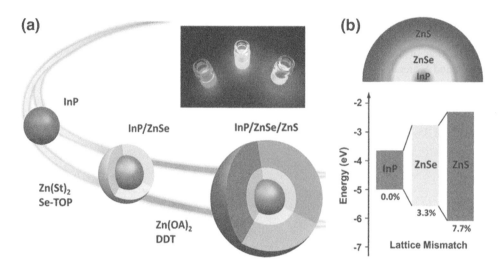

FIGURE 1.13   (a) Schematic of synthesis of InP/ZnSe/ZnS core/shell QDs and (b) the lattice mismatch of InP, ZnSe, and ZnS layers.[142] (From Cao F, Wang S, Wang F, Wu Q, Zhao D, Yang X *Chem. Mater.* 30, 8002, 2018. With permission.)

1.4.3   LEAD HALIDE PEROVSKITE QDS

Recently, ABX_3 that adopts with the $CaTiO_3$-like perovskite structure, where A is an organic or alkali-metal cation, such as $CH_3NH_3^+$ (MA+), $CH(NH_2)_2^+$ (FA+), and Cs+, B is a bivalent cation, such as Pb^{2+} or Sn^{2+}, and X is a monovalent anion, such as Cl−, Br−, and I−, also have attracted more attention. Among them, all inorganic ABX_3 perovskites have a great potential application in micro-LED displays. For example, the $CsPbX_3$ QDs (Figure 1.14) exhibit wide tunable bandgaps (400–700 nm), narrow band emission (FWHM < 35 nm), near-unity PLQY, high defect tolerance, as well as easy synthesis, convenient solution-based processing, and low cost.[143] However, the perovskite QDs

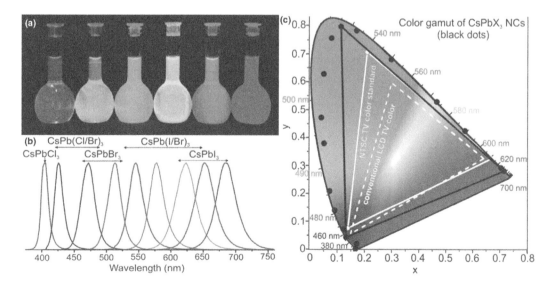

FIGURE 1.14   (a) Photographs of the lead halide perovskite QDs under 365 nm UV light and their (b) corresponding PL spectra and (c) CIE chromaticity coordinates.[121] (From Xuan T, Shi S, Wang L, Kuo HC, Xie RJ *J. Phys. Chem. Lett.* 11, 5184, 2020. With permission.)

suffer from the poor chemical, photo- and thermal stability. They are extremely sensitive to polar solvents due to their inherent ionic nature and are vulnerable to environmental stress like light, oxygen, and heat due to the low formation energy.[144] Many efforts have been made to improve the stability of the perovskite QDs with enhanced performances, which can be classified into three types: compositional engineering, surface engineering, and matrix encapsulation.[145–149]

1.5 CLASSIFICATIONS OF pc-WLEDs

Largely owing to the achievements in the field of UV/blue LEDs and phosphors, pc-WLEDs for general lighting and back-lighting source of traditional LED display and ultrahigh-definition micro-LEDs display have successfully developed in the last decades. For either lighting or back-lighting source for display, there are mainly two most popular approaches as shown in Figure 1.15. (1) A blue LED chip with yellow phosphors or green and red phosphors; (2) A UV LED chip with blue and yellow phosphors (or red, green, and blue phosphors).

1.5.1 pc-WLEDs for General Lighting

For general lighting, there are several key parameters for pc-WLEDs should be considered, mainly including luminous efficacy, CRI, CIE chromaticity coordinates, and CCT.

The pc-WLEDs are widely fabricated by combining a blue LED chip with YAG:Ce yellow phosphor (Figure 1.16(a) and 16(b)). Part of the blue light from the (In,Ga)N LED is absorbed by YAG:Ce and is converted into yellow light. The combination of blue and yellow lights gives a bright white light. Theoretically, this kind of pc-WLEDs has high luminous efficacy up to 350 lm/W,[163] which is more attractive for the creation of a cheap, bright white-lighting source. However, high luminous efficacy comes at the expense of lower CRI (~70%) and higher CCT (>6000K). Therefore, such devices are only suitable for outdoor lighting use.

For general indoor illumination or special lighting application, pc-WLEDs with a CRI of > 90% and the CCT of 2700–4200K is preferable for replacing traditional fluorescent lamps. Such kinds of pc-WLEDs could be produced using a two-phosphor approach by mixing yellow one with red

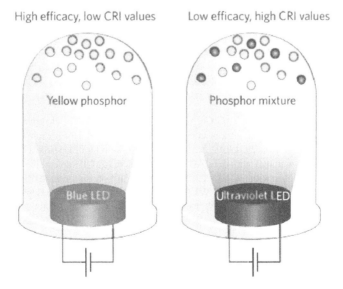

FIGURE 1.15 Two promising ways to produce white light based on LEDs.[162] (From Pimputkar S, Speck JS, DenBaars SP, Nakamura S *Nat. Photonics* 3, 180, 2009. With permission.)

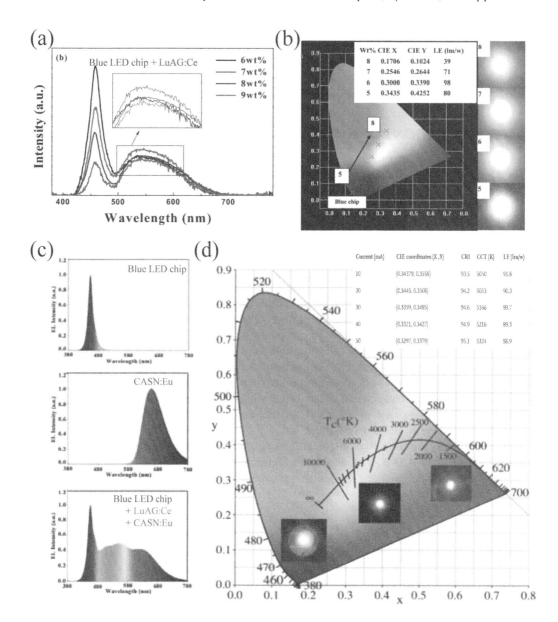

FIGURE 1.16 (a) Emission spectra of YAG:Ce^{3+}-based W-LED module, (b) chromaticity color coordinates and luminous efficiency of W-LEDs with various concentration of YAG:Ce^{3+}.[166] (From Cui S, Chen G, Chen Y, Liu X *J. Mater. Sci. Mater. Electron.* 29, 13019, 2018. With permission.) (c) EL spectra of the blue InGaN chip, CASN:Eu^{2+} and CASN:Eu^{2+} matched with a LuAG:Ce^{3+} phosphor; (d) CIE color coordinates of the WLEDs assembled by coupling a blue chip with a CASN:Eu^{2+} and LuAG:Ce^{3+} phosphor under different driving currents (insets show the photographs of the blue chip, WLED, and red LED, correspondingly).[167] (From Zhang Y, Zhang Z, Liu X, Shao G, Shen L, Liu J, Xiang W, Liang X *Chem. Eng. J.* 401, 125983, 2020. With permission.)

one or green/orange one with red one. For instance, the CRI and CCT values are greatly improved to be ~84% and ~4210K of pc-WLEDs fabricated based on yellow YAG:Ce and red K$_2$SiF$_6$:Mn^{4+}. [164,165] By incorporating CaAlSiN$_3$:Eu^{2+} as a red component and Lu$_3$Al$_5$O$_{12}$:Ce^{3+} as green one, pc-WLEDs have an excellent white-light quality with high luminous efficacy (88.9–91.8 lm/W), high CRI (CRI = 93.5–95.1) and low CCT (5050–5324K), as shown in Figure 1.16(c) and (d).

Beside pc-WLEDs based on blue LED chip, pc-WLEDs could be also fabricated by combining UV LED chip with the mixture of phosphors with different emission color. For instance, as shown in Figure 1.17(a) and (b), pc-WLEDs is fabricated by coupling the phosphors of $BaMgAl_{10}O_{17}:Eu^{2+}$ (BAM:Eu, blue), $Y_3(Al,Ga)_5O_{12}:Ce^{3+}$ (YAGG:Ce, green), and $K_2MgGeO_4:Bi^{3+}$ (KMGO:Bi, orange) with a commercial 310 nm UV LED chip and this prototype device exhibits a high CRI of 93.8 and CCT of 4437K.[168]

For pc-WLEDs based on the mixture of phosphors, it is hard to keep the white-light qualities of pc-WLEDs fully consistent in the process of mass fabrication and longtime usage, due to different specific gravity of different phosphors, the reabsorption between different phosphors, and different aging rates for each phosphor. Theoretically, it is possible to fabricate pc-WLEDs with UV LED chip and single-phase white-emitting phosphor, which could exhibit excellent white-light quality. For instance,

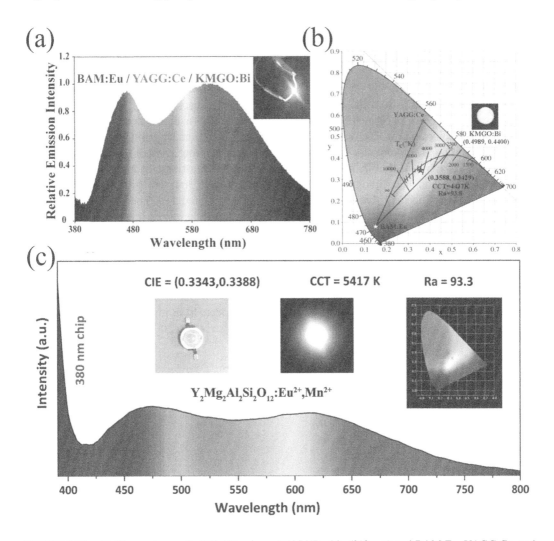

FIGURE 1.17 (a) EL spectrum of a WLED using a UV LED chip (310 nm) and BAM:Eu, YAGG:Ce, and KMGO:Bi phosphors. The insets show photographs of the fabricated white LEDs. (b) The CIE coordinates of BAM:Eu, YAGG:Ce, and KMGO:Bi phosphors and fabricated WLED. (From Li H, Pang R, Liu G, Sun W, Li D, Jiang L, Zhang S, Li C, Feng J, Zhang H *Inorg. Chem.* 57, 12303, 2018. With permission.) (c) EL spectra, relevant parameters, real-scene photos and CIE chromaticity diagram of the $Y_2Mg_2Al_2Si_2O_{12}:Eu^{2+},Mn^{2+}$ phosphor-converted WLED driven by a 30 mA current. (From Zhang X, Zhang D, Zheng Z, Zheng B, Song Y, Zheng K, Sheng Y, Shi Z, Zou H *Dalton Trans.* 49, 17796, 2020. With permission.)

Zhang et al. reported that pc-WLEDs fabricated by using $BaMg_2Al_6Si_9O_{30}$:Eu^{2+},Tb^{3+},Mn^{2+} phosphor, and a 365 nm NUV chip show high CRI of 90.[169] Zou et al. reported pc-WLEDs, fabricated by using a single-phase full-visible-spectrum $Y_2Mg_2Al_2Si_2O_{12}$:Eu^{2+},Mn^{2+}, and UV LED chip, show an ultrahigh CRI of ~93 and a suitable CCT of 5417K, as shown in Figure 1.17(c).[89]

1.5.2 PC-WLEDs FOR BACK-LIGHTING SOURCE FOR DISPLAY

For back-lighting source for display, there are two key parameters, luminous efficacy and color gamut, which should be considered. For back-lighting source in traditional LED display, for instance, Yoshimura et al. successfully developed a high-brightness and thermally robust pc-WLED back-light device, which is mostly able to cover the NTSC (National Television Standards Committee) triangle by using sharp β-Sialon:Eu and sharp K_2SiF_6:Mn^{4+} phosphors (Figure 1.18). This pc-WLED back-light device shows not only wider color gamut but also higher brightness compared to the case using sharp β-Sialon:Eu and broad $CaAlSiN_3$:Eu^{2+} phosphor. Nowadays, the luminous efficacy of this kind of pc-WLED back-light device reaches up to 136 lm/W and its color gamut can cover 96% of the NTSC.[170]

For back-lighting source of ultrahigh-definition micro-LEDs display, the fabrication of the devices is much different from the one for general lighting or back-lighting source in traditional LEDs display. Till now, many methods have been utilized to fabricate QDs color conversion films (CCFs) for full-color micro-LED, such as nanoimprinting, microcontact printing, photolithography, and inkjet printing. The nanoimprinting and microcontact printing technologies suffer from low machining efficiency, uncontrollable transfer printing, strong causticity, or high cost. In contrast, the inkjet printing and photolithography exhibit a simple process, low cost, high degree of automation, and selective patterning. For instance, Kuo et al. demonstrated a method to achieve full-color high-quality micro-LED displays by combining UV micro-LEDs with red-green-blue (RGB) II–VI QD CCFs via the aerosol jet (AJ) technique. Additionally, a photoresist (PR) mold is fabricated to limit the optical cross-talk effect, and the QD CCFs are deposited into the PR mold with clear separation between each other. Furthermore, a distributed Bragg reflector (DBR) covered above the PR mold can significantly enhance both utilization of UV light and the emission intensity of RGB QDs. Lately, the red- and green-emitting CdSe/CdZnS core/shell QD CCFs were combined with the semipolar blue micro-LED chips (50 μm in diameter) to fabricate RGB micro-LED arrays.[172] The RGB micro-LED pixels show bright color individually when driven at 200 A/cm². The EL spectrum

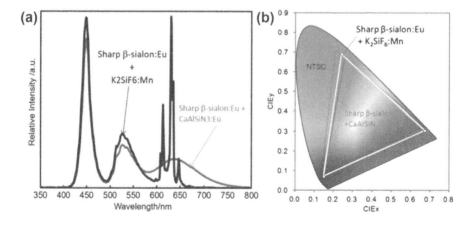

FIGURE 1.18 (a) Emission spectra of the fabricated WLEDs with blue LED, β-Sialon:Eu and K_2SiF_6:Mn^{4+} or $CaAlSiN_3$:Eu^{2+}. (b) The CIE 1931 chromaticity coordinates of the display using fabricated WLEDs.[171] (From Yoshimura K, Fukunaga H, Izumi M, Masuda M, Uemura T, Takahashi K, Xie RJ, Hirosaki N *J. Soc. Inf. Disp.* 24, 449, 2016. With permission.)

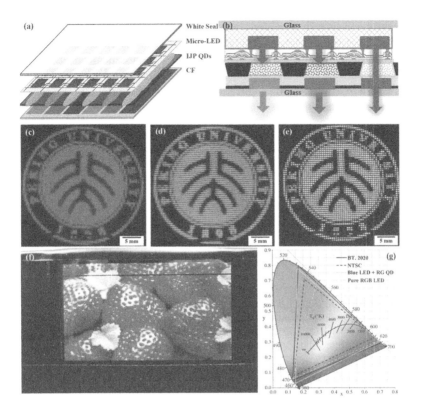

FIGURE 1.19 Schematic structure and display of the fabricated full-color micro-LED by combining green CsPbBr$_3$ QDs and red CdSe QDs with blue micro-LED chip arrays. (a) Schematic illustration and (b) cross-section structure diagram of the displays. (c-e) RGB images of Peaking University logo and (f) red strawberry image manifested by the QDs based micro-LED displays. (g) Color gamut of the micro-LEDs, BT. 2020 and NTSC standard. (From Yin Y, Hu Z, Ali MU, Duan M, Gao L, Liu M, Peng W, Geng J, Pan S, Wu Y, Hou J, Fan J, Li D, Zhang X, Meng H *Adv. Mater. Technol.* 5, 2000251, 2020. With permission.)

of each RGB pixel shows a distinct peak at 630, 536, and 453 nm with the FWHM of 30.5, 24.5, and 24.8 nm, respectively. It thus leads to a wide color gamut of 114.4% of NTSC and 85.4% *Rec.* 2020 in the CIE 1931 space. Simultaneously, the color gamut of the RGB micro-LED pixels remains nearly unchanged under 1–200 A/cm^2 driving conditions because of the excellent stability of blue micro-LEDs and QDs' CCFs. To improve the color gamut of the micro-LED, Meng et al. presented a full-color micro-LED display prototype (Figure 1.19) by combing rationally designed blue micro-LEDs backlight with green CsPbBr$_3$ perovskite QDs, and red CdSe QDs as the CCFs, which shows a wider color gamut (129% NTSC).[173]

1.6 SUMMARY AND OUTLOOK

In the last decades, remarkable progresses and exciting success have been achieved in design, synthesis, and luminescence properties of phosphors, including rare-earth ion or transition-metal ion–activated phosphor in micro- and nanoscale. More importantly, phosphors successfully demonstrated their actual applications in commercial pc-WLEDs for general lighting, traditional LED back-lighting display and ultra-high-definition micro-LED display. There are still many challenges for phosphors that need to be addressed: (1) for traditional LED back-lighting display, synthesis methods of narrow red-emitting fluoride phosphors with high luminescence performance and excellent water resistance are nongreen, which greatly limits its mass production if the care of safety and costs are fully considered. Therefore, new synthesis routes are in great needs. (2) for ultra-high-definition micro-LED display, currently available QDs with excellent luminescence performance almost contain toxic elements, for instance, Cd or Pb. Consequently, new Cd- or Pb-free QDs with high efficient luminescence and excellent stability will certainly be a hot issue in the future. Moreover, motivated by new potential applications constantly emerging, for instance, "smart" light sources, one should pay more attentions to design, synthesis, and luminescence properties of new kinds of phosphors. In summary, it is anticipated that further advances in phosphors and their applications will revolutionize the lighting industry and lead us into a more energy-efficient, bright, colorful, and "smart" future.

REFERENCES

1. Holonyak N Jr, Bevacqua SF (1962) Coherent (visible) light emission from Ga ($As_{1-x}P_x$) junctions. *Appl. Phys. Lett.* 1: 82.
2. Nakamura S, Mukai T, Senoh M (1994) Candela-class high-brightness InGaN/AlGaN double-heterostructure blue-light-emitting diodes. *Appl. Phys. Lett.* 64: 1687.
3. Nakamura S, Senoh M, Mukai T (1993) High-power InGaN/GaN double-heterostructure violet light emitting diodes. *Appl. Phys. Lett.* 62: 2390.
4. Amano H, Sawaki N, Akasaki I, Toyoda Y (1986) Metalorganic vapor phase epitaxial growth of a high quality GaN film using an AlN buffer layer. *Appl. Phys. Lett.* 48: 353.
5. Jang HS, Won YH, Jeon DY (2009) Improvement of electroluminescent property of blue LED coated with highly luminescent yellow-emitting phosphors. *Appl. Phys. B* 95: 715.
6. Du F, Zhuang W, Liu R, Liu Y, Gao W, Zhang X, Xue Y, Hao H (2017) Synthesis, structure and luminescent properties of yellow phosphor $La_3Si_6N_{11}$:Ce^{3+} for high power white-LEDs. *J. Rare Earths* 35: 1059.
7. Kijima N, Seto T, Hirosaki N (2009) A new yellow phosphor $La_3Si_6N_{11}$:Ce^{3+} for white LEDs. *ECS Trans.* 25: 247.
8. Kechele JA, Hecht C, Oeckler O, Schmedt auf der Günne J, Schmidt PJ, Schnick W (2009) $Ba_2AlSi_5N_9$—a new host lattice for Eu^{2+}-doped luminescent materials comprising a nitridoalumosilicate framework with corner- and edge-sharing tetrahedra. *Chem. Mater.* 21: 1288.
9. Zhao Y, Xu H, Zhang X, Zhu G, Yan D, Yu A (2015) Facile synthesis of YAG:Ce^{3+} thick films for phosphor converted white light emitting diodes. *J. Eur. Ceram. Soc.* 35: 3761.
10. Yang YG, Wei L, Xu JH, Yu HJ, Hu YY, Zhang HD, Wang XP, Liu B, Zhang C, Li QG (2019) Luminescence of $Tb_3Al_5O_{12}$ phosphors co-doped with Ce^{3+}/Gd^{3+} for white light-emitting diodes. *Beilstein J. Nanotechnol.* 10: 1237.
11. Park JK, Kim CH, Park SH, Park HD, Choi SY (2004) Application of strontium silicate yellow phosphor for white light-emitting diodes. *Appl. Phys. Lett.* 84: 1647.
12. Saradhi MP, Varadaraju UV (2006) Photoluminescence studies on Eu^{2+}-activated Li_2SrSiO_4 a potential orange-yellow phosphor for solid-state lighting. *Chem. Mater.* 18: 5267.
13. Setlur AA, Heward WJ, Gao Y, Srivastava AM, Chandran RG, Shankar MV (2006) Crystal chemistry and luminescence of Ce^{3+}-doped $Lu_2CaMg_2(Si,Ge)_3O_{12}$ and its use in LED based lighting. *Chem. Mater.* 18: 3314.
14. Long J, Wang Y, Ma C, Yuan X, Dong W, Ma R, Wen Z, Du M, Cao Y (2017) Photoluminescence tuning of $Ca_{8-x}Sr_xMgGd(PO_4)_7$:Eu^{2+}, yMn^{2+} phosphors for applications in white LEDs with excellent color rendering index. *RSC Adv.* 7: 19223.
15. Zhang X, Choi N, Park K, Kim J (2009) Orange emissive $Sr_3Al_2O_5Cl_2$:Eu^{2+} phosphor for warm-white light-emitting diodes. *Solid State Commun.* 149: 1017.
16. Chen Z, Zhang Q, Li Y, Wang H, Xie RJ (2017) A promising orange-yellow-emitting phosphor for high power warm-light white LEDs: pure-phase synthesis and photoluminescence properties. *J. Alloys Compd.* 715: 184.
17. Piao X, Machida K, Horikawa T, Hanzawa H (2007) Self-propagating high temperature synthesis of yellow-emitting $Ba_2Si_5N_8$:Eu^{2+} phosphors for white light-emitting diodes. *Appl. Phys. Lett.* 91: 041908.
18. Li YQ, Delsing ACA, de With G, Hintzen HT (2005) Luminescence properties of Eu^{2+}-activated alkaline-earth silicon-oxynitride $MSi_2O_{2-\delta}N_{2+2/3\delta}$ (M=Ca, Sr, Ba): a promising class of novel LED conversion phosphors. *Chem. Mater.* 17: 3242.

19. Zhou Q, Dolgov L, Srivastava AM, Zhou L, Wang Z, Shi J, Dramićanin MD, Brik MG, Wu M (2018) Mn^{2+} and Mn^{4+} red phosphors: synthesis, luminescence and applications in WLEDs. *J. Mater. Chem. C* 6: 2652.

20. Zhong S, Chen J, Wang S, Liu Q, Wang Y, Wang S (2010) Y_2O_3:Eu^{3+} hexagonal microprisms: fast microwave synthesis and photoluminescence properties. *J. Alloys Compd.* 493: 322.

21. Guo H, Huang X, Zeng Y (2018) Synthesis and photoluminescence properties of novel highly thermal-stable red-emitting $Na_3Sc_2(PO_4)_3$:Eu^{3+} phosphors for UV-excited white-light-emitting diodes. *J. Alloys Compd.* 741: 300.

22. Dabre KV, Dhoble SJ (2014) Synthesis and photoluminescence properties of Eu^{3+}, Sm^{3+} and Pr^{3+} doped Ca_2ZnWO_6 phosphors for phosphor converted LED. *J. Lumin.* 150: 55.

23. Kiran N, Baker AP, Wang GG (2017) Synthesis and luminescence properties of MgO:Sm^{3+} phosphor for white light-emitting diodes. *J. Mol. Struct.* 1129: 211.

24. He X, Zhou J, Lian N, Sun J, Guan M (2010) Sm^{3+}-activated gadolinium molybdate: an intense red-emitting phosphor for solid-state lighting based on InGaN LEDs. *J. Lumin.* 130: 743.

25. Zhang J, Zhang X, Gong M, Shi J, Yu L, Rong C, Lian S (2012) LiSrBO$_3$:Eu^{2+}: a novel broad-band red phosphor under the excitation of a blue light. *Mater. Lett.* 79: 100.

26. Chen D, Zhou Y, Xu W, Zhong J, Ji Z, Xiang W (2016) Enhanced luminescence of Mn^{4+}:$Y_3Al_5O_{12}$ red phosphor via impurity doping. *J. Mater. Chem., C*, 4: 1704.

27. Som S.; Mitra P, Kumar V, Kumar V, Terblans JJ, Swart HC, Sharma SK (2014) The energy transfer phenomena and colour tunability in Y_2O_2S:Eu^{3+}/Dy^{3+} micro-fibers for white emission in solid state lighting applications. *Dalton Trans.* 43: 9860.

28. Uheda K, Hirosaki N, Yamamoto Y, Naito A, Nakajima T, Yamamoto H (2006) Luminescence properties of a red phosphor, CaAlSiN$_3$:Eu^{2+}, for white light-emitting diodes. *Electrochem. Solid-State Lett.* 9: H22.

29. Xianqing P, Takashi H, Hiromasa H, Ken-ichi M (2006) Photoluminescence properties of $Ca_2Si_5N_8$:Eu^{2+} nitride phosphor prepared by carbothermal reduction and nitridation method. *Chem. Lett.* 35: 334.

30. Pust P, Weiler V, Hecht C, Tücks A, Wochnik AS, Henß AK, Wiechert D, Scheu C, Schmidt PJ, Schnick W (2014) Narrow-band red-emitting Sr[LiAl$_3$N$_4$]:Eu^{2+} as a next-generation LED-phosphor material. *Nat. Mater.* 13: 891.

31. Adachi S, Takahashi T (2008) Direct synthesis of K_2SiF_6:Mn^{4+} red phosphor from crushed quartz schist by wet chemical etching. *Electrochem. Solid-State Lett.* 12: J20.

32. Zhou YY, Song EH, Deng TT, Zhang QY (2018) Waterproof narrow-band fluoride red phosphor K_2TiF_6:Mn^{4+} via facile superhydrophobic surface modification. *ACS Appl. Mater. Interfaces* 10: 880.

33. Okamoto S, Yamamoto H (2010) Luminescent-efficiency improvement by alkaline-earth fluorides partially replacing MgO in 3.5MgO·0.5MgF$_2$·GeO$_2$:Mn^{4+} deep-red phosphors for light emitting diodes. *J. Electrochem. Soc.* 157: J59.

34. Kim HS, Machida KI, Horikawa T, Hanzawa H (2015) Luminescence properties of CaAlSiN$_3$:Eu^{2+} phosphor prepared by direct-nitriding method using fine metal hydride powders. *J. Alloys Compd.* 633: 97.

35. Uheda K, Hirosaki N, Yamamoto H (2006) Host lattice materials in the system Ca_3N_2-AlN-Si$_3$N$_4$ for white light emitting diode. *Phys. Status Solidi (A)* 203: 2712.

36. Li S, Liu XJ, Mao R, Huang Z, Xie RJ (2015) Red-emission enhancement of the CaAlSiN$_3$:Eu^{2+} phosphor by partial substitution for Ca_3N_2 by CaCO$_3$ and excess calcium source addition. *RSC Adv.* 5: 76507.

37. Piao X, Machida KI, Horikawa T, Hanzawa H, Shimomura Y, Kijima N (2007) Preparation of CaAlSiN$_3$:Eu^{2+} phosphors by the self-propagating high-temperature synthesis and their luminescent properties. *Chem. Mater.* 19: 4592.

38. Hoerder GJ, Seibald M, Baumann D, Schroder T, Peschke S, Schmid PC, Tyborski T, Pust P, Stoll I, Bergler M, Patzig C, Reissaus S, Krause M, Berthold L, Hoche T, Johrendt D, Huppertz H (2019) Sr[Li$_2$Al$_2$O$_2$N$_2$]:Eu^{2+}—a high performance red phosphor to brighten the future. *Nat. Commun.* 10: 1824.

39. Murphy JE, Garcia-Santamaria F, Setlur AA, Sista S (2015) 62.4: PFS, K_2SiF_6:Mn^{4+}: the red-line emitting LED phosphor behind GE's TriGain Technology™ platform. *SID Symp. Dig. Tech. Pap.* 46: 927.

40. Zhang X, Fang MH, Tsai YT, Lazarowska A, Mahlik S, Lesniewski T, Grinberg M, Pang WK, Pan F, Liang C, Zhou W, Wang J, Lee JF, Cheng BM, Hung TL, Chen YY, Liu RS (2017) Controlling of structural ordering and rigidity of β-SiAlON:Eu through chemical cosubstitution to approach narrow-band-emission for light-emitting diodes application. *Chem. Mater.* 29: 6781.

41. Kim YH, Im WB (2016) A new green phosphor using solid-solution and their structural and optical properties. *Meet. Abstr.* 43: 3190.

42. Praveena R, Shi L, Jang KH, Venkatramu V, Jayasankar CK, Seo HJ (2011) Sol–gel synthesis and thermal stability of luminescence of $Lu_3Al_5O_{12}$:Ce^{3+} nano-garnet. *J. Alloys Compd.* 509: 859.

43. Park WJ, Song YH, Yoon DH (2010) Synthesis and luminescent characteristics of $Ca_{2-x}Sr_xSiO_4$:Eu^{2+} as a potential green-emitting phosphor for near UV-white LED applications. *Mater. Sci. Eng. B* 173: 76.

44. Zhang X, Tang X, Zhang J, Gong M (2010) An efficient and stable green phosphor $SrBaSiO_4$:Eu^{2+} for light-emitting diodes. *J. Lumin.* 130: 2288.

45. Hur S, Song HJ, Roh HS, Kim DW, Hong KS (2013) A novel green-emitting $Ca_{15}(PO_4)_2(SiO_4)_6$:Eu^{2+} phosphor for applications in NUV based W-LEDs. *Mater. Chem. Phys.* 139: 350.

46. Kim JS, Park YH, Kim SM, Choi JC, Park HL (2005) Temperature-dependent emission spectra of M_2SiO_4:Eu^{2+} (M=Ca, Sr, Ba) phosphors for green and greenish white LEDs. *Solid State Commun.* 133: 445.

47. Annadurai G, Kennedy SMM, Sivakumar V (2016) Luminescence properties of a novel green emitting $Ba_2CaZn_2Si_6O_{17}$:Eu^{2+} phosphor for white light – emitting diodes applications. *Superlattices Microstruct.* 93: 57.

48. Zhang X, Zhou F, Shi J, Gong M (2009) $Sr_{3.5}Mg_{0.5}Si_3O_8Cl_4$:$Eu^{2+}$ bluish–green-emitting phosphor for NUV-based LED. *Mater. Lett.* 63: 852.

49. Baginskiy I, Liu RS (2009) Significant improved luminescence intensity of Eu^{2+}-doped $Ca_3SiO_4Cl_2$ green phosphor for white LEDs synthesized through two-stage method. *J. Electrochem. Soc.* 156: G29.

50. Zhang B, Zhang JW, Zhong H, Hao LY, Xu X, Agathopoulos S, Wang CM, Yin LJ (2017) Enhancement of the stability of green-emitting Ba_2SiO_4:Eu^{2+} phosphor by hydrophobic modification. *Mater. Res. Bull.* 92: 46.

51. Koizumi H, Watabe J, Sugiyama S, Hirabayashi H, Homma T (2018) Effect of sintering temperature of Ce^{3+}-doped $Lu_3Al_5O_{12}$ phosphors on light emission and properties of crystal structure for white-light-emitting diodes. *Opt. Rev.* 25: 340.

52. Chen Y, Gong M, Wang G, Su Q (2007) High efficient and low color-temperature white light-emitting diodes with $Tb_3Al_5O_{12}$:Ce^{3+} phosphor. *Appl. Phys. Lett.* 91: 071117.

53. Kang T, Lim S, Lee S, Kang H, Yu Y, Kim J (2019) Luminescent properties of $CaSc_2O_4$:Ce^{3+} green phosphor for white LED and its optical simulation. *Opt. Mater.* 98: 109501.

54. Jang HS, Jeon DY (2007) Yellow-emitting Sr_3SiO_5:Ce^{3+},Li^+ phosphor for white-light-emitting diodes and yellow-light-emitting diodes *Appl. Phys. Lett.* 90: 041906.

55. Im WB, Kim YI, Fellows NN, Masui H, Hirata GA, DenBaars SP, Seshadri R (2008) A yellow-emitting Ce^{3+} phosphor, $La_{1-x}Ce_xSr_2AlO_5$, for white light-emitting diodes. *Appl. Phys. Lett.* 93: 091905.

56. Zhao M, Liao H, Ning L, Zhang Q, Liu Q, Xia Z (2018) Next-generation narrow-band green-emitting $RbLi(Li_3SiO_4)_2$:Eu^{2+} phosphor for backlight display application. *Adv. Mater.* 30: 1802489.

57. Smet PF, Korthout K, Van Haecke JE, Poelman D (2008) Using rare earth doped thiosilicate phosphors in white light emitting LEDs: towards low colour temperature and high colour rendering. *Mater. Sci. Eng. B* 146: 264.

58. Xinmin Z, Hao W, Heping Z, Qiang S (2007) Luminescent properties of $SrGa_2S_4$:Eu^{2+} and its application in green-LEDs. *J. Rare Earths* 25: 701.

59. Fu X, Lü W, Jiao M, You H (2016) Broadband yellowish-green emitting $Ba_4Gd_3Na_3(PO_4)_6F_2$:Eu^{2+} phosphor: structure refinement, energy transfer, and thermal stability. *Inorg. Chem.* 55: 6107.

60. Wu W, Chao K, Liu W, Wei L, Hu D, Wang T, Tegus O (2018) Enhanced orange emission by doping CeB_6 in $CaAlSiN_3$:Ce^{3+} phosphor for application in white LEDs. *J. Rare Earth* 36: 1250.

61. Xie RJ, Hirosaki N, Li HL, Li YQ, Mitomo M (2007) Synthesis and photoluminescence properties of β-sialon:Eu^{2+}($Si_{6-z}Al_zO_zN_{8-z}$:Eu^{2+}): a promising green oxynitride phosphor for white light-emitting diodes. *J. Electrochem. Soc.* 154: J314.

62. Howe B, Diaz AL (2004) Characterization of host-lattice emission and energy transfer in $BaMgAl_{10}O_{17}$:Eu^{2+}. *J. Lumin.* 109: 51.

63. Lee SS, Kim HJ, Byeon SH, Park JC, Kim DK (2005) Thermal-shock-assisted solid-state process for the production of $BaMgAl_{10}O_{17}$:Eu phosphor. *Ind. Eng. Chem. Res.* 44: 4300.

64. Kim KB, Kim YI, Chun HG, Cho TY, Jung JS, Kang JG (2002) Structural and optical properties of $BaMgAl_{10}O_{17}$:Eu^{2+} phosphor. *Chem. Mater.* 14: 5045.

65. Ju H, Wang B, Ma Y, Chen S, Wang H, Yang S (2014) Preparation and luminescence properties of $Na_4CaSi_3O_9$:Ce^{3+} phosphors for solid state lighting. *Ceram. Int.* 40: 11085.

66. Lu FC, Bai LJ, Yang ZP, Han XN (2015) Synthesis and photoluminescence of a novel blue-emitting $Gd_5Si_3O_{12}N$:Ce^{3+} phosphor. *Mater. Lett.* 151: 9.

67. Hao Z, Zhang J, Zhang X, Ren X, Luo Y, Lu S, Wang X (2008) Intense violet-blue emitting ($CaCl_2$/SiO_2):Eu^{2+} phosphor powders for applications in UV-LED based phototherapy illuminators. *J. Phys. D: Appl. Phys.* 41: 182001.

68. Park JK, Jae Choi K, Hae Kim C, Dong Park H, Kon Kim H (2004) Luminescence characteristics of $Sr_3MgSi_2O_8$:Eu blue phosphor for light-emitting diodes. *Electrochem. Solid-State Lett.* 7: H42.

69. Xiao F, Xue YN, Zhang QY (2009) Bluish-green color emitting $Ba_2Si_3O_8$:Eu^{2+} ceramic phosphors for white light-emitting diodes. *Spectrochim. Acta, A* 74: 758.

70. Yuan JL, Zeng XY, Zhao JT, Zhang ZJ, Chen HH, Zhang GB (2007) Rietveld refinement and photoluminescent properties of a new blue-emitting material: Eu^{2+} activated $SrZnP_2O_7$. *J. Solid State Chem.* 180: 3310.

71. Wu ZC, Shi JX, Wang J, Gong ML, Su Q (2006) A novel blue-emitting phosphor $LiSrPO_4$:Eu^{2+} for white LEDs. *J. Solid State Chem.* 179: 2356.

72. Bispo-Jr AG, Lima SAM, Carlos LD, Pires AM, Ferreira RAS (2020) Eu(II)-activated silicates for UV light-emitting diodes tuning into warm white light. *Adv. Eng. Mater.* 22: 2000422.

73. Zeng Q, Tanno H, Egoshi K, Tanamachi N, Zhang S (2006) $Ba_5SiO_4Cl_6$:Eu^{2+}: an intense blue emission phosphor under vacuum ultraviolet and near-ultraviolet excitation. *Appl. Phys. Lett.* 88: 051906.

74. Liao H, Zhao M, Molokeev MS, Liu Q, Xia Z (2018) Learning from a mineral structure toward an ultra-narrow-band blue-emitting silicate phosphor $RbNa_3(Li_3SiO_4)_4$:Eu^{2+}. *Angew. Chem. Int. Ed.* 130: 11902.

75. Zhao M, Liao H, Molokeev MS, Zhou Y, Zhang Q, Liu Q, Xia Z (2019) Emerging ultra-narrow-band cyan-emitting phosphor for white LEDs with enhanced color rendition. *Light: Sci. Appl.* 8: 1.

76. Lan X, Wei Q, Chen Y, Tang W (2012) Luminescence properties of Eu^{2+}-activated $KMg_4(PO_4)_3$ for blue-emitting phosphor. *Opt. Mater.* 34: 1330.

77. Song HJ, Yim DK, Roh HS, Cho IS, Kim SJ, Jin YH, Shim HW, Kim DW, Hong KS (2013) $RbBaPO_4$:Eu^{2+}: a new alternative blue-emitting phosphor for UV-based white light-emitting diodes. *J. Mater. Chem. C* 1: 500.

78. Kohale RL, Dhoble SJ (2013) Eu^{2+} luminescence in $SrCaP_2O_7$ pyrophosphate phosphor. *Luminescence* 28: 656.

79. Kim YH, Arunkumar P, Kim BY, Unithrattil S, Kim E, Moon SH, Hyun JY, Kim KH, Lee D, Lee JS, Im WB (2017) A zero-thermal-quenching phosphor. *Nat. Mater.* 16: 543.

80. Zhang X, Song J, Zhou C, Zhou L, Gong M (2014) High efficiency and broadband blue-emitting $NaCaBO_3$:Ce^{3+} phosphor for NUV light-emitting diodes. *J. Lumin.* 149: 69.

81. Zhang J, Zhang W, Qiu Z, Zhou W, Yu L, Li Z, Lian S (2015) $Li_4SrCa(SiO_4)_2$:Ce^{3+}, a highly efficient near-UV and blue emitting orthosilicate phosphor. *J. Alloys Compd.* 646: 315.

82. Yin LJ, Chen GZ, Zhou ZY, Jian X, Xu B, He JH, Tang H, Luan CH, Xu X, van Ommen JR, Hintzen HT (2015) Improved blue-emitting AlN:Eu^{2+} phosphors by alloying with GaN. *J. Am. Ceram. Soc.* 98: 3897.

83. Strobel P, de Boer T, Weiler V, Schmidt PJ, Moewes A, Schnick W (2018) Luminescence of an oxonitridoberyllate: a study of narrow-band cyan-emitting $Sr[Be_6ON_4]$:Eu^{2+}. *Chem. Mater.* 30: 3122.

84. Liu F, Liu Q, Fang Y, Zhang N, Yang B, Zhao G (2015) White light emission from NaLa $(PO_3)_4$:Dy^{3+} single-phase phosphors for light-emitting diodes. *Ceram. Int.* 41: 1917.

85. Zhang ZW, Song AJ, Ma MZ, Zhang XY, Yue Y, Liu RP (2014) A novel white emission in $Ca_8MgBi(PO_4)_7$:Dy^{3+} single-phase full-color phosphor. *J. Alloys Compd.* 601: 231.

86. Dalal M, Taxak VB, Dalal J, Khatkar A, Chahar S, Devi R, Khatkar SP (2017) Crystal structure and Judd-Ofelt properties of a novel color tunable blue-white-red $Ba_5Zn_4Y_8O_{21}$:Eu^{3+} nanophosphor for near-ultraviolet based WLEDs. *J. Alloys Compd.* 698: 662.

87. Wu ZC, Liu J, Hou WG, Xu J, Gong ML (2010) A new single-host white-light-emitting $BaSrMg(PO_4)_2$:Eu^{2+} phosphor for white-light-emitting diodes. *J. Alloys Compd.* 498: 139.

88. Qiao J, Zhang Z, Zhao J, Xia Z (2019) Tuning of the compositions and multiple activator sites toward single-phased white emission in $(Ca_{9-x}Sr_x)MgK(PO_4)_7$:$Eu^{2+}$ phosphors for solid-state lighting. *Inorg. Chem.* 58: 5006.

89. Zhang X, Zhang D, Zheng Z, Zheng B, Song Y, Zheng K, Sheng Y, Shi Z, Zou H (2020) A single-phase full-visible-spectrum phosphor for white light-emitting diodes with ultra-high color rendering. *Dalton Trans.* 49: 17796.

90. Liu Y, Lan A, Jin Y, Chen G, Zhang X (2015) $Sr_3Bi(PO_4)_3$:Eu^{2+}, Mn^{2+}: single-phase and color-tunable phosphors for white-light LEDs. *Opt. Mater.* 40: 122.

91. Zhang XM, Li WL, Seo HJ (2009) Luminescence and energy transfer in Eu^{2+}, Mn^{2+} co-doped $Li_4SrCa(SiO_4)_2$ for white light-emitting-diodes. *Phys. Lett. A* 373: 3486.

92. Zhu Y, Liang Y, Liu S, Li H, Chen J, Lei W (2018) A strategy for realizing tunable luminescence and full-color emission in $Sr_3Gd_2(Si_3O_9)_2$:Eu phosphors by introducing dual functional Mn^{2+}. *Inorg. Chem. Front.* 5: 2527.

93. Zhang X, Zhu Z, Guo Z, Sun Z, Yang Z, Zhang T, Zhang J, Wu ZC, Wang Z (2019) Dopant preferential site occupation and high efficiency white emission in $K_2BaCa(PO_4)_2$:Eu^{2+}, Mn^{2+} phosphors for high quality white LED applications. *Inorg. Chem. Fron.* 6: 1289.

94. Liu WR, Huang CH, Yeh CW, Tsai JC, Chiu YC, Yeh YT, Liu RS (2012) A study on the luminescence and energy transfer of single-phase and color-tunable $KCaY(PO_4)_2$:Eu^{2+}, Mn^{2+} phosphor for application in white-light LEDs. *Inorg. Chem.* 51: 9636.

95. Xiang J, Zheng J, Zhou Z, Suo H, Zhao X, Zhou X, Zhang N, Molokeev MS, Guo C (2019) Enhancement of red emission and site analysis in Eu^{2+} doped new-type structure $Ba_3CaK(PO_4)_3$ for plant growth white LEDs. *Chem. Eng. J.* 356: 236.

96. Wang Z, Hou X, Liu Y, Hui Z, Huang Z, Fang M, Wu X (2017) Luminescence properties and energy transfer behavior of colour-tunable white-emitting $Sr_4Al_{14}O_{25}$ phosphors with co-doping of Eu^{2+}, Eu^{3+} and Mn^{4+}. *RSC Adv.* 7: 52995.

97. An Z, Liu W, Song Y, Zhang X, Dong R, Zhou X, Zheng K, Sheng Y, Shi Z, Zou H (2019) Color-tunable Eu^{2+}, Eu^{3+} co-doped $Ca_{20}Al_{26}Mg_3Si_3O_{68}$ phosphor for w-LEDs. *J. Mater. Chem. C* 7: 6978.

98. Baran A, Barzowska J, Grinberg M, Mahlik S, Szczodrowski K, Zorenko Y (2013) Binding energies of Eu^{2+} and Eu^{3+} ions in β-Ca_2SiO_4 doped with europium. *Opt. Mater.* 35: 2107.

99. Zhou J, Xia Z (2015) Luminescence color tuning of Ce^{3+}, Tb^{3+} and Eu^{3+} codoped and tri-doped $BaY_2Si_3O_{10}$ phosphors via energy transfer. *J. Mater. Chem. C* 3: 7552.

100. Zhang C, Liu Y, Zhang J, Zhang X, Zhang J, Cheng Z, Jiang J, Jiang H (2016) A single-phase $Ba_9Lu_2Si_6O_{24}$:Eu^{2+}, Ce^{3+}, Mn^{2+} phosphor with tunable full-color emission for NUV-based white LED applications. *Mater. Res. Bull.* 80: 288.

101. Zhang J, Hua Z, Zhang F (2015) Warm white-light generation in $Ca_9MgNa(PO_4)_7$:Sr^{2+}, Mn^{2+}, Ln(Ln=Eu^{2+}, Yb^{3+}, Er^{3+}, Ho^{3+}, and Tm^{3+}) under near-ultraviolet and near-infrared excitation. *Ceram. Int.* 41: 9910.

102. Li C, Zhang C, Hou Z, Wang L, Quan Z, Lian H, Lin J (2009) β-$NaYF_4$ and β-$NaYF_4$:Eu^{3+} microstructures: morphology control and tunable luminescence properties. *J. Phys. Chem. C* 113: 2332.

103. Chen J, Zhang N, Guo C, Pan F, Zhou X, Suo H, Zhao X, Goldys EM (2016) Site-dependent luminescence and thermal stability of Eu^{2+} doped fluorophosphate toward white LEDs for plant growth. *ACS Appl. Mater. Interfaces* 8: 20856.

104. Zhang ZW, Liu L, Zhang XF, Zhang, JP, Zhang WG, Wang DJ (2015) Preparation and investigation of $CaZr_4(PO_4)_6$:Dy^{3+} single-phase full-color phosphor. *Spectrochim. Acta, A* 137: 1.

105. Sun XY, Zhang JC, Liu XG, Lin LW (2012) Enhanced luminescence of novel $Ca_3B_2O_6$:Dy^{3+} phosphors by Li^+ co-doping for LED applications. *Ceram. Int.* 38: 1065.

106. Xu Q, Sun J, Cui D, Di Q, Zeng J (2015) Synthesis and luminescence properties of novel $Sr_3Gd(PO_4)_3$:Dy^{3+} phosphor. *J. Lumin.* 158: 301.

107. Nair GB, Dhoble SJ (2017) White light emission through efficient energy transfer from Ce^{3+} to Dy^{3+} ions in $Ca_3Mg_3(PO_4)_4$ matrix aided by Li^+ charge compensator. *J. Lumin.* 192: 1157.

108. Huang L, Zhu YW, Zhang XJ, Zou R, Pan F, Wang J, Wu MM (2016) HF-free hydrothermal route for synthesis of highly efficient narrow-band red emitting phosphor $K_2Si_{1-x}F_6$:xMn^{4+} for warm white light-emitting diodes. *Chem. Mater.* 28: 1495.

109. Kim M, Park WB, Bang B, Kim CH, Sohn KS (2017) A novel Mn^{4+}-activated red phosphor for use in light emitting diodes, K_3SiF_7:Mn^{4+}. *J. Am. Ceram. Soc.* 100: 1044.

110. Olchowka J, Suta M, Wickleder C (2017) Green synthesis of A_2SiF_6 (A=Li–Cs) nanoparticles using ionic liquids as solvents and as fluorine sources: a simple approach without HF. *Chem. Eur. J.* 23: 12092.

111. Ha J, Novitskaya E, Lam N, Sanchez M, Kim YH, Li Z, Im WB, Graeve OA, McKittrick J (2020) Synthesis of Mn^{4+} activated Na_2SiF_6 red-emitting phosphors using an ionic liquid. *J. Lumin.* 218: 116835.

112. Huang L, Liu Y, Yu J, Zhu Y, Pan F, Xuan T, Brik MG, Wang C, Wang J (2018) Highly stable K_2SiF_6:Mn^{4+}@ K_2SiF_6 composite phosphor with narrow red emission for white LEDs. *ACS Appl. Mater. Interfaces* 10: 18082.

113. Arunkumar P, Kim YH, Kim HJ, Unithrattil S, Im WB (2017) Hydrophobic organic skin as a protective shield for moisture-sensitive phosphor-based optoelectronic devices. *ACS Appl. Mater. Interfaces* 9: 7232.

114. Fang MH, Hsu CS, Su C, Liu W, Wang YH, Liu RS (2018) Integrated surface modification to enhance the luminescence properties of K_2TiF_6:Mn^{4+} phosphor and its application in white-light-emitting diodes. *ACS Appl. Mater. Interfaces* 10: 29233.

115. Nguyen HD, Lin CC, Liu RS (2015) Waterproof alkyl phosphate coated fluoride phosphors for optoelectronic materials. *Angew. Chem. Int. Ed.* 54: 10862.

116. Verstraete R, Rampelberg G, Rijckaert H, Van Driessche I, Coetsee E, Duvenhage MM, Smet PF, Detavernier C, Swart H, Poelman D (2019) Stabilizing fluoride phosphors: surface modification by atomic layer deposition. *Chem. Mater.* 31: 7192.

117. Jang I, Kim J, Kim H, Kim WH, Jeon SW, Kim JP (2017) Enhancement of water resistance and photo-efficiency of K_2SiF_6:Mn^{4+} phosphor through dry-type surface modification. *Colloids Surf. A Physicochem. Eng. Aspect* 520: 850.

118. Murphy JE (2016) Processes for preparing color stable manganese-doped phosphors. U.S. Patent No. 9,399,732. 26.

119. Huang L, Liu Y, Si S, Brik MG, Wang C, Wang J (2018) A new reductive DL-mandelic acid loading approach for moisture-stable Mn^{4+} doped fluorides. *Chem. Commun.* 54: 11857.

120. Lee HE, Shin JH, Park JH, Hong SK, Park SH, Lee SH, Lee JH, Kang IS, Lee KJ (2019) Micro light-emitting diodes for display and flexible biomedical applications. *Adv. Funct. Mater.* 29: 1808075.

121. Xuan T, Shi S, Wang L, Kuo HC, Xie RJ (2020) Inkjet-printed quantum dot color conversion films for high-resolution and full-color micro light-emitting diode displays. *J. Phys. Chem. Lett.* 11: 5184.

122. Yang D, Li X, Zhou W, Zhang S, Meng C, Wu Y, Wang Y, Zeng H (2019) $CsPbBr_3$ quantum dots 2.0: benzenesulfonic acid equivalent ligand awakens complete purification. *Adv. Mater.* 31: 1900767.

123. Moon H, Lee C, Lee W, Kim J, Chae H (2019) Stability of quantum dots, quantum dot films, and quantum dot light-emitting diodes for display applications. *Adv. Mater.* 31: 1804294.

124. Hanifi DA, Bronstein ND, Koscher BA, Nett Z, Swabeck JK, Takano K, Schwartzberg AM, Maserati L, Vandewal K, van de Burgt Y, Salleo A, Alivisatos AP (2019) Redefining near-unity luminescence in quantum dots with photothermal threshold quantum yield. *Science* 363: 1199.

125. Reiss P, Protière M, Li L (2009) Core/shell semiconductor nanocrystals. *Small* 5: 154.

126. Cirillo M, Aubert T, Gomes R, Van Deun R, Emplit P, Biermann A, Lange H, Thomsen C, Brainis E, Hens Z (2014) "Flash" synthesis of CdSe/CdS core–shell quantum dots. *Chem. Mater.* 26: 1154.

127. Silva ACA, da Silva SW, Morais PC, Dantas NO (2014) Shell thickness modulation in ultrasmall CdSe/CdS_xSe_{1-x}/Cds core/shell quantum dots via 1-thioglycerol. *ACS Nano* 8: 1913.

128. Zhu H, Song N, Lian T (2010) Controlling charge separation and recombination rates in CdSe/ZnS type I core–shell quantum dots by shell thicknesses. *J. Am. Chem. Soc.* 132: 15038.

129. Hines MA, Guyot-Sionnest P (1996) Synthesis and characterization of strongly luminescing ZnS-capped CdSe nanocrystals. *J. Phys. Chem.C.* 100: 468.

130. Kim T, Kim KH, Kim S, Choi SM, Jang H, Seo HK, Lee H, Chung DY, Jang E (2020) Efficient and stable blue quantum dot light-emitting diode. *Nature* 586: 385.

131. Zhang W, Chen G, Wang J, Ye BC, Zhong X (2009) Design and synthesis of highly luminescent near-infrared-emitting water-soluble CdTe/CdSe/ZnS core/shell/shell quantum dots. *Inorg. Chem.* 48: 9723.

132. Lim J, Jeong BG, Park M, Kim JK, Pietryga JM, Park YS, Klimov VI, Lee C, Lee DC, Bae WK (2014) Influence of shell thickness on the performance of light-emitting devices based on CdSe/$Zn_{1-x}Cd_xS$ core/shell heterostructured quantum dots. *Adv. Mater.* 26: 8034.

133. Chen O, Zhao J, Chauhan VP, Cui J, Wong C, Harris DK, Wei H, Han HS, Fukumura D, Jain RK, Bawendi MG (2013) Compact high-quality CdSe–CdS core–shell nanocrystals with narrow emission linewidths and suppressed blinking. *Nat. Mater.* 12: 445.

134. Algar WR, Susumu K, Delehanty JB, Medintz IL (2011) Semiconductor quantum dots in bioanalysis: crossing the valley of death. *Anal. Chem.* 83: 8826.

135. Wu Z, Liu P, Zhang W, Wang K, Sun XW (2020) Development of InP quantum dot-based light-emitting diodes. *ACS Energy Lett.* 5: 1095.

136. Tessier MD, Baquero EA, Dupont D, Grigel V, Bladt E, Bals S, Coppel Y, Hens Z, Nayral C, Delpech F (2018) Interfacial oxidation and photoluminescence of InP-based core/shell quantum dots. *Chem. Mater.* 30: 6877.

137. Won YH, Cho O, Kim T, Chung DY, Kim T, Chung H, Jang H, Lee J, Kim D, Jang E (2019) Highly efficient and stable InP/ZnSe/ZnS quantum dot light-emitting diodes. *Nature* 575: 634.

138. Zhang H, Hu N, Zeng Z, Lin Q, Zhang F, Tang A, Jia Y, Li LS, Shen H, Teng F, Du Z (2019) High-efficiency green InP quantum dot-based electroluminescent device comprising thick-shell quantum dots. *Adv. Opt. Mater.* 7: 1801602.

139. Li Y, Hou X, Dai X, Yao Z, Lv L, Jin Y, Peng X (2019) Stoichiometry-controlled InP-based quantum dots: synthesis, photoluminescence, and electroluminescence. *J. Am. Chem. Soc.* 141: 6448.

140. Xu Z, Li Y, Li J, Pu C, Zhou J, Lv L, Peng X (2019) Formation of size-tunable and nearly monodisperse InP nanocrystals: chemical reactions and controlled synthesis. *Chem. Mater.* 31: 5331.

141. Hahm D, Chang JH, Jeong BG, Park P, Kim J, Lee S, Choi J, Kim WD, Rhee S, Lim J, Lee DC, Lee C, Char K, Bae WK (2019) Design principle for bright, robust, and color-pure InP/$ZnSe_xS_{1-x}$/ZnS hetero-structures. *Chem. Mater.* 31: 3476.

142. Cao F, Wang S, Wang F, Wu Q, Zhao D, Yang X (2018) A layer-by-layer growth strategy for large-size InP/ZnSe/ZnS core–shell quantum dots enabling high-efficiency light-emitting diodes. *Chem. Mater.* 30: 8002.

143. Huang H, Bodnarchuk MI, Kershaw SV, Kovalenko MV, Rogach AL (2017) Lead halide perovskite nanocrystals in the research spotlight: stability and defect tolerance. *ACS Energy Lett.* 2: 2071.

144. Wei Y, Cheng Z, Lin J (2019) An overview on enhancing the stability of lead halide perovskite quantum dots and their applications in phosphor-converted LEDs. *Chem. Soc. Rev.* 48: 310.

145. Lou S, Zhou Z, Gan W, Xuan T, Bao Z, Si S, Cao L, Li H, Xia Z, Qiu J, Liu RS, Wang J (2020) In situ synthesis of high-efficiency $CsPbBr_3/CsPb_2Br_5$ composite nanocrystals in aqueous solution of micro-emulsion. *Green Chem.* 22: 5257.

146. Xuan T, Huang J, Liu H, Lou S, Cao L, Gan W, Liu RS, Wang J (2019) Super-hydrophobic cesium lead halide perovskite quantum dot-polymer composites with high stability and luminescent efficiency for wide color gamut white light-emitting diodes. *Chem. Mater.* 31: 1042.

147. Lou S, Xuan T, Wang J (2019) Stability: a desiderated problem for the lead halide perovskites. *Opt. Mater. X* 1: 100023.

148. Xuan T, Lou S, Huang J, Cao L, Yang X, Li H, Wang J (2018) Monodisperse and brightly luminescent $CsPbBr_3/Cs_4PbBr_6$ perovskite composite nanocrystals. *Nanoscale* 10: 9840.

149. Xuan T, Yang X, Lou S, Huang J, Liu Y, Yu J, Li H, Wong KL, Wang C, Wang J (2017) Highly stable $CsPbBr_3$ quantum dots coated with alkyl phosphate for white light-emitting diodes. *Nanoscale* 9: 15286.

150. Wang A, Shen H, Zang S, Lin Q, Wang H, Qian L, Niu J, Song Li L (2015) Bright, efficient, and color-stable violet ZnSe-based quantum dot light-emitting diodes. *Nanoscale* 7: 2951.

151. Wang O, Wang L, Li Z, Xu Q, Lin Q, Wang H, Du Z, Shen H, Li LS (2018) High-efficiency, deep blue $ZnCdS/Cd_xZn_{1-x}S/ZnS$ quantum-dot-light-emitting devices with an EQE exceeding 18%. *Nanoscale* 10: 5650.

152. Shen H, Gao Q, Zhang Y, Lin Y, Lin Q, Li Z, Chen L, Zeng Z, Li X, Jia Y, Wang S, Du Z, Li LS, Zhang Z (2019) Visible quantum dot light-emitting diodes with simultaneous high brightness and efficiency. *Nat. Photonics* 13: 192.

153. Wu Y, Wei C, Li X, Li Y, Qiu S, Shen W, Cai B, Sun Z, Yang D, Deng Z, Zeng H (2018) In situ passivation of $PbBr_6^{4-}$ octahedra toward blue luminescent $CsPbBr_3$ nanoplatelets with near 100% absolute quantum yield. *ACS Energy Lett.* 3: 2030.

154. Zheng X, Yuan S, Liu J, Yin J, Yuan F, Shen WS, Yao K, Wei M, Zhou C, Song K, Zhang BB, Lin Y, Hedhili MN, Wehbe N, Han Y, Sun HT, Lu ZH, Anthopoulos TD, Mohammed OF, Sargent EH, Liao LS, Bakr OM (2020) Chlorine vacancy passivation in mixed halide perovskite quantum dots by organic pseudohalides enables efficient Rec. 2020 blue light-emitting diodes. *ACS Energy Lett.* 5: 793.

155. Cao W, Xiang C, Yang Y, Chen Q, Chen L, Yan X, Qian L (2018) Highly stable QLEDs with improved hole injection via quantum dot structure tailoring. *Nat. Commun.* 9: 2608.

156. Kim Y, Ham S, Jang H, Min JH, Chung H, Lee J, Kim D, Jang E (2019) Bright and uniform green light emitting InP/ZnSe/ZnS quantum dots for wide color gamut displays. *ACS Appl. Nano Mater.* 2: 1496.

157. Chen YM, Zhou Y, Zhao Q, Zhang JY, Ma JP, Xuan TT, Guo SQ, Yong ZJ, Wang J, Kuroiwa Y, Moriyoshi C, Sun HT (2018) $Cs_4PbBr_6/CsPbBr_3$ perovskite composites with near-unity luminescence quantum yield: large-scale synthesis, luminescence and formation mechanism, and white light-emitting diode application. *ACS Appl. Mater. Interfaces* 10: 15905.

158. Dutta A, Behera RK, Pal P, Baitalik S, Pradhan N (2019) Near-unity photoluminescence quantum efficiency for all $CsPbX_3$ (X=Cl, Br, and I) perovskite nanocrystals: a generic synthesis approach. *Angew. Chem. Int. Edit.* 58: 5552.

159. Dai X, Zhang Z, Jin Y, Niu Y, Cao H, Liang X, Chen L, Wang J, Peng X (2014) Solution-processed, high-performance light-emitting diodes based on quantum dots. *Nature* 515: 96.

160. Wang HC, Lin SY, Tang AC, Singh BP, Tong HC, Chen CY, Lee YC, Tsai TL, Liu RS (2016) Mesoporous silica particles integrated with all-inorganic $CsPbBr_3$ perovskite quantum-dot nanocomposites (MP-PQDs) with high stability and wide color gamut used for backlight display. *Angew. Chem. Int. Edit.* 55: 7924.

161. Chiba T, Hayashi Y, Ebe H, Hoshi K, Sato J, Sato S, Pu YJ, Ohisa S, Kido J (2018) Anion-exchange red perovskite quantum dots with ammonium iodine salts for highly efficient light-emitting devices. *Nat. Photonics* 12: 681.

162. Pimputkar S, Speck JS, DenBaars SP, Nakamura S (2009) Prospects for LED lighting. *Nat. Photonics* 3: 180.

163. Withnall R, Silver J, Fern G, Ireland T, Lipman A, Patel B (2008) Experimental and theoretical luminous efficacies of phosphors used in combination with blue-emitting LEDs for lighting and backlighting. *J. Soc. Inf. Disp.* 16: 359.

164. Xu H, Hong F, Liu G, Dong X, Yu W, Wang J (2020) Green route synthesis and optimized luminescence of $K_2SiF_6:Mn^{4+}$ red phosphor for warm WLEDs. *Opt. Mater.* 99: 109500.

165. Zhou Q, Zhou Y, Liu Y, Luo L, Wang Z, Peng J, Yan J, Wu M (2015) A new red phosphor $BaGeF_6$:Mn^{4+}: hydrothermal synthesis, photo-luminescence properties, and its application in warm white LED devices. *J. Mater. Chem. C* 3: 3055.

166. Cui S, Chen G, Chen Y, Liu X (2018) Preparation and luminescent properties of new YAG:Ce^{3+} phosphor in glass (PIG) for white LED applications. *J. Mater. Sci. Mater. Electron.* 29: 13019.

167. Zhang Y, Zhang Z, Liu X, Shao G, Shen L, Liu J, Xiang W, Liang X (2020) A high quantum efficiency $CaAlSiN_3$:Eu^{2+} phosphor-in-glass with excellent optical performance for white light-emitting diodes and blue laser diodes. *Chem. Eng. J.* 401: 125983.

168. Li H, Pang R, Liu G, Sun W, Li D, Jiang L, Zhang S, Li C, Feng J, Zhang H (2018) Synthesis and luminescence properties of Bi^{3+}-activated K_2MgGeO_4: a promising high-brightness orange-emitting phosphor for WLEDs conversion. *Inorg. Chem.* 57: 12303.

169. Lü W, Hao, Z, Zhang X, Luo Y, Wang X, Zhang J (2011) Tunable full-color emitting $BaMg_2Al_6Si_9O_{30}$:Eu^{2+}, Tb^{3+}, Mn^{2+} phosphors based on energy transfer. *Inorg. Chem.* 50: 7846.

170. Li S, Wang L, Tang D, Cho Y, Liu X, Zhou X, Lu L, Zhang L, Takeda T, Hirosaki N, Xie RJ (2018) Achieving high quantum efficiency narrow-band β-sialon:Eu^{2+} phosphors for high-brightness LCD backlights by reducing the Eu^{3+} luminescence killer. *Chem. Mater.* 30: 494.

171. Yoshimura K, Fukunaga H, Izumi M, Masuda M, Uemura T, Takahashi K, Xie RJ, Hirosaki N (2016) White LEDs using the sharp β-sialon:Eu phosphor and Mn-doped red phosphor for wide-color gamut display applications. *J. Soc. Inf. Disp.* 24: 449.

172. Chen SWH, Huang YM, Singh KJ, Hsu YC, Liou FJ, Song J, Choi J, Lee PT, Lin CC, Chen Z, Han J, Wu TZ, Kuo HC (2020) Full-color micro-LED display with high color stability using semipolar (20-21) InGaN LEDs and quantum-dot photoresist. *Photonics Res.* 8: 630.

173. Yin Y, Hu Z, Ali MU, Duan M, Gao L, Liu M, Peng W, Geng J, Pan S, Wu Y, Hou J, Fan J, Li D, Zhang X, Meng H (2020) Full-color micro-LED display with $CsPbBr_3$ perovskite and CdSe quantum dots as color conversion layers. *Adv. Mater. Technol.* 5: 2000251.

2 Phosphor-Converted LED for Backlighting

Yaxin Cao and Yuhua Wang

CONTENTS

2.1 INTRODUCTION

2.1.1 COMPARISON OF DISPLAY APPROACHES

Information constitutes one of the most significant parts of modern society. The generation, transmission, analysis, and feedback of information profoundly influence how the world works, and nowadays, all of these are more dependent on the display instruments. Display technology has gradually but profoundly shaped the lifestyle of human beings, which is widely recognized as an indispensable part of the modern world.[1] The widespread display devices, such as televisions, computers, and smartphones, have opened significant access to the vast information bank of the internet.[2, 3] With the progress in semiconductor technology over the past several decades, display quality has advanced by leaps and bounds toward a wide color gamut, high brightness, and long lifetimes.[4]

Currently, there are two types of display technologies according to the excitation mode. The photoluminescence (PL) type relies on the backlight generated by blue light-emitting diodes (LEDs) combined with down-conversion luminescence materials such as phosphors and quantum dots (QDs), while the electroluminescence (EL) type with self-emission is driven by an electron current. Concretely, they are mainly based on four kinds of device structures as shown in Figure 2.1.[4] For the PL type, the display of colors is based on the backlight. Generally, a beam of backlight would pass through a polarizer, the liquid crystal (LC) layer controlled by thin-film transistors (TFT), the color filters, and another polarizer that is perpendicular to the first one, during which the polarization direction, transmittance, and color of the light are altered.[5, 6] It could be divided mainly into two categories according to the applied luminescent materials, the LC displays (LCDs) based on

DOI: 10.1201/9781003098676-2

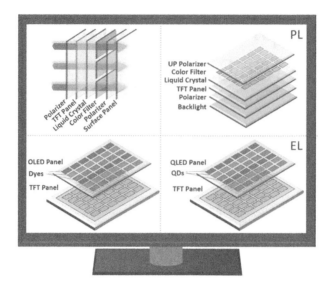

FIGURE 2.1   Schematic of device structures of four different display methods. (Wang, X., Bao, Z., Chang, Y. C. and Liu, R. S., *ACS Energy Lett.* 5, 3374, 2020. With permission.)

phosphors and the QD-adopting LED backlighting LCD devices (QLCDs). For the EL type, the colors that come out from the luminescent materials could be adjusted by the control circuit, for instance, organic LEDs (OLEDs) based on dyes and quantum-dot LEDs (QLEDs) based on QDs are newly developed technologies after LCDs. Some of these devices already have applications in new products, such as televisions, smartphones, and other mobile devices.[7–11] The OLED devices perform better in contrast, thickness, and flexibility but there are still many urgent problems to be solved, especially production problems that need to be solved step by step in the process of mass production line.[12, 13] In the meantime, the QLEDs used in backlighting own excellent wide color gamut but limited by the drawbacks such as low efficiency and poor stability of QDs and the relevant research is still under the laboratory conditions.[14, 15]

LCDs have been well developed since they arrived in the 1960s, and this technology has a huge market in the application of displays in smartphones, tablets, computers, large-screen TVs, data projectors, and other devices in our daily life.[16] The advantages such as modern technology, durability, more reasonable cost make it more competitive at the present stage compared to OLEDs and QLEDs.[17, 18] Thus, attention has been continuously focused on LCDs. Nonetheless, there are still challenges in LCDs, among which the color gamut improvement raises significant concerns these days.

2.1.2 FUNDAMENTALS OF LCDS

To better understand how the LCD works, Figure 2.2 gives the schematic diagram of a popular TFT LCD module in a single pixel. The backlight (light source for the display screen, always the white light) passes through the first polarizer (Polarizer#1), getting polarized light (vertical in the figure), and comes into the TFT-controlled LC layer. The TFT here performs a role of a switch, determining whether the pixel would be charged or not, to avoid the "crosstalk" between a pixel and its neighbors (it has a similar function in OLEDs and QLEDs).[19, 20] The pixel will be charged when the TFT is switched on, and the pixel voltage is supplied by the data lines.[21] When the polarized light passes, the electrically tuned LCs would change their orientations, adjusting the orientations of the polarized light beams and further alter their transmittance through the second polarization filter (Polarizer#2). Because of the transmittance difference caused by LCs, the components of red, green, and blue (RGB) in the exit light in a single pixel is finally determined after the color filters and Polarizer#2.

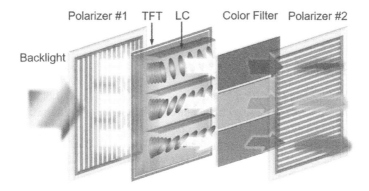

FIGURE 2.2   The schematic diagram of a commonly used TFT LCD module in a single pixel.

Based on the discussed introduction of the LCD working principle, one can see that the LCs change the relative strength of the backlight components which would pass through the RGB color filters while the latter enables the appearance of colors. The category of colors presented by a display device is related to the color gamut. In particular, the color gamut is determined by the position of the RGB fractions of the backlight sources in the Commission International de l'Éclairage (CIE) diagram.[22, 23] It is usually regulated by specific standards and Figure 2.3 depicts several standard color gamut, the standard Red Green and Blue (sRGB) [24]; the National Television Standard Committee (NTSC) standard[25]; the DCI-P3 standard set by Digital Cinema Initiatives (DCI)[26, 27]; and the Rec. 2020 standard defined by the International Telecommunication Union (ITU).[28] The larger the color gamut is, the more the category of colors would be expressed to faithfully reproduce all kinds of natural colors, and colors outside the gamut cannot be displayed.

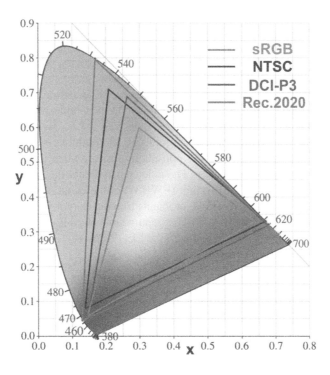

FIGURE 2.3   The CIE 1931 chromaticity coordinate diagram and several standards of color gamut marked in different colors in the diagram.

Among various standards, NTSC is always taken as a reference to illustrate the color gamut that a device could provide.

2.1.3 BACKLIGHTING FOR LCDs

To obtain a larger color gamut for the LCD devices, only red, green, and blue in the highest possible purity are required. Since the initial light source (white light, or rather mixing light) would be re-splited into the primary colors (RGB) by the color filters, it is an effective approach to obtain a comprising narrower band of RGB components by applying additional color filters. However, the luminous flux could be greatly reduced, causing brightness loss of the screen. Besides, the technology of additional color filtration is complex. An alternative approach is to modify the backlight sources. The backlight sources have gone through the development from cold cathode fluorescent lamps (CCFLs), LED arrays that consist of RGB LEDs to the phosphor-converted LEDs (pc-LEDs). [29–31] The pc-LEDs have been widely used in lighting and displays in our daily life nowadays because of their marvelous advantages such as high luminous efficiency, fast switching, low power consumption, durability, and long operational lifetime.[32–34] The color of the backlight, in this case, would be dependent on the emissive part, i.e., the LED chips, as well as the light-converting materials. In pc-LEDs, the LED chips with ultraviolet (UV)/near-ultraviolet (NUV) or blue emission are employed to excite the light-converting materials coated on them. At present, the blue LED chip (based on GaN/GaInN) combined with light-converting materials devices are common used in the LED backlighting for LCD applications.[4, 35–37] For the achievement of a wider color gamut, the light-converting materials are supposed to emit light in an appropriate wavelength range (to match well with the color filters) with a narrow emission band (to reach high color purity and reduce the brightness loss when passing through a filter) and considerable brightness (to ensure the brightness of the screen).[23, 38, 39] From this perspective, the luminescent semiconductor QDs such as CdSe [40, 41] (the II-VI type), InP [42–44] (the III–V type), and $CsPbX_3$[45, 46] (the perovskite-type) [47] which exhibit narrow emission width, saturated color, and tunable emission are attractive candidates for the backlighting.[48] Some of the materials already have the application in QD-adopting LED backlighting LCD devices (QLCDs).[49, 50] Nonetheless, there are still numerous issues to resolve for the application of QDs such as toxicity, poor stability, and the problem in large-scale production.[51–55] Therefore, the mainstream of the current color-converting materials for LCD backlighting are still based on inorganic solid-state phosphors because of their greater stability under chemical/thermal conditions and considerable efficiency with reasonable cost compared to QDs when used in backlighting.[38, 56, 57]

Therefore, we would focus on the progress, problems, and outlooks of the inorganic solid-state phosphors next. What characters should the phosphors possess when they are considered to be used in backlighting? (1) High color purity. The color purity could be determined mainly by the peak position and full-width-at-half-maximum (FWHM) of the emission band. (2) Good stability. Although phosphors are stable on the whole, problems like intensity loss and color drift should be avoided as far as possible. (3) More efficient. This requires the phosphors to perform high quantum yield (QY) themselves and could match well with the LED chips. (4) Costs etc.[58, 59]

2.1.4 CONSTRUCTION AND WORKING OF THE pc-LED BACKLIGHT

A common method to fabricate white light is employing a blue LED chip with the $Y_3Al_5O_{12}:Ce^{3+}$ (YAG:Ce) phosphor, which gives a blue and yellow (B+Y) mixture of light efficiently as shown in Figure 2.4(a).[60, 61] However, the obtained color gamut is limited in this way. This is because the emission of YAG:Ce is broadband without sufficient red component which results in its poor matching with the color filters and the brightness loss. Thus, an alternative approach that combining blue LED chips with red and green phosphors (B+RG) is proposed to avoid the aforementioned issues as much as possible (Figure 2.4(b)). The sulfides-based solution was reported

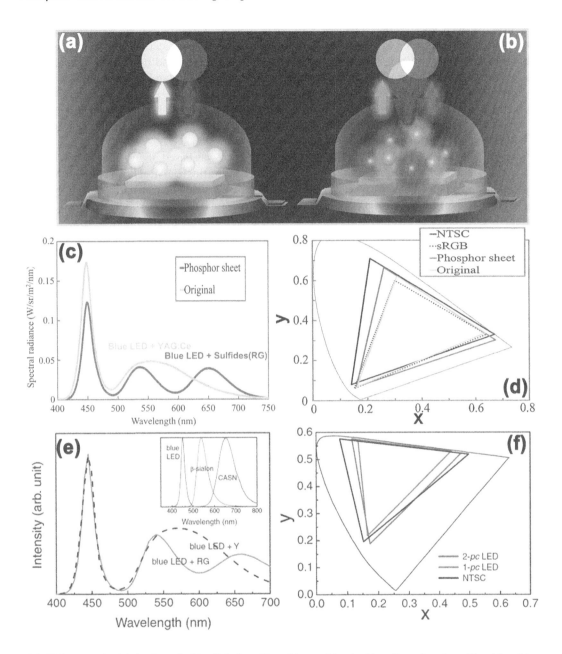

FIGURE 2.4 The fabrication of white light by a blue chip combined with yellow phosphors (a); a blue chip with red and green phosphors (b); the emission spectra (c) and the related CIE 1931 chromaticity diagrams (d) of the backlighting devices for the LCD with the phosphor sheet which contains the $SrGa_2S_4$:Eu and CaS:Eu phosphors compared to the original LCD with the conventional white LEDs which contain the YAG: Ce phosphor; the emission spectra (e) and the related CIE 1976 chromaticity coordinates (f) of the 2-pc (solid) and 1-pc (dotted) LEDs. The insert shows the emission spectra of the blue LED chip, green β-sialon:Eu, and $CaAlSiN_3$:Eu red phosphors. (Ito, Y., Hori, T., Kusunoki, T., Nomura, H. and Kondo, H., *J. Soc. Inf. Disp.* 22, 419, 2015. With permission; Xie, R. J., Hirosaki, N. and Takeda, T., *Appl. Phys. Express 2*, 022401, 2009. With permission.)

by Prof. Yasushi Ito *et al.*[62] The blue LED chip is applied to excite a red phosphor CaS: Eu and a green phosphor $SrGa_2S_4$:Eu. The red/green emission band located at around 650/540 nm, as well as the blue emission located at around 450 nm, could be found in the backlight with the phosphor sheet. The related PL spectra are shown in Figure 2.4(c) and a larger color gamut of ~90% NTSC was realized compared to the original B+Y type (only ~71% NTSC) in CIE1931, which are depicted in Figure 2.4(d). The input electrical power for the LEDs was 7.7 W. However, the sulfides are gradually waived since the employment of sulfides in practice suffers from poor stability against heat, moisture, etc.[63, 64]

Another B+RG-type solution reported by Prof. Xie *et al.* was based on (oxy)nitride phosphors. [35] The green-emitting oxynitride β-sialon:Eu, and red-emitting $CaAlSiN_3$:Eu were utilized for the green and red components, respectively. The Eu-doped β-sialon owns a narrow emission band (FWHM ~55 nm) centered at 535 nm[65], while the emission band of $CaAlSiN_3$:Eu is centered at around 650 nm[66] as displayed in Figure 2.4(e). The corresponding color gamut in a CIE 1976 chromaticity coordinates diagram can be checked in Figure 2.4(f). The color gamut increased from ~71% of the NTSC for the 1-pc (B+Y) type to ~91% for the 2-pc (B+RG) type in a CIE 1976 standard system. CIE 1976 and CIE 1931 systems are both used for the description of the color space but with the different algorithm on the tristimulus values to obtain the chromaticity coordinates.[67, 68] Thus, the color gamut could be found to increase from 68.3 to 82.1% of the NTSC in the CIE 1931 system. $CaAlSiN_3$:Eu is known as an efficient red phosphor with chemical and thermal stability.[66] However, its emission band is quite broad (FWHM ~86 nm) [69] as it can be seen in Figure 2.4(e). With low color purity, the initial brightness of the emission would be reduced when passing through a red filter, making it not appropriate to be a phosphor for backlighting in LCDs. Furthermore, although β-sialon:Eu, the state-of-the-art green phosphor, has been widely used in lighting and display benefiting from the high efficiency and good stability [70], the color purity is not so satisfactory for a greater color gamut (> 90% NTSC). In consequence, many efforts have been paying to develop efficient red and green phosphors for the LCD backlighting applications.

2.1.5 DEVELOPMENT OF PHOSPHORS FOR PC-LED IN BACKLIGHTING

Figure 2.5 gives the timeline of phosphors for the application in pc-LED backlighting. The original considered YAG:Ce^{3+} was first reported as a yellow-emitting phosphor in 1967 [71] and the B+Y-type white pc-LEDs was first commercialized in 1996 by Nichia.[72] Ce^{3+} emits yellow light with a quite broad emission band, which makes it not a valuable candidate for backlighting. And this is very much associated with the doping Ce^{3+} ions in the host. As is well-known, Ce^{3+} performs typical bimodal emission due to its two distinguished levels of the ground state, and consequently, it's much easier for the Ce^{3+}-doped phosphors to get broad emission band rather than the opposite. Therefore, the selection of appropriate activators is of great significance. As we can see, the narrow-band phosphors developed after that have few reports on Ce^{3+}-doped ones. Sulfides have a rich history, CaS and SrS have been recorded as phosphors in the 1700s and 1800s, respectively.[73] The modern application of EL in CaS:Eu^{2+} thin-film devices are reported in 1985, in which red emission band at 663 nm with FWHM of 72 nm (1% mol Eu).[74, 75] Then in the early 1990s, Eu^{2+}-doped $CaGa_2S_4$ and $SrGa_2S_4$ were proposed as a new class of promising thin film electroluminescent (TFEL) phosphors.[76, 77] $SrGa_2S_4$:Eu^{2+} (0.1% mol) had been applied as a commercial phosphor with efficient green emission at around 537–532 nm (FWHM ~35–62 nm from 77 to 500K).[78, 79] The drawbacks of sulfides in chemical/thermal stability for LED using as talked about earlier make them less competitive. In 2005, the oxynitride phosphor β-sialon:Eu^{2+} (~0.003 mol) was reported to perform a green emission peaking at 535 nm with an FWHM of 55 nm, which enables it to be widely used in advanced wide-gamut backlighting devices.[65, 70] Later in 2006, the efficient red nitride phosphor $CaAlSiN_3$:Eu^{2+} (1.6% mol) was reported with robust stability, but it was not suitable for the wide color gamut purpose, considering its broad FWHM of ~90 nm and a great portion

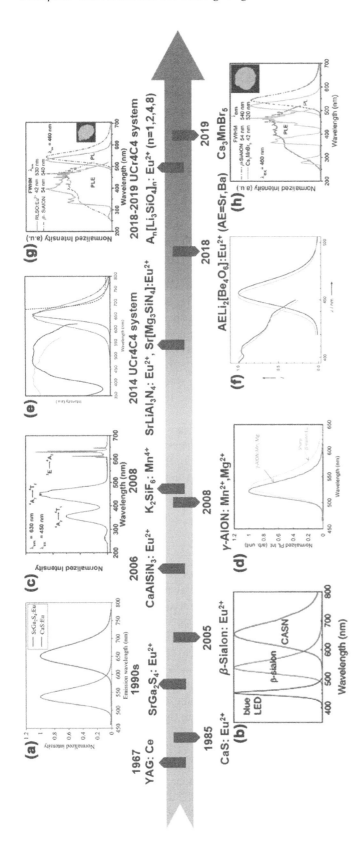

FIGURE 2.5 Progress in narrow band–emitting phosphors. (a) The PL spectra of $SrGa_2S_4:Eu^{2+}$ and $CaS:Eu^{2+}$; (b) the PL spectra of a blue LED chip, β-sialon:Eu^{2+} (β-sialon), and CaAlSiN$_3$:Eu^{2+} (CASN); (c) PL and PLE spectra of $K_2SiF_6:Mn^{4+}$; (d) PL spectra of γ-AlON:Mn^{2+}, Mg^{2+} compared to β-sialon:Eu^{2+}; (e) PL and PLE spectra of $Sr[LiAl_3N_4]:Eu^{2+}$ (blue and red) compared to CaAlSiN$_3$:Eu^{2+} (light, dark gray); (f) PL and PLE spectra of $AELi_2[Be_4O_6]:Eu^{2+}$ (gray for AE = Ba and black for AE = Sr); (g) PL and PLE spectra for RbLi[Li$_3$SiO$_4$]:Eu^{2+} (RLSO: Eu) compared to β-sialon:Eu^{2+}; (h) PL and PLE spectra for Cs$_3$MnBr$_5$ compared to β-sialon:Eu^{2+}. (Ito, Y., Hori, T., Kusunoki, T., Nomura, H. and Kondo, H., *J. Soc. Inf. Disp.* 22, 419, 2015. With permission; Xie, R. J., Hirosaki, N. and Takeda, T., *Appl. Phys. Express* 2, 022401, 2009. With permission; Huang, L., Liu, Y., Liu, J., Zhu, Y., Pan, F., Xuan, T., Brik, M. G., Wang, C. and Wang, J., *ACS Appl. Mater. Interfaces* 10, 18082, 2018. With permission; Yoshimura, K., Fukunaga, H., Izumi, M., Takahashi, K., Xie, R. J. and Hirosaki, N., *Jpn. J. Appl. Phys.* 56, 041701, 2017. With permission; Pust, P., Weiler, V., Hecht, C., Tücks, A., Wochnik, A. S., Henß, A. K., Wiechert, D., Scheu, C., Schmidt, P. J. and Schnick, W., *Nat. Mater.* 9, 891, 2014. With permission; Strobel, P., Maak, C., Weiler, V., Schmidt, P. J. and Schnick, W., *Angew. Chem. Int. Ed.* 57, 8739, 2018. With permission; Zhao, M., Liao, H., Ning, L., Zhang, Q., Liu, Q. and Xia, Z., *Adv. Mater.* 30, 1802489, 2018. With permission; Su, B., Molokeev, M. S. and Xia, Z., *J. Mater. Chem. C* 7, 11220, 2019. With permission.)

of red emission beyond the human eyes' sensitivity.[66, 80, 81] Then, the transition metal Mn-doped phosphors γ-AlON:Mn^{2+}, Mg^{2+}, and KSF:Mn^{4+} are reported in 2008, respectively. The former performs narrow-band green emission at 520 nm with the FWHM of 44 nm while the latter with the emission peak at ~630 nm and the FWHM of ~3 nm [82–84] is regarded as a strong competitor to the commercial CaAlSiN$_3$:Eu^{2+} ever since its first authorized patent in 2011 from GE Company.[85] In 2014, the nitrides Sr[LiAl$_3$N$_4$]:Eu^{2+} (λ_{em} = 654, FWHM ~50 nm) and SrMg$_3$SiN$_4$:Eu^{2+} (λ_{em} = 615, FWHM = 43 nm) were reported in an UCr$_4$C$_4$ structure [86, 87], which is quite important in current narrow-band phosphors and the related (oxy-)nitride phosphors had been reported successively. Except for nitrides, oxide materials with narrow-band emission have also gained development. The oxoberyllate compounds AELi$_2$[Be$_4$O$_6$]:Eu^{2+} (AE = Sr, Ba) reported in 2018, with small stokes shift (~1200 cm^{-1}) and narrow emission band (~1200 cm^{-1}), could be highly excitable in the NUV range and improve the luminous efficacy of an RGB phosphors-converted device.[88] Recently, the oxide phosphors with UCr$_4$C$_4$-type structure have received much attention. The Eu^{2+}-doped A$_x$[Li$_3$SiO$_4$]$_x$ (A = Li, Na, K, Rb, and Cs; x = 1, 2, 4, 8) compounds are NUV or blue light excited with blue to green narrow-band emissions with flexible adjustment.[89–91] In 2019, the narrow-band green-emitting phosphor Cs$_3$MnBr$_5$ with intrinsic Mn^{2+} emission was reported especially with high luminous efficiency (107.66 lm W^{-1}) compared to many other Mn^{2+}-activated phosphors.[92] The development of narrow-band phosphors has greatly expanded the scope of candidates for LCD backlighting application. Nevertheless, can these materials or how can they replace the existing commercial phosphors remain challenging and a lot more research is needed.

Overall, the narrow-band phosphors that already reported with considerable performances are mainly based on three kinds of activators, i.e., Eu^{2+}, Mn^{2+}, and Mn^{4+}. Therefore, a detailed overview of the phosphors with narrow emission band for LCD backlighting would be introduced, revolving around these activators hereinafter to provide possible clues for the further development of LCD backlighting-used phosphors in theoretical and practical aspects.

We have focused on the development of white LED-used luminescence materials, as well as extending of their applications in the field of general lighting, displays, plant growth illumination, and so on. Besides, several potential materials have been obtained for the application in LCD backlighting on the basis of our work, which will also be introduced in the following part.

2.2 PHOSPHORS FOR BACKLIGHTING

2.2.1 Eu^{2+}-Activated Phosphors

Eu^{2+} often gives a band emission due to its 4f^{6}5d^1–4f^7 transition since the energy of the excited state 4f^{6}5d^1 is always lower than that of the lowest excited state ^{6}P of 4f^7 and the participation of the d electron makes it be prominently influenced by the coordination surroundings in the host.[93, 94] Accordingly, when it is doped into compounds, the luminescent properties could be modified by the crystal structures, and the emissions of Eu^{2+} would various form purple to red and even deep-red region.[86, 95–100] Ordinarily, the nephelauxetic effect, crystal field splitting, and the Stokes shift (Figure 2.6) would be considered when discussing Eu^{2+}'s performance in the materials, which will be mainly talked about herein.

For free Eu^{2+} ions, the energy difference between the ground and excited states is reported to be about 4.2–5.4 eV (34,000–43500 cm^{-1}, from the ^{8}S$_{7/2}$ ground state to the high spin states, spin, and dipole allowed).[101]

The basics of crystal field theory (CFT) were proposed by Bathe in 1929, and it was developed in the period of 1930s, 1950s, and after the creation of quantum mechanics. It was used to investigate the electronic structure and properties of transition metal compounds initially and then rare earth–containing compounds, especially in the investigation of the ligands to the d orbital of the central ions.[102, 103] The crystal field splitting of the 5d orbitals describes the energy difference between the highest and lowest 5d energy levels, of which the degree of splitting would be determined by the

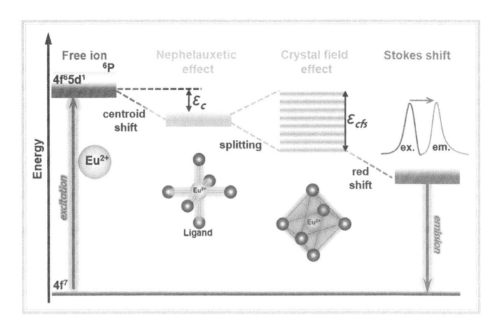

FIGURE 2.6 The luminescence schematic of Eu^{2+} in crystals. The final emission of Eu^{2+} could be influenced by the local environment around it, including the ligands, coordination, etc., in which the nephelauxetic effect, crystal field splitting, and the Stokes shift should be considered.

type of anion coordination polyhedron around the luminescent center.[104] Generally, the crystal field splitting (D_q) can be estimated by the following equation [105–107]:

$$D_q = \frac{1}{6} Ze^2 \frac{r^4}{R^5} \tag{2.1}$$

where Z is the anion charge or valence, e is the charge of an electron, r and R are the radius of the d wavefunction and the bond length, respectively. According to Equation (2.1), the crystal field splitting could be influenced by the categories of the anions (ligand ions), the cations (central ions), and their interaction (related to the bond length). The splitting degree varies in ligand fields of different symmetries, e.g., splitting of a D term in tetrahedral < octahedral < square (planar).[102] When the central ion is fixed, the degree of splitting can be distinguished by the host lattice. It was reported that the general sequence of the host lattice effect is as follows: oxides > aluminates > silicates > borates > phosphates > carbonates > sulfates > fluorides.[108] When the central ion and ligand ion are fixed, the crystal field splitting trends in different compounds could be estimated by the bond lengths. One can speculate that a shorter bond length would result in a larger splitting and, thus, a redshift in the PL spectrum.[109]

The CFT faces a serious problem that the splitting parameters cannot be correctly estimated since the splitting of the d-shell is treated as a pure electrostatic effect in the ionic model formulation of the CFT.[110] "Nephelauxetic" describes the cloud of d-electrons expands when the central ion is surrounded by various other inorganic or organic ions or molecules. The nephelauxetic effect is interpreted as a consequence of the certain increase of the average electron-electron separation in the d shell taking place due to the expansion of the latter (also understood as covalency) in the course of the complex formation. It can be characterized by the nephelauxetic ratio β, which can be written to a good approximation as

$$1 - \beta \approx h(\text{ligands})k(\text{central ion}) \tag{2.2}$$

where h (characterize the ligands) is related to the dielectric constant, which is reproduced by available data on the polarizability of the ligands using the Clausius-Mossotti formula, and k (characterize the central ion) is dependent on the Slater orbital exponent, which describes the distribution of electrons in orbitals of the central ion.[110–112] Qualitatively, one could estimate the centroid shift according to the degree of the nephelauxetic effect. The $4f^65d^1$ state would be centroid shifted when Eu^{2+} is doped into a host lattice and the degree of shift is positively proportional to the covalency between Eu^{2+} and the neighboring ligand ions. The nephelauxetic series for ligands can be presented in the sequence following [112–115]: $F^- < H_2O < NH_3 < Cl^- < Br^- < N^{3-} < I^- < O^{2-} < S^{2-} < Se^{2-} < Te^{2-}$.

Another effect of the host crystal on the position of 5d ($4f^65d^1$ for Eu^{2+}) state is Stokes shift, which is known as the difference between the energy required to excite an electron from the 4f level to the 5d level and the emission energy from the excited state back to the ground state. The Stokes shift describes the energy loss due to coupling of the 5d electron with phonons when the surrounding lattice relaxes to a new equilibrium (from the initial excited status) and results in a redshift in the PL spectrum.[116]

Then how to obtain the narrow-band emission in a Eu^{2+}-doped phosphor? In general, the following aspects should be considered. (1) The structure of the host, for example, the rigid and ordered frameworks toward narrow-band emission and considerable efficiency in phosphors; (2) the limited crystallographic sites, i.e., the crystal should contain a single replaceable cation site or the replaceable cation sites are almost identical; (3) high site symmetry of the activator, including high coordination number, the same bond lengths between the activator and ligands, etc. [38, 117]

The Eu^{2+}-activated materials are introduced according to the category of the host materials, i.e., nitrides, oxynitrides, and oxides here. The structure is, therefore, the determining factor when developing narrow band–emitting phosphors for backlighting. One of the most attractive prototypes is UCr_4C_4. As it can be seen in Figure 2.7(a), UCr_4C_4 crystalized in the tetrahedral system with the space group of $I4/m$ (No. 87). The U, Cr, and C atoms are represented by purple, green, and dark blue spheres, respectively, and the blue tetrahedra are CrC_4 groups. The UCr_4C_4-type compounds could be expressed by the general formula of $AE(A, B)_4X_4$, where AE represents the alkaline earth (AE = Ca, Sr, Ba); A and B are the centers of tetrahedra AX_4 and BX_4; X is the ligand (X = N, O). The tetrahedra connect by corner and edge-sharing to form the channels of *vierer* rings (the term

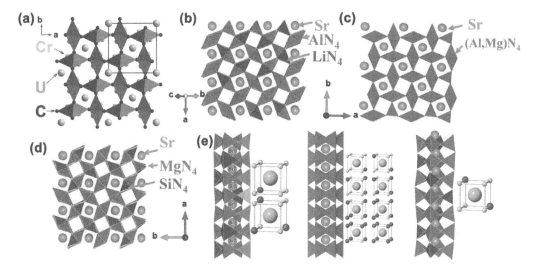

FIGURE 2.7   The crystal structure of (a) UCr_4C_4; the UCr_4C_4-type nitrides (b) $SrLiAl_3N_4$; (c) $SrMg_2Al_2N_4$; (d) $SrMg_3SiN_4$; (e) the channel models and the coordination of Sr in the three nitrides, respectively. (Fang, M. H., Leaño, J. L. and Liu, R. S., *ACS Energy Lett.* 3, 2573–2586, 2018. With permission.)

vierer ring refers to four polyhedra connected to form a ring, and is a nomenclature).[118] AE atoms are located in the tetrahedra-consisted channels, forming the rigid framework with the degree of condensation κ, the atomic ratio (A, B): N, to be 1. Also, the AE^{2+} site is quite appropriate for the Eu^{2+} ion to substitute, enabling extensive research in Eu^{2+}-doped UCr_4C_4-type phosphors.

2.2.1.1 Nitrides

Rare earth–doped nitrides phosphors have attracted much attention since their report in the beginning of the 21st century.[119] Nowadays, they have got applications in the fields of illumination, displays, etc. where there are outstanding phosphors like $CaAlSiN_3$:Eu^{2+}, [66] $Sr_2Si_5N_8$:Eu^{2+}, [120, 121] and $La_3Si_6N_{11}$:Ce^{3+} [122]. Some of our research results in nitride materials have been listed in Table 2.1.

Although we have got some interesting results in white LED-used nitride phosphors, there're very few materials that appropriate for backlighting. However, many excellent nitride-based materials have been reported with narrow bands (typified by the UCr4C4-type nitrides) and they would be introduced in detail in the following.

In 2014, the efficient red-emitting $Sr[LiAl_3N_4]$:Eu^{2+} (SLA) with the narrow band was first reported by Prof. Schnick's group [80], inspiring the subsequent research on phosphors of nitride narrow-band emission, such as $Sr[Mg_2Al_2N_4]$:Eu^{2+} (SMA) and $Sr[Mg_3SiN_4]$:Eu^{2+} (SMS) [87, 133] The structures of SLA, SMA, and SMS are displayed in Figure 2.7(b)–(d), respectively. As can be seen, there are great similarities in their structures. The LiN_4, AlN_4, (for SLA); (Mg, Al)N_4, (for SMA); MgN_4, SiN_4 (for SMS) tetrahedra constitute the channels of *vierer* rings in which each Sr^{2+} ion could coordinate with 8 N^{3-} ions, forming a highly symmetrical cuboid-like polyhedron. These cuboid-like polyhedra would connect by face-sharing, forming endless strands in a particular direction. The ordering property makes the structure highly rigid and simplifies the local coordination environment of the Sr^{2+} ions. The crystal structure of SLA is isotypic to the oxoplumbate $Cs[Na_3PbO_4]$.[134] In SLA, there are two types of Sr^{2+} ions as shown in Figure 2.7(e) (left), of which

---

TABLE 2.1

Some of our Research Results of Eu^{2+}-Doped Nitride Luminescence Materials

Phosphor	S. G.[a]	λ_{ex} (nm)	λ_{em} (nm)	$TQ^{[b]}$ (%)	CCT[d] (K)	CRI[e]	Ref.
$BaLi_2Al_2Si_2N_6$:Eu^{2+}	$P4/ncc$	355	567 (20 GPa)	89.5 (473K)			[123]
$Ca_5Si_2Al_2N_8$:Eu^{2+}	$Pbcn$	450	643	90 (473K)			[124]
$Li_{0.995-x}Mg_xSi_{2-x}Al_xN_3$:$0.005Eu^{2+}$ ($x = 0$–0.04)	$Cmc2_1$	347	565–530	37 (473K)			[125]
$LiSr_4(BN_2)_3$:Eu^{2+}	$Im\bar{3}m$	400	640	<40 (423K)			[126]
$LiSi_2N_3$:Eu^{2+}–xAlN ($0 \le x \le 0.35$)	$Cmc2_1$	355	595	50 (373K)			[127]
$Sr_{1-x}Ca_xYSi_4N_7$:Eu^{2+}($x = 0$–0.5)	$P6_3mc$	347–354	540–564	50 (421K)			[128]
$Ca_2Si_5N_8$:Eu^{2+} (BaF_2 as flux and cation substitution)	Cc	460	617	30 (473K)			[129]
$(Ca_{1-x}Sr_x)_{16}Si_{17}N_{34}$:$Eu^{2+}$	$F\bar{4}3m$	410	653		2642–2817		[130]
Ba^{2+} and Al^{3+}–O^{2-} co-substituted $Sr_2Si_5N_8$:Eu^{2+}	$Pmn2_1$	410	638	>100 (423K)			[131]
$Ca_{1.4-x}Al_{2.8}Si_{9.2}N_{16}$: xEu ($x = 0$–0.3)	$P31c$	400	592	88 (523K)	2056		[132]

[a] S. G.: Space Group, hereinafter inclusive.

[b] the PL intensity at the temperature in the bracket of the initial, e.g., 89.5(473K) means that at 473K, the PL intensity keeps 89.5% of the initial, hereinafter inclusive.

[c] CCT: corresponding color temperature, hereinafter inclusive.

[d] CRI: color rendering index, hereinafter inclusive.

the coordination N^{3-} ions come from LiN_4 and AlN_4 tetrahedra in different proportions (1:7 and 3:5, respectively). This portends two possible luminescence centers in SLA. It is suggested that there are only one kind of Sr site in SMA and SMS. Things are different when the local coordination environment around Sr^{2+} ions is probed. In (Mg, Al)N_4 tetrahedra of SMA, (Mg^{2+}/Al^{3+})-ions are statistically disordered on Wyckoff position 8 h with the site occupancy fraction (SOF) ratio of 0.5:0.5. As it can be seen in Figure 2.7(e) (middle), the numbers of MgN_4 or AlN_4 tetrahedra may range from one to eight statistical permutations and the distances (Mg/Al)-N vary between 1.94 and 2.05 Å, resulting in various local environment around Sr^{2+} ions. But in SMS which is isotypic to the lithosilicate Na[Li_3SiO_4] in Figure 2.7(e) (right),[135] there is only one type of local environment for Sr^{2+} ions, formed by MgN_4 and SiN_4 tetrahedra with a proportion of 6:2. An entire PL spectrum is consistent with the overlapping of the PL spectrum from all of the luminescent centers with various local environments, and accordingly, the spectrum would broaden with the increasing of the categories of local environments.[136, 137] Therefore, the luminescence spectra of Eu^{2+} in the three mentioned materials are distinguishing, which would be discussed later.

The luminescent properties of Eu^{2+}-doped SLA, SMA, and SMS were investigated successively. Figure 2.8(a) shows the broad excitation spectrum (blue) with the maximum at ~466 nm and the narrow emission band (pink) centered at 654 nm with the FWHM of ~1180 cm^{-1} (~50 nm) of SLA: 0.4 mol% Eu^{2+}. Moreover, the Stokes shift of the luminescence is calculated to be as small as 956 cm^{-1}. For comparison, the excitation (gray) and emission (dark gray) spectra of the commercially available $CaAlSiN_3$:Eu^{2+} are showed. The emission band is much broader than the former. It was reported that the emission spectra of (Ca, Sr)$SiAlN_3$:Eu^{2+} are located at ~610–660 nm with the FWHM ~2100–2500 cm^{-1}. In addition, the temperature-dependent emission spectra of SLA: 0.4 mol% are shown in Figure 2.8(b). It has a very high thermal quenching (TQ) temperature, i.e., more than 95% relative quantum efficiency (QE) remains at 200°C compared to the initial at room temperature (RT). Thus, it is an outstanding red-emitting candidate for backlighting applications with high color purity.

AE[$Mg_2Al_2N_4$] (AE = Ca, Sr, Ba) are isotypic, and the structures only show the difference in lattice parameters, i.e., the parameters get larger with the alkaline earth atom size. The luminescence spectra of the AE[$Mg_2Al_2N_4$]: 0.1 mol% Eu^{2+} series are depicted in Figure 2.8(c). The spectra related to Ca[$Mg_2Al_2N_4$]:Eu^{2+}, SMA:Eu^{2+}, and Ba[$Mg_2Al_2N_4$]:Eu^{2+} are depicted in orange, green, and blue, respectively. The excitation spectra are in accordance with the reflectance spectra and the excitation maximum occurs in the range of 450–480 nm. The emission bands show redshift and broadening when AE changes from Ca to Sr and Ba while emission positions are 607 nm (FWHM ~1815 cm^{-1}) for AE = Ca, 612 nm (FWHM ~1823 cm^{-1}) for AE = Sr, and 666 nm (FWHM ~2331 cm^{-1}) for AE = Ba. The significant redshift was concluded to be the result of trapped-excitation emission, which can be caused by the Eu^{2+} 5d band's locating close to the conduction band bottom of the host. [139] The thermal behaviors of the Ca and Sr compounds are complex, whereas the TQ process of the Ba compound is simpler in the whole temperature range which can be fitted by the equation $I_T/I_0 = [1 + A\exp(-E_a/kT)]^{-1}$ (see the dashed curve in Figure 2.8(d)). The PL intensity remains less than 10, 20 and 40% of the initial (7K) for the Ca, Sr, and Ba compounds at 400K, respectively. The performances of AE[$Mg_2Al_2N_4$]:Eu^{2+} are inferior to SLA:Eu^{2+} but provide significant fundamental investigations for the understanding of Eu^{2+}-doped narrowband emitters.

The excitation and emission spectra of SMS: 2 mol% Eu^{2+} in Figure 2.8(e) suggest that it has the excitation and emission maximum at around 450 and 615 nm with the FWHM of ~1170 cm^{-1} (~43 nm), respectively. It is noted that at very low temperature (6K), the QE of SMS: 2 mol% Eu^{2+} reaches about 100% with a narrower FWHM of ~33 nm. It is the narrowest emission for Eu^{2+}-doped phosphors in the red spectral region reported so far. The narrower emission band and higher human eye sensitivity compared to SLA:Eu^{2+} make it attractive. However, the TQ behavior is not so satisfied, as shown in the temperature-dependent PL spectra of SMS: 2 mol% Eu^{2+} (Figure 2.8(f)), the emission intensity remains less than 10% (at RT) of that at 6K under the excitation of 440 nm. The small bandgap (3.9 eV) of SMS results in significant TQ of the luminescence at a higher temperature,

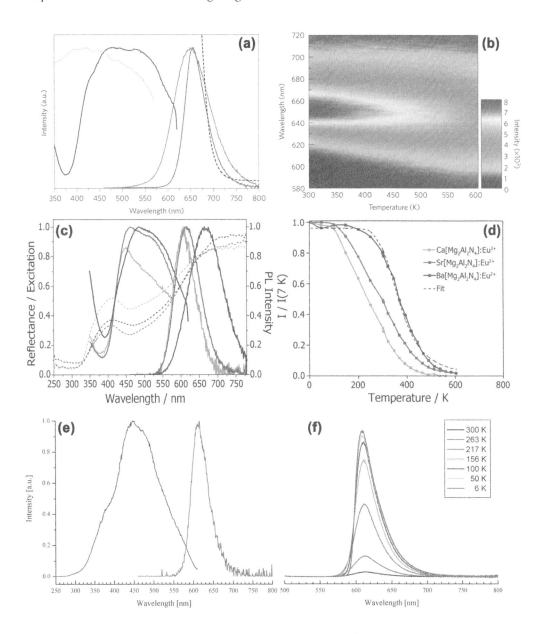

FIGURE 2.8 (a) Excitation (SLA: 0.4 mol% Eu²⁺, blue; CaAlSiN₃:Eu²⁺, light gray) and emission (SLA: 0.4 mol% Eu²⁺, pink; CaAlSiN₃:Eu²⁺, dark gray) spectra for λ_{ex} = 440 nm. The dotted curve represents form the upper limit of sensitivity of the human eye.[138] (b) Temperature-dependent emission spectra of the SLA: 0.4 mol% Eu²⁺ from 300 to 600K. (c) The reflectance (dashed lines), PLE and PL spectra of 0.1 mol% Eu²⁺-doped AE[Mg₂Al₂N₄] (orange for AE = Ca, green for AE = Sr, blue for AE = Ba). The emission spectra are obtained under the excitation of 440 nm. (d) The relative temperature-dependent PL intensity of the related materials. (e) The excitation (blue) and emission (red) spectra of Sr[Mg₃SiN₄]: 2 mol% Eu²⁺ and (b) the related temperature-dependent PL spectra from 6 to 300K under the excitation of 440 nm. (Pust, P., Weiler, V., Hecht, C., Tücks, A., Wochnik, A. S., Henß, A. K., Wiechert, D., Scheu, C., Schmidt, P. J. and Schnick, W., *Nat. Mater.* 9, 891, 2014. With permission; Pust, P., Hintze, F., Hecht, C., Weiler, V., Locher, A., Zitnanska, D., Harm, S., Wiechert, D., Schmidt, P. J. and Schnick, W., *Chem. Mater.* 26, 6113, 2014. With permission. Schmiechen, S., Schneider, H., Wagatha, P., Hecht, C., Schmidt, P. J. and Schnick, W., *Chem. Mater.* 26, 2712, 2014. With permission.)

although the small Stokes shift (~772 cm^{-1}) should lead to good thermal behavior. Later, further investigations on the electronic structures suggested the proximity of the lowest 4f5d states (Eu^{2+}) and the bottom of CB (0.13 eV) is the reason for its lower stability compared to SLA:Eu^{2+} (0.28 eV). Modification on crystal field splitting, bandgap, and the appropriate solid solution was suggested to improve the thermal stability.[140]

Based on the series materials comparable in composition and structure those introduced earlier, a related summary could be given as follow.

1. The bandwidth. In the UCr$_4$C$_4$-type nitrides, in addition to site categories of AE^{2+} ions, the local environment of Eu^{2+} is a significant factor to determine the FWHM of emission bands.
2. The trapped-excitation emission. Large Stokes shift, large FWHM, and nontypical red-shift related to AE^{2+} ions in materials with general structure could indicate anomalous trapped-excitation emission.[139]
3. Thermal stability. The TQ process of these materials could be considered more comprehensively from the aspects of the host bandgap; Stokes shift; trapped-excitation emission, etc.

In addition to the UCr$_4$C$_4$-type nitrides, other nitrides that are sorted by their compositions such as nitridomagnesosilicates, nitridolithoaluminates, and other nitridosilicates doped with Eu^{2+} are also reported with narrow-band emissions.[141–144] Their host lattices provide appropriate positions that are diverse from the UCr$_4$C$_4$-type compounds for Eu^{2+} to substitute. Several of the typical materials would be described later.

The nitridolithoaluminate Ba[Li$_2$(Al$_2$Si$_2$)N$_6$] (BLASN) is crystallized in a tetragonal system and belongs to the space group of *P4/ncc* (no. 130). The lattice parameters are $a = 7.8282(4)$ Å; $c = 9.9557(5)$ Å; cell volume = 610.09(7) Å^3 and the formula units in a single unit cell is 4. It shows similar atomic categories to that of Li$_2$(Ca$_{1.88}$Sr$_{0.12}$)[Mg$_2$Si$_2$N$_6$].[143] There's only one type of Ba site in the lattice, Al and Si share the same site with a half SOF statistically. The degree of condensation κ is found to be 1, which is comparable to the UCr$_4$C$_4$-type ones. The crystal structure is built up from corner- and edge-sharing (Al, Si)N$_4$ tetrahedra and they form two types of *vierer* rings along [0 0 1] direction (Figure 2.9(a)). Every two LiN$_4$ tetrahedra link to each other by edge-sharing and a Li$_4$N$_{12}$ bisphenoid is formed inside the smaller *vierer* ring channels. The larger *vierer* rings are centered by Ba^{2+}, which is coordinated by eight N^{3-} ions, forming a truncated square pyramid (Figure 2.9(b)). The arrangement of the truncated square pyramids are different from the cuboid-like polyhedra in UCr$_4$C$_4$-type structures mentioned earlier, they are staggered along [0 0 1] and connected by common corners in [1 0 0] direction. BLASN: 1 mol% Eu performs a green emission band centered at 532 nm with FWHM = 1962 cm^{-1} (~57 nm). The excitation maximum is located at 395 nm and a strong intensity decrease occurs after that (Figure 2.9(c)). The green emission should be the result of a longer M–N bond length (M = Ba, Sr, Ca or Eu) between 2.93(2) to 3.10(9) Å compared to the UCr$_4$C$_4$-type compounds (red emission) mentioned previously.[80, 87, 133] Moreover, the introduction of a second filled channel leads to an additional degree of freedom in the substitution of elements and, therefore, in luminescence tuning. Recently, the high pressure-induced luminescence properties of BLASN:Eu^{2+} were investigated and the PL spectra (1 mol% Eu^{2+}-doped) are displayed in Figure 2.9(d).[123] The emission intensity drops with a redshift from 532 to 567 nm under the excitation of 355 nm when the hydrostatic pressure reaches ~20 GPa. The gradual compression of host lattice and decrease of bond lengths under increasing pressure lead to a stronger crystal field splitting, therefore, the redshift. The distinguished luminescence centers are proved by decay performance and Gaussian fitting of the sample. As shown in Figure 2.9(e), the peak positions of the two fitted Gaussian peaks are red-shifted to a longer wavelength. In addition to the peak position, the width of the emission band, which is important to the narrow-band phosphor, is also explored under high pressures. The FWHM of BLASN: 1 mol% Eu^{2+} exhibits an interesting variation before and after 7 GPa (Figure 2.9(f)). Compared to it before 7 GPa, the FWHM increases virtually linearly. According to the structure analysis, the value of a/c has a decrease when the pressure exceeds

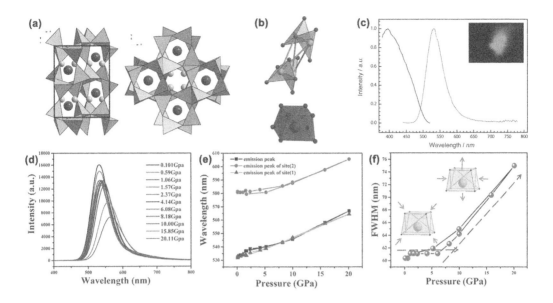

FIGURE 2.9 (a) Crystal structure of $Ba[Li_2(Al_2Si_2)N_6]$ in which (Al, Si)N_4-tetrahedra (orange), Li (violet), Ba (red) are presented in the viewing direction of [0 1 0] and direction [0 0 1], respectively. (b) The bisphenoidal arranged tetrahedra of Li_2N_6 units and the truncated square pyramid of BaN_8. (c) Excitation (blue) and emission (green) spectra of $Ba[Li_2(Al_2Si_2)N_6]$: 1 mol% Eu. (d) The emission spectra of BLASN:0.01Eu^{2+} under varying pressure in the compression process excited at 355 nm. (e) The emission peak positions of the two emission sites and integral spectra. (f) FWHM under varying pressure. (Strobel, P., Schmiechen, S., Siegert, M., Tücks, A., Schmidt, P. J. and Schnick, W., *Chem. Mater.* 27, 6109, 2015. With permission; Wang, Y., Seto, T., Ishigaki, K., Uwatoko, Y., Xiao, G., Zou, B., Li, G., Tang, Z., Li, Z. and Wang, Y., *Adv. Funct. Mater.* 30, 2001384, 2020. With permission.)

6.5 GPa, i.e., the EuN_8 polyhedra would be distorted due to the uneven construction along [0 0 1]. Besides, more defects are generated during this process. As a result, the FWHM of the emission spectra becomes wider.

Another efficient green-emitting nitridolithosilicate $Ba_2LiSi_7AlN_{12}$:Eu^{2+} material was also reported with a narrow emission band and good thermal stability. The $Ba_2LiSi_7AlN_{12}$ crystal belongs to the orthorhombic system in the spaces group of *Pnnm* (no. 58). The lattice parameters are $a = 14.0993(2)$ Å, $b = 4.89670(10)$ Å, $c = 8.07190(10)$ Å, cell volume = 557.28(15) Å^3 and $Z = 2$. There's only one type of Ba site and the Ba^{2+} ions in the lattice would coordinate with eleven N^{3-}. Three types of Si/Al sites with the atom proportion (Si:Al) of 0.875:0.125 and the half occupied Li sites are verified by the crystal analysis. Si^{4+}/Al^{3+} ions occupy the tetrahedral sites (blue tetrahedra), and Li^+ ions occupy the independent tetrahedral sites (red tetrahedra), as shown in Figure 2.10(a). Corner-sharing (Si, Al)N_4 tetrahedra form a corrugated layer (marked A) along the [0 0 1] direction; edge-sharing (Si, Al)N_4 tetrahedra and LiN$_4$ tetrahedra alternately align along the [0 1 0] direction to form a pillar (marked B). Ba^{2+} ions occupy the channels formed by layer A and pillar B along [0 1 0] direction in a zigzag manner. Each Ba^{2+} ion is coordinated by eleven N^{3-}, and the BaN_{11} polyhedra are connected by face sharing. The excitation and emission spectra of $Ba_2LiSi_7AlN_{12}$: ~15 mol% Eu are shown in Figure 2.10(b). The excitation spectrum spans 350 nm to 450 nm. A green-emitting band peaked at ~515 nm with an FWHM of 61 nm (2280 cm^{-1}) is obtained under the excitation of 405 nm while the internal QE was found to be 79%. Although the FWHM is relatively wide compared with that of very narrow green-emitting, it is fairly narrow for a Eu^{2+}-emitting phosphor and is suitable for backlight applications. Besides, the temperature-dependent emission spectra show only a small decrease in luminescence intensity at 200 and 300°C (84 and 76% of the peak intensity

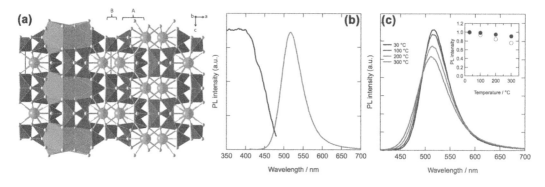

FIGURE 2.10 (a) The crystal structure of $Ba_2LiSi_7AlN_{12}$. (Si, Al)N_4 (Blue), LiN$_4$ (red) tetrahedra, BaN$_{11}$ polyhedra (green), Ba (green), and N (white) are displayed in the viewing direction of [0 1 0]. (b) The excitation (black) and emission (green) spectra of $Ba_2LiSi_7AlN_{12}$: ~15 mol% Eu with the monitoring and stimulating wavelength of 515 and 405 nm, respectively. (c) The related temperature-dependent PL spectra inset shows the peak intensity (open circle) and integrated intensity (filled circle). (Takeda, T., Hirosaki, N., Funahshi, S. and Xie, R. J., *Chem. Mater.* 27, 5892, 2015. With permission.)

observed at RT), as can be seen in Figure 2.10(c). On the whole, many Eu^{2+}-activated nitride luminescence materials can be efficiently excited by blue LED chips with good stability, whereas the preparation conditions are harsh and the preparation cost is high. Besides, the color purity, which is related to the peak position and FWHM of the emission band, still need to be improved.

2.2.1.2 Oxynitrides

Oxynitrides are also good host cadidates for Eu^{2+}, some of our works are listed in Table 2.2.

Among our results, few appropriate oxynitride materials are appropriate for the application in backlighting. However, there are many reported typical and nice oxynitride phosphors with narrowband, which could be seen in the following part.

The Eu^{2+}-doped β-sialon is a well-known phosphor for its favorable properties in emission color, thermal stability, and QE. The chemical composition, $Si_{6-z}Al_zO_zN_{8-z}$ ($0 \leq z \leq 4.2$), is derived from β-Si$_3$N$_4$ by equivalent substitution of Al–O for Si–N pairs, and z represents the substitution

TABLE 2.2
Some of Our Research Results of Eu^{2+}-Doped Oxynitride Luminescence Materials

Phosphor	S. G.	λ_{ex} (nm)	λ_{em} (nm)	TQ (%)	LED devices CCT (K)	CRI	Ref.
$[Mg_{1.25}Si_{1.25}Al_{2.5}]O_3N_3$:$Eu^{2+}$	$R\bar{3}m$	290, 335	460	85 (523K)			[145]
$La_{10-x}Sr_x(Si_{6-x}P_xO_{22}N_2)O_2$:$Eu^{2+}$ ($x = 0$–4)	$P6_3/m$	373–351	520–464	42.6–49.6 (473K)			[146]
$Ba_3Si_6O_9N_4$:Eu^{2+}	$P\bar{3}$	410	494	74 (473K)			[147]
$Li_2SrSiON_2$:Eu^{2+}	$P3_121$	450	586	78 (473K)	4386	90	[148]
$Ca_2(B_{1-y}Si_y)(O_{3-y}N_y)Cl$:0.04$Eu^{2+}$	$P121/c1$	420	525	94 (473K)		~92.8	[149]
$Ca_3Li_{4-y}Si_2N_{6-y}O_y$:$Eu^{2+}$	$C12/m1$	325	702	50 (453K)			[150]
$SrSiAl_2O_3N_2$:Eu^{2+}	$P2_12_12_1$	351	484	56 (423K)			[151]
$Ba_{2.9-x}Mg_xEu_{0.1}Si_6O_{12}N_2$	$P\bar{3}$	368, 400	525	82 (423K)			[152]
$Ba_{3-x}Ca_xSi_6O_{12}N_2$:$Eu^{2+}$	$P\bar{3}$	350, 405	525–536	74 (423K)			[153]
$Sr_2SiN_zO_{4-1.5z}$:Eu^{2+}	$Pmnb$	460	624	87 (523K)			[154]

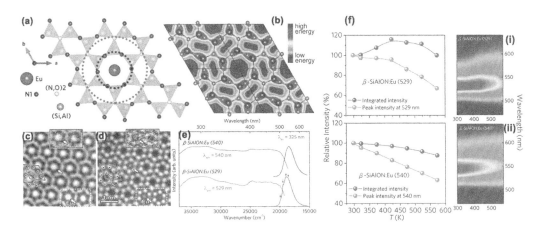

FIGURE 2.11 (a) Crystal structure of $2 \times 2 \times 2$-unit cells of β-sialon:Eu viewed along c-direction, in which Eu (red), (Si, Al) (orange), (O, N)2 (yellow), N1 (blue) are presented. (b) The potential energy surface of the (0 0 1) plane for β-Si_3N_4 calculated using density functional theory (DFT). The low-energy regions shown in violet indicate the most favorable interstitial position for the incorporation of Eu^{2+}. (c) BF and (d) ADF STEM images of β-sialon:Eu observed by our technique. Upper insets in solid-line rectangles show single-scanning images obtained in a conventional manner, resulting in severe quantum noise. Lower insets in dotted rectangles show simulation results. White arrows show Eu atom positions. (e) PLE and PL spectra of β-sialon: Eu (529) and β-sialon:Eu (540) green phosphors (dark arrows show the visible vibronic structure). (f) Thermal quenching behavior of photoluminescence for β-sialon: Eu (529) (i) and β-sialon: Eu (540) (ii); temperature-dependent PL spectra are shown on the right-hand side. (Zhang, X., Fang, M. H., Tsai, Y. T., Lazarowska, A., Mahlik, S., Lesniewski, T., Grinberg, M., Pang, W. K., Pan, F., Liang, C., Zhou, W., Wang, J., Lee, J. F., Cheng, B. M., Hung, T. L., Chen, Y. Y. and Liu, R. S., *Chem. Mater.* 29, 6781, 2017. With permission; Brgoch, J., Gaultois, M. W., Balasubramanian, M., Page, K., Hong, B. C. and Seshadri, R., *Appl. Phys. Lett.*105, 181904, 2014. With permission; Kimoto, K., Xie, R. J., Matsui, Y., Ishizuka, K. and Hirosaki, N., *Appl. Phys. Lett.* 94, 041908, 2009. With permission.)

amount of the Al–O (Si–N) pairs. It belongs to the hexagonal system with the space group of $P6_3/m$ (no. 176) and the Z of 2. Figure 2.11(a) shows the structure of β-sialon:Eu along [0 0 1] direction. There're one type (Si, Al) site and two types of (N, O) sites in the main framework, and the (Si, Al)(N, O)$_4$ tetrahedra connect by corner-sharing to form a one-dimensional channel along [0 0 1] direction.[70, 155] A substitution site is not obvious since Si^{4+} (CN = 6, radius = 0.400 Å) and Al^{3+} (CN = 6, radius = 0.535 Å) are too small for Eu^{2+} (CN = 6, radius = 1.17 Å) to substitute in practical.[156] As a result, the inclusion of the rare earth must occur at an interstitial while the Si/Al ratio could change to maintain electroneutrality. The density functional theory (DFT) calculation of the potential energy surface for β-Si_3N_4 is based on the Vienna *ab initio* Simulation Package (VASP).[157, 158] The result in Figure 2.11(b) reveals two possible interstitial sites, i.e., 12-coordinate sites (Wyckoff position) 2a and 6c. The polyhedral volume of the 6c site (~20 Å^3) is half the size of the 2a site (~40 Å^3), resulting in the greater opportunity for Eu^{2+} to be the center of the 2a site[159], as shown in Figure 2.11(a) (the channel center). More direct evidence for the position of Eu^{2+} in β-sialon is provided by applying the scanning transmission electron microscopy (STEM)-based technique with the calculation of STEM images using a multislice simulation program.[160] Figure 2.11(c) and (d) show the bright-field (BF) and annular dark-field (ADF) imaging of a β-sialon:Eu sample.[161] The inset above and below are single-scanning images obtained in a conventional manner and the simulation results, respectively. The high signal-to-noise (SN) ratio ADF image shows the Eu dopant clearly (pointed by the white arrow). Eu in the interstitial position could coordinate with nine N (the number of the nearest neighbors is always considered to be nine instead of 12 when describing the local environment of Eu^{2+}) to form a symmetric EuN_9 polyhedron and green emission can be obtained with narrow band.

As previously mentioned, the component of β-sialon is adjustable, and therefore, the related luminescent spectra of Eu^{2+} in them are variable. Figure 2.11(e) shows the PLE and PL spectra of β-sialon: Eu (540) ($z = 0.18$, the chemical formula of β-$Si_{5.82}Al_{0.18}O_{0.18}N_{7.82}$:Eu) and β-sialon:Eu (529) ($z = 0.03$, the chemical formula of β-$Si_{5.97}Al_{0.03}O_{0.03}N_{7.97}$:Eu) while the concentration of Eu is set to be 0.02 mol %. The emission of β-sialon:Eu (529) consists of a band centered at 529 nm (FWHM ~49 nm) with several superimposed emission lines. The distinct fine structure observed in the PL and PLE spectra is originated from the ^7F_J states and the lattice phonon in the Eu^{2+} f-d transition, which is indicated by the low-temperature measurements.[162] The fine structure gradually vanished with increasing z value and it almost disappears in β-sialon:Eu (540), of which the band locates at 540 nm with the FWHM of ~54 nm. The temperature-dependent PL spectra of β-sialon:Eu (529) and β-sialon:Eu (540) are displayed in Figure 2.11(f) (i) and (ii), respectively. The emission intensity in both materials decreases continuously when the temperature increase from ambient temperature to 573K. It suggests that β-sialon:Eu (529) has a better thermal performance compared to β-sialon:Eu (540), i.e., smaller redshift (352 vs. 468 cm^{-1}) and smaller intensity drop, due to the more rigid and ordered structure. β-sialon:Eu is one of the most popular phosphors, which is applied as the green component in backlighting due to its outstanding performance.

The crystal structure of $BaSi_2O_2N_2$ was investigated and the structural schematic diagrams are shown in Figure 2.12(a) viewing from the direction of [001]. The structure of $BaSi_2O_2N_2$ exhibited disordered nitrogen atoms and could be described in the space group $Cmcm$ (no. 63). Due to the multiplicity of the single N site in this space group, possible ordered structures can only be obtained by lowering the symmetry. This results in two structure models with fully ordered N positions in the space groups $Pbcn$ (no. 60) and $Cmc2_1$ (no. 36), respectively, which are distinguished by different stacking sequences of the silicate layers. The former is more dominant according to the study.[163] It shows a layer structure, i.e., the flat layers of Ba and $SiON_3$ tetrahedra layers (the silicate layers, Figure 2.12(b)) are arranged alternately. There's only one type of Ba site in the structure and each Ba coordinates with eight O and two farther N atoms with high site symmetry (Figure 2.12(c)) in space group $Pbcn$. $BaSi_2O_2N_2$ was then found to be a good host for Eu^{2+} to generate efficient cyan emission. The PL and PLE spectra of $Ba_{0.9}Eu_{0.1}Si_2O_2N_2$ are displayed in Figure 2.12(d), it shows an emission band centered at 499 nm with the FWHM of ~35 nm, and the QE is 71% at RT. The integrated emission intensity and decay time as a function of temperature is plotted in Figure 2.12(e). The quenching temperature (the temperature for the intensity drops to 50% of the initial at 300K) values for $BaSi_2O_2N_2$: 2 mol% Eu is around 600K. The decay time changes from 0.47 μs (RT) to 0.3 μs (600K). The integrated emission intensities and luminescence decay times could be fitted by the equation $I_T/I_0 = [1 + A\exp(-E_a/kT)]^{-1}$ and $\tau_T/\tau_0 = [1 + B\exp(-E_a/kT)]^{-1}$ as it can be seen in Figure 2.12(e).[164]

The crystal structure of $Sr[Li_2Al_2O_2N_2]$ (SALON) is an ordered variant of the UCr_4C_4-type structure, which belongs to the tetragonal system with the space group of $P4_2/m$ (no. 84). An Al atom is coordinated by three N and one O atoms while a Li atom is coordinated by three O and one N atoms. The $AlON_3$ and LiO_3N tetrahedra consist of the main framework of SALON, including the one-dimensional channels of $vierer$ rings along [0 0], as shown in the left of Figure 2.13(a). Sr atoms could locate in one of the channels in which a $vierer$ ring is comprised by $AlON_3$ and LiO_3N tetrahedra with the proportion of 1:1, forming the condensed network with the degree of condensation κ to be 1. The Sr atom is located at a center of inversion, coordinated by 4 O and 4 N with the Sr–N bond length of 2.760(5) Å, Sr–O bond length of 2.659(4) Å, results in the highly symmetric local environment around Sr (the right of Figure 2.13(a)). The red emission of SALON: 0.7 mol% Eu centered at 614 nm with the FWHM of 48 nm (1286 cm^{-1} or 0.1594 eV) is observed under the excitation of 460 nm. The excitation and emission spectra are shown in Figure 2.13(b). For comparison, the emission spectrum of SLA:Eu is also displayed (the purple curve, $\lambda_{emmax} = 654$ nm, FWHM ~50 nm or 1180 cm^{-1} or 0.1463 eV). Despite the similar FWHM, the emission of SALON: 0.7 mol% Eu locates at a higher energy range and thereby owning a larger overlap in spectra with the human

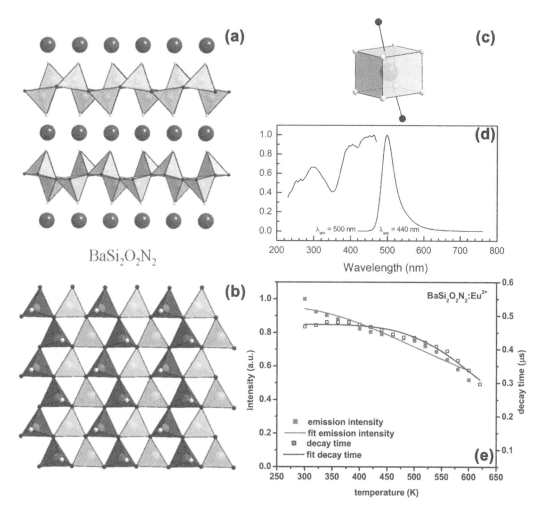

FIGURE 2.12 (a) The crystal structure of $BaSi_2O_2N_2$. The spheres in light gray, black, and dark gray represent O, N, and Ba atoms, respectively. (b) The silicate layers. $SiON_3$ tetrahedra with vertices up are depicted in dark gray, with vertices down in light gray. (c) Coordination of Ba. (d) Excitation and emission spectra of $Ba_{0.9}Eu_{0.1}Si_2O_2N_2$. (e) Temperature dependence of the integrated emission intensity and luminescence decay times of $BaSi_2O_2N_2$: 2 mol% Eu^{2+}. The lines through the data points fit the equation $I_T/I_0 = [1 + A\exp(-E_a/kT)]^{-1}$ and $\tau_T/\tau_0 = [1 + B\exp(-E_a/kT)]^{-1}$, respectively. (Kechele, J. A., Oeckler, O., Stadler, F. and Schnick, W., *Solid State Sci.* 11, 537, 2009. With permission; Li, Y. Q., Delsing, A. C. A., de, G. and Hintzen, H. T., *Chem. Mater.* 17, 3242, 2005. With permission; Bachmann, V., Ronda, C., Oeckler, O., Schnick, W. and Meijerink, A., *Chem. Mater.* 21, 316, 2009. With permission.)

eye sensitivity curve (dotted black). In addition to that, the temperature-dependent (from 298 to 500K) luminescent intensity of SALON: 0.7 mol% Eu in Figure 2.13(c) suggests that the intensity drops by only 4% at 420K compared to that at RT, which surpasses the performance of other narrow-band materials such as SMS:Eu [140] and $Ba[Li_2(Al_2Si_2)N_6]$:Eu [144].

Oxynitrides are attractive candidates for LED-used phosphors because of their favorable properties such as simpler synthesis, lower cost compared to nitrides, good stability, and thus, they are widely investigated in recent years. However, many of them are not satisfactory enough for the practical application in backlighting due to the limitation such as unbefitting emission color and wide FWHM.

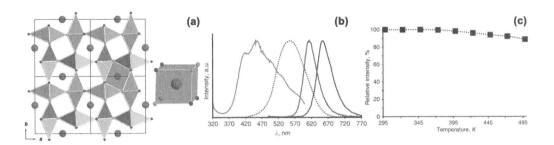

FIGURE 2.13 (a) The crystal structure of $Sr[Li_2Al_2O_2N_2]$ (SALON) along $[0\,0\,\bar{1}]$ direction and the cubic-like coordination of Sr. The spheres in red, green, and blue represent Sr, N, and O atoms, respectively, while the tetrahedra of $AlON_3$ and LiO_3N are depicted in gray and orange, respectively. A Sr could coordinate with 4 O and 4 N to form the cubic-like polyhedron. (b) Normalized excitation (gray, λ_{em} = 614 nm) and emission spectrum (red, excited with λ_{exc} = 460 nm) of SALON: 0.7 mol% Eu and for comparison, the emission spectrum of SLA: 0.4 mol% Eu (purple), as well as the human-eye sensitivity curve (black dotted), are displayed simultaneously.[80] (c) Temperature-dependent PL intensity of SALON: 0.7 mol% Eu from 298 to 500K.(Hoerder, G. J., Seibald, M., Baumann, D., Schröder, T., Peschke, S., Schmid, P. C., Tyborski, T., Pust, P., Stoll, I., Bergler, M., Patzig, C., Reißaus, S., Krause, M., Berthold, L., Höche, T., Johrendt, D. and Huppertz, H., *Nat. Commun.* 10, 1824, 2019. With permission.)

2.2.1.3 Oxides

Oxides are much easier to obtain and cost less than (oxy)nitrides. We have reported many Eu^{2+}-doped oxide materials those are listed in Table 2.3, among which narrowband candidates for the potential application in backlighting are introduced in detail next.

Compared to UCr_4C_4-type nitride phosphors, the isotypic oxides, i.e., the alkali lithosilicate (germanate) phosphors own more flexible composition forms, which allow substitution among alkali atoms (Li, Na, K, Rb, and Cs) in the corresponding site. This also enables the color tunability (green and cyan) in UCr_4C_4-type alkali lithosilicates, which is derived from the structural adjustment. The UCr_4C_4-type alkali lithosilicates could be represented by the formula $A_n[Li_3SiO_4]_n$ (A = one or several from Li, Na, K, Rb, and Cs; n = 1, 2, 4, 8). The structures of these materials are all based on the UCr_4C_4 structure, the LiO_4 and SiO_4 tetrahedra constitute the channels of *vierer* rings with κ = 1 while the A atoms occupy the centers of the *vierer* rings. The difference in structure and the luminescence performance is caused by the composition variety, and therefore, they have different crystallographic characteristics including crystal system, space group, and site symmetry. For a better understanding of the UCr_4C_4-type alkali lithosilicates, several typical materials would be discussed later.

$RbLi[Li_3SiO_4]_2$ (RLSO) has been reported to be a narrow-band emitter but suffers the disadvantages such as low efficiency and chemical stability. The substitution of Li to Na results in the compound $RbNa[Li_3SiO_4]_2$ (RN), which could perform green emission with narrowband as well as enhanced stability. The crystal structure of RN belongs to a monoclinic system in the space group $C2/m$ (no. 12), which is the same as RLSO.[194] The frameworks of the two compounds are distinguished by the coordination and connection type of the LiO_m and NaO_n groups, see Figure 2.14(a). Instead of the 2b (for Li3) and 2a (for Li4) sites in RLSO, Na in RN prefers the 4g site, and the LiO_4 squares along [0 1 0] are replaced by the NaO_8 cubes linking with each other by face sharing. This alteration in the local structure was reported to affect the emission peak position and chemical stability significantly. The PL and PLE spectra of RN: 8 mol% Eu (green solid) are displayed in Figure 2.14(b), as well as that of RLSO: 8 mol% Eu (red dash) and β-sialon: Eu (blue dot). RN: 8 mol% Eu performs green emission band centered at 523 nm with FWHM of 41 nm under the excitation of 455 nm, which is comparable to RLSO: 8 mol% Eu and narrower than β-sialon: Eu.

TABLE 2.3

Some of our Research Results of Eu^{2+}-Doped Oxide Luminescence Materials

Phosphor	S. G.	λ_{ex} (nm)	λ_{em} (nm)	TQ (%)	LED Devices CCT (K)	CRI	Ref.
$BaSc_2Si_3O_{10}:Eu^{2+}$	$P2_1m$	330	445	62 (523K)			[152]
$Ca_{10}(SiO_4)_3(SO_4)_3F_2:Eu^{2+}$	$P2_1m$	330	445	62 (523K)			[165]
$Ba_3P_4O_{13}:Eu^{2+}$	$Pbcm$	365	587	35 (423K)	4354		[166]
$Na_3Sc_2(PO_4)_3:Eu^{2+}$	$R\bar{3}c$	348	458	110 (423K)			[167]
$Sr_8ZnLu(PO_4)_7:Eu^{2+}$	$I12/a1$	400	520	77 (423K)			[168]
$LiBa_{12}(BO_3)_7F_4:Eu^{2+}$	$I4/mcm$	405	644	59 (423K)	4856	84.1	[169]
$Sr_9(Li, Na, K)Mg(PO_4)_7:Eu^{2+}$	$R\bar{3}m$	405	635	<40 (423K)	4287–3131	71.1–94.8	[170]
$K_2ZrSi_3O_9:Eu^{2+}$	$P6_3/m$	400	465	35 (423K)			[171]
$Ba_5(PO_4)_2SiO_4:Eu^{2+}$	$P6_3/m$	405	515	40 (423K)	4561	85.4	[172]
$K_2ZrSi_2O_7:Eu^{2+}$	$P112_1/b$	350	462	24 (423K)	4641	86	[173]
$K_2HfSi_3O_9:Eu^{2+}, Sc^{3+}$	$P\bar{6}$	400	507	100 (423K)	3397– 3527(395ex) 4066(450ex)	84.7– 81.3(395ex) 82.2(405ex)	[174]
$Ba_3ScB_3O_9:Eu^{2+}$	$P6_3cm$	376	735	23 (425K)			[175]
$Ca_{10-x}Sr_x(PO_4)_6F_2:Eu^{2+}$ $(x = 4, 6, 8)$	$P6_3/m$	459–447	343–337	55.01–68.24 (423K)	3728	90.21	[176]
$Ca_6Y_4(SiO_4)_2(PO_4)_4O_2:Eu^{2+}$	$P6_3/m$	346	445	30.7 (398K)			[177]
$Na_2HfSiO_5:Eu^{2+}$	$P121/c1$	369	551	45 (423K)	3875	90	[178]
$Ba_5SiO_4Cl_6:Eu^{2+}$	$C2/c$	345	440	80 (423K)	3956	90.7	[179]
$Na_4Hf_2Si_3O_{12}:Eu^{2+}$	$R\bar{3}c$	365	545	>40 (423K)	3953	87.5	[180]
$Na_2HfSi_2O_7:Eu^{2+}$	$P\bar{1}$	397	460	50 (423K)	4617	92.6	[181]
$Ba_3Ca_4(BO_3)_3(SiO_4)Cl:Eu^{2+}$	$P6_3/m$	350	488	52.5 (423K)	4337	90.1	[182]
$K_3ScSi_2O_7:Eu^{2+}$	$P6_3/mmc$	465	735	70.4 (423K)			[100]
$Sr_{0.9}Ba_{0.1}HfSi_2O_7:Eu^{2+}$	$P121/c1$	352	456	59.5 (423K)	4351	90.2	[183]
$Sr_8CaBi(PO_4)_7:Eu^{2+}$	$I12/a1$	400	606	34 (423K)	7756	75	[184]
$BaSi_2O_5:Eu^{2+}$	$Pcmn$	345	505	46.4 (423K)			[185]
$Sr_{1-x}Ba_xCO_3:Eu^{2+}$	$Pnma$	450	618–750				[186]
$KSrBP_2O_8: Eu^{2+}$	$I\bar{4}2d$	365	462	90.9			[187]
$Sr_2SiN_zO_{4-1.5z}:Eu^{2+}$	$Pmnb$	460	624	87			[188]
$Ca_8Mg(GeO_4)_4Cl_2:Eu^{2+}$	$Fd\bar{3}m$	365, 435	425, 510	76.94 (443K)			[189]
$NaBaScSi_2O_7:Eu^{2+}$	$P2_1m$	427	506	75 (533K)			[190]
$Sr_{2.97-x-y}Eu_{0.03}Mg_xBa_ySiO_5$		400	601	80.99 (413K)			[191]
$CaZr_4(PO_4)_6:Eu^{2+}$	$R\bar{3}$	365	496				[192]
$Ca_7(PO_4)_2(SiO_4)_2:Eu^{2+}$		365	522	34.1 (413K)			[193]

The temperature-dependent luminescence spectra of RN: 8 mol% Eu shown in Figure 2.14(c) suggest a low peak intensity drop of 8.5% (12%) happens under 425K (500K) with the FWHM varies from 41 to 51 nm in the temperature range 300–500K. Besides, RN: 8 mol% Eu is more resistant than RLSO: 8 mol% Eu, e.g., the former keeps 59.6% of that of the pristine sample after 1 hour in an 80°C/80% relative humidity (RH) test while the latter only remains 13% of the pristine sample.

The structure becomes more sophisticated as the compositions of $A_n[Li_3SiO_4]_n$ get more complicated. The crystal structures of $RbNa_3[Li_3SiO_4]_4$ (RNLSO), $RbNa_2K[Li_3SiO_4]_4$ (RNKLSO), and

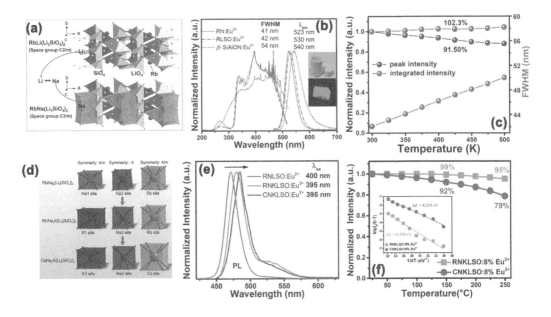

FIGURE 2.14 (a) The crystal structure of RbLi[Li₃SiO₄]₂ (RLSO) and RbNa[Li₃SiO₄]₂ (RN), where purple, white, green, and red spheres represent for Rb, Li, Na, and O atoms, respectively while LiO₄ tetrahedra/square planar, SiO₄ tetrahedra, and NaO₈ cuboid-like polyhedra are depicted in gray, light blue, and green, respectively. (b) PL and PLE spectra for RN: 8 mol% Eu (green solid, λ_{exc} = 455 nm), RLSO: 8 mol% Eu (red dash), and β-sialon: Eu (blue dot). The photographs of RN: 8 mol% Eu sample under daylight and 365 nm lamp are displayed in the inset. (c) Normalized integrated intensity (red), peak intensity (blue), and the FWHM of the emission band (purple) of RN: 8 mol% Eu under varies temperatures from 300 to 500K. (d) The AO₈ (A = Na, K, Rb, and Cs) cuboid-like polyhedra in the crystal structures RbNa₃[Li₃SiO₄]₄ (RNLSO), RbNa₂K[Li₃SiO₄]₄ (RNKLSO) and CsNa₂K[Li₃SiO₄]₄ (CNKLSO) based on the substitution of cations. (e) PL spectra of 8 mol% Eu²⁺-doped RNLSO, RNKLSO, and CNKLSO, respectively. (f) Temperature-dependent integrated PL intensity of RNKLSO: 8 mol% Eu and CNKLSO: 8 mol% Eu. The inset shows the fitted activation energies according to the thermal-related PL intensity. (Liao, H., Zhao, M., Zhou, Y., Molokeev, M. S., Liu, Q., Zhang, Q. and Xia, Z., *Adv. Funct. Mater.* 29, 1901988, 2019. With permission; Zhao, M., Zhou, Y., Molokeev, M. S., Zhang, Q., Liu, Q. and Xia, Z., *Adv. Opt. Mater.* 7, 1801631, 2019. With permission.)

CsNa₂K[Li₃SiO₄]₄ (CNKLSO) are isotypic. Similar to the RLSO and RN talked earlier, the doped Eu would occupy the center of the cuboid-like polyhedra and determining the final performance of the material. Figure 2.14(d) shows the coordination polyhedra in the three kinds of structures. There are three kinds of sites (with the symmetry of 4/*m*, −4, and 4/*m*) in any of the structures. The introduced K in RNLSO would occupy the Na1 site with high symmetry (4/*m*), and the introduced Cs in RNKLSO would occupy the Rb site (4/*m*). The emission spectra of 8 mol% Eu-doped RNLSO, RNKLSO, and CNKLSO are shown in Figure 2.14(e). The PL spectra with asymmetric emission band are observed under the excitation of 400, 395, and 395 nm, respectively and the main peak changes from 471 to 480 and 485 nm. The PL spectra of RNKLSO: 8 mol% Eu, and CNKLSO: 8 mol% Eu could both be fitted by three Gaussian peaks which are corresponding to the three kinds of sites. The fitted peaks can be ascribed by their symmetry difference, namely, when Eu²⁺ occupy the sites with higher symmetry, a narrower emission band would be obtained compared to the lower symmetry sites. The temperature-dependent PL integrated intensity of RNKLSO: 8 mol% Eu and CNKLSO: 8 mol% Eu drops to 99% (95%) and 92% (79%) of that at RT under the temperature of 150 (250)°C (see Figure 2.14(f)). A further study has predicted that the cuboid size would determine the emission color of the Eu²⁺-doped UCr₄C₄-type oxide phosphors, i.e., the cuboid size in the range of 29–34 Å³ would give cyan (blue) emission, whereas in the range of 26–27 Å³, the green emission will be obtained. The cuboid size larger than 40 Å^3 is not suitable for the Eu^{2+} to occupy, which would result in the absence of emission.[195]

Together with these, the recent reported AELi$_2$[Be$_4$O$_6$]:Eu^{2+} (AE = Sr, Ba) is known as a blue emitter with the FWHM of only 1200 cm^{-1} which is comparable to that of Sr[LiAl$_3$N$_4$]:Eu^{2+} (1180 cm^{-1}) and Sr[Mg$_3$SiN$_4$]:Eu^{2+} (1150 cm^{-1}). Both SrLi$_2$[Be$_4$O$_6$] (SLBO) and BaLi$_2$[Be$_4$O$_6$] (BLBO) crystallize in the tetragonal system, belong to the space group of *P4/ncc* (no. 130). The lattice parameters of the compounds show little difference. Therefore, the structure of AELi$_2$[Be$_4$O$_6$] can be discussed using the same model. As displayed in Figure 2.15(a), AELi$_2$[Be$_4$O$_6$] is isotypic with the nitridoalumosilicate BaLi$_2$[(Al$_2$Si$_2$)N$_6$][144] that introduced earlier. The $2 \times 2 \times 1$ supercell along [0 0 1] shows two types of channels of *vierer* rings, consisted of BeO$_4$ tetrahedra (dark gray). They are occupied by AE and Li atoms separately. A view of the channels occupied by AE atoms shows in Figure 2.15(b) suggests the polyhedra of AEO$_8$ connect by corner-sharing without surface contact like that in UCr$_4$C$_4$-type structures. There is only one type of AE site in AELi$_2$[Be$_4$O$_6$] and a single AEO$_8$ polyhedron (best described as a square pyramid trunk) with high symmetry is shown in Figure 2.15(c). The broad excitation band of SLBO: 1 mol% Eu and BLBO: 1 mol% Eu in the UV-blue range (~ 450 nm) and the narrow emission band located at around 456/454 nm with the FWHM of 1200 cm^{-1} also show great similarity in Figure 2.15(d). This is in accordance with their

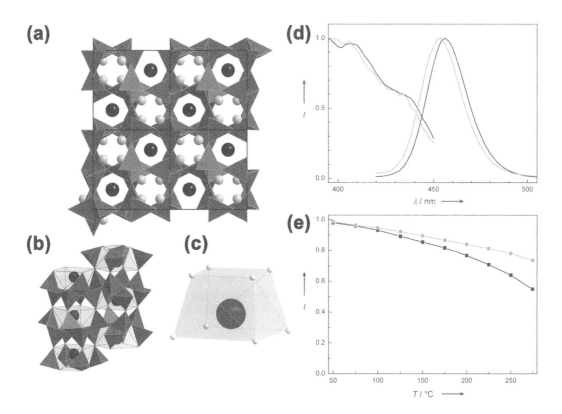

FIGURE 2.15 (a) The crystal structure of AELi$_2$[Be$_4$O$_6$] (AE = Sr, Ba) along [0 0 1]. (b) The channels of *vierer* rings that consist of BeO$_4$ tetrahedra (dark gray) and the corner-sharing AEO$_8$ polyhedra (light gray). (c) The AEO$_8$ polyhedra. The spheres in black, gray, and white represent AE, Li, and O atoms, respectively. (d) The PL and PLE spectra for AELi$_2$[Be$_4$O$_6$]: 1 mol% Eu (AE = Sr, black; AE = Ba, gray) when λ_{exc} = 400 nm. (e) The temperature-dependent integrated PL intensity (AE = Sr, black; AE = Ba, gray), and it remains 76 and 83% at 200°C compared to that at RT for AE = Sr and Ba, respectively. (Strobel, P., Maak, C., Weiler, V., Schmidt, P. J. and Schnick, W., *Angew. Chem. Int. Ed.* 57, 8739, 2018. With permission.)

similarity in lattice parameters. Besides, they exhibit proficient internal (external) QE of 64% (47%) for BLBO: 1 mol% Eu and 47% (16%) for SLBO: 1 mol% Eu under the excitation of 410 nm. The former has better performance under higher temperature, namely, it keeps 83% of the initial (at RT) integrated PL intensity at 200°C while the latter exhibits 76% (see in Figure 2.15(e)). The low quenching temperature is related to the small bandgap of 3.8 eV (for undoped BLBO). $AELi_2[Be_4O_6]$: Eu^{2+} (AE = Sr, Ba) show blue emission centered at around 454–456 nm with narrowband and small Stokes shift (1200 cm^{-1}). It allows the NUV LED-excitation (e.g., 410 nm), enabling the improvement of luminous efficiency and overall conversion efficiency in RGB phosphor-converted light sources.

In addition to the above, the chlorosilicates have demonstrated remarkable potential for use in phosphor-converted LEDs due to their outstanding chemical and physical stabilities, relatively low sintering temperatures, and preferable luminescence properties [196, 197], and they have received extensive concerns.[198–201]

The $Ba_5SiO_4Cl_6$:Eu^{2+} (BSOC) phosphor was first reported in 1979, with high quenching temperature, narrow emission band and comparable emission intensity compared to $BaMgAl_{10}O_{17}$:Eu^{2+} (BAM) especially excited by the NUV light.[202, 203] The host material is derived from the BaO–SiO_2–$BaCl_2$ system and the crystal structure of BSOC is shown in Figure 2.16(a). BSOC belongs to the monoclinic system with a space group of $C2/c$ (no. 15). The framework is formed by the SiO_4 tetrahedra had Ba(O, Cl)$_n$ (n = 9 or 8) polyhedra and only the SiO_4 tetrahedra are displayed for a clearer view. There are three types of Ba sites in the host lattice and the coordination numbers are 9, 9, and 8 for Ba1, Ba2, and Ba3, respectively (Figure 2.16(b)). This results in three kinds of the possible local environment for Eu^{2+} and the emission spectrum of BSOC: Eu^{2+} is, therefore, a composite one as shown in Figure 2.16(c). The PL spectrum peaking at 440 nm (FWHM = 32 nm) could be well fitted by three Gaussian peaks under the excitation of 350 nm. The internal QE of BSOC: 4% mol% Eu^{2+} is about 73%. The peaks I_1–I_3 were ascribed to Eu^{2+}'s emission in Ba1-Ba3 sites, respectively by analyzing their relative D_q values by Equation (2.1). The excitation maximum is located at 345 nm with considerable intensity around 400 nm, indicating its potential use in NUV chip-converted white light devices. In addition to the narrowband emission, it performs comparable thermal stability with the commercial BAM: Eu^{2+}. The peak intensity remains 80% and 90% of the initial (at RT) for BSOC: 4% mol% Eu^{2+} and BAM: Eu^{2+} at 423K, respectively. As seen in Figure 2.16(d), the emission band becomes wider but with a changeless peak position under increasing temperatures, indicating good stability against temperature.

There are still a lot of Eu^{2+}-doped phosphors with application potential, which have not been discussed in detail here. The related information of them is collected and listed in Table 2.4.

Eu^{2+}-doped compounds carry significant weight in narrow-band-emission phosphors, especially in the blue to green range. Eu^{2+}-doped narrow-band-emission phosphors possess the advantages of high efficiency; broad excitation band in UV-blue range which can match well with the LED chips. However, the site variety could significantly influence the FWHM of the emission band and raise the difficulty to realize an extremely narrow band.

2.2.2 Mn^{2+}-Activated Phosphors

The electron configuration of Mn is $1s^2 2s^2 2p^6 3s^2 3p^6 3d^5 4s^2$, and Mn^{2+} (d^5) and Mn^{4+} (d^3) ions are another two activators here to discuss in the narrow-band emission materials.

The $3d$ electrons of Mn^{2+} could be affected by the local environment significantly, and therefore, the luminescence of Mn^{2+}-doped materials is quite sensitive to the crystal field. The crystal field splitting of d orbitals could be qualitatively described by the Tanabe–Sugano (TS) diagram, the energy of the particular levels (E) are plotted as a function of crystal field strength (D_q) in a specific crystal field. Three parameters are necessary to describe an energy level completely in such a diagram, i.e., the Racah parameters B, C, and the crystal field strength D_q, which describes the splitting degree of the free ion terms. In the TS diagram of d^5 configuration in an octahedral crystal

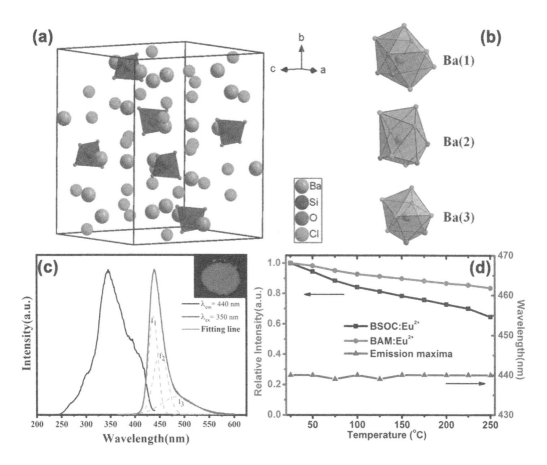

FIGURE 2.16   (a) The crystal structure of $Ba_5SiO_4Cl_6$ (BSOC). (b) The coordination models of the $Ba(O, Cl)_n$ (n = 9 or 8) polyhedra in BSOC. (c) The PL and PLE spectra of BSOC: 4% mol% Eu^{2+} and the Gaussian fittings of the PL curve. Inset is the photograph of the phosphor under the 365 nm UV lamp. (d) The temperature-dependent luminescence intensity of the BSOC: 4% mol% Eu^{2+} sample (black) compared to that of the commercial $BAM:Eu^{2+}$ (red) and the emission peak position variation of the BSOC: 4% mol% Eu^{2+} sample with the temperature (blue). (Wei, Q., Zhou, X., Tang, Z., Wang, X. and Wang, Y., *CrystEngComm* 21, 3660, 2019. With permission.)

field (Figure 2.17), E/B is plotted dependent on D_q/B (with a fixed C/B ratio of 4.5).[216] The solid and dashed lines represent the spin-quartet and spin-doublet states, respectively. It can be seen that the ground state changes from $^6A_1$ to $^2T_2$ in the strong crystal field (beyond D_q/B ~3), indicating that D_q becomes smaller than the spin-paring energy in this case. In addition, the lower excitation energy levels are more dependent on D_q, which is reflected in the considerable variation of emission/absorption spectra of Mn^{2+} in different hosts. The absorption transitions of Mn^{2+} distribute in the UV-blue range with different widths, e.g., the transition band of $^6A_1 \rightarrow ^4A_1$ (^4G) is narrow while that of $^6A_1 \rightarrow ^4T_1$ is wide, which is related to the electron vibronic coupling. The emitting (due to $^4T_1$ (^4G)$\rightarrow ^6A_1$ transition) color of Mn^{2+} emission band could vary from green to deep red, depending on its coordinating environment, e.g., when Mn^{2+} occupies the center of an octahedron, a red emission band can always be obtained while green light can be obtained when it's in a tetrahedral center.[102] Due to the effective influence of the crystal field on the energy levels splitting of Mn^{2+}, a lot of effort have been paid to find appropriate hosts, including many oxynitrides, oxides, and metal halides.

Mn^{2+} is one of the most popular dopants for luminescence materials while the efficiency of the Mn^{2+}-doped materials draws much attention. Some of our research results are listed in Table 2.5.

TABLE 2.4

The Eu^{2+}-Doped Materials with Narrow-Band Emission and their Performances

Compounds	Bandgap (eV)	S.G.[a]	O. C.[b]	CN	λ_{ex} (nm) max	λ_{em} (nm) max	FWHM (nm)	QE (%)	Thermal Stability (%)	Ref.
Sr[LiAl$_3$N$_4$]	4.7	$P\bar{1}$	1	8	~466	654	50	76/52 (I/E)[c]	95 (500K)	[80]
Ca[LiAl$_3$N$_4$]	/	$I4_1/a$	1	8	470	668	60	-	-	[86]
Sr[Mg$_3$SiN$_4$]	~3.9	$I4_1/a$	1	8	450	615	43	-	-	[87]
Ba[Mg$_3$SiN$_4$]	~4.0	$P\bar{1}$	1	8	465	670	88	32	50 (~400K)	[204]
Ca[Mg$_2$Al$_2$N$_4$]	/	$I4/m$	8	8	400	607	62	-	50 (~240K)	[133]
Sr[Mg$_2$Al$_2$N$_4$]	/	$I4/m$	8	8	440	612	62	18	50 (~300K)	[133]
Ba[Mg$_2$Al$_2$N$_4$]	/	$I4/m$	8	8	400	666	77		50 (~360K)	[133]
Ba[Li$_2$Al$_2$Si$_2$N$_6$]	~4.6	$P4/ncc$	1	8	395	532	57		50 (~550K)	[144]
Li$_2$Ca[Mg$_2$Si$_2$N$_6$]	~4.6	$C2/m$	1	6	460	638	~62		-	[143]
Ca$_{18.75}$Li$_{10.5}$[Al$_{39}$N$_{55}$]	/	$Fd\bar{3}m$	3	6	525	647	~54	11(1,450nm[d])	-	[142]
Ba$_2$LiSi$_7$AlN$_{12}$	/	$Pnnm$	1	11	~400	515	61	79(1,405nm)	84 (473K)	[205]
Sr[Li$_2$Al$_2$O$_2$N$_2$]	4.9–5.3	$P4_2/m$	1	8	460	614	41	80	96 (420K)	[206]
BaSi$_2$O$_2$N$_2$	6.6(3)	P_2/m	/	/	460	499	35	>60%	-	[207]
		$Pbcn$ & $Cmcm$	1	8	355	495	0.12eV	71(450 nm)	50 (600K)	[163, 208]
		$Pbcn$	1	7	380	494	48	60(E)	89 (473K)	[209]
β-sialon	5.71, 5.54, 5.47 (z = 0, 0.042, 0.125)	$P6_3/m$	1	9	325	529/540	49/54	64/36 (I/E)/82/65 (I/E)	~100/90 (573K)	[70]
Na$_{0.5}$K$_{0.5}$[Li$_3$SiO$_4$]	/	$I4/m$	3	8	402	486	20.7	76/63 (I/E)	93 (423K)	[210]
RbLi[Li$_3$SiO$_4$]$_2$	4.66	$C2/m$	1	8	395	530	42	80/29 (I/E)	103 (423K)	[194]
RbNa[Li$_3$SiO$_4$]$_2$	/	$C2/m$	1	8	395	523	41	96.2/44 (I/E)	102 (423K)	[211]
RbNa$_3$[Li$_3$SiO$_4$]$_4$	/	$I4/m$	3	8	375	471	22.4	53/13 (I/E)	96 (423K)	[89]
RbNa$_2$K[Li$_3$SiO$_4$]$_4$	/	$I4/m$	3	8	380	480	26	-	99 (423K)	[212]
CsNa$_2$K[Li$_3$SiO$_4$]$_4$	/	$I4/m$	3	8	382	485	26	-	92 (423K)	[212]
CsNa$_2$K[Li$_3$GeO$_4$]$_4$	/	$I4/m$	3	8	398	458	26	10(I)	<10 (423K)	[90]
SrLi$_2$[Be$_4$O$_6$]	/	$P4/ncc$	1	8	410	456	25	47/16(410 nm)	76 (473K)	[88]
BaLi$_2$[Be$_4$O$_6$]	3.8	$P4/ncc$	1	8	410	454	25	64/47(410 nm)	83 (473K)	[88]

(Continued)

TABLE 2.4 (*Continued*)
The Eu^{2+}-Doped Materials with Narrow-Band Emission and their Performances

Compounds	Bandgap (eV)	S.G.[a]	O. C.[b]	CN	λ_{ex} (nm) max	λ_{em} (nm) max	FWHM (nm)	QE (%)	Thermal Stability (%)	Ref.
$Ba_5SiO_4Cl_6$	4.72	$C2/c$	3	9,8	345	440	32(~30)	73(I)	80 (423K)	[202, 203, 213]
$Ba_3Si_6O_9N_4$	6.45,6.79	$P3$	3	9,8	410	493	~46	low	74 (473K)	[147, 214, 215]
$K_2HfSi_3O_9$ (Sc^{3+}-added)	5.5	$P\bar{6}$	2	6	400	507	~59	55.3(I)	~100 (473K)	[174]

[a] S. G. represents for Space Group, hereinafter inclusive.
[b] O. C. represents for Occupancy Categories, hereinafter inclusive.
[c] I represents the Internal Quantum Efficiency; E represents the External Quantum Efficiency, hereinafter inclusive.
[d] Wavelength applied to excite the material when testing the QE, hereinafter inclusive.

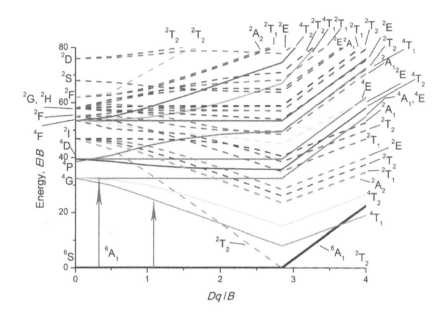

FIGURE 2.17 Tanabe–Sugano diagram for the d^5 electron configuration in the octahedral crystal field. C/B = 4.5. The spin-quartet and spin-doublet states are shown by the solid and dashed lines, respectively. (Zhou, Q., Dolgov, L., Srivastava, A. M., Zhou, L., Wang, Z., Shi, J., Dramićanin, M. D., Brik, M. G. and Wu, M., *J. Mater. Chem. C* 6, 2652, 2018. With permission.)

TABLE 2.5
Some of Our Research Results of Mn^{2+}-Doped White LED-Used Luminescence Materials

Phosphor	S. G.	λ_{ex} (nm)	λ_{em} (nm)	TQ (%)	LED Devices CCT (K)	CRI	Ref.
$Ca_3Hf_2SiAl_2O_{12}$:Ce^{3+}, Mn^{2+}	$Ia\bar{3}d$	400	457–556	<10 (473K)	5379		[217]
$Ca_8NaGd(PO_4)_6F_2$:Ce^{3+}, Mn^{2+}	$P6_3/m$	274/337	577				[218]
$Ca_9La(GeO_4)_{0.75}(PO_4)_6$:$Ce^{3+}$, Mn^{2+}	$R3c$	313	635	70 (473K)			[219]
$Ca_9Al(PO_4)_7$:Ce^{3+},Mn^{2+}	$R3c$	315	654				[220]
$CaSr_2Al_2O_6$:Ce^{3+}, Li^+, Mn^{2+}	$Pa\bar{3}$	358	460–610	22 (473K)	3284K		[221]
$KAlSi_2O_6$:Mn^{2+}	$I4_1/a$	450	513	70 (473K)	4775K		[222]
$K_4CaGe_3O_9$:Mn^{2+}, Yb^{3+}	$Pa\bar{3}$	287	587				[223]
$Na_2BaCa(PO_4)_2$:Eu, Mn	$P3m1$	351	576	90 (473K)	4346K		[224]
$NaBaSc(BO_3)_2$:Ce^{3+}, Mn^{2+}	$R3$	490	630	10 (473K)	3784K	84.36	[225]
$Sr_2P_2O_7$:Eu^{2+}/Eu^{2+}, Mn^{2+}		355	590				[226]
$SrAlSi_4N_7$:Eu^{2+}, Mn^{2+}	$Pna2_1$	370	610	92 (473K)	5467K	83	[227]
$Ca_5La_5(SiO_4)_3(PO_4)_3O_2$:$Ce^{3+}$, Mn^{2+}	$P6_3/m$	355	656				[228]
$Y_3Al_5O_{12}$:Ce^{3+}, Mn^{2+}	$Ia\bar{3}d$	460	593	80.3 (473K)			[229]
$Sr_{10}[(PO_4)_{5.5}(BO_4)_{0.5}](BO_2)$:$Ce^{3+}$, Mn^{2+}	P3	365	650	45 (473K)	3512K		[230]
$NaSrPO_4$:Ce^{3+}, Mn^{2+}	I	365	600				[231]
$Sr_7La_3[(PO_4)_{2.5}(SiO_4)_3(BO_4)_{0.5}](BO_2)$:$Ce^{3+}$, Mn^{2+}	P3	351	628	30 (473K)	4500K		[232]
$Zn_3B_2O_6$:Mn^{2+}	$C2/c$	360	432				[233]

The potential candidates for backlighting, as well as some excellent works reported by other researchers are introduced in the following part.

γ-AlON is a solid solution compound in the Al_2O_3–AlN system with a defective cubic spinel structure in the space group of $Fd\bar{3}m$ (no. 227). As shown in Figure 2.18(a), two types of Al atoms (orange in the 8*a* site, blue in the 16*d* site) occupy the tetrahedral and octahedral centers, respectively, while vacancies occupy the 16*d* site as a regular part in the structure. Mn^{2+} ions are mainly located at the tetrahedral centers when doped, generating the green emission. Figure 2.18(b) shows the PLE and PL spectra of Mn^{2+}-doped γ-AlON. The excitation spectra consist of several bands located at 340, 358, 381, 424, and 445 nm related to the electron transitions from $^6A_1$ to $^4T_2$ (^4P), ^4E (^4G), $^4T_2$, [^4E (^4G), ^4A (^4G)], and $^4T_2$ (^4G), respectively.[234] It shows a green emission band center at 512 (FWHM ~32 nm) and 520 nm (FWHM ~44 nm) for 7 mol% Mn-doped γ-AlON without (dashed) and with 10 mol% Mg (solid), respectively. It suggests the addition of Mg could suppress the formation of corundum during synthesis and at the same time, decrease the defect concentration, enhancing the emission intensity (1.7 times) and QE (from 53% to 62%) of γ-AlON:Mn^{2+}. As is known that an appropriate co-doping of

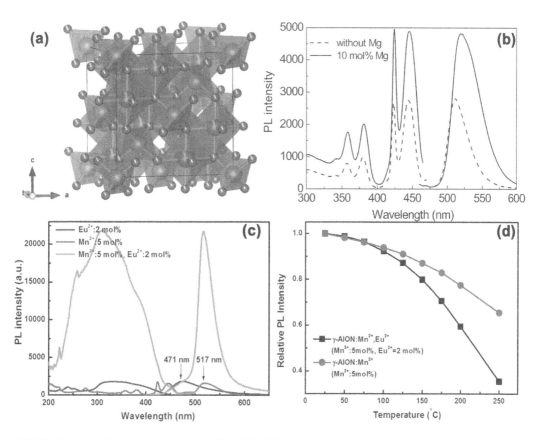

FIGURE 2.18 (a) The crystal structure of γ-AlON. The spheres in orange, blue, and red represent for Al/vacancy in 8*a*, 16*d* sites, O/N in 32*e* site, and the Al(O, N)$_4$ tetrahedra and Al(O, N)$_6$ octahedra are displayed in steel blue. (b) The excitation and emission spectra of γ-AlON: 7 mol% Mn without (dashed) and with 10 mol% Mg (solid). (c) Excitation and emission spectra of γ-AlON: 2 mol% Eu^{2+} (blue), γ-AlON: 5 mol% Mn^{2+} (red), and γ-AlON: 5 mol% Mn^{2+}, 2 mol% Eu^{2+} (green). (d) The temperature-dependent PL intensity of γ-AlON: 5 mol% Mn^{2+} (red) and γ-AlON: 5 mol% Mn^{2+}, 2 mol% Eu^{2+} (black) under the excitation of 405 nm. (Li, S., Xie, R. J., Takeda, T. and Hirosaki, N., *ECS J. Solid State Sci. Technol.* 7, R3064, 2017. With permission; Xie, R. J., Hirosaki, N., Liu, X. J., Takeda, T. and Li, H. L., *Appl. Phys. Lett.* 92, 201905, 2008. With permission; Liu, L., Wang, L., Zhang, C., Cho, Y., Dierre, B., Hirosaki, N., Sekiguchi, T. and Xie, R. J., *Inorg. Chem.* 54, 5556, 2015. With permission.)

sensitizer would also improve the emission intensity of the activators. As shown in Figure 2.18(c), the emission spectrum of γ-AlON:Eu^{2+} (blue) has an appreciable overlap with the excitation spectrum of γ-AlON:Mn^{2+} (red) and the emission of γ-AlON: 5 mol% Mn^{2+} is significantly enhanced by 2 mol% Eu^{2+} co-doping (green) under 365 nm excitation. The emission of Mn^{2+} locates at 517 nm with the FWHM of 45 nm. The intensity is enhanced by ~9 times while the absorption efficiency, internal QE (IQE), and external QE (EQE) are improved from 16 to 65%; 48 to 75%; and 7 to 49%; respectively. [235] The related temperature-dependent PL intensity (Figure 2.18(d)) indicates that it remains 80% of the initial (intensity measured at RT) at 150°C for γ-AlON: 5 mol% Mn^{2+}, 2 mol% Eu^{2+}, which is a little lower than that of γ-AlON: 5 mol% Mn^{2+}. This may be attributed to the lattice distortion caused by the obvious ionic radii mismatch between Eu^{2+} and Al^{3+}.

KAlSi$_2$O$_6$ (KAS) has been reported in the tetragonal system with the space group of $I4_1/a$ (no. 88). A total view of the crystal structure of KAS is shown in Figure 2.19(a) (left). Al (light blue) and Si (blue) share the same lattice site (16f) randomly with an atomic ratio of about 1:2. Each Al or Si atom is coordinated by 4 O atoms. The (Al/Si)O$_4$ tetrahedra are connected by the total angle connection of O[236] as displayed in Figure 2.19(a) (right). The tetrahedra connect with each other to form a six-membered ring unit and these rings constitute the framework of the host lattice by corner-sharing. K atoms (purple) are arranged in the interspace of the framework. The Mn^{2+} ion

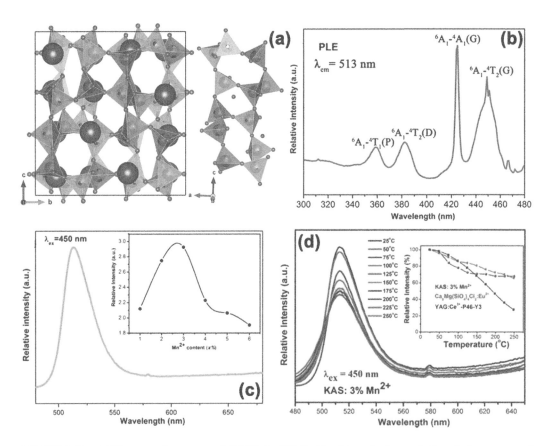

FIGURE 2.19 (a) The crystal structure of KAlSi$_2$O$_6$ (KAS). The (b) excitation and (c) emission spectra of KAS:Mn^{2+}. The inset in the latter shows the peak intensity change with Mn^{2+} doping concentration. (d) The temperature-dependent PL spectra of KAS: 3 mol% Mn^{2+}. The inset shows the peak intensity variation of KAS: 3 mol% Mn^{2+}, YAG: Ce^{3+} (P46-Y3, yellow phosphor), and commercial Ca$_8$Mg(SiO$_4$)$_4$Cl$_2$: Eu^{2+} (green phosphor) with the temperature. (Ding, X., Zhu, G., Wang, Q. and Wang, Y., *RSC Adv.* 5, 30001, 2015. With permission.)

could occupy an Al/Si site due to the comparable ion radii to Al^{3+}/Si^{4+}, locating at the center of the tetrahedra of O. The excitation spectra of KAS:Mn^{2+} in Figure 2.19(b) suggests that when monitored at 513 nm, it has an intensive excitation at ~450 nm, matching with the blue LED chips. A green emission band centered at 513 nm is obtained under the excitation of 450 nm (see Figure 2.19(c)). The Mn^{2+} concentration-dependent peak intensities are displayed in the inset and the intensity maximum occurs when the Mn^{2+} content is about 3 mol% (with the quantum yield of 30.2% under 450 nm excitation). The emission band has a quite narrow width (FWHM = 30 nm), and this results in a higher color purity compared to β-sialon: Eu^{2+}, therefore a wider color gamut. The thermal behavior of KAS: 3 mol% Mn^{2+} (Figure 2.19(d)) is compared with YAG: Ce^{3+} (P46-Y3, yellow phosphor) and commercial $Ca_8Mg(SiO_4)_4Cl_2$:Eu^{2+} (green phosphor) as it can be seen in the inset. The integrated intensity of KAS: 3 mol% Mn^{2+} keeps 67.6% of the initial (298K) at 523K without a shift of the peak position, indicating the considerable thermal stability of the prepared phosphor.[222]

In the aluminate $Sr_2MgAl_{22}O_{36}$ (SMAO), Mn^{2+} could also perform green emission with a narrow band. SMAO is crystallized in the hexagonal system with the space group of $P\bar{6}m2$ (no. 187). The crystal structure of SMAO is shown in Figure 2.20(a). There are two types of Sr atoms (Sr1, Sr2) which are coordinated by nine (yellow) and twelve (medium violet red) O, respectively; eight types

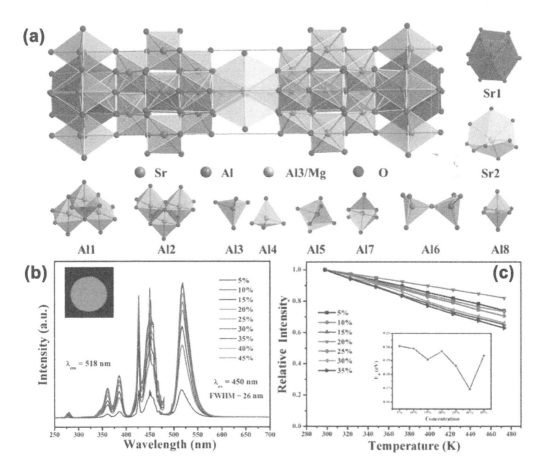

FIGURE 2.20 (a) The crystal structure of $Sr_2MgAl_{22}O_{36}$ (SMAO). Polyhedra in light sea green, yellow, and medium violet red represent $AlO_{4(6)}$ tetrahedra (octahedra), SrO_9 tricapped trigonal prism, and SrO_{12} icosahedron. (b) The excitation and emission spectra of SMAO:xMn^{2+} (5% ≤ x ≤ 45%). The inset is the photograph of SMAO: 30 mol% Mn^{2+} under a 365 nm UV lamp. (c) Temperature-dependent emission intensity of SMAO:xMn^{2+} (5% ≤ x ≤ 35%). The excitation energy (E_a) as a function of Mn^{2+} concentration is shown in the inset. (Zhu, Y., Liang, Y., Liu, S., Li, H. and Chen, J., *Adv. Opt. Mater.* 7, 1801419, 2019. With permission.)

of Al atoms (Al1–Al8) located in the tetrahedral/octahedral (light sea green) centers and Mg occupy the same site as Al3 with the atomic ratio of 1/2. In consideration of the ion radii, Mn^{2+} ions in SMAO are prefer to substitute Mg^{2+} to form MnO_4 tetrahedra. Mn^{2+} in such a coordinated environment could give green emission as talked about earlier. The PLE and PL spectra in Figure 2.20(b) show that the series samples have an excitation band locate at around 450 nm (match well with the blue LED chips) and a narrow-band emission locates at 518 nm with the FWHM of 26 nm, which is narrower than that of β-Sialon:Eu (λ_{em} = 540 nm, FWHM = 60 nm), γ-AlON:Mn^{2+}, Mg^{2+} (λ_{em} = 520 nm, FWHM = 44 nm) introduced earlier. The temperature-dependent PL intensity-change curve in Figure 2.20(c) indicates it could keep 82% of the initial (measured at 298K) intensity at 473K in the SMAO: 20 mol% Mn^{2+} sample. SMAO:Mn^{2+} exhibit comparable performance with the commercialized β-Sialon: Eu, especially its high color purity. Although the aforementioned Mn^{2+}-doped oxides performed good color purity for improved color gamut, the efficiency is not so high and, therefore, limits their applications.

The recently reported bulk metal halides, Cs_3MnBr_5 perform green-emitting with a narrow band. [92] Cs_3MnBr_5 is isostructural to Cs_3CoCl_5, having a tetragonal unit cell with the space group of $I4/mcm$ (no. 140). In the crystal structure of Cs_3MnBr_5 displayed in Figure 2.21(a), two types of Cs atoms distribute randomly while Cs1 atoms occupy the centers (8h) coordinated by eight bromine

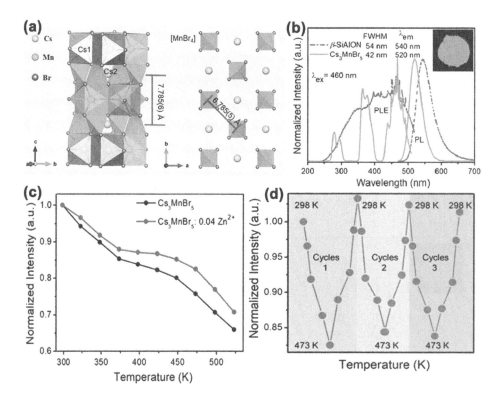

FIGURE 2.21 (a) Crystal structure of Cs_3MnBr_5. The left shows the framework of Cs_3MnBr_5 formed by alternating $CsBr_8$ polyhedra (gray) and $MnBr_4$ tetrahedra (turquoise) layers along [0 0 1], the right is a view of the atom distribution in the *ab* plane and only the $MnBr_4$ tetrahedra are displayed. The distance between neighboring $MnBr_4$ tetrahedra layers along [0 0 1] and that between the nearest Mn atoms in the *ab* plane are marked. (b) The excitation and emission spectra of Cs_3MnBr_5 (green) compared to β-Sialon: Eu (red). (c) The temperature-dependent luminescent intensity of Cs_3MnBr_5 without (black) and with (red) the addition of 4 mol% Zn. (d) The emission intensity of Cs_3MnBr_5: 4 mol% Zn under three times of heating/cooling cycles in the temperature range of 298–473K. (Su, B., Molokeev, M. S. and Xia, Z., *J. Mater. Chem. C* 7, 11220, 2019. With permission.)

TABLE 2.6

The Mn^{2+}-Doped Materials with Narrow-Band Emission and their Performances

Compounds	S.G.	O. C.	CN	λ_{ex} (nm)	λ_{em} (nm)	FWHM (nm)	QE/QY (%)	Thermal Stability[a] (%)	Ref.
γ-AlON:Mg	$Fm\bar{3}m$	1	4	445	520	44	62(I)	88 (423K)	[82, 238, 239]
Cs_3MnBr_5:Zn	$I4/mcm$	1	4	460	520	42	/	82 (423K)	[92]
$SrMgAl_{10}O_{17}$:Eu	$P6_3/mmc$	1	4	432	518	~26	~98(I)	/	[240]
$Sr_2MgAl_{22}O_{36}$	$P\bar{6}m2$	1	4	450	518	26	75/42(I/E)	82 (473K)	[241]
$KAlS_2O_6$	$I4_1/a$	3	4	450	513	30	30.2	67.6 (523K)	[222]
$Zn_3(BO_3)_2$	$Im\bar{3}m$	/	4	426	541	41	61/21(I/E)	70 (523K)	[242]

[a] compared to the intensity at room temperature unless otherwise specified.

atoms to form $CsBr_8$ polyhedra layers and Cs2 coordinates with ten bromine atoms in a $4a$ site. Mn atoms occupy the sites with $4b$ symmetry, forming $MnBr_4$ tetrahedra layers. The layers of $MnBr_4$ tetrahedra and $CsBr_8$ polyhedra layers are arranged layer by layer along [0 0 1] to form the framework of the crystal and the distance between two neighboring $MnBr_4$ tetrahedra layers is about 7.785 (6) Å. It also suggests the distance between the two nearest manganese atoms is around 6.785(5) Å. The layer isolation of the Mn sites restricts the energy migration inside the lattice and, thus, eventually eliminates the concentration quenching in Cs_3MnBr_5.[237] The PL and PLE spectra of Cs_3MnBr_5 (green) in Figure 2.21(b) shows that it could be well excited by blue light and under the excitation of 460 nm, it has higher emitting energy peaking at 520 nm, a smaller FWHM of 42 nm compared to β-Sialon:Eu (red, $\lambda_{em} = 540$ nm, FWHM = 54 nm).[58] The integrated emission intensity of Cs_3MnBr_5 would reduce to 82% of the initial (at 298K) at 473K and the addition of Zn (4 mol%) makes it keep 87% of the initial at 473K (see Figure 2.21(c)). The heating-cooling cycles test result shown in Figure 2.21(d) suggests that after the heating and cooling process three times, the integrated intensity of the emission remains unchanged. The appropriate emission character as well as the considerable thermal/chemical stability makes Cs_3MnBr_5: 4 mol% Zn potential for application in LCD backlighting.

Table 2.6 shows some of the Mn^{2+}-doped materials with narrow-band emission and the related luminescence properties.

It's also worth noting that Mn^{2+}-activated materials always perform low efficiency since the d-d transition of Mn^{2+} belongs to forbidden transition. The efficiency could be improved by targeting an appropriate host to obtain suitable excitation properties (strong excitation band matching well with LED chips); optimization of the sensitization mechanism, enhancing the PL intensity more efficiently. Advanced mechanisms, as well as technics for efficiency improvement of Mn^{2+}-activated materials, are still urgently needed to be addressed in future investigations.

2.2.3 Mn^{4+}-ACTIVATED PHOSPHORS

Mn^{4+} with the electron configuration of d^3 could perform red/deep red emission in some oxide and fluoride materials. To figure out the luminescence properties of Mn^{4+}, the TS diagram displayed in Figure 2.22(a) gives the crystal field-dependent change of energy levels.[245] The ground state is the spin-quartet $^4A_{2g}$ and the energy difference between it and the first excited state $^4T_{2g}$ is always equals to $10D_q$. The first excited state changes to be ^2E_g when D_q/B is greater than 2.1, and it is also seen that the energy of $^2T_{2g}$, especially ^2E_g and $^4T_{2g}$ is almost parallel to the abscissa axis, indicating transitions related to these levels would be less influenced by the crystal field. The strength of crystal field could be distinguished by the luminescence spectra of Mn^{4+}, i.e., it performs band

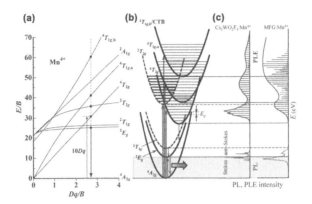

FIGURE 2.22 (a) TS diagram of the d^3 electron configuration for Mn^{4+} in the octahedral field. (b) Configuration coordinate diagram (CCD) for the description of the PL and PLE spectra. (c) The luminescence spectra of Mn^{4+}-doped $Cs_2WO_2F_4$ and magnesium fluorogermanate (MFG). The shadow areas (horizontal bars) in different colors are the theoretical fits[243, 244] of the excitation bands in the configurational-coordinate (CC) model with $^4A_{2g}$ to be the ground state. (Adachi, S. *ECS J. Solid State Sci. Technol.* 9, 016001, 2019. With permission.)

emission originated from $^4T_{2g} \rightarrow {}^4A_{2g}$ spin-allowed transition, whereas sharp peaks attributed to $^2E_g \rightarrow {}^4A_{2g}$ transition, which is parity and spin forbidden but gain intensities with the activation of lattice vibronic modes.[84] The reported Mn^{4+}-doped luminescent materials for backlighting are all sharp peaks with red to deep red emission. Figure 2.22(b) shows the configurational-coordinate (CC) model with $^4A_{2g}$ as the ground state schematically, which can be understood with the luminescence spectra in Figure 2.22(c). The first excited state ^2E_g is responsible for the red emission of the (oxy)fluorides while the emission of the zero phonon line (ZPL) is related to the transition from the parabolic curve bottom of the ^2E_g level. The displacements of the parabolic curves for the excited states ($^4T_{1g}$, $^4T_{2g}$) compared to $^4A_{2g}$ is corresponding to the excitation bands centered in the blue and UV range.

Fluorides activated by Mn^{4+} always have a broad absorption in the UV-blue range and sharp emission peaks located at around 630 nm (related to the $^2E_g \rightarrow {}^4A_{2g}$ transition). Different from fluorides, the sharp-peak emission of Mn^{4+} in oxides always in the spectral range of 630–730 nm, with lower energy than in fluorides. The variation of peak positions of Mn^{4+} spectra is the result of the nephelauxetic effect, i.e., the decrease of Racah parameters B and C compared to that of free ions. Furthermore, the nephelauxetic effect is related to chemical bond properties, interionic distances, angles between the chemical bonds, etc. The nephelauxetic parameter can be introduced to describe the degree of the effect:

$$\beta_1 = \sqrt{\left(\frac{B}{B_0}\right)^2 + \left(\frac{C}{C_0}\right)^2} \qquad (2.3)$$

B ($B_0 = 1160$ cm^{-1}) and C ($C_0 = 4303$ cm^{-1}) are the Racah parameters for Mn^{4+} in the host (free ion). The energy can be shown as a linear function of β_1.

The superiority of Mn^{4+}-doped materials is their narrow emission bands due to its parity and spin forbidden transition from $^2E_g \rightarrow {}^4A_{2g}$. Some of our results related to Mn^{4+}-doped materials are listed in Table 2.7 and other typical and advanced works are introduced next.

The well-known red emission phosphor K_2SiF_6:Mn^{4+} (KSF) has attracted much attention because of its efficient luminescence under UV-blue light excitation, narrow emission band, better stability than organic materials and QDs.[249–252] It is also one of the most popular phosphors applied

TABLE 2.7

Some of our Research Results of Mn^{4+}-Doped White LED-Used Luminescence Materials

Phosphor	S. G.	λ_{ex} (nm)	λ_{em} (nm)	TQ (%)	LED Devices		Ref.
					CCT (K)	CRI	
$K_2Ge_4O_9$:Mn^{4+}	$P3c1$	450	663	10 (473K)	3119	84.1	[246]
$Li_3Mg_2NbO_6$:Mn^{4+}	$Fddd$	470	668	36 (473K)	4649		[247]
$Mg_6ZnGeGa_2O_{12}$:Mn^{4+}	$Cmmm$	420	660	75 (473K)			[248]
$Mg_3Ga_2GeO_8$:Mn^{4+}	$Imma$	419	659	72 (423K)	9332–3440	62.1–80.3	[249]

in the LCD backlighting. KSF is crystallized in the cubic system with the space group of $Fd\bar{3}m$ (no. 227). As shown in Figure 2.23(a), Si atoms are located at six-fold coordinate centers, forming the SiF_6 octahedra and Mn^{4+} ions would occupy the Si sites when doped because of their identical valance and similar ion radii. The PL and PLE spectra of KSF:Mn^{4+} (doped with 0.08 g $KMnO_4$) in Figure 2.23(b) shows the characteristic emission peak ($^2E_g \rightarrow {}^4A_{2g}$ transition) at around 630 nm and the excitation bands centered before 300 nm, at 355 and 450 nm which are ascribed to the transitions from $^4A_{2g}$ to $^4T_{1g,b}$ (^4P), $^4T_{1g,a}$ (^4F), and $^4T_{2g}$, respectively. For comparison, PL and PLE spectra of CaAlSiN$_3$:Eu (dashed red and orange), PL spectrum of β-sialon:Eu (solid blue) are displayed in Figure 2.23(c). The excitation band of KSF:Mn^{4+} is narrower than that of CaAlSiN$_3$:Eu which even covers the green range, having a significant overlap with the emission of β-sialon:Eu and resulting in the efficiency loss when combined with the green phosphor to fabricate white light. On the other hand, the sharp peak emission of KSF:Mn^{4+} has a high color purity and little tail beyond the human-eye sensitivity curve, making it an efficient red-emitting phosphor for the application in LCD backlighting. The thermal/chemical stability of KSF:Mn^{4+} phosphor is not comparable to some (oxy) nitride or oxides in service. Therefore, many efforts have been paid to improve its stability, including coating and single crystal growth.[250, 252, 253] The temperature-related emission intensity (integrated) of the samples uncoated and coated with Al_2O_3 layers are displayed in Figure 2.23(d). It suggests that the samples coated with Al_2O_3 at 120°C for five cycles have a higher quenching temperature (T_{50}, the temperature when the intensity reduced to half of the initial) of 320°C compared to 250°C for the uncoated sample. The coating treatment increase the stability is considered to be the result of surface passivation by the coating layers.

As introduced earlier, the drawbacks of KSF in stability, especially the poor moisture resistance, would limit its further applications. The coating method could improve the moisture resistance to some extent but at the cost of luminescence efficiency and new researches. The Mn^{4+}-activated oxyfluoride Cs_2NbOF_5, reported being with good moisture resistance property, provides one of the new ideas. The crystal structure of the oxyfluoroniobate Cs_2NbOF_5 belongs to the trigonal system with the space group of $P\bar{3}m1$ (no. 164). The crystal structure of Cs_2NbOF_5 is displayed in Figure 2.24(a). The Nb atom could coordinate with six F/O (with a fixed occupation ratio of 5/1) atoms to form the Nb(F/O)$_6$ octahedra (yellow) while the Cs atoms are located in the cell. The PL and PLE spectra of Cs_2NbOF_5: 4.90 at% Mn^{4+} are similar to that of KSF:Mn^{4+} (Figure 2.24(b)). The emission peak locates at around 630 nm with the IQE, EQE of 74.8% and 54.6%, respectively, which are comparable to KSF:Mn^{4+}.[254] The three excitation bands in the range of 250–300 nm are ascribed to the Mn-F charge transfer band (CTB), at 366 nm from $^4A_{2g} \rightarrow {}^4T_{1g,a}$ (^4F) transition, and at 467 nm from $^4A_{2g} \rightarrow {}^4T_{2g}$ transition. Instead of surface treatment of phosphors, Cs_2NbOF_5:Mn^{4+} owns a good moisture-resistant property. As shown in Figure 2.24(c), the PL intensity of the untreated KSF:Mn^{4+} drops quickly after being immersed in deionized water, and it drops to 5.9% of the initial after 240 min. The test results for Cs_2NbOF_5:Mn^{4+} under the same conditions are shown in Figure 2.24(d). It could keep 96.0% of the initial after 240 min immersing in deionized water, indicating the excellent water-resistant feature.

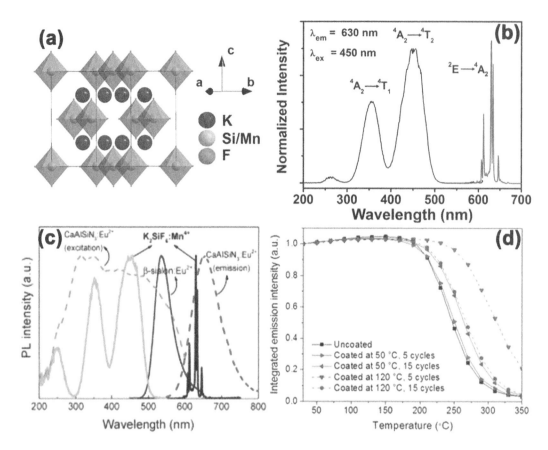

FIGURE 2.23 (a) The crystal structure of K_2SiF_6 (KSF) in a unit cell and Mn occupies the Si site in KSF. (b) The PL ($\lambda_{exc} = 450$ nm) and PLE ($\lambda_{em} = 630$ nm) spectra of KSF:Mn^{4+}. (c) The excitation/emission spectra of KSF:Mn^{4+} (em.: solid black, ex.: solid cyan) compared with CaAlSiN$_3$:Eu (em.: dashed red, ex.: dashed orange) and β-sialon:Eu (em.: solid blue). (d) Temperature-dependent integrated PL intensity of KSF:Mn^{4+} uncoated/coated by Al$_2$O$_3$ with 5/15 cycles. (Huang, L., Liu, Y., Yu, J., Zhu, Y., Pan, F., Xuan, T., Brik, M. G., Wang, C. and Wang, J., *ACS Appl. Mater. Interfaces* 10, 18082, 2018. With permission; Wang, L., Wang, X., Kohsei, T., Yoshimura, K. I., Izumi, M., Hirosaki, N. and Xie, R. J., *Opt. Express* 23, 28707, 2015. With permission; M. ten Kate, O., Zhao, Y., Jansen, K. M. B., van Ommen, J. R. and Hintzen, H. T., *ECS J. Solid State Sci. Technol.* 8, R88, 2019. With permission.)

Mn^{4+} in oxides could also perform red emission with narrow-band peaks, and they always emit with lower energy compared to Mn^{4+}-doped (oxy)fluorides due to the nephelauxetic effect as discussed earlier.

BaMgAl$_{10}$O$_{17}$ (BAM) is familiar to us since the Eu^{2+}-doped BAM has been applied as a commercial blue-emitting phosphor for years. The framework of BAM consists of the orderly-layered MgAl$_{10}$O$_{16}$ spinel blocks and BaO mirror layers (Figure 2.25(a), left part). Eu^{2+} could occupy the Ba site with the coordination number of nine, giving blue emission centered at around 450 nm. When Mn^{4+} is introduced, it is preferred by the Mg/Al site in the center of an octahedron and coordinates with six O^{2-}. As shown in the right part of Figure 2.25(a), the introduction of Mn^{4+}–Mg^{2+} pairs to replace Al^{3+}–Al^{3+} in the lattice leads to the reduction of non-radiative energy transfer. As a result, the BAM:Mn^{4+}, Mg^{2+} samples yield bright peak emission at ~660 nm was obtained. The PLE and PL spectra of BAM: 2 mol% Mn^{4+}, 2 mol% Mg^{2+} are shown in Figure 2.25(b), the PLE spectra perform different intensity distributions compared to the (oxy)fluorides discussed earlier, i.e., the $^4A_{2g} \rightarrow {}^2T_{2g}$ transition occurs because it becomes partially allowed due to the lattice vibrations and

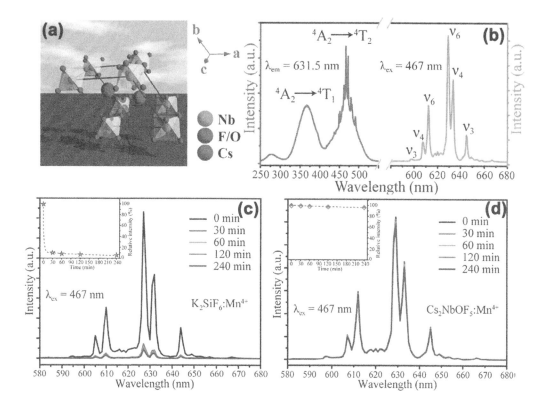

FIGURE 2.24 (a) The crystal structure of Cs_2NbOF_5 in a unit cell. The yellow octahedra represent the $Nb(F/O)_6$ groups. (b) The excitation (red) and emission (green) spectra of Cs_2NbOF_5: 4.90 at% Mn^{4+}. The PL spectra of Cs_2NbOF_5:Mn^{4+} (c) and KSF:Mn^{4+} (d) after immersion in deionized water for 0–240 min. (Zhou, J., Chen, Y., Jiang, C., Milićević, B., Molokeev, M. S., Brik, M. G., Bobrikov, I. A., Yan, J., Li, J. and Wu., M. *Chem. Eng. J.* 405, 126678, 2020. With permission.)

mixes of the $^4A_2$, ^2E, $^2T_2$, and $^4T_2$ states [255]; transitions from $^4A_{2g} \rightarrow {}^4T_{1g}$ dominates rather than the transitions from $^4A_{2g}$ to $^4T_{2g}$, resulting in a lower luminescence efficiency under blue light excitation. The emission that related to $^2E_g \rightarrow {}^4A_{2g}$ transition is located at 660 nm with the FWHM of 30 nm and high color purity of 98.3% under the excitation of 468 nm. The temperature-dependent PL intensity (in both heating and cooling process) of the sample BAM: 2 mol% Mn^{4+}, 2 mol% Mg^{2+} in Figure 2.25(c) shows that the PL intensity (integrated) drops to 45% of the initial (at RT) at 150°C, indicating poor thermal stability.

BAM:Mn^{4+} could emit deep red light with high color purity but in poor thermal stability and without efficient excitation in the NUV-blue range. Nevertheless, the gallium germanate $Mg_3Ga_2GeO_8$:Mn^{4+} (MGG), which could be efficiently excited by deep blue light with better thermal stability was reported earlier. MGG is a result of $MgGa_2O_4$-Mg_2GeO_4 solid solution. The structure of MGG belongs to an orthorhombic system and the space group of *Imma* (no. 74). Mg (Mg1, Mg2, Mg3) and Ga (Ga1, Ga2, Ga3) atoms randomly share the same sites, respectively, and they occupy the centers of octahedra, connecting with each other by edge-sharing (see Figure 2.25(d)). A Ge/Ga4 atom is located at the center of a tetrahedron and connects with (Mg/Ga)O_6 octahedra by edge-sharing. The neutrality can be achieved in the spinel solid solution by the distribution of Mg^{2+}, Ga^{3+}, and Ge^{4+} cations over these sites, allowing some excitability into this geometrically rigid structure. When Mn^{4+} ions are introduced, they get an opportunity to occupy both the octahedra and tetrahedra centers. The excitation spectrum of MGG:Mn^{4+} shows great superiority among lots of Mn^{4+}-doped oxide phosphors, i.e., the excitation band range from 250 to 470 nm with the maximum

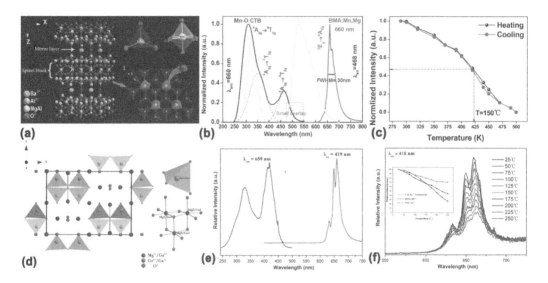

FIGURE 2.25 (a) The crystal structure of $BaMgAl_{10}O_{17}$ (BAM), in which the AlO_6 octahedra (yellow) for Mn^{4+} to occupy and $(Mg/Al)O_6$ tetrahedra (blue) are displayed. (b) The excitation (black) and emission (red) spectra for BAM: 2 mol% Mn^{4+}, 2 mol% Mg^{2+} as well as the emission spectrum of YAG:Ce. (c) The temperature-dependent integrated luminescence intensity of BAM: 2 mol% Mn^{4+}, 2 mol% Mg^{2+} under the excitation of 468 nm. The black and red data represent the results of the heating and cooling process, respectively. (d) The crystal structure of $Mg_3Ga_2GeO_8$ (MGG), the coordination, as well as their connection of $(Mg/Ga)O_6$ octahedra and $(Ge/Ga)O_4$ tetrahedra. (e) The PL and PLE spectra of MGG: 0.5 mol% Mn^{4+}, the excitation, and monitor wavelengths have been marked. (f) Temperature-dependent emission spectra of MGG: 0.5 mol% Mn^{4+} from 25°C (298K) to 250°C (523K). The inset shows the normalized luminescence intensity of MGG: 0.5 mol% Mn^{4+}, as well as the commercial YAG:Ce^{3+} (P46-Y3) and Y_2O_3:Eu^{3+} with increasing temperature. (Wang, B., Lin, H., Huang, F., Xu, J., Chen, H., Lin, Z. and Wang, Y., *Chem. Mater.* 28, 3515, 2016. With permission; Ding, X., Zhu, G., Geng, W., Wang, Q. and Wang, Y., *Inorg. Chem.* 55, 154, 2016. With permission.)

at 419 nm (Figure 2.25(e)).[249] This, on the one hand, could avoid the reabsorption of blue-green light and, on the other hand, could be expected to well match with a NUV (or deep blue) LED chip. The PL spectrum suggests it could emit deep red light peaking at 659 nm with an outstanding QE of 64.7% (0.5 mol% Mn^{4+}-doped). In addition to this, the temperature-dependent emission spectra of MGG: 0.5 mol% Mn^{4+} shown in Figure 2.25(f) indicates that it remains 72% of the initial intensity (at 298K) when the heating temperature is up to 423K, which is comparable with the commercial phosphors YAG:Ce^{3+} (P46-Y3). It demonstrates that MGG:Mn^{4+} is a potential red phosphor matching NUV LED chips to get white light.

Mn^{4+} in oxide compounds always performs a longer emission wavelength (in the range of 630–730 nm) than that in fluorides which is related to the nephelauxetic effect. The emission range is beyond the human eye sensitivity and, thus, takes less remarkable superiority in backlighting but contributes to the low correlated color temperature (CCT) and high color rendering (CRI) in illumination.

A list of Mn^{4+}-doped materials including some of the fluorides, oxyfluorides, and oxides is seen in Table 2.8.

Mn^{4+} as an activator would enable UV-blue excitation band, especially in fluorides. The advantages of the line (narrowband) emissions with high efficiency and color purity as well as the modern synthesis technic make Mn^{4+}-doped fluorides attractive to researchers. However, there's not yet a more effective way to resolve the weakness of chemical stability in consideration of luminescence efficiency and efforts in this aspect are still urgently necessary. Furthermore, the toxicity and pollution during the synthesis of fluorides cannot be effectively avoided at present.

TABLE 2.8

The Mn^{4+}-Doped Materials with Narrow-Band Emission and Their Performances

Compounds	S.G.	λ_{ex} (nm)	λ_{em} (nm)	D_q (cm^{-1})	E (^2E_g) (eV)	ZPL	Ref.
Na$_2$SiF$_6$	$P321$	460	627	1970	2.007	+^{a}	[256]
Na$_2$GeF$_6$	$P321$	465	627	1920	2.008	+	[257]
Na$_2$SnF$_6$	$P4/mnm$	~470	626	1930	2.005	−	[258]
Na$_2$TiF$_6$	$P321$	~460	617	2010	2.006	+	[259]
K$_2$SiF$_6$	$Fm\bar{3}m$	460	630	2030	1.993	−	[84]
K$_2$GeF$_6$	$P\bar{3}m1$	~460	632	1970	1.990	−	[260]
K$_2$SnF$_6$·H$_2$O	$Fddd$	~470	621 (ZPL)	1950	1.998	+	[261]
K$_2$TiF$_6$	$P\bar{3}m1$	~460	~630	2000	1.991	−	[259]
K$_2$ZrF$_6$	$C2/c$	~460^b	~630	Mixture with K$_3$ZrF$_7$			[262]
K$_2$MnF$_6$	$Fm\bar{3}m$	460	630	2075	1.993	−	[263]
	$P6_3mc$	−	630	1910	1.993	+	[264]
Rb$_2$SiF$_6$	$Fm\bar{3}m$	~460	~632	2000	1.992	−	[265]
Cs$_2$SiF$_6$	$Fm\bar{3}m$	~470	~633	1930	1.986	−	[266]
KNaSiF$_6$	$Pnma$	~460	~630	2035	2.000	+	[267]
BaSiF$_6$	$R\bar{3}m$	~470	634	1960	1.987	−	[268]
ZnSiF$_6$·6H$_2$O	$R\bar{3}$	~470	~630	1935	1.991	−	[269]
Na$_3$AlF$_6$	$P2_1/m$	~465	627	1880	1.999	+	[270]
K$_3$AlF$_6$	$I4_1a$	~460	626	1875	2.008	+	[271]
K$_2$LiAlF$_6$	$Fm\bar{3}m$	~460	634	2000	1.984	+	[272]
K$_2$NaAlF$_6$	$Fm\bar{3}m$	~460	630	1975	1.994	+	[272]
Rb$_2$KAlF$_6$	$Fm\bar{3}m$	~460	~630	1930	2.002	+	[273]
Cs$_2$KAlF$_6$	$Fm\bar{3}m$	469	632	1850	1.988	+	[273]
Cs$_2$RbAlF$_6$	$Fm\bar{3}m$	~460	~630	1860	1.994	+	[273]
K$_3$SiF$_7$	$P4/mbm$	~460	~630	1985	2.002	−	[274]
K$_3$ZrF$_7$	$Fm\bar{3}m$	482	629	1850	1.999	+	[275]
K$_2$NbF$_7$	$P2_1/c$	474	627	2131	2.013	+	[276]
Na$_3$TaF$_8$	$C2/c$	462	627	1960	2.002	+	[277]
Ba$_5$AlF$_{13}$	$Fm\bar{3}m$	~460	627	2150	1.999	+	[278]
BaNbOF$_5$	$Pa\bar{3}$	480	628	1750	1.971	/	[279]
BaTiOF$_4$	$Pbcm$	464	631	1920	1.990	−	[280]
Cs$_2$NbOF$_5$	$P\bar{3}m1$	467	632	1775	1.990	−	[281, 282]
Rb$_2$NbOF$_5$	$P3m1$	465	631	1930	1.992	−	[283]
Cs$_2$WO$_2$F$_4$	$P\bar{3}m1$	488	632	2105	1.990^b	−	[284]
Na$_2$WO$_2$F$_4$	$Pbcn$	460	619 (ZPL)	2132	2.003^b	+	[285]
K$_3$TaO$_2$F$_4$	$I4/mmm$	460	630	1930	2.000^b	+	[286]
LiAl$_4$O$_6$F	F	~285	662	2222	1.862^b	/	[287]
Sr$_2$ScO$_3$F	$I4/mmm$	345	697	1640	1.810	/	[288, 289]
α-Al$_2$O$_3$	$R\bar{3}c$	470^b	676^b (R$_1$)	2170	1.832	/	[290]
MgAl$_2$O$_4$	$Fd\bar{3}m$	290	652	1870	1.905	/	[291]
Y$_3$Al$_5$O$_{12}$	$Ia\bar{3}d$	~350	670	1805	1.893	/	[292]
Lu$_3$Al$_5$O$_{12}$	$Ia\bar{3}d$	~300	666	1905	1.902	−	[293]

(Continued)

TABLE 2.8 (*Continued*)

The Mn⁴⁺-Doped Materials with Narrow-Band Emission and Their Performances

Materials	S.G.	λ_{ex} (nm) max	λ_{em} (nm) max	D_q (cm⁻¹)	E (^2E_g) (eV)	ZPL	Ref.
$BaAl_2Ge_2O_8$	$I2/c$	288	668	1945	1.902	/	[294]
$BaGe_4O_9$	$P321$	440	669	1935	1.897	/	[295]
Ba_2GeO_4	$Pnma$	290	667	1985	1.858	/	[296]
$BaMgAl_{10}O_{17}$	$P6_3/mmc$	300	660	1890	1.878	/	[297]
$Ca_{14}Zn_6Ga_{10}O_{35}$	$F23$	313	711	1645	1.790	/	[298]
Li_2MgTiO_4	$Fm\bar{3}m$	~330	676	2101	1.834[b]	/	[299]
$Sr_4Al_{14}O_{25}$	$Pmma$	325	652	2222	1.902[b]	/	[300]
$Mg_3Ga_2GeO_8$	$Imma$	419	659	/	1.882[b]	/	[249]
$K_2Ge_4O_9$	$P\bar{3}c1$	470	663	/	1.870[b]	/	[301]
$Li_3Mg_2NbO_6$	$Fddd$	342	668	2110	1.856[b]	/	[247]
$SrAl_{12}O_{19}$	$P6_3/mmc$	340	655	/	1.893[b]	/	[302]

[a] "+" means the emission of ZPL is strong, whereas "–" means weak.
[b] Recalculated results from the original data (unit conversion).

2.2.4 COLOR GAMUT OF VARIOUS COMBINATION OF PHOSPHORS

One of the most important issues in phosphor-converted LED for backlighting is the improvement of the color gamut. Based on the numerous researches, extending of color gamut and its optimization could be realized by combining LED chips with up-to-date phosphors (previously mentioned). The blue LED chips are applied as the dominant excitation sources in the present studies because of their proven technique, outstanding efficiency, and high compatibility with the current efficient and stable phosphors. Besides, benefiting from its excellent efficiency and modern preparation technology, KSF:Mn⁴⁺ is widely used as the red phosphor in practice. As we know that human eyes are more sensitive to green colors compared to blue and red, while the color gamut could be well extended by improving the color purity in the green range as seen in Figure 2.3. Therefore, there are a lot of research works reported on the color gamut improved by green phosphors. For instance, because of the narrower emission band, the color gamut of β-sialon:Eu (529) contained device could exhibit 89% of the NTSC, which is 7% larger than that of the β-sialon:Eu (540) contained one [70]; the fabrication of LED device using the Mn²⁺, Mg²⁺ co-doped oxynitride γ-AlON and KSF:Mn⁴⁺ achieves a color gamut of 102.4% of the NTSC with excellent reliability comparable to that obtained using sharp β-sialon:Eu [238]; The Mn²⁺-doped narrow-band oxide $Sr_2MgAl_{22}O_{36}$ (SMAO)has a narrow-band emission peaks at 518 nm with a quite small FWHM of 26 nm gives a wide color gamut of 127% NTSC, which is even comparable to the QLCDs and QLEDs [241]; the UCr_4C_4-type $RbNa[Li_3SiO_4]_2$:Eu²⁺ (RN) phosphor exhibits green emission (λ_{em} = 523 nm) with FWHM of 41 nm shows high luminous efficacy and the backlighting device fabricated using it can realize wide color gamut of 113% NTSC.[211] The fabrication of white light by combining blue LED chips and different kinds of red and green phosphors with their color gamut are listed in Table 2.9. The wide color gamut achieved in the LED devices by application of inorganic phosphors with high performances makes LCDs keep competitive in the recent display field.

Red and green phosphors based on blue LED chips have been greatly developed in the past decades while tradeoff among the efficiency, stability, and color gamut that the phosphors could provide have to be carefully considered. Moreover, along with the development of semiconductor technologies, NUV LED chips with RGB phosphors could also have their place in the field of backlighting.

TABLE 2.9

Optical Properties of the Displays with Various Phosphor-Converted White LEDs as LCD Backlighting Devices

No.	LED Chip λ_{em} (nm)	Phosphors λ_{em} (nm)/FWHM(nm)	CCT[a] (K)	Color Gamut (% NTSC)	Ref.
1	~450	$Y_3Al_5O_{12}$:Ce (~550/>100)	4950	68.3	[35]
2	~450	$Y_3Al_5O_{12}$:Ce (~560/>100)	8582[b]	71	[62]
3	~450	β-sialon:Eu (535/~55)	8620	82.1	[35, 66]
		$CaAlSiN_3$:Eu (650/~90)			
4	~450	$SrGa_2S_4$:Eu (540/47)	9319[b]	90	[62]
		CaS:Eu (653/64)			
5	450	β-sialon:Eu (535/54)	8611	85.9	[251]
		K_2SiF_6:Mn^{4+} (631/~3)			
6	460	β-sialon:Eu (540/54)	33,266	82	[70]
		K_2SiF_6:Mn^{4+} (commercial)			
7	460	β-sialon:Eu (529/49)	33,266	89	[70]
		K_2SiF_6:Mn^{4+} (commercial)			
8	~450	$SrGa_2S_4$:Eu (530/55)	8330	86.4	[303]
		K_2SiF_6:Mn^{4+} (~630/10)			
9	445	Sharp β-sialon:Eu (~530/~50)	10,014	95.7	[304]
		K_2SiF_6:Mn^{4+} (GE Corporation)			
10	445	γ-AlON:Mn^{2+}, Mg^{2+} (~520/<45)	10,611	102.4	[238]
		K_2SiF_6:Mn^{4+} (GE TriGain™)			
11	450	β-sialon:Eu (540/–)	–	84.2	[305]
		$Sr[LiAl_3N_4]$:Eu (650/~60)			
12	460	$RbLi[Li_3SiO_4]_2$:Eu (530/42)	6221	107	[194]
		K_2SiF_6:Mn^{4+} (commercial)			
13	455	$RbNa[Li_3SiO_4]_2$:Eu (523/41)	5196	113	[211]
		K_2SiF_6:Mn^{4+} (commercial)			
14	460	$Sr_2MgAl_{22}O_{36}$:Mn^{2+} (518/26)	–	127	[241]
		K_2SiF_6:Mn^{4+} (632/–)			
15	432	$SrMgAl_{10}O_{17}$:Eu^{2+}, Mn^{2+} (518/26)	11,806	114	[240]
		K_2SiF_6:Mn^{4+} (–/–)			

[a] CCT, the abbreviation of correlated color temperature.

[b] Recalculated result, a conversion of the color coordinates.

2.3 SUMMARY AND PERSPECTIVE

Phosphor-based backlighting dominates the current and near-future display field because of its high performance and modern technology. This chapter begins with the device structure, application of phosphors in backlighting and also provides insights into the development of novel materials. The phosphors determine the performance of the display device such as their efficiency, color gamut, and reliability. Therefore, phosphors for backlighting should be with the features including considerable efficiency, narrow-band emission (high color purity), matching well with the RGB color filters, and thermal/chemical stability. There are mainly three kinds of narrow band–emitting phosphors with the activator-host type that could be used for backlighting application, i.e., Eu^{2+}, Mn^{2+}, and Mn^{4+}-activated phosphors, including some typical narrow-band-emitting phosphors such as the commercialized β-sialon:Eu^{2+}, K_2SiF_6:Mn^{4+}, the very famous Eu^{2+}-doped UCr_4C_4-type nitrides/oxides, and other potential materials such as γ-AlON:Mn^{2+}, Ba$_5$SiO$_4$Cl$_6$:Eu^{2+}, and Cs$_2$NbOF$_5$:Mn^{4+}. Their structure, PL properties, as well as the thermal behaviors are introduced in detail. Many other phosphors have also been summarized for the evaluation of their potential applications in backlighting.

In spite of the favorable development of phosphors for backlighting, many scientific and technical issues remain unsolved. Problems like how to improve the performance of existing materials (e.g., the conflict between efficiency and the stability enhancement by coating in KSF:Mn^{4+}), how to begin with the development of a new candidate for backlighting, whether the material can be accurately designed from the available theoretical basis, etc. should be urgently considered and addressed in the near future. To our excitement, with the increasing demands of people for display quality and development of novel LED technologies such as mini-LEDs and micro-LEDs, which promote the display technology forward to a higher quality direction, we believe the application of phosphor-based backlighting will be faced with new opportunities.

ACKNOWLEDGEMENTS

The research work of Prof. Yuhua Wang's Group was financially supported by the National Natural Science Foundation of China (No. U1905213, 51672115, 51372105, 50925206); Research and Development Plan for the Science and Technology Department of Gansu Province (No. 18YF1NA104). We thank Dr. Xicheng Wang and Dr. Yichao Wang (Lanzhou University) for the discussion; Mr. Ruijie Ji, Mr. Haoyang Wang, and Mr. Runtian Kang (Postgraduates in Lanzhou University) for their work in collecting, sorting, and checking of the corresponding literatures and references; Dr. Xin Ding, and Mr. Qi Wei (Lanzhou University) for their corresponding research work.

REFERENCES

1. Castellano JA. 2012. *Handbook of display technology.* Amsterdam, The Netherlands: Elsevier.
2. Yang DK, and Wu ST. 2014. *Fundamentals of liquid crystal devices.* 2nd ed. New York, NY: John Wiley & Sons.
3. Schadt M (2009) Milestone in the history of field-effect liquid crystal displays and materials. *Jpn. J. Appl. Phys.* 48: 03B001.
4. Wang X, Bao Z, Chang YC, and Liu RS (2020) Perovskite quantum dots for application in high color gamut backlighting display of light-emitting diodes. *ACS Energy Lett.* 5: 3374.
5. Saito H. 2014. Structure of liquid crystal display. In *Encyclopedia of polymeric nanomaterials,* edited by Kobayashi, and Müllen. Berlin, Heidelberg: Springer Berlin Heidelberg.
6. Fang MH, Leaño JL, and Liu RS (2018) Control of narrow-band emission in phosphor materials for application in light-emitting diodes. *ACS Energy Lett.* 3: 2573.
7. Görrn P, Sander M, Meyer J, Kröger M, Becker E, Johannes HH, Kowalsky W, and Riedl T (2006) Towards see-through displays: Fully transparent thin-film transistors driving transparent organic light-emitting diodes. *Adv. Mater.* 18: 738.
8. Tsujimura T. 2017. *OLED display fundamentals and applications.* 2nd ed: Hoboken, New Jersey, United States: John Wiley & Sons, Inc.
9. Buckley A. 2013. *Organic light-emitting diodes (OLEDS): Materials, devices and applications.* Amsterdam, The Netherlands: Elsevier.
10. Kim J, Shim HJ, Yang J, Choi MK, Kim DC, Kim J, Hyeon T, and Kim DH (2017) Ultrathin quantum dot display integrated with wearable electronics. *Adv. Mater.* 29: 1700217.
11. Shu Y, Lin X, Qin H, Hu Z, Jin Y, and Peng X (2020) Quantum dots for display applications. *Angew. Chem. Int. Ed.* 59: 22312.
12. Jia H (2018) Who will win the future of display technologies? *Natl. Sci. Rev.* 5: 427.
13. Yeh C, Lo KS, and Lin W (2019) Visual-attention-based pixel dimming technique for OLED displays of mobile devices. *IEEE Trans. Ind. Electron.* 66: 7159.
14. Song J, Fang T, Li J, Xu L, Zhang F, Han B, Shan Q, and Zeng H (2018) Organic–inorganic hybrid passivation enables perovskite QLEDs with an EQE of 16.48%. *Adv. Mater.* 30: 1805409.
15. Won YH, Cho O, Kim T, Chung DY, Kim T, Chung H, Jang H, Lee J, Kim D, and Jang E (2019) Highly efficient and stable InP/ZnSe/ZnS quantum dot light-emitting diodes. *Nature* 575: 634.

16. Huang Y, Hsiang EL, Deng MY, and Wu ST (2020) Mini-LED, Micro-LED and OLED displays: present status and future perspectives. *Light Sci. Appl.* 9: 105.

17. Chen HW, Lee JH, Lin BY, Chen S, and Wu ST (2018) Liquid crystal display and organic light-emitting diode display: Present status and future perspectives. *Light Sci. Appl.* 7: 17168.

18. Kim H-J, Shin M-H, Lee J-Y, Kim J-H, and Kim Y-J (2017) Realization of 95% of the Rec. 2020 color gamut in a highly efficient LCD using a patterned quantum dot film. *Opt. Exp.* 25: 10724.

19. Kagan CR, and Andry P. 2003. *Thin-film transistors.* 1st ed. Boca Raton, FL: CRC Press.

20. Jang HJ, Lee JY, Kwak J, Lee D, Park JH, Lee B, and Noh YY (2019) Progress of display performances: AR, VR, QLED, OLED, and TFT. *J. Inf. Disp.* 20: 1.

21. Street RA (2010) Thin-film transistors. *Adv. Mater.* 21: 2007.

22. Smith AR. 1978. Color gamut transform pairs. Paper read at Siggraph 78: Conference on Computer Graphics & Interactive Techniques.

23. Pust P, Schmidt PJ, and Schnick W (2015) A revolution in lighting. *Nat. Mater.* 14: 454.

24. Commission IE. 1999. Multimedia systems and equipment - colour measurement and management In *Part 2-1: Colour management - Default RGB colour space - sRGB.*

25. EG 27:2004 - SMPTE Engineering Guideline - Supplemental Information for SMPTE 170M and Background on the Development of NTSC Color Standards. 2004. *EG 27:2004*, 1.

26. EG 432-1:2010 - SMPTE Engineering Guideline - Digital Source Processing - Color Processing for D-Cinema. 2010. *EG 432-1:2010*, 1.

27. RP 431-2:2011 - SMPTE Recommended Practice - D-Cinema Quality - Reference Projector and Environment. 2011. *RP 431-2:2011*, 1.

28. Union IT. 2015. Parameter values for ultra-high definition television systems for production and international programme exchange.

29. Holger W, and Thomas J. 2009. LCD backlighting with LED phosphors. Merck Patent GMBH (DE), Winkler Holger (DE), Juestel Thomas (DE).

30. Weindorf PFL, and Zysnarski A. 2007. *Led backlighting system.* US patent, US7193248B2, United States.

31. Kvenvold AM, and Harkavy B. 2005. *High efficiency low power led backlighting system for liquid crystal display.* US patent, US20050231978A1, United States.

32. Ye S, Xiao F, Pan YX, Ma YY, and Zhang QY (2010) Phosphors in phosphor-converted white light-emitting diodes: Recent advances in materials, techniques and properties. *Mater. Sci. Eng. R Rep* 71: 1.

33. Xia Z, and Liu Q (2016) Progress in discovery and structural design of color conversion phosphors for LEDs. *Prog. Mater. Sci.* 84: 59.

34. Nair GB, Swart HC, and Dhoble SJ (2020) A review on the advancements in phosphor-converted light emitting diodes (pc-LEDs): Phosphor synthesis, device fabrication and characterization. *Prog. Mater. Sci.* 109: 100622.

35. Xie RJ, Hirosaki N, and Takeda T (2009) Wide color gamut backlight for liquid crystal displays using three-band phosphor-converted white light-emitting diodes. *Appl. Phys. Express* 2: 022401.

36. Fukuda Y, Matsuda N, Okada A, and Mitsuishi I (2012) White light-emitting diodes for wide-color-gamut backlight using green-emitting Sr-Sialon phosphor. *Jpn. J. Appl. Phys.* 51: 122101.

37. Luo Z, Xu D, and Wu ST (2014) Emerging quantum-dots-enhanced LCDs. *J. Display Technol.* 10: 526.

38. Li S, Xie RJ, Takeda T, and Hirosaki N (2017) Critical review – Narrow-band nitride phosphors for wide color-gamut white LED backlighting. *ECS J. Solid State Sci. Technol.* 7: R3064.

39. Chen L, Chu CI, and Liu RS (2012) Improvement of emission efficiency and color rendering of high-power LED by controlling size of phosphor particles and utilization of different phosphors. *Microelectron. Reliab.* 52: 900.

40. Mattoussi H, Mauro JM, Goldman ER, Anderson GP, Sundar VC, Mikulec FV, and Bawendi MG (2000) Self-assembly of CdSe–ZnS quantum dot bioconjugates using an engineered recombinant protein. *J. Am. Chem. Soc.* 122: 12142.

41. Sagar LK, Bappi G, Johnston A, Chen B, Todorović P, Levina L, Saidaminov MI, García de Arquer FP, Nam DH, Choi MJ, Hoogland S, Voznyy O, and Sargent EH (2020) Suppression of auger recombination by gradient alloying in InAs/CdSe/CdS QDs. *Chem. Mater.* 32: 7703.

42. Achorn OB, Franke D, and Bawendi MG (2020) Seedless continuous injection synthesis of indium phosphide quantum dots as a route to large size and low size dispersity. *Chem. Mater.* 32: 6532.

43. Mićić OI, Sprague J, Lu Z, and Nozik AJ (1996) Highly efficient band-edge emission from InP quantum dots. *Appl. Phys. Lett.* 68: 3150.

44. Zhou X, Ren J, Dong X, Wang X, Seto T, and Wang Y (2020) Controlling the nucleation process of InP/ZnS quantum dots using zeolite as a nucleation site. *CrystEngComm* 22: 3474.

45. Song J, Li J, Li X, Xu L, Dong Y, and Zeng H (2015) Quantum dot light-emitting diodes based on inorganic perovskite cesium lead halides (CsPbX$_3$). *Adv. Mater.* 27: 7162.

46. Ren J, Li T, Zhou X, Dong X, Shorokhov AV, Semenov MB, Krevchik VD, and Wang Y (2019) Encapsulating all-inorganic perovskite quantum dots into mesoporous metal organic frameworks with significantly enhanced stability for optoelectronic applications. *Chem. Eng. J.* 358: 30.

47. Shirasaki Y, Supran GJ, Bawendi MG, and Bulović V (2013) Emergence of colloidal quantum-dot light-emitting technologies. *Nat. Photonics* 7: 13.

48. Yang Z, Gao M, Wu W, Yang X, Sun XW, Zhang J, Wang HC, Liu RS, Han CY, Yang H, and Li W (2019) Recent advances in quantum dot-based light-emitting devices: Challenges and possible solutions. *Mater. Today* 24: 69.

49. Jang E, Jun S, Jang H, Lim J, Kim B, and Kim Y (2010) White-light-emitting diodes with quantum dot color converters for display backlights. *Adv. Mater.* 22: 3076.

50. Chen H, He J, and Wu S (2017) Recent advances on quantum-dot-enhanced liquid-crystal displays. *IEEE J. Sel. Top. Quantum Electron.* 23: 1.

51. Derfus AM, Chan WCW, and Bhatia SN (2004) Probing the cytotoxicity of semiconductor quantum dots. *Nano Lett.* 4: 11.

52. Pietryga JM, Zhuravlev KK, Whitehead M, Klimov VI, and Schaller RD (2008) Evidence for barrierless auger recombination in PbSe nanocrystals: A pressure-dependent study of transient optical absorption. *Phys. Rev. Lett.* 101: 217401.

53. Baldo MA, Lamansky S, Burrows PE, Thompson ME, and Forrest SR (1999) Very high-efficiency green organic light-emitting devices based on electrophosphorescence. *Appl. Phys. Lett.* 75: 4.

54. Park SJ, Link S, Miller WL, Gesquiere A, and Barbara PF (2007) Effect of electric field on the photoluminescence intensity of single CdSe nanocrystals. *Chem. Phys.* 341: 169.

55. Shirasaki Y, Supran GJ, Tisdale WA, and Bulović V (2013) Origin of efficiency roll-off in colloidal quantum-dot light-emitting diodes. *Phys. Rev. Lett.* 110: 217403.

56. Qiao J, Zhao J, Liu Q, and Xia Z (2019) Recent advances in solid-state LED phosphors with thermally stable luminescence. *J. Rare Earths* 37: 565.

57. Zhang Y, Luo L, Chen G, Liu Y, Liu R, and Chen X (2020) Green and red phosphor for LED backlight in wide color gamut LCD. *J. Rare Earths* 38: 1.

58. Li S, Wang L, Tang D, Cho Y, Liu X, Zhou X, Lu L, Zhang L, Takeda T, Hirosaki N, and Xie RJ (2018) Achieving high quantum efficiency narrow-band β-Sialon:Eu^{2+} phosphors for high-brightness LCD backlights by reducing the Eu^{3+} luminescence killer. *Chem. Mater.* 30: 494.

59. Lin H, Hu T, Huang Q, Cheng Y, Wang B, Xu J, Wang J, and Wang Y (2017) Non-rare-earth K$_2$XF$_7$:Mn^{4+} (X = Ta, Nb): A highly-efficient narrow-band red phosphor enabling the application in wide-color-gamut LCD. *Laser Photonics Rev.* 11: 1700148.

60. Mirhosseini R, Schubert MF, Chhajed S, Cho J, and Schubert EF (2009) Improved color rendering and luminous efficacy in phosphor-converted white light-emitting diodes by use of dual-blue emitting active regions. *Opt. Express* 17: 10806.

61. Blasse G, and Bril A (1967) A new phosphor for flying-spot cathode-ray tubes for color television: Yellow-emitting Y$_3$Al$_5$O$_{12}$: Ce^{3+}. *Appl. Phys. Lett.* 11: 53.

62. Ito Y, Hori T, Kusunoki T, Nomura H, and Kondo H (2015) A phosphor sheet and a backlight system providing wider color gamut for LCDs. *J. Soc. Inf. Disp.* 22: 419.

63. Shin HH, Kim JH, Han BY, and Yoo JS (2008) Failure analysis of a phosphor-converted white light-emitting diode due to the CaS:Eu phosphor. *Jpn. J. Appl. Phys.* 47: 3524.

64. Pan Z, and Yan W. 2013. *Near infrared doped phosphors having an alkaline gallate matrix.* US patent, US8894882B2, United States.

65. Hirosaki N, Xie RJ, Kimoto K, Sekiguchi T, Yamamoto Y, Suehiro T, and Mitomo M (2005) Characterization and properties of green-emitting β-SiAlON:Eu^{2+} powder phosphors for white light-emitting diodes. *Appl. Phys. Lett.* 86: 211905.

66. Uheda K, Hirosaki N, Yamamoto Y, Naito A, Nakajima T, and Yamamoto H (2006) Luminescence properties of a red phosphor, CaAlSiN$_3$: Eu^{2+} for white light-emitting diodes. *Electrochem. Solid-State Lett.* 9: H22.

67. Wyszecki G, and Stiles W. 1982. *Color science: Concepts and methods, quantitative data and formulate.* New York, NY: Wiley.

68. Hunt R. 1998. *Measuring colour.* 3rd ed. England: Fountain Press.

69. Wang L, Xie RJ, Suehiro T, Takeda T, and Hirosaki N (2018) Down-conversion nitride materials for solid state lighting: Recent advances and perspectives. *Chem. Rev.* 118: 1951.

70. Zhang X, Fang MH, Tsai YT, Lazarowska A, Mahlik S, Lesniewski T, Grinberg M, Pang WK, Pan F, Liang C, Zhou W, Wang J, Lee JF, Cheng BM, Hung TL, Chen YY, and Liu RS (2017) Controlling of structural ordering and rigidity of β-SiAlON:Eu through chemical cosubstitution to approach narrow-band-emission for light-emitting diodes application. *Chem. Mater.* 29: 6781.

71. Blasse G, and Bril A (1967) A new phosphor for flying-spot cathode-ray tubes for color television: Yellow-emitting $Y_3Al_5O_{12}$-Ce^{3+}. *Appl. Phys. Lett.* 11: 53.

72. Bando K, Sakano K, Noguchi Y, and Shimizu Y (1998) Development of high-bright and pure-white LED lamps. *J. Light Visual Environ.* 22: 1_2.

73. Smet PF, Moreels I, Hens Z, and Poelman D (2010) Luminescence in sulfides: A rich history and a bright future. *Materials* 3: 2834.

74. Tanaka S, Shanker V, Shiiki M, Deguchi H, and Kobayashi H. 1985. Multicolor electroluminescence in alkaline-earth sulfide thin-film devices. Paper read at Proceedings of the SID.

75. Van Haecke JE, Smet PF, De Keyser K, and Poelman D (2007) Single crystal CaS: Eu and SrS: Eu luminescent particles obtained by solvothermal synthesis. *J. Electrochem. Soc.* 154: J278.

76. Nakano F, Uekura N, Nakanishi Y, Hatanaka Y, and Shimaoka G (1997) Preparation of $CaGa_2S_4$: Ce thin films for blue emitting thin-film EL device. *Appl. Surf. Sci.* 121–122: 160.

77. Benalloul P, Barthou C, Benoit J, Eichenauer L, and Zeinert A (1993) IIA-III_2-S_4 ternary compounds: New host matrices for full color thin film electroluminescence displays. *Appl. Phys. Lett.* 63: 1954

78. Bignardi C, Ciulla M, and Iori M (1963) Paramorphism and dysmorphism of the spinal column in school age children. *Boll. Soc. Med. Chir. Cremona.* 17: 63.

79. Chartier C, Barthou C, Benalloul P, and Frigerio JM (2005) Photoluminescence of Eu^{2+} in $SrGa_2S_4$. *J. Lumin.* 111: 147.

80. Pust P, Weiler V, Hecht C, Tücks A, Wochnik AS, Henß A-K, Wiechert D, Scheu C, Schmidt PJ, and Schnick W (2014) Narrow-band red-emitting $Sr[LiAl_3N_4]$:Eu^{2+} as a next-generation LED-phosphor material. *Nat. Mater.* 9: 891.

81. Uheda K, Hirosaki N, and Yamamoto H (2006) Host lattice materials in the system Ca_3N_2–AlN–Si_3N_4 for white light emitting diode. *Phys. Status Solidi A* 203: 2712.

82. Xie RJ, Hirosaki N, Liu XJ, Takeda T, and Li HL (2008) Crystal structure and photoluminescence of Mn^{2+}–Mg^{2+} codoped gamma aluminum oxynitride (γ-AlON): A promising green phosphor for white light-emitting diodes. *Appl. Phys. Lett.* 92: 201905.

83. Adachi S, and Takahashi T (2008) Direct synthesis and properties of K_2SiF_6: Mn^{4+} phosphor by wet chemical etching of Si wafer. *J. Appl. Phys.* 104: 023512.

84. Takahashi T, and Adachi S (2008) Mn^{4+}-activated red photoluminescence in K_2SiF_6 phosphor. *J. Electrochem. Soc.* 155: E183.

85. Setlur AA, Siclovan OP, Lyons RJ, Grigorov LS. 2011. *Moisture-resistant phosphor and associated method.* US patent, US8057706B1, United States.

86. Pust P, Wochnik AS, Baumann E, Schmidt PJ, Wiechert D, Scheu C, and Schnick W (2014) $Ca[LiAl_3N_4]$: Eu^{2+}—A narrow-band red-emitting nitridolithoaluminate. *Chem. Mater.* 26: 3544.

87. Schmiechen S, Schneider H, Wagatha P, Hecht C, Schmidt PJ, and Schnick W (2014) Toward new phosphors for application in illumination-grade white pc-LEDs: The nitridomagnesosilicates $Ca[Mg_3SiN_4]$: Ce^{3+}, $Sr[Mg_3SiN_4]$: Eu^{2+}, and $Eu[Mg_3SiN_4]$. *Chem. Mater.* 26: 2712.

88. Strobel P, Maak C, Weiler V, Schmidt PJ, and Schnick W (2018) Ultra-narrow-band blue-emitting oxoberyllates $AELi_2[Be_4O_6]$: Eu^{2+} (AE = Sr, Ba) paving the way to efficient RGB pc-LEDs. *Angew. Chem. Int. Ed.* 57: 8739.

89. Liao H, Zhao M, Molokeev MS, Liu Q, and Xia Z (2018) Learning from a mineral structure toward an ultra-narrow-band blue-emitting silicate phosphor $RbNa_3(Li_3SiO_4)_4$:Eu^{2+}. *Angew. Chem. Int. Ed.* 57: 11728.

90. Wang W, Tao M, Liu Y, Wei Y, Xing G, Dang P, Lin J, and Li G (2019) Photoluminescence control of UCr_4C_4-type phosphors with superior luminous efficiency and high color purity via controlling site selection of Eu^{2+} activators. *Chem. Mater.* 31: 9200.

91. Dutzler D, Seibald M, Baumann D, Philipp F, Peschke S, and Huppertz H (2019) $RbKLi_2[Li_3SiO_4]_4$: Eu^{2+} an ultra narrow-band phosphor. *Zeitschrift für Naturforschung B* 74: 535.

92. Su B, Molokeev MS, and Xia Z (2019) Mn^{2+}-based narrow-band green-emitting Cs_3MnBr_5 phosphor and the performance optimization by Zn^{2+} alloying. *J. Mater. Chem. C* 7: 11220.

93. Shang M, Li C, and Lin J (2014) How to produce white light in a single-phase host? *Chem. Soc. Rev.* 43: 1372.

94. Li G, Tian Y, Zhao Y, and Lin J (2015) Recent progress in luminescence tuning of Ce^{3+} and Eu^{2+}-activated phosphors for pc-WLEDs. *Chem. Soc. Rev.* 44: 8688.

95. Park WJ, Song YH, Moon JW, Jang DS, and Yoon DH (2009) From blue-purple to red-emitting phosphors, $A_{2-x}B_xP_2O_7$:Eu^{2+}, Mn^{2+} (A And B = Alkaline-Earth Metal) under near-UV pumped white LED applications. *J. Electrochem. Soc.* 156: J148.

96. Singh V, Chakradhar RPS, Rao JL, and Kwak HY (2011) Photoluminescence and EPR studies of $BaMgAl_{10}O_{17}$:Eu^{2+} phosphor with blue-emission synthesized by the solution combustion method. *J. Lumin.* 131: 1714.

97. Wang DY, Huang CH, Wu YC, and Chen TM (2011) $BaZrSi_3O_9$:Eu^{2+}: A cyan-emitting phosphor with high quantum efficiency for white light-emitting diodes. *J. Mater. Chem.* 21: 10818.

98. Liu J, Lian H, Sun J, and Shi C (2005) Characterization and properties of green emitting $Ca_3SiO_4Cl_2$: Eu^{2+} powder phosphor for white light-emitting diodes. *Chem. Lett.* 34: 1340.

99. Saradhi MP, and Varadaraju UV (2006) Photoluminescence studies on Eu^{2+}-activated Li_2SrSiO_4: A potential orange-yellow phosphor for solid-state lighting. *Chem. Mater.* 18: 5267.

100. Zhang Q, Wang X, Tang Z, and Wang Y (2020) A $K_3ScSi_2O_7$:Eu^{2+} based phosphor with broad-band NIR emission and robust thermal stability for NIR pc-LEDs. *Chem. Commun.* 56: 4644.

101. Dorenbos P (2003) Relation between Eu^{2+} and Ce^{3+} f-d-transition energies in inorganic compounds. *J. Phys.: Condens. Matter* 15: 4797.

102. Crystal field theory. 2010. *In Electronic Structure and Properties of Transition Metal Compounds: Introduction to the Theory,* edited by Isaac: John Wiley & Sons, Inc.

103. Morse PM (1932) The theory of electric and magnetic susceptibilities. *Science* 76: 326.

104. Qin X, Liu X, Huang W, Bettinelli M, and Liu X (2017) Lanthanide-activated phosphors based on 4f-5d optical transitions: Theoretical and experimental aspects. *Chem. Rev.* 117: 4488.

105. Gerloch RCS. 1973. *Ligand-field parameters.* London: Cambridge University Press.

106. Rack PD, and Holloway PH (1998) The structure, device physics, and material properties of thin film electroluminescent displays. *Mater. Sci. Eng. R: Rep.* 21: 171.

107. Jang HS, Im WB, Lee DC, Jeon DY, and Kim SS (2007) Enhancement of red spectral emission intensity of $Y_3Al_5O_{12}$: Ce^{3+} phosphor via Pr co-doping and Tb substitution for the application to white LEDs. *J. Lumin.* 126: 371.

108. Lin CC and Liu RS. 2017. Introduction to the basic properties of luminescent materials. In *Phosphor, up conversion nano particles, quantum dots and their applications.* Berlin, Heidelberg: Springer, pp 1–29.

109. Huang CH, Wu PJ, Lee JF, and Chen TM (2011) $(Ca,Mg,Sr)_9Y(PO_4)_7$: Eu^{2+}, Mn^{2+}: Phosphors for white-light near-UV LEDs through crystal field tuning and energy transfer. *J. Mater. Chem.* 21: 10489.

110. Tchougréeff AL, and Dronskowski R (2009) Nephelauxetic effect revisited. *Int. J. Quantum Chem.* 109: 2606.

111. Morrison C, Mason DR, and Kikuchi C (1967) Modified slater integrals for an ion in a solid. *Phys. Lett. A* 24: 607.

112. Jørgensen CK. 1962. The nephelauxetic series. In *Progress in Inorganic Chemistry,* edited by Cotton FA: John Wiley & Sons, Inc, pp 73–124.

113. Lever ABP. 1984. *Inorganic electronic spectroscopy.* 2nd ed. Amsterdam: Elsevier.

114. Bersuker IB. 1976. *The electron structure and properties of coordination compounds.* Moscow: Khimiya.

115. Dorenbos P, Andriessen J, and van Eijk CWE (2003) $4f^{n-1}5d$ centroid shift in lanthanides and relation with anion polarizability, covalency, and cation electronegativity. *J. Solid State Chem.* 171: 133.

116. George NC, Denault KA, and Seshadri R (2013) Phosphors for solid-state white lighting. *Annu. Rev. Mater. Res.* 43: 481.

117. Zhao M, Zhang Q, and Xia Z (2020) Narrow-band emitters in LED backlights for liquid-crystal displays. *Mater. Today* 40: 246.

118. Liebau F. 1985. *Structural chemistry of silicates.* Berlin, Heidelberg: Springer.

119. Xie RJ, Li YQ, Hirosaki N, and Yamamoto H. 2011. *Nitride phosphors and solid-state lighting.* 1st Edition ed. Boca Raton, FL: CRC Press.

120. Krames MR, Mueller GO, Mueller-Mach RB, Bechtel HH, and Schmidt PJ. 2010. *Wavelength conversion for producing white light from high power blue led.* US patent, US20100289044A1, United States.

121. Höppe HA, Lutz H, Morys P, Schnick W, and Seilmeier A (2000) Luminescence in Eu^{2+}-doped $Ba_2Si_5N_8$: Fluorescence, thermoluminescence, and upconversion. *J. Phys. Chem. Solids* 61: 2001.

122. Kijima N, Seto T, and Hirosaki N (2009) A new yellow phosphor $La_3Si_6N_{11}$:Ce^{3+} for white LEDs. *ECS Trans.* 25: 247.

123. Wang Y, Seto T, Ishigaki K, Uwatoko Y, Xiao G, Zou B, Li G, Tang Z, Li Z, and Wang Y (2020) Pressure-driven Eu^{2+}-doped $BaLi_2Al_2Si_2N_6$: A new color tunable narrow-band emission phosphor for spectroscopy and pressure sensor applications. *Adv. Funct. Mater.* 30: 2001384.

124. Ding J, Wu Q, Li Y, Long Q, Wang Y, and Wang Y (2016) Eu^{2+}-activated $Ca_5Si_2Al_2N_8$ – A novel nitrido-alumosilicate red phosphor containing the special polyhedron of separated corner-shared $[Al_2N_6]$ and $[Si_2N_6]$. *Chem. Eng. J.* 302: 466.

125. Ding J, Wu Q, Li Y, Long Q, Wang C, and Wang Y (2015) Synthesis and luminescent properties of the $Li_{0.995-x}Mg_xSi_{2-x}Al_xN_3$: $0.005Eu^{2+}$ phosphors. *J. Am. Ceram. Soc.* 98: 2523.

126. Long Q, Wang C, Ding J, Li Y, Wu Q, and Wang Y (2015) Synthesis and luminescence properties of a novel red-emitting $LiSr_4(BN_2)_3$: Eu^{2+} phosphor. *Dalton Trans.* 44: 14507.

127. Wu Q, Ding J, Wang C, Li Y, Wang X, Mao A, and Wang Y (2015) Effect of a solid solution of AlN on the crystal structure and optical properties of $LiSi_2N_3$:Eu phosphors. *RSC Adv.* 5: 31255.

128. Wang X, Seto T, Zhao Z, Li Y, Wu Q, Li H, and Wang Y (2014) Preparation of $Sr_{1-x}Ca_xYSi_4N_7$:Eu^{2+} solid solutions and their luminescence properties. *J. Mater. Chem. C* 2: 4476.

129. Wang C, Zhao Z, Wang X, Li Y, Wu Q, and Wang Y (2014) Luminescence properties of $Ca_2Si_5N_8$:Eu^{2+} prepared by gas-pressed sintering using baf_2 as flux and cation substitution. *RSC Adv.* 4: 55388.

130. Chen H, Ding J, Ding X, Wang X, Cao Y, Zhao Z, and Wang Y (2017) Synthesis, crystal structure, and luminescence properties of tunable red-emitting nitride solid solutions $(Ca_{1-x}Sr_x)_{16}Si_{17}N_{34}$:$Eu^{2+}$ for white LEDs. *Inorg. Chem.* 56: 10904.

131. Li Z, Seto T, and Wang Y (2020) Enhanced crystallinity and thermal stability of Ba^{2+} and Al^{3+}-O^{2-} co-substituted $Sr_2Si_5N_8$:Eu^{2+}. *J. Mater. Chem. C* 8: 9874.

132. Wu Q, Wang Y, Yang Z, Que M, Li Y, and Wang C (2014) Synthesis and luminescence properties of pure nitride Ca-α-SiAlON with the composition $Ca_{1.4}Al_{2.8}Si_{9.2}N_{16}$ by gas-pressed sintering. *J. Mater. Chem. C* 2: 829.

133. Pust P, Hintze F, Hecht C, Weiler V, Locher A, Zitnanska D, Harm S, Wiechert D, Schmidt PJ, and Schnick W (2014) Group (III) Nitrides $M[Mg_2Al_2N_4]$ (M= Ca, Sr, Ba, Eu) and $Ba[Mg_2Ga_2N_4]$ structural relation and nontypical luminescence properties of Eu^{2+} doped samples. *Chem. Mater.* 26: 6113.

134. Stoll H, and Hoppe R (1987) Ein neues oxoplumbat (IV): $CsNa_3[PbO_4]$. *Rev. Chim. Miner.* 24: 96.

135. Nowitzki B, and Hoppe R (1986) Oxides of type $A[(TO)_n]$: $NaLi_3SiO_4$, $NaLi_3GeO_4$, and $NaLi_3TiO_4$. *Rev. Chim. Miner.* 23: 217.

136. Cao Y, Wang X, Ding J, Zhou X, Seto T, and Wang Y (2020) Constructing a single-white-light emission by finely modulating the occupancy of luminescence centers in europium-doped $(Ca_{1-x}Sr_x)_9Bi(PO_4)_7$ for WLEDs. *J. Mater. Chem. C* 8: 9576.

137. Leaño JL, Lesniewski T, Lazarowska A, Mahlik S, Grinberg M, Sheu HS, and Liu RS (2018) Thermal stabilization and energy transfer in narrow-band red-emitting $Sr[(Mg_2Al_2)_{1-y}(Li_2Si_2)_yN_4]$: Eu^{2+} phosphors. *J. Mater. Chem. C* 6: 5975.

138. Vos JJ (1978) Colorimetric and photometric properties of a 2° fundamental observer. *Color Res. Appl.* 3: 125.

139. Dorenbos P (2003) Anomalous luminescence of Eu^{2+} and Yb^{2+} in inorganic compounds. *J. Phys.: Condens. Matter* 15: 2645.

140. Tolhurst TM, Schmiechen S, Pust P, Schmidt PJ, Schnick W, and Moewes A (2016) Electronic structure, bandgap, and thermal quenching of $Sr[Mg_3SiN_4]$: Eu^{2+} in comparison to $Sr[LiAl_3N_4]$: Eu^{2+}. *Adv. Opt. Mater.* 4: 584.

141. Maak C, Durach D, Martiny C, Schmidt PJ, and Schnick W (2018) Narrow-band yellow-orange emitting $La_{3-x}Ca_{1.5x}Si_6N_{11}$:$Eu^{2+}$ ($x \approx 0.77$): A promising phosphor for next-generation amber pcLEDs. *Chem. Mater.* 30: 3552.

142. Wagatha P, Pust P, Weiler V, Wochnik AS, Schmidt PJ, Scheu C, and Schnick W (2016) $Ca_{18.75}Li_{10.5}$ $[Al_{39}N_{55}]$: Eu^{2+} - supertetrahedron phosphor for solid-state lighting. *Chem. Mater.* 28: 1220.

143. Strobel P, Weiler V, Hecht C, Schmidt PJ, and Schnick W (2017) Luminescence of the narrow-band red emitting nitridomagnesosilicate $Li_2(Ca_{1-x}Sr_x)_2[Mg_2Si_2N_6]$: Eu^{2+} ($x = 0$–0.06). *Chem. Mater.* 29: 1377.

144. Strobel P, Schmiechen S, Siegert M, Tücks A, Schmidt PJ, and Schnick W (2015) Narrow-band green emitting nitridolithoaluminosilicate $Ba[Li_2(Al_2Si_2)N_6]$: Eu^{2+} with framework topology whj for LED/LCD-backlighting applications. *Chem. Mater.* 27: 6109.

145. Li J, Ding J, Cao Y, Zhou X, Ma B, Zhao Z, and Wang Y (2018) Color-tunable phosphor $[Mg_{1.25}Si_{1.25}Al_{2.5}]$ O_3N_3:Eu^{2+} – A new modified polymorph of AlON with double sites related luminescence and low thermal quenching. *ACS Appl. Mater. Interfaces* 10: 37307.

146. Wang Y, Ding J, and Wang Y (2017) Enhancing stability of Eu^{2+} in $La_{10-x}Sr_x(Si_{6-x}P_xO_{22}N_2)O_2$ phosphors by the design of apatite structures with an ($[Si/P][O/N]_4$) framework and tunable luminescence properties. *J. Mater. Chem. C* 5: 985.

147. Mao A, Zhao Z, Wang Y, Wu Q, Wang X, Wang C, and Li Y (2015) Cathodoluminescence properties of $Ba_3Si_6O_9N_4$: Eu^{2+} phosphors. *ECS Solid State Lett.* 4: R17.

148. Mao A, Zhao Z, and Wang Y (2017) Orange phosphor $Li_2SrSiON_2$: Eu^{2+} for blue light chip based warm white LEDs. *RSC Adv.* 7: 42634.

149. Mao A, Zhao Z, Seto T, and Wang Y (2019) Promising color controllable phosphors $CBSOCN_y$: Eu^{2+} with excellent thermal stability. *Mater. Des.* 180: 107865.

150. Wu Q, Ding J, Li Y, Wang X, and Wang Y (2017) Electronic structure and luminescence properties of self-activated and Eu^{2+}/Ce^{3+} doped $Ca_3Li_{4-y}Si_2N_{6-y}O_y$ red-emitting phosphors. *J. Lumin.* 186: 144.

151. Wang X, Zhao Z, Wu Q, Li Y, Wang C, Mao A, and Wang Y (2015) Synthesis, structure, and luminescence properties of $SrSiAl_2O_3N_2$: Eu^{2+} phosphors for light-emitting devices and field emission displays. *Dalton Trans.* 44: 11057.

152. Wang C, Zhao Z, Wu Q, Zhu G, and Wang Y (2015) Enhancing the emission intensity and decreasing the full widths at half maximum of $Ba_3Si_6O_{12}N_2$:Eu^{2+} by Mg^{2+} doping. *Dalton Trans.* 44: 10321.

153. Wang C, Zhao Z, Wu Q, Xin S, and Wang Y (2014) The pure-phase $Ba_{3-x}Ca_xSi_6O_{12}N_2$:$Eu^{2+}$ green phosphor: Synthesis, photoluminescence and thermal properties. *CrystEngComm* 16: 9651.

154. Zhao Z, Yang Z, Shi Y, Wang C, Liu B, Zhu G, and Wang Y (2013) Red-emitting oxonitridosilicate phosphors $Sr_2SiN_zO_{4-1.5z}$: Eu^{2+} for white light-emitting diodes: Structure and luminescence properties. *J. Mater. Chem. C* 1: 1407.

155. Li YQ, Hirosaki N, Xie RJ, Takeda T, and Mitomo M (2008) Crystal and electronic structures, luminescence properties of Eu^{2+}-doped $Si_{6-z}Al_zO_zN_{8-z}$ and $M_ySi_{6-z}Al_{z-y}O_{z+y}N_{8-z-y}$ (M = 2Li, Mg, Ca, Sr, Ba). *J. Solid State Chem.* 181: 3200.

156. Shannon RD (1976) Revised effective ionic radii and systematic studies of interatomic distances in halides and chalcogenides. *Acta Crystallogr., Sect. A: Found. Crystallogr.* 32: 751.

157. Kresse G, and Furthmüller J (1996) Efficient iterative schemes for Ab Initio total-energy calculations using a plane-wave basis set. *Phys. Rev.. B, Condensed Matter* 54: 11169.

158. Kresse G, and Joubert D (1999) From ultrasoft pseudopotentials to the projector augmented-wave method. *Phys. Rev. B* 59: 1758.

159. Brgoch J, Gaultois MW, Balasubramanian M, Page K, Hong BC, and Seshadri R (2014) Local structure and structural rigidity of the green phosphor β-Sialon: Eu^{2+}. *Appl. Phys. Lett.* 105: 181904.

160. Ishizuka K (2002) A practical approach for stem image simulation based on the FFT multislice method. *Ultramicroscopy* 90: 71.

161. Kimoto K, Xie R-J, Matsui Y, Ishizuka K, and Hirosaki N (2009) Direct observation of single dopant atom in light-emitting phosphor of β-SiAlON:Eu^{2+}. *Appl. Phys. Lett.* 94: 041908.

162. Takahashi K, Yoshimura Ki, Harada M, Tomomura Y, Takeda T, Xie RJ, and Hirosaki N (2012) On the origin of fine structure in the photoluminescence spectra of the β-Sialon:Eu^{2+} green phosphor. *Sci. Technol. Adv. Mater.* 13: 015004.

163. Kechele JA, Oeckler O, Stadler F, and Schnick W (2009) Structure elucidation of $BaSi_2O_2N_2$ – A host lattice for rare-earth doped luminescent materials in phosphor-converted (pc)-LEDs. *Solid State Sci.* 11: 537.

164. Bachmann V, Jüstel T, Meijerink A, Ronda C, and Schmidt PJ (2006) Luminescence properties of $SrSi_2O_2N_2$ doped with divalent rare earth ions. *J. Lumin.* 121: 441.

165. Wang Y, Que M, Ci Z, Zhu G, Xin S, Shi Y, and Wang Q (2013) Crystal structure and luminescent properties of a cyan emitting $Ca_{10}(SiO_4)_3(SO_4)_3F_2$: Eu^{2+} phosphor. *CrystEngComm* 15: 6389.

166. Li Y, Chen W, and Wang Y (2016) Synthesis and photoluminescence properties of high-$Ba_3P_4O_{13}$:Eu^{2+} – A broadband yellow-emitting phosphor for near ultraviolet white light-emitting diodes. *Mater. Res. Bull.* 84: 363.

167. Wang X, Zhao Z, Wu Q, Wang C, Wang Q, Yanyan L, and Wang Y (2016) Structure, photoluminescence and abnormal thermal quenching behavior of Eu^{2+}-doped $Na_3Sc_2(PO_4)_3$: A novel blue-emitting phosphor for n-UV LEDs. *J. Mater. Chem. C* 4: 8795.

168. Long Q, Wang C, Li Y, Ding J, and Wang Y (2016) Synthesis and investigation of photo/cathodoluminescence properties of a novel green emission phosphor $Sr_8ZnLu(PO_4)_7$: Eu^{2+}. *J. Alloys Compd.* 671: 372.

169. Ding X, and Wang Y (2017) Commendable Eu^{2+}-doped oxide-matrix-based $LiBa_{12}(BO_3)_7F_4$ red broad emission phosphor excited by NUV light: Electronic and crystal structures, luminescence properties. *ACS Appl. Mater. Interfaces* 9: 23983.

170. Ding X, and Wang Y (2016) Novel orange light emitting phosphor Sr_9(Li, Na, K)$Mg(PO_4)7$:Eu^{2+} excited by NUV light for white LEDs. *Acta Mater.* 120: 281.

171. Ding X, Zhu G, Geng W, Mikami M, and Wang Y (2015) Novel blue and green phosphors obtained from $K_2ZrSi_3O_9$:Eu^{2+} compounds with different charge compensation ions for LEDs under near-UV excitation. *J. Mater. Chem. C* 3: 6676.

172. Ding X, and Wang Y (2017) Structure and photoluminescence properties of a novel apatite green phosphor $Ba_5(PO_4)_2SiO_4:Eu^{2+}$ excited by NUV light. *Phys. Chem. Chem. Phys.* 19: 2449.
173. Tang Z, Wang D, Khan WU, Du S, Wang X, and Wang Y (2016) Novel zirconium silicate phosphor $K_2ZrSi_2O_7:Eu^{2+}$ for white light-emitting diodes and field emission displays. *J. Mater. Chem. C* 4: 5307.
174. Tang Z, Zhang G, and Wang Y (2018) Design and development of a bluish-green luminescent material $(K_2HfSi_3O_9: Eu^{2+})$ with robust thermal stability for white light-emitting diodes. *ACS Photonics* 5: 3801.
175. Tang Z, Zhang Q, Cao Y, Li Y, and Wang Y (2020) Eu^{2+}-doped ultra-broadband VIS-NIR emitting phosphor. *Chem. Eng. J.* 388: 124231.
176. Zhou X, Geng W, Ding J, Wang Y, and Wang Y (2018) Structure, bandgap, photoluminescence evolution and thermal stability improved of Sr replacement apatite phosphors $Ca_{10-x}Sr_x(PO_4)_6F_2: Eu^{2+}$ (x = 4, 6, 8). *Dyes Pigm.* 152: 75.
177. Zhou X, Geng W, and Wang Y (2019) First-principles calculations, structure research and luminescence properties for a novel apatite blue/green phosphor $Ca_6Y_4(SiO_4)_2(PO_4)_4O_2:Eu^{2+}/Tb^{3+}$. *J. Lumin.* 211: 276.
178. Wei Q, Ding J, Chen H, Zhang Q, and Wang Y (2020) A novel yellow-green emitting phosphor with hafnium silicon multiple rings structure for light-emitting diodes and field emission displays. *Chem. Eng. J.* 385: 123392.
179. Feng P, Li G, Guo H, Liu D, and Wang Y (2019) Identifying a cyan ultralong persistent phosphorescence (Ba, Li) $(Si, Ge, P)_2O_5:Eu^{2+}, Pr^{3+}$ via solid solution strategy. *J. Phys. Chem. C* 123: 3102.
180. Wei Q, Ding J, Zhou X, Wang X, and Wang Y (2020) New strategy of designing a novel yellow-emitting phosphor $Na_4Hf_2Si_3O_{12}:Eu^{2+}$ for multifunctional applications. *J. Alloys Compd.* 817: 152762.
181. Wei Q, Ding J, and Wang Y (2020) A novel wide-excitation and narrow-band blue-emitting phosphor with hafnium silicon multiple rings structure for photoluminescence and cathodoluminescence. *J. Alloys Compd.* 831: 154825.
182. Zhang Q, Wang X, and Wang Y (2020) Full-visible-spectrum lighting realized by a novel Eu^{2+}-doped cyan-emitting borosilicate phosphor. *CrystEngComm* 22: 4702.
183. Zhang Q, Ding X, and Wang Y (2019) Novel highly efficient blue-emitting $SrHfSi_2O_7:Eu^{2+}$ phosphor: A potential color converter for WLEDs and FEDs. *Dyes Pigm.* 163: 168.
184. Zhang Q, Wang X, Ding X, and Wang Y (2018) A broad band yellow-emitting $Sr_8CaBi(PO_4)_7: Eu^{2+}$ phosphor for n-UV pumped white light emitting devices. *Dyes Pigm.* 149: 268.
185. Zhang Q, Wang Q, Wang X, Ding X, and Wang Y (2016) Luminescence properties of Eu^{2+}-doped $BaSi_2O_5$ as an efficient green phosphor for light-emitting devices and wide color gamut field emission displays. *New J. Chem.* 40: 8549.
186. Ci Z, and Wang Y (2009) Preparation, electronic structure, and photoluminescence properties of Eu^{2+}-activated carbonate $Sr_{1-x}Ba_xCO_3$ for white light-emitting diodes. *J. Electrochem. Soc.* 156: J267.
187. Wen YAN, Wang Y, Liu B, Zhang F, and Shi Y (2012) Novel blue-emitting $KSrBP_2O_8:Eu^{2+}$ phosphor for near-UV light-emitting diodes. *Funct. Mater. Lett.* 05: 1250048.
188. Zhao Z, Yang Z, Shi Y, Wang C, Liu B, Zhu G, and Wang Y (2013) Red-emitting oxonitridosilicate phosphors $Sr2SiN_xO_{4-1.5x}:Eu^{2+}$ for white light-emitting diodes: Structure and luminescence properties. *J. Mater. Chem. C* 1: 1407.
189. Que M, Ci Z, Wang Y, Zhu G, Liu B, Zhang J, Shi Y, Wen Y, Li Y, and Wang Q (2013) Synthesis and photoluminescence of a new chlorogermanate phosphor $Ca_8Mg(GeO_4)_4Cl_2:Eu^{2+}$. *J. Am. Ceram. Soc.* 96: 223.
190. Zhu G, Shi Y, Mikami M, Shimomura Y, and Wang Y (2014) Electronic structure and photo/cathodoluminescence properties investigation of green emission phosphor $NaBaScSi_2O_7:Eu^{2+}$ with high thermal stability. *CrystEngComm* 16: 6089.
191. Ci Z, Que M, Shi Y, Zhu G, and Wang Y (2014) Enhanced photoluminescence and thermal properties of size mismatch in $Sr_{2.97-x-y}Eu_{0.03}Mg_xBa_ySiO_5$ for high-power white light-emitting diodes. *Inorg. Chem.* 53: 2195.
192. Zhu G, Shi Y, Mikami M, Shimomura Y, and Wang Y (2014) Observation, identification and characterization of strong self-reduction process in a orthophosphate phosphor $CaZr_4(PO_4)_6:Eu$. *Mater. Res. Bull.* 50: 405.
193. Shi Y, Zhu G, Mikami M, Shimomura Y, and Wang Y (2014) Photoluminescence of green-emitting $Ca_7(PO_4)_2(SiO_4)_2: Eu^{2+}$ phosphor for white light emitting diodes. *Opt. Mater. Express* 4: 280.
194. Zhao M, Liao H, Ning L, Zhang Q, Liu Q, and Xia Z (2018) Next-generation narrow-band green-emitting $RbLi(Li_3SiO_4)_2: Eu^{2+}$ phosphor for backlight display application. *Adv. Mater.* 30: e1802489.
195. Fang MH, Mariano COM, Chen PY, Hu SF, and Liu RS (2020) Cuboid-size-controlled color-tunable Eu-doped alkali–lithosilicate phosphors. *Chem. Mater.* 32: 1748.
196. Ding W, Wang J, Zhang M, Zhang Q, and Su Q (2006) A novel orange phosphor of Eu^{2+}-activated calcium chlorosilicate for white light-emitting diodes. *J. Solid State Chem.* 179: 3582.

197. Poort SHM, Reijnhoudt HM, van der Kuip HOT, and Blasse G (1996) Luminescence of Eu^{2+} in Silicate host lattices with alkaline earth ions in a row. *J. Alloys Compd.* 241: 75.

198. Sharma K, Talwar G, Moharil SV, and Ghormare KB (2017) Luminescence of Ce^{3+} in some compounds in the system $CaO-SiO_2-CaCl_2$. *J. Lumin.* 188: 168.

199. Zhu G, Wang Y, Ci Z, Liu B, Shi Y, and Xin S (2012) $Ca_8Mg(SiO_4)_4Cl_2$: Ce^{3+}, Tb^{3+}: A potential single-phased phosphor for white-light-emitting diodes. *J. Lumin.* 132: 531.

200. Hao Z, Zhang X, Luo Y, and Zhang J (2013) Preparation and photoluminescence properties of single-phase $Ca_2SiO_3Cl_2$: Eu^{2+} bluish-green emitting phosphor. *Mater. Lett.* 93: 272.

201. Durach D, Fahrnbauer F, Oeckler O, and Schnick W (2015) $La_6Ba_3[Si_{17}N_{29}O_2]Cl$—An oxonitridosilicate chloride with exceptional structural motifs. *Inorg. Chem.* 54: 8727.

202. Garcia A, Latourrette B, and Fouassier C (1979) $Ba_5SiO_4Cl_6$:Eu, a new blue-emitting photoluminescent material with high quenching temperature. *J. Electrochem. Soc.* 126: 1734.

203. Zeng Q, Tanno H, Egoshi K, Tanamachi N, and Zhang S (2006) $Ba_5SiO_4Cl_6$:Eu^{2+}: An intense blue emission phosphor under vacuum ultraviolet and near-ultraviolet excitation. *Appl. Phys. Lett.* 88: 051906.

204. Schmiechen S, Strobel P, Hecht C, Reith T, Siegert M, Schmidt PJ, Huppertz P, Wiechert D, and Schnick W (2015) Nitridomagnesosilicate $Ba[Mg_3SiN_4]$: Eu^{2+} and structure–property relations of similar narrow-band red nitride phosphors. *Chem. Mater.* 27: 1780.

205. Takeda T, Hirosaki N, Funahshi S, and Xie RJ (2015) Narrow-band green-emitting phosphor $Ba_2LiSi_7AlN_{12}$: Eu^{2+} with high thermal stability discovered by a single particle diagnosis approach. *Chem. Mater.* 27: 5892.

206. Hoerder GJ, Seibald M, Baumann D, Schröder T, Peschke S, Schmid PC, Tyborski T, Pust P, Stoll I, Bergler M, Patzig C, Reißaus S, Krause M, Berthold L, Höche T, Johrendt D, and Huppertz H (2019) $Sr[Li_2Al_2O_2N_2]$:Eu^{2+} – A high performance red phosphor to brighten the future. *Nat. Commun.* 10: 1824.

207. Li YQ, Delsing ACA, de With G, and Hintzen HT (2005) Luminescence properties of Eu^{2+}-activated alkaline-earth silicon-oxynitride $MSi_2O_{2-\delta}N_{2+2/3\delta}$ (M = Ca, Sr, Ba): A promising class of novel LED conversion phosphors. *Chem. Mater.* 17: 3242.

208. Bachmann V, Ronda C, Oeckler O, Schnick W, and Meijerink A (2009) Color point tuning for (Sr, Ca, Ba)$Si_2O_2N_2$:Eu^{2+} for white light LEDs. *Chem. Mater.* 21: 316.

209. Li G, Lin CC, Chen WT, Molokeev MS, Atuchin VV, Chiang CY, Zhou W, Wang CW, Li W-H, Sheu HS, Chan TS, Ma C, and Liu RS (2014) Photoluminescence tuning via cation substitution in oxonitridosilicate phosphors: Dft calculations, different site occupations, and luminescence mechanisms. *Chem. Mater.* 26: 2991.

210. Zhao M, Liao H, Molokeev MS, Zhou Y, Zhang Q, Liu Q, and Xia Z (2019) Emerging ultra-narrow-band cyan-emitting phosphor for white LEDs with enhanced color rendition. *Light Sci. Appl.* 8: 38.

211. Liao H, Zhao M, Zhou Y, Molokeev MS, Liu Q, Zhang Q, and Xia Z (2019) Polyhedron transformation toward stable narrow-band green phosphors for wide-color-gamut liquid crystal display. *Adv. Funct. Mater.* 29: 1901988.

212. Zhao M, Zhou Y, Molokeev MS, Zhang Q, Liu Q, and Xia Z (2019) Discovery of new narrow-band phosphors with the UCr_4C_4-related type structure by alkali cation effect. *Adv. Opt. Mater.* 7: 1801631.

213. Wei Q, Zhou X, Tang Z, Wang X, and Wang Y (2019) A novel blue-emitting Eu^{2+}-doped chlorine silicate phosphor with a narrow band for illumination and displays: Structure and luminescence properties. *CrystEngComm* 21: 3660.

214. Poncé S, Jia Y, Giantomassi M, Mikami M, and Gonze X (2016) Understanding thermal quenching of photoluminescence in oxynitride phosphors from first principles. *J. Phys. Chem. C* 120: 4040.

215. Mikami M (2013) Response function calculations of $Ba_3Si_6O_{12}N_2$ and $Ba_3Si_6O_9N_4$ for the understanding of the optical properties of the Eu-doped phosphors. *Opt. Mater.* 35: 1958.

216. Zhou Q, Dolgov L, Srivastava AM, Zhou L, Wang Z, Shi J, Dramićanin MD, Brik MG, and Wu M (2018) Mn^{2+} and Mn^{4+} red phosphors: Synthesis, luminescence and applications in WLEDs: A review. *J. Mater. Chem. C* 6: 2652.

217. Ding X, Geng W, Wang Q, and Wang Y (2015) Structure, luminescence property and abnormal energy transfer behavior of color-adjustable $Ca_3Hf_2SiAl_2O_{12}$: Ce^{3+}, Mn^{2+} phosphors. *RSC Adv.* 5: 98709.

218. Wang Q, Ci Z, Wang Y, Zhu G, Wen Y, and Shi Y (2013) Crystal structure, photoluminescence properties and energy transfer of Ce^{3+}, Mn^{2+} co-activated $ca_8NaGd(PO_4)_6F_2$ phosphor. *Mater. Res. Bull.* 48: 1065.

219. Xu W, Zhu G, Zhou X, and Wang Y (2015) The structure, photoluminescence and influence of temperature on energy transfer in co-doped $Ca_9La(GeO_4)_{0.75}(PO_4)_6$ red-emission phosphors. *Dalton Trans.* 44: 9241.

220. Wang Q, Ci Z, Zhu G, Que M, Xin S, Wen Y, and Wang Y (2012) Structure and photoluminescence properties of $Ca_9Al(PO_4)_7$:Ce^{3+}, Mn^{2+} phosphors. *ECS J. Solid State Sci. Technol.* 1: R92.

221. Li Y, Shi Y, Zhu G, Wu Q, Li H, Wang X, Wang Q, and Wang Y (2014) A single-component white-emitting $CaSr_2Al_2O_6$:Ce^{3+}, Li^+, Mn^{2+} phosphor via energy transfer. *Inorg. Chem.* 53: 7668.

222. Ding X, Zhu G, Wang Q, and Wang Y (2015) Rare-earth free narrow-band green-emitting $KAlSi_2O_6$:Mn^{2+} phosphor excited by blue light for LED-phosphor material. *RSC Adv.* 5: 30001.

223. Zhou X, Geng W, Guo H, Ding J, and Wang Y (2018) $K_4CaGe_3O_9$:Mn^{2+}, Yb^{3+}: A novel orange-emitting long persistent luminescent phosphor with a special nanostructure. *J. Mater. Chem. C* 6: 7353.

224. Cao Y, Ding J, Ding X, Wang X, and Wang Y (2017) Tunable white light of multi-cation-site $Na_2BaCa(PO_4)_2$:Eu, Mn phosphor: Synthesis, structure and PL/CL properties. *J. Mater. Chem. C* 5: 1184.

225. Geng W, Zhou X, Ding J, and Wang Y (2019) Density-functional theory calculations, luminescence properties and fluorescence ratiometric thermo-sensitivity for a novel borate based red phosphor: $NaBaSc(BO_3)_2$: Ce^{3+}, Mn^{2+}. *J. Mater. Chem. C* 7: 1982.

226. Yu S, Wang D, Wu C, and Wang Y (2017) Synthesis and control the morphology of $Sr_2P_2O_7$: Eu^{2+}/Eu^{2+}, Mn^{2+} phosphors by precipitation method. *Mater. Res. Bull.* 93: 83.

227. Ding J, Seto T, Wang Y, Cao Y, Li H, and Wang Y (2018) A new mode of energy transfer between Mn^{2+} and Eu^{2+} in nitride-based phosphor $SrAlSi_4N_7$ with tunable light and excellent thermal stability. *Chem. Asian J.* 13: 2649.

228. Zhu G, Wang Y, Ci Z, Liu B, Shi Y, and Xin S (2011) $Ca_5La_5(SiO_4)_3(PO_4)_3O_2$: Ce^{3+}, Mn^{2+}: A color-tunable phosphor with efficient energy transfer for white-light-emitting diodes. *J. Electrochem. Soc.* 158: J236.

229. Shi Y, Wang Y, Wen Y, Zhao Z, Liu B, and Yang Z (2012) Tunable luminescence $Y_3Al_5O_{12}$: $0.06Ce^{3+}$, xMn^{2+} phosphors with different charge compensators for warm white light emitting diodes. *Opt. Express* 20: 21656.

230. Zhu G, Xin S, Wen Y, Wang Q, Que M, and Wang Y (2013) Warm white light generation from a single phased phosphor $Sr_{10}[(PO_4)_{5.5}(BO_4)_{0.5}](BO_2)$:$Eu^{2+}$, Mn^{2+}, Tb^{3+} for light emitting diodes. *RSC Adv.* 3: 9311.

231. Li Y, Li H, Liu B, Zhang J, Zhao Z, Yang Z, Wen Y, and Wang Y (2013) Warm-white-light emission from Eu^{2+}/Mn^{2+}-coactivated $NaSrPO_4$ phosphor through energy transfer. *J. Phys. Chem. Solids* 74: 175.

232. Ci Z, Sun Q, Sun M, Jiang X, Qin S, and Wang Y (2014) Structure, photoluminescence and thermal properties of Ce^{3+}, Mn^{2+} co-doped phosphosilicate $Sr_7La_3[(PO_4)_{2.5}(SiO_4)_3(BO_4)_{0.5}](BO_2)$ emission-tunable phosphor. *J. Mater. Chem. C* 2: 5850.

233. Shi Y, Wen Y, Que M, Zhu G, and Wang Y (2014) Structure, photoluminescent and cathodoluminescent properties of a rare-earth free red emitting β-$Zn_3B_2O_6$:Mn^{2+} phosphor. *Dalton Trans.* 43: 2418.

234. Yen W, Shionoya S, and Yamamoto H. 2006. *Phosphor handbook* Boca Raton, FL: CRC.

235. Liu L, Wang L, Zhang C, Cho Y, Dierre B, Hirosaki N, Sekiguchi T, and Xie RJ (2015) Strong energy-transfer-induced enhancement of luminescence efficiency of Eu^{2+}- and Mn^{2+}-codoped gamma-AlON for near-UVLED- pumped solid state lighting. *Inorg. Chem.* 54: 5556.

236. Goryainov S (2005) Pressure-induced amorphization of $Na_2Al_2Si_3O_{10} \cdot 2H_2O$ and $KAlSi_2O_6$ zeolites. *Phys. Status Solidi A* 202: R25.

237. Li J, Liang Q, Cao Y, Yan J, Zhou J, Xu Y, Dolgov L, Meng Y, Shi J, and Wu M (2018) Layered structure produced nonconcentration quenching in a novel Eu^{3+}-doped phosphor. *ACS Appl. Mater. Interfaces* 10: 41479.

238. Yoshimura K, Fukunaga H, Izumi M, Takahashi K, Xie R-J, and Hirosaki N (2017) Achieving super-wide-color-gamut display by using narrow-band green-emitting γ-AlON:Mn, Mg phosphor. *Jpn. J. Appl. Phys.* 56: 041701.

239. Takeda T, Xie RJ, Hirosaki N, Matsushita Y, and Honma T (2012) Manganese valence and coordination structure in Mn, Mg-codoped γ-AlON green phosphor. *J. Solid State Chem.* 194: 71.

240. Kang H, Lee KN, Unithrattil S, Kim HJ, Oh JH, Yoo JS, Im WB, and Do YR (2020) Narrow-band $SrMgAl_{10}O_{17}$:Eu^{2+}, Mn^{2+} green phosphors for wide-color-gamut backlight for LCD displays. *ACS Omega* 5: 19516.

241. Zhu Y, Liang Y, Liu S, Li H, and Chen J (2019) Narrow-band green-emitting $Sr_2MgAl_{22}O_{36}$:Mn^{2+} phosphors with superior thermal stability and wide color gamut for backlighting display applications. *Adv. Opt. Mater.* 7: 1801419.

242. Chen H, and Wang Y (2019) photoluminescence and cathodoluminescence properties of novel rare-earth free narrow-band bright green-emitting ZnB_2O_4:Mn^{2+} phosphor for LEDs and FEDs. *Chem. Eng. J.* 361: 314.

243. Adachi S (2018) Photoluminescence spectra and modeling analyses of Mn^{4+}-activated fluoride phosphors: a review. *J. Lumin.* 197: 119.

244. Adachi S (2018) Photoluminescence properties of Mn^{4+}-activated oxide phosphors for use in white-LED applications: a review. *J. Lumin.* 202: 263.

245. Adachi S (2019) Review—Mn^{4+}-activated red and deep red-emitting phosphors. *ECS J. Solid State Sci. Technol.* 9: 016001.

246. Ding X, Wang Q, and Wang Y (2016) Rare-earth-free red-emitting $K_2Ge_4O_9$:Mn^{4+} phosphor excited by blue light for warm white LEDs. *Phys. Chem. Chem. Phys.* 18: 8088.

247. Wang X, Zhou X, Cao Y, Wei Q, Zhao Z, and Wang Y (2019) Insight into a novel rare-earth-free red-emitting phosphor $Li_3Mg_2NbO_6$:Mn^{4+}: Structure and luminescence properties. *J. Am. Ceram. Soc.* 102: 6724.

248. Ding X, and Wang Y (2017) Structure and photoluminescence properties of rare-earth free narrow-band red-emitting $Mg_6ZnGeGa_2O_{12}$:Mn^{4+} phosphor excited by NUV light. *Opt. Mater.* 64: 445.

249. Ding X, Zhu G, Geng W, Wang Q, and Wang Y (2016) Rare-earth-free high-efficiency narrow-band red-emitting $Mg_3Ga_2GeO_8$:Mn^{4+} phosphor excited by near-UV light for white-light-emitting diodes. *Inorg. Chem.* 55: 154.

250. Huang L, Liu Y, Yu J, Zhu Y, Pan F, Xuan T, Brik MG, Wang C, and Wang J (2018) Highly stable K_2SiF_6:Mn^{4+}@K_2SiF_6 composite phosphor with narrow red emission for white LEDs. *ACS Appl. Mater. Interfaces* 10: 18082.

251. Wang L, Wang X, Kohsei T, Yoshimura K-i, Izumi M, Hirosaki N, and Xie RJ (2015) Highly efficient narrow-band green and red phosphors enabling wider color-gamut LED backlight for more brilliant displays. *Opt. Express* 23: 28707.

252. M. ten Kate O, Zhao Y, Jansen KMB, van Ommen JR, and Hintzen HT (2019) Effects of surface modification on optical properties and thermal stability of K_2SiF_6:Mn^{4+} red phosphors by deposition of an ultrathin Al_2O_3 layer using gas-phase deposition in a fluidized bed reactor. *ECS J. Solid State Sci. Technol.* 8: R88.

253. Wang Z, Yang Z, Wang N, Zhou Q, Zhou J, Ma L, Wang X, Xu Y, Brik MG, Dramićanin MD, and Wu M (2020) Single-crystal red phosphors: Enhanced optical efficiency and improved chemical stability for wLEDs. *Adv. Opt. Mater.* 8: 1901512.

254. Nguyen HD, Lin CC, and Liu RS (2015) Waterproof alkyl phosphate coated fluoride phosphors for optoelectronic materials. *Angew. Chem. Int. Ed.* 54: 10862.

255. Brik MG, and Srivastava AM (2017) Review – A review of the electronic structure and optical properties of ions with d^3 electron configuration (V^{2+}, Cr^{3+}, Mn^{4+}, Fe^{5+}) and main related misconceptions. *ECS J. Solid State Sci. Technol.* 7: R3079.

256. Xu YK, and Adachi S (2009) Properties of Na_2SiF_6:Mn^{4+} and Na_2GeF_6:Mn^{4+} red phosphors synthesized by wet chemical etching. *J. Appl. Phys.* 105: 013525.

257. Wang Z, Liu Y, Zhou Y, Zhou Q, Tan H, Zhang Q, and Peng J (2015) Red-emitting phosphors Na_2XF_6:Mn^{4+} (X = Si, Ge, Ti) with high colour-purity for warm white-light-emitting diodes. *RSC Adv.* 5: 58136.

258. Arai Y, and Adachi S (2011) Optical properties of Mn^{4+}-activated Na_2SnF_6 and Cs_2SnF_6 red phosphors. *J. Lumin.* 131: 2652.

259. Xu YK, and Adachi S (2011) Properties of Mn^{4+}-activated hexafluorotitanate phosphors. *J. Electrochem. Soc.* 158: J58.

260. Adachi S, and Takahashi T (2009) Photoluminescent properties of K_2GeF_6:Mn^{4+} red phosphor synthesized from aqueous HF/$KMnO_4$ solution. *J. Appl. Phys.* 106: 013516.

261. Arai Y, Takahashi T, and Adachi S (2010) Photoluminescent properties of $K_2SnF_6 \cdot H_2O$:Mn^{4+} red phosphor. *Opt. Mater.* 32: 1095.

262. Kasa R, and Adachi S (2012) Mn-activated K_2ZrF_6 and Na_2ZrF_6 phosphors: Sharp red and oscillatory blue-green emissions. *J. Appl. Phys.* 112: 013506.

263. Kasa R, Arai Y, Takahashi T, and Adachi S (2010) Photoluminescent properties of cubic K_2MnF_6 particles synthesized in metal immersed HF/$KMnO_4$ solutions. *J. Appl. Phys.* 108: 113503.

264. Pfeil A (1970) The emission of hexafluoromanganates(IV). *Spectrochim. Acta, A: Mol. Spectrosc.* 26: 1341.

265. Sakurai S, Nakamura T, and Adachi S (2016) Editors' choice—Rb_2SiF_6:Mn^{4+} and Rb_2TiF_6:Mn^{4+} red-emitting phosphors. *ECS J. Solid State Sci. Technol.* 5: R206.

266. Arai Y, and Adachi S (2011) Optical transitions and internal vibronic frequencies of MnF_6^{2-} ions in Cs_2SiF_6 and Cs_2GeF_6. *J. Electrochem. Soc.* 158: J179.

267. Adachi S, Abe H, Kasa R, and Arai T (2011) Synthesis and properties of hetero-dialkaline hexafluoro-silicate red phosphor $KNaSiF_6$:Mn^{4+}. *J. Electrochem. Soc.* 159: J34.

268. Sekiguchi D, Nara JI, and Adachi S (2013) Photoluminescence and Raman scattering spectroscopies of $BaSiF_6$:Mn^{4+} red phosphor. *J. Appl. Phys.* 113: 183516.

269. Hoshino R, and Adachi S (2013) Optical spectroscopy of $ZnSiF_6 \cdot 6H_2O$:Mn^{4+} red phosphor. *J. Appl. Phys.* 114: 213502.

270. Song EH, Wang JQ, Ye S, Jiang XF, Peng MY, and Zhang QY (2016) Room-temperature synthesis and warm-white LED applications of Mn^{4+} ion doped fluoroaluminate red phosphor Na_3AlF_6:Mn^{4+}. *J. Mater. Chem. C* 4: 2480.

271. Song E, Wang J, Shi J, Deng T, Ye S, Peng M, Wang J, Wondraczek L, and Zhang Q (2017) Highly efficient and thermally stable K_3AlF_6:Mn^{4+} as a red phosphor for ultra-high-performance warm white light-emitting diodes. *ACS Appl. Mater. Interfaces* 9: 8805.

272. Zhu Y, Cao L, Brik MG, Zhang X, Huang L, Xuan T, and Wang J (2017) Facile synthesis, morphology and photoluminescence of a novel red fluoride nanophosphor K_2NaAlF_6:Mn^{4+}. *J. Mater. Chem. C* 5: 6420.

273. Deng TT, Song EH, Zhou YY, Wang LY, and Zhang QY (2017) Tailoring photoluminescence stability in double perovskite red phosphors A_2BAlF_6:Mn^{4+} (A = Rb, Cs; B = K, Rb) via neighboring-cation modulation. *J. Mater. Chem. C* 5: 12422.

274. Kim M, Park WB, Bang B, Kim CH, and Sohn KS (2017) A novel Mn^{4+}-activated red phosphor for use in light emitting diodes, K_3SiF_7:Mn^{4+}. *J. Am. Ceram. Soc.* 100: 1044.

275. Tan H, Rong M, Zhou Y, Yang Z, Wang Z, Zhang Q, Wang Q, and Zhou Q (2016) Luminescence behaviour of Mn^{4+} Ions in seven coordination environments of K_3ZrF_7. *Dalton Trans.* 45: 9654.

276. Jansen T, Baur F, and Jüstel T (2017) Red emitting K_2NbF_7:Mn^{4+} and K_2TaF_7:Mn^{4+} for warm-white LED applications. *J. Lumin.* 192: 644.

277. Wang Z, Wang N, Yang Z, Yang Z, Wei Q, Zhou Q, and Liang H (2017) Luminescent properties of novel red-emitting phosphor Na_3TaF_8 with non-equivalent doping of Mn^{4+} for LED backlighting. *J. Lumin.* 192: 690.

278. Qin L, Cai P, Chen C, Wang J, and Seo HJ (2017) Synthesis, structure and optical performance of red-emitting phosphor Ba_5AlF_{13}:Mn^{4+}. *RSC Adv.* 7: 49473.

279. Dong X, Pan Y, Li D, Lian H, and Lin J (2018) A novel red phosphor of Mn^{4+} Ion-doped oxyfluoroniobate $BaNbOF_5$ for warm WLED applications. *CrystEngComm* 20: 5641.

280. Liang Z, Yang Z, Tang H, Guo J, Yang Z, Zhou Q, Tang S, and Wang Z (2019) Synthesis, luminescence properties of a novel oxyfluoride red phosphor $BaTiOF_4$:Mn^{4+} for LED backlighting. *Opt. Mater.* 90: 89.

281. Zhou J, Chen Y, Jiang C, Milićević B, Molokeev MS, Brik MG, Bobrikov IA, Yan J, Li J, and Wu M (2021) High moisture resistance of an efficient Mn^{4+}-activated red phosphor Cs_2NbOF_5:Mn^{4+} for WLEDs. *Chem. Eng. J.* 405: 126678.

282. Wang Q, Yang Z, Wang H, Chen Z, Yang H, Yang J, and Wang Z (2018) Novel Mn^{4+}-activated oxyfluoride Cs_2NbOF_5:Mn^{4+} red phosphor for warm white light-emitting diodes. *Opt. Mater.* 85: 96.

283. Wang Q, Liao J, Kong L, Qiu B, Li J, Huang H, and Wen Hr (2019) Luminescence properties of a non-rare-earth doped oxyfluoride $LiAl_4O_6F$:Mn^{4+} red phosphor for solid-state lighting. *J. Alloys Compd.* 772: 499.

284. Cai P, Qin L, Chen C, Wang J, and Seo HJ (2017) Luminescence, energy transfer and optical thermometry of a novel narrow red emitting phosphor: $Cs_2WO_2F_4$:Mn^{4+}. *Dalton Trans.* 46: 14331.

285. Li L, Pan Y, Huang Y, Huang S, and Wu M (2017) Dual-emissions with energy transfer from the phosphor $Ca_{14}Al_{10}Zn_6O_{35}$: Bi^{3+}, Eu^{3+} for application in agricultural lighting. *J. Alloys Compd.* 724: 735.

286. Zhou Y, Zhang S, Wang X, and Jiao H (2019) Structure and luminescence properties of Mn^{4+}-activated $K_3TaO_2F_4$ red phosphor for white LEDs. *Inorg. Chem.* 58: 4412.

287. Sun L, Devakumar B, Liang J, Wang S, Sun Q, and Huang X (2019) Novel high-efficiency violet-red dual-emitting Lu_2GeO_5: Bi^{3+}, Eu^{3+} phosphors for indoor plant growth lighting. *J. Lumin.* 214: 116544.

288. Kato H, Takeda Y, Kobayashi M, Kobayashi H, and Kakihana M (2018) Photoluminescence properties of layered perovskite-type strontium scandium oxyfluoride activated with Mn^{4+}. *Front. Chem.* 6: 467.

289. Wang Y, Tang K, Zhu B, Wang D, Hao Q, and Wang Y (2015) Synthesis and structure of a new layered oxyfluoride Sr_2ScO_3F with photocatalytic property. *Mater. Res. Bull.* 65: 42.

290. Geschwind S, Kisliuk P, Klein MP, Remeika JP, and Wood DL (1962) Sharp-line fluorescence, electron paramagnetic resonance, and thermoluminescence of Mn^{4+} in alpha-Al_2O_3. *Phys. Rev.* 126: 1684.

291. Peng L, Cao S, Zhao C, Liu B, Han T, Li F, and Li X (2018) Preparation of $Mg_{1+y}Al_{2-x}O_4$: xMn^{4+}, yMg^{2+} deep red phosphor and their optical properties. *Acta Phys. Sin.* 67: 187801.

292. Xu W, Chen D, Yuan S, Zhou Y, and Li S (2017) Tuning excitation and emission of Mn^{4+} emitting center in $Y_3Al_5O_{12}$ by cation substitution. *Chem. Eng. J.* 317: 854.

293. Li F, Cai J, Chi F, Chen Y, Duan C, and Yin M (2017) Investigation of luminescence from LuAG:Mn^{4+} for physiological temperature sensing. *Opt. Mater.* 66: 447.

294. Fu S, and Tian L (2019) A novel deep red emission phosphor $BaAl_2Ge_2O_8$:Mn^{4+} for plant growth LEDs. *Optik* 183: 635.

295. Liang S, Shang M, Lian H, Li K, Zhang Y, and Lin J (2016) Deep red MGe_4O_9:Mn^{4+} (M = Sr, Ba) phosphors: Structure, luminescence properties and application in warm white light emitting diodes. *J. Mater. Chem. C* 4: 6409.

296. Cao R, Luo W, Xiong Q, Jiang S, Luo Z, and Fu J (2015) Synthesis and photoluminescence properties of Ba_2GeO_4:Mn^{4+} novel deep red-emitting phosphor. *Chem. Lett.* 44: 1422.

297. Wang B, Lin H, Huang F, Xu J, Chen H, Lin Z, and Wang Y (2016) Non-rare-earth $BaMgAl_{10-2x}O_{17}$: xMn^{4+}, xMg^{2+}: A narrow-band red phosphor for use as a high-power warm w-LED. *Chem. Mater.* 28: 3515.

298. Yang C, Zhang Z, Hu G, Cao R, Liang X, and Xiang W (2017) A novel deep red phosphor $Ca_{14}Zn_6Ga_{10}O_{35}$:Mn^{4+} as color converter for warm W-LEDs: Structure and luminescence properties. *J. Alloys Compd.* 694: 1201.

299. Jin Y, Hu Y, Wu H, Duan H, Chen L, Fu Y, Ju G, Mu Z, and He M (2016) A deep red phosphor Li_2MgTiO_4:Mn^{4+} exhibiting abnormal emission: Potential application as color converter for warm w-LEDs. *Chem. Eng. J.* 288: 596.

300. Peng M, Yin X, Tanner PA, Brik MG, and Li P (2015) Site occupancy preference, enhancement mechanism, and thermal resistance of Mn^{4+} red luminescence in $Sr_4Al_{14}O_{25}$:Mn^{4+} for warm WLEDs. *Chem. Mater.* 27: 2938.

301. Ding X, Wang Q, and Wang Y (2016) Rare-earth-free red-emitting $K_2Ge_4O_9$:Mn^{4+} phosphor excited by blue light for warm white LEDs. *Phys. Chem. Chem. Phys.: PCCP* 18: 8088.

302. Wang L, Xu Y, Wang D, Zhou R, Ding N, Shi M, Chen Y, Jiang Y, and Wang Y (2013) Deep red phosphors $SrAl_{12}O_{19}$:Mn^{4+}, M (M = Li^+, Na^+, K^+, Mg^{2+}) for high colour rendering white LEDs. *Phys. Status Solidi A* 210: 1433.

303. Oh JH, Kang H, Ko M, and Do YR (2015) Analysis of wide color gamut of green/red bilayered freestanding phosphor film-capped white LEDs for LCD backlight. *Opt. Express* 23: A791.

304. Yoshimura K, Fukunaga H, Izumi M, Masuda M, Uemura T, Takahashi K, Xie RJ, and Hirosaki N (2016) White LEDs using the sharp β-Sialon:Eu phosphor and Mn-doped red phosphor for wide color gamut display applications. *J. Soc. Inf. Disp.* 24: 449.

305. Fang MH, Tsai YT, Sheu HS, Lee JF, and Liu RS (2018) Pressure-controlled synthesis of high-performance $SrLiAl_3N_4$: Eu^{2+} narrow-band red phosphors. *J. Mater. Chem. C* 6: 10174.

3 Phosphor-Converting LED for Broadband IR

Veeramani Rajendran, Ho Chang, and Ru-Shi Liu

CONTENTS

3.1 INTRODUCTION

The electromagnetic spectrum is all kinds of electromagnetic radiation that travel through space at the speed of light in a vacuum. It is very broad and spanning from the radio waves at one end to the gamma rays at the other end with several subdivisions. Infrared radiation is the subdivision that lies between the microwaves and visible light radiations of the electromagnetic spectrum. It is discovered by William Herschel in 1800 and proposed that the infrared can be identified as heat. Infrared radiation is invisible to the naked eye because the sense of the human eye falls after 683 lm/W at 555 nm (bright mode) in photopic vision and 1700 lm/W at 507 nm in scotopic vision (dark mode). Infrared light is significantly utilizing in modern technology for several sectors of application that include the pharmaceutical industry, agriculture, military, academics, and so on. The chief light sources accessible for infrared are the tungsten halogen lamps, laser diodes, light-emitting diode (LED), electrically modulate infrared thermal sources, and black-body systems. Unfortunately, these traditional light sources can't able to fit with the requirements of modern devices in terms of compact size, cost, portability, power consumption, tunable emission spectrum, lesser heat generation, and so on. Hence, a phosphor-converted near-infrared LED (pc-NIR LED) turns into a

DOI: 10.1201/9781003098676-3

FIGURE 3.1 Schematic illustration of the table of content for this chapter (phosphor-converting LED for broadband IR).

favorable light source for modern devices. As it is well-known that the phosphor is the combination of host and activator that will emit light once it is activated with light.

In this chapter, the activators and host systems of the phosphors for the NIR and mid-infrared emission based on a literature survey are disclosed. By following that, it is briefly discussed the emerging applications of broadband phosphor for pc-NIR LED applications with working principles and recent literature progress that includes the NIR spectroscopy, night vision, solar cell, plant growth, persistent luminescence, and anti-counterfeiting. The main theme of this chapter is shown as a schematic illustration in Figure 3.1. Besides, the current bottleneck and future research opportunities of the relevant applications also have been emphasized.

In general, infrared light is subclassified as NIR, mid-infrared, and far-infrared based on the photon energy (International Commission on Illumination) or wavelength (International Organization for Standardization 20473) as per Table 3.1. However, the researchers still adopt their own classification of infrared light in their research reports.

TABLE 3.1
Classification of Infrared Light According to CIE and ISO 20473

	CIE Classification		ISO 20473
	Wavelength (nm)	Photon Energy (THz)	Wavelength (nm)
Near IR, NIR	700–1400	215–430	780–3000
Mid-IR, MIR	1400–3000	100–215	3.0–5.0
Far IR, FIR	3000–0.1 mm	3–100	5.0–10.0

3.2 NEAR-INFRARED WINDOW

NIR light is partly invisible light with a wavelength range of 700–1400 nm. Both the transition metal elements (Mn^{4+}, Cr^{3+}) and rare-earth elements (Yb^{3+}, Pr^{3+}, Er^{3+}, Bi^{3+}, Nd^{3+}, Nb^{3+}) can function as the activator for the obtainment of infrared light in the phosphor materials (light conversion element). However, for the target of the broadband emission spectrum, the most common choice of the activator is Cr^{3+} among others due to the versatile emission characteristics of narrow as well as broad spectral distribution of infrared light. In addition, the greater sensitivity of the $5d$ degenerated orbitals of Cr^{3+} to the coordination environment offers an opportunity to flexibly tune the emission spectrum by modifying the crystal field strength. In an inorganic phosphor system, the combined effects of nephelauxetic effect and crystal field splitting are inevitable phenomenons as shown in Figure 3.2. The nephelauxetic effect means the reduction in the interelectronic repulsion energy of Cr^{3+} due to the formation of a chemical bond with the surrounding anions or ligand. The chemical bond formation results in the delocalization of the outer orbitals due to the sharing of some d electrons with the anion and p- and s-orbitals. After that, based on the coordination environment, the d orbital energies are split further due to the repulsion between d orbitals of the activator and the ligand. In the case of octahedral coordination, $d_{x^2-y^2}$ and d_{z^2} are closer to the activators, so it has higher energy than the other three orbitals d_{xy}, d_{yz}, and d_{xz}. The upper set consists of $d_{x^2-y^2}$ and d_{z^2} orbitals pointing directly toward the central metal ion and is assigned as e_g. On the other hand, the remaining three orbitals d_{xy}, d_{yz}, and d_{xz} are in the lower set and assigned as t_{2g}. The energy value between these two levels is known as crystal field splitting and is commonly denoted as Dq or Δ. The electronic transition between these two sets determines the spectral distribution of light. The crystal field splitting is sensitive to the coordination environment. In the case of tetrahedral coordination, the pattern of splitting is inverse (e_g gains lower energy, while t_{2g} gains higher energy with a different energy value of Dq or Δ).

3.2.1 TANABE-SUGANO DIAGRAM

Tanabe-Sugano (T-S) is a powerful tool to understand the electronic transition of transition metal ions in more detail. There is a dedicated diagram for each electronic configuration (d^2 to d^8). One can refer to any chemistry textbook for a better understanding of the construction of T-S diagrams and its energy term symbol notations. However, the T-S diagram of d^3 electronic configuration is

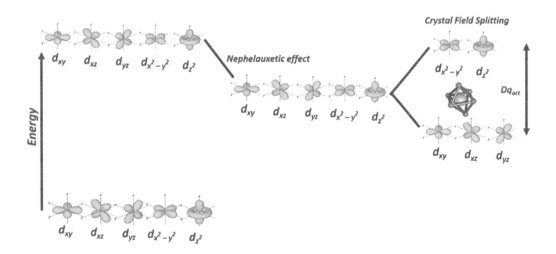

FIGURE 3.2 Schematic illustration of nephelauxetic effect and crystal field splitting for d–d transition in the octahedral coordination.

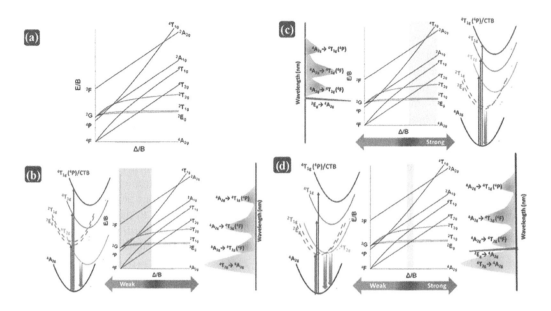

FIGURE 3.3 (a) Tanabe-Sugano diagram of d^3 electronic configuration with the schematic illustration of transition bands in the region of (b) weak crystal field, (c) strong crystal field, and (d) intermediate crystal field.

explained slightly in this chapter to understand the tuning mechanism of crystal field engineering in Cr^{3+} for the broadband emission. The ground state is $^4A_{2g}$, which is the element of the free-ion state ^4F of d^3 ion. At the small crystal field value (on the left-hand side) $Dq<0.4$ as shown in Figure 3.3(a), $^4T_{2g}$ and $^4T_{1g}$ are the first and second excited states, respectively, which are also the elements of the free-ion state ^4F. The slopes of these higher energy states $^4T_{2g}$ and $^4T_{1g}$ (^4P) are strongly dependent on the value of Dq. On the other hand, at $0.4<Dq<1.4$ as shown in Figure 3.3(b), the first and second excited states are $^4T_{2g}$ and $^4T_{1g}$ as consistent as $Dq<0.4$. However, the third excited stated is turned as ^2E_g. Since the optical transition is allowed to $^4A_{2g}\rightarrow^4T_{2g}$, $^4T_{1g}$ (^4F), and $^4T_{1g}$ (^4P), the three optical transition bands can be observable in the excitation spectrum. On the other hand, the optical transitions of $^4A_{2g}\rightarrow^2E_g$, $^2T_{1g}$, and $^2T_{2g}$ are both parity and spin forbidden. Therefore, the optical transition peak can't be observable in the excitation spectrum. Nevertheless, the lowest excited state is $^4T_{2g}$ till $Dq<1.4$. So the optical transition from $^4T_{2g}\rightarrow^4A_{2g}$ results in the broader emission spectral distribution of infrared light. In contrast at the higher crystal field value of $Dq>2.4$ as shown in Figure 3.3(c), the first lowest excited state is turned as ^2E_g followed by $^4T_{2g}$, $^4T_{1g}$ (^4F), and $^4T_{1g}$ (^4P) as the subsequent excited stated. Aforementioned, the optical transition of $^4A_{2g}\rightarrow^2E_g$, $^2T_{1g}$ is parity and spin forbidden. Hence, there will be no difference in the excitation spectrum as compared with the weak crystal field case. In contrast, ^2E_g can gain energy from the local lattice vibrations of the host and still emit the infrared light like a sharp line or restricted narrow spectrum. On the other hand, in some phosphors systems, this optical transition can be observable as sharp tips in the excitation band that is commonly pronounced as Fano-antiresonance lines. The crystal field value range of $1.4<Dq<2.3$ is the region of interest in the optical spectroscopy of Cr^{3+} because the higher lying energy states intersect together around this region as shown in Figure 3.3(d). It is often difficult to locate the exact position of the lower excited states. In all cases, the lowest excited states ^2E_g and $^4T_{2g}$ are located very close to each other. Hence, the resultant emission spectrum exhibits the sharp line embedded on the broadband spectral distribution of infrared light as the optical transitions from ^2E_g and $^4T_{2g}$ to $^4A_{2g}$. On the other hand, it is also important to consider that the T-S diagram is constructed on the assumption of ideal octahedral symmetry (Cr^{3+} in O_h) and no spin-orbit interactions. However, it is not always being the ideal O_h symmetry in the real crystal, so the deviation in the symmetry can further split these energy states. In those situations, the resultant emission spectrum can contain several emission peaks.

3.2.2 HOST SYSTEMS FOR NEAR-INFRARED WINDOW

Several chemical compositions activated with Cr^{3+} are known for NIR emissions. Table 3.2 summarizes the list of the phosphors reported for the NIR emission in the various host (categorized based on the crystal system). On the other hand, NIR emission is also accountable by the other activators of transition metal elements and rare-earth elements. However, on the considerations of the spectral distribution of NIR light (full-width-half-maximum, FWHM), only a few chemical system with Cr^{3+} as an activator can be considerable.

3.2.2.1 Cubic, Spinel, and Inverse Spinel

The cubic crystal system is the simplest and common conceptual model among the seven crystal systems with a unit cell resembling the cube. The phosphors reported for the infrared light with the cubic crystal system are in three kinds, namely garnet, spinel, and inverse spinel structure. The garnet structured compounds are in the space group of $Ia\bar{3}d$ with the generalized chemical formulae of $A_3B_5O_{12}$ or $A_3B_2C_3O_{12}$, where A, B, and C are cations. In the crystal, cation A is in dodecahedral coordination with oxygen atoms $[AO_8]$, while the other cations B and C are in the octahedral $[BO_6]$ and tetrahedral $[CO_4]$ coordination, respectively, with oxygen atoms. Moreover, cation A in dodecahedral coordination sharing one common edge with cation B in octahedral and with the cation C in tetrahedral coordination. On the other hand, the cations B and C are sharing the common-edge oxygen atom. When Cr^{3+} is doped into the crystal lattice, it will occupy the octahedral coordination in the cation B site than the tetrahedral. As per the point charge model, the crystal field splitting Dq is inversely proportional to the bond length between the central ion (B/Cr) and the anion (oxygen). In simple words, the longer bond length can lead to the weaker crystal field splitting, while the shorter bond length results in the stronger crystal field. Hence, the structural modifications in the octahedral site can affect the crystal field splitting that in turn can tune the spectral distribution of the infrared light. Cation modification is the simplest and straightforward method to alter Dq values. By this strategy, the Dq/B values are tuned from ~2.39 to 2.62 and even more. At the same time, the error bar in the calculation of Racah parameters is highly variable among the reported values due to misleading in considering the energy position of the transitions. Hence, it can impact the predicted values of Dq/B in the T-S diagram. However, the significant differences can be notable in the cation substitution strategies especially in terms of centered emission wavelength. For example, the garnet system of $Y_3Al_2Ga_3O_{12}$ exhibits the NIR– luminescence in the range of 650–950 nm with the centered emission wavelength of ~690 nm.[6] The substitution of Sc^{3+} in the position of cation B causes the local lattice and volume expansion of the unit cell that in turn decreases the crystal field strength of Cr^{3+}. Hence, the centered emission wavelength of $Y_3Sc_2Ga_3O_{12}$ is red shifted to ~740 nm.[14] Technically, the increased distance between the central metal ions and anion length causes the redshift to decrease the crystal field splitting energy between t_{2g} and e_g states. This in turn encourages the possibility of high spin with broadband emission.

On the other hand, the structural modifications on the cation A site results in the indirect crystal field engineering effects in cation B. It is mentioned earlier, cation A is sharing a common edge separately with cations B and C, respectively. Hence, the expansion of cation A local lattice indirectly induces the local lattice expansion in the cation B, while the cation C local lattice is shrunk due to the sharing of a common apex oxygen atom with cation B and to maintain the structural rigidity. In the garnet system of $A_3Sc_2Ga_3O_{12}$ (A=Lu^{3+}, Gd^{3+}, Y^{3+}, La^{3+}), the Shannon ionic size is increased as follows: $Lu^{3+}<Y^{3+}<Gd^{3+}<La^{3+}$. With increasing the ionic size of atoms in cation A, Sc–O bond length is also increased as follows: 1.993 Å<2.018 Å<2.041 Å<2.086 Å. The increasing bond length modifies the values of crystal field strength that in turn shifts the spectrum toward the higher wavelength. Cr^{3+}-doped $Lu_3Sc_2Ga_3O_{12}$ garnet shows the emission range of 700–1000 nm with a centered

TABLE 3.2

List of the Phosphors Reported for the NIR Emission in the Various Hosts

Chemical Composition	Activator	Emission Range (nm)	λ_{em} (nm)	λ_{exc} (nm)	Crystal System, Space Group	Application	Ref.
$Y_3Al_2Ga_3O_{12}$	Cr^{3+}	600–800	690	455	Cubic, $Ia\overline{3}d$	N/A	[1]
$Ca_3Ga_2Ge_3O_{12}$	Cr^{3+}	650–1100	749, 803, 907	254	Cubic, $Ia\overline{3}d$	N/A	[2]
$Zn_3Al_2Ge_3O_{12}$	Cr^{3+}	650–750	697	397	Cubic, $Ia\overline{3}d$	Persistent luminescence	[3]
$Y_3Ga_5O_{12}$	Cr^{3+}	650–950	715	450	Cubic, $Ia\overline{3}d$	N/A	[4]
$Gd_3Sc_2Ga_3O_{12}$	Cr^{3+}	650–950	750	450	Cubic, $Ia\overline{3}d$	N/A	[4]
$La_3Sc_2Ga_3O_{12}$	Cr^{3+}	700–1000	722	440	Cubic, $Ia\overline{3}d$	Spectroscopy	[5]
$Y_3Sc_2Ga_3O_{12}$	Cr^{3+}	640–850	740	450	Cubic, $Ia\overline{3}d$	Spectroscopy	[5]
$Gd_3Sc_2Ga_3O_{12}$	Cr^{3+}	650–860	754	460	Cubic, $Ia\overline{3}d$	Spectroscopy	[5]
$Lu_3Sc_2Ga_3O_{12}$	Cr^{3+}	620–800	818	480	Cubic, $Ia\overline{3}d$	Spectroscopy	[5]
$Y_3Al_2Ga_3O_{12}$	Cr^{3+}	650–950	690	254	Cubic, $Ia\overline{3}d$	Persistent luminescence	[6]
$Ca_2LuZr_2Al_3O_{12}$	Cr^{3+}	650–850	754, 813	460	Cubic, $Ia\overline{3}d$	Spectroscopy	[7]
$Na_2CaSn_2Ge_3O_{12}$	Cr^{3+}	700–850	810	470	Cubic, $Ia\overline{3}d$	Persistent luminescence	[8]
$Na_2CaTi_2Ge_3O_{12}$	Cr^{3+}	700–850	780	470	Cubic, $Ia\overline{3}d$	Persistent luminescence	[8]
$Ca_2LuHf_2Al_3O_{12}$	Cr^{3+}	700–1000	855–785	460	Cubic, $Ia\overline{3}d$	Bio-imaging	[9]
$Gd_3Sc_2Ga_3O_{12}$	Cr^{3+}, Al^{3+}	650–850	761–715	460	Cubic, $Ia\overline{3}d$	Spectroscopy	[10]
$Y_2CaAl_4SiO_{12}$	Cr^{3+}	600–1100	689, 738–768	440	Cubic, $Ia\overline{3}d$	Night vision	[11]
$Ca_3Sc_2Si_3O_{12}$	Cr^{3+}	650–1000	770	460	Cubic, $Ia\overline{3}d$	Night vision	[12]
$Mg_3Y_2Ge_3O_{12}$	Cr^{3+}	650–1250	771–811	436	Cubic, $Ia\overline{3}d$	Spectroscopy	[13]
$Y_3Sc_2Ga_3O_{12}$	Cr^{3+}	650–950	740–750	450	Cubic, $Ia\overline{3}d$	Persistent luminescence and night vision	[14]
$Ca_2LuHf_2Al_3O_{12}$	Cr^{3+}	650–1100	775	460	Cubic, $Ia\overline{3}d$	N/A	[15]
$Ca_2LuHf_2Al_3O_{12}$	Cr^{3+}, Ce^{3+}	650–1100	769–797	460	Cubic, $Ia\overline{3}d$	N/A	[15]

(Continued)

TABLE 3.2 (*Continued*)
List of the Phosphors Reported for the NIR Emission in the Various Hosts

Chemical Composition	Activator	Emission Range (nm)	λ_{em} (nm)	λ_{exc} (nm)	Crystal System, Space Group	Application	Ref.
$Ca_3LuScGa_2Ge_2O_{12}$	Cr^{3+}	650–1100	800	465	Cubic, $Ia\overline{3}d$	Spectroscopy	[16]
$MgY_2Al_4SiO_{12}$	Cr^{3+}	660–720	689	580	Cubic, $Ia\overline{3}d$	N/A	[17]
$Y_3Ga_5O_{12}$	Cr^{3+}, Yb^{3+}	600–850, 950–1050	688, 1032	451	Cubic, $Ia\overline{3}d$	N/A	[18]
$Ca_3Ga_2Ge_3O_{12}$	Cr^{3+}, Bi^{3+}	600–850	739	440	Cubic, $Ia\overline{3}d$	N/A	[19]
$Ca_3Sc_2Si_3O_{12}$	$Cr^{3+}, Nd^{3+}, Yb^{3+}, Ce^{3+}$	700–1100	770, 1060, 975	450	Cubic, $Ia\overline{3}d$	Spectroscopy	[20]
$Ca_3Ga_2Ge_3O_{12}$	Yb^{3+}, Pr^{3+}	700–1100	975	254	Cubic, $Ia\overline{3}d$	Persistent luminescence	[21]
$Y_3Ga_5O_{12}$	$Ce^{3+}, Pr^{3+}, Yb^{3+}$	500–650, 900–1100	1029	450	Cubic, $Ia\overline{3}d$	N/A	[22]
$Ca_2LuZr_2Al_3O_{12}$	Cr^{3+}, Yb^{3+}	730–880, 900–1100	780, 1032	455	Cubic, $Ia\overline{3}d$	Spectroscopy	[23]
$Ca_3Ga_2Ge_3O_{12}$	Yb^{3+}, Pr^{3+}	700–1100	975	254	Cubic, $Ia\overline{3}d$	Persistent luminescence	[21]
$Li_6CaLa_2Sb_2O_{12}$	Mn^{4+}, Al^{3+}	660–750	715	320	Cubic, $Ia\overline{3}d$	Plant growth	[24]
$Li_3Zn_8Al_5Ge_9O_{36}$	Cr^{3+}	650–800	700	258	Cubic	Persistent luminescence	[25]
$Ca_{14}Zn_6Al_{10}O_{35}$	Cr^{3+}	650–750	711	415	Cubic, $F23$	Persistent luminescence	[26]
$Zn_3Ga_2Ge_2O_{10}$	Cr^{3+}	450–850	713, 520	254	Cubic, $Fd\overline{3}m$	Persistent luminescence	[27]
$Zn_3(Al_{0.8}Ga_{0.2})_2Ge_2O_{10}$	Cr^{3+}	600–900	705, 721	410	Cubic, $Fd\overline{3}m$	N/A	[28]
$Zn_3(Al_{0.9}In_{0.12})_2Ge_2O_{10}$	Cr^{3+}	600–900	705	410	Cubic, $Fd\overline{3}m$	N/A	[28]
Zn_2SnO_4	Cr^{3+}	650–1200	800	320	Cubic, $Fd\overline{3}m$	Persistent luminescence	[29]
$Zn_{1.8}Al_{0.4}Sn_{0.8}O_4$	Cr^{3+}	650–1200	722	320	Cubic, $Fd\overline{3}m$	Persistent luminescence	[29]
$Zn_{0.8}Al_2Ge_{0.2}O_4$	Cr^{3+}	600–750	706, 715, 720, 668, 678	442	Cubic, $Fd\overline{3}m$	Persistent luminescence	[30]
$Zn_{1.25}Ga_{1.5}Ge_{0.25}O_4$	Cr^{3+}	650–800	698, 720	280	Cubic, $Fd\overline{3}m$	Persistent luminescence	[31]
$InMgGaO_4$	Cr^{3+}	650–1200	712, 736, 915	581	Cubic, $Fd\overline{3}m$	N/A	[32]
$In_{0.9}Al_{0.1}MgGaO_4$	Cr^{3+}	650–1200	712, 736, −915	581	Cubic, $Fd\overline{3}m$	N/A	[32]
K_3AlF_6, K_3GaF_6	Cr^{3+}	650–1000	750	442	Cubic, $Fd\overline{3}m$	Spectroscopy	[33]

(*Continued*)

TABLE 3.2 (Continued)

List of the Phosphors Reported for the NIR Emission in the Various Hosts

Chemical Composition	Activator	Emission Range (nm)	λ_{em} (nm)	λ_{exc} (nm)	Crystal System, Space Group	Application	Ref.
$MgAl_2O_4$	Mn^{4+}	700–950	528, 825	450	Cubic, $Fd\bar{3}m$	Night vision	[34]
$LiGa_5O_8$	Cr^{3+}, Sn^{4+}	650–800	717, 727, 736, 703	419	Cubic, $P4_332$	Persistent luminescence	[35]
$LiGa_4(Ge, Si)O_8$	Cr^{3+}	650–850	718	416	Cubic, $P4_332$	Persistent luminescence	[36]
$Ca_{14}Zn_6Ga_{10}O_{35}$	Mn^{4+}–Mn^{5+}	650–800, 1100–1300	677, 687, 695, 704, 713, 1152	330	Cubic, $P23$	Night vision	[37]
$MgTa_2O_6$	Cr^{3+}	700–1150	910, 834	460	Tetragonal, $P4_2/mnm$	Bio-imaging	[38]
Sr_2MgWO_6	Cr^{3+}	700–1000	763, 755, 800	371	Tetragonal, $I4/m$	Persistent luminescence	[39]
$K_2Ga_2Sn_6O_{16}$	Cr^{3+}	650–1300	830	430	Tetragonal, $I4/m$	Persistent luminescence and bio-imaging	[40]
$Sr_{2.95}Si_{0.95}Al_{0.05}O_5$	Ce^{3+}, Nd^{3+}	900–1200	1093	410	Tetragonal, $P4/nnc$	N/A	[41]
$CaMoO_4$	Tb^{3+}, Yb^{3+}	950–1050	489, 545, 585, 622, 669, 997	306	Tetragonal, $I4_1/a$	Solar cell	[42]
$CaGdAlO_4$	Mn^{4+}, Yb^{3+}	650–750, 925–1100	712, 981	300	Tetragonal, $I4/mmm$	Solar cells	[43]
$CaTiO_3$	Yb^{3+}, Bi^{3+}	900–1100	978, 998	374	Tetragonal, $P4mm$	Persistent luminescence	[44]
$Bi_2Ga_4O_9$	Cr^{3+}	650–900	800	442	Orthorhombic, $Pbam$	Optical thermometry	[45]
$Bi_2Al_4O_9$	Cr^{3+}	650–900	705.5, 699.0	442	Orthorhombic, $Pbam$	Optical thermometry	[46]
$KTiOPO_4$	Cr^{3+}	700–1000	833	–	Orthorhombic, $Pna2_1$	N/A	[47]
$CaGa_2O_4$	Yb^{3+}, Cr^{3+}	650–1100	980	266	Orthorhombic, $Pna2_1$	Persistent luminescence	[48]
$CaSc_2O_4$	Ce^{3+}, Yb^{3+}	450–700, 900–1100	511, 563, 976, 1000	466	Orthorhombic, $Pna2_1$	Solar cell	[49]
$NaScSi_2O_6$	Cr^{3+}	700–1000	800	450	Monoclinic, $C2/c$	Spectroscopy	[50]
$LiInSi_2O_6$	Cr^{3+}	700–1100	840	460	Monoclinic, $C2/c$	Spectroscopy	[51]
$NaScGe_2O_6$	Cr^{3+}	700–1250	895	490	Monoclinic, $C2/c$	Persistent luminescence	[52]
$Mg_3Ga_2Ge_3O_{16}$	Cr^{3+}	600–850	693, 715	267	Monoclinic, $C2/c$	Optical information storage and read out	[53]
Lu_2GeO_5	Bi^{3+}, Yb^{3+}	300–450, 950–1100	370, 976	276	Monoclinic, $C2/c$	N/A	[54]
La_2MgZrO_6	Cr^{3+}	650–1200	825	460	Monoclinic, $P2_1/n$	N/A	[55]
Ca_2MgWO_6	Cr^{3+}	650–1000	803	460	Monoclinic, $P2_1/n$	Persistent luminescence	[56]
$Ca_4O(PO_4)_2$	Eu^{2+}, Lu^{3+}, La^{3+}, Gd^{3+}, Ce^{3+}, Tm^{3+}, Y^{3+}	550–850	690	467	Monoclinic, $P2_1$	Persistent luminescence	[57]

(Continued)

TABLE 3.2 (Continued)

List of the Phosphors Reported for the NIR Emission in the Various Hosts

Chemical Composition	Activator	Emission Range (nm)	λ_{em} (nm)	λ_{exc} (nm)	Crystal System, Space Group	Application	Ref.
$(Mg_{0.5}Zn_{0.5})_{2.97}(PO_4)_2$	Mn^{2+}	550–850	630, 730	402, 417	Monoclinic, $P2_1/c$	N/A	[58]
$Ba_2Y(BO_3)_2Cl$	Ce^{3+}, Yb^{3+}	900–1100	471, 972	355	Monoclinic, $P2_1/m$	Solar cell	[59]
Gd_2ZnTiO_6	Mn^{4+}, Yb^{3+}	650–720, 925–1075	670, 685, 690,704, 980	365	Monoclinic, $P2_1/c$	N/A	[60]
$TbZn(B_5O_{10})$	Yb^{3+}	920–1060	974	365	Monoclinic, $P2_1/c$	N/A	[61]
$SrGa_{12}O_{19}$	Cr^{3+}	650–950	750	425	Hexagonal, $P6_3/mmc$	Persistent luminescence	[62]
$LaMgGa_{11}O_{19}$	Cr^{3+}	650–900	775	–	Hexagonal, $P6_3/mmc$	Near-infrared spectroscopy	[63]
$K_3LuSi_2O_7$	Eu^{2+}	550–850	760	460	Hexagonal, $P6_3/mmc$	Night vision	[64]
$BaZrSi_3O_9$	Cr^{3+}	700–1000	800	455	Hexagonal, $P\bar{6}c2$	Solar cell	[65]
YBO_3	Ce^{3+}, Yb^{3+}	950–1050	971	360	Hexagonal, $P6_3/m$	Solar cell	[66]
$LaMgAl_{11}O_{19}$	Cr^{3+}, Yb^{3+}	650–1090	698, 980	455	Hexagonal, $P6_3/mmc$	Solar cell	[67]
$GdAl_3(BO_3)_4$	Ce^{3+}, Yb^{3+}	900–1100	975	320	Rhombohedral, $R32$	c-Si solar cells	[68]
$ScBO_3$	Cr^{3+}	700–1000	800	470	Rhombohedral, $R\bar{3}c$	Bio-imaging	[69]
$ScBO_3$	Cr^{3+}	700–950	800	450	Rhombohedral, $R3m$	Spectroscopy	[70]
$GdAl_3(BO_3)_4$	Ce^{3+}, Yb^{3+}	900–1100	975	320	Rhombohedral, $R32$	c-Si solar cells	[68]
$La_3GaGe_5O_{16}$	Cr^{3+}	650–1050	700, 780	470	Triclinic, $P\bar{1}$	Spectroscopy	[71]
$Ca_3Ga_2Ge_4O_{14}$	Cr^{3+}	650–900	745	470	Trigonal, $P321$	N/A	[72]
$LaAlO_3$	Mn^{4+}	650–800	697, 704, 710, 718, 724, 731	335	Trigonal, $R\bar{3}c$	Persistent luminescence	[73]
$La_3Ga_5GeO_{14}$	Cr^{3+}	600–1200	850, 920	473	Trigonal, $P321$	Spectroscopy	[74]
$YAl_3(BO_3)_4$	Cr^{3+}, Yb^{3+}	670–800, 950–1050	710, 980	450	Trigonal, $R32$	Spectroscopy	[50]
La_2MgGeO_6	Mn^{4+}	650–750	678, 684, 695, 705, 709.5	450	Trigonal, $R3$	Persistent luminescence	[75]
La_2MgGeO_6	Cr^{3+}	640–750	710	310	Trigonal, $R3$	Persistent luminescence	[76]

emission at 722 nm. On replacing Lu^{3+} with Y^{3+}, the centered emission is shifted around ~740 nm with an increase in *fwhm*. By following this, the centered emission is redshifted to ~754 and ~818 nm for the replacements of Gd^{3+} and La^{3+}, respectively.[5]

Even though the cation substitution strategy offers greater tuning of crystal field engineering and wider selectivity of atoms, the FWHM is always restricted within the range of 650–950 nm in most garnet systems. Hence, the co-unit substitution strategy is proposed in the garnet system for an enhancement of *fwhm* and luminescence intensity. In the co-unit substitution strategy, the cations A and C are simultaneously substituted with the bigger ionic-sized divalent cation and smaller ionic-sized tetravalent cation, respectively. This concurrent substitution allows much space for cation B in octahedral coordination to expand its local lattice due to the shrinking of tetrahedral lattice in cation C site and only slighter expansion of dodecahedral lattice in cation A. Xie *et al.* [11] reported that the substitution of Ca^{2+} for Y^{3+} and Si^{4+} for Al^{3+} in the garnet structure of $Y_3Al_5O_{12}$ can result in the broader NIR light spectrum in the range of 600–1100 nm. The chemical formula reported for the co-unit substitution is $Y_{3-x}Ca_xAl_{5-x}Si_xO_{12}:Cr^{3+}$. At $x=0$, there are several emission peaks with a restricted FWHM of 40 nm. At the value of $x=1$, there are intense R line and broadband emissions centered at 689 and 744 nm, respectively, with an enhanced FWHM of 160 nm. On the other hand, Dq/B values also fall from 2.62 ($x=0$) to 2.43 ($x=1$) that further supports the spectrum changes and effectiveness of the co-unit substitution strategy.

Zhang *et al.* [9] also reported that the co-unit substitution technique in the garnet structure with the chemical composition of $Ca_{3-x}Lu_xHf_2Al_{2+x}Si_{1-x}O_{12}:Cr^{3+}$. It is reported that the substitution units of $[Lu^{3+}–Al^{3+}]$ for $[Ca^{2+}–Si^{4+}]$ alter the local lattice units in the crystal structure that in turn enhance the luminescence intensity and thermal stability. The introduction of Lu^{3+} disrupts the ordered arrangement of CaO_8 into disordered by the broadening of the Raman peaks at 176 and 216 cm^{-1}. In contrast, the disordered arrangements in tetrahedral units of $[Si/AlO_4]$ are restructured into ordered arrangements by the emergence of a sharp peak at 764 cm^{-1} in the series of 690–900 cm^{-1}. As the effect, these local structural modifications create two luminescence Cr^{3+} centers (Cr1 and Cr2) that are responsible for the enhancement of luminescence intensity by about 81.5 times. The Cr1 center is located at 785 nm even at $x=0$ excitable by 280 nm, while the Cr2 is at 855 nm that is a more prevailing emission for 480 nm excitation. With increasing values of x, the ratio of Cr1/Cr2 is also increased that accounts for the blue shift from 855 to 785 nm. Moreover, the chemical unit substitution of $[Lu^{3+}–Al^{3+}]$ for $[Ca^{2+}–Si^{4+}]$ also increases the thermal stability of the phosphor. With a respective to the room temperature luminescence intensity, the co-unit substituted samples ($x=1$) still maintain the luminescence intensity of 87.1% even at higher temperatures due to the slower non-radiative process from Cr1, while, on the other hand, $x=0$ samples show only 40.8%.

Other than the co-unit substitution approach, the modest way to extend the FWHM value of the NIR spectrum is the addition of rare-earth elements as co-dopants. In general, the rare-earth elements can be excitable by a wavelength of more than 600 nm. Hence, the introduction of rare-earth elements into the garnet system results in the energy transfer from Cr^{3+} to the rare-earth element that in turn extends the FWHM. Among the rare earths, Yb^{3+} is the most preferable candidate as co-dopant and sensitizers due to the efficient superimposition of Cr^{3+} emission and Yb^{3+} excitation band. Under the excitation of 440 nm, Yb^{3+} un-doped $Ca_2LuZr_2Al_3O_{12}:0.08\ Cr^{3+}$ garnet shows the NIR emission in the range of 730–880 nm with the centered emission and FWHM of 780 and 150 nm, respectively. Yb^{3+} substitution in $Ca_2LuZr_2Al_3O_{12}:0.08\ Cr^{3+}$ results in emissions in the range of 900–1100 nm. As the whole, $Ca_2LuZr_2Al_3O_{12}:0.08\ Cr^{3+}, 0.01\ Yb^{3+}$ covers the super broadband NIR emission in the range of 700–1100 nm due to the energy transfer from Cr^{3+} to Yb^{3+} with super FWHM value of 330 nm and internal quantum efficiency of ~77.2%.[23] Even though the emission range is extended beyond 900 nm, the spectral distribution of infrared light is not homogenous. Another advantage of Yb^{3+} substitution is the enhancement of thermal stability property. Yb^{3+} is excitable by Cr^{3+} and thermally stable than Cr^{3+} due to the competition between energy transfer and thermal de-excitation in Cr^{3+}-emitting states. The energy transfer process leads to the presence of Yb^{3+} ions as the closed pairs of $Cr^{3+}–Yb^{3+}$. Hence, energy can be easily transferred to Yb^{3+} toward thermal de-excitation in Cr^{3+} emissions. Yb^{3+} additions can overwhelm the non-radiative emissions with an emission intensity greater than the room temperature in the range of 100–400 K.

The spinel and inverse spinel structure also belong to the cubic crystal system, but it differs from the garnet structure of its distinct coordinating features. Unlike the garnet structure, there are tetrahedral and octahedral environments only due to the possession of one divalent and trivalent cations, respectively. It is generally represented by the chemical formula AB_2O_4, where A and B are the divalent and trivalent cations, respectively. In spinel, the divalent cation occupies 1/8 of tetrahedral space, while the trivalent cations occupy 1/2 of octahedral spaces. On the other hand, the inverse spinel also has an exact chemical composition as spinel. However, in inverse spinel, one part of the trivalent cation will occupy the tetrahedral sites, while the other part will be in the octahedral site. It can be denoted as $(A, B)_{tetra}B_{octa}O_4$. The crystal compounds can be classified as spinel and inverse structures based on their chemical metal elements. There is a greater tendency for inverse spinel structure formation if the divalent cation (A) is a transition metal element while the trivalent cation (B) is non-transition metal or transition metal element with d^0, d^5, d^{10} configurations. The space group of $Fd\bar{3}m$ is typically assigned to the spinel system for all inversion degrees, on the assumption that all cations are randomly distributed within each sub-lattice. However, in the inverse system, the cation arrangement is affected by the inversion degree.

The coexistence occupation by trivalent cations in tetrahedral and octahedral sites in inverse spinel leads to the generation of intrinsic and antisite defects in the system. This makes the inverse spinel structure the potential candidate for the NIR persistent luminescence. As the interest for persistent luminescence, the general method is the generation of point defect (vacancy defect) or oxygen deficiency by the substitution of trivalent or tetravalent cations in the elemental position of tetravalent or trivalent cations, respectively. As it is previously discussed, these substitutions are not only generating defects but also affect the crystal field strength of the activator. Zn_2SnO_4 shows the NIR luminescence in the range of 650–1200 nm with the centered emission at 800 nm. The crystal field strength (Dq/B) was calculated as 2.292. On the substitution of the Al^{3+} for Sn^{4+} ions, Dq/B value is enhanced to 2.925 for $Zn_{1.8}Al_{0.4}Sn_{0.8}O_4$ with the central emission shifted to 722 nm with the emergence of R line at 702 nm.[29] In addition to the Zn vacancies and Zn interstitials defects, these nonequivalent substitutions deepen the trap concentrations especially at higher concentrations of Al^{3+}. Hence, the crystal field engineering in the inverse structure for the stronger crystal field is advantageous for the persistent luminescence applications. Similarly, Pan et al. [77] investigated the solid solubility between the spinel and inverse spinel systems of $ZnGa_2O_4$ and Zn_2GeO_4 and developed the zinc gallogermanates with the generalized chemical formula of $Zn_xGa_yGe_zO_{(x+(3y/2)+2z)}:Cr^{3+}$ where x, y, and z are integers from 1 to 5. In the series of the zinc gallogermanates, $Zn_3Ga_2Ge_2O_{10}:0.5$ Cr^{3+} shows the emissions at 650–1000 nm with the centered emission at 715 nm. Unfortunately, the crystal structure is not solved yet due to the similar X-ray scattering factors of Zn, Ga, and Ge. By following that, the series of samples was reported by several researchers in the zinc gallogermanates, for example, $Zn_3(Al_{0.8}Ga_{0.2})_2Ge_2O_{10}$, [28] and $Zn_3Al_2Ge_3O_{12}$.[3]

Li et al. [32] reported the isostructural transformation of $InMgGaO_4$ from rhombohedral (space group, $R\bar{3}m$) to the spinel (space group, $Fd\bar{3}m$) on the incorporation of Cr^{3+} due to the breakage of one Mg/Ga–O bond. The ionic size difference between Cr^{3+} and In^{3+} in octahedral coordination is almost 23%. Hence, the local lattice of the octahedral will shrink, while the hexahedron of [Mg/GaO$_5$] can be attracted toward the octahedron, which contributes to the rupture of the longer Mg/Ga–O bonds to form the spinel structure. The increasing of Cr^{3+} ions concentrations leads to the sequential transforming from the stronger crystal field (~2.55) to intermediate crystal field (~2.22) followed by the weaker values with the systematic changes of hexahedral [Mg/GaO$_5$] to tetrahedral [Mg/GaO$_4$]. On the other hand, under an excitation wavelength of 580 nm, it shows the broadband emission 650–1200 nm with R line at 713 nm, two broad bands at 736 and 915 nm, respectively. It is postulated that the broadband emission at 915 nm should from inverted [Mg/GaO$_6$], while the emission at 715 and 736 nm can assign to the intermediate crystal field the regular octahedron [InO$_6$]. The assignment is based on the fact that the inverted octahedron shows a much weaker crystal field than the regular octahedron. Then, the redshift can be due to the increased nephelauxetic effect. It is calculated that the nephelauxetic ratio of Cr^{3+} (1.66) is smaller than In^{3+} (1.78). Hence, the position of excited states decreased with the increase of covalence that can result in a redshift of the R line. Moreover, the structural distortion in the regular and inverted octahedron due to Cr^{3+} amplifies the redshift at the broadband emission due to the reduction in the crystal field.

3.2.2.2 Tetragonal

Structurally, the tetragonal unit cell is identical to a cubic crystal with three axes that meet at 90°. However, the length of one axis is longer than the other two axes that are the same length. This makes the unit cell a rectangular prism. There are two basic types of Bravais lattice: primitive (stretching from a simple cube) and body-centered (stretching either the face-centered or body-centered cubic lattice). It is also important to note that the face-centered tetragonal is equivalent to the body-centered tetragonal. It means that the Cr^{3+} behavior in body-centered tetragonal can be more identified as a garnet structure.

Sr_2MgWO_6 crystallizes in the space group of $I4/m$ consists of two octahedral sites of $[MgO_6]$ and $[WO_6]$, and one 12-coordinated polyhedron site of $[SrO_{12}]$. Cr^{3+} is considered to be a substitute at $[MgO_6]$ due to the nearly identical ionic size and feasibility of a six-coordinate environment. Sr_2MgWO_6:Cr^{3+} shows the intense broad emission at 650–1050 nm range with the distinct spectral features of several narrow lines around 800 nm, R lines, and $N2$ lines at 763 and 765 nm, respectively.[39] In contrast, another system of Ca_2MgWO_6:Cr^{3+} shows the single-centered broadband emission at 800 nm in the range of 650–1050 nm.[56] These two systems can be distinguished only in terms of emission intensity and spectral feature that is due to the different values of Dq/B (Ca_2MgWO_6:Cr^{3+}=2.006; Sr_2MgWO_6:Cr^{3+}=2.924). It is a clear insight that Dq/B is controllable by the crystal system and coordination symmetry. Even though Cr^{3+} occupies the Mg^{2+} site in both systems along the c axis, the slightly higher ionic size of Sr^{2+} results in the tetragonal lattice with higher symmetry than the monoclinic system of Ca_2MgWO_6:Cr^{3+}. Hence, the higher symmetry of the tetragonal system results in a stronger crystal system with intensified emission. Also, the broadening can be related to the electron-phonon coupling that requires further analysis in both systems.

The tetragonal host lattice with one-dimensional chain structure is also interesting kind for its luminescent properties due to the positive effect of carrier mobility. The chemical formula of $A_2B_8O_{16}$ (A=K^+, Cs^+; B=Ga^{3+}, Mn^{4+}, Sn^{4+}, Ti^{4+}) is generally called hollandite-type compounds, which is allotropic from tetragonal ($I4/m$) to the monoclinic ($I2/m$) with an increase of ionic size B. $K_2Ga_2Sn_6O_{16}$ belongs to the tetragonal class with one-dimensional framework of double-chained (Ga, Sn)O_6 octahedrons edge shared with adjacent ones along the c axis. $K_2Ga_2Sn_6O_{16}$:Cr^{3+} shows the broadband emission around 650–1300 nm centered at 830 nm.[40] Besides, the nephelauxetic effect is increased when the higher electronegativity element of Ga^{3+} (1.81) and Sn^{4+} (1.96) is replaced by the lower electronegativity element of Cr^{3+} (1.66) and thus makes the centered emission position redshifted on increasing the concentrations of Cr^{3+}. On the other hand, the Dq/B value is increased due to the shrinkage of local lattice, when bigger Sn^{4+} is replaced by smaller Cr^{3+} ions. To overcome this, the Dq/B value can be lowered from ~2.71–3.00 to ~2.96–2.78 by substitution of Gd^{3+} for the same concentration of Cr^{3+}.[40] The Gd^{3+} can decrease the lattice relaxation process, and the probability of non-radiative transitions by reducing the energy transfer from Cr^{3+} to luminescence killer center. This in turn affects the degree of electron-lattice coupling that can result in reduced FWHM.

3.2.2.3 Orthorhombic

The orthorhombic crystal system is also the one-direction stretched cubic lattice as like the tetragonal with the rectangular prism base. However, the length of the three crystallographic axes is different and interests at a 90° angle. It is also a double-fold crystal symmetry structure with a rotation of 180° down the c axis. The structure consists of tetrahedral cation units in the direction of the chain, and two tetrahedral units are bound together by a single mutual anion. The octahedral cation unit is located between these tetrahedral units and shares the anion atom with the tetrahedral units. The host types of phosphates, tungstate, and very few oxides are coming under this system. The orthorhombic lattice exhibits the intermediate crystal field values around approximately ~2.3. The mullite-type structure with the chemical composition of $Bi_2Ga_2O_9$ is characterized by the chains of GaO_6 linked with the tetrahedral dimers $[Ga_2O_7]$ and highly asymmetric $[BiO_3]$ groups. On the substitution of Cr^{3+}, $Bi_2Ga_2O_9$:Cr^{3+} shows the broadband emission in the range of 650–950 nm centered at 800 nm with two R lines at 702 and 710 nm.[45] It is previously mentioned that the T-S diagram is designed on the assumption that Cr^{3+} is at O_h symmetry. Hence, the splitting of R lines with a difference of ~170 cm^{-1} should be due to distortion of host lattice on the incorporation of Cr^{3+}. In this case, the distortion is chiefly due to stereochemically active lone electron pairs of Bi^{3+}. The formation of antibonding orbital due to the s–p interactions (Bi–$6s$ and O–$2p$ electrons) has been stabilized through interaction with the unoccupied Bi–$6p$ orbital that causes distortions on the octahedral sites. This distortion on the octahedral causes the splitting of R lines in both isostructural compounds of $Bi_2Ga_4O_9$ and $Bi_2Al_4O_9$.[46] However, the relative intensity of the R lines can be finely controllable based on the nature of chemical elements.

$KTiOPO_4$ single crystal doped with Cr^{3+} impurity crystallized in $Pna2_1$ space group. The framework of the crystal structure is the helical chains of distorted TiO_6 octahedral units along c axis sharing two corner oxygen and a bridge with PO_4 tetrahedrons. K^+ is located in the channels of the structure with dodecahedral or ninefold coordination.[47] The local lattice of distorted TiO_6 consists of anomalous shorter and longer bond length of 1.74 and 2.10 Å. This arrangement offers great diversity and versatile structural modifications in $KTiOPO_4$ for tuning the luminescence. For example, the chemical formula of $KTiOPO_4$ can be rewritten as $MM'OXO_4$, where M=K, Na, Ag, Tl, Rb, or Cs; M'=Ti, Sn, Zr, Ge, Al, Cr, Fe, V, Nb, Ta, Sc, Ga, or Ce; and X=P, As, or Si. However, the selection of M, M', and X has a drastic effect on the nonlinear optical properties of the crystal. Despite the charge compensation requirement for the substitution of Cr^{3+} in Ti^{4+} ions, Cr^{3+} ions are yet in the valence of +3 at four equivalent sites in the unit cell with C_1 symmetry. Under 532 nm, Cr^{3+}-doped $KTiOPO_4$ shows the intensified broadband around 730–1030 nm with centered emission at 833 nm.

$CaSc_2O_4$ and $SrSc_2O_4$ are isostructural and interesting compounds for infrared luminescence. Both compounds are crystallized in the $Pnam$ space group. The structure is made up of edge-sharing two octahedral ScO_6 moieties that are linked through corner oxygen atoms to form a tunnel structured framework with large trigonal prismatic cavities for the occupation of Sr^{2+} or Ca^{2+} ions. There are two active crystallographic sites for Sc^{3+} in the crystal structure. Sr^{2+}-containing sample shows maximum emission at 860 nm, [49] while the emission is centered at 819 nm for Ca^{2+}-containing sample.[48] Moreover, the excitation is also shifted as like emission case. The crystal field calculations also evidence a much weaker crystal field of 1.95 (Dq/B) for $SrSc_2O_4$:Cr^{3+} than $CaSc_2O_4$:Cr^{3+} (Dq/B=2.1). As with garnet structure, the cation substitution in the nearby environment weakens the nephelauxetic effect or reducing the Racah parameter value (β=0.7625 for $CaSc_2O_4$ and β=0.7745 for $SrSc_2O_4$), which enhances the covalent character and reduces the bond length of the Cr–O bonds. The low-temperature emission spectra locate $^4T_2 \rightarrow ^4A_2$ transition zero-phonon line at 750 nm, which further evidence that the $^4T_2$ state is well below the ^2E state that is typically situated around 700 nm as like typical weak crystal field environment.

3.2.2.4 Monoclinic

The monoclinic crystal system consists of three unequal length axes, of which one unequal axis is perpendicular to the other two axes. It is typically a double symmetry crystal structure with a rotation of 180° around the perpendicular axis. There are two types of Bravais lattice in the monoclinic system, one is the primitive monoclinic (P), while the other one is body-centered (I). Most of the oxide compounds reported for the monoclinic system are in the space group of $C2/c$. In the case of $C2/c$, the crystal structure is made up of two polyhedral layers. One layer consists of corner-sharing tetrahedral polyhedrons and another polyhedral layer consists of edge-sharing octahedral polyhedrons and embedded monovalent ions. Hence, it offers one crystallographic site for the integration of Cr^{3+} ions with the strong electron-phonon coupling and weaker crystal field, which results in the greater FWHM. Cr^{3+}-doped $NaScGe_2O_6$ exhibits the emission range of 700–1250 nm with the centered emission at 895 nm due to the lesser Dq/B of 1.96.[52] Cr^{3+}-doped $NaScSi_2O_6$ exhibits the emission range of 700–1000 nm with the centered emission at 800 nm due to the lesser Dq/B of 1.82.[50] Similarly, Cr^{3+}-doped $LiInSi_2O_6$ also shows the emission range of 700–1100 nm with the centered emission and Dq/B value of 840 nm and 1.75, respectively.[51]

Apart from this above, there are some tungstates, phosphates, and chlorides are also reported in the different space group for solar cell and persistent luminescence applications.

3.2.2.5 Hexagonal

The hexagonal family is so unique from the other systems, due to the possession of two structures (hexagonal [1 sixfold axis] and trigonal [1 threefold axis]) based on the axis of rotation, and two Bravais lattice systems (rhombohedral and hexagonal system). The typical hexagonal structure is defined by the concurrent angle of 120° between the three equilateral or horizontal axes and the perpendicular vertical axes of one to the other. It's one of the complex crystal families with no generalized chemical formula. Hence, we discussed only some of the interesting compounds for the hexagonal family. $SrGa_{12}O_{19}$, [62] $LaMgGa_{11}O_{19}$, [63] and $LaMgAl_{11}O_{19}$ [67] belong to the same group of $P6_3/mmc$ and are renowned for the possession of five polyhedral environments for the trivalent cations (Ga^{3+} and Al^{3+}). Out of the five polyhedral, one polyhedral is in regular high symmetric octahedral, two are in distorted octahedral, while the other two are in tetrahedral and hexahedral, respectively. On the incorporation of Cr^{3+} ions, $SrGa_{12}O_{19}$ shows the emission in the range of 650–950 nm with several narrow emission bands, while the higher intensity is at 750 nm. Dq/B value is calculated as 2.51.[62] However, the actual substitution site of Cr^{3+} in $SrGa_{12}O_{19}$ is unclear till now. Liu and coworkers introduce Mg^{2+} ions into this complex structure with the modified chemical composition as $LaMgGa_{11}O_{19}$. Interestingly, Mg^{2+} coexists with the Ga^{3+} in the hexahedral environment. As with $SrGa_{12}O_{19}$, $LaMgGa_{11}O_{19}$:Cr^{3+} also has emission in the range of 650–1100 nm with several narrow bands, which is attributed to the existence of multiple Cr^{3+} luminescent centers and thus, in turn, causes the redshift on increasing the concentration of Cr^{3+} due to the energy transfer among the sites.[63]

Perovskite and borates are known for the rhombohedral structure with the space group of $R\bar{3}c$. $ScBO_3$:Cr^{3+} is a well-studied compound for its higher quantum efficiency of 72.8% for the maximized emission at 800 nm in the range of 700–1000 nm.[69, 70]

3.3 MID-INFRARED WINDOW

3.3.1 Activators and Host Selections

The selection of activators for the useful range of infrared emission in the second window is shown in Figure 3.4 and Table 3.3. Lanthanide ions are considered as the ideal participants for the emission from NIR to mid-infrared due to the abundant energy levels of $4f$ electron configurations. Among the lanthanide ions, Yb^{3+}, Nd^{3+}, Pr^{3+}, Er^{3+}, and Tm^{3+} have a capability for mid-infrared emissions. Since the $4f$ electrons of lanthanide ions are strongly shielded from the outer filled $5d$ and $5p$ orbitals, the intra-configurational transition of $4f$–$4f$ is hardly affected by the crystal field. Hence, the emission wavelength and spectral distribution of Yb^{3+}, Nd^{3+}, Pr^{3+}, Er^{3+}, and Tm^{3+} in the mid-infrared window is most constrained. In some instances, lanthanide ions can also adopt $5d$–$4f$ inter-configurational transition that depends on the valence state and emits broadband.

For lanthanide ions, there are several choices of a host that include molybdates, vanadates, niobates, oxides, oxy-carbonates, phosphates, borates, and titanates. Aforementioned, $4f$–$4f$ spin-forbidden transition is independent of crystal field and the emission wavelength is almost fixed despite the host. Yb^{3+} ($^4f_{13}$) has two energy levels with the ground state of $^2F_{7/2}$. Due to stark splitting between the manifolds of $^2F_{7/2}$ (ground state) and $^2F_{5/2}$ (excited state), it always has a standard

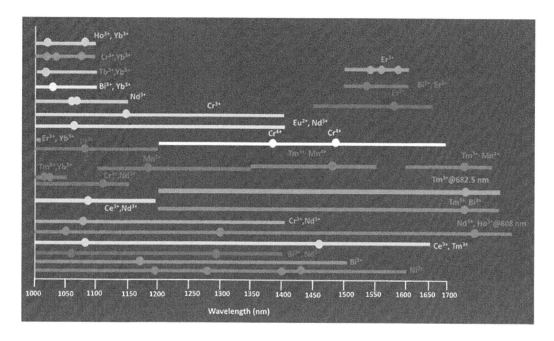

FIGURE 3.4 Schematic illustration of the selection of activators in mid-infrared wavelengths.

emission peak at 1000 nm in the range of 950–1050 nm.[59, 78] Yb^{3+} ion is greatly employing sensitizers in solar cell applications. Therefore, it is always co-doped with other lanthanide ions (Tm^{3+}, Ho^{3+}, Nd^{3+}, Tb^{3+}) to transfer the excitation energy in the solar cell.[85, 86]

Nd^{3+} ($^4f_3$) and Er^{3+} ($^4f_{11}$) possess a similar electronic structure, but Nd^{3+} has only one pair of similar energy levels, whereas Er^{3+} has several pairs of similar energy levels. Thus, the emission spectrum of Nd^{3+}-doped compounds contains three sharp peaks at ~900, ~1060, and ~1340 nm due to the respective transition of $^4F_{3/2}{\rightarrow}^4I_{9/2}$, $^4F_{3/2}{\rightarrow}^4I_{11/2}$, $^4F_{3/2}{\rightarrow}^4I_{13/2}$.[81] On the other hand, Er^{3+}-doped compounds also several peaks 980 ($^4I_{11/2}{\rightarrow}^4I_{15/2}$), 1011 ($^4I_{11/2}{\rightarrow}^4I_{15/2}$), 1029 ($^4I_{11/2}{\rightarrow}^4I_{15/2}$), 1513 ($^4I_{13/2}{\rightarrow}^4I_{15/2}$), 1535 ($^4I_{13/2}{\rightarrow}^4I_{15/2}$), 1545 ($^4I_{13/2}{\rightarrow}^4I_{15/2}$), 1552 ($^4I_{13/2}{\rightarrow}^4I_{15/2}$), 1575 nm ($^4I_{13/2}{\rightarrow}^4I_{15/2}$).[107] Nd^{3+} is most famous for laser material due to its longer wavelength with shorter decay time, but it is greatly utilizing as a biomedicine probe due to the greater bio-capability and lesser scattering by tissues.

Tm^{3+} is a unique and second least common member of lanthanides with only trace amounts on Earth. The free ion of Tm^{3+} is a promising candidate for laser application in both visible and infrared regions due to the stronger absorption at 800 nm. The incorporation of Tm^{3+} ions in the oxide host results in different kinds of emission channels, which make it a promising material in several fields of applications. In general, Tm^{3+} emission relates to the transition from the excited states of $^3F_4$, $^3H_5$, $^3H_4$, $^3F_3$, $^3F_2$, and $^1G_4$ in the manifold of $^4F_{12}$ to the $^3H_6$ ground state. These excited states can be further split into two levels with 22–30 cm^{-1}, which results in the Tm–Tm interaction and optical hyperfine structure.[84]

Pr^{3+} is the renowned luminescent center for red-emitting persistent luminescence, especially in the biomedicine field. Majority of Pr^{3+}-doped compound shows emission at 610 nm due to pronounced strong $^1D_2{\rightarrow}^3H_4$ transition. However, it can also emit NIR emission at 720 and 900 nm due to $^1D_2{\rightarrow}^3H_5$ transition and $^1D_2{\rightarrow}^3H_6$, respectively. Importantly, Pr^{3+} is capable of mid-infrared emission due to the transition of $^1D_2{\rightarrow}^3F_3$, $^3F_4$.[83]

On the other hand, the transition metal elements of Ni^{2+}, Co^{2+}, Mn^{4+}, Cr^{4+}, and Bi^{3+} display emission in the mid-infrared range for the wider spectral distribution. Moreover, the sensitivity of transition metal elements for the surrounding environments due to the unfilled d orbitals offers the

TABLE 3.3

List of the Phosphors Reported for the Mid-Infrared Emission in the Various Hosts

Chemical Composition	Activator	Emission Range (nm)	λ_{em} (nm)	λ_{exc} (nm)	Ref.
$SrMoO_4$	Yb^{3+}	Host=400–600, Yb^{3+}=950–1050	493, 1000	290	[78]
$Ba_2LaV_3O_{11}$	Yb^{3+}	Host=400–700, 900–1100	502, 1004	338	[79]
Lu_3NbO_7	Nd^{3+}	850–1100	1060	800	[80]
$Sr_3(VO_4)_2$	Nd^{3+}	1050–1100	1075, 1064	506	[81]
$La_2O_2CO_3$	Nd^{3+}	850–1400	870, 1064, 1336	780	[82]
$Gd_2O_2CO_3$	Nd^{3+}	850–1400	870, 1064, 1336	823	[82]
$Y_2O_2CO_3$	Nd^{3+}	850–1400	870, 1064, 1336	823	[82]
$MgGeO_3$, $CdSiO_3$	Pr^{3+}	570–1200	625, 900, 1085	254	[83]
$YNbO_4$	Tm^{3+}	750–1020, 1200–2500	Tm^{3+}=802, 1820	682.5	[84]
$LuPO_4$	Tm^{3+}, Yb^{3+}	Tm^{3+}=625–675, Yb^{3+}=925–1050	649, 778, 1003	468	[85]
$SrMoO_4$	Tm^{3+}, Yb^{3+}	900–1050	648, 998	283	[86]
$Ca_3(PO_4)_2$	Tm^{3+}, Ce^{3+}	1000–1650	1181, 1461	356	[87]
$YNbO_4$	Tm^{3+}, Bi^{3+}	N/A, 1200–2800	648, 791, 1820	461	[88]
$Y_3Al_5O_{12}$	Bi^{3+}, Yb^{3+}	900–1100, 300–700	Bi^{3+}=304, 460, Yb^{3+}=1028	275	[89]
$BaGd_2ZnO_5$	Nd^{3+}, Yb^{3+}	900–1150	Yb^{3+}=423, 473, 978, Nd^{3+}=1071	359	[90]
$LuPO_4$	Tb^{3+}, Yb^{3+}	900–1100	Tb^{3+}=545, Yb^{3+}=1003	489	[91]
$LuPO_4$	Tb^{3+}, Yb^{3+}	880–1100	Yb^{3+}=980, 1018	484	[92]
$GdBa_3B_9O_{18}$	Mn^{4+}, Tm^{3+}	650–750, 775–825, 1350–1550, 1600–2100	Tm^{3+}=800, 1488, 1800, Mn^{4+}=698	311	[93]
Y_2MgTiO_6	Ce^{3+}, Nd^{3+}	800–1200	Nd^{3+}=916, 1093	440	[94]
$(Sr_{0.6}Ca_{0.4})_3(Al_{0.6}Si_{0.4})O_{4.4}F_{0.6}$	Cr^{3+}, Nd^{3+}, Yb^{3+}	650–1150	Cr^{3+}=698, Nd^{3+}=915, 1103, Yb^{3+}=958, 980, 1011, 1031, 1087	413	[95]
Al_2O_3	Yb^{3+}, Nd^{3+}	850–1150	Yb^{3+}=972, Nd^{3+}=1060	520	[96]
$Na_2GdMg_2V_3O_{12}$	Cr^{3+}, Nd^{3+}	650–850, 850–1400	691, 1060	430, 808	[97]
$GdY_2Al_3Ga_2O_{12}$	Cr^{3+}, Nd^{3+}	700–900, 850–1400	Nd^{3+}=1062, Cr^{3+}=750 nm	808, 267	[98]
$Ca_3Ga_2Ge_3O_{12}$	Nd^{3+}, Ho^{3+}	1000–1900, 1000–3500	Ho^{3+}=2100, 2700, Nd^{3+}=900, 1050, 1300, 1800	808	[99]
$NaLa_9(GeO_4)_6O_2$	Bi^{3+}, Nd^{3+}	375–600, 850–1400	Bi^{3+}=480, Nd^{3+}=815, 876, 895, 925, 1058, 1328	316	[100]
La_2GeO_5	Ho^{3+}, Yb^{3+}	500–800, 900–1100	Ho^{3+}=551, 756, Ho^{3+}, Yb^{3+}=1208, 1033, Yb^{3+}=980, 1080	400	[101]
Lu_2O_3	Nd^{3+}, Ho^{3+}, Yb^{3+}	900–1100, 850–1450	Nd^{3+}=502, 909, 1061, 1339, 980, 1189, Yb^{3+}=977, 1004, 1069	340	[102]
$Ba_2YV_3O_{11}$	Eu^{2+}, Nd^{3+}	800–1400	1069	400	[103]
$Ca_5(PO_4)_3Cl$					

(Continued)

TABLE 3.3 (*Continued*)

List of the Phosphors Reported for the Mid-Infrared Emission in the Various Hosts

Chemical Composition	Activator	Emission Range (nm)	λ_{em} (nm)	λ_{exc} (nm)	Ref.
Li_2SrSiO_4	Ce^{3+}, Pr^{3+}	850–1100, 400–500	Pr^{3+}=428, 611, 635, 650, 1039	450	[104]
$LaAlO_3$	Er^{3+}	930–1050, 1450–1650	Er^{3+}=986, 1553	405	[105]
$NaBaPO_4$	Er^{3+}, Bi^{3+}	1500–1600	1534	377	[106]
Y_2O_3	Er^{3+}, Li^+, Yb^{3+}	950–1010, 1500–1600	Er^{3+}=980, 1011, 1029, 1513, 1535, 1545, 1552, 1575, Yb^{3+}=975, 1011, 1029		[107]
$Zn_3Ga_2Ge_2O_{10}$	Ni^{2+}	1050–1600	1290	254	[108]
$La_3Ga_5GeO_{14}$	Ni^{2+}	1050–1600	1430		[109]
$Y_3Al_5GaO_{12}$	Ni^{2+}	1200–1600	1400	400	[109]
$SrGa_4O_7$·Gd_zZnTiO_6	Cr^{3+}, Cr^{4+}	650–1200, 1200–1700	770, 1395, 1476	440, 450	[110]
Mg_2SiO_4	Cr^{3+}, Cr^{4+}	650–1400	745–813, 1150	450	[111]
$BaBPO_5$	Bi^{3+}	600–720, 900–1500	642, 1162	478	[112]
M_2SiO_4 (M=Ba, Sr, Ca)	Mn^{5+}	1100–1400	1190	320	[113]

advantage of spectral tuning of the emission as, unlike lanthanide ions. For the transition metal ions, very limited chemical compositions are reported till now. This could be mainly due to the limitations of the instrumentation. That is, InGaAS photodetector or special detector is needed for measuring the wavelength of more than 700 nm. Additionally, advanced instrumentations are needed to explore the luminescence mechanism and other properties of the phosphors. In general, Ni^{2+} can ideally substitute for Ga^{3+}, so the gallates an excellent host system for mid-infrared emission too. Pan *et al.* [108] reported three different gallate host systems of Ni^{2+}-doped $Zn_3Ga_2Ge_2O_{10}$, $LiGa_5O_8$, $La_3Ga_5GeO_{14}$ for the infrared emission centered at 1290, 1220, and 1430 nm, respectively, in the second window of 1050–1600 nm that corresponds to the spin-allowed transitions $^3T_2$ (^3F)→$^3A_2$ (^3F) of Ni^{2+}. Similarly, Ni^{2+}-doped garnet structure ($Y_3Al_5GaO_{12}$) was reported by Hu and coworkers for the mid-infrared emission at 1400 nm in the range of 1200–1600 nm.[109] Unfortunately, these systems are mainly reported for persistent luminescence applications. This denotes plenty of research opportunities available in designing the mid-infrared phosphors for pc-NIR LED applications. On the other hand, the presence of Cr^{4+} and Mn^{5+} ions in the tetrahedral environment can give mid-infrared emission in the range of 1200–1700 and 1100–1400 nm, respectively. Cr^{4+} ions in the tetrahedral coordination of $SrGa_4O_7$ host leads to respective emission at ~1395 and ~1476 nm due to $^1E(^1D)$→$^3A_2(^3F)$ and $^3T_2(^3F)$→$^3A_2(^3F)$ transitions as per d^8 T-S diagram.[110] Similarly, Mn^{5+} in the tetrahedral coordination of M_2SiO_4 (M=Ba, Sr, Ca) leads to emission at 1190 nm due to $^1E(^3F)$→$^3A_2(^3F)$ transition as per d^8 transition.[113] The mid-infrared emission of Mn^{5+} and Cr^{4+} is considered as special cases due to the challenges of controlling the valence state in the conventional solid-state reaction.

3.4 APPLICATIONS OF BROADBAND NEAR-INFRARED PHOSPHOR FOR LED

3.4.1 NEAR-INFRARED SPECTROSCOPY

NIR spectroscopy is the most versatile tool for the quantitative and qualitative detections of more than 20 decades in various fields, including pharmaceutical, agriculture, astronomy, and food industry. In principle, each molecule is composed of several atoms that are connected by bonds. The connected bonds can be both primary (ionic, metallic, and covalent) and secondary bonds (van der Waals). However, based on the bond strength and atom weight, each bond will absorb the defined wavelength of light and endures the vibrations for the shorter times (as shown in Figure 3.5(a)). Moreover, it is distinctive for each molecule and turns into a "spectral fingerprint". Hence, it is possible to distinguish the two molecules by monitoring light behaviors in terms of absorption, reflection, and transmission. By coupling with the statistical data analysis tool, it is more convenient to extract the features of the spectra fingerprint details and compare them with the standard database.[114]

The instrumentation and application areas of NIR spectroscopy extend gradually over the period. After the discovery of the infrared light in 1800, Charles Wheatstone constructed the infrared spectrometer for commercial usage in the year 1835s. It is extensively employed in academic and space research for studying the material structure, solar system, and natural vegetation, etc. Around the 1900s, it is almost an inevitable tool in academics especially for investigating the organic compounds after the significant contributions of spectroscopists, Mr. Hertzberg, Mr. Coblenz, and Mr. Angstrom. In the later 2000s, the NIR spectrometer turns into a significant tool in agricultural products for quality control activities.[115]

In general, the spectrometer devices comprise several components like a light source, monochromatic, detector, and optical accessories. However, the most crucial component in the spectrometer device is the light source. In traditional, the light source is usually a tungsten halogen light bulb that is bulky and also possesses the constraints of limited lifetime, a requirement of glass envelope to contain the generation of heat, longer waiting time to attain spectral stability. In some cases, the light source can be Xenon lamps and duplex lamps (deuterium and tungsten combination lamps). But then again, these light sources don't have smooth spectral distribution and also suffer drawbacks

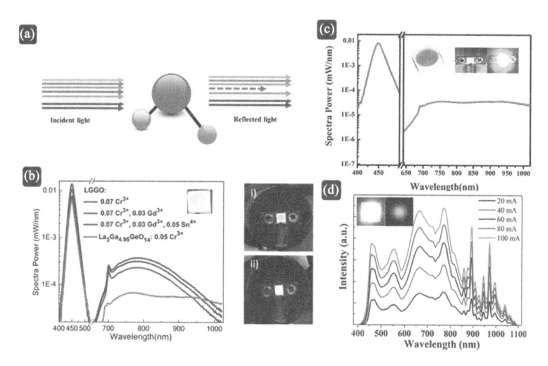

FIGURE 3.5 Emission spectra of pc-NIR LED for different NIR phosphors (a) schematic illustration of light interaction with the molecule, (b) blue LED chip with $La_3GaGe_5O_{16}$:Cr^{3+}, (c) blue LED chip with $La_3Ga_5GeO_{14}$:Cr^{3+}, and (d) blue LED chip with the different combination of phosphors such as $Y_2CaAl_4SiO_{12}$:Cr^{3+}, $CaAlSiN_3$:Eu^{2+}, and $Y_3Al_5O_{12}$:Ce^{3+}. (Rajendran, V., Fang, M.H., Guzman, G.N.D., Lesniewski, T., Mahlik, S., Grinberg, M., Leniec, G., Kaczmarek, S.M., Lin, Y.S., Lu, K.M. and Lin, C.M., *ACS Energy Lett.* 3, 2679, 2018. With permission; Rajendran, V., Lesniewski, T., Mahlik, S., Grinberg, M., Leniec, G., Kaczmarek, S.M., Pang, W.K., Lin, Y.S., Lu, K.M., Lin, C.M. and Chang, H., *ACS Photonics* 6, 3215, 2019. With permission; Yao, L., Shao, Q., Xu, X., Dong, Y., Liang, C., He, J. and Jiang, J., *Ceram. Int.* 45, 14249, 2020. With permission)

as like tungsten halogen lamp. This is the major obstacle when swapping the infrared spectrometer devices from the laboratory to the real-time nondestructive investigations for the nonscientific community as well.[115, 116]

For portable or handheld type spectrometers, a compact light source is urgently needed with the possessed features of the broader spectral distribution of light like a tungsten halogen lamp. In the consideration of LED technology, GaAs (Gallium Arsenide) is the well-known and commercially available LED with the peak emission at ~870 nm for various applications, including sensors. However, the versatility and temperature characteristic of GaAs-based LED is very poor. Furthermore, the manufacturing conditions always fluctuate the obtainment of predetermined peak emission wavelength that results in poor yield and increased production cost. In contrast, pc-NIR LED is an emergent technology with the benefits of longer lifetimes, better compactness, lower consumption of electricity, higher spectral stability, fast modulation capability, lower cost, wide operating temperature, and customized smooth spectrum.

In the design of photo-spectrometers with pc-NIR LED as a light source, the potential of the fabricated pc-NIR LED device was evaluated in terms of radiant flux and spectral distribution range. Radiant flux or radiant power is defined as the amount of radiant energy that emerged from the radiant power source per unit time. It is generally described in units of W, mW, and μW. As a general rule, the radiant power is directly proportional to the operating current. Hence, it is also important to consider the operating current for the radiant flux. As a whole, the ideal light source of spectrometers should have a broader distribution of light with higher radiant flux as much as possible.

To achieve this, there are two strategies commonly adopted in fabricating pc-NIR LED devices for spectroscopy. One is the pc-NIR LED device with single broadband NIR phosphors, while the other is the pc-NIR device with the combinations of two or more NIR phosphors.

Table 3.4 lists the NIR phosphors reported for the spectroscopy applications along with photo-electric performance. It is interesting to note that all of these research articles have been published after 2017 that suggests that pc-NIR LED light source for spectroscopy is still in the early stages of development. In other words, there are plenty of research scope opportunities that are available in pc-NIR LED for the spectrometers. Very few chemical systems are qualified for single-phosphor-based broadband NIR light source.

Rajendran *et al.* [74] reported a lanthanum gallogermanate system of $La_3Ga_5GeO_{14}:Cr^{3+}$ for a NIR light range of 600–1200 nm with the FWHM of 330 nm. Such super broadband is mainly due to the emissions from two emitting centers at 750 and 920 nm (Figure 3.5(c)). Besides, these two emitting centers are in octahedral and tetrahedral coordination with the different valence of Cr^{3+} activators. Similarly, Nanai *et al.* [111] also reported a $MgSiO_4:Cr^{3+}$ system for super broadband NIR light in the range of 650–1400 nm from two distinct luminescence centers at 800 and 1150 nm. It is also proposed that the activator ions are incorporated in the two different coordination geometry of octahedral and tetrahedral with different valences of Cr^{3+} and Cr^{4+}, respectively. At the operating current of 350 mA, the output powers of 18.2 and ~170 mW were obtained for the fabricated pc-NIR LED device of blue LED with $La_3Ga_5GeO_{14}:Cr^{3+}$ and $MgSiO_4:Cr^{3+}$. It is again dispute in the principal governing factors of the radiant flux because it is directed to the several factors like

TABLE 3.4
List of the Phosphors Reported for the Near-Infrared Spectroscopy along with Photoelectric Performance

Phosphor	Emission (nm)	FWHM (nm)	Input Current (mA)	NIR Power (mW)	Efficiency (%)	Year	Ref.
$Lu_3Al_5O_{12}:0.05Ce^{3+}, 0.5\%$ Cr^{3+}+Bi-doped GeO_2 glass	500–850, 1000–1600	N/A	100	47	N/A	2017	[117]
$La_3Ga_5GeO_{14}:Cr^{3+}$	600–1200	330	350	18.2	76.59	2018	[74]
$Ca_2LuZr_2Al_3O_{12}:Cr^{3+}$	650–850	117	20	2.448	4.1	2018	[7]
$ScBO_3:Cr^{3+}$	700–950	120	120	26	7	2018	[70]
$YAl_3(BO_3)_4:Cr^{3+}, Yb^{3+}$+$NaScSi_2$ $O_6:Cr^{3+}$	650–850, 980, 750–950	N/A	100	26	8.6	2018	[50]
$Ca_2LuHf_2Al_3O_{12}:Cr^{3+}$	700–1000	N/A	100	46.09	15.75	2019	[9]
$MgSiO_4:Cr^{3+}, Cr^{4+}$	650–1400	N/A	800	367.5	N/A	2019	[111]
$La_3GaGe_5O_{16}:Cr^{3+}$	650–1050	185	350	43.1	52.5	2019	[71]
$La_3GaGe_5O_{16}:Cr^{3+}, Gd^{3+}$	650–1050	185	350	56.3	60.6	2019	[71]
$La_3GaGe_5O_{16}:Cr^{3+}, Gd^{3+}, Sn^{4+}$	650–1050	186	350	65.2	68.27	2019	[71]
$K_3AlF_6:Cr^{3+}$	650–1000	N/A	350	7	N/A	2019	[33]
$K_3GaF_6:Cr^{3+}$	650–1000	N/A	350	8.4	N/A	2019	[33]
$LiInSi_2O_6:Cr^{3+}$	700–1100	143	210	51.6	17.2	2019	[51]
$Ca_3Sc_2Si_3O_{12}:Cr^{3+}, Ce^{3+},$ Yb^{3+}, Nd^{3+}	700–900, 900–1100	440	100	14.6	4.9	2019	[20]
$Ca_3Sc_2Si_3O_{12}:Cr^{3+}, Ce^{3+}, Yb^{3+},$ Nd^{3+}+$CaAlSiN_3:Eu^{2+}$	700–1100	522	100	15.6	5.2	2019	[20]
$ScBO_3:Cr^{3+}$	700–1000	120	350	39.11	N/A	2020	[69]
$Gd_3Sc_2Ga_3O_{12}:Cr^{3+}, Al^{3+}$	650–1000	~99	1000	750.8	18.1	2020	[10]
$NaScGe_2O_6:Cr^{3+}$	700–1250	162	350	12.07	N/A	2020	[52]
$Y_2CaAl_4SiO_{12}:Cr^{3+}$	650–900	160	150	800	N/A	2020	[11]

injecting current, excitation chip size, LED size, and the ratio between powder and resin. However, the obtainment of satisfactorily radiant flux with the broadest spectral distribution of light is the minimum criterion for spectroscopy light source. On the other hand, several chemical systems are reported for single-phosphor-based broadband NIR light source. From Table 3.4, it is also interesting that the different chemical systems exhibit the different spectral distribution of infrared light with varying performance. For example, the lanthanum gallogermanate system with the chemical composition of $La_3GaGe_5O_{16}:Cr^{3+}$ exhibits the NIR in the range of 600–1050 nm with the FWHM value of 185 nm (Figure 3.5(b)).[71] The borate system of $ScBO_3$ exhibits the emission range of 600–950 nm with the FWHM value of 120 nm.[70] It is suggested that the presence of multiple luminescence centers in the single phosphor system is advantageous for the multiple elemental detections in NIR spectroscopy applications.

In contrast, the pc-NIR LED source can also have fabricated with two or more infrared phosphors, red phosphors to achieve the broadest spectral distribution as much as possible. For example, Yao et al. [20] combined the multiple-element doped $Ca_3Sc_2Si_3O_{12}:Cr^{3+}$, Ce^{3+}, Yb^{3+}, Nd^{3+} NIR phosphor and commercial red phosphor ($CaAlSiN_3:Eu^{2+}$) along with a blue LED chip to achieve the highest FWHM of 522 nm. Similarly, Mao et al. [11] also combined the $Y_2CaAl_4SiO_{12}:Cr^{3+}$ and commercial red phosphor ($CaAlSiN_3:Eu^{2+}$) for the boarded spectral distribution of NIR light as shown in Figure 3.5(d).

3.4.2 NIGHT VISION

The eye vision mechanism for human and animals are entirely different. The retina in the human eye consists of two photoreceptors: cones and rod to identify the wavelength and intensity of color, respectively. On the other hand, each animal has a unique color perception mechanism. For example, the cattle have dichromatic vision, while the fowl have a tetrachromatic vision. However, most animals have larger pupils to receive more light and help them to view at wider angles and in a dark environment. Moreover, some animals have an extra thinner layer of tissues in the retina called tapetum that acts as a natural mirror and reflects the light to the eye. This thinner layer bounces the light almost twice in the retina that provides the night vision for hunting. However, the human eye doesn't possess such an arrangement for vision in a dark environment. Hence, humans need electro-optical devices to see at night.

Several types of imaging technology and tools are available for night vision purposes, for example, night vision goggles, thermal imaging, shortwave infrared imagers, and infrared cameras. Despite their difference, the night vision devices share the simplest three blocks (optical objective, image intensifier tube, and optical ocular) of architectural design as shown in Figure 3.6(a). The purpose of the optical objective is to receive the light reflected from the object, while the optical ocular is for human vision. The image intensifier tube of the night vision device has a photocathode followed by an internal anode and phosphors screen. The application of high voltage creates a strong electrostatic field between the photocathode and the anode. Once the infrared light reflected from the optical objective reaches the photocathode, the electrons are emitted and accelerated toward the phosphor screen. This in turns the visible image in the optical ocular. Even the structure of night vision devices looks simpler, the process of creating an output image is quite complicated and involves several design rules.[118]

It is also important to review the history of the night vision devices development for understanding the scope and opportunities of research. Before the end of the Second World War, night vision devices are developed and used solely for military activities. With the advancements of modern solid-state technologies, the product was commercialized around 1980. Other than military activities, night vision devices are greatly employing in law enforcement, fire and rescue activities, natural resource agencies, security, medical, and engineering applications.[119] However, the intellectual property right of image intensifier tubes by big manufacturers and other competing surveillance imaging technology made night vision devices as old-fashioned research for several decades. Since

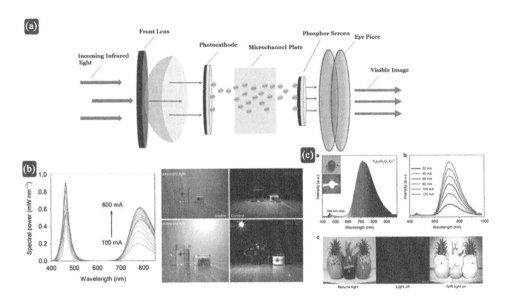

FIGURE 3.6 (a) Working principle of night vision goggles. The photographs of the objects capture under the fabricated pc-NIR LEDs as the light source of night vision applications – (b) $Ca_3Sc_2Si_3O_{12}:Cr^{3+}$, and (c) $K_3LuSi_2O_7:Eu^{2+}$. (Jia, Z., Yuan, C., Liu, Y., Wang, X.J., Sun, P., Wang, L., Jiang, H. and Jiang, J., *Light Sci. Appl.* 9, 86, 2020. With permission; Qiao, J., Zhou, G., Zhou, Y., Zhang, Q. and Xia, Z., *Nat. Comm.* 10, 5267, 2019. With permission)

the sun is not present in the nighttime, the source of infrared light to the object for the reflection and absorption is from the moon. It is well-known that the climate is unpredictable and also impossible to ensure the moonlight can reach all the surfaces of the Earth at all times. Hence, the infrared light-emitting source has to be included in the night vision devices.[120]

For a longer time, the night vision devices relied on industrial halogen and tungsten bulbs along with special filters as the light source for the NIR illumination. As aforementioned, the traditional light sources are short-lived and will generate a larger amount of heat. On the other hand, laser illuminators are cost-savers than LED due to the higher luminance and also eco-friendly. However, laser illuminators are single wavelength emitters. For the broader spectrum, it is necessary to employ more laser illuminators in the series which further adds complication to the design. Similarly, LED technology is current-savers with the lack of spectral tuning features. Therefore, the recent developments of pc-NIR LED turn an alternative rising light source for night vision goggles. The technical constraints as the light source in the night vision devices are the sensitivity around the red to infrared light in the emission range of 630–930 nm. Table 3.2 lists the NIR phosphors reported for night vision application along with the photoelectric performance. It is very clear that the concept of pc-NIR LED is booming up and the technology is at the initial stage of development. The potential of the proposed NIR phosphor system is commonly demonstrating by a simple experiment of capturing the photograph of the object under the NIR LED source in a dark environment.

Jia *et al.* [12] demonstrated the effectiveness of pc-NIR LED for night vision applications by simply capturing the photographs of the objects such as water, and milk, in the cups under the fluorescent light and fabricated pc-NIR device (460 nm blue LED for excitation and $Ca_3Sc_2Si_3O_{12}:Cr^{3+}$ as the conversion layer). As shown in Figure 3.6(b), the logo on the milk and water cups is clear under the lights of a fluorescent and fabricated pc-NIR device in ON Mode. However, the logo on the water cup is unclear in the image captured in OFF mode by using the NIR camera, but the logo on both water and milk cup is more clear under the fabricated NIR LED even in OFF mode. Similarly, Qiao *et al.* [64] demonstrated the helpfulness of pc-NIR LED for night vision application by using rare-earth ion (Eu^{2+}) activated phosphors as an alternative to the conventional Cr^{3+}-doped system. As shown in Figure 3.6(c), the fruits placed under the pc-NIR LED (blue LED as excitation source and $K_3LuSi_2O_7:Eu^{2+}$ as the conversion layer) light source can be detected as like natural light.

3.4.3 SOLAR CELL

To save the earth from the threatening of global warming, most of the countries revised their energy plan from nonrenewable to renewable energy as much as possible. Among many renewable technologies, solar energy has proved to be one of the most promising solutions to the energy shortage in the world. To harness the energies of the sun for electricity production, a solar cell must be efficient and cost-effective to cope with traditional sources. Thanks to its advantages, the overall deployed photovoltaic (PV) capacity at the end of 2014 amounted to less than 177 gigawatts (GW) as shown by the Photovoltaic Power System Program of the International Energy Agency (IEA PVPS).[121] The maximum theoretical efficiency of 1.1 eV silicon solar cell was approximately 30%. However, the recent solar technologies that include wafer, thin-film, organic, and nanostructured devices already pushed these limits. In general, the solar cell efficiency is hindered by the factors of non-absorption of low-energy photons, thermalization losses (absorption of high-energy photons), extraction losses (unavoidable charge carrier recombination), and optical loss (incomplete absorption, reflection, and shading).[122]

The spectral mismatch between the incident solar spectrum and solar cell absorption contributed to the greater loss of solar cell efficiency. As straightforward, if the incident photon energy (E) is greater than or equal to the bandgap energy (E_g), then the photon can be absorbed in the solar cell and generate photocurrent. In contrast, if $E>E_g$ and $E>E_g$, then the photons can be transmitted or lost due to thermalization in the solar cell, respectively.[123] Hence, the chief key that determines the solar cell efficiency is the spectral mismatch. In principle, the solar cell can harness energy through any one of the three modes. (1) concentrating photovoltaics (The lens concentrates incoming light that increases solar cell efficiency and decreases material demand, while the cooling sink prevents the cell from overheating), (2) upconversion (UC) layer-solar cell (The energies of two or more are merged into a single photon through UC, then this new photon is transmitted to the cell for utilization, (3) down-conversion layer-solar cell (The absorbed high-energy photon is split into longer wavelength for better absorption of the solar cell). In opposition, luminescent downshifting is a mechanism that transfers the energy of a short wavelength photon to lower energy.[124]

The main idea behind the usage of the down-converting layer is that the solar cell also has an ideal external quantum efficiency (EQE) that varies with wavelength such that, ideally, having as much spectrum as possible will be an excellent way to increase performance. The mechanism of down-converting is opposite to the UC as shown in Figure 3.7. The electrons of single ions absorb the incident UV photon and travel to the second excited state. Before decaying, the two NIR photons will be generated through any one of the three routes for resonant energy transfer between two distinctions. At last, the reverse cooperative energy transfer takes place as the ion receives the photon and transfers the energy to two other ions promoting it to a state of an excited state. At the same time, these states decay releasing two low-energy photons. As compare with UC, down-converting is a linear process and independent of light intensity. Downshifting is the mechanism in which a high-energy photon is converted into a photon with lower energy as more like the down-conversion process. However, one absorbed photon emits two photons of lower energy in the down-conversion process, while the downshifting process emits the highest of one photon for each absorbed photon. Hence, the quantum efficiency of the downshifting process is less than unity and also encounterable thermalization loss.

Tables 3.2 and 3.3 list down-conversion phosphors reported for the solar cell. Due to the intra-configurational $4f$–$4f$ transition characteristic with different energies, lanthanide ions are believed as the probable elements for down-converting the radiations in a solar cell because $4f$ electrons

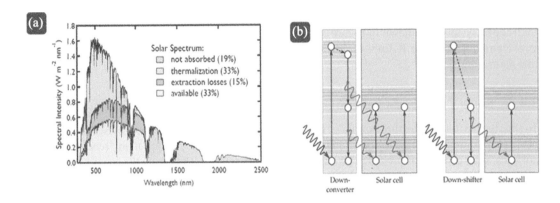

FIGURE 3.7 (a) Solar spectrum represented with the energy losses that occur in a silicon cell, (b) working principle of down-conversion and downshifting in the solar cell. (Day, J., Senthilarasu, S. and Mallick, T.K., *Renew. Energy* 132, 186, 2019. With permission; De La Mora, M.B., Amelines-Sarria, O., Monroy, B.M., Hernández-Pérez, C.D. and Lugo, J.E., *Sol. Energy Mater. Sol. Cells* 165, 59, 2017. With permission)

are largely insulated from adjacent crystal fields due to the outer filled $5s$ and $5p$ shell. Hence, the emission spectrum is unlikely affected by the crystal field ($4f$–$4f$ spin-forbidden transition is not related to the chemical bonding) and it more behaves like free ions. Among the lanthanide ions for down-converting, Pr^{3+}, Tm^{3+}, Er^{3+}, and Gd^{3+} are capable of single ion-dependent cascade emission. Energy transfer from the lanthanide ion to another happens through several processes, such as (1) electron migration and hole migration; (2) excitons migration; (3) resonance between atoms with spectrum overlaps; and (4) photon reabsorption by another activator ion or sensitizer.[125] However, the major obstacle encountering by down-converting material as the top layer of the solar cell is the absorption and scattering losses. Hence, photoluminescence quantum yield more than 115% is mandatorily required.[126] Several lanthanide ion systems exhibiting a down-converting process with energy transfer greater than 100% (theoretically photoluminescence quantum yields can more than 150%) have been reported. One of the effective strategies to enhance the absorption is the addition of sensitizers such as Eu^{2+} and Yb^{2+} or transition metal ions such as Mn^{2+}. For example, the quantum efficiency of $SrAl_2O_4$:Yb^{3+} can attain 147.36% by the addition of Eu^{2+} as sensitizers. [127] Thus, the visible energy at 515 nm is effectively converted to 980 nm for silicon solar cells. Yb^{3+}-doped Na_2YMg_2 $(VO_4)_3$ exhibits an intensive NIR emission at 974 nm under UV illumination of 240–400 nm.[128] Also, the visible photon from the host is converted to NIR photons with a quantum efficiency of nearly 160%. Thus, it demonstrates that Yb^{3+}-sensitized material is the optimal candidate for solar cell applications.

Thanks to its performance, PV is currently an excellent choice to have energy services that mitigate the effect on the atmosphere. Concerning the use of down-converters in solar cells to improve the performance of PV, certain considerations must be considered for commercial applications, such as the availability of down-converters, the deposition methods that allow the wide surface area to be protected, as well as the durability of atmospheric conditions. Other than the selective use of lanthanide ions in various hosts, the narrowing the size of the host in nanoscales such as quantum dots, nano-phosphor, quantum dot graphene, and silicon nanoparticles are believed to be emerging research areas to enhance the performance of solar cells.

3.4.4 PLANT GROWTH

The population growth, rapid globalization, industrialization, and urban densification limit the space of cultivation for food. Besides, global warming and climatic changes threaten the productivity of agriculture. According to the news of the United Nations, the world will require food production of more than 70% to keep pace with a population of 9.6 billion in the year 2050.[129] Hence, the controlled environment agriculture or indoor plant farming can be an effective rising choice that can meet the food demands of the future. A plant needs light, nutrients, humidity, and carbon dioxide for its growth. Among them, sunlight is the natural key element for plant development, although it does not only supply energy for plant growth, series of chemical photosynthesis processes but also influences their increasing morphology by various phytochromes in plant cells. Others are the environmental factors that are mostly produced or controlled by humans. By providing an artificial light source with the ideal light spectrum requisite for the growth of the plant, it is possible to grow crops efficiently all over the year despite climatic conditions.[130]

It is the basic understanding that the reflection of green light from the sun by the plant's leaves makes its color green to the human eyes. This is also one of the reasons that the photoscopic spectrum of the human eye is maximized at 555 nm. Other than the green light, the blue light (400–520 nm), red light (600–700 nm), infrared light (700–740 nm) are absorbed by the photoreceptors in the plant for its nutrient, growth, and development.[131] *Chlorophyll a* and *Chlorophyll b* are the primary photoreceptors responsible for the photosynthesis activity in the plant, which absorb the blue light at 425 and 470 nm, and red light at 660 and 640 nm, respectively. During the photosynthesis process, the chlorophyll pigments *a* and *b* in the plant absorb the light energy from the sun and convert the water, carbon dioxide, minerals to oxygen and carbohydrates. Carbohydrates are the sugars that are the fuel for the plant's functioning and metabolism. On the other hand, the plant also contains some other photoreceptors such as carotenoids (470–500 nm), cryptochromes (350 and 450 nm), phototropins (350 and 450 nm), and phytochromes. Phytochromes are the interesting photoreceptors that regulate the plant morphology and photomorphogenic functions. Phytochromes have two chemical forms of *Pr* and *Pfr* that are interconvertible. *Pr* is inactive and exists in the dark time, while *Pfr* is the active form and regulates the physiological responses of the plant. As a general rule, *Pr* (inactive form) absorbs the red photon of 670 nm and transforms into the active form of *Pfr*. As the exchange, *Pfr* absorbs the infrared photon of 730 nm and transfers back to the inactive form of *Pr* as shown in Figure 3.8(a). These interconvertible chemical forms are inevitable because it controls leaf size, stem growth, seed germination, and flowering, etc.[132, 133] In designing the artificial light source for plant growth, the ratio proportion between red and infrared light should be the customizing value based on the type of interesting plant. Because excessive exposure to infrared photon can reverse the effect of a red photon is a short-day plant or vice versa.

In addition to the photosynthetic active region (PAR, 400–700 nm), the vegetation can be greatly altered with the ratio of R (photon irradiance between 655 and 665 nm) and NR (photon irradiance between 725 and 735 nm). On the other hand, each plant has its defense mechanism to its neighbor's vegetation. Based on the response to the PAR and R/NR ratio, a plant can be classified into shade avoidance and shade tolerance plants. The monocotyledon plants possess the character of shade avoidance effect and require enriched R light (low R/NR ratio, short-day plant).[134] In the case of a flowering plant like *Arabidopsis* and *Brassica*, when grown in low R:FR ratios, Arabidopsis exhibitions a reduction of cotyledon, leaf expansion, an increased elongation of hypocotyl, and petioles. The shade avoidance effect changes are rapid and reversible. In contrast, the dicotyledonous plant possesses the character of shade tolerance and requires enriched NR light (high R/NR ratio, long day). Photomorphogenic changes associated with shade tolerance are decreased leaf growth, reduced branching, and increased apical dominance.[135]

The total photon output in terms of photosynthetic photon flux (PPF) and PPF density (PPFD) has also to be considered when fabricating the light source for plant growth in addition to PAR and ratio of R/NR. PF measures the overall amount of photosynthetic photons (those within the PAR wavelength) produced by the light source (regardless of where the photons are sent), while PPFD measures how many of these photons fall within the given surface area per second. Both PPF and PPFD are expressed in units of μmol. However, PPFD values decrease as the distance increases.

The higher efficiency of InGaN LEDs and its accessibility for blue light has inspired the researchers to design the downshifting phosphors that can convert the blue photon from LED to the red and

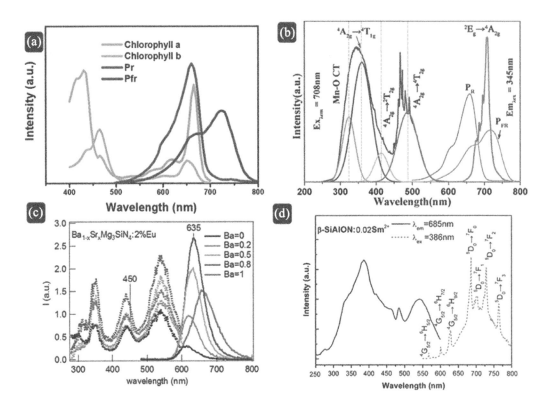

FIGURE 3.8 (a) Absorption spectrum of plant photoreceptors, excitation, and emission spectrum – (b) La(Mg, Ti)$_{0.5}$O$_3$:Mn^{4+}, (c) Ba$_{0.8}$Sr$_{0.2}$Mg$_3$SiN$_4$:Eu^{2+}, and (d) β-SiAlON:Sm^{2+}. (Zhou, Z., Zheng, J., Shi, R., Zhang, N., Chen, J., Zhang, R., Suo, H., Goldys, E.M. and Guo, C., *ACS Appl. Mater. Interfaces* 9, 6177, 2017. With permission; Osborne, R.A., Cherepy, N.J., Åberg, D., Zhou, F., Seeley, Z.M., Payne, S.A., Drobshoff, A.D., Comanzo, H.A. and Srivastava, A.M., *Opt. Mater.* 84, 130, 2018. With permission; Yang, Z., Zhao, Z., Que, M. and Wang, Y., *Opt. Mater.* 35, 1348, 2013. With permission)

infrared photons to match the absorption spectrum of the photoreceptors. Again, the choice of luminescence centers ultimately influences the output photon type in the design of inorganic phosphors. In-plant growth, the current bottleneck is the inorganic phosphors (or downshifting phosphors) for the red and infrared photons to activate the phytochrome photoreceptors since there are wider choices of inorganics phosphors for the red emission. For example: Mn^{4+} activated (K, Mg)$_2$SiF$_6$, Eu^{2+} activated Sr[LiAl$_3$N$_4$], and CaAlSiN$_3$. However, the research is mostly accelerated toward the infrared emission. The transition metal ions of Mn^{4+} and Cr^{3+} and rare-earth ions of Eu^{2+} and Sm^{2+} are the commonly choosing luminescence centers for infrared emission.

Mn^{4+}-activated phosphors are known for their line emission (^{2}E→^{4}A$_2$) in the red to a NIR range of 630–740 nm. Depending on the host selection, the emission wavelength could be shifted, but the emission range is always restricted due to its inherent nature of a strong crystal field. In recent times, publications reporting for the plant growth by using Mn^{4+} are also gradually rising. Mn^{4+}-activated La(Mg, Ti)$_{0.5}$O$_3$ prepared through sol-gel route possess the space group of $Pm\overline{3}m$.[136] There are two possible crystallographic sites (Mg^{2+} or Ti^{4+}) for the incorporation of the activator, but the energy band and density of state (DOS) calculation proved out that Mn^{4+} is substituted for Ti^{4+}. Since the electronegativity of Mn^{4+} is greater than Ti^{4+} and Mg^{2+}, the substitution of Mn^{4+} for Ti^{4+} leads to a stronger attraction with the surrounding O atoms than the interactions of Mg with O atoms. Hence, the bond length of Mn/Ti–O will decrease with the concurrent increase of Mg–O bond length. Under the excitation wavelength of 345 or 487 nm, Mn^{4+}-activated $La(Mg,Ti)_{0.5}O_3$ exhibits an intensified NIR emission at 708 nm with the anti-stokes phonon sidebands (AS-PSB) at 675, 680, 685, 690, and 695 nm due to the multiple vibrational modes of $^2E \rightarrow {}^4A_2$ in the $[MnO_6]$ environment as shown in Figure 3.8(b).[136] Even though the emission band at 708 nm overlaps effectively with the absorption band of phytochrome *Pfr*, it is hardly excitable by the blue light, especially at 450 nm. For the real application, it is necessary criteria that the prepared phosphors should absorb blue emissions of InGaN LED strongly. Moreover, the phosphors should capable of retaining their luminescence even at high operating LED temperatures. In this case, the luminescence intensity is only 53% at the temperature of 423 K that is comparable with other Mn^{4+} systems of 50% for $Gd_2ZnTiO_6:Mn^{4+}$ [60] and $Sr_4Al_{14}O_{25}:Mn^{4+}$.[137]

CaGdAlO_4$:Mn^{4+}$ prepared by solid-state reaction also exhibits an intense sharp NIR emission at 715 nm due to the spin-forbidden $^2E_g \rightarrow {}^4A_{2g}$ transition of Mn^{4+} under the excitation of 349 nm. Again it is well clear that the obtained emission matches with the absorption of phytochrome *Pfr*, but excitation bands are maximized only at 352, 394, and 489 nm due to $^4A_2 \rightarrow {}^4T_1$, $^4A_2 \rightarrow {}^2T_2$, $^4A_2 \rightarrow {}^4T_2$, respectively.[138] Aforementioned, the absorption of blue light from InGaN LED is a mandatory requirement for the visualization of plant growth applications in real time. In short, most of Mn^{4+}-activated oxide, borates, tungstate, perovskites, and oxyfluorides are efficient in emitting the NIR light that coincides well with the absorption of phytochrome *Pfr*. For example, $Ca_3La_2W_2O_{12}:Mn^{4+}$ ($\lambda_{emission}$=711 nm, $\lambda_{excitation}$=360 and 469 nm), [139] $KLaMgWO_6:Mn^{4+}$ ($\lambda_{emission}$=696 nm, $\lambda_{excitation}$=365 and 486nm), [140] $La_2LiSbO_6:Mn^{4+}$ ($\lambda_{emission}$=695, 706, 713, 716, 722, 729 nm, $\lambda_{excitation}$=314, 342, and 470nm).[141] However, the inefficient absorption of blue light and poor temperature stability turns Mn^{4+}-activated system into an inefficient candidate for plant growth application.

Cr^{3+} is another popular transition metal element to obtain NIR light. However, in the case of Cr^{3+}-doped materials, there is the probability of both spin-forbidden and spin-allowed transitions based on the strength of the crystal field. As a recap, the spin-forbidden transition will always occur due to the stronger crystal field and results in the line or narrow emission. On the other hand, broadband is caused due to weaker crystal fields and spin-allowed transition. The line or narrow emission spectrum of Cr^{3+}-doped material is more recommendable over the broadband emission for plant growth because the emission wavelength over 750 nm is considered as the energy spillover and unnecessary for the plant growth. Moreover, it also reduces the efficiency of pc-NIR LED devices. Hence, Cr^{3+}-doped materials with the crystal field strength (Dq/B) more than 2.3 is considered as the potential candidate for plant growth. Even though the excitation band is not maximized at 450 nm, yet it is excitable by blue light effectively.

One of the best examples is Cr^{3+}-doped $Y_3Ga_5O_{12}$. It exhibits distinct three excitation bands located at 284, 447, and 619 nm due to $^4A_2 \rightarrow {}^4T_1(^4P)$, $^4A_2 \rightarrow {}^4T_1(^4F)$, and $^4A_2 \rightarrow {}^4T_2(^4F)$ transitions, respectively. Under 254 nm excitation, the emission spectrum is composed of a sharp *R* line at 690 nm superimposed on the broadband emission in the range of 650–800 nm.[1] Hence, it is very clear that the Cr^{3+}-doped $Y_3Ga_5O_{12}$ can absorb the blue light, and also the emission spectrum can overlap with the phytochrome photoreceptors.

Fang *et al.* [142] reported the NIR phosphor $Ga_2O_3:Cr^{3+}$ with the highest internal quantum efficiency of 92.4% through solid-state reaction. Aforementioned, the excitation bands of Cr^{3+} doped can't be maximized at 450 nm. As like, the photoluminescence excitation spectrum of $Ga_2O_3:Cr^{3+}$ is characterized by the two bands at 440 and 608 nm, which is very close to the absorption of blue light especially at 450 nm. On the other hand, the NIR emission consists of two sharp *R* lines at 695.5 and 688.8 nm that correspond to the $^2E \rightarrow {}^4A_2$ transitions. The existence of *R* lines superimposed on the broadband of 650–950 nm indicates the existence of a stronger crystal field character. Besides, the calculation of crystal field strength as ~2.6 evidences the observation. In our opinion, broadband can be due to electron-phonon coupling. Although the emission range is over 750 nm in $Ga_2O_3:Cr^{3+}$, the emission is maximized within 650–720 nm. It means that the spillover is still there, but it will be minimum as compared with the pure broadband emission of phosphors. The potential of $Ga_2O_3:Cr^{3+}$ is demonstrated by the growing two kinds of the plant of *Aglaonema* and *Plectranthus amboinicus* under the fabricated LED. They demonstrated by exposing both plants to the sunlight in the day time, but only one group of the plant is exposed to fabricated NIR LED for 12 h. As the result, *Aglaonema* and *P. amboinicus* plants exposed to NIR grew 5 and 8%, respectively, higher than the control plant group. In short, Cr^{3+}-doped materials with the higher crystal field strength can also be suitable candidates for plant growth applications.

The rare-earth ions are attractive activators for plant growth applications in current times due to their unique feature of broader excitation spectrum span from UV to visible. Also, the quantum efficiency and temperature stability of the rare-earth ions' activator compounds are considerably high. Among the rare-earth ions, Eu^{2+}, Eu^{3+}, Sm^{2+} ions can generate both red and NIR photons on excitation. Due to the inter-configurational electronic transition of Eu^{2+} ($4f^7(^8S_{7/2} \rightarrow {}^4F_6(^8S_{7/2})5d^1$), it can generate red photons only for phytochrome *Pr* than *Pfr*. Eu^{2+}-activated $Li_2Ca_2Mg_2Si_2N_6$ is reported with the red emission and FWHM of 638 and 62 nm, respectively, at the excitation wavelength of 460 nm.[143] As expected, the photoexcitation spectrum of $Li_2Ca_2Mg_2Si_2N_6$:Eu^{2+} shows broad absorption in the blue region as like diffuse reflectance spectrum. Besides, $Li_2Ca_2Mg_2Si_2N_6$:Eu^{2+} still retains the luminescence intensity of 64% at 473 K. Overall, the *Pak-choi* seedlings grown under the sunlight for 14 days were exposed to white LED and fabricated LED with 455 nm for 10 days. The plant exposed to the fabricated LED shows increased fresh weight, total chlorophyll content, soluble protein content, and total soluble sugar than the daylight and white LED. Similarly, $Ba_{0.8}Sr_{0.2}Mg_3SiN_4$:$Eu^{2+}$ also emits only red photons at 635 nm under the excitation wavelength of 450 nm. Even though the excitation spectrum is not broad, there are three bands maximized at 360, 450, and 550 nm. Besides, the phosphors also show quantum efficiency of ~50% as shown in Figure 3.8(c).[144] Thus, it can conclude that the Eu^{2+}-doped compounds have strong broad blue absorption, good quantum efficiency, and convincing thermal stability, but it can't generate a NIR photon. Hence, the application of Eu^{2+}-doped compounds for plant growth application is limited.

On the other hand, the Eu^{3+} ion can emit light in the range of 570–720 nm due to the intra-configurational electronic transition ($4f \rightarrow 4f$). The emission spectrum of the Eu^{3+}-doped compound is characterized as the presence of several narrow lines in both red and NIR regions with the respective electronic transition from $^5D_0$ to ^4F_J (J=6). However, the emission at 615 nm is strongest among the others and assigned as $^5D_0 \rightarrow {}^7F_2$ transition, while the infrared emission at 657 ($^5D_0 \rightarrow {}^7F_3$) and 703 nm ($^5D_0 \rightarrow {}^7F_4$) is slightly weaker. The luminescence intensity of Eu^{3+} is highly sensitive to its coordination environments in the structure especially $^5D_0 \rightarrow {}^7F_2$. Hence, the luminescence intensity ratio between 615 and 702 nm can be varied greatly either by selecting the host or crystal structure tuning. For example, in the crystal system of $Sr_2ZnGe_2O_7$ (tetragonal, $P\bar{4}21m$), Eu^{3+} is considered to be in Sr^{2+} site with the coordination number (CN) of 8. In this case, the NIR emission at 704 nm ($^5D_0 \rightarrow {}^7F_4$) is higher than the red emission at 615 nm ($^5D_0 \rightarrow {}^7F_3$) under 394 nm.[145] In $Ca_{14}Al_{10}Zn_6O_{35}$:Eu^{3+} (Cubic, $F23$) system, Eu^{3+} replaced Ca^{2+} ions in the octahedral coordination shows the predominant red emission at 612 nm.[146] Similarly, in $NaYb(MoO_4)_2$:Eu^{3+} system, the red emission at 615 nm (electric dipole transitions) takes the highest luminescence than 655 and 592 nm.[147] Lu_2GeO_5:Eu^{3+} also shows higher luminescence intensity at 612 nm than the other typical emission lines of 594, 655, and 703 nm.[148] Based on the host, Eu^{3+} ions can emit selective red and NIR photons to match the absorption spectrum of both *Pr* and *Pfr* which is advantageous for plant growth. However, the major drawback is the excitation by UV light. It is hardly excitable by blue light. To enhance the absorption of blue light, it is mandatory to co-substitute the sensitizer like Bi^{3+}.

Sm^{2+} with $4f^6$ electronic configuration is more beneficial than other rare-earth ions and transition metal ions for plant growth applications. Since the nuclear charge of Sm^{2+} is lesser than Eu^{3+}, the energy of Sm^{2+} ($4f^6$) is decreased by about 15–20% relatively. This energy difference causes Sm^{2+} ions to emit photons effectively in the range of 590–700 nm. As a concern with the emission spectrum, Sm^{2+} can encounter both inter-configurational and intra-configurational emissions transition as compare with Eu^{2+} and Eu^{3+}. Again, the type of emission transition depends on the relative positions of $4f^55d$ and $^5D_0$ ($4f^6$) excited states, and also the Stokes shift of emission. For example,

β-SiAlON:Sm^{2+} exhibits intra-configurational emission as the typical four main sharp emission lines at 685 ($^5D_0 \rightarrow ^7F_0$), 690 ($^5D_0 \rightarrow ^7F_1$), 728 ($^5D_0 \rightarrow ^7F_2$), and 763 nm ($^5D_0 \rightarrow ^7F_3$) as shown in Figure 3.8(d).[149] At the same time, it can also show broadband around 600–800 nm due to the inter-configurational emission ($4f^5 5d \rightarrow 4f^6$). At 250°C, the luminescence intensity is still 84% as compared with the room temperature. On the other hand, SrAlF$_5$:Sm^{2+} exhibits intra-configurational emission only around 650–750 nm with the three sharp peaks at 680 ($^5D_0 \rightarrow ^7F_0$), 690 ($^5D_0 \rightarrow ^7F_1$), and 720 nm ($^5D_0 \rightarrow ^7F_2$).[150] This is so clear that the intra-configurational ($4f^6 - 4f^6$) emission transition has a good overlap with the absorption spectra of the phytochromes (*Pr* and *Pfr*). In contrast, Sr$_2$B$_5$O$_9$Cl:Sm^{2+} exhibits pure inter-configurational emission at 700 nm, which also overlaps with the phytochrome absorption spectra.[151] Another advantage of the Sm^{2+} compound is the excitation band, it can span from UV to visible due to the inter-configurational $4f^6 \rightarrow 4f^5 5d$ transition. Furthermore, the stronger coupling between $5d$ orbitals and $4f^6$ cores leads to the arising of several absorption components that make the Sm^{2+}-doped compounds as good absorbing toward blue photons. The thermal stability is also good for Sm^{2+}-doped compounds. For example, in Ca$_{0.91}$Y$_{0.09}$F$_{2.09}$:Sm^{2+}, the lowest Sm^{2+} $4f^5 5d$ state is very close to the conduction band. Hence, the increase of temperature leads to enhancement of ionization rate from $4f^5 5d$ to the conduction band and decreases the population of ions at the $^5D_0$ level. Thus, non-radiative relaxation is slowed. In short, Sm^{2+} is the promising rare-earth activator ion to stimulate the phytochromes photoreceptors (*Pr* and *Pfr*) due to its distinct broader absorption spectrum, better quantum efficiency, good temperature stability, and less spillover of energy for plant growth. However, the significant disadvantage of Sm^{2+} is the material selectively. Only fluoride compounds can sustain the Sm ions in the valence of +2. In oxide and nitride compounds, Sm ions will always exist as Sm^{3+} instead of Sm^{2+} ions.

3.4.5 PHOSPHORESCENCE

Persistent luminescence/phosphorescence (Pers PL) is the fascinating optical process, where the emission of light from the luminescent material persists for the longest time from several seconds to a few days even after ceasing the source of excitation. It is also called phosphorescence, and afterglow or long-lasting phosphorescence. In general, phosphor for persistent luminescence phosphors consists of two active components. One is the emitter, while the other is a trap. The emitters are the active luminescent centers and emit the light photons, while the traps are crystal lattice defects or co-dopants and perform the function of capturing the excitation energy in terms of electrons and gradually transfer to the emitter for the photoemission under the thermal or physical irradiation. To add further, the emitter decides the wavelength of photons, while the photoemission intensity and duration are related to the trap type (shallow or deep) and its level. The solid-state reaction is the traditional method to synthesize persistent luminescence phosphors. Currently, the persistent luminescence phosphors in nanometer size for the bio-imaging application was also synthesized through the hydro/solvothermal method, template method, wax-sealed method, and so on.[152]

For NIR Pers PL, the best choices of activators can be transition metal elements such as Cr^{3+}, Mn^{4+}, and Mn^{2+}. To explain the Pers PL mechanism in detail, let us consider Cr^{3+} in a strong crystal field as a typical example. Figure 3.9 shows the schematic diagram for the energy level of Cr^{3+} between the valence and conduction band of the host. Moreover, it is also assumed that Cr^{3+} containing host in this typical example possesses both shallow and deep trap. In general, the shallow traps are due to the intrinsic defects in the compound such as anion vacancies, antisite defects due to the inequivalent chemical substitutions and exist the trap depth of 0.5–1 eV, while the deep traps are due to the addition of co-dopants with the trap depth of more than 1 eV. For Cr^{3+}, $^4A_2$ is the ground state, while the first and second excited states are $^4T_1$ and $^4T_2$, respectively. When the light is irradiated for a defined time, the electrons in the ground state ($^4A_2$) of Cr^{3+} will be photoionized to the conduction band (step 1). After that, the electrons in the conduction band will be consequently trapped by both shallow trap (step 2) and deep trap (step 3). After the process of trapping, the de-trapping process will start by the release of electrons thermally from shallow trap via conduction

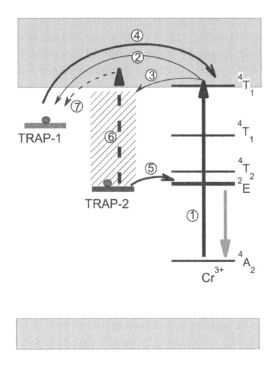

FIGURE 3.9 Persistent luminescence mechanism of Cr^{3+}. (Liu, F., Yan, W., Chuang, Y.J., Zhen, Z., Xie, J. and Pan, Z., *Sci. Rep.* 3, 1554, 2013. With permission)

band and recombine with the ionized Cr^{3+} ions for the initial NIR Pers PL (step 4). The de-trapping of electrons from a shallow trap will continue till the electron can be emptied. Once the electrons were emptied in the shallow trap, the electrons will be de-trapped from the deep traps for the Pers PL (step 5).[153] It should be important to note that the physics of the Pers PL phenomenon is quite complex and yet a lot of research has been going for it. And, the trap type and its number can vary for different materials.

As discussed in the previous sections of crystal field splitting, Cr^{3+}-doped compounds can be a tune for both narrowband and broadband emission of NIR light by simply modifying the value of crystal field strength. Hence, Cr^{3+}-doped NIR PL materials have created increasing interest among the researchers. Table 3.1 lists Cr^{3+}-activated PL materials reported for the past decade. The most widely common and promising host for Cr^{3+} NIR persistent luminescence is the spinel and inverse structure due to the inherent generation of antisite defects and oxygen vacancy on the incorporation of Cr^{3+} ions. Besides, the probability of a stronger crystal field for Cr^{3+} in the spinel and inverse structure further adds the advantages. Cr^{3+}-doped $ZnGa_2O_4$ is known for spinel structure, but it also possesses a few percentages of inverse spinel symmetry too (nearly 3%). Hence, Cr^{3+} ions can present in the regular octahedral (Cr_n^{3+}) and slightly distorted octahedral (Cr_{dis}^{3+}) in $ZnGa_2O_4$. Thus, Cr^{3+} ions in unperturbed octahedral and slightly distorted octahedral (neighboring antisite defect) lead to the respective emission of R line and $N2$ line at 688 and 695 nm in the range of 650–750 nm. At the same time, the R line and $N2$ lines are accompanied by the Stokes phonon sidebands (S-PSB) at 708 and 715 nm and AS-PSB at 663, 670, and 680 nm as just like the typical ruby. By varying the nominal ratio of Zn/Ga, the antisite defects of ZnGa′ (negatively charged) and GaZn° (positively charged) that in turn can vary the intensity ratio of $N2/R$ due to the distortion of the local polyhedron of Cr^{3+}. Furthermore, the persistent luminescence decay time is also improved for lowering the nominal ratio of Zn/Ga (higher antisite defects). As a whole, it is observed that Cr^{3+} ions in the neighboring antisite defect play a critical role in the trapping process of electrons, and the direct excitation of the $N2$ line can trigger the performance of persistent luminescence.[154]

Pan *et al.* [77] included the tetravalent cation Ge^{4+} in the typical spinel structure formula and designed the system of Cr^{3+}-doped zinc gallogermanates with the standard chemical formula of $Zn_xGa_yGe_zO_{(x+(3y/2)+2z)}:tCr^{3+}$, mR, where R is a co-dopant (alkaline earth ions, lanthanide ions, and Li^+ ions), x, y, and z are integers from 1 to 5; t is 0.01–5 mol%, and m is 0–5 mol%. Every mixture of these variables will still yield materials with a remarkable NIR of continuous luminescence. It is reported that the $Zn_3Ga_2Ge_2O_{10}:0.5\%$ Cr^{3+} can exhibit the NIR persistent luminescence decay time for more than 360 h in the range of 650–1000 nm centered at 696 nm ($^2E{\rightarrow}^4A_2$). It is further believed that the Ge^{4+} vacancies (V_{Ge}) and O^{2-} vacancies (V_O) generated at higher sintering temperatures can be found in the neighborhood of Cr^{3+} ions and form the defect cluster of $V_{Ge}-Cr^{3+}-V_O$. Thus, it acts as the electron trapping centers to store energy in the zinc gallogermanates. In contrast, the emission spectrum is blue-shifted to 694 nm in the system of Cr^{3+}-doped $Zn_3Al_2Ge_2O_{10}$.[3] However, the persistent luminescence activity is still due to the intrinsic defects of Zn^{2-} vacancies and O^{2-} vacancies. The addition of Ca^{2+} can stabilize the O^{2-} vacancies and intensifies the persistent time. The addition of alkaline earth ions, lanthanide ions, and Li^+ ions as co-dopants can intensify the NIR emission and afterglow time.

The lanthanum gallogermanates system of $La_3Ga_5GeO_{14}:Cr^{3+}$ can emit broadband NIR luminescence in the range of 600–1500 nm with a persistent time of more than 1 h under UV excitation (~240–360 nm).[155] However, the addition of co-dopants such as Li^+, Zn^{2+}, Ca^{2+}, Mg^{2+}, and Dy^{3+} will create more trapping centers for carriers, which in turn can boost up the persistent time and intensity of phosphorescence.[156] As compared with zinc gallogermanates, the phosphorescence luminescence activity in $La_3Ga_5GeO_{14}:Cr^{3+}$ is mainly of photogenerated Cr^{4+} ions instead of defect clusters. In brief, the electrons from the ground state of Cr^{3+} are stimulated to $^4T_1(te^2)$ or some energy levels with higher energies under UV irradiations. Then, the electrons are captured by the oxygen vacancies and generate Cr^{4+} ions. The thermal release of photogenerated Cr^{4+} ions leads to the phosphorescence of Cr^{3+}. Thus, it can infer that the persistent luminescence activity is independent of active luminescent centers, and purely influential by the host structure and chemical constituents. However, the FWHM of the emission spectrum can determine by the crystal field strength of the active luminescent center.

Cr^{3+}-doped garnet structure is also an exciting material for the persistent luminescence studies due to the flexibility of cation substitution. Xu *et al.* [6] reported two different series of Cr^{3+}-doped garnets with the generalized chemical formulae of $A_3Ga_{4.99}Cr_{0.01}O_{12}$ (A=Y, Gd, Lu) and $A_3Sc_{1.99}Cr_{0.01}Ga_3O_{12}$ (A=Y, Gd, Lu) for the NIR luminescence in the range of 650–950 nm. On considering the ionic size order of $Gd^{3+}>Y^{3+}>Lu^{3+}$ and $Sc^{3+}>Ga^{3+}$, the crystal field strength of Cr^{3+} is gradually weakened. Hence, the spin type of Cr^{3+} transition changes from spin-forbidden in $Lu_3Ga_{4.99}Cr_{0.01}O_{12}$ to spin-allowed in $Gd_3Sc_{1.99}Cr_{0.01}Ga_3O_{12}$. In other words, $Lu_3Ga_{4.99}Cr_{0.01}O_{12}$ shows the spectral features of R line at 690 nm and phonon-side bands, while $Gd_3Sc_{1.99}Cr_{0.01}Ga_3O_{12}$ has only broadband at 770 nm. Thus, the change of Cr^{3+} spin-type affects the fluorescence decay time and also the persistent intensity. However, all six garnets show the consistent two types of traps. One of them is due to an electron trap connected to Cr^{3+} ions, and the other is due to an intrinsic in the garnet hosts. In general, the blue light (460 nm) irradiation can't able to assist the electron trapping-de-trapping channel through the conduction band due to the facility of Cr^{3+} ions to $^4T_1$ (^4F) only. In contrast, $Gd_3Sc_{1.99}Cr_{0.01}Ga_3O_{12}$ can persist luminescence even at 20 K. Thus, it suggests that the visible light irradiation is yet possible in the garnet structure through the tunneling process. As per the host-referred binding energy diagram, Yb^{3+} could introduce a new electron trap, which would lead to an increase in Cr^{3+} at room temperate in comparison with the other lanthanides ions of Sm^{3+}, Eu^{3+}, Tm^{3+}, Yb^{3+}.

Mn^{4+} also possesses the same electronic configuration as Cr^{3+}, but it will always experience a strong crystal field in the octahedral environment despite any host. However, the position of sharp emission lines ($^2E{\rightarrow}^4A_2$) can be variable based on the host in the region of 650–800 nm. Mn^{4+} has a wider choice of substitutability than Cr^{3+}. The ideal substitution elements for Cr^{3+} are Ga^{3+}, Al^{3+}, or Sc^{3+}, but Mn^{4+} can replace Ga^{3+}, Al^{3+}, Ge^{4+}, Si^{4+}, Ti^{4+}, and Zr^{4+} ions due to the similar valence and ionic size. Perovskites are considered as the ideal host for Mn^{4+} persistent luminescence due to the ability to generate several intrinsic defects on the substitution of nonequivalent elements in the structure. Perovskite belongs to the rhombohedral crystal system and can describe with the generalized formula of ABO_3, where $A=La^{3+}$, Gd^{3+}, Y^{3+}, and $B=Sb^{2+}/Zn^{2+}–Ti^{4+}$ pair. Let us consider $LaAlO_3$ perovskite as a typical example.[157] In $LaAlO_3$ structure, Mn^{4+} is believed to substitute for the central cation Al^{3+} is in octahedral coordination. $LaAlO_3$:Mn^{4+} can show persistent luminescence in the range of 650–750 nm with a persistent time of more than 4 h. Even though the ionic sizes of Al^{3+} and Mn^{4+} ($Mn^{4+}=0.53$ Å and $Al^{3+}=0.535$Å) are nearly identical, the inequivalent valence leads to the generation of one positive defect for $[Mn_{Al}]$. To neutralize this positive defect, the structure needs equivalent negative charges or annihilation of positive effective charge. Hence, the shallow and deep traps for the electrons storage are formed due to the positive effective charges of $[Mn_{Al}]$ and negative effective charge of $[V_{Al}]$. Other than these two, the system can also generate the defects of positive effective charges of $[Al^I]$, $[La^I]$, $[V_O]$, and negative effective charges of $[O^I]$, $[V_{Al}]$, $[V_{La}]$, $[Mn_{Al}^{(II)}]$. Interestingly, Mn^{4+} is not the only emitting center, but also the defects. Hence, the defect concentrations can be increased by simplifying increasing the concentration, but the intensity of fluorescence will be quenched. To counteract this, Li *et al.* [157] co-substitute the tetravalent Ge^{4+} to enrichment the defect density. As a result, $LaAlO_3$:0.1% Mn^{4+}, 0.9% Ge^{4+} shows the afterglow time duration more than 20 h after the irradiation with UV for 10 min. Similarly, $GdAlO_3$:0.1% Mn^{4+}, 0.9% Ge^{4+} also show emission in the range of 650–750 nm with the six sharp peaks at 680, 685, 693, 698, 704, and 719 nm under the excitation wavelength of 326 nm. It also shows persistent luminescence of more than 20 h, but still, it is inferior as compared with $LaAlO_3$. On the other hand, the persistent luminescence intensity can enhance in $LaAlO_3$ by engineering the trap depth. Du *et al.* [158] engineered the trap depth by the substitution of Na^+, Ca^{2+}, Sr^{2+}, or Ba^{2+} in the polyhedral units of $[LaO_{12}]$.

Mn^{2+} is in d^5 electron electronic configuration and also has an unfilled d orbital as like Cr^{3+}, and Mn^{4+}. The energy transition of Mn^{2+} is comparatively sensitive to the crystal field because of the involvement of the outermost d shells. As Mn^{2+} ions are located in various crystallographic conditions, they can produce a broad emission band ranging from blue-green to far-red (parity-forbidden transition, $^4T_1$ [^4G] to $^6A_1$ [^6S]). In general, Mn^{2+} in the tetrahedral environment produces green emission, while Mn^{2+} in the octahedral environment can emit red to NIR emission based on the host chemical composition. In contrast to Cr^{3+} and Mn^{4+}, the emission wavelength of Mn^{2+} ions is chiefly decided by the CN in the hosts.

With the primary model of $CaAl_2O_4$:Mn^{2+}, Ce^{3+} for green persistent luminescence at 525 nm for over 10 h, it is extended to the yellow (550 nm) and red (660 nm) persistent luminescence by the modification of the host system as $Ca_2Al_2SiO_7$:Mn^{2+}, Ce^{3+}, and $MgSiO_3$:Mn^{2+}, Eu^{2+}, Dy^{3+}, respectively. The purpose of Dy^{3+} is to build hole-trapped centers in $MgSiO_3$:Mn^{2+}, Eu^{2+}, Dy^{3+}, where Mn^{2+} obtained energy from the excitation of electron-hole pairs. Inspired by the role of Mn^{2+} for red persistent luminescence in $MgSiO_3$, Scherman and coworkers further developed this model to the potential of NIR emission at 690 nm by the incorporation of Ca and Zn.[159] Besides, it is experimentally evidenced that Mn^{2+} is the final emitting center in $Ca_{0.2}Zn_{0.9}Mg_{0.9}Si_2O_6$:$Eu^{2+}$, Mn^{2+}, Dy^{3+}, whereas Dy^{3+} forms the trap centers. Quite different from other transition metal ion, Mn^{2+} emits NIR emission for several hours due to the persistent energy transfer (PET) from the rare-earth ions. [160] After multiple surface modification phases, the persistent luminescence nanoparticles of the same system were employed for in vivo imaging and opened up a wide variety of possible applications for biomedicine.

Besides the silicate system, the germanate hosts have also been well studied to induce Mn^{2+}-activated NIR Pers PL, such as $CaMgGe_2O_6$ and $MgGeO_3$.[161] Unfortunately, a few research works are dedicated to the design and discovery of Mn^{2+}-doped NIR persistent luminescence. Because, the majority of Mn^{2+}-doped persistent phosphors emit a green or orange, which can meet the criteria for bio-imaging application.

The rare-earth ions with the states of Eu^{2+}, Yb^{2+}, Pr^{4+}, Ce^{4+}, Tb^{4+}, and Sm^{2+} can also emit NIR emission based on the host material and sintering conditions. The unique feature of rare-earth ions is the denser energy level that makes the persistent luminescence excitation from UV visible. However, the $5d$ state of RE is too high to be charged and there are no carrier traps with sufficient depths or trap distributions at room temperature. Thus, the emission is restricted from orange to red for most of the rare-earth ions activated. A few host materials have been reported to provide RE ions-activated NIR PersL luminescence. Besides that, the persistent luminescence duration of rare-earth ion-activated compound is much inferior as compare with transition metal ions. For example, Yb-activated $(Ba_{1-x}Sr_x)AlSi_5O_2N_7$ Pers PL exhibits broadband NIR persistent luminescence centered at 664 nm and sharp line at 980 nm due to Yb^{2+} and Yb^{3+}, respectively. However, the persistent luminescence time is only for 60 min that is very low as compare with Cr^{3+} and Mn^{4+}.[162] In short, the rare-earth ion-activated compounds are still lacking and still more research has to come out. For the persistent luminescence in mid-infrared and far-infrared windows, the two strategies are commonly available. One is the doping of a single activator, while the other is a PET process between two or three different emitting centers.

The rare-earth ions of Yb^{3+}, Nd^{3+}, Er^{3+}, and Ho^{3+} are considered as the possible candidates for the persistent luminescence in mid-infrared and far-infrared due to the $4f$–$4f$ intra-configuration transitions. In other words, the emission wavelength for these rare-earth ions is preferably designed instead of a host system and crystal strength.

Yb^{3+} ions always have standard emission at 1000 nm due to the Stark splitting between the manifolds of $^2F_{7/2}$ (ground state) and $^2F_{5/2}$ (excited state). In general, it is greatly used in silicon photovoltaics along with lanthanide ions. Pan et $al.$ [163] demonstrated successfully that Yb^{3+}-doped $MgGeO_3$ can exhibits persistent luminescence at 1000 nm for more than 100 h. It is proposed out that the $MgGeO_3$:Yb^{3+} has shallow traps with continuous trap distribution and the high energy photochromic-center-type deep traps. Both traps can receive an electron from the host under UV excitation of 254 nm and forms a charge-transfer state. After that, the trapped electrons recombined with the charge transfer band (CTB) and subsequently populate the excited state of $^2F_{5/2}$ through the intersystem crossing process. Later, the persistent luminescence occurs at 1000 nm due to non-radiative relaxation from $^2F_{5/2}$ to $^2F_{7/2}$. On the other hand, the emission can be extended from 900–1700 nm by the incorporation of trivalent lanthanide ions such as Er^{3+}, Yb^{3+}, Pr^{3+}, Nd^{3+}, Ho^{3+}, and Tm^{3+} in the host structure of $CaSn_2O_4$.[164]

Pr^{3+} is known for red emission at 608 nm due to the $^1D_2 \rightarrow ^3H_4$ transition. However, Pr^{3+} ions can also emit infrared emission in the second window from the states of $^3H_5$ (~720 nm), $^3H_6$ (~900 nm), or $^3F_{3,4}$ (~1080 nm) to $^1D_2$. Hence, Pr^{3+} is also one of the desired candidates to achieve mid-infrared emission. For example, $MgGeO_3$:Pr^{3+} and $CdSiO_3$:Pr^{3+} emit strong afterglow emission in NIR-I and NIR-II in addition to conventional red emission. The chicken breast model tissue examination pieces of evidence that the mid-infrared emission at 1085 nm had better penetration strength.[83]

Nd^{3+} ions can show emission at ~900, ~1060, and ~1340 nm due to the energy transitions from the $^4F_{3/2}$ state to other states of $^4I_{9/2}$, $^4I_{11/2}$, $^4I_{13/2}$, respectively. The parity-forbidden transition of $4f$–$4f$ makes Nd^{3+} ions a low extinction coefficient. Besides, Nd^{3+} can create traps with appropriate depth to facilitate the mid-infrared persistent luminescence of other rare-earth ions. For example, $Sr_3Sn_2O_7$:Nd^{3+} emits intense mid-infrared Pers PL between 800 and 1400 nm due to the $4f$–$4f$ transitions of Nd^{3+} ions.[165] The unusual configuration of oxygen octahedral tilting and rotation and intra-layer plane defect in $Sr_3Sn_2O_7$:Nd^{3+} led to a higher Pers PL strength than the standard perovskite structure of $SrSnO_3$. As a proof-of-concept, the image of chicken wings recorded by the InGaAs-CCD camera shows a high image contrast. Unfortunately, the persistent duration is only ~1 h that doesn't meet the requirements for biomedical applications.

The photoionization processes of the rare-earth ions are difficult to accomplish under atmospheric conditions, which hinder their ability as efficient emitting centers for the development of NIR Pers PL. Hence, the PET between two or three different emitting centers is required to visualize the infrared emission in mid-infrared and far-infrared. Tanabe and coworkers reported plenty of research works for the PET process with the pairs of Er^{3+}–Cr^{3+}, Ce^{3+}–Cr^{3+}, Nd^{3+}–Ce^{3+}–Cr^{3+}, Ho^{3+}–Cr^{3+}, Cr^{3+}–Sm^{3+}, Ho^{3+}–Cr^{3+}–Sm^{3+}.[6, 105, 166–168] Then, we consider the examination of garnet structured $Y_3Al_2Ga_3O_{12}$ co-doped with Er^{3+} and Cr^{3+} ions as the typical example. $Y_3Al_2Ga_3O_{12}:Er^{3+}$ and $Y_3Al_2Ga_3O_{12}:Cr^{3+}$ show respective intense emission at 700 nm due to $^2E \rightarrow ^4A_2$ transition of Cr^{3+}, and peak emission at ~1532 nm due to $^4I_{13/2} \rightarrow ^4I_{15/2}$ transition of Er^{3+}.[6] Moreover, both Cr^{3+} and Er^{3+} show a persistent duration of more than 10 h. Due to possible energy transfer from $Cr^{3+} \rightarrow Er^{3+}$, the emission spectrum of $Y_3Al_2Ga_3O_{12}:Er^{3+}-Cr^{3+}$ shows obvious splitting at 966 and 1532 nm due to the transition of Er^{3+} from $^4I_{11/2}/^4I_{13/2} \rightarrow ^4I_{15/2}$. Under UV irradiations, the excited electron of Cr^{3+} was caught by the electron trapping centers with two separate trap depths, and subsequently, electrons are released from the traps to the excited state $(Cr^{3+})^*$ with the aid of thermal activation at room temperature. Then the standard persistent luminescence process of Cr^{3+} has happened. All at once, the Cr^{3+} to Er^{3+} PET was followed by the Er^{3+}-related persistent luminescence at 966 and 1532 nm. Furthermore, owing to the significantly decreased light dispersion of Er^{3+} Pers PL around 1500 nm falling into the far-infrared range, the optical imagery of 1-cm thick raw pork tissues shows that the Er^{3+} PersL with a longer wavelength can obtain better image contrast and higher spatial resolution than the Cr^{3+} Pers PL in the NIR frame.

Transition metal ions Ni^{2+} and Co^{2+} ions have been big choices in the selection of NIR-persistent luminescence over 1000 nm. $Zn_3Ga_2Ge_2O_{10}:Ni^{2+}$, $LiGa_5O_8:Ni^{2+}$, and $La_3Ga_5GeO_{14}:Ni^{2+}$ show the broadband infrared emission in the range of 1000–1600 centered at ~1290, ~1220, and ~1430 nm, respectively, due to the spin-allowed transitions $^3T_2$ (^3F) $\rightarrow ^3A_2$ (^3F) of Ni^{2+}.[108] $Zn_3Ga_2Ge_2O_{10}:Ni^{2+}$ can persist for more than 12 h. By following that, Pellerin et al. [169] reported that the local crystal field around Ni^{2+} ions can be tailored by the element substitution of Sn and Zn for Ga in the $ZnGa_2O_4$ host. Thus, the centered emission is effectively adjusted from 1270–1430 nm in the mid-infrared window.

Due to the advantage of background free, the persistent luminescent material in mid-infrared and far-infrared is currently targeted for anti-counterfeiting technology. $Zn_2Ga_2Sn_{0.5}O_6:Ni^{2+}$ has the strongest emission at 1350 nm, while the doping of Yb^{3+} intensifies the emission.[170] As the proof-of-concept, $Zn_2Ga_2Sn_{0.5}O_6:Ni^{2+}$, Yb^{3+} is integrated with visible persistent luminescence for multichannel light anti-fake effects can be accomplished at various sources of excitation and at different periods, which demonstrated a higher camouflage, flexibility, and signal-to-noise ratio relative to current phosphors.

3.4.6 Anti-Counterfeiting

Globalization connects the world in different aspects that include technology, trade, economy, culture, media, and politics. However, counterfeiting activities threaten the integrity of globalization and also challenge the governments, industries, and customers. Several anti-technologies have been developed in the past decades to prevent unlawful activities in the modern world such as magnetic response, plasmonic security labels, and taggants. Among them, the usage of taggants is considered to be ideal techniques due to their low cost, better coding/decoding capacity, nontoxic, simplicity, and eco-friendly.[171] Based on the application purpose, the taggants materials can be classified into four categories, property marketing, anti-counterfeiting, tracking, and monitoring. To meet the abovementioned purposes, the taggants can be classified into three types based on their forms such as physical, spectroscopy, DNA, and chemical taggants.[171–173] For anti-counterfeiting purposes, the chemical as well as spectroscopic taggants are applicable. In both methods, more than two organic or inorganic luminescence material possessing different optical properties are combined to form the single mixture with a unique spectral signature. However, in the spectroscopic method, the emission profile of the applied taggant is measured by simple spectrophotometry techniques, while the chemical coding in terms of binary digits or letters is accomplished by the fixed mixture that can be used either be present or be absent from a mixture in the chemical method.[172]

Under the excitation of UV or NIR light, the conventional inorganic luminescence materials used for anti-counterfeiting purposes may exhibit single- or double-mode fluorescence in different luminescence techniques such as photoluminescence, persistent luminescence, and UC luminescence.

However, it can be counterfeited sometimes with ease by selecting the other materials that also have similar emissions. Hence, the designing of inorganic luminescence material with multimodal luminescence is essential and in greater demand for effective anti-counterfeiting activities.

The hydrothermal methods synthesized Y_2O_3:Yb^{3+}/Er^{3+} possess the morphology of nanorods and can emit green and red light at 562 ($^2H_{11/2}$, $^4S_{3/2}$/$^4I_{15/2}$) and 660 nm ($^4F_{9/2}$/$^4I_{15/2}$), respectively, under the excitations of 980 nm.[171] The green emission is more intensified than red light due to the energy transfer and excited-state absorption from Yb^{3+}→Er^{3+}. Besides, there is also a strong emission at 1127 nm for 980 nm excitation. Interestingly, Y_2O_3:Yb^{3+}/Er^{3+} nanorods also exhibit downshifting luminescence in the range of 1300–1700 nm for the same 980 nm excitation. Furthermore, Y_2O_3:Yb^{3+}/Er^{3+} nanorods exhibit only green emission at 563 nm under the conventional downshifting excitation wavelength of 453 nm. Thus, 980 nm excitation plays a dual role of UC by emitting green light at 563 nm and also downshifting by the emission of infrared light in the range of 1300–1700 nm. Conversely, it is also excitable by 453 nm for green emission. Such a feature of multiple excitations Y_2O_3:Yb^{3+}/Er^{3+} on is advantageous for anti-counterfeiting. At last, the luminescence ink was prepared by mixing Y_2O_3:Yb^{3+}/Er^{3+} powders with an inexpensive PVC gold medium and printed the security codes on the black paper by a standard screen printing technique. Under normal light, the printed pattern is not visible to the naked eye, but the pattern glows green and red color under the excitation of 980 nm. However, further excitation with 379 nm inhibits the green color. At last, the more realistic proof of multiple excitations on the Indian currency as a legitimate proof of its application.

Shi *et al.* [174] proposed the concept of trichromatic persistent luminescence for anti-counterfeiting applications. In the host structure of $Na_2CaGe_2O_6$, the substitution of Pb^{2+} and Y^{3+} results in persistent deep-blue emission due to the creation of photo chargeable energy trap and trapping level by Pb^{2+} and Y^{3+}, respectively. Similarly, the co-substitution pairs of Pb^{2+} and Tb^{3+} lead to the green emission, while the pair of Pb^{2+}, Mn^{2+}, and Yb^{3+} results in the red emission. Interestingly, under the emission of 400 nm, the excitation spectrum spans from 250 to 320 nm, while the excitation spectrum diminishes to 230–320 and 230–260 nm for the 550 and 600 nm, respectively. This demonstrates that the emission color change from white-blue to deep-blue, green-blue to deep-blue, and red-violet to blue-violet by simply varying the excitation wavelength from 250 to 320 nm. The time-resolved and color-coded multiple informational images with three distinct features were designed with persistent luminescence materials and cotton fabric. The butterfly pattern, the flower pattern, and the leaf are made up of $Na_2CaGe_2O_6$:Pb^{2+}, Y^{3+}; $Na_2CaGe_2O_6$:Pb^{2+}, Tb^{3+}; and $Na_2CaGe_2O_6$:Pb^{2+}, Mn^{3+}, Yb^{3+}, respectively. Under normal light, the printed image on common cotton fabric is not visible, but the white-blue butterfly, the red-violet flower, and the green-blue leaf can be observed under 24 nm. Unlike conventional anti-counterfeiting materials, the green and red light fade immediately after shutting off the UV light, but only blue fluorescence glows. After the excitation of 295 nm, the whole image appears again.

As an alternative to excitation-dependent dual-mode luminescence, Zhang *et al.* [175] proposed a multimode emission technique for anti-counterfeiting and encryption applications. The nanoparticles of $ZnAl_{1.4}Ge_{0.3}O_{3.7}$ and $ZnAl_{1.4}Ge_{0.3}O_{3.7}$:1% Cr^{3+} prepared through the hydrothermal method exhibit the dual emissions of blue light and NIR light centered at 690 and 475 nm. However, the blue emission is excitable only by UV less than 300 nm, while NIR emission is excitable by both visible light and UV light. Hence, the blue and NIR emission was assigned due to the Zn vacancies of a host, and GeZn antisite defects close to Cr^{3+}, respectively. Thus, the proposed system $ZnAl_{1.4}Ge_{0.3}O_{3.7}$:1% Cr^{3+} itself possesses the strong NIR persistent luminescence of more than 10 h. Besides, the photo-stimulated thermoluminescence of 980 nm intensifies the NIR emission of 690 nm due to the release of more electrons from the deep trap. Thus, the combined features of dual photoluminescence at 475 and 690 nm for different excitations, NIR persistent luminescence, and photo-stimulated luminescence by 980 nm are the advantages for the modern anti-counterfeiting applications. In the end, the effectiveness of proposed triple mode emissions for anti-counterfeiting is demonstrated by printing the number pattern (100 and 738) and QR code on the paper with the prepared luminescence inks of $ZnAl_{1.4}Ge_{0.3}O_{3.7}$ and $ZnAl_{1.4}Ge_{0.3}O_{3.7}$:1% Cr^{3+} as cyan and magenta. Under the excitation of 254 nm, the number codes of 100 and 738 show different colors initially. At the same time, the shutting off of 254 nm light fades the bluish-white and reveals the NIR emission of $ZnAl_{1.4}Ge_{0.3}O_{3.7}$:1% Cr^{3+} strongly. Consequently, the photostimulation by using 980 nm light can intensify the NIR persistent luminescence.

The anti-counterfeiting security level can be increased further by designing the persistent luminescent phosphors that provide the emission in the second window of infrared light. Ma *et al.* [170] reported that the substitution of Yb^{3+} and Ni^{2+} ions in the spinel structure of $Zn_2Ga_2Sn_{0.5}O_6$ can result in the emission in the range of 1100–1600 nm centered at 1349. The emission arises due to spin-allowed transition $^3T_2(^3F) \rightarrow {}^3A_2(^3F)$ of Ni^{2+} in octahedral coordination. On the other hand, $Zn_2Ga_2Sn_{0.5}O_6$:Yb^{3+}, Ni^{2+} can be versatility excitable by both UV and red light of 396 and 640 nm. Besides, the characteristics of photoluminescence and persistent luminescence with stronger glow times are critical and advantageous for anti-counterfeiting. As per that condition, $Zn_2Ga_2Sn_{0.5}O_6$:Yb^{3+}, Ni^{2+} has the stronger fluorescence intensity at 1349 nm for nearly 70 times of excitation-emission cycles, and also stable persistent luminescence intensity with the decay time of ~5 min for one cycle. As the proof-of-concept, the "U" pattern is constructed independently with $Zn_2Ga_2Sn_{0.5}O_6$:Yb^{3+}, Ni^{2+} and visible persistent luminescent commercial phosphors. It is anticipated that the U pattern consisting of visible persistent phosphors is observable for 15 s after UV shut off, while the pattern made up of $Zn_2Ga_2Sn_{0.5}O_6$:Yb^{3+}, Ni^{2+} is indistinguishable. However, the usage of the NIR detector reveals the pattern clearly, which evidence the second infrared phosphor advantaged for security information and anti-counterfeiting. Not limited to, the application of different excitation sources such as UV and 808 nm laser reveals the hidden number separately which further the information storage.

3.5 CONCLUSIONS

We have summarized the activators and host systems of phosphors reported for the NIR and mid-infared emissions. For the broadband NIR emission, as Cr^{3+} as the potential emitter, the effect of crystal system selection on the crystal field strength, which in turn affects the emission spectrum and tuning strategies of centered emission are discussed briefly. Apart from that, the working principle, current, and future opportunities of research are predominantly explained in the fields of NIR spectroscopy, night vision, solar cell, phosphoresce, and anti-counterfeiting applications. On the whole, it is found out that the mid-infrared phosphors excitable by a blue LED for pc-LED applications are deficient.

3.6 PERSPECTIVE

The shrinkage of LED size in terms of micro-LED and mini-LED already is pioneering the back-lighting technology for the feature of fine dimming. This directly impacts the particle size of the infrared phosphor and also influences the forthcoming researches toward the investigations of novel techniques for synthesizing the infrared emitting phosphors in nanometer scale. The infrared emitting nanoparticles are great in demand for in vivo biomedical applications due to the significant advantage of the lesser signal-to-noise ratio. This breaks the existing difficulty of visualizing the deep organs of the human body in real time. Not only the bio-applications, but the infrared nanoparticles also unwrap the possibility in energy applications such as solar cells, anti-counterfeiting, sensing technology due to the greater surface area. We hope that the infrared mini-LED can uncover a wide range of theranostics, photo-bio-modulation, and bio-sensing applications.

On the other hand, the pc-NIR LED technology is not adequately fully-fledged for practical applications due to lesser radiant flux and also the constrained spectral distribution of light. Additionally, the broadband mid-infrared phosphors excitable by blue light are still lacking. Hence, more innovative research methods are required to enhance the performances of the pc-NIR LED and also find out the potential emitters and host system.

ACKNOWLEDGMENTS

This work was supported by the Ministry of Science and Technology of Taiwan (Contract Nos. MOST 109-2112-M-003-011, MOST 109-2113-M-002-020-MY3, MOST 107-2113-M-002-008-MY3, MOST 107-2923-M-002-004-MY3, and MOST 106-2112-M-003-007-MY3).

REFERENCES

1. Katayama Y, Viana B, Gourier D, Xu J, and Tanabe S (2016) Photostimulation induced persistent luminescence in $Y_3Al_2Ga_3O_{12}$: Cr^{3+}. *Opt. Mater. Express* 6: 1405.
2. Lin H, Bai G, Yu T, Tsang MK, Zhang Q, and Hao J (2017) Site occupancy and near-infrared luminescence in $Ca_3Ga_2Ge_3O_{12}$: Cr^{3+} persistent phosphor. *Adv. Opt. Mater.* 5: 1700227.
3. Huang W, Wu D, Gong X, and Deng C (2018) Cr^{3+} activated $Zn_3Al_2Ge_3O_{12}$: a novel near-infrared long persistent phosphor. *J. Mater. Sci.: Mater. Electron.* 29: 5275.
4. Butkute S, Gaigalas E, Beganskiene A, Ivanauskas F, Ramanauskas R, and Kareiva A (2018) Sol-gel combustion synthesis of high-quality chromium-doped mixed-metal garnets $Y_3Ga_5O_{12}$ and $Gd_3Sc_2Ga_3O_{12}$. *J. Alloys Compd.* 739: 504.
5. Malysa B, Meijerink A, and Jüstel T (2018) Temperature dependent Cr^{3+} photoluminescence in garnets of the type $X_3Sc_2Ga_3O_{12}$ (X = Lu, Y, Gd, La). *J. Lumin.* 202: 523.
6. Xu J, Murata D, Ueda J, Viana B, and Tanabe S (2018) Toward rechargeable persistent luminescence for the first and third biological windows via persistent energy transfer and electron trap redistribution. *Inorg. Chem.* 57: 5194.
7. Zhang L, Zhang S, Hao Z, Zhang X, Pan G, Luo Y, Wu H, and Zhang J (2018) A high efficiency broadband near-infrared $Ca_2LuZr_2Al_3O_{12}$: Cr^{3+} garnet phosphor for blue LED chips. *J. Mater. Chem. C* 6: 4967.
8. Zhou X, Ju G, Dai T, and Hu Y (2019) Endowing Cr^{3+}-doped non-gallate garnet phosphors with near-infrared long-persistent luminescence in weak fields. *Opt. Mater.* 96: 109322.
9. Zhang L, Wang D, Hao Z, Zhang X, Pan G, Wu H, and Zhang J (2019) Cr^{3+}-doped broadband NIR garnet phosphor with enhanced luminescence and its application in NIR spectroscopy. *Adv. Opt. Mater.* 7: 1900185.
10. Basore ET, Xiao W, Liu X, Wu J, and Qiu J (2020) Broadband near-infrared garnet phosphors with near-unity internal quantum efficiency. *Adv. Opt. Mater.* 8: 2000296.
11. Mao M, Zhou T, Zeng H, Wang L, Huang F, Tang X, and Xie RJ (2020) Broadband near-infrared (NIR) emission realized by the crystal-field engineering of $Y_{3-x}Ca_xAl_{5-x}Si_xO_{12}$: Cr^{3+} (x = 0–2.0) garnet phosphors. *J. Mater. Chem. C* 8: 1981.
12. Jia Z, Yuan C, Liu Y, Wang XJ, Sun P, Wang L, Jiang H, and Jiang J (2020) Strategies to approach high performance in Cr^{3+}-doped phosphors for high-power NIR-LED light sources. *Light Sci. Appl.* 9: 86.
13. Meng X, Zhang X, Shi X, Qiu K, Wang Z, Wang D, Zhao J, Li X, Yang Z, and Li P (2020) Designing a super broadband near infrared material $Mg_3Y_2Ge_3O_{12}$: Cr^{3+} using cation inversion for future light sources. *RSC Adv.* 10: 19106.
14. Yang X, Chen W, Wang D, Chai X, Xie G, Xia Z, Molokeev MS, Liu Y, and Lei B (2020) Near-infrared photoluminescence and phosphorescence properties of Cr^{3+}-doped garnet-type $Y_3Sc_2Ga_3O_{12}$. *J. Lumin.* 225: 117392.
15. Wu J, Zhuang W, Liu R, Liu Y, Gao T, Yan C, Cao M, Tian J, and Chen X (2020) Broadband near-infrared luminescence and energy transfer of Cr^{3+}, Ce^{3+} co-doped $Ca_2LuHf_2Al_3O_{12}$ phosphors. *J. Rare Earths* 39: 269.
16. Bai B, Dang P, Huang D, Lian H, and Lin J (2020) Broadband near-infrared emitting $Ca_2LuScGa_2Ge_2O_{12}$:Cr^{3+} phosphors: luminescence properties and application in light-emitting diodes. *Inorg. Chem.* 59: 13481.
17. Kim IW, Kaur S, Yadav A, Rao AS, Saravanakumar S, Rao JL, and Singh V (2020) Structural, luminescence and EPR properties of deep red emitting $MgY_2Al_4SiO_{12}$: Cr^{3+} garnet phosphor. *J. Lumin.* 220: 116975.
18. Ye S, Zhou J, Wang S, Hu R, Wang D, and Qiu J (2013) Broadband downshifting luminescence in Cr^{3+}-Yb^{3+} codoped garnet for efficient photovoltaic generation. *Opt. Express* 21: 4167.
19. Liu C, Xia Z, Molokeev MS, and Liu Q (2015) Synthesis, crystal structure, and enhanced luminescence of garnet-type $Ca_3Ga_2Ge_3O_{12}$: Cr^{3+} by codoping Bi^{3+}. *J. Am. Ceram. Soc.* 98: 1870.
20. Yao L, Shao Q, Xu X, Dong Y, Liang C, He J, and Jiang J (2019) Broadband emission of single-phase $Ca_3Sc_2Si_3O_{12}$: Cr^{3+}/Ln^{3+} (Ln = Nd, Yb, Ce) phosphors for novel solid-state light sources with visible to near-infrared light output. *Ceram. Int.* 45: 14249.

21. Dai WB, Lei YF, Zhou J, Xu M, Chu LL, Li L, Zhao P, and Zhang ZH (2017) Near-infrared quantum-cutting and long-persistent phosphor $Ca_3Ga_2Ge_3O_{12}$: Pr^{3+}, Yb^{3+} for application in *in vivo* bioimaging and dye-sensitized solar Cells. *J. Alloys Compd.* 726: 230.

22. Han C, Luo L, Dong G, Zhang W, and Wang Y (2017) Enhanced visible to near infrared downshifting in Ce^{3+}/Yb^{3+}-co-doped yttrium aluminum garnet through Pr^{3+} doping. *Mater. Res. Bull.* 96: 286.

23. He S, Zhang L, Wu H, Wu H, Pan G, Hao Z, Zhang X, Zhang L, Zhang H, and Zhang J (2020) Efficient super broadband NIR $Ca_2LuZr_2Al_3O_{12}$: Cr^{3+},Yb^{3+} garnet phosphor for pc-LED light source toward NIR spectroscopy applications. *Adv. Opt. Mater.* 8: 1901684.

24. Han YJ, Wang S, Liu H, Shi L, Dong XZ, Fan RR, Liu C, Mao ZY, Wang DJ, Zhang ZW, and Zhao Y (2020) A novel Al^{3+} modified $Li_6CaLa_2Sb_2O_{12}$: Mn^{4+} far-red-emitting phosphor with garnet structure for plant cultivation. *J. Lumin.* 221: 117031.

25. Xue F, Hu Y, Fan L, Ju G, Lv Y, and Li Y (2017) Cr^{3+}-activated $Li_5Zn_8Al_5Ge_9O_{36}$: a near-infrared long-afterglow phosphor. *J. Am. Ceram. Soc.* 100: 3070.

26. Sun F, Xie R, Guan L, and Zhang C (2016) The near-infrared long-persistent phosphorescence of Cr^{3+}-activated non-gallate phosphor. *Mater. Lett.* 164: 39.

27. Wu Y, Li Y, Qin X, Chen R, Wu D, Liu S, and Qiu J (2015) Near-infrared long-persistent phosphor of $Zn_3Ga_2Ge_2O_{10}$: Cr^{3+} sintered in different atmosphere. *Spectrochim. Acta A Mol. Biomol. Spectrosc.* 151: 385.

28. Bai Q, Zhao S, Guan L, Wang Z, Li P, and Xu Z (2018) Design and control of the luminescence of Cr^{3+}-doped phosphors in the near-infrared I region by fitting the crystal field. *Cryst. Growth Des.* 18: 3178.

29. Li Y, Li Y, Chen R, Sharafudeen K, Zhou S, Gecevicius M, Wang H, Dong G, Wu Y, Qin X, and Qiu J (2015) Tailoring of the trap distribution and crystal field in Cr^{3+}-doped non-gallate phosphors with near-infrared long-persistence phosphorescence. *NPG Asia Mater.* 7: e180.

30. Xi L, Wang Y, Yin L, and Townsend PD (2020) The influence of germanium ions on the infrared long persistent phosphor $Zn_{1+x}Al_{2-2x}Ge_xO_4$: Cr. *J. Alloys Compd.* 820: 153094.

31. Wang B, Chen Z, Li X, Zhou J, and Zeng Q (2020) Photostimulated near-infrared persistent luminescence Cr^{3+}-doped Zn-Ga-Ge-O phosphor with high QE for optical information storage. *J. Alloys Compd.* 812: 152119.

32. Meng X, Wang Z, Qiu K, Li Y, Liu J, Wang Z, Liu S, Li X, Yang Z, and Li P (2018) Design of a novel near-infrared phosphor by controlling the cationic coordination environment. *Cryst. Growth Des.* 18: 4691.

33. Lee C, Bao Z, Fang MH, Lesniewski T, Mahlik S, Grinberg M, Leniec G, Kaczmarek SM, Brik MG, Tsai YT, Tsai TL, and Liu RS (2020) Chromium(III)-doped fluoride phosphors with broadband infrared emission for light-emitting diodes. *Inorg. Chem.* 59: 376.

34. Song E, Jiang X, Zhou Y, Lin Z, Ye S, Xia Z, and Zhang Q (2019) Heavy Mn^{2+} doped $MgAl_2O_4$ phosphor for high-efficient near-infrared light-emitting diode and the night-vision application. *Adv. Opt. Mater.* 7: 1901105.

35. Huang W, Gong X, Cui R, Li X, Li L, Wang X, and Deng C (2018) Enhanced persistent luminescence of $LiGa_5O_8$: Cr^{3+} near-infrared phosphors by codoping Sn^{4+}. *J. Mater. Sci.: Mater. Electron.* 29: 10535.

36. De Clercq OQ, Martin LIDJ, Korthout K, Kusakovskij J, Vrielinck H, and Poelman D (2017) Probing the local structure of the near-infrared emitting persistent phosphor $LiGa_5O_8$: Cr^{3+}. *J. Mater. Chem. C* 5: 10861.

37. Liao Z, Xu H, Zhao W, Yang H, Zhong J, Zhang H, Nie Z, and Zhou ZK (2020) Energy transfer from Mn^{4+} to Mn^{5+} and near infrared emission with wide excitation band in $Ca_{14}Zn_6Ga_{10}O_{35}$:Mn phosphors. *Chem. Eng. J.* 395: 125060.

38. Liu G, Molokeev MS, Lei B, and Xia Z (2020) Two-site Cr^{3+} occupation in the $MgTa_2O_6$:Cr^{3+} phosphor toward broad-band near-infrared emission for vessel visualization. *J. Mater. Chem. C* 8: 9322.

39. Xu DD, Qiu ZC, Zhang Q, Huang LJ, Ye YY, Cao LW, and Meng JX (2019) Sr_2MgWO_6: Cr^{3+} phosphors with effective near-infrared fluorescence and long-lasting phosphorescence. *J. Alloys Compd.* 781: 473.

40. Lai J, Shen W, Qiu J, Zhou D, Long Z, Yang Y, Zhang K, Khan I, and Wang Q (2020) Broadband near-infrared emission enhancement in $K_2Ga_2Sn_6O_{16}$: Cr^{3+} phosphor by electron-lattice coupling regulation. *J. Am. Ceram. Soc.* 103: 5067.

41. Chen J, Liu J, Yao H, Li X, Yao Z, Yue H, and Yu X (2015) Preparation and application of strong near-infrared emission phosphor Sr_3SiO_5: Ce^{3+}, Al^{3+}, Nd^{3+}. *J. Am. Ceram. Soc.* 98: 1836.

42. Bai W, Liu Y, Wang Y, Qiang X, and Feng L (2015) Cooperative down-conversion and near-infrared luminescence of Tb^{3+}–Yb^{3+} co-doped $CaMoO_4$ broadband phosphor. *Ceram. Int.* 41: 12896.

43. Hu J, Zhang Y, Lu B, Xia H, Ye H, and Chen B (2019) Efficient conversion of broad UV–visible light to near-infrared emission in Mn^{4+}/Yb^{3+} co-doped $CaGdAlO_4$ phosphors. *J. Lumin.* 210: 189.

44. Zou Z, Wu C, Li X, Zhang J, Li H, Wang D, and Wang Y (2017) Near-infrared persistent luminescence of Yb^{3+} in perovskite phosphor. *Opt. Lett.* 42: 4510.

45. Back M, Ueda J, Brik MG, Lesniewski T, Grinberg M, and Tanabe S (2018) Revisiting Cr^{3+}-doped $Bi_2Ga_4O_9$ spectroscopy: crystal field effect and optical thermometric behavior of near-infrared-emitting singly-activated phosphors. *ACS Appl. Mater. Interfaces* 10: 41512.

46. Back M, Ueda J, Xu J, Asami K, Brik MG, and Tanabe S (2020) Effective ratiometric luminescent thermal sensor by Cr^{3+}-doped mullite Bi2Al4O9 with robust and reliable performances. *Adv. Opt. Mater.* 8: 2000124.

47. Pestryakov E, Petrov V, Volkov A, and Maslov V (2002) Spectroscopic properties of Cr^{3+} ions in KTP single crystals. In *XI Feofilov Symposium on Spectroscopy of Crystals Activated by Rare-Earth and Transition Metal Ions 4766.*

48. Rai M, Mishra K, Rai SB, and Morthekai P (2018) Tailoring UV-blue sensitization effect in enhancing near infrared emission in X, Yb^{3+}: $CaGa_2O_4$ (X = 0, Eu^{3+}, Bi^{3+}, Cr^{3+}) phosphor for solar energy conversion. *Mater. Res. Bull.* 105: 192.

49. Li J, Chen L, Hao Z, Zhang X, Zhang L, Luo Y, and Zhang J (2015) Efficient near-infrared downconversion and energy transfer mechanism of Ce^{3+}/Yb^{3+} codoped calcium scandate phosphor. *Inorg. Chem.* 54: 4806.

50. Shao Q, Ding H, Yao L, Xu J, Liang C, Li Z, Dong Y, and Jiang J (2018) Broadband near-infrared light source derived from Cr^{3+}-doped phosphors and a blue LED chip. *Opt. Lett.* 43: 5251.

51. Xu X, Shao Q, Yao L, Dong Y, and Jiang J (2020) Highly efficient and thermally stable Cr^{3+}-activated silicate phosphors for broadband near-infrared LED applications. *Chem. Eng. J.* 383: 123108.

52. Zhou X, Geng W, Li J, Wang Y, Ding J, and Wang Y (2020) An ultraviolet–visible and near-infrared-responded broadband NIR phosphor and its NIR spectroscopy application. *Adv. Opt. Mater.* 8: 1902003.

53. Zhan Y, Jin Y, Wu H, Yuan L, Ju G, Lv Y, and Hu Y (2019) Cr^{3+}-doped $Mg_4Ga_4Ge_3O_{16}$ near-infrared phosphor membrane for optical information storage and recording. *J. Alloys Compd.* 777: 991.

54. Luo H, Zhang S, Mu Z, Wu F, Nie Z, Zhu D, Feng X, and Zhang Q (2019) Near-infrared quantum cutting via energy transfer in Bi^{3+}, Yb^{3+} co-doped Lu_2GeO_5 down-converting phosphor. *J. Alloys Compd.* 784: 611.

55. Zeng H, Zhou T, Wang L, and Xie RJ (2019) Two-site occupation for exploring ultra-broadband near-infrared phosphor – double-perovskite La_2MgZrO_6: Cr^{3+}. *Chem. Mater.* 31: 5245.

56. Xu D, Wu X, Zhang Q, Li W, Wang T, Cao L, and Meng J (2018) Fluorescence property of novel near-infrared phosphor Ca_2MgWO_6: Cr^{3+}. *J. Alloys Compd.* 731: 156.

57. Chen W, Wang Y, Zeng W, Li G, and Guo H (2016) Design, synthesis and characterization of near-infrared long persistent phosphors $Ca_4(PO_4)_2O$: Eu^{2+}, R^{3+} (R = Lu, La, Gd, Ce, Tm, Y). *RSC Adv.* 6: 331.

58. Ma YY, Hu JQ, Song EH, Ye S, and Zhang QY (2015) Regulation of red to near-infrared emission in Mn^{2+} single doped magnesium zinc phosphate solid-solution phosphors by modification of the crystal field. *J. Mater. Chem. C* 3: 12443.

59. Zhao J, Guo C, and Li T (2015) Near-infrared down-conversion and energy transfer mechanism of Ce^{3+}-Yb^{3+} co-doped $Ba_2Y(BO_3)_2Cl$ phosphors. *ECS J. Solid State Sci. Technol.* 5: R3055.

60. Xiang J, Chen J, Zhang N, Yao H, and Guo C (2018) Far red and near infrared double-wavelength emitting phosphor Gd_2ZnTiO_6: Mn^{4+}, Yb^{3+} for plant cultivation LEDs. *Dyes Pigm.* 154: 257.

61. Li H, Lu Y, Li C, Deng D, Rong K, Jing X, Wang L, and Xu S (2019) Near-infrared down-conversion luminescence in Yb^{3+} doped self-activated TbZn (B_5O_{10}) phosphor. *J. Optoelectron. Adv. Mater.* 21: 373.

62. Xu J, Chen D, Yu Y, Zhu W, Zhou J, and Wang Y (2014) Cr^{3+}:$SrGa_{12}O_{19}$: a broadband near-infrared long-persistent phosphor. *Chem. Asian J.* 9: 1020.

63. Liu S, Wang Z, Cai H, Song Z, and Liu Q (2020) Highly efficient near-infrared phosphor $LaMgGa_{11}O_{19}$: Cr^{3+}. *Inorg. Chem. Front.* 7: 1467.

64. Qiao J, Zhou G, Zhou Y, Zhang Q, and Xia Z (2019) Divalent europium-doped near-infrared-emitting phosphor for light-emitting diodes. *Nat. Commun.* 10: 5267.

65. Wu X, Xu D, Li W, Wang T, Cao L, and Meng J (2017) Synthesis and luminescence of novel near-infrared emitting $BaZrSi_3O_9$: Cr^{3+} phosphors. *Lumin* 32: 1554.

66. Hao Y, Wang Y, Hu X, Liu X, Liu E, Fan J, Miao H, and Sun Q (2016) YBO_3: Ce^{3+}, Yb^{3+} based near-infrared quantum cutting phosphors: synthesis and application to solar cells. *Ceram. Int.* 42: 9396.

67. Luo X, Yang X, and Xiao S (2018) Conversion of broadband UV-visible to near infrared emission by $LaMgAl_{11}O_{19}$: Cr^{3+}, Yb^{3+} phosphors. *Mater. Res. Bull.* 101: 73.

68. Zhang Q, Ni H, Lin L, Ding J, Ding J, and Li X (2019) Communication – an intense broadband sensitized near-infrared emitting $GdAl_3(BO_3)_4$: Ce^{3+},Yb^{3+} phosphor. *ECS J. Solid State Sci. Technol.* 8: R47.

69. Fang MH, Huang PY, Bao Z, Majewska N, Leśniewski T, Mahlik S, Grinberg M, Leniec G, Kaczmarek SM, Yang CW, Lu KM, Sheu HS, and Liu RS (2020) Penetrating biological tissue using light-emitting diodes with a highly efficient near-infrared ScBO$_3$: Cr^{3+} phosphor. *Chem. Mater.* 32: 2166.

70. Shao Q, Ding H, Yao L, Xu J, Liang C, and Jiang J (2018) Photoluminescence properties of a ScBO$_3$: Cr^{3+} phosphor and its applications for broadband near-infrared LEDs. *RSC Adv.* 8: 12035.

71. Rajendran V, Lesniewski T, Mahlik S, Grinberg M, Leniec G, Kaczmarek SM, Pang WK, Lin YS, Lu KM, Lin CM, Chang H, Hu SF, and Liu RS (2019) Ultra-broadband phosphors converted near-infrared light emitting diode with efficient radiant power for spectroscopy applications. *ACS Photonics* 6: 3215.

72. Qiu K, Zhang H, Yuan T, Shi J, Wu S, Meng X, Liu J, Li P, Yang Z, and Wang Z (2019) Design and improve of the near-infrared phosphor by adjusting the energy levels or constructing new defects. *Spectrochim. Acta A Mol. Biomol. Spectrosc.* 219: 401.

73. Du J, De Clercq OQ, Korthout K, and Poelman D (2017) LaAlO$_3$: Mn^{4+} as near-infrared emitting persistent luminescence phosphor for medical imaging: a charge compensation study. *Materials* 10: 1422.

74. Rajendran V, Fang MH, Guzman GND, Lesniewski T, Mahlik S, Grinberg M, Leniec G, Kaczmarek SM, Lin YS, Lu KM, Lin CM, Chang H, Hu SF, and Liu RS (2018) Super broadband near-infrared phosphors with high radiant flux as future light sources for spectroscopy applications. *ACS Energy Lett.* 3: 2679.

75. Du J and Poelman D (2019) Facile Synthesis of Mn^{4+}-activated double perovskite germanate phosphors with near-infrared persistent luminescence. *Nanomaterials* 9: 1759.

76. Wang Q, Mu Z, Zhang S, Zhang Q, Zhu D, Feng J, Du Q, and Wu F (2019) A novel near infrared long-persistent phosphor La$_2$MgGeO$_6$: Cr^{3+}, RE^{3+} (RE = Dy, Sm). *J. Lumin.* 206: 618.

77. Pan Z, Lu Y-Y, and Liu F (2012) Sunlight-activated long-persistent luminescence in the near-infrared from Cr^{3+}-doped zinc gallogermanates. *Nat. Mater.* 11: 58.

78. Luo X, Shen J, Huang H, Xu L, Wang Z, Chen Y, and Li L (2016) Near-infrared quantum cutting in Yb^{3+} doped SrMoO$_4$ phosphors. *J. Nanosci. Nanotechnol.* 16: 3494.

79. Zhao J, Guo C, Li T, Song D, and Su X (2015) Near-infrared down-conversion and energy transfer mechanism in Yb^{3+}-doped Ba$_2$LaV$_3$O$_{11}$ phosphors. *Phys. Chem. Chem. Phys.* 17: 26330.

80. İlhan M, Ekmekçi MK, Oraltay RG, and Başak AS (2017) Structural and near-infrared properties of Nd^{3+} activated Lu$_3$NbO$_7$ phosphor. *J. Fluoresc.* 27: 199.

81. Sawala N, Koparkar K, Bajaj N, and Omanwar S (2016) Near-infrared spectral downshifting in Sr$_{(3-x)}$(VO$_4$)$_2$: xNd^{3+} phosphor. *Bull. Mater. Sci.* 39: 1625.

82. Watanabe M, Sejima Y, Oka R, Ida S, and Masui T (2019) Submicron-sized phosphors based on hexagonal rare earth oxycarbonate for near-infrared excitation and emission. *J. Asian Ceram. Soc.* 7: 502.

83. Liang Y, Liu F, Chen Y, Wang X, Sun K, and Pan Z (2017) Red/near-infrared/short-wave infrared multiband persistent luminescence in Pr^{3+}-doped persistent phosphors. *Dalton Trans.* 46: 11149.

84. Chen X, Salamo GJ, Li S, Wang J, Guo Y, Gao Y, He L, Ma H, Tao J, Sun P, Lin W, and Liu Q (2015) Two-photon, three-photon, and four-photon excellent near-infrared quantum cutting luminescence of Tm^{3+} ion activator emerged in Tm^{3+}: YNbO$_4$ powder phosphor one material simultaneously. *Phys. B: Condens. Matter* 479: 159.

85. Li L, Wang Y, Shen J, Chang W, Jin T, Wei X, and Tian Y (2016) Near-infrared downconversion in LuPO$_4$: Tm^{3+}, Yb^{3+} phosphors. *J. Nanosci. Nanotechnol.* 16: 3511.

86. Li L, Pan Y, Chang W, Feng Z, Chen P, Li C, Zeng Z, and Zhou X (2017) Near-infrared downconversion luminescence of SrMoO$_4$: Tm^{3+}, Yb^{3+} phosphors. *Mater. Res. Bull.* 93: 144.

87. Dong SL, Lin HH, Yu T, and Zhang QY (2014) Near-infrared quantum-cutting luminescence and energy transfer properties of Ca$_3$(PO$_4$)$_2$:Tm^{3+},Ce^{3+} phosphors. *J. Appl. Phys.* 116: 023517.

88. Chen X, Li S, Salamo GJ, Li Y, He L, Yang G, Gao Y, and Liu Q (2015) Sensitized intense near-infrared downconversion quantum cutting three-photon luminescence phenomena of the Tm^{3+} ion activator in Tm^{3+}, Bi^{3+}: YNbO$_4$ powder phosphor. *Opt. Express* 23: A51.

89. Kumar KS, Lou C, Manohari AG, Cao H, and Pribat D (2018) Broadband down-conversion of near-infrared emission in Bi^{3+}-Yb^{3+} co-doped Y$_3$Al$_5$O$_{12}$ phosphors. *Optik* 157: 492.

90. Liu L, Li M, Cai S, Yang Y, and Mai Y (2015) Near-infrared quantum cutting in Nd^{3+} and Yb^{3+} doped BaGd$_2$ZnO$_5$ phosphors. *Opt. Mater. Express* 5: 756.

91. Pan Y, Li L, Chang W, Chen W, Li C, Chen P, and Zeng Z (2017) Efficient near-infrared quantum cutting in Tb^{3+}, Yb^{3+} codoped LuPO$_4$ phosphors. *J. Rare Earths* 35: 235.

92. Liu T, Cai J, Cui Y, Song Z, He M, Zhao H, Zhang Z, and Tao H (2018) Synthesis and near-infrared quantum cutting of Tb^{3+}, Yb^{3+} codoped GdBa$_3$B$_9$O$_{18}$ phosphors. *Micro Nano Lett.* 13: 108.

93. Li J, Liao J, Wen HR, Kong L, Wang M, and Chen J (2019) Multiwavelength near infrared downshift and downconversion emission of Tm^{3+} in double perovskite Y$_2$MgTiO$_6$: Mn^{4+}/Tm^{3+} phosphors via resonance energy transfer. *J. Lumin.* 213: 356.

94. Chen J, Liu J, Yin H, Jiang S, Yao H, and Yu X (2016) Efficient near-infrared emission of Ce^{3+}–Nd^{3+} codoped $(Sr_{0.6}Ca_{0.4})_3(Al_{0.6}Si_{0.4})O_{4.4}F_{0.6}$ phosphors for c- Si solar cell. *J. Am. Ceram. Soc.* 99: 141.

95. Tawalare PK, Bhatkar VB, Omanwar SK, and Moharil SV (2019) Cr^{3+} sensitized near infrared emission in Al_2O_3:Cr,Nd/Yb phosphors. *J. Alloys Compd.* 790: 1192.

96. Song D, Guo C, Zhao J, Suo H, Zhao X, Zhou X, and Liu G (2016) Host sensitized near-infrared emission in Nd^{3+}-Yb^{3+} co-doped $Na_2GdMg_2V_3O_{12}$ phosphor. *Ceram. Int.* 42: 12988.

97. Hou D, Zhang Y, Li JY, Li H, Lin H, Lin Z, Dong J, Huang R, and Song J (2020) Discovery of near-infrared persistent phosphorescence and stokes luminescence in Cr^{3+} and Nd^{3+} doped $GdY_2Al_3Ga_2O_{12}$ dual mode phosphors. *J. Lumin.* 221: 117053.

98. Lin H, Yu T, Bai G, Tsang MK, Zhang Q, and Hao J (2016) Enhanced energy transfer in Nd^{3+}/Cr^{3+} co-doped $Ca_3Ga_2Ge_3O_{12}$ phosphors with near-infrared and long-lasting luminescence properties. *J. Mater. Chem. C* 4: 3396.

99. Baklanova YV, Lipina OA, Enyashin AN, Surat LL, Tyutyunnik AP, Tarakina NV, Fortes AD, Chufarov AY, Gorbatov EV, and Zubkov VG (2018) Nd^{3+}, Ho^{3+}-codoped apatite-related $NaLa_9(GeO_4)_6O_2$ phosphors for the near- and middle-infrared region. *Dalton Trans.* 47: 14041.

100. Li J, Zhang S, Luo H, Mu Z, Li Z, Du Q, Feng J, and Wu F (2018) Efficient near ultraviolet to near infrared downconversion photoluminescence of La_2GeO_5: Bi^{3+}, Nd^{3+} phosphor for silicon-based solar cells. *Opt. Mater.* 85: 523.

101. Xiang G, Ma Y, Zhou X, Jiang S, Li L, Luo X, Hao Z, Zhang X, Pan G-H, Luo Y, and Zhang J (2017) Investigation of the energy-transfer mechanism in Ho^{3+}- and Yb^{3+}-codoped Lu_2O_3 phosphor with efficient near-infrared downconversion. *Inorg. Chem.* 56: 1498.

102. Zhao J, Wang X, Li L, and Guo C (2020) Near-infrared emissions in host sensitized $Ba_2YV_3O_{11}$: RE^{3+} (RE = Nd, Ho, Yb) down-converting phosphors. *Ceram. Int.* 46: 5015.

103. Tawalare PK, Bhatkar VB, Omanwar SK, and Moharil SV (2018) Near-infrared emitting $Ca_5(PO_4)_3Cl$: Eu^{2+}, Nd^{3+} phosphor for modification of the solar spectrum. *Lumin* 33: 1288.

104. Chen J, Guo C, Yang Z, Li T, and Zhao J (2016) Li_2SrSiO_4: Ce^{3+}, Pr^{3+} phosphor with blue, red, and near-infrared emissions used for plant growth LED. *J. Am. Ceram. Soc.* 99: 218.

105. Xu J, Murata D, Katayama Y, Ueda J, and Tanabe S (2017) Cr^{3+}/Er^{3+} co-doped $LaAlO_3$ perovskite phosphor: a near-infrared persistent luminescence probe covering the first and third biological windows. *J. Mater. Chem. B* 5: 6385.

106. Hong J, Lin L, Li X, Xie J, Qin Q, and Zheng Z (2019) Enhancement of near-infrared quantum-cutting luminescence in $NaBaPO_4$: Er^{3+} phosphors by Bi^{3+}. *Opt. Mater.* 98: 109471.

107. Huerta EF, De Anda J, Martínez-Merlin I, Caldiño U, and Falcony C (2020) Near-infrared luminescence spectroscopy in yttrium oxide phosphor activated with Er^{3+}, Li^+ and Yb^{3+} ions for application in photovoltaic systems. *J. Lumin.* 224: 117271.

108. Liu F, Liang Y, Chen Y, and Pan Z (2016) Divalent nickel-activated gallate-based persistent phosphors in the short-wave infrared. *Adv. Opt. Mater.* 4: 562.

109. Yuan L, Jin Y, Zhu D, Mou Z, Xie G, and Hu Y (2020) Ni^{2+}-doped yttrium aluminum gallium garnet phosphors: bandgap engineering for broad-band wavelength-tunable shortwave-infrared long-persistent luminescence and photochromism. *ACS Sustainable Chem. Eng.* 8: 6543.

110. Lai J, Qiu J, Wang Q, Zhou D, Long Z, Yang Y, Hu S, Li X, Pi J, and Wang J (2020) Disentangling site occupancy, cation regulation, and oxidation state regulation of the broadband near infrared emission in a chromium-doped $SrGa_4O_7$ phosphor. *Inorg. Chem. Front.* 7: 2313.

111. Nanai Y, Ishida R, Urabe Y, Nishimura S, and Fuchi S (2019) Octave-spanning broad luminescence of Cr^{3+}, Cr^{4+}-codoped Mg_2SiO_4 phosphor for ultra-wideband near-infrared LEDs. *Jpn. J. Appl. Phys.* 58: SFFD02.

112. Liu BM, Yong ZJ, Zhou Y, Zhou DD, Zheng LR, Li LN, Yu HM, and Sun HT (2016) Creation of near-infrared luminescent phosphors enabled by topotactic reduction of bismuth-activated red-emitting crystals. *J. Mater. Chem. C* 4: 9489.

113. Zhang X, Nie J, Liu S, and Qiu J (2018) Structural variation and near infrared luminescence in Mn^{5+}-doped M_2SiO_4 (M=Ba, Sr, Ca) phosphors by cation substitution. *J. Mater. Sci.: Mater. Electron.* 29: 6419.

114. De Guzman GNA, Fang MH, Liang CH, Bao Z, Hu SF, and Liu RS (2020) Near-infrared phosphors and their full potential: a review on practical applications and future perspectives. *J. Lumin.* 219: 116944.

115. Rajendran V, Chang H, and Liu RS (2019) Recent progress on broadband near-infrared phosphors-converted light emitting diodes for future miniature spectrometers. *Opt. Mater. X* 1: 100011.

116. De Guzman GNA, Hu SF, and Liu RS (2020) Enticing applications of near-infrared phosphors: review and future perspectives. *J. Chin. Chem. Soc.* 68: 206.

117. Hayashi D, van Dongen AM, Boerekamp J, Spoor S, Lucassen G, and Schleipen J (2017) A broadband LED source in visible to short-wave-infrared wavelengths for spectral tumor diagnostics. *Appl. Phys. Lett.* 110: 233701.

118. Nwazor NO and Orakwue SI (2020) A review of night vision technology. *AJST* 4: 265.

119. Chrzanowski K (2013) Review of night vision technology. *Opto-Electron. Rev.* 21: 153.

120. Perić D and Livada B (2020) Night vision technology breakthroughts. In 9th International Scientific Conference on Defensive Technologies OTEH, Belgrade, Serbia.

121. IEA-PVPS P (2014) PVPS report snapshot of global PV 1992-2013. In *Report IEA-PVPS T1-24.*

122. de la Mora MB, Amelines Sarria O, Monroy BM, Hernández Pérez CD, and Lugo JE (2017) Materials for downconversion in solar cells: perspectives and challenges. *Sol. Energy Mater. Sol. Cells* 165: 59.

123. Day J, Senthilarasu S, and Mallick TK (2019) Improving spectral modification for applications in solar cells: a review. *Renew. Energy* 132: 186.

124. Lee TD and Ebong AU (2017) A review of thin film solar cell technologies and challenges. *Renew. Sust. Energ. Rev.* 70: 1286.

125. Richards BS (2006) Luminescent layers for enhanced silicon solar cell performance: down-conversion. *Sol. Energy Mater. Sol. Cells* 90: 1189.

126. Huang XY and Zhang QY (2010) Near-infrared quantum cutting via cooperative energy transfer in Gd_2O_3:Bi^{3+},Yb^{3+} phosphors. *J. Appl. Phys.* 107: 063505.

127. Tai Y, Zheng G, Wang H, and Bai J (2015) Broadband down-conversion based near infrared quantum cutting in Eu^{2+}–Yb^{3+} co-doped $SrAl_2O_4$ for crystalline silicon solar cells. *J. Solid State Chem.* 226: 250.

128. Li Y, Wei X, Chen H, Pan Y, and Ji Y (2015) Near-infrared downconversion through host sensitized energy transfer in Yb^{3+}-doped $Na_2YMg_2(VO_4)_3$. *Phys. B: Condens. Matter* 478: 95.

129. Fleischmann WBM (2013) World must sustainably produce 70 percent more food by mid-century – UN report. https://news.un.org/en/story/2013/12/456912 (Accessed on 8 December 2020).

130. Massa GD, Kim HH, Wheeler RM, and Mitchell CA (2008) Plant productivity in response to LED lighting. *HortScience* 43: 1951.

131. Pocock T (2015) Light-emitting diodes and the modulation of specialty crops: light sensing and signaling networks in plants. *HortScience* 50: 1281.

132. Pocock T (2017) Influence of light-emitting diodes (LEDs) on light sensing and signaling networks in plants. In *Light Emitting Diodes for Agriculture: Smart Lighting,* edited by Dutta Gupta. Singapore: Springer Singapore.

133. Kreslavski VD, Los DA, Schmitt FJ, Zharmukhamedov SK, Kuznetsov VV, and Allakhverdiev SI (2018) The impact of the phytochromes on photosynthetic processes. *Biochim. Biophys. Acta* 1859: 400.

134. Casal JJ (2012) Shade avoidance. In *The Arabidopsis Book 2012.* e0157.

135. Valladares F and Niinemets Ü (2008) Shade tolerance, a key plant feature of complex nature and consequences. *Annu. Rev. Ecol. Evol. Syst.* 39: 237.

136. Zhou Z, Zheng J, Shi R, Zhang N, Chen J, Zhang R, Suo H, Goldys EM, and Guo C (2017) Ab initio site occupancy and far-red emission of Mn^{4+} in cubic-phase $La(MgTi)_{1/2}O_3$ for plant cultivation. *ACS Appl. Mater. Interfaces* 9: 6177.

137. Peng M, Yin X, Tanner PA, Brik MG, and Li P (2015) Site occupancy preference, enhancement mechanism, and thermal resistance of Mn^{4+} red luminescence in $Sr_4Al_{14}O_{25}$: Mn^{4+} for warm WLEDs. *Chem. Mater.* 27: 2938.

138. Sun Q, Wang S, Li B, Guo H, and Huang X (2018) Synthesis and photoluminescence properties of deep red-emitting $CaGdAlO_4$: Mn^{4+} phosphors for plant growth LEDs. *J. Lumin.* 203: 371.

139. Huang X and Guo H (2018) Finding a novel highly efficient Mn^{4+}-activated $Ca_3La_2W_2O_{12}$ far-red emitting phosphor with excellent responsiveness to phytochrome Pfr: towards indoor plant cultivation application. *Dyes Pigm.* 152: 36.

140. Liang J, Devakumar B, Sun L, Sun Q, Wang S, Li B, Chen D, and Huang X (2019) Mn^{4+}-activated $KLaMgWO_6$: a new high-efficiency far-red phosphor for indoor plant growth LEDs. *Ceram. Int.* 45: 4564.

141. Srivastava AM, Brik MG, Comanzo HA, Beers WW, Cohen WE, and Pocock T (2017) Spectroscopy of Mn^{4+} in double perovskites, La_2LiSbO_6 and La_2MgTiO_6: deep red photon generators for agriculture LEDs. *ECS J. Solid State Sci. Technol.* 7: R3158.

142. Fang MH, De Guzman GNA, Bao Z, Majewska N, Mahlik S, Grinberg M, Leniec G, Kaczmarek SM, Yang CW, Lu KM, Sheu HS, Hu SF, and Liu RS (2020) Ultra-high-efficiency near-infrared Ga_2O_3: Cr^{3+} phosphor and controlling of phytochrome. *J. Mater. Chem. C* 8: 11013.

143. Yang X, Zhang Y, Zhang X, Chen J, Huang H, Wang D, Chai X, Xie G, Molokeev MS, Zhang H, Liu Y, and Lei B (2020) Facile synthesis of the desired red phosphor $Li_2Ca_2Mg_2Si_2N_6$: Eu^{2+} for high CRI white LEDs and plant growth LED device. *J. Am. Ceram. Soc.* 103: 1773.

144. Osborne RA, Cherepy NJ, Åberg D, Zhou F, Seeley ZM, Payne SA, Drobshoff AD, Comanzo HA, and Srivastava AM (2018) $Ba_{(1-x)}Sr_xMg_3SiN_4$: Eu narrowband red phosphor. *Opt. Mater.* 84: 130.

145. Chen YL, Wang J, and Wang GQ (2020) Photoluminescence properties of near-UV pumped deep red emitting $Sr_2ZnGe_2O_7$: Eu^{3+} phosphors for plant growth LEDs. *Opt. Mater.* 106: 110022.

146. Li L, Pan Y, Huang Y, Huang S, and Wu M (2017) Dual-emissions with energy transfer from the phosphor $Ca_{14}Al_{10}Zn_6O_{35}$: Bi^{3+}, Eu^{3+} for application in agricultural lighting. *J. Alloys Compd.* 724: 735.

147. Zhang A, Jia M, Sun Z, Liu G, Fu Z, Sheng T, Li P, and Lin F (2019) High concentration Eu^{3+}-doped $NaYb(MoO_4)_2$ multifunctional material: thermometer and plant growth lamp matching phytochrome PR. *J. Alloys Compd.* 782: 203.

148. Sun L, Devakumar B, Liang J, Wang S, Sun Q, and Huang X (2019) Novel high-efficiency violet-red dual-emitting Lu_2GeO_5: Bi^{3+}, Eu^{3+} phosphors for indoor plant growth lighting. *J. Lumin.* 214: 116544.

149. Yang Z, Zhao Z, Que M, and Wang Y (2013) Photoluminescence and thermal stability of β-SiAlON:Re (Re = Sm, Dy) phosphors. *Opt. Mater.* 35: 1348.

150. Hagemann H, Kubel F, Bill H, and Gingl F (2004) $^5D_0 \rightarrow {}^7F_0$ transitions of Sm^{2+} in $SrMgF_4$:Sm^{2+}. *J. Alloys Compd.* 374: 194.

151. Dotsenko VP, Radionov VN, and Voloshinovskii AS (1998) Luminescence of Sm^{2+} in strontium haloborates. *Mater. Chem. Phys.* 57: 134.

152. Zhou Z, Li Y, and Peng M (2020) Near-infrared persistent phosphors: synthesis, design, and applications. *Chem. Eng. J.* 399: 125688.

153. Liu F, Yan W, Chuang YJ, Zhen Z, Xie J, and Pan Z (2013) Photostimulated near-infrared persistent luminescence as a new optical read-out from Cr^{3+}-doped $LiGa_5O_8$. *Sci. Rep.* 3: 1554.

154. Bessière A, Sharma SK, Basavaraju N, Priolkar KR, Binet L, Viana B, Bos AJJ, Maldiney T, Richard C, Scherman D, and Gourier D (2014) Storage of visible light for long-lasting phosphorescence in chromium-doped zinc gallate. *Chem. Mater.* 26: 1365.

155. Yan W, Liu F, Lu Y-Y, Wang XJ, Yin M, and Pan Z (2010) Near infrared long-persistent phosphorescence in $La_3Ga_5GeO_{14}$: Cr^{3+} phosphor. *Opt. Express* 18: 20215.

156. Jia D, Lewis LA, and Wang XJ (2010) Cr^{3+}-doped lanthanum gallogermanate phosphors with long persistent IR emission. *Electrochem. Solid-State Lett.* 13: J32.

157. Li Y, Li YY, Sharafudeen K, Dong GP, Zhou SF, Ma ZJ, Peng MY, and Qiu JR (2014) A strategy for developing near infrared long-persistent phosphors: taking $MAlO_3$:Mn^{4+}, Ge^{4+} (M = La, Gd) as an example. *J. Mater. Chem. C* 2: 2019.

158. Du J, De Clercq OQ, and Poelman D (2018) Thermoluminescence and near-infrared persistent luminescence in $LaAlO_3$: Mn^{4+},R (R= Na^+, Ca^{2+}, Sr^{2+}, Ba^{2+}) ceramics. *Ceram. Int.* 44: 21613.

159. Maldiney T, Lecointre A, Viana B, Bessière A, Bessodes M, Gourier D, Richard C, and Scherman D (2011) Controlling electron trap depth to enhance optical properties of persistent luminescence nanoparticles for *in vivo* imaging. *J. Am. Chem. Soc.* 133: 11810.

160. le Masne de Chermont Q, Chanéac C, Seguin J, Pellé F, Maîtrejean S, Jolivet JP, Gourier D, Bessodes M, and Scherman D (2007) Nanoprobes with near-infrared persistent luminescence for *in vivo* imaging. *Proc. Natl. Acad. Sci. U.S.A.* 104: 9266.

161. Katayama Y, Ueda J, and Tanabe S (2014) Effect of Bi_2O_3 doping on persistent luminescence of $MgGeO_3$: Mn^{2+} phosphor. *Opt. Mater. Express* 4: 613.

162. Lv Y, Wang L, Zhuang Y, Zhou TL, and Xie RJ (2017) Discovery of the Yb^{2+}–Yb^{3+} couple as red-to-NIR persistent luminescence emitters in Yb-activated $(Ba_{1-x}Sr_x)AlSi_5O_2N_7$ phosphors. *J. Mater. Chem. C* 5: 7095.

163. Liang YJ, Liu F, Chen YF, Wang XJ, Sun KN, and Pan Z (2016) New function of the Yb^{3+} ion as an efficient emitter of persistent luminescence in the short-wave infrared. *Light Sci. Appl.* 5: e16124.

164. Liang Y, Liu F, Chen Y, Wang X, Sun K, and Pan Z (2017) Extending the applications for lanthanide ions: efficient emitters in short-wave infrared persistent luminescence. *J. Mater. Chem. C* 5: 6488.

165. Kamimura S, Xu CN, Yamada H, Marriott G, Hyodo K, and Ohno T (2017) Near-infrared luminescence from double-perovskite $Sr_3Sn_2O_7$: Nd^{3+}: a new class of probe for *in vivo* imaging in the second optical window of biological tissue. *J. Ceram. Soc. Jpn.* 125: 591.

166. Xu J, Murata D, Ueda J, and Tanabe S (2016) Near-infrared long persistent luminescence of Er^{3+} in garnet for the third bio-imaging window. *J. Mater. Chem. C* 4: 11096.

167. Xu J, Tanabe S, Sontakke AD, and Ueda J (2015) Near-infrared multi-wavelengths long persistent luminescence of Nd^{3+} ion through persistent energy transfer in Ce^{3+}, Cr^{3+} co-doped $Y_3Al_2Ga_3O_{12}$ for the first and second bio-imaging windows. *Appl. Phys. Lett.* 107: 081903.

168. Xu J, Murata D, So B, Asami K, Ueda J, Heo J, and Tanabe S (2018) 1.2 μm persistent luminescence of Ho^{3+} in $LaAlO_3$ and $LaGaO_3$ perovskites. *J. Mater. Chem. C* 6: 11374.

169. Pellerin M, Castaing V, Gourier D, Chanéac C, and Viana B. (2018). *Persistent Luminescence of Transition Metal (Co, Ni...)-Doped ZnGa2O4 Phosphors for Applications in the Near-Infrared Range.* Vol. 10533. SPIE OPTO: SPIE: San Francisco, California.

170. Ma C, Liu H, Ren F, Liu Z, Sun Q, Zhao C, and Li Z (2020) The second near-infrared window persistent luminescence for anti-counterfeiting application. *Cryst. Growth Des.* 20: 1859.

171. Kumar P, Dwivedi J, and Gupta BK (2014) Highly luminescent dual mode rare-earth nanorod assisted multi-stage excitable security ink for anti-counterfeiting applications. *J. Mater. Chem. C* 2: 10468.

172. Gooch J, Daniel B, Abbate V, and Frascione N (2016) Taggant materials in forensic science: a review. *TrAC, Trends Anal. Chem.* 83: 49.

173. Kumar P, Singh S, and Gupta BK (2016) Future prospects of luminescent nanomaterial based security inks: from synthesis to anti-counterfeiting applications. *Nanoscale* 8: 14297.

174. Shi C, Shen X, Zhu Y, Li X, Pang Z, and Ge M (2019) Excitation wavelength-dependent dual-mode luminescence emission for dynamic multicolor anticounterfeiting. *ACS Appl. Mater. Interfaces* 11: 18548.

175. Zhang Y, Huang R, Li H, Lin Z, Hou D, Guo Y, Song J, Song C, Lin Z, Zhang W, Wang J, Chu PK, and Zhu C (2020) Triple-mode emissions with invisible near-infrared after-glow from Cr^{3+}-doped zinc aluminum germanium nanoparticles for advanced anti-counterfeiting applications. *Small* 16: 2003121.

4 Laser Phosphors

Shihai You and Rong-Jun Xie

CONTENTS

4.1 INTRODUCTION

InGaN light-emitting diode (LED)-based solid-state lighting (SSL) has been already considered the fourth-generation green light sources because of its high electricity-to-light conversion efficiency, low power consumption, long lifetime, compactness, and environmental friendliness. It, therefore, finds extensive applications in general lighting, liquid crystal display (LCD) backlights, vehicles headlamps, street lighting, and decorated lighting. However, there is a significant challenge in the development of white LEDs (wLEDs) with simultaneously high luminance and high luminous flux. Taking automotive headlamps for an example, the current density of LEDs used is on the order of 1 A mm^{-2}, producing white light with luminance of ~130 cd mm^{-2} and a total luminous flux in the range of 1–10,000 lm. These brightness and fluxes are not high enough for some applications, such as auxiliary high-beam headlamps which project light to 600 m, fiber-based endoscopy, and cinema projectors. In these cases, the white light should have brightness larger than 500 cd mm^{-2}, much higher than the state-of-the-art wLEDs. The challenge is caused by the well-known "efficiency droop" of the blue InGaN LED chip (Figure 4.1).[1] When the input power density reaches 100 W mm^{-2}, the power-conversion efficiency (PCE) of a state-of-the-art 450 nm InGaN LED drops to an only ~10%. It is much lower than 70% at an input powder density of 5 W mm^{-2}. Till now, there are no solutions for such a bottleneck problem.

In contrast, laser diodes (LDs) show a direct proportionality between the input power and resulting PCE, and they, therefore, have a much higher threshold of "efficiency droop" and high efficiencies can be achieved at very high current (or power) densities. For example, the state-of-the-art 450 nm LDs have a PCE of ~30% at an input power density of 200 W mm^{-2}, which is approximately three times higher than that of InGaN LEDs driven under the same input power density (Figure 4.1). Therefore, using LDs instead of LEDs as the primary source to pump phosphors can generate light with both superbrightness and high luminous flux. Furthermore, laser-driven lighting (or laser

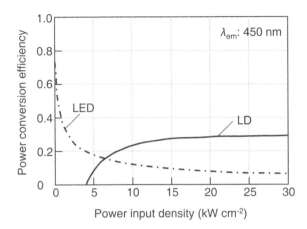

FIGURE 4.1 Power-conversion efficiencies versus input power density of a state-of-the-art blue LED and a state-of-the-art blue laser diode (LD). (Wierer, J.J., Tsao, J.Y. and Sizov, D.S., *Laser Photonics Rev.*, 7, 963–993, 2013. With permission)

lighting) promises smaller emitting area, higher efficiency, lower beam divergence, lower power consumption, faster switching, and compactness, allowing applications in projectors, automotive headlighting, and medical industries.

To create white light, a generally accepted option is to combine phosphor(s) with an ultraviolet- or blue-LED chip, which is also named as phosphor-converted *w*LEDs (pc-*w*LEDs). The phosphors used in *w*LEDs, *i.e.*, LED phosphors, play an indispensable role in determining the luminous efficacy (LE), color rendition, color gamut, color temperature, and longevity of lighting sources. Similarly, phosphors are also key wavelength-conversion materials in laser lighting, but they are in a remote-phosphor configuration.[2] In the case of laser lighting, we, therefore, name them "laser phosphors". However, traditional LED phosphors embedded in organic resin are prone to aging and damage under high power density laser irradiation, due to their poor thermal stability and low thermal conductivity. Moreover, luminance saturation usually occurs for LED phosphors when the power density of the laser light exceeds a threshold. Both thermal quenching and luminance saturation will lead to loss in luminance or luminous flux, shift in chromaticity, and reduction in reliability of lighting sources. Therefore, traditional phosphor-in-silicone (PiS) color converters are not suitable for laser lighting, and it is necessary to develop and design laser phosphors for both super-high luminance and luminous flux laser lighting.

In this chapter, requirements for laser phosphors will be introduced, followed by the classification of laser phosphors in terms of single crystals, phosphor-in-glass (PiG), phosphor films, and phosphor ceramics. Finally, some perspectives for laser phosphors are demonstrated.

4.2   REQUIREMENTS FOR LASER PHOSPHORS

Solid-state laser lighting is an exciting development that requires new phosphor geometries to handle greater light fluxes involved. To pursue better performance of lighting devices, laser phosphors must be carefully selected or designed. Although there are a huge number of luminescent materials, only those matching well with the selection rules can be considered candidates for laser lighting. In addition, laser phosphors are utilized in a remote configuration and usually have a composite microstructure (except single crystals), their geometry needs to be carefully considered or built. In the following list, some requirements are given for laser phosphors.

1. *Small thermal quenching.* Localized heat generation will occur when phosphors are irradiated by high power density laser light, owing to the quantum defect or spectral Stokes shift

during color/wavelength conversion. The thermal quenching behavior of a phosphor indicates its capability of maintaining the luminescence at high temperatures. A small thermal quenching means high thermal stability of the phosphor.

2. *Small luminance saturation.* Luminance saturation is often observed when phosphors are excited under high light power or flux density, of which the output luminous flux remains unchanged or declines when the pump power exceeds a critical value. The saturation mechanism is still argued but can be roughly ascribed to thermal saturation and/or optical saturation. The thermal saturation is linked to thermal quenching, and the optical saturation is due to the depletion of ground state population or excited state absorption (ESA). The luminance saturation is the obstacle to realize super-high luminance or super-high power laser lighting, and thus, small saturation or high saturation threshold is required for laser phosphors.

3. *High quantum efficiency.* Luminescence loss, owing to non-radiative transitions, is a dominant origin of heat generation. To reduce the luminescence loss, laser phosphors should have a high internal quantum efficiency (IQE). In addition, high absorption of light will generate more heat owing to the Stokes shift loss. Therefore, the concentration of activator needs to be carefully considered.

4. *Good thermal management.* Thermal issue is a big challenge for laser phosphors, and thermal management is, therefore, of great significance for improving the light output, chromaticity stability and device reliability. Rapid heat dissipation is required when heat is generated and accumulated, otherwise the temperature of phosphors under the laser irradiation will rise quickly and finally thermal quenching occurs. A higher thermal conductivity of the color converter means a higher capability of heat dissipation. It is, thus, an important topic to design heat sink structures or prepare thermally conductive laser phosphors for maintaining a sufficient low temperature.

5. *Short decay time.* It is generally accepted that luminance saturation is a combination of thermally and optically induced saturation. Thermal saturation is resulted from luminescence quenching, whereas optical saturation is caused by ESA or the depletion of ground state population under a large excitation density.[3,4] Optical saturation usually occurs when the fluorescence decay time of the activator is not short enough for the recombination of electron and hole. A longer decay time implies a higher possibility of electrons being pumped into the conduction band of host, leading to the optical saturation. Therefore, phosphors with short decay times are chosen for laser lighting, for example, the Ce^{3+}-doped one.

6. *Appropriate scattering effect.* The laser source has a quite narrow beam and strong light intensity, and in a transmissive mode, the laser light hence easily penetrates to the color converter and comes out from the back side, leaving the phosphors unirradiated. This will result in the nonuniformity of light, such as yellow ring. In a reflective mode, the laser light will be almost reflected from the front side, and the converted light is, thus, less leading to the nonuniformity of light. Therefore, scattering centers, such as pores and secondary phases, are introduced for (i) increasing the pump absorption, (ii) providing an efficient light extraction and a high luminous flux, and (iii) achieving uniform light by mixing the pumped and converted light well.

4.3 CLASSIFICATION OF LASER PHOSPHORS

4.3.1 SINGLE CRYSTALS

Luminescent single crystals have a perfect crystal structure without any grain boundaries, thus exhibiting outstanding thermal stability, high IQE, and comparatively high thermal conductivity when compared to their counterparts, such as phosphor powders or phosphor ceramics. Due to their thermal robustness, luminescent single crystals, *i.e.*, the $Y_3Al_5O_{12}:Ce^{3+}$ (YAG:Ce) represented garnet single crystals, attract great attentions as color converters for high-brightness white laser lighting. A series of $(Y_{1-x}Lu_x)_3Al_5O_{12}:Ce^{3+}$ and $(Y_{1-x}Gd_x)_3Al_5O_{12}:Ce^{3+}$ single crystals were grown by the Czochralski method for laser excitation (Figure 4.2).[5–7] These luminescent single crystals possess a high IQE over 95% and exhibit an excellent thermal stability up to at least 300°C. Furthermore, they have two orders of magnitude higher thermal conductivity compared to organic resins–encapsulated phosphors (*i.e.*, PiS) and can, therefore, facilitate the dissipation of heat generated during the light conversion process. Apart from yellow-emitting YAG:Ce single crystals, green-emitting $Lu_3Al_5O_{12}:Ce^{3+}$ (LuAG:Ce) and red-emitting $K_2SiF_6:Mn^{4+}$ (KSF:Mn) single crystals have also been developed for laser lighting.[8,9] Under a 5-W blue LD excitation, LuAG:Ce single crystal can maintain a low work temperature of 62°C, achieving extremely stable emission without any efficiency loss. But, the luminance saturation threshold of KSF:Mn is only around 0.3 W mm^{-2}, hardly fulfilling the requirements of laser lighting. This can be ascribed to the long fluorescence decay time of Mn^{4+} activators, which causes serious optical saturation.

Recently, the optical performance of a series of garnet single crystals, including LuAG:Ce, YAG:Ce, $Gd_3(Ga,Al)_5O_{12}:Ce^{3+}$ (GAGG:Ce), and $(Gd,Y)_3Al_5O_{12}:Ce^{3+}$ (GdYAG:Ce), for laser lighting has been systematically investigated (Figure 4.3).[10] The substitution of Y^{3+} by Lu^{3+} leads to a blue-shifted emission, whereas the incorporation of Gd^{3+} induces a red-shifted emission, finally obtaining single crystals with tunable emissions (*i.e.*, green, yellow, and orange). When coupled to blue LDs, they can create white laser light with different correlated color temperatures (CCTs) and color rendering indices (*R*a). For example, the GdYAG:Ce single crystal converted white laser light has a CCT of 3500–4000 K and a *R*a of 80, while that based on YAG:Ce single crystals possesses a high CCT of 4500–5500 K and a low *R*a of around 70. Both YAG:Ce and LuAG:Ce single crystals have a high IQE of about 95% and excellent thermal stability; hence, they can withstand high-power-density blue laser excitation and achieve high-brightness white laser lighting. The red emission–enhanced GdYAG:Ce single crystal also has a relatively high IQE of ~90%, but its IQE decreases sharply at a high temperature (*i.e.*, 300°C). Therefore, it really improves the optical quality of white laser lighting, but its saturation threshold is relatively low. Among these single crystals, GAGG:Ce has the lowest IQE (~80%) and largest thermal quenching, making it unsuitable for high-power laser excitation.

The superior performance over organic resins–encapsulated phosphors makes luminescent single crystal an attractive color converter in high-power white laser lighting. However, it should also be noted that the activator concentration in single crystals is one order of magnitude lower than that

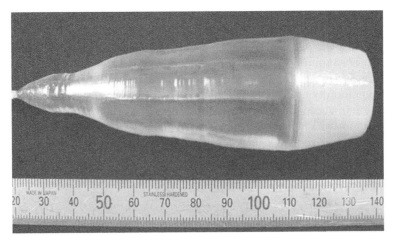

FIGURE 4.2 Photograph of $(Y_{0.8}Lu_{0.2})_3Al_5O_{12}:Ce^{3+}$ single crystal. (Arjoca, S., Víllora, E.G., Inomata, D., Aoki, K., Sugahara, Y. and Shimamura, K., *Mater. Res. Express*, 1, 025041, 2014. With permission)

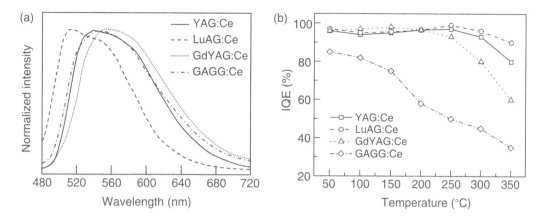

FIGURE 4.3 (a) Emission spectra (λ_{ex} = 445 nm) and (b) temperature-dependent internal quantum efficiency (IQE) of $Y_3Al_5O_{12}$:Ce^{3+} (YAG:Ce), $Lu_3Al_5O_{12}$:Ce^{3+} (LuAG:Ce), $Gd_3(Ga,Al)_5O_{12}$:Ce^{3+} (GAGG:Ce), and $(Gd,Y)_3Al_5O_{12}$:Ce^{3+} (GdYAG:Ce) luminescent single crystals. (Balci, M.H., Chen, F., Cunbul, A.B., Svensen, Ø., Akram, M.N. and Chen, X., *Opt. Rev.*, 25, 166–174, 2018. With permission)

of phosphor powders, and thus, the absorbance and external quantum efficiency (EQE) of single crystals are considerably low, which further leads to limited or insufficient luminance. Usually, the EQE of YAG:Ce single crystals is only about 50%, although its powder form has a high value of more than 90%.[11] When combined with an incident laser, it does withstand a high excitation powder density but only produces a limited luminous flux as well as a relatively low LE. Moreover, due to the absence of light-scattering factors, *i.e.*, grain boundaries, single crystals suffer from a low light extraction efficiency and poor light uniformity. Some patterns, *i.e.*, cone-shaped microwell arrays, have, therefore, been made on the surface of single crystal to enhance the light scattering and finally alleviate the problems mentioned above. Unfortunately, the alleviation is not significant. Furthermore, the growth of single crystals is complicated, costly, and time-consuming, thus severely hindering their large-scale fabrication. And the phosphor materials that can be grown into single crystals are almost limited to the garnet system so far. Because of these limitations, luminescent single crystal converted white laser lighting is still at its laboratory stage and not commercially available.

4.3.2 PHOSPHOR-IN-GLASS (PIG)

Glass materials own a relatively higher thermal stability and thermal conductivity (~1 W m⁻¹ K⁻¹) than that of organic resins (0.1–0.4 W m⁻¹ K⁻¹); therefore, they have been selected as alternative matrices for phosphors encapsulation in recent years. This selection of stable all-inorganic encapsulation, thus, well avoids the aging of organic resins and consequently the performance degradation of *w*LEDs. PiG, where phosphor particles are uniformly dispersed in a glass matrix, can inherit excellent luminescent properties of phosphor as well as all the advantages of glass, making it a promising color converter for solid-state lighting sources.

Usually, PiG is prepared by melting or co-sintering the mixture of phosphor powders and glass frits. This is much more feasible and economical in preparation than other all-inorganic color converters, like single crystals and phosphor ceramics. In 2010, Segawa *et al.* designed and prepared yellow-emitting Ca-α-SiAlON:Eu^{2+}-PiGs for applications in *w*LEDs by using borate and tellurite glasses as matrices, respectively.[12] Then, PiGs based on various phosphors, such as YAG:Ce, LuAG:Ce, $SrAl_2O_4$:Eu^{2+}, β-SiAlON:Eu^{2+} (β-SiAlON:Eu), $CaAlSiN_3$:Eu^{2+} (CASN:Eu), $SrMgAl_{10}O_{17}$:Mn^{4+}, and even some nanocrystals, have been developed. In fact, almost all kinds of phosphors can be introduced into a glass matrix. And the great versatility for various phosphors is exactly one of the most attractive merits of PiG color converters. To obtain excellent optical performance of PiG color converters, two key factors must be highly considered: (i) the chemical erosion to the phosphor powders from the glass matrix must be effectively avoided to retain the luminescent properties of the raw phosphor; (ii) the reflective index of the glass should be close to that of the phosphor to relieve the interface scattering loss and keep PiGs transparent.[13] As the former one is more significant, various strategies have been developed to suppress the interfacial reaction between phosphor particles and the glass matrix, such as adopting low-melting glasses (*i.e.*, tellurite, phosphate, borate glasses) or pure SiO_2 glass, and applying quick and high-pressure consolidation techniques, *i.e.*, spark plasma sintering (SPS) and hot isostatic pressing. Considering the high reactivity and diffusivity of low-valence ions (Na^+, K^+, Zn^{2+}, Ca^{2+}, *etc.*) in multicomponent glasses, the fused silica glass was used as the matrix and a series of efficient and stable PiGs with various phosphors were successfully fabricated.[14]

Initially, PiGs were used to replace organic color converters (*i.e.*, PiS) in ordinary *w*LEDs, and then they proved excellent performance in high-power *w*LEDs.[15,16] By co-sintering YAG:Ce phosphor and tellurite glass frits, a transparent YAG:Ce-PiG was fabricated as a new-generation color converter for high-power *w*LEDs.[15] Benefiting from the high reliability of inorganic glass, the YAG:Ce-PiG plates exhibit excellent heat and humidity resistance and, thus, gain longevity: only 7.6% LE loss is observed after aging for 600 h at 150°C, much superior to that of the conventional organic resin encapsulation; and only 5.6% LE degradation is detected after boiling in water for 24 h. Apart from the superb stability, the YAG:Ce-PiG converted *w*LED present outstanding optical performances, yielding a LE of 124 lm W^{-1}, CCT of 6674 K, and *R*a of 70, under an operating current of 350 mA. In addition to YAG:Ce-PiGs, there are a large number of PiGs based on various phosphors being developed for *w*LED applications.[16] However, they are beyond the topic of this chapter and not discussed here in detail.

In recent years, phosphor-converted laser lighting is springing up in a diverse range of applications, resulting in urgent demands for reliable all-inorganic color converters. Hence, some bulk PiGs, such as YAG:Ce, *β*-SiAlON:Eu, and CASN:Eu, have been investigated under laser excitation.[17–19] Some PiG-based white LDs (*w*LDs) have also been demonstrated, and unfortunately, all of them have not achieved a really high luminance till now. This is mainly due to the low thermal conductivity (~1 W m^{-1} K^{-1}) of the glass matrix, which cannot dissipate heat quickly and, thus, encounters luminance saturation at a very low incident laser power density (usually <2 W mm^{-2}). Besides, to achieve a good transparency in PiGs, the phosphor content remains low (usually <10 wt%), which also hinders the luminance improvement. So, even though PiG color converters possess competitive advantages in easy fabrication and great versatility, their application in laser lighting is limited due to their small saturation thresholds and low luminance output. To overcome these drawbacks, phosphor films (equals to "PiG films" in this chapter) have been developed to enable PiGs to be applied in high-power laser-driven lighting.[20] They will be discussed systematically in the following section.

4.3.2.1   YAG:Ce-PiGs

Current investigations on PiGs mainly focus on the dispersion of oxide phosphors (especially YAG:Ce) in the glass matrix. In 2017, a YAG:Ce-PiG plate was prepared and then coupled with a 445 nm LD, which generated a bright white light with color coordinates of 0.329 and 0.333, CCT of 5649 K, and LE of 110 lm W^{-1}.[13] Then, Yu *et al.* schematically compared the optical performance of YAG:Ce-PiG and YAG:Ce-PiS under high-power blue laser excitation.[17] As shown in Figure 4.4(a), the YAG:Ce-PiG plate is highly transparent, with phosphor particles being uniformly dispersed in the glass matrix. Under a 0.75-W blue laser excitation, the temperature of YAG:Ce-PiS quickly rises to above 706°C and the surface of the sample is burned down in a short time of about 60 s. By contrast, the temperature of YAG:Ce-PiG plate increases in a small range (<100°C), and the surface of the sample remains unchanged (Figure 4.4(b)). With the extension of irradiation time, the PiG plate can maintain 85% of its original emission intensity, the intensity of which is 10 times higher than

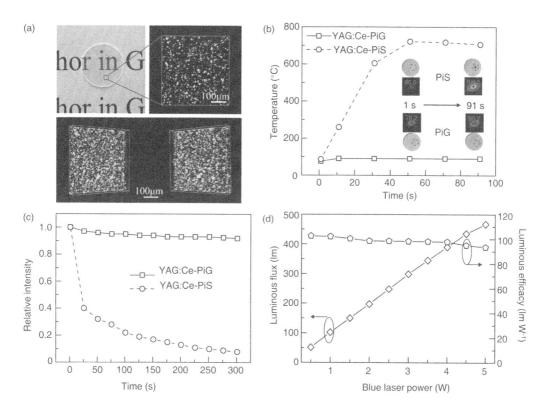

FIGURE 4.4 (a) Photograph, 2D surface fluorescence distribution image, and 3D reconstructed image of YAG:Ce-PiG plate; (b) temperature and (c) relative emission intensity change of YAG:Ce PiG and PiS samples excited by a blue LD at a constant power (0.75 W) over time, the insets of (b) are corresponding photographs and thermal infrared images; (d) luminous flux and luminous efficacy of YAG:Ce-PiG under excitation with varying laser powers. (Yu, J., Si, S., Liu, Y., Zhang, X., Cho, Y., Tian, Z., Xie, R., Zhang, H., Li, Y. and Wang, J., *J. Mater. Chem. C*, 6, 8212–8218, 2018. With permission)

that of the PiS sample (Figure 4.4(c)). Since YAG:Ce-PiG has superior heat-resistance performance over YAG:Ce-PiS, it can withstand an incident blue laser power of 5 W and give a luminous flux of 467 lm as well as a LE (the ratio of luminous flux to incident laser power, LE_{opt}) of 93.4 lm W^{-1} (Figure 4.4(d)).

4.3.2.2 β-SiAlON:Eu^{2+}- and CaAlSiN$_3$:Eu^{2+}-PiGs

Benefiting from the compositional and structural versatility, nitride phosphors show compelling photoluminescent properties, such as abundant emission colors, tunable emission spectra, high quantum efficiency, and small thermal quenching.[21] Thus, PiGs based on nitride phosphors, such as Ca-α-SiAlON:Eu^{2+}, La$_3$Si$_6$N$_{11}$:Ce^{3+}, β-SiAlON:Eu, and CASN:Eu, also attract great attentions for laser lighting. Zhu *et al.* fabricated β-SiAlON:Eu- and CASN:Eu-PiG plates by the co-sintering method, and studied their optical performance under blue laser excitation.[18,19] Under blue laser irradiation, β-SiAlON:Eu- and CASN:Eu-PiG plates, respectively, emitted intense green and red colors, but they both showed a decreased EQE, suggesting the emission degradation induced by the chemical erosion of phosphors from the glass matrix. As shown in Figure 4.5, the luminance saturation thresholds of β-SiAlON:Eu- and CASN:Eu-PiG plates are 0.98 and 0.5 W mm^{-2}, respectively. The corresponding luminance fluxes are 275 and 39 lm. The low saturation threshold and small luminance flux are caused by the general drawbacks of PiGs, *i.e.*, low thermal conductivity and low phosphor content, as well as the nonthermal optical saturation resulted from the intrinsic long fluorescence decay time of the Eu^{2+} activator (*i.e.*, several hundred nanoseconds).

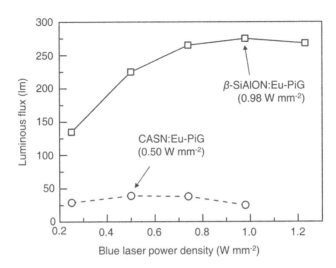

FIGURE 4.5 Effect of incident laser power on luminous flux of β-SiAlON:Eu^{2+} (β-SiAlON:Eu) and CaAlSiN$_3$:Eu^{2+} (CASN:Eu) PiG plates. (Zhu, Q.-Q., Wang, X.-J., Wang, L., Hirosaki, N., Nishimura, T., Tian, Z.-F., Li, Q., Xu, Y.-Z., Xu, X. and Xie, R.-J., *J. Mater. Chem. C*, 3, 10761, 2015; Zhu, Q.-Q., Xu, X., Wang, L., Tian, Z.-F., Xu, Y.-Z., Hirosaki, N. and Xie, R.-J., *J. Alloy. Compd.*, 702, 193–198, 2017. With permission)

4.3.3 PHOSPHOR FILMS

Phosphor film is a composite color converter, where a thin PiG film with high phosphor contents is tightly co-sintered on a high thermally conductive substrate. Phosphor films can be facilely prepared by some well-established techniques, such as screen printing, blade coating, and spin coating. Compared to PiG plates, phosphor films fully inherit the advantages (*i.e.*, easy fabrication, versatility for various phosphors) but perfectly avoid the disadvantages of PiG plates (*i.e.*, low thermal conductivity, low phosphor content), thus attracting great attentions from both the academic and industrial fields. The merits of phosphor films include the following aspects: (i) the thickness of most PiG films is only in several tens of micrometers, thus effectively reducing the heat transfer distance and finally increasing the whole capability of heat dissipation; (ii) the phosphor content in PiG films is much larger than that in PiG plates (*i.e.*, sometimes >80 wt%), which increases the conversion efficiency as well as the luminous flux of the device; (iii) the substrates with high thermal conductivities act as not only the carrier for heat dissipation but also the support of excellent mechanical performance, leading to robust color converters; (iv) phosphor films can be structured by alternating or stacking phosphor layers of different composition or thickness, so they offer great freedom in optical performance tuning.

Due to successful improvements in heat dissipation and phosphor content, PiG films have been intensively studied for high-power laser lighting. Various PiG films based on different phosphors (*i.e.*, oxide phosphors, nitride phosphors) sintered on different substrates (*i.e.*, glasses, ceramics, metals) have been demonstrated for applications in laser lighting and displays, as listed in Table 4.1.

4.3.3.1 Phosphor Films on Glass Substrates

Owing to their similar properties, phosphor films were first co-sintered on glass substrates. In 2016, Yoshimura *et al.* prepared β-SiAlON:Eu and Ca-α-SiAlON:Eu^{2+} PiG (SiO$_2$ glass matrix) films on a glass substrate and studied their luminescent properties as color converters for laser lighting.[22] The optical performance was evaluated in a transmissive mode, and a 440 nm laser beam was used as the excitation light (Figure 4.6(a)). The cross-sectional SEM image (Figure 4.6(b)) confirms that the PiG film is tightly bonded to the glass substrate. Under UV light irradiation, the β-SiAlON:Eu and

TABLE 4.1
Some Typical Phosphor Films for Laser Lighting

Substrate	Phosphor	Glass Matrix	IQE (EQE)[a] ex. 450 nm	ST[b] (W mm^{-2})	LF[c] (lm)	L.E.[d] (lm mm^{-2})	LE$_{opt}$[e] (lm W^{-1})	Ref.
Glass	β-SiAlON:Eu^{2+}	SiO$_2$	78% (56%)	–	–		–	22
	Ca-α-SiAlON:Eu^{2+}	SiO$_2$	93% (81%)	–	–		–	22
	Y$_3$Al$_5$O$_{12}$:Ce^{3+}	SiO$_2$	90% (–)	5 W[f]	–		–	23
Alumina	Lu$_3$Al$_5$O$_{12}$:Ce^{3+}	B$_2$O$_3$–Al$_2$O$_3$–ZnO–SiO$_2$	86% (–)	1.7	5496	320	199	24
Sapphire	Y$_3$Al$_5$O$_{12}$:Ce^{3+}	SiO$_2$	95% (52%)	14.3	905	3480	234	25
	Y$_3$Al$_5$O$_{12}$:Ce^{3+} + CaAlSiN$_3$:Eu^{2+} (mass ratio of 12:1)	SiO$_2$	81% (36%)	12.9	404	1554	137	25
	CaAlSiN$_3$:Eu^{2+}	SiO$_2$–Al$_2$O$_3$–B$_2$O$_3$–CaO	79% (–)	3.2	192	48	15	26
1DPCs-sapphire[g]	Y$_3$Al$_5$O$_{12}$:Ce^{3+}	B$_2$O$_3$–SiO$_2$–Al$_2$O$_3$–BaO–ZnO	86% (77%)	11.2	1839	2343	210	27
	Y$_3$Al$_5$O$_{12}$:Ce^{3+} + Ca-α-SiAlON:Eu^{2+} (dual layer)	B$_2$O$_3$–SiO$_2$–Al$_2$O$_3$–BaO–ZnO	–	8.7	1330	1694	194	27
	Y$_3$Al$_5$O$_{12}$:Ce^{3+} + CaAlSiN$_3$:Eu^{2+} (dual layer, CaAlSiN$_3$-in-silicone)	B$_2$O$_3$–SiO$_2$–Al$_2$O$_3$–BaO–ZnO	–	7.8	1259	1603	205	27
	La$_3$Si$_6$N$_{11}$:Ce^{3+}	SiO$_2$–Al$_2$O$_3$–Na$_2$O–CaO–TiO$_2$	72% (–)	12.91	1076	2143	166	28
	CaAlSiN$_3$:Ce^{3+}	SiO$_2$–Al$_2$O$_3$–BaO–ZnO–B$_2$O$_3$	–	12.9	233	464	36	29
	Ca$_{0.75}$Li$_{0.15}$Al$_{0.75}$Si$_{1.25}$N$_{2.9}$O$_{0.1}$:Ce^{3+}	SiO$_2$–Al$_2$O$_3$–BaO–ZnO–B$_2$O$_3$	–	17.8	297	592	33	29
	β-SiAlON:Eu^{2+}	SiO$_2$–BaO–Al$_2$O$_3$–ZnO	61% (41%)	6.09	675	860	141	30
Aluminum (Al)	Y$_3$Al$_5$O$_{12}$:Ce^{3+}	B$_2$O$_3$–SiO$_2$–Al$_2$O$_3$	–	>4 W[h]	430[i]	–	107	31

Note: [a]IQE (EQE), internal (external) quantum efficiency; [b]ST, luminance saturation threshold; [c]LF, luminous flux; [d]LE, luminous emittance, luminous flux per unit area; [e]LE$_{opt}$, LE at ST, defined as the ratio of luminous flux to incident laser optical power; [f]5 W, incident laser power when luminance saturated, no laser spot size was given; [g]1DPCs-sapphire, one-dimensional photonic crystals–coated sapphire; [h]4 W, the largest incident laser power given (no saturation); [i]430, luminous flux at an incident laser power of 4 W (no saturation).

Ca-α-SiAlON:Eu^{2+} phosphor films show characteristic green and orange emissions, respectively. Their combination (β-SiAlON:Eu + Ca-α-SiAlON:Eu^{2+}, denoted as SiAlON later) gives an intensive yellow emission, corresponding to a broadband spectrum covering the green to orange region (Figure 4.6(c)). The SiAlON-PiG film can withstand a high-power blue laser light of >4 W and give an illuminance of more than 1000 lx (Figure 4.6(d)). Similarly, a YAG:Ce-PiG film on a quartz glass substrate has also been prepared, which can endure a blue laser power of 4.1 W.[23]

Unfortunately, due to the intrinsic low thermal conductivity of glass substrates (~1 W m^{-1} K^{-1}), phosphor films sintered on glass substrates are difficult to be applied for high-power-density laser excitation. Researchers, therefore, shift their attentions to substrates with a much higher thermal conductivity, such as ceramic and metallic substrates.

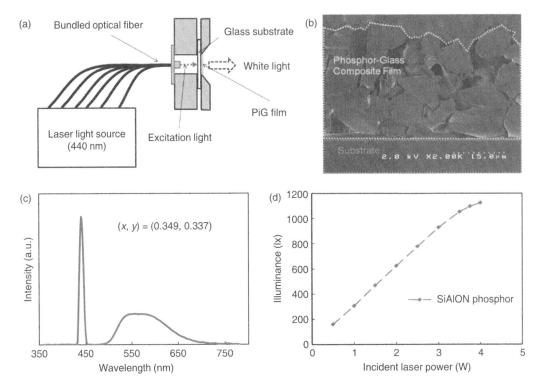

FIGURE 4.6　(a) Schematic diagram of the transmissive measurement mode; (b) cross-sectional SEM image of the SiAlON-PiG film; (c) emission spectrum and chromaticity coordinates of white light converted by the SiAlON-PiG film under 0.5-W blue laser excitation (λ_{ex} = 440 nm); (d) excitation light power-dependent illuminance of the light source. (Yoshimura, K., Annen, K., Fukunaga, H., Harada, M., Izumu, M., Takahashi, K., Uchikoshi, T., Xie, R.-J. and Hirosaki, N., *Jpn. J. Appl. Phys.*, 55, 042102, 2016. With permission)

4.3.3.2 Phosphor Films on Ceramic Substrates

Ceramic plates with high thermal conductivity, *i.e.*, Al_2O_3, sapphire, and AlN, were chosen as substrates for phosphor films. Among them, sapphire is most widely used because of its high thermal conductivity (~30 W m^{-1} K^{-1}), excellent mechanical properties, and high in-line transmittance (~86%), especially suitable for the transmissive mode. Wei *et al.* sintered the YAG:Ce-PiG film on a sapphire substrate and investigated effects of the phosphor content and film thickness on luminance saturation.[32] They found that both the threshold value of luminance saturation and the luminous flux were reduced with increasing the phosphor content and film thickness. An optimal phosphor film (mass ratio of phosphor to glass of 1:1 and thickness of 50 μm) can bear a maximum blue laser power density of 10.16 W mm^{-2} (saturation threshold) and create a high-brightness white laser light with a luminous flux of 1048 lm. To further enhance the heat dissipation of the outside surface of phosphor films, a dual sapphire plate structure was designed for laser-driven white lighting.[33] The YAG:Ce-PiG film is settled between two pieces of sapphire plates, thus providing two effective heat transfer channels. As a result, it achieves considerably low working temperature (<50°C) comparing to PiG films sintered on a single sapphire plate (>150°C) when excited by high-power laser.

To fully utilize the potential of sapphire, the sapphire coated with one-dimensional photonic crystals (1DPCs-sapphire) was selected as a substrate, namely, the backward side is covered by an antireflection (AR) layer to enhance the transmittance of the incident blue light, and the forward side is coated by a blue-pass (BP) filter to reflect the backward yellow emission.[27,28,34] Both the AR layer and BP filter consist of thin alternating high and low refractive index layers, *i.e.*, TiO_2 and SiO_2. The AR layer on the front side enhances the transmittance from 85 to 92%, while the BP filter

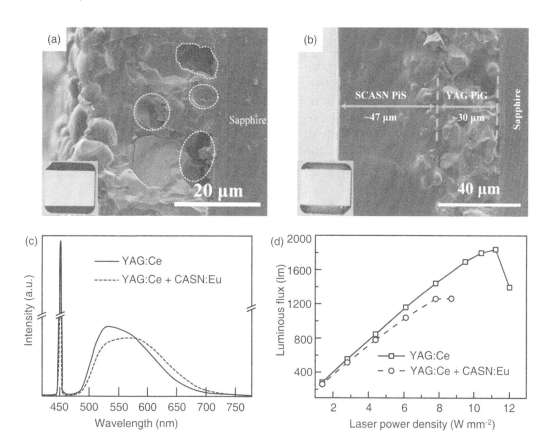

FIGURE 4.7 Cross-sectional SEM images of (a) YAG:Ce-PiG film and (b) CASN:Eu-PiS stacked YAG:Ce-PiG film, the insets are corresponding photographs; (c) emission spectra of two phosphor films under blue laser excitation (λ_{ex} = 450 nm); (d) changes in luminous flux of two films with increasing incident laser power. (Zheng, P., Li, S., Wang, L., Zhou, T.-L., You, S., Takeda, T., Hirosaki, N. and Xie, R.-J., *ACS Appl. Mater. Interfaces*, 10, 14930–14940, 2018. With permission)

on the back side further increases it up to 97%, and almost 100% reflects the backward yellow light. Zheng *et al.* prepared the YAG:Ce-PiG film on a 1DPCs-sapphire using the blade coating method (Figure 4.7).[27] It can withstand a high-power-density blue laser excitation of 11.20 W mm⁻² and emit high brightness white light with a luminous flux of 1839 lm (luminous emittance, luminous flux per unit area, of 2343 lm mm⁻²), LE_{opt} of 210 lm W⁻¹, Ra of 68 and CCT of 6504 K (Figure 4.7(d)). Moreover, a red-emitting CASN:Eu-PiS film was stacked on the YAG:Ce-PiG film to improve the white light quality (Figure 4.7(b)). Then, the Ra can be increased to 74 and the CCT can be lowered to 5649 K, but the saturation threshold and luminous flux were also sacrificed, the values of which decreased to 7.80 W mm⁻² and 1259 lm, respectively. Inspired by this report, PiG films based on other phosphors, *i.e.*, LuAG:Ce, co-sintered on sapphire substrates have been also developed.[20,25]

Differing from oxide phosphors, nitride phosphors are prone to reacting with oxide glass matrices, thus, may severely damage their luminescent properties. In 2019, You *et al.* co-sintered an efficient and thermally robust $La_3Si_6N_{11}$:Ce^{3+} (LSN:Ce)-PiG film on the 1DPCs-sapphire, successfully achieving high-brightness blue laser–driven white lighting for the first time by using a nitride phosphor (Figure 4.8).[28] The LSN:Ce-PiG film can bear a blue laser power density of up to 12.90 W mm⁻² and show a high luminous flux of 1076 lm (luminous emittance of 2143 lm mm⁻²). This interesting result is ascribed to the chemical inertness of LSN:Ce phosphor, which, thus, minimizes the erosion of phosphor particles by the glass matrix and almost remains its original luminescent properties.

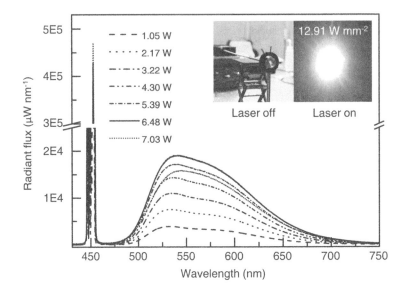

FIGURE 4.8 Emission spectra of $La_3Si_6N_{11}$:Ce^{3+}-PiG films under irradiations with increasing incident blue laser power from 1.05 to 7.03 W, the insets are the photograph of the laser-driven white lighting device and the corresponding luminescent image. (You, S., Li, S., Zheng, P., Zhou, T., Wang, L., Liu, L., Hirosaki, N., Xu, F. and Xie, R.-J. *Laser Photonics Rev.*, 13, 1800216, 2019. With permission)

After this, some efficient PiG films based on other nitride phosphors, including β-SiAlON:Eu and CASN:Eu, have been developed for high-power laser lighting and displays, and their saturation thresholds can reach as high as 6.09 and 3.2 W mm^{-2}, respectively.[26,30]

Apart from the transparent sapphire substrate, phosphor films have also been sintered on some opaque ceramic plates, such as Al_2O_3 ceramics, Al-coated sapphire, and AlN, and used as reflective color converters in laser-driven lighting and displays.[24,35]

4.3.3.3 Phosphor Films on Metallic Substrates

Metallic materials have a very high thermal conductivity, which are usually used as heat sinks for high-power LED or LD chips. To further improve the heat dissipation of phosphor films, aluminum (Al) with a thermal conductivity of >200 W m^{-1} K^{-1} was adopted as the substrate.[31,36] In 2017, Park *et al.* designed and fabricated a thermally robust phosphor-aluminum composite (PAC) color converter *via* the SPS method.[31] The color converter was composed of three parts: (i) a YAG:Ce-PiG film (50 vol% YAG:Ce + 50 vol% glass) with a thickness of about 500 μm, (ii) an Al substrate, and (iii) a thermally graded layer (75 vol% Al + 25 vol% glass) sandwiched between the PiG film and Al substrate (Figure 4.9(a)). With such a design, the PAC can own a thermal conductivity of 31.6 W m^{-1} K^{-1}, enabling it to be excited under a high-power blue laser of 4 W without any saturation and generate a white light with a luminous flux of 430 lm (Figure 4.9(b)). Moreover, a thermoelectric module was attached at the Al side, and it can collect the waste heat and produce electric energy with an output voltage and current of 289 mV and 77 mA, respectively (under 4-W blue laser excitation). To better manipulate the heat flow from YAG:Ce-PiG film to the Al substrate, they also carefully monitored the temperature distribution of PAC to optimize the thermally graded layers design (the inset of Figure 4.9(b)).[37] This research not only provides a promising color converter type for high-power laser lighting, but also demonstrates an interesting idea for energy recycling.

In summary, preparing phosphor films on high thermally conductive substrates successfully broadens the practical applications of PiGs, making them a kind of very promising color converters applied in high-power laser lighting and displays.

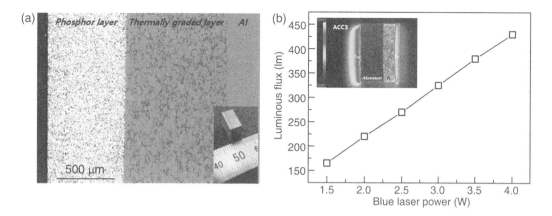

FIGURE 4.9 (a) Cross-sectional SEM image and photograph (inset) of the YAG:Ce-PiG layer on the Al substrate; (b) luminous flux of the sample with increasing incident laser power, the inset is temperature distribution of sample under laser excitation monitored by an infrared camera. (Park, J., Kim, J. and Kwon, H., *Adv. Opt. Mater.*, 5, 1700317, 2017; Park, J., Cho, S. and Kwon, H., *Sci. Rep.*, 8, 17852, 2018. With permission)

4.3.4 PHOSPHOR CERAMICS

Phosphor ceramics are important optical polycrystalline materials. They can be used as scintillators for X-ray detector, *i.e.*, $(Y,Gd)_2O_3:Eu^{3+}$, Pr^{3+} and $Gd_2O_2S:Pr^{3+},Ce^{3+}$, or as laser ceramics for laser generation, *i.e.*, Nd:YAG, or as remote phosphor layer in *w*LEDs, *i.e.*, YAG:Ce. In 2008, a (Y,Gd)AG:Ce phosphor ceramic plate (PCP) was used to combine with a blue LED chip to produce white light.[38] In 2014, Lenef *et al.* proposed laser activation of a remote YAG:Ce phosphor ceramic for producing very high luminance solid-state light sources, thus starting the journey of phosphor ceramics applied in high-power laser lighting and displays.[2]

Generally, phosphor ceramics for laser lighting can be classified into two groups: (i) monolithic phosphor ceramics, obtained by the densification of phosphor powders or in-situ synthesis from raw materials, *i.e.*, the garnet phosphor ceramics; (ii) composite phosphor ceramics, where phosphor particles are embedded in the transparent ceramic matrix, *i.e.*, $YAG:Ce-in-Al_2O_3$ ($YAG:Ce/Al_2O_3$). To achieve high densification and excellent luminescent properties, phosphor ceramics are usually prepared by pressure-assisted sintering methods, such as vacuum sintering, hot pressing sintering, hot isostatic pressing, and SPS. Compared to single crystals, polycrystalline phosphor ceramics allow for a great versatility for various phosphors and a controllable microstructure (such as pores, secondary phases, grain boundaries, and grain sizes), thus offering greater flexibility in designing phosphor compositions and controlling the light-scattering and absorption processes. Besides, it is much easier to prepare phosphor ceramics than single crystals. Compared to PiGs, phosphor ceramics show a higher thermal conductivity, better mechanical properties, and superior thermal and chemical stabilities. These merits, therefore, make phosphor ceramics a very important type of color converters in laser-driven lighting and displays. Table 4.2 summarizes phosphor ceramics developed in recent years for laser excitation. It can be found that phosphor ceramics are very limited to the garnet system, especially YAG:Ce. Owing to the intrinsic low diffusion rate of nitrides, most nitride phosphors with excellent luminescent properties are difficult to be fully densified. In addition, even if nitride phosphor ceramics can be densified with sintering aids, their photoluminescence properties are hardly maintained. Hence, it is still a big challenge to fabricate high-quality phosphor ceramics from phosphor powders with varying emission colors.

TABLE 4.2
Phosphor Ceramics Proposed for Laser Lighting

Group	Emission Colors	Phosphor Ceramics	Preparation Method	EQE[a] (ex. 450 nm)	k[b] (W m^{-1} K^{-1})	ST[c] (W mm^{-2})	L.E.[e] (lm mm^{-2})	LF[d] (lm)	LE$_{opt}$[f] (lm W^{-1})	Ref.
Monolithic	Blue	$Ba_{0.985}MgAl_{10}O_{17}$:0.015Eu^{2+} h[g] = 1 mm	Spark plasma sintering (SPS)	66% (ex. 340 nm)	–	–	–	–	–	39
	Green	$Lu_{2.999}Al_5O_{12}$:0.001Ce^{3+} h = 1 mm	In-situ vacuum sintering	73%	–	49	3976	7934	162	40
		$Lu_{2.985}Al_5O_{12}$:0.015Ce^{3+} h = 1.2 mm	Vacuum sintering	–	–	2.15	1540	218	101	41
		$Lu_3Al_5O_{12}$:Ce^{3+} h = 0.1 mm	SPS	77%	6.3	24	472	1302	54	42
		$Lu_{2.997}Al_5O_{12}$:0.003Ce^{3+} (2.88% porosity) h = 0.28 mm	In-situ vacuum sintering	–	–	–	~600[h]	–	~200	43
	Yellow	$Y_{2.999}Al_5O_{12}$:0.001Ce^{3+} h = 1 mm	In-situ vacuum sintering	55%	–	25.98	2227	4454	171	40
		$Y_{2.985}Al_5O_{12}$:0.015Ce^{3+} (nano-sized)	Vacuum sintering	–	–	19.1	1424	2733	120	44
		$Y_{2.994}Al_5O_{12}$:0.006Ce^{3+} (15% porosity) h = 0.5 mm	In-situ vacuum sintering	58%	7.05	7.92	850	1700	215	45
		$Y_{2.985}Al_5O_{12}$:0.015Ce^{3+} (24% Al$_2$O$_3$ grains) h = 0.3 mm	In-situ vacuum sintering	75%	13.75	14	992	1976	141	46
	Orange	$Y_{1.649}Gd_{1.35}Al_5O_{12}$:0.001Ce^{3+} h = 1 mm	In-situ vacuum sintering	70.5%	–	7.04	393	786	112	40
	Red	$CaAlSiN_3$:Eu^{2+} h = 0.15 mm	SPS	60%	4	1.5	~200	~63	42.2	47,48
		Y_2O_3:Eu^{3+} h = 0.5 mm	Hot isostatic pressing (HIP)	–	10.9	–	–	–	–	49
	White	AlN:Ce^{3+}	Current activated pressure assisted densification	–	90	–	–	–	–	50

(Continued)

TABLE 4.2 (Continued)
Phosphor Ceramics Proposed for Laser Lighting

Group	Emission Colors	Phosphor Ceramics	Preparation Method	EQE[a] (ex. 450 nm)	k[b] (W m^{-1} K^{-1})	ST[c] (W mm^{-2})	LF[d] (lm)	L.E.[e] (lm mm^{-2})	LE$_{opt}$[f] (lm W^{-1})	Ref.
Composite	Green	$Y_{2.92}Al_3Ga_2O_{12}:0.08Ce^{3+}/Y_3Al_5O_{12}$ h = 0.5 mm	Vacuum sintering	68.3%	9.4	8.32	–	700	84	51
	Yellow	$Y_3Al_5O_{12}:Ce^{3+}/Al_2O_3$ h = 0.1 mm	SPS	76%	18.5	>50	~2000[i]	–	44.4	52
		$Y_3Al_5O_{12}:Ce^{3+}/Al_2O_3$ h = 0.15 mm	SPS	–	~18	3.88	631	315.5	81.3	53
		$Y_{2.94}Al_3Ga_2O_{12}:0.06Ce^{3+}/Y_3Al_5O_{12}$ h = 0.5 mm	HIP	70.2%	8.9	9.6	–	1220	127	54

Note: [a]EQE, external quantum efficiency; [b]k, thermal conductivity at room temperature; [c]ST, luminance saturation threshold; [d]LF, luminous flux; [e]LE, luminous emittance, luminous flux per unit area; [f]LE$_{opt}$, LE at ST, the ratio of luminous flux to incident laser optical power; [g]h, ceramic plate thickness; [h]600, luminous flux at an incident laser power of 45 W (no saturation); [i]2000, luminous flux at an incident laser power of 3 W (no saturation).

4.3.4.1 Monolithic Phosphor Ceramics

4.3.4.1.1 Yellow-Emitting YAG:Ce

As a golden partner of blue LEDs/LDs, yellow-emitting YAG:Ce phosphor ceramics have been intensively investigated as color converters for high-power wLEDs and wLDs. In 2016, Song *et al.* fabricated YAG:Ce PCPs *via* vacuum sintering, by using micro- and nano-sized YAG:Ce powder as starting materials, and investigated their optical performance under high-power blue laser excitation.[44] They found that PCPs prepared from nano-YAG:Ce powders had a finer and more uniform microstructure than those prepared from micro-YAG:Ce powders, hence showing better optical properties, *including* higher luminance saturation threshold, luminous flux (luminous emittance) and LE. As presented in Figure 4.10(a), the nano-structured PCP can withstand a higher blue laser power density (14.77 versus 7.75 W mm^{-2}) and generate luminous emittance three times higher than the microstructured one. By combining the optimized nano-YAG:Ce PCP with a blue LD chip, high-brightness white light can be created with a luminous flux of 1424.6 lm (luminous emittance of 2733 lm mm^{-2}), LE$_{opt}$ of 218 lm W^{-1}, Ra of 54.2, and CCT of 5994 K, under a blue laser power density of 19.1 W mm^{-2} (Figure 4.10(b)). These results suggest that the YAG:Ce phosphor ceramic could serve as a promising color converter for laser lighting. Henceforth, many investigations on YAG:Ce phosphor ceramics for high-power laser lighting have been carried out.[40,55] By gluing a YAG:Ce phosphor ceramic ring to an aluminum wheel, the phosphor ceramic wheel could work under a super-high blue laser power density of 187 W mm^{-2} excitation and emit a luminance of 9000 cd mm^{-2}, when the wheel speed was 4000 rpm.[55]

The reflective index of the YAG:Ce PCP is about 1.82, and this high value will produce a quite small critical angle and, thus, strong total internal reflection, which seriously deteriorates the light extraction efficiency.[45] To solve this problem, light-scattering centers should be introduced into PCPs. Moreover, LDs are a kind of point light sources and have a much higher powder density than LEDs, so light-scattering centers are also necessary to (i) increase the absorption of the incident laser and (ii) make the laser beam uniform. To this end, Al$_2$O$_3$ is usually introduced as a secondary phase into YAG:Ce PCPs.[46,56] With 24 wt% Al$_2$O$_3$ particles introducing into the YAG:Ce ceramics (Figure 4.11(a)), it effectively increases the blue light absorption and EQE, leading to a 27.3% enhancement in LE$_{opt}$ (157 versus 123 lm W^{-1}) compared to the Al$_2$O$_3$-free YAG:Ce ceramic plate.[46] However, considering the quite small reflective index difference between Al$_2$O$_3$ (~1.76) and YAG (~1.82), Al$_2$O$_3$ grains do not have a significant scattering effect. Then, pores are recognized as more efficient scattering centers in phosphor ceramics due to the following two reasons: (i) large reflective index difference between air (~1) and YAG, and (ii) no absorption of the emitted light by air.

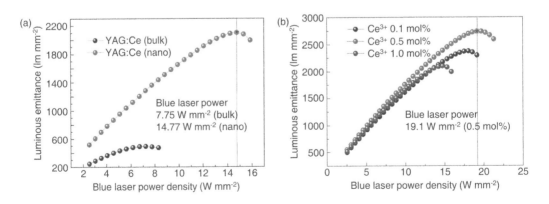

FIGURE 4.10 (a) Luminous emittance of YAG:Ce phosphor ceramics prepared from bulk- and nano-YAG:Ce powders with increasing blue laser power density; (b) laser power density-dependent luminous emittance of nano-YAG:Ce phosphor ceramics with different Ce^{3+} concentrations. (Song, Y.H., Ji, E.K., Jeong, B.W., Jung, M.K., Kim, E.Y. and Yoon, D.H., *Sci. Rep.*, 6, 31206, 2016. With permission)

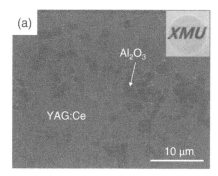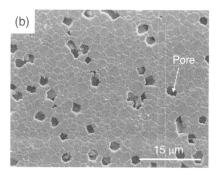

FIGURE 4.11 SEM images of YAG:Ce phosphor ceramics with (a) 24 wt% Al_2O_3 and (b) a porosity of 15%. Inset in (a) is a photograph of a ceramic plate. (Wang, J., Tang, X., Zheng, P., Li, S., Zhou, T. and Xie, R.-J., *J. Mater. Chem. C*, 7, 3901–3908, 2019; Zheng, P., Li, S., Wei, R., Wang, L., Zhou, T.-L., Xu, Y.-R., Takeda, T., Hirosaki, N. and Xie, R.-J. *Laser Photonics Rev.*, 13, 1900147, 2019. With permission)

Zheng *et al.* fabricated a series of YAG:Ce ceramics with a controlled size and content of pores by using poly(methyl methacrylate) microspheres as a pore-forming agent and schematically investigated their optical performance under high-power laser excitation (Figure 4.11(b)).[45] A high porosity of 15% can effectively reduce the light beam size, yielding super-high luminance and excellent light uniformity. The results suggest that high-scattering phosphor ceramics would be a promising color converter for high-quality laser lighting, which also provides a new idea for designing laser phosphors.

To improve the light quality (*i.e.*, color rendition and color temperature) of YAG:Ce converted *w*LDs, some red emission–enhanced YAG:Ce phosphor ceramics, *i.e.*, (Y,Gd)AG:Ce,[57] $Y_3Mg_xAl_{5-2x}Si_xO_{12}$:Ce[3+],[58] have been developed. But the improvement in light quality is limited. To enrich the preparation technology, three-dimensional (3D) printing technique has been applied to fabricate YAG:Ce phosphor ceramics.[59] Besides, a new strategy for preparing YAG:Ce ceramic films on a sapphire substrate has also been proposed.[60] So, as an excellent color conversion material for SSL, YAG:Ce ceramics have been extensively investigated, but it is still a tough mission to further improve the light quality and conversion efficiency.

4.3.4.1.2 Green-Emitting $Lu_3Al_5O_{12}$:Ce[3+]

$Lu_3Al_5O_{12}$:Ce[3+] (LuAG:Ce) is an excellent green component with high quantum efficiency and thermal stability. It also has been investigated as a green color converter for laser lighting. Application of LuAG:Ce phosphor ceramics in *w*LEDs showed a high LE of 223.4 lm W[-1].[61] Xu *et al.* prepared LuAG:Ce transparent ceramics *via* the SPS technique by using commercially available LuAG:Ce powder as starting material and LiF as a sintering additive (Figure 4.12(a)).[42] The as-prepared LuAG:Ce phosphor ceramic emitted a typical broadband spectrum centered at 523 nm under 450 nm excitation with a superior EQE of about 77%. More importantly, it showed excellent thermal stability with only 4.1% loss in luminescence at 200°C, as well as a very high reliability with a small drop (1.9%) in emission intensity after 1000 h aging at 85°C and 85% relative humidity. The ceramic sample also owned a good thermal conductivity of 6.3 W m[-1] K[-1], which is about 13 times higher than the powder sample (0.5 W m[-1] K[-1]). These merits, thus, allow the LuAG:Ce transparent ceramic to be a promising color converter for high-power laser lighting. Under an 8.7-W blue laser excitation, it yields a luminous flux of 472 lm with a wall-plug efficacy of 54.3 lm W[-1] and maintains a relatively low surface temperature of ~126°C (Figure 4.12(b)).

Similar to YAG:Ce phosphor ceramics, pores have been also introduced into LuAG:Ce phosphor ceramics to modulate their scattering effects. A LE_{opt} of over 200 lm W[-1] was obtained when the LuAG:Ce ceramic plate had a porosity of 2.88% under the 450 nm blue laser excitation.[43] Li *et al.* used a water-cooled heat sink to enhance the heat dissipation of LuAG:Ce phosphor ceramics, and

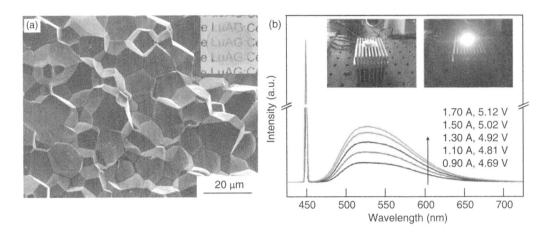

FIGURE 4.12 (a) Photograph and cross-sectional SEM image of LuAG:Ce phosphor ceramics; (b) emission spectra driven by a blue LD under varying operating currents and voltages, and insets are photos of lighting device. (Xu, J., Wang, J., Gong, Y., Ruan, X., Liu, Z., Hu, B., Liu, B., Li, H., Wang, X. and Du, B., *J. Eur. Ceram. Soc.*, 38, 343–347, 2018. With permission)

the temperature of the spot where laser light irradiates can remain as low as 65°C.[41] It effectively compensates the negative impact of thermal quenching. They also proposed the patterning of the surface of ceramic plates to mitigate the total internal reflection induced light loss and, thus, further improved the luminous flux and LE of LuAG:Ce phosphor ceramics.

4.3.4.1.3 Red-Emitting CaAlSiN$_3$:Eu^{2+}

As one of best red phosphors, CASN:Eu plays an important role in achieving high color rendering and wide color gamut. To be applied in high-power laser lighting and displays, a bulk CASN:Eu ceramic is demanded to survive from high-power-density laser excitation. But it is a great challenge to sinter fully densified CASN:Eu phosphor ceramics with acceptable optical properties because of the intrinsic low diffusion rate of nitrides. In 2015, Pricha *et al.* first fabricated CASN:Eu phosphor ceramics *via* pressureless sintering. Unfortunately, only a low relative density of 80% was obtained, which cannot well meet the requirements for SSL.[62]

Considering that CASN and Si$_2$N$_2$O are isostructural and can form a solid solution to a certain degree, Li *et al.* adopted Si$_3$N$_4$ and SiO$_2$ as dual sintering additives and successfully prepared a translucent CASN:Eu phosphor ceramic *via* the SPS technique for the first time.[47,48] The CASN:Eu phosphor ceramic had a high relative density of >99%, the microstructure of which consisting of core-shell structured red-emitting CASN:Eu phosphor particles uniformly dispersed in a nonluminescent Ca-α-SiAlON matrix, as shown in Figure 4.13(a). Compared to its powder form, the CASN:Eu ceramic showed an enhanced thermal stability (15% increase) and a high thermal conductivity of 4 W m^{-1} K^{-1}. It also maintained a high EQE of 60% upon 450 nm excitation, which is 87% of the powder. Under blue laser excitation (450 nm), the CASN:Eu ceramic gives an intense and broadband red emission peaking at 647 nm (Figure 4.13(b)). With increasing the blue laser power density from 21 to 150 W cm^{-2}, the output luminous flux of the CASN:Eu ceramic showed a linear increase from 30 to 200 lm with a stable LE$_{opt}$ of around 42 lm W^{-1}. The encouraging optical performance of the CASN:Eu phosphor ceramic enables it to be a potential red color converter for use in laser lighting and displays, although it has obvious luminance saturation and cannot be applied with high power/flux density laser.

4.3.4.1.4 Others

Some other phosphor ceramics, such as blue-emitting BaMgAl$_{10}$O$_{17}$:Eu^{2+} and white-emitting AlN:Ce^{3+}, have also been developed and demonstrated their potential applications in laser-driven SSL.

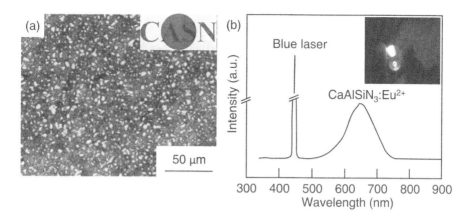

FIGURE 4.13 (a) Photograph and cross-sectional SEM image of the CaAlSiN$_3$:Eu^{2+} (CASN:Eu) red phosphor ceramic; (b) emission spectrum excited by a blue LD and emitting picture. (Li, S., Zhu, Q., Wang, L., Tang, D., Cho, Y., Liu, X., Hirosaki, N., Nishimura, T., Sekiguchi, T., Huang, Z. and Xie, R.-J., *J. Mater. Chem. C*, 4, 8197–8205, 2016; Li, S., Tang, D., Tian, Z., Liu, X., Takeda, T., Hirosaki, N., Xu, F., Huang, Z. and Xie, R.-J., *J. Mater. Chem. C*, 5, 1042–1051, 2017. With permission)

A monolithic translucent BaMgAl$_{10}$O$_{17}$:Eu^{2+} ceramic can convert the ultraviolet laser light to blue light with the same efficiency as the starting powder and exhibit superior thermal management in comparison with the silicone encapsulation.[39] The white-emitting AlN:Ce^{3+} phosphor ceramic (λ_{ex} = 375 nm), fabricated by a current-activated pressure-assisted densification technique, has a thermal conductivity of 90 W m^{-1} K^{-1}, holding a significant promise for producing high-brightness UV-laser-driven white lighting.[50]

4.3.4.2 Composite Phosphor Ceramics

4.3.4.2.1 YAG:Ce/Al$_2$O$_3$

The theoretical thermal conductivity of YAG is 9–14 W m^{-1} K^{-1}, which still needs to be greatly enhanced to quickly dissipate the heat generated upon strong laser irradiation. This need drives the design of composite phosphor ceramics with a much higher thermal conductivity, *i.e.*, YAG:Ce/Al$_2$O$_3$, where YAG:Ce phosphor particles are embedded in a nonluminescent Al$_2$O$_3$ matrix. Al$_2$O$_3$ is selected as the matrix owing to: (i) high thermal conductivity (32–35 W m^{-1} K^{-1}); (ii) small thermal expansion coefficient difference between YAG and Al$_2$O$_3$ (8.0 × 10^{-6} versus 8.4 × 10^{-6} K^{-1}); (iii) large bandgap of 7–8 eV; and (iv) facile fabrication of transparent Al$_2$O$_3$ ceramics.

In 2016, Li *et al.* used commercial YAG:Ce phosphor powders (5–20 μm) and α-Al$_2$O$_3$ (0.25 μm) as staring materials and fabricated a series of YAG:Ce/Al$_2$O$_3$ composite phosphor ceramics *via* the SPS method.[52] Under a sintering temperature of 1360°C and a pressure of 80 MPa, the composite ceramics can be fully densified with a relative density of 99.7%. As shown in Figure 4.14, YAG:Ce phosphor particles are uniformly dispersed in the Al$_2$O$_3$ matrix and the interfaces between phosphor particles and matrix are very clear. Large YAG:Ce particles show a good crystallinity and a high EQE of 76% (under 460 nm excitation) while the fine Al$_2$O$_3$ grains (0.5–2 μm) contribute to the superior in-line transmittance of 55% at 800 nm by minimizing the birefringence-related scattering. More importantly, the composite phosphor ceramic exhibits a high thermal conductivity of 18.5 W m^{-1} K^{-1} and a significant improvement in thermal stability with only 8% emission reduction at 200°C. Because of these excellent optical and thermal properties, the YAG:Ce/Al$_2$O$_3$ ceramic shows no luminance saturation even at a high blue laser power density of 50 W mm^{-2}. When coupled with a 45-W blue LD, it gives a high luminous flux of 2000 lm with a CCT of 5200 K. The high-brightness white light source validates that the YAG:Ce/Al$_2$O$_3$ ceramic is highly suitable for high-power laser lighting. Inspired by this work, a series of YAG:Ce/Al$_2$O$_3$ composite phosphor ceramics have

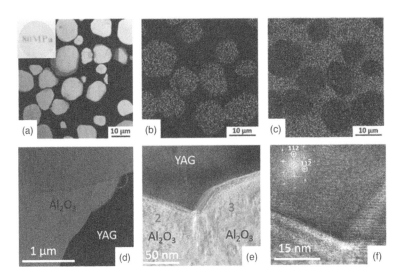

FIGURE 4.14 (a) Cathodoluminescence mapping taken by monitoring the 540 nm emission of the YAG:Ce/Al$_2$O$_3$ composite ceramic, showing bright YAG:Ce particles dispersed in a nonluminescent Al$_2$O$_3$ matrix; (b and c) EDS mappings for Y and Al elements; (d) bright field, and (e and f) high-resolution TEM images of the interfacial area between YAG:Ce and Al$_2$O$_3$. (Li, S., Zhu, Q., Tang, D., Liu, X., Ouyang, G., Cao, L., Hirosaki, N., Nishimura, T., Huang, Z. and Xie, R.-J., *J. Mater. Chem. C*, 4, 8648–8654, 2016. With permission)

been designed and prepared.[20,63] Apart from the heat dissipation enhancement, the highly scattering nature of ceramic composites also benefits to the light uniformity.

4.3.4.2.2 YAG:Ce/YAG

Given that the difference in the refractive index of two distinct phases, *i.e.*, Al$_2$O$_3$ and YAG, will greatly reduce the transparency of composites phosphor ceramics, Ce^{3+}-free YAG phase was selected as ceramic matrix for dispersing the YAG:Ce phosphor. Zhu *et al.* fabricated YAG:Ce/YAG composite phosphor ceramics by hot isostatic pressing sintering, which possessed a high thermal conductivity of 8.9 W m^{-1} K^{-1} at room temperature and a high in-line transmittance of ~65%.[54] The transmittance was much higher than that of the YAG:Ce/Al$_2$O$_3$ composite (~55%), which allows for the reduction of the light-scattering loss. The YAG:Ce/YAG phosphor ceramic can withstand a high blue laser power density of 9.6 W mm^{-2} and have a luminous emittance of 1220 lm mm^{-2}. Similarly, a green-emitting Y$_3$Al$_{5-x}$Ga$_x$O$_{12}$:Ce^{3+}-in-YAG (YAGG:Ce/YAG) PCP was also designed, which showed a high luminous emittance of 700 lm mm^{-2} under a maximum blue laser power density of 8.2 W mm^{-2} excitation.[51]

Besides, a YAG:Ce/CaF$_2$ composite phosphor ceramic was designed and successfully prepared at a ultralow sintering temperature of 700°C by a hot-pressing sintering.[64]

Figure 4.15 summarizes the advantages and disadvantages of different laser phosphors. Single crystals have superior thermal robustness and high IQE owing to its ideal structure but suffer from low light extraction efficiency and EQE. Moreover, their growth is quite complicated and costly, and only a small number of materials, *i.e.*, the garnet system, can be grown into large single crystals. PiGs possess competitive advantages in easy fabrication and great versatility for various phosphors; however, their applications in laser lighting are hindered because of the poor thermal conductivity of glass matrices and low phosphor content. Phosphors films co-sintered on high thermally conductive substrates not only inherit the merits of PiGs but also do not have their drawbacks, whereas phosphor ceramics have advantages of single crystals and their microstructures and optical properties can be easily tailored. These merits make them very promising color converters for high-power laser excitation. Of course, phosphor films and phosphor ceramics with better performance, *i.e.*,

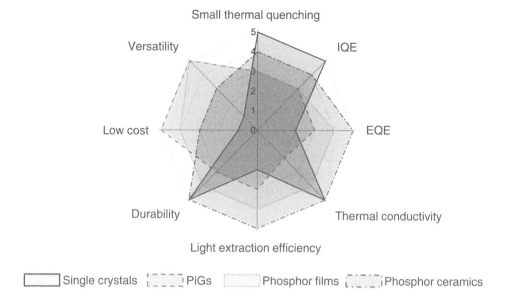

FIGURE 4.15   Radar graph highlighting the advantages and disadvantages of different laser phosphors.

higher efficiency, higher saturation threshold, and varying emission colors, are still urgently needed for efficient and high-quality laser lighting and displays.

4.4   PERSPECTIVES

As key materials in laser lighting and displays, laser phosphors have been emerged as a new type of color converters and receive increasing attentions from both academic and industrial communities. Differing from LED phosphors, laser phosphors are challenging with super-high power/flux density excitation, leading to some problems that have not been encountered before, such as luminance saturation. As mentioned in Section 4.3, to fulfill the requirements for laser lighting and display applications, laser phosphors are designed and used in different forms, including single crystals, PiG plates, phosphor ceramics, and phosphor films. When selecting laser phosphors, luminance saturation and thermal quenching should be first considered. In addition, the following issues also need to be concerned about.

1. *Development of novel laser phosphors.* Till now, a very few phosphors have been proposed for use in combination with laser light and are almost limited to yellow-emitting YAG:Ce and green-emitting LuAG:Ce. Although the commercialized β-SiAlON:Eu^{2+} (green), CaAlSiN$_3$:Eu^{2+} (red), and K$_2$SiF$_6$:Mn^{4+} (red) phosphors are investigated as laser phosphors, they have quite low luminance saturation and luminance flux due to their longer decay time of activators. This makes them hardly used for superbrightness projection. Therefore, it is of urgent to explore new green or red phosphors with high saturation thresholds. By considering the decay time as a key factor affecting the luminance saturation, Ce^{3+} would be an excellent activator for laser phosphors as it has a decay time about 5% of Eu^{2+} and 0.001% of Mn^{2+} or Mn^{4+}. In fact, La$_3$Si$_6$N$_{11}$:Ce^{3+} and Ca$_{1-x-y}$Li$_x$Al$_{1-x-y}$Si$_{1+x+y}$N$_{3-y}$O$_y$:Ce^{3+} were reported to be promising orange-emitting laser phosphors, which have the saturation threshold of 12.9 and 17.4 W mm^{-2}, respectively.[28],[29] Furthermore, a red-emitting HP-CaSiN$_2$:Ce^{3+} (λ_{em} = 620 nm) was discovered by calculating the bandgap, structure distortion, and bond length of 46 Ce^{3+}-doped nitride compounds.[65] This red phosphor had

a saturation threshold of 10.89 W mm^{-2} when used in a phosphor wheel. On other hand, there are no solid descriptors for screening appropriate host of laser phosphors, and it is still a big challenge for chemists and materials scientists.

2. *Realization of high color rendition and wide color gamut.* As the laser light beam is very narrow, it is difficult to reach color rendition as high as the case of using LED chips. As to laser displays, a wider color gamut is necessary for realizing vivid images and maintaining higher competition than advanced panel displays such as LCDs, quantum dot LEDs (QLEDs), and organic LEDs (OLEDs). However, the color quality of phosphors-converted laser lighting is limited by the lack of robust red phosphors, especially the narrow-band red and green phosphors. Currently available Eu^{2+}-, Mn^{2+}-, or Mn^{4+}-doped red and green phosphors cannot be used for superbrightness laser lighting as they are easily saturated or damaged under high power/flux laser irradiation. Although the Ce^{3+}-doped phosphors can be survived from laser light excitation with higher power/flux densities, there are quite few green- or red-emitting materials. $La_3(Si,Al)_6N_{11}:Ce^{3+}$ was developed as an interesting red-emitting laser phosphor, but it had a very low quantum efficiency and hence a large saturation.[66,67] Furthermore, the Ce^{3+}-doped phosphors usually exhibit broad emission bands, which are not suitable for display applications. Therefore, it is a tough mission to discover narrow-band red and green laser phosphors.

3. *Enhancement of light uniformity and extraction efficiency.* The uniformity and brightness of the output light largely depends on luminescent materials and packaging configurations. Discussion on the packaging configurations is beyond the scope of this chapter, but it is extremely important for practical applications of laser lighting. As to laser phosphors, to search for excellent hosts is a must, but the microstructure or geometry plays a key role in determine the uniformity, size, and efficiency of the output white light. For example, the introduction of pores into the color converters would enhance the scattering effect and, therefore, improve the conversion efficiency of phosphor particles and the homogenous mixing of the emitted (LDs) and converted (laser phosphors) light. Both the light uniformity and extraction efficiency of the lighting device are finally enhanced. An alternative option for creating light-scattering centers is to build composite microstructures containing secondary phases, such as Al_2O_3 and SiO_2. However, the presence of pores or secondary phases usually reduces the transparency or thermal conductivity (except Al_2O_3) of the whole color converter, which decreases the luminance saturation threshold of laser phosphors and the luminous flux of the device. A good balance between the light uniformity and luminous flux needs to be considered by optimizing the microstructure of laser phosphors. In addition, the microstructure and geometry will be differently designed or built for laser phosphors used in the reflective and transmissive modes.

REFERENCES

1. Wierer, J.J., Tsao, J.Y. and Sizov, D.S., Comparison between blue lasers and light-emitting diodes for future solid-state lighting, *Laser Photonics Rev.*, 7, 963–993, 2013.
2. Kane, M.H., Jiao, J., Dietz, N., Huang, J.-J., Lenef, A., Kelso, J., Tchoul, M., Mehl, O., Sorg, J. and Zheng, Y., Laser-activated remote phosphor conversion with ceramic phosphors, *Proc. SPIE*, 9190, 91900C, 2014.
3. Lenef, A., Raukas, M., Wang, J. and Li, C., Phosphor performance under high intensity excitation by InGaN laser diodes, ECS J. Solid State Sci. Technol., 9, 016019, 2020.
4. Shchekin, O.B., Schmidt, P.J., Jin, F., Lawrence, N., Vampola, K.J., Bechtel, H., Chamberlin, D.R., Mueller-Mach, R. and Mueller, G.O., Excitation dependent quenching of luminescence in LED phosphors, *Phys. Status Solidi RRL*, 10, 310–314, 2016.
5. Arjoca, S., Víllora, E.G., Inomata, D., Aoki, K., Sugahara, Y. and Shimamura, K., Ce:$(Y_{1-x}Lu_x)_3Al_5O_{12}$ single-crystal phosphor plates for high-brightness white LEDs/LDs with high-color rendering ($Ra > 90$) and temperature stability, *Mater. Res. Express*, 1, 025041, 2014.

6. Arjoca, S., Víllora, E.G., Inomata, D., Aoki, K., Sugahara, Y. and Shimamura, K., Temperature dependence of Ce:YAG single-crystal phosphors for high-brightness white LEDs/LDs, *Mater. Res. Express*, 2, 055503, 2015.

7. Arjoca, S., Inomata, D., Matsushita, Y. and Shimamura, K., Growth and optical properties of $(Y_{1-x}Gd_x)_3Al_5O_{12}$:Ce single crystal phosphors for high-brightness neutral white LEDs and LDs, *CrystEngComm*, 18, 4799–4806, 2016.

8. Kang, T.W., Park, K.W., Ryu, J.H., Lim, S.G., Yu, Y. M. and Kim, J.S., Strong thermal stability of $Lu_3Al_5O_{12}$:Ce^{3+} single crystal phosphor for laser lighting, *J. Lumin.*, 191, 35–39, 2017.

9. Zhou, Y., Yu, C., Song, E., Wang, Y., Ming, H., Xia, Z. and Zhang, Q., Three birds with one stone: K_2SiF_6:Mn^{4+} single crystal phosphors for high-power and laser-driven lighting, *Adv. Opt. Mater.*, 8, 2000976, 2020.

10. Balci, M.H., Chen, F., Cunbul, A.B., Svensen, Ø., Akram, M.N. and Chen, X., Comparative study of blue laser diode driven cerium-doped single crystal phosphors in application of high-power lighting and display technologies, *Opt. Rev.*, 25, 166–174, 2018.

11. Park, K.W., Lim, S.G., Deressa, G., Kim, J.S., Kang, T.W., Choi, H.L., Yu, Y.M., Kim, Y.S., Ryu, J.G., Lee, S.H. and Kim, T.H., High power and temperature luminescence of $Y_3Al_5O_{12}$:Ce^{3+} bulky and pulverized single crystal phosphors by a floating-zone method, *J. Lumin.*, 168, 334–338, 2015.

12. Segawa, H., Ogata, S., Hirosaki, N., Inoue, S., Shimizu, T., Tansho, M., Ohki, S. and Deguchi, K., Fabrication of glasses of dispersed yellow oxynitride phosphor for white light-emitting diodes, *Opt. Mater.*, 33, 170–175, 2010.

13. Zhang, X., Yu, J., Wang, J., Lei, B., Liu, Y., Cho, Y., Xie, R.-J., Zhang, H.-W., Li, Y., Tian, Z., Li, Y. and Su, Q., All-inorganic light convertor based on phosphor-in-glass engineering for next-generation modular high-brightness white LEDs/LDs, *ACS Photonics*, 4, 986–995, 2017.

14. Zhang, D., Xiao, W., Liu, C., Liu, X., Ren, J., Xu, B. and Qiu, J., Highly efficient phosphor-glass composites by pressureless sintering, *Nat. Commun.*, 11, 2805, 2020.

15. Zhang, R., Lin, H., Yu, Y., Chen, D., Xu, J. and Wang, Y., A new-generation color converter for high-power white LED: transparent Ce^{3+}:YAG phosphor-in-glass, *Laser Photonics Rev.*, 8, 158–164, 2014.

16. Lin, H., Hu, T., Cheng, Y., Chen, M. and Wang, Y., Glass ceramic phosphors: towards long-lifetime high-power white light-emitting-diode applications-a review, *Laser Photonics Rev.*, 12, 1700344, 2018.

17. Yu, J., Si, S., Liu, Y., Zhang, X., Cho, Y., Tian, Z., Xie, R., Zhang, H., Li, Y. and Wang, J., High-power laser-driven phosphor-in-glass for excellently high conversion efficiency white light generation for special illumination or display backlighting, *J. Mater. Chem. C*, 6, 8212–8218, 2018.

18. Zhu, Q.-Q., Wang, X.-J., Wang, L., Hirosaki, N., Nishimura, T., Tian, Z.-F., Li, Q., Xu, Y.-Z., Xu, X. and Xie, R.-J., β-Sialon:Eu phosphor-in-glass: a robust green color converter for high power blue laser lighting, *J. Mater. Chem. C*, 3, 10761–10766, 2015.

19. Zhu, Q.-Q., Xu, X., Wang, L., Tian, Z.-F., Xu, Y.-Z., Hirosaki, N. and Xie, R.-J., A robust red-emitting phosphor-in-glass (PiG) for use in white lighting sources pumped by blue laser diodes, *J. Alloy. Compd.*, 702, 193–198, 2017.

20. Li, S., Wang, L., Hirosaki, N. and Xie, R.-J., Color conversion materials for high-brightness laser-driven solid-state lighting, *Laser Photonics Rev.*, 12, 1800173, 2018.

21. Wang, L., Xie, R.-J., Suehiro, T., Takeda, T. and Hirosaki, N., Down-conversion nitride materials for solid state lighting: recent advances and perspectives, *Chem. Rev.*, 118, 1951–2009, 2018.

22. Yoshimura, K., Annen, K., Fukunaga, H., Harada, M., Izumi, M., Takahashi, K., Uchikoshi, T., Xie, R.-J. and Hirosaki, N., Optical properties of solid-state laser lighting devices using SiAlON phosphor–glass composite films as wavelength converters, *Jpn. J. Appl. Phys.*, 55, 042102, 2016.

23. Xu, J., Hu, B., Xu, C., Wang, J., Liu, B., Li, H., Wang, X., Du, B. and Gong, Y., Carbon-free synthesis and luminescence saturation in a thick YAG:Ce film for laser-driven white lighting, *J. Eur. Ceram. Soc.*, 39, 631–634, 2019.

24. Xu, J., Yang, Y., Wang, J., Du, B., Santamaría, A.A., Hu, B., Liu, B., Ji, H., Dam-Hansen, C. and Jensen, O.B., Industry-friendly synthesis and high saturation threshold of a LuAG:Ce/glass composite film realizing high-brightness laser lighting, *J. Eur. Ceram. Soc.*, 40, 6031–6036, 2020.

25. Wu, H., Hao, Z., Pan, G.-H., Zhang, L., Wu, H., Zhang, X., Zhang, L. and Zhang, J., Phosphor-SiO_2 composite films suitable for white laser lighting with excellent color rendering, *J. Eur. Ceram. Soc.*, 40, 2439–2444, 2020.

26. Xu, J., Yang, Y., Guo, Z., Hu, B., Wang, J., Du, B., Liu, B., Ji, H., Dam-Hansen, C. and Jensen, O.B., Design of a $CaAlSiN_3$:Eu/glass composite film: facile synthesis, high saturation-threshold and application in high-power laser lighting, *J. Eur. Ceram. Soc.*, 40, 4704–4708, 2020.

27. Zheng, P., Li, S., Wang, L., Zhou, T.-L., You, S., Takeda, T., Hirosaki, N. and Xie, R.-J., Unique color converter architecture enabling phosphor-in-glass (PiG) films suitable for high-power and high-luminance laser-driven white lighting, *ACS Appl. Mater. Interfaces*, 10, 14930–14940, 2018.

28. You, S., Li, S., Zheng, P., Zhou, T., Wang, L., Liu, L., Horisaki, N., Xu, F. and Xie, R.-J., A thermally robust $La_3Si_6N_{11}$:Ce-in-glass film for high-brightness blue-laser-driven solid state lighting, *Laser Photonics Rev.*, 13, 1800216, 2019.

29. You, S., Li, S., Wang, L., Takeda, T., Hirosaki, N. and Xie, R.-J., Ternary solid solution phosphors $Ca_{1-x-y}Li_xAl_{1-x-y}Si_{1+x+y}N_{3-y}O_y$:$Ce^{3+}$ with enhanced thermal stability for high-power laser lighting, *Chem. Eng. J.*, 404, 126575, 2021.

30. Wang, L., Wei, R., Zheng, P., You, S., Zhou, T., Yi, W., Takeda, T., Hirosaki, N. and Xie, R.-J., Realizing high-brightness and ultra-wide color gamut laser-driven backlighting by using laminated phosphor-in-glass (PiG) films, *J. Mater. Chem. C*, 8, 1746–1754, 2020.

31. Park, J., Kim, J. and Kwon, H., Phosphor-aluminum composite for energy recycling with high-power white lighting, *Adv. Opt. Mater.*, 5, 1700347, 2017.

32. Wei, R., Wang, L., Zheng, P., Zeng, H., Pan, G., Zhang, H., Liang, P., Zhou, T. and Xie, R.-J., On the luminance saturation of phosphor-in-glass (PiG) films for blue-laser-driven white lighting: effects of the phosphor content and the film thickness, *J. Eur. Ceram. Soc.*, 39, 1909–1917, 2019.

33. Peng, Y., Mou, Y., Sun, Q., Cheng, H., Chen, M. and Luo, X., Facile fabrication of heat-conducting phosphor-in-glass with dual-sapphire plates for laser-driven white lighting, *J. Alloy. Compd.*, 790, 744–749, 2019.

34. Yang, Y., Zhuang, S. and Kai, B., High brightness laser-driven white emitter for etendue-limited applications, *Appl. Opt.*, 56, 8321–8325, 2017.

35. Peng, Y., Sun, Q., Liu, J., Mou, Y., Wang, X., Chen, M. and Luo, X., Reflective phosphor-in-glass color converter for laser-driven white lighting, *IEEE Photon. Technol. Lett.*, 32, 983–986, 2020.

36. Wang, H., Mou, Y., Peng, Y., Zhang, Y., Wang, A., Xu, L., Long, H., Chen, M., Dai, J. and Chen, C., Fabrication of phosphor glass film on aluminum plate by using lead-free tellurite glass for laser-driven white lighting, *J. Alloy. Compd.*, 814, 152321, 2020.

37. Park, J., Cho, S. and Kwon, H., Aluminum-ceramic composites for thermal management in energy-conversion systems, *Sci. Rep.*, 8, 17852, 2018.

38. Bechtel, H., Schmidt, P., Busselt, W. and Schreinemacher, B.S., Lumiramic: a new phosphor technology for high performance solid state light sources, *Proc. SPIE*, 7058, 70580E-1, 2008.

39. Cozzan, C., Brady, M.J., O'Dea, N., Levin, E.E., Nakamura, S., DenBaars, S.P. and Seshadri, R., Monolithic translucent $BaMgAl_{10}O_{17}$:Eu^{2+} phosphors for laser-driven solid state lighting, *AIP Adv.*, 6, 105005, 2016.

40. Xu, Y., Li, S., Zheng, P., Wang, L., You, S., Takeda, T., Hirosaki, N. and Xie, R.-J., A search for extra-high brightness laser-driven color converters by investigating thermally-induced luminance saturation, *J. Mater. Chem. C*, 7, 11449–11456, 2019.

41. Li, K., Shi, Y., Jia, F., Price, C., Gong, Y., Huang, J., Copner, N., Cao, H., Yang, L., Chen, S., Chen, H. and Li, J., Low etendue yellow-green solid-state light generation by laser-pumped LuAG:Ce ceramic, *IEEE Photon. Technol. Lett.*, 30, 939–942, 2018.

42. Xu, J., Wang, J., Gong, Y., Ruan, X., Liu, Z., Hu, B., Liu, B., Li, H., Wang, X. and Du, B., Investigation of an LuAG:Ce translucent ceramic synthesized via spark plasma sintering: towards a facile synthetic route, robust thermal performance, and high-power solid state laser lighting, *J. Eur. Ceram. Soc.*, 38, 343–347, 2018.

43. Zhang, Y., Hu, S., Wang, Z., Zhou, G. and Wang, S., Pore-existing $Lu_3Al_5O_{12}$:Ce ceramic phosphor: an efficient green color converter for laser light source, *J. Lumin.*, 197, 331–334, 2018.

44. Song, Y.H., Ji, E.K., Jeong, B.W., Jung, M.K., Kim, E.Y. and Yoon, D.H., High power laser-driven ceramic phosphor plate for outstanding efficient white light conversion in application of automotive lighting, *Sci. Rep.*, 6, 31206, 2016.

45. Zheng, P., Li, S., Wei, R., Wang, L., Zhou, T.-L., Xu, Y.-R., Takeda, T., Hirosaki, N. and Xie, R.-J., Unique design strategy for laser-driven color converters enabling superhigh-luminance and high-directionality white light, *Laser Photonics Rev.*, 13, 1900147, 2019.

46. Wang, J., Tang, X., Zheng, P., Li, S., Zhou, T. and Xie, R.-J., Self-thermal management YAG:Ce-Al_2O_3 color converters enabling high-brightness laser-driven solid state lighting in a transmissive configuration, *J. Mater. Chem. C*, 7, 3901–3908, 2019.

47. Li, S., Zhu, Q., Wang, L., Tang, D., Cho, Y., Liu, X., Hirosaki, N., Nishimura, T., Sekiguchi, T., Huang, Z. and Xie, R.-J., $CaAlSiN_3$:Eu^{2+} translucent ceramic: a promising robust and efficient red color converter for solid state laser displays and lighting, *J. Mater. Chem. C*, 4, 8197–8205, 2016.

48. Li, S., Tang, D., Tian, Z., Liu, X., Takeda, T., Hirosaki, N., Xu, F., Huang, Z. and Xie, R.-J., New insights into the microstructure of translucent $CaAlSiN_3:Eu^{2+}$ phosphor ceramics for solid-state laser lighting, *J. Mater. Chem. C*, 5, 1042–1051, 2017.

49. Zhu, Q.-Q., Yang, P.-F., Wang, Z.-Y. and Hu, P.-C., Additive-free $Y_2O_3:Eu^{3+}$ red-emitting transparent ceramic with superior thermal conductivity for high-power UV LEDs and UV LDs, *J. Eur. Ceram. Soc.*, 40, 2426–2431, 2020.

50. Wieg, A.T., Penilla, E.H., Hardin, C.L., Kodera, Y. and Garay, J.E., Broadband white light emission from Ce:AlN ceramics: high thermal conductivity down-converters for LED and laser-driven solid state lighting, *APL Mater.*, 4, 126105, 2016.

51. Zhu, Q.-Q., Meng, Y., Zhang, H., Li, S., Wang, L. and Xie, R.-J., YAGG:Ce phosphor-in-YAG ceramic: an efficient green color converter suitable for high-power blue laser lighting, *ACS Appl. Electron. Mater.*, 2, 2644–2650, 2020.

52. Li, S., Zhu, Q., Tang, D., Liu, X., Ouyang, G., Cao, L., Hirosaki, N., Nishimura, T., Huang, Z. and Xie, R.-J., Al_2O_3-YAG:Ce composite phosphor ceramic: a thermally robust and efficient color converter for solid state laser lighting, *J. Mater. Chem. C*, 4, 8648–8654, 2016.

53. Xu, J., Yang, Y., Guo, Z., Corell, D.D., Du, B., Liu, B., Ji, H., Dam-Hansen, C. and Jensen, O.B., Comparative study of Al_2O_3-YAG:Ce composite ceramic and single crystal YAG:Ce phosphors for high-power laser lighting, *Ceram. Int.*, 46, 17923–17928, 2020.

54. Zhu, Q.-Q., Li, S., Yuan, Q., Zhang, H. and Wang, L., Transparent YAG:Ce ceramic with designed low light scattering for high-power blue LED and LD applications, *J. Eur. Ceram. Soc.*, 41, 735–740, 2021.

55. Hagemann, V., Seidl, A. and Weidmann, G., Ceramic phosphor wheels for high luminance SSL-light sources with >500W of laser power for digital projection, *Proc. SPIE*, 10940, 1094017, 2019.

56. Song, Y.H., Ji, E.K., Jeong, B.W., Jung, M.K., Kim, E.Y., Lee, C.W. and Yoon, D.H., Design of laser-driven high-efficiency Al_2O_3/YAG:Ce^{3+} ceramic converter for automotive lighting: fabrication, luminous emittance, and tunable color space, *Dyes Pigments*, 139, 688–692, 2017.

57. Liu, X., Zhou, H., Hu, Z., Chen, X., Shi, Y., Zou, J. and Li, J., Transparent Ce:GdYAG ceramic color converters for high-brightness white LEDs and LDs, *Opt. Mater.*, 88, 97–102, 2019.

58. Yao, Q., Hu, P., Sun, P., Liu, M., Dong, R., Chao, K., Liu, Y., Jiang, J. and Jiang, H., YAG:Ce^{3+} transparent ceramic phosphors brighten the next-generation laser-driven lighting, *Adv. Mater.*, 32, 1907888, 2020.

59. Hu, S., Liu, Y., Zhang, Y., Xue, Z., Wang, Z., Zhou, G., Lu, C., Li, H. and Wang, S., 3D printed ceramic phosphor and the photoluminescence property under blue laser excitation, *J. Eur. Ceram. Soc.*, 39, 2731–2738, 2019.

60. Zhu, Q., Ding, S., Xiahou, J., Li, S., Sun, X. and Li, J.G., A groundbreaking strategy for fabricating YAG:Ce^{3+} transparent ceramic films via sintering of LRH nanosheets on a sapphire substrate, *Chem. Commun.*, 56, 12761–12764, 2020.

61. Ma, C., Tang, F., Chen, J., Ma, R., Yuan, X., Wen, Z., Long, J., Li, J., Du, M., Zhang, J. and Cao, Y., Spectral, energy resolution properties and green-yellow LEDs applications of transparent $Ce^{3+}:Lu_3Al_5O_{12}$ ceramics, *J. Eur. Ceram. Soc.*, 36, 4205–4213, 2016.

62. Pricha, I., Rossner, W., Moos, R. and McKittrick, J., Layered ceramic phosphors based on $CaAlSiN_3$:Eu and YAG:Ce for white light-emitting diodes, *J. Am. Ceram. Soc.*, 99, 211–217, 2016.

63. Cozzan, C., Lheureux, G., O'Dea, N., Levin, E.E., Graser, J., Sparks, T.D., Nakamura, S., DenBaars, S.P., Weisbuch, C. and Seshadri, R., Stable, heat-conducting phosphor composites for high-power laser lighting, *ACS Appl. Mater. Interfaces*, 10, 5673–5681, 2018.

64. Gu, C., Wang, X.-J., Xia, C., Li, S., Liu, P., Li, D., Li, H., Zhou, G., Zhang, J. and Xie, R.-J., A new CaF_2-YAG:Ce composite phosphor ceramic for high-power and high-color-rendering WLEDs, *J. Mater. Chem. C*, 7, 8569–8574, 2019.

65. Xia, Y., Li, S., Zhang, Y., Takeda, T., Hirosaki, N. and Xie, R.-J., Discovery of a Ce^{3+}-activated red nitride phosphor for high-brightness solid-state lighting, *J. Mater. Chem. C*, 8, 14402–14408, 2020.

66. You, S., Li, S., Jia, Y. and Xie, R.-J., Interstitial site engineering for creating unusual red emission in $La_3Si_6N_{11}:Ce^{3+}$, *Chem. Mater.*, 32, 3631–3640, 2020.

67. Nitta, M., Nagao, N., Nomura, Y., Hirasawa, T., Sakai, Y., Ogata, T., Azuma, M., Torii, S., Ishigaki, T. and Inada, Y., High-brightness red-emitting phosphor $La_3(Si,Al)_6(O,N)_{11}:Ce^{3+}$ for next-generation solid-state light sources, *ACS Appl. Mater. Interfaces*, 12, 31652–31658, 2020.

5 Smart Phosphors

Man-Chung Wong, Yang Zhang, and Jianhua Hao

CONTENTS

5.1 INTRODUCTION

The new era demands more intelligent materials. A subgroup of these intelligent materials termed smart materials have the capability to sense, process and respond in a controllable and reversible way to one or more environmental stimuli (e.g., electric or magnetic field, mechanical stress, temperature, chemical or light signals). The research and developmental of smart material have witness extensive growth and smart material have already been applied in all aspects of modern life. For example, electrochromic materials have been widely applied in smart windows, in which the transmittance can be modulated through controlling the charge carriers inside. Likewise, luminescence is an emission of photons from the matter when it is excited by an external energy source. Luminescence can be categorized into different types in accordance to the type of excitation energy, such as photoluminescence (PL), electroluminescence (EL), cathodoluminescence (CL) and mechanoluminescence (ML). A significant group of luminescent materials termed phosphors are composed of an inorganic crystalline host incorporated with intentionally doped impurities that function as luminescent activators, as shown in Figure 5.1(a).[1] If the absorption cross-section of the activators is small, other kinds of impurities with larger absorption cross-section can be co-doped to enhance the luminescence efficiency. This particular kind of dopant ions is defined as sensitizers

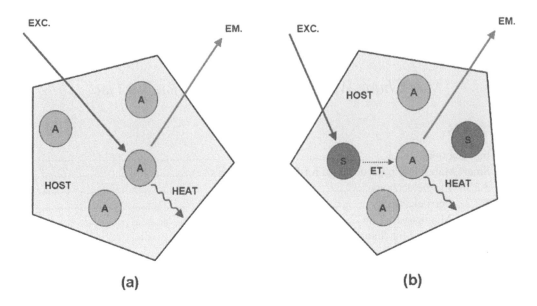

FIGURE 5.1 Schematic diagram of the general mechanism of luminescence from phosphors. (a) Emission from luminescent center or activator in the host lattice. (b) Energy transfer from a sensitizer to an activator to commence luminescence emission. Prefix: A: activator; S: sensitizer; EXC: excitation; EM: emission; ET: energy transfer. (Zhang, Y. and Hao, J., *J. Mater. Chem. C* 1, 5607, 2013. With permission.)

that can absorb the excitation energy and subsequently transfer the absorbed energy to the activators (Figure 5.1(b)). So far, metal ions, including lanthanide (Ln) ions and transition metal (TM) ions, are mostly utilized as dopants. In particular the intra-shell d–d transitions are usually exploited in TM ions, while both parity allowed $5d$–$4f$ (e.g., Ce^{3+}) and parity forbidden $4f$–$4f$ (e.g., Tb^{3+}) are favorable for Ln ions. Metal ion–doped phosphors have shown abundant applications from optoelectronics to biomedicine, and the scope of applications have already covered wide aspects of human life with novel and vital application under innovate and develop. The past decades have witnessed unprecedented achievements in the development of these types of phosphor materials. However, further investigation and development of these phosphors remain a pressing issue as modern society emerges into the era of intelligence. Especially the types of luminescence of a particular phosphor capable of emitting are generally preordained upon its synthesization, which limited the scope of application. Furthermore, it can be argued that the number of phosphors that could absorb different forms of energy and utilizes these energy to luminance in a comparable manner is rather limited, let alone *in situ* modulation of the emission characteristic (bandwidth, spectra, emission intensity etc.). Therefore, smart phosphor capable of being excited by a variety of physical stimuli and providing a real-time on-demanded modulation of luminescence characteristic is highly desired, which is a hotspot for the research field of optical functional material. Obviously, luminescence material capable of being integrated or incorporated into any smart phosphor system must have a superior performance in both emission and susceptibility to different types of excitation. The selection of both activators and host materials determines the luminescent properties of phosphors. However, there are limited cases in which the emission properties can be *in-situ* modulated. For instance, the jellyfish green fluorescent protein can be optically switched between the 'on' or 'off' emission state, such that it has been utilized for cellular biological labeling. Nevertheless, this type of switching or tuning in solid-state phosphors is scarce, even though this capability is highly desirable for both fundamental research and practical applications.

Spectroscopic tuning and enhancement of luminescence are the core concern for all kinds of phosphors. Modifying the luminescence in phosphors excited by a given excitation source can be achieved through conventional chemical approaches, such as changing the composition of host materials and/or doping ions, surface modification and core-shell structure.[2–4] However, these chemical methods are essentially an irreversible and *ex-situ* process. Therefore, it is unlikely to understand the kinetic process of how the luminescence changes with structural symmetry through this conventional approach. Moreover, it is almost impossible to isolate the pure crystal-field effect from other extrinsic effects present in different samples such as chemical inhomogeneities and defects. Thereby, the development of stimuli-responsive luminescence material has attracted much attention. Particularly, novel phosphor with emissions characteristics such as the emission color and intensity or switching between bright and dark state can be controlled on demand. Herein, to facilitate a more practical, controllable and efficient modulation of luminescence properties in a dynamic and real-time manner, we introduce the concept of smart phosphors. Smart phosphors are designed materials with luminescent properties that can be modulated in a controlled manner by external stimuli, such as electric or magnetic fields, stress, light, temperature, moisture or chemical signals. In principle, smart phosphors can be achieved through the coupling between smart and luminescent materials properties. In a broad sense, smart phosphors enable the dynamical control and modulation of luminescence characteristic in one single compound or composite structure through various strategic and systemic coupling of functional properties with the luminescence. Based on this principle, various types of smart phosphors can be conceived, promote and enhance a much greater degree of freedom for designing and utilizing smart phosphors in an unprecedented manner. Such achievement will improve the phosphor's performance and deepen the understanding of the mechanism underneath the luminescent processes. In light of these merits, our group has been pioneering on the forefront of smart phosphors. Our general strategy adapted to develop smart phosphor is shown in Figure 5.2. Selected smart materials, including ferroelectric, piezoelectric and magnetostrictive materials, integrated with predefined metal ion–doped phosphors, in terms of bulk, composite, thin-film and 2D materials, embark a tuning or stimulating the luminescence in an *in-situ*, dynamical and reversible manner. Smart phosphors have great potential for future optoelectronic and photonic devices due to the intrinsic functionality coupled with light emission. Moreover, compared to extensive coupling studies of smart materials such as multiferroics and magnetoelectric (ME) sensor and memory device,[5,6] the smart material coupling effects on the luminescence have been overlooked. In the past decade, our group has explored the coupling of various smart material properties such as ferroelectric and magnetostrictive characteristics with luminescence and triggered the emergence of this new research field of smart phosphors.

This chapter will provide an introduction of smart phosphor and the exciting breakthroughs in a variety of smart phosphors. It will first briefly discuss various properties of selected smart materials that have been used in the developmentation of novel smart phosphors. Then, we will introduce the design strategies in the formulation and fabrication of smart phosphors, especially the coupling between the aforementioned smart material with metal ions doped phosphor. Afterwards, some representative works on developing different types of smart phosphors will be highlighted. Lastly, the summary and perspectives on smart phosphors will be presented.

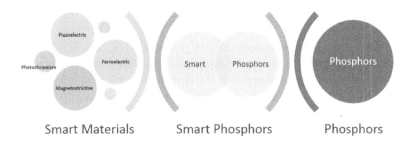

FIGURE 5.2 The constitute strategy of realizing smart phosphors.

5.2 SMART MATERIALS PROPERTIES USED FOR SMART PHOSPHORS

In general, smart materials are divided into two main categories. Passive smart materials, like optical fiber, which can transfer a portion of certain types of energy. For instance, wavelengths inside the predetermined bandwidth will be reflected by fiber Bragg grating, which is sensitive to external stimuli such as inducing strain or variations in the surrounding temperature or gaseous. These processes can be regarded as the passive optical process based on the light-object interaction such as reflection, absorption and transmission. This resultant emission of this passive process is generally preordained and cannot be moderated on site. On the contrary, active smart materials can moderate their characteristics or structures through stimulation with appropriate kinds of energy source. Smart materials that respond to physical stimuli constitute a major category of active smart material. Piezoelectric, ferroelectric, magnetostrictive materials and shape memory occupy an important position among the family of smart materials and have been the preferred choices to compose smart phosphor.

5.2.1 Piezoelectricity

Piezoelectric materials generate an electric potential under an applied strain. Or conversely, an applied electric field will induce a mechanical deformation through the converse piezoelectric effect. Piezoelectricity as an intrinsic property, can be attributed to the non-centrosymmetric crystal structures. The direct piezoelectric effect of a piezoelectric material defines the relationship between the polarization of the material and the strain or stress the material can produces. Converse piezoelectric response modes are typically employed to tailor the properties of materials via piezoelectric-induced strain. Nowadays, the most technologically advanced piezoelectric materials include ceramics or single crystals of perovskite structure (ABO_3), such as $Pb(Zr,Ti)O_3$ (PZT) and $BaTiO_3$ (BTO). PZT ceramics are the binary solid solutions of $PbZrO_3$ (PZO) and $PbTiO_3$ (PTO). The piezoelectric coefficients strongly depend on the chemical compositions and generally achieve the largest values at the vicinity of the morphotropic phase boundary (MPB). The piezoelectric coefficients of PZT drastically increase near the MPB located at the composition with the Zr/Ti ratio of 53/47. After decades of development, single-crystal relaxor ferroelectric $Pb(Mg_{1/3}Nb_{2/3})O_3–PbTiO_3$ (PMN-PT) offers the currently largest available piezoelectric response and has been extensively utilized in numerous applications, especially in the field of strain engineering. The piezoelectric coefficients d_{33} of PMN-PT can be up to ~2500 pC/N. Like PZT, the piezoelectric coefficients of PMN-PT are strongly enhanced at the MPB between the rhombohedral and tetragonal phases. The giant piezoelectric effect of PMN-PT can be ascribed to an electric field–induced phase transition from rhombohedral to a tetragonal structure.[7] The advantages of strain engineering on phosphors using piezoelectric actuators can be summarized as the following points. First, the strain generation and modulation from the piezoelectric materials can be realized on demand through applying an electric field. Second, piezoelectric actuators can work in an electric-controllable linear motion with high frequency in contrast to conventional mechanical impact sources. Lastly, inorganic piezoelectric materials can be used under low-temperature condition.

5.2.2 Ferroelectricity

Ferroelectric materials possess a spontaneous electric polarization that can be reversed by an applied electric field. Ferroelectric materials are an attractive class of essential components for various functional devices, including nonvolatile memories, tunnel junctions, field-effect transistors (FET), photovoltaics, high dielectric and tunable microwave devices. The most important class of ferroelectric materials shares a perovskite structure with ABO_3 formula. Ferroelectric materials have unique characteristics, including spontaneous electric polarization, polarization reversal and phase transition at Curie temperature T_c, where a cubic paraelectric state is evolved to a lower symmetry ferroelectric state. BTO is often regarded as an archetypical ferroelectric material. Below 393K, cubic BTO distorts to a ferroelectric tetragonal structure. Upon cooling to 278K, it transforms into an orthorhombic structure. A further phase transition of BTO from orthorhombic to rhombohedral structure occurs around 183K. The paraelectric-to-ferroelectric phase transition gives rise to the ferroelectricity in BTO. To be precise, during phase transition, the centers of the positive and negative charge may not coincide even without the application of an external electric field, resulting in a spontaneous polarization.[8] The polarization vector can be switched under an applied electric field. Ferroelectric crystals consist of domains, i.e., regions of aligned polarization vectors. Initially, a ferroelectric may exhibit zero macroscopic polarization due to randomly oriented domains. During the poling process, ferroelectric materials show the hysteresis curves of polarization versus electric field (P-E loop). The macroscopic polarization of the ferroelectric material will increase and reach a saturation value with increasing applied electric field. A remnant polarization P_r will exist and define the value of the polarization after removing the electric field. As the applied electric field decreases and reaches the coercive field E_c, the net polarization then becomes zero, meaning that the volume fractions of the domains with opposite directions are equal. Along with the electric field sweeps between the positive and negative directions, the relationship between the applied electric field and polarization is formed, and a hysteresis loop can be observed. Moreover, Landau free energy curve indicates a bistable nature of the ferroelectric polarization, which lays the physical basis for numerous ferroelectric-based applications, such as nonvolatile ferroelectric random access memory (FeRAM).

Ferroelectric materials are characterized by their asymmetry crystalline structures and their responsiveness to various physical stimuli including electric field, temperature and strain. For example, epitaxial strain emerges when ferroelectric films are deposited on the substrates with lattice mismatch, the induced in-plane strain can dramatically affect the polarization properties of the ferroelectric films.[9] Besides strain, an applied electric field can also modulate the crystal structure. Taking tetragonal BTO as an example, the unit cell elongates along with the spontaneous polarization direction. An applied electric field along the direction of spontaneous polarization will further elongate the c-axis of the lattice, and the in-plane a-axis of the lattice will contract at the same time. Such ionic crystal will then be further electrically polarized. Three primary sources will contribute these electrical polarizations, namely, electronic, ionic and dipole reorientation with each source's contribution depended on the frequency of the applied field. Dipole reorientation can follow the applied electric field up to MHz even GHz, which implies that ferroelectric materials with permanent dipoles are functional in switching application within such frequency range.

5.2.3 MAGNETOSTRICTION

Magnetic materials have been exploited for more than millennia. Nowadays, magnetic materials used in actuation and transducers have been a research hotspot of smart materials. Magnetic responsive materials are the topic of intense research due to their breakthrough applications in the biomedical, microfluidics and microelectronics fields. These applications are mainly dependent on the magneto-mechanical coupling, i.e., the magnetostrictive effect. The magnetostrictive materials exhibit a strain due to the reorientation of magnetic moment in response to an applied magnetic field. In case of zero external magnetic field, the magnetic moments in a paramagnetic material are randomly oriented. Upon an external magnetic field, these magnetic moments rotate to minimize the magnetization energy. This rotation of magnetic domains in the microscopic view cause the magnetostrictive materials to expand or contract following the external magnetic field's direction and, thus, converting electromagnetic energy into mechanical energy. The parameter for measuring the degree of magnetostriction is the magnetostriction coefficient λ. Similar to the definition of normal strain in elastic mechanics, the magnetostriction coefficient is defined as the change per unit length of a magnetostrictive material under an external magnetic field's action. In principle, all magnetic materials possess magnetostriction to some degree, but giant magnetostriction occurs in a small fraction of materials such as rare-earth iron alloy Terfenol-D (terbium-iron-dysprosium alloy) and galfenol (gallium-iron alloy).[10] However, these giant magnetostrictive materials (GMMs) suffer from the limitations associated with the bulkiness, fragile and the requirement of various pretreatments. These weaknesses severely hinder their utilization in smart phosphors. On the other hand, polymer-based magnetic composite materials have been extensively studied due to their high versatility and sensitivity to subtle environmental changes. They are adopting a more essential role in a wide range of magnetic related applications. These magnetic responsive polymer composites offer tailored, exceptional magnetic responsive feature by merging magnetic and polymer materials. These magnetic materials are generally composed of inorganic magnetic particles, such as a super-paramagnetic iron oxide or alloy, hard magnetic materials (Co, Ni, FePt). Given the wide variety of magnetic materials, magnetic material selection will depend on how the smart phosphor functions. One significant merit of the polymer-based magnetic composite materials is that the design and arranged magnetic moment allows them to respond in a pre-ordinated manner to an alternating magnetic field.

5.3 PHOSPHORS USED FOR SMART PHOSPHOR

Smart phosphors can respond to a variety of physical stimuli and provide a real-time, on demanded modulation of luminescence characteristics. Noticeably, the luminescence properties of a phosphor are generally preordained upon its synthesis. In a broad sense, phosphors are composed of a host and an activation system. The chemical neutrality of the host materials has a significant influence on the emission duration and serves as a framework of the phosphor. In addition, the host dictated the types of activator that can be incorporated into the phosphor and, ultimately, the types of luminescence that the phosphor is capable of emitting.[11] Moreover, owning to the lattice parameter of the host and the efficiency of interaction between the external stimulation and phonon energy,[12] it can be argued that the number of phosphors that could effectively respond to physical stimuli is somewhat limited, let alone be able to moderate the emission characteristic *in-situ* (bandwidth, spectra, emission intensity, etc.). Luminescence material capable of integrated or incorporated into the smart phosphor system must possess a superior performance in both emission and susceptibility to different types of excitation. As a major subcategory of luminescent materials, metal ions–doped phosphor material, host matrix is composed of crystalline semiconductor or perovskite, and in incorporated with impurities ions in low concentrations, typically TM ions or Ln series ions functioning as luminescent center. Such a kind of metal ion–doped phosphor is an excellent candidate for integrating and developing smart phosphor owning to its high versatility and excellent luminescence performance, which provides us great opportunities to couple with various physical fields. Herein, we give a brief introduction to them.

5.3.1 LANTHANIDE IONS

The Ln series comprise the 15 metallic chemical elements, from lanthanum through lutetium (Figure 5.3). The Ln ions are characterized by an incompletely filled $4f$ shell and shielded from the surrounding by the filled $5s^2$ and $5p^6$ orbitals, this results in the sharp and narrow lines of the emission spectra from the transitions within the $4f$ configuration. While the emission spectra arising from $5d$–$4f$ transitions (*e.g.*, Ce^{3+}) are broader as the $5d$ electrons are unshielded and heavily influenced by their surroundings. The lifetime of the emissions due to $4f$–$4f$ transitions are substantially long-lived, mostly in the range of milliseconds because of the forbidden characteristic of f–f transitions in free $4f$ ions. In the case where luminescence originated from parity-allowed transitions, such as $5d$–$4f$ transitions, a much faster lifetime ($\sim10^{-5}$ s) can be observed. Because of the existence of their abundant ladder-like energy levels, trivalent Ln ions, such as Er^{3+}, Tm^{3+}, Eu^{3+} and Pr^{3+}, have been extensively studied as activator ions for luminescent materials. Recent studies of trivalent Ln ions–doped phosphors had been extensively demonstrated excellent downconversion and/or upconversion luminescence covering from ultraviolet (UV) to near-infrared (NIR) region.

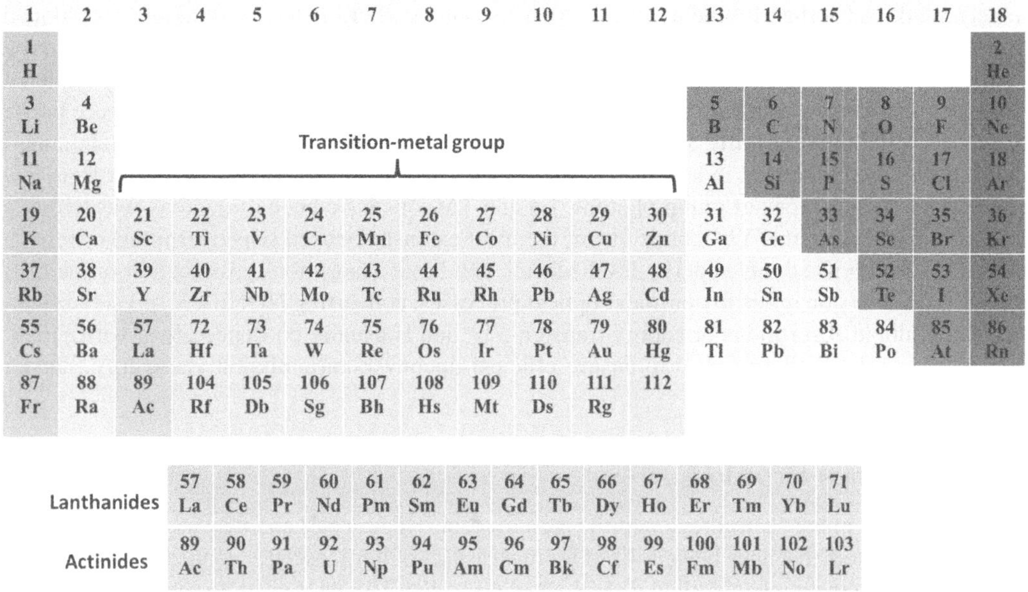

1	2	3	4	5	6	7	8	9	10	11	12	13	14	15	16	17	18
1 H																	2 He
3 Li	4 Be				Transition-metal group							5 B	6 C	7 N	8 O	9 F	10 Ne
11 Na	12 Mg											13 Al	14 Si	15 P	16 S	17 Cl	18 Ar
19 K	20 Ca	21 Sc	22 Ti	23 V	24 Cr	25 Mn	26 Fe	27 Co	28 Ni	29 Cu	30 Zn	31 Ga	32 Ge	33 As	34 Se	35 Br	36 Kr
37 Rb	38 Sr	39 Y	40 Zr	41 Nb	42 Mo	43 Tc	44 Ru	45 Rh	46 Pb	47 Ag	48 Cd	49 In	50 Sn	51 Sb	52 Te	53 I	54 Xe
55 Cs	56 Ba	57 La	72 Hf	73 Ta	74 W	75 Re	76 Os	77 Ir	78 Pt	79 Au	80 Hg	81 Tl	82 Pb	83 Bi	84 Po	85 At	86 Rn
87 Fr	88 Ra	89 Ac	104 Rf	105 Db	106 Sg	107 Bh	108 Hs	109 Mt	110 Ds	111 Rg	112						

Lanthanides	57 La	58 Ce	59 Pr	60 Nd	61 Pm	62 Sm	63 Eu	64 Gd	65 Tb	66 Dy	67 Ho	68 Er	69 Tm	70 Yb	71 Lu
Actinides	89 Ac	90 Th	91 Pa	92 U	93 Np	94 Pu	95 Am	96 Cm	97 Bk	98 Cf	99 Es	100 Fm	101 Mb	102 No	103 Lr

FIGURE 5.3 Element that utilized in metal ions–doped phosphor, the transition elements (d-block elements) are located in the central region of the periodic table. The metallic elements with atomic numbers from 57 (lanthanum) to 71 (lutetium) inclusive are the lanthanides. (Zhang, Y. and Hao, J., *J. Mater. Chem. C* 1, 5607, 2013. With permission.)

By precise control of the dopant ions' combinations and concentration, one can obtain single-phase white light-emitting phosphors.

Er^{3+} ion has drawn broad attention as activation ion. Its characteristic emission at 1.55 μm matches the minimum attenuation of silica optical fibers commonly used in optical fibers communication. Moreover, it is highly suitable to be utilized in upconversion phosphors. The efficiency of the upconversion process of Er^{3+} is exceptionally high, owing to its ladder-like arranged energy level structure. The energy gaps between the $^4I_{11/2}$ and $^4I_{15/2}$ states (~10,350 cm^{-1}) and between the $^4I_{11/2}$ and $^4F_{7/2}$ states (~10,370 cm^{-1}) are almost the same. Besides, the energy difference between $^4F_{9/2}$ and $^4I_{13/2}$ states is also in the identical region. Thus, under 980 nm laser excitation, upconversion emission may occur via energy transitions of these levels. The red and green emissions correspond to $4f$ transitions of $^4F_{9/2} \rightarrow {}^4I_{15/2}$ and $^2H_{11/2}/{}^4S_{3/2} \rightarrow {}^4I_{15/2}$ of the Er^{3+} ions, respectively.

Tm^{3+} ions can emit blue light located at 460 and 476 nm, corresponding to $^1D_2 \rightarrow {}^3F_4$ and $^1G_4 \rightarrow {}^3H_6$ transitions. Achieving such a short wavelength emission is of significant to obtain blue color for full-color display applications. Recently, incorporating Tm^{3+} to realizes white light sources have attracted intensive interest. The blue component from the Tm^{3+} ion can be employed to form single-phased white light–emitting phosphors. In addition to the blue emissions, Tm^{3+} ions can also emit luminescence in the red (640–710 nm) and NIR region (807 nm), which can be applied in biological imaging.

Phosphors contained Eu^{3+} ions show red emissions around 616 nm. The characteristic spectra bands of Eu^{3+} ions in hosts locate at around 579, 593, 614, 652 and 705 nm, which correspond to $^5D_0 \rightarrow {}^7F_j$ transitions ($j = 0$, 1, 2, 3 and 4). It is well known that the luminescence of Eu^{3+} is sensitive to the host crystal symmetry around Eu^{3+} ions. For instance, the electric dipole transition $^5D_0 \rightarrow {}^7F_2$ is forbidden when Eu^{3+} ion is located at aversion symmetry site. While the magnetic dipole transition $^5D_0 \rightarrow {}^7F_1$ is insensitive to the site symmetry and hardly varies with the crystal filed around Eu^{3+} ions. Hence, the Eu^{3+} ions are regarded as a probe to detect the site symmetry around Eu^{3+} in the host materials.

In many single-doped luminescent materials, the excitation radiation is either incapable of or merely weakly absorbed by the activator. As mentioned earlier, different ion is often co-doped to the host lattice to enhance emission efficiency. The second ion can absorb the exciting energy more effectively and, subsequently, transfer it to the activator. Therby this type of ion is called a sensitizer. Emissions from Ln ions often originate from the intra-4f transitions which are parity forbidden while only partially allowed through the mixing of opposite parity wave functions if only Ln ions are embedded in a host. In light of this, most lanthanide activators suffer from low absorption cross-sections of pump photons. Ln and TM ions have been utilized as sensitizers in different doping systems. Yb^{3+} ion is one of the most extensively used sensitizer in upconversion phosphors due to its suitable energy level scheme and large absorption cross-section in the NIR region. Upconversion refers to nonlinear optical processes, converting NIR to UV or visible emission via multiple absorptions or energy transfer. Yb^{3+} ion is usually co-doped along with activators such as Er^{3+}, Tm^{3+} and Ho^{3+} to ensure efficient upconversion emission via energy transfer process.

5.3.2 Transition Metal Ions

TM ions located in the central region of the periodic table (Figure 5.3) and they are another type of luminescent centers that have been utilized for most aspect of luminescence application. In general, TM ions have an incompletely filled d-shell (d^n, $0 < n < 10$). The 3d transitions metal ions utilized in phosphors have three electrons (Cr^{3+} and Mn^{4+}) or five electrons (Mn^{2+} and Fe^{3+}) occupying the outermost 3d orbitals. In strong contrast with lanthanide ions, the 3d orbitals are located in an outer orbit, which is not shielded from the host lattice by any occupied orbitals. Thus, the emission properties are strongly crystal-field dependent and modified by changing the host lattice. Broadly speaking, undefined features characterize the TMs due to the strong coupling of electronic transitions to vibrational transitions of the host lattice.

The luminescence of Cr^{3+} in Al_2O_3 was utilized in the first solid-state laser. The emission consists of two strong lines at 694.3 and 692.9 nm (the so-called R1 and R2 lines). Spectroscopic properties of Cr^{3+} ions depend strongly on the crystal field, demonstrated in the lowest excited states. $^4T_2$ level is sensitive to the crystal field, so luminescence of Ce^{3+} arises from the spin-allowed transition $^4T_2 \rightarrow {}^4A_2$ characterized by broadband with a short lifetime. In contrast, the emission spectra related to the spitted high energy level is characterized by sharp lines and relatively long lifetime. The emissions of Ce^{3+} can be varied from red into a green emission with the decrease of the crystal field strength.

The 3d electrons of Mn^{2+} ion are not shielded from the host lattice by any outer orbitals. Thus, the energy difference between the excited state and the ground state is effectively modified by the host lattice. Like most TM ions, the Mn^{2+} ion has a broad emission band, the position and width are crystal-field dependent. The emission between $^4T_1$ and $^6A_1$ can shift from green to deep red with increasing the crystal field strength. ZnS-doped Mn^{2+} is one of the most successful phosphors used in PL, ML and EL. Typically, ZnS:Mn^{2+}-based thin-film EL (TFEL) devices have shown great applications in flat panel display and lighting source.

The Ni^{2+} ions show a complicated emission spectrum due to the appearance of emission transitions from multiple levels. One of the exciting research topics is the observation of broadband NIR emissions from Ni^{2+}-doped phosphors. Ni^{2+} in an appropriate octahedral crystal has been found to exhibit ultra-broadband near-infrared luminescence. The Ni^{2+}-doped phosphors have long been attracting attention. It is demonstrated that Ni^{2+} can act as a single emission species in multiple octahedral local environments. The ultra-broadband luminescence property can be attributed to the nature of electronic transitions of Ni^{2+}, i.e., extremely strong electron-phonon coupling. The Ni^{2+}-doped phosphors have been considered as promising candidates for tunable laser and broadband optical amplifier.

5.3.3 HOST MATERIALS

In a sense, the host material can be regarded as the vital part in determining the ultimate luminescent properties. It is necessary to effectively optimize the distribution of the dopant ions functioning as luminescent centers inside the host lattice and prohibit any unintended non-radiative processes. Dopant ions in a solid-state host can be considered as impurities embedded in the host lattice. For instance, when the dopant ions replace the host ions substitutionally, the dimension of the host lattice and the distance between the dopant ions and their relative spatial position varies. Thereby, the dopant materials generally require close lattice or ionic radius matching to render the desired formation of crystal defects and lattice stress. Thus, an ideal host materials should also have low phonon energies to minimize non-radiative relaxation loss. For display and LED applications, host materials must have a large bandgap, thus it will be transparent to visible light. Overall, hosts are usually insulators, such as oxide, sulfide, fluoride, aluminate and silicate. Some semiconductors can also serve as hosts for optically active ions, as long as the luminescent excited state does not overlap with the conduction band (CB) leading to quenching. Nonetheless, the host materials must also be mechanically and chemically stable in their operating environment.

Another aspect of significant influence on the optical properties of metal ion–doped phosphors arsied from the host materials can be observed from the luminescence originated from $4f$–$4f$ transitions of Ln ions and $3d$–$3d$ transitions of TM metal ions, since they are both partially forbidden. However, when the doped ion is introduced in a non-centrosymmetric host lattice, the parity selection rules are relaxed. Low symmetry hosts exert a crystal-field containing more uneven components around the dopant ions. The perturbation caused by the odd-order term of crystal-field, forcing these parity-forbidden transitions to be allowed. In principle, the lower symmetry at the dopant ion site corresponds to the higher transition probabilities.[13] Therefore, the host should provide suitable site symmetry for the dopant ions.

5.4 COUPLING STRATEGIES BETWEEN SMART MATERIALS AND PHOSPHORS

After introducing several essential kinds of smart materials and the basic of metal ion–doped phosphors, we will introduce some strategies to formulate the smart phosphors.

5.4.1 STRAIN TUNING BASED ON PIEZOELECTRICS

Strain engineering based on piezoelectrics has been utilized to modulate the physical properties of various materials, such as ferromagnets, superconductors, quantum dots and 2D semiconductor materials. As mentioned earlier, single-crystal relaxor ferroelectric PMN-PT, endowed with giant piezoelectric strain and with piezoelectric tunability at cryogenic temperatures, has advanced to the forefront of strain engineering. The (0 0 1)-oriented PMN-PT single crystal is generally cut into rectangular plates with thicknesses between 200 and 500 μm., the PMN-PT then subjugated to polarization with a large poling electric field along the (0 0 1) direction prior to utilization. Thereafter, once an electric field was applied on the PMN-PT in the poling direction, the lattice of the PMN-PT will elongate along the direction of the electric field, resulting in anisotropic biaxial strain perpendicular to the electric field.

To obtain a larger lattice mismatch–induced strain, both thin-film growth and layer material transfer should be meticulously processed to form excel coherency between the thin-film or 2D material with the PMN-PT substrate. The integration of graphene-like 2D atomic-thick crystals onto a PMN-PT actuator always involves the interaction with the supporting substrate by vdW forces. Single-crystal PMN-PT has a good lattice match with conventional dielectrics, such as $SrTiO_3$ (STO), BTO and BFO. Some most utilized conductive oxides, including $SrRuO_3$ (SRO), $La_xSr_{1-x}MnO_3$ (LSMO) and $La_xCa_{1-x}MnO_3$ (LCMO), also have small lattice mismatch with PMN-PT. Therefore,

PMN-PT is an ideal piezoelectric substrate for epitaxial growth oxide heterostructures. The past few years have witnessed significant advances in the methodology of high-quality piezoelectric thin films fabrication, as well as in their integration into MEMS. The merit of this integration is that a giant strain can be generated with a reduced applied voltage. For instance, the breakthroughs of integration of epitaxial PMN-PT films with Si platform enable the realization of monolithic MEMS devices rendered a dramatically reduced operation voltages and physical dimensions.

5.4.2 METHODS BASED ON PIEZOPHOTONICS

The piezophotonics effect coined by Wang is the coupling between piezoelectric properties and photoexcitation, where strain-induced piezopotential modulates and controls the photoexcitation properties.[14] Specifically, ML can respond to mechanical stimuli and subsequent emission of light, in general, also called piezoluminescence specifically for piezoelectrics.[15] An archetypical ML material ZnS:Mn shows an intense ML through a characteristic piezoluminescence emission mechanism. As stress is imposed onto the phosphor, the wurtzite-structured ZnS gives rise to a piezopotential, which tilts the energy bands of ZnS. A first principle calculation confirms that Mn^{2+} ion doping introduces both electron and hole defect states between the conduction and valence bands (VBs).[16] The corresponding electron in the upper electron defect state detraps and escapes into the CB. Then the detrapped electrons recombine with the holes non-radiatively and transfer the energy to the Mn^{2+} ions. When the Mn^{2+} ion returns from the excited state ($^4T_1$) to the ground state ($^6A_1$), a photon is emitted.

Currently, ML materials have been fabricated in different forms, including crystals, powders, nanoparticles and thin films, and can be stimulated under different forms of mechanical energy stimuli, such as ultrasonic-, impact-, tribo- and compression excitation. Particularly, piezoluminescence is a peculiar type of ML. It is distinctive from other types of ML, since light emission will occur upon an elastic deformation was imposed onto a piezoluminescence phosphor. This reproducible, intriguing mechano-optical conversion effect has attracted considerable attention and offered promising opportunities for numerous applications.[15] The underlining mechanisms that govern the transfer of mechanical energy to light emission in ML materials are complex and varied significantly between each ML phosphor.[17] The most common and controllable piezophotonic luminescence is achieved by composing ML phosphors coated on the top of piezoelectric actuators. As mentioned earlier, relaxor ferroelectric single-crystal PMN-PT has superior piezoelectric coefficients and can generate large linear forces under applied voltages with fast actuation, which has been used for piezophotonics. For the ML phosphors integrated with the piezoelectric actuator, the piezoluminescence intensity is a function of the frequency and magnitude of the applied voltage. The luminescence intensity (I) of ML is proportional to the strain (σ) and its change rate ($\partial\sigma/\partial t$), which can be written in the form[18]

$$I = \alpha\sigma\frac{\partial\sigma}{\partial t}\left(1 - \exp\left(-\frac{t}{\tau}\right)\right) \tag{5.1}$$

where α is a constant describing the strain coupling to ML; τ is the lifetime of the interaction associated with the ML. The strain (σ) created by the converse piezoelectric effect is expressed as

$$\lambda = \frac{\partial\sigma}{\partial E} = d\frac{\partial\sigma}{\partial V} \tag{5.2}$$

where λ is the piezoelectric coefficient. E and V are the electric field and voltage, respectively; d is the thickness of the piezoelectric film. Equation (5.1) can be expressed as

$$I = \alpha\lambda\frac{\sigma}{d}\frac{\partial V}{\partial t}\left(1 - \exp\left(-\frac{t}{\tau}\right)\right) \tag{5.3}$$

According to Equation (5.3), it is known that a static electric field is unable to excite ML due to $\partial V / \partial t = 0$. A periodic electric signal is applied to the piezoelectric actuator. We average the luminescence intensity I during one time period. This averaged ML intensity (I_a) can be written as

$$I_a = \frac{1}{T} \int_0^T I dt = f * \int_0^T \alpha \lambda \frac{\sigma}{d} \frac{\partial V}{\partial t} \left(1 - \exp\left(-\frac{t}{\tau} \right) \right) dt \tag{5.4}$$

T and f are the period and frequency of the external electric voltage, respectively. For $t \gg \tau$, I_a can be expressed as

$$I_a = f * \int_0^{V_0} \alpha \lambda \frac{\sigma}{d} dV = \frac{1}{2} f \alpha \lambda^2 \left(\frac{V_0}{d} \right)^2 \tag{5.5}$$

where V_0 is the maximum value of the voltage applied to the piezoelectric actuator. From Equation (5.5), we can see that the total ML intensity has a positive linear relationship with f and V_0^2.

5.4.3 Ferroelectric Host Coupling

In inorganic lanthanide-doped phosphors, the optical properties are determined by doped lanthanide ions and the host's crystal structure. Thereby, selection of appropriate host materials is essential to endow lanthanide-doped phosphors with desirable luminescent spectra and enhanced light emission. The variation of the host symmetry can modulate the light emission from the lanthanide-doped phosphors. According to Laporte's parity selection rule of electrical dipole (ED) radiation, f-f ED transitions of free Ln^{3+} ions are not allowed. Once the Ln^{3+} ions are embedded in a non-centrosymmetric host lattice, the mixing of odd-parity configurations from the host crystal-field somewhat relaxes the selection rules. Therefore, a lower symmetry host in principle can exert more uneven parity into the $4f$ electronic wavefunctions, giving rise to the ED transitions of Ln^{3+} ions with a higher probability. The Judd-Ofelt (J-O) theory is devoted to the theoretical prediction of the spectroscopic properties of Ln^{3+} ions,[19, 20] in which the spontaneous emission probability A_{ed} for an ED transition between the initial J manifold $|[S,L]J>$ and the final J manifold $|[S',L']J'>$ is determined by

$$A_{ed} = \frac{64\pi^4 e^2}{3h(2J+1)\lambda^3} \left[\frac{n(n^2+2)^2}{9} \right] S_{ed} \tag{5.6}$$

where e is the electron charge, λ is the mean wavelength of the transition, n is the refractive index and h is plank constant. The ED line strength S_{ed}:

$$S_{ed} = \sum_{t=2,4,6} \Omega_t \left| \left\langle 4f^n [S,L]J \left\| U^{(t)} \right\| 4f^n [S',L']J' \right\rangle \right|^2 \tag{5.7}$$

where $<\|U^{(t)}\|>$ are reduced matrix elements of the unit tensor operators, and Ω_t (t=2,4,6) are the J-O intensity parameters, containing the parity perturbations of the crystal-field effect. It is worth noticing a particular portions of ED transitions of the lanthanide ions that are peculiarly sensitive to changes in the surrounding environment. It was concluded that the inhomogeneity of the surrounding dielectric contributes to the so-called hypersensitive transitions. Among the three J-O parameters, the Ω_2 term is closely associated with the hypersensitive transitions. The magnitude of Ω_2 term is found to correlate with the host lattice's asymmetry positively.

The most straightforward method to vary the host symmetry is to change the host material. Nevertheless, it is an *ex-situ* and irreversible manner. Host materials made from ferroelectrics are able to respond to external physical stimuli. Under an electric field, charged ions of ferroelectric materials demonstrated asymmetry displacement and result in a small distortion in the crystal dimension. In a ferroelectric host coupling with the doped lanthanide ions, a system of Er^{3+} ions–doped BTO is used as an example to investigate the electric field–induced modification on the luminescence of Er^{3+} ions. Since its discovery, BTO is considered as the archtypical perovskite ferroelectrics. It possesses a tetragonal structure at room temperature, where the unit cell is elongated along the c-axis. It is found that Er^{3+} ion can replace Ba^{2+} or Ti^{4+} ion depending on the local Ba/Ti ratio in BTO. If Er^{3+} ion substitutes Ti^{4+} ion, Er^{3+} ion is located at the B-site and octahedrally coordinated by oxygen ions. Under an electric field, Ti^{4+} ions move along the direction of the electric field, while the O^{2-} ions are shifted against the electric field. The BTO lattices undergo electric field–induced asymmetric structural transformation. The distortion of the host symmetry is proportional to the applied electric field, which gives rise to the modulation on the optical properties of Er^{3+} ions.

5.4.4 MAGNETIC-INDUCED LUMINESCENCE

Compared to the luminescence emission by the excitation of photon and electric field, little work is reported on the light emission by another common source, i.e., magnetic field. The correlation between light emission and magnetic field stimulation has often been ignored. In most reports, the term of magnetoluminescence with a prefix of 'magneto' is about PL or EL under the influence of externally applied magnetic field based on interchange spin-spin interaction, spin-dependent exciting processes and Lorentz effects, which does not implies magnetic-induced luminescence (MIL). Notably, the magnetoluminescence effects typically occur under extreme conditions of a high magnetic field and low temperature. By comparison, the realization of light-emission stimulated by magnetic field faces an even more significant challenge, and hence, the phenomenon of intrinsic MIL is extremely rare.

As mentioned earlier, the piezophotonic effect is a two-way coupling between piezoelectricity and photoexcitation process. This two-way coupling effect between other external stimuli and the piezophotonic effect can lead to an intriguing properties. We first proposed a strategy to observe MIL by coupling the magnetostrictive effect to the piezophotonic effect, as shown in Figure 5.4(a).[21] In our design, the MIL composite is composed of two parts, the magnetic phase which is a magnetic elastomer layer formed with paramagnetic Fe–Co–Ni alloy particles embedded in the poly(dimethylsiloxane) (PDMS), while the other part is the ML phosphors phase composed of ZnS doped with metal ions (e.g., Al, Cu) mixed with PDMS. Under an alternative magnetic field, the magnetic elastomer can impose strain on the ML phosphors to generate light emission (Figure 5.4(b)).

Such MIL is a result of the product of the magnetostrictive (magnetic/mechanical) effect in the magnetic phase and ML (mechanical/luminescent) effect in the phosphor, namely,

$$\text{MIL} = \frac{\text{magnetic}}{\text{mechanical}} \times \frac{\text{mechanical}}{\text{luminescent}} \tag{5.8}$$

Assume that an external magnetic field (H) is applied to the strain (σ) coupled with the two-phase system. Thus, magnetostrictive coefficient λ in a magnetic-phase can be described by

$$\lambda = \frac{\partial \sigma}{\partial H} \tag{5.9}$$

On the other hand, ML intensity (I) is proportional to the strain and its rate of change ($\partial \sigma / \partial t$), which can be written in the form

$$I = \alpha \sigma \frac{\partial \sigma}{\partial t} (1 - \exp(-t/\tau)) \tag{5.10}$$

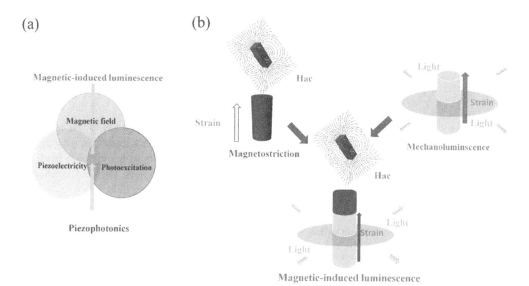

FIGURE 5.4 (a) Schematic diagram illustrating the field of piezophotonics and its integration with magnetic field to introduce magnetic-induced luminescence (MIL). (b) Implementation of MIL effect via strain-mediated magneto-luminescent coupling. (Wong, M. C., Chen, L., Tsang, M. K., Zhang, Y., and Hao, J., *Adv. Mater.* 27, 4488, 2015. With permission.)

where α is a constant of coupling strain to luminescence; τ is the lifetime of the interaction associated with ML event. Therefore, the MIL intensity in the strain-mediated two phases triggered by the magnetic field can be expressed as

$$I = \alpha\lambda\beta\sigma \frac{\partial H}{\partial t}(1 - \exp(-t/\tau)) \tag{5.11}$$

where β is a coupling factor between the magnetic and phosphor phases ($\beta \leq 1$). For some magnetic actuators, maintaining the linear relationship of σ and H with correlation coefficient c under specific operation range, Equation (5.4) can be further described as

$$I = \alpha\lambda\beta cH \frac{\partial H}{\partial t}(1 - \exp(-t/\tau)) \tag{5.12}$$

For $t \gg \tau$, Equation (5.12) can be simplified as,

$$I = \alpha\sigma \frac{\partial \sigma}{\partial t} \tag{5.13}$$

By integrating Equation (5.13) when taking $\sigma = 0$ at $t = 0$, we obtain

$$\int_0^t I dt = \int_0^{\sigma_0} \alpha\sigma d\sigma \tag{5.14}$$

where σ_0 is the maximum value of the applied strain. From Equation (5.14), the total luminescence intensity I_T is

$$I_T = \frac{1}{2}\alpha\sigma_0^2 \tag{5.15}$$

When substituting the integral result of Equation (5.9) into Equation (5.15) and taking into account the coupling factor between the two phases, we find

$$I_T = \frac{1}{2}\alpha\lambda\beta H_0^2 \tag{5.16}$$

where H_0 is the maximum value of the magnetic field.

Hence, the total luminescence intensity of MIL is proportional to the product of the three parameters (α, λ, β) and square of H_0. According to Equation (5.11), it is apparent that static dc magnetic field is unable to excite the luminescence because of $\partial H / \partial t = 0$. For alternating magnetic field, e.g., $H = H_0 \sin \omega t = \sqrt{2} H_{rms} \sin \omega t$, where H_{rms} is the root-mean-square magnetic field, MIL intensity will therefore depend on the amplitude and frequency of the input magnetic field.

5.4.5 LIGHT-DRIVEN PHOTOCHROMISM

Smart phosphors responding to light stimulus have numerous advantages over other stimuli, such as contactless and ease accessibility. Photochromism refers to a reversible photo-induced transformation of a chemical species between different forms having different absorption spectra. Photochromic host materials doped with luminescence activator lanthanide ions simultaneously possess the capability of photochromic reaction and light emission behavior in one single compound. In recent years, various kinds of lanthanide-doped bismuth layer-structure ferroelectrics, including $Na_{0.5}Bi_{2.5}Nb_2O_9$ (NBN), and Sr_2SnO_4 exhibit superior photochromic performance.[22–24] During the high-temperature sintering process, the volatilization of sodium and potassium elements introduces consequent defects (e.g., Na, K and O vacancies), which can trap electrons or holes resulting in color centers upon UV or visible light irradiation. When the absorption band of the color centers overlaps the emission bands of doped lanthanide ions, the luminescence emission can be consciously modulated through energy migration between them. Among various readout manners, optical readout based on the luminescence contrast has been regarded as the most convenient and efficient way because of its high sensitivity, resolution, contrast and fast response. Various lanthanide ions can generate upconversion emission under long wavelength (e.g., 980 nm), which cannot trigger a photochromic reaction in the lanthanide ion–doped photochromic materials. Thus, readout and write-in operations can eliminate undesired interference, which is highly essential for a nondestructive readout. Therefore, lanthanide-doped photochromics promises enormous potential for photomemories owing to the added luminescence switching provided by the intrinsic photochromism coupled with the lanthanide luminescence.

5.5 PROGRESS IN DEVELOPING SMART PHOSPHORS

With the fast development of smart materials, control and coupling of the luminescence processes from smart phosphors are attracting extensive interest. This section will demonstrate some recent progress in developing smart phosphors with diverse designs and configurations and their distinctive applications characteristics.

5.5.1 STRAIN ENGINEERING ON THE PL OF METAL ION–DOPED PHOSPHORS

Mechanical stimuli, including friction, impact, compression and ultrasonic vibration, are often encountered in research and engineering designs. It is noteworthy that ML materials can convert various mechanical energies into light emission, and we will discuss this issue in the next section. By simply utilizing a stress, tuning of the luminescence of smart phosphors can be achieved. More importantly, through tailoring the substrate-induced strain of thin-film heterostructures, the luminescence properties of phosphors can be finely tuned. PMN-PT is capable of providing sizeable biaxial strain arising from the converse piezoelectric effect. The electric field–controlled strain based on PMN-PT has previously been applied to modulate transport, magnetic and optical behaviors of different material systems. Benefiting from the piezoelectric-induced strain of PMN-PT, the luminescence of phosphors can be modulated in a real-time and reversible way.[25] Epitaxial Yb^{3+}/Er^{3+}-doped BTO thin films were grown on PMN-PT substrate. Upconversion PL in Yb^{3+}/Er^{3+}-doped BTO thin films was effectively tuned by the elastic strain arising from PMN-PT. After poling the PMN-PT substrate, the upconversion intensity decreases over the whole wavelength range. When the electric field is increased from 0 to 10 kV/cm, it is clear that the upconversion is modified. Figure 5.5(a) shows that the modification ratios I/I_0 for $^2H_{11/2}/^4S_{2/3} \rightarrow {}^4I_{15/2}$ and $^4F_{9/2} \rightarrow {}^4I_{15/2}$ are ~25% and ~21%, respectively. It could be found in Figure 5.5(b) that the change in intensities is almost linearly dependent on the electric field applied to the PMN-PT substrate. To clarify and gain an understanding of the mechanism responsible for the observed PL emission modification in BTO:Yb/Er thin films, the lattice deformation of BTO:Yb/Er thin film under dc bias voltage should be considered.

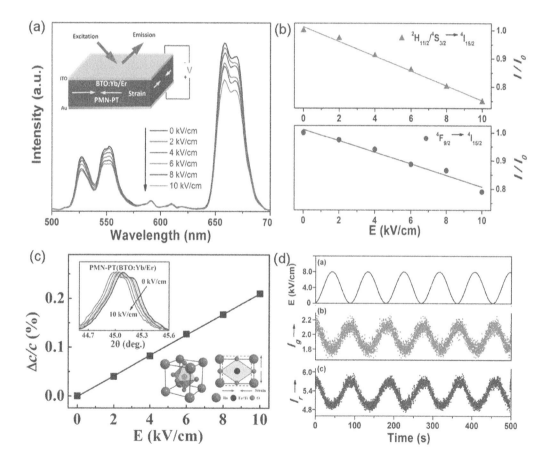

FIGURE 5.5 (a) The upconversion emission spectra of the (BTO:Yb/Er)/PMN-PT heterostructure under dc bias ranging from 0 to 10 kV/cm. Inset shows the schematic of the corresponding smart phosphor. (b) The change ratios for green and red emission bands as a function of the applied electric field. (c) The relative changes in the lattice constant $\Delta c/c$ of PMN-PT as a function of the electric field. Up left inset shows the *in-situ* XRD patterns in the vicinity of reflections for (BTO:Yb/Er)/PMN-PT heterostructure excited by different electric fields strength. Bottom right inset shows the schematic of the compressive BTO:Yb/Er thin film biaxially strained to match the substrate PMN-PT. (d) The kinetic PL response at the wavelength of 523 and 656 nm of BTO:Yb/Er smart phosphor film under a sinusoidal ac bias. (Zheng, M., Sun, H., Chan, M. K., and Kwok, K. W., *Nano Energy* 55, 22, 2019. With permission.)

The obtained BTO:Yb/Er thin film has a cubic structure with an in-plane parameter of $a = 4.010$ Å, which is nearly 0.35% smaller than that of PMN-PT. Therefore, the BTO:Yb/Er films are subjected to lateral restraint from the PMN-PT substrates. *In-situ* X-ray diffraction analysis in Figure 5.5(c) indicates a strain-induced lattice change of the BTO host under different electric fields. The systematic shift of PMN-PT (0 0 2) and BTO:Yb/Er (0 0 2) reflection peaks demonstrate that the electric field successfully modified lattice distortion of PMN-PT and BTO:Yb/Er. The calculated out-of-plane lattice constant c of PMN-PT increases from 4.019 to 4.027 Å. The linear dependence of relative changes $\Delta c/c$ of PMN-PT changes with increasing E, indicating the piezoelectric nature of induced strain in PMN-PT. The maximum values of $\Delta c/c$ for PMN-PT is 0.21% when $E = 10$ kV/cm. A decreased in-plane lattice constant would accompany volume-preserving distortion results an increase in out-of-plane lattice constant. We could conclude that a decrease in the in-plane lattice dimension of PMN-PT will release the lateral restraint in BTO:Yb/Er film. It is widely accepted that the crystal symmetry of the host materials plays an essential role in modifying the PL of dopant ions. Our previous results indicated that the Ti^{4+} ions in BTO could be replaced by the substitution ions Er^{3+}/Yb^{3+}, resulting in a non-centrosymmetric lattice.[26] When the BTO:Yb/Er film grown on PMN-PT substrate, the lattice misfit induced large lateral strain induces lower symmetry around Er^{3+} ions, leading to the observed PL emission. When the E field is applied, the strain and the crystal field around Er^{3+} ions are reduced. It is generally accepted that $4f$–$4f$ electric dipole transition probabilities will be suppressed in higher symmetry structure. Thus, the recovery of symmetry in the BTO host will lead to the modulation of PL intensity. More intriguingly, integrating lanthanide-doped ferroelectric films with giant piezoelectric actuators provide an *in-situ* and dynamic modulation of the light emissions as shown in Figure 5.5(d).

As mentioned earlier, the crystal field has a strong influence on the luminescence of TM ions. TM ion Ni^{2+} has been chosen as the dopant ion since the fluctuation of crystal-field strength may result in the modulation of the energy level between the $^3T_2$ and $^3A_2$ levels of Ni^{2+} ion. One can expect that the luminescence of Ni^{2+} ions can be tuned by fine-tuning the crystal field strength under strain stimulus. Furthermore, Ni^{2+}-doped phosphors are considered as one of the promising candidates for tunable laser and broadband NIR emission. STO is known to be an incipient ferroelectric and fundamental material of oxide-based heterostructures. To realize tunable luminescence from TM ion–doped phosphors corporated smart phosphor, TM ion Ni^{2+}-doped STO thin films were grown on PMN-PT.[27] When increasing dc bias voltage on PMN-PT from 0 V to 500 V, the emission peak position shows a blue shift from 1326 to 1313 nm. Meanwhile, the PL intensity gradually increases from 8.3 to 9.13 (a.u. intensity). This intensity slightly reduced as the voltage increases from zero bias to 500 V. XRD results confirmed that the in-plane compressive strain produced by the PMN-PT substrate were transferred to the STO:Ni thin film. The strain causes compression in the $[NiO]_6$ octahedron phase. This compression causes the crystal field around Ni^{2+} ion enhanced and weaken the lattice relaxation since the Ni^{2+}–O^{2-} bond distance decreases as the applied bias voltage increased. The splittering of energy thereby become more significant results in an increase in the transition energy from the excited state to the ground state, rendering in a blue-shifted emission peak. The reduced non-radiative transition enhances the emission intensity and elongates the emission lifetime.

5.5.2 Piezophotonic-Induced Light Emission

Piezophotonic light-emitting devices have great potential for future micro- and nanoscale systems due to the added functionality provided by the light emission through the electromechanical transduction. The piezophotonic effect is a two-way coupling effect between piezoelectricity and photoexcitation properties, where the strain-induced piezoelectric potential modulates the band structure within piezoelectric phosphors and, thus, tunes/controls the relevant optical process. The realization of light emission stimulated by the piezophotonic effect is to initiate the ML process replacing p-n junction-based light-emitting diodes (LEDs) for general lighting purposes.

In earlier study, an intriguing series of multifunctional material, Pr^{3+}-doped $BaTiO_3$–$CaTiO_3$ ceramics was fabricated, and electro-mechano-optical coupling was realized.[28] XRD analysis shows that Pr^{3+}-doped $[(1-x)BaTiO_3-xCaTiO_3]$ ceramics belongs to diphasic materials for $x \geq$ 0.25, in which tetragonal $Ba_{0.77}Ca_{0.23}TiO_3$ and orthorhombic $Ba_{0.1}Ca_{0.9}TiO_3$ coexist. A significant enhancement in the magnitude of the induced electrostrictive strain were observed in the ceramics via an electric field–induced non-180° domain rotation. Upon an application of stress, a strong red ML emission can be seen by naked eyes. The interactions between the ionic polarization in Pr^{3+}-doped $Ba_{0.77}Ca_{0.23}TiO_3$ and the ferroelectric behavior in $Ba_{0.77}Ca_{0.23}TiO_3$ are responsible for the strong ML. The three way coupling achieved in a single compound has aroused a new surge of interest in multifunctional materials. Later, Xu et al. further proposed a high-performance piezo-multifunctional material named 'multi-piezo' that simultaneously exhibits piezoelectricity and efficient luminescence.[29] The study was performed by doping lanthanide Pr^{3+} ions into a $LiNbO_3$ host by precisely tuning the Li/Nb ratio in nonstoichiometric $LiNbO_3$:Pr^{3+}, a material that exhibits a high luminescence intensity. In particular, $LiNbO_3$:Pr^{3+} shows excellent strain sensitivity at the lowest strain levels, with no threshold for stress sensing. These multiple piezo-properties are useful for multifunctional device application.

In light of this work, an agile and controllable piezophotonic light emission can be realized by integrating the ML phosphors layer with piezoelectric actuators. Our group has demonstrated the fabrication and characteristics of strain-induced piezoelectric potential stimulated luminescence from ZnS:Mn film grown on piezoelectric $Pb(Mg_{1/3}Nb_{2/3})O_3$–$xPbTiO_3$ (PMN-PT) substrate.[30] Compared with 1D nanostructures, the thin-film structure would increase the device reproducibility and reduce the fabrication cost because of the mature thin-film fabrication technologies. Additionally, a thin-film structure facilitates the integration and patterning of the device. More importantly, the simultaneous generation of light and ultrasound wave is first demonstrated using a single system in this work (Figure 5.6(a)). We can tune the luminescence of the ZnS:Mn films and generate ultrasound signal via a converse piezoelectric effect in PMN-PT upon applying an electric-field. Such a multifunctional source may find applications in a variety of systems. For instance, various techniques such as high-frequency ultrasound imaging and fluorescence spectroscopy have been widely used for medical diagnosis. However, each of these techniques has intrinsic advantages and deficiencies. There is a need to develop a combined approach capable of evaluating structural characteristics and biochemical properties, which may provide higher predictivity and complementary information than a single approach, either by optical or ultrasound method. Therefore, the integration of novel dual-modal sources combining light-emission and ultrasound generation on a single wafer with detecting system will aid the development of a hybrid system in tissue diagnosis. The wurtzite-type Mn^{2+}-doped ZnS can also possess a noncentral symmetric structure and inherently exhibit the piezoluminescence characteristics. Single crystal PMN-PT was chosen as a substrate. PMN-PT single-crystal features large piezoelectric coefficients and extraordinary high electromechanical coupling factors, and these merits translate into the high performance of the ultrasound transducer fabricated with PMN-PT single-crystal. ZnS:Mn thin films were grown on (0 0 1) cut PMN-PT single-crystal substrate by pulsed laser deposition. We applied a positive poling voltage of 500 V across the ZnS:Mn/PMN-PT structure to make the PMN-PT substrate positively polarized. The luminescence spectra of ZnS:Mn film by electric field operating at 500 Hz and 200 V_{pp}. The emission peak is found at ~588 nm in the electric field–induced emission spectrum of ZnS:Mn, similar to the PL spectrum of ZnS:Mn. This indicates that the electric field–induced luminescence also results from the $^4T_1 \rightarrow {}^6A_1$ transition of Mn^{2+}. The PMN-PT substrate has been positively polarized prior to the measurements of the luminescent properties of the ZnS:Mn film. When an ac electric field with a smaller field strength than the coercive field of the PMN-PT was applied to the polarized PMN-PT substrate, the PMN-PT substrate's lattice will expand or contract along the electric field direction. Our previous studies have shown that the PMN-PT substrate's lattice can expand and contract along the c axis at the same frequency as that of the applied electric field. The electric field–induced lattice displacements originate from the converse piezoelectric effect of the

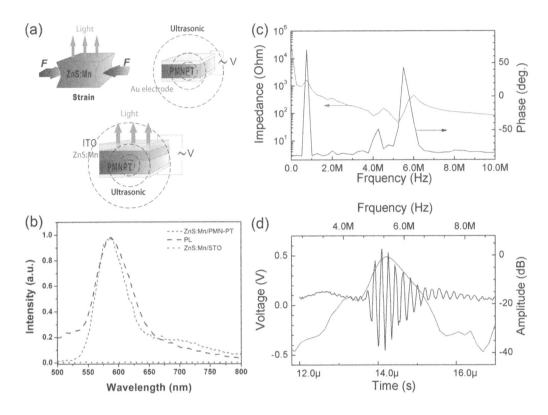

FIGURE 5.6 (a) The setup used to measure the dual-mode light and ultrasound emission of ZnS:Mn/ PMN-PT based smart phosphor. (b) The luminescence spectra from ZnS:Mn/PMN-PT smart phosphor operating under an alternative electric field. (c) The thickness mode impedance and phase spectra for the PMN-PT single-crystal that cultivated in the smart phosphor. (d) Pulse-echo waveform and frequency spectrum from ZnS:Mn/PMN-PT structures. (Zhan, Y., Gao, G., Chan, H. L., Dai, J., Wang, Y., and Hao, J., *Adv. Mater.* 24, 1729, 2012. With permission.)

PMN-PT substrate. Such a modulation in the lattice parameter *c* will cause changes in the in-plane lattice parameters *a* and *b* due to the Poisson effect, which can subsequently impose an in-plane strain on the ZnS:Mn film. XRD and TEM results indicate that the ZnS:Mn film is of good quality and firmly adheres to the underlying substrate. Wurtzite ZnS:Mn possesses an asymmetrical structure. Hence, the substrate-imposed strain can give rise to a piezoelectric potential and result in the piezoluminescence. The earlier result indicates that it is possible to realize on-chip multimode source integrated thin-film phosphors with piezoelectric materials. Various imaging technologies, including optical imaging and ultrasonic imaging, can provide various anatomical structure or composition information. Nevertheless, each imaging modality has its own merits and application limitations. Great efforts have been focused on the development of a multimodal imaging system. There have been reports on dual-modal diagnostic devices combining optical and ultrasonic imaging, i.e., a hybrid system combines optical fiber and an ultrasonic transducer. It would be attractive if a single source combined with light and ultrasound could be produced and integrated into a single detecting system. To validate the novel signal source's feasibility, the ZnS:Mn/PMN-PT structure was poled along the thickness direction and the test signal was also applied along the same direction. Figure 5.6(c) shows the sample's impedance and phase spectra with PMN-PT thickness of 0.5 mm as a function of frequency in a frequency range from 0.5 to 30 MHz. It can be seen that the plate mode resonance is found at ~750 kHz. The thickness mode resonance frequency (f_r) and the antiresonance frequency (f_a) can be observed at 5.25 and 6.05 MHz, respectively. The thickness electromechanical coupling coefficient k_t can be derived from

$$k_t^2 = \frac{\pi}{2} \frac{f_r}{f_a} \tan\left(\frac{\pi}{2} \frac{f_a - f_r}{f_a} \right) \tag{5.17}$$

k_t is found to be 0.53. Figure 5.6(d) shows a pulse-echo waveform and frequency spectrum of the PMN-PT coated with ZnS:Mn film. The structure's center frequency was about 5.48 MHz, which is in the ultrasonic frequency range well suited for medical diagnostics. The fractional bandwidth at −6 dB was measured to be around 16.6%. At any rate, the results are indicative of the dual-functions, i.e., light source and ultrasound transducer in the device structure of ZnS:Mn thin film grown on PMN-PT.

Nevertheless, the PMN-PT provides unprecedented control over the EL emission from the ZnS:Mn. During a typical operation, ac electric field applied onto the smart phosphor as in conventional alternative EL device. At the same time, the emission peak at 588 nm stems from the transition of the luminescent center of Mn^{2+} from $^4T_1 \rightarrow {}^6A_1$. However, the intensity of luminescence can be controlled and varied significantly. Before the measurement, the PMN-PT was polarized along its c-axis. When an ac electric field is applied across the PMN-PT, a strain will be induced to expand lattice in PMN-PT along the electric field's direction. This strain will generate mechanical stress onto the ZnS:Mn phosphor, which provoked a sudden burst of luminescence emission. As the electric field remains steady, the luminescence intensity gradually faded away. Once the electric field is changed into the opposite direction, their conversed piezoelectric ability of PMN-PT may take effect and then strain is induced onto the ZnS:Mn and luminescence may commence as a short burst. Intriguingly, the luminescence is ceased until the electric field pointed in the same direction with the polarization again. Therefore, a dynamic and on-demand modulation of the EL characteristic could be realized by controlling the applied electric field's direction.

Furthermore, our group also conducted a resonance investigation on this smart phosphor. During the investigation, the phosphor was subjected to a sinusoidal voltage of varying frequency in two frequencies ranges. At a frequency below 1 kHz, the luminescence intensity varies linearly with the electrical signal frequency, whereas as a high-frequency range, <1 MHz, the measured luminescence intensity demonstrated a maximized intensity at 650 kHz. A dual-mode operation of ultrasound generation, as emission and modulation of luminescence, can be realized in this smart phosphor. This novel smart phosphor coupled strain engineering, piezoelectric potential and photonic characteristic to drive and modulate EL. This simultaneous generation and modulation of emission characteristics render our smart phosphor system with unprecedented simplicity, which offers great insight into future development in a novel smart phosphor that manifests multifunctional and response to various physical stimuli.

One severe hindrance on developing chip-integrated ZnS:Mn/PMN-PT smart phosphor devices is the thickness of the PMN-PT bulks. The PMN-PT actuators used for piezophotonic luminescence were made from single-crystal bulk materials ranging from 0.5- to 1-mm thickness. Piezophotonic light-emitting sources based on PMN-PT bulk are severely restricted by some challenges, such as a high voltage burden (up to hundreds of volts), low integration density and micro-manufacturing difficulties. Also, it is difficult to integrate many piezoelectric elements with different patterns together on a single chip. To fabricate a large-area array with each element being addressable would promise dramatically greater design freedom to realize smaller light-emitting elements, facilitating its applications as light sources and displays. Developing chip-integrated devices or incorporating such photonic components onto a Si platform is even more highly sought after in this field. Utilizing piezoelectric thin films as central active elements is an appealing alternative way to overcome all these problems. The utilization of piezoelectric thin films strongly reduces the voltage burden and allows us to take advantage of mature micro-manufacturing techniques.[31] Figure 5.7(a) schematically shows the device fabrication process. The (0 0 1)-oriented single-crystal PMN-PT was bonded

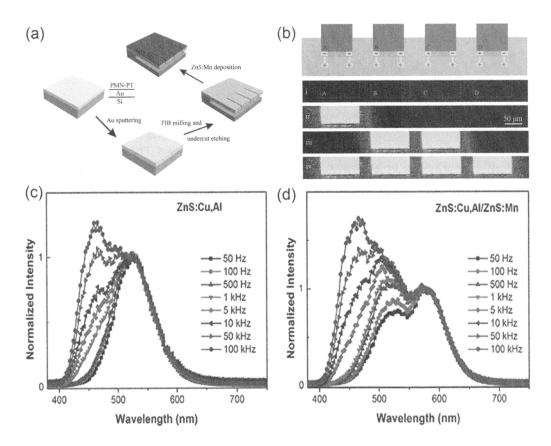

FIGURE 5.7 (a) Schematic illustration of the fabrication process for the smart phosphor. (b) Visualization of addressable piezophotonic smart phosphor. Up: Schematic illustration of the prototype device. Low: Demonstration of four representative addressable light emission states. (c) The normalized luminescence spectra of piezophotonic thin film. (d) The normalized luminescence spectra of smart phosphor thin film under various electric field strength and frequency excitation. (Chen, Y., Zhang, Y., Karnaushenko, D., Chen, L., Hao, J., Ding, F., and Schmidt, O.G., *Adv. Mater.* 29, 1605165, 2017. With permission.)

on Si. Then it was mechanically ground down to tens of microns. However, to pattern PMN-PT films with such thickness remains challenging. To further thin down the films, we turn to inductively coupled plasma reactive ion etching (ICP-RIE) instead of mechanical grinding. Compared with other etching methods such as wet chemical etching, the ICP-RIE can produce a very smooth surface.. We etched the PMN-PT film down to 7-μm thickness, thin enough for further patterning. X-ray θ–2θ scans display only (0 0 l) diffraction peaks of a pure perovskite phase of PMN-PT, confirming the microprocessing does not introduce any secondary phase or impurities into the thin film. This is essential to guarantee that the PMN-PT thin film preserves the giant piezoelectric properties comparable to its bulk counterpart. An array of actuators has been produced from the single-crystal PMN-PT thin film, with each active element having a footprint of 120 μm in length and 100 μm in width. Each PMN-PT thin-film actuator can be individually addressed, generating local deformation to trigger piezoluminescence. When an alternating current (AC) peak-to-peak voltage (V_{pp}) of 20 V at 150 Hz is applied to the PMN-PT film in a vertical direction, ML emission can be observed from the ZnS:Mn thin film, resulting from the $^4T_1 \rightarrow {}^6A_1$ transition of the Mn^{2+} ion. The in-plane strain of the ZnS:Mn film can be reversibly and quasi- linearly controlled by the voltage applied vertically across the PMN-PT thin film. First principle calculations confirm Mn^{2+} ion doping introducing electron and hole defect states between the conduction and VBs. The electron in the upper electron defect state detraps and escapes into the CB. Then the detrapped electrons recombine with the holes non-radiatively and transfer the energy to the Mn^{2+} ions, allowing a photon emission. The magnitude of strain determines the CB's tilt angle, which medicate the amount of detrapped electrons. Hence, an immense strain results in more intense piezoluminescence. Furthermore, it is interesting to find that pulsed emission of luminescence only appears when the bias is switched on and off. If the bias remains unchanged, the piezoluminescence will fade away. The attenuation of the piezoluminescence may be ascribed to the depletion of detrapped electrons and the absence of piezoelectric field. Under a fixed magnitude of strain, only a certain number of detrapped electrons escape into the CB. Hence, the piezoluminescence will extinguish as soon as the detrapped electrons exhausted. Once the strain is released, the tilt CB becomes flat. The electrons will redistribute, while the traps will be refilled. Thereby a constantly oscillating strain results in a sustaining piezoluminescence.

The ability to individually control the chip-integrated piezophotonic components is highly desirable. Incorporating such components into a Si platform should be appealing for developing on-chip piezophotonic devices. The integration of such devices on PMN-PT bulk has been challenging because of the large footprint of individual light-emitting elements, high voltage burden and high production costs. Hence, we had further demonstrated a prototype piezophotonic smart phosphor to circumvent these challenges. Figure 5.7(b) shows the sketch of such a device. Four light-emitting units are encoded from A to D, which can be electrically triggered independently. Each unit can produce local deformation not influenced by others. Figure 5.7(b) shows the addressable characteristics of the device. When all external voltages are switched off, no light-emission is observed from all the four units (situation i). To individually address each unit, we first triggered unit A with 24 V_{pp} at 150 Hz and bright light can be observed in unit A only (situation ii). In situation iii, we switch off unit A, and excite units B and C, only units B and C glow. Situation iv shows that four elements are triggered simultaneously. Here, the addressability promises more flexibility for many intriguing applications, especially when used as light sources or displays with each unit as active pixel.

To realize the modulation of the emission wavelength of piezophotonic luminescence, it has been demonstrated that, by regulating the mixing ratio of two or more ML materials, the emission color thereby can be tuned. However, it is essentially an irreversible and *ex-situ* method. Previous research reported that the ML spectrum of ZnS:Al, Cu shifted to shorter wavelength as the strain rate increased, due to the increasing recombination of the electrons in the CB (or shallow donor level) and holes in the VB (or the *e* state of Cu). The typical mechanical stretching-releasing system can only provide the strain rate up to several hundreds of hertz. Thus, shifts of only several nanometers were observed. Here, our electrical-triggered PMN-PT-based smart phosphor can be stimulated the phosphor up to megahertz, which is suitable for realizing color manipulation of ML from ZnS:Al, Cu-contained phosphor layers. Figure 5.7(c) shows the spectral shape of ZnS:Al, Cu under the frequencies increased from 50 Hz to 100 kHz, the applied voltage was kept at 20 V_{pp}. The spectra are normalized to the peak of ZnS:Al, Cu at 522 nm for intuitively showing the wavelength shift. When the strain rate is increased, the light emission of ZnS:Al, Cu around 460 nm enhances and gradually dominates the emission. The calculated Commission Internationale de L'Eclairage (CIE) coordinates suggest the shift from (0.27, 0.57) at 50 Hz to (0.20, 0.30) at 100 kHz. To obtain a patterned multicolored device, a bilayer film composed of ZnS:Cu,Al and ZnS:Mn was deposited on the PMN-PT. Figure 5.7(d) shows the normalized spectra of ZnS:Cu,Al/ZnS:Mn bilayer. The calculated spectra is normalized by the peak wavelength of the ZnS:Mn. The results have shown that the spectral shape of ZnS:Mn is unchanged with increasing frequency, which is consistent with previous reports.[32] On the other hand, the intensity of ZnS:Cu,Al clearly increases. The calculated CIE coordinates shift from (0.39, 0.50) to (0.26, 0.31) with the frequency increasing from 50 Hz to 100 kHz for the ZnS:Cu,Al/ZnS:Mn bilayer emission. As a result, a color-tunable light emission from orange to blue-green is obtained. These results imply that continuous and reversible controllable color manipulation can be achieved through real-time regulating the strain rate.

Consequently, compared to its bulk counterparts, the devices incorporating single-crystal PMN-PT thin films have several improved figures of merit. First, the utilization of PMN-PT thin film strongly reduces the voltage burden for the application. Suspended PMN-PT thin films generate the required strain under reduced voltages of only 10–20 V. Second, thin-film structures benefit from efficient micro-manufacturing processes, which promises great design freedom for on-chip integration. The realization of addressable light-emitting units with small footprint enables scalable on-chip light sources. A continuous and reversible controllable color manipulation has been demonstrated through real-time regulating the strain rate of the PMN-PT film. Our proof-of-concept smart phosphor solves the main problems hampering the applications of piezophotonic luminescence devices (e.g., high voltage burden, not addressable and low integration density) and offers the possibility to develop more compact and colorful piezophotonic light sources and displays. ZnS:Mn film grown onto PMN-PT by epitaxy method is rigid, limiting the spread applications of ZnS-based phosphors. We have further developed flexible light-emitting devices made from the composite ML phosphor-based hybrid structures, which can meet the easy and low-cost requirements.[33] The piezophototronic luminescence device composed of polymer phosphor layer coated on the top of piezoelectric actuator. Such composite phosphors and the related hybrid devices are capable of responding to different types of external stimuli, including electric field, uniaxial strains of stretch and mechanical writing and piezoelectric biaxial strain, resulting in the observed white and green light emissions by the naked eyes.

5.5.3 Ferroelectric Tuning of Lanthanide-Doped Phosphors

It is well-known that BTO as a prototype ferroelectric will undergo successive phase transitions with changing the temperature. Meanwhile, Er^{3+} ion is very sensitive to the change of the surrounding environment. Our group has investigated the PL response of the Er^{3+}-doped BTO ceramics within the temperature range from 15 to 300K. Er^{3+} ion can substitute Ba^{2+} (A-site) or Ti^{4+} ion (B-site) depending on Ba/Ti ration.[26] When the temperature passed through 150 and 250K, two sudden changes in the PL intensity of the Er^{3+} ion located at A-site was observed, which may arise from the phase transition–induced host lattice changes. The PL intensity of the Er^{3+} ion at B-site presented a different response. With the temperature decreasing from 300K to 15K, the electrons at the upper $^2H_{11/2}$ level were thermally quenched to the nearby lower $^4S_{3/2}$ level. Therefore, the light emission from $^2H_{11/2}$ level was strongly suppressed, while the emission from $^4S_{3/2}$ level increases remarkably.

Besides thermal quenching, the enhanced light emission is also due to the lower symmetry around the Er^{3+} ion resulting from the phase transitions. Thus, Er^{3+} ion can be used as a spectroscopic probe for the occurrence of ferroelectric phase transition. Besides Er^{3+} ion, Pr^{3+} is another most utilized lanthanide dopant ion for many applications. Shen *et al.* have used Pr^{3+} ions to probe the polarization and phase transition of $(Ba_{0.77}Ca_{0.23})TiO_3$. After poling, red light emission from the $^1D_2 \rightarrow {}^3H_4$ transition of Pr^{3+} ion is enhanced around 30%. The variation of the crystal field during poling leads to the enhancement of the red emission. Both the red and blue emissions of Pr^{3+} ion show the peaks at around 100K, which is coincided with the phase transition temperature (from orthorhombic to tetragonal) of $(Ba_{0.77}Ca_{0.23})TiO_3$ ceramic. Lanthanide ions can be used to probe the ferroelectric behavior. Conversely, ferroelectric polarization can also modulate the luminescent properties of lanthanide-doped ferroelectrics. Chu *et al.* have studied the relationship between PL spectra of Er^{3+}-doped KNO with the poling electric field.[34] The polarization of KNO effectively enhanced the fine stark splitting components of $^2H_{11/2}/^4S_{3/2} \rightarrow {}^4I_{15/2}$ transitions of Er^{3+} ions.

Our group is the first to realize dynamic modulation of upconversion PL via an electric field in epitaxial Yb^{3+}/Er^{3+}-doped BTO thin films (Figure 5.8(a)).[35] BTO:Yb/Er film was deposited on conductive SRO-coated STO as in typical parallel-plate capacitor structure. Top transparent electrode of 200 nm-thick indium tin oxide (ITO) deposited at about 250°C was utilized, so that the excitation and emission light can pass through the ITO layer. Compared to bulk materials, the use of thin film of BTO:Yb/Er here would facilitate applying a higher electric field to the sample

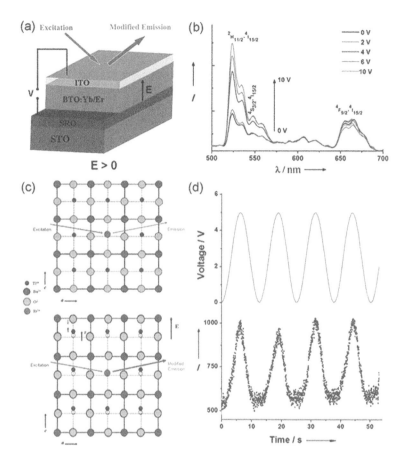

FIGURE 5.8 (a) The smart phosphor that capable to modulated the emission properties (upconversion emission) under an electric field. (b) The upconversion emission spectra of the BTO:Yb/Er smart phosphor thin film under dc bias voltages ranging from 0 to 10 V. (c) The tetragonal lattice of Er-doped BTO smart phosphor and the mechanism of the subsequent modulation of emission without and with an external electric field. (d) PL emission at 523 nm as a function of time while the sinusoidal ac electric field is applied to the BTO:Yb/Er smart phosphor thin film. (Hao, J., Zhang, Y., and Wei, X., *Angew. Chem.* 50, 6876, 2011. With permission.)

under a low bias voltage. The overall green emission from the $^{2}H_{11/2}/^{4}S_{3/2} \rightarrow {}^{4}I_{15/2}$ transitions of Er^{3+} ion is amplified 2.7 times with a bias of only 10 V as shown in Figure 5.8(b). While, external electric field has hardly any influence on the red emission from the $^{4}F_{9/2} \rightarrow {}^{4}I_{15/2}$ transition. Numerous studies indicated that the host materials' crystal symmetry could significantly influence the PL of the dopant ions indeed. The observed results are mainly ascribed to the unique crystal structure of ferroelectric materials, allowing it to be responsive to an external electric field. In a sense, the change of symmetry-related variants in a single ferroelectric host by an external electric field is in analogy to the variation of symmetry in different hosts made by the conventional chemical route. Specifically, the prototypical perovskite ferroelectric BTO used in this work has a ferroelectric tetragonal structure with the point group 4mm at room temperature, and Ti is shifted related to the negatively charged oxygens, producing a polarization. BTO is non-centrosymmetric in the tetragonal phase. Our previous study showed that Er^{3+} ions can substitute Ti^{4+} ions; hence, Er^{3+} ions in the lattice are non-centrosymmetric, breaking inversion symmetry and mixing states of opposite parity even in the absence of electric field. Therefore, the perturbation caused by the odd-order term of the crystal field forces the parity forbidden 4f–4f electric dipole transitions allowed. When applying an electric field along the direction of spontaneous polarization, the c-axis of the lattice elongates and promotes the structural asymmetry of the host materials, approaching lower symmetry around Er^{3+} ions (Figure 5.8(c)). Typically, the lower symmetry at the site of lanthanide ions, the more uneven crystal field components can mix opposite-parity into $4f$ configurational levels and subsequently increase the transition probabilities of the dopant ions. So we suggest the observed modification of PL emission in BTO:Yb/Er thin film is attributed to an increase of distortion of Er^{3+} site symmetry caused by external electric field applying on the host material. The wavelength-dependent enhancement can be ascribed to the fact that the green emission originated from a hypersensitive transition of Er^{3+} ion is more sensitive to the local field changes. Considering the ferroelectrics' structural symmetry is switchable, it offers an opportunity to tune the light emission reversibly and dynamically. Figure 5.8(d) shows the synchronous control of the PL intensity with a sinusoidal voltage. It clearly demonstrated a real time enhancement and modulation of upconversion PL can be realized by applying relatively low voltage to Yb/Er co-doped BTO thin films. Our results will aid further investigation of luminescence and widespread applications because of additional degree of freedom in the design of smart phosphor. In addition, it should be pointed out that current upconversion phosphors are still limited by poor upconversion efficiencies. The lanthanide-based upconversion materials depends on the occurrence of radiative transitions (both absorptive and emissive), which are nevertheless theoretically parity forbidden. The conventional method to address this issue is to distort the host lattice symmetry synthetically. Herein, an electric field–enhanced upconversion emission was demonstrated. In principle, we have achieved significant enhancement of upconversion quantum yield electrically.

The electric field not only introduces polarization switching and crystal deformation in ferroelectric material but also gives rise to phase transitions. The electric field–induced phase transition is irreversible and can put a dominant effect on the PL emissions. Kwok et al. have investigated the electric field–induced modulation of PL in Pr^{3+}-doped $Ba_{0.85}Ca_{0.15}Ti_{0.90}Zr_{0.10}O_3$ (BCTZ:Pr) ceramic (Figure 5.9(a)).[36] The PL modulation in BCTZ:Pr was ascribed to the electric field–induced polarization switching, as well as the tetragonal-to-rhombohedral transformation. After the phase transition is completed, the PL tuning is mainly controlled by reversing the polarization. Nonvolatile characteristic of ferroelectrics is the basis of the ferroelectric random access memory (FeRAM). Reversible and nonvolatile tuning of PL by an electric field is high desirable for developing optical read-out ferroelectric memory. It has been reported that PMN-PT possesses a metastable polarization state, where the polarization vectors point to the lateral direction. Inspired by this, they have realized reversible and nonvolatile tuning of PL from BCTZ:Pr thin-film combined with PMN-PT single crystal.[37] Figure 5.9(b) shows the electrically induced lateral strain in the PMN-PT (1 1 1) under different electric fields. When the electric field is larger than the coercive field, asymmetrical butterfly-like strain curve can be observed due to the vertical polarization (P_r^+) 180° switching. If the electric field is smaller than the coercive field, a hysteretic response of the lateral strain is found arising from non-180° ferroelastic polarization switching. A stable lateral polarization ($P_r^{\parallel}$) can be obtained. Herein, the lateral strain of the PMNPT single crystal was measured by a strain gauge as shown in Figure 5.9(c). As shown in Figure 5.9(d), the reversible and nonvolatile modulation of PL was achieved by imposing a series of electric pulses on the PMN-PT. Two kinds of reversible and nonvolatile residual strains can be obtained via tuning the ferroelectric domains between the vertical (upward or downward) and lateral directions by external electric fields (Figure 5.9(e)). Based on these strain states of the PMN-PT substrate, the reversible and nonvolatile tuning of PL has achieved in the BCTZ:Pr thin films.

5.5.4 MIL

Luminescence from phosphor material directly excited by magnetic field have manifested merely limited report, though, it can be argue that magnetic field induced luminescence is extremely difficult to implement in a single compound. It is worthy to notice that the availability of various types of laminate structures with great flexibility of the design makes it possible to tailor their properties

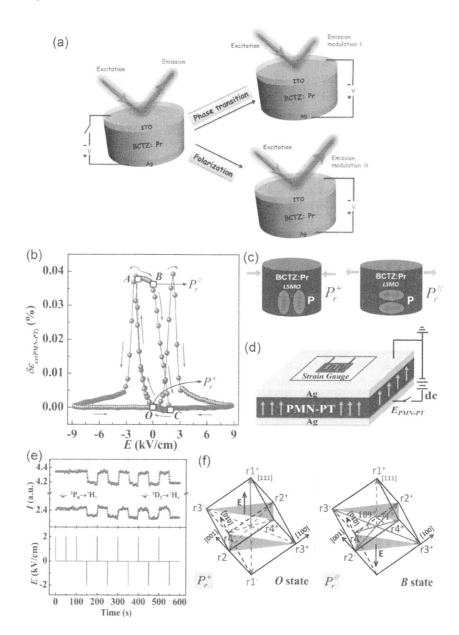

FIGURE 5.9 (a) Schematic illustration for electric field–controllable PL emissions of the BCTZ:Pr ceramic-type smart phosphor. (b) The lateral strain in the PMN-PT (1 1 1) under different electric fields sweeping. The green and red lines illustrate the sweeping fields larger and smaller than the coercive field of PMN-PT. (c) Schematic illustrations of two ferroelastic strain states. (d) The schematic for measuring the lateral strain–induced by the PMN-PT. (e) Nonvolatile PL modulation by the electric pulses. (f) Schematic illustrations for different polarization states (P_r^+ and $P_r^{//}$) of the PMN-PT.(Sun, H., Wu, X., Chung, T. H., Kwok, K. W., *Sci. Rep.* 6, 28677, 2016. With permission; Zheng, M., Sun, H., Chan, M-K., and Kwok, K. W., *Nano Energy* 55, 22, 2019. With permission.)

via interfacial coupling. Hence, by considering the direct magnetic-luminescence coupling is exceptionally difficult to realize in a single-phased material, the development of composite laminates may be an alternative solution to tackle the major problem at this stage, particularly a practical approach based on composite laminates should be favorable for various magnetic applications. In the real-world, magnetic fields exist in numerous systems, and, therefore, the detection of magnetic field is essential for environmental surveillance, mineral exploring and safety monitoring. For example, the large magnetic flux generated from a grid-connected power wire can be used to monitor power consumption of electric appliances in power industry. It is known that conventional magnetic sensing materials and devices are restricted to employing the conversion from magnetic field input to electric signal output, such as Hall, magnetoresistive and ME effects, as well as recently developed nanogenerator. These sensor nodes are usually mounted on the power wire of each appliance. Nonetheless, it is essential to explore more approach beyond the ME process. It would be very attractive if MIL-based laminate materials and devices are conceived, which are capable of showing the ability to respond and harvesting energy from magnetic fields. In contrast to conventional magnetic sensors, the realization of MIL-based smart phosphor will enjoy competitive advantages, including real-time visualization, remote sensing without making electric contact, and the realization of nondestructive and noninvasive detection.

In order to realize a direct magnetic field–induced luminescence (MIL), in which light emission can be remotely triggered by a magnetic-field with relatively weak field strength via strain-mediated coupling with piezophotonic phosphors to realize MIL.[21] we proposed a multiphase composite composed of metal ion–doped wurtzite ZnS phosphor mixed in polymer matrix forming the phosphor phase, while superparamagnetic Fe–Co–Ni alloy was used in the magnetic phase as a magnetic field sensitizer and actuator. These two phases were interlinked via the polymer matrix such that the strain induced by the actuation of the magnetic phase can be coupled into the phosphor phase and provoked magnetic-induced luminescence. To observe and maximize MIL, the design of the three components, polymer, magnetic particle and phosphor, should be optimized to reach a compromise between elastic module, magnetic permeability, luminescence and other requirements in the hybrid structure. Figure 5.10(a) shows the SEM image of the microstructure of the resultant composite. The MIL smart phosphor shows no significant hysteresis because of negligible magnetic remnant and coercive field ($H_c \sim 9$ Oe) when applying magnetic field within the range of several kOe.

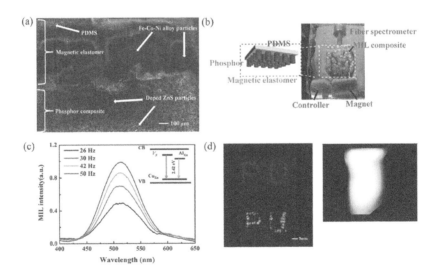

FIGURE 5.10 (a) SEM image demonstrating the interface of the fabricated multiphased laminated MIL composite consisted of superparamagnetic phase and piezoluminescence phase smart phosphor. (b) The simple setup for realizing room temperature MIL (c) The emission spectra of MIL from the multiphased smart phosphor activated by an alternative magnetic field with a frequency of 26 to 50 Hz with fixed field strength. The inset demonstrating the corresponding energy transition inside the piezoluminescence phosphor for the emission (d) Left: Photograph of logo the MIL smart phosphor illuminating 'PU' representing the abbreviation for the Hong Kong Polytechnic University driven by low magnetic field without (upper) and with magnetic field (lower). Right: White color emission of MIL light source under magnetic field. (Wong, M. C., Chen, L., Tsang, M. K., Zhang, Y., and Hao, J., *Adv. Mater.* 27, 4488, 2015. With permission.)

The magnetic phase of the smart phosphor under an alternative magnetic field will stretch along the axis of the cylindrical sample. Both strain and stress increase monotonically when the dc magnetic field increases from 2 to 3.5 kOe. Figure 5.10(b) shows the setup for measuring the MIL. A sinusoidal magnetic field strength of root means square magnitude as low as 1.4 kOe with a frequency 15 Hz are primed to trigger the MIL from our smart phosphor during a typical operation. The obtained MIL emission profiles of Al, Cu-doped ZnS and Al, Cu, Mn-doped ZnS phosphors are similar to the previously observed PL, EL and ML spectra from the corresponding phosphors (Figure 5.10(c)). The dominant peak at 509 nm can be explained from the donor-acceptor (D-A) recombination of $Al_{Zn} \rightarrow Cu_{Zn}$. The phosphor phases of metal ion–doped ZnS used in this work possess a wurtzite-type asymmetric structure and inherently exhibit piezoelectric characteristics. The strain generated by the magnetic phase gives rise to an inner piezoelectric potential in the metal ion–doped ZnS particles situated in the phosphor phase. It is noted that the piezoelectric potential created by the strain in the metal ion–doped ZnS has a polarity. Since alternating magnetic fields are employed, the structure should experience tensile and compressive strains in one cycle, which would induce piezoelectric potential with opposite polarity in the metal ion–doped ZnS. Such effect may result in different response of luminescence at the rising and falling edges. The MIL intensity peaked at 509 nm from Al, Cu-doped ZnS can be tuned by changing the magnetic field's frequency f and the root mean square field strength H_{rms}. It is noticeable that the MIL intensity increases linearly when increasing either f or H_{rms}^2, which is consistent with the proposed MIL model as earlier. This phenomenon can be explained by the fact that the increases in magnetic field strength enhance the magnitude of the induced strain. Hence, more charge carriers in the phosphor are detrapped, resulting in the increased MIL intensity. Notably, the linear relationship of MIL intensity versus f and H_{rms}^2 should benefit to developing novel MIL based magnetic sensors.

The transient characteristic of MIL from this type of smart phosphor is also investigated. In this measurement, alternative magnetic field with the frequency of 50 Hz was initially switched on and retained at $H_{rms} = 3.5$ kOe for 30 s and subsequently switched off. According to the fitting curves, the rise time (τ_r) and the fall time (τ_f) are 2.85 and 3.15 s, respectively. It should be noticed the rise time serves as the response time of the MIL phosphor composite in the device applications. Moreover, the figure-of merits of the novel type of device will highly depend on the constant of coupling strain to luminescence (α), magnetostrictive coefficient (λ), coupling factor between the magnetic and phosphor phases (β) based on our proposed MIL model.

To show the MIL's potential applications, we have demonstrated several MIL smart phosphor devices such as display and white lighting, as shown in Figure 5.10(d). Furthermore, the ability of this intriguing emission triggering properties rendered a spectroscopic tuning methodology to demonstrated the practicality of MIL smart phosphor.

Undoubtedly, the traditional methods for modulating the emission color from phosphor rely heavily on the chemical routes, which provide limited opportunity for *in-situ* and on-demand control over the emission color once they were formulated and fabricated. Novel MIL phenomenon has been observed from the phosphor composites via strain-mediated coupling. Later, remote and temporal tuning of MIL and color gamut via modulating magnetic field was also demonstrated.[38] Metal ion-doped ZnS has led to implementing a platform for achieving multicolor emissions, which features ladder-like arranged energy levels, producing a set of emission wavelengths. The magnetostrictive strain triggers the adjacent piezophosphor to produce light emission (Figure 5.11(a)). When the excitation frequencies are set at no higher than 50 Hz, the composite produces a dominant green emission band peaked at 503 nm. Intriguingly, when increasing the frequency of the magnetic field while fixing the magnetic field strength, the emission shows a trend of an intensity increase in the shoulder peak at 472 nm of blue emission band, while the intensity of green emission at 503 nm exhibits a gradual suppression simultaneously as shown in Figure 5.11(b). Remarkably, as the modulation frequency further increases, the suppression of emission intensity at 503 nm continues, while the blue emission intensity continuously increases until the main spectrum peaked at 472 nm is apparent when $f = 470$ Hz. Figure 5.11(c) shows the integrated emission intensities as a function

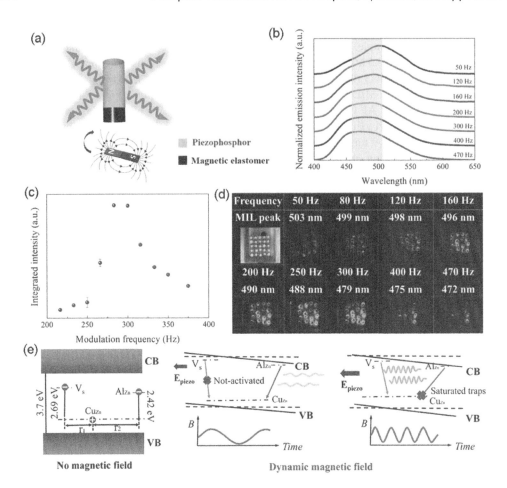

FIGURE 5.11 (a) Schematic illustration of the multiphase laminated MIL composite consisted of superparamagnetic elastomer (magnetic phase) and piezophosphor (piezoluminescence phase) smart phosphor, excited and modulated by an alternative magnetic field. (b) Normalized emission intensity of the smart phosphor composite under various frequencies, showing the shift in emission peak wavelength. (c) Integrated MIL emission intensity as a function of the frequency of magnetic field at fixed strength (2.0 kOe). (d) Photographs of the smart phosphor showing from green to blue color variation by changing the frequency of the magnetic field. (e) The mechanism of color modulation of MIL smart phosphor by changing magnetic field frequency from the smart phosphor. (Wong, M. C., Chen, L., Bai, G., Huang, L. B., and Hao, J., *Adv. Mater.* 29, 1701945, 2017. With permission.)

of f when the strength of the applied magnetic field was fixed. Therefore, we achieved a dynamic and *in-situ* control over the piezophotonic emission wavelength by merely adjusting the magnetic field excitation frequency. Figure 5.11(d) illustrates the photographs presenting the emission color varied from green to blue and intensity variation when f was modulated from 50 to 470 Hz. Figure 5.10(e) shows the proposed temporal tuning mechanism of MIL from ZnS:Al, Cu piezophosphor. The underlying mechanism can be briefly described as the follows. When Al and Cu ions are doped into ZnS host lattice, non-equivalent substitutions take place along with introducing various kinds of defects. Several kinds of luminescent centers stem from these defects, which can give rise to different color emission. Among them, sulfide vacancies (V_S) are generally attributed to the lattice mismatch in the host crystal, forming a shallow donor state below the CB. Meanwhile, Al^{3+} related defects Al_{Zn} create deeper donor levels under the CB which act as trapping centers for electrons. Cu^+-related defects Cu_{Zn} carry one negative charge acting as the acceptor levels, which can trap holes above the VB. These charge traps in different energy levels are bounded together by Coulomb interaction to form luminescent centers in the form of donor-acceptor (D-A) pairs with discrete spatial intra-pair separation (r). The relationship between the emitted photon energy and the intra-pair separation can be expressed as

$$E(r) = E_g - (E_A + E_D) + \frac{q^2}{4\pi\varepsilon_r r} \tag{5.18}$$

where E_g is the energy band gap, E_A and E_D represent the trap depth of acceptor and donor, respectively; q is the electron charge, and ε_r is dielectric constant of ZnS phosphor.

When a magnetic field (H) is applied to the composites, the induced magnetostrictive strain (ε) will generate an inner piezoelectric potential in ZnS. This piezoelectric potential induces an increase of electrostatic energy of electron in a charge trap, which can be expressed as,

$$\Delta E = \frac{1}{k}\left(\frac{\lambda q H d_0}{\varepsilon_r}\right)^2 \tag{5.19}$$

where d_0 is the local piezoelectric constant, k is the force constant of host lattice. Since piezophotonic emission is essentially a dynamic process, the relationship between the change rate of the electrostatic energy and the applied magnetic field excitation can be written as

$$\frac{dE}{dt} = \frac{2H\dot{H}}{k}\left(\frac{\lambda q d_0}{\varepsilon_r}\right)^2 \tag{5.20}$$

where $\dot{H} = dH/dt$ is the change rate of magnetic field which is related to the strain rate of the sample. When the induced strain sustains for a time t, the instantaneous trap depth E_D' of the charge trap would decrease to,

$$E_D' = E_D - \frac{2H\dot{H}}{k}\left(\frac{\lambda q d_0}{\varepsilon_r}\right)^2 t \tag{5.21}$$

From Equations (5.18) and (5.21), it is conceivable that the binding energy of trapped electron decreases with increasing modulation frequency of magnetic excitation.

Hence, the temporal tuning mechanism is proposed as follows. When applying an alternating magnetic field to the composites, magnetostrictive strain–induced piezopotential causes tilting band structure of ZnS and the detrapping of electrons in the donor states takes place. At low magnetic field modulation frequency (middle panel of Figure 5.11(e)), the deep charge traps Al_{Zn} sites are excited, leading to a green emission. As the frequency increases further (right panel of Figure 5.11(e)), the trap depths of these donor states started to decrease according to Equation (5.21). Therefore, the shallower charge traps V_s could be triggered and ionized even the magnitude of magnetic field remains constant, leading to an increase in the emission intensity at 472 nm. In addition, a small r between the V_s–Cu_{Zn} pair leads to a large overlapping of the pair's wavefunction, resulting a fast decay time and high e^-–h^+ recombination rate. In the meantime, the Al_{Zn}–Cu_{Zn} pairs with a larger r value will be saturated readily due to its small overlapping of wavefunction under high frequency of excitation modulation. Therefore, the green emission intensity saturates and then declines as the excitation frequency increases, while blue emission will subsequently dominate in the luminescence spectrum. These new findings of the luminescent materials with the ability to be accessed and modulated temporally will offer opportunities for applications in the fields of magnetic optical sensing, piezophotonics, energy harvester, nondestructive environmental surveillance, novel light sources and displays.

5.5.5 INORGANIC LANTHANIDE-DOPED PHOTOCHROMICS

Photochromism refers to a reversible photo-induced transformation of a chemical species between different forms arises from the diverse absorption spectra of the species. Photochromic materials enable reversible manipulation and possess erasable/rewritable capability, ensuring them to be adequate candidates as optical memories. Many organic photochromic molecules, including spiropyrans and diarylethenes, have been considered for optical memory applications. However, few of them can meet the stringent thermo- and photostable requirements. Despite being effective photoswitching, organic materials demonstrate photochemical fatigue upon many photo-irradiation cycles. Compared with organic photochromics, inorganic photochromic materials display superior thermal and chemical stability. Of the inorganic photochromic materials studied so far, TM oxides such as WO_3 have been extensively investigated for smart windows and ophthalmic glasses. However, intrinsic WO_3 also faces some challenges, including poor reversibility, low fatigability and selection of irradiation light, which definitely hamper its application for optical memory. In recent years, some kinds of lanthanide-doped bismuth layer-structure ferroelectrics exhibit superior photochromic performance. During high-temperature sintering process, the volatilization of sodium and potassium elements introduces consequent defects (e.g., Na, K and O vacancies), which can trap electrons or holes resulting in color centers upon light irradiation. Among various readout manners, optical readout using the luminescence contrast has been regarded the most convenient and efficient way because of its high sensitivity, resolution, contrast and fast response. Therefore, lanthanide-doped photochromics promises great potential for photomemories owing to the added luminescence switching provided by the intrinsic photochromism coupled with the lanthanide luminescence. If the absorption bands of color centers overlap the emission bands of dopant lanthanide ions, the luminescence emission can be consciously modulated through energy migration between them. So far, some researchers have proposed several kinds of lanthanide-doped photochromic materials endowed with dramatically luminescence switching contrast. Zhang et al. have reported that reversible luminescence photoswitching upon photochromic action in lanthanide-doped ferroelectric material, $Na_{0.5}Bi_{2.5}Nb_2O_9$:Pr^{3+} (NBN:Pr).[22] Upon exposure to visible light or sunlight, the material exhibited a reversible color change between green to dark grey by alternating visible light irradiation and thermal stimulus. More importantly, the PL intensity from Pr^{3+} ion can be effectively tuned, and the emission intensity has no significant degradation after several periods. Later, they utilized Er^{3+} ion to replace Pr^{3+} as the activator ion.[23] The purpose of doping Er^{3+} ions into photochromic NBN film is to realize a nondestructive optical readout. Er^{3+} ions can generate upconversion emission under near-infrared 980 nm laser irradiation, eliminating the interference between the optical write-in (407 nm) and readout (980 nm) operations. In undoped NBN sample, there is no upconversion emission from Er^{3+} ion under 980 nm laser excitation. The bandgap of undoped NBN film is about 3.2 eV (corresponding to 387.5 nm), which is much shorter than the readout wavelength (980 nm). Because of the inability of triggering photochromism through the incident near-infrared 980 nm photon, the undesired interference between read and write operations can eliminated.

Lanthanide-doped inorganic photochromic ceramics face several problems and plights, such as low storage density, less portable and processing inconvenience, not to mention integration with wearable electronics. Memories incorporating lanthanide-doped photochromic ferroelectric thin films as the central active element might be an appealing option to overcome abovementioned problems. Generally, oxide ceramics are brittle under mechanical deformation. Nevertheless, their thin-film counterparts can tolerate much larger biaxial strain up to ~ 6%, which promises significantly improvement on the design freedom in order to realize flexible optical memory. Recent breakthroughs in growing high-quality complex oxides films on flexible substrates have offered great opportunities to fabricate photochromic material-based flexible photomemories. We have demonstrated a flexible and rewritable nonvolatile photomemory, in which inorganic photochromic material Er^{3+}-doped NBN (NBN:Er) was chosen as the active component. NBN:Er thin films were fabricated using PLD under moderate temperature on a flexible PI substrate.[39] Figure 5.12(a)

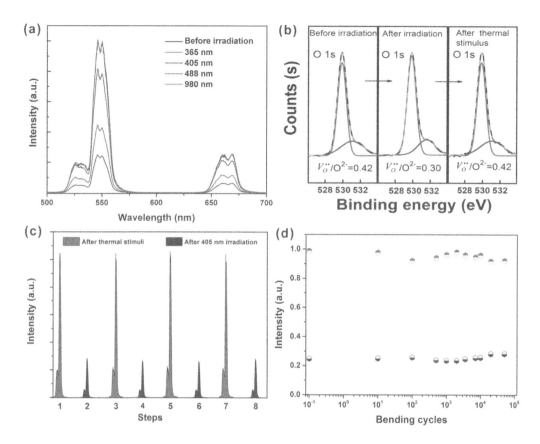

FIGURE 5.12 (a) PL spectra of the NBN:Er smart phosphor film under different wavelengths light irradiation. (b) The O 1s XPS spectra in three states: before irradiation, after irradiation and after the thermal stimulus. (c) Upconversion spectra of the NBN:Er film upon alternating 405 nm light irradiation and thermal stimuli. (d) Two data phases endurance characteristics of the NBN:Er/PI film after 10^5 bending cycles. (Chen, H., Dong, Z., Chen, W., Sun, L., Du, X., Zhao, Y., Chen, P., Wu, Z., Liu,. W., and Zhang, Y., *Adv. Opt. Mater.* 8, 1902125, 2020. With permission.)

shows the optical switching behavior of the UC luminescence from NBN:Er/PI structures illuminated with different wavelength light irradiations (365, 405, 488 and 980 nm) for 30 s. The photochromic effect and PL modulation under photostimulation are closely related to the volatilization of Bi and Na elements. Accordingly, some oxygen vacancies ($V_O^{\cdot\cdot}$) were also formed as charge compensation in the NBN:Er film. Defects V_{Na}', V_{Bi}''' and $V_O^{\cdot\cdot}$ introduce defect levels within the forbidden gap. Under photo-stimulation, the electrons in the VB can jump into to the defect levels or the conductive band and then trapped by $V_O^{\cdot\cdot}$. The left holes are trapped by vacancies of A-sites (mainly containing V_{Na}' and V_{Bi}''') (Figure 5.12(b)). Through energy transfer between Er^{3+} and NBN host, the overlapping between the absorption band of the color centers and the doped Er^{3+} ion's emission bands gives rise to the PL modulations. As shown in Figure 5.12(c), the upconversion spectra assigned to the $^2H_{11/2}/^4S_{3/2} \rightarrow {}^4I_{15/2}$ transitions suffer PL intensities switching, while their profiles and peak positions remain almost unchanged during repeated write/erase operations. The NBN:Er thin film on PI exhibits reversible, high contrast photoswitching characteristics upon photo-stimulation. Such memories show comprehensive robust characteristics that sustain well up to 10^5 bending cycles and maintain stability over many write-read-erase cycles (Figure 5.12(d)). The proof-of-concept flexible photomemory smart phosphor simultaneously enables thermal and photostability, nondestructive readout capability and fatigue resistance during cyclic write and erase processes.

5.6 CONCLUSION AND PERSPECTIVE

In this chapter, we presented the definition, the constitution of smart phosphors and the diverse mechanisms behind them. Smart phosphors provide an ideal platform for exploring novel luminescence phenomena and fundamental study on the interaction between smart materials and luminescence. Some of the examples presented earlier indicate that the dynamical process of the luminescence responses to the structural changes can be deduced while isolating from other interference factors. Smart phosphors provide unusual routes to control and modulate the luminescence in *in-situ*, dynamical and reversible manners. The coupling between smart materials properties to phosphors promises unprecedented degree of freedom in designing novel optoelectronic and photonic devices, with enhanced performance or even new functions, primed for the era of intelligence.

Despite these encouraging progress, there are still many issues to be addressed. Development in this field needs a further understanding of fundamental physics. Some mentioned interactions between ferroelectric and optical processes are still open questions. For instance, the explanation of modulation of the Ln^{3+}-doped ferroelectrics is based on the semi-empirical J-O theory. Theoretical model deduced using first-principles density-functional theory should be created for Ln^{3+}-doped ferroelectrics. Moreover, the fundamental issues of stimuli responses of this luminescence are rarely addressed, resulting in the formulation and development of new smart phosphor countered a significant difficulty. Secondly, the improvement associated with the coupling efficiency, response speed and durability are still in urgent needs of the investigation. Regarding polymer phosphor composites, it is promising to incorporate highly durable and more smart material properties such as self-healing and shape memory already used in other research fields[40–42] Moreover, it can be noticed that since the discovery of unique luminescent properties in various 2D materials.[43–47] Distinctive properties can be induced by interlayered coupling, novel physical properties have been generated and reported by the interlayer coupling. Thereby, this interlayer coupling factor can be further exploited for enhancing the coupling efficiency in smart phosphor. Thirdly, the brittle nature of inorganic smart materials is unsuitable for use in flexible and transparent optoelectronics. Considering their superior flexibility and mechanical strength, recent observed piezoelectricity and ferroelectricity in 2D materials are expected to become increasingly important in developing piezo-phototronic devices. Remarkable spectroscopic shifts on PL emission have been realized by various piezoelectric substrates, providing new and exciting opportunities to study the physical mechanism of piezoelectric properties of these vdW heterostructures.[48–51] Hybrid perovskites have risen up as a class of important optoelectronic materials due to their exceptional attributes, especially confirming these novel perovskites possessing piezoelectricity and ferroelectricity properties. The recent breakthrough on the integration of millimeter scale perovskites thin film with other functional material provides valuable opportunities for realizing hybrid multilayer optoelectronic system and devices.[52] These emerging materials offer dramatically greater degree of freedom to develop novel smart phosphors. It is anticipated that a convergence of smart materials coupled with emerging phosphors will lead to exciting and unprecedented smart phosphors.

ACKNOWLEDGMENTS

This work was supported by the PolyU Postdoctoral Fellowships Scheme (YW4N) and Research Grant Council (RGC) of Hong Kong (RGC GRF No. PolyU 153025/19P). Y. Zhang acknowledges support from NSFC under project No. 11874230.

REFERENCES

1. Zhang Y, Hao J (2013) Metal-ion doped luminescent thin films for optoelectronic applications. *J. Mater. Chem. C* 1: 5607
2. Wang F, Liu X (2008) Upconversion multicolor fine-tuning: Visible to near-infrared emission from lanthanide-doped NaYF4 nanoparticles. *J. Am. Chem. Soc.* 130: 5642

3. Sedlmeier A, Gorris HH (2015) Surface modification and characterization of photon-upconverting nanoparticles for bioanalytical applications. *Chem. Soc. Rev.* 44: 1526

4. Wang F, Deng R, Wang J, Wang Q, Han Y, Zhu H, Chen X, Liu X (2011) Tuning upconversion through energy migration in core–shell nanoparticles. *Nat. Mater.* 10: 968

5. Scott JF (2007) Multiferroic memories. *Nat. Mater.* 6: 256

6. Eerenstein W, Mathur ND, Scott JF (2006) Multiferroic and magnetoelectric materials. *Nature* 442: 759

7. Baek SH, Rzchowski MS, Aksyuk VA (2012) Giant piezoelectricity in PMN-PT thin films: Beyond PZT. *MRS Bull.* 37: 1022

8. Martin LW, Rappe AM (2016) Thin-film ferroelectric materials and their applications. *Rev. Mater.* 2: 16087

9. Suo Z (1998) Stress and strain in ferroelectrics. *Curr. Opin. Solid State Mater. Sci.* 3: 486

10. Liu J, Jiang C, Xu H (2012) Giant magnetostrictive materials. *Sci. China Technol. Sci.* 55: 1319

11. De Guzman GNA, Fang MH, Liang CH, Bao Z, Hu SF, Liu RS (2020) [INVITED] Near-infrared phosphors and their full potential: A review on practical applications and future perspectives. *J. Lumin.* 219: 116944

12. Xiang G, Xia Q, Xu S, Liu X, Jiang S, Wang Y, Zhou X, Li L, Ma L, Wang X (2021) Multipath optical thermometry realized in CaSc2O4: Yb3+/Er3+ with high sensitivity and superior resolution. *J. Am. Ceram. Soc.* 104: 2711

13. Wang F, Liu X (2009) Recent advances in the chemistry of lanthanide-doped upconversion nanocrystals. *Chem. Soc. Rev.* 38: 976

14. Wang X, Peng D, Huang B, Pan C, Wang ZL (2019) Piezophotonic effect based on mechanoluminescent materials for advanced flexible optoelectronic applications. *Nano Energy* 55: 389

15. Hao J, Xu C-N (2018) Piezophotonics: From fundamentals and materials to applications. *MRS Bull.* 43: 965

16. Wang X, Zhang H, Yu R, Dong L, Peng D, Zhang A, Zhang Y, Liu H, Pan C, Wang ZL (2015) Dynamic pressure mapping of personalized handwriting by a flexible sensor matrix based on the mechanoluminescence process. *Adv. Mater.* 27: 2324

17. Zhang JC, Wang X, Marriott G, Xu CN (2019) Trap-controlled mechanoluminescent materials. *Prog. Mater Sci.* 103: 678

18. Chandra BP, Xu CN, Yamada H, Zheng XG (2010) Luminescence induced by elastic deformation of ZnS:Mn nanoparticles. *J. Lumin.* 130: 442

19. Judd BR (1962) Optical absorption intensities of rare-earth ions. *Phys. Rev.* 127: 750

20. Weber MJ (1967) Probabilities for radiative and nonradiative decay of Er^{3+} in LaF_3 *Phys. Rev.* 157: 262

21. Wong MC, Chen L, Tsang MK, Zhang Y, Hao J (2015) Magnetic-induced luminescence from flexible composite laminates by coupling magnetic field to piezophotonic effect. *Adv. Mater.* 27: 4488

22. Zhang Q, Sun H, Wang X, Hao X, An S (2015) Reversible luminescence modulation upon photochromic reactions in rare-earth doped ferroelectric oxides by in situ photoluminescence spectroscopy. *ACS Appl. Mater. Interfaces* 7: 25289

23. Zhang Q, Zheng X, Sun H, Li W, Wang X, Hao X, An S (2016) Dual-mode luminescence modulation upon visible-light-driven photochromism with high contrast for inorganic luminescence ferroelectrics. *ACS Appl. Mater. Interfaces* 8: 4789

24. Zhang Y, Luo L, Li K, Li W, Hou Y (2018) Reversible up-conversion luminescence modulation based on UV-VIS light-controlled photochromism in Er^{3+} doped Sr_2SnO_4. *J. Mater. Chem. C* 6: 13148

25. Wu Z, Zhang Y, Bai G, Tang W, Gao J, Hao J (2014) Effect of biaxial strain induced by piezoelectric PMN-PT on the upconversion photoluminescence of $BaTiO_3$:Yb/Er thin films. *Opt. Express* 22: 29014

26. Zhang Y, Hao J, Mak CL, Wei X (2011) Effects of site substitutions and concentration on upconversion luminescence of Er^{3+}-doped perovskite titanate. *Opt. Express* 19: 1824

27. Bai G, Zhang Y, Hao J (2014) Tuning of near-infrared luminescence of $SrTiO_3$:Ni^{2+} thin films grown on piezoelectric PMN-PT via strain engineering. *Sci. Rep.* 4: 5724

28. Wang X, Xu CN, Yamada H, Nishikubo K, Zheng XG (2005) Electro-mechano-optical conversions in Pr^{3+}-doped $BaTiO_3$–$CaTiO_3$ ceramics. *Adv. Mater.* 17: 1254

29. Tu D, Xu CN, Yoshida A, Fujihala M, Hirotsu J, Zheng XG (2017) $LiNbO_3$:Pr^{3+}: A Multipiezo material with simultaneous piezoelectricity and sensitive piezoluminescence. *Adv. Mater.* 29: 1606914

30. Zhang Y, Gao G, Chan HL, Dai J, Wang Y, Hao J (2012) Piezo-phototronic effect-induced dual-mode light and ultrasound emissions from ZnS:Mn/PMN-PT thin-film structures. *Adv. Mater.* 24: 1729

31. Chen Y, Zhang Y, Karnaushenko D, Chen L, Hao J, Ding F, Schmidt OG (2017) Addressable and color-tunable piezophotonic light-emitting stripes. *Adv. Mater.* 29: 1605165

32. Jeong SM, Song S, Lee SK, Ha NY (2013) Color manipulation of mechanoluminescence from stress-activated composite films. *Adv. Mater.* 25: 6194

33. Chen L, Wong MC, Bai G, Jie W, Hao J (2015) White and green light emissions of flexible polymer composites under electric field and multiple strains. *Nano Energy* 14: 372

34. Chu SY, Wen CH, Tyan SL, Lin YG, Juang YD, Wen CK (2004) Polarization tuning the Stokes photoluminescence spectra of erbium doped $KNbO_3$ ceramics. *J. Appl. Phys.* 96: 2552

35. Hao J, Zhang Y, Wei X (2011) Electric-induced enhancement and modulation of upconversion photoluminescence in epitaxial $BaTiO_3$:Yb/Er thin films. *Angew. Chem.* 50: 6876

36. Sun HL, Wu X, Chung TH, Kwok KW (2016) In-situ electric field-induced modulation of photoluminescence in Pr-doped $Ba_{0.85}Ca_{0.15}Ti_{0.90}Zr_{0.10}O_3$ lead-free ceramics. *Sci. Rep.* 6: 28677

37. Zheng M, Sun H, Chan MK, Kwok K (2019) Reversible and nonvolatile tuning of photoluminescence response by electric field for reconfigurable luminescent memory devices. *Nano Energy* 55: 22

38. Wong MC, Chen L, Bai G, Huang L-B, Hao J (2017) Temporal and remote tuning of piezophotonic-effect-induced luminescence and color gamut via modulating magnetic field. *Adv. Mater.* 29: 1701945

39. Chen H, Dong Z, Chen W, Sun L, Du X, Zhao Y, Chen P, Wu Z, Liu W, Zhang Y (2020) Flexible and rewritable non-volatile photomemory based on inorganic lanthanide-doped photochromic thin films. *Adv. Opt. Mater.* 8: 1902125

40. Xu W, Huang LB, Hao J (2017) Fully self-healing and shape-tailorable triboelectric nanogenerators based on healable polymer and magnetic-assisted electrode. *Nano Energy* 40: 399

41. Xu W, Wong MC, Hao J (2019) Strategies and progress on improving robustness and reliability of triboelectric nanogenerators. *Nano Energy* 55: 203

42. Xu W, Wong MC, Guo Q, Jia T, Hao J (2019) Healable and shape-memory dual functional polymers for reliable and multipurpose mechanical energy harvesting devices. *J. Mater. Chem. A* 7: 16267

43. Jie W, Yang Z, Bai G, Hao J (2018) Luminescence in 2D materials and van der Waals heterostructures. *Adv. Opt. Mater.* 6: 1701296

44. Bai G, Yuan S, Zhao Y, Yang Z, Choi SY, Chai Y, Yu SF, Lau SP, Hao J (2016) 2D layered materials of rare-earth Er-doped MoS_2 with NIR-to-NIR down- and up-conversion photoluminescence. *Adv. Mater.* 28: 7472

45. Lyu Y, Wu Z, Io WF, Hao J (2019) Observation and theoretical analysis of near-infrared luminescence from CVD grown lanthanide Er doped monolayer MoS_2 triangles. *Appl. Phys. Lett.* 115: 153105

46. Bai G, Lyu Y, Wu Z, Xu S, Hao J (2020) Lanthanide near-infrared emission and energy transfer in layered WS_2/MoS_2 heterostructure. *Sci. China Mater.* 63: 575

47. Liu Y, Bai G, Lyu Y, Hua Y, Ye R, Zhang J, Chen L, Xu S, Hao J (2020) Ultrabroadband tuning and fine structure of emission spectra in lanthanide Er-doped ZnSe nanosheets for display and temperature sensing. *ACS Nano* 14: 16003

48. Guo F, Lyu Y, Jedrzejczyk MB, Zhao Y, Io WF, Bai G, Wu W, Hao J (2020) Piezoelectric biaxial strain effects on the optical and photoluminescence spectra of 2D III–VI compound α-In_2Se_3 nanosheets. *Appl. Phys. Lett.* 116: 113101

49. Yuan S, Luo X, Chan HL, Xiao C, Dai Y, Xie M, Hao J (2019) Room-temperature ferroelectricity in $MoTe_2$ down to the atomic monolayer limit. *Nat. Commun.* 10: 1

50. Yuan S, Io WF, Mao J, Chen Y, Luo X, Hao J (2020) Enhanced piezoelectric response of layered In_2Se_3/MoS_2 nanosheet-based van der Waals heterostructures. *ACS Appl. Nano Mater.* 3: 11979

51. Io WF, Yuan S, Pang SY, Wong LW, Zhao J, Hao J (2020) Temperature- and thickness-dependence of robust out-of-plane ferroelectricity in CVD grown ultrathin van der Waals α-In_2Se_3 layers. *Nano Res.* 13: 1897

52. Ding R, Liu CK, Wu Z, Guo F, Pang SY, Wong LW, Io WF, Yuan S, Wong MC, Jedrzejczyk MB (2020) A general wet transferring approach for diffusion-facilitated space-confined grown perovskite single-crystalline optoelectronic thin films. *Nano Lett.* 20: 2747

6 Quantum Dots for Display Applications

Zhaobing Tang, Shuo Ding, Xuanyu Zhang, Chunyan Wu,
Nathan Evan Stott, Chaoyu Xiang, and Lei Qian

CONTENTS

6.1 INTRODUCTION

In 1984, Louis Brus was the first one to show the modified bandgap energy of semiconductor crystal with its radius below the bulk Bohr exciton radius.[1] After that, quantum dots (QDs) have been gaining considerable attentions among the optoelectronic industry, especially in luminous applications, due to obtain any high-purity color emission by varying the QDs' size of the same material system. Employing solution chemistry, between ~1 and ~10 nm, the growth of QDs material is finely controlled in the special solution of reactor, which leads to the excellent dispersibility in organic solvent that is the crucial step of effective cost manufacturing method. Specially, the surfaces of QDs are dominated, which are composed of cations and anions with amounts of dangling bands that are easily oxidized to generate traps, influencing significantly the efficiency and stability. To eliminate those traps, the core/shell structure is extensively adopted, promoted the commercial application in display, including LCD, microLED and so on, which also brought the enormously improvement of electroluminescence (EL) application that is comparable with commercial OLED in terms of efficiency, after optimized the device structure. Notably, QDs recently have attracted lots of attentions on scientific interest and commercialized application in the field of display.

DOI: 10.1201/9781003098676-6

In this chapter, we focus mainly on QDs material and its applications, including the brief introduction and synthetic method of QDs, and display applications that contain photoluminescence and electroluminescence, presented their recent progress of correlational researches. Finally, the future of QDs for display applications are expressed, which is still prospective.

6.2 QUANTUM CONFINEMENT AND QUANTUM DOTS

6.2.1 QUANTUM CONFINEMENT EFFECT AND QUANTUM PROPERTIES

The first thing to understand when defining QDs is the quantum confinement effect. Generally speaking, when the size of semiconductor materials is small enough (typically at nanometer levels, below the classical exciton Bohr radius of the given material), electrons will be confined by the boundary of materials and the electronic band structure will be split, or "quantized," into discrete electronic energy states, and thus, the "bandgap" or band-edge transition energy gap of quantum-confined materials is larger than that of the same bulk material. This phenomenon is an underlying result of the aforementioned quantum confinement effect. Thus, QDs may more simply be defined as nanoscale semiconductor materials that are small enough to be experiencing the quantum confinement effect. As a result of the quantum confinement effect, as shown in Figure 6.1, the band-edge energy photoemission wavelength can be tuned continuously through tuning the size of the QDs. For the same given material composition, the smaller the QD size is, the larger the band-edge energy transition will be and, thus, the bluer the photoluminescence (PL) emission wavelength will become.

Fundamentally, during the process of light emission, it is excitons that are confined under the quantum confinement effect. An exciton is a quasi-particle formed by an electron-hole pair, which

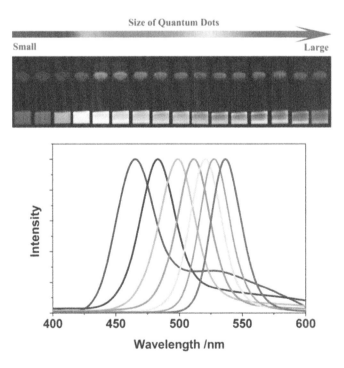

FIGURE 6.1 Top-sixteen emission colors from small (blue) to large (red) CdSe QDs excited by a near-ultraviolet lamp; size of QDs can be from ~1 to ~10 nm. Bottom-photoluminescence emission spectra of a size series of CdSe QDs. (Bera, D., Qian, L., Tseng, T. K., Holloway, P. H., *Mater.* 3, 2260, 2010. With permission.)

can be simplified down to the classical model of a negative electron revolving around a positive hole (absence of an electron), and the radius of the electron orbit of revolution (the distance between the electron and hole) is called the Bohr radius in accordance with the Bohr model proposed by Niels Bohr in 1931. Generally speaking, when the size of a certain material is comparable to or smaller than its classical exciton Bohr radius, the exciton inside will be confined, resulting in significant quantum size properties.

Apart from the tunable band-edge emission, QDs also have other unique properties. For example, since the size of QD is relatively small, the surface area-to-volume ratio of QDs is much larger than that of the same bulk material. The relatively large surface area of QDs induces deficient bonding or coordination to surface atoms, leading to a large ratio of dangling bonds at the surfaces of QD materials. As a consequence, the chemical reactivity of bare QDs is high, expanding its application into the realm of high-performance catalysts.[2] After binding with ligands, the surfaces of QDs are believed to be passivated such that the electronic energy structure as well as spin states are also changed significantly, allowing for other powerful strategies to alter and exploit the properties of QDs. Additionally, the nanoscale nature of QDs results in other interesting physical phenomena, such as the dielectric confinement effect, the coulomb blockade effect, and quantum tunneling to greatly broaden the potential applications of QDs.

6.2.2 EMISSIVE QUANTUM DOTS

Up until the present time, the most important applications of QDs are still based upon their emissive optical properties. First, QDs can have extremely high luminescent efficiency. The PL quantum yield (PLQY) of QDs can easily reach up to 100%,[3] indicating negligible non-radiation losses inside QDs of given structures under the right conditions. Second, the emissions spectra of QDs can be narrow if the distribution of sizes are tight, enabling the potential for fabrication of QD light-emitting diodes (QLEDs) with high color purity. Third, the tunable band-edge emission wavelengths of QDs greatly reduce the obstacle of full-color QLED displays. Fourth, QDs can show good stability to water and air given appropriately engineered materials structures and packaging such that the operational lifetimes of QLED devices has already exceeded thousands of hours.[4] Fifth, the synthesis of QDs has become more facile and scalable as research and development on these materials has progressed, showing a bright future for mass-production and commercialization. Sixth, and finally, QDs are solution processable. QDs can be easily dispersed into multiple kinds of solvents and media to form stable QD inks such that large-scale manufacturing processes, such as inkjet and slot-die printing, can be applied.

6.3 DEVELOPMENT OF QDs

The history of QDs dates back to 1980 when Russian physicist Ekimov discovered QDs in semiconductor crystals.[5] Later, in 1984, Brus at Bell Labs derived the relationship between size and the band-edge energies of semiconductor QD nanoparticles.[1] In 1993, Murray and Bawendi introduced high-temperature injection of organometallic precursor methods for cadmium-based (Cd-based) QD synthesis,[6] which greatly improved narrowing of the band-edge emission through tighter size-distributions and allowed for greater solution processability. Higher PLQYs, narrower band-edge emissions without the necessity of size-selective precipitation, greatly reduced production costs, and improved fabrication efficiencies were achieved by Stott and Bawendi through development of synthetic processes making use of safer, air-stable cadmium precursors and continuous flow reactor (CFR) technology.[7, 8] In the early 21st century, some companies, like QD vision and Nanosys, licensed the inventions of Stott and Bawendi to pioneer the commercialization of QD display technologies while focusing on solving the problems of mass-production and cost-control of QDs. In 2011, Qian *et al.* [9] developed an all-solution process method for QLED fabrication, greatly improving the efficiencies and lifetimes of such devices. Since then, large companies, such as Samsung and Sony, have founded their own QLED research teams, and a vast range of QLED devices and display televisions have ushered in a brand new era of QD device development.

Currently, the best performance of QDs for light emission and display applications are (core) shell structured QDs. For (core)shell QDs, the core and shell materials should have good correlations between electronic band structures (bandgaps and band offsets) and minimal lattice constant mismatch, such as in the case of CdSe core and ZnSe shell materials. Compared to bare QD cores, (core)shell structures can better passivate the surface states of the cores and reduce surface defects, suppressing non-radiative decay loss during the light emission process. Furthermore, the shell structure can confine excitons better and increase the probability of exciton recombination, thereby increases the PLQY of such QDs. Finally, the (core)shell structure is more resistant to oxidation and degradation processes, causing impressive enhancement of the stability and reliability of QDs. The synthesis of (core)shell structures will be discussed in greater detail below.

The history of QDs can be traced back to 1980 in which the three-dimensional quantum size effect was first discovered in glass by Russian physicist Ekimov.[10] However, the colloidal QDs in use today were first systematically investigated by Brus and his coworkers.[11] He established the concept of colloidal QDs and derived a model to quantitatively evaluate the correlation between particle size and the quantum confinement effect.[1, 12] The equation used to determine the "bandgap," or band-edge energy gap, of QDs is derived from this model proposed by Brus: [13]

$$E_g^* = E_g + \frac{\hbar^2 \pi^2}{2R^2} \left[\frac{1}{m_e^*} + \frac{1}{m_h^*} \right] - \frac{1.786e^2}{\varepsilon R} - 0.248 E_R^* \tag{6.1}$$

where E_g^* is the "bandgap" of the QD, E_g is the bandgap of the corresponding bulk material, $\hbar$ is the reduced Planck's Constant, R is the radius of the QD, m_e & m_h are effective mass of electrons and holes in the material, respectively, ε is dielectric constant of the bulk material, and E_R^* is the Rydberg energy (which is equal to $e^4/(2\varepsilon^2\hbar^2(m_e^{-1} + m_h^{-1}))$).[14]

6.3.1 DEVELOPMENT OF SOLUTION SYNTHESIS OF QUANTUM DOTS

With development of the organometallic synthetic method, organometallic precursors rapidly injected into high-temperature "coordinating" solvents was introduced for synthesizing CdSe QDs by Brus and Bawendi.[6, 15] Then, highly monodispersed CdSe QDs were synthesized by Bawendi in 1993,[16] which was a milestone for high-quality nanocrystal QD synthesis.[17] The concept of "focusing of size-distribution," which was proposed by Alivistos in 1998, was based on the study of the high-quality CdSe QD synthetic processes developed at the time.[18] Although it was difficult for scientists to understand the mechanisms of forming narrowly size-distributed and shape-controlled nanocrystals in the early 1990s, the stage for the rational synthesis of colloidal QDs was conceived at the end of 1990s, and this system of rational synthesis of colloidal nanocrystals has been developing ever since.[19] The evolution of the development of the system of colloidal QD synthesis can be roughly generalized by a recipe for CdSe QDs exhibiting following:

$$TOPO(solvent)TOP - Me_2Cd + TOP - Se \xrightarrow{TOPO, 300°C} CdSe \tag{6.2}$$

Tri-n-octylphosphine oxide (TOPO) is a high-boiling solvent that contains trace ligand impurities. Tri-n-octylphosphine (TOP) is a coordinating solvent for elemental selenium. Dimethylcadmium (Me$_2$Cd) is cadmium precursor. Se-TOP mixed with TOP-Me$_2$Cd was injected into TOPO solvent at 300°C. [7]

$$TOPO(solvent)HPA \text{ or } TDPA(ligands)Cd(TDPA)_2 + TOP - Se \xrightarrow{300°C} CdSe \tag{6.3}$$

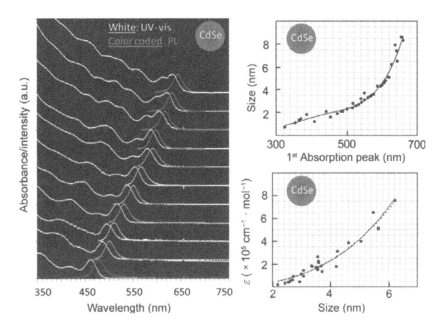

FIGURE 6.2 Absorption and photoluminescence spectrum of monodispersed, high-quality CdSe QDs of different particle sizes. (Peng, X., *Nano Res.* **2**, 425, 2009. With permission.)

CdO, Cd(Ac)₂, and CdCO₃ was used as Cd source, which was dissolved in the TOPO solvent with ligands hexylphosphonic acid (HPA) or tetradecylphosphonic acid (TDPA), 2001, Peng's recipe.[20, 21]

$$ODE(solvent)OA(ligands)Cd(OA)_2 + TOP - Se \xrightarrow{250°C} CdSe \tag{6.4}$$

Octadecene (ODE) is a noncoordinating solvent, oleic acid (OA) is ligand for dissolving CdO, Cd(Ac)₂, or CdCO₃ in ODE, 2002, Peng's recipe.[22, 23]

Nowadays, ODE is widely used as a solvent for synthesizing various kinds of QDs. Generally, ligands used for dissolving metal precursors in the field of QD synthesis include fatty acids with different chain lengths, amines, phosphines, and phosphonic acids.[23] For cadmium-based QD synthesis, the extremely toxic and expensive dimethyl cadmium (Me₂Cd) has been replaced by air stable and relatively safer cadmium sources, such as CdO, Cd(Ac)₂ Cd(acac)₂, and so on. Thus, the so-called greener method for synthesizing high-quality, monodispersed CdSe QDs has been widely used up until the present time (Figure 6.2).[18]

6.3.2 (CORE)SHELL STRUCTURED QUANTUM DOTS

Bare core CdSe QDs suffer from non-radiative recombination, resulting in lower PLQYs. However, with high-quality shells to passivate the surfaces of CdSe cores, non-radiative decay losses are suppressed, and the shell can also confine excitons to increase radiative recombination rates. As a (core)shell structure of QDs can significantly increase PLQY and the stability of QDs, the method of growing high-quality shells upon CdSe cores was proposed in 2003 (Figure 6.3).[24] The "successive ion layer adsorption and reaction" (SILAR) method was developed as a significant concept to grow highly crystalline shells in successive steps of cation addition and reaction followed by anion addition and reaction to build shells layer by layer. The performance of (core)shell QDs is affected by the quality of shell growth. Narrow size-distribution and emission line width, high PLQY, and blinking suppression Cd-based QDs were reported in 2013,[25] which significantly pushed the commercial application of QDs (Figure 6.4).

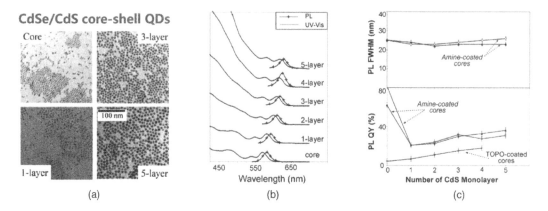

(a)   (b)   (c)

FIGURE 6.3 (CdSe)CdS (core)shell QDs, (a) TEM images of different shell layers, (b) absorption and photoluminescence spectrum, and (c) emission line width and PLQY. (Li, J. J., Wang, Y. A., Guo, W., Keay, J. C., Mishima, T. D., Johnson, M. B., Peng, X., *J. Am. Chem. Soc.* **125**, 12567, 2003. With permission.)

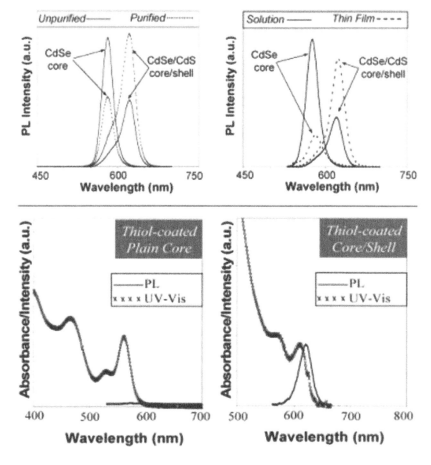

FIGURE 6.4 (CdSe)CdS (core)shell QDs (a–d) absorption and photoluminescence spectrum emission line width and PLQY. (Chen, O., Zhao, J., Chauhan, V.P., Cui, J., Wong, C., Harris, D.K., Wei, H., Han, H.S., Fukumura, D., Jain, R.K., Bawendi, M.G., *Nat. Mater.* **12**, 445, 2013. With permission.)

6.3.3 Composition Gradient Quantum Dots

In addition to the binary (core)shell structure QDs, like (CdSe)CdS, (CdSe)ZnS and (CdSe)ZnSe, composition gradient (core)shell QDs with $((CdSe)Cd_{1-x}Zn_xSe_{1-y}S_y)ZnS$ structures were also synthesized in 2008 (Figure 6.5).[26] This was achieved through adjusting the precursor reactivities between cation precursors and anion precursors. The advantages of this approach can be summarized as follows: A facile one-pot method producing QDs with PLQYs up to 80% in a wide emission range from 500 to 610 nm. The wide emission range was achieved by controlling the precursor ratios. Moreover, the lattice mismatch between core and shell was minimized by the composition gradient shells, reducing lattice strain to allow for thicker shell coatings. In 2015, RGB QLED performance was significantly increased through introducing ZnSe as an intermediate shell instead of CdS (Figure 6.6).[27] With the optimization of composition gradient Cd-based (core)shell structured QDs, stable high-efficiency QLED devices with high emission external quantum efficiencies (EQEs) were fabricated.

6.4 CADMIUM-FREE QUANTUM DOTS

Although Cd-based QDs have been greatly developed, the environmental conservation is an increasingly important topic, which requires making use of environmental QDs but not toxicant QDs. Consequently, cadmium-free QDs (CFQDs) are an attractive hotspot in the researching luminance materials. During recent decades, the most studied CFQDs have included InP, ZnSe, $CuInS_2$, and group IV semiconductors (carbon QDs and silicon QDs). With a 1.35 eV bulk bandgap and 10 nm exciton Bohr radius, InP QDs can be tuned to emit blue, green, and red light; and their emission character is also competitive. Therefore, InP QLEDs are regarded as the most promising alternative Cd-free QLEDs for display and lighting applications. Hence, the synthesis history of luminescent InP QDs is relatively early, which can be traced back to the 1990s. However, it's difficult to obtain high-quality InP nanocrystals because of the "focusing of size-distribution", which involves burst nucleation followed by diffusion-controlled growth. The challenge for the synthesis of InP nanocrystals is how to control the nucleation and growth processes to regulate the size-distribution.[27] Therefore, the progress of InP QLEDs still lags far behind that of the state-of-the-art Cd-based QLEDs. Hence, to overcome this problem, many novel synthetic techniques and precursor reagents have been invented, which has offered improved control over the reaction kinetics to yield higher quality InP QDs to improve the performance of InP QDs. To understand progress about the synthesis of high-quality InP QDs, this chapter will introduce several commonly used methods for the synthesis InP QDs.

6.4.1 Hot-Injection Method

The hot-injection method, also known as the high-temperature oil phase method, organic liquid phase method, and organometallic method, is accomplished by adding a high-boiling organic solvent, organometallic precursor, and organic ligand compound into the reactor. The organometallic precursor is mixed with a nonmetal precursor, which are injected to instantly decompose to form QD nuclei, which are then added upon by controlled focused growth. The organic ligands coordinate onto the surfaces of the nanocrystalline nuclei through electrostatic binding to prevent the aggregation of the nuclei and allow for controlled growth by diffusion through the resulting steric barrier. In the synthesis of QDs, the size, morphology, and growth kinetics of QDs are altered by adjusting the concentrations and types of organic ligands.[28] In general, ligands with too low of an adsorption force will easily lead to larger sizes of QDs or aggregation, while ligands with too strong of an adsorption force will inhibit the growth of QDs. Commonly used ligands are surfactant

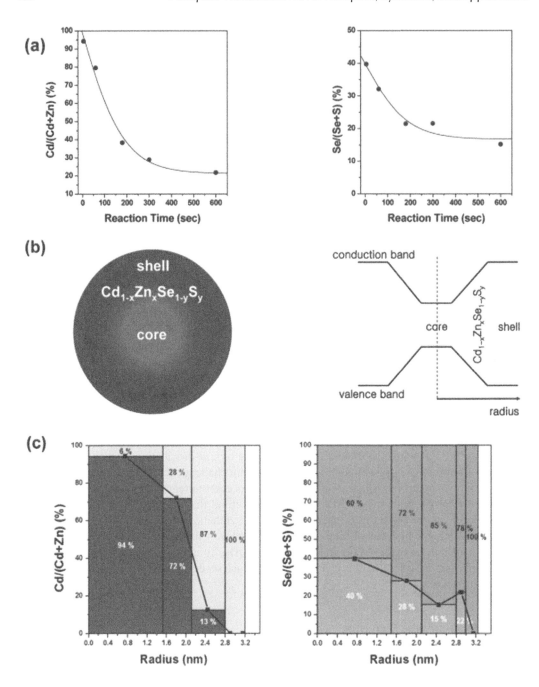

FIGURE 6.5 Composition gradient QDs (a) correlation of precursor ratio to reactions time, (b) band diagram of composition gradient QDs, (c) element distribution of composition gradient QDs. (Bae, W. K., Char, K., Hur, H., Lee, S., *Chem. Mater.* **20**, 531, 2008. With permission.)

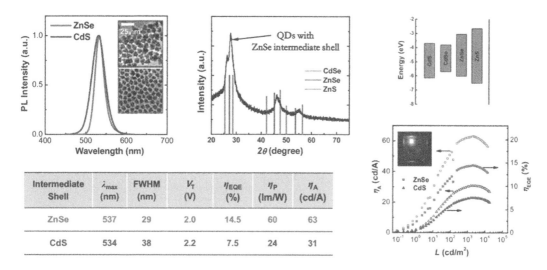

Intermediate Shell	λ_{max} (nm)	FWHM (nm)	V_T (V)	η_{EQE} (%)	η_P (lm/W)	η_A (cd/A)
ZnSe	537	29	2.0	14.5	60	63
CdS	534	38	2.2	7.5	24	31

FIGURE 6.6 Composition gradient QDs with ZnSe as an intermediate shell and fabricated QLED device. (Yang, Y., Zheng, Y., Cao, W., Titov, A., Hyvonen, J., Manders, J.R., Xue, J., Holloway, P.H., Qian, L. *Nat. Photonics.* **9**, 259, 2015. With permission.)

chemicals, such as long-chain alkyl carboxylic acids, amines, thiols, phosphines, phosphonic acids, alkyl phosphoric acids, and other organic surfactants. The hot-injection method is derived from the synthesis method reported by Murray *et al.* [6] in 1993. The organometallic compound dimethyl cadmium was used as the cadmium source, the sulfur source was bis(trimethylsilyl) sulfide, and the selenium source was trioctylphosphine selenide (TOPSe). Based on TOPO as the solvent, mono-disperse CdS and CdSe semiconductor QDs were obtained. This method was subsequently and gradually improved and promoted toward the synthesis of II–VI semiconductor QDs. The synthesis of III-V semiconductor QDs also refers to the method of II–VI QDs. However, the synthesis of high-quality III–V QDs was not successful at the beginning. This mainly manifested in the broad size-distributions of the synthesized III–V QDs and the absence of the characteristic QD absorption peaks, poor crystalline quality, and difficulty dispersing into solvents. Nozik *et al.* [29] made a major breakthrough in 1994. They used the oxalate complex of $InCl_3$ and tris(trimethylsilyl)phosphine ($P(Si(CH_3)_3)_3$) reacted in TOPO solvent when synthesizing InP QDs to successfully synthesize relatively high-quality InP QDs (size dispersion 20%) compared to prior methods, promoting the development of III–V QD synthetic technologies based upon InP. Later, Peng *et al.* [30, 31] reported the use of indium acetate and long-chain fatty acids of various chain lengths (such as myristic acid, palmitic acid, and OA) to generate indium precursors which can react with organic phosphorus sources in the non-coordinating solvent ODE at intermediate to high temperatures to form InP and InAs QDs. They found that InP QD quality is much more dependent on fatty acid ligands than CdSe QDs, and the window of fatty acid ligand concentration is very narrow. Too much or too little ligand can easily lead to poor size-distributions of InP QDs. Adding an appropriate amount of octylamine can control the growth rate of InP QDs and reduce the nucleation temperature from 270°C to about 200°C to obtain high-quality InP QDs whose fluorescence emission range is extended to 750 nm with the PLQY of InP QDs increased to 40% after overcoating with a ZnS or ZnSe shell. [32] Subsequently, Prasad *et al.* [33] systematically studied and optimized the parameters of different fatty acid chain lengths, concentrations, and nucleation temperatures. The synthesized InP QDs had a PLQYs of 0.5%–2% and full-width-at-half-maximum (FWHM) of the emission peaks at approximately 110 nm. Reiss *et al.* [34] reported that the PLQY of (InP)ZnS QDs synthesized by the one-pot method reached 50%–70%, which further improved the potential application capabilities of InP QDs to make QD-based biological cell imaging and photoelectric applications possible (Figure 6.7(a)–(c)).

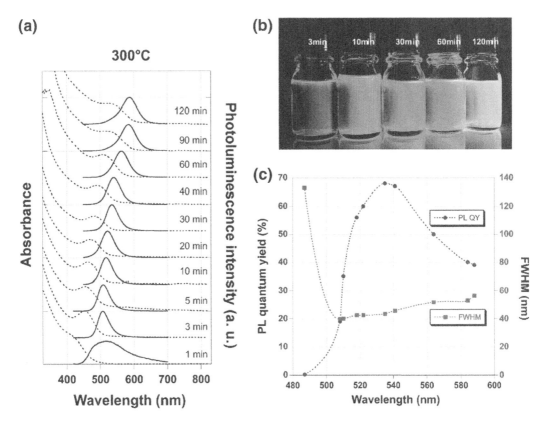

FIGURE 6.7 (a) Evolution of the PL (λ_{ex} = 400 nm) and absorption spectra with reaction time (vertically shifted for clarity). (b) Photograph of selected samples under UV light. (c) Evolution of the fluorescence QY and of the PL line width. (Li, L. and Reiss, P., *J. Am. Chem. Soc.* 130, 11588, 2008. With permission.)

6.4.2 HYDROTHERMAL METHOD

The hydrothermal method is a method in which water or organic solvent is used as the reaction medium in which high pressures induce reaction of precursors with each other to form QDs in a stainless steel autoclave at a particular temperature. The hydrothermal method using non-aqueous solvents is called the solvothermal method. It has the advantages of simplicity, high efficiency, low reaction temperatures (generally less than 200°C), no need to remove oxygen and water, and no flammable organic phosphorus source. Qian *et al.* [35, 36] reported a simple and efficient solvothermal synthesis method to produce InP QDs. Initially, a reducing metal salt was added to xylene solvent, and the reactor was sealed and heated to 150°C to grow InP nanocrystals with a size of 15 nm. InP nanocrystals of 9 nm can be obtained by mixing indium chloride, potassium borohydride, and yellow phosphorus in methanol-benzene solvent through a solvothermal reaction for 4 hours at temperature. However, the size distribution of the synthesized InP QDs is relatively large, and its quality cannot be compared with that of InP QDs synthesized by the thermal injection method. In 2008, Li *et al.* [37, 38] mixed $InCl_3$ with dodecylamine and tris(dimethylamino)phosphine ($P(N(CH_3)_2)_3$) in toluene to synthesize InP QDs at 180°C for 24 hours. However, these kinds of InP QDs had large size-distributions because the nucleation and growth occurred almost simultaneously. Therefore, it was necessary to further use size screening methods to separate QDs of different particle sizes, and the PLQY of InP QDs could only reach 58% after hydrofluoric acid etching treatment. Yang *et al.* [39, 40] also used this method to synthesize InP QD cores and then overcoated them using zinc

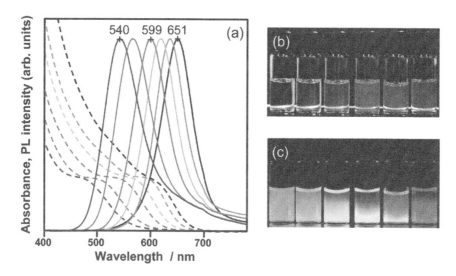

FIGURE 6.8 The absorption and fluorescence spectra of InP QDs synthesized by (a) solvothermal method, (b) visible light, and (c) fluorescence pictures. (Li, C., Ando, M., Enomoto, H., Murase, N., *J. Phys. Chem. C* 112, 20190, 2008. With permission.)

stearate and dodecanethiol to form (InP)ZnS (core)shell QDs. After size sorting, the (InP)ZnS QD PLQY achieved 24%–60% as shown in Figure 6.8.

At present, InP QDs synthesized by hydrothermal methods still have the disadvantages of large size-distributions, low PLQYs, and difficulty in controlling the direct synthesis of (core)shell QDs. Additionally, particle sizes cannot be monitored in real time during the hydrothermal synthesis of QDs.

6.4.3 MICROWAVE-ASSISTED SYNTHESIS

The microwave-assisted synthesis method is widely used to prepare high-quality nanomaterials because of its rapid heating rate and uniform temperature distribution, which provides a scalable platform for industrial applications.[41, 42] By judicious choice of the solvents, passivating ligands, and reactants; the nanomaterial precursors can be selectively heated preferentially with regards to the solvent or passivating ligand. Selective heating in the microwave cavity is advantageous in organic synthesis, and, in general, these microwave synthetic methodologies are quite adaptable to reactions that have high energies of activation and slow reaction rates.[43, 44]

The first microwave-assisted synthesis of InP QDs was reported in 2005 by Strouse and co-workers, who found that InGaP and InP were rapidly formed at 280°C within minutes, yielding clean reactions and highly monodisperse size-distributions that required no size-selective precipitation and resulted in the highest out of batch quantum efficiency reported to date of 15% prior to chemical etching, as shown in Figure 6.9.[41] Then, the group found that using ionic liquids as the solvent can enhance QYs because of the balance of the growth and the surface etching of the QDs by the function of ionic liquids.[45] Additionally, through changing of the microwave power and reaction temperatures, InP QDs with tunable emissions (544–630 nm) were reported with the assistance of dodecylamine, which facilitates QD formation by destabilizing InP clusters and increasing the nucleation rate.[46, 47]

6.4.4 MICROFLUIDIC SYNTHESIS

The microfluidic synthesis of colloidal QDs is a more promising technique than conventional batch synthesis techniques. The benefits of microfluidic synthesis are as follows: (i) efficient mixing, (ii) high heat and mass transfer, (iii) high surface-to-volume ratio, (iv) temperature control,

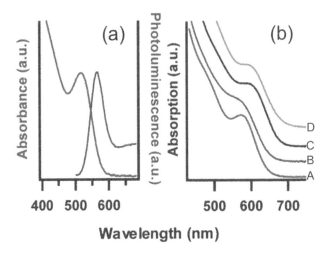

FIGURE 6.9 Absorbance and photoluminescence of InP in toluene (a). Absorbance of a series of InP nanoparticles is formed in the presence of ionic liquids at 280°C for 15 min at 280 W (b): (A) as-prepared InP nanoparticles with no ionic liquid present, (B) InP with trihexyltetradecylphosphonium decanoate, (C) InP with trihexyltetradecylphosphonium bromide, and (D) InP with trioctylphosphine oxide. (Gerbec, J.A., Magana, D., Washington, A., Strouse, G.F., *J. Am. Chem. Soc.* 127, 15791, 2005. With permission.)

(iv) continuous production, and (v) low reagent consumption. As such, microfluidic synthesis is an ideal technique for large-scale production.[48, 49] A schematic of a multistage microfluidic platform for the synthesis of multilayered QDs is shown in Figure 6.10.[50] After the initial attempt of the preparation of QDs based on a microfluidic system by Edel and Fortt in 2002,[51] many studies have reported the synthesis of various types of QDs.

Nightingale and de Mello first synthesized InP QDs by single-capillary and Y-shaped microfluidic devices in 2009.[52] In this study, the premixed precursor solutions were injected into a glass

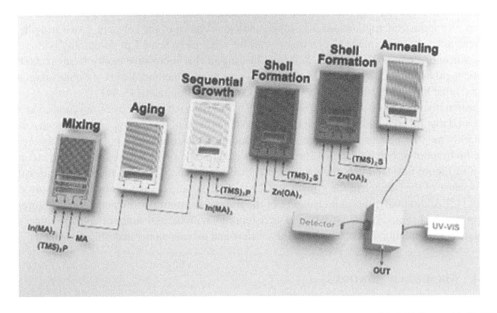

FIGURE 6.10 Schematic of a multistage microfluidic platform for synthesis of InP/ZnS core/shell QDs. (Edel, J.B., Fortt, R., DeMello, J.C., DeMello, A.J., *Chem. Commun.* 10, 1136, 2002. With permission.)

capillary submerged in an oil bath at controlled temperature. To overcome the drawbacks that the precursor concentrations cannot be controlled or varied to synthesize QDs with good quality, the authors designed a Y-shaped microfluidic device. Here, a two-in/one-out Y-shaped glass microfluidic chip reactor was designed (~300°C) to enable variation of the indium and phosphorus concentrations. They found that the precursor injection rate can influence the emission spectra as well as the QY.

Two years later, three-stage microfluidic reactors, namely, mixing, aging, and sequential injection, were invented by Baek *et al.* [53] During synthesis, octane as a supercritical solvent was selected to provide excellent mixing and fast diffusivity for producing homogeneous reaction conditions. The temperature can be increased to 130–175°C, which was the crystallization temperature of the precursors at the mixing stage. Then the temperature can be further increased to 200–340°C at the annealing stage. Using this device, high-quality InP QDs were produced within 2 minutes. However, (core)shell structure QDs cannot be realized by this microfluidic system.

In 2011, to synthesize (core)shell structures QDs, Kim *et al.* [54] invented a device to synthesize (core)shell-structured (InP)ZnS QDs. By optimizing the flow rate and temperature, the scholars successfully synthesized (InP)ZnS QDs of four different colors, including bluish green, yellow, orange, and red, and the PLQYs of the corresponding QDs were approximately 20%, 42%, 34%, and 37%, respectively. Later, Ippen *et al.* [55] reported a synthetic technique for producing high-quality InP QDs in a CFR using toluene as the solvent. However, the PLQYs of the resultant InP QD products did not increase compared with those obtained from batch systems and continuous-flow devices. Jensen and co-workers synthesized (InP)ZnS, (InP)ZnSe, (InP)CdS, and (InAs)InP QDs using six chip reactors for mixing, aging, sequential growth of the core and two shells, and annealing.[56] By varying the size of InP core QDs, a set of (InP)ZnS QDs with sizes from 3.8 to 4.9 nm were prepared, with tunable band-edge emissions (554–681 nm) and high PLQYs (32%–40%).

Despite the remarkable advances of microfluidic synthesis, it has not been widely adopted to synthesize nanomaterials because of certain problems. For example, it is difficult to make use of microfluidic chips in a conventional chemical laboratory due to crystalline-scale formation and solidification of high melting point solvents, which makes the chip life short.

In conclusion, QDs material have been on the rise in commercialization, with realized high stability, excellent emissive characteristics and superior solubility, which exhibit a variety of applications, especially in new display. Next, we briefly introduce the display applications of QDs.

6.5 THE DISPLAY APPLICATIONS OF QUANTUM DOTS

For the unique optoelectronic characteristics of tunable spectra, narrow emission peaks, solution processability, and high quantum efficiencies; QDs are prospective materials for display technology to improve color gamut. Now, in the LCD displays within the last five years, QDs are extensively applied in the higher priced market, which have enhanced its color saturation. Moreover, through combining QDs for color conversion, microLED display technology is a new paradigm for flat-panel televisions by introducing micron-sized LEDs as the back-light. In active display technologies with superior color purity, brightness, viewing angle, and high efficiencies, QLED is the ultimate target of QD applications. All in all, QD applications in the optoelectronic field are fairly widespread; however, it can be summed up into two major categories: PL and EL technologies. In this segment, reviews of PL and EL QDs display technologies are presented. To realize QD deposition, low-cost printable methods are employed since QDs can be completely dispersed into organic solvents, which will be briefly introduced.

6.5.1 QUANTUM DOTS FOR PHOTOLUMINESCENT APPLICATIONS

Due to the high exciton binding energy for rigorous quantum confinement, the emissive characteristics of QDs are excellent by optical excitation, especially in color purity and quantum yield. After employing a (core)shell structure, the thermal stability and photochemical resistibility of QDs are enormously improved, leading to the possibility of commercial applications in display technologies. About five years ago, to enhance the color performance of LCD displays, QDs materials were extensively applied within the backlights of LCD televisions, including the methods of on-chip, remote composite films, quantum rail technology, *etc.* More recently, as a color conversion material, PL applications of QDs are becoming more ubiquitous in display technologies, including future new-generation displays such as miniLED/microLED and QD-OLED. We summarize the PL applications of QDs in the sections below.

6.5.1.1   Quantum Dots for LCD Backlight Units

Thin-film transistor (TFT) LCD is the dominant flat panel display technology used in our daily lives, after several decades of extensive development in advanced manufacturing technology, which has the advantage of effective cost, longevity, and high resolution. When comparing to active emission displays with high contrast ratios, wide viewing angles, high color gamuts, and flexibility, the performance metrics of LCD are still far behind in exhibiting color (as shown in Figure 6.11).[57] At present, by local dimming of the backlight, dual cell, and ultrathin backlight module, the contrast ratio and thickness of LCD are visibly improved. However, on color reproduction, LCD with white LED backlights can achieve only a color gamut of about 80% AdobeRGB, which cannot satisfy the requirements for a higher quality display.[58] Several methods have been applied to enhance color presentation, such as covering the color filter (CF) and LED backlight with a different light source, but they lead to significantly reducing the efficiency of the backlight. To narrow the emissive spectrum for richer color, a CF is used for enhancing the three optical primary colors, but its transmittance is only about 30% such that the light efficiency is significantly reduced.[59] Improving the LED backlight by developing a new white LED with RGB colors is an attractive option since the EL technology is more complex due to the requirements a driving current and that the three primary colors must be separated. Moreover, the efficiency of green LED is far behind blue and red LED, presenting serious challenges.

In the context of those challenges in LCD backlight, employing red and green QDs as optical downconverters of some of the blue LED light, a promising backlight technology has emerged from high stability QDs. The white backlight spectrum consists of the blue emission peak from the LED

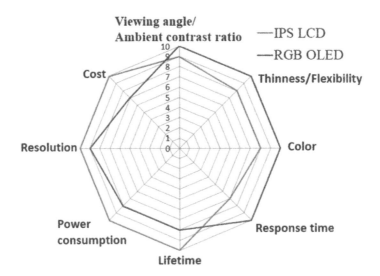

FIGURE 6.11   Performance comparisons of LCD and OLED. (Chen, H. W., Zhu, R. D., He, J., Duan, W., Hu, W., Lu, Y. Q., Li, M. C., Lee, S. L., Dong, Y. J., Wu, S. T., *Light-Sci. Appl.* 2017, 6. With permission.)

and red/green PL emission peaks emitted from QDs excited by the blue LED. Accordingly, forming three separate emission peaks, the color gamut of LCD is above 100% NTSC 1976 (national television system committee) after enhancing the color purity of the backlight by QDs.[60] Therefore, many companies have actively stepped into the QD business, including QD material suppliers, device developers, and display manufacturers, toward on-chip, remote composite film, and quantum rail applications to promote QDs toward commercial applications in high-end LCD displays. Up until now, according to the QD position within the backlight, there are three types of QD LCD displays: on-chip, remote composite film, and quantum rail.[58] The first one, on-chip, is made by employing encapsulated QDs directly on top of the blue LED, leading to more effective costs resulting from the smaller amount of QD materials required. The QD enhancement film (QDEF) is a QD polymer composite used as a remote technique that is extensively employed in backlights for reducing the requirements of QDs stability but requiring more materials for added cost. Quantum rail is another scheme for QD application in display technology. In 2013, Sony was the first company to sell televisions with backlights employing QD Vision quantum rail technology, which displayed the advantage of QDs for improved color gamut but for a much higher cost. Then, some TV companies, such as Samsung, TCL, and Hisense, began to join successively with the industry of QDs for LCD TV technologies. QDs for display applications has been becoming a focused area of high interest. In the following parts, we explain in detail the manners in which QDs are employed for PL applications in LCD display technology.

Among those three configurations, the method of QDs on-chip (as shown in Figure 6.12(a)) is the most effective cost for consuming the minimum amount of QD materials because red and green QDs are deposited directly drop on the LED chip to form a QD lens or film to significantly simplify the optical design of LCD backlight by drop-in replacement. Due to the high sensitivity of QDs to water and oxygen, a hermetic encapsulation is necessary for prolonging the operative lifetime, including two approaches of designing an appropriate QD shell structure and hermetic sealing of the QD film over the LED chip, which increases the total design complexity and expense. Nevertheless, the surface temperature of the LED is very high (~150°C) when it is working at temperature, which can lead to the rapid degradation of QDs.[61] Furthermore, direct exposure to the intense high-energy light flux of the blue LED, the QDs on the LED chip will degrade very rapidly. [62] Therefore, the thermal and optical reliabilities of QDs must be dramatically improved before this approach of QDs on-chip is extensively applied.

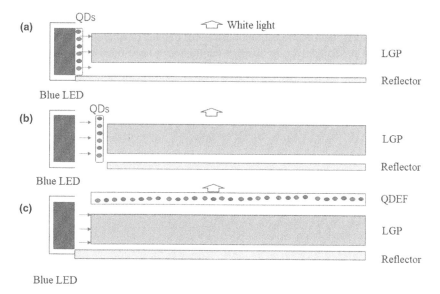

FIGURE 6.12   The configurations of QD applications in LCD: (a) on-chip, (b) quantum rail, and (c) remote film.

To avoid the degradation of QDs at the high junction temperature and intense light excitation of the LED, quantum rail is an alternative geometry, especially in large-size TVs, which will obviously improve the lifetimes of the QDs and which require more acceptable amounts QDs materials. In fact, Sony adopted this configuration in some models of their 55-in. TVs in 2013.[63] As shown in Figure 6.12(b), the mixing of red and green QDs, encapsulated within a tube, sandwich between the LED and the light guide plate (LGP), emitting uniform white light. The blue light arising from driven LED penetrates the quantum rail QD-composite tube, forming the RGB three primary colors backlight toward LGP. After careful designs of the concentration ratio of QDs and scattering nanoparticles, and accurate calculations of optical distance, the light efficiency and color uniformity can be achieved with this method of backlight. However, there is a sticky issue in the assembly of the display because the fragile tube is damaged easily and the rigorous requirements for optical distance greatly increase the difficulty of assembly. After 2016, a general consensus of abandoning quantum rail technology has been reached within the display industry.

At present, QDs used in remote application composite films is the most commonly employed configuration, as shown in Figure 6.12(c), which comprises a QD-composite film between and coated on both sides by two barrier films, to form a sandwich structure, known commonly as QDEF. Due to the remote location far away from the LED chip, QD films placed above the LGP are decoupled spatially from the source of heat and intense light to enhance significantly the reliability and stability of the QDs. Simultaneously, the two plastic barrier films can strictly prevent oxygen and moisture from diffusing into the QDEF to degrade the QDs and with excellent flexibility and high transparent capacity. It greatly benefits from long operation lifetimes and a simplified assembly process, promoting the commercial application of QDEFs in LCD display technology. Since the QD materials consumption is high, especially in large-scale displays, the price of LCD displays with QDEFs is significantly increased. In 2016, accelerated aging tests revealed an operative lifetime of QDEFs at over 30,000 hours.[64] Thus, with the development of such optical designs, the approach of QDs in remote applications has been adopted extensively in high-end products. Many domestic and overseas manufacturers, including Samsung, BOE, and TCL, have embraced this configuration in commercial LCD products. As market demands keep growing, the price should reduce gradually over time.

However, LCD technology has been advancing continuously in color saturation, color gamut, contrast ratio, response time, and view angle as technologies such as local dimming, dual cell, miniLED backlight, and QDs enhancement are applied maturely in flat panel displays. For the high complexity of LCD geometry (Figure 6.13), the capacities of portability and flexibility are limited, which are the most crucial specifications in modern smart display technologies such as smartphones, intelligent wear, *etc.* Consequently, new display technologies have been suggested and developed

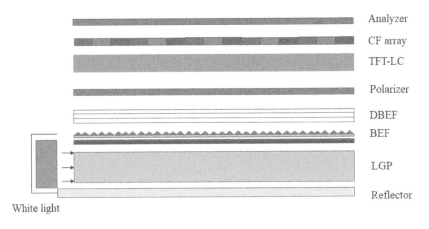

FIGURE 6.13 The typical geometry of an LCD display.

vigorously, including OLED, Mini/MicroLED, and QLED; however, LCD will still remain appealing for the coming decades in some display areas due to low cost, particularly with the introduction of QD materials. Excluding QD-based LCD displays in the application of down-conversion PL, QD materials have tremendous application potential in QD-OLED and Mini/MicroLED.

6.5.1.2 Quantum Dots for Mini/MicroLED

In 2012, a 55-in. microLED TV panel was showed by Sony with over 6 million microLED chips to achieve 1920 × 1080 resolution, which received lots of attention. Since the continued development of LED chips and assemblies, Samsung exhibited the first modular 146-in. LED display in 2018 named "The Wall." After this, mini/microLED display technologies have achieved much encouraging progress in resolution and luminance, especially in the rapidly advancing of LED chip manufacturing, but ponysize LED chips are faced with the difficulty of industrial manufacturing. Today, miniLED chips have been applied widely in LCD backlights for improved display capability, such as contrast ratio, to achieve the ultimate three primary color display, which must still confront many difficulties that are mostly unsolved now. First, toward full color mini/microLED displays comprising millions of chips, every pixel contains three different color chips as subpixels, which requires a method that is suitable for multiple mass transfer as one of the most difficult problems. Second, for employed three color LED chips calling for unequable driving currents, complex driving is necessary, which greatly increases the complexity of designing and preparing backplane, driving control, *etc*. Third, the efficiency of mini/microLED displays puts them out of reach for consumers, especially under the limitation of the low efficiency of the green chip called the "green gap," with the device dimension shrink leading to more serious reduction of efficiency, which obviously appears in the red chip.[65] Therefore, a new scheme is put forward to solve those problems called QD color conversion (QDCC) after combining the advantages of QDs, which could substantially reduce the manufacturing complexity by integrating photoetching processes and printing techniques, leading to the development trend of mini/microLED displays.[66]

By making use of the compelling characteristics of QDs, mini/microLED displays could offer those potential advantages of wide color gamut, high contrast ratio, fast response time, and wide viewing angle to play a crucial role in new active displays and illustration.[67] In Figure 6.14, the geometry of a mini/microLED display with QDCC is exhibited, pumping every subpixel of patterned QDCC by blue LED pixels. To suppress the blue excitation light from the LED, two approaches are used, including a distributed Bragg reflector (DBR) above the QDs layer and a CF that LCD adopts, leading to require more complex fabrication processes and increased production expenses. Based upon the high absorption cross section of the QD layer, QDs themselves

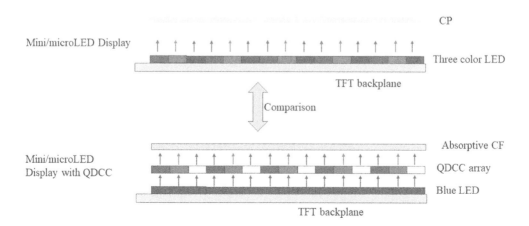

FIGURE 6.14 The geometry of a mini/microLED display with QDCC compared to without QDCC.

absorbing is the best way to completely remove excess blue light through employing the extinction coefficient and volume fraction of the QDs that are used for adjusting the color balance. Due to the demands of pixel density in QDCC, manufacturing processes of inkjet printing (IJP) and photolithography are utilized to form patterns, mixing QDs with the photoresist polymer and UV-curing adhesive.

With the improving stability of QDs and making progress in device fabrication of LEDs, mini/microLED display possesses an optimistic perspective in next generation of display technologies, such as smart wear, vehicle monitors, and digital displays. Although the commercial application has not been realized yet, mini/microLED emissive displays will step gradually onto the stage in the not-too-distance future.

6.5.2 QUANTUM DOTS FOR EL APPLICATIONS

EL QLED technology has attracted a great deal of attention because there are many excellent EL properties of narrow emissive FWHM, tunable EL spectrum, solution processability, and high efficiency.[68] Compared with PL applications, QDs have some special requirements for injection of electrons and holes from two separate directions. With the (core)shell structure of QDs, good interface contact must be formed between the outer shell and organic ligands to allow transport accessibly of charge carriers while maintaining electrochemical and photochemical stability, which could affect the efficiency and operative lifetime of the device.

Up to now, the EQE of green and red QLED are over 20%, making a giant leap forward in the past decade, which is comparable with commercial phosphorescent OLED. In 2011, Qian *et al.* [69] exhibited an organic-inorganic hybrid structure employing an n-type metal oxide semiconductor of zinc oxide (ZnO) nanoparticles as the electron transport layer (ETL) for alternating traditional organic ETL, realizing the temporal highest brightness of QLED. Then, using this hybrid structure (in Figure 6.15), QLED devices have been further developed by optimizing the synthetic process of ZnO nanoparticles and perfecting this hybrid structure to balance the injection of electrons and holes, the EQE of which has achieved to over 20% in 2014.[70] Soon after, some QLED panels are being exhibited one after another by display manufacturers at home and abroad. Recently, the EQE of QLED devices has exceeded 30% after adopting alloyed QDs, and its operational lifetime is over 1.8 million hours at 100 cd/m^2.[71] Moreover, researchers at Samsung have reported achievements of an EQE of blue QLED above 20% with operating lifetimes over 15,000 hours at 100 cd/m^2 (T$_{50}$), after passivating QD materials, which indicates an exciting prospect in industrial application.[72]

To boost the performance in QLED, immense amounts of solid research has been carried out to explore the operational mechanism of such devices. As a general rule, the EL mechanism of QLED

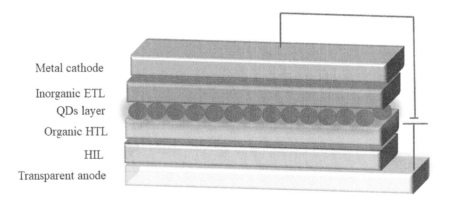

Metal cathode
Inorganic ETL
QDs layer
Organic HTL
HIL
Transparent anode

FIGURE 6.15 The structure diagram of a typical hybrid QLED.

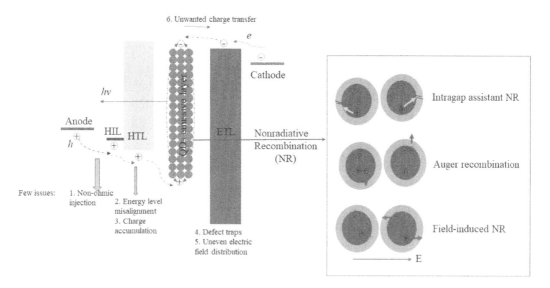

FIGURE 6.16 A schematic diagram of QLED device physics, including the non-radiative recombination in EL.

is based upon direct charge injection through injecting an electron and a hole from the charge transport layers (CTLs) into a given QD to form an exciton from which subsequently occurs radiative recombination to release energy with a photon, as shown in the Figure 6.16. There are several sequentially elementary processes in QLED EL: (i) The injection of charge carrier from electrode into CTLs, (ii) the transportation of charge carriers in CTLs, (iii) the carrier injection into QDs, (iv) the exciton formation of electron and hole pair in QDs, (v) the radiative recombination of excitons in QDs, and (vi) light emitting externally from the device inside out.[73] All of these physical processes could be affected by the optoelectrical characteristics of every layer in the device, including energy band levels, electrical conductivity, work function, defect states, and optical properties. Consequently, such issues as cover energy level misalignment, nonohmic contact, material defects, and disequilibria conductivity could decrease the device efficiency and lifetime, giving rise to unwanted charge transfer, space charge accumulation, and uneven electric field distribution. To improve the EQE of QLED, the balance injection of charge carriers and the suppression of non-radiative recombination are crucial by exploiting appropriate energy band CTL materials, removing QDs defects, and optimizing the device structure.[26]

In general, while hole injection into neutral QDs is considered to be inefficient, researchers have found that the intermediate negatively charged state of QDs triggers confinement-enhanced Coulomb interactions, which simultaneously accelerate hole injection and hinder excessive electron injection.[74] By re-optimizing the (core)shell electronic structure of QDs, the working lifetime of QD devices was increased to more than 2,000 hours of T_{95} (at 1,000 nits) for the first time, which is fully sufficient for industrialization.[75] Because a major problem of QLED is the lifetime of blue devices in commercial applications, Chen *et al.* explored the failure modes of blue QLED devices to discover it is mainly due to the rapid degradation of the QD-ETL knot, not the degradation of the HTL layer. To solve the problem of device lifetime, it is necessary to design the CBM position to be more suitable for electron injection from the ETL and QD materials into the EML layer.[76] When the mechanism of device EL mechanism is understood more and more clearly, the issues of stability and efficiency of QLED will be solved quickly.

Despite some subsistent problems in QLED, AM (active matrix) QLED is promising based upon the excellent dispersibility of QDs in organic solvents, which has the enormous potential for achieving full color with a low-cost solution process to precisely pattern the emissive layers employing IJP, transfer printing, blade coating, photolithography, and so on. Among those technologies, IJP is

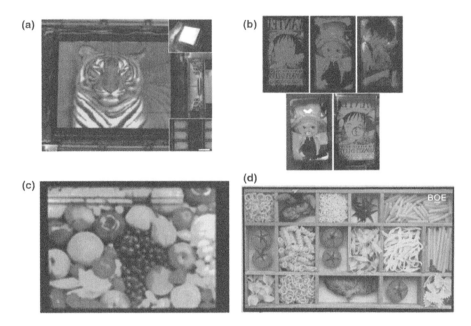

FIGURE 6.17 Demo of QLED display: (a) Samsung, (b) Nanophotonica, (c) South China University of Technology, and (d) BOE.

a most ideal approach for fabricating full color QLED displays due to the advantages of being mask-free, non-contact, and easy for patterning; which has been extensively applied in depositing solvable materials that include organic polymers and small molecules, colloidal nanomaterials, and carbon materials.[77] All of the display panel manufacturers, for example, Samsung, BOE, and TCL, are planning for the developing layout of QLED display based upon IJP, and even some upstream suppliers are starting to develop their own QLED technological capabilities, such as Nanophotonica and Nanosys. In China, the technology of printable QLED is a key technical program that is vigorously supported by government to lead the novel display industry.

As early as in 2009, (as shown in Figure 6.17a) Samsung has exhibited a QLED display of a 320×240-pixel array with an image of a 4-in. crosslink QLED device using an α-Si TFT backplane for the AM drive. Nanophotonica also showed a 4.3-in. monocolor AM display of 480×800 pixels driven by a poly-Si TFT backplane, taking advantage of nozzle flow printing to realize 150-ppi resolution of 50×160-μm pixel size (figure 6.17b). Using IJP from a mixed solvent system on a modified ZnO transparent layer to form coffee-ring-free QD thin films for QLED, a 2-in. full color AM QLED display of 120 ppi was shown by South China University of Technology with a brightness of 400 cd/m^2, high color gamut of 109%, and high contrast ratio of 50,000:1 (figure 6.17c). This year, BOE has unveiled a 55-in. 4K AM-QLED that is an upgrade to the 14-in. AM-QLED display in 2017, based on IJP, with a resolution of 3840×2160 and a color gamut of 119% NTSC (figure 6.17d). Regarding the operating lifetime, passivating the trap states induced by anions and cations on QD surfaces, the half lifetime of IJP QLED was more than 1,721,000 hours when adopting dual ionic passivation and a proper ligand, which greatly opened the industrial application of QLED technology.[78]

6.6　CONCLUSION AND THE FUTURE OF QUANTUM DOTS FOR DISPLAY APPLICATIONS

The PL applications of QDs have achieved rapid success in the past five years, benefiting from the maturity of QD synthetic chemistry and their superior performances of electrical, chemical, and optical properties, led to a great step forward for LCD, which holds a tremendous potential for mini/

microLED displays in new generation display technology. Even so, the technology of EL-QLED is of unprecedented importance for the future of display, especially based on IJP process, cost effectiveness, wide color gamut, foldable flexibility, and ultrathin form factors. However, there are some crucial aspects to be worked out for speeding up the commercialization of QLED, for example, the high desire for CFQDs, long operating lifetimes of blue QLED, and a high-resolution patterning technique. Presently, Cd-based QDs are commonly used in QLED for superior properties, which is strictly controlled in some key consumer electronics markets. A number of CFQDs, such as InP, $CuInS_2$, ZnSe, have been applied in QLED devices to show a maximum EQE of 21.4%, which represents a comparable performance to Cd-based QLED. On the other hand, although the company JOLED has announced that the industrial production of IJP OLED has been realized, a higher resolution of patterning technology is imminently necessary for wide application, for which IJP and photolithography are the most promising. At present, the IJP technique could achieve the highest resolution of 200 ppi in an industrial process, which could be able to reach 400 ppi by optimal design of pixel definition layers. Moreover, the photolithography technique is maturely applied in the traditional semiconductor industry, which can be employed in the patterning process of QLED to reach very high resolution. However, there are crucial issues to device degradation within the requisite technological processes, such as realizing a three colors sub-pixel, the compatibility of QDs and photoresist, the coexistence between the process of preparing the backplane, and the technique of photolithographic patterning.[79]

In spite of the subsistent challenges and competitors, there is still a prospective future for QLED to be extensively applied in display technology and even be a strong contender to LCD, OLED, and mini/microLED, and we have an expectation of the successful commercialization of full color QLED display in the near future.

REFERENCES

1. Brus L. E. (1984) Electron-electron and electron-hole interactions in small semiconductor crystallites: The size dependence of the lowest excited electronic state. *J. Chem. Phys.* 80: 4403.
2. Mohanty B., Ghorbani-Asl M, Kretschmer S, Ghosh A, Guha P, Panda SK, Jena B, Krasheninnikov AV, Jena BK. (2018) MoS2 quantum dots as efficient catalyst materials for the oxygen evolution reaction. *ACS Catal.* 8: 1683.
3. Nasilowski M., Spinicelli P., Patriarche G., Dubertret B. (2015) Gradient CdSe/CdS quantum dots with room temperature biexciton unity quantum yield. *Nano Lett.* 15: 3953.
4. Cao W, Xiang C, Yang Y, Chen Q, Chen L, Yan X, Qian L. (2018) Highly stable QLEDs with improved hole injection via quantum dot structure tailoring. *Nat. Commun.* 9: 2608.
5. Ekimov A. I., Onushchenko, A. A. (1981) Quantum size effect in three-dimensional microscopic semiconductor crystals. *ZhETF Pis ma Redaktsiiu.* 34: 363 34.
6. Murray C, Norris DJ, Bawendi MG, (1993) Synthesis and characterization of nearly monodisperse CdE (E = S, Se, Te) semiconductor nanocrystallites. *J. Am. Chem. Soc.* 115: 8706.
7. Bawendi M. G., Stott N. E., Preparation of nanocrystallites. U.S. Patents 6,576,291; 6,821,337; and 7,138,098.
8. Stott N. E., Jensen K. F., Bawendi M. G., Yen B. K. H. Method of preparing nanocrystals. U.S. Patent 7,229,497.
9. Qian L, Zheng Y, Xue J, Holloway P.H. (2011) Stable and efficient quantum-dot light-emitting diodes based on solution-processed multilayer structures. *Nat. Photonics* 5: 543.
10. Ekimov AI. (1981) Quantum size effect in three-dimensional microscopic semiconductor crystals. *JETP. Lett.* 34: 345.
11. Rossetti R, Nakahara S, Brus LE. (1983) Quantum size effects in the redox potentials, resonance Raman spectra, and electronic spectra of CdS crystallites in aqueous solution. *J. Chem. Phys.* 79: 1086.
12. Brus L. (1986) Electronic wave functions in semiconductor clusters: Experiment and theory. *J. Phys. Chem.* 90: 2555.
13. Wang Y, Herron N. (1991) Nanometer-sized semiconductor clusters: Materials synthesis, quantum size effects, and photophysical properties. *J. Phys. Chem.* 95: 525.
14. Li T. L., Teng H. (2010) Solution synthesis of high-quality CuInS2 quantum dots as sensitizers for TiO2 photoelectrodes. *J. Mater. Chem.* 20: 3656.

15. Bawendi M. G., Kortan A.R., Steigerwald, M. L., Brus, L. E. (1989) X-ray structural characterization of larger CdSe semiconductor clusters. *J. Chem. Phys.* 91: 7282.

16. Steigerwald M. L., Brus L. E. (1990) Semiconductor crystallites: A class of large molecules. *Acc. Chem. Res.* 23: 183.

17. Peng X. G. (2009) An essay on synthetic chemistry of colloidal nanocrystals. *Nano Res.* 2: 425.

18. Peng X, Wickham J, Alivisatos AP. (1998) Kinetics of II-VI and III-V colloidal semiconductor nanocrystal growth: 'Focusing' of size distributions. *J. Am. Chem. Soc.* 120: 5343.

19. Qu L, Peng ZA, Peng X. (2001) Alternative routes toward high quality CdSe nanocrystals. *Nano Lett.* 1: 333.

20. Peng Z. A., Peng X. (2001) Formation of high-quality CdTe, CdSe, and CdS nanocrystals using CdO as precursor. *J. Am. Chem. Soc.* 123: 183.

21. Yu WW, Peng X. (2002) Noncoordinating solvents : Tunable reactivity. *Communications* 114: 2368.

22. Battaglia D, Peng X. (2002) Formation of high quality InP and InAs nanocrystals in a noncoordinating solvent. *Nano Lett.* **2**: 1027.

23. Li JJ, Wang YA, Guo W, Keay JC, Mishima TD, Johnson MB, Peng *X.* (2003) Large-scale synthesis of nearly monodisperse CdSe/CdS (core)shell nanocrystals using air-stable reagents via successive ion layer adsorption and reaction. *J. Am. Chem. Soc.* 125: 12567.

24. Chen O, Zhao J, Chauhan VP, Cui J, Wong C, Harris DK, Wei H, Han HS, Fukumura D, Jain RK, Bawendi *MG.* (2013) Compact high-quality CdSe-CdS (core)shell nanocrystals with narrow emission linewidths and suppressed blinking. *Nat. Mater.* **12**: 445.

25. Bae WK, Char K, Hur H, Lee S. (2008) Single-step synthesis of quantum dots with chemical composition gradients. *Chem. Mater.* 20: 531.

26. Yang Y, Zheng Y, Cao W, Titov A, Hyvonen J, Manders JR, Xue J, Holloway PH, Qian L. (2015) High-efficiency light-emitting devices based on quantum dots with tailored nanostructures. *Nat. Photonics* 9: 259.

27. Chen B, Li D, Wang F. (2020) InP quantum dots: Synthesis and lighting applications. *Small* 16: e2002454.

28. Peng X, Thessing J. (2005) Controlled synthesis of high-quality semiconductor nanocrystals. In *Semiconductor Nanocrystals and Silicate Nanoparticles*, edited by Peng, and Mingos. Berlin, Heidelberg: Springer Berlin Heidelberg.

29. Micic OI, Curtis CJ, Jones KM, Sprague JR, Nozik AJ. (1994) Synthesis and characterization of InP quantum dots. *J. Phys. Chem.* 98: 4966.

30. Battaglia D, Peng X. (2002) Formation of high quality InP and InAs nanocrystals in a noncoordinating solvent. *Nano Lett.* 2: 1027.

31. Xie R, Peng X. (2009) Synthesis of Cu-doped InP nanocrystals (d-dots) with ZnSe diffusion barrier as efficient and color-tunable NIR emitters. *J. Am. Chem. Soc.* 131: 10645.

32. Mushonga P, Onani MO, Madiehe AM, Meyer M. (2013) One-pot synthesis and characterization of InP/ZnSe semiconductor nanocrystals. *Mater. Lett.* 95: 37.

33. Lucey DW, MacRae DJ, Furis M, Sahoo Y, Cartwright AN, Prasad PN. (2005) Monodispersed InP quantum dots prepared by colloidal chemistry in a noncoordinating solvent. *Chem. Mater.* 17: 3754.

34. Li L, Reiss P. (2008) One-pot synthesis of highly luminescent InP/ZnS nanocrystals without precursor injection. *J. Am. Chem. Soc.* 130: 11588.

35. Li B, Xie Y, Huang J, Liu Y, Qian Y. (2001) A novel method for the preparation of III–V semiconductors: Sonochemical synthesis of InP nanocrystals. *Ultrason. Sonochem.* 8: 331.

36. Wei S, Lu J, Yu W, Qian Y. (2004) InP nanocrystals via surfactant-aided hydrothermal synthesis. *J. Appl. Phys.* 95: 3683.

37. Li C, Ando M, Enomoto H, Murase N. (2008) Highly luminescent water-soluble InP/ZnS nanocrystals prepared via reactive phase transfer and photochemical processing. *J. Phys. Chem. C* 112: 20190.

38. Murase N, Li C. (2008) Highly luminescent water-soluble InP/ZnS nanocrystals prepared via reactive phase transfer and photochemical processing. *J. Lumin.* 128: 1896.

39. Lee JC, Jang EP, Jang DS, Choi Y, Choi M, Yang H. (2013) Solvothermal preparation and fluorescent properties of color-tunable InP/ZnS quantum dots. *J. Lumin.* 134: 798.

40. Byun HJ, Song WS, Yang H. (2011) Facile consecutive solvothermal growth of highly fluorescent InP/ZnS (core)shell quantum dots using a safer phosphorus source. *Nanotechnology* 22: 235605.

41. Gerbec JA, Magana D, Washington A, Strouse GF. (2005) Microwave-enhanced reaction rates for nanoparticle synthesis. *J. Am. Chem. Soc.* 127: 15791.

42. Ashley B, Vakil PN, Lynch BB, Dyer CM, Tracy JB, Owens J, Strouse GF. (2017) Microwave enhancement of autocatalytic growth of nanometals. *ACS Nano* 11: 9957.

43. Guzelian AA, Katari JB, Kadavanich AV, Banin U, Hamad K, Juban E, Alivisatos AP, Wolters RH, Arnold CC, Heath JR. (1996) Synthesis of size-selected, surface-passivated InP nanocrystals. *J. Phys. Chem.* 100: 7212.

44. Heath JR. (1998) Covalency in semiconductor quantum dots. *Chem. Soc. Rev.* 27: 65.

45. Lovingood DD, Strouse GF. (2008) Microwave induced in-situ active ion etching of growing InP nanocrystals. *Nano Lett.* 8: 3394.

46. Siramdas R, McLaurin EJ. (2015) InP nanocrystals with color-tunable luminescence by microwave-assisted ionic-liquid etching. *Chem. Mater.* 27: 1432.

47. Kubendhiran S, Bao Z, Dave K, Liu RS. (2017) InP nanocrystals with color-tunable luminescence by microwave-assisted ionic-liquid etching. *Chem. Mater.* 29: 2101.

48. Kubendhiran S, Bao Z, Dave, K, Liu RS. (2019) Microfluidic synthesis of semiconducting colloidal quantum dots and their applications. *ACS Appl. Nano Mater.* 2: 1773.

49. Kwon BH, Lee KG, Park TJ, Kim H, Lee TJ, Lee SJ, Jeon DY. (2012) Continuous in situ synthesis of ZnSe/ZnS (core)shell quantum dots in a microfluidic reaction system and its application for light-emitting diodes. *Small* 8: 3257.

50. Volk AA, Epps RW, Abolhasani M. (2021) Accelerated development of colloidal nanomaterials enabled by modular microfluidic reactors: Toward autonomous robotic experimentation. *Adv. Mater.* 33: 2004495.

51. Edel JB, Fortt R, DeMello JC, DeMello AJ. (2002) Microfluidic routes to the controlled production of nanoparticles. *Chem Commun (Camb)* 10(2002): 1136.

52. Nightingale AM, de Mello JC. (2009) Controlled synthesis of III-V quantum dots in microfluidic reactors. *Chemphyschem* 10: 2612.

53. Baek J, Allen PM, Bawendi MG, Jensen KF. (2011) Investigation of indium phosphide nanocrystal synthesis using a high-temperature and high-pressure continuous flow microreactor. *Angew. Chem. Int. Ed.* 50: 627.

54. Kim K, Jeong S, Woo JY, Han CS. (2012) Successive and large-scale synthesis of InP/ZnS quantum dots in a hybrid reactor and their application to white LEDs. *Nanotechnology* 23: 065602.

55. Ippen C, Schneider B, Pries C, Kröpke S, Greco T, Holländer A. (2015) Large-scale synthesis of high quality InP quantum dots in a continuous flow-reactor under supercritical conditions. *Nanotechnology* 26: 085604.

56. Baek J, Shen Y, Lignos I, Bawendi MG, Jensen KF. (2018) Multistage microfluidic platform for the continuous synthesis of III–V core/shell quantum dots. *Angew. Chem. Int. Ed.* 57: 10915.

57. Dai X, Deng Y, Peng X, Jin Y. (2017) Quantum-Dot Light-Emitting Diodes for Large-Area Displays: Towards the Dawn of Commercialization. *Adv. Mater.* 29: 1607022.

58. Chen HW, Zhu RD, He J, Duan W, Hu W, Lu YQ, Li MC, Lee SL, Dong YJ, Wu ST. (2017) Going beyond the limit of an LCD's color gamut. *Light Sci. Appl.* 6: e17043.

59. Luo Z, Xu D, Wu ST. (2014) Emerging quantum-dots-enhanced LCDs. *J. Disp. Technol.* 10: 526.

60. Luo Z, Chen Y, Wu S. (2013) Wide color gamut LCD with a quantum dot backlight. *Opt. Express.* 21: 26269.

61. Zhao Y, Riemersma C, Pietra F, Koole R, de Mello Donegá C, Meijerink A. (2012) High-temperature luminescence quenching of colloidal quantum dots. *ACS Nano* 6: 9058.

62. Srivastava AK, Zhang W, Schneider J, Halpert JE, Rogach AL. (2019) Luminescent down-conversion semiconductor quantum dots and aligned quantum rods for liquid crystal displays. *Adv. Sci.* 6: 1901345.

63. Coe-Sullivan S. (2016) 20-1: invited paper: the quantum dot revolution: marching towards the mainstream. *SID Symp. Digest Tech. Pap.* 47: 239.

64. Thielen J, Lamb D, Lemon A, Tibbits J, Derlofske JV, Nelson. (2016) 27-2: invited paper: correlation of accelerated aging to in-device lifetime of quantum dot enhancement film. *SID Symp. Dig. Tech. Pap.* 47: 336.

65. Hwang D, Mughal A, Pynn CD, Nakamura S, DenBaars SP. (2017) Sustained high external quantum efficiency in ultrasmall blue III–nitride micro-LEDs. *Appl. Phys. Exp.* 10: 032101.

66. Huang Y, Hsiang EL, Deng MY, Wu ST. (2020) Mini-LED, Micro-LED and OLED displays: present status and future perspectives. *Light Sci. Appl.* 9: 16.

67. Liu Z, Lin CH, Hyun BR, Sher CW, Lv Z, Luo B, Jiang F, Wu T, Ho CH, Kuo HC, He JH. (2020) Micro-light-emitting diodes with quantum dots in display technology. *Light Sci. Appl.* 9: 83.

68. Shirasaki Y, Supran GJ, Bawendi MG, Bulović V. (2012) Emergence of colloidal quantum-dot light-emitting technologies. *Nat. Photonics* 7: 13.

69. Qian L, Zheng Y, Xue J, Holloway PH. (2011) Stable and efficient quantum-dot light-emitting diodes based on solution-processed multilayer structures. *Nat. Photonics* 5: 543.

70. Dai X, Zhang Z, Jin Y, Niu Y, Cao H, Liang X, Chen L, Wang J, Peng X. (2014) Solution-processed, high-performance light-emitting diodes based on quantum dots. *Nature* 515: 96.

71. Sadeghi S, Abkenar SK, Ow-Yang CW, Nizamoglu S. (2019) Efficient white LEDs using liquid-state magic-sized CdSe quantum dots. *Sci. Rep.* 9: 10061.

72. Kim T, Kim KH, Kim S, Choi SM, Jang H, Seo HK, Lee H, Chung DY, Jang E. (2020) Efficient and stable blue quantum dot light-emitting diode. *Nature* 586: 385–389.

73. Yuan Q, Wang T, Yu P, Zhang H, Zhang H, Ji W. (2021) A review on the electroluminescence properties of quantum-dot light-emitting diodes. *Org. Electron.* 90: 106086.

74. Deng Y, Lin X, Fang W, Di D, Wang L, Friend RH, Peng X, Jin Y. (2020) Eciphering exciton-generation processes in quantum-dot electroluminescence. *Nat. Commun.* 11: 2309.

75. Cao W, Xiang C, Yang Y, Chen Q, Chen L, Yan X, Qian L. (2018) Highly stable QLEDs with improved hole injection via quantum dot structure tailoring. *Nat. Commun.* 9: 2608.

76. Chen S, Cao W, Liu T, Tsang SW, Yang Y, Yan X, Qian L. (2019) On the degradation mechanisms of quantum-dot light-emitting diodes. *Nat. Commun.* 10: 765.

77. Sun Y, Jiang Y, Sun XW, Zhang S, Chen S. (2019) Beyond OLED: efficient quantum dot light-emitting diodes for display and lighting application. *Chem. Rec.* 19: 1729–1752.

78. Xiang C, Wu L, Lu Z, Li M, Wen Y, Yang Y, Liu W, Zhang T, Cao W, Tsang SW, Shan B. (2020) High efficiency and stability of ink-jet printed quantum dot light emitting diodes. *Nat. Commun.* 11: 1646.

79. Yang J, Choi MK, Yang UJ, Kim SY, Kim YS, Kim JH, Kim DH, Hyeon T. (2021) Toward full-color electroluminescent quantum dot displays. *Nano Lett.* 21: 26.

7 Colloidal Quantum Dots and Their Applications

Sheng Cao and Jialong Zhao

CONTENTS

7.1 INTRODUCTION

Quantum dots (QDs) are composed of a small number of atoms, which are quasi-zero-dimensional semiconductor nanocrystals that confine excitons in three-dimensional space. The carrier motion is limited in three-dimensional space in the QD, so sometimes QD is also called "artificial atom", "superatom" or "quantum dot atom", which is a concept put forward clearly in 1980s–1990s.[1] Modern QD technology can be traced back to the 1970s, which was developed to solve the global energy crisis. The initial research began in the early 1980s with the scientists from two laboratories: Louis Brus of Bell Laboratories and Alexander Efros of Yoffe Institute. Brus and his colleagues found that the different sizes of CdS QDs could produce different colors, and they proposed the theory of "quantum confinement effect".[2] Subsequently, the research on the luminescence characteristics and mechanism of colloidal CdS QDs has gradually become a hot topic in the world. These works are helpful to understand the quantum confinement effect of QDs, reveal the relationship between the size and color of QDs and lay a foundation for the application of QDs.

The size of QDs is generally 1–10 nm. Due to the quantum confinement of electrons and holes, the conduction and valence bands (VBs) of QDs are not continuous but become discrete energy levels with molecular characteristics, which can emit fluorescence that is higher than the bandgap of the bulk material after excitation. After nearly 40 years of development, QDs have gone through the process from a single structure to a complex system with different components and structures, which mainly benefits from the continuous development and optimization of chemical synthesis methods as well as in-depth understanding of the growth mechanism of QDs. So far, chemistry and material scientists have played a leading role in this field. Since the development of hot-injection synthesis technology with the continuous improvement and maturity of QD preparation technology in 1993, research have started for the application of QDs in biological and electronic devices.[3-5] In particular, Alivisatos group and Nie group published papers on QDs as biological probes in 1998,[6,7] which solved the problem of QDs dissolving in aqueous solution for the first time and proved how to combine QDs with biological macromolecules through surface groups, which sets off a research upsurge of QDs in biomedical application. With the help of various physical effects of QDs, such as quantum size, dielectric confinement, surface, quantum tunneling and Coulomb blocking effects, it can be predicted that QDs will be widely used in solar energy conversion, luminescence and display, biolabeling and ultrasensitive detection in the near future.

In this chapter, we summarize the development of colloidal QDs around their preparation, structure, optical properties and applications. From the perspective of QD preparation and luminescence mechanism, we first introduce the basic optical properties of QDs, including the quantum confinement effect, the luminescence principle, the overview of the preparation and the structure of QDs. Then, we review the research history and progress of several important QDs, such as II-VI, III-V, I-III-V, IV, lead halide perovskite and the corresponding doped QDs. The optical properties of QDs associated with their potential applications in field of light-emitting devices, bioluminescence imaging and optical temperature sensors are further discussed. Finally, we put forward our personal viewpoints on the future research interest and direction of QDs.

7.2 THE FUNDAMENTAL OPTICAL PROPERTIES OF QDs

The structure of high-quality optical colloidal QDs is shown in Figure 7.1, which is mainly composed of inorganic nanocrystal core, shell and organic surface ligands. Among them, the nanocrystal core usually determines the luminescent color of the material to a large extent. The shell structure is mainly used to control the optical properties and stability of QD. The introduction of surface ligands makes the as-synthesized QDs have a good solubility and play a role in passivating the surface. Some basic optical properties of QDs will be introduced in detail in the following content.

7.2.1 QUANTUM CONFINEMENT EFFECT

Colloidal QDs are nanoscaled single-crystal semiconductor materials synthesized by solution method. When a semiconductor QD absorbs photons, the electrons in its VB are excited to the conduction band (CB), and holes corresponding to the excited electrons are generated in the VB. Such photogenerated electron and hole are attracted to each other by strong Coulomb interaction in the QD and behave as an exciton.

In the early 1980s, Brus et al. discovered that QDs could absorb and emit light with shorter wavelength than their bulk ones.[2] In addition, the optical properties of QDs are found related to their

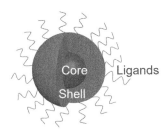

FIGURE 7.1   Schematic of the typical structure of a high-performance colloidal QD.

own size.[8] To explain this experimental phenomenon, the quantum confinement effect is proposed. According to this theory, when the geometric size of semiconductor is gradually reduced to less than the exciton Bohr radius of bulk materials, the energy levels of VB and CB will change from continuous to discrete, as shown in Figure 7.2(a). The relationship between the energy level of the first excited state and its size can be expressed quantitatively. The ground state energy level is the reference zero point, and the specific expression of the first transition state energy level is as follows[8]:

$$E = E_g + \frac{h^2\pi^2}{2R^2}\left(\frac{1}{m_e} + \frac{1}{m_h}\right) - \frac{1.8e^2}{\varepsilon R} \tag{7.1}$$

where R is the radius of the QD, m_e and m_h are the effective masses of electrons and holes respectively, and ε is the dielectric constant of the material. The first term E_g is the bandgap of semiconductor bulk materials, the second term is the energy change brought by quantum confinement effect and the third term is the Coulomb interaction energy of electrons and holes. It can be seen that the smaller size of the QDs is, the more significant influence of quantum confinement on the photogenerated electrons and holes carry, and thus increase the bandgap. The results show that with the decrease of QD size, the corresponding energy of photon absorption and emission increases and the exciton absorption and photoluminescence (PL) peaks gradually shift to blue.

The quantum confinement effect makes the luminescence of QDs have the characteristics of continuously tunable wavelength, as shown in Figure 7.2(b). The absorption spectra of QDs reflect the absorption of different energy levels. For the band edge or the first exciton level, there is an obvious absorption peak, usually called the first exciton absorption peak, in the absorption spectrum due to the determination of the bandgap. The first exciton absorption peak usually shifts to red with the increase of the size of QDs. While for the higher exciton energy states, the exciton absorption peak becomes smooth in a shorter wavelength range. Compared with organic dye molecules, QDs have wider absorption band and stronger absorption capacity.

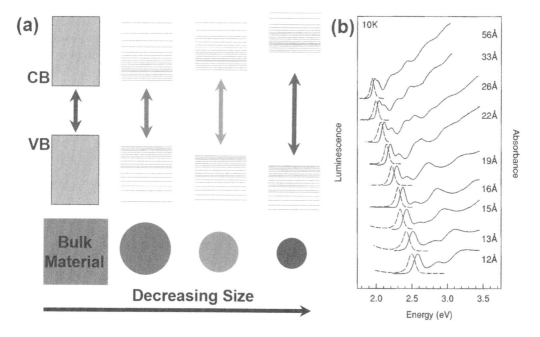

FIGURE 7.2 (a) Schematic diagram of quantum confinement effect on the energy levels of semiconductor material. (b) Normalized absorption and PL spectra of CdSe QDs at 10K with various sizes. (Efros, A. et al. *Phys. Rev. B* 54, 4843–4856, 1996. With permission)

In general, compared with other luminescent materials, QDs with quantum confinement effect have the advantages of enhanced exciton binding energy, wide absorption, narrow emission, high PL quantum efficiency, high thermal and chemical stability and so on. In addition, the modification of organic ligands on the surface of QDs can prevent the aggregation of QDs and make them stable in the solution, so that they have better solution processability. With their excellent optical and structural properties, QDs have wide application prospects in lighting, display, photovoltaic devices, biomedical science and other fields.

7.2.2 THE LUMINESCENCE PRINCIPLE OF QDS

The luminescence mechanism of QDs is similar to that of bulk semiconductors. Figure 7.3(a) describes the basic process of exciton state luminescence in QDs. When the QD absorbs the excited photons, the electrons in the VB will be excited to the energy level of the CB (I process). Then, the electron and hole reach the bottom of CB and the top of VB, respectively, through the vibration relaxation process, forming the electron-hole pair by Coulomb interaction, which is called exciton (II process). If the exciton is recombined by radiation, it will release photons corresponding to its band edge energy (III process). Since the exciton emission of QDs mainly comes from the photons emitted by the first excited state, the band edge emission process is the intrinsic emission of the first excited state of QDs by default.

Beside the intrinsic exciton emission, as shown in Figure 7.3(b), the defect state emission may also exist in the QDs. In the process of QD growth, the hanging bond of anion and cation is easy to appear on the surface, which is generally considered the surface defect states. The surface defects of QDs can capture photo-generated carriers (IV process) and make electrons or holes enter the defect level directly in the form of vibration relaxation and/or electron tunneling (V process). These trapped electrons or holes have radiative recombination (defect state luminescence, VI process) or non-radiative recombination (VII process) with remaining holes or electrons in the QD or in the same defect level. Compared with the exciton state, the defect state luminescence is not only inefficient but also a red-shifted in the emission wavelength. To avoid the defect states, it is needed to find an effective way to eliminate the surface defects of QDs, so that they cannot capture the carriers generated in the QDs. The organic ligands on the surface of QDs can not only prevent the aggregation effect among QDs but also participate in the elimination of the anion and cation suspension bonds. Therefore, organic ligands are important factors to ensure the high PL quantum yield (QY) of QDs and also an essential part of the colloidal QD system.[9]

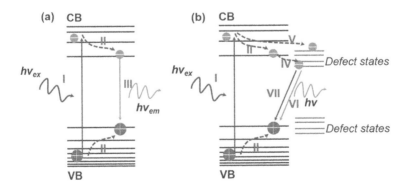

FIGURE 7.3 Schematic diagram of exciton state (a) and defects (b) luminescence process of QD. I refer to the process of QDs absorbing photons. II is the process of vibrational relaxation of hot carriers. III is the process of exciton radiative recombination. IV is the process in which band edge excitons are trapped by defect states. V is the process of thermal electron vibration relaxation to the defect energy level on the surface of QD. VI is the process of defect state luminescence. VII is the process of electron hole non-radiative recombination in the defect state.

7.2.3   OVERVIEW OF THE PREPARATION OF QDs

The preparation of QDs plays a key role in the research and development of QDs. Only stable and reliable QD materials can be used to further study the basic properties and industrial applications of QDs. In the 1980s, due to the imperfect synthesis technology, the optical properties of QDs were not improved significantly. Until the 1990s, Bawendi group[5] proposed a new synthesis method based on the pyrolysis of organometallic reagents in thermal coordinating solvent and synthesized high-quality CdSe QDs. This is a milestone breakthrough because they solved the two major problems that had not been solved by the previous synthesis methods, namely, the uneven particle size distribution and the low luminous efficiency caused by surface defects. However, there are also limitations, such as the materials used are toxic and high cost. This is also the reason for the slow development of QD synthesis technology in the past ten years. Until about 2000, Peng group[10–12] proposed a simple green synthesis route. They used cadmium oxide instead of dimethylcadmium and synthesized high-quality QDs by adjusting the solvent type, precursor reaction activity and other conditions. Since then, a large number of synthesis methods have emerged, and the research of QDs has stepped into a new stage.

In recent years, through the study of the mechanism of colloidal QDs, it is found that the two key factors for the synthesis of high-quality QDs are rapid nucleation and slow growth.[4,13] The main factors affecting nucleation are temperature and precursor concentration.[9,14,15] In the QD synthesis reaction system, the nucleation and growth of QDs follow the "Lamer diffusion and defocusing model",[14,16] as shown in Figure 7.4. With the deepening of research, researchers found that when the monomer concentration in the reaction exceeds the critical value (C^*), a large number of small nuclei will be produced explosively. As the monomer is consumed, the concentration of the reactants decreases. This is because the particle size distribution is inversely proportional to the growth rate in this model. So, when the concentration is below the critical value, no new nuclei are produced. In contrast, larger particles grow more slowly than smaller ones. The particles with small size have large chemical potential. The defocusing effect causes the small particle to break up and melt, which is swallowed up by the QDs with large size and keeps the QDs growing slowly. Chemical equilibrium is a dynamic equilibrium process, and the growth of QDs is no exception. After nucleation, the smaller nuclei will dissolve due to the higher surface free energy and grow again into monomers on the larger nuclei. This process of large particles "engulfing" the smaller particles is called "Ostwald ripening". Ostwald ripening can cause the core size to lose focus and widen the full width at half maximum (fwhm) of the luminescence.[17] Therefore, we should pay attention to the control of reaction time to avoid Ostwald ripening under high temperature for a long time.

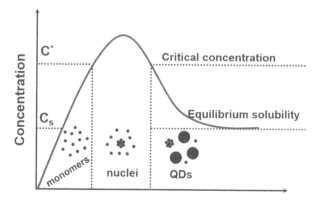

FIGURE 7.4   Schematic diagram of nucleation and growth process of nanocrystals in solution. The precursor is dissolved in the solvent to form monomer, then nucleated and finally promoted the growth of QDs through the aggregation of nuclei.

At present, QD synthesis can be divided into two categories according to the reaction solvent. One is aqueous synthesis, which is a direct method of synthesis of QDs in aqueous solution. The preparation of QDs by aqueous phase method is simple, low cost and low toxicity and can be applied to biomedical research without further surface modification.[18] It should be realized that the low boiling point of aqueous solvent is not conducive to the growth of QD crystal. Generally, there are many defects in the synthesized products, and the resulting fwhm of fluorescence emission is wide and the PL QY is not high. After a lot of efforts by researchers, some high-quality QDs, such as CdTe-based water QDs, have been reported, with PL QY in the range of 60–85%.[19,20]

Organic phase synthesis, usually referred to high-temperature solution pyrolysis, is another way to prepare QDs. This method is usually carried out in high-temperature organic solvents, through the injection of various precursors and thermal decomposition to synthesize a series of QDs.[5,14,16,21] Due to the high temperature in the synthesis of QDs, the QDs obtained have good crystallinity and monodispersity, and the fluorescence QY is high, usually more than 60%, and can be well dissolved in a variety of organic solvents. After proper surface modification, these oil-soluble QDs can also be transferred to aqueous phase for further application in the field of biological detection.

High-temperature solution pyrolysis is mainly divided into hot-injection[22–24] and one-pot[22] method, which have become the main method to synthesize high-quality QDs. The hot-injection method is to inject precursor solution (such as metal precursor and sulfur group precursor) into organic reaction solvent rapidly at high temperature, and the reaction solution nucleates rapidly at high temperature. As the reaction proceeds, the concentration of monomers in the solution decreases. When a critical value is reached, the QDs enter into a slow growth process, and the size increases gradually. After Ostwald ripening process, the QDs needed are finally synthesized. One-pot method is to mix all reagents into the reaction vessel in advance, control the growth of QDs by heating the reaction solution and adjusting the proportion and temperature of components. This method can be used to prepare high-quality QDs at low temperature, which is easy to operate and convenient for large-scale growth. For example, Zhang et al.[25] developed a scalable, reproducible and low-cost synthesis method for preparing high-quality CdS/Zn$_x$Cd$_{1-x}$S, CdSe/Zn$_x$Cd$_{1-x}$S and CdTe/Zn$_x$Cd$_{1-x}$S QDs with gradient alloy shell structure by directly heating commercial available CdO, Zn(NO$_3$)$_2$ and chalcogenide elements in octadecene media at air. As shown in Figure 7.5, the luminescence color of the obtained QDs can be easily adjusted from ultraviolet (UV) to near infrared (NIR) by simply changing the reactant and feed ratio. In addition, the PL efficiency of the as-prepared QDs can reach up to 80%.

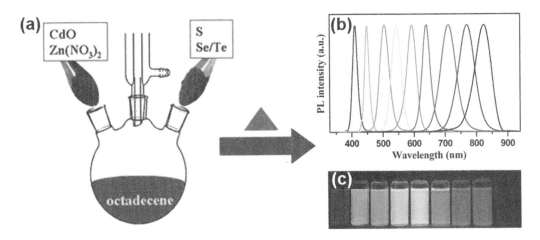

FIGURE 7.5 (a) Schematic single-step one-pot synthesis of QDs. (b) PL spectra of QDs with emission wavelength ranging from ultraviolet to NIR window. (c) Typical photos of QDs irradiated by ultraviolet light. (Zhang, W. et al. *ACS Nano* 6, 11066–11073, 2012. With permission)

7.2.4 QD STRUCTURES

As mentioned previously, QD consists of nanocrystal core, shell and organic ligands. Nanocrystal core is the emission center of QD that largely determines the emission wavelength and fwhm of QD. Because the average diameter of QDs is only a few nanometers, this high surface-to-volume ratio will lead to the optical properties of QDs affected by the surface trap states caused by dangling bonds and surface defects. Generally, such trap states in the electronic energy level structure of QDs capture the excited electrons or holes, which leads to the non-radiative recombination of QDs and the deterioration of their optical properties.[26–28] Therefore, the bare core QDs are very sensitive to the changes of the surrounding environment, showing low luminous efficiency and stability. Although organic ligands can greatly improve the luminous efficiency of QDs, it is not enough for QDs to maintain high fluorescence efficiency and photochemical stability. Generally speaking, to obtain an ideal exciton state luminescence, inorganic shell materials must be grown on the QD core materials.[29] In addition, the growth of an appropriate inorganic shell on the QD core material can not only adjust the exciton state luminescence behavior of the QD but also make the excited state luminescence center of the QD far away from the dangling bonds (defects) effect. Therefore, core-shell QDs have more stable optical properties and higher fluorescence quantum efficiency, which are also the most widely used structure in the field of application.

In the core-shell QDs, the shell plays different roles according to the bandgaps and relative position of electronic energy levels of the involved materials. The energy band arrangement of common materials is shown in Figure 7.6(a). According to the energy-level alignment of core and shell materials, the core-shell QD can be divided into three types, denominated type-I, reverse type-I and type-II band alignment (Figure 7.6(b)).[30–33] Type-I means that the bandgap of the shell material is larger than that of the core and both electrons and holes are confined in the core. Reverse type-I means that the bandgap of the shell material is smaller than that of the core. In this case, the holes

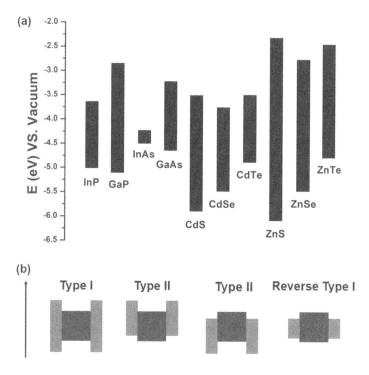

FIGURE 7.6 (a) Electronic energy levels of typical semiconductors. (b) Diagram of energy level alignment in QD core-shell system. The upper and lower edges of the rectangle correspond to the positions of the conduction and valence bands of the core (blue color) and shell material (green color), respectively.

and electrons are partially or completely confined in the shell according to the thickness of the shell. Type-II means that the VB edge or CB edge of the shell material is located in the bandgap of the core, and the energy band positions of the core and shell materials are staggered. When QDs are excited, the type II structure will lead to the spatial separation of holes and electrons in different regions of the core-shell structure.

In type-I structure QDs, the main purpose of shell is to passivate the core surface. To improve the luminescence intensity and photochemical stability of QDs, the core with optical activity is physically separated from the surrounding medium by shell. This is due to the fact that the surface of nanocrystal core is usually acted as the carrier trapped state energy level, which leads to the decrease of PL efficiency.[34] The typical core-shell QDs of type-I prototype structure are CdSe/ZnS, in which the bandgap of ZnS (3.61 eV) is larger than that of CdSe core (1.74 eV), and the CB minima (CBMs) (the VB maxima [VBMs]) of shell is significantly higher (lower) than that of core, respectively. This makes the exciton subject to strong spatial confinement effect, and the wave functions of electrons and holes are mainly confined in the CdSe core. The experimental results at the level of solution aggregation and single particle show that the stability of CdSe/ZnS core-shell QDs is much better than that with single structure of CdSe.[35,36] However, it should be noted that although the early literature considered that the luminescence performance of CdSe/ZnS core-shell QDs is better than that of CdSe single structure QDs. Recent literature showed that CdSe single structure QDs could approach the near-unity PL QY.[37] As mentioned earlier, the main advantage of type-I core-shell QDs is to improve their photophysical and photochemical stability.

In reverse type-I structure QDs, since the carriers are dispersed in the shell materials to some extent, the PL peak of the reverse type-I QDs shows a significant red shift. By using this property, the PL emission wavelength can be adjusted by the thickness of shell. These structures are mainly studied in CdS/HgS,[38] CdS/CdSe[39,40] and ZnSe/CdSe.[40–42] The photobleaching resistance and PL QY of these structures will be greatly improved by coating a wider bandgap shell on their surface.

In the type-II structure QDs, the main purpose of coating the shell is to make the emission wavelength of the QD red-shift obviously. The staggered energy level will have a smaller effective energy level for core and shell materials. Research on the structure to control the thickness of the shell to obtain the emission color is required, which is difficult to be achieved by other materials. Type II has been applied to the preparation of NIR QDs, such as CdSe/CdTe,[43,44] ZnSe/CdS[45,46] and CdSe/ZnTe.[47,48] Compared with type-I structure QDs, the fluorescence decay lifetime of type-II QDs is significantly longer. In this type of core-shell QDs, one of the carriers (electrons or holes) is located in the shell, and the QDs with appropriate shell materials have the similar effect to that of type-I system in improving the luminous efficiency and light stability.

Core-shell QDs have become the focus of scientists. It is found that the high properties of core-shell QDs generally require shell epitaxial growth. Therefore, the energy level structure of core-shell material is not the only standard of structure design, and the key principle of its design is that the core-shell material needs the same crystal structure and small lattice mismatch degree. On the contrary, the growth of the shell leads to strain, which will form defect states at the core-shell interface and/or in the shell. These defects can be act as trap states of photogenerated carriers and reduce the fluorescence quantum efficiency.

7.3 SELECTED QD SYSTEMS

In recent years, the preparation of QDs has received extensive attention and development. Figure 7.7(a) shows different separated fluorescent QDs developed to date and their tunable emission spectrum windows cover from UV to NIR region. In particular, almost 100% fluorescent quantum efficiency has been realized in CdSe, InP and perovskite-based systems. Meanwhile, a very active subfield in this area involves modification of properties of QDs by incorporating optically active dopants, typically transition ions of Mn and Cu, into the lattice of the host QDs. The range of fluorescent colors can also cover the whole visible region with the use of two typical transition methods are shown

(a) *Tunable emission colors from QDs*

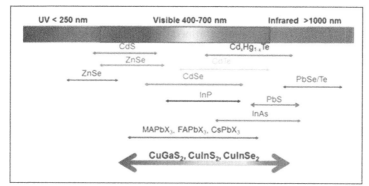

(b) *Tunable emission colors from doped QDs*

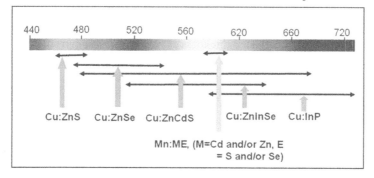

FIGURE 7.7 Schematic presentation of tunable emission range of different (a) intrinsic QD (Chen, B., Pradhan, N., and Zhong, H. *J. Phys. Chem. Lett.* 9, 435–445, 2018. With permission) and (b) doped QD (Pradhan, N. and Sarma, D. D. *J. Phys. Chem. Lett.* 2, 2818–2826, 2011. With permission) systems.

in Figure 7.7(b). This section discusses progress in the synthesis of different families of highly luminescent semiconductor QDs, which aims to discuss some common problems in the preparation process of QDs, hoping to stimulate further thinking and discussion of the readers.

7.3.1 Group II-VI QDs

The representative of group II-VI QDs is binary metal chalcogenide, which can be expressed by the general formula ME (M=Cd, Hg, Zn etc., E=S, Se and Te). To date, there are many examples of colloidal QD synthesis in this kind of system. II-VI semiconductor materials have generally direct bandgaps (for example, 2.5, 1.74 and 1.45 eV for CdS, CdSe and CdTe, respectively) and good extinction coefficients, which lead to a huge demand for applications.[17,18,49] CdE (E=S, Se, Te) is one of the most deeply studied colloidal QDs. Most of them are prepared by hot-injection method, which can accurately control the particle size, shape and particle size distribution. After nearly 40 years of development, the luminous efficiency of CdE QDs after surface passivation is probably near to unity. With the application of QDs becoming more and more mature, the demand for nontoxic compounds without Cd is increasing. Due to the chemical similarity between various transition metals, it is expected that other types of metal chalcogenide QDs can be prepared by using a synthesis route similar to that of Cd-based QDs.

In the development of high efficiency Cd-based QDs, researchers have devoted a lot of efforts to optimize the luminescent properties of core-shell QDs. The basic idea of core-shell structure of QD is to insert an insulating region between the excited state and the localized trapped state on the outer surface of core, so as to reduce or even eliminate the influence of surface trap states on the luminescence properties. In the CdSe/ZnS type-I core-shell structure, the excited state electron and hole wave functions of QD are limited to the core due to the sharp interface between ZnS shell and CdSe core; thus, the luminesce color is determined by the core size. However, due to the lattice mismatch between ZnS and CdSe, the interface is in the state of stress, and structural defects are formed, which leads to the degradation of the core luminescent properties.[50–52] It is found that the lattice strain can be reduced by using the buffer layer of alloying and/or composition gradient. For example, CdSe/CdS/Zn$_x$Cd$_{1-x}$S/ZnS QDs were prepared by introducing an alloying intermediate layer into the shell, and Zn$_x$Cd$_{1-x}$S alloy layer was prepared by using the successive ion layer adhesion and reaction (SILAR) technique, which resulted in the change of shell composition from CdS to ZnS in radial direction, and the luminous efficiency of the as-prepared QDs increased by about 80%.[51] In addition, proper shell modification can also suppress the "blinking" phenomenon of single QD. The so-called single QD blinking phenomenon refers to the "on" and "off" PL of single QD under the continuous excitation of high flux. It is found that when the surface of CdSe QDs is covered with 19 monolayers of CdS shell, the nonblinking behavior of the QDs is observed under the excitation of continuous light.[53] Especially in 2013, Qin et al.[54] reported a series of phase pure zinc blende CdSe-based QDs with CdS shell thickness of 4–16 monolayers. As shown in Figure 7.8(a), these CdSe/CdS QDs with different shell thickness exhibit nonblinking in single-exciton region. In addition, these core-shell QDs possess well-controlled single excited-state decay dynamics at both single-dot and ensemble levels. These results indicate that these unique bright nonblinking QDs have a good application prospect in both single-dot and ensemble levels.

The luminescent color of group II-VI QDs can be tuned by adjusting the composition of QD core, so that all colors in the visible spectrum can be realized. For example, Zhong et al.[55] reported an effective high-temperature synthetic strategy. A series of high-quality Zn$_x$Cd$_{1-x}$Se alloy QDs with emission wavelength of 460–630 nm were prepared by adding Zn and Se into prepared CdSe QD cores. Yuan et al.[56] reported the synthesis of CdZnSe/CdZnS core-shell alloy QDs. As shown in Figure 7.8(b), these QDs show PL emission covering a broad wavelength range of 470–650 nm and maintaining high PL QYs. It is worth noting that large-scale synthesis of number of high-efficient QDs have been reported,[22,57] which will greatly promote the practical application of QDs.

With the application of QDs becoming more and more mature, the demand for nontoxic compounds without Cd is increasing. The ZnSe QD is a promising alternative "green" blue luminescent nanocrystal material, in which ZnSe has a large bandgap of 2.7 eV and theoretically can achieve 460 nm emission wavelength.[15,30,58,59] To obtain high-quality ZnSe QDs with type-I core-shell structure, ZnS or ZnSe shell can be overcoated on the ZnSe core according to the energy band alignment requirements. ZnSe/ZnS QDs with tunable emission color are synthesized by one-pot, hot-injection and aqueous solution synthesis. These QDs have a tunable emission range of 390–450 nm, narrow fwhm and high fluorescence efficiency.[60] For example, Han et al.[61] reported a ZnSe/ZnSe/ZnS QDs via prepared ZnSeTe core and then coated with multi-shell. The resulting QDs exhibit a pure blue emission with a PL peak of 445 nm, an fwhm of 25 nm and a high PL QY of 84%. These ZnSe-based QDs with high luminous efficiency bring hope for the large-scale commercial application of QDs.

7.3.2 Group III-V QDs

Group III-V semiconductors are binary compounds formed by the combination of IIIA group metal and group VA nonmetal elements. Because of the unique properties such as high electron mobility, direct bandgap and small exciton binding energy, III-V semiconductors are widely used in high-performance optoelectronic devices.[62] Except for some nitrides, most III-V semiconductors have smaller lattice ionic properties than II-VI semiconductors. Compared with ion reactions, the formation of more covalent bonds usually requires more stringent reaction conditions, such as high reaction temperature, long reaction time and higher reactive precursors. These conditions are usually

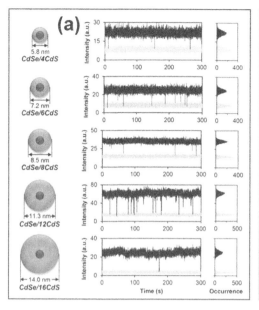
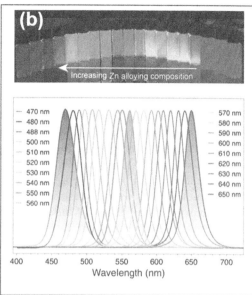

FIGURE 7.8 (a) PL intensity time traces of single zinc blende CdSe/CdS QDs. The core radius of these QDs is 1.55 nm, and the shell thickness varied from 1.36 to 5.44 nm, corresponding to 4–16 monolayers of CdS shell (Qin, H. et al. *J. Am. Chem. Soc.* 136, 179–187, 2014. With permission) (b) photograph (top) and PL spectra (bottom) of the CdZnSe/CdZnS alloy QDs, the PL emission profiles cover the full visible range from 470 to 650 nm (Yuan, Y. et al. *Chem. Mater.* 31, 2635–2643, 2019. With permission).

not conducive to the precise control of the shape and size of QDs. To obtain high-quality materials, it is necessary to optimize the activity and reaction conditions of precursors. In addition, III-V semiconductor such as InP QDs are more sensitive to air than II-VI QDs (such as CdSe), which makes most of the synthesis require oxygen free conditions.[61,63] At present, compared with II-VI QDs, the development of the III-V QDs is still in the exploratory stage. In this section, the development of InP QDs with high PL QY in the typical group III-V is briefly introduced.

The synthesis of InP QDs can be traced back to the 1990s.[64,65] The early InP QDs have no shell structure, and their PL spectra usually show wide emission, asymmetry and sometimes tailing. These problems are closely related to the surface traps, in which the non-radiative recombination takes place, and the PL is suppressed to a large extent. The wide fwhm emission is sometimes due to the wide particle size distribution. In 2002, InP QDs with symmetric PL spectra were synthesized by using uncoordinated solvents and strictly controlled reaction conditions.[66] It is found that the key to solve the problems of wide emission, long tail in PL spectrum is to fine-tune the synthesis conditions of the reaction. However, the PL QY of all InP QDs in solution is still less than 10%.

To improve the luminescent efficiency of InP QDs, the InP/ZnCdSe$_2$ core-shell QDs were designed and prepared.[67] However, due to the small offset between the minimum CB of InP and ZnCdSe$_2$, the electron confinement effect of the shell is weak, and the luminescent efficiency of as-synthesized QDs is not high. For this reason, researchers began to focus on wider bandgap semiconductors, such as ZnS, as potential shell materials. Due to the large lattice mismatch between InP and ZnS, the epitaxial growth of ZnS shell on InP core is not easy to achieve.[68] After years of efforts, a new method to synthesize InP/ZnS core-shell QDs at relatively low temperatures was developed. By this method, the emission color of QDs can be adjusted in the range of 480–750 nm, and their luminescent efficiency can be increased to 40%. The PL dynamics study shows that the poor PL of InP QDs is mainly caused by the electron capture on the InP surface.[69] To improve the optical properties of InP/ZnS QDs, it is necessary to reduce the defects on the surface of InP core.

It is found that InZnP/ZnS and InPZnS/ZnS QDs with alloy core can better limit electrons,[70,71] which inspires the concept of core/shell/shell QDs. The ZnS is considered the best material for the outermost layer of InP QDs because ZnS has good optical stability and air stability, and wider bandgap. To reduce the stress defects at the core-shell interface, two promising intermediate shell candidates, GaP and ZnSe, were designed. Kim et al.[72] reported highly stable and luminescent InP/GaP/ZnS QDs. The use of GaP shell passivates the surface and removes the traps, resulting in the as-obtained QDs with a maximum PL QY of 85%. Park et al.[73] introduced the same structure of InP/GaP/ZnS core/shell/shell QDs and obtained a full color range from blue to red, with high luminous performance (blue QD QY: ~40%, fwhm: 50 nm; green QD QY: ~85%, fwhm: 41 nm; red QD QY: ~60%, fwhm: 65 nm). In 2019, Peng group[74] used stoichiometric ratio to strictly control the proportion of anions and cations in the core-shell growth of QDs and eliminated the influence of acetic acid produced in the reaction process. Green and red InP/ZnSe/ZnS QDs with 90 and 100% luminescent QYs were synthesized. In the same year, Won et al.[75] used hydrofluoric acid to remove the oxide on the surface of InP nucleus in the synthesis of QDs and coated double shell at high temperature. The obtained InP/ZnSe/ZnS QDs exhibit luminescent peak of 630 nm with fwhm of 35 nm and unity luminescent QY. At present, the preparation of high-quality InP-based QDs is developing rapidly toward large-scale applications.

7.3.3 I-III-V QDs

Ternary I-III-VI (I=Cu, Ag; III=In, Al, Ga; VI=S, Te, Se) QDs have been received grown interest due to the low toxic and environment friendly elements as compared to binary II-VI QDs.[76,77] Ternary metal chalcogenides can be derived from binary compounds by replacing two divalent metals with one monovalent and one trivalent cation, or two trivalent cations with one divalent and one tetravalent cation.[78] It should be pointed out that ternary semiconductors are certainly not limited to metal chalcogenides. Due to the similarity of basic synthesis concepts and precursors of metals and sulfur elements, the preparation methods and growth mechanism of ternary and multivariate QDs are similar to those of binary metal sulfur compounds. The main challenge in the synthesis of multicomponent QDs is to balance the reactivity of different metal precursors, which is also a prerequisite for controlling their composition and crystal phase.[76] So far, there have been a large number of reports on the research of I-III-VI QDs, especially $CuInS_2$, $CuInSe_2$, $AgInS_2$ and $AgInSe_2$.

Compared with the groups II-VI and III-V QDs, I-III-VI QDs have three remarkable optical properties, that is longer fluorescence lifetime (100–300 ns), larger Stokes shift (300–500 meV) and wider fwhm (>300 meV).[76,78] Among them, the longer fluorescence lifetime is related to the larger Stokes shift and the recombination of the defect energy levels, while the broad fwhm is mainly due to the chemical component and/or size dispersion of the multiple QDs.[79] Taking Cu–In–S QDs as an example, Yang et al.[80] thought that the large luminescent linewidth of $CuInS_2$ QDs was mainly related to the distribution of the distance between the donor and the acceptor and the interaction between photons and phonons. To reduce their size distribution, the size separation of $CuInS_2$ QDs with different particle size ranges was carried out by selective precipitation. The results show that the fwhm of QDs is still as high as 250 meV. However, Klimov et al.[81] found that the fwhm of single QDs is significantly reduced from 300 meV of ensemble samples to 60 meV. Therefore, the main reason for the larger fwhm of $CuInS_2$ QDs is that there are randomly distributed emission centers in the QD host.

The luminescent color of I-III-VI QDs is controlled by the size and chemical composition of QDs. As shown in Figure 7.9(a), the PL spectra obtained from $CuInS_2$ and $CuInSe_2$ QDs can be tuned in the visible and NIR spectral window.[76–78,82] In addition, the synthesis of I-III-VI QDs has also been extended to gram level, which is expected to be used as commercial materials.[76,83] Figure 7.9(b) shows the typical reaction devices and products in the large-scale synthesis of Cu-In-S-based QDs. The availability of gram high-quality QDs is expected to promote the practical application of QDs. The commission internationale de l'éclairage (CIE) chromaticity coordinates and possible color windows of $CuGaS_2$, $CuInS_2$ and $AgInS_2$ QDs are shown in Figure 7.9(c). The wide emission of most

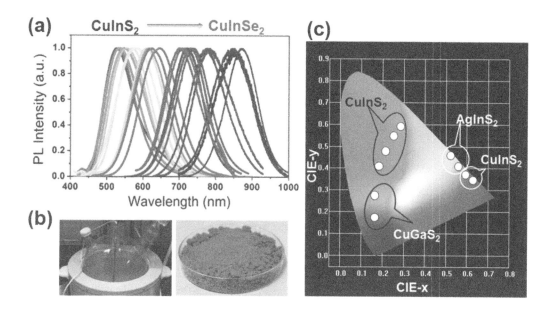

FIGURE 7.9 (a) PL spectra of CuInS$_2$ and CuInSe$_2$ QDs tuned by chemical composition and ZnS shell. (b) Digital photos of gram scale synthetic reaction flask and purified Cu-In-S-based QDs. (c) CIE chromaticity coordinates of CuInS$_2$, AgInS$_2$ and CuGaS$_2$ QDs. (Chen, B., Pradhan, N., and Zhong, H. *J. Phys. Chem. Lett. 9*, 435-445, 2018., with permission).

visible light regions and simple solution processing features make these QDs as ideal candidates for high-quality illumination.[77]

7.3.4 IV QDs

Group IV element QDs include carbon, silicon and germanium colloidal QDs, and their research is much less than compound QDs. One of the biggest differences between group IV QDs and the widely studied II-VI QDs is that the covalent properties of the bonds in silicon and germanium require high temperature to form crystals. It is reported that the bond dissociation energies of silicon and germanium are 327 and 274 kJ/mol, respectively, indicating that the two elements are easy to form stable amorphous phase.[17] In addition, the dissociation energies of Si–O and Ge–O are 798 and 662 kJ/mol, respectively,[17] which indicates that the two elements have a strong tendency to combine with oxygen. Compared with the surface ligand chemistry of compound semiconductor QDs, the chemical synthesis of carbon, silicon and germanium QDs is still challenging.

The synthesis method of group IV element QDs is not mature enough. Taking the synthesis of carbon QDs (CQDs) as an example, it mainly includes two strategies: top-down and bottom-up. The top-down strategy for the synthesis of CQDs includes the decomposition of large carbon structures such as nanodiamond, graphite, carbon smoke, activated carbon and graphite oxide by arc discharge, laser ablation and electrochemical oxidation. The bottom-up synthesis of CQDs can be realized from molecular precursors (such as citrate and carbohydrate) through hydrothermal/solvothermal, load synthesis, electrochemistry and microwave synthesis routes. At present, the CQDs with high luminous efficiency mainly come from the latter. Liu et al.[84] prepared CQDs with fluorescence QY of 75% by hydrothermal method using citric acid and trimethylaminomethane as precursors and water as reaction medium. Yuan et al.[85] synthesized red, green and blue CQDs as shown in Figure 7.10(a) by solvothermal method with threefold symmetric phloroglucinol (PG) as the reagent and ethanol as reaction medium. The as-synthesized CQDs exhibit high PL QY up to 54–72% and

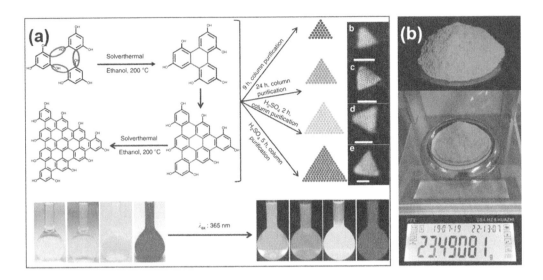

FIGURE 7.10 (a) Preparation of CQDs with full colors by solvothermal treatment, and photographs of CQDs under daylight and fluorescence images under UV light (Yuan, F. et al. *Nat. Commun.* 9, 2249, 2018. With permission) (b) photographs of a CQDs@g-C₃N₄ composite exposed to daylight and UV light (Meng, L. et al. *Mater. Chem. Front.* 4, 517–523, 2020. With permission).

narrow PL emission with fwhm of 29–30 nm. Qu et al.[86] used citric acid as carbon source and urea as nitrogen source to prepare green light CQDs with fluorescence quantum efficiency of 14% by microwave method. Meng et al.[87] reported an in-situ synthesis CDs@g-C$_3$N$_4$ composite phosphor via two-step microwave-assisted heating of citric acid and urea precursors. The as-obtained composite phosphors possess PLQY reaching 62%. A batch of synthesized luminescent CQDs is more than 20 g as shown in Figure 7.10(b), which provides the possibility for its potential application.

In general, the preparation of colloidal QDs of group IV elements, such as carbon, silicon and germanium, has not been able to summarize clear rules and related principles. However, in any case, it can be found that the synthesis of group IV element QDs is quite different from that of classical group II-VI compound QDs. It should be noted that although a lot of related research work has been reported, the size distribution, surface structure and optical properties of related QDs are still not comparable to those of classical II-VI QDs, and the in-depth study of synthesis mechanism is still lacking. For more details of group IV QDs, please refer to the latest review articles.[88–91]

7.3.5 Perovskite QDs

Perovskite QDs are a new type of luminescent materials developed in recent years.[92] They have shown impressive luminescent properties despite the publication only for a few years. For example, up to 90% of the PL quantum efficiency and high color purity (fwhm of 12–40 nm). The structure of perovskite is generally represented by ABX_3, where A refers to univalent inorganic or organic cations, such as Cs^+ and methylammonium (MA^+). B is a smaller cation, including Pb^{2+} and Sn^{2+}, and X is Cl^-, Br^- or I^-, or a mixture of them. In the perovskite structure, the luminescent property depends on the elements occupying A, M and X positions.

In 2015, Protesescu et al.[93] synthesized all inorganic $CsPbX_3$ QDs (X=Cl^-, Br^- or I^-, or halide mixture) for the first time by hot-injection method. It is found that the perovskite has rapid nucleation and growth kinetics, and the reaction time is only a few seconds. The as-prepared perovskite QDs have good monodispersity, and the color of QDs can be controlled by adjusting the type and content of halogen elements. As shown in Figure 7.11, the fluorescence emission of $CsPbX_3$ QDs can cover the whole visible region. The hot-injection synthesis method for $CsPbX_3$ QDs needs high

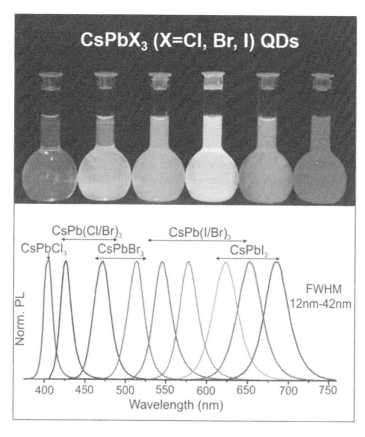

FIGURE 7.11 Photo of CsPbX$_3$ NCs (X=Cl, Br, I) solutions under UV lamp and representative PL spectra. (Protesescu, L. et al., *Nano Lett.* 15, 3692–3696, 2015. With permission)

reaction temperature, and the reaction needs to end quickly after injecting precursor (with the help of ice bath conditions). The injection process is relatively complex, and the quality of the synthesized QDs is highly dependent on the reaction environment. Therefore, room-temperature synthesis strategies such as ligand-assisted precipitation (LARP) and supersaturated recrystallization (SR) are proposed. In 2015, Zhong group[94] established a method of LARP of MAPbX$_3$ (X=Cl, Br, I) QDs at room temperature. Based on the fact that dimethylformamide (DMF) is a good solvent for dissolving small molecules and inorganic salts, PbBr$_2$, MABr, *n*-octylamine and oleic acid are first dissolved in DMF, and then formed a transparent and free layered precursor solution. LARP technique is realized by simply mixing MAPbX$_3$ precursor solution dissolved in DMF with long-chain organic ligands into a bad solvent such as toluene under strong agitation. The role of long-chain ligands is to control the crystallization process and promote the formation of stable colloidal MAPbX$_3$ QDs. The quantum efficiency of colloidal MAPbX$_3$ QDs solution obtained by this method is as high as 70%. In 2016, Zeng group[95] reported another room-temperature approach to synthesis all inorganic CsPbX$_3$ (X=Cl, Br, I) QDs, namely SR. CsPbX$_3$ QDs are synthesized by transferring Cs$^+$, Pb^{2+} and X from DMF to toluene at room temperature. Because the solubility of these ions in toluene is lower than that in DMF, they precipitate in the form of CsPbX$_3$ QDs under strong agitation. Although these CsPbX$_3$ QDs are formed at room temperature, they have blue, green and red emission with PL QY of 70, 80, 95% and fwhm of 18, 35 and 20 nm, respectively.

To date, high-temperature hot-injection and room-temperature synthesis have become the two most important strategies for preparing perovskite QDs.[92,96,97] In these methods, the nucleation and growth process of QDs can be analyzed according to the classical Lamer theory.[92] Although the colloidal perovskite QDs show excellent luminescent properties, there are still facing some key problems such as poor stability.[98] The perovskite QDs are less stable to high temperature, humidity and light. When polar water molecules interact with them, perovskite will decompose or dissolve. To address this issues, various strategies have been studied, including the introduction of a new surface capping agent or nanocomposite coating on perovskite QDs. Recently, core-shell (pseudo II type) $CsPbBr_3$/ZnS QDs have been prepared by Ravi et al.[99] Compared with pristine $CsPbBr_3$ QDs, the PL lifetime of core-shell QDs is increased by about 15 times, which provides the possibility to solve the stability of perovskite. In addition, most of the perovskite QDs contain lead, which is a heavy metal–limiting practical application. The lead-free perovskite QDs prepared by substituting Sn or Mn for lead are reported, but their properties are poor.[92,97,98,100] Therefore, to apply perovskite QDs to practical applications, further research is needed to solve these obstacles.

7.3.6 DOPED QDS

Doped QDs mainly refer to the introduction of other impurities in the semiconductor lattice, generally a small amount of transition metal ions or rare earth ions, forming semiconductor QDs with new properties. Doping can significantly change the optical and electrical properties of the QDs. High-quality doped QDs have not only the intrinsic luminescent properties of host QDs but also some other unique properties. For example, (i) the excitation wavelength of doped QDs depends on the host material, but the luminescent wavelength mainly depends on the nature of doped ions; (ii) the large Stokes shift; (iii) the luminescence stability and excellent photochemistry stability at high temperature; (iv) the doped QDs with specific emission wavelength can be obtained by properly selecting host materials and impurities.

In general, doping of colloidal semiconductor nanocrystals is more difficult to realize compared to their bulk counterparts. After years of exploration, people have made achievements in some systems to insert dopant ions into the lattice of colloidal QDs. Doped QDs can be synthesized by common hot-injection method (injecting dopant precursor into mixed cationic precursor), which can adjust the size, shape, composition and the corresponding or resulting optical properties. Dopants such as Mn or Cu ions can be doped in-situ during synthesis or after the formation of host QDs. In the synthesis of doped QDs, a large number of different doping systems have been successfully synthesized, although some materials and impurity ions are still not matched, which makes the understanding of doping growth mechanism need further study. The early view was that the solubility of impurities in QDs was similar to that in bulk materials, but this model could not explain the high content of Mn in II-VI group semiconductor materials.[101] Another view is "self-cleaning theory".[102] Compared with bulk materials, Mn doping content in QDs is lower. This view points out that when the QD is in thermal equilibrium, impurity ions can be easily ejected, thus maintaining the lattice integrity of the host. Erwin and Norris et al.[103] proposed that the successful doping of impurity ions can be divided into two processes: adsorption and growth. The impurity ions first adsorb on the surface of the host material and then grow new shells to make the impurity ions in the host material. According to this model, Mn-doped CdSe QDs were successfully synthesized. All of the three models have their own application scope, but they cannot fully explain the experimental results of a large number of doped QDs. In 2005, Pradhan and Peng et al.[104] proposed the synthesis route of "nucleation doping", separated the nucleation process and growth process of QDs and successfully obtained a series of high-quality doped QDs. At present, there are two main methods for the preparation of doped QDs, that is nucleation doping and growth doping, as shown in Figure 7.12(a). Nucleation-doping refers to the introduction of impurity ions into the host QD material when it grows and nucleates, so that the impurity ions are mainly doped in the center of the host material. Growth-doping is a doping method that makes the host material nucleate first, add impurity ions to adsorb on the surface of the host material and then continue to grow the host material on its surface.

Broadly, the luminescence of dopant ions in QDs can be divided into discrete center luminescence and composite center luminescence. For discrete center luminescence, such as Mn^{2+}, only the dopant

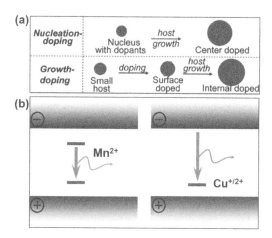

FIGURE 7.12 (a) Schematic presentation of nucleation- and growth-doping (Pradhan, N., *J. Am. Chem. Soc.* 127, 17586–17587, 2005. With permission) (b) recombination process of two types of dopant emissions, where one is dependent on the *d–d* states of Mn dopant (left), other uses both dopant and host energy states (right).

d–d states are involved. While for composite center luminescence, such as $Cu^{+/2+}$, both the induced energy states of the host state and the dopant state are involved.[105] Figure 7.12(b) presents the schematic diagram of recombination process of dopant emissions with two different types. As shown that the emission the *d–d* state emission of Mn-doped QDs is independent of the host bandgap, while the luminescent mechanism of Cu-doped QDs is only one dopant state in the recombination process.

The luminescence of Mn^{2+} is generally believed to be caused by spin forbidden relaxation between $^4T_1$ and $^6A_1$ states in tetrahedral coordination Mn^{2+} ions.[101,105–107] Under the light excitation, the exciton produced on the host transfers its energy to the Mn^{2+} states, which makes the spins of the Mn^{2+} *d* orbital electrons reverse and produces the *d–d* transition emission. In principle, this emission normally appears at approximately 2.12 eV (585 nm), and the fwhm of PL peak remains about 60–80 nm with an excited state lifetime in the range of several milliseconds.[108–111] Generally, strong crystal field and/or large crystal lattice torsional strain around Mn^{2+} sites can cause large crystal-field splitting of Mn^{2+} 3d orbitals, which correspondingly leads to gap contraction between $^4T_1$ and $^6A_1$ states and results in a red shift of Mn^{2+} emission. On this basis, by adjusting the position or lattice pressure of Mn^{2+} dopant ions, Mn^{2+} emission can be partially tuned in the orange-red part of the visible spectrum.[112] It is noted that the color of Mn^{2+} emission can even be adjusted to cover the whole visible spectrum by finely adjusting the strain produced through epitaxial growth of the upper shell of QD core.[113] Compared with the traditional Mn^{2+}-doped II-VI QDs, the crystal structure of multicomponent QDs is more complex, which will exacerbate the crystal field splitting of Mn^{2+} 3d orbits and endow Mn^{2+} ions with more abundant luminescence behavior.[37,109,112,114] To date, a series of Mn^{2+}-doped multicomponent QDs, such as Mn^{2+}-doped Cu–In–S,[108] Zn–In–S[109,110] and Cd–In–S[115] have been synthesized. It is worth noting that the near-unity QY and single-exponential decay of red Mn^{2+} emission peaking at 627 nm in doped $CsPbCl_3$ QDs have been reported recently,[114] which will greatly promote the application of Mn^{2+}-doped QDs in solid state lighting.

The color of the doped QDs with composite luminescent centers is tunable, among which Cu-doped QDs are the most representative. These doped ions have high quantum efficiency and wide color tunable window that mainly come from the recombination process between the host band and Cu-induced impurity state. Therefore, the tunable color of doped QDs is mainly related to tunable bandgap, either directly or through impurity or trap states. The luminescence mechanism of Cu-doped QDs has been reported in some previous publications.[106,116,117] It is generally accepted that the luminescence of the Cu dopant is related to host bandgap and dopant energy level, which is originated from the recombination relaxation between the electron in the CB of the host QD and the hole in the Cu T_2 state.[118] The non-radiative process in Cu-doped QDs is related to surface defects.[118] Because the band edge emission of Cu-doped QDs is several orders of magnitude faster than that of Cu dopant–related emission, the emission of QDs is, therefore, controlled by band edge transition. The Cu dopant–related emission will be produced in the presence of surface hole traps in the host QDs, and the intensity of Cu ions emission could be increase with the increase of the ratio of surface hole traps. Therefore, intrinsic impurity bands including bulk and surface defects jointly affect the luminescence properties of Cu-doped QDs, and these emission properties can be tuned by carefully controlling these parameters.

The color of Cu ion emission can be extended from visible window to NIR window by changing the host bandgap, size and chemical composition of QDs and controlling the shape of QDs based on quantum confinement effect (as shown in Figure 7.7(b)). In general, Cu-doped ZnS QDs are mainly green emission,[119] while Cu-doped ZnSe QDs have a wider luminescent range.[120] Cu-doped CdS QDs mainly show the emission range from orange to red light,[121] and Cu-doped InP QDs show NIR luminescence.[122] However, due to the change of the position of Cu d state in different host and the position of doping environment in the lattice, the exact position of Cu state and the origin of tunable doping emission have not been clearly revealed to date,[118] which needs further study.

Apart from Mn and Cu dopants mentioned earlier, other dopants such as Ag, Ni, Pb and lanthanide elements have also been reported for doping QDs, but they are significantly less popular than those of Mn and Cu doping. From the viewpoint of future application of luminescent materials, dopants of Ag, Cr and Ni ions may be potentially since dopant emission has been observed in these doped QDs. For example, Jana et al.[123] reported a series of Ag, Cr, Ni and Cu-doped CdZnS QDs which emit efficient, tunable, stable dopant state PL, and these doped QDs represent four different prominent colors in the visible spectral window. However, doping these transition metal ions still faces great challenging due to the size and charge mismatch between the doped ions and the matrix. Based on the previous research results of Mn and Cu doping, it will have some reference values for the further preparation of other high-quality doped QDs and the study of their optical properties.

7.4 APPLICATIONS

QDs have many physical and chemical properties different from macroscopic bulk materials due to the quantum confinement effect, which make them show a very broad application prospect in many fields, such as solar energy conversion, display and lighting, catalysis and biomedicine. To date, they also have a profound impact on the sustainable development of life science and information technology and the basic research in the field of matter. From the perspective of luminescent applications, the following section focus on the application of QDs in light-emitting diodes (LEDs), bioimaging and ratiometric temperature sensors.

7.4.1 LEDs

Colloidal QD-based LEDs (QLEDs) have been a hot topic in academic and industrial research in recent years. This is because of their unique features such as size- and chemical composition-tunable emission, high PL QY, high saturation color, easy fabrication, which have potential applications in next-generation lighting and display. In general, QLEDs can be simply classified as down-conversion and electrically driven LED devices. The schematic structure of down-conversion QLEDs is shown in Figure 7.13(a). This type QLEDs used the QDs as phosphors fabricated under the light excitation of a blue GaN LED chip. For example, Yuan et al.[124] used Cu and Mn–codoped Zn-In-S QDs as a single light conversion layer to construct high-quality white LEDs. As shown in Figure 7.13(b), the white LED shows bright natural white light with a color rendering index of 95 and the luminous efficiency is 73.2 lm/W. To date, many companies such as Samsung, LG and TCL have been developing these devices as backlight display units in liquid crystal display (LCD) incorporates. As compared to the traditional white-LED backlight used phosphor as the down-converter, the QD-based LCD

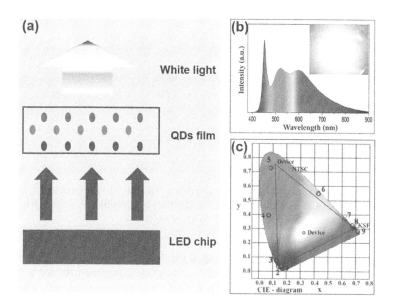

FIGURE 7.13   (a) Schematic structure of down-conversion type QLEDs. (b) EL spectrum of white LED based on Cu and Mn–codoped Zn–In–S QDs, inset shows the corresponding illumination photo (Yuan, X. et al. *ACS Appl. Mater. Interfaces* 7, 8659–8666, 2015. With permission) (c) CIE color coordinates corresponding to the $CH_3NH_3PbX_3$ QDs and their white LED devices and NTSC standard (bright area) (Zhang, F. et al. *ACS Nano* 9, 4533–4542, 2015. With permission).

contributes to a broader color gamut of >100% area of the National Television System Committee (NTSC) standard. It is worth noting that for the novel perovskite QDs, the color gamut coverage area reaches 130% of NTSC 1931 (Figure 7.13(c)). This QD enhanced backlight can present more vivid colors for LCDs and may endow the LCDs industry enter the second life.

The principle of electrically driven QLEDs is similar to the organic LED (OLED) technology, where the emitting layer (EML) of QDs is sandwiched between the electron transport layer (ETL) and the hole transport layer (HTL), as shown in Figure 7.14(a). The electrically driven QLEDs are more attractive due to the advantage of high saturation colors inherited from the narrow emission of QDs as comprising OLED, thus making QLED technology to be one of the promising candidates for next-generation lighting and display. The operating mechanism of electrically driven QLEDs is to inject electrons and holes into the EML through cathode anode, respectively, and then form excitons and emit light through their radiative recombination. Therefore, to obtain high brightness and high efficiency devices, it is necessary to inject carriers into EML effectively. To achieve this, the multilayered structures similar to the development of OLEDs is designed by inserting charge transport layers (CTLs) between the electrode and the EML, namely hole injection layer (HIL), electron injection layer (EIL), HTL and ETL, as shown in Figure 7.14(a). Various parameters, such as brightness, power efficiency and current efficiency, are used to evaluate the performance of QLEDs. However, since these values depend on the color sensitivity of the human eye, it is difficult to compare absolute values at different wavelengths. To estimate the efficiency of QLEDs with different color-emitting arcading to the same standard, the external quantum efficiency (EQE) defined as the ratio of the total number of photons emitted by QLEDs to the number of electrons injected has been used and is expressed as:

$$EQE = \gamma \cdot \eta_{rad} \cdot \eta_{out} \qquad (7.2)$$

where γ is the percentage of injected charges used to form excitons, η_{rad} is the ratio of the excitons recombined by radiative recombination and η_{out} is the out-coupling efficiency, which determines the

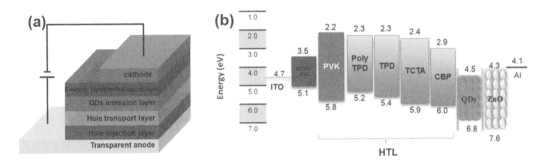

FIGURE 7.14 (a) Schematic structure of electrically driven QLEDs. (b) Schematic diagram illustrates the bandgap and band position of several common CTLs and different emission color alloy QDs. (Ho, M. D. et al. *ACS Appl. Mater. Interfaces* 5, 12369–12374, 2013. With permission)

number of photons generated that can escape out of the device. Therefore, it is necessary to improve η_{rad}, γ and η_{out} to achieve highly efficient QLEDs, and the specific measures are as follows.

The simplest way to improve η_{rad} of QLEDs is to suppress the non-radiative exciton decay processes occurring in QD emitters. It has been reported that the defects on the core surface of QDs lead to the existence of trap states, which make them become non-radiative decay centers. By designing QDs with core-shell structure, it has been proved this is the best scheme to suppress the exciton quenching process. As mentioned earlier, PL QYs is reinforced by coating the core with a shell material to reduce surface defects. As the results, the QDs with high PL QY are used to construct QLEDs which usually show excellent electroluminescence performance.[125–127]

In addition, charge transport modification related to γ is considered an effective way to improve EQE of QLEDs. The charge transfer mainly depends on the energy level structure and mobility of CTL.[128] Figure 7.15(b) summarized the band energy levels of some common CTLs. Due to the high mobility of metal oxides and the small injection barrier between QD and ZnO, electrons can be easily transferred to QD emitters, thus making ZnO as the most commonly used ETL material. However, the hole transport capacity of QLEDs is relatively insufficient because the transport

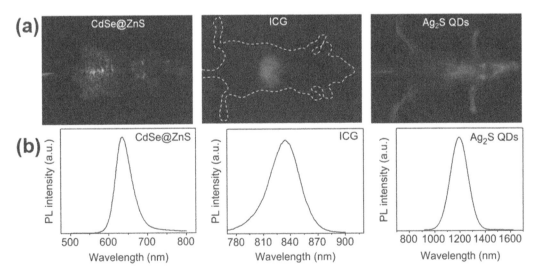

FIGURE 7.15 (a) In vivo fluorescence images and (b) PL spectra of CdSe/ZnS QDs, ICG (NIR-I) and Ag₂S QDs (NIR-II). The fluorescent materials were injected into nude mice for 5 minutes and then the fluorescence imaging was performed. (Li, C. et al. *Biomaterials* 35, 393–400, 2014. With permission)

capacity of organic HTLs is lower than that of ZnO-based ETLs. In addition, the VBMs of common QDs such as CdSe are much deeper than the highest occupied molecular orbital (HOMO) level of organic HTL materials currently available, which results in a large hole injection barrier. The results show that the number of electrons and holes in EML is not equal, which leads to serious charge imbalance. The excess electrons in QDs will lead to Auger recombination, which significantly reduces the device efficiency. One way to solve this issue is to find suitable HTL materials such as surface-modified NiO_x.[129] Another way is to design QD with a favorable energy level alignment for charge injection.[130,131] These problems can also be solved by adjusting the energy level structure and mobility of ETL to prevent excessive electron injection. The existing research results have shown that the performance of QLEDs can also be improved by passivating with wide bandgap semiconductors such as polyethylenimine (PEI),[132] poly(methyl methacrylate) (PMMA)[133] and alumina,[134] or doping ZnO ETL with other metals such as Ga,[135] Li[136] and Mg.[137]

The η_{out} associated with out-coupling technology is another starting point to be considered for enhancing EQE of QLEDs. It is reported that the optical estimation of the out-coupling efficiency for QLEDs with planar structures is only 20%, which means that almost 80% of the emitted photons are trapped inside the device. This low out-coupling efficiency of planar QLEDs is mainly due to the total internal reflection (TIR) of the air/substrate, substrate/ITO interfaces and the plasma loss on the metal electrode.[128] Therefore, increasing η_{out} is considered to be another important way to enhance the performance of QLEDs. To date, there are many works have been done on the extraction of trapped photons, which can be divided into external extraction structure (EES) and internal extraction structure (IES). EES is built on the outside surface of the substrate to adjust the TIR at the air/substrate interface, while IES is built into device to modulate the TIR at the substrate/ITO interface. At present, both methods are used to improve the electroluminescence performance of QLEDs.[138–140]

Based on the earlier three aspects, QLEDs technology has experienced tremendous development in the past few years. The EQE of QLEDs has been developed from less than 0.1% of the first prototype device[141] to the present ~20% for red, green and blue colors with corresponding brightness of more than 10,000 cd/m^2.[142,143] It is worth noting that the EQE and operation lifetime of red, green and blue QLED devices can almost met the requirements for displays even comparable to the state of art OLEDs.[144] In particular, the recently reported red-QLEDs fall to 95% of the initial brightness at 1000 cd/m^2 has reached 7668 hours,[144] which provide hope for the application of QLEDs in outdoor displays and lightings.

7.4.2 Bioluminescence Imaging

Bioluminescence imaging has the advantages of high sensitivity, low cost, high frame rate, strong portability and great potential for multiple imaging. It is a very useful biological research tool, which can provide high-resolution biological tissue information in vivo and in vitro. In particular, bioluminescence imaging has high sensitivity and real-time performance, which can meet the needs of surgery and endoscopic surgery. In addition, bioluminescence imaging plays an important role in early diagnosis and image-guided tumor resection. In normal in vitro and in vivo optical imaging, cells or tissues are usually labeled with organic dyes and fluorescence is collected from these dyes using appropriate imaging tools. Optical probes for these applications should have high PL QY and strong resistance to photobleaching. Due to their excellent optical properties and robust stability, QDs have become one of the most ideal biological fluorescence imaging materials. Furthermore, QDs have large specific surface area and abundant surface chemical properties, which can be combined with biomolecules for targeted imaging. Unlike most dyes, QDs with different emission wavelengths can be excited by a single light source, which is helpful for the multiple imaging of QDs labeled samples.

QDs-based bioluminescence imaging has been widely studied since 1998.[6,145,146] Among them, CdSe QDs and their core-shell structures have been the most active luminescent materials in the field of bioluminescence imaging.[146] The imaging of QDs in animals has made great progress in the last decade. In particular, it is noted that the excited-state lifetime of doped QDs is much longer than pristine QDs. From the analysis point of view, such long-lived emission of QDs has obvious advantages in time-resolved fluorescence mode. In the time-resolved bioluminescence imaging mode, the analyte signal of biological tissue can be collected and analyzed after the background attenuation of tissue spontaneous fluorescence and photon scattering, so as to improve the imaging accuracy.[147,148]

It is found that the absorption of tissue is low in the range of 650–1450 nm, and the light scattering and autofluorescence of tissue components decrease with the increase of imaging wavelength in this range. Compared with visible light and NIR-I region (NIR-I, 750–900 nm), NIR-II region (1000–1700 nm) has been proved to be able to better avoid the interference of biological background signals such as tissue spontaneous fluorescence and scattering. For example, Li et al.[149] injected red CdSe/ZnS QDs, small molecule fluorescence dye indocyanine green (ICG, NIR-I) and Ag_2S QDs (NIR-II) into mice. After 5 minutes, the mice were imaged with 455, 704 and 808 nm excitation light sources, respectively. As shown in Figure 7.15, it can be seen that mice injected with CdTe/CdSe QDs showed a strong green yellow signal in the visible light emission window, indicating that the signal contained tissue autofluorescence. The red signal concentration in the liver of mice injected with ICG indicated that the half-life of ICG in blood circulation is short. While the red signal widely distributed in the whole body of mice injected with Ag_2S QDs indicates that its half-life in blood circulation is long, it proves that NIR-II QDs can be used for visualization of blood vessels in vivo. These studies provide the possibility for early disease diagnosis and imaging guidance treatment. To date, the NIR-II QDs has been more and more used in biomedical imaging field.

7.4.3 Ratiometric Temperature Sensors

Temperature is one of the key parameters to evaluate physical, chemical and biological characteristics. With the rapid development of nanotechnology, nanotechnology, microelectronics and integrated photonics, the accurate measurement of local temperature on the nanoscaled materials have aroused more and more interest in the scientific community. Optical temperature sensor can measure the temperature on the nanometer scale by recording the change of intrinsic optical properties of the probe material caused by temperature. For example, according to the change of PL peak position, fwhm, PL intensity and PL lifetime with temperature, it is used to monitor the change of local temperature. Single emission QD is the most widely used temperature sensor material, but the optical properties of QDs are affected by many factors other than temperature. So far, the sensitivity of optical temperature sensor based on single-emission QD is still relatively weak.

Compared with the single emission QD temperature sensor, the ratiometric optical temperature sensor allows self-calibration of the system and has better robustness and reliability. The temperature sensor based on dual emission QDs shows two different PL channels, which come from two different radiative deexcitation paths of the same QD. In most cases, the dual emission from the same QD can be used for self-calibration temperature measurement, so the accurate temperature measurement can be realized. The temperature response range and sensitivity are two important indexes to qualify the performance of ratiometric temperature sensor.[150–152] Among them, the thermal sensitivity S_m is defined as:

$$S_m = \frac{d_R}{d_T} \tag{7.3}$$

$$R = \frac{I_{A1}}{I_{A2}} \tag{7.4}$$

in which R is the PL ratio at a given temperature (T), and I_{A1} and I_{A2} are the integrated PL intensities of peaks 1 and 2. A good ratiometric temperature sensor generally possesses both high values S_m and wide temperature response range.

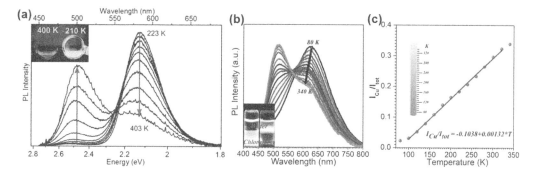

FIGURE 7.16 (a) Temperature-dependent PL spectra of colloidal $Zn_{1-x}Mn_xSe/ZnCdSe$ QDs, inset shows color photo of the PL of QDs at 210 and 400K under UV irradiation (Vlaskin, V. A. et al. *Nano Lett.* 10, 3670–3674, 2010. With permission) (b) variable-temperature PL spectra of Cu and Mn–codoped Zn–In–S QDs, the PL spectra were normalized by the total integrated PL intensities. The inset shows the photo of QDs dispersed in chloroform and water under UV irradiation. (c) The ratio of I_{Cu}/I_{tot} versus temperature (Cao, S. et al. *Adv. Funct. Mater.* 26, 7224–7233, 2016. With permission).

So far, the approaches for realizing dual emission of QDs used in ratiometric temperature sensors are mainly divided into three ways: (i) to control band edge and surface trap state emissions by controlling the synthesis process of plain QDs. For example, Jethi et al.[153] reported that CdSe QDs with dual emission from the band edge and surface trap states can be used as a good candidate material toward optical ratiometric temperature sensors. (ii) The second way is controlling the chemical composition and crystal structures of core-shell interface in "giant" QDs systems to obtain dual emissions. By this way, ultrasensitive and self-calibrating thermometers were reported by dual emission of PbS/CdS "giant" QDs.[154] (iii) The third way is by doping transition metal ions such as Mn ions in QDs to realize the dual PL emission from the band edge excitonic and the dopant ions.[150] Gamelin group has done a lot of systematic work in this area,[154,155] and they reported that the dual emission from two thermodynamically coupling of Mn dopant ions and host excitons in Mn-doped Cd-based QDs is conducive to fabricate stable and repeatable highly sensitive optical temperature sensors. Specially, Vlaskin et al.[156] reported $Zn_{1-x}Mn_xSe/ZnCdSe$ QDs and measured the PL at the temperature range of 223–403K. The normalized PL spectra relative to the total integrated PL intensity at various temperatures as shown in Figure 7.16(a). The sensitivity of this optical ratiometric temperature sensor is 0.9%/K and is independent of the chemical composition of QDs. In addition to single-ion doping, double-ion codoping can also realize dual fluorescence emission in QDs. Cao et al.[150] reported that the preparation of Cu and Mn–codoped Zn–In–S QDs with dual emissions at 512 and 612 nm from Cu and Mn dopants, respectively. The emission intensity ratio of 512/612 nm can be adjusted by changing the concentration of Cu and Mn ions. As shown in Figure 7.16(b) and (c), the temperature-dependent PL spectra of Cu and Mn–codoped Zn–In–S QDs show that the relative PL intensity of Cu ions increases with the increase of temperature, and the sensitivity of temperature detection S_m is 0.132%/K at the broad temperature window of 100–320K. The advantage of this kind of codoped QDs is that it not only provides dual emission form the double doped ions in a single QD but also gives the corresponding temperature sensor good detection repeatability with the help of stable PL properties of doped ions, which shows great hope for real temperature measurement in the future.

7.5 SUMMARY AND OUTLOOK

The research of colloidal QDs began in the early 1980s. The early research of QDs focused on synthesis and basic optical and structural characterization. Now, the research scope and field of QDs are gradually expanding. In-situ characterization and direct imaging technology are also introduced, and rich information about the QD growth is gradually obtained. In recent decades, new phenomena have been found in the photophysical processes and electronic properties of QDs. Photovoltaic and light-emitting devices based on QDs are also developing vigorously and can compete with other existing technologies step by step. The research of QDs has been applied to many fields, such as electronics, biomedicine, catalysis and new energy, and is expected to lead to a series of new technologies. However, there are still a series of challenges in the research of QDs, which are listed as follows.

1. Large-scale and precise synthesis of QDs. Accurate control of the size, shape and composition of QDs is the basic requirement of QD applications. Although the synthesis of QDs has been relatively mature, there are also a lot of literature reports. However, the large-scale and accurate synthesis of QDs is still facing a huge challenge. At present, researchers are trying to deepen the understanding of the nucleation and growth mechanism of QDs and try to associate them with the transformation of precursors and monomers. The nucleation and early growth stage of QDs are very important for the growth control of QDs. Therefore, in-situ observation of this process is very necessary. In-situ X-ray diffraction, in-situ absorption and emission spectrum monitoring, in-situ electron microscope observation of QD growth can promote our understanding of the nucleation stage and early growth of QDs. On the basis of further understanding of QD growth mechanism, it is expected to develop a new green synthesis method of colloidal QDs suitable for mass production.

2. Applications of QDs. In QD-based devices, the surface of QD must be combined with other materials to introduce an interface that can control the characteristics and performance of the device. However, the interface engineering between the QD layer and other device components is still in its infancy, which is of great significance for using new materials and new methods to design the QD surface and optimize the device performance. It should be pointed out that for the research of optoelectronic devices, first of all, efforts should be made to prepare high-quality QD materials, construct perfect QD films and then design the most effective device structure. With the continuous improvement and optimization of QD materials and device structures, high efficiency QD-based devices can be created.

3. Toxicity and environmental issues of QDs. As we discussed earlier, the application of QDs involves LEDs, biological imaging, temperature sensor and other fields. At present, most of the reported QD applications mainly focus on cadmium- and lead-based QDs. Because of their toxicity, these QDs have great potential danger in the long-term application in real life, and the prospect of commercialization is not widely recognized. In recent years, some environmentally friendly alternative materials have been reported, such as group IV element QDs and group III-V compound QDs. In addition to toxicity and environmental problems, the contents of some rare metal resources, such as indium and tellurium, involved in QDs must also be considered. Efficient chemical synthesis of nontoxic and high abundance element-based QDs with excellent optical properties of cadmium- and lead-based QDs is still a challenging task.

QD-related technologies and products are gradually entering our lives. Although QD materials and their applications still face variety of great challenges at present, we believe that the development of QD and its related research will bring revolutionary technological breakthroughs and illuminate our future.

REFERENCES

1. Klimov, V. I., *Nanocrystal quantum dots*. CRC Press: Boca Raton, FL, 2010.
2. Brus, L. E., Electron-electron and electron-hole interactions in small semiconductor crystallites: the size dependence of the lowest excited electronic state. *J. Chem. Phys.* 80, 4403–4409, 1984.

3. Pietryga, J. M.; Park, Y.-S.; Lim, J.; Fidler, A. F.; Bae, W. K.; Brovelli, S.; Klimov, V. I., Spectroscopic and device aspects of nanocrystal quantum dots. *Chem. Rev.* 116, 10513–10622, 2016.

4. Wegner, K. D.; Hildebrandt, N., Quantum dots: bright and versatile in vitro and in vivo fluorescence imaging biosensors. *Chem. Soc. Rev.* 44, 4792–4834, 2015.

5. Murray, C. B.; Norris, D. J.; Bawendi, M. G., Synthesis and characterization of nearly monodisperse CdE (E = sulfur, selenium, tellurium) semiconductor nanocrystallites. *J. Am. Chem. Soc.* 115, 8706–8715, 1993.

6. Bruchez, M.; Moronne, M.; Gin, P.; Weiss, S.; Alivisatos, A. P., Semiconductor nanocrystals as fluorescent biological labels. *Science* 281, 2013, 1998.

7. Chan, W. C. W.; Nie, S., Quantum dot bioconjugates for ultrasensitive nonisotopic detection. *Science* 281, 2016, 1998.

8. Brus, L., Electronic wave functions in semiconductor clusters: experiment and theory. *J. Phys. Chem.* 90, 2555–2560, 1986.

9. Qu, L.; Peng, X., Control of photoluminescence properties of CdSe nanocrystals in growth. *J. Am. Chem. Soc.* 124, 2049–2055, 2002.

10. Peng, X.; Manna, L.; Yang, W.; Wickham, J.; Scher, E.; Kadavanich, A.; Alivisatos, A. P., Shape control of CdSe nanocrystals. *Nature* 404, 59–61, 2000.

11. Peng, Z. A.; Peng, X., Formation of high-quality CdTe, CdSe, and CdS nanocrystals using CdO as precursor. *J. Am. Chem. Soc.* 123, 183–184, 2001.

12. Peng, X., Green chemical approaches toward high-quality semiconductor nanocrystals. *Chem. Eur. J.* 8, 334–339, 2002.

13. Mashford, B. S.; Stevenson, M.; Popovic, Z.; Hamilton, C.; Zhou, Z.; Breen, C.; Steckel, J.; Bulovic, V.; Bawendi, M.; Coe-Sullivan, S.; Kazlas, P. T., High-efficiency quantum-dot light-emitting devices with enhanced charge injection. *Nat. Photonics* 7, 407–412, 2013.

14. Thanh, N. T. K.; Maclean, N.; Mahiddine, S., Mechanisms of nucleation and growth of nanoparticles in solution. *Chem. Rev.* 114, 7610–7630, 2014.

15. Xu, G.; Zeng, S.; Zhang, B.; Swihart, M. T.; Yong, K.-T.; Prasad, P. N., New generation cadmium-free quantum dots for biophotonics and nanomedicine. *Chem. Rev.* 116, 12234–12327, 2016.

16. LaMer, V. K.; Dinegar, R. H., Theory, production and mechanism of formation of monodispersed hydrosols. *J. Am. Chem. Soc.* 72, 4847–4854, 1950.

17. Reiss, P.; Carrière, M.; Lincheneau, C.; Vaure, L.; Tamang, S., Synthesis of semiconductor nanocrystals, focusing on nontoxic and earth-abundant materials. *Chem. Rev.* 116, 10731–10819, 2016.

18. Jing, L.; Kershaw, S. V.; Li, Y.; Huang, X.; Li, Y.; Rogach, A. L.; Gao, M., Aqueous based semiconductor nanocrystals. *Chem. Rev.* 116, 10623–10730, 2016.

19. Rogach, A. L.; Franzl, T.; Klar, T. A.; Feldmann, J.; Gaponik, N.; Lesnyak, V.; Shavel, A.; Eychmüller, A.; Rakovich, Y. P.; Donegan, J. F., Aqueous synthesis of thiol-capped CdTe nanocrystals: state-of-the-art. *J. Phys. Chem. C* 111, 14628–14637, 2007.

20. Bao, H.; Gong, Y.; Li, Z.; Gao, M., Enhancement effect of illumination on the photoluminescence of water-soluble CdTe nanocrystals: toward highly fluorescent CdTe/CdS core-shell structure. *Chem. Mater.* 16, 3853–3859, 2004.

21. Park, J.; Joo, J.; Kwon, S. G.; Jang, Y.; Hyeon, T., Synthesis of monodisperse spherical nanocrystals. *Angew. Chem. Int. Edit.* 46, 4630–4660, 2007.

22. Bae, W. K.; Nam, M. K.; Char, K.; Lee, S., Gram-scale one-pot synthesis of highly luminescent blue emitting $Cd_{1-x}Zn_xS/ZnS$ nanocrystals. *Chem. Mater.* 20, 5307–5313, 2008.

23. Blackman, B.; Battaglia, D. M.; Mishima, T. D.; Johnson, M. B.; Peng, X., Control of the morphology of complex semiconductor nanocrystals with a type II heterojunction, dots vs peanuts, by thermal cycling. *Chem. Mater.* 19, 3815–3821, 2007.

24. Li, J. J.; Wang, Y. A.; Guo, W.; Keay, J. C.; Mishima, T. D.; Johnson, M. B.; Peng, X., Large-scale synthesis of nearly monodisperse CdSe/CdS core/shell nanocrystals using air-stable reagents via successive ion layer adsorption and reaction. *J. Am. Chem. Soc.* 125, 12567–12575, 2003.

25. Zhang, W.; Zhang, H.; Feng, Y.; Zhong, X., Scalable single-step noninjection synthesis of high-quality core/shell quantum dots with emission tunable from violet to near infrared. *ACS Nano* 6, 11066–11073, 2012.

26. Bawendi, M. G.; Carroll, P. J.; Wilson, W. L.; Brus, L. E., Luminescence properties of CdSe quantum crystallites: resonance between interior and surface localized states. *J. Chem. Phys.* 96, 946–954, 1992.

27. Wang, X.; Qu, L.; Zhang, J.; Peng, X.; Xiao, M., Surface-related emission in highly luminescent CdSe quantum dots. *Nano Lett.* 3, 1103–1106, 2003.

28. Klimov, V. I.; McBranch, D. W.; Leatherdale, C. A.; Bawendi, M. G., Electron and hole relaxation pathways in semiconductor quantum dots. *Phys. Rev. B* 60, 13740–13749, 1999.

29. Hines, M. A.; Guyot-Sionnest, P., Synthesis and characterization of strongly luminescing ZnS-capped CdSe nanocrystals. *J. Phys. Chem.* 100, 468–471, 1996.

30. Yoffe, A. D., Semiconductor quantum dots and related systems: electronic, optical, luminescence and related properties of low dimensional systems. *Adv. Phys.* 50, 1–208, 2001.

31. Reiss, P.; Protière, M.; Li, L., Core/shell semiconductor nanocrystals. *Small* 5, 154–168, 2009.

32. Vasudevan, D.; Gaddam, R. R.; Trinchi, A.; Cole, I., Core-shell quantum dots: properties and applications. *J. Alloys Compd.* 636, 395–404, 2015.

33. Navarro-Pardo, F.; Zhao, H.; Wang, Z. M.; Rosei, F., Structure/property relations in "giant" semiconductor nanocrystals: opportunities in photonics and electronics. *Acc. Chem. Res.* 51, 609–618, 2018.

34. Wang, C.; Barba, D.; Selopal, G. S.; Zhao, H.; Liu, J.; Zhang, H.; Sun, S.; Rosei, F., Enhanced photocurrent generation in proton-irradiated "giant" CdSe/CdS core/shell quantum dots. *Adv. Funct. Mater.* 29, 1904501, 2019.

35. Peng, X.; Schlamp, M. C.; Kadavanich, A. V.; Alivisatos, A. P., Epitaxial growth of highly luminescent CdSe/CdS core/shell nanocrystals with photostability and electronic accessibility. *J. Am. Chem. Soc.* 119, 7019–7029, 1997.

36. Nirmal, M.; Dabbousi, B. O.; Bawendi, M. G.; Macklin, J. J.; Trautman, J. K.; Harris, T. D.; Brus, L. E., Fluorescence intermittency in single cadmium selenide nanocrystals. *Nature* 383, 802–804, 1996.

37. Gao, Y.; Peng, X., Photogenerated excitons in plain core CdSe nanocrystals with unity radiative decay in single channel: the effects of surface and ligands. *J. Am. Chem. Soc.* 137, 4230–4235, 2015.

38. Mews, A.; Eychmueller, A.; Giersig, M.; Schooss, D.; Weller, H., Preparation, characterization, and photophysics of the quantum dot quantum well system cadmium sulfide/mercury sulfide/cadmium sulfide. *J. Phys. Chem.* 98, 934–941, 1994.

39. Pan, Z.; Zhang, H.; Cheng, K.; Hou, Y.; Hua, J.; Zhong, X., Highly efficient inverted type-I CdS/CdSe core/shell structure QD-sensitized solar cells. *ACS Nano* 6, 3982–3991, 2012.

40. Balet, L. P.; Ivanov, S. A.; Piryatinski, A.; Achermann, M.; Klimov, V. I., Inverted core/shell nanocrystals continuously tunable between type-I and type-II localization regimes. *Nano Lett.* 4, 1485–1488, 2004.

41. Groeneveld, E.; Witteman, L.; Lefferts, M.; Ke, X.; Bals, S.; Van Tendeloo, G.; de Mello Donega, C., Tailoring ZnSe-CdSe colloidal quantum dots via cation exchange: from core/shell to alloy nanocrystals. *ACS Nano* 7, 7913–7930, 2013.

42. Zhong, X.; Xie, R.; Zhang, Y.; Basché, T.; Knoll, W., High-quality violet- to red-emitting ZnSe/CdSe core/shell nanocrystals. *Chem. Mater.* 17, 4038–4042, 2005.

43. Smith, C. T.; Tyrrell, E. J.; Leontiadou, M. A.; Miloszewski, J.; Walsh, T.; Cadirci, M.; Page, R.; O'Brien, P.; Binks, D.; Tomić, S., Energy structure of CdSe/CdTe type II colloidal quantum dots—do phonon bottlenecks remain for thick shells? *Sol. Energy Mater. Sol. Cells* 158, 160–167, 2016.

44. Chen, C.-Y.; Cheng, C.-T.; Lai, C.-W.; Hu, Y.-H.; Chou, P.-T.; Chou, Y.-H.; Chiu, H.-T., Type-II CdSe/CdTe/ZnTe (core–shell–shell) quantum dots with cascade band edges: the separation of electron (at CdSe) and hole (at ZnTe) by the CdTe layer. *Small* 1, 1215–1220, 2005.

45. Boldt, K.; Schwarz, K. N.; Kirkwood, N.; Smith, T. A.; Mulvaney, P., Electronic structure engineering in ZnSe/CdS type-II nanoparticles by interface alloying. *J. Phys. Chem. C* 118, 13276–13284, 2014.

46. Gui, R.; An, X., Layer-by-layer aqueous synthesis, characterization and fluorescence properties of type-II CdTe/CdS core/shell quantum dots with near-infrared emission. *RSC Adv.* 3, 20959–20969, 2013.

47. Naifar, A.; Zeiri, N.; Nasrallah, S. A.-B.; Said, M., Optical properties of CdSe/ZnTe type II core shell nanostructures. *Optik* 146, 90–97, 2017.

48. Kim, S.; Fisher, B.; Eisler, H.-J.; Bawendi, M., Type-II quantum dots: CdTe/CdSe(core/shell) and CdSe/ZnTe(core/shell) heterostructures. *J. Am. Chem. Soc.* 125, 11466–11467, 2003.

49. Regulacio, M. D.; Han, M.-Y., Composition-tunable alloyed semiconductor nanocrystals. *Acc. Chem. Res.* 43, 621–630, 2010.

50. Xu, S.; Shen, H.; Zhou, C.; Yuan, H.; Liu, C.; Wang, H.; Ma, L.; Li, L. S., Effect of shell thickness on the optical properties in CdSe/CdS/Zn$_{0.5}$Cd$_{0.5}$S/ZnS and CdSe/CdS/Zn$_x$Cd$_{1-x}$S/ZnS core/multishell nanocrystals. *J. Phys. Chem. C* 115, 20876–20881, 2011.

51. Xie, R.; Kolb, U.; Li, J.; Basché, T.; Mews, A., Synthesis and characterization of highly luminescent CdSe-Core CdS/Zn$_{0.5}$Cd$_{0.5}$S/ZnS multishell nanocrystals. *J. Am. Chem. Soc.* 127, 7480–7488, 2005.

52. Talapin, D. V.; Mekis, I.; Götzinger, S.; Kornowski, A.; Benson, O.; Weller, H., CdSe/CdS/ZnS and CdSe/ZnSe/ZnS core-shell-shell nanocrystals. *J. Phys. Chem. B* 108, 18826–18831, 2004.

53. Chen, Y.; Vela, J.; Htoon, H.; Casson, J. L.; Werder, D. J.; Bussian, D. A.; Klimov, V. I.; Hollingsworth, J. A., "Giant" multishell CdSe nanocrystal quantum dots with suppressed blinking. *J. Am. Chem. Soc.* 130, 5026–5027, 2008.

54. Qin, H.; Niu, Y.; Meng, R.; Lin, X.; Lai, R.; Fang, W.; Peng, X., Single-dot spectroscopy of zinc-blende CdSe/CdS core/shell nanocrystals: nonblinking and correlation with ensemble measurements. *J. Am. Chem. Soc.* 136, 179–187, 2014.

55. Zhong, X.; Han, M.; Dong, Z.; White, T. J.; Knoll, W., Composition-tunable $Zn_xCd_{1-x}Se$ nanocrystals with high luminescence and stability. *J. Am. Chem. Soc.* 125, 8589–8594, 2003.

56. Yuan, Y.; Zhu, H.; Wang, X.; Cui, D.; Gao, Z.; Su, D.; Zhao, J.; Chen, O., Cu-catalyzed synthesis of CdZnSe-CdZnS alloy quantum dots with highly tunable emission. *Chem. Mater.* 31, 2635–2643, 2019.

57. Pu, Y.; Cai, F.; Wang, D.; Wang, J.-X.; Chen, J.-F., Colloidal synthesis of semiconductor quantum dots toward large-scale production: a review. *Ind. Eng. Chem. Res.* 57, 1790–1802, 2018.

58. Wang, A.; Shen, H.; Zang, S.; Lin, Q.; Wang, H.; Qian, L.; Niu, J.; Li, L. S., Bright, efficient, and color-stable violet ZnSe-based quantum dot light-emitting diodes. *Nanoscale* 7, 2951–2959, 2015.

59. Wang, S.; Li, J. J.; Lv, Y.; Wu, R.; Xing, M.; Shen, H.; Wang, H.; Li, L. S.; Chen, X., Synthesis of reabsorption-suppressed type-II/type-I ZnSe/CdS/ZnS core/shell quantum dots and their application for immunosorbent assay. *Nanoscale Res. Lett.* 12, 380–380, 2017.

60. Han, C.-Y.; Yang, H., Development of colloidal quantum dots for electrically driven light-emitting devices. *J. Korean Ceram. Soc.* 54, 449–469, 2017.

61. Han, C.-Y.; Lee, S.-H.; Song, S.-W.; Yoon, S.-Y.; Jo, J.-H.; Jo, D.-Y.; Kim, H.-M.; Lee, B.-J.; Kim, H.-S.; Yang, H., More than 9% efficient ZnSeTe quantum dot-based blue electroluminescent devices. *ACS Energy Lett.*, 1568–1576, 2020.

62. Kuech, T. F., III-V compound semiconductors: growth and structures. *Prog. Cryst. Growth Charact. Mater.* 62, 352–370, 2016.

63. Wu, Z.; Liu, P.; Zhang, W.; Wang, K.; Sun, X. W., Development of InP quantum dot-based light-emitting Diodes. *ACS Energy Lett.* 5, 1095–1106, 2020.

64. Guzelian, A. A.; Katari, J. E. B.; Kadavanich, A. V.; Banin, U.; Hamad, K.; Juban, E.; Alivisatos, A. P.; Wolters, R. H.; Arnold, C. C.; Heath, J. R., Synthesis of size-selected, surface-passivated InP nanocrystals. *J. Phys. Chem.* 100, 7212–7219, 1996.

65. Micic, O. I.; Sprague, J. R.; Curtis, C. J.; Jones, K. M.; Machol, J. L.; Nozik, A. J.; Giessen, H.; Fluegel, B.; Mohs, G.; Peyghambarian, N., Synthesis and characterization of InP, GaP, and $GaInP_2$ quantum dots. *J. Phys. Chem.* 99, 7754–7759, 1995.

66. Battaglia, D.; Peng, X., Formation of high quality InP and InAs nanocrystals in a noncoordinating solvent. *Nano Lett.* 2, 1027–1030, 2002.

67. Mićić, O. I.; Smith, B. B.; Nozik, A. J., Core-shell quantum dots of lattice-matched $ZnCdSe_2$ shells on InP cores: experiment and theory. *J. Phys. Chem. B* 104, 12149–12156, 2000.

68. Haubold, S.; Haase, M.; Kornowski, A.; Weller, H., Strongly luminescent InP/ZnS core–shell nanoparticles. *ChemPhysChem* 2, 331–334, 2001.

69. Kim, M. R.; Chung, J. H.; Lee, M.; Lee, S.; Jang, D.-J., Fabrication, spectroscopy, and dynamics of highly luminescent core-shell InP@ZnSe quantum dots. *J. Colloid Interface Sci.* 350, 5–9, 2010.

70. Thuy, U. T. D.; Reiss, P.; Liem, N. Q., Luminescence properties of In(Zn)P alloy core/ZnS shell quantum dots. *Appl. Phys. Lett.* 97, 193104, 2010.

71. Kim, T.; Kim, S. W.; Kang, M.; Kim, S.-W., Large-scale synthesis of InPZnS alloy quantum dots with dodecanethiol as a composition controller. *J. Phys. Chem. Lett.* 3, 214–218, 2012.

72. Kim, S.; Kim, T.; Kang, M.; Kwak, S. K.; Yoo, T. W.; Park, L. S.; Yang, I.; Hwang, S.; Lee, J. E.; Kim, S. K.; Kim, S.-W., Highly luminescent InP/GaP/ZnS nanocrystals and their application to white light-emitting diodes. *J. Am. Chem. Soc.* 134, 3804–3809, 2012.

73. Park, J. P.; Lee, J.-J.; Kim, S.-W., Highly luminescent InP/GaP/ZnS QDs emitting in the entire color range via a heating up process. *Sci. Rep.* 6, 30094, 2016.

74. Li, Y.; Hou, X.; Dai, X.; Yao, Z.; Lv, L.; Jin, Y.; Peng, X., Stoichiometry-controlled InP-based quantum dots: synthesis, photoluminescence, and electroluminescence. *J. Am. Chem. Soc.* 141, 6448–6452, 2019.

75. Won, Y.-H.; Cho, O.; Kim, T.; Chung, D.-Y.; Kim, T.; Chung, H.; Jang, H.; Lee, J.; Kim, D.; Jang, E., Highly efficient and stable InP/ZnSe/ZnS quantum dot light-emitting diodes. *Nature* 575, 634–638, 2019.

76. Zhong, H.; Bai, Z.; Zou, B., Tuning the luminescence properties of colloidal I–III–VI semiconductor nanocrystals for optoelectronics and biotechnology applications. *J. Phys. Chem. Lett.* 3, 3167–3175, 2012.

77. Chen, B.; Pradhan, N.; Zhong, H., From large-scale synthesis to lighting device applications of ternary I–III–VI semiconductor nanocrystals: inspiring greener material emitters. *J. Phys. Chem. Lett.* 9, 435–445, 2018.

78. Aldakov, D.; Lefrançois, A.; Reiss, P., Ternary and quaternary metal chalcogenide nanocrystals: synthesis, properties and applications. *J. Mater. Chem. C* 1, 3756–3776, 2013.

79. Hamanaka, Y.; Ogawa, T.; Tsuzuki, M.; Kuzuya, T., Photoluminescence properties and its origin of AgInS$_2$ quantum dots with chalcopyrite structure. *J. Phys. Chem. C* 115, 1786–1792, 2011.

80. Nam, D.-E.; Song, W.-S.; Yang, H., Facile, air-insensitive solvothermal synthesis of emission-tunable CuInS$_2$/ZnS quantum dots with high quantum yields. *J. Mater. Chem.* 21, 18220–18226, 2011.

81. Zang, H.; Li, H.; Makarov, N. S.; Velizhanin, K. A.; Wu, K.; Park, Y.-S.; Klimov, V. I., Thick-shell CuInS$_2$/ZnS quantum dots with suppressed "blinking" and narrow single-particle emission line widths. *Nano Lett.* 17, 1787–1795, 2017.

82. Chen, B.; Zhong, H.; Zhang, W.; Tan, Z. A.; Li, Y.; Yu, C.; Zhai, T.; Bando, Y.; Yang, S.; Zou, B., Highly emissive and color-tunable CuInS$_2$-based colloidal semiconductor nanocrystals: off-stoichiometry effects and improved electroluminescence performance. *Adv. Funct. Mater.* 22, 2081–2088, 2012.

83. Kang, X.; Yang, Y.; Huang, L.; Tao, Y.; Wang, L.; Pan, D., Large-scale synthesis of water-soluble CuInSe$_2$/ZnS and AgInSe$_2$/ZnS core/shell quantum dots. *Green Chem.* 17, 4482–4488, 2015.

84. Liu, Y.; Zhou, L.; Li, Y.; Deng, R.; Zhang, H., Highly fluorescent nitrogen-doped carbon dots with excellent thermal and photo stability applied as invisible ink for loading important information and anti-counterfeiting. *Nanoscale* 9, 491–496, 2017.

85. Yuan, F.; Yuan, T.; Sui, L.; Wang, Z.; Xi, Z.; Li, Y.; Li, X.; Fan, L.; Tan, Z. a.; Chen, A.; Jin, M.; Yang, S., Engineering triangular carbon quantum dots with unprecedented narrow bandwidth emission for multicolored LEDs. *Nat. Commun.* 9, 2249, 2018.

86. Qu, S.; Liu, X.; Guo, X.; Chu, M.; Zhang, L.; Shen, D., Amplified spontaneous green emission and lasing emission from carbon nanoparticles. *Adv. Funct. Mater.* 24, 2689–2695, 2014.

87. Meng, L.; Ushakova, E. V.; Zhou, Z.; Liu, E.; Li, D.; Zhou, D.; Tan, Z.; Qu, S.; Rogach, A. L., Microwave-assisted in situ large scale synthesis of a carbon dots@g-C$_3$N$_4$ composite phosphor for white light-emitting devices. *Mater. Chem. Front.* 4, 517–523, 2020.

88. Gayen, B.; Palchoudhury, S.; Chowdhury, J., Carbon dots: a mystic star in the world of nanoscience. *J. Nanomater.* 2019, 3451307, 2019.

89. Morozova, S.; Alikina, M.; Vinogradov, A.; Pagliaro, M., Silicon quantum dots: synthesis, encapsulation, and application in light-emitting diodes. *Front. Chem.* 8, 191, 2020.

90. Zhu, Z.; Zhai, Y.; Li, Z.; Zhu, P.; Mao, S.; Zhu, C.; Du, D.; Belfiore, L. A.; Tang, J.; Lin, Y., Red carbon dots: optical property regulations and applications. *Mater. Today* 30, 52–79, 2019.

91. Carolan, D., Recent advances in germanium nanocrystals: synthesis, optical properties and applications. *Prog. Mater. Sci.* 90, 128–158, 2017.

92. Shamsi, J.; Urban, A. S.; Imran, M.; De Trizio, L.; Manna, L., Metal halide perovskite nanocrystals: synthesis, post-synthesis modifications, and their optical properties. *Chem. Rev.* 119, 3296–3348, 2019.

93. Protesescu, L.; Yakunin, S.; Bodnarchuk, M. I.; Krieg, F.; Caputo, R.; Hendon, C. H.; Yang, R. X.; Walsh, A.; Kovalenko, M. V., Nanocrystals of cesium lead halide perovskites (CsPbX$_3$, X = Cl, Br, and I): novel optoelectronic materials showing bright emission with wide color gamut. *Nano Lett.* 15, 3692–3696, 2015.

94. Zhang, F.; Zhong, H.; Chen, C.; Wu, X.-g.; Hu, X.; Huang, H.; Han, J.; Zou, B.; Dong, Y., Brightly luminescent and color-tunable colloidal CH$_3$NH$_3$PbX$_3$ (X = Br, I, Cl) quantum dots: potential alternatives for display technology. *ACS Nano* 9, 4533–4542, 2015.

95. Li, X.; Wu, Y.; Zhang, S.; Cai, B.; Gu, Y.; Song, J.; Zeng, H., CsPbX$_3$ quantum dots for lighting and displays: room-temperature synthesis, photoluminescence superiorities, underlying origins and white light-emitting diodes. *Adv. Funct. Mater.* 26, 2435–2445, 2016.

96. Bai, K.; Zeng, R.; Ke, B.; Cao, S.; Xue, X.; Tan, R.; Zou, B., Synthesis of high-efficient Mn^{2+} doped CsPbCl$_3$ perovskite nanocrystals in toluene and surprised lattice ejection of dopants at mild temperature. *J. Alloys Compd.* 806, 858–863, 2019.

97. Wang, Y.; Cao, S.; Li, J.; Li, H.; Yuan, X.; Zhao, J., Improved ultraviolet radiation stability of Mn^{2+}-doped CsPbCl$_3$ nanocrystals via B-site Sn doping. *CrystEngComm* 21, 6238–6245, 2019.

98. Zhou, Y.; Zhao, Y., Chemical stability and instability of inorganic halide perovskites. *Energy Environ. Sci.* 12, 1495–1511, 2019.

99. Ravi, V. K.; Saikia, S.; Yadav, S.; Nawale, V.; Nag, A., CsPbBr$_3$/ZnS core/shell type nanocrystals for enhancing luminescence lifetime and water stability. *ACS Energy Lett.* 5, 1794–1796, 2020.

100. Zhao, Y.; Li, J.; Dong, Y.; Song, J., Synthesis of colloidal halide perovskite quantum dots/nanocrystals: progresses and advances. *Isr. J. Chem.* 59, 649–660, 2019.

101. Nag, A.; Chakraborty, S.; Sarma, D. D., To dope Mn^{2+} in a semiconducting nanocrystal. *J. Am. Chem. Soc.* 130, 10605–10611, 2008.

102. Dalpian, G. M.; Chelikowsky, J. R., Self-purification in semiconductor nanocrystals. *Phys. Rev. Lett.* 96, 226802, 2006.

103. Erwin, S. C.; Zu, L.; Haftel, M. I.; Efros, A. L.; Kennedy, T. A.; Norris, D. J., Doping semiconductor nanocrystals. *Nature* 436, 91–94, 2005.

104. Pradhan, N.; Goorskey, D.; Thessing, J.; Peng, X., An alternative of CdSe nanocrystal emitters: pure and tunable impurity emissions in ZnSe nanocrystals. *J. Am. Chem. Soc.* 127, 17586–17587, 2005.

105. Pradhan, N.; Das Adhikari, S.; Nag, A.; Sarma, D. D., Luminescence, plasmonic, and magnetic properties of doped semiconductor nanocrystals. *Angew. Chem. Int. Edit.* 56, 7038–7054, 2017.

106. Pradhan, N.; Sarma, D. D., Advances in light-emitting doped semiconductor nanocrystals. *J. Phys. Chem. Lett.* 2, 2818–2826, 2011.

107. Cao, S.; Jia, L.; Wang, L.; Gao, F.; Wei, G.; Zheng, J.; Yang, W., Efficient energy transfer from 1, 3, 5-tris (N-phenylbenzimidazol-2, yl) benzene to Mn: CdS quantum dots. *Jpn. J. Appl. Phys.* 53, 04EG07, 2014.

108. Cao, S.; Li, C.; Wang, L.; Shang, M.; Wei, G.; Zheng, J.; Yang, W., Long-lived and well-resolved Mn^{2+} ion emissions in CuInS-ZnS quantum dots. *Sci. Rep.* 4, 7510, 2014.

109. Cao, S.; Zhao, J.; Yang, W.; Li, C.; Zheng, J., Mn^{2+}-doped Zn-In-S quantum dots with tunable bandgaps and high photoluminescence properties. *J. Mater. Chem. C* 3, 8844–8851, 2015.

110. Cao, S.; Zheng, J.; Dai, C.; Wang, L.; Li, C.; Yang, W.; Shang, M., Doping concentration-dependent photoluminescence properties of Mn-doped Zn–In–S quantum dots. *J. Mater. Sci.* 53, 1286–1296, 2018.

111. Cao, S.; Zheng, J.; Zhao, J.; Wang, L.; Gao, F.; Wei, G.; Zeng, R.; Tian, L.; Yang, W., Highly efficient and well-resolved Mn^{2+} ion emission in MnS/ZnS/CdS quantum dots. *J. Mater. Chem. C* 1, 2540–2547, 2013.

112. Pradhan, N., Red-tuned Mn d-d emission in doped semiconductor nanocrystals. *ChemPhysChem* 17, 1087–1094, 2016.

113. Hazarika, A.; Pandey, A.; Sarma, D. D., Rainbow emission from an atomic transition in doped quantum dots. *J. Phys. Chem. Lett.* 5, 2208–2213, 2014.

114. Ji, S.; Yuan, X.; Cao, S.; Ji, W.; Zhang, H.; Wang, Y.; Li, H.; Zhao, J.; Zou, B., Near-unity red Mn^{2+} photoluminescence quantum yield of doped CsPbCl$_3$ nanocrystals with Cd incorporation. *J. Phys. Chem. Lett.* 11, 2142–2149, 2020.

115. Cao, S.; Dai, C.; Yao, S.; Zou, B.; Zhao, J., Synthesis and optical properties of Mn^{2+}-doped Cd-In-S colloidal nanocrystals. *J. Mater. Sci.* 55, 12801–12810, 2020.

116. Knowles, K. E.; Hartstein, K. H.; Kilburn, T. B.; Marchioro, A.; Nelson, H. D.; Whitham, P. J.; Gamelin, D. R., Luminescent colloidal semiconductor nanocrystals containing copper: synthesis, photophysics, and applications. *Chem. Rev.* 116, 10820–10851, 2016.

117. Cao, S.; Ji, W.; Zhao, J.; Yang, W.; Li, C.; Zheng, J., Color-tunable photoluminescence of Cu-doped Zn-In-Se quantum dots and their electroluminescence properties. *J. Mater. Chem. C* 4, 581–588, 2016.

118. Srivastava, B. B.; Jana, S.; Pradhan, N., Doping Cu in semiconductor nanocrystals: some old and some new physical insights. *J. Am. Chem. Soc.* 133, 1007–1015, 2011.

119. Bol, A. A.; Ferwerda, J.; Bergwerff, J. A.; Meijerink, A., Luminescence of nanocrystalline ZnS:Cu^{2+}. *J. Lumin.* 99, 325–334, 2002.

120. Wang, C.; Xu, S.; Wang, Z.; Cui, Y., Key roles of impurities in the stability of internally doped Cu:ZnSe nanocrystals in aqueous solution. *J. Phys. Chem. C* 115, 18486–18493, 2011.

121. Lien, V. T. K.; Tan, P. M.; Hien, N. T.; Hoa, V. X.; Chi, T. T. K.; Truong, N. X.; Oanh, V. T. K.; Thuy, N. T. M.; Ca, N. X., Tunable photoluminescent Cu-doped CdS/ZnSe type-II core/shell quantum dots. *J. Lumin.* 215, 116627, 2019.

122. Xie, R.; Battaglia, D.; Peng, X., Colloidal InP nanocrystals as efficient emitters covering blue to near-infrared. *J. Am. Chem. Soc.* 129, 15432–15433, 2007.

123. Jana, S.; Manna, G.; Srivastava, B. B.; Pradhan, N., Tuning the emission colors of semiconductor nanocrystals beyond their bandgap tunability: all in the dope. *Small* 9, 3753–3758, 2013.

124. Yuan, X.; Ma, R.; Zhang, W.; Hua, J.; Meng, X.; Zhong, X.; Zhang, J.; Zhao, J.; Li, H., Dual emissive manganese and copper co-doped Zn-In-S quantum dots as a single color-converter for high color rendering white-light-emitting diodes. *ACS Appl. Mater. Interfaces* 7, 8659–8666, 2015.

125. Bae, W. K.; Park, Y.-S.; Lim, J.; Lee, D.; Padilha, L. A.; McDaniel, H.; Robel, I.; Lee, C.; Pietryga, J. M.; Klimov, V. I., Controlling the influence of auger recombination on the performance of quantum-dot light-emitting diodes. *Nat. Commun.* 4, 2661, 2013.

126. Lee, K.-H.; Lee, J.-H.; Song, W.-S.; Ko, H.; Lee, C.; Lee, J.-H.; Yang, H., Highly efficient, color-pure, color-stable blue quantum dot light-emitting devices. *ACS Nano* 7, 7295–7302, 2013.

127. Shirasaki, Y.; Supran, G. J.; Bawendi, M. G.; Bulović, V., Emergence of colloidal quantum-dot light-emitting technologies. *Nat. Photonics* 7, 13–23, 2013.

128. Sun, Y.; Jiang, Y.; Sun, X. W.; Zhang, S.; Chen, S., Beyond OLED: efficient quantum dot light-emitting diodes for display and lighting application. *Chem. Rec.* 19, 1729–1752, 2019.

129. Lin, J.; Dai, X.; Liang, X.; Chen, D.; Zheng, X.; Li, Y.; Deng, Y.; Du, H.; Ye, Y.; Chen, D.; Lin, C.; Ma, L.; Bao, Q.; Zhang, H.; Wang, L.; Peng, X.; Jin, Y., High-performance quantum-dot light-emitting diodes using NiO_x hole-injection layers with a high and stable work function. *Adv. Funct. Mater.* 30, 1907265, 2020.

130. Cao, W.; Xiang, C.; Yang, Y.; Chen, Q.; Chen, L.; Yan, X.; Qian, L., Highly stable QLEDs with improved hole injection via quantum dot structure tailoring. *Nat. Commun.* 9, 2608, 2018.

131. Pu, C.; Dai, X.; Shu, Y.; Zhu, M.; Deng, Y.; Jin, Y.; Peng, X., Electrochemically-stable ligands bridge the photoluminescence-electroluminescence gap of quantum dots. *Nat. Commun.* 11, 937, 2020.

132. Davidson-Hall, T.; Aziz, H., Perspective: toward highly stable electroluminescent quantum dot light-emitting devices in the visible range. *Appl. Phys. Lett.* 116, 010502, 2020.

133. Dai, X.; Zhang, Z.; Jin, Y.; Niu, Y.; Cao, H.; Liang, X.; Chen, L.; Wang, J.; Peng, X., Solution-processed, high-performance light-emitting diodes based on quantum dots. *Nature* 515, 96–99, 2014.

134. Ji, W.; Shen, H.; Zhang, H.; Kang, Z.; Zhang, H., Over 800% efficiency enhancement of all-inorganic quantum-dot light emitting diodes with an ultrathin alumina passivating layer. *Nanoscale* 10, 11103–11109, 2018.

135. Cao, S.; Zheng, J.; Zhao, J.; Yang, Z.; Li, C.; Guan, X.; Yang, W.; Shang, M.; Wu, T., Enhancing the performance of quantum dot light-emitting diodes using room-temperature-processed Ga-doped ZnO nanoparticles as the electron transport layer. *ACS Appl. Mater. Interfaces* 9, 15605–15614, 2017.

136. Kim, H.-M.; Cho, S.; Kim, J.; Shin, H.; Jang, J., Li and Mg co-doped zinc oxide electron transporting layer for highly efficient quantum dot light-emitting diodes. *ACS Appl. Mater. Interfaces* 10, 24028–24036, 2018.

137. Zhang, H.; Sun, X.; Chen, S., Over 100 cd A^{-1} efficient quantum dot light-emitting diodes with inverted tandem structure. *Adv. Funct. Mater.* 27, 1700610, 2017.

138. Yu, R.; Yin, F.; Huang, X.; Ji, W., Molding hemispherical microlens arrays on flexible substrates for highly efficient inverted quantum dot light emitting diodes. *J. Mater. Chem. C* 5, 6682–6687, 2017.

139. Yang, X.; Dev, K.; Wang, J.; Mutlugun, E.; Dang, C.; Zhao, Y.; Liu, S.; Tang, Y.; Tan, S. T.; Sun, X. W.; Demir, H. V., Light extraction efficiency enhancement of colloidal quantum dot light-emitting diodes using large-scale nanopillar arrays. *Adv. Funct. Mater.* 24, 5977–5984, 2014.

140. Ding, K.; Fang, Y.; Dong, S.; Chen, H.; Luo, B.; Jiang, K.; Gu, H.; Fan, L.; Liu, S.; Hu, B.; Wang, L., 24.1% external quantum efficiency of flexible quantum dot light-emitting diodes by light extraction of silver nanowire transparent electrodes. *Adv. Opt. Mater.* 6, 1800347, 2018.

141. Colvin, V. L.; Schlamp, M. C.; Alivisatos, A. P., Light-emitting diodes made from cadmium selenide nanocrystals and a semiconducting polymer. *Nature* 370, 354–357, 1994.

142. Shen, H.; Gao, Q.; Zhang, Y.; Lin, Y.; Lin, Q.; Li, Z.; Chen, L.; Zeng, Z.; Li, X.; Jia, Y.; Wang, S.; Du, Z.; Li, L. S.; Zhang, Z., Visible quantum dot light-emitting diodes with simultaneous high brightness and efficiency. *Nat. Photonics* 13, 192–197, 2019.

143. Wang, L.; Lin, J.; Hu, Y.; Guo, X.; Lv, Y.; Tang, Z.; Zhao, J.; Fan, Y.; Zhang, N.; Wang, Y.; Liu, X., Blue quantum dot light-emitting diodes with high electroluminescent efficiency. *ACS Appl. Mater. Interfaces* 9, 38755–38760, 2017.

144. Liu, D.; Cao, S.; Wang, S.; Wang, H.; Dai, W.; Zou, B.; Zhao, J.; Wang, Y., Highly stable red quantum dot light-emitting diodes with long T_{95} operation lifetimes. *J. Phys. Chem. Lett.* 11, 3111–3115, 2020.

145. Michalet, X.; Pinaud, F. F.; Bentolila, L. A.; Tsay, J. M.; Doose, S.; Li, J. J.; Sundaresan, G.; Wu, A. M.; Gambhir, S. S.; Weiss, S., Quantum dots for live cells, in vivo imaging, and diagnostics. *Science* 307, 538, 2005.

146. Petryayeva, E.; Algar, W. R.; Medintz, I. L., Quantum dots in bioanalysis: a review of applications across various platforms for fluorescence spectroscopy and imaging. *Appl. Spectrosc.* 67, 215–252, 2013.

147. Wu, P.; Yan, X.-P., Doped quantum dots for chemo/biosensing and bioimaging. *Chem. Soc. Rev.* 42, 5489–5521, 2013.

148. Li, C.; Wu, P., Cu-doped quantum dots: a new class of near-infrared emitting fluorophores for bioanalysis and bioimaging. *Luminescence* 34, 782–789, 2019.

149. Li, C.; Zhang, Y.; Wang, M.; Zhang, Y.; Chen, G.; Li, L.; Wu, D.; Wang, Q., In vivo real-time visualization of tissue blood flow and angiogenesis using Ag_2S quantum dots in the NIR-II window. *Biomaterials* 35, 393–400, 2014.

150. Cao, S.; Zheng, J.; Zhao, J.; Yang, Z.; Shang, M.; Li, C.; Yang, W.; Fang, X., Robust and stable ratiometric temperature sensor based on Zn-In-S quantum dots with intrinsic dual-dopant ion emissions. *Adv. Funct. Mater.* 26, 7224–7233, 2016.

151. Cao, S.; Dai, C.; Zhao, J.; Zou, B., Synthesis of dual-emission Ag- and Mn-codoped Zn-In-S nanocrystals and their optical radiometric temperature sensors. J. Nanopart. Res. 21, 242, 2019.

152. Zhao, H.; Vomiero, A.; Rosei, F., Tailoring the heterostructure of colloidal quantum dots for ratiometric optical nanothermometry. *Small* 16, 2000804, 2020.

153. Jethi, L.; Krause, M. M.; Kambhampati, P., Toward ratiometric nanothermometry via intrinsic dual emission from semiconductor nanocrystals. *J. Phys. Chem. Lett.* 6, 718–721, 2015.

154. Zhao, H.; Vomiero, A.; Rosei, F., Ultrasensitive, biocompatible, self-calibrating, multiparametric temperature sensors. *Small* 11, 5741–5746, 2015.

155. McLaurin, E. J.; Bradshaw, L. R.; Gamelin, D. R., Dual-emitting nanoscale temperature sensors. *Chem. Mater.* 25, 1283–1292, 2013.

156. Vlaskin, V. A.; Janssen, N.; van Rijssel, J.; Beaulac, R.; Gamelin, D. R., Tunable dual emission in doped semiconductor nanocrystals. *Nano Lett.* 10, 3670–3674, 2010.

8 Lanthanide-Doped Upconversion Nanoparticles for Super-Resolution Imaging

Chaohao Chen, Jiayan Liao, and Dayong Jin

CONTENTS

DOI: 10.1201/9781003098676-8

8.1   FUNDAMENTALS AND DEVELOPMENT OF PHOTON UPCONVERSION

Luminescent probes, including both organic days and inorganic nanoparticles, have become an essential tool required in biological and biomedical imaging studies. Organic dyes and fluorescent proteins are prone to photobleaching, and quantum dots (QDots) contain toxic heavy metals that are harmful to biological samples. Conventional luminescent probes only emit lower energy photons under excitation from ultraviolet (UV) or visible light of higher energy that can cause photodamage and autofluorescence of biomolecules, resulting in a low signal-to-noise ratio of the collected imaging data. Photon upconversion has attracted enormous attentions as the nonlinear anti-Stokes' optical process can convert long-wavelength irradiance excitation into the short-wavelength spectrum region.[1, 2]

There are some critical requirements for the photon upconversion process, for instance, the long lifetimes of the excited states and the ladder-like arrangement of the energy levels with similar spacing. Suitable hosts doped with transition metal ions have been reported to display upconversion emissions.[3, 4] Actinide-doped materials have also been investigated for their upconversion abilities.[5, 6] Upconversion processes have also been realized in the two-photon absorption of QDots and organic dyes, second-harmonic generation and anti-Stokes Raman scattering.[7–11] However, these systems require challenging conditions, such as low temperatures, high-power-density photon flux or expensive ultrashort pulsed lasers.

Much higher upconversion efficiencies have been achieved by using lanthanide-doped solids. Lanthanide ions are suitable elements for photon upconversion because their energy level spacings of 4f-valence electrons exhibit ladder-like electronic structures that correspond to the emission states of the near-infrared (NIR) and visible photon energies. Accordingly, unique nonlinear processes, in which two or more NIR pump photons are absorbed through real intermediate energy levels of lanthanide ions, lead to the emission of light at a wavelength shorter than incident radiation. This was independently formulated in bulk crystals in the mid-1960s by Auzel, Ovsyankin and Feofilov.[12–14] Since that time, much research has been done in lanthanide-doped upconversion luminescence, and their remarkable optical properties, including massive anti-Stokes shift, excellent photostability, sharp and tunable multi-peak line emission and a long and widely tunable domain of the emission lifetime, as well as high resistance to optical blinking and photobleaching, have been discovered and used. Such intrinsic features make frequency upconversion by lanthanide-based upconversion materials appealing for photonic and biological applications.

8.1.1   Lanthanide-Based Upconversion Nanoparticles

Lanthanide-doped upconversion nanoparticles (UCNPs) consist of a suitable inorganic and photostable host embedded with various trivalent lanthanide ions (Figure 8.1(a)).[15] A crystalline host lattice featuring a low phonon energy environment is beneficial to the photon upconversion process, where the non-radiative decay rate of lanthanides via multiphoton relaxation can effectively be suppressed. Photon upconversion in lanthanide-doped crystals relies on the physically existing intermediary energy level. Consequently, it possesses higher frequency conversion efficiency (Figure 8.1(b)). The lanthanide family (from La to Lu) exhibits similar physical and chemical properties due to the significant similarity present in their electron configuration.

Upconversion luminescence generally originates from electronic transitions within the $4f^N$ configuration of lanthanides. Although these 4f electronic transitions in principle are electric dipoles

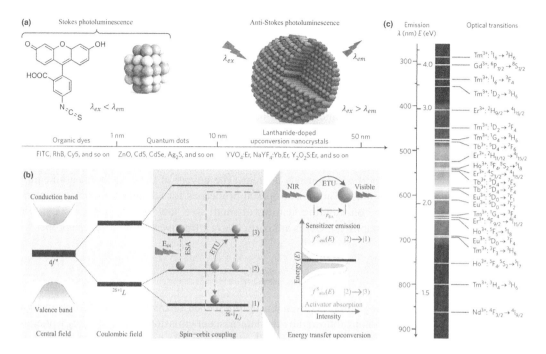

FIGURE 8.1 Lanthanide-doped nanoparticles and photon upconversion. (a) A comparison of three lumines-cent nanomaterials: organic dyes, quantum dots and UCNPs. (b) Electronic energy level diagrams of trivalent Ln^{3+} ions in relation to upconversion processes. (c) Typical Ln^{3+}-based upconversion emission bands covering a broad range of wavelengths. (Zhou B, Shi B, Jin D, and Liu X, *Nat. Nano.* 10, 924, 2015. With permission.)

prohibited in quantum mechanical selection rules, they are relaxed through the local crystal field-induced intermixing of higher electronic configurations with *f* states. Moreover, the lanthanide's $4f^N$ electronic configuration can split into an abundance of energy sublevels resulting from appreciable electronic repulsion from the Coulombic interaction of the electrons. The spin-orbit coupling among f-electrons also can split the $4f^N$ electronic configuration of lanthanides, leading to a rich energy-level pattern as well as weak perturbations in the crystal field (Figure 8.1(b)).[16, 17]

Moreover, the partially filled 4f intra-configurational transitions of lanthanides are effectively shielded by outer complete 5s and 5p shells, resulting in a weak electron-phonon coupling and, thus, line-like sharp emissions as a consequence of electronic transitions.[18] Upconversion emissions are rarely influenced by the environment due to optical transitions being localized and generally not interfered with by quantum confinement – this leads to unique optical properties such as a sharp emission band, reliable photostability and long excited-state lifetime. It benefits the sequential exci-tations and ion-ion interactions in the excited states of lanthanide ions. Such intrinsic features make frequency upconversion by lanthanide-based upconversion materials appealing for photonic and biological applications. The upconversion luminescence intensity generally has a nonlinear depen-dence on the excitation light density:[19]

$$I = kP^n \tag{8.1}$$

where I is the luminescence intensity, P is the pump laser power, k is material-related coefficient and n is the required number of the excitation photons for the upconversion luminescence. When the I is plotted in a double-logarithmical representation versus P, a n-photon emission process will have a slope of n used in the low power regime. However, the slope will be smaller than n in the high power regime and generate the phenomenon of saturation. This is an effect because of the competi-tive process between the decay rate and upconverted rate at the intermediate states.

Such unique features make lanthanide-doped upconversion materials ideal candidates for a variety of applications, ranging from bioassay and photocatalysis to super-resolution imaging and three-dimensional (3D) volumetric displays.[20–23] By properly selecting the type of lanthanide dopants in a nanoparticle, the upconversion emission wavelength can be tuned precisely to span the UV and NIR regions (Figure 8.1(c)).

Until the late 1990s, when great strides were being made in nanoscience and nanotechnology in materials science, upconversion nanomaterials have been promoted. The high-quality UCNPs were designed and routinely synthesized with controllable composition, crystalline phase, size and shape through different chemical techniques.[24, 25] The upconversion process in nanomaterials is similar in bulk materials, but, due to the size confinement present in nanoparticles, ion-ion and ion-lattice interactions are modified to various degrees. The advent of these UCNPs significantly expands the potential applications of nonlinear optical properties for advanced nanophotonic and biological research. Due to the outstanding features of small size, low toxicity and excellent biocompatibility and high photostability, UCNPs are particularly attractive as luminescence probes for bioimaging applications.

8.1.2 Mechanism of Upconverting Luminescence

As a nonlinear optical phenomenon, upconversion luminescence forms the sequential absorption of two or more low-energy photons followed by the luminescent emission of a high-energy photon. Auzel first described the concept of "upconversion", summarized its mechanism and reported on upconverted visible emission based on energy transfer by using Yb^{3+} to sensitize Er^{3+} and Tm^{3+} ions.[26] After decades of development, photon upconverting processes involving lanthanide ions are currently predominantly categorized into five classes of mechanisms: excited-state absorption (ESA), energy-transfer upconversion (ETU), photon avalanche (PA), cooperative energy transfer (CET) and energy migration-mediated upconversion (EMU).[27, 28] All of these processes involve the sequential absorption of two or more photons,[29] as shown in Figure 8.2.

8.1.2.1 Excited-State Absorption

ESA is an appropriate pumping mechanism made possible by the successive absorption of pumping photons, which most likely happens in the singly doped upconversion materials. In a typical ESA process, excitation takes the form of successive absorption of pumping photons by a single ion. If excitation energy is resonant with the transition from the ground state to the excited metastable intermediate state, phonon absorption occurs through the ESA process. A second pumping

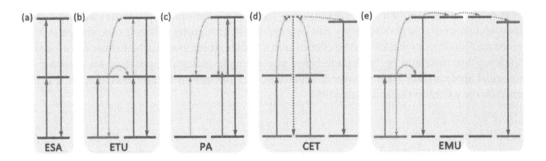

FIGURE 8.2 Principal upconversion mechanisms of lanthanide-doped nanomaterials. Red arrows stand for direct excitation processes, blue arrows represent radiative emission processes and dashed arrows represent energy transfer processes. (a) Excited-state absorption (ESA). (b) Energy-transfer upconversion (ETU). (c) Photon avalanche (PA). (d) Cooperative energy transfer (CET). (e) Energy migration-mediated upconversion (EMU). (Zheng W, Huang P, Tu D, Ma E, Zhu H, and Chen X, *Chem. Soc. Rev.* 44, 1379, 2015. With permission.)

photon promotes the excited electrons from excited metastable level to higher lying state. Finally, the upconverted emission will occur when the photon drops from higher excited state back to the ground state. The ladder-like energy states of Ln^{3+} ions are required to achieve an efficient ESA process. Such rigorous requirements make only a small set of lanthanides such as Ho^{3+}, Er^{3+}, Nd^{3+} and Tm^{3+} have such energy level structures for the ESA process.[30–32]

8.1.2.2 Energy Transfer Upconversion

Both the ETU and ESA processes utilize sequential absorption of excitation photons to populate the intermediate levels of Ln^{3+} dopant. However, ETU is by far the most efficient upconversion process through energy transfer involving the participation of two types of neighboring ions: the sensitizer and activator. In a sensitizer-activator system, the sensitizer is pumped to the excited states upon absorbing a pump photon. A non-radiative energy transfer process successively promotes the sensitizer's ability to donate its harvested energy to the activator via a dipole-dipole resonant interaction. Meanwhile, the activator relaxes back to the ground state. Yb^{3+} ions are frequently used as sensitizers in the ETU process, conveniently resonant with many f–f transitions of Er^{3+}, Tm^{3+} and Ho^{3+}. Yb^{3+}–Er^{3+} and Yb^{3+}–Tm^{3+} and Yb^{3+}–Ho^{3+} doped ion-pairs in $NaYF_4$ materials are typical and efficient dual ions-activated upconversion systems (Figure 8.3).[33–36]

8.1.2.3 Photon Avalanche

PA is an infrequent upconversion process that requires an excitation power above a specific threshold value. Chivian and coworkers first discovered the PA process in a Pr^{3+}-doped infrared quantum counter in 1979.[37] The ground state is populated to the intermediate excited state by nonresonant weak ESA because of the mismatched energy. The luminescent centers may populate to the upper visible-emitting state through the resonant ESA process. After the metastable level population is finished, the occupation of the intermediate states occurs through efficient cross-relaxation energy transfer between the excited ion and neighboring ground-state ion. The weak ground-state absorption and strong feedback looping of ESA followed by further initiated cross-relaxation exponentially increase exponentially intermediate states population above the excitation threshold, producing strong upconverting emissions as an avalanche process. PA occurs when the consumption of intermediate states ions is less than that of the ground state ions, which is desirable for upconverting lasers. Recently, except for compact lasers application, the PA process serves to realize super-resolution imaging.[23, 38]

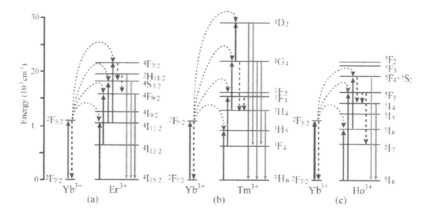

FIGURE 8.3 Schematic energy level diagrams showing typical UC processes for (a) Yb^{3+} and Er^{3+}, (b) Yb^{3+} and Tm^{3+} and (c) Yb^{3+} and Ho^{3+}. (Zhou J, Liu Q, Feng W, Sun Y, and Li F, *Chem. Rev.* 115, 395–465, 2015. With permission.)

8.1.2.4　Cooperative Energy Transfer

CET process involves the interaction of three ions and both sensitizers and activators are essential. The population of emission levels usually arises from the CET of adjacent ions. Two energy donors are cooperatively activated to a virtual excited state by absorbing excitation photons, and they fulfill a simultaneous energy transfer to a neighboring acceptor.[39] Upconverting involves two processes: cooperative sensitization (e.g., $Yb^{3+}-Tb^{3+}$, $Yb^{3+}-Eu^{3+}$ and $Yb^{3+}-Pr^{3+}$ ion pairs,) [40, 41] and cooperative luminescence (e.g., $Yb^{3+}-Yb^{3+}$ ion pair [42]). The upconversion efficiency of CET is relatively poor due to the absence of a long-lived intermediate energy state of the activator, especially in nanomaterials. Most of the relevant studies of CET have focused on bulk materials. Due to the virtual pair levels during electronic transitions leading to CET being less efficient, a few research studies demonstrated the feasibility of this process for nanomaterials.[43]

8.1.2.5　Energy Migration-Mediated Upconversion

Except for the mechanisms mentioned earlier, a novel upconverting mechanism known as EMU was proposed.[44] Efficient tunable upconversion emissions in $NaGdF_4$:Tm/Yb@$NaGdF_4$:Ln core-shell nanoparticles were realized through Gd sublattice-mediated energy migration. The phenomenon of EMU is essentially a type of energy transfer from the upconversion process. It proposes a solution for enhancing the efficiency of upconversion emissions from activators without proper intermediate energy levels and builds a bridge between sensitized ions and the activated ion of mismatched energy states. In the EMU method, several types of lanthanide dopants are incorporated into separate layers in the core-shell structure nanomaterials. Sensitizer ions absorb NIR excitation photons, and the higher excited state of the accumulator ions is populated via an ETU process from the sensitizer. Some of the accumulator ions at high-lying excited state do not generate radiation relaxation via their transition but instead move the energy through the accumulator to the migrator. The migrator ions conduct the random energy hopping throughout the core-shell interface, resulting in a radiation transition that produces upconverted luminescence by new activators. The rational core-shell structure and arrangement are vital to diminish luminescence quenching and facilitate the energy transfer at the core-shell interface so that an efficient EMU process can be generated.

In the past, great efforts were made to measure and evaluate the efficiency of energy transfer pathways. In studies by Auzel et al., the quantum efficiency of the ETU process was found to be two or three orders of magnitude higher than those of the ESA and CET processes. ETU is a universal process in the multi-doped upconversion system. To some extent, EMU is an extension of ETU benefiting from developing the core-shell synthetic technique. Lanthanide ions with exceptional energy levels were introduced into separate layers. Photon upconversion in core-shell nanoparticles opens up a new door for regulating dopant interactions. Furthermore, in core-shell nanoparticles, incompatible dopant ions can be spatially confined in separate layers to eliminate the quenching effect caused by short-range cross-relaxation between ions. Moreover, a long-range energy migration process from the outside layer to the inner core could be further harnessed to manipulate the photon dynamic route.[45]

8.1.3　Advances of UCNPs in Biological Imaging

Stepwise optical excitation, energy transfer and various nonlinear and collective light-matter interaction processes act together to convert low-energy excitation photons into visible or UV regions. The upconversion photoluminescence of UCNPs is not only an interesting optical phenomenon but also has great potential for biological imaging.[46] Imaging techniques can benefit from the extreme photophysical advantages of UCNPs, such as low autofluorescence background, deep penetration depth, extreme photostability and low toxicity.

8.1.3.1　Low Autofluorescence Background

The first significant advantage of UCNPs as biological imaging probes is that UCNPs use the NIR light as the exciting source. Most of the biomolecules cannot absorb the NIR photons well, alleviating the autofluorescence problem. Typically, autofluorescence is a kind of fluorescence resulting from the excitation of cellular molecules, and it contributes to the background and competes with the signal.[46] Background autofluorescence will hinder the collection of the expected signal, particularly when the signal intensity is not strong enough. With NIR excitation, the background noise can be reduced to a negligible level as biosamples have no NIR range absorption. Consequently, the obtained images with UCNPs' probes display a very high level of signal-to-noise ratio. With such a low autofluorescence background, it is possible to detect a single UCNP.[47]

8.1.3.2 Deep Penetration Depth

Another advantage of UCNPs by using the NIR excitation source is the ability of the deep penetration depth through tissue.[48] For deep biological tissue imaging, absorption and scattering are the main factors that cause the signal attenuation proportional to the depth, particularly severe in the visible light range. To overcome this challenge, the development of efficient and biocompatible probes with excitation or emission located within the "biological transparency window" of the NIR range has attracted significant interests for optical imaging through deep tissues. With the upconverting process by converting NIR photons to visible/NIR light, UCNPs have several unique advantages for optical deep tissue imaging.

8.1.3.3 High Photostability

Apart from the low autofluorescence and high penetration depth, UCNPs have been known to exhibit high photostability without photoblinking and photobleaching at the single-particle level. [49] Previous experiments have indicated that the emission intensity of UCNPs can remain constant when under a fixed excitation laser power.[50] The higher excitation intensity alleviates concentration quenching in upconverting luminescence,[47] which opens a new door to achieving single UCNP imaging with high brightness of the light by using a focused excitation irradiance, enabling continuous imaging with microseconds per dwell time.[51] Photobleaching is another indicator when evaluating the photostability of the probes. Due to chemical degradation, the conventional fluorescent probes may irreversibly not emit photons after a finite period of excitation,[46] limiting the time window for imaging of the living cells. By contrast, UCNPs that assisted live-cell imaging has been demonstrated with the ability to resist photobleaching events, even when subjected to hours of continuous intense excitation, providing rich, dynamic information useful for biological research.[52]

8.1.3.4 Low Toxicity

The low toxicity of UCNPs is also an essential advantage for the biological imaging application. [53] Previous research has reported that this nanoparticle creates minimal toxic outcomes in cells or organisms. For investigating the toxicity of UCNPs on whole organisms, the distribution of UCNPs was analyzed in the nematode by imaging the upconverting emissions.[54] Moreover, their cytotoxicity depends on their surface properties. Other investigations[55, 56] indicated that accumulation in organs is an essential parameter for the toxicity of UCNPs. As major elimination organs of the circulatory system, the liver and spleen are occupied by larger amounts of UCNPs coated with citrate or 6-hexanoic acid.[57]

8.2 SUPER-RESOLVED FLUORESCENCE MICROSCOPY

Optical microscopy has played a critical role in the biological and medical fields as it helps us to investigate and discover the fine structures that are invisible to the naked eye. With the first microscope, the British physicist Robert Hook discovered the plant cell wall structure in 1665. A contemporary Dutch scientist, A. Leeuwenhoek, utilized a homemade microscope to observe the body's inner workings on a microscopic level, seeing bacteria, sperm and even blood cells flowing through capillaries.[58] With further developments in optical microscope technology in the next few centuries, imaging of cells and their biological structures has enabled us to understand how cells function. During the last three decades, continuous advances in technology and modern research methods have promoted the imaging field via a broad, interdisciplinary effort, leading to noteworthy improvements like better labeling, more in-depth observation and increased spatial and temporal resolution. The significance of imaging modalities was recognized as the essential breakthrough invention when in 2014 the Nobel Prize in Chemistry was awarded for the development of super-resolved fluorescence microscopy. The full potential capabilities of different super-resolution microscopy methods have yet to be explored, so the best approach to the advanced applications remains uncertain. For this reason, super-resolution imaging becomes an active field as nanometric resolution imaging is providing valuable insights into various biological systems, which are important developments for future research.[59]

8.2.1 Diffraction Limit in Microscopy

Based on experimental evidence and principles of physics, Abbe proposed a famous and vital conclusion that the resolution of the microsystem could not bypass the diffraction limit.[60] Specifically, the structure features smaller than 200 nm in size are not resolvable using a conventional light microscope due to optical diffraction. Abbe described the transverse dimension of the far-field microscopic focal spot with the following equation:

$$\Delta r \approx \frac{\lambda}{2n\sin\alpha} = \frac{0.61\lambda}{NA} \tag{8.2}$$

where λ represents the wavelength, n is the refractive index of the medium between the sample and the objective, α is the semi-objective aperture angle and NA represents the numerical aperture. The size of the refractive index n and the aperture half-angle α are constrained by objective technical conditions. Equation (8.2) represents the full-width-at-half-maximum (FWHM) of the point spread function (PSF) and is commonly used as a quantitative definition of resolution in far-field light microscopy.

Confronted with the resolution limit imposed by diffraction described in Equation (8.2), scientists first opted for the obvious solution by increasing the numerical aperture or reducing the wavelength of the excitation laser. Nevertheless, the resolution of the optical microscopy was limited to approximately half wavelength of the light, even with perfect lenses and optimal alignment. Instead of visible light, the researcher ultimately employed an electron beam as a powerful way to probe the sample to obtain more detailed information after realizing the wave nature of electrons. Following the same physical principle but with a smaller wavelength down to one nanometer, electron microscopes such as scanning electron microscopy (SEM) and transmission electron microscopy (TEM)[61] reveal the subcellular structure and atomic organization of the materials with accuracy approaching the Angstrom range. However, the use of those electron imaging approaches is significantly restricted in life science research, as the vacuum operation requirement precludes the observation of living systems. Moreover, electron microscopes impose a limitation on the specific label and are not suitable for imaging multiple cellular structures in one organic sample. Furthermore, protocols and procedures in transmission and scanning electron microscopes are relatively costly and time-consuming.

Compared with electron microscopy, light (fluorescent) microscopy play an essential role in the life sciences for cell imaging, which enabled minimally invasive optical-based observation of events over long periods of time. By taking advantage of evanescent waves, total internal reflection fluorescence (TIRF) microscopy [62] first achieved high axial resolution, but its lateral resolution remained at around the diffraction limit. Another near-field technique – scanning near-field optical microscopy (SNOM) [63] – employs the evanescent light wave at the end of a fiber tip. The fluorescent signal on the surface is collected by the fiber tip to form an optical image, whose resolution depends on the aperture size instead of the wavelength, while the interior of the cell is not visible, because the intensity of the evanescent wave exhibits an exponential decay as a function of the distance from the fiber tip surface. Thus, SNOM only collects high-resolution information near the surface of the sample, remaining limited to the throughput.

It would be advantageous to devise optical far-field methods that can image the subcellular structures and proteins within living cells.[64] Over the past several decades, far-field fluorescence microscopy became an essential tool in the disciplines of physical, chemical, material and biomedical research, especially in studying the living cells and organisms. However, the light-imposed diffraction limitation on the spatial resolution restricts the applications of optical microscopes, as it remains a pivotal challenge to improve the resolution of far-field light microscopy.

8.2.2 SUPER-RESOLUTION TECHNOLOGY

Recent years have witnessed the development of super-resolution far-field optical microscopy techniques. To break the diffraction barrier in far-field microscopy, Stefan Hell introduced the concept of reversible saturable or switchable optical fluorescence transitions (RESOLFT). This involved switchable or reversibly saturable fluorophore transitions, where the diffraction barrier can be overcome by a saturated optical transition (depletion) between two optically distinguishable states (on/off) in fluorescent molecules. To distinguish the fluorescent molecule from the adjacent ones, it technically requires the control of the fluorescence molecules switching between the bright and dark states. For this consideration, optical super-resolution technologies employ the photophysical and photochemical processes of fluorescent emitting, making possible the optical switching of fluorescent molecules.

According to the fundamental principles, super-resolution technology can be divided into two categories to deliver the nanolevel resolution. The first type of super-resolution technology mainly utilizes sub-diffraction-limit pattern excitation methods. Among the methods that improve resolution by pattern excitation modification, the most important techniques are referred to as stimulated emission depletion (STED) microscopy,[65, 66] ground-state depletion (GSD) microscopy [67] and structured illumination microscopy (SIM).[68] The second technique relies on stochastically detecting the precise localization of single molecules, including photoactivated localization microscopy (PALM) [69] and stochastic optical reconstruction microscopy (STORM).[70] There are many variations based on these schemes, as well as advanced imaging process methods, which could be combined to improve the performance of those existing imaging modalities. Due to the optical stability of UCNPs, we only discuss the application in the first type of super-resolution technology in this chapter.

8.3 NANOPARTICLES IN SUPER-RESOLUTION TECHNIQUES

8.3.1 ADVANTAGES OF NANOPARTICLES IN SUPER-RESOLUTION IMAGING

With the advantages of small size and biocompatibility, organic dyes and fluorescent proteins are now widespread imaging probes in the above far-field super-resolution microscopies. However, the disadvantages of current fluorescent proteins and molecular dyes are also apparent in the super-resolution applications, particularly for the long-term tracking of single molecular and real-time imaging of subcellular structures. Typically, they are too dim to offer enough signal in fast imaging and can be rapidly photobleached under the excitation light. Therefore, the super-resolution community seeks alternative probes with intense brightness and resistance to photobleaching to revolutionize microscopy by revealing the events inside the cells.

Although nanoparticles are larger in size than dye molecules and proteins, several groups have successfully induced fluorescent nanoparticles into the imaging communities to explore their capability in functional subcellular imaging.[71] Nanoparticles have attracted a great of attention because they are excellent candidates for biolabeling and biosensing,[72] drug delivery,[73] cancer therapy,[48] long-term molecular tracking,[52] solar cell application [74] and triggering chemical reaction.[75] These uses are all due to their remarkable photophysical properties. For the potential applications of nanoparticles, mapping of the centers within individual nanoparticles has become significantly important. Recently, a variety of super-resolution modalities have employed nanoparticles as fluorescent nanoprobes, including QDots,[76–82] carbon dots (CDots),[83–87] polymer dots (PDots),[88–93] aggregation-induced emission (AIE) dots,[94–98] and UCNPs.[99–101]

Various kinds of nanoparticles have demonstrated considerable advantages in super-resolution imaging for biological research; however, only a few analyses have conducted long-term tracking and super-resolution imaging in deep biological tissues. The main reason is that the excitation and emission of these nanoparticles are located at the range of visible and UV wavelength with a low penetration ability in the deep tissue.[102] Moreover, these fluorescent nanoparticles are relatively photostable compared with the dyes and proteins, but it is still a significant challenge for them to resist the photobleaching under excitation over several hours for long-term observation.[29]

Addressing the issues mentioned earlier, UCNPs serve as a new type of multiphoton probe with high-density emitters in small volumes. Each UCNP contains thousands of co-doped lanthanide ions that form a network of photon sensitizers and activators, which have the unique optical property of converting NIR to shorter wavelength NIR, visible and UV emission.[103] UCNPs have attracted considerable attention due to the superior physicochemical features, such as significant anti-Stokes spectral shifts, background-free, photostable, low toxicity and high-imaging penetration ability.

8.3.2 UCNPs in STED Microscopy

8.3.2.1 Principle of STED Microscopy

In 1990, Stephen Hell proposed STED microscopy and the first experiment with this was conducted with his colleagues four years later.[65] STED also was the first technique successfully applied to super-resolution biological imaging of fixed cells.[66] As stated above, STED employed spatially modulated and saturable transitions between two fluorescent molecular states to engineer the PSF (Figure 8.4). The principle of this technique relied on a time-sequential readout of fluorescent probe photoswitching.

A conventional STED optical setup configuration is presented in Figure 8.4(a) showing the phase plate, excitation and depletion lasers (green and red beams, respectively), dichromatic mirrors, objective and tube lens. The probe is illuminated under two synchronized ultrafast tiny aligned sources consisting of an excitation pulse laser followed by a doughnut-shaped depletion pulse laser which is also named the STED beam. Generally, the pulse width of the excitation laser is shorter than that of the counterpart. By taking advantage of the timescales for molecular relaxation, pulsed lasers create radially symmetric depletion regions. Figure 8.4(b) demonstrates that fluorophores occur when a molecule encounters a photon that matches the energy gap between the ground (S_0) and excited (S_1) states. Upon interaction between the photon and excited fluorophore, the excited fluorophores instantaneously return to the ground state through stimulated emission before fluorescence emission can occur. Consequently, the STED beam can effectively deplete fluorescence in the selected zones near the focal spot of the excited fluorophores (Figure 8.4(c)). The deactivation of the fluorophores occurs in the whole focus volume, excluding the center of focus.

STED microscopy is one type of PSF engineering technique that sharpens the size of the focal spot, equivalent to expanding the microscope spatial frequency passband. Although both laser beams remain diffraction-limited, the final achievable resolution can easily bypass the diffraction limit since the STED beam is shaped to feature a near-zero-intensity point at the center of focus with the exponentially growing intensity toward the periphery. The nonlinear depletion of the excited

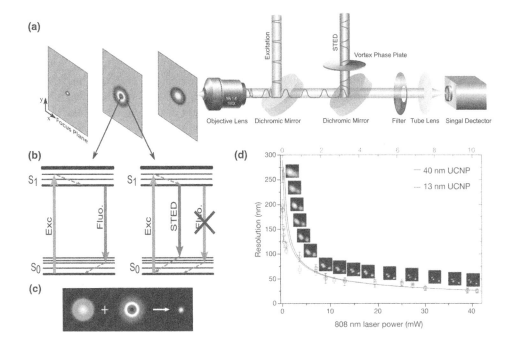

FIGURE 8.4 The principle of a STED microscope. (a) Experimental setup for STED microscopy. (b) The simplified energy levels of the fluorescent switching in STED. (c) A schematic illustration of a STED micros-copy. (d) The nonlinear depletion curve is shown as the normalized fluorescent signal curve is plotted as a function of the STED laser power. The insets show simulated depletion efficiency profiles for different values of maximum STED intensity. (Modified from Liu Y, Lu Y, Yang X, Zheng X, Wen S, Wang F, Vidal X, Zhao J, Liu D, Zhou Z, Ma C, Zhou J, Piper J, Xi P, and Jin D, *Nature* 543, 229, 2017. With permission.)

fluorescent state by the STED beam constitutes the basis for achieving images with a resolution under the diffraction limit (Figure 8.4(d)).[23] Most spontaneous fluorescence can be depleted once the power of the STED beam is above the statured threshold. With the increased STED beam power, the "nonfluorescent area" is expanding, but with little impact on the fluorescence in the central area. When two laser pulses are superimposed, only probes that sit in the center of the STED beam can have the fluorescence emission, significantly restricting the emission and narrowing the size of the effective PSF. The FWHM of the remaining effective PSF is well approximated by an expanded form of Equation (8.3).

$$\Delta r_{STED} \approx \frac{\lambda}{2n \sin \alpha \sqrt{1 + \left(\frac{I_{max}}{I_{sat}}\right)}} \tag{8.3}$$

where $I_{sat} = h\upsilon / \tau_{fl} \sigma$ is defined as the characteristic threshold intensity at which half of the maximal fluorescent signal is elicited, I_{max} is the maximum STED beam power, $h\upsilon$ is the photon energy, σ is the cross-section of excitation and τ_{fl} is the fluorescence lifetime of the excited state.

By increasing the power of the depletion laser, we can obtain an arbitrarily small Δr_{STED} to overcome the diffraction limit. Underlying Equation (8.3), the effective resolution increase with STED is proportional to the power of the depletion laser. The resolution is theoretically not fundamentally limited by diffraction anymore but rather by the residual intensity in the dough-nut minimum in practice and the material properties, like the largest I_{max} that may be tolerated by the specimen. It becomes problematic at extremely intense laser powers that are likely to cause destruction and rapid photobleaching of the fluorescent probe. By taking the non-bleaching feature of the nitrogen-vacancy (NV) center in diamond,[104] the highest resolution of 5.8 nm has been reported under the laser intensity of 3.7 GW/cm^2. For biological structure imaging, the STED system can achieve a resolution of 20 nm with organic dyes [105] and a resolution of 50 nm with fluorescent proteins.[106]

For recording a complete image, both lasers are raster-scanned across the samples, similar to the conventional confocal microscope. One of the advantages of STED microscopy is that effective resolution increase is entirely dictated by the experimental configuration and the laser beam powers applied to the sample. Moreover, the image is acquired as the beam scans along with the sample without additional image processing for the first generation of STED microscopy. Meanwhile, the imaging speed of STED has reached up to 200 fps [66] and up to 28 fps for the biological samples.[105]

It has been three decades since the STED concept was first proposed, and much work has documented the enhanced performance of this imaging system, especially for those advanced versions to address the improvements in axial resolution for 3D imaging. By coupling two STED depletion beams with two opposing objective lenses as a 4Pi-style configuration, researchers have developed iso-STED [107] to significantly improve resolution in both the lateral and axial directions. This technique yields an isotropic PSF with a resolution reaching 40 nm along the axes. Such an asymmetrical focal spot has benefits in recording high-resolution images from deep within biological cells and tissues, such as visualizing the essential structures in the mitochondria of intact cells.[108] Moreover, digital image processing can be employed to reduce the effect of side lobes and other artifacts that compromise the PSF. By exploiting the saturated fluorescence excitation, a compressed sensing approach [109] has been recently reported as demonstrating the potential to obtain 3D super-resolved imaging, principally through a single 2D raster scan with tightly focused PSF.

8.3.2.2 Low-Power STED Microscopy

Recent studies have indicated that UCNPs with thousands of emitters per nanoparticle can serve as suitable single-molecule probes, enabling bright emission in super-resolution imaging. The concept of UCNP-assisted STED nanoscopy (UCNP-STED) was proposed and demonstrated to achieve sub-30 nm optical resolution in resolving the cluster of single 13 nm UCNPs (Figure 8.5), with an excitation of 980 nm laser at 1-mW power and 808 nm laser fixed at 40 mW as the depletion. [23] Specifically, the metastable $^3H_4$ level was populated due to the PA-like effect that was caused by cross-relaxation when high thulium ions (Tm^{3+})-doped UCNPs excited at 980 nm laser beam, resulting in population inversion relative to the $^3H_6$ ground level. The depletion 808 nm laser corresponding to the energy gap ($^3H_4 \rightarrow {}^3H_6$) triggered stimulated emission to inhibit blue emission (Figure 8.5(b)). Taking advantage of reducing the required power, UCNP-STED has the potential to generate the multiple-doughnut beam to realize rapid parallelized nanoscope.[110] Conducting UCNP-STED with a depletion beam at lower power, barely any autofluorescence could be generated to obtain the superior signal-to-background ratio (Figure 8.5(c) and (d)). Moreover, the continued work further presents a resolution of 80 nm in resolving cellular cytoskeleton protein structures (Figure 8.5(g)–(j)).[38]

Although UCNP-assisted STED reduces two orders of magnitude power than conventional dye-based STED, several restrictions still exist. First, like the other STED-like modalities, the depletion beam requires much higher power than that of the excitation laser. Second, this STED scheme calls for a proficient skill in aligning dual laser beams, which confines the technique to specialized laboratories and applications. Third, the low-efficiency four-photon emission in these UCNP-assisted STED methods is not suitable for high-sensitivity super-resolution imaging. Fourth, the specific requirement of the depletion beam to match the targeted bandgap of the energy levels in UCNPs limits the preponderance in various emission wavelengths. In this regard, a more convenient and efficient approach is imperative to utilize UCNPs in super-resolution microscope instead of a stimulated emission procedure.

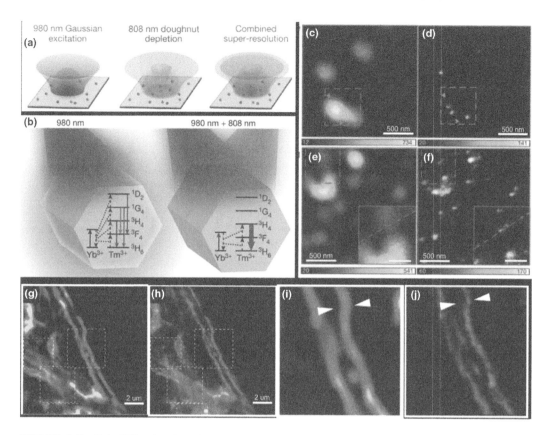

FIGURE 8.5 Sub-diffraction imaging with UCNPs. (a) Diagrams of the UCNP-assisted STED super-resolution imaging. (b) Energy level diagrams of Yb/Tm-co-doped UCNPs under 980 nm illumination (left), and under both 980 and 808 nm illumination (right). (c) and (d) Confocal (left) and super-resolution (right) images of the 40 nm 8% Tm-doped UCNPs. (e) and (f) Confocal (left) and super-resolution (right) images of the 13 nm 8% Tm-doped UCNPs. (g) Diffraction-limited image of UCNP-stained cytoskeleton structures. (h) STED super-resolution image of (g). (e) The confocal image from the panel in (g). (j) STED image of (i). (Modified from Liu Y, Lu Y, Yang X, Zheng X, Wen S, Wang F, Vidal X, Zhao J, Liu D, Zhou Z, Ma C, Zhou J, Piper J, Xi P, and Jin D, *Nature* 543, 229, 2017. With permission; Modified from Zhan Q, Liu H, Wang B, Wu Q, Pu R, Zhou C, Huang B, Peng X, Ågren H, and He S, *Nat. Commun.* 8, 1058, 2017. With permission.)

8.3.3 UCNPs in FED Microscopy

8.3.3.1 Principle of FED Microscopy

As a derivative of STED, fluorescence emission difference (FED) microscopy [111, 112] is a promising approach that addresses the abovementioned challenges in the commercialized laser scanning microscopy system without requiring specific dyes and emission spectra. FED microscopy (Figure 8.6) is an effective method that can improve the resolution of laser scanning microscopy based on two sequential scans to obtain two images with a Gaussian PSF (PSF_{Gau}) and a doughnut PSF (PSF_{Dou}), respectively. The super-resolution image is achieved by subtracting the image obtained by the doughnut beam scan from the image obtained from the Gaussian beam scan. An appropriate normalizing coefficient (r), typically between 0.7 and 1, is used to adjust the imaging quality. The subtraction process follows the equations:

$$PSF_{FED} = PSF_{Gau} - r * PSF_{Dou} \tag{8.4}$$

$$F_{FED} = F_{Gau} - r * F_{Dou} \tag{8.5}$$

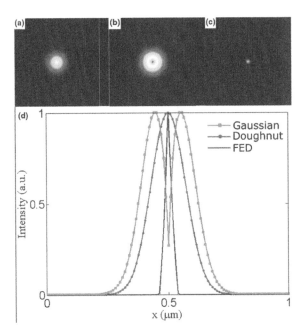

FIGURE 8.6 The principle of a FED image subtraction microscope. (a) The Gaussian PSF. (b) The doughnut PSF. (c) The processed FED PSF by subtracting (b) from (a). (d) The corresponding cross-section profiles in (a)–(c).

where PSF_{FED} is the processed effective PSF; PSF_{Gau} is Gaussian PSF; PSF_{Dou} is doughnut PSF; r is normalizing coefficient; F_{FED} is the processed subtraction image; F_{Gau} is the image taken with PSF_{Gau}; and F_{Dou} is the image taken with PSF_{Dou}.

The details of the system and the imaging procedure for the point-scanning microscopy have been described.[113] Briefly, the emission is collected by an objective lens with a high numerical aperture and focused by the tube lens onto the single-photon detector, so that the effective PSF $(h_{\text{eff}}(x,y))$ can be described as:

$$\begin{cases} h_{\text{eff}}(x,y) = h_{\text{em}}(x,y) \times h_c(x,y) \\ h_{\text{em}}(x,y) = \eta(i) \times h_{\text{exc}}(x,y) \end{cases} \tag{8.6}$$

where $h_{\text{em}}(x,y)$ is the PSF of emission; $h_c(x,y)$ is the PSF of the confocal collection system; $h_{\text{exc}}(x,y)$ is the PSF of the excitation beam (doughnut beam); and $\eta(i)$ is the excitation power dependent emission intensity curve; The FWHM of the intensity dip in $h_{\text{eff}}(x,y)$ represents the nanoscopy resolution.

The experimentally measured intensity distribution PSF $(h_{\text{exp}}(x,y))$ on the image plane in the system is the convolution between the $h_{\text{eff}}(x,y)$ and the spatial distribution profile $(h_{\text{fluo}}(x,y))$ of fluorescent signals as written below:

$$h_{\text{exp}}(x,y) = h_{\text{eff}}(x,y) * h_{\text{fluo}}(x,y) \tag{8.7}$$

For generating a super-resolution image of the single nanoparticle by FED, the researchers employ a tightly focused doughnut-shaped excitation beam to scan across a sample containing UCNPs. Only when a single UCNP sits in the middle of the doughnut beam does FED generate a negative contrast. Using the definition of resolution in GSD microscopy,[114] the FWHM of the dip at the measured

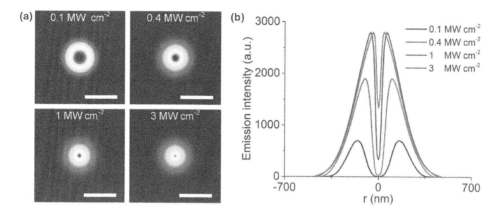

FIGURE 8.7 The principle of FED nanoscopy using UCNP as a multiphoton probe for deep tissue imaging. (a) The simulated "negative" contrast images of the saturated upconversion emission of a single UCNP at four different excitation powers of 0.1, 0.4, 1 and 3 MW/cm². (b) Cross-section profiles in (a). Pixel size, 10 nm. Scale bar is 500 nm. (Modified from Chen C, Wang F, Wen S, Su QP, Wu MC, Liu Y, Wang B, Li D, Shan X, Kianinia M, Aharonovich I, Toth M, Jackson S, Xi P and Jin D. *Nat. Commun.* 9, 3290, 2018. With permission.)

PSF of a single UCNP defines the experimental resolution achieved by FED (Figure 8.7(b)). By taking advantage of the uniform size of monodispersed UCNPs, the actual resolution of FED can be calculated through deconvolution with the size of a single nanoparticle. With the nonlinear excitation process, the optical resolution of FED is limited by diffraction limitation under low excitation power (Figure 8.7(a)). However, this limitation can be easily overturned with high power.

The resolution of FED at a specific excitation power density is primarily determined by the emission saturation curve of UCNPs. There are three features from the curve affecting the resolution. The first feature is the saturation intensity point (I_S) to achieve the half value of the maximum emission intensity.[115] The second feature is the power point (I_{MAX}) which sets out to achieve the maximum emission intensity with fixed I_S. In other words, smaller I_{MAX} indicates the more superlinear shape of the curvature between I_S and I_{MAX}. The third feature is the onset value of the curve, which is defined as the power point (in the unit of I_S) to achieve e^{-2} of the maximum emission intensity. Tremendous onset indicates a more underlinear shape of the curvature between 0 and I_S. When excitation power is smaller than I_S, the nanoparticle has not reached the non-saturation state. The acquired image of the UCNPs reflects the PSF of the doughnut beam because the emission signal is proportional to the excitation intensities. When the applied laser intensity is over I_S, the acquired pattern will be affected due to the optical nonlinear response of the emitter. The emission saturation should result in the expansion of the obtained image and sharpened the dip at the center of the doughnut pattern, as shown in Figure 8.7(a). Precisely, the increased excitation power elevates the fluorescence at the center region, while the fluorescence signals away from the center retain the same values since they have already reached the maximum. Typically, it is easy to overcome the diffraction limitation by employing a relatively high laser much over I_S. Lower values of I_S and/or I_{MAX} shrink the size of a dark spot in the doughnut emission PSF, thereby enhancing the resolution. The more significant onset of the curve offers a lower depth of the PSF and thus yields a better resolution.

For demonstrating the effect of the PSF on the FED image subtraction, a numerical experiment is considered in a series of patterns that comprised emitters with varying distances. In the center region, the distance between these points is closer to each other. The points move further away as they spread out, and this design allows us to obtain a distribution of points with different spatial frequencies in a plane. Figure 8.8 demonstrates the simulation results of the FED image subtraction. Compared with the image obtained from PSF$_{Gau}$ (Figure 8.8(a) and (d)), the resolution of the processed image by subtracting the emission PSF$_{Dou}$ from the PSF$_{Gau}$ has been significantly enhanced (Figure 8.8(c) and 8.8(e)).

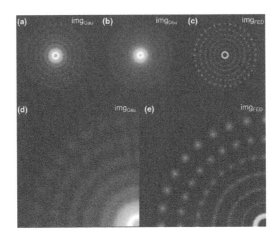

FIGURE 8.8 Simulation result by FED image subtraction. (a) The scanning image by Gaussian PSF and (b) Doughnut PSF. (c) The processed result is achieved by subtracting the image in (b) from that in (a). (d) The magnified quarter image in (a). (e) The magnified quarter image in (c). In this simulation, the appropriate normalizing coefficient r is 0.9. (Modified from Chen C, Liu B, Liu Y, Liao J, Shan X, Wang F, and Jin D, *Adv.Mater.* 17:2008847, 2021. With permission.)

Although FED microscopy has improved the resolution, the main shortcomings of the subtraction-based modality suffer from the image distortion of artifacts and critical information loss. The subtraction balances the initial image's intensities, and the direct subtraction of normalized data sets often generates areas with negative intensity values, which is a potential source of data loss. To eliminate this effect, researchers introduce an appropriate normalizing coefficient r to adjust the strength of the subtraction. However, the r as a constant for the entire imaging area will cause over-subtracting and the resultant data loss. A few years ago, researchers further employed a pixel assignment [116] approach to adjust the contrast factor in each pixel, but at the expense of the overall resolution. To date, many efforts have been made to improve the performance of FED in several aspects, e.g., employing a spatial light modulator for laser switching, eliminating deformations via beam modulation and enhancing resolution by using a saturation effect. These features have made FED a more powerful super-resolution microscopy imaging method.

8.3.3.2   Non-Bleaching FED Microscopy

As a facile super-resolution method, FED microscopy releases the requirements of the intense depletion laser and is independent of the species of agents, which marks out FED microscopy as having great potential. However, conventional FED microscopy employs fluorescent dyes and proteins, which suffer from the effect of photobleaching and are not suitable for long-time tracking and imaging. Moreover, the saturated FED microscopy requires heightened illumination intensities to achieve saturated excitation effect. This will further increase the photobleaching effect more severely and also cause photodamage issues to the biological samples.

By introducing $Nd^{3+}/Yb^{3+}/Er^{3+}$-co-doped UCNPs with red emission (660 nm), the non-bleaching FED microscopy achieves 172 nm resolution by using 800 nm 10-MW/cm^2 excitations.[50] Figure 8.9 demonstrates the experimental imaging study of a single nanoparticle under different excitation intensities (0.1, 1 and 10 MW/cm^2, respectively). When under the intense excitation intensity (1 or 10 MW/cm^2), the signal-to-noise ratio of image was increased, facilitating the imaging resolution quantification. When comparing the FWHM of solid PSFs in Figure 8.9(b) and (c), the PSFs widen due to the emission saturation effect. For the doughnut PSFs, higher intensity results in a narrower dark center (Figure 8.9(e) and (f)). With the subtraction process (Figure 8.9(h) and (i)), the obtained FED PSFs' resolution has been significantly enhanced.

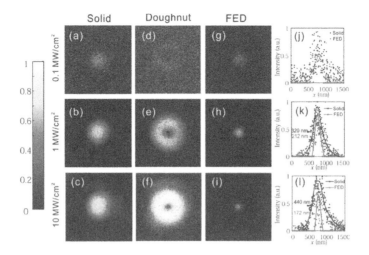

FIGURE 8.9 Imaging of single UCNP at corresponding excitation intensities. (a)–(c) Solid laser-excited PSFs at 0.1, 1 and 10 MW/cm², respectively. (d)–(f) Doughnut laser excited PSFs at the corresponding excitation intensity. (g)–(i) FED PSFs at corresponding excitation intensity with subtraction factors set to 0.83. (j)–(l) The intensity profile along the image center of solid and FED PSFs. (Wu Q, Huang B, Peng X, He S, and Zhan Q, *Opt. Express* 25, 30885, 2017. With permission.)

It also presents that the saturated emission of UCNPs under the intense excitation laser intensity can further improve the resolution. Two adjacent points in the sub-diffraction limited volume can be differentiated (Figure 8.10(a)–(c)), demonstrating the superior resolving ability of FED is better than conventional confocal images. Moreover, images of monodispersed UCNPs show no intensity decrease (Figure 8.10(d) and (e)), even under after half-hour laser scanning, suggesting they do not suffer from photobleaching.

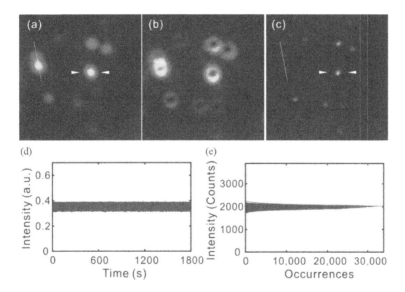

FIGURE 8.10 Imaging of monodispersed UCNPs under 10-MW/cm² irradiation. (a) Image scanned by Gaussian laser, and (b) by doughnut laser. (c) Subtraction image of (a) and (b), $r = 0.83$. (d) The time trace of emission intensity from UCNPs under 10-MW/cm² continuous laser illumination for 30 minutes with 100-μs dwell time. (e) The stability of time traces and the single-peak Gaussian distributions of corresponding histograms in (d). (Wu Q, Huang B, Peng X, He S, and Zhan Q, *Opt. Express* 25, 25, 30885–30894, 2017. With permission.)

Compared to the STED microscopy, the simplified optical system with a polarization-maintaining single-mode fiber-coupled 808 nm diode laser was designed to perform the non-bleaching FED imaging. The prepared UCNPs solution was highly diluted and dispersed on a coverslip using a spin-coating method for imaging experiment. An image of the Gaussian PSF and an image of Doughnut PSF would be recorded by switching the two laser beams.

8.3.3.3 Near-Infrared Emission Saturation Microscopy for Deep Tissue Imaging

Through the transparent biological window,[102] NIR light does not excite autofluorescence background and has minimal phototoxicity. With the advantage of the long wavelength, multiphoton laser scanning microscopy is an alternative method for imaging the biological samples with tens to hundreds of microns thickness.[117] Nevertheless, it requires high-power and expensive femtosecond lasers to achieve super-resolution. This is due to the poor efficiency of the nonlinear multiphoton process, for instance, small multiphoton absorption cross-sections of the probes.[118]

As a NIR-in and NIR-out configuration, UCNPs represent an ideal fluorescent probe that would enable upconversion nanoscopy imaging in deep tissue. Parallel to the non-bleaching FED, the researchers employ UCNPs to unlock a novel mode of NIR emission saturation (NIRES) microscopy for super-resolution imaging inside the deep tissue, with much lower excitation intensity than that required in conventional dyes.[113] NIRES nanoscopy takes advantage of multi-intermediate ladder-like energy levels of UCNPs (Figure 8.11(a)), easily converting 980 nm photons into 800 nm photons. The emission saturation curve of the two-photon state $^3H_4$ (800 nm) (Figure 8.11(b)) shows the early onset of upconversion emissions at low-excitation power density and sharp rising-up slope, reflecting its nonlinear energy transfer assisted the photon upconversion process. The decline in the saturation curve under high excitation power (larger than 2 mW) is due to the energy being further upconverted from $^3H_4$ to the higher energy states.

The researchers further examine the penetration depth and resolution of NIRES imaging through deep tissue (Figure 8.12). In this experiment, 4% Tm^{3+} 40% Yb^{3+}-co-doped UCNPs are attached behind a 93-μm thick slice of mouse liver tissue, which allows UCNPs to diffuse into the tissue slice for super-resolution imaging of single UCNPs from different depths (Figure 8.12). Due to the aforementioned strong attenuation for visible emissions (Figure 8.12(b)), there is only 11.3% of

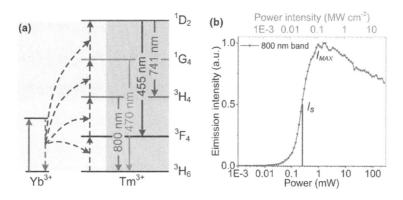

FIGURE 8.11 The simplified energy levels and 800 nm emission power dependence of Yb^{3+} and Tm^{3+}-co-doped UCNPs. (a) The sensitizer Yb^{3+} ions initiate the photon upconversion process by a linear and sequential absorption of 980 nm excitation. Due to the multiple long-lived intermediate states, the energy is stepwise transferred onto the scaffold energy levels of emitters Tm^{3+}. This eventually facilitates multiphoton upconversion emissions, including those from the four-photon upconversion excited state $^1D_2$ (455 and 741 nm), three-photon state $^1G_4$ (470 nm) and two-photon excited state $^3H_4$ (800 nm). (b) The saturation intensity curve of the 800 nm emissions derives from UCNPs (40 nm $NaYF_4$: 20% Yb^{3+}, 4% Tm^{3+}) under 980 nm excitation. (Modified from Chen C, Wang F, Wen S, Su QP, Wu MC, Liu Y, Wang B, Li D, Shan X, Kianinia M, Aharonovich I, Toth M, Jackson S, Xi P and Jin D. *Nat. Commun.* 9, 3290, 2018. With permission.)

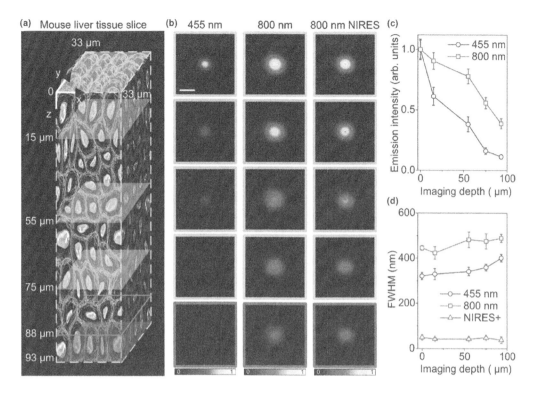

FIGURE 8.12 The penetration depth of different emission bands and optical resolution of different imaging modalities at different depth of a liver tissue slice. (a) An illustration of a mouse liver tissue slice with 93-μm thickness. (b) Single-particle imaging at different depth in liver tissue. Left, confocal images from 455 nm emission; middle, confocal images from 800 nm emission; right, the corresponding NIRES images. (c) The normalized emission attenuation at different depths through the liver tissue. (d) The corresponding FWHM in (b); The resolutions of NIRES in (d) are 49.6 ± 11.1 nm (0 μm), 42.4 ± 6.2 nm (15 μm), 42.4 ± 7.2 nm (55 μm), 48.0 ± 7.3 nm (75 μm), 38.2 ± 14.3 nm (93 μm). Benefiting from the controlled synthesis of intensity-monodispersed UCNPs, every single nanoparticle can be differentiated from each cluster by comparing their emission intensity to statistical averaged value. The pixel dwell time for confocal and NIRES is 3 ms. The pixel size for confocal and NIRES is 10 nm. The scale bar is 500 nm. (Modified from Chen C, Wang F, Wen S, Su QP, Wu MC, Liu Y, Wang B, Li D, Shan X, Kiania M, Aharonovich I, Toth M, Jackson S, Xi P and Jin D. *Nat. Commun.* 9, 3290, 2018. With permission.)

455 nm emission left in confocal imaging (Figure 8.12(c)). More encouragingly, by increasing excitation power to compensate for the aberration-induced distortion on excitation PSF,[119] a relatively consistent resolution of sub-50 nm has been maintained without any aberration correction through tissue as deep as 93 μm.

The saturation curve of UCNP can be optimized by tuning the doping concentration of emitters. This work further demonstrates that the material design of UCNPs can improve their optical properties to enhance resolution, suggesting a new topic for the material sciences community to investigate, specifically more efficient super-resolution probes through deep tissue.[120] Combined with the self-healing Bessel beam, the researchers explored this sub-diffraction imaging modality to map single nanoparticles inside a strong scattering tumor cells spheroid.[121]

8.3.3.4 One-Scan FED Microscopy for Increasing Imaging Speed

For the purposes of biological research, imaging speed is another essential parameter. However, the conventional FED microscopy requires two different modulated images for the subtraction process, restricting the imaging speed. Moreover, the additional data acquisition may compromise the performance of FED due to the sample drifting issue, especially in a dynamic environment. With the excitation-orthogonalized UCNPs, one-scan FED microscopy eliminates the issues mentioned earlier by using two synchronized Gaussian and doughnut beams.[122]

One-scan FED microscopy is achieved by employing UCNPs with a facile structure $NaYF_4:2\%Er^{3+}@NaYF_4@NaYF_4:20\%Yb^{3+}/2\%Tm^{3+}$ nanoparticles (Figure 8.13(a)), which emit two-photon green luminescence under 808 nm laser excitation and three/five-photon blue one under 940 nm laser excitation, respectively. An inert $NaYF_4$ isolation shell was designed to block energy transfer between the segment of Yb^{3+}/Tm^{3+} and that of Er^{3+}. The proposed luminescence mechanism of the core and shell is demonstrated in Figure 8.13(b), respectively. Er^{3+} singly doped $NaYF_4$ can absorb 808 nm photon energy and generate two-photon green emissions at 525 and 550 nm, which are assigned to the transitions $^2H_{11/2} \rightarrow {}^4I_{15/2}$ and $^4S_{3/2} \rightarrow {}^4I_{15/2}$ of Er^{3+}, respectively. The outer layer Yb^{3+}/Tm^{3+}-co-doped $NaYF_4$ can emit three/five-photon blue emissions at 475 and 455 nm under the excitation of a 940 nm laser, originating from the transitions $^1D_2 \rightarrow {}^3F_4$ and $^1G_4 \rightarrow {}^3H_6$ of Tm^{3+}, respectively. The core/shell/shell structure can be excited orthogonally under the selective excitation

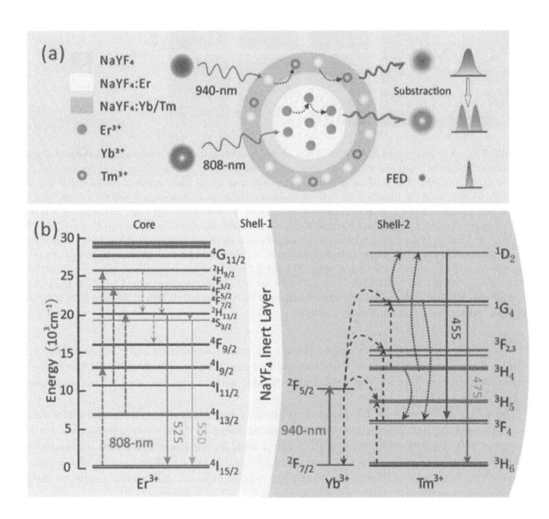

FIGURE 8.13 Design layout of UCNP nanoprobes with orthogonalized dual Ex/Em. (a) Schematic design of $NaYF_4:2\%Er^{3+}@NaYF_4@NaYF_4:20\%Yb^{3+}/2\%Tm^{3+}$ nanoparticles and schematic illustration of one-scan FED theory. (b) The proposed upconversion luminescence mechanism of dual excitation and emission. (Huang B, Wu Q, Peng X, Yao L, Peng D, and Zhan Q, *Nanoscale*. 10, 21025, 2018. With permission.)

lasers, and emissions can be detected by two separate detecting channels, ensuring the performance of FED with a high signal-to-noise ratio. Through the one-scan method, an image with Gaussian spots and another with doughnut spots can be obtained simultaneously in two different channels. Based on the earlier consideration, a super-resolution image can be obtained by simply implementing image subtraction.

Under the two orthogonalized excitations of NIR lasers, spectrally nonoverlapped emission bands and images were investigated. The proposed one-scan FED nanoscopy reduces the laser power compared with the previously reported STED imaging using UCNPs. Therefore, one-scan FED nanoscopy offers the potential for low-power super-resolution imaging. It is even feasible to implement immediate super-resolution microscopy imaging when combined with real-time imaging processing. This can be a good choice for long-term monitoring of live cells because of drift-free and quick scanning, the non-bleaching properties of UCNPs and low-power NIR excitation lasers. Apart from this, the proposed optically orthogonalized nanoprobes also have considerable potential in applications of high throughput, multichannel optical imaging and sensing. Intriguingly, due to its low excitation power requirement and simple control system, it is viable to further accelerate the image acquisition through parallel excitation and a detection method like parallelized STED nanoscopy.

8.3.3.5 Single Beam One-Scan FED for Simplifying the Configuration

However, the complexity of the coupling of two lasers hinders the development of this one-scan FED technology. By taking advantage of the apparent contrast in saturation intensity curves of upconversion emissions from the multi-intermediate states [31], researchers scan the sample of single UCNPs using a tiny focused doughnut illumination beam and detect from multiple emission channels, including 800 and 740 nm (shown in Figure 8.14(c)).[124] Benefiting from the multiple long-lived intermediate states (Figure 8.14(a)), the sensitized photon energy can be stepwise transferred onto the scaffold energy levels of Tm^{3+} emitters, which eventually facilitates the multiphoton upconversion emissions from the two-photon excited state (800 nm, $^3H_4 \rightarrow {}^3H_6$) and four-photon excited state (740 nm, $^1D_2 \rightarrow {}^3H_4$). The upconversion emission at 800 nm from the lower intermediate excited level has a lower saturation threshold compared with emissions from the higher levels, as confirmed by the nonlinear fluorescence saturation curves shown in Figure 8.14(b).

The researchers found that under low excitation power, both channels display the doughnut-pattern PSFs.[123] The difference in heterochromatic saturated emission PSFs becomes significant with an increased excitation power. Because of the non-zero feature at the center dip of doughnut beam (around 1.4%), at an intense power density, the 800 nm saturated emission PSF eventually becomes a "Gaussian PSF" with two-photon upconversion emission at center reaching the maxima (Figure 8.14(c)). The center dip rises sharply with the increased excitation power. This can be explained by the increased excitation power elevating the two-photon fluorescence at the center to reach the maxima, while the fluorescence signals away from the center retain the same values since they have reached saturation. Eventually, the doughnut-shaped profile turns into a Gaussian one when fully saturated power is evident. In contrast, the 740 nm saturated emission PSF remains in a doughnut shape. Moreover, the research works also indicate to reduce the excitation power by designing the optimal synthesis of UCNPs, including the activator and sensitizer doping concentration, core-shell structures, and choose the optimal emission bands.[120]

8.3.3.6 Fourier Fusion for Enhancing the Image Quality

Although the one-scan FED method has simplified the two scans-based modalities of FED, the main shortcomings of both subtraction-based modalities suffer from the image distortion of artifacts and critical information loss. The subtraction balances the intensities of the initial images, and the direct subtraction of normalized data sets often generates areas with negative intensity values, which is a potential source of data loss. When an appropriate normalizing coefficient (r) is introduced to adjust the strength of the subtraction, the r, as a constant for the entire imaging area, will

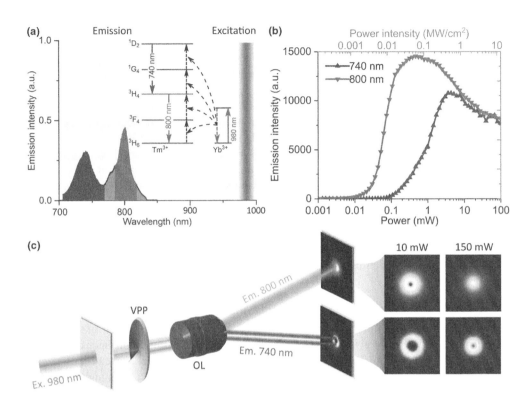

FIGURE 8.14 Heterochromatic emission saturation contrast produced by UCNPs and power-dependent emission PSF patterns under a tightly focused doughnut beam illumination. (a) The upconversion emission spectrum of a typical 40 nm single nanoparticle (NaYF$_4$: 40% Yb^{3+}, 4% Tm^{3+}) upon 980 nm excitation laser at a power density of 10 MW/cm^2. Inset is a simplified energy level and upconversion process of Yb^{3+}- and Tm^{3+}-co-doped UCNPs. The multiphoton NIR upconversion emissions mainly from two-photon excited state (800 nm, $^3H_4 \rightarrow {}^3H_6$) and four-photon excited state (740 nm, $^1D_2 \rightarrow {}^3H_4$). (b) The two distinct power-dependent saturation intensity curves of the 800 and 740 nm emissions. (c) The power-dependent PSF patterns of two emission bands from a single UCNP under a doughnut-shaped excitation beam. Under the low excitation power (10 mW), both channels display the doughnut PSFs. The 800 nm emission converts to Gaussian PSF when increasing the laser power to 150 mW. (Modified from Chen C, Liu B, Liu Y, Liao J, Shan X, Wang F, and Jin D, *Adv.Mater.* 17:2008847, 2021. With permission.)

cause over-subtracting and the resultant data loss. As the processed image (Figure 8.8(e)) shows, the subtraction method creates an absence in critical spatial frequencies. Moreover, the processed image depicts the dots with different sizes, and these are apparently the artifacts.

Alternatively, the researchers transfer the heterochromatic saturated emission PSFs into Fourier domain and achieve the super-resolution imaging by conceptualizing the frequency shifting mechanism.[123] This strategy can effectively resolve any issues of information loss and image distortion associated with the spatial domain subtraction approach since Fourier domain optical transfer functions (OTF) fusion can maximize the overall coverage of emission PSF patterns. By using a doughnut beam illumination, the spatial frequency components are encoded into both the doughnut excitation patterns, and the doughnut saturated emission patterns (Gaussian), including the high spatial frequency components being captured in the broader range of the detection OTF.[124] Figure 8.15 compares the difference in frequency components between OTFs obtained by two emissions' PSFs. Obviously, PSF$_{Dou}$ generates more content at a high spatial frequency than that generated by PSF$_{Gau}$. More significant content of high frequencies in OTF helps recover high spatial frequency components in the image through the deconvolution algorithm.

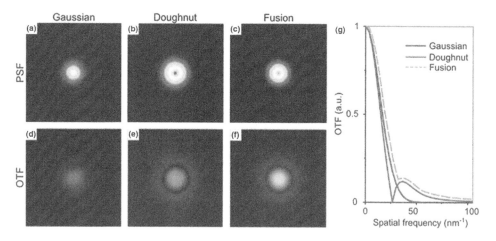

FIGURE 8.15 Schematics of Fourier fusion for super-resolution microscopy. (a) Gaussian PSF. (b) Doughnut PSF. (c) Fourier fusion PSF. (d) OTF of Gaussian. (e) OTF of Doughnut. (f) OTF of Fourier fusion. (g) The corresponding cross-section profiles of OTF in (d)–(f). (Modified from Chen C, Liu B, Liu Y, Liao J, Shan X, Wang F, and Jin D, *Adv.Mater.* 17:2008847, 2021. With permission.)

However, the emission PSF_{Dou} has a gap in the intermediate frequency range, resulting in a deficiency at the mediate to low spatial frequencies (Figure 8.15(e)). Meanwhile, this loss can be compensated for by the OTF components generated from PSF_{Gau} (Figure 8.15(d)). Therefore, heterochromatic OTF fusion of saturated emission PSFs in Fourier domain takes advantage of both the PSF_{Dou} resolving power in providing high frequency content and the compensation effect from PSF_{Gau} in covering the medium frequency range (Figure 8.15(f)). Following a similar study,[124] the researchers develop a heterochromatic Fourier spectrum fusion method to alleviate the frequency deficiency at specific frequency points as well as the partial deficiency induced by the saturation effect. The procedure is given by:

$$OTF_{eff} = OTF_{Gau} \times mask_{Gau} + OTF_{Dou} \times mask_{Dou} \times r_1 \tag{8.8}$$

$$I_{eff} = I_{Gau} \times mask_{Gau} + I_{Dou} \times mask_{Dou} \times r_1 \tag{8.9}$$

where OTF_{eff} and I_{eff} are the final processed effective OTF and image, respectively. OTF_{Gau} and OTF_{Dou} are the Fourier transform of the corresponding PSF_{Gau} and PSF_{Dou}, respectively. The $mask_{Gau}$ (Gaussian low pass) and $mask_{Dou}$ (Gaussian high pass) are the cut-off Fourier frequency filters, respectively. r_1 is a variable that modifies the ratio of these fused components for achieving the optimal synthetic system OTF_{eff}. I_{Gau} and I_{Dou} are the acquired images with the corresponding PSF_{Gau} and PSF_{Dou}, respectively.

Using the approach of Fourier domain heterochromatic fusion of the image by the Gaussian-like (oversaturated doughnut) 800 nm emission PSF (Figure 8.16(a), inset OTF) and the image by the saturating doughnut 740 nm emission PSF (Figure 8.16(b), inset OTF), as shown in Figure 8.16(c), demonstrate here is the ability to resolve the discrete nanoparticles with a discernible distance of 120 nm.[124] Furthermore, the magnified comparison images and the corresponding line profiles in Figure 8.16(f) quantitatively illustrate the enhanced high imaging quality achieved by our Fourier domain fusion method, compared with the image deconvoluted by Gaussian-like PSF and the one processed by subtraction. As a result of the obviously different nonlinear emission responses from the multiple intermediate excited states, the varying degrees of saturating emission PSFs provide the key for the Fourier domain heterochromatic fusion of the images simultaneously obtained from the multiple emission color bands. The line profiles of the processed image of a single nanoparticle, shown

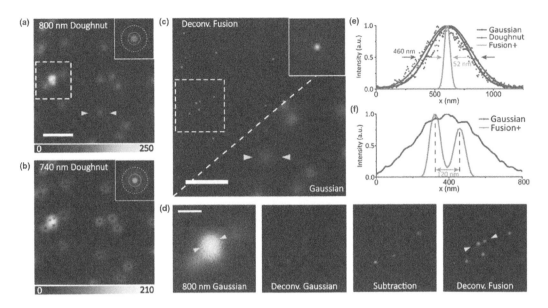

FIGURE 8.16 Super-resolution imaging of UCNPs using Fourier domain heterochromatic fusion. (a) The 800 nm emission band image of UCNPs under a 980 nm intense doughnut beam excitation. With UCNPs' 800 nm emission being oversaturated, the emission PSF shows a Gaussian-like profile. Inset is the corresponding OTF. (b) The 740 nm emission band image of UCNPs under the same 980 nm doughnut beam excitation, showing the doughnut emission PSF. Inset shows the corresponding OTF. (c) The left part is the super-resolution imaging result by Fourier domain fusing the OTFs of (a) with (b). In contrast, the right part is the 800 nm confocal image using a 980 nm Gaussian beam excitation. Inset is the fused OTF. (d) The magnified area of interest to illustrate the comparison imaging results using the various image process algorithms, including Richardson-Lucy deconvolution with Gaussian PSF (Deconv. Gaussian), subtraction of the doughnut image from the Gaussian image (subtraction) and the Fourier domain fusion (Deconv. Fusion), respectively. (e) Line profiles of a single UCNP from (a) and (c). (f) Line profiles of two nearby UCNPs in confocal and fusion images from (d). The excitation power used in this experiment is 150 mW. Pixel dwell time, 1 ms. Pixel size, 10 nm. The scale bars are 1.5 μm in (a) and (c), and 500 nm in (d). (Modified from Chen C, Liu B, Liu Y, Liao J, Shan X, Wang F, and Jin D, *Adv.Mater.* 17:2008847, 2021. With permission.)

in Figure 8.16(e), further demonstrate the quantified result of the significantly enhanced FWHM of 52 nm ($\sim\lambda/20$) from 460 nm.

This is the first study to explore the multicolor nonlinear responses and Fourier domain fusion for enhancing spatial resolution beyond the diffraction limit. It maximizes the capabilities of Fourier domain OTF fusion of multiple saturated emission PSFs in the spectral regime. Compared with the temporal domain modulation of excitation modality that requires switching illumination pattern [124] or laser mode [125] with dual excitations procedure, the single scan method is simple, fast and stable. It can avoid the use of additional optical components and procedures in correcting the sample drifts between multiple sequential recordings. The single-beam scanning mode using a simplified optics setup is compatible with the standard commercial or lab-based laser scanning microscopes; therefore, they may overcome the current bottleneck issue associated with the system's complexity and stability.

8.3.4 UCNPs in Photon-Avalanche Single-Beam Super-Resolution Imaging

8.3.4.1 Principle of Photon-Avalanche Single-Beam Microscopy

Combining the doughnut beam-based super-resolution techniques with UCNPs is advantageous because they can be imaged in the background-free and without photobleaching at biologically safe under the NIR excitation. However, the discrepancy between currently available super-resolution

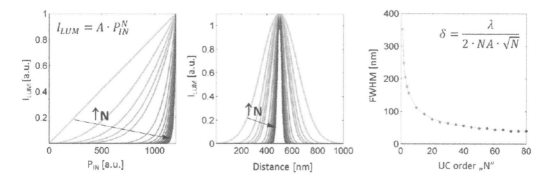

FIGURE 8.17 Introduction to photon avalanche emission and multiphoton super-resolution imaging. The impact of nonlinearity of the luminescent label on the photoexcitation power density dependence of emission. (Bednarkiewicz A, Chan EM, Kotulska A, Marciniak L, and Prorok K, *Nanoscale Horiz.* 4, 881, 2019. With permission.)

techniques and what scientists require is still large. What is needed are more straightforward and inexpensive approaches to manipulate light below the diffraction limit.

Recently, PA in UCNPs has been employed to bypass the diffraction limit by using a single Gaussian beam.[126] This imaging approach requires no complex instrumentation, excitation beam shaping or patterning, image post-processing, or alignment procedures.[127] This simple modality relies on the highly nonlinear dependence of the emission intensity on the excitation power (Figure 8.17). The deeply sub-diffraction resolution will be achieved automatically during the scanning confocal microscopy since the size of the emission PSFs is scaled inversely with the square root of the degree of nonlinearity. According to Equation (8.10), the resolution is inversely proportional to the number of photons involved in the emission process:[128]

$$\Delta r = \frac{\Delta r_0}{\sqrt{N}} = \frac{\lambda}{2\mathrm{NA}\sqrt{N}} \tag{8.10}$$

where Δr is the resolution, Δr_0 is the single-photon diffraction-limited resolution, N is the number of photons involved in the emission process or the order of nonlinearity, λ is the excitation wavelength and NA is the numerical aperture of the objective lens.

Thus, it is possible to realize sub-diffraction imaging is possible when $N > 1$ since the intensity of multiphoton luminescence near the center of the excitation beam spot will be enhanced relative to the intensity near its edges. This is due to the multiphoton luminescence intensity scaling as $I_{\mathrm{lum}} = P_{\mathrm{in}}^N$, where P_{in} is the incident power density.

For the PA in the lanthanide-based nanostructure, a single ground-state absorption will trigger a chain reaction of ESA and cross-relaxation between surrounding ions, resulting in upconverted emissions. The PA mechanism amplifies the population of excited states, like the 800 nm emission from $^3H_4$ level. Specifically, the intermediate state ($^3F_4$) can be repopulated through a positive feedback loop of ESA, following by the cross-relaxation back down to the same intermediate state. This process can effectively double the population of $^3F_4$ state, amplifying the looping in excited-state populations.

By combining the superlinear concept with conventional UCNPs, the researchers achieve a more than several folds' improvement in resolution during 2-photon, 3-photon or 4-photon upconversion imaging.[128–132] Based on the positive optical feedback in each nanocrystal, PA enables the emission scales to be nonlinear with the 22nd power of pump intensity, achieving sub-70 nm spatial resolution using only simple scanning confocal microscopy.[127] Further enhancement in resolution will be challenging because the higher order upconverted emissions will further compromise efficiency. Also, it is still a technical challenge to detect the higher order UV emission by using the

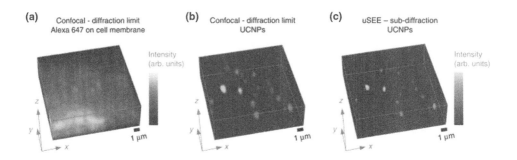

FIGURE 8.18 3D sub-diffraction microscopy imaging in neuronal cells. (a) WGA-Alexa Fluor 647, highlighting the cell plasma membrane. (b) Confocal images of the UCNPs inside the cell taken at excitation power density corresponding to the lateral diffraction limit (11.8 mW/μm^2). (c) Corresponding sub-diffraction images of the UCNPs inside the cell taken at low power (1.7 mW/μm^2). (Denkova D, Ploschner M, Das M, Parker LM, Zheng X, Lu Y, Orth A, Packer NH, and Piper JA, *Nat. Commun.* 10, 3695, 2019. With permission.)

high NA objective lens. Nonetheless, according to theoretical simulation modeling, it can evaluate the feasibility of resolving 20 nm features when nonlinearity exceeds 80.[126]

According to the proposed modeling, UCNP-assisted superlinear approach confines the emission PSF to less than half of conventional confocal PSF without using a depleting beam. This method can be easily conducted without complicated beam overlap adjustments and temporal synchronization of the pump and depletion beams.[126] Any confocal microscope with an appropriate excitation wavelength and spectral filters can perform super-resolution imaging with this superlinear scheme, achievable without extensive modification.

8.3.4.2 Conventional Confocal for 3D Sub-Diffraction Imaging

Furthermore, a similar approach named superlinear excitation-emission microscopy provides experimental evidence by realizing 3D sub-diffraction imaging in a biological sample on a conventional confocal configuration (Figure 8.18).[133] The researchers employed NaYF$_4$ nanocrystals with comparatively higher doping concentration (8%) of Tm^{3+} ions. The design makes it possible to reach a superlinear slope of 6.2, which is almost double the steepness reported in the current literature. At convenient experimental conditions, the resolution was doubly enhanced compared to the diffraction limit both in the axial and lateral direction. Furthermore, it verifies that the UCNPs can be continuously visualized in 3D sub-diffraction imaging conditions for more than 5 hours in the fixed neuronal cells. Also, the superlinear excitation-emission can combine with STED imaging modality can further improve the lateral resolution of approximately 80 nm under lower excitation power, exacting the advantages of both modalities, making it more applicable in a biological environment.[134]

8.3.5 UCNPs in Nonlinear SIM

8.3.5.1 Principle of SIM

Unlike the earlier point-scanning super-resolution modalities, SIM is a wide-field technique that bypasses the diffraction barrier.[68] It extracts a higher resolution by exploiting the interference (moiré patterns) of diffraction orders when two grid patterns are overlaid at an angle (Figure 8.19(a)). As shown in Figure 8.19(b), SIM typically requires closely spaced periodic patterns to down-modulate the high spatial frequency information in the sample so that with the support of OTF the high-frequency information can be reconstructed. Since the resolution is only enhanced in the direction perpendicular to the line-shaped zero nodes, the pattern must be rotated in several directions to cover many angles in the focal plane. With a specialized image-processing algorithm, high-frequency information can be extracted from the raw data to produce a reconstructed image, doubling the resolution of traditional

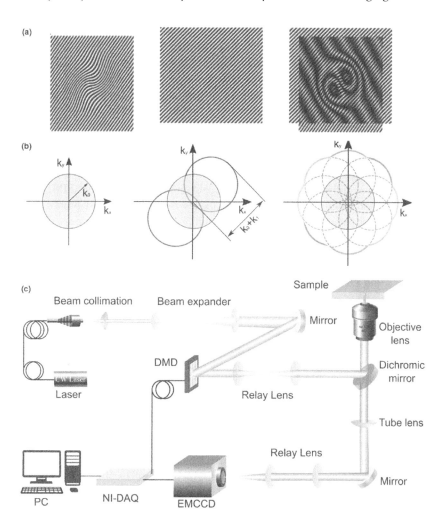

FIGURE 8.19 The principle of a SIM microscope. (a) An unknown sample structure is multiplied by a known regular illumination pattern to generate a better moiré fringes pattern. (b) The corresponding Fourier components. The moiré fringes occur at the spatial difference frequencies between the pattern frequency and each spatial frequency component of the sample structure. (c) Experimental setup for SIM microscopy. (Modified from Gustafsson MG, *Proc. Natl. Acad. Sci. USA* 102, 13081, 2005. With permission; Modified from Liu B, Chen C, Di X, Liao J, Wen S, Su QP, Shan X, Xu ZQ, Ju LA, Mi C, Wang F and Jin D, *Nano Lett.* 20, 4775, 2020. With permission.)

diffraction-limited light microscopy. This method uses a grid array of line-shaped intensity maxima and minima instead of the conventional doughnut-shaped excitation beam that is employed by other PSF engineering techniques of STED and GSD described earlier. The mechanism of the enhanced resolution can be described in terms of spatial frequencies and Fourier components.

The nonlinear structured illumination schemes, for example, saturated pattern excitation microscopy (SPEM) [135] and saturated SIM (SSIM) [68] have demonstrated their ability to take advantage of emission saturation where high-resolution information is acquired in the system. In SPEM and SSIM, two powerful interfering beams from the same light source generate a standing wave grid pattern to form the structured illumination. Due to exposure to the intense illumination, most of the exposed fluorescent probes within the specimen saturated leave only narrow dark line-shaped volumes, lasting approximately 200 nm at the edges of the interference pattern. By increasing the excitation power, the nonlinear saturated photo-response can help to further improve the resolution of SIM in the regime of

50 nm and resolve subcellular structures. New advances made in the denoising process and modified excitation conditions have been applied to SIM, for example, Hessian-SIM and grazing incidence SIM were recently devised with high imaging speed. They are applied to observations of ultrastructures of cellular organelles and their structural dynamics, such as mitochondrial cristae.

The next challenge is to explore the possibilities of using these techniques for thick tissue neuroscience imaging and nanomedicine tracking, as the strong scattering and absorption aberrate the structured illumination patterns and introduce unwanted out-of-focus light, both deteriorating the imaging resolution. To address this challenge, NIR excitation has been implemented to mitigate the absorption. Two-photon or multiphoton excitation in conjugation with spot scanning SIM has been reported to improve the imaging depth through tissue but at the price of low speed caused by the spot-scanning structure. Organic fluorescent dyes and proteins are the most common imaging probes for SIM, because of their outstanding staining and specific ability to target organelles. Nevertheless, these probes require a tightly focused and high-power pulsed laser to activate the multiphoton absorption process, due to their small multiphoton absorption cross-section. The required high excitation power, especially by nonlinear SIM, restricts long-duration visualization of subcellular structure in living cells.

8.3.5.2　Wide-Field Deep Tissue Super-Resolution Imaging

Very recently, the nonlinear upconversion SIM induces a strategy to create video-rate super-resolution imaging through thick biological tissues by using UCNPs.[136] The unique nonlinear photoresponse of UCNPs makes it possible for an efficient nonlinear mode on SIM to obtain high-frequency harmonics in the Fourier domain of the imaging plane. In this way, nonlinear SIM can be achieved with gentle excitation power. It further demonstrates that fine-tuning of the doping concentrations in UCNPs can modify the nonlinearity of the photon response to further enhance the resolution to 1/7th of the excitation wavelength (Figure 8.20).

Based on time-wavelength-space conversion, the multiphoton upconversion time-encoded SIM surpasses the speed limitation on image acquisition with modulation and scanning rate of 50 MHz while maintaining low excitation power.[137] The progress made in nonlinear SIM using UCNPs lies in the improved resolution under a low excitation power, particularly for single nanoparticles dynamic tracking through deep tissue. Most recently, based on the tailored lifetime profiles of UCNPs, time-resolved SIM (TR-SIM) was proposed for high-throughput multiplexing super-resolution imaging.[138]

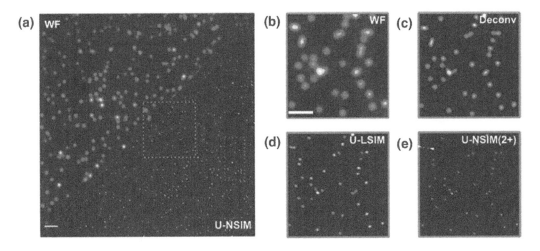

FIGURE 8.20 Super-resolution imaging reconstructions of upconversion nanocrystals. (a) Wide-field (left) and super-resolution (right) images of the 4% Tm-doped UCNPs. Scale bar: 2 μm. (b)–(e) Comparison imaging results of a selected area (orange frame) with different imaging modalities: (b) wide-field microscopy; (c) Wiener deconvolution; (d) linear SIM; (e) nonlinear SIM. Scale bar: 2 μm. (Liu B, Chen C, Di X, Liao J, Wen S, Su,QP, Shan X, Xu ZQ, Ju LA, Mi C, Wang F and Jin D, *Nano Lett.* 20, 4775, 2020. With permission.)

8.4   CHALLENGES

With the nonlinear photon response property, UCNPs have successfully emerged as a powerful potential fluorescent probe for super-resolution imaging. To achieve a low photo-toxicity sub-diffraction imaging with lower excitation power, controllable synthesis procedure of UCNPs will further benefit the energy transfer process and resultant saturation intensity properties. Compared to conventional fluorescent dyes or proteins, the relatively large size of nanoparticles and their complex surface chemistry give rise to difficulties in specific labeling of subcellular targets.

8.4.1   Designing Smaller and Brighter UCNPs

For UCNP-assisted subcellular imaging techniques, designing and synthesizing smaller and brighter nanoparticles continues to be a significant challenge. Ideal fluorescent nanoparticles as probes are expected to interfere minimally with biological applications. Biological applications have been successfully demonstrated using fluorescent nanoparticles with a diameter of less than 12 nm.[38, 139] For instance, Frangioni and coworkers reported that QDots < 5 nm can be easily metabolized by renal clearance while the QDots >15 nm trend accumulates in the liver and spleen in vivo behavior. [140] UCNPs of smaller size are less likely to affect the function of their label target sterically. Also, the nanoparticles may move passively through the membranes of subcellular organelles. Conjugation of smaller UCNPs with translocation dendrimers, peptides and other pharmacophore ligands will confer further targeting specificity to probes.[71] Typically, the basic requirement for fluorescent probes in super-resolution microscopy is to be smaller than the subcellular structures. According to the Nyquist-Shannon sampling theorem, the label density has to be high enough to ensure the average distance between labeled adjacent probes is half that of the desired resolution.[141]

Apart from size, the brightness of UCNPs is another critical issue for the sub-diffraction applications. However, there are unavoidable trade-offs between brightness and size.[142] It can be challenging to reduce the size of UCNPs without affecting their remarkable performance. Specifically, the brightness of UCNPs decreases proportionally to the number of doping ions that they contain. Also, a smaller size increases the ratio of surface area to volume, thus leading to enhanced surface quenching and reduced brightness.[143] Moreover, the brightness of UCNPs has been limited by the energy transfer and relaxation within individual nanoparticles. Due to distinct energy transfer pathways resulting from the dopant-host interactions, different host materials doped, and varied doped concentrations of activators display diverse emission profiles ratio and emission intensity.[144]

Strategies involving iterative rounds of kinetic modeling, to develop UCNPs for super-resolution imaging, such as optimization of crystal host and dopant concentrations [145] and coating of nanocrystals with thin shells, have been used to generate UCNPs in the sub-10 nm range.[146] Doing so maintains optical stability, high uniformity, brightness and monodispersity. These new rules for designing smaller and brighter UCNPs address key obstacles for optimizing UCNPs as sub-diffraction imaging probes. Although much progress has been made in the past decade, research for new strategies and new methods for better performance is not stopping, especially in smaller and uniform UCNPs with intense brightness.

8.4.2   Surface Modification of UCNPs

Surface functionalization and bioconjugation strategies determine the stability of UCNPs in physiological environments and their specificity in labeling subcellular structures. As the highly monodispersed UCNPs are mainly synthesized in high-boiling organic solvents, oleic acid is distributed on the surface of the nanoparticles to form a hydrophobic layer of surface ligands. Subsequently, it is necessary to conduct a surface modification stage to yield a hydrophilic surface composition before applying the UCNPs as the fluorescent probe. One of the fast surface modifications is to induce an additional surface modification step to replace the hydrophobic ligands by hydrophilic ligands as the growth-controlling reagent. However, it is still a challenge to synthesize UCNPs with hydrophilic ligands directly.

Currently, silanization of silica-coated UCNPs, ligand interaction and ligand exchange reactions are the most widely applied methods for generating UCNPs that form stable dispersions and can be further functionalized.[53] Various biomolecules can be conjugated with the functional groups yielded from the silanization of silica-coated UCNPs, but it may increase the nanoparticle volume. For the non-covalent intermolecular interactions of new ligands with the UCNP surface, we can achieve a hydrophilic surface functionalization by partially oxidizing the hydrophobic surface ligands. Alternatively, the ligand exchange reaction can be an excellent choice to replace the hydrophobic surface ligands by hydrophilic ligands.[147, 148] In a ligand exchange reaction, hydrophilic ligands need to compete with the densely oleic acid which is distributed on the UCNP surface.[149] However, hydrophilic ligands have higher surface energy and can be more easily packed compared to the hydrophobic counterpart. As a result, ligand exchange reactions depend on strong coordination with lanthanide ions to efficiently compete for the lanthanide binding sites on the UCNPs' surface. Also, several parameters should be considered when choosing the suitable surface ligand, like reducing the surface quenching of the upconversion emissions, cytotoxicity and biocompatibility.

A subsequent modification step is to bind biomolecules to the surface of UCNPs. Due to the highly positively charged surface, trivalent lanthanide-ion-doped UCNPs are less stable than other fluorescent nanoparticles, like PDots and CDots, in physiological environments. For this reason, we need to pay more attention to using ligand molecules with more negative charges to form more robust anchoring to UCNPs for efficient bioconjugation.[150] Moreover, the dispersed UCNPs must remain stable in the long term in aqueous media, avoiding aggregation and precipitation during the labeling process. Most sub-diffraction applications require a modification of the UCNP surface by antibodies, proteins or lectins to specifically label cellular target structures. These modifications should be available under mild conditions without leading UCNPs aggregation issues or altering the functionality of the biomolecule.

Although there remain several challenges for specifically labeling UCNPs to the subcellular structures in live cells, recent progress was achieved in the functionalization of UCNPs. Bioconjugation,[148–152] cell optogenetics,[153, 154] and long-term tracking in cells [155, 156] are empowering UCNPs to track more biological events. The tuning of multiple emission colors [157, 158] and lifetimes [36] of UCNPs will permit multiplexed subcellular structures imaging and long-term single-molecules tracking.[71]

8.5   CONCLUSION AND FUTURE OPPORTUNITIES

In this chapter, the focus has been on reviewing newly developed methods to improve the performance of super-resolution microscopes by using UCNPs. With the nonlinear photon response of UCNPs, the novel approaches make several concrete advances for sub-diffraction applications in terms of laser power, probe's optical stability, image depth, speed, overall quality, 3D and wide-field imaging.

By inducing the PA-like effect, a UCNPs-assisted STED nanoscopy offers two orders of magnitude lower saturation intensity than those of conventional fluorescent probes, achieving sub-30 nm optical resolution in resolving the cluster of single 13 nm UCNPs. For releasing the proficient skills in aligning dual laser beam, non-bleaching FED microscopy achieves sub-diffraction resolution by introducing $Nd^{3+}/Yb^{3+}/Er^{3+}$-co-doped UCNPs with red emission. By taking advantage of NIR-in and NIR-out optical nonlinear response curve from the single nanoparticle, researcher pushes this technique to unlock a new mode of deep tissue super-resolution imaging. With the excitation-orthogonalized UCNPs, one-scan FED microscopy further eliminates the sample drifting issues and increases imaging speed. Taking advantage of the heterochromatic optical nonlinear response curves in a single UCNPs, a multicolor Fourier fusion algorithm can achieve the maximum spatial frequency information, yielding an overall enhanced image quality. Conversely, PA in UCNPs is employed as the superlinear concept to release 3D sub-diffraction imaging in a biological sample on a conventional confocal configuration. Recently, the nonlinear upconversion SIM induces a strategy to create video-rate super-resolution imaging through thick biological tissues utilizing UCNPs.

Significant progress has been made in applying the nonlinear photon response in UCNPs for super-resolution imaging in recent years. The next step with the UCNP-assisted nanoscopy modalities would be to apply them to inspect biological samples. For future analyses, two key potential research directions are briefly discussed here: first, designing smaller UCNPs with intense brightness; and second, releasing surface bioconjugation strategies to specific labeling UCNPs to the subcellular structures. Thus, researchers working in different fields and the industry must collaborate to apply UCNPs because they are potent probes in the sub-diffraction imaging for biological studies. We expect that all such future research directions can enhance the performance of developed super-resolution imaging systems.

REFERENCES

1. Liu G (2015) Advances in the theoretical understanding of photon upconversion in rare-earth activated nanophosphors. *Chem. Soc. Rev.* 44: 1635.
2. Zhang F (2015) *Photon upconversion nanomaterials.* Vol. 416: Berlin, Heidelberg: Springer.
3. Xiao F, Xie S, Yi R, Peng S, and Xie C (2020) Room temperature broadband upconversion luminescence in Yb^{3+} and Mn^{2+} codoped $Sr_5(PO_4)_3Cl$ phosphors. *J. Lumin.* 219: 116943.
4. Song EH, Ding S, Wu M, Ye S, Xiao F, Dong GP, and Zhang QY (2013) Temperature-tunable upconversion luminescence of perovskite nanocrystals $KZnF_3:Yb^{3+},Mn^{2+}$. *J. Mater. Chem. C* 1: 4209.
5. Karbowiak M, Mech A, Drożdżyński J, and Edelstein NM (2003) Crystal-field analysis, upconversion, and excited-state dynamics for $(U^{4+},U^{3+}):Ba_2YCl_7$ single crystals. *Phys. Rev. B* 67: 195108.
6. Yin M, Joubert MF, and Krupa JC (1997) Infrared to green up-conversion in $LaCl_3:U^{3+}$. *J. Lumin.* 75: 221.
7. Xu L, Xu C, Sun H, Zeng T, Zhang J, and Zhao H (2018) Self-induced second-harmonic and sum-frequency generation from interfacial engineered Er^{3+}/Fe^{3+} doped $LiNbO_3$ single crystal via femtosecond laser ablation. *ACS Photonics* 5: 4463.
8. Venkatakrishnarao D, Narayana YS, Mohaiddon MA, Mamonov EA, Mitetelo N, Kolmychek IA, Maydykovskiy AI, Novikov VB, Murzina TV, and Chandrasekar R (2017) Two-photon luminescence and second-harmonic generation in organic nonlinear surface comprised of self-assembled frustum shaped organic microlasers. *Adv. Mater.* 29: 1605260.
9. Xu L, Zhang J, Yin L, Long X, Zhang W, and Zhang Q (2020) Recent progress in efficient organic two-photon dyes for fluorescence imaging and photodynamic therapy. *J. Mater. Chem. C* 8: 6342.
10. Meiling TT, Cywiński PJ, and Löhmannsröben HG (2018) Two-photon excitation fluorescence spectroscopy of quantum dots: photophysical properties and application in bioassays. *J. Phys. Chem. C* 122: 9641.
11. Liu Y, Gou H, Huang X, Zhang G, Xi K, and Jia X (2020) Rational synthesis of highly efficient ultra-narrow red-emitting carbon quantum dots for NIR-II two-photon bioimaging. *Nanoscale* 12: 1589.
12. Auzel F (1976) Multiphonon-assisted anti-Stokes and Stokes fluorescence of triply ionized rare-earth ions. *Phys. Rev. B* 13: 2809.
13. Ovsyakin V, and Feofilov P (1966) Cooperative sensitization of luminescence in crystals activated with rare earth ions. *J. Exp. Theor. Phys.* 4: 317.
14. Auzel FE (1973) Materials and devices using double-pumped-phosphors with energy transfer. *Proc. IEEE* 61: 758.
15. Zhou B, Shi B, Jin D, and Liu X (2015) Controlling upconversion nanocrystals for emerging applications. *Nat. Nanotechnol.* 10: 924.
16. Qin X, Liu X, Huang W, Bettinelli M, and Liu X (2017) Lanthanide-activated phosphors based on 4f-5d optical transitions: theoretical and experimental aspects. *Chem. Rev.* 117: 4488.
17. Huang B (2016) 4f fine-structure levels as the dominant error in the electronic structures of binary lanthanide oxides. *J. Comput. Chem.* 37: 825.
18. Binnemans K (2009) Lanthanide-based luminescent hybrid materials. *Chem. Rev.* 109: 4283.
19. Suyver J, Aebischer A, García-Revilla S, Gerner P, and Güdel H (2005) Anomalous power dependence of sensitized upconversion luminescence. *Phys. Rev. B* 71: 125123.

20. He H, Liu B, Wen S, Liao J, Lin G, Zhou J, and Jin D (2018) Quantitative lateral flow strip sensor using highly doped upconversion nanoparticles. *Anal. Chem.* 90: 12356.
21. Yang W, Li X, Chi D, Zhang H, and Liu X (2014) Lanthanide-doped upconversion materials: emerging applications for photovoltaics and photocatalysis. *Nanotechnology* 25: 482001.
22. Deng R, Qin F, Chen R, Huang W, Hong M, and Liu X (2015) Temporal full-colour tuning through non-steady-state upconversion. *Nat. Nanotechnol.* 10: 237.
23. Liu Y, Lu Y, Yang X, Zheng X, Wen S, Wang F, Vidal X, Zhao J, Liu D, and Zhou Z (2017) Amplified stimulated emission in upconversion nanoparticles for super-resolution nanoscopy. *Nature* 543: 229.
24. Liu D, Xu X, Du Y, Qin X, Zhang Y, Ma C, Wen S, Ren W, Goldys EM, and Piper JA (2016) Three-dimensional controlled growth of monodisperse sub-50 nm heterogeneous nanocrystals. *Nat. Commun.* 7: 10254.
25. Zheng K, Loh KY, Wang Y, Chen Q, Fan J, Jung T, Nam SH, Suh YD, and Liu X (2019) Recent advances in upconversion nanocrystals: expanding the kaleidoscopic toolbox for emerging applications. *Nano Today* 29: 100797.
26. Auzel F (2004) Upconversion and anti-Stokes processes with f and d ions in solids. *Chem. Rev.* 104: 139.
27. Dong H, Sun LD, and Yan CH (2015) Energy transfer in lanthanide upconversion studies for extended optical applications. *Chem. Soc. Rev.* 44: 1608.
28. Liao J (2020) Optical fingerprints of upconversion nanoparticles for super-capacity multiplexing.
29. Zheng W, Huang P, Tu D, Ma E, Zhu H, and Chen X (2015) Lanthanide-doped upconversion nano-bioprobes: electronic structures, optical properties, and biodetection. *Chem. Soc. Rev.* 44: 1379.
30. Sun T, Li Y, Ho WL, Zhu Q, Chen X, Jin L, Zhu H, Huang B, Lin J, and Little BE (2019) Integrating temporal and spatial control of electronic transitions for bright multiphoton upconversion. *Nat. Commun.* 10: 1811.
31. Liao J, Jin D, Chen C, Li Y, and Zhou J (2020) Helix shape power-dependent properties of single upconversion nanoparticles. *J. Phys. Chem. Lett.* 11: 2883.
32. Liu L, Wang S, Zhao B, Pei P, Fan Y, Li X, and Zhang F (2018) Er^{3+} sensitized 1530 nm to 1180 nm second near-infrared window upconversion nanocrystals for in vivo biosensing. *Angew. Chem. Int. Ed.* 130: 7640.
33. Zhou J, Liu Q, Feng W, Sun Y, and Li F (2015) Upconversion luminescent materials: advances and applications. *Chem. Rev.* 115: 395.
34. Liao J, Yang Z, Wu H, Yan D, Qiu J, Song Z, Yang Y, Zhou D, and Yin Z (2013) Enhancement of the up-conversion luminescence of Yb^{3+}/Er^{3+} or Yb^{3+}/Tm^{3+} co-doped $NaYF_4$ nanoparticles by photonic crystals. *J. Mater. Chem. C* 1: 6541.
35. Li W, Xu J, He Q, Sun Y, Sun S, Chen W, Guzik M, Boulon G, and Hu L (2020) Highly stable green and red up-conversion of $LiYF_4:Yb^{3+},Ho^{3+}$ for potential application in fluorescent labeling. *J. Alloys Compd.* 845: 155820.
36. Lu Y, Zhao J, Zhang R, Liu Y, Liu D, Goldys EM, Yang X, Xi P, Sunna A, and Lu J (2014) Tunable lifetime multiplexing using luminescent nanocrystals. *Nat. Photonics* 8: 32.
37. Chivian JS, Case W, and Eden D (1979) The photon avalanche: a new phenomenon in Pr^{3+}-based infra-red quantum counters. *Appl. Phys. Lett.* 35: 124.
38. Zhan Q, Liu H, Wang B, Wu Q, Pu R, Zhou C, Huang B, Peng X, Ågren H, and He S (2017) Achieving high-efficiency emission depletion nanoscopy by employing cross relaxation in upconversion nanoparticles. *Nat. Commun.* 8: 1058.
39. Strek W, Deren P, and Bednarkiewicz A (2000) Cooperative processes in $KYb(WO_4)_2$ crystal doped with Eu^{3+} and Tb^{3+} ions. *J. Lumin.* 87: 999.
40. Hao S, Shao W, Qiu H, Shang Y, Fan R, Guo X, Zhao L, Chen G, and Yang C (2014) Tuning the size and upconversion emission of $NaYF_4:Yb^{3+}/Pr^{3+}$ nanoparticles through Yb^{3+} doping. *RSC Adv.* 4: 56302.
41. Dong H, Sun LD, Wang YF, Xiao JW, Tu D, Chen X, and Yan CH (2016) Photon upconversion in $Yb^{3+}–Tb^{3+}$ and $Yb^{3+}–Eu^{3+}$ activated core/shell nanoparticles with dual-band excitation. *J. Mater. Chem. C* 4: 4186.
42. Qin WP, Liu ZY, Sin CN, Wu CF, Qin GS, Chen Z, and Zheng KZ (2014) Multi-ion cooperative processes in Yb^{3+} clusters. *Light Sci. Appl.* 3: e193.
43. Zhou B, Yang W, Han S, Sun Q, and Liu X (2015) Photon upconversion through Tb^{3+}-mediated interfacial energy transfer. *Adv. Mater.* 27: 6208.
44. Su Q, Han S, Xie X, Zhu H, Chen H, Chen CK, Liu RS, Chen X, Wang F, and Liu X (2012) The effect of surface coating on energy migration-mediated upconversion. *J. Am. Chem. Soc.* 134: 20849.
45. Zuo J, Sun D, Tu L, Wu Y, Cao Y, Xue B, Zhang Y, Chang Y, Liu X, and Kong X (2018) Precisely tailoring upconversion dynamics via energy migration in core–shell nanostructures. *Angew. Chem. Int. Ed.* 57: 3054.

46. Park YI, Lee KT, Suh YD, and Hyeon T (2015) Upconverting nanoparticles: a versatile platform for wide-field two-photon microscopy and multi-modal in vivo imaging. *Chem. Soc. Rev.* 44: 1302.

47. Zhao J, Jin D, Schartner EP, Lu Y, Liu Y, Zvyagin AV, Zhang L, Dawes JM, Xi P, and Piper JA (2013) Single-nanocrystal sensitivity achieved by enhanced upconversion luminescence. *Nat. Nanotechnol.* 8: 729.

48. Tian G, Zhang X, Gu Z, and Zhao Y (2015) Recent advances in upconversion nanoparticles-based multifunctional nanocomposites for combined cancer therapy. *Adv. Mater.* 27: 7692.

49. Wu S, Han G, Milliron DJ, Aloni S, Altoe V, Talapin DV, Cohen BE, and Schuck PJ (2009) Non-blinking and photostable upconverted luminescence from single lanthanide-doped nanocrystals. *Proc. Natl. Acad. Sci. U.S.A.* 106: 10917.

50. Wu Q, Huang B, Peng X, He S, and Zhan Q (2017) Non-bleaching fluorescence emission difference microscopy using single 808 nm laser excited red upconversion emission. *Opt. Express* 25: 30885.

51. Peng X, Huang B, Pu R, Liu H, Zhang T, Widengren J, Zhan Q, and Ågren H (2019) Fast upconversion super-resolution microscopy with 10 μs per pixel dwell times. *Nanoscale* 11: 1563.

52. Nam SH, Bae YM, Park YI, Kim JH, Kim HM, Choi JS, Lee KT, Hyeon T, and Suh YD (2011) Long-term real-time tracking of lanthanide ion doped upconverting nanoparticles in living cells. *Angew. Chem. Int. Ed.* 123: 6217.

53. Sedlmeier A, and Gorris HH (2015) Surface modification and characterization of photon-upconverting nanoparticles for bioanalytical applications. *Chem. Soc. Rev.* 44: 1526.

54. Wu Z, Guo C, Liang S, Zhang H, Wang L, Sun H, and Yang B (2012) A pluronic F127 coating strategy to produce stable up-conversion $NaYF_4$:Yb,Er(Tm) nanoparticles in culture media for bioimaging. *J. Mater. Chem.* 22: 18596.

55. Xiong L, Yang T, Yang Y, Xu C, and Li F (2010) Long-term in vivo biodistribution imaging and toxicity of polyacrylic acid-coated upconversion nanophosphors. *Biomaterials* 31: 7078.

56. Gnach A, Lipinski T, Bednarkiewicz A, Rybka J, and Capobianco JA (2015) Upconverting nanoparticles: assessing the toxicity. *Chem. Soc. Rev.* 44: 1561.

57. Zhou J, Yu M, Sun Y, Zhang X, Zhu X, Wu Z, Wu D, and Li F (2011) Fluorine-18-labeled Gd^{3+}/Yb^{3+}/Er^{3+} co-doped $NaYF_4$ nanophosphors for multimodality PET/MR/UCL imaging. *Biomaterials* 32: 1148.

58. Uluç K, Kujoth GC, and Başkaya MK (2009) Operating microscopes: past, present, and future. *Neurosurgical Focus* 27: E4.

59. Hell SW, Sahl SJ, Bates M, Zhuang X, Heintzmann R, Booth MJ, Bewersdorf J, Shtengel G, Hess H, and Tinnefeld P (2015) The 2015 super-resolution microscopy roadmap. *J. Phys. D: Appl. Phys.* 48: 443001.

60. Klar TA, Engel E, and Hell SW (2001) Breaking Abbe's diffraction resolution limit in fluorescence microscopy with stimulated emission depletion beams of various shapes. *Phys. Rev. E* 64: 066613.

61. Effersø F, Auken E, and Sørensen KI (1999) Inversion of band-limited TEM responses. *Geophys. Prospect.* 47: 551.

62. Axelrod D (2001) Total internal reflection fluorescence microscopy in cell biology. *Traffic* 2: 764.

63. Vobornik D, and Vobornik S (2008) Scanning near-field optical microscopy. *Bosnian J. Basic Med. Sci.* 8: 63.

64. Tressler C (2013) Characterization and alignment of the STED Doughnut using fluorescence correlation spectroscopy.

65. Hell SW, and Wichmann J (1994) Breaking the diffraction resolution limit by stimulated emission: stimulated-emission-depletion fluorescence microscopy. *Opt. Lett.* 19: 780.

66. Willig KI, Harke B, Medda R, and Hell SW (2007) STED microscopy with continuous wave beams. *Nat. Methods* 4: 915.

67. Hell SW, and Kroug M (1995) Ground-state-depletion fluorscence microscopy: a concept for breaking the diffraction resolution limit. *Appl. Phys. B* 60: 495.

68. Gustafsson MG (2005) Nonlinear structured-illumination microscopy: wide-field fluorescence imaging with theoretically unlimited resolution. *Proc. Natl. Acad. Sci. U.S.A.* 102: 13081.

69. Betzig E, Patterson GH, Sougrat R, Lindwasser OW, Olenych S, Bonifacino JS, Davidson MW, Lippincott-Schwartz J, and Hess HF (2006) Imaging intracellular fluorescent proteins at nanometer resolution. *Science* 313: 1642.

70. Rust MJ, Bates M, and Zhuang X (2006) Sub-diffraction-limit imaging by stochastic optical reconstruction microscopy (STORM). *Nat. Methods* 3: 793.

71. Jin D, Xi P, Wang B, Zhang L, Enderlein J, and van Oijen AM (2018) Nanoparticles for super-resolution microscopy and single-molecule tracking. *Nat. Methods* 15: 415.

72. Drees C, Raj AN, Kurre R, Busch KB, Haase M, and Piehler J (2016) Engineered upconversion nanoparticles for resolving protein interactions inside living cells. *Angew. Chem. Int. Ed.* 55: 11668.

73. Yao C, Wang P, Li X, Hu X, Hou J, Wang L, and Zhang F (2016) Near-infrared-triggered azobenzene-liposome/upconversion nanoparticle hybrid vesicles for remotely controlled drug delivery to overcome cancer multidrug resistance. *Adv. Mater.* 28: 9341.

74. He M, Pang X, Liu X, Jiang B, He Y, Snaith H, and Lin Z (2016) Monodisperse dual-functional upconversion nanoparticles enabled near-infrared organolead halide perovskite solar cells. *Angew. Chem. Int. Ed.* 128: 4352.

75. Wu S, and Butt HJ (2016) Near-infrared-sensitive materials based on upconverting nanoparticles. *Adv. Mater.* 28: 1208.

76. Dertinger T, Colyer R, Iyer G, Weiss S, and Enderlein J (2009) Fast, background-free, 3D super-resolution optical fluctuation imaging (SOFI). *Proc. Natl. Acad. Sci. U.S.A.* 106: 22287.

77. Wang Y, Fruhwirth G, Cai E, Ng T, and Selvin PR (2013) 3D super-resolution imaging with blinking quantum dots. *Nano Lett.* 13: 5233.

78. Zeng Z, Chen X, Wang H, Huang N, Shan C, Zhang H, Teng J, and Xi P (2015) Fast super-resolution imaging with ultra-high labeling density achieved by joint tagging super-resolution optical fluctuation imaging. *Sci. Rep.* 5: 1.

79. Li R, Chen X, Lin Z, Wang Y, and Sun Y (2018) Expansion enhanced nanoscopy. *Nanoscale* 10: 17552.

80. Yang X, Zhanghao K, Wang H, Liu Y, Wang F, Zhang X, Shi K, Gao J, Jin D, and Xi P (2016) Versatile application of fluorescent quantum dot labels in super-resolution fluorescence microscopy. *ACS Photonics* 3: 1611.

81. Zhao M, Ye S, Peng X, Song J, and Qu J (2019) Green emitted CdSe@ ZnS quantum dots for FLIM and STED imaging applications. *J. Innovative Opt. Health Sci.* 12: 1940003.

82. Hanne J, Falk HJ, Görlitz F, Hoyer P, Engelhardt J, Sahl SJ, and Hell SW (2015) STED nanoscopy with fluorescent quantum dots. *Nat. Commun.* 6: 7127.

83. Zhi B, Cui Y, Wang S, Frank BP, Williams DN, Brown RP, Melby ES, Hamers RJ, Rosenzweig Z, and Fairbrother DH (2018) Malic acid carbon dots: from super-resolution live-cell imaging to highly efficient separation. *ACS Nano* 12: 5741.

84. Chizhik AM, Stein S, Dekaliuk MO, Battle C, Li W, Huss A, Platen M, Schaap IA, Gregor I, and Demchenko AP (2016) Super-resolution optical fluctuation bio-imaging with dual-color carbon nanodots. *Nano Lett.* 16: 237.

85. Leménager G, De Luca E, Sun YP, and Pompa PP (2014) Super-resolution fluorescence imaging of biocompatible carbon dots. *Nanoscale* 6: 8617.

86. Bu L, Luo T, Peng H, Li L, Long D, Peng J, and Huang J (2019) One-step synthesis of N-doped carbon dots, and their applications in curcumin sensing, fluorescent inks, and super-resolution nanoscopy. *Microchim. Acta* 186: 675.

87. He H, Liu X, Li S, Wang X, Wang Q, Li J, Wang J, Ren H, Ge B, and Wang S (2017) High-density super-resolution localization imaging with blinking carbon dots. *Anal. Chem.* 89: 11831.

88. Chen X, Li R, Liu Z, Sun K, Sun Z, Chen D, Xu G, Xi P, Wu C, and Sun Y (2017) Small photoblinking semiconductor polymer dots for fluorescence nanoscopy. *Adv. Mater.* 29: 1604850.

89. Chen X, Liu Z, Li R, Shan C, Zeng Z, Xue B, Yuan W, Mo C, Xi P, and Wu C (2017) Multicolor super-resolution fluorescence microscopy with blue and carmine small photoblinking polymer dots. *ACS Nano* 11: 8084.

90. Liu Z, Liu J, Zhang Z, Sun Z, Shao X, Guo J, Xi L, Yuan Z, Zhang X, and Chiu DT (2020) Narrowband polymer dots with pronounced fluorescence fluctuations for dual-color super-resolution imaging. *Nanoscale* 12: 7522.

91. Sun Z, Liu Z, Chen H, Li R, Sun Y, Chen D, Xu G, Liu L, and Wu C (2019) Semiconducting polymer dots with modulated photoblinking for high-order super-resolution optical fluctuation imaging. *Adv. Opt. Mater.* 7: 1900007.

92. Cox S, Rosten E, Monypenny J, Jovanovic-Talisman T, Burnette DT, Lippincott-Schwartz J, Jones GE, and Heintzmann R (2012) Bayesian localization microscopy reveals nanoscale podosome dynamics. *Nat. Methods* 9: 195.

93. Burnette DT, Sengupta P, Dai Y, Lippincott-Schwartz J, and Kachar B (2011) Bleaching/blinking assisted localization microscopy for superresolution imaging using standard fluorescent molecules. *Proc. Natl. Acad. Sci. U.S.A.* 108: 21081.

94. Lo CYW, Chen S, Creed SJ, Kang M, Zhao N, Tang BZ, and Elgass KD (2016) Novel super-resolution capable mitochondrial probe, MitoRed AIE, enables assessment of real-time molecular mitochondrial dynamics. *Sci. Rep.* 6: 30855.

95. Li D, Ni X, Zhang X, Liu L, Qu J, Ding D, and Qian J (2018) Aggregation-induced emission luminogen-assisted stimulated emission depletion nanoscopy for super-resolution mitochondrial visualization in live cells. *Nano Res.* 11: 6023.

96. Zhou J, Yu G, and Huang F (2016) AIE opens new applications in super-resolution imaging. *J. Mater. Chem. B* 4: 7761.

97. Gu X, Zhao E, Zhao T, Kang M, Gui C, Lam JW, Du S, Loy MM, and Tang BZ (2016) A mitochondrion-specific photoactivatable fluorescence turn-on AIE-based bioprobe for localization super-resolution microscope. *Adv. Mater.* 28: 5064.

98. Wang YL, Fan C, Xin B, Zhang JP, Luo T, Chen ZQ, Zhou QY, Yu Q, Li XN, and Huang ZL (2018) AIE-based super-resolution imaging probes for β-amyloid plaques in mouse brains. *Mater. Chem. Front.* 2: 1554.

99. Wu R, Zhan Q, Liu H, Wen X, Wang B, and He S (2015) Optical depletion mechanism of upconverting luminescence and its potential for multi-photon STED-like microscopy. *Opt. Express* 23: 32401.

100. Dong H, Sun LD, and Yan CH (2021) Lanthanide-doped upconversion nanoparticles for super-resolution microscopy. *Front. Chem.* 8: 619377.

101. Chen C (2020) Multi-photon nonlinear fluorescence emission in upconversion nanoparticles for super-resolution imaging.

102. Hong G, Lee JC, Robinson JT, Raaz U, Xie L, Huang NF, Cooke JP, and Dai H (2012) Multifunctional in vivo vascular imaging using near-infrared II fluorescence. *Nat. Med.* 18: 1841.

103. Han S, Deng R, Xie X, and Liu X (2014) Enhancing luminescence in lanthanide-doped upconversion nanoparticles. *Angew. Chem. Int. Ed.* 53: 11702.

104. Rittweger E, Han KY, Irvine SE, Eggeling C, and Hell SW (2009) STED microscopy reveals crystal colour centres with nanometric resolution. *Nat. Photonics* 3: 144.

105. Westphal V, Rizzoli SO, Lauterbach MA, Kamin D, Jahn R, and Hell SW (2008) Video-rate far-field optical nanoscopy dissects synaptic vesicle movement. *Science* 320: 246.

106. Hein B, Willig KI, and Hell SW (2008) Stimulated emission depletion (STED) nanoscopy of a fluorescent protein-labeled organelle inside a living cell. *Proc. Natl. Acad. Sci. USA* 105: 14271.

107. Schmidt R, Wurm CA, Jakobs S, Engelhardt J, Egner A, and Hell SW (2008) Spherical nanosized focal spot unravels the interior of cells. *Nat. Methods* 5: 539.

108. Schmidt R, Wurm CA, Punge A, Egner A, Jakobs S, and Hell SW (2009) Mitochondrial cristae revealed with focused light. *Nano Lett.* 9: 2508.

109. Pascucci M, Ganesan S, Tripathi A, Katz O, Emiliani V, and Guillon M (2019) Compressive three-dimensional super-resolution microscopy with speckle-saturated fluorescence excitation. *Nat. Commun.* 10: 1327.

110. Chmyrov A, Keller J, Grotjohann T, Ratz M, d'Este E, Jakobs S, Eggeling C, and Hell SW (2013) Nanoscopy with more than 100,000'doughnuts'. *Nat. Methods* 10: 737.

111. Kuang C, Li S, Liu W, Hao X, Gu Z, Wang Y, Ge J, Li H, and Liu X (2013) Breaking the diffraction barrier using fluorescence emission difference microscopy. *Sci. Rep.* 3: 1441.

112. Ge B, Zhu L, Kuang C, Zhang D, Fang Y, Ma Y, and Liu X (2015) Fluorescence emission difference with defocused surface plasmon-coupled emission microscopy. *Opt. Express* 23: 32561.

113. Chen C, Wang F, Wen S, Su QP, Wu MC, Liu Y, Wang B, Li D, Shan X, Kianinia M, Aharonovich I, Toth M, Jackson S, Xi P and Jin D (2018) Multi-photon near-infrared emission saturation nanoscopy using upconversion nanoparticles. *Nat. Commun.* 9: 3290.

114. Rittweger E, Wildanger D, and Hell S (2009) Far-field fluorescence nanoscopy of diamond color centers by ground state depletion. *EPL* 86: 14001.

115. Han KY, Kim SK, Eggeling C, and Hell SW (2010) Metastable dark states enable ground state depletion microscopy of nitrogen vacancy centers in diamond with diffraction-unlimited resolution. *Nano Lett.* 10: 3199.

116. Korobchevskaya K, Peres C, Li Z, Antipov A, Sheppard CJ, Diaspro A, and Bianchini P (2016) Intensity weighted subtraction microscopy approach for image contrast and resolution enhancement. *Sci. Rep.* 6: 25816.

117. Zheng W, Wu Y, Winter P, Fischer R, Dalle Nogare D, Hong A, McCormick C, Christensen R, Dempsey WP, and Arnold DB (2017) Adaptive optics improves multiphoton super-resolution imaging. *Nat. Methods* 14: 869.

118. Bianchini P, Harke B, Galiani S, Vicidomini G, and Diaspro A (2012) Single-wavelength two-photon excitation–stimulated emission depletion (SW2PE-STED) superresolution imaging. *Proc. Natl. Acad. Sci. USA* 109: 6390.

119. Urban NT, Willig KI, Hell SW, and Nägerl UV (2011) STED nanoscopy of actin dynamics in synapses deep inside living brain slices. *Biophys. J.* 101: 1277.

120. Chen C, Wang F, Wen S, Liu Y, Shan X, and Jin D (2019) Upconversion nanoparticles assisted multiphoton fluorescence saturation microscopy. Paper read at Nanoscale Imaging, Sensing, and Actuation for Biomedical Applications XVI.

121. Liu Y, Wang F, Lu H, Fang G, Wen S, Chen C, Shan X, Xu X, Zhang L, and Stenzel M (2020) Super-resolution mapping of single nanoparticles inside tumor spheroids. *Small* 16: 1905572.

122. Huang B, Wu Q, Peng X, Yao L, Peng D, and Zhan Q (2018) One-scan fluorescence emission difference nanoscopy developed with excitation orthogonalized upconversion nanoparticles. *Nanoscale* 10: 21025.

123. Voronin DV, Kozlova AA, Verkhovskii RA, Ermakov AV, Makarkin MA, Inozemtseva OA, and Bratashov DN (2020) Detection of rare objects by flow cytometry: imaging, cell sorting, and deep learning approaches. *Int. J. Mol. Sci.* 21: 2323.

124. Chen C, Liu B, Liu Y, Liao J, Shan X, Wang F, and Jin D (2021) Heterochromatic nonlinear optical responses in upconversion nanoparticles for super-resolution nanoscopy. *Adv.Mater.* 17:2008847.

125. Thibon L, Piché M, and De Koninck Y (2018) Resolution enhancement in laser scanning microscopy with deconvolution switching laser modes (D-SLAM). *Opt. Express* 26: 24881.

126. Bednarkiewicz A, Chan EM, Kotulska A, Marciniak L, and Prorok K (2019) Photon avalanche in lanthanide doped nanoparticles for biomedical applications: super-resolution imaging. *Nanoscale Horiz.* 4: 881.

127. Lee C, Xu EZ, Liu Y, Teitelboim A, Yao K, Fernandez-Bravo A, Kotulska AM, Nam SH, Suh YD, and Bednarkiewicz A (2021) Giant nonlinear optical responses from photon-avalanching nanoparticles. *Nature* 589: 230.

128. Wang B, Zhan Q, Zhao Y, Wu R, Liu J, and He S (2016) Visible-to-visible four-photon ultrahigh resolution microscopic imaging with 730 nm diode laser excited nanocrystals. *Opt. Express* 24: A302.

129. Liu H, Xu CT, and Andersson-Engels S (2014) Potential of multi-photon upconversion emissions for fluorescence diffuse optical imaging. *Opt. Express* 22: 17782.

130. Caillat L, Hajj B, Shynkar V, Michely L, Chauvat D, Zyss J, and Pellé F (2013) Multiphoton upconversion in rare earth doped nanocrystals for sub-diffractive microscopy. *Appl. Phys. Lett.* 102: 143114.

131. Kostyuk AB, Vorotnov AD, Ivanov AV, Volovetskiy AB, Kruglov AV, Sencha LM, Liang L, Guryev EL, Vodeneev VA, and Deyev SM (2019) Resolution and contrast enhancement of laser-scanning multiphoton microscopy using thulium-doped upconversion nanoparticles. *Nano Res.* 12: 2933.

132. De Camillis S, Ren P, Cao Y, Plöschner M, Denkova D, Zheng X, Lu Y, and Piper JA (2020) Controlling the non-linear emission of upconversion nanoparticles to enhance super-resolution imaging performance. *Nanoscale* 12: 20347.

133. Denkova D, Ploschner M, Das M, Parker LM, Zheng X, Lu Y, Orth A, Packer NH, and Piper JA (2019) 3D sub-diffraction imaging in a conventional confocal configuration by exploiting super-linear emitters. *Nat. Commun.* 10: 3695.

134. Plöschner M, Denkova D, De Camillis S, Das M, Parker LM, Zheng X, Lu Y, Ojosnegros S, and Piper JA (2020) Simultaneous super-linear excitation-emission and emission depletion allows imaging of upconversion nanoparticles with higher sub-diffraction resolution. *Opt. Express* 28: 24308.

135. Gustafsson MG (2000) Surpassing the lateral resolution limit by a factor of two using structured illumination microscopy. *J. Microsc.* 198: 82.

136. Liu B, Chen C, Di X, Liao J, Wen S, Su QP, Shan X, Xu Z-Q, Ju LA, Mi C, Wang F, and Jin D (2020) Upconversion nonlinear structured illumination microscopy. *Nano Lett.* 20: 4775.

137. Hu C, Wu Z, Yang X, Zhao W, Ma C, Chen M, Xi P, and Chen H (2020) MUTE-SIM: multiphoton upconversion time-encoded structured illumination microscopy. *OSA Continuum* 3: 594.

138. Liu B, Liao J, Song Y, Lu J, Zhou J, and Wang F (2021) Multiplexed structured illumination super-resolution imaging with time-domain upconversion nanoparticles. arXiv preprint arXiv:2101.07911.

139. Shi F, and Zhao Y (2014) Sub-10 nm and monodisperse β-NaYF$_4$:Yb,Tm,Gd nanocrystals with intense ultraviolet upconversion luminescence. *J. Mater. Chem. C* 2: 2198.

140. Choi HS, Liu W, Misra P, Tanaka E, Zimmer JP, Ipe BI, Bawendi MG, and Frangioni JV (2007) Renal clearance of quantum dots. *Nat. Biotechnol.* 25: 1165.

141. Shroff H, Galbraith CG, Galbraith JA, and Betzig E (2008) Live-cell photoactivated localization microscopy of nanoscale adhesion dynamics. *Nat. Methods* 5: 417.

142. Gargas DJ, Chan EM, Ostrowski AD, Aloni S, Altoe MVP, Barnard ES, Sanii B, Urban JJ, Milliron DJ, and Cohen BE (2014) Engineering bright sub-10 nm upconverting nanocrystals for single-molecule imaging. *Nat. Nanotechnol.* 9: 300.

143. Kraft M, Würth C, Muhr V, Hirsch T, and Resch-Genger U (2018) Particle-size-dependent upconversion luminescence of NaYF$_4$:Yb,Er nanoparticles in organic solvents and water at different excitation power densities. *Nano Res.* 11: 6360.

144. Wen S, Zhou J, Zheng K, Bednarkiewicz A, Liu X, and Jin D (2018) Advances in highly doped upconversion nanoparticles. *Nat. Commun.* 9: 2415.

145. Ma C, Xu X, Wang F, Zhou Z, Liu D, Zhao J, Guan M, Lang CI, and Jin D (2017) Optimal sensitizer concentration in single upconversion nanocrystals. *Nano Lett.* 17: 2858.

146. Liu X, Kong X, Zhang Y, Tu L, Wang Y, Zeng Q, Li C, Shi Z, and Zhang H (2011) Breakthrough in concentration quenching threshold of upconversion luminescence via spatial separation of the emitter doping area for bio-applications. *Chem. Commun.* 47: 11957.

147. Kostiv U, Farka Zk, Mickert MJ, Gorris HH, Velychkivska N, Pop-Georgievski O, Pastucha Mj, Odstrčiliková Ek, Skládal P, and Horák D (2020) Versatile bioconjugation strategies of PEG-modified upconversion nanoparticles for bioanalytical applications. *Biomacromolecules* 21: 4502.

148. Duong HT, Chen Y, Tawfik SA, Wen S, Parviz M, Shimoni O, and Jin D (2018) Systematic investigation of functional ligands for colloidal stable upconversion nanoparticles. *RSC Adv.* 8: 4842.

149. Ren W, Wen S, Tawfik SA, Su QP, Lin G, Ju LA, Ford MJ, Ghodke H, van Oijen AM, and Jin D (2018) Anisotropic functionalization of upconversion nanoparticles. *Chem. Sci.* 9: 4352.

150. Li X, Tang Y, Xu L, Kong X, Zhang L, Chang Y, Zhao H, Zhang H, and Liu X (2017) Dependence between cytotoxicity and dynamic subcellular localization of up-conversion nanoparticles with different surface charges. *RSC Adv.* 7: 33502.

151. He H, Howard CB, Chen Y, Wen S, Lin G, Zhou J, Thurecht KJ, and Jin D (2018) Bispecific antibody-functionalized upconversion nanoprobe. *Anal. Chem.* 90: 3024.

152. Sun Y, Zhang W, Wang B, Xu X, Chou J, Shimoni O, Ung AT, and Jin D (2018) A supramolecular self-assembly strategy for upconversion nanoparticle bioconjugation. *Chem. Commun.* 54: 3851.

153. Chen S, Weitemier AZ, Zeng X, He L, Wang X, Tao Y, Huang AJ, Hashimotodani Y, Kano M, and Iwasaki H (2018) Near-infrared deep brain stimulation via upconversion nanoparticle–mediated optogenetics. *Science* 359: 679.

154. Yadav K, Chou AC, Ulaganathan RK, Gao HD, Lee HM, Pan CY, and Chen YT (2017) Targeted and efficient activation of channelrhodopsins expressed in living cells via specifically-bound upconversion nanoparticles. *Nanoscale* 9: 9457.

155. Wang F, Wen S, He H, Wang B, Zhou Z, Shimoni O, and Jin D (2018) Microscopic inspection and tracking of single upconversion nanoparticles in living cells. *Light Sci. Appl.* 7: 18007.

156. Bae YM, Park YI, Nam SH, Kim JH, Lee K, Kim HM, Yoo B, Choi JS, Lee KT, and Hyeon T (2012) Endocytosis, intracellular transport, and exocytosis of lanthanide-doped upconverting nanoparticles in single living cells. *Biomaterials* 33: 9080.

157. Lin G, Baker MA, Hong M, and Jin D (2018) The quest for optical multiplexing in bio-discoveries. *Chem* 4: 997.

158. Zhang Y, Zhang L, Deng R, Tian J, Zong Y, Jin D, and Liu X (2014) Multicolor barcoding in a single upconversion crystal. *J. Am. Chem. Soc.* 136: 4893.

9 Upconversion Nanophosphors for Photonic Application

Qi Zhu and Feng Wang

CONTENTS

9.1 INTRODUCTION

The concept of "upconversion" was first put forward in 1959 by Nicolaas Bloembergen as a theoretical assumption, which was experimentally verified by François Auzel in 1966 through the observation of visible emission under near-infrared (NIR) excitation.[1, 2] Typically, the upconversion phenomenon refers to nonlinear optical processes that generate high-energy emissions by successive absorption of multiple lower energy photons. Upconversion phosphors are generally composed of inorganic host crystals incorporated with lanthanide dopant ions.[3] Figure 9.1(a) and (b) shows the upconversion luminescence of lanthanide-doped phosphors produced through the physically existing intermediary energy states of lanthanide ions. In an upconversion process, the host lattice provides a strong crystal field surrounding the dopants along with lattice vibrations that impose manifold effects on the optical performance of the dopant ions.[4–6]

Owing to the advances in chemical syntheses, lanthanide-doped upconversion nanophosphors (UCNPs) have emerged in the 2000s and largely expanded the scope of upconversion research.[7, 8] Compared with the bulk counterparts, nanocrystalline host materials allow for precise control of dopant distribution at the nanometer length scale through core–shell nanostructural engineering. By applying an undoped inert shell to reduce surface quenching and separating undesired energy migration processes, the upconversion emission intensity can be largely enhanced even at unusually high dopant concentrations.[9] Furthermore, a set of lanthanide dopants can be integrated into separate layers of a core–shell particle to generate novel photon conversion processes that are inaccessible by conventional bulk materials, which is revealed in Figure 9.1(c).[10] Specifically, upconversion

DOI: 10.1201/9781003098676-9

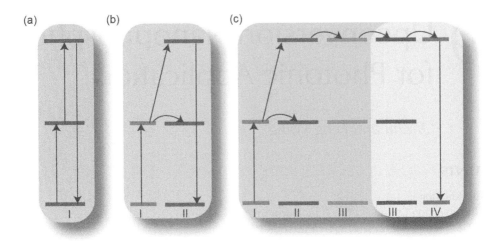

FIGURE 9.1 Common upconversion processes in lanthanide-doped phosphors. (a) Excited-state absorption (ESA) in a single lanthanide ion that emits one photon via absorption of two photons. (b) Energy transfer upconversion (ETU), including two lanthanide ions, where one ion transfers energy to the adjacent ion for upconversion emission. (c) Energy migration–mediated upconversion (EMU) involving four types of lanthanide ions in core–shell nanophosphors. They are sensitizer (type I) that harvests excitation energy, accumulator (type II) that accepts energy from surrounding sensitizer to establish a population in the high-lying excited state, migrator (type III) that extracts energy from excited accumulators and delivers the energy through the host lattice, and activator (type IV) that releases upconversion emission. (Wang, F., Deng, R., Wang, J., Wang, Q., Han, Y., Zhu, H., Chen, X. and Liu, X., *Nat. Mater.* 10, 968, 2011. With permission.)

emissions ranging from ultraviolet to infrared have been realized in a series of lanthanide (Er^{3+}, Tm^{3+}, Ho^{3+}, Eu^{3+}, Tb^{3+}, Sm^{3+}, Dy^{3+}, Pr^{3+}, and Ce^{3+}) and Mn^{2+} ions following multiwavelength excitation at 808, 980, and 1550 nm.[10–17]

UCNPs are also superior for practical applications due to their tunable shape, size, and surface chemistry. To date, upconversion nanocrystals of diverse shapes (e.g., sphere, rod, and disk) and sizes (5–50 nm) have been prepared with high uniformity. Besides, several surface modifications approaches have been established to control the surface wetting property and reactivity.[18, 19] As a result, UCNPs can be readily integrated with other types of optical nanostructures and components. For example, UCNPs have been coupled with organic dyes or plasmonic nanostructures for enhanced control over upconversion properties.[20–22] UCNPs have also been incorporated into polymeric cavities for modulating the excitation and emission processes.[23, 24] Notably, with the development of laser technology and optimization of characterization instrumentation, the potential of lanthanide-doped UCNPs has been greatly released, giving rise to promising applications in the fields of biology, energy, and information technology.[3, 25, 26]

In this chapter, we focus on emerging photonic applications that take advantage of lanthanide-doped UCNPs, which include luminescence nanothermometry (LNT), upconversion lasing, super-resolution microscopy, and solar energy conversion. We discuss the working principles and recent progress of the relevant researches. We also attempt to highlight the challenges and opportunities for future investigations.

9.2 UPCONVERSION LUMINESCENCE NANOTHERMOMETRY

LNT permits robust, rapid, and noninvasive temperature sensing with high thermal sensitivity and temporal–spatial resolution.[27, 28] Lanthanide-doped UCNPs demonstrate potentials in LNT owing to their appropriate optical characters, namely thermal stability, narrow emission profiles,

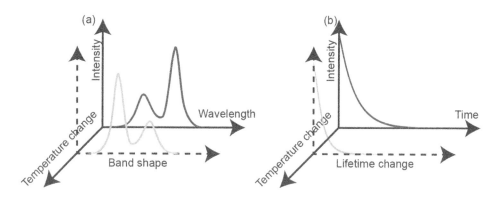

FIGURE 9.2 Major working principles for upconversion nanothermometry. (a) Emission profile alteration manifested as relative intensity changes of two emission peaks. (b) Emission lifetime change, as decay processes are sensitive to the thermal field for precise temperature reading.

and tunable emission wavelength across the full spectrum.[29] As seen in Figure 9.2, two main optical factors of UCNPs can realize temperature reading, which are emission profile (or relative emission intensity) and excited-state lifetime.[30] UCNPs typically display multipeak emission profiles, the relative intensities of which are well correlated with the temperature. Besides, decay processes in UCNPs are sensitive to temperature, providing possibilities to extract temperature reading through the luminescence lifetime.

9.2.1 TEMPERATURE SENSING BY EMISSION PROFILE

LNT based on relative peak intensity is called ratiometric thermometry. The two optical peaks for temperature reading could come from one type of optical center or from two distinct emission centers, which are classified as single-center LNT and multicenter LNT, respectively.[30]

Lanthanide ions are characterized by abundant energy levels. Two adjacent electronic states in a lanthanide ion are thermally coupled when the energy separation is less than several hundred wavenumbers. The population distributions in two closely spacing energy states obey the Boltzmann statistics, which can be represented according to Equation (9.1).[30]

$$N_2 = N_1 \cdot \exp\left(-\frac{\Delta E}{k_B \cdot T}\right) \tag{9.1}$$

where N_1 stands for population of the low energy level, N_2 stands for population of the high energy level, ΔE stands for energy difference between these two energy levels, k_B stands for Boltzmann constant, and T is the absolute temperature.

The corresponding luminescence intensity deriving from these two energy levels is manifested as:

$$I_1 = \phi_1 \cdot N_1 \tag{9.2}$$

$$I_2 = \phi_2 \cdot N_2 \tag{9.3}$$

where ϕ_1 and ϕ_2 are constants related to intrinsic properties of the energy levels. As a result, the fluorescence intensity ratio is listed as:

$$\frac{I_1}{I_2} = \phi_1/\phi_2 \cdot \exp(\Delta E / k_B \cdot T) \tag{9.4}$$

According to Equation (9.4), when the energy difference between two excited energy levels is small enough, a tiny temperature change can induce a large alteration of emission intensity ratio. In this circumstance, population redistribution occurs under temperature stimulation. Several lanthanide ions can show sensitive change in relative emission intensity as a function of temperature, including Er^{3+} ($^2H_{11/2} \rightarrow ^4I_{15/2}/^4S_{3/2} \rightarrow ^4I_{15/2}$), Nd^{3+} ($^4F_{5/2} \rightarrow ^4I_{9/2}/^4F_{3/2} \rightarrow ^4I_{9/2}$), Tm^{3+} (Stark sublevels $^3H_4|0 \rightarrow ^3H_6/^3H_4|1 \rightarrow ^3H_6$), Pr^{3+} ($^3P_1 \rightarrow ^3H_5/^3P_0 \rightarrow ^3H_5$), and Dy^{3+} ($^4I_{15/2} \rightarrow ^6H_{15/2}/^4F_{9/2} \rightarrow ^6H_{15/2}$).[31–35]

Multicenter UCNP is another well-behaved luminescence nanothermometer, which is normally based on the emission intensity ratio between two distinct emitters. Compared with the thermally coupled energy level in a single ion, the separated emission centers exhibit strong temperature-dependent optical behavior to produce high thermal sensitivity and fast luminescence intensity ratio response.[36] As a representative example, Lis and co-workers reported a $NaYF_4$:Yb/Er@ SiO_2 nanoparticle, in which they compared the intensity ratio of thermal-coupled energy level for Er^{3+} ions (550/640 nm) and non-thermalized emission between Yb^{3+} ions and Er^{3+} ions (I_{1010}/I_{810} and I_{1010}/I_{640}). Interestingly, the intensity ratio of Yb/Er (I_{1010}/I_{810}) shows the highest sensitivity of 1.64%/K.[37]

Organic dye molecules contribute to high sensitivity of upconversion LNT by forming hybrid systems with UCNPs and imposing additional effects on the upconversion emissions. Vetrone and co-workers reported a $NaGdF_4$:Er/Yb-pNIPAM-FluoProbe532A system, in which the dye molecule (FluoProbe532A) quenched green emission from the $NaGdF_4$:Er/Yb nanophosphors. The quenching process was affected by collapsing or stretching of the polymer chain (pNIPAM) during temperature change. Accordingly, the intensity ratio of green (from 496 to 580 nm) and total emissions (from 300 to 840 nm) was applied for temperature reading and the sensitivity reached the maximum value of 0.89%/K at 45°C.[38] In another example, as demonstrated in Figure 9.3(a), Li and co-workers designed a hybrid system consisting of triplet–triplet annihilation (TTA) and $NaYF_4$:Nd (5%) UCNPs. The emission of the TTA system was temperature sensitive and used as an indicator signal. By contrast, the Nd^{3+} emission was almost stable against temperature change and used as a reference signal. This hybrid system demonstrated a high thermal sensitivity up to ~7.1%/K that holds the potential for precise temperature detection in tissue or inflammation-induced temperature change in mice.[39]

Owing to the sensitive perception of the surrounding environment and excellent thermal stability, upconversion LNT is proved to be a simple and powerful technique to probe temperature and thermal transportation at the nanoscale, which is difficult to achieve using common detection methods. Typically, upconversion LNT functions as a high sensitivity primary thermometry to reveal temperatures in physical processes. In 2016, Carlos and co-workers measured instantaneous ballistic velocity of suspended Brownian nanocrystals with a high relative sensitivity of 1.15%/K at 296K by employing $NaYF_4$:Yb/Er@$NaYF_4$ nanocrystals as LNT. As shown in Figure 9.3(b), after turning on the heater, the heat flux in the nanofluid changed, leading to variation of the emission intensity ratio of Er^{3+} ions ($I_{525}/I_{545\,nm}$, $^2H_{11/2} \rightarrow ^4I_{15/2}/^4S_{3/2} \rightarrow ^4I_{15/2}$). At each position x_i on the x-axis, a set of the critical time t_{0i} that represents the time needed to record a variation in emission in response to temperature change was collected. By compiling a different position x_i and the corresponding critical time t_{0i}, the measured instantaneous ballistic velocity of nanoparticles was obtained. Due to the much shorter timescale (~10^{-12} s) for upconversion phosphors to thermalize in comparison with its relaxation time as well as short Boltzmann redistribution time, the UCNPs acted as a wonderful nanoplatform to investigate Brownian motion for suspending nanoparticles.[40]

In a subsequent study, Carlos and co-workers successfully determined the thermal conductivity (0.20 ± 0.02 W/(m K) at 300K) of lipid bilayers that were coated around Yb/Er-doped $LiYF_4$ nanophosphors as temperature probes. Figure 9.3(c) discusses the temperature gradient across the bilayer measured with a sensitivity of 1.27%/K and a temperature uncertainty of 0.11K.[41] In this work, the $LiYF_4$:Yb/Er nanophosphors were used to measure the actual temperature of nanoparticles, while an immersed thermal couple was employed to reveal the temperature of the surrounding nanofluids. For the nanophosphors coated with lipid bilayers, the temperature gradient as a function of laser

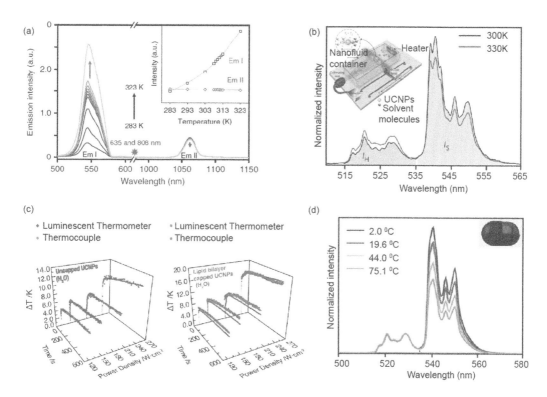

FIGURE 9.3 (a) The emission spectra of TTA-Nd nanophosphors in aqueous solution under simultaneous 635- and 808 nm excitations. Inset: the integral emission intensity of TTA (green line) and Nd emission (red line) as a function of temperature. (b) Emission spectra of nanofluid at 300K and 330K. Inset: the schematic of the experimental setup. (c) Temperature profiles of lipid bilayer coated UCNPs and uncoated counterparts dispersed in water under different excitation power densities. The temperatures detected by an immersed thermometer were also presented for comparison. (d) Emission spectra of NaLuF$_4$:Yb/Er@NaLuF$_4$@Carbon UCNPs collected at different temperatures. Inset: the schematic diagram of the nanophosphors. (Xu, M., Zou, X., Su, Q., Yuan, W., Cao, C., Wang, Q., Zhu, X., Feng, W. and Li, F., *Nat. Commun.* 9, 2698, 2018. With permission; Brites, C. D. S., Xie, X., Debasu, M. L., Qin, X., Chen, R., Huang, W., Rocha, J., Liu, X. and Carlos, L. D., *Nat. Nanotechnol.* 11, 851, 2016. With permission; Bastos, A. R. N., Brites, C. D. S., Rojas Gutierrez, P. A., DeWolf, C., Ferreira, R. A. S., Capobianco, J. A. and Carlos, L. D. *Adv. Funct. Mater.* 29, 1905474, 2019. With permission; Zhu, X., Feng, W., Chang, J., Tan, Y. W., Li, J., Chen, M., Sun, Y. and Li, F., *Nat. Commun.* 7, 10437, 2016. With permission.)

power density was observed in either H$_2$O or D$_2$O. Under low power density excitation, the lipid bilayers acted as the thermal barrier to block thermal transfer from UCNPs to the external environment. This effect narrowed with the enhancement of excitation power density. Beyond a critical power density of around 150 W/cm^2, the temperature difference between the two thermometers decreased and thermal equilibrium was reached.

Moreover, the combination of upconversion LNT and heat generation materials brings about synergetic effects for accurate temperature control in photothermal therapy and optoelectronic devices. Figure 9.3(d) discusses the realization of simultaneous photothermal therapy and accurate temperature feedback with a sensitivity of 1%/K by employing NaLuF$_4$:Yb/Er@NaLuF$_4$@Carbon upconversion nanocomposite, which was proposed by Li and co-workers in 2016. The carbon shell acted as a photothermal agent to ablate cancer cells and the NaLuF$_4$:Yb/Er@NaLuF$_4$ nanophosphors served as a nanothermometer for monitoring the real-time temperature and evading tissue lesion due to hyperthermia.[42]

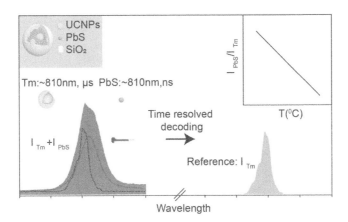

FIGURE 9.4 Schematic illustration of PbS-QD/Tm-UCNP hybrids for lifetime-based nanothermometry. Inset: schematic diagram of the hybrid structure (upper left) and the intensity ratio of PbS-QDs and Tm-UCNPs as a function of temperature (upper right). (Qiu, X., Zhou, Q., Zhu, X., Wu, Z., Feng, W. and Li, F., *Nat. Commun.* 11, 4, 2020. With permission.)

9.2.2 Temperature Sensing by Lifetime

Luminescence lifetime refers to the time needed for emission intensity decaying to 1/e of its original level.[30] Thermal field is an important factor that influences decay processes between electronic levels due to an increase of nonradiative relaxation rates with an increase in temperature.[43] Importantly, characterizations of lifetime are insensitive to the influence of ambient environments such as electromagnetic interference, which makes lifetime an accurate temperature indicator.[44] Till now, lifetime-based LNT has been realized in a set of lanthanide activators (e.g., Eu^{3+}, Tm^{3+}, and Er^{3+}) that are doped in different hosts.[44–46]

Recently, composite structures comprising multiple luminescent centers that share similar emission peak positions with significantly different lifetimes appear as well-behaved lifetime-based nanothermometers. In 2020, Li and co-workers designed a class of hybrid lanthanide nanoclusters that combined PbS quantum dots (QDs) and $NaYbF_4$:Tm@$NaYF_4$:Yb@$NaYF_4$:Nd UCNPs, as shown in Figure 9.4, which featured dual emission lifetimes at ~810 nm and rendered a high thermal sensitivity of 5.6%/K. Under 865 nm NIR excitation, the PbS QDs functioned as temperature-sensitive parts with upconversion emission at 814 nm (ns lifetime scale) and the Tm-doped UCNPs functioned as the referential part with emission at 804 nm (μs lifetime scale). Notably, both excitation and emission wavelengths are located in the NIR region, which showed outstanding performance in temperature monitor inside the tumor region during photothermal therapy.[47]

9.3 UPCONVERSION LASING

Optical microcavities are important components in the field of photonics. Their excellent ability for light confinement significantly enhances light-matter interaction, making them preeminent candidates for lasing and spontaneous amplification emission.[48] Owing to their long lifetime, sharp emission bandwidth, and tunable optical properties, lanthanide-doped UCNPs are appropriate gain media in photonic devices. In recent decades, great advances have been achieved in the preparation of microcavities integrated with UCNPs. The combination of UCNPs and microcavities not only boosts the upconversion efficiency but also broadens the spectral range of microlasers.

Figure 9.5 depicts three major types of cavities that support lasing or light amplification from UCNPs, Fabry–Perot (FP) cavity, whispering gallery mode (WGM) cavity, and integrated plasmonic cavity. The FP and WGM cavities belong to optical microresonators where the UCNPs are

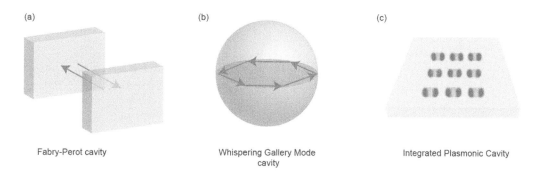

(a)   (b)   (c)

Fabry-Perot cavity   Whispering Gallery Mode   Integrated Plasmonic Cavity
                                      cavity

FIGURE 9.5   The schematic diagram of (a) Fabry–Perot cavity, (b) whispering gallery mode cavity, and (c) integrated plasmonic cavity.

mixed with a dielectric medium of a high reactive index to promote the formation of total internal reflection at the cavity sidewalls.[49] The emission wavelengths can be amplified under resonance conditions:

$$n_{\text{eff}} \cdot L = m \cdot \lambda_m \tag{9.5}$$

where n_{eff} is the effective refractive index of the cavity, L is cavity length for each round-trip, λ_m is the mth-order resonance wavelength, and m is an integer. When the resonators have a good quality factor, the stimulated radiation energy is capable of exceeding stimulated absorption to form population inversion of gain medium (UCNP), which is the prerequisite to produce lasing emission. Plasma cavities are recently burgeoning for upconversion lasing based on the surface plasmon amplification effect. The surface plasmon polaritons generated from noble metals facilitate remarkably optical mode volume shrink of the surrounding gain media, producing stimulated emission of radiation.[50] Notably, the isolated plasmonic cavities suffer from low Q value due to the considerable ohmic losses associated with the dissipation of pump energy into heat.[51, 52] Thus, integrated plasmonic cavity arrays emerge as the appropriate choice to restrain radiative loss because of the electromagnetic interactions between individual units.[53]

9.3.1   FP CAVITY

An FP cavity is composed of two parallel reflecting surfaces for the resonant lights repeatedly reflected back and forth to pass through the whole interspace.[54] Especially, FP cavities have open-access geometry, which is in favor of high-power laser output.[55] The cooperation of UCNPs with FP cavities exhibits valid emission amplification or lasing, which mitigates the low efficiency of UCNPs and facilitates their use in biomedical therapy and information encryption.

Figure 9.6(a) displays a typical example of the amplified spontaneous emission of NaYF$_4$:Yb/Er@NaYF$_4$ core–shell nanophosphors in FP cavity reported by Yu and co-workers. The full width at half maximum of the emission peaks at around 410, 540, and 655 nm showed narrowing tendency when the excitation power attained the threshold value, indicating amplified spontaneous emission from the UCNPs as the gain medium in the FP cavity.[56] In 2018, Ruan and co-workers reported random lasing in the ultraviolet and blue spectral region from NaYF$_4$:Yb/Tm nanophosphor film that sandwiched between Al and quartz reflectors under excitation of a 980 nm nanosecond laser.[57] As seen in Figure 9.6(b), when the excitation power exceeded the thresholds (450 kW/cm^2 for 345 nm and 62 kW/cm^2 for 474 nm), the lasing peaks at around 345 and 474 nm were detected. Nevertheless, the lasing spectra indicated the existence of random lasing because the spectra varied at different detection angles. This phenomenon can be attributed to the inhomogeneous distribution of nanophosphors in the thin film due to solvent evaporation–induced aggregation.

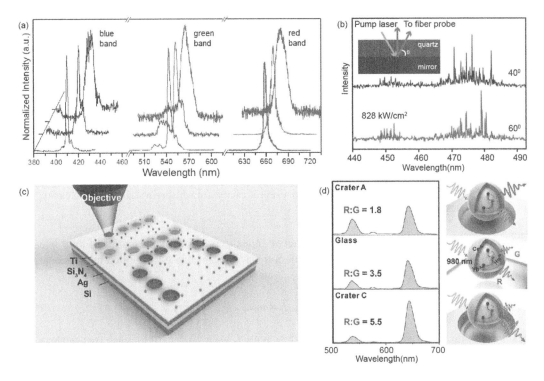

FIGURE 9.6 (a) Normalized emission spectra of FP cavity under different excitation powers. (b) Emission spectra of random lasing at observation angles of 40° and 60°. Inset: schematic diagram of sandwich structure incorporated with NaYF$_4$:Yb/Tm (20/2%) UCNPs. (c) Sketch of MIM-structured nanocraters. (d) Ratiometric upconversion emission tuning based on various nanocraters and nanoparticle compositions. (Zhu, H., Chen, X., Jin, L. M., Wang, Q. J., Wang, F. and Yu, S. F., *ACS Nano* 7, 11420, 2013. With permission; Peng, Y. P., Lu, W., Ren, P., Ni, Y., Wang, Y., Zhang, L., Zeng, Y.J., Zhang, W. and Ruan, S., *Photonics Res.* 6, 947, 2018. With permission; Feng, Z., Hu, D., Liang, L., Xu, J., Cao, Y., Zhan, Q., Guan, B. O., Liu, X. and Li, X., Adv. *Opt. Mater.* 7, 1900610, 2019. With permission.)

Intriguingly, FP cavity combined with plasmonic effect brings about a selective enhancement of emission peaks of UCNPs, demonstrating potentials for encryption.[58] Figure 9.6(c) and (d) shows a Ti–Si$_3$N$_4$–Ag metal–insulator–metal (MIM) structure to tune upconversion emission of NaYF$_4$:Yb/Ho/Ce nanophosphors designed by Li and co-workers. Femtosecond-pulsed laser beams were applied to create two ablated nanocraters in the MIM structure, which exhibited resonance peaks at 538 and 650 nm. When illuminated by a 980 nm laser, FP resonance modes, which were in the form of longitudinal gap plasmons, were generated to obtain increased field confinement inside the craters, leading to obvious luminescence enhancement.

9.3.2 WGM CAVITY

WGM microcavities are characterized by high Q factors (usually surpass 10^{11}) and facile fabrication processes, making them excellent supports for high-performance optical devices.[59] UCNPs are appealing candidates as the gain medium in WGM cavities to realize emission regulation or lasing that covers the spectral range from deep ultraviolet to NIR.[56, 60, 61] Benefiting from the high dispersibility of UCNPs in polymers such as silicon resin and polystyrene (PS), a collection of microcavities with diverse geometries and configurations (spheres, bottles, cylinders, and disks) can be readily fabricated for lasing through the formation of WGMs.[16, 62–66]

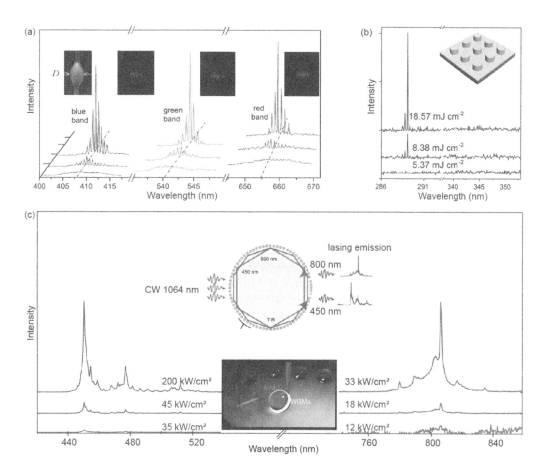

FIGURE 9.7 (a) Lasing emission spectra from NaYbF$_4$:Yb/Er@NaYF$_4$ UCNPs in a bottle-like microcavity (D = 80 μm) under three-pulse excitation scheme at 980 nm of different powers. (b) Ultraviolet B (298 nm) lasing from LiYbF$_4$:Tm (1%)@LiYbF$_4$@LiLuF$_4$ UCNPs in patterned SiO$_2$ substrate under 980 nm excitation. Inset: schematic diagram of the SiO$_2$ substrate structure. (c) Lasing emission at 450 and 807 nm under 1064 nm excitation at various power densities. Inset: schematic diagram of excitation as well as lasing processes in the PS bead. (Zhu, H., Chen, X., Jin, L. M., Wang, Q. J., Wang, F. and Yu, S. F., *ACS Nano* 7, 11420, 2013. With permission; Jin, L., Wu, Y., Wang, Y., Liu, S., Zhang, Y., Li, Z., Chen, X., Zhang, W., Xiao, S. and Song, Q., *Adv. Mater.* 31, 1807079, 2019. With permission; Fernandez-Bravo, A., Yao, K., Barnard, E. S., Borys, N. J., Levy, E. S., Tian, B., Tajon, C. A., Moretti, L., Altoe, M. V., Aloni, S., Beketayev, K., Scotognella, F., Cohen, B. E., Chan, E. M. and Schuck, P. J., *Nat. Nanotechnol.* 13, 572, 2018. With permission.)

In 2013, Yu and Wang et al. first reported a bottle-like WGM microcavity laser using NaYF$_4$:Yb/Er@NaYF$_4$ nanophosphors as the gain medium as illustrated in Figure 9.7(a). A three-pulse pump scheme was devised to improve the excitation efficiency, which reduced optical damage induced by high pumping energy of a single pulse and maximized the optical gain of the two-photon (540 nm Er^{3+}:^{4}S$_{3/2}$ → ^{4}I$_{15/2}$, 655 nm Er^{3+}:^{4}F$_{9/2}$ → ^{4}I$_{15/2}$) or three-photon (410 nm Er^{3+}:^{2}H$_{9/2}$ → ^{4}I$_{15/2}$) upconversion processes.[56]. Based on this work, the same team reported the deep ultraviolet lasing at 311 nm (Gd^{3+}:^{6}P$_{7/2}$ → ^{8}S$_{7/2}$) using NaYF$_4$@NaYbF$_4$:Tm/Gd (1/y%)@NaYF$_4$ nanophosphors under a five-pulse pump scheme at 980 nm. The UCNPs demonstrated enhancement of upconversion emission due to the confinement of energy migration in the inner shell that resulted in reduced energy loss. This optimizing strategy endowed UCNPs with two orders of magnitude higher optical gain than the conventional three-photon upconversion process of a Yb/Tm system. Particularly, single-mode lasing was obtained by employing a thin microresonator (D = 20 μm), which is unique in lanthanide nanophosphors due to their narrow gain bandwidth.[64] In addition to NaYF$_4$-based hosts, upconversion lasing emissions from bottle-like cavities have also been obtained in other materials, such as ~311 nm (Gd^{3+}) from Lu$_6$O$_5$F$_8$ nanocrystals by 447 nm excitation[16] and ~548 nm (Er^{3+}) from KLu$_2$F$_7$ nanocrystals by 980 nm excitation,[67] indicating the versatility of UCNPs as gain media for WGM lasing.

Recently, periodic structure of resonant cavities has emerged with superiorities in terms of controllability and repeatability. Figure 9.7(b) demonstrates 289 nm lasing from LiYbF$_4$:Tm (1%)@ LiYbF$_4$@LiLuF$_4$ core–shell–shell nanophosphors under 980 nm excitation through the use of on-chip integrated SiO$_2$ pillars proposed by Song and co-workers.[60] The lasing emission was extended to the ultraviolet B region owing to the highly Yb-doped inner shell that promoted the absorption of excitation energy to boost the intensity of 289 nm emission from Tm^{3+} ions ($^1I_6 \rightarrow ^3H_6$). By changing the pillar size, controllable mode numbers of lasing emissions were realized on the UCNPs-on-SiO$_2$ microdisks. Particularly, single-mode lasing emission was observed from microdisks of 10 μm in diameter at a threshold of ~121 mJ/cm^2.

Apart from lasing emission in the ultraviolet and visible region, NIR lasing is also accomplished in spherical cavities composed of PS beads. As shown in Figure 9.7(c), Schuck and co-workers reported continuous-wave (CW) excited lasing at 450 and 808 nm using 5 μm PS beads with high Q factors of 10^3~10^4 and low threshold power of 14 kW/cm^2. Energy-looping NaYF$_4$:Tm^{3+} nanophosphors were chosen as gain media due to their efficient upconversion emission under 1064 nm excitation. The UCNPs were attached to the surface of PS beads through a partially swelling and deswelling process.[62] Consequently, the surface condition of PS beads took effect on the excitation threshold of the spherical cavities.[63] Uniform sub-monolayer UCNPs on the PS beads' surface were realized through precise tuning of the surface charge, contributing to around a 25-fold reduction of threshold to 1.7 ± 0.7 kW/cm^2.

9.3.3 PLASMA CAVITY

Upconversion lasing from plasmonic cavities has appeared recently owing to their low threshold, high stability, and directional lasing output.[61] Particularly, the integrated plasmonic arrays can produce strong near-field enhancements due to collective, coherent couplings between the intense optical field and oscillating charges on the metal surface. In general, the light amplification of plasmonic cavities stems from their large local density of optical states (LDOS).[53] The enhancement of the spontaneous radiation rate is described by the Purcell factor:

$$F = 3/4\pi^2 \cdot (Q/V) \cdot (\lambda/2n)^3 \quad (9.6)$$

where Q is the quality factor, V is the mode volume, λ is the resonant wavelength, and n is the refractive index of the medium. For integrated plasmonic cavities, they have ultrasmall mode volume and high Q factor thus brings about ultra-large LDOS. If the LDOS is large enough, stimulated emission processes can also be increased for laser production. Figure 9.8(a) denotes the special ability of plasmonic cavities to selectively enhance upconversion intensity for lasing emission. In 2019, Odom and co-workers reported single-mode upconverting lasing at around 664 nm from NaYF$_4$:Yb/Er@ NaYF$_4$ nanoparticles coupled with Ag nanopillar arrays by excitation at 980 nm with a threshold as low as 29 W/cm^2.[61] As seen in Figure 9.8(b), the lattice spacing a_0 between each pillar was adjustable (450–460 nm) to selectively match the red emission of Er^{3+} ions ($^4F_{9/2} \rightarrow ^4I_{15/2}$) for the generation of single-mode lasing emission.

9.4 SUPER-RESOLUTION MICROSCOPY

Super-resolution microscopy (SRM) that breaks the diffraction limits (λ/2NA) has been developed as a powerful method for exploring biological structure at the nanoscale.[68] UCNPs are typically excited by NIR light (980 nm) that is located within the biological transparent window. Besides,

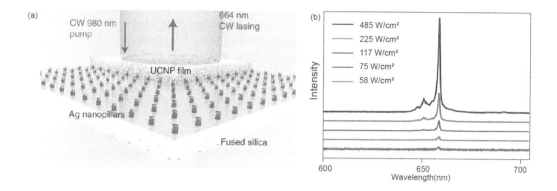

FIGURE 9.8 (a) Schematic diagram of Ag arrays with UCNPs coating on the top. The Ag nanopillars were 80 nm in diameter and 50 nm in height. The spacing (a_0) between Ag nanopillars was 450 nm. (b) Power-dependent lasing spectra from Ag nanopillar arrays at $\lambda = 664$ nm. (Fernandez-Bravo, A., Wang, D., Barnard, E. S., Teitelboim, A., Tajon, C., Guan, J., Schatz, G. C., Cohen, B. E., Chan, E. M., Schuck, P. J. and Odom, T. W., *Nat. Mater.* 18, 1172, 2019. With permission.)

UCNPs also show tunable emission peaks, good photostability, and strong emission intensity.[69] These distinctive optical properties make UCNPs beneficial for bioimaging.

To date, UCNPs as luminescent probes are involved in two main types of SRM techniques, stimulated emission depletion (STED) microscopy and structured illumination microscopy (SIM), which are illustrated in Figure 9.9. The key to breaking the diffraction limitation is to increase the distinguished molecules in the diffraction-limited volume. Generally, SRM employs patterned illuminations to modulate upconversion emissions inside the diffraction-limited volume.[70] It's worth noting that the spatial modulation in STED is in a "negative" manner, where the additive light

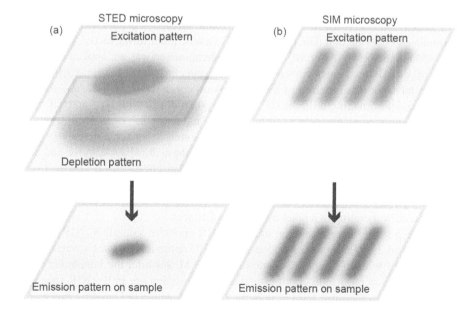

FIGURE 9.9 The patterned illumination for super-resolution optical microscopy. (a) In STED microscopy, samples are illuminated by an excitation light spot and a "doughnut"-shaped depletion pattern. (b) In SIM, samples are excited by a set of periodic light patterns.

field is applied to suppress the emission of UCNPs. In contrast, illumination modulation in SIM is in a "positive" manner, that is, the sample is directly excited to generate luminescent patterns.[71] Besides, the acquisition of super-resolution imaging through these two methods is discrepant. For the STED microscopy, images are obtained directly from the original data, while computational treatment is indispensable to get final images for SIM.[70]

9.4.1 UPCONVERSION-BASED STED MICROSCOPY

A classical STED microscope system consists of two synchronized ultrafast illuminance beams: excitation and depletion beams. In general, the excitation beam pumps electrons to high-lying excited energy levels, which emits through spontaneous radiation in the absence of the depletion beam. Once the wavelength-matched depletion beam is applied, which is typically "doughnut" shaped, the excited state is depopulated through stimulated radiation before the spontaneous radiation process happens. Theoretically, if the depletion intensity is strong enough, only the UCNPs distributed in the center of the "doughnut" beam can emit, where the depletion light field is nearly zero. As a result, a much smaller light spot can be obtained in comparison with the focal spot of a conventional microscope.[70] Particularly, the saturation response of UCNPs promotes resolution enhancement. When the intensity of the depletion beam is beyond the saturation level, the depletion efficiency is largely boosted owing to the nonexistent excited state in the depletion region.[71]

According to the characteristics of STED microscopy, the luminescent probes should simultaneously respond to the Gaussian excitation and "doughnut" depletion, with strong emission intensity and high depletion efficiency. Highly doped UCNPs are shown to be decent candidates for STED with good spatial resolution, fast pixel dwell time, and satisfactory intensity. The high doping level of activators is vital to facilitate cross-relaxation and realize population inversion that triggers amplification of stimulated emission, leading to depopulation of metastable energy levels for emission depletion. Figure 9.10(a) demonstrates an upconversion-STED technique using $NaYF_4$:Yb/Tm (20/8%) UCNPs, registering a resolution of 28 nm and a pixel dwell time of 4 ms under 9.75 MW/cm^2 excitation by Jin's group in 2017.[72] Under 980 nm Gaussian excitation profile, the typical upconversion emissions from $^3H_4$, $^1G_4$, and $^1D_2$ levels of Tm^{3+} ions were observed. However, the addition of an 808 nm Gauss–Laguerre mode "doughnut" depletion profile (matching $^3H_4 \rightarrow {}^3H_6$) induced stimulated emission that depopulated the $^3H_4$ levels, manifesting as the inhibition of Tm^{3+} emission at 455 nm as shown in Figure 9.10(b). This phenomenon was not applicable for low Tm^{3+} doping (1%), which exhibited enhancement of upconversion emission under synergetic 980/808 nm excitation. Similar work was reported by He and co-workers in the same year, and they additionally realized two-color imaging utilizing the Yb/Tm system doped with Eu^{3+} and Tb^{3+} ions under 975/810 nm excitation, demonstrating a resolution of 66 nm and a fast pixel dwell time of 100 μs. [73] Unfortunately, the long lifetime and weak emission intensity of the aforementioned samples restrict the scanning speed in STED imaging. To overcome these drawbacks, Ågren and co-workers designed a highly Yb^{3+} doping strategy ($NaYF_4$@$NaYbF_4$:Tm (10%)) to boost the absorption of excitation energy with retained depletion efficiency.[74] The Yb-based UCNPs revealed dramatically enhanced luminescence intensity and accelerated emission kinetics, contributing to improved imaging speeds, corresponding to a pixel dwell time of 10 μs.

To achieve deep tissue super-resolution imaging, NIR emission saturation (NIRES) nanoscopy was recently devised. NIRES nanoscopy adopts the STED approach but uses a single "doughnut"-like excitation beam to enable super-resolution imaging. Meanwhile, the lanthanide dopant ions are carefully selected to ensure that both excitation and emission are in the NIR spectral region. NIRES approach retains the same level of imaging resolution with STED microscopy and provides a remarkable light penetration depth as well as a simplified experimental setup. In 2018, Jin's group demonstrated NIRES nanoscopy using $NaYF_4$:Yb/Tm nanophosphors to realize single-particle imaging through 93 μm thick liver tissue with a resolution of 50 nm.[75] As shown in Figure 9.10(c), the 980 nm excitation laser (larger than 1 MW/cm^2) was shaped as a "doughnut" beam to saturate

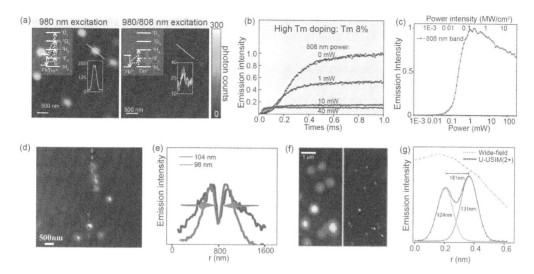

FIGURE 9.10 (a) Confocal images of NaYF$_4$:Yb/Tm(20/8%) under continuous-wave 980 nm laser excitation (left panel) and simultaneous 980-and 808 nm excitation (right panel). Inset: energy level diagrams of the 8% Tm^{3+}-doped UCNPs under corresponding excitations. (b) Transient response of the 455 nm emission from 8% Tm-doped UCNPs under synchronous 980 nm (power fixed at 1 mW) and 808 nm pulses (1 ms duration, power varied from 0 to 40 mW). (c) The saturation intensity curves of 800 nm emissions from NaYF$_4$:Yb/Tm (20/4%) UCNPs under 980 nm excitation. (d) The zoomed-in super-resolution images of UNCPs at 55.9 μm depth inside the spheroid under 8.9 MW/cm^2 980 nm excitation. (e) Line profiles of the UCNPs labeled in (d), revealing full width at half maximum (FWHM) of 104 and 98 nm, respectively. (f) Imaging reconstructions from wide-field (left panel) and super-resolution (right panel) images of NaYF$_4$: Yb/Tm (40/4%) nanophosphors. (g) Line profiles of two UCNPS at the lower right corner in (e). (Liu, Y., Lu, Y., Yang, X., Zheng, X., Wen, S., Wang, F., Vidal, X., Zhao, J., Liu, D., Zhou, Z., Ma, C., Zhou, J., Piper, J. A., Xi, P. and Jin, D. *Nature* 543, 229, 2017. With permission; Chen, C., Wang, F., Wen, S., Su, Q. P., Wu, M. C. L., Liu, Y., Wang, B., Li, D., Shan, X., Kianinia, M., Aharonovich, I., Toth, M., Jackson, S. P., Xi, P. and Jin, D., *Nat. Commun.* 9, 3290, 2018. With permission; Liu, Y., Wang, F., Lu, H., Fang, G., Wen, S., Chen, C., Shan, X., Xu, X., Zhang, L., Stenzel, M. and Jin, D., *Small* 16, 1905572, 2020. With permission; Liu, B., Chen, C., Di, X., Liao, J., Wen, S., Su, Q. P., Shan, X., Xu, Z. Q., Ju, L. A., Mi, C., Wang, F. and Jin, D., *Nano Lett.* 20, 4775, 2020. With permission.)

metastable level ($^3H_4$) of the Tm^{3+} ions, resulting in bright NIR emission at 800 nm for detection. Compared with the 455 nm emission that decayed to 11.3% after passing through the 93 μm thick tissue, the 800 nm emission preserved 38.7% of the initial intensity. In the following development, Jin and co-workers optimized the 980 nm excitation by applying Bessel–Laguerre–Gaussian beam to alleviate light absorption and scattering in high cell density cancer spheroids.[76] Figure 9.10(d) and (e) demonstrates 55.9 μm mapping inside a spheroid with a resolution of 98 nm by adjusting the excitation beam to minimize the loss of both excitation and emission lights.

9.4.2 UPCONVERSION-BASED SIM

SIM adopts spatially varying illumination to excite the labeled sample in the diffraction-limited region to overcome the diffraction limit.[77] The excitation light is typically a set of periodic patterns consisting of standing waves of different orientations or phases. Under the structured illumination, the microscopy can detect increased spatial frequency that conveys different information of the sample, thereby giving rise to a high resolution.[70] Each of the obtained image snapshots is composed of sample structure information and excitation patterns. Therefore, computationally reconstruction is necessary to get the final sample structure by separating the different regions to their proper positions in the frequency space through scanning and rotating the patterns of multiple snapshots.[71, 78] Similar to STED microscopy, the saturation of UCNPs enables high image resolution.[71]

Compared with STED microscopy, one of the conspicuous merits of SIM is fast scan speed because the focused depletion beam is not required in this system. In 2020, Jin's group demonstrated an upconversion SIM using $NaYF_4$:Yb/Tm (40/4%) nanophosphors as the luminescent probe. As seen in Figure 9.10(f) and (g), under 975 nm excitation, NIR imaging at 800 nm was demonstrated with a resolution below 131 nm and an imaging rate of 1 Hz.[79] A high doping strategy was applied in designing the UCNPs, which were 4% for Tm^{3+} and 40% for Yb^{3+}. The high Tm^{3+} concentration promoted cross-relaxation between adjacent ions and enhances the nonlinearity of the photoresponse, which was critical to obtain high resolution. The high Yb^{3+} concentration facilitated the absorption of excitation energy to compensate for the sacrifice of emission intensity due to concentration quenching in Tm^{3+} ions. Besides, the high brightness lowered the needed excitation power density to 4 kW/cm^2, which effectively prevented the photothermal effect.

9.5 SOLAR ENERGY CONVERSION

Sunlight is a kind of free and abundant green energy that has a wide spectral wavelength ranging from ultraviolet to infrared (280–2500 nm). The contained solar energy can be harvested and converted into electric energy through photovoltaic (PV) technologies to meet the enormous consumption need in the future.[80] However, the existing solar cells suffer from limited energy conversion efficiency (less than 30%) due to the inherent restriction of PV materials, which only absorb the photons with energy above their bandgaps. UCNPs can enhance the solar energy conversion efficiency by converting the unabsorbed sub-bandgap photons into over-bandgap photons that can be utilized by PV devices, which is described in Figure 9.11.[81] To date, upconversion has boosted the efficiency of several types of solar cells, including dye-sensitized solar cells (DSSCs) and perovskite solar cells (PSCs).

9.5.1 UPCONVERSION-BOOSTED DSSCs

DSSCs have been received popularity owing to their cost-effective and eco-friendly properties. Nevertheless, they are faced with severe problems that limit their power conversion efficiency (PCE) due to the absorption mismatch between sensitizers in cells and solar irradiation spectrum. [82] The improvement of light capture ability is necessary to boost the efficiency of DSSCs. Novel photoelectrode materials containing UCNPs are appropriate candidates for optimizing the performance of DSSCs.

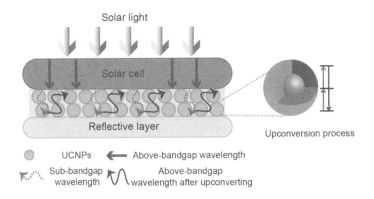

FIGURE 9.11 The schematic diagram showing the working principle of UCNPs in solar cells.

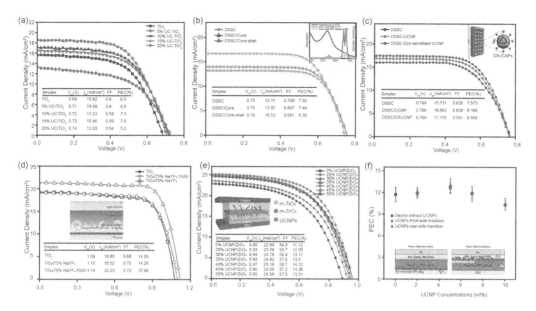

FIGURE 9.12 (a) Photocurrent–voltage curves of YbF$_3$-Ho/TiO$_2$ heterostructures under AM 1.5G. (b) Photocurrent–voltage curves of untreated DSSC, NaYF$_4$:Yb/Er incorporated DSSC, and NaYF$_4$:Yb/Er@ YOF:Yb/Er incorporated DSSC under AM 1.5G. Inset: absorption spectrum of N719 and emission spectra of UCNPs-incorporated TiO$_2$. (c) Photocurrent–voltage curves of untreated DSSC, UCNPs-incorporated DSSC, and dye-sensitized UCNPs (DSUCNP)-incorporated DSSC under AM 1.5G. Inset: schematic diagram of the DSUCNP-incorporated DSSC device. (d) Photocurrent–voltage curves of untreated TiO$_2$, 75% of NaYF$_4$-incorporated TiO$_2$ and 75% of NaYF$_4$:Yb/Er-incorporated TiO$_2$ under AM 1.5G. Inset: schematic configuration of the PSC device. (e) Photocurrent–voltage curves of UCNPs-treated ZrO$_2$ PSCs under AM 1.5G. Inset: schematic configuration of the PSC device. (f) PCE of UCNPs-incorporated PSCs at the front or rear side. Inset: schematic configuration of the PSC devices. (Yu, J., Yang, Y., Fan, R., Wang, P., Dong, Y., *Nanoscale* 8, 4173, 2016. With permission; Tian, L., Shang, Y., Hao, S., Han, Q., Chen, T., Lv, W. and Yang, C., *Adv. Funct. Mater.* 28, 1803946, 2018. With permission; Hao, S., Shang, Y., Li, D., Ågren, H., Yang, C., Chen, G., *Nanoscale* 9, 6711, 2017. With permission; Roh, J., Yu, H. and Jang J. *ACS Appl. Mater. Interfaces* 8, 19847, 2016. With permission; Li, Y., Zhao, L., Xiao, M., Huang, Y., Dong, B., Xu, Z., Wan, L., Li, W. and Wang, S., *Nanoscale* 10, 22003, 2018. With permission; Schoenauer Sebag, M., Hu, Z., de Oliveira Lima, K., Xiang, H., Gredin, P., Mortier, M., Billot, L., Aigouy, L. and Chen, Z., *ACS Appl. Energ. Mater.* 1, 3537, 2018. With permission.)

Figure 9.12(a) represents a typical example reported by Dong and co-workers in 2016, describing the DSSCs performance enhancement (8% overall conversion efficiency and 19% increase of photocurrents) with optimized photoelectrodes composed of an YbF$_3$-Ho/TiO$_2$ nanoheterostructure. This hybrid promoted the PCE by synergistically improving the utilization of NIR light and suppressing the photoinduced charge recombination for prolonging the lifetime of excited electrons.[82]

A high upconversion emission intensity is essential for improving the performance of PV devices. The use of protective shell coating is a viable way for enhancing upconversion emissions. In 2018, Yang and co-workers reported a class of NaYF$_4$:Yb/Er@YOF:Yb/Er core–shell UCNPs and achieved 25-fold enhancement of upconversion emission intensity due to the suppression of surface quenching by the YOF:Yb/Er outmost shell layer. Under 980 nm excitation, the regular dye (N719) in DSSCs showed 100% absorption of green emission and 80% red emission from the Er^{3+} ions. Figure 9.12(b) indicates 17.1% increase of the DSSC efficiency owing to boosted utilization of sub-bandgap NIR light after coating with the oxyfluoride shell layer.[83]

Dye-sensitization is another powerful method to enhance upconversion. In 2017, Chen and co-workers reported an organic dye (IR783)-sensitized NaYF$_4$:Yb/Er (10/2%)@NaYF$_4$:Nd (30%) UCNP. The IR 783 were attached to the surface of the UCNPs for wide spectral absorption from 670 to 860 nm, followed by energy transfer to Nd^{3+} ions and subsequently to Yb^{3+}/Er^{3+} ions for intense upconversion emission.[84] The upconversion emission peaks at 540/650 nm matched well with the absorption spectrum of regular light-absorbing dyes (N719) in DSSCs. Figure 9.12(c) describes a PCE enhancement by 13.1% by incorporating the UCNPs into the TiO_2 photoanode of the DSSC.

9.5.2 UPCONVERSION-BOOSTED PSCs

Organometal halide-based PSCs emerge as the promising PV technology in recent decades. The PSCs originate from DSSCs, in which perovskite materials are applied as sensitizers to replace dye molecules in DSSCs.[85] Similar to DSSCs, the PSCs suffer from spectral mismatch losses between sunlight and perovskite nanocrystals limiting the efficiency of solar cells. To reduce spectral mismatch losses and promote photocurrent generation, UCNPs are usually combined with mesoporous semiconductor oxide nanoparticle films.

Yb/Er-codoped UCNPs are appropriate candidates as efficient upconverters in PSCs. In 2016, Jang and co-workers combined $NaYF_4$:Yb/Er nanoprisms with TiO_2 nanoparticles in $CH_3NH_3PbI_3$ PSCs as an upconverting mesoporous layer to capture NIR light. As demonstrated in Figure 9.12(d), the upconversion mixed layer demonstrated 13.74% enhancement of PCE in comparison with the bare TiO_2 layer.[86] In another development, Mao and co-workers directly doped Yb/Er (13/6%) ions into TiO_2 nanorods as the electron transfer materials in $MAPbI_{3-x}Cl_x$ PSCs, which demonstrated a 20.8% enhancement of PCE.[87]

Figure 9.12(e) exhibits a 28.8% PCE enhancement of $FA_{0.4}MA_{0.6}PbI_3$ PSCs in a report of Wang and co-workers in 2018, based on the use of $NaYbF_4$:Ho UCNPs and ZrO_2 as the scaffold layer. ZrO_2 has a large bandgap and can isolate TiO_2 with the counter electrode, producing expanded absorption and higher photovoltage. Meanwhile, the incorporated $NaYbF_4$:Ho^{3+} UCNPs helped to increase NIR light harvest. Under the optimal ratio (40 wt%) of UCNPs and ZrO_2, the PSC exhibited a high PCE of 14.32% ($J_{sc} = 25.16$ mA/cm^2, $V_{oc} = 0.975$ V, and FF = 58.7%) compared with the ZrO_2-based device with a PCE of 11.12%.[88]

In a recent study, the contributions of UCNPs to PCE enhancement were macroscopically and microscopically proved by Chen and co-workers. Figure 9.12(f) shows 6.1% and 6.5% of PCE enhancements of $FA_{0.83}Cs_{0.17}Pb(I_{0.6}Br_{0.4})_3$ (FA = formamidinium [$HC(NH_2)_2$]) PSCs when KY_7F_{22}:Yb/Er (20/5%) UCNPs were inserted into the front side (fluorine doped tin oxide [FTO]/perovskite interface) or the rear side (perovskite/HTL interface), respectively.[89] Particularly, a device with a half upconversion layer was designed and observed under light beam-induced current (LBIC)/fluorescence mapping setup. When the excitation spot moved from the part without UCNPs to that with UCNPs, the short-circuit current revealed threefold enhancement. The mapping experiments confirmed the contribution of UCNPs to efficiency enhancement.

9.6 CONCLUSION

In this chapter, we have summarized the basic properties of lanthanide-based UCNPs with an emphasis on the emerging applications that take advantage of these nanophosphors. Despite the prominent performance of UCNPs such as tunable emission and excellent stability, they still face severe challenges: (1) low quantum efficiency; although high excitation power density can remit this problem to some extent, the exorbitant optical energy normally causes device damages; (2) limited absorption band and cross section. This problem restricts the choice of excitation light sources, leading to finite available NIR laser excitation wavelength, which impedes the practical applications of UCNPs.

ACKNOWLEDGMENTS

F. Wang acknowledges the National Natural Science Foundation of China (21773200 and 21573185) and the Research Grants Council of Hong Kong (CityU 11204717 and 11205219) for supporting this work.

REFERENCES

1. Zhou B, Shi B, Jin D, Liu X (2015) Controlling upconversion nanocrystals for emerging applications. *Nat. Nanotechnol.* 10: 924.
2. Auzel F (2004) Upconversion and anti-stokes processes with f and d ions in solids. *Chem. Rev.* 104: 139.
3. Wang F, Banerjee D, Liu Y, Chen X, Liu X (2010) Upconversion nanoparticles in biological labeling, imaging, and therapy. *Analyst* 135: 1839.
4. Patra A, Friend CS, Kapoor R, Prasad PN (2003) Effect of crystal nature on upconversion luminescence in Er^{3+}:ZrO_2 nanocrystals. *Appl. Phys. Lett.* 83: 284.
5. Yi G, Lu H, Zhao S, Ge Y, Yang W, Chen D, Guo LH (2004) Synthesis, characterization, and biological application of size-controlled nanocrystalline $NaYF_4$:Yb,Er infrared-to-visible up-conversion phosphors. *Nano Lett.* 4: 2191.
6. Krämer KW, Biner D, Frei G, Güdel HU, Hehlen MP, Lüthi SR (2004) Hexagonal sodium yttrium fluoride based green and blue emitting upconversion phosphors. *Chem. Mater.* 16: 1244.
7. Bünzli JCG, Piguet C (2005) Taking advantage of luminescent lanthanide ions. *Chem. Soc. Rev.* 34: 1048.
8. Chen G, Qiu H, Prasad PN, Chen X (2014) Upconversion nanoparticles: Design, nanochemistry, and applications in theranostics. *Chem. Rev.* 114: 5161.
9. Chen B, Wang F (2020) Combating concentration quenching in upconversion nanoparticles. *Acc. Chem. Res.* 53: 358.
10. Wang F, Deng R, Wang J, Wang Q, Han Y, Zhu H, Chen X, Liu X (2011) Tuning upconversion through energy migration in core–shell nanoparticles. *Nat. Mater.* 10: 968.
11. Su Q, Han S, Xie X, Zhu H, Chen H, Chen CK, Liu RS, Chen X, Wang F, Liu X (2012) The effect of surface coating on energy migration-mediated upconversion. *J. Am. Chem. Soc.* 134: 20849.
12. Li X, Liu X, Chevrier DM, Qin X, Xie X, Song S, Zhang H, Zhang P, Liu X (2015) Energy migration upconversion in manganese(II)-doped nanoparticles. *Angew. Chem. Int. Edit.* 54: 13312.
13. Zhou B, Tao L, Chai Y, Lau SP, Zhang Q, Tsang YH (2016) Constructing interfacial energy transfer for photon up- and down-conversion from lanthanides in a core–shell nanostructure. *Angew. Chem. Int. Edit.* 55: 12356.
14. Liu Y, Zhou S, Zhuo Z, Li R, Chen Z, Hong M, Chen X (2016) In vitro upconverting/downshifting luminescent detection of tumor markers based on Eu^{3+}-activated core–shell–shell lanthanide nanoprobes. *Chem. Sci.* 7: 5013.
15. Chen X, Jin L, Sun T, Kong W, Yu SF, Wang F (2017) Energy migration upconversion in Ce(III)-doped heterogeneous core–shell–shell nanoparticles. *Small* 13: 1701479.
16. Du Y, Wang Y, Deng Z, Chen X, Yang X, Sun T, Zhang X, Zhu G, Yu SF, Wang F (2020) Blue-pumped deep ultraviolet lasing from lanthanide-doped $Lu_6O_5F_8$ upconversion nanocrystals. *Adv. Opt. Mater.* 8: 1900968.
17. Liu X, Wang Y, Li X, Yi Z, Deng R, Liang L, Xie X, Loong DB, Song S, Fan D, All AH, Zhang H, Huang L, Liu X (2017) Binary temporal upconversion codes of Mn^{2+}-activated nanoparticles for multilevel anti-counterfeiting. *Nat. Commun.* 8: 899.
18. Liu D, Xu X, Du Y, Qin X, Zhang Y, Ma C, Wen S, Ren W, Goldys EM, Piper JA, Dou S, Liu X, Jin D (2016) Three-dimensional controlled growth of monodisperse sub-50 nm heterogeneous nanocrystals. *Nat. Commun.* 7: 10254.
19. Zheng K, Loh KY, Wang Y, Chen Q, Fan J, Jung T, Nam SH, Suh YD, Liu X (2019) Recent advances in upconversion nanocrystals: Expanding the kaleidoscopic toolbox for emerging applications. *Nano Today* 29: 100797.
20. Wang X, Valiev RR, Ohulchanskyy TY, Ågren H, Yang C, Chen G (2017) Dye-sensitized lanthanide-doped upconversion nanoparticles. *Chem. Soc. Rev.* 46: 4150.
21. Dong J, Gao W, Han Q, Wang Y, Qi J, Yan X, Sun M (2019) Plasmon-enhanced upconversion photoluminescence: Mechanism and application. *Rev. Phys.* 4: 100026.
22. Zhu Q, Sun T, Wang F (2019) Optical tuning in lanthanide-based nanostructures. *J. Phys. D: Appl. Phys.* 53: 053002.
23. Chen X, Sun T, Wang F (2020) Lanthanide-based luminescent materials for waveguide and lasing. *Chem. Asian J.* 15: 21.
24. Venkatakrishnarao D, Mamonov EA, Murzina TV, Chandrasekar R (2018) Advanced organic and polymer whispering-gallery-mode microresonators for enhanced nonlinear optical light. *Adv. Opt. Mater.* 6: 1800343.
25. Goldschmidt JC, Fischer S (2015) Upconversion for photovoltaics – A review of materials, devices and concepts for performance enhancement. *Adv. Opt. Mater.* 3: 510.

26. Yao W, Tian Q, Wu W (2019) Tunable emissions of upconversion fluorescence for security applications. *Adv. Opt. Mater.* 7: 1801171.

27. Rosal BD, Ximendes E, Rocha U, Jaque D (2017) In vivo luminescence nanothermometry: From materials to applications. *Adv. Opt. Mater.* 5: 1600508.

28. Brites CDS, Balabhadra S, Carlos D (2019) Lanthanide-based thermometers: At the cutting-edge of luminescence thermometry. *Adv. Opt. Mater.* 7: 1801239.

29. Li Z, Yuan H, Yuan W, Su Q, Li F (2018) Upconversion nanoprobes for biodetections. *Coord. Chem. Rev.* 354: 155.

30. Jaque D, Vetrone F (2012) Luminescence nanothermometry. *Nanoscale* 4: 4301.

31. Geitenbeek RG, Prins PT, Albrecht W, van Blaaderen A, Weckhuysen BM, Meijerink A (2017) NaYF$_4$:Er^{3+},Yb^{3+}/SiO$_2$ core/shell upconverting nanocrystals for luminescence thermometry up to 900 K. *J. Phys. Chem. C* 121: 3503.

32. Shang Y, Han Q, Hao S, Chen T, Zhu Y, Wang Z, Yang C (2019) Dual-mode upconversion nanoprobe enables broad-range thermometry from cryogenic to room temperature. *ACS Appl. Mater. Interfaces* 11: 42455.

33. Balabhadra S, Debasu ML, Brites CDS, Nunes LAO, Malta OL, Rocha J, Bettinelli M, Carlos LD (2015) Boosting the sensitivity of Nd^{3+}-based luminescent nanothermometers. *Nanoscale* 7: 17261.

34. Brites CDS, Fiaczyk K, Ramalho JFCB, Sójka M, Carlos LD, Zych E (2018) Widening the temperature range of luminescent thermometers through the intra- and interconfigurational transitions of Pr^{3+}. *Adv. Opt. Mater.* 6: 1701318.

35. Xia T, Cui Y, Yang Y, Qian G (2017) A luminescent ratiometric thermometer based on thermally coupled levels of a Dy-MOF. *J. Mater. Chem. C* 5: 5044.

36. Mi C, Zhou J, Wang F, Lin G, Jin D (2019) Ultrasensitive ratiometric nanothermometer with large dynamic range and photostability. *Chem. Mater.* 31: 9480.

37. Runowski M, Stopikowska N, Szeremeta D, Goderski S, Skwierczyńska M, Lis S (2019) Upconverting lanthanide fluoride core@shell nanorods for luminescent thermometry in the first and second biological windows: β-NaYF$_4$:Yb^{3+}–Er^{3+}@SiO$_2$ temperature sensor. *ACS Appl. Mater. Interfaces* 11: 13389.

38. Hemmer E, Quintanilla M, Légaré F, Vetrone F (2015) Temperature-induced energy transfer in dye-conjugated upconverting nanoparticles: A new candidate for nanothermometry. *Chem. Mater.* 27: 235.

39. Xu M, Zou X, Su Q, Yuan W, Cao C, Wang Q, Zhu X, Feng W, Li F (2018) Ratiometric nanothermometer in vivo based on triplet sensitized upconversion. *Nat. Commun.* 9: 2698.

40. Brites CDS, Xie X, Debasu ML, Qin X, Chen R, Huang W, Rocha J, Liu X, Carlos LD (2016) Instantaneous ballistic velocity of suspended Brownian nanocrystals measured by upconversion nanothermometry. *Nat. Nanotechnol.* 11: 851.

41. Bastos ARN, Brites CDS, Rojas Gutierrez PA, DeWolf C, Ferreira RAS, Capobianco JA, Carlos LD (2019) Thermal properties of lipid bilayers determined using upconversion nanothermometry. *Adv. Funct. Mater.* 29: 1905474.

42. Zhu X, Feng W, Chang J, Tan YW, Li J, Chen M, Sun Y, Li F (2016) Temperature-feedback upconversion nanocomposite for accurate photothermal therapy at facile temperature. *Nat. Commun.* 7: 10437.

43. Yu W, Xu W, Song H, Zhang S (2014) Temperature-dependent upconversion luminescence and dynamics of NaYF$_4$:Yb^{3+}/Er^{3+} nanocrystals: Influence of particle size and crystalline phase. *Dalton Trans.* 43: 6139.

44. Fu Y, Zhao L, Guo Y, Yu H (2019) Up-conversion luminescence lifetime thermometry based on the $^1G_4$ state of Tm^{3+} modulated by cross relaxation processes. *Dalton Trans.* 48: 16034.

45. Shi R, Li B, Liu C, Liang H (2016) On doping Eu^{3+} in Sr$_{0.99}$La$_{1.01}$Zn$_{0.99}$O$_{3.495}$: The photoluminescence, population pathway, de-excitation mechanism, and decay dynamics. *J. Phys. Chem. C* 120: 19365.

46. Yao L, Li Y, Xu D, Lin H, Peng Y, Yang S, Zhang Y (2019) Upconversion luminescence enhancement and lifetime based thermometry of Na(Gd/Lu)F4 solid solutions. *New J. Chem.* 43: 3848.

47. Qiu X, Zhou Q, Zhu X, Wu Z, Feng W, Li F (2020) Ratiometric upconversion nanothermometry with dual emission at the same wavelength decoded via a time-resolved technique. *Nat. Commun.* 11: 4.

48. He L, Özdemir ŞK, Yang L (2013) Whispering gallery microcavity lasers. *Laser Photon. Rev.* 7: 60.

49. Feng S, Lei T, Chen H, Cai H, Luo X, Poon AW (2012) Silicon photonics: From a microresonator perspective. *Laser Photon. Rev.* 6: 145.

50. Lu YJ, Kim J, Chen HY, Wu C, Dabidian N, Sanders CE, Wang CY, Lu MY, Li BH, Qiu X, Chang WH, Chen LJ, Shvets G, Shih CK, Gwo S (2012) Plasmonic nanolaser using epitaxially grown silver film. *Science* 337: 450.

51. Oulton RF, Sorger VJ, Zentgraf T, Ma RM, Gladden C, Dai L, Bartal G, Zhang X (2009) Plasmon lasers at deep subwavelength scale. *Nature* 461: 629.

52. Ambati M, Nam SH, Ulin-Avila E, Genov DA, Bartal G, Zhang X (2008) Observation of stimulated emission of surface plasmon polaritons. *Nano Lett.* 8: 3998.

53. Zhou W, Dridi M, Suh JY, Kim CH, Co DT, Wasielewski MR, Schatz GC, Odom TW (2013) Lasing action in strongly coupled plasmonic nanocavity arrays. *Nat. Nanotechnol.* 8: 506.

54. Heylman KD, Knapper KA, Horak EH, Rea MT, Vanga SK, Goldsmith RH (2017) Optical microresonators for sensing and transduction: A materials perspective. *Adv. Mater.* 29: 1700037.

55. Vallance C, Trichet AAP, James D, Dolan PR, Smith JM (2016) Open-access microcavities for chemical sensing. *Nanotechnology* 27: 274003.

56. Zhu H, Chen X, Jin LM, Wang QJ, Wang F, Yu SF (2013) Amplified spontaneous emission and lasing from lanthanide-doped up-conversion nanocrystals. *ACS Nano* 7: 11420.

57. Peng YP, Lu W, Ren P, Ni Y, Wang Y, Zhang L, Zeng YJ, Zhang W, Ruan S (2018) Integration of nanoscale light emitters: An efficient ultraviolet and blue random lasing from NaYF$_4$:Yb/Tm hexagonal nanocrystals. *Photonics Res.* 6: 947.

58. Feng Z, Hu D, Liang L, Xu J, Cao Y, Zhan Q, Guan BO, Liu X, Li X (2019) Laser-splashed plasmonic nanocrater for ratiometric upconversion regulation and encryption. *Adv. Opt. Mater.* 7: 1900610.

59. Savchenkov AA, Matsko AB, Ilchenko VS, Maleki L (2007) Optical resonators with ten million finesse. *Opt. Express* 15: 6768.

60. Jin L, Wu Y, Wang Y, Liu S, Zhang Y, Li Z, Chen X, Zhang W, Xiao S, Song Q (2019) Mass-manufactural lanthanide-based ultraviolet b microlasers. *Adv. Mater.* 31: 1807079.

61. Fernandez-Bravo A, Wang D, Barnard ES, Teitelboim A, Tajon C, Guan J, Schatz GC, Cohen BE, Chan EM, Schuck PJ, Odom TW (2019) Ultralow-threshold, continuous-wave upconverting lasing from subwavelength plasmons. *Nat. Mater.* 18: 1172.

62. Fernandez-Bravo A, Yao K, Barnard ES, Borys NJ, Levy ES, Tian B, Tajon CA, Moretti L, Altoe MV, Aloni S, Beketayev K, Scotognella F, Cohen BE, Chan EM, Schuck PJ (2018) Continuous-wave upconverting nanoparticle microlasers. *Nat. Nanotechnol.* 13: 572.

63. Liu Y, Teitelboim A, Fernandez-Bravo A, Yao K, Altoe MVP, Aloni S, Zhang C, Cohen BE, Schuck PJ, Chan EM (2020) Controlled assembly of upconverting nanoparticles for low-threshold microlasers and their imaging in scattering media. *ACS Nano* 14: 1508.

64. Chen X, Jin L, Kong W, Sun T, Zhang W, Liu X, Fan J, Yu SF, Wang F (2016) Confining energy migration in upconversion nanoparticles towards deep ultraviolet lasing. *Nat. Commun.* 7: 10304.

65. Jin LM, Chen X, Siu CK, Wang F, Yu SF (2017) Enhancing multiphoton upconversion from NaYF$_4$:Yb/Tm@NaYF$_4$ core–shell nanoparticles via the use of laser cavity. *ACS Nano* 11: 843.

66. Sun T, Li Y, Ho WL, Zhu Q, Chen X, Jin L, Zhu H, Huang B, Lin J, Little BE, Chu ST, Wang F (2019) Integrating temporal and spatial control of electronic transitions for bright multiphoton upconversion. *Nat. Commun.* 10: 1811.

67. Bian W, Lin Y, Wang T, Yu X, Qiu J, Zhou M, Luo H, Yu SF, Xu X (2018) Direct identification of surface defects and their influence on the optical characteristics of upconversion nanoparticles. *ACS Nano* 12: 3623.

68. Hoover EE, Squier JA (2013) Advances in multiphoton microscopy technology. *Nat. Photonics* 7: 93.

69. Jin D, Xi P, Wang B, Zhang L, Enderlein J, van Oijen AM (2018) Nanoparticles for super-resolution microscopy and single-molecule tracking. *Nat. Methods* 15: 415.

70. Sigal YM, Zhou R, Zhuang X (2018) Visualizing and discovering cellular structures with super-resolution microscopy. *Science* 361: 880.

71. Huang B, Babcock H, Zhuang X (2010) Breaking the diffraction barrier: Super-resolution imaging of cells. *Cell* 143: 1047.

72. Liu Y, Lu Y, Yang X, Zheng X, Wen S, Wang F, Vidal X, Zhao J, Liu D, Zhou Z, Ma C, Zhou J, Piper JA, Xi P, Jin D (2017) Amplified stimulated emission in upconversion nanoparticles for super-resolution nanoscopy. *Nature* 543: 229.

73. Zhan Q, Liu H, Wang B, Wu Q, Pu R, Zhou C, Huang B, Peng X, Ågren H, He S (2017) Achieving high-efficiency emission depletion nanoscopy by employing cross relaxation in upconversion nanoparticles. *Nat. Commun.* 8: 1058.

74. Peng X, Huang B, Pu R, Liu H, Zhang T, Widengren J, Zhan Q, Ågren H (2019) Fast upconversion super-resolution microscopy with 10 μs per pixel dwell times. *Nanoscale* 11: 1563.

75. Chen C, Wang F, Wen S, Su QP, Wu MCL, Liu Y, Wang B, Li D, Shan X, Kianinia M, Aharonovich I, Toth M, Jackson SP, Xi P, Jin D (2018) Multi-photon near-infrared emission saturation nanoscopy using upconversion nanoparticles. *Nat. Commun.* 9: 3290.

76. Liu Y, Wang F, Lu H, Fang G, Wen S, Chen C, Shan X, Xu X, Zhang L, Stenzel M, Jin D (2020) Super-resolution mapping of single nanoparticles inside tumor spheroids. *Small* 16: 1905572.

77. Wu Y, Shroff H (2018) Faster, sharper, and deeper: Structured illumination microscopy for biological imaging. *Nat. Methods* 15: 1011.

78. Heintzmann R, Gustafsson MGL (2009) Subdiffraction resolution in continuous samples. *Nat. Photonics* 3: 362.

79. Liu B, Chen C, Di X, Liao J, Wen S, Su QP, Shan X, Xu ZQ, Ju LA, Mi C, Wang F, Jin D (2020) Upconversion nonlinear structured illumination microscopy. *Nano Lett.* 20: 4775.

80. Huang X, Han S, Huang W, Liu X (2013) Enhancing solar cell efficiency: The search for luminescent materials as spectral converters. *Chem. Soc. Rev.* 42: 173.

81. Ferreira RAS, Correia SFH, Monguzzi A, Liu X, Meinardi F (2020) Spectral converters for photovoltaics – What's ahead. *Mater. Today* 33: 105.

82. Yu J, Yang Y, Fan R, Wang P, Dong Y (2016) Enhanced photovoltaic performance of dye-sensitized solar cells using a new photoelectrode material: Upconversion YbF_3-Ho/TiO_2 nanoheterostructures. *Nanoscale* 8: 4173.

83. Tian L, Shang Y, Hao S, Han Q, Chen T, Lv W, Yang C (2018) Constructing a "native" oxyfluoride layer on fluoride particles for enhanced upconversion luminescence. *Adv. Funct. Mater.* 28: 1803946.

84. Hao S, Shang Y, Li D, Ågren H, Yang C, Chen G (2017) Enhancing dye-sensitized solar cell efficiency through broadband near-infrared upconverting nanoparticles. *Nanoscale* 9: 6711.

85. Park NG (2015) Perovskite solar cells: An emerging photovoltaic technology. *Mater. Today* 18: 65.

86. Roh J, Yu H, Jang J (2016) Hexagonal β-$NaYF_4$:Yb^{3+}, Er^{3+} nanoprism-incorporated upconverting layer in perovskite solar cells for near-infrared sunlight harvesting. *ACS Appl. Mater. Interfaces* 8: 19847.

87. Wang X, Zhang Z, Qin J, Shi W, Liu Y, Gao H, Mao Y (2017) Enhanced photovoltaic performance of perovskite solar cells based on Er-Yb co-doped TiO_2 nanorod arrays. *Electrochim. Acta* 245: 839.

88. Li Y, Zhao L, Xiao M, Huang Y, Dong B, Xu Z, Wan L, Li W, Wang S (2018) Synergic effects of upconversion nanoparticles NaYbF4:Ho^{3+} and ZrO_2 enhanced the efficiency in hole-conductor-free perovskite solar cells. *Nanoscale* 10: 22003.

89. Schoenauer Sebag M, Hu Z, de Oliveira Lima K, Xiang H, Gredin P, Mortier M, Billot L, Aigouy L, Chen Z (2018) Microscopic evidence of upconversion-induced near-infrared light harvest in hybrid perovskite solar cells. *ACS Appl. Energ. Mater.* 1: 3537.

10 Upconversion Nanophosphors for Bioimaging

Jiating Xu and Jun Lin

CONTENTS

10.1 INTRODUCTION

Modern biomedical imaging techniques are extremely important tools in clinical diagnosis and therapy assessment, allowing the noninvasive, highly sensitive and specific observation of pathological and physiological events associated with human diseases in living subjects [1–3]. In the past several decades, a diversity of imaging modalities, such as magnetic resonance (MR) imaging (MRI), X-ray computed tomography (CT) imaging, fluorescence imaging, photoacoustic (PA) imaging and positron emission tomography (PET), combined with appropriate probes, have been well developed and greatly expedited the prosperity of bioimaging in preclinical and clinical practice. Among these modalities, fluorescence imaging has received considerable attention in biomedical research owing to its nonhazardous radiation, quick feedback, high sensitivity, low cost, and so forth. Significantly, fluorescence imaging is highly promising for realizing accurate information feedback at the molecular level. Moreover, this technique enables the final images to be obtained within milliseconds and with a resolution of up to tens of nanometers [4–7].

In the technique of optical bioimaging, an excitation light is used to radiate the specimen and another beam of light emitted from the specimen is collected by a camera. When the excitation photons penetrate the biotissues, four processes of interface reflection, in-tissue scattering, in-tissue

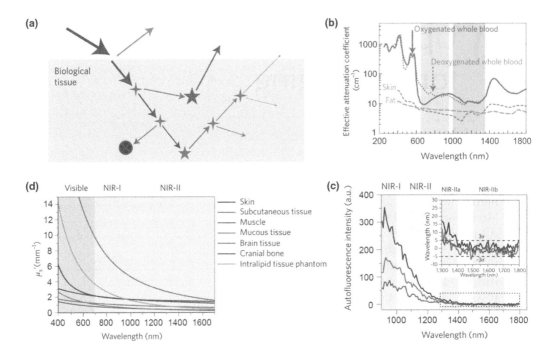

FIGURE 10.1 The general path of excitation photons in fluorescence bioimaging: Light-tissue interactions resulting from impinging excitation light (blue). Interface reflection (cyan), scattering (green), absorption (black circle with purple cross) and autofluorescence (brown) all contribute to the loss of signal (fluorescence, red) and the gain of noise (a). Plots of the effective attenuation coefficient (on a log scale) versus wavelength show absorption and scattering from different biocomponents (b). Autofluorescence spectra of *ex vivo* mouse liver (black), spleen (red) and heart tissue (blue) under 808 nm excitation light. Inset: Autofluorescence spectra at high wavelengths (c). Reduced scattering coefficients of different biotissues and of an intralipid scattering tissue phantom as a function of wavelength in the 400- to 1700 nm region, which covers the VIS, NIR-I and NIR-II ranges (d). (a), (c) and (d) Reprinted with the permission from Ref. [8]. Copyright 2017 Nature Publishing Group. (b) Reprinted with the permission from Ref. [9]. Copyright 2009 Nature Publishing Group.

absorption and tissue autofluorescence related to light-tissue interactions, should be considered (Figure 10.1(a)). For conventional fluorescence imaging, the excitation light is usually visible (VIS) light, which has shallow tissue penetration, thus conventional optical imaging techniques mainly focus on superficial lesions and often require tumor-bearing regions to be exposed surgically. Due to the existent absorption and scattering, the excitation light will be greatly attenuated when it penetrates biotissues. Biological molecules show strong absorptions in the VIS and ultraviolet (UV) regions (Figure 10.1(b)); therefore, tissue absorption occurs mainly in these spectral ranges. Thus, the near-infrared (NIR) window is named biological transparency window. Besides, for almost all biotissues, it has been demonstrated that there are decreased tissues scattering at longer wavelengths (Figure 10.1(c)). Taken together, the deeper tissue penetration can be achieved using NIR excitation light. Moreover, when the NIR excitation is used, the tissue autofluorescence can be also greatly alleviated, allowing for a much higher image contrast (Figure 10.1(d)). In conclusion, for NIR fluorescence imaging, both excitation and emission light with wavelengths > 700 nm facilitate deep tissue imaging with a high signal-to-background ratio (SBR).

From the perspective of optimizing excitation light source, the rare-earth (RE) upconversion nanophosphors (UCNPs) with NIR (700–1700 nm) excitation laser show the best performance. Interestingly, as for UCNPs, the abundant f-orbital configurations coupled with their unique electronic structures endow the RE ions in the solid materials rich energy levels varying from UV to VIS and NIR (Figure 10.2). Due to the effective shielding of the 4f orbitals caused by the filled

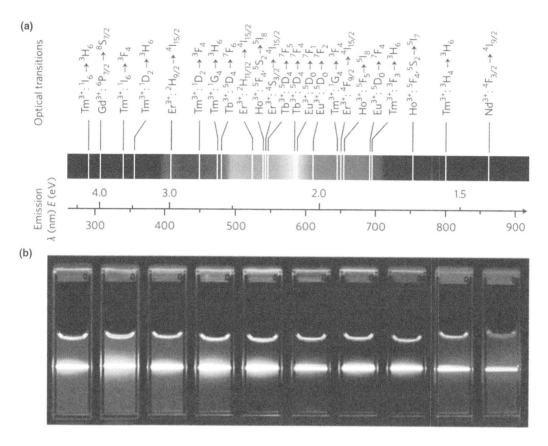

FIGURE 10.2 RE UC emission bands covering a broad range of wavelengths from UV to NIR and their corresponding main optical transitions (a). Compiled luminescent photos showing fine-tuned VIS to NIR emitting UCNPs (b). (a) Reprinted with the permission from Ref. [15]. Copyright 2015 Nature Publishing Group. (b) Reprinted with the permission from Ref. [16]. Copyright 2014 American Chemical Society.

5s and 5p orbitals, the energy levels of the RE ions are well defined, and the spectra derived from inner 4f–4f transitions are acuminate and insensitive to outer environment; all of these features enable sharp and tunable emissions of UCNPs with high photostability [10, 11]. It is worth noting that the emissions of RE ions involve only atomic transitions, thus they are extremely resistant to photobleaching [10]. What is more, the emission line width of UCNPs is only ~10–20 nm (full width at half maximum), which is obviously narrower than that of transition metal ions (~100 nm) and quantum dots (QDs) (~25–40 nm) [12–14]. These attractive characteristics make UCNPs promising as optical bioimaging probes.

Recently, the focus of UC fluorescence imaging has shifted from the traditional NIR-I-excited UC-emissive nanoconstructs to the NIR-II window (1000–1700 nm) excitation [17–19]. The NIR-II excitation outperforms the NIR-I excitation due to the further decreased photon scattering, absorption and tissue autofluorescence [20, 21]. Thus, in this chapter, we focus on the recent advancement of UCNPs that can be excited by the NIR laser and put an emphasis on that with NIR-II excitation. First, the advantages of NIR-excited UCNPs for fluorescence imaging will be introduced. After that, recent reports regarding the UCNPs-based bioimaging including the applications of UCNPs for single-modal upconversion luminescence (UCL) optical imaging, UCL-based multimodal imaging, nanoscopy imaging, multiplexed imaging, ratiometric imaging, hypoxia imaging, potassium imaging, myeloperoxidase (MPO) imaging, mRNA imaging and neuronal activity imaging will be summarized. Finally, the challenges and the future development directions of the UCNPs-based bioimaging will be discussed, and an emphasis will be put on how to use UCNPs to realize the integration of disease diagnosis and treatment.

10.2 MECHANISMS OF RE UCL

With abundant energy levels of 4f electron configurations, RE ions can undergo various UC processes, and the corresponding UC mechanisms can be summarized as excited state absorption (ESA), energy transfer upconversion (ETU), photon avalanche (PA), cooperative energy transfer (CET) upconversion, and energy migration-mediated upconversion (EMU) [22]. In an ESA process (Figure 10.3(a)), RE ions with multiple energy levels can undergo the successive absorption of two or more low-energy photons, resulting in the transition from the ground to excited state, and further to a higher excited state. High-energy photons could be released within such transitions. ETU process includes two types of luminescent centers, a sensitizer and an activator (Figure 10.3(b)). The absorption cross section of the sensitizer is usually larger than that of the activator. Upon excitation with the pump photons, the excited sensitizer transfers energy to adjacent activators resonantly. UC emission is generated from the activator when electrons drop back to the ground state.

PA process was discovered using Pr^{3+} ion-based infrared quantum counters (Figure 10.3(c)). The energy gap between the intermediate state and the ground state is in a mismatch with the energy of the pump photon. Once electrons are excited to the intermediate state, an ESA process is likely to occur to populate the higher excited state. Subsequently, resonant cross-relaxation (CR) takes place

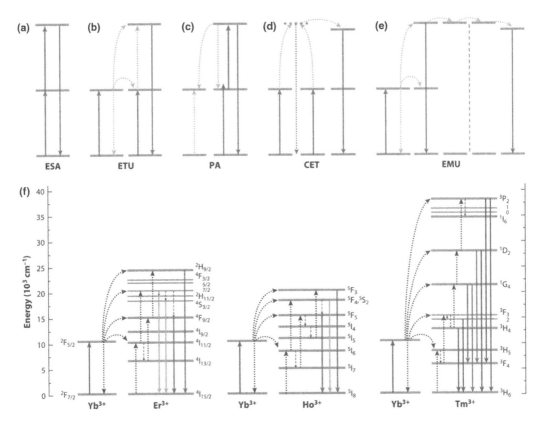

FIGURE 10.3 Basic energy transfer mechanisms of RE-based UC emissions (a)–(e): ESA (a), ETU (b), PA (c), CET UC (d), and EMU UC (e). Energy-level diagrams and proposed UC energy transfer pathways in the Yb^{3+}–Er^{3+}, Yb^{3+}–Ho^{3+} and Yb^{3+}–Tm^{3+} pairs (f). Adapted with permission form Ref. [22]. Copyright 2015 Annual Reviews Incorporated.

between the superexcited ion and adjacent ground state ion, yielding two ions in the intermediate state. Repeating the CR process, exponential population of the intermediate state is sure to occur, along with excitation above the threshold. In this case, PA-induced UC emissions are readily produced as long as the consumption of superexcited ions is less than that of ground state ions.

Similar to ETU, two types of luminescent centers are required in the CET process (Figure 10.3(d)): a cooperative sensitizer and activator. The main difference between the two processes is the absence of adequate long-lived intermediate energy levels in the activators in CET. In CET, UC emission results from simultaneous energy transfer from two sensitizers to one activator. EMU mechanism was proposed based on energy transfer within core-shell nanostructures [23]. An EMU process incorporates four types of luminescent centers with defined concentrations into separated layers: a sensitizer, accumulator, migrator and activator (Figure 10.3(e)). Upon excitation with low-energy photons, an ETU process occurs, populating the higher excited state of the accumulator. Then the energy is donated to an adjacent migrator in the same region, followed by further energy transfer through the core-shell interface to a migrator in the neighboring region. Finally, the energy is trapped by an activator, giving out UC emissions as electrons drop back to the ground state. Meanwhile, UC emissions from the accumulator ions can also occur.

Among five types of UC emission mechanisms, the emission efficiency of ETU-motivated UCNPs is relatively high and overwhelmingly dominate correlative studies. Among the RE ions, the Yb^{3+} ion is the best choice for the sensitizer. In addition, the absorption cross section of the Yb^{3+} ion is $9.11 \times 10-21$ cm^{-2} (976 nm; $^2F_{7/2} \rightarrow 2F_{5/2}$), larger than most of the RE ions. More importantly, the energy-level diagram of the Yb^{3+} ion is quite simple, with only one excited state of $^2F_{5/2}$, which matches well with those of many RE ions. In Yb^{3+}–Ln^{3+}-codoped ETU-based UCNPs, Yb^{3+} absorbs NIR light ($^2F_{7/2} \rightarrow ^2F_{5/2}$) and then donates the obtained energy to adjacent Ln^{3+}. As for the activators, the Er^{3+}, Ho^{3+} and Tm^{3+} ions are three typical activators for the ETU process owing to their ladder-like arrangement of energy levels and excellent level matching with the Yb^{3+} ion (Figure 10.3(f)). The doping content of the Yb^{3+} ion is usually kept at 20% or higher, whereas that of the activators is lower than 2%.

10.3 ADVANTAGES OF UCNPS FOR OPTICAL BIOIMAGING

10.3.1 HIGH PENETRATION DEPTH

During the past several decades, the 980- to 800 nm laser-excited UCNPs have been extensively studied for *in vivo* UCL bioimaging. The excitation wavelengths of UC process are located within the optical transmission window of biotissues, thus the UCL *in vivo* imaging based on RE UCNPs affords a higher penetration depth. As previously reported, Chen et al. reported that UC optical images from a suspension of $NaYbF_4$:Yb,Tm@CaF_2 core-shell nanocrystals can be successfully acquired through 3.2 cm of pork tissue [24]. Though great progress has been made, the penetration depths need to be improved for some *in vivo* bioimaging applications. Note that, the recent attention has been shifting to develop NIR-II-to-NIR-II fluorescence imaging based on RE UCNPs for further improving penetration depth in biotissues [25].

10.3.2 LOW PHOTOBLEACHING

The 4f orbitals of RE ions are shielded by the 6s, 5p and 5d orbitals, thus their spectra caused by f–f transitions are narrow and insensitive to the surrounding environment. In addition, the energy levels of RE ions enable large shifts between the excitation and emission bands, which can be hundreds of nanometers, containing discrete gaps without absorption. Furthermore, the RE emissions involve only atomic transitions, thus they are highly resistant to photobleaching. The nature of the inorganic host lattice of RE UCNPs brought about its high photostability. Overall, the low vibrational energy losses, independence of host material and the large variety of absorption and emission wavelengths make RE ions ideal for UC spectral conversion. When longer wavelength excitation of NIR laser is used for RE UCNPs-based nanosystems, higher photostability is observed with the fluorescence of the annihilator as the detection signal [26]. By contrast, for some organic UC nanoprobes, upon the excitation of shorter wavelength (higher energy), laser will damage the structure of the corresponding annihilator molecules and, hence, give rise to the photobleaching phenomenon.

10.3.3 LOW DETECTION LIMITATIONS

The minimum number of cells detected in living animals by optical imaging is an important parameter in the evaluation of the sensitivity of an *in vivo* imaging technique. When the RE UCNPs were utilized for optical bioimaging, there is nearly no autofluorescence of biological samples, thereby realizing the high signal-to-noise ratio (SNR), which is the root cause of low detection limitations. Xu et al. reported that around ten mesenchymal stem cells labeled with $NaYF_4$:Yb,Er UCNPs could be detected by *in vivo* imaging with 50 nm particles [27]. Sun et al. have validated that when human nasopharyngeal epidermal carcinoma cell labeled with sub-10 nm β-$NaLuF_4$:Yb,Tm NPs were injected into a mouse subcutaneously or intravenously, low detection limits were achieved for *in vivo* whole-body bioimaging with UC emission SNR of 3 and 10, respectively [28].

10.3.4 MULTIPLEX UC BIOIMAGING

Considering that the currently developed RE UCNPs have the optical features of diversity of excitation laser source and emission color, tunability of fluorescence intensity and temporal lifetime, thus the UC *in vivo* multiplexed bioimaging can be easily achieved. Note here, these optical parameters can not only work by themselves to be multiplexing encoders but also ally with each other to remarkably enrich the encoding capability of UCNPs. Chatterjee et al. demonstrated the multiplexed lymphangiography of three groups of lymph nodes using three different RE UCNPs with multiwavelength emissions [29]. Along with the rapid development of optical multiplexing, more and more research interests have been already shifted to develop RE UCNPs-based optical encoding vehicles.

10.3.5 HIGH BIOSAFETY

The toxicity of the highly crystalized RE-based to biosome is extremely low especially for those having proper surface modifications [30]. Here, we briefly clarify the suitability of NIR-excited UCNPs for optical bioimaging. First, the UCNPs with proper surface modification pose high biocompatibility. Second, due to the high degree of crystallization, UCNPs perform a good chemical stability, to some extent, which contributes to the biocompatibility as well. Furthermore, the chemical stability also ensures the particles' high photostability with non-photobleaching. The last but not the least, the biotissues have ignorable NIR absorption, and that's why the issues of phototoxicity and overheating can be obviated.

10.4 UCNPS FOR BIOIMAGING

Table 10.1 summarizes the recent works about RE UCNPs for bioimaging study. In the following section, the typical works about using UCNPs for single-mode UCL imaging, UCL-participated multimodal imaging, nanoscopy imaging, multiplexed imaging, ratiometric imaging, hypoxia imaging, potassium imaging, MPO imaging, mRNA imaging and neuronal activity imaging will be introduced in detail.

10.4.1 SINGLE-MODE OPTICAL IMAGING

As early as in 2008, Zhang et al. demonstrated that polyethyleneimine (PEI)-modified $NaYF_4$:Yb/Er UCNPs with the diameter of ~50 nm could be employed for *in vivo* imaging in anesthetized

TABLE 10.1
Typical Examples of UCNPs Studied for Bioimaging

UCNPs	Ex/Em Wavelength	Application	Reference
PEI-NaYF$_4$:Yb,Er	980/540, 653	VIS imaging	[29]
NaYF$_4$:2%Tm,20%Yb	975/800	NIR-I imaging	[31]
PEG-Y$_2$O$_3$:Yb,Er	980/660	VIS imaging	[32]
NaYF$_4$:Gd,Er/Yb/Eu	975/649, 653, 667	VIS and MR imaging	[33]
NaYF$_4$:20%Yb,1.8%Er,0.2%Tm-PEG-RGD	980/530, 650	VIS imaging	[34]
NaYF$_4$:Yb,Tm	980/800	NIR-I imaging	[27]
RGD-PEG-Er-Y$_2$O$_3$	980/660–740	VIS imaging	[35]
Citrate-NaYF$_4$:Yb,Er	980/520, 540, 654	VIS imaging	[36]
OA-PAA-PEG-NaYF$_4$:Yb,Er(Tm)	980/550, 650	VIS multiplexed imaging	[37]
NaYF$_4$:18%Yb,2%Er-AD	980/521, 540	VIS imaging	[38]
PAA-NaYF$_4$:20%Yb,1%Tm	980/475, 695, 800	NIR-I imaging	[39]
AA-NaGdF$_4$:Tm,Er,Yb	980/800	NIR-I and MR imaging	[40]
NaYF$_4$:Yb,Er-FA	980/640–680	VIS imaging	[41]
FA-LaF$_3$:Yb,Ho/Tm-OAAA	980/540–580	VIS imaging	[42]
CaF$_2$:Tm,Yb	920/800	NIR-I imaging	[43]
NaYF$_4$:Yb,Tm	980/800	NIR-I multiplexed imaging	[44]
NaYF$_4$:Yb,Er-PEI	980/531–561, 633–685	VIS imaging	[45]
NaYF$_4$:20%Yb,2%Er-OA-CD	980/540, 660	VIS, FL and PET imaging	[46]
N719-NaYF$_4$:20%Yb,1.6%Er,0.4%Tm	980/520–560	Ratiometric imaging	[47]
^{18}F/FA-aAA-Gd-NaYF$_4$:Yb,Er	980/521, 540, 654	VIS CT, MR and PET imaging	[48]
β-NaLuF$_4$:24%Gd,20%Yb,1%Tm	980/800	NIR-I imaging	[49]
NaYF$_4$:Yb,Tm@NaYF$_4$:Yb,Tm and NaYF$_4$:Yb,Tm@ NaYF$_4$:Yb,Tm@NaYF$_4$	980/800	NIR-I imaging	[50]
^{18}F-Y$_2$O$_3$,^{18}F-NaYF$_4$,^{18}F-Y(OH)$_3$	980/475, 800	NIR-I and PET imaging	[51]
KGdF$_4$:Tm,Yb@KGdF$_4$@PEG-phospholipid	980/803	NIR-I and MR imaging	[52]
NaYF$_4$:Yb,Tm@Fe$_x$O$_y$	980/800	NIR-I and MR imaging	[53]
YOF:Yb,Er@YOF	980/669	VIS imaging	[54]
NaYbF$_4$:Tm,Er,Ho	915/800	NIR-I imaging	[55]
NaYF$_4$:Yb,Tm	980/800	NIR-I imaging	[56]
α-NaYbF$_4$:Tm@CaF$_2$	980/800	NIR-I imaging	[24]
NaYF$_4$:Yb,Tm	980/291, 350, 362, 450, 479, and 648	VIS imaging	[57]
NaYF$_4$:Yb/Er@SiO$_2$@P(NIPAM-co-MAA)	980/523, 541	VIS imaging	[58]
Mn^{2+}-doped NaYF$_4$:Yb,Er	980/650–670	VIS imaging	[59]
β-NaYF$_4$:Yb,Er(Tm)-transferrin	980/543	VIS imaging	[60]
NaLuF$_4$:Yb,Tm@SiO$_2$-GdDTPA	980/800	NIR-I, MR and CT imaging	[61]
PEG-NaYbF$_4$:Tm	980/800	NIR-I and CT imaging	[62]
PAA-NaLuF$_4$:Yb,Tm	980/800	NIR-I imaging	[63]
GdF$_3$:Yb,Er,Li@silica	980/671	VIS and MR imaging	[64]
GdVO$_4$:Yb,Ho@Si	980/648, 658, 540	VIS and MR imaging	[65]
Amine-BaGdF$_5$:Yb/Er	980/500–600, 600–700	VIS and MR imaging	[66]
PEG-BaGdF$_5$:Yb,Er	980/500–600, 600–700	VIS, MR and CT imaging	[67]
NaLuF$_4$:Gd,Yb,Er/Tm	980/520–560, 640–680	VIS, MR and CT imaging	[68]
Gd$_2$O$_3$:Yb,Er/Tm	980/800	NIR-I and MR imaging	[69]

(Continued)

TABLE 10.1 (*Continued*)
Typical Examples of UCNPs Studied for Bioimaging

UCNPs	Ex/Em Wavelength	Application	Reference
Y_2O_3:Er,Yb@SiO_2@LDH-5FU	980/640–700	VIS imaging and chemotherapy	[70]
$NaYF_4$:Yb,Tm@$NaGdF_4$:Yb	980/800	NIR-I, MR, CT imaging and PDT	[71]
$NaYF_4$:Yb,Tm@$NaYbF_4$:Er@$NaYF_4$:Yb,Tm	980/653	VIS imaging	[72]
PEG-$NaGdF_4$:Yb,Er	980/521, 541, 655	VIS and MR imaging	[73]
$BaYF_5$:Yb,Er	980/515–555	VIS and CT imaging	[74]
$NaYF_4$:Tm,Yb@$NaYF_4$@$mSiO_2$	980/800	NIR-I imaging and drug delivery	[75]
$NaYF_4$@$NaLuF_4$:Yb,Er/Tm	980/800	NIR-I imaging	[76]
hCy7-$NaYF_4$:Yb,Er,Tm	980/534–560, 635–680 and 800	Ratiometric imaging	[77]
PAA-$BaYbF_5$:Tm	980/800	NIR-I, X-ray and CT imaging	[78]
$NaYF_4$:Yb,Tm@$NaLuF_4$@$NaYF_4$@$NaGdF_4$	980/800	NIR-I, MR and CT imaging	[79]
Cit-[153]Sm-$NaLuF_4$:Yb,Gd,Tm	980/800	NIR-I and SPECT imaging	[80]
$NaLuF_4$:Yb,Tm	980/800	NIR-I and CT imaging	[28]
$NaGdF_4$:Yb,Er@$NaGdF_4$:Nd,Yb	808/540, 650	VIS imaging	[81]
$NaLuF_4$:[153]Sm,Yb,Tm	980/400–500	VIS and SPECT imaging	[82]
$NaYF_4$:Yb,Er (Ho or Tm)	920/800	NIR-I imaging and PDT	[83]
Phospholipid-polyethyleneglycol (DSPE-PEG)-$NaYbF_4$:Gd,Tm	980/800	NIR-I imaging	[84]
$NaYF_4$:Tm,Yb	975/476, 650, 700	VIS imaging	[85]
Ad-RGD@βPCD@$NaYF_4$:Yb,Er	980/540	VIS imaging	[86]
$NaYF_4$:59%Yb^{3+},0.5%Tm^{3+}@$NaYF_4$:20%Gd^{3+}-PAA	980/800	NIR-I, MR and CT imaging	[87]
NaErF4:Yb	980/664	VIS, MR and CT imaging	[88]
$NaYF_4$:Yb,Er-Cys and $NaYF_4$:Yb,Tm-Cys	980/520–560, 620–670 and 800	VIS imaging NIR imaging	[89]
$NaYF_4$:Yb^{3+},Er^{3+}-PASP	915/670	VIS imaging	[90]
$NaYF_4$:Yb^{3+},Tm^{3+},Co^{2+}-OCC	980/800	NIR-I and MR imaging	[91]
$BaGdF_5$:20%Yb^{3+}/2%Tm^{3+}@$BaGdF_5$:2%Yb^{3+}-$NHNH_2$-DOX	980/800	NIR-I, CT, MRI and drug delivery	[92]
BSA-dextran-$NaYF_4$:Yb,Er,Tm	980/540	VIS imaging	[93]
Zwitterionic phospholipids modified $NaGdF_4$:Yb^{3+}/Er^{3+}	980/514–560, 635–680	VIS imaging	[94]
PEG-$NaLuF_4$:Yb/Er	980/515–555	VIS and X-ray imaging	[95]
PEI-$NaYbF_4$:Er	980/500–600, 600–700	VIS and X-ray imaging	[96]
Ag@SiO_2@Lu_2O_3:Gd/Yb/Er	980/520–560, 625–700	VIS imaging	[97]
α-CD-CaF_2:Yb,Er@CaF_2	980/540	VIS, CT imaging and drug delivery	[98]
$NaLuF_4$:Gd/Yb/Er	980/500–600, 600–700	VIS and X-ray imaging	[99]
$NaLuF_4$:Mn/Yb/Er	980/600–700	VIS imaging	[100]
$NaYF_4$:Yb/Er-g-HPG-RB	980/542, 588	VIS imaging	[101]
CHCl-$NaYF_4$:Yb/Er/Tm	980/541, 654, 800	H_2S imaging	[102]
POSS-$NaYF_4$:Yb,Er@$NaGdF_4$	980/500–600	VIS and MR imaging	[103]
$NaYF_4$:Yb,Er@$NaYF_4$:Yb@$NaNdF_4$:Yb@$NaYF_4$:Yb@PAA	808/540	VIS imaging and drug delivery	[104]

(Continued)

TABLE 10.1 (*Continued*)
Typical Examples of UCNPs Studied for Bioimaging

UCNPs	Ex/Em Wavelength	Application	Reference
GdOF:Yb/Er/Mn@mSiO$_2$-ZnPC-CDs	980/540, 650	VIS, CT, MR and PT imaging-guided PDT, PTT and chemotherapy	[105]
NaYF$_4$:Yb/Tm@NaYF$_4$@mSiO$_2$-Ir	980/477, 600	Hypoxia imaging	[106]
NaGdF$_4$:Yb^{3+}/Er^{3+}@NaGdF$_4$:Nd^{3+}@ Sodium-Gluconate	980/525, 540, 600, 656	VIS, MR and CT imaging	[107]
Porphyrin-phospholipid-NaYbF$_4$:Tm-NaYF$_4$	980/800	NIR-I, FL, CT, PET, CL and PA imaging	[108]
NaYF$_4$:Yb/Ho@NaYF$_4$:Nd@NaYF$_4$-RB	980/650	VIS imaging and PDT	[109]
PAA-NaYF$_4$:Gd/Yb/Er	980/520, 540, 660	VIS, CT and MR imaging	[110]
PEG/RB-Pt(IV)-NaGdF$_4$:Yb/Nd@ NaGdF$_4$:Yb/Er@NaGdF$_4$	808/550, 650	VIS imaging, PDT and chemotherapy	[111]
LiYF$_4$:Yb^{3+}/Tm^{3+}@SiO$_2$@GPS@CH/PhL/ FITC-BSA/PEGBA	980/792	NIR imaging and macromolecular delivery	[112]
NaYF$_4$:Yb^{3+},Er^{3+}-anti-hIgG and NaYF$_4$:Yb^{3+},Tm^{3+}-anti-hIgM	980/520–560 and 450–480	VIS multiplexing	[113]
Citrate-coated NaGdF$_4$:Yb^{3+},Tm^{3+},Bi^{3+}	980/800	NIR-I and CT imaging	[114]
NaYF$_4$:Gd^{3+},Tm^{3+}@NaGdF$_4$	1064/800	NIR-I imaging	[115]
NaYF$_4$:Yb^{3+}/Ho^{3+}/Mn^{2+}	980/645	VIS imaging	[116]
NaGdF$_4$:Yb^{3+},Tm^{3+}@NaGdF$_4$@Azo-Lipo	980/800	NIR-I imaging and drug delivery	[117]
RGDS(p)-NaGdF$_4$:Yb,Er@NaGdF$_4$	980/500–600, 600–700	VIS imaging	[21]
BaYbF$_5$:Gd/Er	980/520, 545, 660	VIS, CT and MR imaging	[118]
Silica-coated NaYF$_4$:30%Yb,2%Tm (20 nm), NaYF$_4$:30%Yb,2%Tm,30%Lu (34 nm) and NaYF$_4$:25%Yb,0.5%Tm (23 nm)	980/800	NIR-I multiplexing	[119]
NaYF$_4$:20%Yb,8%Tm	980/455	Nanoscopy imaging	[120]
NaYF$_4$:Yb^{3+},Er^{3+}-pHrodo Red	980/550, 590	Ratiometric imaging	[121]
PEG-Zn$_{1.1}$Ga$_{1.8}$Ge$_{0.1}$O$_4$:Cr-β-NaYbF$_4$:Tm@ NaYF$_4$	980/700	VIS imaging	[122]
NaYF$_4$:Yb/Er	980/651	VIS imaging	[123]
Li(Gd,Y)F$_4$:Yb,Er/LiGdF$_4$-PAA	980/485–585	VIS imaging	[124]
NaYF$_4$:Gd^{3+},Yb^{3+},Er^{3+}@TRF-PpIX	980/522, 543, 654	VIS imaging	[125]
K$_{0.3}$Bi$_{0.7}$F$_{2.4}$:20%Yb^{3+}/0.5%Tm^{3+}	980/740–840	NIR-I imaging	[126]
NaYF$_4$:20%Yb^{3+},4%Tm^{3+}	980/800	NIR-I nanoscopy imaging	[127]
Ag@PVP-Ba$_2$GdF$_7$:Yb^{3+},Ho^{3+}	980/540, 650	VIS, CT and PTT	[128]
RB-FC-NaYF$_4$:Yb^{3+},Er^{3+},Tm^{3+}	980/500–560, 600–700	Ratiometric imaging	[129]
TPAMC-NaYF$_4$:Yb,Er,Tm@PEG	980/600–680	Ratiometric imaging	[130]
NaErF$_4$:2%Ho@NaYF$_4$	1530/1180	Ratiometric imaging	[25]
NaYF$_4$:Yb^{3+},Tm^{3+}	980/450–500	VIS multiplexing	[131]
NaGdF$_4$:Yb,Er@NaGdF$_4$:Yb @Ce6@mSiO$_2$@ mMnO$_2$-PEG	980/530–570, 630–680	VIS, MR and CT imaging-guided PDT, CDT and chemotherapy	[132]
NaYF$_4$:Yb^{3+},Er^{3+}@NaGdF$_4$@SiO$_2$-COOH	980/522, 542, 654	VIS imaging	[133]
NaYF$_4$:Yb,Er@NaYF$_4$-Tz/FA-PEG	980/654	VIS imaging and PDT	[134]
NaYF$_4$:Yb^{3+},Er^{3+}@silica-N=FA	980/650	VIS imaging	[135]

(Continued)

TABLE 10.1 (*Continued*)

Typical Examples of UCNPs Studied for Bioimaging

UCNPs	Ex/Em Wavelength	Application	Reference
$NaErF_4$:10%Yb@$NaYF_4$:40%Yb@ $NaNdF_4$:10%Yb@$NaGdF_4$:20%Yb@ SiO_2-UEA-I	808/650	VIS imaging	[136]
PAAO-$NaYF_4$:Yb/Nd/Er@ $NaYF_4$:Nd-DCM-H_2O_2	808/515–560, 640–675	Ratiomereic imaging	[137]
FeC-$NaGdF_4$:Yb^{3+},Tm^{3+}@$NaGdF_4$-PEG	980/800	NIR-I, MRI and PTT	[138]
Yb/Tm/GZO@SiO_2	980/640–660	VIS imaging	[139]
Yb^{3+}/Tm^{3+} coupled UCNPs	980/365	mRNA imaging	[140]
$NaYF_4$:Er,Yb@$NaYF_4$@RB	980/540, 654	VIS imaging and PDT	[141]
$NaErF_4$@$NaYF_4$	1540/540, 654	VIS multiplexing	[142]
$NaYF_4$@$NaYbF_4$@$NaYF_4$:Yb^{3+}/Tm^{3+}@$NaYF_4$	980/808	NIR-I multipexing	[143]
$NaYF_4$:Yb,Tm@$NaYF_4$-[Ru(dpp)$_3$]$^{2+}$Cl$_2$	980/600–625	Hypoxia imaging	[144]
$NaYF_4$:Yb/Tm@$NaYF_4$:Yb/Nd@ hmSiO$_2$-PBFI	980/500–600	Potassium imaging	[145]
$NaYF_4$:Yb/Gd/Tm-DPA	980/425–500	Neuronal activities imaging	[146]
PC-$NaGdF_4$:Yb,Er	980/652	Myeloperoxidase imaging	[147]

Wistar rats (Figure 10.4) after the rats were subjected to direct injection of nanoparticles under the skin in the groin and upper leg regions [29]. Upon 980 nm NIR laser irradiation, the nanoparticles could be detected up to 10 mm beneath the skin, far deeper than depth through the use of QDs. These results provided promising examples for *in vivo* optical imaging at a deeper tissue level for performing minimal invasive imaging in real time. From then on, more and more UC VIS-emissive UCNPs were invented for single-modality optical imaging [32, 58, 81, 83, 123, 124, 133, 134, 141]. Especially, with the superiority of the deep penetration of the emitted fluorescence, considerable attention has been paid on developing red-emissive UCNPs for optical imaging study [54, 59, 64, 116, 136, 139, 148, 149].

Also in 2008, Prasad et al. synthesized UC NIR-I emissive nanoparticles (NPs) of $NaYF_4$:Yb^{3+},Tm^{3+} and studied their optical imaging properties [31]. Figure 10.5(a) and 10.5(b), respectively, showed the emission spectrum and the TEM image of the obtained NPs. Because of the filled 5s2 and 5p6 electrons in outer shells, the 4f electrons of the RE ions are effectively shielded from the surrounding crystal field; therefore, the spectrum shows the characteristic multiple emission peaks of the Tm^{3+} ion (Figure 10.5(c)). Thereafter, the *in vitro* cellular imaging assay was conducted on human pancreatic cancer cells to determine whether NPs can be used for cellular imaging. Figure 10.5(d) shows the transmission and PL images of cells treated with the NPs following 975 nm light excitation. Interestingly, the localized emission spectrum acquired from the cells showed the characteristic emission peak of the Tm^{3+} ion at 800 nm (inset of Figure 10.5(d)). No autofluorescence was detected, confirming the capability of the Tm^{3+}-activated upconversion nanoparticles (UCNPs) for high-contrast *in vitro* cellular imaging. Figure 10.5(e) presents the *in vivo* whole-body images of a BALB/c mouse injected with the prepared UCNPs. The high-contrast images clearly demonstrate the feasibility to image the NIR-emitting NPs (shown as red). In the right image of Figure 10.5(e), a scan in the range 700–850 nm shows an intense NIR emission peaking at 800 nm. Notably, the signal was readily detectable both through the skin (without hair removal) and after the dissection of the mouse. From Figure 10.5(e), the saturated levels of emission in the spleen and liver can be observed, indicating a high uptake of the developed UCNPs by these two organs. The NIR UC fluorescence imaging results in a high contrast between the emission signal of the UCNPs and the

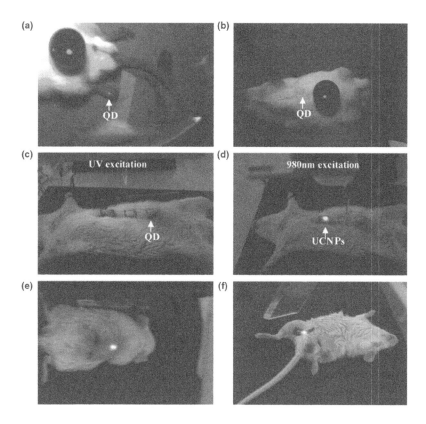

FIGURE 10.4 *In vivo* imaging of rat: Quantum dots (QDs) injected into translucent skin of foot (a) show fluorescence, but not through thicker skin of back (b) or abdomen (c); PEI/NaYF₄:Yb/Er nanoparticles injected below abdominal skin (d), thigh muscles (e) or below skin of back (f) show luminescence. QDs on a black disk in (a) and (b) are used as the control. Reprinted from Ref. [29]. Copyright 2008, with the permission from Elsevier.

background, with light penetration depth up to ~20 mm in mouse. It is worth nothing that the developed UCNPs showed no obvious *in vivo* toxicity after 48-h injection.

The suitability of the α-NaYbF₄:Tm³⁺@CaF₂ NPs for *in vivo* UC NIR-I imaging was examined on mice. The hair on the back of the mouse was shaved, while the hair on the belly remained unshaved. The hyaluronic acid (HA)-coated NPs were injected to a BALB/c mouse intravenously *via* the tail vein, then the BALB/c mouse was imaged under 980 nm laser excitation for *in vivo* emission at 3-h postinjection. The scattered excitation light was cut off by an emission filter in front of the imaging camera objective. As displayed in Figure 10.6(a)–(f), a high-contrast image of the mouse injected with the core-shell NPs was obtained, demonstrating the feasibility to spectrally distinguish the characteristic emission of the NPs. The NIR-I UC fluorescence signal was readily detectable through the skin of unshaved (Figure 10.6(a)–(c)) and shaved (Figure 10.6(d)–(f)) parts of the mouse. Significantly, an intense photoluminescence with an emission peak at 800 nm was easily detected (Figure 10.6(f)). Therein, the SBR is calculated to be 310, and it is of big significance to note that after the mice injected with the HA-coated core-shell NPs were sacrificed, their main organs were excised and imaged without any UCL signal being detected. The biocompatible CaF₂ shell coating brings a 35-fold increase to the total UCL intensity. Also, the high-contrast UCL imaging of deep tissues is achieved using a cuvette with a NP aqueous dispersion covered with 3.2-cm thick pork tissue and a NPs-loaded synthetic fibrous mesh wrapped around rat femoral bone [24]. Since then, Tm³⁺-activated NaYF₄, NaGdF₄, NaLuF₄ and NaYbF₄ UCNPs have been widely studied for NIR-I UC bioimaging [39, 42, 43, 49, 50, 52, 56, 63, 76, 84, 85, 122].

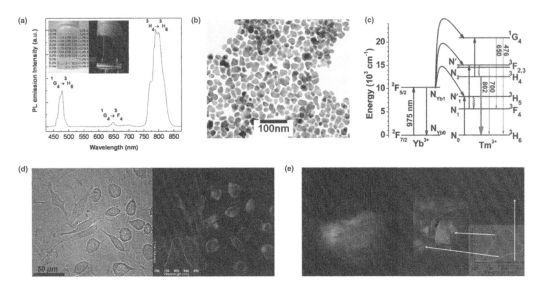

FIGURE 10.5 UC emission spectrum (a), TEM image (b) and proposed emission mechanism (c) of NaYF$_4$:Yb^{3+},Tm^{3+} NPs. *In vitro* transmission (left) and luminescence (right) images of Panc 1 cells treated with NaYF$_4$:Yb^{3+},Tm^{3+} NPs. Inset shows localized emission spectra taken from cells (red) and background (black) (d). Whole-body images of a mouse injected i.v. with NaYF$_4$:Yb^{3+},Tm^{3+} NPs; intact mouse (left), same mouse after dissection (right). The red color indicates emission from UCNPs, green and black show background as indicated by the arrows. The inset presents the PL spectra corresponding to the spectrally unmixed components of the multispectral image obtained with the Maestro system (e). Reproduced with permission from Ref. [31]. Copyright 2008 American Chemical Society.

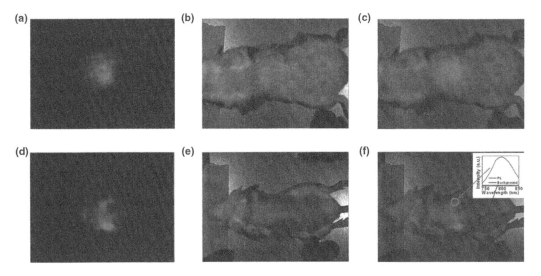

FIGURE 10.6 Whole-animal imaging of a BALB/c mouse injected *via* the tail vein with HA-coated α-NaYbF$_4$:Tm^{3+}@CaF$_2$ core-shell NPs. UCL images (a) and (d), bright-field images (b) and (e) and merged bright-field and UCL images (f). The mouse was imaged in the belly (a)–(c) and the back (d)–(f) positions. Inset in (f) shows the spectra of the NIR UCL and background taken from the circled area. Reproduced with permission from Ref. [24]. Copyright 2012 American Chemical Society.

Actually, it should be pointed out that excitation at ~980 nm poses some intrinsic disadvantages, such as unavoidable heat generation and limited tissues penetration depth. As accepted, water is the main component of the human body and possesses a strong absorption of 980 nm photons. As a result, the 980 nm excitation photon is greatly attenuated when penetrating biotissues; therefore, its penetration depth is significantly decreased. Moreover, nearly all of the excitation energy absorbed by bio-bodies would be transformed into thermal energy, which inevitably causes a detrimental effect to cells and tissues. To compromise this issue, Andersson-Engels et al. proposed an alternative approach by exciting the $NaYbF_4$:Yb^{3+},Er^{3+}/Yb^{3+},Ho^{3+}/Yb^{3+},Tm^{3+} NPs with 915 nm light, and the more efficient NIR-I UC emission was detected [55]. They also demonstrated that a 915 nm excitation laser allows a larger imaging depth and drastically less heating in animals or tissues due to the minimal absorption by water molecules. Through a detailed and reasonable comparison with a 980 nm laser, it can be concluded that less heating of the biological specimen and deeper tissue penetration were achieved in NIR-I fluorescence imaging by using a 915 nm laser. Similarly, Adolfo Speghini et al. validated that 920 nm laser-excited CaF_2:Tm^{3+},Yb^{3+} NPs result in a much deeper penetration imaging power than that of CaF_2:Er^{3+},Yb^{3+} [43].

In 2016, there was a rare work about NIR-II laser (1064 nm)-excited NIR-I UC brain imaging [115]. Therein, the energy-looping NPs (ELNPs) of $NaYF_4$:Gd^{3+},Tm^{3+}@$NaGdF_4$ were studied for a NIR-I optical imaging (Figure 10.7(a)). Upon 1064 nm laser excitation, the $NaYF_4$:Gd^{3+},Tm^{3+} core NPs produced UC emission at 800 nm (Figure 10.7(b)). As demonstrated in Figure 10.7(c), the 1064 nm photon has a higher transmission in brain tissues, which enables the imaging of micron-scale features through millimeters. To validate that the ELNPs could be applied and unambiguously visualized inside cells under 1064 nm excitation, ELNPs with a diameter of 11 nm and a 2 nm inert shell were fed to HeLa cells. After the cell fixation, the confocal microscopy and spectral images acquired with 1064 nm excitation and 800 nm emission present spotted patterns corresponding to endosomal staining with an undetectable background (Figure 10.7(d1)–(d3)). As a control group, an identical procedure that excluded ELNPs was conducted to validate that this emission originated from the ELNPs and not tissue autofluorescence (Figure 10.7(d4)).

To validate that these ELNPs could be imaged systematically through different tissues and phantoms, polystyrene beads were used to cloak them. The optical imaging of ELNPs under 1064 nm excitation was performed on fixed coronal slices of an adult mouse brain (Figure 10.7(e1)). The ELNP-loaded beads were imaged through the frontal cortex of brain slices with thicknesses of 0.5 and 1.0 mm (Figure 10.7(e2)). As shown, the beads were clearly visualized with 1064 nm confocal microscopy through the 1.0-mm brain slice. The line cuts in Figure 10.7(e3) show no resolution loss through the brain slice when imaging with a 1064 nm laser. The energy-looping mechanism employed in this study enables the ELNPs to bypass weak GSA transitions in favor of strong ESA transitions, a concept that has not been realized in optical imaging. Significantly, the ELNPs can be imaged in cells and deep tissue using commonly available continuous wave NIR-II laser excitation and standard silicon detectors These experimental results demonstrate that ELNPs are a promising class of NIR fluorescent probes for high-fidelity visualization in cells and tissue.

During the rapid development period of the UCNPs-based optical imaging, there are some works that aim at surface modification surface modifying/optical engineering of UCNPs [34–36, 38, 41, 45, 57, 60, 86, 89, 90, 93, 94, 97, 101, 104, 135, 150, 151] and followed that the UC optical imaging-guided therapeutic nanotechnology achieved rapid development [70, 75, 109, 111, 112, 117, 125].

10.4.2 UC Optical Imaging-Based Multimodal Imaging

From the aspect of imaging information, biomedical images can be divided into two modes: (i) one mode provides structure information, such as US, MRI and CT; (ii) another mode provides function and molecular information, such as single photon emission computed tomography (SPECT), PET, and florescence. These different imaging methods have their own advantages and disadvantages, for example, CT imaging can picture the internal bone and tissue of the human body, but its imaging

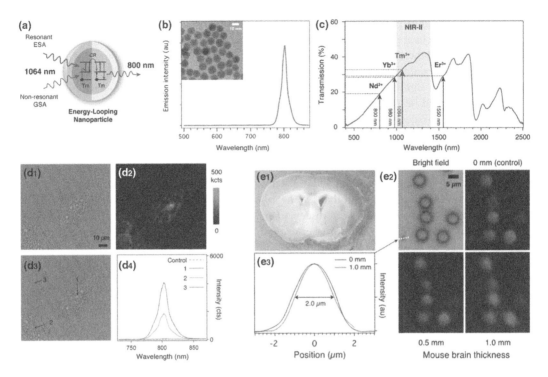

FIGURE 10.7 Schematic illustration of the energy transfer mechanism in the energy-looping NP when excited by a 1064 nm laser (a). Experimental UC emission spectra of $NaYF_4$:Gd^{3+},Tm^{3+} NPs excited at 1064 nm. Inset: TEM of the measured NPs (b). Transmission spectrum through the frontal cortex of a 1-mm-thick, fixed coronal brain slice from an adult mouse. Common NIR excitation wavelengths of Nd^{3+}, Yb^{3+} and Er^{3+} ions are super-imposed, along with the 1064 nm looping excitation line of the Tm^{3+} ion and the NIR-II window between 1000 and 1400 nm (c). Confocal imaging of HeLa cells fixed after incubation with 1.5%Tm^{3+} ELNPs: Bright-field (d1), integrated emission between 740 and 870 nm (d2), and overlaid micrographs (d3) under 1064 nm excitation. Emission spectra (d4) from numbered points in (d3). Note, the absence of a measurable 800 nm emission outside of the cell boundaries (3) and in a vehicle control lacking ELNPs. A representative mouse brain slice through which ELNP beads were imaged (e1). Bright-field and confocal luminescence micrographs of Tm^{3+}-doped ELNP beads imaged through 0, 0.5 and 1 mm-thick brain slices (e2). Intensity line cuts of unbound ELNPs as measured along the white dashed line shown in (e2) and (e3), illustrating two 1 m resolution through 1-mm brain slices. Reproduced with permission from Ref. [115]. Copyright 2016 American Chemical Society.

contrast on soft tissue is relatively low; MRI has high resolution, but low sensitivity [152]. Single-mode imaging often has inherent disadvantages and is unable to provide comprehensive information, which makes it difficult for achieving precise diagnosis. Thus, the integration of different image modes is a nice way to realize their complementary advantages and further to provide more comprehensive diagnostic information. With the development of RE UC optical imaging technology, increasing attention has been paid to inventing optical imaging-based multimodal imaging.

In the field of exploiting UCNPs-based multimodal biomedical imaging contrast agents, the researchers have paid great efforts. For example, as early as in the year of 2009, Prasad's group prepared biorecognition biomolecules-modified UCNPs for combined optical and MRI [33]. Overall, several strategies have been developed to fabricate optical and MRI dual-modal imaging nanostructures based on UCNPs. The first method involves introducing Gd^{3+}/Co^{2+} into the lattice of UCNPs to provide an MRI effect [40, 65, 66, 69, 73, 91, 103]. The second method is to combine UCNPs with other magnetic nanostructures by sequential growth or coating. For example, superparamagnetic iron oxide nanoparticles have been widely used in MRI due to their special magnetic properties and good biocompatibility [53].

In 2012, PEGylated $NaYbF_4:Tm^{3+}$ nanoprobe was reported for CT/optical dual-modal imaging *in vivo* 100. Since then, this type of CT/optical dual-modal imaging agents have gained an increasing exploitation [28, 74, 80, 82, 95, 96, 99, 100, 114, 126]. For example, ^{18}F is the most widely used radionuclide for PET imaging in clinical diagnostics. The most widely used host material of the UCNPs is fluoride, and it is easy to introduce ^{18}F into the UCNPs to realize PET and UCL imaging functions. Nevertheless, the half-life of ^{18}F is too short (1.829 h), which greatly limits its application in long-term imaging. To overcome this problem, Li's group prepared [153]Sm-labeled $NaLuF_4:Yb,Tm$, which was used for UCL and SPECT bimodal imaging [82]. The half-life (46.3 h) of [153]Sm is much longer than that of [18]F. In addition, it emits medium-energy xenon rays for long-term SPECT imaging. As shown in Figure 10.8(a), a significant UCL signal was observed 1 h after intravenous injection of the $NaLuF_4:Yb,Tm$ nanoparticles. Furthermore, *in situ* and *ex vivo* UCL images were also conducted in order to investigate the distribution of the particles in various organs. As it can be seen, the UCL signal was clearly VIS in the liver and spleen, whereas virtually no signal was seen in the other organs of the mouse. As the liver and spleen are the main organs for eliminating foreign nanoparticles, early accumulation by the liver and spleen was expected and is related to the clearance of nanoparticles from the blood by cells of the mononuclear phagocytic system.

Further, the application of $NaLuF_4:^{153}Sm,Yb,Tm$ NPs as radiotracers was investigated, including *in vivo* SPECT imaging and *ex vivo* biodistribution studies. As displayed in Figure 10.8(b), intense radioactive signals were detected exclusively in the liver and spleen, confirming that the $NaLuF_4:^{153}Sm,Yb,Tm$ nanoparticles were rapidly taken up by these two organs, which are consistent with data obtained from the *in vivo* UCL images (Figure 10.8(a)). At 24 h, a much larger uptake by the spleen compared to the liver was observed (Figure 10.8(b5)), which is partially due to the spleen being the largest organ of the immune system. The abovementioned findings indicate that the [153]Sm-doping and *in vivo* SPECT imaging strategy described herein provides a facile method for real-time tracking with high sensitivity for three-dimensional (3D) bioimaging *in vivo* and easy quantification of the distribution of the nanoparticles.

In pursuing more precise bioimaging, increasing studies have been focused on developing UCNPs for (PET/MRI/UCL or CT/MRI/UCL) trimodal imaging [46, 48, 51, 67, 68, 78, 79, 87, 88, 92, 107, 108, 110, 118]. The limited planar resolution of MRI imaging technology makes it unsuitable for cellular level imaging. CT can provide high-resolution 3D structural details of tissue based on the differences in X-ray absorption from different tissues. Nevertheless, due to its low sensitivity to soft tissues, its application in disease diagnosis is greatly limited. Fortunately, these issues can be solved with optical imaging.

A typical work about UCNPs-based optical, CT and MRI trimodal imaging contrast, in which the $NaLuF_4:Yb^{3+},Tm^{3+}$ NPs were used to provide the nanosystem X-ray attenuation and UCL capabilities; the SiO_2 shell was grown on the surface of the NPs as the bridge to conjugate the NPs with the DTPA–Gd complex, which enabled the magnetic property of the final sample [61]. As displayed in Figure 10.9(a1), ROI 1 is on the top right of the mouse abdomen and from it a strong 800 nm UC emission was detected. The mean emission intensity of the UC emission from ROI 1 was more than 330, and those from ROI 2 and ROI 3 are about 28 and 7, respectively, thus the SBR was calculated to be ~15 for the NIR-I UC imaging of the nude mouse. Figure 10.9(a2) presents an *ex vivo* NIR-I UCL image for the nude mouse and it can be seen that the 800 nm signal comes from the viscera region. The intensities of 800 nm UC signal in ROI 4, ROI 5 and ROI 6 are 470, 8 and 35, respectively, thus the SBR for the *ex vivo* imaging is bigger than 17. The histological luminescence image of the viscera is shown in Figure 10.9(a3). In the liver, lung, and spleen, the 800 nm UC signal can be easily detected, while intensities of the UCL signal in heart, kidney, intestines and stomach are low. These results imply that the $UCNP@SiO_2$-GdDTPA NPs are enriched in the liver, lung and spleen. Subsequently, a Kunming mouse was intravenously injected with $UCNP@SiO_2$-GdDTPA to demonstrate its *in vivo* MRI capability. As shown in upper images of Figure 10.9(b), the brightness of the liver increased as the time prolonged, which indicates that the MR signal in the liver was positively enhanced after intravenous injection of NPs. The middle images in Figure 10.9(b)

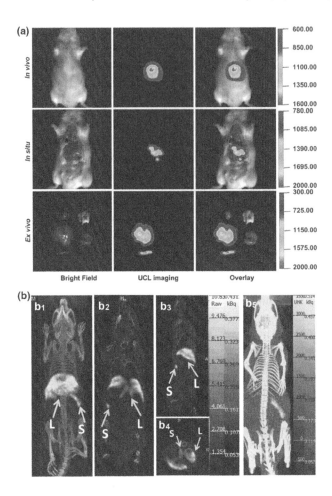

FIGURE 10.8　*In vivo, in situ* and *ex vivo* imaging of the Kunming mouse 1 h after tail vein injection of the Sm-UCNPs (20 mg kg⁻¹) (a). *In vivo* SPECT images after intravenous injection of Sm-UCNPs (b). Whole-body three-dimensional projection (b1), coronal (b2), sagittal (b3) and transversal images acquired at 1 h (b4) and whole-body three-dimensional projection images acquired at 24 h (b5) are shown, respectively. The arrows inset point to the liver (L) and spleen (S). Reproduced with permission from Ref. [82]. Copyright 2013 Elsevier.

demonstrate the T_1 distribution of the liver, in which the region of the liver becomes dark blue with prolonged time. In the bottom images of Figure 10.9(b), the colored images of the liver imply a variation of the T_1 MR signal. Obviously, the MR signal of the colorful region is enhanced positively with prolonged time. The dependence of the T_1 enhancement rate of the MR signal with time was also found. As for the spleen, a positive enhancement of the MR signal was also detected after intravenous injection of the sample. The upper images of Figure 10.9(c) show the T_1-weighted imaging result of the spleen. Unlike that observed for the liver, the brightness of the spleen region did not increase with time. The T_1 distribution of the spleen is shown in the middle images Figure 10.9(c), where the region of the spleen became darker with prolonged time. At last, in the bottom images of Figure 10.9(c), the colored T_1-weighted images of the spleen as a function of time are displayed. The green region shrank gradually as time increased, and the white and orange regions of the spleen became larger. The abovementioned results imply that UCNP@SiO$_2$-GdDTPA could be potentially utilized as an MRI positive-contrast agent.

In that study, the CT value for UCNP@SiO$_2$-GdDTPA was tested to be ~220 Hounsfield units (HU), which is 50% higher than that of iopromide. To demonstrate that the prepared UCNP@SiO$_2$-GdDTPA NPs can be used as a long-term CT imaging contrast agent *in vivo*, 3 mg day⁻¹ UCNP@

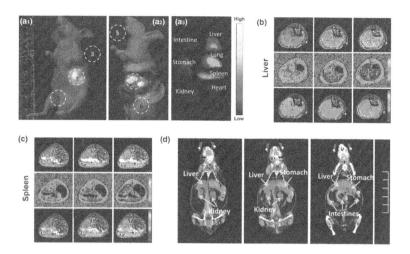

FIGURE 10.9 *In vivo* NIR-I fluorescence imaging of a nude mouse after injection with UCNP@SiO$_2$-GdDTPA for 10 min (a1). *Ex vivo* PL image of a nude mouse sacrificed (a2). *Ex vivo* PL image of viscera (a3). T_1-weighted MR images (upper), T_1 distribution images (middle) and local colorized T_1-weighted MR images (bottom) of a liver after injection with UCNP@SiO$_2$-GdDTPA for 0, 30 and 120 min (b). T_1-weighted MR images (upper), T_1 distribution images (middle) and local colorized T_1-weighted MR images (bottom) of a spleen after injection with UCNP@SiO$_2$-GdDTPA for 0, 30 and 120 min (c). Serials coronal CT images of a Kunming mouse at different layers after injection with UCNP@SiO$_2$-GdDTPA (d). Reproduced with permission from Ref. [61]. Copyright 2012 Elsevier.

SiO$_2$-GdDTPA was intravenously injected into a Kunming mouse for 1 week to obtain the viscera imaging of the abdomen. As a result, after 1 week of injections of UCNP@SiO$_2$-GdDTPA, the viscera in the abdomen can be distinguished clearly by three orthographic views and 3D reconstruction. Also, serial coronal images of the abdomen were scanned for further observation of the viscera. As shown in Figure 10.9(d), the stomach, liver, kidneys and intestines were imaged in the order of decreasing depth of the coronal planes. These results indicate that the UCNP@SiO$_2$-GdDTPA NPs could be gradually enriched in the liver and spleen. In conclusion, the developed nanosystem could be potentially utilized as a CT imaging-contrast agent. This report has enlightening significance in the field of multimode imaging based on UC NIR-I emitting LDNCs. With the advancement of multimodal imaging research, more and more UCNPs have been exploited as multiple imaging-guided therapeutic nanomedicines [71, 98, 105, 128, 132, 138].

10.4.3 NANOSCOPY IMAGING

Nanoscopy imaging is an emerging field of optical imaging, and the study of UCNPs in such an area is very rare. In 2017, Jin and his cooperators demonstrated the application of UCNPs doped with high concentrations (8%) of Tm^{3+} ions in low-power super-resolution stimulated emission depletion (STED) microscopy and achieve nanometerscale optical resolution (nanoscopy) imaging [120]. As illustrated in Figure 10.10(a), when the highly doped UCNPs were excited at a wavelength of 980 nm, the photon UC process was sensitized by Yb^{3+} stepwise transfer of that energy onto the scaffold energy levels of the Tm^{3+} emitters and eventually upconverted emission from the two-photon $^3H_4$, three-photon $^1G_4$ or four-photon $^1D_2$ levels of Tm^{3+}. The reduced inter-emitter distance at high Tm^{3+} doping concentration leads to intense CR, inducing a PA-like effect that rapidly populates the metastable $^3H_4$ level, resulting in population inversion relative to the $^3H_6$ ground level within a single nanoparticle. As a result, illumination by a laser at 808 nm, matching the UC band of the $^3H_4 \rightarrow {}^3H_6$ transition, can trigger amplified stimulated emission to discharge the $^3H_4$ intermediate level, so that the UC pathway to generate blue luminescence can be optically inhibited. Figure 10.10(b) displays

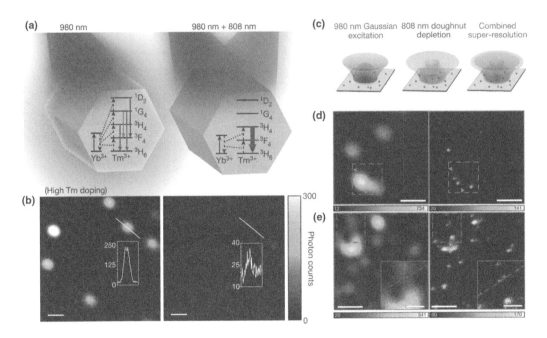

FIGURE 10.10 Energy-level diagrams of Yb/Tm-codoped UCNPs under 980 nm illumination (left), and under both 980- and 808 nm illumination (right) (a). Confocal images in 455 nm UC emission of the 8% Tm-doped UCNPs under continuous wave 980 nm laser (left) and under both 980 and 808 nm dual laser (right) illumination (b). Diagrams of the UC-STED super-resolution imaging, in which a Gaussian excitation profile (980 nm) and a Gauss-Laguerre mode "doughnut" depletion profile (808 nm) at far field are employed (c). Confocal (left) and super-resolution (right) images of the 40 nm 8% Tm-doped UCNPs. The 980 and 808 nm intensities were 0.66 and 9.75 MW cm^{-2}, respectively. Pixel dwell time, 4 ms; scale bars, 500 nm. Dashed boxes mark an area containing closely spaced 13 nm UCNPs that can be resolved in UC-STED but not in confocal imaging (d). As (e) but for the 13 nm 8% Tm-doped UCNPs. The 980 and 808 nm intensities were 0.66 and 7.5 MW cm^{-2}, respectively. Pixel dwell time, 6 ms; scale bars, 500 nm for the main images and 200 nm for the insets. Dashed boxes mark an area containing closely spaced 13 nm UCNPs that can be resolved in UC-STED but not in confocal imaging and are shown enlarged in the insets marked by solid boxes (e). Reproduced with permission from Ref. [120]. Copyright 2017 Nature Publishing Group.

the confocal images of single UCNPs doped with 8% Tm^{3+} ions (and 20% Yb^{3+}). Obviously, the UC emission under continuous-wave (CW) 980 nm excitation was inhibited once a CW 808 nm probe beam was applied. By contrast, the UCNPs with low doping concentration (1% Tm^{3+} and 20% Yb^{3+}) showed negligible optical switching effects under the same experiment conditions (data not shown).

To further explore the use of highly doped UCNPs for optical super-resolution imaging; the 808 nm beam was spatially modulated to produce a doughnut-shaped point spread function (PSF) that overlapped with the Gaussian PSF of the 980 nm excitation beam at the focal plane (illustrated in Figure 10.10(c)). The nanocrystals on the periphery of the 980 nm PSF would, therefore, be expected to be optically switched off by the 808 nm beam, leading to an effective excitation spot smaller than the optical diffraction limit, similar to the situation in STED microscopy. To prove that concept, two samples of monodispersed UCNPs, both doped with 8%Tm and 20%Yb, were characterized with average sizes of 39.8 and 12.9 nm, respectively. A region with a size comparable to the optical diffraction limit but containing three 40 nm UCNPs was selected, and a sequence of far-field optical super-resolution images that clearly resolve the adjacent UCNPs were recorded (Figure 10.10(d)). Spot sizes of 48.3 nm were obtained by UC-STED microscopy at an 808 nm intensity of 9.75 mW cm^{-2} (Figure 10.10(d)). Deconvolution based on the simple Pythagorean equation shows this result corresponds to a markedly improved resolution of 27.4 nm, representing a 13-fold improvement over the optical diffraction limit, or 1/36 of the excitation wavelength. Also, the authors examined 12.9 nm UCNPs and calculated a resolution of 28.4 nm (Figure 10.10(e)), which confirms the resolution for the present UC-STED system of ~28 nm. These engineered UCNPs offer saturation intensity two orders of magnitude lower than those of fluorescent probes currently employed in STED microscopy, suggesting a new way of alleviating the square-root law that typically limits the resolution that can be practically achieved by such techniques.

In 2018, the UCNPs of 40 nm $NaYF_4$:20%Yb^{3+},4%Tm^{3+} were investigated to unlock a new mode of NIR emission saturation nanoscopy for deep tissue super-resolution imaging with excitation intensity several orders of magnitude lower than that required by conventional multiphoton fluorescence microscopy dyes [127]. Using a doughnut beam excitation from a 980 nm diode laser and detecting at 800 nm, they achieve a resolution of sub 50 nm, 1/20th of the excitation wavelength, in imaging of single UCNP through 93-μm thick liver tissue. This method offers a simple solution for deep tissue super resolution imaging and single-molecule tracking. NIR emission saturation offers a great deal of simplicity and stability, as an advanced tool to achieve super resolution in deep tissue.

In another paralleled work, Wu et al. demonstrated a fluorescence emission difference nanoscopy using UCNPs, with 172 nm resolution achieved using 800 nm 10-MW cm^{-2} excitation and 660 nm emission detection [153]. With the goal to achieve super-resolution imaging using low laser power and longer wavelength, i.e., to reduce photo toxicity applied to live cells, controlled synthesis of RE UCNPs will further optimize energy transfer process and resultant saturation intensity properties for biophotonics to surpass the diffraction limit.

10.4.4 MULTIPLEXED IMAGING

Multiplexed imaging technology is an effective strategy for simultaneously identifying different biological organisms. The utilization of UCNPs for optical multiplexing introduces unconventional parameters in both the spatial and temporal dimensions, which greatly enhances the optical encoding capabilities. As reported, with the rapid development of UCNPs design and preparation, the emission properties of UCNPs were adjusted and applied to multiplexed UCL imaging [37, 44, 72, 113, 119, 131, 142, 143].

In a very latest work, the UC emission of Tm^{3+} at 808 nm was investigated by Chen et al. for optical multiplexing [143]. Therein, the lifetime of the core-multishell $NaYF_4$@$NaYbF_4$@$NaYF_4$:Yb^{3+}/Tm^{3+}@$NaYF_4$ NPs was tuned in a wide range of 78–2157 μs through controlling the doping content of Yb^{3+} in $NaYF_4$:Yb^{3+}/Tm^{3+} layer and the thickness of the $NaYbF_4$ layer. As shown in Figure 10.11(a), the tetradomain core-multishell UCNPs with a set of lifetimes of $\tau_1-\tau_{10}$ can be easily distinguished with their home-built lifetime imaging system.

To demonstrate the potent of earlier obtained samples for temporal optical multiplexed UC imaging, a scientific-grade silicon camera was utilized to capture temporal UCL images, which was synchronized with a square-wave pulsed excitation laser but triggered at a precisely defined delay time (Figure 10.11(b)). The polyacrylic acid (PAA)-coated core-multishell UCNPs show a slightly shorter lifetime than the corresponding ones dispersed in hexane (Figure 10.11(c)), possibly attributed to the existence of long-distance coupling between Yb^{3+} ions and the vibronic −OH group of water molecules. The PAA-coated core-multishell UCNPs with a lifetime of $\tau_5 = 1158$ μs were administrated into a Kunming mouse through tail vein injection and implemented subcutaneous injection of PAA-coated UCNPs with lifetimes of $\tau_2 = 920$ and $\tau_9 = 1528$ μs into the left and right of the abdomen (Figure 10.11(d), bottom row). Indeed, temporal optical multiplexed UC can be clearly seen in the liver and two abdomen subcutis with distinct lifetime-hued colors. Moreover, the lifetime values estimated through a time-delayed imaging process are reproducible (τ_2 and τ_9 revealed in two mice shown in the upper and bottom rows of Figure 10.11(d)) and agree well with those measured in solutions in Figure 10.11(c). Notably, the position and shape of the lifetime-multicolored areas correspond well with that achieved through UCL intensity imaging. It can be predicted in the near future, in the field of optical multiplexing based on UCNPs, more and more attention will be paid on this kind of emission lifetime-participated multiparameter encoding.

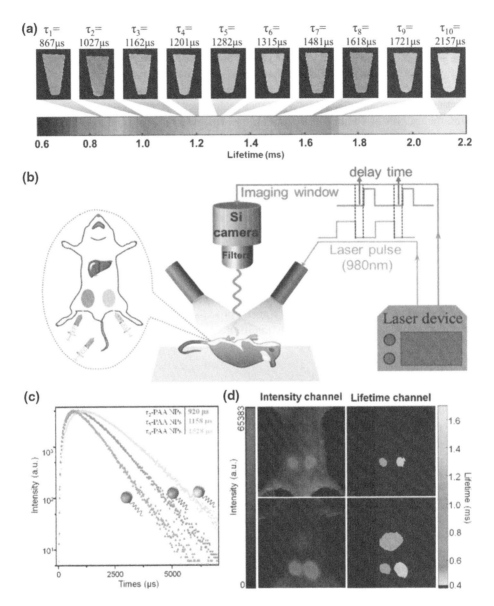

FIGURE 10.11 Pseudocolor-mapped luminescence lifetime images of our tetradomain core-multishell NPs in centrifuge tubes using a home-built time-resolved UC imaging system (a). Schematic illustration of NPs administration into a mouse, which was then imaged using a home-built time-resolved UCL imaging system. Deep-cooling silicon camera is synchronized with square-wave excitation laser pulse but triggered at a precisely defined delay time. Camera keeps integrating until arrival of the next laser pulse (b). Measured UCL decay profiles of PAA-coated core-multishell NPs with lifetimes of τ_2, τ_5 and τ_9 (aqueous dispersion) (c). UCL intensity (top left) and lifetime (top right) imaging of PAA-coated core-multishell nanoparticles with lifetimes of τ_2 and τ_9 in a Kunming mouse and also in a second Kunming mouse (bottom) (subcutaneous injection of these nanoparticles into abdomen). Intravenous administration of PAA-coated core-multishell nanoparticles with a lifetime of τ_5 was implemented for the second mouse, enabling us to light up the internal organs for both luminescence intensity (bottom left) and lifetime (bottom right) UC imaging (d). τ values correspond to those of nanoparticles in Figure 10.11(a). Reproduced with permission from Ref. [143]. Copyright 2020 American Chemical Society.

10.4.5 RATIOMETRIC IMAGING

Ratiometric imaging is a technique based on the relative changes of the emission intensity between different emission peaks of UCNPs-based nanosysytem to selectively reflect the concentration of a certain component in the living body. The RE UCNPs can be used for optical ratiometric imaging due to its basic property of multiple emission peaks. In recent one decade, the UCNPs-based nanoarchitectures have been attracted a certain degree of attention in the point of ratiometric bioimaging.

In the aspect of bioimaging of methylmercury (MeHg$^+$)-using UCNPs, Li et al. finished a significant work in 2013 [77]. Therein, the authors demonstrated an UC luminescence resonance energy transfer (LRET) nanosystem composed of the NPs NaYF$_4$:Yb,Er,Tm and MeHg$^+$-responsible NIR cyanine dye hCy7 (hCy7-UCNPs) for UCL monitoring of MeHg$^+$ within living small animal for the first time. Using the ratiometric UCL emission at 800–660 nm as a detection signal, the detection limit of hCy7-UCNPs for MeHg$^+$ in aqueous solution was as low as 0.18 ppb (Figure 10.12(a)).

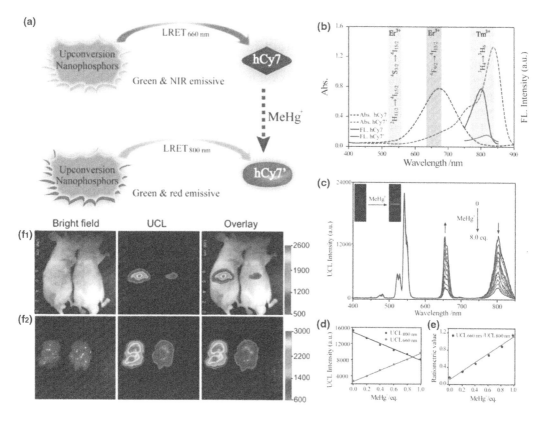

FIGURE 10.12 Schematic illustration of the UC LRET process from the UC emission of UCNPs to the absorption of the dye hCy7 or hCy7′ (a). UV-VIS absorption (dashes) and emission (solid) spectra of hCy7 and in the absence and presence of MeHg$^+$ (excitation with 730 nm). The range of main UCL emission bands of Er^{3+} and Tm^{3+} is also shown with different color (b). UCL spectra of 0.05-mg mL^{-1} h Cy7-UCNPs in the aqueous solution upon gradual addition of MeHg$^+$ (from 0 to 8 equiv). Inset: the photos showing change in the red UCL emissions (c). The UCL emissions at 660 ± 2 and 800 ± 2 nm as a function of MeHg$^+$ concentration (d). The ratio of the UCL emission at 660 ± 2 to 800 ± 2 nm as a function of MeHg$^+$ concentration (e). *In vivo* UCL images of 40-μg hCy7-UCNPs-pretreated living mice injected intravenously with 0.2-mL normal saline (left mouse) or 0.1-mM MeHg$^+$ solution (right mouse) (f1). The corresponding UCL images of the livers that were isolated from the earlier dissected mice. The UCL emission was collected at 800 ± 12 nm upon irradiation at 980 nm (f2). Reproduced with permission from Ref. [77]. Copyright 2013 American Chemical Society.

The absorption and fluorescence emission spectra of hCy7 in the absence and presence of MeHg$^+$ are shown in Figure 10.12(b). Moreover, this interaction between hCy7-UCNPs and MeHg+ also affected the upconversion emission signals by the LRET process. Following the addition of MeHg$^+$, UCL intensity at 660 nm increased significantly, and UCL emission at 800 nm decreased for hCy7-UCNPs under excitation at 980 nm (Figure 10.12(c)). The UCL emission change of hCy7-UCNPs following addition of MeHg$^+$ was attributed to the LRET process. It can be seen from Figure 10.12(c) that following the addition of MeHg$^+$, the spectral overlap between the red UCL emission ($\lambda_{em} = 635$–680 nm) of UCNPs and absorption ($\lambda_{abs} = 670$ nm) of hCy7 was reduced, causing a decrease in the LRET from the red UCL of UCNPs to hCy7. Consequently, the red UCL emissions of hCy7-UCNPs increased in the presence of MeHg$^+$ (Figure 10.12(c)). Nevertheless, spectral overlap between the NIR UCL emission ($\lambda_{em} = 800$ nm) of UCNPs and absorption ($\lambda_{abs} = 845$ nm) of hCy7′ was increased, leading to an increase in the LRET from the NIR UCL of UCNPs to hCy7′. As a result, the NIR UCL emissions of hCy7-UCNPs stopped following the addition of MeHg$^+$ (Figure 10.12(c)). Using UCL emission at 800 and 660 nm as detection signals (Figure 10.12(d)), the detection limits for MeHg$^+$ in aqueous solution were tested to be ~0.80 and 0.82 ppb, respectively. Moreover, the variation in UCL ratio UCL660/UCL800 nm versus the addition of MeHg$^+$ showed a good linear relationship (Figure 10.12I).

The *in vivo* accumulation of MeHg$^+$ is related to the development of visceral injury, especially liver injury in mammals. Herein, the hCy7-UCNPs nanosystem was investigated to monitor MeHg$^+$ in a mouse model. Male Kunming mice were injected intravenously with hCy7-UCNPs and divided into two equal groups. As shown in Figure 10.12(f), the left group (control group) was injected intravenously with physiological saline, and the right group was injected intravenously with MeHg$^+$ in saline. UCL imaging *in vivo* indicated that the hCy7-UCNPs mainly accumulated in the livers of mice (Figure 10.12(f)). Compared with control mice treated with normal saline, the UCL at 800 nm decreased to 50% of the original value after treatment with MeHg$^+$ (Figure 10.12(f1)). The mice in the two groups were subsequently dissected to isolate the livers. The UCL emission in the liver of MeHg$^+$-pretreated mice was weaker than that in the control mice (Figure 10.12(f2)). Such a successful ratiometric UCL-based nanosystem for bioimaging of MeHg$^+$ provides a new design strategy for further novel probes for highly sensitive *in vivo* bioimaging studies.

In 2018, Zhang et al. invented a core-shell NPs NaErF$_4$Ho@NaYF$_4$ for upconverting 1530 nm excitation laser to 650-, 980- and 1180 nm emissions (Figure 10.13(a)) [25]. Taking advantage of the multiple emissions and large anti-Stokes shifting (>350 nm), the obtained NIR-II UCNPs can be used for the ratiometric bioimaging of H$_2$O$_2$. In comparison with the emission at 650 nm, emissions at 980 and 1180 nm exhibit similar attenuation kinetics in the simulation tissue, which is very important for the *in vivo* ratiometric fluorescent sensor (Figure 10.13(b)). Combined with the selective absorption property at 980 nm and the H$_2$O$_2$ sensitivity of IR1061 in Fenton reaction, the ratiometric fluorescent (I_{980}/I_{1180}) H$_2$O$_2$ sensor can be readily fabricated by encapsulating NaErF$_4$:Ho@NaYF$_4$ NIR-II UCNPs with UC emission peaks at 1180 and 980 nm, IR1061 with strong absorption at 800–1100 nm and Fe^{2+} with Fenton reaction activity in polycaprolactone. As shown in Figure 10.13(b), the NIR-II ratiometric fluorescent sensor can respond to localized OH radicals generated from H$_2$O$_2$, which can cause the destruction of IR1061 dye and subsequently weaken its absorption band at 800–1100 nm. The 980 nm emission (I_{980}) is greatly suppressed because of the strong secondary absorption (the non-LRET process) of IR1061 at 980 nm in the absence of H$_2$O$_2$, which gradually recovered with increasing of the concentration of H$_2$O$_2$. In stark contrast, intensity of the 1180 nm emission (I_{1180}) is very stable. In accordance with the Lambert-Beer law, the log(I_{980}/I_{1180}) is a linear function of the H$_2$O$_2$ concentration. The NIR-II UC features of NaErF$_4$:Ho@NaYF$_4$ with large Stokes shift is the key factor to realize the NIR-II sensing. Since the wide absorption bands of the IR1061 organic dye, it is difficult to realize the NIR-II sensor with regular NIR-II lanthanide probes having the excitation wavelengths during 700–980 nm, which are overlapping with the absorption band of the IR1061 organic probes.

The H$_2$O$_2$-sensitive microneedle patch was further used for *in vivo* bioimaging of H$_2$O$_2$ in the inflammation site (Figure 10.13(c)). Owing to low autofluorescence and tissue scattering of the

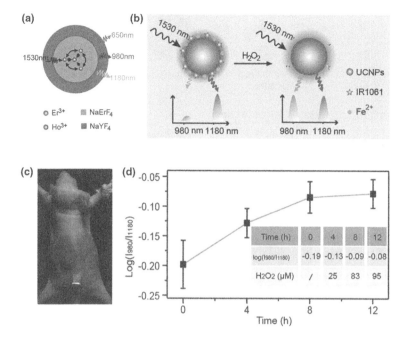

FIGURE 10.13 Energy transfer mechanism in the NaErF$_4$:2%Ho@NaYF$_4$ core-shell UCNPs (a). Illustration of ratiometric fluorescent sensor for H$_2$O$_2$ (b). Photograph of a mouse treated with the microneedle patch (c). Ratiometric fluorescence (I$_{980}$/I$_{1180}$) of the microneedle patches and corresponding H$_2$O$_2$ concentration at different time (d). Reproduced with permission from Ref. [25]. Copyright 2018 Wiley.

NIR-II luminescence, the luminescence images of the microneedle array were still very clear under the skin tissue. The signal from single needle (200 × 200 μm^2) could be easily distinguished. Through the linear correlation between log(I$_{980}$/I$_{1180}$) and H$_2$O$_2$ concentration (Figure 10.13(d)), the concentration of H$_2$O$_2$ in the inflammatory site could reach as high as 95 μM in 12 h, which was consistent with previous reports in the similar inflammation model (50–300 μM). In such a point, Wang et al. had reported fluorescence resonance energy transfer (FRET)-based UCNPs sensitized by Nd^{3+} for the ratiometric imaging of H$_2$O$_2$ *in vivo* [137].

Based on a similar mechanism, there are some other works that were reported to ratiometric imaging: the living components of metal ions [47, 129], hydrogen sulfide [102, 130], H$^+$ [121], et al. In the future, more and more ratiometric imaging nanosystems will be constructed through designing cascade chemical or physical reaction based on UCNPs.

10.4.6 Hypoxia Imaging

In a recent work, Li et al. designed and fabricated a unique core-satellite nanostructure, where many UCNPs were attached onto the surfaces of bio-MOFs containing rich O$_2$ indicators, which are expected to maximize the FRET efficiency under NIR excitation [144]. The genetically engineered murine model that may accurately mimic TME and progression of human lung cancer is introduced to provide reliable preclinical evaluation on the core-satellite nanosensors for *in vivo* tracking of non-small-cell lung cancer lesion *via* hypoxia imaging.

Synthesis process of the core-satellite nanostructures is schemed in Figure 10.14(a). In view of the characteristic emission bands of NaYF$_4$:Yb,Tm@NaYF$_4$ core-shell UCNPs at 450 and 477 nm, the molecule tris(4,7-diphenyl-1,10-phenanthroline) ruthenium(II) dichloride ([Ru(dpp)$_3$]$^{2+}$Cl$_2$) with strong and well-matched absorption in the wavelength region of 350–550 nm is chosen as the

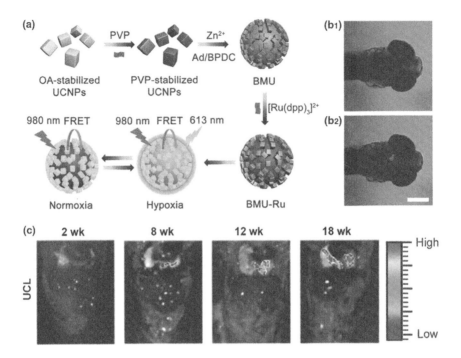

FIGURE 10.14 Preparation and characterization of BMU core-satellite nanostructures. Strategy for fabrication of sensors based on UCNPs, bio-MOF-100 and $[Ru(dpp)_3]^{2+}Cl_2$ for cycling hypoxia response under NIR excitation (a). Zebrafish after incubation with BMU-Ru nanosensors for 1 h was followed by addition of 15-mM BDM at 300 s. After BDM is washed out, red emission in the brain of zebrafish disappears (b1). After adding BDM again, red emission in the brain is recovered (b2). The process was repeated three times at least. λ_{ex} = 980 nm, λ_{em} = 600–625 nm; scale bar: 300 μm. UCL images of the mice lung after intravenous injection of BMU-Ru nanosensors at different intervals (2, 8, 12 and 16 weeks) (c). Reproduced with permission from Ref. [144]. Copyright 2020 Wiley.

quenchable indicator for O_2 thanks to the occurrence of effective FRET between them. The *in vivo* hypoxia imaging of BMU-Ru nanosensors was checked with the zebrafish system. As demonstrated in Figure 10.14(b), disappearance of emission is discerned when 2,3-butanedione is washed out with PBS, demonstrating that the O_2 content in the brain of zebrafish is recovered. Importantly, the process of emission recovery and quenching combined with normoxic and hypoxic conditions is well reversible in zebrafish system.

The prepared BMU-Ru nanosensors were also employed to track the non-small-cell lung cancer lesion progression under NIR laser excitation. As manifested in Figure 10.14(c), the UCL signal of BMU-Ru nanosensors begins to appear within 8 weeks, presenting an early stage in tumor development. As time prolongs to 12 weeks, the signal accounts for 40–50% area of lung; meanwhile the signal intensity becomes stronger, illustrating that the lesion of non-small-cell lung cancer worsens and causes more hypoxia state; while at 16 weeks, observation of 70–80% signal coverage accompanied by more and larger red color signal area in lung indicates that lung hypoxia further increases and non-small-cell lung cancer deteriorates into the late stage.

In another intriguing work, a phosphorescent iridium(III) complex-modified UCNPs is developed, which can monitor O_2 concentration and also reduce autofluorescence under both downconversion (DC) and UC channels [106]. Through such a well-defined nanostructure, gradient O_2 concentration can be detected clearly through confocal microscopy luminescence intensity imaging, phosphorescence lifetime imaging microscopy and time-gated imaging, which is meaningful to O_2 sensing in biological tissues with nonuniform O_2 distribution.

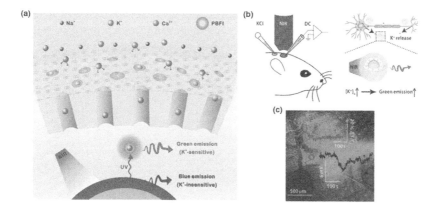

FIGURE 10.15 Schematics showing a magnified view of the nanosensor and its K⁺ sensing mechanism (a). Schematic of the electrophysiological recording (left) and optical imaging (right) of CSD in the intact mouse brain. Left: The mouse skull was thinned to visualize the cortical surface. CSD was evoked by either KCl incubation or pinprick. A single electrode was inserted to detect CSD by measuring local field potential. Right: CSD-associated K⁺ release was visualized by detecting the increased fluorescence from the nanosensor (b). Surface fluorescence image showing nanosensor-stained brain parenchyma. Insets show the time courses of the nanosensor's fluorescence signal (red line) at the red region of interest (ROI) and the field potential (black line) recorded near the red ROI (c). Adapted with permission from Ref. [145]. Copyright 2020 American Association for the Advancement of Science.

10.4.7 POTASSIUM IMAGING

In a recently published work, Shi and coworkers reported a K⁺ nanosensor based on UCNPs of NaYF$_4$:Yb/Tm@NaYF$_4$:Yb/Nd that can be excited by 808 nm light and used for *in vivo* K⁺ imaging [145]. As depicted in Figure 10.15(a), the UCNPs convert NIR light to UV light that excites the acceptor potassium binding benzofuran isophthalate (PBFI) *via* LRET. The outer surface of the silica nanoparticles was subsequently shielded with a thin layer of K⁺-selective filter membrane containing micropores. Therein, the filter membrane layer allows only K⁺ to diffuse into and out of the nanosensor, thus excluding the interference from other cations. Once diffused into the nanosensor, K⁺ will bind to PBFI immediately. Upon NIR irradiation, the upconverted UV photons from the UCNPs excite PBFI, leading to the emission of K⁺-bonded PBFI.

Afterwards, the shielded nanosensors were applied to image cortical spreading depression (CSD) in the mouse brain, which is a wave-like propagation of neural activity. After injecting the PEG-modified shielded nanosensors into the mouse cerebral cortex in a cranial window, CSD events were triggered by KCl incubation in a craniotomy ~3 mm away to avoid interference from such externally applied K⁺. Simultaneously, the local field potential and the optical signal through the cranial window were recorded (Figure 10.15(b)). The increases in nanosensor fluorescence agreed well with electrophysiological CSD events (Figure 10.15(c)). Moreover, simultaneous monitoring of both K⁺ and Ca²⁺ in the zebrafish brain was also achieved using the developed UCNPs-based nanosensor in such work. It is anticipated that the shielded nanosensor will have broad applications in brain research and improve our understanding of abnormal K⁺-related diseases.

10.4.8 MYELOPEROXIDASE IMAGING

In the aspect of bioimaging of MPO, Xian et al. selected phycocyanin (PC) as an energy acceptor to develop a UCNPs-based nanoprobe for the first time [147]. Specifically, the NIR-responsive nanoprobe based on PC-UCNPs was fabricated by covalent conjugated commercial PC with NaGdF$_4$:Yb,Er UCNPs. As depicted in Figure 10.16, the UCL of UCNPs was quenched by PC

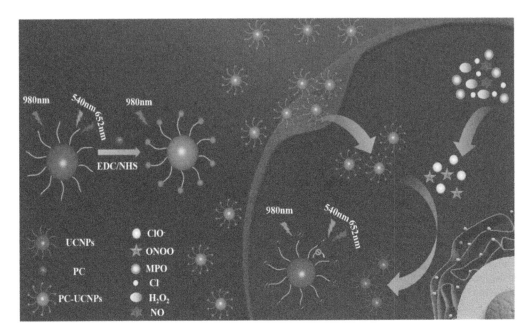

FIGURE 10.16 Schematic illustration of the rational modulation of PC-UCNPs NIR nanoprobe to detect the bioactivity of MPO in cells. Adapted with permission from Ref. [147]. Copyright 2020 American Chemical Society.

quencher after formation of the PC-UCNPs. When the ClO^-/MPO–H_2O_2–Cl^- or $ONOO^-$/MPO–NO cycle was presented in the system, the UCL was recovered because the structure of PC was destroyed by the reaction between $ONOO^-$/ClO^- and PC. On the basis of the "turn-on" luminescence, the nanoprobe is used to evaluate the bioactivity of MPO. Significantly, it has been successfully applied in bioimaging toward living cells and tissues during an inflammatory process based on the advantages of superior biocompatibility and the deep light penetration of NIR light. Hence, PC can act as the recognition element of MPO as well as the energy acceptor of UCNPs in this work.

10.4.9 mRNA Imaging

In the field of UCNPs-participated mRNA bioimaging, Xia's group reported an interesting work. Figure 10.17 shows the principle and whole process of NIR light triggered nucleic acid cascade recycling amplification reaction in living cancer cells. The photoactivatable DNA hairpin with quencher (H) was hybridized with single strand linker DNA labeled with TAMRA fluorophore (L), which was immobilized on the surface of UCNPs, forming the probe complex with quenched fluorescence. This probe complex can go inside cancer cells *via* endocytosis and proceed nucleic acid displacement reaction and trigger the exonuclease III assisted nucleic acid cascade recycling amplification under NIR irradiation and in the present of target mRNA and exonuclease III, generating amplified fluorescence signal. As demonstrated, MnSOD mRNA imaging in living HeLa cancer cells *via* this developed NIR light activated nucleic acid cascade recycling amplification strategy achieves about 5-fold fluorescence intensity than that without signal amplification. The developed method shows good spatiotemporal controllability that enables on-demand controlling of the when and where of the signal amplification process in living cancer cells.

10.4.10 Neuronal Activity Imaging

In a very latest report, Shi and his cooperators invented NIR voltage nanosensors to enable real-time *in vivo* bioimaging of neuronal activities [146]. As shown in Figure 10.18(a), due to

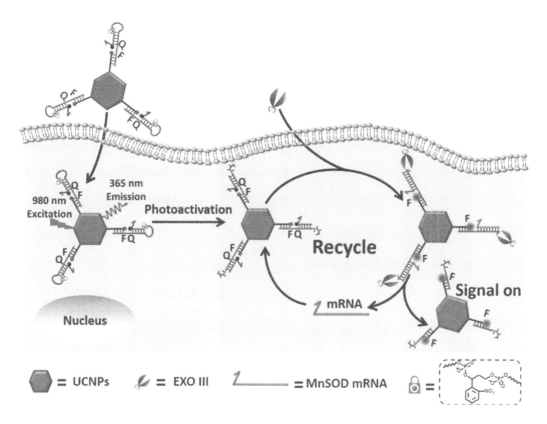

FIGURE 10.17 Schematic overview of NIR light activated nucleic acid cascade recycling amplification reaction in living cancer cells for photocontrollable signal-amplified mRNA imaging. Adapted with permission from Ref. [140]. Copyright 2020 American Chemical Society.

the distance-dependent changes in the FRET efficiency between the UCNPs and dipicrylamine (DPA), UCNPs upconvert NIR to UV-VIS luminescence emissions of different intensities under varied membrane potentials. Under the hyperpolarized state, DPA molecules prefer to locate close to the outer leaflet of the plasma membrane, leading to increased efficiency of FRET and decreased luminescence intensity of UCNPs (top of Figure 10.18(a)). Under the depolarization state, the translocation of DPA to the inner leaflet of the plasma membrane decreases the efficiency of FRET due to the increased distance between the two moieties, resulting in increased UCNPs luminescence (bottom of Figure 10.18(a)).

At the cellular level, single neurons reflect the low-frequency (<1 Hz) fluctuation as a subthreshold membrane potential (Figure 10.18(b)). To check the performance of the voltage nanosensor in monitoring subthreshold fluctuations in intact animals, we labeled mouse cortical neurons by injecting a mixture of both GSH-modified UCNPs and DPA into the primary somatosensory cortex of mice under pentobarbiturate anesthesia. Dozens of neurons were labeled across the tiny durotomy (Figure 10.18(c)). Subsequently, we checked whether our nanosensors can detect differential oscillations of subthreshold neuronal activities before and after a tail pinch, a stimulus applied to the mouse to increase its vigilance level (Figure 10.18(b)).

Because of the high stability of the emission of UCNPs, the nanosensor luminescence can be recorded for as long as 20 min without significant decay. All these results demonstrate that the designed nanosensor can report the brain state-related subthreshold membrane potential fluctuations in the brain of intact mice. Overall, the developed voltage nanosensors exhibit great potential in brain activity mapping, helping to efficiently dissect the functional operations of neuronal circuits.

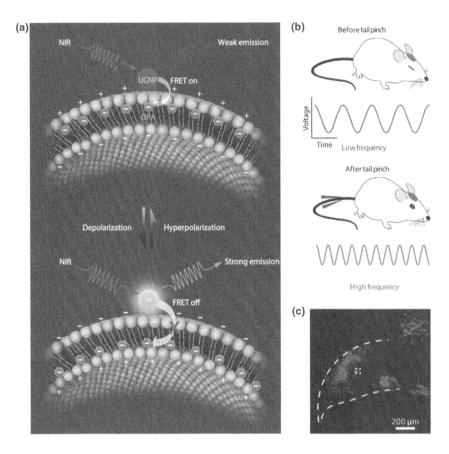

FIGURE 10.18 Schematic illustration showing the basic mechanism of the voltage nanosensor (a). Schematic illustration of the mouse brain state before and after the tail pinch (b). The mouse underlying pentobarbiturate anesthesia was loaded with the voltage nanosensor at the neocortex through the durotomy (white dash line) and received tail pinch. Inset, a typical labeled neuron, marked by the white dash square in the middle (c). Adapted with permission from Ref. [146]. Copyright 2020 American Chemical Society.

10.4.11 SUMMARY

This chapter has summarized the recent progress of RE UCNPs in the field of bioimaging. NIR excitation laser source affords deep tissue penetration, high temporal and spatial resolution and minimal autofluorescence, thus diversifying the bioimaging applications of RE UCNPs. It has been validated that NIR-II exciting photon possess up to a 1000-fold reduction in scattering losses when compared with the NIR-I photon [8]. Namely, UCNPs with both excitation and emission light located in the NIR window (especially in the NIR-II window) are highly suitable for improving the sensitivity and resolution of optical imaging. Significantly, when used for *in vivo* bioimaging, the NIR-II-excited UCNPs pose great advantages in obviating overheating and phototoxicity issues caused by excitation photon. Moreover, it has been fully demonstrated that RE UC optical imaging having great potency in combining multimodal bioimaging and disease treatment. Not only that, the investigation of RE UCNPs in other emerging optical imaging fields (hypoxia imaging, living component imaging, ions imaging, neuron imaging, *etc.*) have been developing swiftly. Overall, the achievements have enriched our understanding about UCNPs and their potential applications in bioimaging. Nevertheless, there is still a long way to go before clinical translation of UCNPs due to the existence of several challenges.

The first challenge should be confronted is the biocompatibility and long-term toxicity, although UCNPs have been proved to be low toxic in *in vitro* and *in vivo* experiment. However, these results were obtained from small animals such as mice and zebra fish, which are probably different from human body. Despite the negligible toxicity for *in vitro* and *in vivo* in a short-term study after modifying the surface of the particles with biocompatible ligand moieties, its long-term immune effect is still a task that requires further investigation. Moreover, the uptake of UCNPs in reticuloendothelial system (liver, spleen, *etc.*) should be seriously considered, as it would greatly influence its application toward biomedical regions.

The second challenge is that the quantum yield (QY) of the UCNPs is still not very high. In fact, the QY is mainly limited by the low NIR absorption cross section of the RE ions. Till now, some NIR dye molecules have been employed recently as antennas to enhance and broaden the excitation of UCNPs, which help in improving the conversion efficacy [154]. This improvement in efficacy is a result of the increased absorptivity of the nanostructure complexes. To improve the UCL efficiency, another crucial factor that we need to look at it is the size distribution of the UCNPs as it has been known that smaller size nanostructures improve the overall cellular uptake, while at the same time it generally induces an inferior emission. Overall, the abovementioned challenges will compel researchers to exploit alternative UCNPs with ultrasmall size, higher QY, longer excitation wavelength and ideal surface modification to address the potential issues that existed in most current RE UCNPs.

Currently, most of the invented UCNPs have no specific interactions to targets of interest and cannot accurately image biological events of interest. Thus, it is highly significant to design intelligent probes in response to specific biological events or targets, thereby achieving a higher SBR. It is predicted that, in the field of UCNPs-based optical bioimaging, more and more new technology and applications will be developed. In the followed years, the UC emission lifetime-participated optical multiplexing will play a more and more important role in bioimaging field. Ultimately, through designing bioresponsive UC nanosystems, the multimodal imaging-guided therapy will have a great progress in the area of disease theranostic. All in all, we believe that the continuous innovations and new discoveries made in the area of material chemistry, biology, and photophysics will enable the functionalized UCNPs to expand its applicability.

REFERENCES

1. Kim H, Beack S, Han S, Shin M, Lee T, Park Y, Kim KS, Yetisen AK, Yun SH, Kwon W, Hahn SK (2018) Multifunctional photonic nanomaterials for diagnostic, therapeutic, and theranostic applications. *Adv Mater* 30: 1701460.
2. Chen G, Qiu H, Prasad PN, Chen X (2014) Upconversion nanoparticles: Design, nanochemistry, and applications in theranostics. *Chem Rev* 114: 5161–5214.
3. Choi JS, Kim S, Yoo D, Shin TH, Kim H, Gomes MD, Kim SH, Pines A, Cheon J (2017) Distance-dependent magnetic resonance tuning as a versatile MRI sensing platform for biological targets. *Nat Mater* 16: 537–542.
4. Yuan L, Lin W, Zheng K, Zhu S (2012) FRET-based small-molecule fluorescent probes: Rational design and bioimaging applications. *Acc Chem Res* 46: 1462–1473.
5. Li X, Gao X, Shi W, Ma H (2014) Design strategies for water-soluble small molecular chromogenic and fluorogenic probes. *Chem Rev* 114: 590–659.
6. Dong H, Sun L-D, Yan C-H (2015) Energy transfer in lanthanide upconversion studies for extended optical applications. *Chem Soc Rev* 44: 1608–1634.
7. Wei Y, Yang X, Ma Y, Wang S, Yuan Q (2016) Lanthanide-doped nanoparticles with near-infrared-to-near-infrared luminescence for bioimaging. *Chin J Chem* 34: 558–569.
8. Hong G, Antaris AL, Dai H (2017) Near-infrared fluorophores for biomedical imaging. *Nat Biomed Eng* 1: 0010.
9. Smith AM, Mancini MC, Nie SM (2009) BIOIMAGING Second window for *in vivo* imaging. *Nat Nanotechnol* 4: 710–711.
10. Qin X, Liu X, Huang W, Bettinelli M, Liu X (2017) Lanthanide-activated phosphors based on 4f-5d optical transitions: Theoretical and experimental aspects. *Chem Rev* 117: 4488–4527.

11. Liu Y, Tu D, Zhu H, Chen X (2013) Lanthanide-doped luminescent nanoprobes: Controlled synthesis, optical spectroscopy, and bioapplications. *Chem Soc Rev* 42: 6924–6958.

12. Zhang H, Chen Z-H, Liu X, Zhang F (2020) A mini-review on recent progress of new sensitizers for luminescence of lanthanide doped nanomaterials. *Nano Res* 13: 1795–1809.

13. Xu S, Cui J, Wang L (2016) Recent developments of low-toxicity NIR II quantum dots for sensing and bioimaging. *Trends Analyt Chem* 80: 149–155.

14. Chan WCW, Nie S (1998) Quantum dot bioconjugates for ultrasensitive nonisotopic detection. *Science* 281: 2016–2018.

15. Zhou B, Shi BY, Jin DY, Liu XG (2015) Controlling upconversion nanocrystals for emerging applications. *Nat Nanotechnol* 10: 924–936.

16. Wang F, Liu X (2014) Multicolor tuning of lanthanide-doped nanoparticles by single wavelength excitation. *Acc Chem Res* 47: 1378–1385.

17. Zhang Z, Fang X, Liu Z, Liu H, Chen D, He S, Zheng J, Yang B, Qin W, Zhang X, Wu C (2020) Semiconducting polymer dots with dual-enhanced NIR-IIa fluorescence for through-skull mouse-brain imaging. *Angew Chem Int Ed* 59: 3691–3698.

18. Kenry, Duan Y, Liu B (2018) Recent advances of optical imaging in the second near-infrared window. *Adv Mater* 30: 1802394.

19. Smith AM, Mancini MC, Nie S (2009) Second window for *in vivo* imaging. *Nat Nanotechnol* 4: 710–711.

20. He S, Song J, Qu J, Cheng Z (2018) Crucial breakthrough of second near-infrared biological window fluorophores: Design and synthesis toward multimodal imaging and theranostics. *Chem Soc Rev* 47: 4258–4278.

21. Yao C, Wang P, Wang R, Zhou L, El-Toni AM, Lu Y, Li X, Zhang F (2016) Facile peptides functionalization of lanthanide-based nanocrystals through phosphorylation tethering for efficient *in vivo* NIR-to-NIR bioimaging. *Anal Chem* 88: 1930–1936.

22. Sun LD, Dong H, Zhang PZ, Yan CH (2015) Upconversion of rare earth nanomaterials. *Annu Rev Phys Chem* 66: 619–642.

23. Wang F, Deng R, Wang J, Wang Q, Han Y, Zhu H, Chen X, Liu X (2011) Tuning upconversion through energy migration in core-shell nanoparticles. *Nat Mater* 10: 968–973.

24. Chen G, Shen J, Ohulchanskyy TY, Patel NJ, Kutikov A, Li Z, Song J, Pandey RK, Agren H, Prasad PN, Han G (2012) (alpha-NaYbF$_4$:Tm^{3+})/CaF$_2$ core/shell nanoparticles with efficient near-infrared to near-infrared upconversion for high-contrast deep tissue bioimaging. *ACS Nano* 6: 8280–8287.

25. Liu L, Wang S, Zhao B, Pei P, Fan Y, Li X, Zhang F (2018) Er(3+) Sensitized 1530 nm to 1180 nm second near-infrared window upconversion nanocrystals for *in vivo* biosensing. *Angew Chem Int Ed* 57: 7518–7522.

26. Song CX, Zhang SB, Zhou QA, Hai H, Zhao DF, Hui YZ (2017) Upconversion nanoparticles for bioimaging. *Nanotechnol Rev* 6: 233–242.

27. Xu CT, Axelsson J, Andersson-Engels S (2009) Fluorescence diffuse optical tomography using upconverting nanoparticles. *Appl Phys Lett* 94: 251107.

28. Sun Y, Peng JJ, Feng W, Li FY (2013) Upconversion nanophosphors NaLuF(4):Yb,Tm for lymphatic imaging *in vivo* by real-time upconversion luminescence imaging under ambient light and high-resolution X-ray CT. *Theranostics* 3: 346–353.

29. Chatterjee DK, Rufaihah AJ, Zhang Y (2008) Upconversion fluorescence imaging of cells and small animals using lanthanide doped nanocrystals. *Biomaterials* 29: 937–943.

30. Dong H, Du SR, Zheng XY, Lyu GM, Sun LD, Li LD, Zhang PZ, Zhang C, Yan CH (2015) Lanthanide nanoparticles: From design toward bioimaging and therapy. *Chem Rev* 115: 10725–10815.

31. Nyk M, Kumar R, Ohulchanskyy TY, Bergey EJ, Prasad PN (2008) High contrast *in vitro* and *in vivo* photoluminescence bioimaging using near infrared to near infrared up-conversion in Tm^{3+} and Yb^{3+} doped fluoride nanophosphors. *Nano Lett* 8: 3834–3838.

32. Hilderbrand SA, Shao FW, Salthouse C, Mahmood U, Weissleder R (2009) Upconverting luminescent nanomaterials: Application to *in vivo* bioimaging. *Chem Commun*: 4188–4190. DOI: 10.1039/B905927J.

33. Kumar R, Nyk M, Ohulchanskyy TY, Flask CA, Prasad PN (2009) Combined optical and MR bioimaging using rare earth ion doped NaYF$_4$ nanocrystals. *Adv Funct Mater* 19: 853–859.

34. Xiong L, Chen Z, Tian Q, Cao T, Xu C, Li F (2009) High contrast upconversion luminescence targeted imaging *in vivo* using peptide-labeled nanophosphors. *Anal Chem* 81: 8687–8694.

35. Zako T, Nagata H, Terada N, Utsumi A, Sakono M, Yohda M, Ueda H, Soga K, Maeda M (2009) Cyclic RGD peptide-labeled upconversion nanophosphors for tumor cell-targeted imaging. *Biochem Biophys Res Commun* 381: 54–58.

36. Cao TY, Yang TS, Gao Y, Yang Y, Hu H, Li FY (2010) Water-soluble NaYF$_4$:Yb/Er upconversion nanophosphors: Synthesis, characteristics and application in bioimaging. *Inorg Chem Commun* 13: 392–394.

37. Cheng LA, Yang K, Zhang SA, Shao MW, Lee ST, Liu ZA (2010) Highly-sensitive multiplexed *in vivo* imaging using pegylated upconversion nanoparticles. *Nano Res* 3: 722–732.

38. Liu QA, Li CY, Yang TS, Yi T, Li FY (2010) "Drawing" upconversion nanophosphors into water through host-guest interaction. *Chem Commun* 46: 5551–5553.

39. Xiong LQ, Yang TS, Yang Y, Xu CJ, Li FY (2010) Long-term *in vivo* biodistribution imaging and toxicity of polyacrylic acid-coated upconversion nanophosphors. *Biomaterials* 31: 7078–7085.

40. Zhou J, Sun Y, Du XX, Xiong LQ, Hu H, Li FY (2010) Dual-modality *in vivo* imaging using rare-earth nanocrystals with near-infrared to near-infrared (NIR-to-NIR) upconversion luminescence and magnetic resonance properties. *Biomaterials* 31: 3287–3295.

41. Zhou J, Yao LM, Li CY, Li FY (2010) A versatile fabrication of upconversion nanophosphors with functional-surface tunable ligands. *J Mater Chem* 20: 8078–8085.

42. Cao TY, Yang Y, Gao YA, Zhou J, Li ZQ, Li FY (2011) High-quality water-soluble and surface-functionalized upconversion nanocrystals as luminescent probes for bioimaging. *Biomaterials* 32: 2959–2968.

43. Dong N-N, Pedroni M, Piccinelli F, Conti G, Sbarbati A, Ramírez-Hernández JE, Maestro LM, Iglesias-de la Cruz MC, Sanz-Rodriguez F, Juarranz A, Chen F, Vetrone F, Capobianco JA, Solé JG, Bettinelli M, Jaque D, Speghini A (2011) NIR-to-NIR two-photon excited CaF_2:Tm^{3+},Yb^{3+} nanoparticles: Multifunctional nanoprobes for highly penetrating fluorescence bio-imaging. *ACS Nano* 5: 8665–8671.

44. Jeong S, Won N, Lee J, Bang J, Yoo J, Kim SG, Chang JA, Kim J, Kim S (2011) Multiplexed near-infrared *in vivo* imaging complementarily using quantum dots and upconverting $NaYF_4$:Yb^{3+},Tm^{3+} nanoparticles. *Chem Commun* 47: 8022–8024.

45. Jin JF, Gu YJ, Man CWY, Cheng JP, Xu ZH, Zhang Y, Wang HS, Lee VHY, Cheng SH, Wong WT (2011) Polymer-coated $NaYF_4$:Yb^{3+},Er^{3+} upconversion nanoparticles for charge-dependent cellular imaging. *ACS Nano* 5: 7838–7847.

46. Liu Q, Chen M, Sun Y, Chen GY, Yang TS, Gao Y, Zhang XZ, Li FY (2011) Multifunctional rare-earth self-assembled nanosystem for tri-modal upconversion luminescence/fluorescence/positron emission tomography imaging. *Biomaterials* 32: 8243–8253.

47. Liu Q, Peng JJ, Sun LN, Li FY (2011) High-efficiency upconversion luminescent sensing and bioimaging of Hg(II) by chromophoric ruthenium complex-assembled nanophosphors. *ACS Nano* 5: 8040–8048.

48. Liu Q, Sun Y, Li CG, Zhou J, Li CY, Yang TS, Zhang XZ, Yi T, Wu DM, Li FY (2011) F-18-labeled magnetic-upconversion nanophosphors *via* rare-earth cation-assisted ligand assembly. *ACS Nano* 5: 3146–3157.

49. Liu Q, Sun Y, Yang TS, Feng W, Li CG, Li FY (2011) Sub-10 nm hexagonal lanthanide-doped $NaLuF_4$ upconversion nanocrystals for sensitive bioimaging *in vivo*. *J Am Chem Soc* 133: 17122–17125.

50. Pichaandi J, Boyer JC, Delaney KR, van Veggel F (2011) Two-photon upconversion laser (scanning and wide-field) microscopy using Ln(3+)-doped $NaYF_4$ upconverting nanocrystals: A critical evaluation of their performance and potential in bioimaging. *J Phys Chem C* 115: 19054–19064.

51. Sun Y, Yu MX, Liang S, Zhang YJ, Li CG, Mou TT, Yang WJ, Zhang XZ, Li BA, Huang CH, Li FY (2011) Fluorine-18 labeled rare-earth nanoparticles for positron emission tomography (PET) imaging of sentinel lymph node. *Biomaterials* 32: 2999–3007.

52. Wong H-T, Vetrone F, Naccache R, Chan HLW, Hao J, Capobianco JA (2011) Water dispersible ultra-small multifunctional $KGdF_4$:Tm^{3+},Yb^{3+} nanoparticles with near-infrared to near-infrared upconversion. *J Mater Chem* 21: 16589.

53. Xia A, Gao Y, Zhou J, Li C, Yang T, Wu D, Wu L, Li F (2011) Core-shell $NaYF_4$:Yb^{3+},Tm^{3+}@Fe_xO_y nanocrystals for dual-modality T_2-enhanced magnetic resonance and NIR-to-NIR upconversion luminescent imaging of small-animal lymphatic node. *Biomaterials* 32: 7200–7208.

54. Yi GS, Peng YF, Gao ZQ (2011) Strong red-emitting near-infrared-to-visible upconversion fluorescent nanoparticles. *Chem Mater* 23: 2729–2734.

55. Zhan Q, Qian J, Liang H, Somesfalean G, Wang D, He S, Zhang Z, Andersson-Engels S (2011) Using 915 nm laser excited Tm^{3+}/Er^{3+}/Ho^{3+}-doped $NaYbF_4$ upconversion nanoparticles for *in vitro* and deeper *in vivo* bioimaging without overheating irradiation. *ACS Nano* 5: 3744–3757.

56. Zhou JC, Yang ZL, Dong W, Tang RJ, Sun LD, Yan CH (2011) Bioimaging and toxicity assessments of near-infrared upconversion luminescent $NaYF_4$:Yb,Tm nanocrystals. *Biomaterials* 32: 9059–9067.

57. Chen H, Zhai XS, Li D, Wang LL, Zhao D, Qin WP (2012) Water-soluble Yb^{3+},Tm^{3+} codoped $NaYF_4$ nanoparticles: Synthesis, characteristics and bioimaging. *J Alloys Compd* 511: 70–73.

58. Dai Y, Ma P, Cheng Z, Kang X, Zhang X, Hou Z, Li C, Yang D, Zhai X, Lin J (2012) Up-conversion cell imaging and pH-induced thermally controlled drug release from $NaYF_4$:Yb^{3+}/Er^{3+}@hydrogel core-shell hybrid microspheres. *ACS Nano* 6: 3327–3338.

59. Tian G, Gu Z, Zhou L, Yin W, Liu X, Yan L, Jin S, Ren W, Xing G, Li S, Zhao Y (2012) Mn^{2+} dopant-controlled synthesis of $NaYF_4$:Yb/Er upconversion nanoparticles for *in vivo* imaging and drug delivery. *Adv Mater* 24: 1226–1231.

60. Wang Z, Liu CH, Chang LJ, Li ZP (2012) One-pot synthesis of water-soluble and carboxyl-functionalized beta-$NaYF_4$:Yb,Er(Tm) upconversion nanocrystals and their application for bioimaging. *J Mater Chem* 22: 12186–12192.

61. Xia A, Chen M, Gao Y, Wu D, Feng W, Li F (2012) Gd^{3+} complex-modified $NaLuF_4$-based upconversion nanophosphors for trimodality imaging of NIR-to-NIR upconversion luminescence, X-ray computed tomography and magnetic resonance. *Biomaterials* 33: 5394–5405.

62. Xing H, Bu W, Ren Q, Zheng X, Li M, Zhang S, Qu H, Wang Z, Hua Y, Zhao K, Zhou L, Peng W, Shi J (2012) A $NaYbF_4$:Tm^{3+} nanoprobe for CT and NIR-to-NIR fluorescent bimodal imaging. *Biomaterials* 33: 5384–5393.

63. Yang TS, Sun Y, Liu Q, Feng W, Yang PY, Li FY (2012) Cubic sub-20 nm $NaLuF_4$-based upconversion nanophosphors for high-contrast bioimaging in different animal species. *Biomaterials* 33: 3733–3742.

64. Yin WY, Zhao LN, Zhou LJ, Gu ZJ, Liu XX, Tian G, Jin S, Yan L, Ren WL, Xing GM, Zhao YL (2012) Enhanced red emission from GdF_3:Yb^{3+},Er^{3+} upconversion nanocrystals by Li^+ doping and their application for bioimaging. *Chem-Eur J* 18: 9239–9245.

65. Yin WY, Zhou LJ, Gu ZJ, Tian G, Jin S, Yan L, Liu XX, Xing GM, Ren WL, Liu F, Pan ZW, Zhao YL (2012) Lanthanide-doped $GdVO_4$ upconversion nanophosphors with tunable emissions and their applications for biomedical imaging. *J Mater Chem* 22: 6974–6981.

66. Zeng SJ, Tsang MK, Chan CF, Wong KL, Fei B, Hao JH (2012) Dual-modal fluorescent/magnetic bioprobes based on small sized upconversion nanoparticles of amine-functionalized $BaGdF_5$:Yb/Er. *Nanoscale* 4: 5118–5124.

67. Zeng SJ, Tsang MK, Chan CF, Wong KL, Hao JH (2012) PEG modified BaGdF5:Yb/Er nanoprobes for multi-modal upconversion fluorescent, *in vivo* X-ray computed tomography and biomagnetic imaging. *Biomaterials* 33: 9232–9238.

68. Zhou J, Zhu X, Chen M, Sun Y, Li F (2012) Water-stable $NaLuF_4$-based upconversion nanophosphors with long-term validity for multimodal lymphatic imaging. *Biomaterials* 33: 6201–6210.

69. Zhou L, Gu Z, Liu X, Yin W, Tian G, Yan L, Jin S, Ren W, Xing G, Li W, Chang X, Hu Z, Zhao Y (2012) Size-tunable synthesis of lanthanide-doped Gd_2O_3 nanoparticles and their applications for optical and magnetic resonance imaging. *J Mater Chem* 22: 966–974.

70. Chen CP, Yee LK, Gong H, Zhang Y, Xu R (2013) A facile synthesis of strong near infrared fluorescent layered double hydroxide nanovehicles with an anticancer drug for tumor optical imaging and therapy. *Nanoscale* 5: 4314–4320.

71. Dai Y, Xiao H, Liu J, Yuan Q, Ma Pa, Yang D, Li C, Cheng Z, Hou Z, Yang P, Lin J (2013) *In vivo* multimodality imaging and cancer therapy by near-infrared light-triggered trans-platinum pro-drug-conjugated upconverison nanoparticles. *J Am Chem Soc* 135: 18920–18929.

72. Dou Q, Idris NM, Zhang Y (2013) Sandwich-structured upconversion nanoparticles with tunable color for multiplexed cell labeling. *Biomaterials* 34: 1722–1731.

73. Liu C, Gao Z, Zeng J, Hou Y, Fang F, Li Y, Qiao R, Shen L, Lei H, Yang W, Gao M (2013) Magnetic/upconversion fluorescent $NaGdF_4$:Yb,Er nanoparticle-based dual-modal molecular probes for imaging tiny tumors *in vivo*. *ACS Nano* 7: 7227–7240.

74. Liu HR, Lu W, Wang HB, Rao L, Yi ZG, Zeng SJ, Hao JH (2013) Simultaneous synthesis and amine-functionalization of single-phase $BaYF_5$:Yb/Er nanoprobe for dual-modal *in vivo* upconversion fluorescence and long-lasting X-ray computed tomography imaging. *Nanoscale* 5: 6023–6029.

75. Liu J, Bu W, Pan L, Shi J (2013) NIR-triggered anticancer drug delivery by upconverting nanoparticles with integrated azobenzene-modified mesoporous silica. *Angew Chem Int Ed* 52: 4375–4379.

76. Liu Q, Feng W, Yang TS, Yi T, Li FY (2013) Upconversion luminescence imaging of cells and small animals. *Nat Protoc* 8: 2033–2044.

77. Liu Y, Chen M, Cao TY, Sun Y, Li CY, Liu Q, Yang TS, Yao LM, Feng W, Li FY (2013) A cyanine-modified nanosystem for *in vivo* upconversion luminescence bioimaging of methylmercury. *J Am Chem Soc* 135: 9869–9876.

78. Liu Z, Ju E, Liu J, Du Y, Li Z, Yuan Q, Ren J, Qu X (2013) Direct visualization of gastrointestinal tract with lanthanide-doped $BaYbF_5$ upconversion nanoprobes. *Biomaterials* 34: 7444–7452.

79. Shen JW, Yang CX, Dong LX, Sun HR, Gao K, Yan XP (2013) Incorporation of computed tomography and magnetic resonance imaging function into NaYF4:Yb/Tm upconversion nanoparticles for *in vivo* trimodal bioimaging. *Anal Chem* 85: 12166–12172.

80. Sun Y, Liu Q, Peng JJ, Feng W, Zhang YJ, Yang PY, Li FY (2013) Radioisotope post-labeling upconversion nanophosphors for *in vivo* quantitative tracking. *Biomaterials* 34: 2289–2295.

81. Wang YF, Liu GY, Sun LD, Xiao JW, Zhou JC, Yan CH (2013) Nd^{3+}-sensitized upconversion nanophosphors: Efficient *in vivo* bioimaging probes with minimized heating effect. *ACS Nano* 7: 7200–7206.

82. Yang Y, Sun Y, Cao TY, Peng JJ, Liu Y, Wu YQ, Feng W, Zhang YJ, Li FY (2013) Hydrothermal synthesis of $NaLuF_4$: ^{153}Sm,Yb,Tm nanoparticles and their application in dual-modality upconversion luminescence and SPECT bioimaging. *Biomaterials* 34: 774–783.

83. Zhan QQ, He SL, Qian J, Cheng H, Cai FH (2013) Optimization of optical excitation of upconversion nanoparticles for rapid microscopy and deeper tissue imaging with higher quantum yield. *Theranostics* 3: 306–316.

84. Damasco JA, Chen G, Shao W, Agren H, Huang H, Song W, Lovell JF, Prasad PN (2014) Size-tunable and monodisperse Tm^{3+}/Gd^{3+}-doped hexagonal $NaYbF_4$ nanoparticles with engineered efficient near infrared-to-near infrared upconversion for *in vivo* imaging. *ACS Appl Mater Interfaces* 6: 13884–13893.

85. Gnach A, Prorok K, Misiak M, Cichy B, Bednarkiewicz A (2014) Up-converting $NaYF_4$:0.1%Tm^{3+}, 20%Yb^{3+} nanoparticles as luminescent labels for deep-tissue optical imaging. *J Rare Earth* 32: 207–212.

86. Ma C, Bian T, Yang S, Liu CH, Zhang TR, Yang JF, Li YH, Li JS, Yang RH, Tan WH (2014) Fabrication of versatile cyclodextrin-functionalized upconversion luminescence nanoplatform for biomedical imaging. *Anal Chem* 86: 6508–6515.

87. Tang SH, Wang JN, Yang CX, Dong LX, Kong DL, Yan XP (2014) Ultrasonic assisted preparation of lanthanide-oleate complexes for the synthesis of multifunctional monodisperse upconversion nanoparticles for multimodal imaging. *Nanoscale* 6: 8037–8044.

88. Wang HB, Lu W, Zeng TM, Yi ZG, Rao L, Liu HR, Zeng SJ (2014) Multi-functional $NaErF_4$:Yb nanorods: Enhanced red upconversion emission, *in vitro* cell, *in vivo* X-ray, and T_2-weighted magnetic resonance imaging. *Nanoscale* 6: 2855–2860.

89. Wei ZW, Sun LN, Liu JL, Zhang JZ, Yang HR, Yang Y, Shi LY (2014) Cysteine modified rare-earth up-converting nanoparticles for *in vitro* and *in vivo* bioimaging. *Biomaterials* 35: 387–392.

90. Xia A, Deng YY, Shi H, Hu J, Zhang J, Wu SS, Chen Q, Huang XH, Shen J (2014) Polypeptide-functionalized $NaYF_4$:Yb^{3+},Er^{3+} nanoparticles: Red-emission biomarkers for high quality bioimaging using a 915 nm laser. *ACS Appl Mater Interfaces* 6: 18329–18336.

91. Xia A, Zhang X, Zhang J, Deng Y, Chen Q, Wu S, Huang X, Shen J (2014) Enhanced dual contrast agent, Co^{2+}-doped $NaYF_4$:Yb^{3+},Tm^{3+} nanorods, for near infrared-to-near infrared upconversion luminescence and magnetic resonance imaging. *Biomaterials* 35: 9167–9176.

92. Yang DM, Dai YL, Liu JH, Zhou Y, Chen YY, Li CX, Ma PA, Lin J (2014) Ultra-small $BaGdF_5$-based upconversion nanoparticles as drug carriers and multimodal imaging probes. *Biomaterials* 35: 2011–2023.

93. Yang TS, Liu Q, Li JC, Pu SZ, Yang PY, Li FY (2014) Photoswitchable upconversion nanophosphors for small animal imaging *in vivo*. *RSC Adv* 4: 15613–15619.

94. Yao C, Wang PY, Zhou L, Wang R, Li XM, Zhao DY, Zhang F (2014) Highly biocompatible zwitterionic phospholipids coated upconversion nanoparticles for efficient bioimaging. *Anal Chem* 86: 9749–9757.

95. Yi ZG, Lu W, Xu YR, Yang J, Deng L, Qian C, Zeng TM, Wang HB, Rao L, Liu HR, Zeng SJ (2014) PEGylated $NaLuF_4$:Yb/Er upconversion nanophosphors for *in vivo* synergistic fluorescence/X-ray bioimaging and long-lasting, real-time tracking. *Biomaterials* 35: 9689–9697.

96. Yi ZG, Zeng SJ, Lu W, Wang HB, Rao L, Liu HR, Hao JH (2014) Synergistic dual-modality *in vivo* upconversion luminescence/X-ray imaging and tracking of amine-functionalized $NaYbF_4$:Er nanoprobes. *ACS Appl Mater Interfaces* 6: 3839–3846.

97. Yin DG, Wang CC, Ouyang J, Zhang XY, Jiao Z, Feng Y, Song KL, Liu B, Cao XZ, Zhang L, Han YL, Wu MH (2014) Synthesis of a novel core-shell nanocomposite $Ag@SiO_2@Lu_2O_3$:Gd/Yb/Er for large enhancing upconversion luminescence and bioimaging. *ACS Appl Mater Interfaces* 6: 18480–18488.

98. Yin WY, Tian G, Ren WL, Yan L, Jin S, Gu ZJ, Zhou LJ, Li J, Zhao YL (2014) Design of multifunctional alkali ion doped CaF_2 upconversion nanoparticles for simultaneous bioimaging and therapy. *Dalton Trans* 43: 3861–3870.

99. Zeng SJ, Wang HB, Lu W, Yi ZG, Rao L, Liu HR, Hao JH (2014) Dual-modal upconversion fluorescent/X-ray imaging using ligand-free hexagonal phase $NaLuF_4$:Gd/Yb/Er nanorods for blood vessel visualization. *Biomaterials* 35: 2934–2941.

100. Zeng SJ, Yi ZG, Lu W, Qian C, Wang HB, Rao L, Zeng TM, Liu HR, Liu HJ, Fei B, Hao JH (2014) Simultaneous realization of phase/size manipulation, upconversion luminescence enhancement, and blood vessel imaging in multifunctional nanoprobes through transition metal Mn^{2+} doping. *Adv Funct Mater* 24: 4051–4059.

101. Zhou L, He BZ, Huang JC, Cheng ZH, Xu X, Wei C (2014) Multihydroxy dendritic upconversion nanoparticles with enhanced water dispersibility and surface functionality for bioimaging. *ACS Appl Mater Interfaces* 6: 7719–7727.

102. Zhou Y, Chen WQ, Zhu JX, Pei WB, Wang CY, Huang L, Yao C, Yan QY, Huang W, Loo JSC, Zhang QC (2014) Inorganic-organic hybrid nanoprobe for NIR-excited imaging of hydrogen sulfide in cell cultures and inflammation in a mouse model. *Small* 10: 4874–4885.

103. Ge XQ, Dong L, Sun LN, Song ZM, Wei RY, Shi LY, Chen HG (2015) New nanoplatforms based on UCNPs linking with polyhedral oligomeric silsesquioxane (POSS) for multimodal bioimaging. *Nanoscale* 7: 7206–7215.

104. Liu B, Chen YY, Li CX, He F, Hou ZY, Huang SS, Zhu HM, Chen XY, Lin J (2015) Poly(acrylic acid) modification of Nd^{3+}-sensitized upconversion nanophosphors for highly efficient UCL imaging and pH-responsive drug delivery. *Adv Funct Mater* 25: 4717–4729.

105. Lv R, Yang P, He F, Gai S, Li C, Dai Y, Yang G, Lin J (2015) A yolk-like multifunctional platform for multimodal imaging and synergistic therapy triggered by a single near-infrared light. *ACS Nano* 9: 1630–1647.

106. Lv W, Yang TS, Yu Q, Zhao Q, Zhang KY, Liang H, Liu SJ, Li FY, Huang W (2015) A phosphorescent iridium(III) complex-modified nanoprobe for hypoxia bioimaging via time-resolved luminescence microscopy. *Adv Sci* 2: 1500107.

107. Ma DD, Meng LJ, Chen YZ, Hu M, Chen YK, Huang C, Shang J, Wang RF, Guo YM, Yang J (2015) $NaGdF_4$:Yb^{3+}/Er^{3+}@$NaGdF_4$:Nd^{3+}@sodium-gluconate: Multifunctional and biocompatible ultrasmall core-shell nanohybrids for UCL/MR/CT multimodal imaging. *ACS Appl Mater Interfaces* 7: 16257–16265.

108. Rieffel J, Chen F, Kim J, Chen GY, Shao W, Shao S, Chitgupi U, Hernandez R, Graves SA, Nickles RJ, Prasad PN, Kim C, Cai WB, Lovell JF (2015) Hexamodal imaging with porphyrin-phospholipid-coated upconversion nanoparticles. *Adv Mater* 27: 1785–1790.

109. Wang D, Xue B, Kong X, Tu L, Liu X, Zhang Y, Chang Y, Luo Y, Zhao H, Zhang H (2015) 808 nm driven Nd^{3+}-sensitized upconversion nanostructures for photodynamic therapy and simultaneous fluorescence imaging. *Nanoscale* 7: 190–197.

110. Yi ZG, Lu W, Liu HR, Zeng SJ (2015) High quality polyacrylic acid modified multifunction luminescent nanorods for tri-modality bioimaging, *in vivo* long-lasting tracking and biodistribution. *Nanoscale* 7: 542–550.

111. Ai F, Sun T, Xu Z, Wang Z, Kong W, To MW, Wang F, Zhu G (2016) An upconversion nanoplatform for simultaneous photodynamic therapy and Pt chemotherapy to combat cisplatin resistance. *Dalton Trans* 45: 13052–13060.

112. Jalani G, Naccache R, Rosenzweig DH, Haglund L, Vetrone F, Cerruti M (2016) Photocleavable hydrogel-coated upconverting nanoparticles: A multifunctional theranostic platform for NIR imaging and on-demand macromolecular delivery. *J Am Chem Soc* 138: 1078–1083.

113. Kale V, Pakkila H, Vainio J, Ahomaa A, Sirkka N, Lyytikainen A, Talha SM, Kutsaya A, Waris M, Julkunen I, Soukka T (2016) Spectrally and spatially multiplexed serological array-in-well assay utilizing two-color upconversion luminescence imaging. *Anal Chem* 88: 4470–4477.

114. Lei P, Zhang P, Yao S, Song S, Dong L, Xu X, Liu X, Du K, Feng J, Zhang H (2016) Optimization of Bi^{3+} in upconversion nanoparticles induced simultaneous enhancement of near-infrared optical and X-ray computed tomography imaging capability. *ACS Appl Mater Interfaces* 8: 27490–27497.

115. Levy ES, Tajon CA, Bischof TS, Iafrati J, Fernandez-Bravo A, Garfield DJ, Chamanzar M, Maharbiz MM, Sohal VS, Schuck PJ, Cohen BE, Chan EM (2016) Energy-looping nanoparticles: Harnessing excited-state absorption for deep-tissue imaging. *ACS Nano* 10: 8423–8433.

116. Reddy KL, Rai M, Prabhakar N, Arppe R, Rai SB, Singh SK, Rosenholm JM, Krishnan V (2016) Controlled synthesis, bioimaging and toxicity assessments in strong red emitting Mn^{2+} doped $NaYF_4$:Yb^{3+}/Ho^{3+} nanophosphors. *RSC Adv* 6: 53698–53704.

117. Yao C, Wang P, Li X, Hu X, Hou J, Wang L, Zhang F (2016) Near-infrared-triggered azobenzene-liposome/upconversion nanoparticle hybrid vesicles for remotely controlled drug delivery to overcome cancer multidrug resistance. *Adv Mater* 28: 9341–9348.

118. Li XL, Yi ZG, Xue ZL, Zeng SJ, Liu HR (2017) Multifunctional $BaYbF_5$:Gd/Er upconversion nanoparticles for *in vivo* tri-modal upconversion optical, X-ray computed tomography and magnetic resonance imaging. *Mater Sci Eng C Mater Biol Appl* 75: 510–516.

119. Liu H, Jayakumar MK, Huang K, Wang Z, Zheng X, Agren H, Zhang Y (2017) Phase angle encoded upconversion luminescent nanocrystals for multiplexing applications. *Nanoscale* 9: 1676–1686.

120. Liu Y, Lu Y, Yang X, Zheng X, Wen S, Wang F, Vidal X, Zhao J, Liu D, Zhou Z, Ma C, Zhou J, Piper JA, Xi P, Jin D (2017) Amplified stimulated emission in upconversion nanoparticles for super-resolution nanoscopy. *Nature* 543: 229–233.

121. Nareoja T, Deguchi T, Christ S, Peltomaa R, Prabhakar N, Fazeli E, Perala N, Rosenholm JM, Arppe R, Soukka T, Schaferling M (2017) Ratiometric sensing and imaging of intracellular pH using polyethylenimine-coated photon upconversion nanoprobes. *Anal Chem* 89: 1501–1508.

122. Qiu XC, Zhu XJ, Xu M, Yuan W, Feng W, Li FY (2017) Hybrid nanoclusters for near-infrared to near-infrared upconverted persistent luminescence bioimaging. *ACS Appl Mater Interfaces* 9: 32583–32590.

123. Reddy KL, Prabhakar N, Arppe R, Rosenholm JM, Krishnan V (2017) Microwave-assisted one-step synthesis of acetate-capped NaYF$_4$:Yb/Er upconversion nanocrystals and their application in bioimaging. *J Mater Sci* 52: 5738–5750.

124. Shin J, Kim Y, Lee J, Kim S, Jang HS (2017) Highly bright and photostable Li(Gd,Y)F$_4$:Yb,Er/LiGdF$_4$ core/shell upconversion nanophosphors for bioimaging applications. *Part Part Syst Charact* 34: 1600183.

125. Wang D, Zhu L, Pu Y, Wang JX, Chen JF, Dai LM (2017) Transferrin-coated magnetic upconversion nanoparticles for efficient photodynamic therapy with near-infrared irradiation and luminescence bioimaging. *Nanoscale* 9: 11214–11221.

126. An R, Lei P, Zhang P, Xu X, Feng J, Zhang H (2018) Near-infrared optical and X-ray computed tomography dual-modal imaging probe based on novel lanthanide-doped K$_{0.3}$Bi$_{0.7}$F$_{2.4}$ upconversion nanoparticles. *Nanoscale* 10: 1394–1402.

127. Chen C, Wang F, Wen S, Su QP, Wu MCL, Liu Y, Wang B, Li D, Shan X, Kianinia M, Aharonovich I, Toth M, Jackson SP, Xi P, Jin D (2018) Multi-photon near-infrared emission saturation nanoscopy using upconversion nanoparticles. *Nat Commun* 9: 3290.

128. Chen ZY, Liu GX, Sui JT, Li D, Song Y, Hong F, Dong XT, Wang JX, Yu WS (2018) Multifunctional PVP-Ba(2)GdF$_7$:Yb^{3+},Ho^{3+} coated on Ag nanospheres for bioimaging and tumor photothermal therapy. *Appl Surf Sci* 458: 931–939.

129. Gu B, Ye MN, Nie LN, Fang Y, Wang ZL, Zhang X, Zhang H, Zhou Y, Zhang QC (2018) Organic-dye-modified upconversion nanoparticle as a multichannel probe to detect Cu^{2+} in living cells. *ACS Appl Mater Interfaces* 10: 1028–1032.

130. Li X, Zhao H, Ji Y, Yin C, Li J, Yang Z, Tang YF, Zhang QC, Fan QL, Huang W (2018) Lysosome-assisted mitochondrial targeting nanoprobe based on dye-modified upconversion nanophosphors for ratiometric imaging of mitochondrial hydrogen sulfide. *ACS Appl Mater Interfaces* 10: 39544–39556.

131. Wang F, Wen S, He H, Wang B, Zhou Z, Shimoni O, Jin D (2018) Microscopic inspection and tracking of single upconversion nanoparticles in living cells. *Light Sci Appl* 7: 18007.

132. Xu J, Han W, Yang P, Jia T, Dong S, Bi H, Gulzar A, Yang D, Gai S, He F, Lin J, Li C (2018) Tumor microenvironment-responsive mesoporous MnO$_2$-coated upconversion nanoplatform for self-enhanced tumor theranostics. *Adv Funct Mater* 28: 1803804.

133. Abualrejal MMA, Eid K, Tian RR, Liu L, Chen HD, Abdullah AM, Wang ZX (2019) Rational synthesis of three-dimensional core-double shell upconversion nanodendrites with ultrabright luminescence for bioimaging application. *Chem Sci* 10: 7591–7599.

134. Feng Y, Wu Y, Zuo J, Tu L, Que I, Chang Y, Cruz LJ, Chan A, Zhang H (2019) Assembly of upconversion nanophotosensitizer *in vivo* to achieve scatheless real-time imaging and selective photodynamic therapy. *Biomaterials* 201: 33–41.

135. Phuong HT, Huong TT, Vinh LT, Khuyen HT, Thao DT, Huong NT, Lien PT, Minh LQ (2019) Synthesis and characterization of NaYF$_4$:Yb^{3+},Er^{3+}@silica-N=folic acid nanocomplex for bioimaginable detecting MCF-7 breast cancer cells. *J Rare Earth* 37: 1183–1187.

136. Tian RR, Zhao S, Liu GF, Chen HD, Ma LN, You HP, Liu CM, Wang ZX (2019) Construction of lanthanide-doped upconversion nanoparticle-Uelx Europaeus Agglutinin-I bioconjugates with brightness red emission for ultrasensitive *in vivo* imaging of colorectal tumor. *Biomaterials* 212: 64–72.

137. Wang H, Li YK, Yang M, Wang P, Gu YQ (2019) FRET-based upconversion nanoprobe sensitized by Nd^{3+} for the ratiometric detection of hydrogen peroxide *in vivo*. *ACS Appl Mater Interfaces* 11: 7441–7449.

138. Wang JX, Yao CJ, Shen B, Zhu XH, Li Y, Shi LY, Zhang Y, Liu JL, Wang YL, Sun LN (2019) Upconversion-magnetic carbon sphere for near infrared light-triggered bioimaging and photothermal therapy. *Theranostics* 9: 608–619.

139. Bai YD, Li YM, Wang R, Li YM (2020) Low toxicity, high resolution, and red tissue imaging in the vivo of Yb/Tm/GZO@SiO$_2$ core-shell upconversion nanoparticles. *Acs Omega* 5: 5346–5355.

140. Duan R, Li T, Duan Z, Huang F, Xia F (2020) Near-infrared light activated nucleic acid cascade recycling amplification for spatiotemporally controllable signal amplified mRNA imaging. *Anal Chem* 92: 5846–5854.

141. Feng Y, Chen H, Wu Y, Que I, Tamburini F, Baldazzi F, Chang Y, Zhang H (2020) Optical imaging and pH-awakening therapy of deep tissue cancer based on specific upconversion nanophotosensitizers. *Biomaterials* 230: 119637.

142. Huang J, Li J, Zhang X, Zhang W, Yu Z, Ling B, Yang X, Zhang Y (2020) Artificial atomic vacancies tailor near-infrared II excited multiplexing upconversion in core-shell lanthanide nanoparticles. *Nano Lett* 20: 5236–5242.
143. Li H, Tan M, Wang X, Li F, Zhang Y, Zhao L, Yang C, Chen G (2020) Temporal multiplexed *in vivo* upconversion imaging. *J Am Chem Soc* 142: 2023–2030.
144. Li YT, Liu JM, Wang ZC, Jin J, Liu YL, Chen CY, Tang ZY (2020) Optimizing energy transfer in nanostructures enables *in vivo* cancer lesion tracking *via* near-infrared excited hypoxia imaging. *Adv Mater* 32: 1907718.
145. Liu J, Pan L, Shang C, Lu B, Wu R, Feng Y, Chen W, Zhang R, Bu J, Xiong Z, Bu W, Du J, Shi J (2020) A highly sensitive and selective nanosensor for near-infrared potassium imaging. *Sci Adv* 6: eaax9757.
146. Liu J, Zhang R, Shang C, Zhang Y, Feng Y, Pan L, Xu B, Hyeon T, Bu W, Shi J, Du J (2020) Near-infrared voltage nanosensors enable real-time imaging of neuronal activities in mice and zebrafish. *J Am Chem Soc* 142: 7858–7867.
147. You Y, Cheng S, Zhang L, Zhu Y, Zhang C, Xian Y (2020) Rational modulation of the luminescence of upconversion nanomaterials with phycocyanin for the sensing and imaging of myeloperoxidase during an inflammatory process. *Anal Chem* 92: 5091–5099.
148. Fukushima S, Furukawa T, Niioka H, Ichimiya M, Sannomiya T, Miyake J, Ashida M, Araki T, Hashimoto M (2016) Synthesis of Y_2O_3 nanophosphors by homogeneous precipitation method using excessive urea for cathodoluminescence and upconversion luminescence bioimaging. *Opt Mater Express* 6: 831–843.
149. Wang J, Wang F, Wang C, Liu Z, Liu X (2011) Single-band upconversion emission in lanthanide-doped $KMnF_3$ nanocrystals. *Angew Chem Int Ed* 50: 10369–10372.
150. Li RB, Ji ZX, Dong JY, Chang CH, Wang X, Sun BB, Wang MY, Liao YP, Zink JI, Nel AE, Xia T (2015) Enhancing the imaging and biosafety of upconversion nanoparticles through phosphonate coating. *ACS Nano* 9: 3293–3306.
151. Park YI, Nam SH, Kim JH, Bae YM, Yoo B, Kim HM, Jeon KS, Park HS, Choi JS, Lee KT, Suh YD, Hyeon T (2013) Comparative study of upconverting nanoparticles with various crystal structures, core/shell structures, and surface characteristics. *J Phys Chem C* 117: 2239–2244.
152. Xu J, Gulzar A, Yang P, Bi H, Yang D, Gai S, He F, Lin J, Xing B, Jin D (2019) Recent advances in near-infrared emitting lanthanide-doped nanoconstructs: Mechanism, design and application for bioimaging. *Coord Chem Rev* 381: 104–134.
153. Wu Q, Huang B, Peng X, He S, Zhan Q (2017) Non-bleaching fluorescence emission difference microscopy using single 808 nm laser excited red upconversion emission. *Opt Express* 25: 30885–30894.
154. Zou W, Visser C, Maduro JA, Pshenichnikov MS, Hummelen JC (2012) Broadband dye-sensitized upconversion of near-infrared light. *Nat Photonics* 6: 560–564.

11 Multiphoton Downconversion *Quantum Cutting*

Dechao Yu and Qinyuan Zhang

CONTENTS

11.1 CONCEPT OF QUANTUM CUTTING

With the comprehensive understanding of basic scientific issues about luminescence and luminescent materials, as well as the intense drive of their widespread and irreplaceable applications, luminescence and luminescent materials have been developed fast over the past 100 years.[1,2] At present, commercial lanthanide (Ln) ions-based phosphors that are employed in different fluorescent tubes, X-ray imaging and color televisions have quantum efficiencies (QEs) close to the theoretical limit (100%).[3,4] However, such optimal phosphors do not exhibit good luminescent performance with excitation in vacuum ultraviolet (VUV) region (energy more than 50,000 cm^{-1}, corresponding to wavelength shorter than 200 nm).[5] It is known that the Xe/Ne discharge produces VUV photons spanning 147–190 nm, energy of which is two more times as that of visible photon (400–700 nm). Hence, for devices of plasma display panel (PDP) and mercury-free fluorescent lamp (MFFL), energies of incident VUV photons are mostly dissipated as heat during luminescence. In practical applications, screen efficiency of current PDP phosphors is only half that of cathode-ray tube used in television sets.[6–8] As a consequence, to prepare high-performance VUV-excited luminescent materials and even to develop phosphors with QE greater than unity are considerably necessary in enhancing the luminescent properties of PDP and MFFL devices.

Quantum cutting refers to a luminescent process of cutting one high-energy photon into two or more low-energy photons, which, also known as photon cascade emission, photon (quantum) splitting or downconversion, is just reverse to upconversion luminescence. Through such absorption and reemission, QE exceeding 100% is realized. In 1957, Dexter theoretically treated the possibilities

DOI: 10.1201/9781003098676-11

of QE more than unity[9]: (i) for a single activator having three energy levels, if the separation between adjacent levels is large enough to match visible light, can yield two visible photons by two-step consecutive radiative transitions; (ii) for a couple of ions, a VUV-excited sensitizer may simultaneously transfer its energy to two nearby activators, and each activator accepts half energy of the VUV photon for visible photon reemission. In 1974, Sommerdijk *et al.* and Piper *et al.* experimentally recorded the stepwise transitions of $^1S_0 \rightarrow {}^3P_j, {}^1I_6$ and $^3P_0 \rightarrow {}^3F_j, {}^3H_j$ in Pr^{3+}-doped fluorides with VUV excitation of 185 nm, which emit photons at about 407 and 488 nm, respectively, having internal QE of about 140%.[10,11] In 1999, Meijerink *et al.* demonstrated that under excitation of one VUV photon around 172 nm $LiGdF_4:Eu^{3+}$ phosphors can yield two red photons by Eu^{3+} activators following stepwise resonant energy transfer (ET) from Gd^{3+} to Eu^{3+}.[12] A value of about 200% was achieved for the QE of Gd^{3+}–Eu^{3+} couple. These research outcomes extremely stimulate scientists' interests especially in the fields related to PDP, MFFL, etc. To obtain highly efficient VUV-to-visible quantum cutting, some other Ln^{3+} ions like Er^{3+}, Tm^{3+}, Pr^{3+} and Nd^{3+} were added as co-dopants to further improve the involved ET efficiency.[13-23] For example, in the $Gd^{3+}/Er^{3+}/Tb^{3+}$ ternary ions, multiphoton cutting was efficiently realized for green emissions.[19] However, because the first-step transition of Pr^{3+} splitter gives rise to photon emission in UV ~340 nm or near-UV ~400 nm, such photons are invisible to human eyes and cannot play an effective role in display and lighting.[3,24,25] Thus, numerous attempts were thoroughly performed by co-doping suitable acceptor (A) ions like Eu^{3+}, Er^{3+}, Tm^{3+}, Mn^{2+} and Cr^{3+} to efficiently convert the first-step (near-)UV part to visible photon emission by resonant cross-relaxation (CR) from Pr^{3+} to A ions.[26-32]

In recent years, intensive investigations and explorations have been devoted to efficiently convert sunlight to electricity using photovoltaic cells. However, relatively low photon conversion efficiency of solar cells has become one of shortcomings to limit their mass production and extensive applications in long term. Nowadays, the commercial single crystalline Silicon (*c*-Si)-based solar cells (band gap, E_g, ~1.12 eV) only have average conversion efficiency around 15%–18%,[33] while the Shockley-Queisser efficiency limit is about 30% for a single-junction solar cell at E_g of 1.1 eV.[34] How to enhance the conversion efficiency of solar cell becomes an international frontier subject. Besides altering the structure design of solar cells, sunlight spectral modification through near-infrared (NIR)-to-visible upconversion and that through UV/blue-to-NIR downconversion have been proposed as feasible schemes to significantly improve the efficiency of solar cells.[35-43] In theory, downconversion of an incident high-energy sunlight photon to two low-energy photons that are just absorbed by solar cells will greatly reduce the thermalization losses caused upon direct absorption of high-energy photon. Trupke *et al.* in 2002 established a model utilizing downconversion layer on the top of single-junction solar cells with E_g ~1.1 eV,[44] where each sunlight photon with energy more than 2.2 eV is ideally downconverted to two photons, each of which has energy just above the E_g, and absorption of which creates two pairs of "electron-hole". Through detailed-balance calculation, the efficiency of such new-generation solar cell will reach up to 40%, far more than the Shockley-Queisser limit.[34,44] In experiments, Vergeer *et al.* and Zhang *et al.* first demonstrated NIR quantum cutting for the Tb^{3+}/Yb^{3+} and Ln^{3+}/Yb^{3+} (Ln = Tb, Tm) couples under blue excitation, respectively.[45-48] Basically, Ln^{3+} (Ln = Tb, Tm) donor ion efficiently absorb one blue photon (~470 nm) to excite its long-lived energy state at about 20,000 cm^{-1}, and the energy can be simultaneously transferred to two nearby Yb^{3+} ions onto the $^2F_{5/2}$ state at about 10,000 cm^{-1}, which efficiently emit two NIR photons at about 1000 nm. Through such cooperative ET process, QE close to 2 was realized. Because the reemitted photon ~1000 nm has energy just above the E_g of *c*-Si, in principle, such efficient NIR quantum cutting material is suited as downconverting layer for *c*-Si solar cells, thereby greatly increasing conversion efficiency. Then immense amounts of researches were carried out to develop new type of NIR quantum cutting for Ln^{3+}/Yb^{3+} couples (Ln = Pr, Dy, Er, Ho, Nd) and to significantly optimize their luminescence properties.[49-57] On the other hand, the recent interesting reports on NIR multiphoton photon splitting and/or NIR multiphoton cascade emission in Ln^{3+} (Ln = Er, Dy, Ho, Tm, Pr, Nd) single ion greatly enrich the investigations of NIR downconversion systems[58–65] and stimulated the development of novel efficient NIR quantum cutting materials as well as the exploration of their application potentials.

11.2 VUV-TO-VISIBLE QUANTUM CUTTING

Investigation onto the energy-level structures of Ln ions as well as their colorful electronic transitions in near-UV, visible and NIR regions has been lasting for more than 100 years. However, clear recognition to highly excited levels of Ln ions in VUV region was almost achieved in the recent 30 years. With the fast development of modern testing techniques, studies of highly excited levels of Ln ions as well as the visible quantum cutting-involved VUV levels and transitions become more and more in-depth and comprehensive. All the related factors to influence luminescence and QE of Ln ions, such as photon absorption, ET paths, concentration quenching, temperature quenching and nonradiative relaxation (NR) were studied systematically with numerous progresses.

Typical VUV-to-visible downconversion can be categorized into four types: (i) photon cascade emission of Ln^{3+} (Ln = Pr, Tm, Er, Gd) single ion; (ii) CR processes-induced quantum cutting in Pr^{3+}/A (A = Eu^{3+}, Er^{3+}, Mn^{2+}, Cr^{3+}) couple; (iii) quantum cutting through stepwise ET scheme in dual and ternary ions such as VUV-to-visible photon cutting of Gd^{3+}/Eu^{3+} couple; and (iv) quantum cutting via cooperative ET process such as VUV-to-visible downconversion in $CaSO_4$:Tb^{3+}, Na^+ representative.

11.2.1 Photon Cascade Emission of Ln^{3+} (Ln = Pr, Tm, Er, Gd) Single Ion

In principle, radiative transitions from Pr^{3+} ion under VUV excitation are divided into two categories: inter-configurational $4f5d \rightarrow 4f^2$ and intra-configurational $4f^2 \rightarrow 4f^2$.[2,66,67] As shown in Figure 11.1(a), the $^1S_0$ level of Pr^{3+} is located at about 47,000 cm^{-1}, while the energetic location of the $4f5d$ states of Pr^{3+} directly depends on properties of host lattice. As the lowest $4f5d$ state lies below the $^1S_0$ state of Pr^{3+}, the high energetic excitation will lead to broadband emission due to parity-allowed transitions from the lowest $4f5d$ state to $4f^2$ states, as illustrated in Figure 11.1(a). Thus, no photon cascade emission is observed. Here, the $4f5d \rightarrow 4f^2$ emissions are dominantly situated in UV region, which may be useful as materials for fast-response scintillator. Whereas, hosts with weak crystal fields, low phonon energies, large band gap energies, large cation-anion distances and large coordination numbers for the substitution enable Pr^{3+} $4f5d$ states to be above Pr^{3+} $^1S_0$ level.[67]

In Pr^{3+}-doped YF_3 and cubic (α-)$NaYF_4$ phosphors, the relative weak crystal fields of Pr–F bonds make the lowest $4f5d$ energy state above Pr^{3+} $^1S_0$ state (Figure 11.1(a)), where with VUV excitation photon cascade emission was first experimentally observed by Sommerdijk *et al.* and Piper *et al.*[10,11]

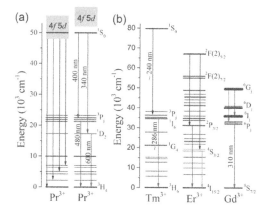

FIGURE 11.1 (a) Energy-level diagram of Pr^{3+} identifying scheme of photon cascade emission, and (b) that of Tm^{3+}, Er^{3+} and Gd^{3+} illustrating the mechanism of involved photon cascade emission, respectively.

Once Pr^{3+} $^1S_0$ state is populated by direct excitation of a deep-UV photon (~215 nm) or by fast NR from the intra-configurational $4f5d$ state excited by a VUV photon (Figure 11.1(a)), a first-step transition of $^1S_0 \rightarrow {}^3P_2$, $^1I_6$ will dominantly occur for a photon emission at about 400 nm, and then either $^3P_{0,1} \rightarrow {}^3H_j$ or $^1D_2 \rightarrow {}^3H_4$ takes place for a second photon emission in visible range (450–700 nm). In practice, the most intense second-step emission can either be $^3P_0 \rightarrow {}^3H_4$ (~480 nm) or $^1D_2 \rightarrow {}^3H_4$ (~600 nm), depending upon the (non-)radiative decay possibilities from $^3P_0$ to $^1D_2$ state, which is generally determined by the maximum vibration frequency (i.e., phonon energy) of host lattice. According to energy gap law, the band gap of $^3P_0 \rightarrow {}^1D_2$ of Pr^{3+}, approximately 3400 cm^{-1}, can be easily bridged by emitting 3–4 phonons in borates, phosphates, silicates, etc.[68–74] Besides, from the $^3P_j/^1D_2$ excited state, 95% of radiations are situated in the visible range,[63] which suggests that only the large luminescence branching ratios from $^1S_0$ to $^3P_2$,$^1I_6$ can satisfy the requirement of efficient visible emissions from Pr^{3+} splitters. Moreover, by using fluorescence branching ratios of the $^1S_0$ and $^3P_0$ states of Pr^{3+}, QE of such photon cascade emission is approximated to be 140%.[10,11] These results stimulated significant interests to prepare efficient visible quantum cutting materials for new generation of lighting and display. Various fluoride-/oxide-based hosts such as aluminates, borates and sulfates were successfully employed to realize visible quantum cutting of Pr^{3+} ions.[30–32,67,70,75–78]

Tm^{3+} ion is a possible candidate for photon cascade emission due to the favorable energy distribution of its excited electronic levels, as schematically depicted in Figure 11.1(b).[79] Because Tm^{3+} ion and Pr^{3+} ion lie in the same Ln series, the general electronic energy-level structure of Tm^{3+} can be roughly viewed as that of Pr^{3+} but expanded in energy by two times due to the doubled strength of the spin-orbit interaction changing from Pr^{3+} to Tm^{3+}.[80] The $^1S_0$ level of Tm^{3+}, located at 75,000 cm^{-1}, however, is not available for quantum cutting because of inevitable severe interference by the lower $5d$ levels. On the other hand, the energy-level scheme of Tm^{3+} offers the possibility to radiate from the $^3P_2$ excited state through the $^1G_4$ intermediate state to the ground state, thereby leading to emission of two visible photons. Such quantum cutting phenomenon has been reported in $LaF_3:Tm^{3+}$ but with visible QE less than 50%.[81] The lowering of visible QE is caused by infrared emissions due to the multiphonon relaxation of $^3F_{2,3} \rightarrow {}^3F_4$ of Tm^{3+}.

As shown in Figure 11.1(b), Er^{3+} ion has also got an energy-level structure with abundant excited states in visible region, which could meet with the basic requirement of visible photon cascade emission. Under VUV excitation, photon cascade emission has been observed in Er^{3+}-doped $LiYF_4$ and LaF_3 from the $^2F(2)_{5/2}$ state with $^2P_{3/2}$ or $^4S_{3/2}$ acting as an intermediate level,[13] as illustrated in Figure 11.1(b); however, the most intense emissions were located in the UV region. By means of luminescence branching ratios calculated on the basis of Judd-Ofelt theory, the maximum visible QE for $LaF_3:Er^{3+}$ was found to be 112%,[82] which is not sufficient for a Xe-discharge lamp phosphor. Nevertheless, if certain co-dopant(s) can efficiently convert the UV emissions into visible within the photon cascade emission of Er^{3+}, efficient quantum cutting phosphors with Er^{3+} ion will be quite promising.

In 1997, Wegh *et al.* made a systematic spectroscopic study of the $4f^7$ energy levels of Gd^{3+} in $LiYF_4$ in the VUV region (50,000–70,000 cm^{-1}).[14] Energy-level diagram of Gd^{3+} in Figure 11.1(b) schematically shows the possibilities for photon cascade emission.[16] Energy gap between the $^6G_{7/2}$ level and the lower lying ^6D_j levels is about 8000 cm^{-1}, which significantly suppresses multiphonon relaxation but benefits radiative transitions. Following a VUV photon excites Gd^{3+} into the ^6G_j states, photon cascade emission is indeed detected in $LiYF_4:5\%$ Gd^{3+} by a first-step transition of $^6G_j \rightarrow {}^6P_j$ for a red photon emission at about 600 nm, and a second-step $^6P_j \rightarrow {}^8S_{7/2}$ transition for a UV photon around 313 nm.[14] Notably, similar to the case of Er^{3+} splitter, the dominant UV emission makes low visible QE for the Gd^{3+} splitter. Moreover, from the Gd^{3+} ^6G_j excited state, there exist several decay pathways, among which only the $^6G_j \rightarrow {}^6P_j$ transition corresponds to photon emission in visible region.

Disregarding the emission wavelengths (colors) from Ln^{3+} splitters, up to now, demonstration of such interesting photon cascade emissions as well as that of quantum cutting luminescence for all Ln^{3+}-couples are always extracted from excitation and emission spectra, which is actually indirect and ambiguous. De Jong *et al.* recently designed a new experimental setup to directly test the occurrence of photon bunching in the emission from this macroscopic photon splitting Pr^{3+}-doped $NaLaF_4$ phosphors.[83] In case the photon splitting happened, the non-Poissonian photon emission statics is possibly observed as an unambiguous proof. Ultimately, the two-photon cascade emission of Pr^{3+} was successfully validated through such "improved" photon correction technique (single-photon detector with ultrahigh efficiencies, high time resolution and low dark counts in visible region is necessary to this setup).

11.2.2 Quantum Cutting Via Cross-Relaxation in Pr^{3+}/A $(A = Eu^{3+}, Er^{3+}, Mn^{2+}, Cr^{3+})$ Couple

Intense level of research activities has provided a special focus on Pr^{3+}-doped phosphors for efficient VUV-to-visible quantum cutting since its first demonstration of photon cascade emission in the early 1970s.[10,11] However, in all different hosts, the parity forbidden intra-$4f$ transition of $^1S_0 \rightarrow {}^3P_2$, $^1I_6$ can only yield violet photon around 400 nm, which comprises 60%–80% transitions from Pr^{3+} $^1S_0$ state but invisible to human eyes and useless to lighting and display.[24,84] So that optimization to efficiently convert the near-UV color of Pr^{3+} splitter to visible was performed by adding a second appropriate dopant as acceptor (A) ion. In principle, the requirements for such A ion are: (i) strong absorption around 400 nm; (ii) predominant emission situated in visible region with high sensitivity of the human eyes; and (iii) the A co-dopant should not have any energy levels negatively interfering with the Pr^{3+}-involved photon cascade emission.

Mn^{2+} ion has been considered a suitable A ion for Pr^{3+} splitter because its main absorption transitions within the $3d^5$ configuration well coincide with the violet emission of Pr^{3+} and its emissions efficiently span green-red range.[85] As schematically depicted in Figure 11.2, transition of $^1S_0 \rightarrow {}^3P_2$, $^1I_6$ of Pr^{3+} at about 405 nm and that of $^1S_0 \rightarrow {}^1D_2$ at 340 nm match well with the absorption of $^6A_{1g} \rightarrow {}^4E_g, {}^4A_{1g}$ of Mn^{2+} at 410 nm and that of $^6A_{1g} \rightarrow {}^4T_{2g}$ at 350 nm, respectively.[84] Under VUV excitation, CR ET (cross-ET) processes will efficiently take place from Pr^{3+} to Mn^{2+}, through which the violet part of Pr^{3+} splitter will finally excite the $^4T_{1g}$ state of Mn^{2+} following fast NR from the upper $^4E_g, {}^4A_{1g}$

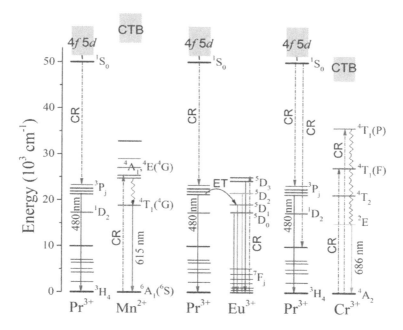

FIGURE 11.2 Energy-level diagrams of Pr^{3+}, Mn^{2+}, Eu^{3+} and Cr^{3+} ions schematically depicting VUV-to-visible quantum cutting through CR processes in Pr^{3+}/Mn^{2+}, Pr^{3+}/Eu^{3+} and Pr^{3+}/Cr^{3+} couples, respectively.

and $^4T_{2g}$ states. Hence, a first green/red photon will be obtained from the $^4T_{1g} \rightarrow ^6A_{1g}$ transition of Mn^{2+}. Subsequently, the residual energy in the $^3P_0$ or $^1D_2$ state of Pr^{3+} results in a second photon emission in blue or red color by occupying the $^3H_4$ ground state of Pr^{3+}. Ideally visible QE greater than unity is realized for the Pr^{3+}/Mn^{2+} couple if the involved cross-ET process works efficiently and the luminescence efficiencies of Pr^{3+} and Mn^{2+} are high enough.

In practice, visible quantum cutting in the Pr^{3+}/Mn^{2+}-co-doped $SrAl_{12}O_{19}$ was stated by a US patent,[86] while van der Kolk *et al.* and Kück *et al.* revealed no efficient cross-ET from Pr^{3+} to Mn^{2+} in $SrAlF_5$, $CaAlF_5$, $NaMgF_3$, $KMgF_3$, SrY_2F_8, YF_3, CaF_2, $LiBaF_3$, etc.[28,29] These results may be due to the selection rules, or the total spin of the system change or there is no empty excited state with suitable symmetry near the $^1S_0$ state of Pr^{3+}. Whereas, visible quantum cutting for the Pr^{3+}/Mn^{2+} couple was successfully identified via the abovementioned cross-ET process in SrB_4O_7 and $LaMgB_5O_{10}$ hosts.[30,87,88] Taking SrB_4O_7 as an example, the expected spectral overlaps between the emission spectra of SrB_4O_7:Pr^{3+} sample and the excitation spectra of SrB_4O_7:Mn^{2+} sample have clearly been detected in 330–430 nm, which promotes the ET process from Pr^{3+} to Mn^{2+} for the Pr^{3+}/Mn^{2+}-co-doped SrB_4O_7 sample.[30,88] Although SrB_4O_7:Mn^{2+} shows almost no emission from Mn^{2+}, in SrB_4O_7:Pr^{3+}, Mn^{2+}, besides the line emissions of Pr^{3+}, a broadband emission at 615 nm from Mn^{2+} was observed due to efficient occurrence of cross-ET from Pr^{3+} to Mn^{2+}. By comparing the intensities of Pr^{3+} $^1S_0$ emission in SrB_4O_7:Pr^{3+} with that in SrB_4O_7:Pr^{3+}, Mn^{2+}, the cross-ET efficiency was evaluated to be 43%, resulting in an optimal QE of 143%.[30,88] Similarly, in $LaMgB_5O_{10}$:5% Pr^{3+}, 5% Mn^{2+} sample the cross-ET efficiency of $Pr^{3+} \rightarrow Mn^{2+}$ was calculated to be about 30%.[87,89]

A shown in Figure 11.2, efficient CR between Pr^{3+}: $^1S_0 \rightarrow ^1I_6$ and Eu^{3+}: $^7F_{0,1} \rightarrow ^5D_3$, $^5L_6$ is expected for the Pr^{3+}/Eu^{3+} couple because of the spectral overlap around 400 nm between Pr^{3+} emission and Eu^{3+} absorption.[26,90] However, experimentally the red/orange emissions from Eu^{3+} were not observed in YF_3:Pr^{3+}, Eu^{3+} upon VUV excitation into Pr^{3+} $4f5d$ bands.[91] To determine whether the absence of Eu^{3+} emission is caused by a low ET rate or not, the critical distance between Pr and Eu ions in YF_3:Pr^{3+}, Eu^{3+} was calculated to be 0.40 nm, quite similar to the distance (0.359 nm) between nearest neighbors in YF_3 host lattice.[26] This result indicates that for nearest neighboring pairs of Pr^{3+}–Eu^{3+}, the ET rate will be comparable to the radiative decay rate from the $^1S_0$ level of Pr^{3+}. That is, the cross-ET process and the following emissions from Eu^{3+} should occur. To have an insight into why the Eu^{3+} emissions were absent, luminescence spectra as well as time-resolved luminescence spectra were measured upon excitation into Pr^{3+} $^1S_0$ level for YF_3:1% Pr^{3+}, x% Eu^{3+} ($x = 0, 5, 10$). All analysis reveals that the addition of Eu^{3+} ion will unexpectedly introduce quenching center for the Pr^{3+} $^1S_0$ emissions. This quenching process is attributed to fast NR through a Eu^{2+}–Pr^{4+} metal-to-metal charge-transfer state (CTS). As a result, the interaction between Eu^{3+} and Pr^{3+} does not lead to any efficient Eu^{3+} emission.[25,26]

Cr^{3+} ion was chosen as a promising A ion for Pr^{3+} splitter due to its absorption transitions well resonant to Pr^{3+} $^1S_0$ emissions.[92] Nie *et al.* recently reported the cross-ET processes of the Pr^{3+}/Cr^{3+} couple for efficient visible quantum cutting in $SrAl_{12}O_{19}$ and $CaAl_{12}O_{19}$ phosphors.[76] In the hosts, Pr^{3+} and Cr^{3+} dopants will replace Ca^{2+} (Sr^{2+}) and Al^{3+} sub-lattice sites, respectively. The average Ca-to-Al distance (4.64 Å) is much shorter than the Ca-to-Ca distance (10.95 Å), which may facilitate efficient ET from an excited Pr^{3+} ion to a nearby Cr^{3+} ion. As shown in Figure 11.2, upon VUV excitation onto Pr^{3+} $4f5d$ states, cross-ET process from Pr^{3+} to Cr^{3+} will first occur through either Pr^{3+}: $^1S_0 \rightarrow ^3P_2$, $^1I_6$ + Cr^{3+}: $^4A_2 \rightarrow ^4T_1(F)$ or Pr^{3+}: $^1S_0 \rightarrow ^1G_4$ + Cr^{3+}: $^4A_2 \rightarrow ^4T_1(P)$, and then following fast NR, the populated ^2E state of Cr^{3+} emits red photon around 686 nm to occupy the $^4A_2$ ground state. In the second step, energy in the $^3P_2$, $^1I_6$ excited states quickly populating the $^3P_0$ state of Pr^{3+} is mainly transferred to another nearby Cr^{3+} ion for the second red photon emission, whereas in the other case of Pr^{3+} $^1G_4$ populated, the second emission is located in NIR region, invisible to human eyes. Due to the occurrence of quantum cutting in the Pr^{3+}/Cr^{3+} couple, an increase of the relative emission intensity of $^3P_0$ or $^1G_4$ is expected in comparison with Pr^{3+} single ion, which is just a spectral proof about the occurrence of cross-ET process. Through a series of calculations, visible QE for $CaAl_{12}O_{19}$:1% Pr^{3+}, 5% Cr^{3+} and $SrAl_{12}O_{19}$:2% Pr^{3+}, 5% Cr^{3+} is approximated to 143% and 147%, respectively.[76]

11.2.3 Quantum Cutting Via Stepwise Energy Transfer in Dual and Ternary Ions

Distinct from the photon cascade emission of Pr^{3+} splitters and the Pr^{3+}/A couple to convert near-UV part of Pr^{3+}, Wegh *et al.* in 1999 innovatively achieved visible quantum cutting for red emissions in the Gd^{3+}/Eu^{3+}-co-doped $LiGdF_4$ phosphors through a two-step ET process,[12] as schematically shown in Figure 11.3. As a Gd^{3+} ion was excited by a VUV photon around 200 nm onto the ^6G_j state, following fast energy migration over Gd^{3+} sub-lattice sites, the first-step cross-ET between Gd^{3+}: $^6G_j \rightarrow ^6P_j$ (~16,950 cm^{-1}) and Eu^{3+}: $^7F_0 \rightarrow ^5D_0$ (17,000 cm^{-1}) will make Eu^{3+} $^5D_0$ occupied for a red photon emission at about 610 nm, and subsequently energy in the populated Gd^{3+} ^6P_j state resonantly excites another neighboring Eu^{3+} ion into the ^5H_j state, from which energy will nonradiatively occupy Eu^{3+} ^5D_j ($j = 0, 1, 2, 3$) state to finally bring about a second photon emission mainly at 590/610 nm. By taking into account the two-step ET processes involved in the quantum cutting for two photon emissions and the process that emits only one photon by direct ET approach, Wegh *et al.* determined the efficiency of cross-ET process by,

$$\frac{P_{CR}}{P_{CR} + P_{DT}} = \frac{R(^5D_0/^5D_{1,2,3})_{6_{G_j}} - R(^5D_0/^5D_{1,2,3})_{6_{I_j}}}{R(^5D_0/^5D_{1,2,3})_{6_{I_j}} + 1} \qquad (11.1)$$

where P_{CR} (P_{DT}) is the probability for cross-ET (direct ET) from Gd^{3+} to Eu^{3+}, $R(^5D_0/^5D_{1,2,3})$ is the ratio of Eu^{3+} $^5D_0$ emission intensity to Eu^{3+} $^5D_{1,2,3}$ emission intensities and the subscript (^6G_j or ^6I_j) indicates the excitation level for which the ratio is obtained.[12,18] A value is calculated to be ~0.9, showing that nine of ten Gd^{3+} ions in the ^6G_j excited state relax by the two-step ET process to Eu^{3+}. That is, on the assumption of no energy losses caused by impurities and defects, a visible QE of 190% can be achieved. Because the radiative emissions from Eu^{3+} activators are mainly located at about 600 nm, rather useful to lighting and display, such results have been significantly stimulating the development of Gd^{3+}/Eu^{3+}-co-doped system.[93–98]

Moreover, attempts of measuring external QE of a quantum cutting system were performed mainly in two approaches: (i) comparison of emission spectra of $LiGdF_4$:Eu^{3+} phosphors with that of commercial Y_2O_3:Eu^{3+} phosphors, and (ii) evaluation based on diffuse reflectance spectra.[99]

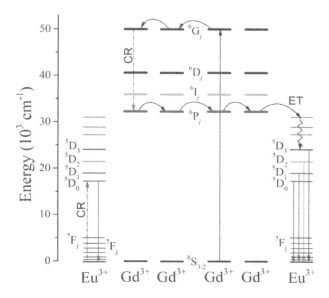

FIGURE 11.3 Schematic energy-level diagrams of Gd^{3+} and Eu^{3+} ions showing mechanism of quantum cutting via two-step ET processes from Gd^{3+} to Eu^{3+} ions.

The both approaches obtained similar value of QE, ~32%, for $LiGdF_4:Eu^{3+}$ phosphors under VUV ~202 nm excitation, which, however, is quite lower than the internal QE ~190%. In practice, the parity-forbidden intra-$4f$ transitions of Gd^{3+} can only absorb a small part of VUV excitation energy, while most of VUV excitation energy is dissipated as heat by defects and/or impurities in the crystal lattice. Such effects will definitely lead to much lower external QE. Besides, QE value of the Gd^{3+}/Eu^{3+} quantum cutter shows clear dependence on host matrix and concentration of Eu^{3+} ions. For example, $BaF_2:Gd^{3+}$, Eu^{3+} has QE ~194%, much more than that ~140% for $KLiGdF_5:Gd^{3+}$, Eu^{3+}.[95,100] Some studies indicate that optimization of Eu^{3+} concentration can expectedly adjust the radii distance of Gd^{3+}-to-Eu^{3+}, which thereby maximizes the cross-ET rate from Gd^{3+} to Eu^{3+} to achieve an enhanced QE value.[98,100]

Apart from quantum cutting in the Gd^{3+}/Eu^{3+} couple, visible quantum cutting was also observed in $K_2GdF_5:Tb^{3+}$ and $BaGdF_5:Tb^{3+}$ phosphors.[21,101] In contrast, in the Gd^{3+}/Tb^{3+}-co-doped phosphors, quantum cutting only happens upon VUV excitation onto Tb^{3+} ion. $K_2GdF_5:Tb^{3+}$ and/or $BaGdF_5:Tb^{3+}$ phosphors feature broad excitation lines centered at 212 and 172 nm because of the spin-allowed transitions from the $^7F_6$ ground state to low-spin $4f^75d$ state of Tb^{3+}. As schematically illustrated in Figure 11.4(a), upon excitation of 212 nm into $4f^75d$ state of Tb^{3+} ion, a first-step CR process happens between Tb^{3+}: $4f5d \rightarrow {}^5D_2$ and Tb^{3+}: $^3F_j \rightarrow {}^5D_4$ or between Tb^{3+}: $4f5d \rightarrow {}^5D_3$ and Gd^{3+}: $^8S_{7/2} \rightarrow {}^6I_j$, which enables a first emission of green photon (~542 nm) from Tb^{3+} or UV photon (~315 nm) from Gd^{3+}, and then the remaining energy in the $^5D_2$ state will nonradiatively decay to the $^5D_3$ or $^5D_4$ level of Tb^{3+} for a second emission of blue or green photon. Similarly, under excitation of 172 nm (Figure 11.4(b)), a first-step CR process will happen between Tb^{3+} ions to emit one green photon, and then the residual energy will be de-excited in two probable paths: (i) through fast NR the populated $^5D_3$ or $^5D_4$ level of Tb^{3+} emit a second photon in blue/green range; (ii) following resonant ET from Tb^{3+} to Gd^{3+}, transition of $^6P_j \rightarrow {}^8S_{7/2}$ of Gd^{3+} will lead to a second UV photon emission. Here, it should be noted that once a UV photon ~274 nm directly excites $K_2GdF_5:Tb^{3+}$ sample the

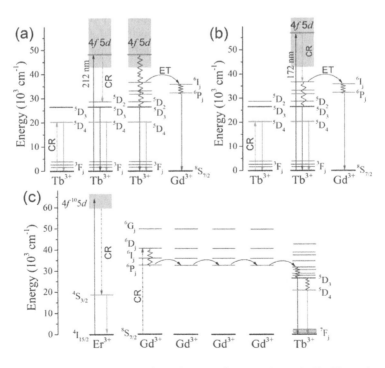

FIGURE 11.4 Energy-level diagrams of Gd^{3+}, Tb^{3+} and Er^{3+} ions schematically illustrating the stepwise ET-involved quantum cutting in the Tb^{3+}/Gd^{3+} couple under excitation of (a) 212 and (b) 172 nm, and (c) that in the Er^{3+}/Tb^{3+}/Gd^{3+} ternary system.

above-mentioned quantum cutting will not happen, and a direct ET from Gd^{3+} ^6P_j excited state to Tb^{3+} is observed in experiments.

Efficiency of cross-ET between Tb^{3+} and a nearby Gd^{3+} for the Gd^{3+}/Tb^{3+} couple can be regularly calculated as,[21,101]

$$\frac{P_{CR}}{P_{CR}+P_{DT}} = \frac{R(^5D_4/\text{rest})_{Tb^{3+}} - R(^5D_4/\text{rest})_{Gd^{3+}}}{R(^5D_4/\text{rest})_{Tb^{3+}}+1} \tag{11.2}$$

where P_{CR} represents the probability of CR process and P_{DT} is the probability of direct ET process. $R(^5D_4/\text{rest})$ is the ratio of photoluminescence intensity of $^5D_4$ to that attributed to the $^5D_3$ of Tb^{3+} and the $^6P_{7/2}$ of Gd^{3+}, and the subscript indicates excitation from Tb^{3+} or Gd^{3+}. For the sample of K_2GdF_5:11% Tb^{3+}, a calculated value of QE was found to be 189% and 187% under excitation of 212 and 172 nm, respectively.[21] In the case of $BaGdF_5$:15% Tb^{3+}, QE was evaluated to be 168% and 180% under excitation of 215 and 187 nm, respectively.[101] Moreover, dependence of cross-ET efficiency on Tb^{3+} concentration was investigated in K_2GdF_5:x% Tb^{3+} (x from 1 to 11) samples under VUV excitation at 172 and 212 nm, respectively. The results indicate that the CR efficiency increase monotonically by increasing Tb^{3+} concentration upon both excitation cases.

In 2000, Wegh et al. reported visible quantum cutting in the $Gd^{3+}/Tb^{3+}/Er^{3+}$ triply doped $LiGdF_4$ phosphors, as schematically shown in Figure 11.4(c).[19] In such system, quantum cutting will take place upon VUV excitation of Er^{3+} onto $4f^{10}5d$ levels: energy in the lowest $4f^{10}5d$ state will excite a nearby Gd^{3+} ion via CR process between Er^{3+}: $4f^{10}5d \rightarrow {^4S_{3/2}}$ and Gd^{3+}: $^8S_{7/2} \rightarrow {^6D_j}$, in which the occupied $^4S_{3/2}$ state of Er^{3+} radiate one green photon to populate the $^4I_{15/2}$ ground state, and following fast NR from the ^6D_j to ^6P_j states of Gd^{3+}, the excited energy will hop among the Gd^{3+} sub-lattice sites and finally to excite an encountered Tb^{3+} ion, thereby emitting a second blue or green photon due to the transition of $^5D_3 \rightarrow {^7F_j}$ or $^5D_4 \rightarrow {^7F_j}$, respectively. Similarly, efficiency of such quantum cutting system can be estimated by comparing the emission intensity of Er^{3+} $^4S_{3/2}$ excited state under VUV excitation (quantum cutting happened) with that under UV excitation (no quantum cutting happened for "extra" $^4S_{3/2}$ emission). A calculation expression is typically presented as[19]:

$$\frac{P_{ET}}{P_{ET+}+P_{ET-}} = \frac{R(^4S_{3/2}/\text{rest})_{Er^{3+}} - R(^4S_{3/2}/\text{rest})_{Gd^{3+}}}{R(^4S_{3/2}/\text{rest})_{Gd^{3+}}+1} \tag{11.3}$$

where P_{ET+} and P_{ET-} are the probabilities of the desired and the undesired ET possibilities from the $4f^{10}5d$ states of Er^{3+} to Gd^{3+}, respectively. $R(^4S_{3/2}/\text{rest})_{Er^{3+}}$ and $R(^4S_{3/2}/\text{rest})_{Gd^{3+}}$ are the intensity ratios of Er^{3+} $^4S_{3/2}$ emission to all remaining emissions upon Er^{3+} $f \rightarrow d$ excitation and upon the Gd^{3+} ^6I_j excitation, respectively. If the nonradiative losses due to energy migration at the defects and impurities are ignored, the QE can be given by $P_{ET}/(P_{ET+}+P_{ET-})+1$. Finally, efficiency for the CR steps of $LiGdF_4$:1.5% Er^{3+}, 0.3% Tb^{3+} was calculated to be 30%, resulting in a maximum QE of 130%.[19]

11.2.4 Quantum Cutting Via Cooperative Energy Transfer in $CaSO_4$:Tb^{3+}, Na^+

Similar to fluorides, $CaSO_4$ has a large bandgap (>10 eV) and a weak crystal field, which is considered as a promising VUV-stimulated quantum cutting candidate. Within SO_4^{2-} complexes, the CTS-mediated anion exciton was reported to create the self-trapped exciton (STE)-like state. In 2007, Lakshmanan et al. reported a second-order cooperative ET from one SO_4^{2-} STE to two nearby Tb^{3+} ions for green emissions in $CaSO_4$:Tb^{3+}, Na^+ under VUV excitation of 147 nm.[102] Using Na^+ as a charge compensator, $CaSO_4$:Tb^{3+}, Na^+ phosphors exhibit two prominent excitation bands centered at 147 (~8.44 eV) and 216 nm (~5.74 eV), which are from the charge-transfer excitations within SO_4^{2-} complexes and the $4f^8 \rightarrow 4f^75d$ transitions of Tb^{3+} ions, respectively. Hence, a direct CTS-mediated VUV absorption is far off resonance in $CaSO_4$:Tb^{3+}, Na^+. The mobility of self-trapped anion excitons is restricted, unlike that of host-relaxed free excitons. This makes the Stokes-shifted luminescence upon 147 nm excitation a highly unlikely and inefficient process in $CaSO_4$ host. On the other hand, the STE states created within SO_4^{2-} complexes feature maximum emission energy of 8.44 eV, which is energetically possible to trigger a second-order quantum cutting to two nearby Tb^{3+} ions.[102] In such quantum cutting, the upper limit of energy absorption by the Tb^{3+} acceptors will be 4.22 eV, just matching with the $^5H_5$ level, and the lower limit is determined to be 3.26 eV, specifically corresponding to the $^5D_3$ level of Tb^{3+} ion. Between the $^5H_5$ and $^5D_3$ levels, there are nine operative levels (some of which are wide absorption) to satisfy the resonance condition of a second-order quantum cutting.

By comparing integrated intensity of emissions in 320–720 nm of $CaSO_4$:Tb^{3+}, Na^+ excited at 147 nm with that of YBO_3:Tb^{3+} standard, QE was evaluated to be 117% ± 8% for the $CaSO_4$:4% Tb^{3+}, 12% Na^+ sample.[102] It is believed that a reduction in the grain size as well as an improvement in its morphology would further increase its luminescence efficiency under VUV excitation for real applications. However, negative factors such as spectral overlap of host-anion excitons and charge compensator, luminescence quenching by flux and concentration quenching effects operating at energy states higher than $^5D_4$ or $^5D_3$ levels of Tb^{3+} may prevent the $CaSO_4$:Tb^{3+}, Na^+ to reach up to a theoretical maximum QE of 200%.

11.3 UV/BLUE-TO-NIR QUANTUM CUTTING IN LN³⁺/YB³⁺ COUPLE

With the in-depth studies of quantum cutting luminescence as well as the emerging needs for efficient NIR fluorescence in optoelectronic applications, the research has not only been focused on VUV-excited visible quantum cutting but also been extended to NIR region.[25,42,43] Usually energy of a UV/visible photon is two or more times as that of a NIR photon, it is energetically possible to realize quantum cutting for efficient NIR emission with QE more than unity.[9,38] Moreover, Ln^{3+} ions naturally have abundant energy levels spanning UV-visible-NIR regions, where absorption/excitation, mutual ET processes and emissions can be readily tuned and fitted according to the application requirements. Overall NIR quantum cutting for the Ln^{3+}/Yb^{3+} couples can be typically divided into three categories: (i) second-order cooperative quantum cutting such as the blue-excited Tb^{3+}/Yb^{3+} representative couple; (ii) first-order resonant quantum cutting such as the blue-excited Pr^{3+}/Yb^{3+} couple and the UV-/blue-excited Ho^{3+}/Yb^{3+} couple; and (iii) phonon energy-mediated quantum cutting in the Tm^{3+}/Yb^{3+} couple.

11.3.1 Second-Order Cooperative Quantum Cutting in Ln³⁺/Yb³⁺ (Ln = Tb, Tm) Couple

On the basis of a second-order cooperative ET model, Vergeer *et al.* demonstrated a NIR quantum cutting for 1000 nm emission in the Tb^{3+}/Yb^{3+}-co-doped YPO_4 phosphors under blue excitation of 489 nm,[45] and meanwhile Zhang *et al.* systematically explored the quantum cutting in the Ln^{3+}/Yb^{3+} (Ln = Tb, Tm)-co-doped $GdBO_3$ and $GdAl_3(BO_3)_4$ phosphors under blue excitation of 470–490 nm.[46–48] As illustrated in Figure 11.5(a), Tb^{3+} and Tm^{3+} ions both have blue-emitting states (i.e., Tb^{3+} $^5D_4$ and Tm^{3+} $^1G_4$) that are located at about 20,000 cm^{-1}, while Yb^{3+} ions is blank in visible region but has unique $^2F_{5/2}$ excited state at about 10,000 cm^{-1}. Moreover, Tb^{3+} and Tm^{3+} ions do not have intermediate levels well resonant to the $^2F_{5/2}$ excited state of Yb^{3+}. Correspondingly, as blue light excites the Ln^{3+}/Yb^{3+} (Ln = Tb, Tm) couple, energy of one blue photon absorbed by a Ln^{3+} ion can be cooperatively transferred to two nearby Yb^{3+} ions, each of which finally emits one NIR photon around 1000 nm via efficient $^2F_{5/2} \rightarrow ^2F_{7/2}$ transition of Yb^{3+} (Figure 11.5(b) and (c)). As a result of efficient cooperative ET, intensities of emissions from the blue-emitting state of Ln^{3+} as well as its fluorescence lifetime (Figure 11.5(d)) decrease obviously with increasing Yb^{3+} concentration, while Yb^{3+} emission at about 1000 nm first increases to a maximum and then decreases due to concentration quenching.[48] Moreover, Vergeer *et al.* theoretically modeled the time-resolved signals of Tb^{3+} $^5D_4 \rightarrow ^7F_6$ versus Yb^{3+} concentration by Monte Carlo methods and proved that only the cooperative

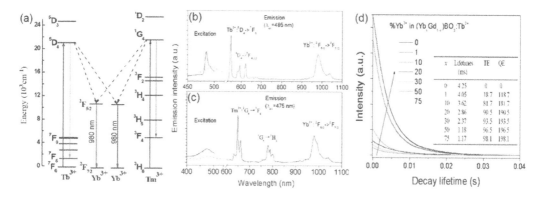

FIGURE 11.5 (a) Energy-level diagrams of Tb^{3+}, Tm^{3+} and Yb^{3+} schematically showing the blue-to-NIR quantum cutting via a second-order cooperative ET model for the Tb^{3+}/Yb^{3+} and Tm^{3+}/Yb^{3+} couples, respectively, (b and c) corresponding photoexcitation and visible-NIR emission spectra of (b) Tb^{3+}/Yb^{3+} and (c) $Tm^{3+}/$ Yb^{3+} couples and (d) decay curves of Tb^{3+}: $^5D_4 \rightarrow {}^7F_5$ fluorescence as a function of Yb^{3+} concentration.

ET of 1 Tb^{3+}-to-2 Yb^{3+} works well in YPO_4:1% Tb^{3+}, x% Yb^{3+}.[45] Based on the ET efficiency (η_{ET}) from Ln^{3+} to Yb^{3+}, QE of a cooperative quantum cutting can be calculated as:

$$\eta = \eta_{Ln}\left(1 - \eta_{x\%\ Yb}\right) + 2\eta_{x\%\ Yb} \tag{11.4}$$

where η_{Ln} is the QE of Ln^{3+} donor ions, which is usually set to be 1. Through calculation, an optimal value of 188% was reported for the $YbPO_4$:1% Tb^{3+} sample,[45] and a comparative value of 198.1% was got for the $GdBO_3$:1% Tb^{3+}, 75% Yb^{3+} sample.[46] Because energy of photon around 1000 nm perfectly matches with the E_g of c-Si (~1.12 eV), ideal Ln^{3+}/Yb^{3+} quantum cutting materials were expected to serve as an effective downconverting layer on the top of commercial Si-based solar cells.[35,36,44] Driven by efficient NIR quantum cutting as well as its promising applications in photovoltaics, numerous studies have been performed to optimize the Ln^{3+}/Yb^{3+}-co-doped systems and to develop some other new NIR quantum cutting systems even for various application purposes.[103–116] Thereinto, the Ln^{3+}/Yb^{3+}-co-doped oxyfluoride glass ceramics with QE close to 2 seems to be promising candidates as downconverting layer due to their excellent chemical and mechanical properties as well as high transparency to visible light.[106–108]

11.3.2 First-Order Resonant Quantum Cutting in $Ln^{3+}/$ Yb^{3+} (Ln = Pr, Dy, Er, Nd, Ho) Couple

In the earlier stage of NIR quantum cutting research, the second-order cooperative ET mechanism was also proposed for the Pr^{3+}/Yb^{3+} couple under blue excitation, while distinct phenomena were experimentally observed in comparison with the well-demonstrated Tb^{3+}/Yb^{3+} cooperative downconverter: (i) Yb^{3+} concentration was optimized to be about 2% in the Pr^{3+}/Yb^{3+} co-doping, which is roughly 2 times lower than the case in the Tb^{3+}/Yb^{3+} couple; (ii) at the optimal Yb^{3+} concentration ET efficiency of Pr^{3+} $^3P_0 \rightarrow Yb^{3+}$ $^2F_{5/2}$ was about 1.5–2 times higher than that of Tb^{3+} $^5D_4 \rightarrow Yb^{3+}$ $^2F_{5/2}$.[45–48,117] In 2009, van der Ende *et al.* proposed a first-order resonant ET model for NIR quantum cutting in the Pr^{3+}/Yb^{3+}-co-doped fluorides (SrF_2 and $LiYF_4$) under blue excitation of 440 nm.[49,52] As illustrated in Figure 11.6(a), Pr^{3+} $^1G_4$ state is located at about 9800 cm^{-1} and has energy difference ~500 cm^{-1} to Yb^{3+} $^2F_{5/2}$ excited state, requiring absorption of one to two phonons for compensation, which easily happens according to energy gap law. Following the first-step cross-ET between Pr^{3+}: $^3P_0 \rightarrow {}^1G_4$ and Yb^{3+}: $^2F_{7/2} \rightarrow {}^2F_{5/2}$, the occupied Pr^{3+} $^1G_4$ state will nearly resonantly transfer its energy to another nearby Yb^{3+} ion. The 1 blue-to-2 NIR downconversion will be efficiently achieved for the Pr^{3+}/Yb^{3+} couple. Similarly intense NIR emission at 1000 nm was observed at quite low Yb^{3+} concentration ~5%.[49] Moreover, comparison of Monte Carlo modeling to a second-order cooperative ET with that to a first-order resonant ET clearly reveals that the first-order resonant ET works better for quantum cutting in the Pr^{3+}/Yb^{3+} couple.[52] Because a first-order ET rate is generally 1000-fold faster than a second-order ET rate, creation of NIR resonant quantum cutting materials has become a hot research topic. Theoretically, amongst all Ln^{3+} ions, the Dy^{3+}, Er^{3+}, Nd^{3+} and Ho^{3+} ions all have energy levels lying at about 20,000 cm^{-1} and meanwhile have intermediate levels located at ~10,000 cm^{-1}, as illustrated in Figure 11.6(b)–(e), which are preconditions to serve as resonance donor to Yb^{3+} acceptor in the Ln^{3+}/Yb^{3+}-co-doped system.[38] In practice, many research groups all over the world reported efficient NIR quantum cutting in the Ln^{3+}/Yb^{3+} (Ln = Dy, Er, Nd, Ho) couple by a first-order resonant ET mechanism.[50,51,53–57,118–129] Bai *et al.* reported resonant quantum cutting in the Dy^{3+}/Yb^{3+}-co-doped zeolites under excitation of 430 nm (Figure 11.6(b)), but the slight enhancement of NIR emission at 1000 nm by co-doping Yb^{3+} ions suggests a low efficiency mainly caused by severe multiphonon relaxation from the unstable $^6H_{5/2}$ intermediate state.[53] In contrast, Er^{3+} has a long-lived $^4I_{11/2}$ intermediate state (lying at 10,130 cm^{-1}), but its blue-excited $^4F_{7/2}$ state is quite unstable due to its smaller energy gap to the following $^2H_{11/2}$ state, ΔE ~1250 cm^{-1} (Figure 11.6(c)). In practice, only bromide host like $Cs_3Y_2Br_9$ with low phonon energy around 180 cm^{-1} can effectively suppress multiphonon relaxation from the $^4F_{7/2}$ to $^2H_{11/2}$ state and enable a first-order quantum cutting in the Er^{3+}/Yb^{3+} couple with photon-cutting efficiency close to 200%.[51] Similarly, Nd^{3+} ion has a well-known $^4F_{5/2}$ state (lying at 12,500 cm^{-1}), but its blue-excited states around $^2G_{9/2}$ are unstable and nonradiative at all (Figure 11.6(d)).[50,125] An expected first-order quantum cutting from Nd^{3+}

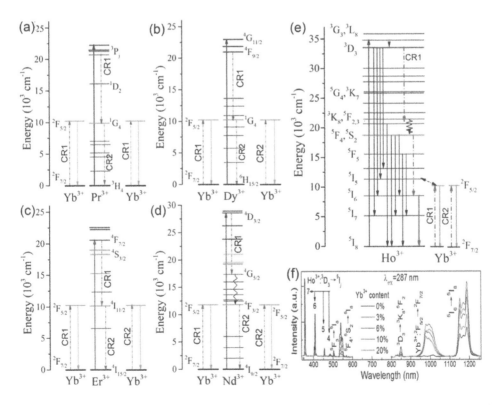

FIGURE 11.6 (a–e) Schematic energy-level diagrams illustrating NIR quantum cutting by means of a first-order resonant ET model in (a) Pr^{3+}/Yb^{3+}, (b) Dy^{3+}/Yb^{3+}, (c) Er^{3+}/Yb^{3+}, (d) Nd^{3+}/Yb^{3+} and (e) Ho^{3+}/Yb^{3+} couples, respectively and (f) Yb^{3+} concentration-dependent emission spectra of the Ho^{3+}/Yb^{3+}-co-doped β-NaYF$_4$ phosphors under excitation of 287 nm.

$^2G_{9/2}$ state maybe inefficient or ignorable, but fortunately a long-lived $^4D_{3/2}$ state (situated around 28,000 cm^{-1}) above $^2G_{9/2}$ state has been proved by Meijer *et al.* to make efficient two-step resonant quantum cutting in the Nd^{3+}/Yb^{3+} couple.[50] Here, we may notice that the abundant energy-level structures of Ln^{3+} ions provide endless possibilities to design efficient NIR quantum cutting materials, but we should be cautious enough to avoid mistakes or controversies.[123,125] The Ho^{3+}/Yb^{3+} couple will be an appropriate example: Ho^{3+} ion has metastable levels of $^5I_5$ at 11,200 cm^{-1}, $^5F_3$ at 20,600 cm^{-1}, $^5G_4$ at 25,800 cm^{-1}, $^3D_3$ at 33,100 cm^{-1}, etc., as well as so many energy gaps resonant to the Yb^{3+} absorption (Figure 11.6(e)).[55] Lin *et al.* and Yu *et al.* experimentally revealed that energy in the $^5F_3$ excited state will nonradiatively populate the next $^5S_2$, $^5F_4$ state of Ho^{3+}, from which CR process between Ho^{3+}: $^5S_2$, $^5F_4 \rightarrow {}^5I_6$ and Yb^{3+}: $^2F_{7/2} \rightarrow {}^2F_{5/2}$ first occurs efficiently for one photon emission at 1000 nm, and then energy in the $^5I_6$ excited state decays to ground state by emitting another NIR photon around 1180 nm.[54,57] At the same time, Deng *et al.* reported two-step resonant quantum cutting in the Ho^{3+}/Yb^{3+} couple from the UV-excited $^5G_4$ state of Ho^{3+} with $^5F_5$ serving as an intermediate level.[56] Furthermore, under excitation of 287 nm, Yu *et al.* demonstrated that from the UV-excited $^3D_3$ state of Ho^{3+} (Figure 11.6(e) and (f)) a three-photon NIR quantum cutting occurs first by two-step cross-ET from the $^3D_3$ and $^5S_2$,$^5F_4$ states of Ho^{3+} to the Yb^{3+} $^2F_{5/2}$ state for two photon emissions at 1000 nm and a third-step transition of $^5I_6 \rightarrow {}^5I_8$ of Ho^{3+} for one more photon emission at 1180 nm.[55] Correspondingly, with increasing Yb^{3+} concentration (Figure 11.6(f)), intensities of emissions from the $^3D_3$ and $^5S_2$, $^5F_4$ states of Ho^{3+} decrease markedly, while the intensity of Yb^{3+} emission solely increases to a maximum at 10% Yb^{3+} co-dopant, and that of Ho^{3+} emission from the $^5I_6$ state monotonously rises up without obvious quenching effects. Based on the experimental data and theoretical speculations, an internal QE was estimated to be about 246% for the hexagonal (β-) NaYF$_4$:1% Ho^{3+}, 20% Yb^{3+} sample.

11.3.3 Phonon Energy-Mediated Quantum Cutting in Tm^{3+}/Yb^{3+} Couple

NIR quantum cutting for the Ln^{3+}/Yb^{3+} couple not only can be designed and optimized by the type and concentration of co-dopants and by varying excitation lights, but actually can also be mediated by fluorescence environments surrounding the Ln^{3+}/Yb^{3+} couple.[130,131] Most recently Yu *et al.* demonstrated an interesting phenomenon of phonon energy-mediated NIR quantum cutting in the Tm^{3+}/Yb^{3+} couple under blue excitation of about 470 nm.[131] Tm^{3+} ion has a long-lived $^1G_4$ blue-emitting state but no intermediate level resonant with the $^2F_{5/2}$ excited state of Yb^{3+} at 10,000 cm^{-1}, and the well-known $^1G_4 \rightarrow {}^3H_5$ transition of Tm^{3+} has a large mismatch about 2000–3000 cm^{-1} with the $^2F_{7/2} \rightarrow {}^2F_{5/2}$ absorption of Yb^{3+} (Figure 11.7(a) and (b)). Over a quite long time, the Tm^{3+}/Yb^{3+} couple has been widely studied for its second-order cooperative downconversion for 1000 nm emission under excitation of 470 nm (Figure 11.7(c) and (d)).[48,108,132–135] Whereas it can be predicted that, in certain host lattices with maximum phonon energy more than 500 cm^{-1}, the energy mismatch between Tm^{3+}: $^1G_4 \rightarrow {}^3H_5$ and Yb^{3+}: $^2F_{7/2} \rightarrow {}^2F_{5/2}$ can be readily compensated for efficient cross-ET process, which, finally emitting one NIR photon around 1000 nm per absorbing one blue photon, will compete with or even to disable the cooperative ET of 1 Tm^{3+}-to-2 Yb^{3+} (Figure 11.7(e) and (f)). Four kinds of host matrixes, YBO$_3$, Y$_3$Al$_5$O$_{12}$ (YAG), Y$_2$O$_3$ and NaYF$_4$, are employed to tune the maximum phonon energy from 1050, to 860, to 600 and finally to 370 cm^{-1}.[131,136] In principle, the energy gap ~2000–3000 cm^{-1} can be bridged by emitting at least two, three, four and six phonons in YBO$_3$, YAG, Y$_2$O$_3$ and NaYF$_4$, respectively. Simply judged from the energy gap law (multiphonon relaxation involved less than five phonons would be efficient), phonon energy "surrounding" the Tm^{3+}/Yb^{3+} couple essentially determine the ET pathways for the blue-to-NIR conversion luminescence. Most importantly, on the basis of a second-order cooperative ET mechanism and a first-order resonant ET mechanism, through Monte Carlo methods, Yu *et al.* modeled the time-resolved luminescence of Tm^{3+} $^1G_4$ excited state as a function of Yb^{3+} concentration in YBO$_3$ (Figure 11.7(g) and (h)), YAG (Figure 11.7(i) and (j)), Y$_2$O$_3$ (Figure 11.7(k) and (l)) and NaYF$_4$ (Figure 11.7(m) and (n)), respectively.[131] The cooperative quantum cutting within the Tm^{3+}/Yb^{3+} couple can only be well fit

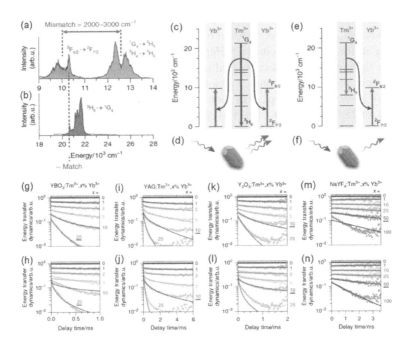

FIGURE 11.7 (a) NIR emission spectrum of YBO_3:0.1% Tm^{3+}, 2% Yb^{3+} under excitation of 465 nm, and (b) corresponding excitation spectrum monitoring Tm^{3+} $^1G_4$ emission. (c) Cooperative ET from one Tm^{3+} ion to two nearby Yb^{3+} ions, and (d) one-to-two photon cutting eventually achieved with Yb^{3+} reemitting two 1000 nm photons. (e) Phonon-assisted CR of Tm^{3+}: $^1G_4 \rightarrow Yb^{3+}$: $^2F_{5/2}$, and (f) one-to-one photon conversion by Yb^{3+} reemitting at 1000 nm. (g–n) Dependence of Tm^{3+}-to-Yb^{3+} ET dynamics on Yb^{3+} concentration in (g and h) YBO_3, (i and j) YAG, (k and l) Y_2O_3 and (m and n) $NaYF_4$. Panels (g, i, k, m) show results of a fit to model of phonon-assisted CR, whereas panels (h, j, l, n) show those to a model of cooperative ET process.

in $NaYF_4$, while the multiphonon-assisted cross-ET model works perfect for the Tm^{3+}/Yb^{3+} couple in other three hosts. Specifically value of the Tm^{3+}-to-Yb^{3+} CR strength in $NaYF_4$ is $(7 \pm 6) \times 10^1$ $Å^6$ ms^{-1}, which is thus two orders of magnitude slower than that in the other three high-phonon-energy hosts. Through calculations, the overall QE up to 132% can be rationally obtained for the sample of $NaYbF_4$:Tm^{3+}. Besides, it should be noted that, for the high-phonon-energy hosts co-activated by the Tm^{3+}/Yb^{3+} couple, following the phonon-assisted cross-ET process energy in the occupied $^3H_5$ state of Tm^{3+} will fast relax to the lower lying $^3H_4$ state, which further radiates a second mid-infrared photon around 1800 nm to populate the $^3H_6$ ground state.[62,64,130]

11.4 NIR PHOTON SPLITTING IN LN^{3+} SINGLY DOPED SYSTEMS

Over the past ten more years, systematic investigations had been performed in the Ln^{3+}/Yb^{3+} couple mainly for the sake of cutting one incident UV/blue photon into two or more NIR ~1000 nm photons. Such ideal photon cutting for an enhanced 1000 nm photon emission would significantly satisfy the perspectives to improve the photo-response of c-Si solar cells. However, current popular pursuit of preferable quantum cutting effects for the Ln^{3+}/Yb^{3+} couple always introduces many controversies and even erroneous conclusions about the involved ET mechanisms.[39,48–52,124–135,137–139] Therefore, discarding Yb^{3+} acceptor to well investigate the electronic transitions of Ln^{3+}-self in NIR region as well as the ET interaction between Ln^{3+} ions would be crucial to elucidate the mechanisms of efficient quantum cutting for novel downconverting materials.[64,140,141] Moreover, some other fields like solar cells with smaller E_g (E_g of Ge ~0.67 eV, E_g of $CuInSe_2$ ~1.0 eV, E_g of GaSb ~0.7 eV),[142] new NIR laser, NIR photon counter and bioimaging through NIR window (780–1700 nm) also urgently call for novel NIR photon-splitting materials for efficient versatile fluorescence in a much broad region.

11.4.1 NIR Photon Cascade Emission of Ln³⁺ (Ln = Dy, Pr, Ho, Tm) Single Ion

On the basis of the theoretical discussion about photon cascade emission by Dexter,[9] splitting the energy of one UV/visible photon into two or more "small" parts, each corresponding to one NIR photon emission, is completely probable for Ln^{3+} single ion. Such NIR photon cascade emission was first reported in Dy^{3+} single ion, which has a long-lived $^4F_{9/2}$ blue-emitting level at about 21,000 cm⁻¹ and several intermediate levels of ^6H_j around 10,000 cm⁻¹, as schematically shown in Figure 11.8(a).[60] Once the $^4F_{9/2}$ level is excited, one NIR photon at about 834 (991) nm will be emitted to make energy occupy the $^6H_{7/2}$, $^6F_{9/2}$ ($^6H_{5/2}$) level. Subsequently the populated $^6H_{5/2}$ state will give out a second NIR photon to occupy the $^6H_{15/2}$ ground state. Such consecutive transitions from the $^4F_{9/2}$ level with ^6H_j acting as intermediate level constitute the two-photon cascade emission, but its efficiency is rather low because of severe multiphonon relaxation from the ^6H_j intermediate levels as well as severe CR between Dy^{3+} ions. Similarly, the two-step NIR transitions from the $^3P_0$ blue-emitting state of Pr^{3+} take place efficiently with the $^1G_4$ state serving as an effective intermediate level, as illustrated in Figure 11.8(b).[63] An absorbed blue photon around 440 nm is split into two NIR photons reemitted at 915 and 990 nm, respectively. Because the multiple NIR emission bands around 1000 nm overlap completely, such stepwise photon cascade emission is hard to be spectroscopically identified. To prove the involved ET dynamics of Pr^{3+}, Yu et al. innovatively compared the time-resolved emission spectra of $GCPr_{0.1}$ in 770–1150 nm upon pulsed light respectively exciting the $^1G_4$ and $^1D_2$ states of Pr^{3+}. As shown in Figure 11.8(c), only NIR emission band emerges initially (~7.5 μs) at about 915 nm, directly visualizing the preferential occurrence of $^3P_0 \rightarrow {}^1G_4$ transition of Pr^{3+}. With prolonging delay time, the 915 nm emission increases rapidly in intensity and broadens asymmetrically because of contribution of additional NIR fluorescence around 900 nm. Another emission peak at 1040 nm appears weakly at delay time ~8.0 μs, and then its intensity rises markedly with the presence of an extra shoulder at ~990 nm from delay time around 19.5 μs (Figure 11.8(c)). However, in Figure 11.8(d), there does not present any emission at 915 nm and other NIR region at ~7.5 μs, which further validates that the 915 nm emission is only from the $^3P_0 \rightarrow {}^1G_4$ transition. From earlier delay time ~8.0 μs, the

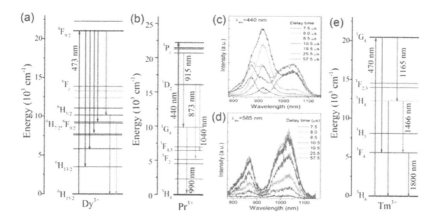

FIGURE 11.8 (a) Energy-level diagram of Dy^{3+} ion illustrating its NIR photon cascade emission by stepwise transitions. (b) Similar to part (a) but for Pr^{3+} ion, and (c and d) corresponding time-resolved emission spectra under pulsed light excitation of (c) 440 and (d) 585 nm, respectively. (e) Energy-level diagram of Tm^{3+} ion schematically elucidating the 1 blue-to-3 NIR photon cascade emission by stepwise transitions.

two NIR emission bands at 873 and 1040 nm emerge simultaneously and rise up with an almost identical trend. Moreover, in good consistence with Figure 11.8(c), the shoulder around 990 nm becomes more obvious in Figure 11.8(d) as delay time increases. These results reveal that the later appearance of 990 nm emission does originate from the $^1G_4$ excited state of Pr^{3+}.

On the basis of Judd-Ofelt theory, Yu *et al.* proposed a feasible method to evaluate the QE for NIR photon cascade emission by considering luminescence branching ratios.[60,63] For a specific Ln^{3+} singly doped material, luminescence branching ratios of the original blue-emitting levels of photon splitter and that of the intermediate level can be rationally calculated. By taking into account the stepwise NIR transitions, an overall QE, η_{all}, can be expressed as:

$$\eta_{all} = \eta_{vis} + \eta_{near-IR} \approx \eta_{orig} + \left(\beta_{first-step} + \beta_{second-step}\right)\eta_{orig}\eta_{inter} \qquad (11.5)$$

where $\beta_{first-step}$ and $\beta_{second-step}$ are luminescence branching ratio of first-step NIR transition and that of second-step NIR transition of the Ln^{3+} photon splitter, and η_{orig} (η_{inter}) is luminescence efficiency of original (intermediate) level, which is typically set to be 1 by ignoring all luminescence-quenching effects. Through this, QE of $GdVO_4$:Dy^{3+} photon-splitting material was estimated to be about 111%,[60] and that of $GCPr_{0.1}$ was reported to be 104%,[63] both of which are quite low. However, such feeble stepwise NIR transitions just corresponding to emission at 1000 nm would undoubtedly become the resonant ET routes once Yb^{3+} co-doped with such Ln^{3+} splitter. For instance, efficiency close to 200% for NIR 1000 nm emission was obtained for the 5% Yb^{3+} co-doped with SrF_2:0.1% Pr^{3+} phosphors.[49]

Following the in-depth understanding of stepwise transitions, a sequential three-step NIR photon cascade emission of Tm^{3+} single ion was demonstrated in various systems.[62,143-149] As schematically illustrated in Figure 11.8(e), upon a blue photon exciting Tm^{3+} into the $^1G_4$ state, the first NIR photon can be emitted through the $^1G_4 \rightarrow ^3H_4$ transition at about 1165 nm. Then energy in the populated $^3H_4$ state efficiently decays to the $^3F_4$ state, yielding the second NIR photon around 1466 nm. Finally, the occupied $^3F_4$ state relaxes to the $^3H_6$ ground state by giving out the third NIR photon at about 1800 nm. The energy gaps of $^1G_4 \rightarrow ^3F_{2,3}$, $^3H_4 \rightarrow ^3H_5$ and $^3F_4 \rightarrow ^3H_6$ are about 6000, 4000 and 5600 cm^{-1}, respectively, while the maximum phonon energy of fluorides is about 400 cm^{-1}.[136] According to the energy gap law, radiative decay would dominate over multiphonon relaxation for the $^1G_4$, $^3H_4$ and $^3F_4$ states of Tm^{3+}, thereby facilitating the occurrence of three-step photon cascade emission. Notably for the sake of designing such effective three-step photon cascade emission, a high Tm^{3+}-doping concentration such as >0.5%–1% should be avoided to hamper the efficient CR process between Tm^{3+} ions.[81] On the basis of luminescence branching ratios, a theoretical maximum QE was calculated to be about 158% for Tm^{3+} NIR splitter. However, in terms of an integrating sphere and technical calibrations of visible R928, NIR R5509-72 and middle-IR PbSe detectors, the total QE was experimentally measured to be about 32% for the β-$NaYF_4$:1% Tm^{3+} phosphors.[62]

Ho^{3+} ion has abundant energy-level structure ranging from deep-UV, to visible and to NIR areas, as schematically illustrated in Figure 11.9(a) according to the Russell-Saunders Stark level structure of activator ions. Under UV excitation of 287 nm, the $^3D_3$ state of Ho^{3+} will be nonradiatively populated to efficiently emit UV/blue/green photons (inset of Figure 11.9(b)) and/or a first NIR photon around 850 nm through the $^3D_3 \rightarrow ^3K_8$,$^5F_2$ transition of Ho^{3+} (Figure 11.9(b)).[61,150] Simply judged by the energy gap law, the large energy gap of $^3D_3 \rightarrow ^5G_2$ (~2800 cm^{-1}) will need at least seven phonons for effective compensation in the β-$NaYF_4$ host lattice (maximum phonon energy ~400 cm^{-1}), which will efficiently hamper the multiphonon relaxation from the $^3D_3$ excited state, through its following states to the blue-/green-emitting states of Ho^{3+}. Following the $^5S_2$, $^5F_4$ green-emitting state nonradiatively populated from the $^3K_8$, $^5F_2$ blue-emitting state, additional relaxation to the $^5I_6$ intermediate level will radiatively take place to give out a second NIR photon around 1015 nm photon (Figure 11.9(a) and (b)).[61,150-152] Finally, energy in the populated $^5I_6$ state will decay to the $^5I_8$ ground state by emitting a third photon at about 1180 nm. Similar to the case of $^3D_3$ excited level, the large energy gap from the $^5S_2$, $^5F_4$ level to the next lower lying $^5F_5$ level (about 2900 cm^{-1}) and that

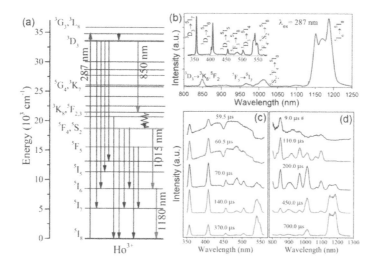

FIGURE 11.9 (a) Energy-level diagram showing stepwise transitions of 1 UV-to-3 NIR photon cascade emission from the $^3D_3$ excited state of Ho^{3+}, (b) corresponding UV-NIR luminescence spectra and (c and d) time-resolved emission spectra of Ho^{3+} in (c) UV-visible region and (d) NIR region, respectively.

from the $^5I_6$ to $^5I_7$ level (about 3500 cm^{-1}) enable radiative transitions to dominate over multiphonon relaxation from the $^5S_2$, $^5F_4$ and $^5I_6$ states. Hence, NIR three-photon cascade emission of Ho^{3+} single ion will take place efficiently in hosts with low-phonon-energy.[61,150] To prove the cascade NIR transitions, time-resolved luminescence spectra of β-NaYF$_4$:1% Ho^{3+} were measured under pulsed light excitation of 287 nm.[61] As expected in Figure 11.9(c), the emission bands at 360 and 410 nm from the $^3D_3$ state dominate initially at delay time around 59.5 μs, that at 485 nm from the lower lying $^5F_3$ state emerges after a delay time of 60.5 μs, and that at 540 nm from the $^5S_2$, $^5F_4$ state rises up clearly after 70 μs. These ET dynamics reveal that the $^3D_3$ excited state might be faster depopulated to the $^3K_8$, $^5F_2$ states through one-step NIR transition at 850 nm rather than by stepwise nonradiative decay, and after going through NR in longer decay time, the energy will populate the $^5S_2$, $^5F_4$ state, part of which yields emission of NIR photons. A second set of time-resolved emission spectra in NIR region is shown in Figure 11.9(d), at the early stage around 9 μs, only the NIR emission at about 850 nm is detected due to the $^3D_3 \rightarrow$ $^3K_8$, $^5F_2$ transition of Ho^{3+}. After 110 μs, the NIR emission at 1015 nm from the $^5S_2$, $^5F_4 \rightarrow$ $^5I_6$ transition and that of 965 nm from the $^5F_5$ state appear clearly. Finally, around delay time of 200 μs, the emission band around 1180 nm occurs because of the $^5I_6 \rightarrow$ $^5I_8$ transition and then monotonically increases in intensity. These results further directly evidence the efficient occurrence of three-step NIR photon cascade emission of Ho^{3+} from the $^3D_3$ excited state with $^3K_8$, $^5F_2$ and $^5I_6$ serving as intermediate levels. Based on the luminescence branching ratios of Ho^{3+} in NaYF$_4$, an optimal QE is obtained to be 124%.[61]

11.4.2 MULTIPHOTON SPLITTING BY CROSS-RELAXATION OF LN^{3+} ($LN = ER$, TM) SINGLE ION

Photon splitting by multiple cross-ET processes was first proposed by Chen *et al.* in $Er_{0.3}Gd_{0.7}VO_4$ crystal, where an absorbed UV/green photon is split into two or more NIR photons reemitted at about 1530 nm, respectively.[59] For a process of three-photon splitting, two consecutive CR processes are required to occur with high efficiency. As schematically shown in Figure 11.10(a), upon absorption of a green photon around 522 nm, and/or fast NR from upper UV-/blue-excited states, the $^2H_{11/2}$, $^4S_{3/2}$ state of Er^{3+} will be occupied with relatively long-lived status. At an elevated Er^{3+} concentration, such as 30% Er^{3+} in $GdVO_4$ crystal, first-step CR process of $^2H_{11/2}$, $^4S_{3/2} \rightarrow$ $^4I_{9/2} + ^4I_{15/2} \rightarrow$ $^4I_{13/2}$ (CR1) and second-step CR process of $^4I_{9/2} \rightarrow$ $^4I_{13/2} + ^4I_{15/2} \rightarrow$ $^4I_{13/2}$ (CR2) take place efficiently to make energy triply populate the $^4I_{13/2}$ state of Er^{3+}, each of which thereby emits NIR photon around 1530 nm.[153–155] Chen et al. evaluated an approximate QE of 178.55% for 1532.5 nm emission in the $Er_{0.3}Gd_{0.7}VO_4$ crystal.[59] However, it has to be noted that the CR2 process from Er^{3+} $^4I_{9/2}$ state has energy mismatch of negative 800 cm^{-1} and belongs to anti-Stokes process. And more the $^4I_{9/2}$ excited state, located at ~12,450 cm^{-1}, only lies at 2200 cm^{-1} above the next lower lying $^4I_{11/2}$ state, which would be easily de-excited by multiphonon relaxation in host lattice with large phonon energy.

Yu et al. systematically studied the green-to-NIR photon splitting in YVO_4:x% Er^{3+} ($x = 1$–20) phosphors.[58] Figure 11.10(b) shows dependence of the integrated NIR emission intensities on Er^{3+}-doping concentration. Interestingly, with increasing Er^{3+} concentration, the emission of 856/1236 nm increases to reach a maximum at 2% Er^{3+}, while that of 1000 and 1530 nm reach a maximum at 20% Er^{3+} and 5% Er^{3+}, respectively. Moreover, emission from the $^4I_{9/2}$ excited state of Er^{3+}, corresponding to NIR photons around 805 nm, is not detectable at all just because the energy difference of $^4I_{9/2} \rightarrow {^4I_{11/2}}$, ~2200 cm^{-1}, can be bridged by only three phonons in YVO_4 (maximum phonon energy ~890 cm^{-1}).[58,60] To have an insight into the possibility of (weak) emission from the $^4I_{9/2}$ excited level, low-temperature (4 and 50K) emission spectra were further measured for YVO_4:0.5% Er^{3+} and YVO_4:2% Er^{3+} under 522 nm excitation (Figure 11.10(c)). Low temperatures will slow down multiphonon relaxation and possibly allow for the observation of weak $^4I_{9/2}$ emission at 4K.[156] The 0.5% Er^{3+} doping level will hamper efficient CR, and above 2% Er^{3+}, efficient cross-ET from the $^4S_{3/2}$ excited level is expected to populate the $^4I_{9/2}$ state and possibly give rise to the $^4I_{9/2}$ emission. However, none of the samples yield any detectable emission around 805 nm, indicating that multiphonon quenching is severe for Er^{3+} in the vanadate. Evidence that the anti-Stokes CR2 is inefficient is obtained from the trends shown in Figure 11.10(b): efficient CR2 process would deplete energy in the $^4I_{9/2}$ state to the $^4I_{13/2}$ state directly but not the $^4I_{11/2}$ state of Er^{3+}, which would suppress the emission intensity of $^4I_{11/2} \rightarrow {^4I_{15/2}}$ around 1000 nm, in stark contrast to the monotonic enhancement of 1000 nm emission with increasing Er^{3+} concentration in Figure 11.10(b). Actually the situation for CR2 process differs from that for CR1 process. The $^4S_{3/2}$ level has the $^2H_{11/2}$ level at ~700 cm^{-1} higher energies, while for the $^4I_{9/2}$ level, there is no higher energy level that can be thermally populated to

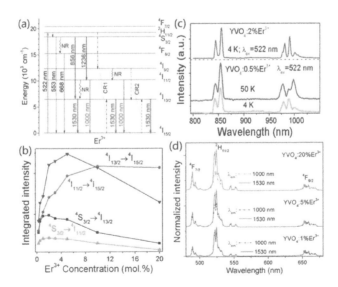

FIGURE 11.10 (a) Energy-level diagram of Er^{3+} illustrating ET mechanisms of two- or three-photon splitting by multistep CR processes, (b) integrated intensities of NIR emissions versus Er^{3+} concentration in YVO_4, (c) low-temperature NIR emission spectra of YVO_4:2% Er^{3+} (4K) and that of YVO_4:0.5% Er^{3+} (4 and 50K) under excitation of 522 nm, and (d) excitation spectra normalized at the $^4F_{9/2}$ excited state, while monitoring 1000 nm (broken lines) or 1530 nm (solid lines) emissions of YVO_4:x% Er^{3+} ($x = 1, 5, 20$).

allow for resonant ET and the transfer relies on coupling with weaker vibronic transitions. Also the reduced matrix elements U_λ^2 for the "donor" transition on Er^{3+} ($^4I_{9/2} \rightarrow {}^4I_{13/2}$) are smaller (0.0003, 0.0081, 0.64).[157]

From Figure 11.10(d), Yu et al. further determined if and how efficient the downconversion for 1530 nm is from the $^2H_{11/2}/^4S_{3/2}$ state via the double-splitting processes.[58] In the case of photon splitting, the ratio of the integral intensity of green excitation band over that of red excitation band by monitoring the emission wavelength at 1530 nm ($I_{1530\ nm}$) should be much greater than that by monitoring the emission at 1000 nm ($I_{1000\ nm}$): CR from the excited $^2H_{11/2}/^4S_{3/2}$ state at higher Er^{3+} concentrations will promote emission of more than one 1530 nm photon but excitation in the $^4F_{9/2}$ state cannot, while a single 1000 nm photon follows excitation in both the $^2H_{11/2}/^4S_{3/2}$ and $^4F_{9/2}$ states, as depicted in Figure 11.10(a). The ratio of $I_{1530\ nm}/I_{1000\ nm}$ is about 1.03, 1.71 and 1.91 for the samples doped with 1%, 5% and 20% Er^{3+}, respectively. These results validate that the photon splitting for 1530 nm does occur from the $^2H_{11/2}/^4S_{3/2}$ state and its efficiency strongly increases with increasing Er^{3+} content, to around 190% at 20%Er^{3+}. At this high Er^{3+} concentration, there is, however, considerable concentration quenching that lowers the actual efficiency. The highest overall QE is thus realized in the ~5% Er^{3+}-doped material where the photon-splitting efficiency is around 70% and the concentration quenching is still limited.

It is well known that cross-ET interaction between dopants will become efficient at an elevated concentration in the case of existing energy resonance gaps between ions. Tm^{3+} ion has rich energy-level structure spanning UV-to-IR region, where the several nearly resonant energy gaps between Tm^{3+} ions make the luminescence properties of Tm^{3+} sensitive to its concentration in host.[81,149] Under excitation of UV photons around 360 nm, Chen et al. briefly presented a four-photon splitting in $(Y_{1-x}Tm_x)_3Al_5O_{12}$ ($x = 0.005$ and 0.2) through CR process of $^1D_2 \rightarrow {}^3H_4 + {}^3H_6 \rightarrow {}^3F_{2,3}$ and that of $^3H_4 \rightarrow {}^3F_4 + {}^3H_6 \rightarrow {}^3F_4$ for an enhanced $^3F_4 \rightarrow {}^3H_6$ emission at 1788 nm.[158,159] A theoretical up-limit of QE is approximated to be 282.12% for such NIR four-photon splitting. Meanwhile, Yu et al. experimentally demonstrated the efficient occurrence of multistep cross-ET processes between Tm^{3+} ions and modeled the involved ET dynamics as a function of Tm^{3+} concentration in Gd_2O_2S host.[65] Upon UV ~365 nm photons exciting Tm^{3+} onto the $^1D_2$ state, a serious of emissions from $^1D_2$, $^1G_4$, $^3H_4$ and $^3F_4$ states were recorded at diluted 0.1% Tm^{3+} in Gd_2O_2S (Figure 11.11(a)). With increasing Tm^{3+} concentration, all emissions from the $^1D_2$, $^1G_4$ and $^3H_4$ excited levels decrease obviously, while that from the $^3F_4$ excited level increases greatly till 5% Tm^{3+}. These phenomena are resulted from the dependence of cross-ET process on Tm^{3+} concentration (Figure 11.11(b)), which completely changes the colorful spontaneous emissions of Tm^{3+} into the enhanced $^3F_4 \rightarrow {}^3H_6$ emission at about 1850 nm. Despite the theoretical speculations to the multistep working cross-ET paths on the basis of the almost equivalent energy transitions as well as their reduced matrix elements, Yu et al. experimentally detected an additional emerging NIR emission around 1215 nm from the $^3H_5 \rightarrow {}^3H_6$ transition at higher Tm^{3+} concentration around 10% (Figure 11.11(a)), which directly convinces the involved cross-ET pathways for the four-photon splitting, i.e., mainly via an initial CR of $^1D_2 \rightarrow {}^3H_4 + {}^3H_6 \rightarrow {}^3F_{2,3}$ and a following CR of $^3F_{2,3} \rightarrow {}^3F_4 + {}^3H_6 \rightarrow {}^3H_5$ and/or $^3H_4 \rightarrow {}^3F_4 + {}^3H_6 \rightarrow {}^3F_4$ between Tm^{3+} ions.[65] Moreover, Yu et al. developed a cross-ET model by dipole-dipole interaction,

$$I(t) = I(0)e^{-t/\tau_0} \prod_{i}^{shells} (1 - x + xe^{-C_{cr}t/r_i^6})^{n_i} \qquad (11.6)$$

where x is Tm^{3+} acceptor concentration, $\Gamma_{xr} = C_{xr}r^{-6}$ is the cross-ET rate, C_{xr} (the strength of cross-ET) can alternatively be converted to a critical radius $R_0 = \sqrt[6]{\tau_0 C_{xr}}$, and ($r_i$, n_i) represents the "neighbor list" given by the crystal structure of Gd_2O_2S. The best fit is obtained for an intrinsic lifetime $\tau_0 = 4.0\ \mu s$, and $C_{xr} = 32.8\ nm^6\ ms^{-1}$ (corresponding to $R_0 = 7.1$ Å, and a nearest-neighbor cross-ET rate of $1/0.06\ \mu s$, 69× faster than the intrinsic decay). These results validated that the cross-ET from the $^1D_2$ level competes strongly with radiative decay (see similar demonstration of cross-ET from the $^3F_{2,3}$ and

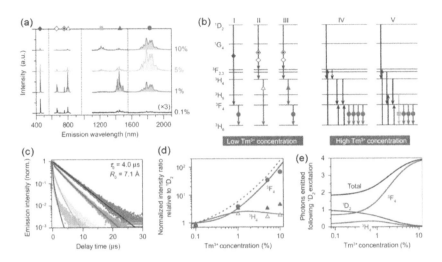

FIGURE 11.11 (a) Emission spectra of Gd_2O_2S:$x\%$ Tm^{3+} ($x = 0.1–10$) excited at 365 nm into $^1D_2$ state, (b) radiation paths from the $^1D_2$ state, (c) decay curves of $^1D_2 \to {}^3F_4$ at 465 nm, (d) emission intensity from $^3F_4$ (red circles) and $^3H_4$ (green filled triangles: $^3H_4 \to {}^3F_4$; green open triangles: $^3H_4 \to {}^3H_6$) relative to that from $^1D_2$ state and (e) total QE of Gd_2O_2S:Tm^{3+} photon splitter as Tm^{3+} $^1D_2$ state excited.

$^3H_4$ states in Ref. 65). Also, the comparison of the relative emission intensities in the spectra with the predications of "light yield" model exhibited a good correspondence in Figure 11.11(d), confirming the occurrence of one-to-four NIR photon splitting in Gd_2O_2S:Tm^{3+}. Ultimately, theoretical photon yield for the $^1D_2$ excited level in Figure 11.11(e) indicated that, as the cross-ET processes become more efficient at 10% Tm^{3+} in Gd_2O_2S, total QE of the four-photon splitting increases up to 388%.

ACKNOWLEDGMENTS

Q.Z. acknowledge financial support from the National Science Foundation of China (U1601205 and 51472088 and 51125005).

REFERENCES

1. Blasse, G. and Grabmaier, B.C., *Luminescent Materials*, Berlin: Springer, 1994.
2. Feldmann, C., Justel, T., Ronda, C.R., and Schmidt, P.J., *Adv. Funct. Mater.*, 13, 511, 2003.
3. Ronda, C.R., *J. Alloys Compd.*, 225, 534, 1995.
4. Henderson, B. and Imbusch, G.F., *Optical Spectroscopy in Inorganic Solids*, Oxford: Clarendon, 1998.
5. Moine, B. and Bizarri, G., *Opt. Mater.*, 28, 58, 2006.
6. Ronda, C.R., Justel, T., and Nikol, H., *J. Alloys Compd.*, 275, 669, 1998.
7. Sommerer, T.J., *J. Phys. D: Appl. Phys.*, 29, 769, 1996.
8. Justel, T. and Nikol, H., *Adv. Mater.*, 12, 527, 2000.
9. Dexter, D.L., *Phys. Rev.*, 108, 630, 1957.
10. Sommerdijk, J.L., Bril, A., and de Jager, A.W., *J. Lumin.*, 8, 341, 1974.
11. Piper, W.W., de Luca, J.A., and Ham, F.D., *J. Lumin.*, 8, 344, 1974.
12. Wegh, R.T., Donker, H., Oskam, K.D., and Meijerink, A., *Science*, 283, 663, 1999.
13. Wegh, R.T., van Loef, E.V.D., Burdick, G.W., and Meijerink, A., *Mol. Phys.*, 101, 1047, 2003.
14. Wegh, R.T., Donker, H, Meijerink, A., Lamminmäki, R.J., and Hölsä, J., *Phys. Rev. B*, 56, 13841, 1997.
15. Yang, Z., Lin, J.H., Su, M.Z., Tao, Y., and Wang, W., *J. Alloys Compd.*, 308, 94, 2000.
16. Feofilov, S.P., Zhou, Y., Seo, H.J., Jeong, J.Y., Keszler, D.A., and Meltzer, R.S., *Phys. Rev. B*, 74, 085101, 2006.
17. Zhou, Y., Feofilov, S.P., Seo, H.J., Jeong, J.Y., Keszler, D.A., and Meltzer, R.S., *Phys. Rev. B*, 77, 075129, 2008.
18. Wegh, R.T., Donker, H., Oskam, K.D., and Meijerink, A., *J. Lumin.*, 82, 93, 1999.

19. Wegh, R.T., van Loef, E.V.D., and Meijerink, A., *J. Lumin.*, 90, 111, 2000.
20. Peijzel, P.S., Schrama, W.J.M., and Meijerink, A., *Mol. Phys.*, 102, 1285, 2004.
21. Lee, T.J., Luo, L.Y., Diau, E.W.-G., and Chen, T.-M., *Appl. Phys. Lett.*, 89, 131121, 2006.
22. Jia, W., Zhou, Y., Feofilov, S.P., Meltzer, R.S., Jeong, J.Y., and Keszler, D., *Phys. Rev. B*, 72, 075114, 2005.
23. Babin, V., Oskam, K.D., Vergeer, P., and Meijerink, A., *Radiat. Meas.*, 38, 767, 2004.
24. Justel, T., Nikol, H., and Ronda, C., *Angew. Chem. Int. Ed.*, 37, 3084, 1998.
25. Zhang, Q.Y. and Huang, X.Y., *Prog. Mater. Sci.*, 55, 353, 2010.
26. Vergeer, P., Babin, V., and Meijerink, A., *J. Lumin.*, 114, 267, 2005.
27. Wang, X.-J., Huang, S., Lu, L., Yen, W.M., Srivastava, A.M., and Setlur, A.A., *Opt. Commun.*, 195, 405, 2001.
28. van der Kolk, E., Dorenbos, P., van Eijk, C.W.E., Vink, A.P., Weil, M., and Chaminade, J.P., *J. Appl. Phys.*, 95, 7867, 2004.
29. Kück, S. and Sokolska, I., *J. Phys. Condens. Matter.*, 18, 5447, 2006.
30. Chen, Y.H., Shi, C.S., Yan, W.Z., Qi, Z.M., and Fu, Y.B., *Appl. Phys. Lett.*, 88, 061906, 2006.
31. Nie, Z.G., Zhang, J.H., Zhang, X., Luo, Y.S., Lu, S.Z, and Wang, X.J., *J. Lumin.*, 119–120, 332, 2006.
32. Nie, Z.G., Zhang, J.H., Zhang, X., Ren, X.G., Zhang, G.B., and Wang, X.-J., *Opt. Lett.*, 32, 991, 2007.
33. van der Zwaan, B. and Rabl, A., *Solar Energy*, 74, 19, 2003.
34. Shockley, W. and Queisser, H.J., *J. Appl. Phys.*, 32, 510, 1961.
35. Richards, B.S., *Sol. Energy Mater. Sol. Cells*, 90, 2329, 2006.
36. Richards, B.S., *Sol. Energy Mater. Sol. Cells*, 90, 1189, 2006.
37. Strümpel, C., McCann, M., Beaucarne, G., Arkhipov, V., Slaoui, A., del Canizo, C., and Tobias, T., *Sol. Energy Mater. Sol. Cells*, 91, 238, 2007.
38. van der Ende, B.M., Aarts, L., and Meijerink, A., *Phys. Chem. Chem. Phys.*, 11, 11081, 2009.
39. Yu, D.C., Rabouw, F.T., Boon, W.Q., Kieboom, T., Ye, S., Zhang, Q.Y., and Meijerink, A., *Phys. Rev. B*, 90, 165126, 2014.
40. Chen, D.Q., Wang, Y.S., and Hong, M.C., *Nano Energy*, 1, 73, 2012.
41. Lian, H.Z., Hou, Z.Y., Shang, M.M., Geng, D.L., Zhang, Y., and Lin, J., *Energy*, 57, 270, 2013.
42. Yu, D.C. and Zhang, Q.Y., *Sci. Sin. Chim.*, 11, 1431, 2013.
43. Huang, X.Y., Han, S.Y., Huang, W., and Liu, X.G., *Chem. Soc. Rev.*, 42, 173, 2013.
44. Trupke, T., Green, M.A., and Würfel, P., *J. Appl. Phys.*, 92, 1668, 2002.
45. Vergeer, P., Vlugt, T.J.H., Kox, M.H.F., den Hertog, M.I., van der Eerden, J.P.J.M., and Meijerink, A., *Phys. Rev. B*, 71, 014119, 2005.
46. Zhang, Q.Y., Yang, C.H., and Pan, Y.X., *Appl. Phys. Lett.*, 90, 021107, 2007.
47. Zhang, Q.Y., Yang, C.H., Jiang, Z.H., Ji, X.H., *Appl. Phys. Lett.*, 90, 061914, 2007.
48. Zhang, Q.Y., Yang, G.F., and Jiang, Z.H, *Appl. Phys. Lett.*, 91, 051903, 2007.
49. van der Ende, B.M., Aarts, L., and Meijerink, A., *Adv. Mater.*, 21, 3073, 2009.
50. Meijer, J.M., Aarts, L., van der Ende, B.M., Vlugt, T.J.H., and Meijerink, A., *Phys. Rev. B*, 81, 035107, 2010.
51. Eilers, J.J., Biner, D., van Wijngaarden, J.T., Krämer, K., Güdel, H.-U., and Meijerink, A., *Appl. Phys. Lett.*, 96, 151106, 2010.
52. van Wijngaarden, J.T., Scheidelaar, S., Vlugt, T.J.H., Reid, M.F., and Meijerink, A., *Phys. Rev. B*, 81, 155112, 2010.
53. Bai, Z.H., Fujii, M., Hasegawa, T., Imakita, K., Mizuhata, M., and Hayashi, S., *J. Phys. D: Appl. Phys.*, 44, 455301, 2011.
54. Yu, D.C., Huang, X.Y., and Zhang, Q.Y., *J. Alloys Compd.*, 509, 9919, 2011.
55. Yu, D.C., Ye, S., Huang, X.Y., and Zhang, Q.Y., *AIP Adv.*, 2, 022124, 2012.
56. Deng, K.M., Gong, T., Hu, L.X., Wei, X.T., Chen, Y.H., and Yin, M., *Opt. Express*, 19, 1749, 2011.
57. Lin, H., Chen, D.Q., Yu, Y.L., Yang, A.P., and Wang, Y.S., *Opt. Lett.*, 36, 876, 2011.
58. Yu, D.C., Yu, T., Wang, Y.Z., Zhang, Q.Y., and Meijerink, A., *Phys. Rev. Appl.*, 13, 024076, 2020.
59. Chen, X.B., Wu, J.G., Xu, X.L., Zhang, Y.Z., Sawanobori, N., Zhang, C.L., Pan, Q.H., and Salamo, G.J., *Opt. Lett.*, 34, 887, 2009.
60. Yu, D.C., Ye, S., Peng, M.Y., Zhang, Q.Y., Qiu, J.R., Wang, J., and Wondraczek, L., *Sol. Energy Matter. Sol. Cells*, 95, 1590, 2011.
61. Yu, D.C., Huang, X.Y., Ye, S., Peng, M.Y., Zhang, Q.Y., and Wondraczek, L., *Appl. Phys. Lett.*, 99, 161904, 2011.
62. Yu, D.C., Ye, S., Peng, M.Y., Zhang, Q.Y., and Wondraczek, L., *Appl. Phys. Lett.*, 100, 191911, 2012.
63. Yu, D.C., Chen, Q.J., Lin, H.H., Wang, Y.Z., and Zhang, Q.Y., *Opt. Mater. Express*, 6, 197, 2016.
64. Shahi, P.K., Singh, P., Rai, S.B., and Bahadur, A., *Inorg. Chem.*, 55, 1535, 2016.

65. Yu, D.-C., Martín-Rodríguez, R., Zhagn, Q.-Y., Meijerink, A., and Rabouw, F.T., *Light Sci. Appl.*, 4, e344, 2015.

66. Dieke, G.H., *Spectra and Energy Levels of Rare Earth Ions in Crystals*. New York, NY: Interscience, 1968.

67. van der Kolk, E., Dorenbos, P., Vink, A.P., Perego, R.C., van Eijk, C.W.E., and Lakshmanan, A.R., *Phys. Rev. B*, 64, 195129, 2001.

68. van Dijk, J.M.F. and Schuurmans, M.F.H., *J. Chem. Phys.*, 78, 5317, 1983.

69. Schuurmans, M.F.H. and van Dijk, J.M.F., *Physica B + C*, 23, 131, 1984.

70. Srivastava, A.M., Doughty, D.A., and Beers, W.W., *J. Electrochem. Soc.*, 144, L190, 1997.

71. You, F.T., Huang, S.H., Meng, C.X., Wang, D.W., Xu, J.H., Huang, Y., and Zhang, G.B., *J. Lumin.*, 122–123, 58, 2007.

72. Zheng, W., Zhu, H., Li, R., Tu, D., Liu, Y., Luo, W., and Chen, X., *Phys. Chem. Chem. Phys.*, 14, 6974, 2012.

73. Dorenbos, P., Marsman, M., van Eijk, C.W.E., Korzhik, M.V., and Minkov, B.I., *Radiat. Eff. Defects Solids*, 135, 325, 1995.

74. Kim, J.S., Park, Y.H., Choi, J.C., and Park, H.L., *J. Electrochem. Soc.*, 152, H135, 2005.

75. van der Kolk, E., Dorenbos, P., van Eijk, C.W.E., *J. Phys. Condens. Matter.*, 13, 5471, 2001.

76. Nie, Z.G., Zhang, J.H., Zhang, X., Ren, X.G., Di, W.H., Zhang, G.B., Zhang, D.H., and Wang, X-J., *J. Phys. Condens. Matter.*, 19, 076204, 2007.

77. Srivastava, A.M., Doughty, D.A., and Beers, W.W., *J. Electrochem. Soc.*, 143, 4113, 1996.

78. Vink, A.P., Dorenbos, P., and van Eijk, C.W.E., *Phys. Rev. B*, 66, 075118, 2002.

79. Hölsä, J., Lastusaari, M., Maryško, M., and Tukia, M., *J. Solid State Chem.*, 178, 435, 2005.

80. Pappalardo, R., *J. Lumin.*, 14, 159, 1976.

81. Tanner, P.A., Mak, C.S.K., Kwok, W.M., Phillips, D.L., and Joubert, A.-F., *J. Phys. Chem. B*, 106, 3606, 2002.

82. Peijzel, P.S. and Meijerink, A., *Chem. Phys. Lett.*, 401, 241, 2005.

83. de Jong, M., Meijerink, A., and Rabouw, F.T., *Nat. Commun.*, 8, 15537, 2017.

84. Meijerink, A., Wegh, R.T., Vergeer, P., and Vlugt, T., *Opt. Mater.*, 28, 575, 2006.

85. Song, E.H., Zhou, Y.Y., Wei, Y., Han, X.X.,Tao, Z.R., Qiu, R.L., Xia, Z.G., and Zhang, Q.Y., *J. Mater. Chem. C*, 7, 8192, 2019.

86. Park, W., Summers, C.J., Do, Y., Park, D., and Yang, H., US Patent, 6669867 B2, December 30; 2003.

87. Fu, Y.B., Zhang, G.B., Qi, Z.M., Wu, W.Q., and Shi, C.S., *J. Lumin.*, 124, 370, 2007.

88. Chen, Y.H., Yan, W.Z., and Shi, C.S., *J. Lumin.*, 122–123, 21–24, 2007.

89. Fu, Y.B., Zhang, G.B., Wu, W.Q., Qi, Z.M., Chen, Y.H., Wang, D.W., and Shi, C.S., *J. Solid State Chem.*, 68, 1779, 2007.

90. Lee, T.J., Luo, L.Y., Cheng, B.M., Diau, W.-G., and Chen, T.-M., *Appl. Phys. Lett.*, 92, 081106, 2008.

91. Zachau, M., Zwaschka, F., and Kummer, F., In: Ronda, C.R., Welker, T., eds. *Proceedings of the Sixth International Conference on Luminescent Materials*. Pennington, NJ: The Electrochemical Society Inc., 1998, 314.

92. Yu, D.C., Zhou, Y.S., Ma, C.S., Melman, J.H., Baroudi, K.M., LaCapra, M., and Riman, R.E., *ACS Appl. Electron. Mater.*, 1, 2325, 2019.

93. Lepoutre, S., Boyer, D., and Mahiou, R., *J. Lumin.*, 128, 635, 2008.

94. You, F.T., Wang, Y., Lin, J.H., and Tao, Y., *J. Alloys Compd.*, 343, 151, 2002.

95. Liu, B., Chen, Y.H., Shi, C.S., Tang, H.G., and Tao, Y., *J. Lumin.*, 101, 155, 2003.

96. Kodama, N. and Watanabe, Y., *Appl. Phys. Lett.*, 84, 4141, 2004.

97. Karbowiaka, M., Mecha, A., and Romanowski, W.R., *J. Lumin.*, 114, 65, 2005.

98. Takeuchi, N., Ishida, S., Matsumura, A., and Ishikawa, Y.-I., *J. Phys. Chem. B*, 108, 12397, 2004.

99. Feldmann, C., Justel, T., Ronda, C.R., and Wiechert, D.U., *J. Lumin.*, 92, 245, 2001.

100. Kodama, N. and Oishi, S., *J. Appl. Phys.*, 98, 103515, 2005.

101. Tzeng, H.Y., Cheng, B.M., and Chen, T.M., *J. Lumin.*, 122–123, 917, 2007.

102. Lakshmanan, A.R., Kim, S.B., Jang, H.M., Kum, B.G., Kang, B.K., Heo, S., and Seo, D., *Adv. Funct. Mater.*, 17, 212, 2007.

103. Huang, X.Y., Yu, D.C., and Zhang, Q.Y., *J. Appl. Phys.*, 106, 113521, 2009.

104. Zhang, Q.H., Wang, J., Zhang, G.G., and Su, Q., *J. Mater. Chem.*, 19, 7088, 2009.

105. Huang, X.Y. and Zhang, Q.Y., *J. Appl. Phys.*, 105, 053521, 2009.

106. Chen, D.Q., Wang, Y.S., Yu, Y.L., Huang, P., and Weng, F.Y., *Opt. Lett.*, 33, 1884, 2008.

107. Ye, S., Zhu, B., Chen, J.X., Luo, J., and Qiu, J.R., *Appl. Phys. Lett.*, 92, 141112, 2008.

108. Ye, S., Zhu, B., Luo, J., Chen, J.X., Lakshminarayana, G., and Qiu, J.R., *Opt. Express*, 16, 8989, 2008.

109. Duan, Q.Q., Qin, F., Wang, D., Xu, W., Cheng, J.M., Zhang, Z.G., and Cao, W.W., *J. Appl. Phys.*, 110, 113503, 2011.
110. Zhou, J.J., Teng, Y., Ye, S., Zhuang, Y.X., and Qiu, J.R., *Chem. Phys. Lett.*, 486, 116, 2010.
111. An, Y.-T., Labbé, C., Cardin, J., Morales, M., and Gourbilleau, F., *Adv. Opt. Mater.*, 1, 855, 2013.
112. Enrichi, F., Armellini, C., Belmokhtar, S., Bouajaj, A., Chiappini, A., Ferrari, M., Quandt, A., Righini, G.C., Vomiero, A., and Zur, L., *J. Lumin.*, 193, 44, 2018.
113. Shao, W., Lim, C.-K., Li, Q., Swihart, M.T., and Prasad, P.N., *Nano Lett.*, 18, 4922, 2018.
114. Wang, Z.J. and Meijerink, A., *J. Phys. Chem. Lett.*, 9, 4522, 2018.
115. Dumont, L., Cardin, J., Benzo, P., Carrada, M., Labbé, C., Richard, A.L., Ingram, D.C., Jadwisienczak, W.W., and Gourbilleau, F., *Sol. Energy Mater. Sol. Cells*, 145, 84, 2016.
116. Guo, L.Y., Yu, H., Liu, J.J., Wu, B., Guo, Y., Fu, Y.T., and Zhao, L.J., *J. Alloys Compd.*, 784, 739, 2019.
117. Chen, X.P., Huang, X.Y., and Zhang, Q.Y., *J. Appl. Phys.*, 106, 063518, 2009.
118. Aarts, L., van der Ende, B.M., and Meijerink, A., *J. Appl. Phys.*, 106, 023522, 2009.
119. Aarts, L., Jaeqx, S., van der Ende, B.M., and Meijerink, A., *J. Lumin.*, 131, 608, 2011.
120. Fan, B., Point, C., Adam, J.-L., Zhang, X.H., Fan, X.P., and Ma, H.L., *J. Appl. Phys.*, 110, 113107, 2011.
121. Rakov, N. and Maciel, G.S., *J. Appl. Phys.*, 110, 083519, 2011.
122. Rodríguez, V.D., Tikhomirov, V.K., Méndez-Ramos, J., Yanes, A.C., and Moshchalkov, V.V., *Sol. Energy Mater. Sol. Cells*, 94, 1612, 2010.
123. Borrero-González, L.J. and Nunes, L.A.O., *J. Phys.: Condens. Matter.*, 24, 385501, 2012.
124. Zhang, J.-X., Hou, D.J., Li, J.-Y., Lin, H.H., Huang, R., Zhang, Y., and Hao, J.H., *J. Phys. Chem. C*, 124, 19774, 2020.
125. Chen, D.Q., Yu, Y.L., Lin, H., Huang, P., Shan, Z.F., and Wang, Y.S., *Opt. Lett.*, 35, 220, 2010.
126. Yu, T., Yu, D.C., Lin, H.H., and Zhang, Q.Y., *J. Alloys Compd.*, 695, 1154, 2017.
127. Xiang, G.T., Ma, Y., Zhou, X.J., Jiang, S., Li, L., Luo, X.B., Hao, Z.D., Zhang, X., Pan, G.-H., Luo, Y.S., and Zhang, J.H., *Inorg. Chem.*, 56, 1498, 2017.
128. Chen, T.J., Yang, X.L., Xia, W.B., Gao, X.J., Li, W., and Xiao, S.G., *Mater. Sci. Eng. B*, 211, 20, 2016.
129. Suresh, K. and Jayasankar, C.K., *J. Alloys Compd.*, 788, 1048, 2019.
130. Zheng, W., Zhu, H.M., Li, R.F., Tu, D.T., Liu, Y.S., Luo, W.Q., and Chen, X.Y., *Phys. Chem. Chem. Phys.*, 14, 6974, 2012.
131. Yu, D.C., Yu, T., van Bunningen, A.J., Zhang, Q.Y., Meijerink, A., and Rabouw, F.T., *Light Sci. Appl.*, 9, 107, 2020.
132. Xie, L.C., Wang, Y.H., and Zhang, H.J., *Appl. Phys. Lett.*, 94, 061905, 2009.
133. Zhang, Q., Zhu, B., Zhuang, Y.X., Chen, G.R., Liu, X.F., Zhang, G., Qiu, J.R., and Chen, D.P., *J. Am. Ceram. Soc.*, 93, 654, 2010.
134. Jiang, G.C., Wei, X.T., Chen, Y.H., Duan, C.K., and Yin, M., *J. Rare Earth*, 31, 27, 2013.
135. Fu, L., Xia, H.-P., Dong, Y.-M., Li, S.-S., Gu, X.-M., Zhang, J.-L., Wang, D.-J., Jiang, H.-C., and Chen, B.-J., *Chin. J. Chem. Phys.*, 28, 73, 2015.
136. Gao, G.J., Turshatov, A., Howard, I.A., Busko, D., Joseph, R., Hudry, D., and Richards, B.S., *Adv. Sustainable Syst.*, 1, 1600033, 2017.
137. Yu, T., Yu, D.C., Zhang, Q.Y., and Meijerink, A., *Phys. Rev. B*, 98, 134308, 2018.
138. Ueda, J. and Tanabe, S., *J. Appl. Phys.*, 106, 043101, 2009.
139. Chen, D.Q., Wang, Y.S., Yu, Y.L., Huang, P., and Weng, F.Y., *J. Appl. Phys.*, 104, 116105, 2008.
140. Huang, L.L., Lin, L., Zheng, B., Huang, H., Feng, Z.H., Wang, Z.Z., Li, X.Y., and Zheng, Z.Q., *Opt. Commun.*, 441, 170, 2019.
141. Sun, T.Y., Chen, X., Jin, L.M., Li, H.-W., Chen, B., Fan, B., Moine, B., Qiao, X.S., Fan, X.P., Tsang, S.-W., Yu, S.F., and Wang, F., *J. Phys. Chem. Lett.*, 8, 5099, 2017.
142. Ten Kate, O.M., de Jong, M., Hintzen, H.T., and van der Kolk, E., *J. Appl. Phys.*, 114, 084502, 2013.
143. Yu, D.C., Zhang, J.P., Chen, Q.J., Zhang, W.J., Yang, Z.M., and Zhang, Q.Y., *Appl. Phys. Lett.*, 101, 171108, 2012.
144. Wang, Y.Z., Yu, D.C., Lin, H.H., Ye, S., Peng, M.Y., and Zhang, J., *Appl. Phys.*, 114, 203510, 2013.
145. Jaffrès, A., Viana, B., and van der Kolk, E., *Chem. Phys. Lett.*, 527, 42, 2012.
146. Zhang, J.P., Yu, D.C., Zhang, F.F., Peng, M.Y., and Zhang, Q.Y., *Opt. Mater. Express*, 4, 111, 2013.
147. Yu, T., Lin, H.H., Yu, D.C., Ye, S., and Zhang, Q.Y., *J. Phys. Chem. C*, 119, 26643, 2015.
148. Dong, S.L., Lin, H.H., Yu, T., and Zhang, Q.Y., *J. Appl. Phys.*, 116, 023517, 2014.
149. ten Kate, O.M. and van der Kolk, E., *J. Lumin.*, 156, 262, 2014.
150. Song, P. and Jiang, C., *IEEE Photon. J.*, 5, 8400209, 2013.
151. Yu, D.C, Huang, X.Y., Ye, S., Zhang, Q.Y., and Wang, J., *AIP Adv.*, 1, 042161, 2011.

152. Zhang, W.J., Yu, D.C., Zhang, J.P., Qian, Q., Xu, S.H., Yang, Z.M., and Zhang, Q.Y., *Opt. Mater. Express*, 2, 636, 2012.
153. Chen, X.-B., Yang, G.-J., Song, L., Yang, X.-D., Liu, D.-H., Chen, Y., Ding, F.-L., and Wu, Z.-L., *Acta Phys. Sin.*, 61, 037804, 2012.
154. Miritello, M., Savio, R.L., Cardile, P., and Priolo, F., *Phys. Rev. B*, 81, 041411(R), 2010.
155. Zhou, J.J., Teng, Y., Liu, X.F., Ye, S., Xu, X.Q., Ma, Z.J., and Qiu, J.R., *Opt. Express*, 18, 21663, 2010.
156. Yu, D.C., Ballato, J., and Riman, R.E., *J. Phys. Chem. C*, 120, 9958, 2016.
157. Dexter, D.L., *J. Chem. Phys.*, 21, 836, 1953.
158. Chen, X.B., Salamo, G.J., Yang, G.J., Li, Y.L., Ding, X.L., Gao, Y., Liu, Q.L., and Guo, J.H., *Opt. Express*, 21, A829, 2013.
159. Chen, X.B., Li, S., Salamo, G.J., Li, Y.L., He, L.Z., Yang, G.J., Gao, Y., and Liu, Q.L., *Opt. Express*, 23, A51, 2013.

12 Mechanoluminescent Phosphors

Yixi Zhuang

CONTENTS

12.1 INTRODUCTION

Mechanoluminescent (ML) phosphors are solid-state compounds that give photon emissions upon mechanical stimuli. The unique force-responsive luminescence characteristics make ML phosphors greatly different from the conventional photoluminescent (PL) phosphors since their first discovery. Over the last few decades, ML phosphors have been developed into an important subgroup of phosphors that show charming optical functionalities and promising applications as sensing materials and energy conversion materials.

This chapter aims to give a brief introduction to ML phosphors, their structures, ML performances, characterization methods, and applications. The brief history of ML phosphors is reviewed and the general classification of mechanoluminescence is given in Section 12.1. A list of representative ML phosphors is presented in Section 12.2 and their crystal structures and microstructures are discussed. The characterization methods and instruments used for studying ML phosphors that are quite different from those for PL phosphors are introduced in Section 12.3. Several important applications of ML phosphors especially in the emerging fields of information technologies are highlighted in Section 12.4. Finally, a short summary and some reference review papers for further study are given in Section 12.5.

12.1.1 BRIEF HISTORY OF ML PHOSPHORS

The first report on ML can be traced back to 1605. When a cube sugar was scrapped with a knife in the dark, Bacon F. observed a sparkling light. He recorded this amazing phenomenon in his writing *The Advancement of Learning*.[1] Although the ML of sugar or other solid materials must have been observed before Bacon F., the year 1605 is generally regarded as the beginning of scientific research on ML phosphors. Since then, similar ML phenomena originating from fracture or breaking of solid particles were reported in a large number of compounds. In fact, the reported luminescence from fracture belongs to a type of ML, generally known as *fractoluminescence*. In 1664, Boyle R. documented another important event on ML. According to his description, a white light flashed in the dark when he rubbed the surface of a diamond with a finger or a steel needle.[2] Obviously, the observed ML on the diamond surface was not caused by fracture. It should be related to contact friction between two different solid materials and thus is categorized as *triboluminescence*. After the diamond, more materials showing emission when rubbed against another object have been found. The *triboluminescence* can be further subcategorized into *triboelectrically*, *tribochemically*, and *tribothermally induced ML*.

In 1999, Xu C.-N. et al. reported an important type of ML in the typical persistent luminescent phosphors, such as $SrAl_2O_4:Eu^{2+}$ and $Ca_2Al_2SiO_7:Ce^{3+}$.[3, 4] After being charged with ultraviolet (UV) light, these phosphors gave intense ML upon mechanical stimuli. Amazingly, the ML occurred in the elastic deformation range and the ML could be completely recovered after sufficient UV light recharging. This nondestructive and recoverable ML, namely *elasticoluminescence*, has greatly enhanced the application potential of ML phosphors and advanced them into practical applications in the structural health diagnosis of large buildings. In 2013, Jeong S. M. et al. reported self-recoverable ML in a composite film of ZnS:Cu particles embedded in polydimethylsiloxane (PDMS).[5] Without any preexcitation, the composite film exhibited repeatable ML upon mechanical stimuli over 100,000 times. Although the mechanism is still controversial, the self-recoverable ML phosphors without the requirement of preexcitation are considered highly promising for dynamic stress sensing applications.

The milestones and some important events in the history of ML phosphors are summarized in a timeline in Figure 12.1. The development profile of ML phosphors shows that the major achievements and contributions are focused on discovering new phenomena, reporting new materials and revealing the possible mechanisms before 2010. Undoubtedly, the exploration of ML phosphors with improved performances and the efforts to achieve a better understanding of the ML mechanism

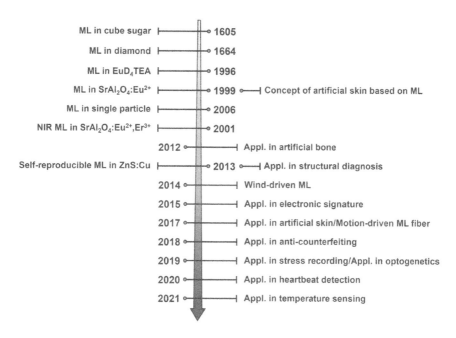

FIGURE 12.1 Overview of the research on ML phosphors. The left side shows the discovery of typical ML phenomena or ML phosphors. The right side summarizes the ground-breaking work reporting on new concepts or applications of ML phosphors.

have never stopped in the whole history; however, it should be noticed that after 2010, more attentions have been paid to the application research of ML phosphors. As the research focus is shifting from basic materials science to multidisciplinary technical applications, many innovative concepts and novel applications have been proposed in recent years. The emerging concepts and applications of ML phosphors, such as artificial skin, electronic signature, biomechanical engineering, and stress recording, have stimulated professional technicians in other fields to join in the research of ML phosphors, making ML phosphors and ML-related sensing technology a significant research hotspot. Some representative applications of ML phosphors will be introduced in Section 12.4.

12.1.2 GENERAL CLASSIFICATION OF ML

Since the first report, thousands of ML phosphors have been discovered. According to the action mode of mechanical stimulus on the phosphors, the ML in these phosphors can be generally classified into four categories (see Figure 12.2). *Elasticoluminescence* (*E*), *plasticoluminescence* (*P*), and *fractoluminescence* (*F*) can be understood as stress-induced luminescence from solids during three different strain stages, including elastic deformation, plastic deformation, and fracture stages, respectively. Notably, the three types of ML have different responses of emission to the applied stress or strain. The intensity of elasticoluminescence exhibits a linear relationship to the magnitude of stress or strain, usually over a large range of stress or strain. *Elasticoluminescence* is therefore very useful for sensing the relative size of stress or imaging the stress distribution. *Plasticoluminescence* is also a type of deformational ML. However, the plasticoluminescence usually does not show good recoverability under repeated mechanical stimuli. It is investigated in only a few compounds. *Fractoluminescence* occurs in the final stage of mechanical action; thus, it is easy to understand that fractoluminescence could be accompanied by elasticoluminescence and plasticoluminescence. Since the fracture of solid may release a lot of energy by breaking the chemical bonding, the fractoluminescence intensity of many compounds is high, which is observable even

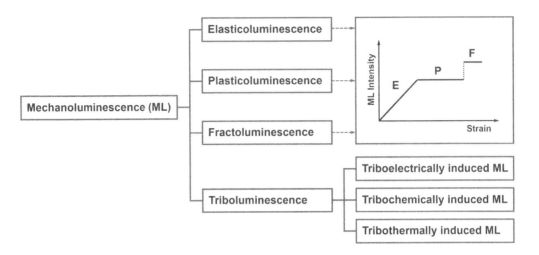

FIGURE 12.2 General classification of ML.

in daylight. However, fractoluminescence is unrecoverable and its application is limited to the fields such as early warning and failure monitoring of engineering structures.

Regarding *triboluminescence*, the frictional contact and separation of two different materials on the surface is the energy source of the luminescence. The friction of solids induces luminescence either by establishing an internal electric field, triggering a chemical reaction, or generating a thermal effect, and it can be further divided into three subcategories according to the different intermediate processes.

It should be noted that the terms of ML given in Figure 12.2 are those widely accepted in the field of inorganic chemistry. In the long history of ML, other terms on mechanically induced luminescence have been proposed and used in publications. The different terms are possibly derived from different cultural backgrounds, technical fields, and various experimental methods. They may cause misleading and misunderstanding to the researchers from different fields to some context. For example, the term *triboluminescence* was originally used to refer to the light emission when rubbing a solid material (came from the Greek *tribein*). However, rubbing a solid also may result in fractures or breaks and induce fractoluminescence. Therefore, some materials exactly giving fractoluminescence are also named triboluminescence materials especially in the field of organic chemistry. *Piezoluminescence* was originally used for light emission when pressing a solid (came from the Greek *piezein*). The term piezoluminescence has been subsequently extended to all kinds of mechanoluminescence possibly because pressing is the most common way to apply stress on solids. This term has been in less use recently and replaced by the common term mechanoluminescence. Furthermore, another term *piezophotonics* was proposed by Wang Z. L. et al. to refer to the light emission originated from a piezoelectric effect.[6] The piezophotonics opens up a new disciplinary field but it is only used for the phosphors showing a verified piezoelectric effect.

For a material under a given mechanical stimulus, in some cases, it is difficult to determine the specific ML type. The difficulty partially comes from the fact that a seemingly single type of mechanical action may exactly give multiple mechanical stimuli to the materials. This is particularly evident in the widely used and studied composite materials with inorganic ML particles embedded in soft organic matrix. For example, stretching or compressing a $ZnS:Cu@PDMS$ composite may generate at least (i) frictions between the ML phosphors and the organic matrix and (ii) compression forces on the phosphors. Even for the well-known ML phosphors (such as $SrAl_2O_4:Eu^{2+}$, $ZnS:Cu$, and related ML composites), the type and mechanism of ML still need further investigation.

12.2 ML PHOSPHORS

Exactly, ML phosphors are a subgroup of PL phosphors, which means an efficient emitting center is indispensable in ML phosphors. The reason why an ML phosphor is different from conventional PL phosphors may lie in a specific structure with the ability either to convert mechanical energy to photonic energy or to release energy as photon emission upon a mechanical stimulus. The analysis of the structure (including microstructure) is an important subject in the study of ML phosphors. However, although some interesting results have been found, a universal and convincing conclusion on the relationship between the structure and ML performance has not been reached.

It is roughly estimated that approximately 50% of inorganic solids and over 30% of organic crystals show fractoluminescence. The recent efforts on fractoluminescent materials are focused on the organic compounds that show high-brightness fractoluminescence up to a level visible in daylight, which are largely related to aggregation-induced emission or intermolecular interaction adjustment. [7] On the other hand, the research interests on inorganic ML phosphors are mostly shifted to nondestructive elasticoluminescence. In this section, we will briefly discuss the known ML phosphors (mostly elasticoluminescent phosphors), their structures (or microstructures), the methods to explore new ML phosphors, and efforts to enhance the ML performance.

12.2.1 REPORTED ML PHOSPHORS AND THEIR STRUCTURES

Table 12.1 lists some typical ML phosphors reported in the last few decades (fractoluminescent phosphors are not included here). These phosphors are grouped according to the type of crystal structure in an attempt to find a possible relationship between the structure and ML performance. The dopant that gives efficient ML for each crystal is also provided in the table and the emission wavelengths are presented. In this section, we briefly introduce the ML phosphors with representative crystal structures and those that have been considered promising for future applications. For further information on the ML compounds and their structures, interested readers may refer to the papers in refs. [8, 9]

12.2.1.1 Rock Salt

ML has been reported in the compounds with a rock salt structure since the beginning of the last century. These rock salt compounds without an artificially introduced dopant (such as KCl and KBr) show symmetric crystal structure but gave intense ML under deformation.[10] The emission came from color centers that could be created by X-ray or γ-ray irradiation. The rock salt compounds showed a low stress threshold for dislocation movements, and thus the reported ML in the irradiated rock salt compounds usually included both elasticoluminescence and plasticoluminescence.[11] The ML intensity is proportional to the stress size within the elastic range and is linear to the mechanical energy during the plastic deformation.

12.2.1.2 Wurtzite

ZnS:Mn^{2+}/Cu with a wurtzite structure is one of the most remarkable and well-studied ML phosphors. The discovery of ML in ZnS:Mn^{2+}/Cu could be traced back at least to 1986 by B. P. Chandra,[12] but self-recoverable ML in a ZnS:Cu@PDMS composite film over 100,000 load cycles in the absence of light preexcitation was reported in 2013 by Jeong S. M. et al.[5] The high-brightness and recoverable ML without the need for preexcitation has ignited widespread interests on the ZnS:Mn^{2+}/Cu, especially in the research of the self-powered stress sensing applications.[13] Since wurtzite is a typical piezoelectric crystal without an inversion symmetric center, it may be reasonable to attribute the origin of ML in the ZnS:Mn^{2+}/Cu to the interaction between the piezoelectricity and the electrons in shallow traps. However, the piezoelectric effect is difficult to explain why the ZnS:Mn^{2+}/Cu composite films give bright ML even at a relatively small stress (< 0.1 MPa) and why self-recoverable ML is only observed during continuous deformation of large strains. Although its

TABLE 12.1

Part of Reported ML Phosphors, Their Structures, and Emission Wavelengths

Structure	Host	Space Group	Symm[1]	Dopant[2]	Wavelength[3] (nm)	Refs.
Rock salt	KCl	$F\bar{m}3m$	Yes	–	455	[10]
				Eu^{2+}	420	[36]
	KBr	$F\bar{m}3m$	Yes	–	463	[10]
	KI	$F\bar{m}3m$	Yes	–	472	[10]
	NaCl	$F\bar{m}3m$	Yes	–	450	[10]
	MgO	$F\bar{m}3m$	Yes	–	420, 520	[37]
Wurtzite	ZnS	$P6_3mc$	No	Mn^{2+}	585	[13]
				Cu	520	[5]
	CaZnOS	$P6_3mc$	No	Mn^{2+}	580	[38]
				Ln^{3+}	Multicolor	[14]
	SrZnOS	$P6_3mc$	No	Mn^{2+}	600	[15]
				Ln^{3+}	Multicolor	[15]
	BaZnOS	$Cmcm$	Yes	Mn^{2+}	630	[39]
Perovskite	$(Ba,Ca)TiO_3$	$P4mm$	No	Pr^{3+}	613	[16]
	$LiNbO_3$	$R3c$	No	Pr^{3+}	613	[17]
	$NaNbO_3$	$Pbma$	No	Pr^{3+}	613	[18]
	$CaNb_2O_6$	$Pbcn$	Yes	Pr^{3+}	613	[19]
	$Ca_2Nb_2O_7$	$P2_1$	No	Pr^{3+}	613	[19]
	$Ca_3Nb_2O_8$	$Fm3m, P4/nnc$	Yes	Pr^{3+}	613	[19]
	$Ca_3Ti_2O_7$	$Cmc2_1$	No	Pr^{3+}	613	[40]
Spinel	$ZnAl_2O_4$	$Fd\bar{3}m$	Yes	Mn^{2+}	560	[20]
	$ZnGa_2O_4$	$Fd\bar{3}m$	Yes	Mn^{2+}	505	[21]
	$MgGa_2O_4$	$Fd\bar{3}m$	Yes	Mn^{2+}	506	[21]
Tridymite	$SrAl_2O_4$	$P2_1$	No	Eu^{2+}	520	[24]
	$(Ba,Sr)Al_2O_4$	$P2_1, P6_3$	No	Eu^{2+}	520	[25]
	$Zn_2(Ge,Si)O_4$	$R3$	No	Mn^{2+}	535	[41]
Melilite	$Ca_2MgSi_2O_7$	$P\bar{4}2_1m$	No	Eu^{2+}	530	[27]
	$Sr_2MgSi_2O_7$	$P\bar{4}2_1m$	No	Eu^{2+}	500	[28]
	$SrBaMgSi_2O_7$	$P\bar{4}2_1m$	No	Eu^{2+}	440	[28]
	$Ca_2Al_2SiO_7$	$P\bar{4}2_1m$	No	Ce^{3+}	417	[4]
	$CaYAl_3O_7$	$P\bar{4}2_1m$	No	Eu^{2+}	440	[29]
Anorthite	$CaAl_2Si_2O_8$	$P\bar{1}$	Yes	Eu^{2+}	430	[30]
	$(Ca,Sr)Al_2Si_2O_8$	$I\bar{1}$	No	Eu^{2+}	420	[31]
	$SrMg_2(PO_4)_2$	$P\bar{3}m$	No	Eu^{2+}	412	[42]
Oxynitride	$SrSi_2O_2N_2$	$P1$	No	Eu^{2+}	540	[35]
	$(Sr,Ba)Si_2O_2N_2$	$P1$	No	Eu^{2+}	570	[35]
	$BaSi_2O_2N_2$	$Cmc2_1$	No	Eu^{2+}	498	[33]
Others	ZrO_2	$P2_1/c$	Yes	Ti^{4+}	470	[43]
	$CaZr(PO_4)_2$	$Pna2_1$	No	Eu^{2+}	474	[44]
	$SrZn_2OS_2$	$Pmn2_1$	No	Mn^{2+}	580	[15]

[1] **Symm**: Yes = Centrosymmetric structure; No = Non-centrosymmetric structure.

[2] **Dopant**: – = No dopant; Ln^{3+} = Various species of trivalent lanthanides.

[3] **Wavelength**: Multicolor = Multicolor emission depending on the species of Ln^{3+}.

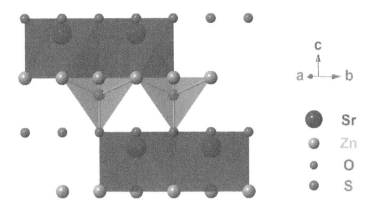

FIGURE 12.3 Crystal structure of SrZnOS.

application has been highly anticipated, the detailed ML mechanism in $ZnS:Mn^{2+}/Cu$ phosphors and $ZnS:Mn^{2+}/Cu$-based composites remains to be further clarified.

CaZnOS and SrZnOS exhibit the same wurtzite structure as ZnS. In the crystal, the Ca^{2+}/Sr^{2+} and Zn^{2+} cations are coordinated with mixed anions, forming $[CaS_3O_3]/[SrS_3O_3]$ octahedra and $[ZnS_3O]$ tetrahedra, respectively (Figure 12.3). These polar mixed-anion building units have the same crystalline orientation (*e.g.*, all ZnO bonds are parallel to the *c* axis) and thus construct large piezoelectricity in crystal through periodic arrangement of the building units. Importantly, the CaZnOS and SrZnOS hosts provide suitable sites for the substitutions of trivalent lanthanides (Ln^{3+}), which makes it possible to tune the emission wavelength of ML in a wide range by selecting different Ln^{3+} activators.[14, 15] It turns out the ML in Mn^{2+}/Ln^{3+}-doped CaZnOS or SrZnOS, similar to the $ZnS:Mn^{2+}/Cu$, could be self-recovered in cycles of mechanical actions without the need for light preexcitation. The wavelength-tunable and self-recoverable ML makes the CaZnOS- and SrZnOS-related ML phosphors widely applicable for the uses in different sensing applications, especially in those fields requiring ML in the near-infrared (NIR) or UV region.

12.2.1.3 Perovskite

Perovskite compounds such as $(Ba,Ca)TiO_3$ are one of the best-known piezoelectric crystals. Because of the large piezoelectric constant, it is not surprising that the perovskite compounds have been extensively studied for ML. Xu C.-N. et al. investigated electro-mechano-optical conversions in $(Ba_{1-x}Ca_x)TiO_3:Pr^{3+}$ ceramics and reported strong ML in a diphasic region ($x = 0.25–0.90$).[16] They concluded that the ML of the $(Ba_{1-x}Ca_x)TiO_3:Pr^{3+}$ could be attributed to the strong interaction between the large ionic polarization in a Ca-rich phase (to provide surface charge upon mechanical pressure by the piezoelectric effect) with the domains in a Ba-rich phase (to act as luminescent body).

Lanthanide-doped niobates are also important ML phosphors with the perovskite-type structure, such as $LiNdO_3:Pr^{3+}$ (Figure 12.4), $NaNdO_3:Pr^{3+}$, and $Ca_xNd_2O_{5+x}:Pr^{3+}$ ($x = 1, 2, 3$).[17–19] In the $Li_xNdO_3:Pr^{3+}$, a slightly Li-rich nonstoichiometric composition exhibits unusually high ML intensity and strain sensitivity, which can be attributed to the excellent piezoelectric and optoelectronic properties. The series of $Ca_xNd_2O_{5+x}:Pr^{3+}$ ($x = 1, 2, 3$) compounds is an interesting example to study the correlation between the ML performance and the crystal structure. Although all the three compounds show ML and piezoelectricity, only the $Ca_2Nd_2O_7:Pr^{3+}$ exactly belongs to a non-centrosymmetric crystal and it possesses the most intense ML. In contrast, the piezoelectricity in the $CaNd_2O_6:Pr^{3+}$ and $Ca_3Nd_2O_8:Pr^{3+}$ is possibly due to the anisotropy and cation vacancies of the hosts.

12.2.1.4 Spinel

Ideally, spinel belongs to cubic crystals without centrosymmetry. Somewhat unexpectedly, intense ML from $ZnAl_2O_4:Mn^{2+}$, $ZnGa_2O_4:Mn^{2+}$, and $MgGa_2O_4:Mn^{2+}$ after UV irradiation was discovered

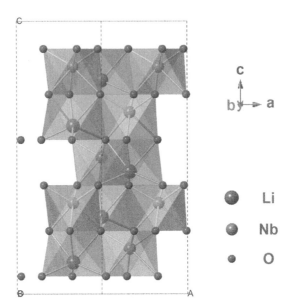

FIGURE 12.4 Crystal structure of LiNbO$_3$ along [0 1 0].

by Xu C.-N. et al.[20, 21] It turned out the ML should be related to the anti-site defects or cationic vacancies in these spinels (partially reverse spinel), which was well supported by the significant enhancement of ML after a reduction treatment. On the other hand, in the normal spinel MgAl$_2$O$_4$:Mn^{2+} where anti-site defects were almost impossible, ML and persistent luminescence of Mn^{2+} could hardly be observed.

12.2.1.5 Tridymite

Tridymite is a high-temperature phase of silica. The crystal structure of tridymite is featured with vertex-shared [SiO$_4$] tetrahedra, ring-like nets in planes, and a periodic layer-stacked three-dimensional (3D) network (Figure 12.5). Due to the high-brightness ML, α-SrAl$_2$O$_4$:Eu^{2+} in the tridymite family has become the most-studied and the best-known elasticoluminescent (ML) phosphor since the first report in 1999.[3] The α-SrAl$_2$O$_4$:Eu^{2+} belongs to a non-centrosymmetric crystal with space group of $P2_1$. It is generally accepted that the ML in the α-SrAl$_2$O$_4$:Eu^{2+} is attributed to

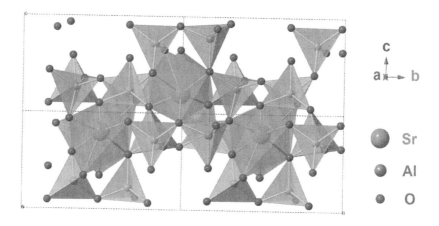

FIGURE 12.5 Crystal structure of α-SrAl$_2$O$_4$ along [1 0 0].

an induced piezoelectric field during the elastic deformation. It turns out that the α-SrAl$_2$O$_4$:Eu^{2+} lattice is strongly anisotropic because the linear thermal expansion coefficient in the c axis is one order of magnitude smaller than those in the a and b axes.[22] Dynamic microstructural observation also indicates that the deformation of the c axis is smaller than those of a and b axes when the crystal was subjected to hydrostatic pressure.[23, 24] The extraordinary anisotropy of the α-SrAl$_2$O$_4$ is thus considered a possible reason for the outstanding ML performance.

In another study, the influence of structural evolution on the ML properties in (Sr$_{1-x}$Ba$_x$)Al$_2$O$_4$:Eu^{2+},Eu^{3+} is examined.[25] While intense ML is observed in the composition range of x = 0–0.4, it totally disappears when $0 > 0.5$. The variation of ML coincides with the phase transformation of the (Sr$_{1-x}$Ba$_x$)Al$_2$O$_4$ from monoclinic to hexagonal around x = 0.3–0.43. The monoclinic crystal structure is considered critical to obtain the strong ML. On the other hand, it has been well proved that the ML can be significantly improved by cation co-doping (*e.g.*, Zr^{4+}).[26] The trap states can be also a significant factor for the superior ML in the α-SrAl$_2$O$_4$.

12.2.1.6 Melilite

Melilite is a group of polycationic compounds with a typical composition of $X_2M_AM_{B2}O_7$, in which X is a large-size cation from Na, Ca, Ba, or Ln, M_A is a small-size cation from Mg, Zn, Al, or Y, and M_B is from Si, Ge, Al, or B. Melilite compounds generally crystalize in a tetragonal structure with space group of $P\bar{4}2_1m$. Taking Ca$_2$MgAl$_2$O$_7$ as an example, both Mg and Al cations are tetrahedrally coordinated (Figure 12.6). These [MgO$_4$] and [AlO$_4$] tetrahedra are bridged by O atoms, forming tetrahedra layers on (0 0 1) and leaving large channels along [0 0 1] to host Ca^{2+} atoms. The unique channel structure creates significant anisotropy, which could be the origin of the observed ML when an efficient luminescent center such of Eu^{2+} or Ce^{3+} is introduced into the Ca$_2$MgAl$_2$O$_7$ or other Melilite-type hosts.[27–29]

12.2.1.7 Anorthite

ML in CaAl$_2$Si$_2$O$_8$:Eu^{2+} with an anorthite structure was first reported by Xu C.-N. et al.[30] Although the CaAl$_2$Si$_2$O$_8$ (triclinic, space group of $P\bar{1}$) shows centrosymmetry, elasticoluminescence indeed appeared upon compressive loads. The ML could be attributed to spontaneous strains and ferroelasticity from intrinsic disorders of [SiO$_4$] and [AlO$_4$] tetrahedra in the CaAl$_2$Si$_2$O$_8$. Furthermore, the ML performance can be greatly enhanced in the (Ca$_{1-x}$Sr$_x$)Al$_2$Si$_2$O$_8$:Eu^{2+} solid solution by increasing the Sr content from 0 to 0.4. When x continues to increase to 0.8, the ML completely disappears.[31] Structural examination reveals that phase transformation from $P\bar{1}$ to $I\bar{1}$ occurs at around 0.4 and from $I\bar{1}$ to $I2/c$ at around 0.7, respectively.[32] The anorthite solid solution gives a good example to control the ML performance by structural evolution (such as phase variation of the host, trap density, electron population function).

12.2.1.8 Oxynitride

ASi$_2$O$_2$N$_2$:Eu^{2+} (A = Sr and Ba) is an oxynitride phosphor with high quantum efficiency for lighting. The first report on ML in the BaSi$_2$O$_2$N$_2$:Eu^{2+} (blue emission) is contributed by Smet P. F. et al. in

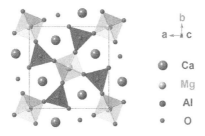

FIGURE 12.6 Crystal structure of Ca$_2$MgAl$_2$O$_7$ along [0 0 1].

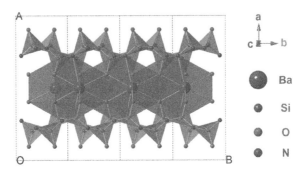

FIGURE 12.7   Crystal structure of $BaSi_2O_2N_2$ along [0 0 1].

2012.[33] Although there are three possible structure models in space groups of *Cmcm*, *Cmc2₁*, and *Pbcn* for $BaSi_2O_2N_2$ according to Schnick W. et al., [34] only the *Cmc2₁* shows non-centrosymmetry. The observed intense ML in $BaSi_2O_2N_2$:Eu^{2+} is thus possibly derived from the model in the *Cmc2₁* space group. Zhuang Y. et al. also found that the $SrSi_2O_2N_2$:Eu^{2+} (green) and (Sr,Ba)$Si_2O_2N_2$:Eu^{2+} (yellow) compounds in the space group of *P1* exhibited bright ML, although the intensity was relatively weaker than the Ba compound.[35] In terms of the coordination states, each cation in this group is bonded with two species of anions, forming [BaO_6N_2]/[SrO_6N_2] and [$SiON_3$] polyhedra, respectively (Figure 12.7). These mixed-anion polyhedra create a highly asymmetric environment in a near range and may contribute to the formation of piezoelectric fields upon mechanical stimuli. Nevertheless, the structural origin of ML in the $ASi_2O_2N_2$:Eu^{2+} (A = Ca, Sr, and Ba) family is still unclear and deserves further study.

Throughout the compounds listed in Table 12.1, one can find that most of the ML phosphors (more than half) show non-centrosymmetry in space group. Therefore, non-centrosymmetry in crystal structure is considered highly favorable for the formation of an electric field, which could be constructed either by a piezoelectric effect or triboelectric effect. On the other hand, there are still many compounds exhibiting centrosymmetric space groups. Since the centrosymmetric crystal structure cannot explain the origin of ML based on the widely used mechanism, some researchers have studied the microstructures of these centrosymmetric compounds and found that point defects, defect clusters, mixed-ion coordination states, domain structures, *etc.* could lead to the destruction of local symmetry. The local asymmetry could further create a built-in electric field under mechanical stimulus and finally produce ML.

12.2.2   METHODS TO EXPLORE NEW ML PHOSPHORS

In the early days of ML research without a clear understanding on the ML origin, most of the ML compounds were either discovered by accident in experiments or developed on the basis of simple synthetic experience. This not only leads to slow progress in the discovery of new materials but also makes it difficult to improve the ML performance of the existing materials. Obviously, a rational material design method can significantly speed up the development process and provide a useful guideline for performance improvement. In the field of ML materials, some efforts have been devoted to an effective method for the development of new materials recently. Three possible routes for ML materials' exploration have been reported.

The first one is to develop ML phosphors from the pool of reported persistent luminescent phosphors. Xu C.-N. et al. have proposed an energy-level model to clarify the mechanism of elasticoluminescence based on a large number of ML phosphors. According to the energy-level model as depicted in Figure 12.8(a), charge carriers are excited from emitting centers by light irradiation and then captured by some traps. When the phosphor is subjected to mechanical stress, the captured charge carriers can be released from traps at a much accelerated rate and give ML. This

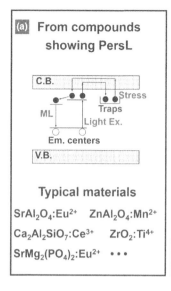

(a) From compounds showing PersL

Typical materials

$SrAl_2O_4$:Eu^{2+} $ZnAl_2O_4$:Mn^{2+}

$Ca_2Al_2SiO_7$:Ce^{3+} ZrO_2:Ti^{4+}

$SrMg_2(PO_4)_2$:Eu^{2+} • • •

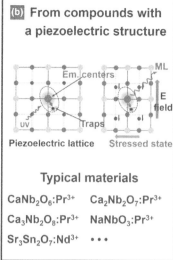

(b) From compounds with a piezoelectric structure

Typical materials

$CaNb_2O_6$:Pr^{3+} $Ca_2Nb_2O_7$:Pr^{3+}

$Ca_3Nb_2O_8$:Pr^{3+} $NaNbO_3$:Pr^{3+}

$Sr_3Sn_2O_7$:Nd^{3+} • • •

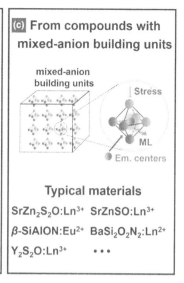

(c) From compounds with mixed-anion building units

Typical materials

$SrZn_2S_2O$:Ln^{3+} $SrZnSO$:Ln^{3+}

β-SiAlON:Eu^{2+} $BaSi_2O_2N_2$:Ln^{2+}

Y_2S_2O:Ln^{3+} • • •

FIGURE 12.8 Three possible routes to explore new ML phosphors. (a) From compounds showing persistent luminescence (PersL). (b) From compounds with a piezoelectric structure. (c) From compounds with mixed-anion building units. Typical materials found based on the three routes are given on the bottom. (Akiyama M., Xu C.-N., Nonaka K., *Applied Physics Letters* 81: 457–459, 2002. With permission; Zhang J.-C., Long Y.-Z., Yan X., Wang X., Wang F., *Chemistry of Materials* 28: 4052–4057, 2016. With permission; Chen C., Zhuang Y., Tu D., Wang X., Pan C., Xie R.-J., *Nano Energy* 68: 104329, 2020. With permission.)

process is possible due to an electric field generated by a piezoelectric effect or a triboelectric effect. Since traps are a precondition to realize elasticoluminescence, it is easy to understand that many elasticoluminescent phosphors are found from the typical compounds showing bright persistent luminescence (*e.g.*, $SrAl_2O_4$:Eu^{2+}, $ZnAl_2O_4$:Mn^{2+}). Nevertheless, some studies also suggest that the traps that contribute to persistent luminescence may be different from those that contribute to ML, further complicating the role of traps in ML phosphors. In this sense, to control the trap depth, its population, distribution, and the interaction with emitting centers in phosphors are important for the development of high-performance ML phosphors.

The second one is to create ML phosphors from the compounds with a piezoelectric crystal structure. Zhang J.-C. and coworkers have proposed an idea of incorporating trivalent lanthanides into classical piezoelectric compounds and accordingly developed a series of ML phosphors such as $Ca_2Nb_2O_7$:Pr^{3+} and $NaNbO_3$:Pr^{3+} (see Figure 12.8(b)). This route in fact verifies that the piezoelectric crystal structure is generally valid to generate ML in stress-applied phosphors by generating a piezoelectric electric field that promotes the charge carriers released from traps. Thus, the ML phosphors based on this approach should also have appropriate traps to capture the charge carriers under light irradiation.

The third one is to explore ML phosphors from the compounds with mixed-anion building units. This approach is proposed by Y. Zhuang et al. on the basis of a study on $SrZn_2OS_2$:Ln^{3+}, $SrZnOS$:Ln^{3+}, β-SiAlON:Eu^{2+}, and $SrSi_2O_2N_2$:Ln^{2+}.[15] It is further used to develop more mixed-anion ML compounds, such as Y_2O_2S:Ln^{3+} and $Y_2Si_3O_3N_4$:Ln^{3+}. In the mixed-anion building units $[AX_mY_n]$ (*A* is a central cation; *X* and *Y* are different species of anions surrounding the cations), the local symmetry of cation could be broken or the asymmetry of the cation could be further enlarged due to the differences of charge, ionic radius, electronegativity, and polarity of different anions. The local asymmetry caused by the mixed-anion building units and crystalline non-centrosymmetry formed by the periodic arrangement of the building units may become the structural origin of the ML (Figure 12.8(c)). Although the detailed mechanism has not been fully understood, the observed

ML in the above mixed-anion compounds generally show excellent self-recoverability without the need for light preexcitation. Compared with the ZnS:Cu/Mn^{2+} (the most important self-recoverable ML phosphor), these mixed-anion compounds extend the wavelength of ML to a much wider range, including UV (CaZnOS:Pb^{2+}) and NIR (SrZnOS:Er^{3+}) regions, and thus may greatly expand the application fields of ML mechanical sensing.

It should be noted that due to the diversity and complexity of ML mechanism as well as the lack of in-depth understanding of the relationship between its structure and ML properties, the methods mentioned above still have great limitation.

12.2.3 Methods to Improve ML Intensity

The intensity of ML under a certain mechanical stimulus is the fundamental property of an ML phosphor, which determines the sensing sensitivity if the phosphor is to be used for stress sensing. In recent years, great efforts have been made to improve the ML intensity.

12.2.3.1 Modifying the Microstructure or Crystal Structure of ML Phosphors

As the traps in the ML phosphors provide room for the storage of charge carriers, defect engineering (such as to boost the trap density in elasticoluminescence materials) should be an effective method to improve the ML intensity. For example, Zhang H. et al. reported that the ML intensity of SrAl$_2$O$_4$:Ce^{3+} could be enhanced by ~100 times after co-doping with 1.5% Ho^{3+}. The enhancement is majorly attributed to the formation of abundant shallow traps by aliovalent substitution of Ho^{3+} for Sr^{2+}, which greatly enlarged the storage capacity for charge carriers during light irradiation (Figure 12.9(a)).[45] The co-doping strategy is further applied to other materials and brought in obvious enhancement of ML intensity. Typical examples can be found in co-doping Dy^{3+} into Ca$_2$MgSi$_2$O$_7$:Eu^{2+},[27] Sr^{2+} into CaAl$_2$Si$_2$O$_8$:Eu^{2+},[31] Nd^{3+} into CaZnOS:Mn^{2+},[46] Gd^{3+} into (Ba,Ti)TiO$_3$:Pr^{3+},[47] Gd^{3+} into LiNbO$_3$:Pr^{3+},[48] La^{3+} into (Sr,Ca,Ba)$_2$SnO$_4$:Sm^{3+},[49] Zr^{4+} into SrAl$_2$O$_4$:Eu^{2+},[26] and Ge^{4+} into Sr$_3$Sn$_2$O$_7$:Sm^{3+}[50]. Reduction treatment is also a feasible way to increase the trap density. An approximately 100-fold enhancement in the ML intensity was reported in ZnAl$_2$O$_4$:Mn^{2+} after heat treated in a H$_2$-N$_2$ atmosphere at 1300°C.[20] Also, by applying an appropriate hydrogenation treatment, the ML intensity of ZnS:Cu could be significantly enhanced due to the increased concentration of sulfur vacancies.[51] In addition, Xu C.-N. and coworkers reported bright reddish-orange ML in Sr$_{n+1}$Sn$_n$O$_{3n+1}$:Sm^{3+} ($n = 1–\infty$) with designable layer units. The ML intensity of Sr$_3$Sn$_2$O$_7$:Sm^{3+} was improved ~1000 times than that of SrSnO$_3$:Sm^{3+} due to the layered structure and efficient charge transfer between the host and the emitting centers Sm^{3+} (Figure 12.9(b)).[52]

Regarding the ZnS:Cu/Mn^{2+}, promoting the migration and recombination rates of charge carriers under certain stress loading could be a key factor to improve the ML intensity. Sang Y. and coworkers fabricated a multilayer composite material with ZnS:Cu,Mn^{2+} ML phosphors sandwiched between two ferroelectric polyvinylidene fluoride (PVDF) layers.[53] The study showed that the additional external electric field induced by the PVDF layers might reduce the trap depth of ZnS:Cu,Mn^{2+} under stress loading, which drove more charge carriers to be released and recombine with emitting centers (Figure 12.9(c)). Moreover, Peng D. et al. reported a ZnS/CaZnOS heterojunction that exhibited recoverable ML with an extraordinary intensity of ~2 times higher than that of commercial ZnS:Mn^{2+} phosphors.[54] The high-performance ML might originate from the efficient charge transfer and promoted recombination arising from the conduction band offset in the heterojunction interface region, which was well supported by density functional theory calculations (Figure 12.9(d)).

12.2.3.2 Designing Specific Structures in ML Composite Films

In many cases, ML phosphors are incorporated into soft organic matrix to form composite films (or bulks) for the stress sensing/imaging applications. The structural features of the composite films also show significant influence on the ML intensity. These structural features include but are not

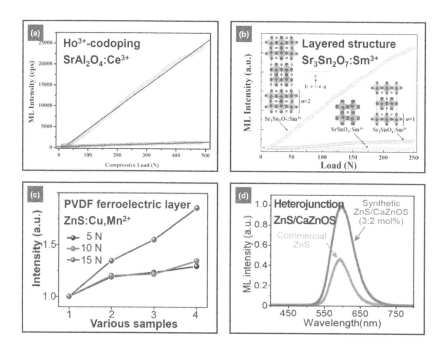

FIGURE 12.9 Improving the ML intensity of phosphors by modifying the microstructure or crystal structure. (a) ML intensity as a function of compressive load in $SrAl_2O_4:Ce^{3+}$ and $SrAl_2O_4:Ce^{3+},Ho^{3+}$. (b) ML intensity as a function of compressive load in $SrSnO_3:Sm^{3+}$, $Sr_2SnO_4:Sm^{3+}$, and $Sr_3Sn_2O_7:Sm^{3+}$. (c) ML enhancement in $ZnS:Cu,Mn^{2+}$ due to ferroelectric PVDF layers. (d) ML spectra of Cu-doped $ZnS/CaZnOS$ heterojunctions and commercial $ZnS:Mn^{2+}$ phosphors. (Zhang H., Yamada H., Terasaki N., Xu C.-N., *Applied Physics Letters* 91: 081905, 2007. With permission; Kamimura S., Yamada H., Xu C.-N., *Applied Physics Letters* 101: 091113, 2012. With permission; Wang F., Wang F., Wang X., Wang S., Jiang J., Liu Q., Hao X., Han L., Wang J., Pan C., Liu H., Sang Y., *Nano Energy* 63: 103861, 2019. With permission; Peng D., Jiang Y., Huang B., Du Y., Zhao J., Zhang X., Ma R., Golovynskyi S., Chen B., Wang F., *Advanced Materials* 32: 1907747, 2020. With permission.)

limited to the mechanical properties of the organic matrix, the size and distribution of introduced pores, organic-inorganic interfaces, and deliberately design array structures. They may result in variation of stress conduction efficiency, charge carrier mobility, or photon scattering inside the composite films and finally affect the output of ML. For example, Vinogradov D. A. et al. investigated the ML intensity of $SrAl_2O_4:Eu^{2+},Dy^{3+}$ nanoparticles in different organic matrices, such as boehmite and PDMS.[55] They found that the boehmite that was formed through polymerization of alumina sol-gel could concentrate applied stress and increased the tension of each ML particle by 2–3 orders of magnitude compared to $SrAl_2O_4:Eu^{2+},Dy^{3+}$ in PDMS. Song Y. and coworkers modified the PDMS matrix and adjusted the elastic modulus by adding SiO_2 nanoparticles.[56] The rigid SiO_2 nanoparticles concentrated the applied stress surrounding $ZnS:Cu/Mn^{2+}@Al_2O_3$ microparticles and thus improved the ML intensity by ~5 times.

Alternatively, some researchers designed specific array structures, *e.g.*, square, pyramidal, and cylindrical arrays, to concentrate the applied stress on specific areas in a two-dimensional plane. Wang Z. L. et al. fabricated a square array structure with a pixel size of 50×50 μm^2 in area and 100×100 μm^2 in center-to-center distance by spin-coating mixtures of $ZnS:Mn^{2+}$/photoresists on a polyethylene terephthalate (PET) substrate and curing square patterns. The array structure enabled clear imaging of the stress distribution with no cross talk between adjacent pixels and achieved good resolution and distinguishability.[6] The same group also designed a pyramidal array of $ZnS:Mn^{2+}@$ PDMS composites with 40×40 μm^2 in the area and 100×100 μm^2 by a silicon molding process. [57] The pyramid architecture allowed the applied stress to be effectively concentrated near the center of the pyramids, which accordingly reduced the threshold of the stress sensor to ~0.6 MPa. In addition, Zhu G. et al. adopted a separate layer consisting of aligned PDMS pillars.[58] The pillars with 90 μm in diameter and 60 μm in height could concentrate stress, making the intensity of the ML layer twice than that with a plane-surface stamp. Peng H. et al. used an advanced 3D printing technology to fabricate rigid nylon templates with similar cylindrical arrays.[59] ZnS:Cu@ PDMS composite films superimposed on the cylindrical array architecture could effectively harvest mechanical energy into light even under low external stress.

The ML intensity enhancement due to the structural design of composite films can usually be superimposed with that by the optimization of crystal structure. Consequently, in addition to studying the relationship between the crystal structure and the ML performance, the investigation of the structure of composite sensing film and sensor is also a significant research topic to achieve high-sensitivity ML stress sensing.

12.3 CHARACTERIZATION INSTRUMENTS AND CHARACTERIZATIONS OF ML PHOSPHORS

Generally, ML phosphors give transient light emission at the moment of mechanical action. The excitation method and the signal collection mode for the ML phosphors should be largely different from those used for the conventional PL phosphors. In this section, the instruments and methods for the characterization of ML phosphors and the evaluation of their potential applications are overviewed. The emphasis is put on those devices and methods exclusively used for the ML studies.

Because phosphor particles tend to move under mechanical stimuli, the ML phosphors are often mixed with organic polymers to prepare composite films or blocks (pellets) for ML characterizations and applications. In addition to fixing the phosphor particles, another important function of the soft organic matrix is to make the brittle phosphor particles not be broken even under large mechanical impact. Organic polymers that have been used as the matrix for ML phosphors to prepare composite films or blocks include PDMS,[60] epoxy resin,[61] polymethyl methacrylate (PMMA),[62] polyurethanes (PU),[63] polystyrene (PS),[64] PVDF,[53] polyvinyl chloride (PVC),[58] silicone,[65] and hydrogel.[66] In this section, most of the characterization instruments for ML phosphors are actually for composite films containing ML phosphors.

12.3.1 Characterization Instruments for ML Phosphors

12.3.1.1 Excitation Unit

Different from PL, the excitation of ML requires a unit that is able to impose controllable mechanical action to the studied phosphors. Universal testing machine is the most commonly used instrument to output compressive, tensile, or shear forces in a controllable mode. Typically, universal testing machine is used to compress a disk of ML phosphors@epoxy resin composites along the cross section under a fixed increasing rate of stress or strain.[45] This test is particularly useful to research the linear relationship between the ML intensity and the stress size. Alternatively, universal testing machine can be applied to stretch a thin sheet or a fiber that is covered with ML phosphors, which shows the stress distribution on the surface of the sheet.[67] Linear motor is another common mechanical unit to output repeated and high-frequency reciprocating motions.[56] It is widely used to test the recoverability of ML by stretching or bending soft composite ML films. Linear or rotation motor also can be integrated with a load that is perpendicular to the moving plane to generate frictional movement on the surface of a moving object. The former provides periodic reciprocating motion,[45] while the latter gives directional rotating motion.[20]

In addition, some researchers have tried to combine an alternating magnetic field with magnetostrictive body to produce alternating deformation or stress, which is further transferred to ML phosphors.[68, 69] Another feasible remote excitation source for ML is an ultrasonic wave. The ultrasonic wave can easily penetrate through biological tissues to excite the ML phosphors inside the living body, which may be widely used in future biological applications.[70, 71] Also, ML can be excited by loads that have random or ever-changing sizes, such as handwriting, skin movement, wind movement, or free falling.[6, 57, 72] Generally, the above excitation methods are designed to explore specific applications, not just to investigate the basic ML properties of materials.

12.3.1.2 Detection Unit

In principle, photodetectors are the detection units for ML. Since ML is a type of photoemission with relatively weaker intensity compared with PL, it requires a photodetector working in a highly sensitive mode to collect the weak light, although this is not a big problem for modern optoelectronics technology. Two types of photodetectors have been applied in the ML research: single-point detectors and two-dimensional array sensors. Single-point detectors include photomultiplier tubes (PMTs),[4] photodiode (PDs),[73] and charge-coupled device (CCD) spectrometers. [54] Two-dimensional array sensors include CCD and complementary metal oxide semiconductor (CMOS) image sensors.[74, 75]

PMTs are one of the most important photodetectors due to their superior detection sensitivity. PMTs with different detection wavelength ranges are available, which are majorly dependent on the species of semiconductor materials to fabricate the PMT. Si-type PMTs are the most common photodetectors for collecting visible photons; while with the growing attention of NIR ML phosphors, the use of InGaAs-type PMTs is also increasing.[76] Compared with PMTs, PDs have the advantages of good linearity, low cost, small size, long life, high quantum efficiency, and low driving voltage, which is suitable for the detection of rapidly changing optical signals. However, the disadvantages of PDs are small detection area and low sensitivity. CCD spectrometers are significant multichannel detectors. When connected with an optical fiber, CCD spectrometers provide a convenient way to monitor the spectral response of an object (usually from a single point) under mechanical stimuli in real time. Finally, CCD or CMOS array sensors have become indispensable instruments as image detectors to image the stress distribution on the surface of an ML film.[6] The CCD or CMOS sensors used in the ML study may include consumer-grade digital cameras, digital video recorders (for dynamic monitoring), and professional optical imagers.

12.3.1.3 Combination Configuration

Except for the remote excitation methods such as through a magnetic field or ultrasonic waves, the mechanical excitation unit is generally in contact with the ML composite films (or ML phosphors in some cases) and imposes stress directly on them. According to the contact mode between the ML film and photodetector (either remote or near-field) as well as the direction of photon acquisition (either forward or backward), there are four different combination configurations of the excitation (the same to ML composite films) and detection units (summarized in Figure 12.10).

Taking advantage of the instantaneous transmission of photons through space, the widely adopted configurations are to separate the photodetector from the excitation unit (ML composite film) to realize remote sensing (Figure 12.10(a) and (b)). To avoid blocking the ML by the excitation unit located on the upper surface, a transparent substrate is placed under the ML film. The ML generated on the bottom surface passes through the substrate and is finally collected by a remote photodetector (Figure 12.10(a)).[6, 72] When applying stress in a direction parallel to the surface of ML films, photodetectors can be installed above the upper surface. This configuration can be used to monitor the stress distribution on the surface of an object coated with a thin ML film (Figure 12.10(b)).[77] In addition, the space between the ML film and photodetector can be removed to build near-field configurations (Figure 12.10(c) and (d)). Among them, the near-field & forward (N&F) configuration by integrating ML film directly on optoelectronic image sensor is a promising scheme to construct a high-resolution stress sensor (Figure 12.10(c)). This scheme is particularly valuable for the development of flexible stress sensors for the advanced sensing applications such as wearable devices and electronic skins. Finally, a near-field & backward (N&B) configuration that the excitation unit exerts

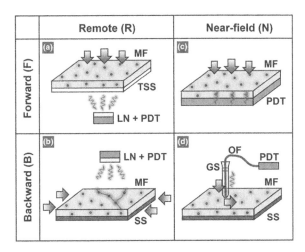

FIGURE 12.10 Combination configurations of the excitation and detection units for ML. The stress provided by the mechanical excitation unit is indicated by red arrows. The combination configurations are classified into (a–d) four types primarily according to the contact mode between the ML films and photodetector (remote or near-field) as well as the direction of photon acquisition (forward or backward). Abbreviations in the figure: MF (ML film), TSS (transparent substrate), LN (lens), PDT (photodetector), SS (substrate), GS (glass sleeve), OF (optical fiber). (Zhuang Y., Xie R.-J., *Advanced Materials* 33: 2005925, 2021. With permission.)

stress directly on the upper surface of the ML film and the emitted photons are also collected from the upper surface has also been designed (Figure 12.10(d)).[15] This configuration can be achieved by a delicately designed glass sleeve encapsulating an optical fiber. It can be used as a probe to scan the surface mechanical state of the ML films point by point.

12.3.1.4　Characterization Instruments for ML

In the last few decades, different instruments for the characterization of ML phosphors have been constructed by several research groups. Xu C.-N. et al. developed a modular testing system to record the ML intensity of a composite film under tensile load by using a CCD camera (Figure 12.11(a)). [78] The CCD camera can be simply replaced with a PMT, PD, or spectrometer to realize high-sensitivity detection or acquire spectral information as needed. Compressive force or other types of loads also can be given by a universal testing machine. Obviously, this testing system belongs to the remote & backward (R&B) configuration and it is widely used to monitor the variation of ML from the same point under different mechanical loads. Zhu G. and coworkers developed an ML recording system with the remote & forward (R&F) configuration as shown in Figure 12.11(b).[72] In this system, ML is driven by vertical force such as handwriting from the upper surface. The ML generated on the lower surface of the thin film passes through space and is collected by a remote detector under the film. When an image sensor is installed, this system can easily obtain the imaging of ML. The group of Xu C.-N. reported another ML recording system with the N&F configuration (Figure 12.11(c)).[79] A photosensitive layer is placed directly under the ML film to record all the ML signals generated from the film under friction or other mechanical loads. Undoubtedly, the photosensitive layer can be replaced by an array photodetector to obtain the real-time ML signals. Recently, Zhuang Y. and coworkers reported an ML probing system with the N&B configuration (Figure 12.11(d)).[15] In this system, the excitation and detection units are integrated into one by using an optical fiber encapsulated into a glass sleeve. As a result, the tip of the glass sleeve is able to collect ML from the phosphors or films at the moment when applying stress on them.

In addition to the representative characterization instruments described above, other instruments for specific ML study have been developed. Fontenot R. S. et al. reported a drop-tower apparatus to record the transient light from ML phosphors, mostly for fractoluminescent phosphors when

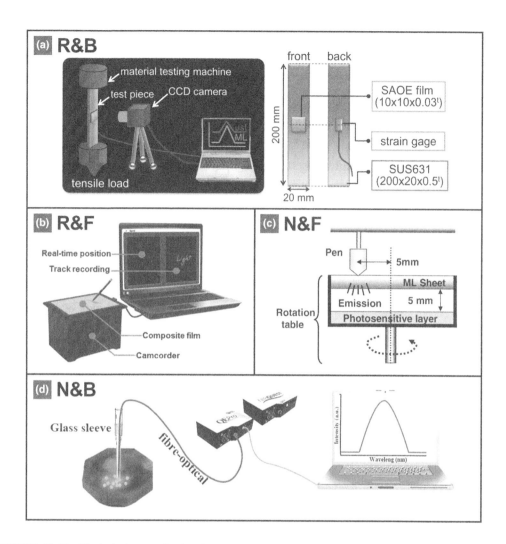

FIGURE 12.11 Typical characterization instruments for ML. (a) An ML testing system with the remote & backward (R&B) configuration that is driven with a material testing machine and monitored with a CCD camera. (b) An ML recording system with the remote & forward (R&F) configuration that can be driven by handwriting and recorded with a remote camera. (c) An ML recording system with the near-field & forward (N&F) configuration that is driven with a rotation motor and recorded by a photosensitive layer. (d) An ML probing system with the near-field & backward (N&B) configuration in which the ML is driven by grinding and simultaneously monitored with a fiber spectrometer. (Fujio Y., Xu C.-N., Nishibori M., Teraoka Y., Kamitani K., Terasaki N., Ueno N., *Journal of Advanced Dielectrics* 4: 1450016, 2014. With permission; Wei X.Y., Wang X., Kuang S.Y., Su L., Li H.Y., Wang Y., Pan C., Wang Z.L., Zhu G., *Advanced Materials* 28: 6656–6664, 2016. With permission; Terasaki N., Xu C.-N., *Japanese Journal of Applied Physics* 48: 04C150, 2009. With permission; Chen C., Zhuang Y., Tu D., Wang X., Pan C., Xie R.-J., *Nano Energy* 68: 104329, 2020. With permission.)

impacted by a free-falling ball.[73] The impact energy can be controlled by the height and weight of the ball. Notably, the ML intensity as a function of time can be precisely recorded by using an oscilloscope and an amplifier. Xu C.-N. and coworkers have invented a device based on atomic force microscopy (AFM) that is able to measure the ML from a single microparticle by using the AFM probe.[80] This device is important to observe the ML in different regions of elastic, plastic, and destructive deformation as well as study the ML mechanism of single particles.

The ML characterization instruments reported so far are all developed or set up by individual research groups and are oriented to laboratory research. A standardized and universal ML characterization instrument capable of quantitative excitation (by mechanical actions) and quantitative detection (by a correctly calibrated photodetector) has not been successfully developed.

12.3.2 Characterization of ML Phosphors

12.3.2.1 ML Spectra

The emission spectrum is the basic property of ML phosphors, which is related to their application field. Since ML is generally transient emission, the acquisition of ML spectra requires multichannel photodetectors such as CCD-type fiber spectrometer. In most cases, the ML spectra of phosphors are almost identical to their PL spectra. For example, the PL spectra of $CaZnOS:Mn^{2+}$ show a single broad emission band at 610 nm, and similar spectra can be recorded when the $CaZnOS:Mn^{2+}$ is subjected to impact, friction, or compression. However, the ML of some phosphors may show a little spectral shift or emission band loss rather than the PL spectra. For example, Smet P. F. et al. found that the ML spectra of $BaSi_2O_2N_2:Eu^{2+}$ redshifted by ~4 nm compared to the PL spectra.[33] After excluding the influence of instrument-related spectral response and resolution on the experimental results, the authors suggested that the redshift might be caused by the variation of crystal field strength during the mechanical action.

As shown in Table 12.1, the emission wavelength in the reported ML phosphors can be obtained in a wide range, covering from UV, visible, to NIR regions. The emission wavelength of ML phosphors can be controlled by doping different species of emitting centers into a suitable host, such as the CaZnOS or SrZnOS.[14] Particularly, due to the promising applications in the field of biomedical engineering, the ML phosphors giving NIR emission are getting more and more attention.[76]

12.3.2.2 ML Intensity

As mentioned before, high ML intensity of a phosphor under certain stress is the basic requirement for its application. In actual characterizations, the ML intensity can be affected by the species of ML phosphors, deformation size, deformational speed, the nature of mechanical stimulation, *etc.* The ML intensity can be defined in terms of $cd \cdot m^{-2}$. According to the literature, the ML intensity in the following typical phosphors (exactly inorganic@organic composites) are ZnS:Cu/Mn^{2+}@ PDMS (83.3–125.5 $cd \cdot m^{-2}$),[60] ZnS:Cu/Mn^{2+}@PDMS (21 $cd \cdot m^{-2}$),[81] $CaZr(PO_4)_2:Eu^{2+}$@epoxy resin (~15 $mcd \cdot m^{-2}$),[44] $CaZnOS:Mn^{2+}$@epoxy resin (30–40 $mcd \cdot m^{-2}$),[38] and $Ca_2Nb_2O_7:Pr^{3+}$@ epoxy resin (35 $mcd \cdot m^{-2}$).[19] It should be noted that even for the same material, the ML intensity measured by different researchers can be greatly different due to the difference of the mode of applied mechanical stimulation and the used characterization instrument. Consequently, it is of high significance to standardize the test conditions, including the preparation of ML materials (composition of the composites), the light excitation condition, the mode of mechanical stimulation, and the calibration of characterization instruments.

12.3.2.3 Linearity

A good linearity between the ML intensity and the stress amplitude is also an important prerequisite for the stress sensing applications. Xu C.-N. et al. first reported the linear increases of ML (elasticoluminescence) intensity in $SrAl_2O_4:Eu^{2+}$ with the amplitude of compressive force in 1999. [3] The linear relationship was also proven valid in $SrAl_2O_4:Eu^{2+}$ under various types of mechanical actions, such as compression, shear deformation, torsion, and ultrasonic vibration.[24, 82–84] For the composite films containing ZnS:Mn^{2+} or ZnS:Cu particles, the ML intensity was also found to be linear with several types of mechanical parameters, such as the sliding stress, sliding velocity, stretching rate, tensile strain, torque, gas flow velocity, magnetic field frequency, and square of the magnetic field strength.[5, 60, 67, 68, 72, 85] Although the underlying mechanism is still unclear, it has been proven that linearity is a common feature of most reported ML materials (except for those that only show fractoluminescence). In particular, the linearity of ML ensures high reliability and accuracy in stress distribution imaging.

12.3.2.4 Recoverability

Considering the need for sensing repeatability in sensing applications, it is necessary to characterize the emission recoverability of ML materials. Except for the fractoluminescence and plasticolumines-cence, most ML phosphors show recoverability under multiple cycles of mechanical actions. However, most of them can only recover the initial ML intensity after sufficient charging by light irradiation, while only a small part (*e.g.,* ZnS:Cu, ZnS:Mn^{2+}, CaZnOS:Ln^{3+}, SrZnOS:Ln^{3+}) keeps the same ML intensity under the multiple cycles of mechanical actions even without preexcitation of light (*i.e.,* self-reproducible). Till now, the reported species and properties of self-reproducible ML materials seem to have not yet met the great expectations in stress sensing applications. Further efforts should be made to develop more self-reproducible ML materials with enhanced sensitivity and intensity.

12.4 APPLICATIONS TO STRESS SENSING

Over the last few decades, thousands of ML phosphors, including the well-known SrAl$_2$O$_4$:Eu^{2+} and ZnS:Cu/Mn^{2+}, have been developed and investigated. Along with the development of the materials showing high ML intensity, excellent recoverability, and reliable linearity, the application of ML in sensing technology has received more and more attention. Due to the outstanding features of remoteness, stress distribution visualization, flexibility, easy integration, and self-powering, the ML-based sensing technology has exhibited significant advantages over the conventional sensing technologies. Although it has progressed rapidly in the last decades, the ML-based sensing technology is still in its infancy compared with the conventional technologies. The coming years must be an important period for the fast development of ML-based sensing technology and its applications. In this section, we will review the applications of ML in several important and promising fields and give several typical examples of the applications.

12.4.1 Structural Health Diagnosis and Damage Early Warning

ML-based sensing technology has been applied to visualize the cracks and defects on or inside engineering structures such as large buildings, bridges, highways, and other metal structures. This is because the cracks and defects on or inside structures may significantly affect the stress distribution on the surface. Although X-ray or ultrasound imaging is widely used to inspect big cracks or defects in engineering structures, it is difficult to detect those "invisible" cracks and defects in the early and developing stages. In contrast, the ML-based sensing technology is much sensitive to these early cracks and defects that create an unbalanced stress distribution around them. The ML-based sensing technology is particularly useful for early warning of damage or failure in engineering structures. More importantly, the ML-based sensing technology provides a simple, effective, and low-cost solution for remote, real-time, and large-area visualization of engineering structures. Consequently, it shows great potential in a variety of industrial applications.

The application of ML phosphors in structural health diagnosis is to prepare ML composite films and attach them as sensing sheets on the monitored structures like their artificial skin, which presents a real-time stress/strain distribution on the surface through the ML intensity imaging. Typically, the group of Xu C.-N. prepared stress/strain sensor sheets from the SrAl$_2$O$_4$:Eu^{2+} ML phosphors and attached them onto concrete structures of a highway bridge.[26] When a heavy vehicle passed through the bridge, the obtained ML image under stress was able to reflect the stress/strain distribution on the surface of concretes and the inner shallow layer (Figure 12.12). Therefore, the high-risk areas with overstress or excessive fatigue in the concrete structure can be identified by the ML-based health diagnosis method. Using ML imaging as a monitor to perform accurate

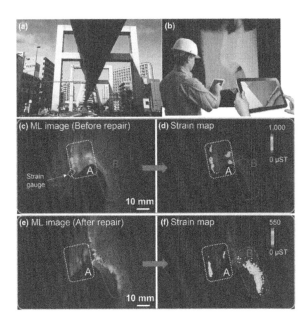

FIGURE 12.12 ML-based stress/strain sensor sheets containing $SrAl_2O_4:Eu^{2+}$ ML phosphors for on-site quantitative strain imaging and structural health diagnosis of a highway bridge. (a, b) Outside view and onsite inspection of the highway bridge. (c, d) ML Images and stress distribution before repair. (e, f) ML Images and stress distribution after repair. (Liu L., Xu C.-N., Yoshida A., Tu D., Ueno N., Kainuma S., *Advanced Materials Technologies* 4: 1800336, 2019. With permission.)

maintenance work in the high-risk areas can significantly extend the service life of the engineering structure at a low cost.

In addition, the ML-based stress/strain sensors can also be used for structural health diagnosis of high-pressure vessels, such as liquid hydrogen tanks or pipelines.[86] In particular, the ML sensors attached on external surfaces can visualize the cracks inside the high-pressure vessels, thus providing an effective nondestructive assessment technique.

Aircraft and spacecraft are subjected to various types of mechanical actions during flights. The structural health diagnosis and surficial stress monitoring are highly important for flight safety. However, monitoring the structural health and stress state of aircraft and spacecraft is still challenging due to the large-area surface and complex working environment. Korman V. et al. have reported a sensor sheet based on the ML phosphors $ZnS:Mn^{2+}$ for real-time monitoring of the surficial stress of high-speed moving objects, which showed great application potential in aircraft and spacecraft.[87] They used an air gun at speeds of 2–6 km s^{-1} to simulate mechanical impacts in space and adopted a silicon photodetector to collect ML signals. The sensor sheet attached to the back of the target plate produced yellow ML with a decay time of ~0.3 ms during the ultrahigh-speed impact, which ensured an ultrafast response to the impact. Due to the limited energy in space, this self-powered stress sensor shows a promising application in the structural health diagnosis of spacecraft.

12.4.2 BIOMECHANICAL ENGINEERING

Biomechanical engineering applies the principles and methods of mechanics to quantitatively analyze the mechanical problems in biology. Notably, biomechanical engineering is critical to understand the dynamic processes of movement, growth, and disease development in biological systems. Since most of the biological tissues such as bones and teeth are located inside the body, a noncontact and real-time

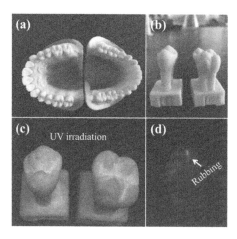

FIGURE 12.13 Artificial teeth fabricated from composites of $SrAl_2O_4:Eu^{2+},Dy^{3+}$ ML phosphors and denture base resin. (a) and (b) Photograph image under natural light, (c) PL image under UV excitation, and (d) ML image under mechanical rubbing. (Jiang Y., Wang F., Zhou H., Fan Z., Wu C., Zhang J., Liu B., Wang Z., *Materials Science & Engineering C* 92: 374–380, 2018. With permission.)

stress sensing technology is of high value in biological research. However, an effective and low-cost stress sensing technology that can be used in a living body has not been fully developed.

In 2012, Xu C.-N. et al. proposed an idea of using ML phosphors to image the surficial stress distribution of artificial bones.[88] They compared the ML imaging method with the thermoelastic stress analysis and concluded that the ML imaging method might exhibit advantages in temporal response and spatial resolution. Furthermore, Wang Z. et al. applied the ML imaging method to stress analysis of artificial teeth and developed a new approach to visualize the stress distribution on teeth with a super-high resolution (Figure 12.13).[89]

Due to the excellent penetrability of NIR light through biological tissues, NIR-emitting ML phosphors are in high demand in the biomechanical applications. Accordingly, several groups have devoted to the development of NIR-emitting ML phosphors, such as $SrAl_2O_4:Eu^{2+},Er^{3+}$ [90] and $Sr_3Sn_2O_7:Nd^{3+}$.[91] They also demonstrated the feasibility of ML imaging through deep biological tissues by *in vitro* experiments. These preliminary results indicate that the ML imaging technology may provide a noncontact, real-time, and tissue-penetrating mechanical sensing approach and thus will be a very promising mechanical probe for biomechanical engineering applications. However, there is still a long way to go for the practical application of ML-based biomechanical sensing technology.

12.4.3 Electronic Skin and Wearable Devices

A sensing layer composed of ML phosphors can be used as electronic skin (artificial skin) of wearable devices, robots, or prosthetics to provide a tactile sensing function and monitor the health state. In addition to being easy to use for low-cost, large-area sensing, the ML-based sensing layer with flexible organic matrix may also exhibit superior stretchability and self-healing ability compared to conventional stress sensors.

The idea of using ML films as artificial skin was first proposed by Xu C.-N. et al. in 1999.[92] In that pioneering work, they chose the $SrAl_2O_4:Eu^{2+}$ phosphors to prepare a flexible stress sensing layer that, when stressed, would generate ML with intensity proportional to the magnitude of the stress and would be collected by a remote photodetector. This exciting work has inspired extensive research on reproducible ML phosphors and the active exploration of the applications of ML-based sensing in the following 20 years. Recently, Song Y. et al. used SiO_2 nanoparticles–modified

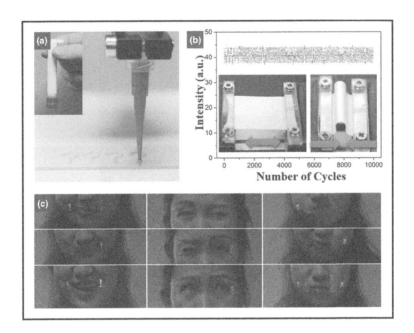

FIGURE 12.14 Skin-attachable sensor to monitor the motion of facial skin. (a) 3D printing process, (b) durability tests, and (c) visualized ML response to facial skin motion. (Qian X., Cai Z., Su M., Li F., Fang W., Li Y., Zhou X., Li Q., Feng X., Li W., Hu X., Wang X., Pan C., Song Y., *Advanced Materials* 30: 1800291, 2018. With permission.)

ZnS:Cu/Mn^{2+}@Al$_2$O$_3$@PDMS composites and 3D printing technology to prepare a skin-attachable stress sensing layer (Figure 12.14).[56] This sensing layer enabled visual facial expression management by monitoring the ML generated by the movement of facial skin. In addition, several research groups are working on integrating the ML-based sensing technology with traditional capacitive or resistance technology to design novel multimodal stress sensors with an extremely wide sensing range for the electronic skin applications.[57, 93, 94]

12.4.4 STRESS RECORDING

In general, ML phosphors emit photons at the moment when subjected to mechanical action. Real-time detection is a typical feature for the ML-based stress sensing technology. Nevertheless, some researchers have found that some ML phosphors have the ability to record the history of stress without the on-site use of photodetectors. Since no power supply is required for the entire process of mechanical action, stress recording (non-real-time stress sensing) is particularly beneficial, especially in cases when mechanical action is long or unpredictable.

In 2009, Xu C.-N. et al. proposed the concept of stress recording by combining the high-brightness ML phosphors SrAl$_2$O$_4$:Eu^{2+} with a photosensitive material (Polaroid 669) to prepare a two-layer stress recording medium.[79] In this two-layer medium, the photons given out from the ML phosphors can be accumulatively absorbed by the photosensitive material, resulting in a quantitative and measurable change in its absorbance. Furthermore, Zhuang Y. and coworkers reported a new force-induced charge carrier storage (FICS) effect for stress recording in deep trap–containing ML phosphors such as BaSi$_2$O$_2$N$_2$:Eu^{2+},Dy^{3+} (Figure 12.15).[35] They demonstrated three applications of the FICS effect in electronic signature recording, falling point monitoring, and vehicle collision recording, which exhibited outstanding advantages of distributed recording, long-term storage, and no need for continuous power supplies.

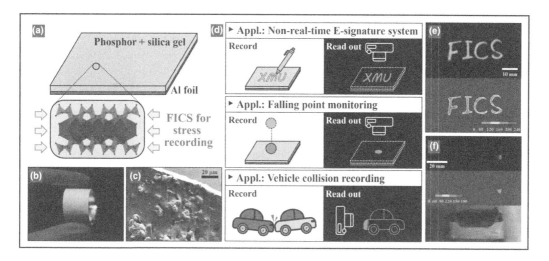

FIGURE 12.15 Applications of the FICS effect in stress recording. (a) Schematic diagram of a composite film containing the deep-trap ML phosphor $BaSi_2O_2N_2:Eu^{2+},Dy^{3+}$. (b) and (c) Photographic and SEM image of the composite film. (d) Schematic diagrams for three stress recording applications. (e) and (f) TL images and TL intensity maps. (Zhuang Y., Tu D., Chen C., Wang L., Zhang H., Xue H., Yuan C., Chen G., Pan C., Dai L., Xie R.J., *Light-Science & Applications* 9: 182, 2020. With permission.)

12.5 SUMMARY

The history of recorded scientific research on ML phosphors has been more than 500 years, while the rapid development of recoverable ML phosphors is in the last 20 years. Among them, ten years after 2010 has witnessed many researchers pushing ML phosphors and ML-based sensing technology into more new applications.

For the classification of ML, it has been basically agreed in the inorganic chemistry field that it can be categorized into four groups, *i.e.*, elasticoluminescence, plasticoluminescence, fractoluminescence, and triboluminescence. However, the understanding of the mechanism of each type of ML remains largely controversial. It could be hopeful to solve the difficulty of the ML mechanism by combining advanced microstructure characterization, dynamic mechanical simulation, and accurate electronic structure calculation.

Regarding the species of recoverable ML phosphors, some researchers have tried to establish the relationship between the crystal structure and ML. There is only weak evidence that non-centrosymmetry may contribute to the production of ML; ML nevertheless is also found in many crystals with a centrosymmetric structure. Although point defects, defect clusters, mixed-ion coordination states, domain structures, *etc.* have been proposed to be the origin of the destruction of local symmetry in the literature, the study of microstructure and its relationship with ML is still insufficient.

The instrument used to characterize ML phosphors requires a specific mechanical excitation unit and a highly sensitive detector for weak light acquisition or an array of detectors for luminescent imaging, which is very different from that used for the conventional PL phosphors. Although ML characterization instruments have developed greatly in the last 20 years, a standardized instrument has been lacking for a long time. To some extent, this hinders the rapid development and application of ML phosphors.

Finally, with the emergence of new technological fields such as wearable devices and biomechanical engineering, the application of stress luminescence must become increasingly important. This requires not only to further improve the performance of ML phosphors but also to consider how to match and optimize ML phosphors with sensors.

This chapter is only a rough introduction to ML phosphors. Readers with further interests on ML phosphors and their applications can be referred to the following review papers.[7–9, 76, 95–100]

REFERENCES

1. Bacon F., 1605. *The advancement of learning*. London, UK: Novum Organum.
2. Boyle R., 1664. *Experiments and considerations touching colours*. New York, NY: Johnson Reprint Corporation.
3. Xu C.-N., Watanabe T., Akiyama M., Zheng X.-G., 1999. Direct view of stress distribution in solid by mechanoluminescence. *Applied Physics Letters* 74: 2414–2416.
4. Akiyama M., Xu C.-N., Matsui H., Nonaka K., Watanabe T., 1999. Recovery phenomenon of mechanoluminescence from $Ca_2Al_2SiO_7$:Ce by irradiation with ultraviolet light. *Applied Physics Letters* 75: 2548–2550.
5. Moon Jeong S., Song S., Lee S.-K., Choi B., 2013. Mechanically driven light-generator with high durability. *Applied Physics Letters* 102: 051110.
6. Wang X., Zhang H., Yu R., Dong L., Peng D., Zhang A., Zhang Y., Liu H., Pan C., Wang Z.L., 2015. Dynamic pressure mapping of personalized handwriting by a flexible sensor matrix based on the mechanoluminescence process. *Advanced Materials* 27: 2324–2331.
7. Xie Y., Li Z., 2018. Triboluminescence: Recalling interest and new aspects. *Chem* 4: 943–971.
8. Zhang J.-C., Wang X., Marriott G., Xu C.-N., 2019. Trap-controlled mechanoluminescent materials. *Progress in Materials Science* 103: 678–742.
9. Feng A., Smet P.F., 2018. A review of mechanoluminescence in inorganic solids: Compounds, mechanisms, models and applications. *Materials* 11: 484.
10. Chandra B.P., Baghel R.N., Singh P.K., Luka A.K., 2009. Deformation-induced excitation of the luminescence centres in coloured alkali halide crystals. *Radiation Effects and Defects in Solids* 164: 500–507.
11. Kruglov A.S., Ei-Shanshoury I.A., Matta M.K., 1966. On the luminescence of γ-rayed KCl crystals induced by plastic deformation. *Journal of the Physical Society of Japan* 21: 2147–2153.
12. Chandra B.P., Dubey P.K., Datt S.C., 1986. Mechano- and electroluminescence of heavily doped ZnS:Mn phosphors. *Physica Status Solidi A* 97: K59–K61.
13. Xu C.-N., Zheng X.-G., Watanabe T., Akiyama M., Usui I., 1999. Enhancement of adhesion and triboluminescence of ZnS:Mn films by annealing technique. *Thin Solid Films* 352: 273–277.
14. Du Y., Jiang Y., Sun T., Zhao J., Huang B., Peng D., Wang F., 2019. Mechanically excited multicolor luminescence in lanthanide ions. *Advanced Materials* 31: 1807062.
15. Chen C., Zhuang Y., Tu D., Wang X., Pan C., Xie R.-J., 2020. Creating visible-to-near-infrared mechanoluminescence in mixed-anion compounds $SrZn_2S_2O$ and SrZnSO. *Nano Energy* 68: 104329.
16. Wang X., Xu C.N., Yamada H., Nishikubo K., Zheng X.G., 2005. Electro-mechano-optical conversions in Pr^{3+}-doped $BaTiO_3$-$CaTiO_3$ ceramics. *Advanced Materials* 17: 1254–1258.
17. Tu D., Xu C.N., Yoshida A., Fujihala M., Hirotsu J., Zheng X.G., 2017. $LiNbO_3$:Pr^{3+}: A multipiezo material with simultaneous piezoelectricity and sensitive piezoluminescence. *Advanced Materials* 29: 1606914.
18. Zhang J.C., Pan C., Zhu Y.F., Zhao L.Z., He H.W., Liu X., Qiu J., 2018. Achieving thermo-mechano-opto-responsive bitemporal colorful luminescence via multiplexing of dual lanthanides in piezoelectric particles and its multidimensional anticounterfeiting. *Advanced Materials* 30: 1804644.
19. Zhang J.-C., Long Y.-Z., Yan X., Wang X., Wang F., 2016. Creating recoverable mechanoluminescence in piezoelectric calcium niobates through Pr^{3+} doping. *Chemistry of Materials* 28: 4052–4057.
20. Matsui H., Xu C.-N., Tateyama H., 2001. Stress-stimulated luminescence from $ZnAl_2O_4$:Mn. *Applied Physics Letters* 78: 1068–1070.
21. Matsui H., Xu C.-N., Akiyama M., Watanabe T., 2000. Strong mechanoluminescence from UV-irradiated spinels of $ZnGa_2O_4$:Mn and $MgGa_2O_4$:Mn. *Japanese Journal of Applied Physics* 39: 6582–6586.
22. Yamada H., Kusaba H., Xu C.-N., 2008. Anisotropic lattice behavior in elasticoluminescent material $SrAl_2O_4$:Eu^{2+}. *Applied Physics Letters* 92: 101909.
23. Yamada H., Xu C.-N., 2007. *Ab initio* calculations of the mechanical properties of $SrAl_2O_4$ stuffed tridymite. *Journal of Applied Physics* 102: 126103.
24. Xu C.-N., Yamada H., Wang X., Zheng X.-G., 2004. Strong elasticoluminescence from monoclinic-structure $SrAl_2O_4$. *Applied Physics Letters* 84: 3040–3042.
25. Sakaihara I., Tanaka K., Wakasugi T., Ota R., Fujita K., Hirao K., Ishihara T., 2002. Triboluminescence of (Sr,Ba)Al_2O_4 polycrystals doped with Eu^{3+} and Eu^{2+}. *Japanese Journal of Applied Physics* 41: 1419–1423.

26. Liu L., Xu C.-N., Yoshida A., Tu D., Ueno N., Kainuma S., 2019. Scalable elasticoluminescent strain sensor for precise dynamic stress imaging and onsite infrastructure diagnosis. *Advanced Materials Technologies* 4: 1800336.

27. Zhang H., Yamada H., Terasaki N., Xu C.-N., 2008. Green mechanoluminescence of $Ca_2MgSi_2O_7$:Eu and $Ca_2MgSi_2O_7$:Eu,Dy. *Journal of The Electrochemical Society* 155: J55–J57.

28. Zhang H., Terasaki N., Yamada H., Xu C.-N., 2009. Mechanoluminescence of europium-doped $SrAMgSi_2O_7$ (A=Ca, Sr, Ba). *Japanese Journal of Applied Physics* 48: 04C109.

29. Zhang H., Yamada H., Terasaki N., Xu C.-N., 2008. Blue light emission from stress-activated $CaYAl_3O_7$:Eu. *Journal of The Electrochemical Society* 155: J128–J131.

30. Zhang L., Yamada H., Imai Y., Xu C.-N., 2008. Observation of elasticoluminescence from $CaAl_2Si_2O_8$:Eu^{2+} and its water resistance behavior. *Journal of The Electrochemical Society* 155: J63–J65.

31. Zhang L., Xu C.-N., Yamada H., Bu N., 2010. Enhancement of mechanoluminescence in $CaAl_2Si_2O_8$:Eu^{2+} by partial Sr^{2+} substitution for Ca^{2+}. *Journal of The Electrochemical Society* 157: J50–J53.

32. Feng A., Michels S., Lamberti A., Van Paepegem W., Smet P.F., 2020. Relating structural phase transitions to mechanoluminescence: The case of the $ca_{1-x}sr_xal_2si_2o_8$:$1\%Eu^{2+}$,$1\%Pr^{3+}$ anorthite. *Acta Materialia* 183: 493–503.

33. Botterman J., Eeckhout K.V.d., Baere I.D., Poelman D., Smet P.F., 2012. Mechanoluminescence in $BaSi_2O_2N_2$:Eu. *Acta Materialia* 60: 5494–5500.

34. Kechele J.A., Oeckler O., Stadler F., Schnick W., 2009. Structure elucidation of $BaSi_2O_2N_2$ – A host lattice for rare-earth doped luminescent materials in phosphor-converted (pc)-LEDs. *Solid State Sciences* 11: 537–543.

35. Zhuang Y., Tu D., Chen C., Wang L., Zhang H., Xue H., Yuan C., Chen G., Pan C., Dai L., Xie R.J., 2020. Force-induced charge carrier storage: A new route for stress recording. *Light-Science & Applications* 9: 182.

36. Nakamura S., Kawaguchi K., Ohgaku T., 2009. Deformation luminescence of X-irradiated KCl:Eu^{2+} by bending test. *IOP Conference Series: Materials Science and Engineering* 3: 012022.

37. Dickinson J.T., Scudiero L., Yasuda K., Kim M.W., Langford S.C., 1997. Dynamic tribological probes particle emission andtransient electrical measurements. *Tribology Letters* 3: 53–67.

38. Zhang J.C., Xu C.N., Kamimura S., Terasawa Y., Yamada H., Wang X., 2013. An intense elastico-mechanoluminescence material CaZnOS:Mn^{2+} for sensing and imaging multiple mechanical stresses. *Optics Express* 21: 12976–12986.

39. Li L., Wong K.-L., Li P., Peng M., 2016. Mechanoluminescence properties of Mn^{2+}-doped baznos phosphor. *Journal of Materials Chemistry C* 4: 8166–8170.

40. Fan X.H., Zhang J.C., Zhang M., Pan C., Yan X., Han W.P., Zhang H.D., Long Y.Z., Wang X., 2017. Piezoluminescence from ferroelectric $Ca_3Ti_2O_7$:Pr^{3+} long-persistent phosphor. *Optics Express* 25: 14238–14246.

41. Zhao H., Wang X., Li J., Li Y., Yao X., 2016. Strong mechanoluminescence of $Zn_2(Ge_{0.9}Si_{0.1})O_4$:Mn with weak persistent luminescence. *Applied Physics Express* 9: 012104.

42. Kamimura S., Yamada H., Xu C.-N., 2012. Development of new elasticoluminescent material $SrMg_2(PO_4)_2$:Eu. *Journal of Luminescence* 132: 526–530.

43. Akiyama M., Xu C.-N., Nonaka K., 2002. Intense visible light emission from stress-activated ZrO_2:Ti. *Applied Physics Letters* 81: 457–459.

44. Zhang J.C., Xu C.N., Long Y.Z., 2013. Elastico-mechanoluminescence in $CaZr(PO_4)_2$:Eu^{2+} with multiple trap levels. *Optics Express* 21: 13699–13709.

45. Zhang H., Yamada H., Terasaki N., Xu C.-N., 2007. Ultraviolet mechanoluminescence from $SrAl_2O_4$:Ce and $SrAl_2O_4$:Ce,Ho. *Applied Physics Letters* 91: 081905.

46. Su M., Li P., Zheng S., Wang X., Shi J., Sun X., Zhang H., 2020. Largely enhanced elastico-mechanoluminescence of CaZnOS: Mn^{2+} by co-doping with Nd^{3+} ions. *Journal of Luminescence* 217: 116777.

47. Zhang J.-C., Wan Y., Xin X., Han W.-P., Zhang H.-D., Sun B., Long Y.-Z., Wang X., 2014. Elastico-mechanoluminescent enhancement with Gd^{3+} codoping in diphase (Ba,Ca)TiO_3:Pr^{3+}. *Optical Materials Express* 4: 2300–2309.

48. Qiu G., Fang H., Wang X., Li Y., 2018. Largely enhanced mechanoluminescence properties in Pr^{3+}/Gd^{3+} co-doped $LiNbO_3$ phosphors. *Ceramics International* 44: 15411–15417.

49. Zhao H., Chai X., Wang X., Li Y., Yao X., 2016. Mechanoluminescence in (Sr,Ca,Ba)$_2SnO_4$: Sm^{3+},La^{3+} ceramics. *Journal of Alloys and Compounds* 656: 94–97.

50. Li J., Xu C.-N., Tu D., Chai X., Wang X., Liu L., Kawasaki E., 2018. Tailoring bandgap and trap distribution via Si or Ge substitution for Sn to improve mechanoluminescence in $Sr_3Sn_2O_7$:Sm^{3+} layered perovskite oxide. *Acta Materialia* 145: 462–469.

51. Gan J., Kang M.G., Meeker M.A., Khodaparast G.A., Bodnar R.J., Mahaney J.E., Maurya D., Priya S., 2017. Enhanced piezoluminescence in non-stoichiometric ZnS:Cu microparticle based light emitting elastomers. *Journal of Materials Chemistry C* 5: 5387–5394.

52. Kamimura S., Yamada H., Xu C.-N., 2012. Strong reddish-orange light emission from stress-activated $Sr_{n+1}Sn_nO_{3n+1}$:Sm^{3+} (n = 1, 2, ∞) with perovskite-related structures. *Applied Physics Letters* 101: 091113.

53. Wang F., Wang F., Wang X., Wang S., Jiang J., Liu Q., Hao X., Han L., Wang J., Pan C., Liu H., Sang Y., 2019. Mechanoluminescence enhancement of ZnS:Cu,Mn with piezotronic effect induced trap-depth reduction originated from PVDF ferroelectric film. *Nano Energy* 63: 103861.

54. Peng D., Jiang Y., Huang B., Du Y., Zhao J., Zhang X., Ma R., Golovynskyi S., Chen B., Wang F., 2020. A ZnS/CaZnOS heterojunction for efficient mechanical-to-optical energy conversion by conduction band offset. *Advanced Materials* 32: 1907747.

55. Ilatovskii D.A., Tyutkov N.A., Vinogradov V.V., Vinogradov A.V., 2018. Stimuli-responsive mechanoluminescence in different matrices. *ACS Omega* 3: 18803–18810.

56. Qian X., Cai Z., Su M., Li F., Fang W., Li Y., Zhou X., Li Q., Feng X., Li W., Hu X., Wang X., Pan C., Song Y., 2018. Printable skin-driven mechanoluminescence devices via nanodoped matrix modification. *Advanced Materirals* 30: 1800291.

57. Wang X., Que M., Chen M., Han X., Li X., Pan C., Wang Z.L., 2017. Full dynamic-range pressure sensor matrix based on optical and electrical dual-mode sensing. *Advanced Materials* 29: 1605817.

58. Wei X.Y., Liu L., Wang H.L., Kuang S.Y., Zhu X., Wang Z.L., Zhang Y., Zhu G., 2018. High-intensity triboelectrification-induced electroluminescence by microsized contacts for self-powered display and illumination. *Advanced Materials Interfaces* 5: 1701063.

59. Bao L., Xu X., Zuo Y., Zhang J., Liu F., Yang Y., Xu F., Sun X., Peng H., 2019. Piezoluminescent devices by designing array structures. *Science Bulletin* 64: 151–157.

60. Jeong S.M., Song S., Joo K.-I., Kim J., Hwang S.-H., Jeong J., Kim H., 2014. Bright, wind-driven white mechanoluminescence from zinc sulphide microparticles embedded in a polydimethylsiloxane elastomer. *Energy & Environmental Science* 7: 3338–3346.

61. Sohn K.-S., Seo S.Y., Kwon Y.N., Park H.D., 2002. Direct observation of crack tip stress field using the mechanoluminescence of $SrAl_2O_4$:(Eu,Dy,Nd). *Journal of the American Ceramic Society* 85: 712–714.

62. Zhang H., Peng D., Wang W., Dong L., Pan C., 2015. Mechanically induced light emission and infrared-laser-induced upconversion in the Er-doped CaZnOS multifunctional piezoelectric semiconductor for optical pressure and temperature sensing. *The Journal of Physical Chemistry C* 119: 28136–28142.

63. Persits N., Aharoni A., Tur M., 2017. Quantitative characterization of ZnS:Mn embedded polyurethane optical emission in three mechanoluminescent regimes. *Journal of Luminescence* 181: 467–476.

64. Akintola T.M., Tran P., Lucien C., Dickens T., 2020. Additive manufacturing of functional polymer-based composite with enhanced mechanoluminescence (ZnS:Mn) performance. *Journal of Composite Materials* 54: 3181–3188.

65. Larson C., Peele B., Li S., Robinson S., Totaro M., Beccai L., Mazzolai B., Shepherd R., 2016. Highly stretchable electroluminescent skin for optical signaling and tactile sensing. *Science* 351: 1071–1074.

66. İncel A., Reddy S.M., Demir M.M., 2017. A new method to extend the stress response of triboluminescent crystals by using hydrogels. *Materials Letters* 186: 210–213.

67. Jeong S.M., Song S., Seo H.-J., Choi W.M., Hwang S.-H., Lee S.G., Lim S.K., 2017. Battery-free, human-motion-powered light-emitting fabric: Mechanoluminescent textile. *Advanced Sustainable Systems* 1: 1700126.

68. Wong M.-C., Chen L., Tsang M.-K., Zhang Y., Hao J., 2015. Magnetic-induced luminescence from flexible composite laminates by coupling magnetic field to piezophotonic effect. *Advanced Materials* 27: 4488–4495.

69. Wong M.C., Chen L., Bai G., Huang L.B., Hao J., 2017. Temporal and remote tuning of piezophotonic-effect-induced luminescence and color gamut via modulating magnetic field. *Advanced Materials* 29: 1701945.

70. Wu X., Zhu X., Chong P., Liu J., Andre L.N., Ong K.S., Brinson K., Jr., Mahdi A.I., Li J., Fenno L.E., Wang H., Hong G., 2019. Sono-optogenetics facilitated by a circulation-delivered rechargeable light source for minimally invasive optogenetics. *Proceedings of the National Academy of the Sciences of the United States of America* 116: 26332–26342.

71. Terasaki N., Yamada H., Xu C.-N., 2013. Ultrasonic wave induced mechanoluminescence and its application for photocatalysis as ubiquitous light source. *Catalysis Today* 201: 203–208.

72. Wei X.Y., Wang X., Kuang S.Y., Su L., Li H.Y., Wang Y., Pan C., Wang Z.L., Zhu G., 2016. Dynamic triboelectrification-induced electroluminescence and its use in visualized sensing. *Advanced Materials* 28: 6656–6664.

73. Fontenot R.S., Hollerman W.A., Aggarwal M.D., Bhat K.N., Goedeke S.M., 2012. A versatile low-cost laboratory apparatus for testing triboluminescent materials. *Measurement* 45: 431–436.

74. Xu C.-N., Zheng X.-G., Akiyama M., Nonaka K., Watanabe T., 2000. Dynamic visualization of stress distribution by mechanoluminescence image. *Applied Physics Letters* 76: 179–181.

75. Wu C., Zeng S., Wang Z., Wang F., Zhou H., Zhang J., Ci Z., Sun L., 2018. Efficient mechanoluminescent elastomers for dual-responsive anticounterfeiting device and stretching/strain sensor with multi-mode sensibility. *Advanced Functional Materials* 28: 1803168.

76. Xiong P., Peng M., Yang Z., 2021. Near-infrared mechanoluminescence crystals: A review. *iScience* 24: 101944.

77. Terasaki N., Xu C.-N., 2013. Historical-log recording system for crack opening and growth based on mechanoluminescent flexible sensor. *IEEE Sensors Journal* 13: 3999–4004.

78. Fujio Y., Xu C.-N., Nishibori M., Teraoka Y., Kamitani K., Terasaki N., Ueno N., 2014. Development of highly sensitive mechanoluminescent sensor aiming at small strain measurement. *Journal of Advanced Dielectrics* 4: 1450016.

79. Terasaki N., Xu C.-N., 2009. Mechanoluminescence recording device integrated with photosensitive material and europium-doped SrAl2O4. *Japanese Journal of Applied Physics* 48: 04C150.

80. Sakai K., Koga T., Imai Y., Maehara S., Xu C.N., 2006. Observation of mechanically induced luminescence from microparticles. *Physical Chemistry Chemical Physics* 8: 2819–2822.

81. Chen L., Wong M.-C., Bai G., Jie W., Hao J., 2015. White and green light emissions of flexible polymer composites under electric field and multiple strains. *Nano Energy* 14: 372–381.

82. Li C.S., Xu C.N., Zhang L., Yamada H., Imai Y., Wang W.X., 2008. Dynamic visualization of stress distribution by mechanoluminescence image. *Key Engineering Materials* 388: 265–268.

83. Kim G.-W., Kim J.-S., 2014. Dynamic torsional response analysis of mechanoluminescent paint and its application to non-contacting automotive torque transducers. *Measurement Science and Technology* 25: 015009.

84. Zhan T., Xu C.-N., Fukuda O., Yamada H., Li C., 2011. Direct visualization of ultrasonic power distribution using mechanoluminescent film. *Ultrasonics Sonochemistry* 18: 436–439.

85. Kim J.S., Kim G.-W., 2014. New non-contacting torque sensor based on the mechanoluminescence of ZnS:Cu microparticles. *Sensors and Actuators A: Physical* 218: 125–131.

86. Fujio Y., Xu C.-N., Terasawa Y., Sakata Y., Yamabe J., Ueno N., Terasaki N., Yoshida A., Watanabe S., Murakami Y., 2016. Sheet sensor using SrAl2O4:Eu mechanoluminescent material for visualizing inner crack of high-pressure hydrogen vessel. *International Journal of Hydrogen Energy* 41: 1333–1340.

87. Korman V., Goedeke S.M., Hollerman W.A., Bergeron N.P., Allison S.W., Moore R.J., 2006. Developing a phosphor-based health monitoring sensor suite for future spacecraft. *Proceeding of SPIE: Sensors for Propulsion Measurement Applications* 6222: 62220B.

88. Hyodo K., Terasawa Y., Xu C.-N., Sugaya H., Mishima H., Miyakawa S., 2012. Mechanoluminescent stress imaging for hard tissue biomechanics. *Journal of Biomechanics* 45: S263.

89. Jiang Y., Wang F., Zhou H., Fan Z., Wu C., Zhang J., Liu B., Wang Z., 2018. Optimization of strontium aluminate-based mechanoluminescence materials for occlusal examination of artificial tooth. *Materials Science & Engineering C* 92: 374–380.

90. Terasawa Y., Xu C.N., Yamada H., Kubo M., 2011. Near infra-red mechanoluminescence from strontium aluminate doped with rare-earth ions. *IOP Conference Series: Materials Science and Engineering* 18: 212013.

91. Xiong P., Peng M., 2019. Near infrared mechanoluminescence from the Nd^{3+} doped perovskite LiNbO3:Nd^{3+} for stress sensors. *Journal of Materials Chemistry C* 7: 6301–6307.

92. Xu C.N., Watanabe T., Akiyama M., Zheng X.G., 1999. Artificial skin to sense mechanical stress by visible light emission. *Applied Physics Letters* 74: 1236–1238.

93. Zhang Y., Fang Y., Li J., Zhou Q., Xiao Y., Zhang K., Luo B., Zhou J., Hu B., 2017. Dual-mode electronic skin with integrated tactile sensing and visualized injury warning. *ACS Applied Materials & Interfaces* 9: 37493–37500.

94. Jang J., Kim H., Ji S., Kim H.J., Kang M.S., Kim T.S., Won J.-e., Lee J.-H., Cheon J., Kang K., Im W.B., Park J.-U., 2019. Mechanoluminescent, air-dielectric MoS$_2$ transistors as active-matrix pressure sensors for wide detection ranges from footsteps to cellular motions. *Nano Letters* 20: 66–74.

95. Zhuang Y., Xie R.-J., 2021. Mechanoluminescence rebrightening the prospects of stress sensing: A review. *Advanced Materials* 33: 2005925.

96. Bünzli J.-C.G., Wong K.-L., 2018. Lanthanide mechanoluminescence. *Journal of Rare Earths* 36: 1–41.

97. Zhang H., Wei Y., Huang X., Huang W., 2019. Recent development of elastico-mechanoluminescent phosphors. *Journal of Luminescence* 207: 137–148.

98. Peng D., Chen B., Wang F., 2015. Recent advances in doped mechanoluminescent phosphors. *Chempluschem* 80: 1209–1215.

99. Wang C., Peng D., Pan C., 2020. Mechanoluminescence materials for advanced artificial skin. *Science Bulletin* 65: 1147–1149.

100. Mukherjee S., Thilagar P., 2019. Renaissance of organic triboluminescent materials. *Angewandte Chemie International Edition* 58: 7922–7932.

13 Long Persistent Phosphors

Feng Liu and Xiao-Jun Wang

CONTENTS

13.1 INTRODUCTION

Persistent phosphor as a kind of light-emitting material can store excitation energy in so-called traps and then persistently releases the energy in the form of light emission after ceasing the excitation. This delayed emission after the end of the excitation is called long persistent luminescence, also called afterglow, which relates to the trapping/detrapping of electrons or holes in the persistent phosphors.

The phenomenon of persistent luminescence has been known for centuries. However, only in the last few decades, the technologies and mechanisms investigating the persistent luminescence properties are developed. In particular, since the milestone material $SrAl_2O_4$:Eu^{2+},Dy^{3+} was reported by Matsuzawa *et al.* in 1996,[1] the field of persistent phosphors has been rapidly developed. [2–4] Today, blue-green-emitting persistent phosphors (e.g., $SrAl_2O_4$:Eu^{2+},Dy^{3+} and $CaAl_2O_4$:Eu^{2+},Nd^{3+}) have achieved good commercial success for various civil applications (e.g., safety signs, watch dials and toys). Besides the commercial applications, persistent phosphors also have found potential in other fields. For example, a promising application for persistent phosphors (nano-sized particles) is on *in vivo* optical imaging in modern biology and medical diagnostics.[2–4] Accordingly, novel

DOI: 10.1201/9781003098676-13

materials have been developed, especially the Cr^{3+}-doped persistent phosphors.[5,6] Moreover, to the bioimaging applications using persistent luminescence technology, there is a challenge on the *in vivo* excitation of the phosphors. That is, since persistent phosphors are generally capable of being charged by ultraviolet illumination, which has short tissue penetration and is harmful to normal tissues, it is not possible to recharge the persistent nanoprobes using ultraviolet light once the probes have been inside a living subject. Thus, limited luminescence longevity is the major obstacle to the use of persistent luminescence technology in bioimaging. To date, persistent luminescence has been extensively studied, first on visible persistent luminescence, followed by infrared and recently on ultraviolet. Hundreds of persistent phosphors have been reported.[2–4]

In this chapter, recent developments of persistent phosphor research are surveyed. More background information may be found in the second version of the handbook containing a chapter/section of the same title.[7] The chapter deals with the following aspects.

- First, the trapping/detrapping of electrons in the persistent luminescence will be demonstrated on the basis of thermoluminescence spectroscopy (Section 13.1).
- Then the excitation process of persistent luminescence (Section 13.2) will be discussed, involving the up-conversion charging technique developed recently.
- Finally, we give an introduction to emission (Section 13.3) of persistent luminescence. A novel glow-in-the-bright phenomenon will be introduced, which advances the understanding of persistent luminescence.

13.2 TRAPPING AND DETRAPPING

Persistent luminescence research has been well developed in the last decades, largely stimulated by the famous glow-in-the-dark phosphor, $SrAl_2O_4$:Eu^{2+},Dy^{3+}. The generally accepted picture of persistent luminescence has its origins in the energy band theory of solids. The persistent luminescence process can be qualitatively demonstrated by a simple schematic model based on electron transfer assumption, as shown in Figure 13.1. Upon irradiation, electrons are promoted to the delocalized states (conduction band continuums or excited states associated with delocalization properties). Whenever the lattice defects occur, or if there are impurities within the lattice, there is a breakdown in the periodicity of the crystalline structure and it becomes possible for electrons to possess energies in the energy gap. That is, the existence of structure imperfection in the host lattice or the incorporation of impurities can give rise to lattice distortion, resulting in the localized energy levels within the energy gap. The introduced energy levels may be discrete or distributed, depending on the nature of defect and host lattice. The charges may be trapped at the defect energy levels. On the contrary, the previously trapped charges may be freed by the externally thermal stimulation, resulting in long-term emissions from the emitting centers in the phosphors. The delay corresponds to the time the electrons spend in the traps.

13.2.1 THERMOLUMINESCENCE

For persistent phosphors, persistent luminescence performance generally depends on the properties of traps, while the information related to the traps can be provided by thermoluminescence measurement. Thermoluminescence refers to a luminescence phenomenon induced by thermally stimulated recombination of trapped electrons and holes in phosphors, which have been subjected to prior irradiation. The resulting curve of recombination luminescence intensity versus temperature is usually called a glow curve. Consequently, thermoluminescence involves a metastable state consisting of trapped electron or hole, which can be triggered to recombine by heating stimulation; the recombination energy is transferred to an emitting center. During the thermoluminescence, the irradiated phosphor is heated to a temperature at which the energy barriers of traps can be overcome thermally. The basic processes and spectroscopic techniques about the thermoluminescence

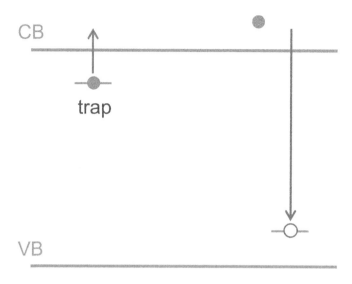

FIGURE 13.1 Scheme representation of persistent luminescence. Up- and down-arrows represent thermal releasing of electron and recombination radiation, respectively.

have been well introduced before.[8,9] Here, we would not go into the details of thermoluminescence but present two remarks on the measurements.

13.2.1.1 Effect of Black-Body Radiation

Black-body radiation in the thermoluminescence measurements usually originates from the sample, as well as the heating platform. Since the background radiation generally locates in the infrared region, one can deduce that the black-body radiation may affect significantly the infrared-emitting thermoluminescence measurement. A nice example using an infrared persistent phosphor, $Zn_3Ga_2Ge_2O_{10}:Ni^{2+}$, is presented here. Upon excitation, the phosphor exhibits infrared persistent luminescence featuring a broad spectral band with the maximum at 1290 nm.[10] Accordingly, thermoluminescence measurements are carried out by monitoring the infrared emission. In Figure 13.2, the purple solid-line curve is the measured thermoluminescence curve of $Zn_3Ga_2Ge_2O_{10}:Ni^{2+}$ phosphor by monitoring the emission of Ni^{2+}. Before the measurement, the sample was pre-irradiated with 300 nm light for 20 min to fill the traps. The curve consists of a broad emission band peaking at about 80°C and a steep rise at a high-temperature region. To understand the origin of the rise at the high-temperature region, we then thermally cleaned the same sample at 420°C for 30 min (the traps were thermally emptied) and recorded the heating read-out curve using the same procedure, as the black dashed-line curve shown in Figure 13.2. The cleaned sample still exhibits a remarkable rise at the high-temperature region starting at ~100°C. Since the traps in the cleaned sample were thermally emptied, this high-temperature emission is unrelated to the electron traps and should be attributed to black-body radiation from both the sample and the heating system. Therefore, the real thermoluminescence output of the $Zn_3Ga_2Ge_2O_{10}:Ni^{2+}$ phosphor can be obtained by subtracting the black-body radiation contribution obtained on the cleaned sample (dashed-line curve) from the thermoluminescence curve obtained on the pre-irradiated sample (solid-line curve), as the pink dot-dashed-line curve shown in Figure 13.2. The thus obtained corrected thermoluminescence curve exhibits a single broad band peaking at 80°C.

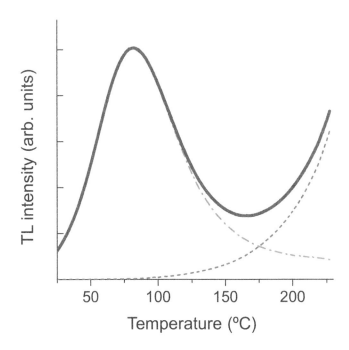

FIGURE 13.2 Thermoluminescence (TL) measurement and the effect of black-body radiation on the TL measurement in $Zn_3Ga_2Ge_2O_{10}$:Ni^{2+} infrared persistent phosphor. These curves are recorded by monitoring at 1290 nm.

13.2.1.2 Effect of Thermal Quenching

Effect of thermal quenching on the steady-state photoluminescence of phosphors is a common phenomenon. For a storage phosphor, if its emitting level is sensitive to ambient temperature, care should be taken on the thermoluminescence analysis. Here, we demonstrate such effect by comparing isostructural $Y_3Ga_5O_{12}$:Bi^{3+} and $Y_3Al_5O_{12}$:Bi^{3+} phosphors.[11] Figure 13.3(a) gives the thermoluminescence curves for the $Y_3Ga_5O_{12}$:Bi^{3+} and $Y_3Al_5O_{12}$:Bi^{3+}. The two curves share similar spectral structures in the low-temperature region (30–180°C). Whereas, compared with the thermoluminescence curve of $Y_3Al_5O_{12}$:Bi^{3+}, the high-temperature band of $Y_3Ga_5O_{12}$:Bi^{3+} is nearly disappeared. At first sight on the curves in Figure 13.3(a), one may simply think there is no population in the deep traps of $Y_3Ga_5O_{12}$:Bi^{3+}, since the thermoluminescence intensity in the high-temperature region is pretty low. However, it should be realized that, compared to the case of visible or infrared luminescence, the thermal quenching of ultraviolet luminescence is generally more remarkable and may affect the thermoluminescence measurement.

The apparent difference between the thermoluminescence spectral shapes of $Y_3Ga_5O_{12}$:Bi^{3+} and $Y_3Al_5O_{12}$:Bi^{3+} inspires us to learn the emission capability of the ultraviolet emitting levels at different temperatures. For each phosphor, the steady-state photoluminescence intensities are recorded as a function of excitation temperatures. As shown in Figure 13.3(b), from 30 to 270°C, the emission intensities decrease gradually upon 254 nm excitation. While the emission intensity of $Y_3Ga_5O_{12}$:Bi^{3+} decreases very steeply along the temperature, the intensity of $Y_3Al_5O_{12}$:Bi^{3+} decreases slightly. By combining the observation in thermoluminescence measurements of the two phosphors (Figure 13.3(a)), one can deduce that the difference of thermoluminescence spectral distributions may stem from the different luminescence thermal quenching properties.

To further illustrate the effect of luminescence thermal quenching on the thermoluminescence measurements in $Y_3Ga_5O_{12}$:Bi^{3+} and $Y_3Al_5O_{12}$:Bi^{3+}, a scheme on the competition between the thermal quenching and the thermoluminescence emission is proposed. Figure 13.3(c) shows that the $^3P_1$ emitting level in $Y_3Ga_5O_{12}$:Bi^{3+} is situated closer to the bottom of conduction band than that in

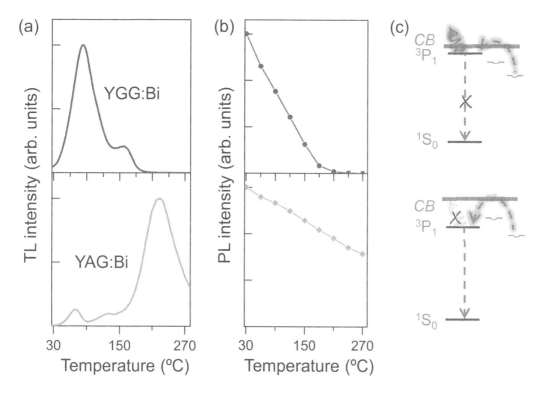

FIGURE 13.3 Thermoluminescence (TL) and photoluminescence (PL) thermal quenching of $Y_3Ga_5O_{12}:Bi^{3+}$ (YGG:Bi) and $Y_3Al_5O_{12}:Bi^{3+}$ (YAG:Bi). (a) TL curves monitored ultraviolet emissions of YGG:Bi (316 nm) and YAG:Bi (303 nm). (b) Temperature dependence of steady-state PL intensity for each phosphor under 254 nm excitation. (c) Schematic illustrations of TL and thermal ionization quenching at high temperature for YGG:Bi and YAG:Bi.

$Y_3Al_5O_{12}:Bi^{3+}$. The smaller energy difference can explain the lower temperature at which quenching by thermal ionization starts. Accordingly, even though the deep traps of $Y_3Ga_5O_{12}:Bi^{3+}$ are released effectively at high temperature, the thermal ionization disables the thermoluminescence emission from the $^3P_1$ level. On the contrary, in $Y_3Al_5O_{12}:Bi^{3+}$, the large energy difference between the $^3P_1$ emitting level and the bottom of conduction band inactivates the thermal ionization, so that the deep traps of $Y_3Al_5O_{12}:Bi^{3+}$ can be evaluated by the thermoluminescence measurement.

To gain more insight into the deep traps in $Y_3Ga_5O_{12}:Bi^{3+}$, an extended thermoluminescence experiment is presented. The $Y_3Ga_5O_{12}:Bi^{3+}$ phosphor is initially illuminated at 200°C by a 254 nm ultraviolet lamp for 60 s, followed by a fast cooling down step (in 5 s). Subsequently, the thermoluminescence measurement starts at 180 s after stoppage of the illumination. Figure 13.4(a) gives the resulting thermoluminescence curve (dashed-line curve). Compared with the curve of $Y_3Ga_5O_{12}:Bi^{3+}$ shown in Figure 13.4(a), a majority of the low-temperature thermoluminescence band disappears, indicating that the shallow traps are inactive (i.e., emptied) during the illumination due to the thermal energy available at 200°C. That is, the shallow traps do not contribute to the thermoluminescence under such an excitation condition. For the deep traps, as mentioned above, their storage capability cannot be directly evaluated by the corresponding thermoluminescence intensity due to the effect of remarkable thermal quenching.

Subsequently, another extended thermoluminescence experiment is conduced, in which photostimulation approach is introduced. First, we illuminate the $Y_3Ga_5O_{12}:Bi^{3+}$ phosphor for 60 s at 200°C using the 254 nm ultraviolet lamp to fill the deep traps (note: shallow traps have been thermally emptied during the excitation). After cooling down to room temperature, the charged

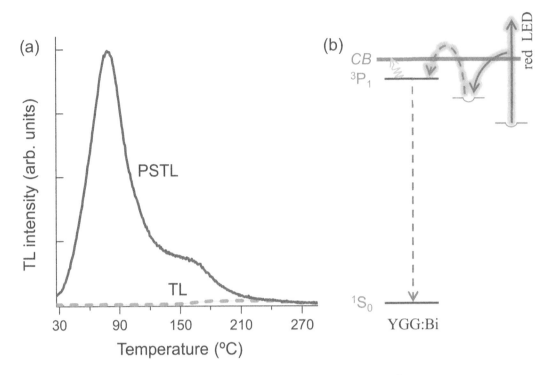

FIGURE 13.4 Photostimulated thermoluminescence (PSTL) of $Y_3Ga_5O_{12}$:Bi^{3+}. (a) TL (dashed-line) and PSTL (solid-line) curves. (b) Schematic illustration of the PSTL.

phosphor is excited by using a 630 nm red LED. Compared with the dashed-line curve shown in Figure 13.4(a), the red LED illumination significantly changes the thermoluminescence curve profile (solid-line curve presented in Figure 13.4(a)), i.e., the low-temperature thermoluminescence band reappears. It indicates that the shallow traps are refilled upon the LED illumination, and thus a photostimulated thermoluminescence (PSTL) phenomenon occurs. The PSTL curve exhibits a similar spectral shape to that in Figure 13.3(a). Such a result fits well-prediction. That is, there is a remarkable population of deep traps in $Y_3Ga_5O_{12}$:Bi^{3+}, even though the high-temperature thermoluminescence intensity looks very weak. Accordingly, a schematic illustration accounting for the reappeared PSTL intensity of $Y_3Ga_5O_{12}$:Bi^{3+} in the low-temperature region is put forward. As illustrated in Figure 13.4(b), upon the red-light illumination at room temperature, some electrons in the deep traps are photo-released and the thermally emptied shallow traps are refilled. Subsequently, all the traps are released during heating, followed by the PSTL emission from the $^3P_1$ energy level in $Y_3Ga_5O_{12}$:Bi^{3+}.

13.2.2 Photostimulated Persistent Luminescence (PSPL)

Figure 13.4 has shown that, for persistent phosphors, the distribution of traps can be manipulated by photostimulation approach. Accordingly, one expects that the persistent luminescence emission performances can be improved by taking advantage of photostimulation technique. Here, for illustrating the effect of photostimulation, we present thermoluminescence measurements on $LiGa_5O_8$:Cr^{3+} infrared persistent phosphor.[12] The measurements are conducted undergoing different delay times, as shown in Figure 13.5(a). When delay time increases to 120 h (the solid-line curve), a majority of the shallow-trap band disappears and the deep-trap band still exists, indicating that it is the deep traps that are responsible for the superlong persistent

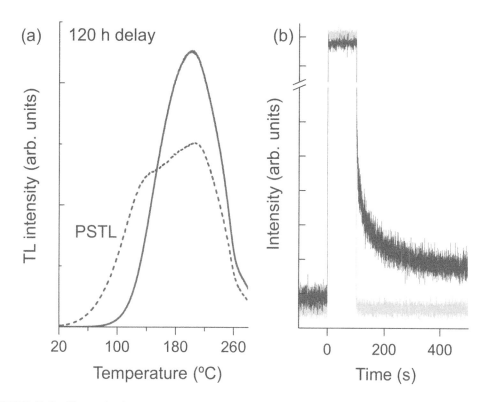

FIGURE 13.5 Thermoluminescence spectra and photostimulated persistent luminescence decay curves of LiGa$_5$O$_8$:Cr^{3+} phosphor. (a) The solid-line curve is acquired at a delay time of 120 h. The dashed-line curve was acquired on a 120 h-decayed phosphor after stimulation by 400 nm light for 100 s. (b) Photostimulated persistent luminescence decay curves monitored at 716 nm. The brown curve was acquired on a 120 h-decayed disc (pre-irradiated by 300 nm light), while the gray curve was recorded on a bleached sample (without ultraviolet pre-irradiation).

luminescence at room temperature. Subsequently, we conducted PSTL measurements on the decayed LiGa$_5$O$_8$:Cr^{3+} samples. The dashed-line curve in Figure 13.5(a) shows the thermoluminescence curve of a 120 h-decayed LiGa$_5$O$_8$:Cr^{3+} disc after being exposed to 400 nm illumination for 100 s. Compared with the 120 h-decayed sample without photostimulation, the 400 nm light stimulation significantly changes the thermoluminescence curve profile, i.e., the deep-trap band intensity decreases, while the shallow-trap band reappears. This means that after the 400 nm light photostimulation, some of the electrons in the deep traps are photo-released and the emptied shallow traps are refilled.

The photostimulation-induced electron trap redistribution, especially the refill of the shallow traps, suggests that photostimulation can affect the persistent luminescence behaviors of the pre-irradiated LiGa$_5$O$_8$:Cr^{3+} phosphor. To verify this assumption, we illuminated a 120 h-decayed LiGa$_5$O$_8$:Cr^{3+} disc with 400 nm light for 100 s and measured its persistent luminescence decay curve (brown curve). Figure 13.5(b) clearly shows that the 400 nm light illumination increases the persistent luminescence intensity and thus a PSPL phenomenon occurs.

The above results on PSPL phenomenon in LiGa$_5$O$_8$:Cr^{3+} at room temperature indicate that the PSPL write-in process fills the deep traps with electrons (i.e., forms the photochromic centers), while the PSPL read-out process releases the captured electrons from the filled deep traps (i.e., the photochromic centers) to the conduction band, followed by the refill of the emptied shallow traps.

13.3 EXCITATION

13.3.1 Charging by Ionizing Radiation

Based on an electron transfer assumption, for a persistent phosphor to maximally exhibit its persistent luminescence potential, the electron traps in the material, which are usually located close to the bottom of the conduction band, need to be fully filled upon an ionizing excitation via the conduction band state.

Because of the high energy of trap level, the persistent luminescence process requires high excitation energy (usually ultraviolet light), which is higher than the energy of the emission light. According to previous studies, the photoionization energy of phosphors can be determined from a persistent luminescence or thermoluminescence excitation spectrum (i.e., a plot of integrated emission intensity versus the illumination wavelength). In addition, the charging efficiency can be investigated using the time-dependent charging curve, its stored energy before reaching the equilibrium and the integrated intensity of the released phosphorescence after excitation ceased.[13]

13.3.1.1 Excitation Spectrum of Persistent Luminescence

Figure 13.6 shows an example of the persistent luminescence excitation spectrum of $Lu_2Pr_{0.01}Gd_{0.99}Al_2Ga_3O_{12}$ phosphor.[14] Although the steady-state photoluminescence can be effectively excited by visible-light wavelengths (440–490 nm; Figure 13.6), the situation for persistent luminescence is different because of the different activation mechanisms between them. To understand the effectiveness of different excitation wavelengths (energies) to the persistent luminescence, the afterglow decay curves monitored the ultraviolet emission under the excitation of different wavelengths between 200 and 500 nm are measured. Figure 13.6 gives the afterglow intensity of the phosphor as the function of the excitation wavelengths over 200–500 nm spectral range. The spectrum clearly shows that the persistent luminescence can be effectively achieved under ultraviolet

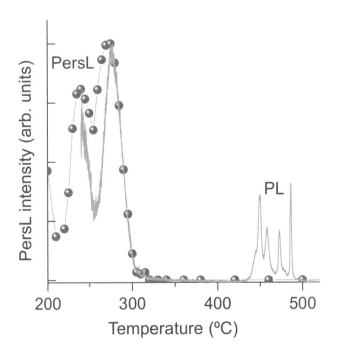

FIGURE 13.6 Trap filling spectrum obtained by plotting the persistent luminescence (PersL) intensities as a function of excitation wavelengths. For comparison, the photoluminescence (PL) excitation spectrum obtained is also presented as the solid-line curve.

illumination (200–320 nm) but cannot be achieved under visible-light illumination, even though the visible-light excitation is effective to the steady-state photoluminescence. Such spectrum is an indication of delocalization energy for filling the traps. The comparison between the two excitation spectra shows that the onset of the electron delocalization energy (~300 nm) coincides with the onset of the $4f^2 \rightarrow 4f5d$ excitation energy, revealing the delocalized character of the 4f5d excited state at room temperature.

13.3.1.2 Identification of Delocalized State

In a photoionization measurement, which is based on persistent luminescence or thermoluminescence excitation spectroscopy, if the excitation energy is high enough to promote the ground state electron to a delocalized state, the delocalized electron possibly will be captured by electron traps [15] and subsequently released during heating, followed by the recombination with the ionized luminescent center. Accordingly, the persistent luminescence or thermoluminescence excitation spectroscopy may be used to identify the delocalized state.

Figure 13.7(a) gives an emission spectrum of $MgGeO_3:Pr^{3+}$ upon 254 nm excitation at room temperature.[16] Besides the f→f transition peaking at 626 nm, the emission spectrum consists of a weak emission band peaking at around 420 nm. According to Blasse's study,[17] there is generally no d→f transition emission of Pr^{3+} at room temperature in phosphors containing d^{10} ions (e.g., In^{3+}, Ga^{3+} or Ge^{4+}). A similar effect has been observed when the Al^{3+} ion in $Y_3Al_5O_{12}:Pr^{3+}$ is replaced by Ga^{3+} with $3d^{10}$ configuration. While the $Y_3Al_5O_{12}:Pr^{3+}$ shows efficient d→f transition emission, the $Y_3Ga_5O_{12}:Pr^{3+}$ does not. The absence of d→f transition in $Y_3Ga_5O_{12}:Pr^{3+}$ has been ascribed to an electron transfer quenching. That is, the lowest level of the 4f5d configuration is situated in the conduction band of $Y_3Ga_5O_{12}$, so that a photoionization of the Pr^{3+} occurs. Accordingly, a low-lying impurity-trapped exciton state is proposed to account for the quenching of the lowest 4f5d level in $Y_3Ga_5O_{12}:Pr^{3+}$. The hole of the impurity-trapped exciton is on the ionized Pr^{3+} ion, while the

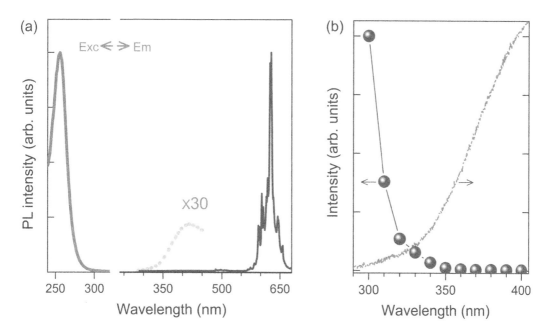

FIGURE 13.7   (a) Photoluminescence (PL) emission and excitation spectra of $MgGeO_3:Pr^{3+}$ recorded at room temperature. (b) Thermoluminescence excitation spectrum. The onset of electron delocalization energy (~340 nm) is an indication of photoionization threshold of the phosphor at room temperature. A room-temperature PL emission spectrum of the phosphor is also presented.

electron is delocalized over the neighboring d^{10} ions. Like in $Y_3Ga_5O_{12}$:Pr^{3+}, we consider that the $MgGeO_3$:Pr^{3+} system has somewhere a low-lying impurity-trapped exciton state, which may quench the d→f transition emission.

To gain insight on the impurity-trapped exciton state in $MgGeO_3$:Pr^{3+}, a photoionization experiment based on thermoluminescence excitation spectroscopy is conducted. The measurements have been carried out by illuminating the $MgGeO_3$:Pr^{3+} phosphor at room temperature with monochromatic light, the wavelength of which is tuned between 300 and 400 nm. Figure 13.7(b) shows that, at wavelengths shorter than 340 nm, the trap filling occurs and the thermoluminescence intensity increases rapidly with decreasing wavelength. It means that 340 nm is the onset of photoionization threshold at room temperature, revealing the delocalized character of excited state in $MgGeO_3$:Pr^{3+}. Notably, the thermoluminescence excitation spectrum conforms to the "mirror image" of the weak emission band (Figure 13.7(b)), providing strong evidence that the weak emission band in $MgGeO_3$:Pr^{3+} originates from an impurity-trapped exciton composition associated with ionization properties. Thus, the thermoluminescence excitation spectroscopy reveals that an impurity-trapped exciton state quenches the 4f5d state of Pr^{3+} in $MgGeO_3$.

13.3.2 Up-Conversion Charging (UCC)

In research and applications of persistent phosphors, it is a general knowledge that high-energy ionizing radiation (e.g., ultraviolet light from mercury lamps, xenon arc lamps and the Sun, as well as X-ray irradiation) is consistently used to charge persistent phosphors. For some emerging applications, such as persistent luminescence probe for bioimaging, however, the development has been largely hindered by the constraint of high-energy excitability of persistent luminescence.[18] Taking into account the fact that low-energy illumination (e.g., visible or infrared light) is much more suitable and less harmful than ultraviolet light for practical applications, taking advantage of the low-energy light excitation is therefore an urgent issue to be solved in the persistent luminescence area.

Several low-energy excitation approaches have been reported,[12, 18–21] in which up-conversion charging (UCC) is a promising candidate for charging phosphors using low-energy excitation light sources.[20] In a typical UCC process, the low-energy incident photons can promote the phosphor system from ground state to high-energy delocalized state via a two-step up-conversion excitation channel, followed by the fill of the traps. When the stored excitation energy is gradually released, persistent luminescence signal can be generated. To achieve UCC, two prerequisites generally need to be met: (1) the activator ion possesses a long-lifetime intermediate state and has a tendency to be oxidized in appropriate phosphor systems. The candidates for such activator ions may be Pr^{3+}, Tb^{3+}, Sm^{2+}, Cr^{3+} or Mn^{2+}. (2) The excitation light features a relatively high intensity, as well as an appropriate wavelength output in the visible or infrared region. The choice of excitation light source may be monochromatic light-emitting diode, laser or white light sources. According to these two prerequisites, one can expect that some other activator ions, such as rare-earth Pr^{3+} ion and transition metal Mn^{2+} ion, should also exhibit the UCC phenomenon when doped in appropriate hosts.

Trivalent chromium (Cr^{3+}) ion with d^3 electron configuration is well known for its broadband absorption in the ultraviolet-to-visible spectral region in phosphors. Crystals doped with Cr^{3+} ion have numerous optical and spectral applications, with the most famous one being the ruby crystal (Al_2O_3:Cr^{3+}) – on which Maiman realized the first working laser in 1960 through exciting the crystal using a high-intensity green-light (~560 nm) flash tube.[22]

In recent years, Cr^{3+}-activated gallate-based NIR persistent phosphors (e.g., $LiGa_5O_8$:Cr^{3+}, $Zn_3Ga_2Ge_2O_{10}$:Cr^{3+}, and $ZnGa_2O_4$:Cr^{3+}) have attracted considerable attention because of their promising applications in bioimaging and night-vision applications.[2] While the conventional photoluminescence of these Cr^{3+}-activated phosphors can be effectively excited by both the ultraviolet and visible light (~250–680 nm) of a xenon lamp, their persistent luminescence can only be effectively produced by the high-energy ultraviolet light. This phenomenon is understandable because individual visible-light photon from the xenon lamp cannot directly reach the high-energy delocalization state (i.e., an excited state associated with delocalization properties) of Cr^{3+} to fill the electron traps.

However, the knowledge obtained from the low-intensity xenon lamp excitation is one-sided. In Maiman's pioneering work, he also observed an up-conversion excitation phenomenon in ruby.[22] That is, the green-flashlight excitation first populated the ^2E level of Cr^{3+} ion, and during the long lifetime (~5 ms) of the ^2E metastable state, the Cr^{3+} ion is further raised to a higher lying charge transfer state. This observation clearly indicated the occurrence of transitions between two excited optical states of Cr^{3+} under intense green-light excitation. However, this two-photon excitation process in ruby did not receive much attention and further study, because often this type of up-conversion excitation prevents a material from becoming a good laser material; that is, such up-conversion excitation step decreases the population inversion that is essential for achieving laser radiation. Although the up-conversion excitation phenomenon in ruby is undesirable for laser applications, it inspires us to speculate that it may provide a promising solution to effectively charge the Cr^{3+}-activated persistent phosphors using low-energy, high-intensity visible-light sources.

UCC approach is proposed to effectively charge Cr^{3+}-activated persistent phosphors using a visible-light laser or light-emitting diode, as schematically depicted in Figure 13.8(a). In this section, we use $LaMgG_{11}O_{19}:Cr^{3+}$ persistent phosphor as a representative material to demonstrate the UCC process. [23] When a high-intensity visible-light laser is used as the excitation source, according to the UCC concept in Figure 13.8(a), the occurrence of the UCC process of Cr^{3+} ion should consist of two steps. The first absorption of a visible-light photon excites the system to the $^4T_2$ state of Cr^{3+} ion, followed by a fast non-radiative relaxation to the ^2E emitting state. The ^2E level can be regarded as the up-conversion intermediate state because of its long stable-state luminescence lifetime. Subsequently, under an up-conversion scheme, the system is further promoted to the high-energy $^4T_1(^4P)$ state, which is associated with a delocalization property at room temperature. The excited electrons will delocalize with a significant probability from the $^4T_1(^4P)$ state and then fill the electron traps.

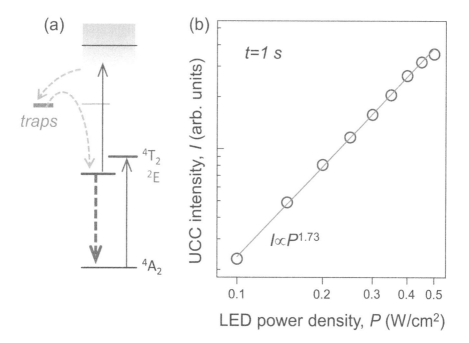

FIGURE 13.8 (a) Schematic representation of the UCC-induced persistent luminescence process, which consists of two-step ionization, trapping, detrapping and persistent luminescence emission. (b) Double-logarithmic plot of the UCC-induced persistent luminescence intensity (I) versus the power density (P) of the excitation source.

After ceasing the excitation, the gradual emptying of traps yields a long-lasting emission from the Cr^{3+} ion at room temperature.

From the UCC concept and the UCC properties in $LaMgGa_{11}O_{19}:Cr^{3+}$ persistent phosphor, we found that in order to achieve UCC in a persistent phosphor, there are requirements on both the excitation source and the activator ion. For the excitation source, its output power intensity needs to be relatively high (e.g., >5 mW/cm^2 for $LaMgGa_{11}O_{19}:Cr^{3+}$ phosphor using a 450 nm laser), and its photon energy should be higher than the metastable-state energy of the activator ion (e.g., the ^2E level in Cr^{3+}). For the activator ion, two prerequisites are simultaneously required: a long-lifetime intermediate state and a high-energy delocalization state. The lifetime of the intermediate state should be sufficiently long, so that the excited electron in the intermediate state can be further promoted during its lifetime. In regard to the delocalization state, the activator ion should have a tendency to be oxidized in solid (e.g., $Cr^{3+} \rightarrow Cr^{4+}$), so that electron transfer process can take place between the activator ion and the electron traps.

The nonlinear excitation nature of the persistent luminescence in the $LaMgGa_{11}O_{19}:Cr^{3+}$ phosphor has been verified by measuring the dependence of thermoluminescence intensity on excitation power density. We have recorded the thermoluminescence curves for the Cr^{3+} emission after the blue illumination with fixed illumination duration of 1 s and varied power densities from 0.1 to 0.5 W/cm^2. Figure 13.8(b) depicts the integrated thermoluminescence intensity (I) as a function of the illumination power density (P) at double-logarithmic coordination. For the illumination with exposure duration of 1 s, the I–P curve is consistent well with a quadratic relationship ($I \propto P^{1.73}$), providing strong evidence that two-photon excitation is involved in the up-conversion process.

The low-energy visible-light excitability offered by the UCC is expected to have significant impacts on both fundamental luminescence research and practical applications of persistent phosphors. For instance, in biomedical research, the UCC makes effective *in vivo* charging persistent optical probes using tissue-friendly visible or infrared light sources possible. Moreover, since the UCC appears to be a common phenomenon in persistent phosphors containing UCC-enabling activator ions, many existing persistent phosphors that were previously well-studied using X-ray or ultraviolet excitation can now be revisited using visible or infrared light excitation, enabling new luminescence properties to be discovered and new applications to be developed.

13.3.3 UP-CONVERSION CHARGING DYNAMICS

For persistent phosphors, the integrated thermoluminescence intensity is an indication of electron population in the traps. Thus, thermoluminescence measurement may reveal excitation power dependence of the charging performance of phosphors.

In a charging process of a persistent phosphor, on one hand, the traps can be filled through a two-step photoionization of activator ion (i.e., the electron trapping process). On the other hand, the external illumination can release some trapped electrons to the delocalized continuum state of the phosphor (i.e., the photostimulated detrapping process). That is, the excitation light may release some trapped electrons while filling the traps (i.e., excitation light–stimulated detrapping).

To demonstrate the UCC dynamics, a simple schematic illustration is presented in Figure 13.9(a). For a phosphor system, while the traps are being filled under illumination, the trapped electrons may be simultaneously liberated by the illumination energy (i.e., excitation light–stimulated detrapping) and ambient thermal energy (i.e., ambient temperature–stimulated detrapping). During the illumination, rate equation describing the time evolution of electron population in traps (N) is

$$\frac{dN}{dt} = n \cdot A_{tr} - N \cdot A_{d1} - N \cdot A_{d2}, \tag{13.1}$$

where n is the electron population of excited state (associated with delocalization property), A_{tr} the trapping rate, A_{d1} and A_{d2} the detrapping rates corresponding to excitation light–stimulated and

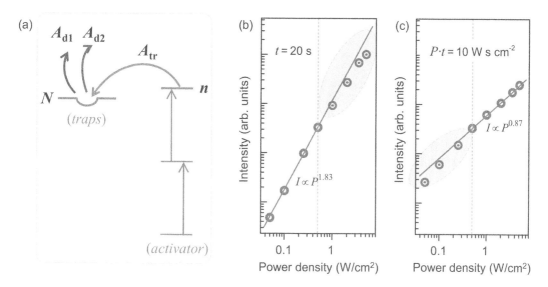

FIGURE 13.9 (a) Schematic illustration of up-conversion charging (UCC) dynamics. N is the population in traps, n is the population of excited state, A_{tr} is the trapping rate and A_{d1} and A_{d2} are the excitation light–stimulated detrapping rate and ambient temperature–stimulated detrapping rate, respectively. (b), (c) The integrated thermoluminescence intensity (I) is plotted on double-logarithmic scale against the excitation power density (P). The straight line is a quadratic fit of the data.

ambient temperature–stimulated detrapping, respectively. Since the filling or emptying of traps is generally considered to be much slower than that of the inner-atom transition, we assume that n is independent of the present illumination duration. Subsequently, we solve Equation (13.1) by taking into account the initial condition (i.e., $N=0$ at $t=0$ for the emptied traps), obtaining the solution:

$$N = \frac{n \cdot A_{tr}}{A_{d1} + A_{d2}} \left[1 - e^{-(A_{d1}+A_{d2})\cdot t} \right]. \tag{13.2}$$

The power $(A_{d1}+A_{d2})\cdot t$ stands for the effect of detrapping in the UCC process. The term $(A_{d1}\cdot t)$ represents detrapping effect caused by the excitation light stimulation, which is proportional to illumination dose. While the term $(A_{d2}\cdot t)$ represents detrapping effect caused by the ambient-temperature stimulation.

For an up-conversion process, the electron population of excited state (n) depends on the illumination power density (P) and the number of pump photons (m),[24] or $n \propto P^m$. While for the detrapping, we assume that $A_{d1} \propto P$ and A_{d2} is a constant at a certain temperature (temperature dependent). Thus, according to Equation (13.2), the competition between the trapping and detrapping should depend on the illumination power and duration. In the following, we will further discuss Equation (13.2) under two different conditions.

13.3.3.1 Effect of Detrapping Is not Remarkable

If the effect of detrapping is not remarkable under a certain illumination condition, the first-order Taylor series approximation of the exponential function in Equation (13.2) is reasonable, namely,

$$e^{-(A_{d1}+A_{d2})\cdot t} \approx 1 - (A_{d1} + A_{d2}) \cdot t. \tag{13.3}$$

Thus, Equation (13.2) may be written as

$$N \approx n \cdot A_{tr} \cdot t. \tag{13.4}$$

When the illumination duration t is fixed, we have

$$N \propto P^m \tag{13.5}$$

Alternatively, when the illumination dose $P \cdot t$ is fixed, we have

$$N \propto P^{m-1}. \tag{13.6}$$

Equations (13.5) and (13.6) may be used as rules of thumb to determine the involvement of multiphoton excitation in the UCC process.

13.3.3.2 Effect of Detrapping Is Remarkable

If the effect of detrapping is remarkable during illumination, the N–P function will deviate from Equation (13.5) or (13.6). That is, when t is fixed, $(A_{d2} \cdot t)$ in Equation (13.2) is a constant at a certain temperature, while $(A_{d1} \cdot t)$ increases with the power density P. Equation (13.3), as well as Equation (13.5), will be invalid upon illumination at a high-power density. For the same reason, when $P \cdot t$ is fixed, $(A_{d1} \cdot t)$ in Equation (13.2) is a constant and $(A_{d2} \cdot t)$ increases with the illumination duration. As a consequence, Equation (13.3), as well as Equation (13.6), will no longer be valid upon illumination with a low-power density.

For a persistent phosphor, the population in traps can be evaluated from the area under thermoluminescence curve.[25] Accordingly, we record the thermoluminescence of $LaMgGa_{11}O_{19}:Pr^{3+}$ and plot the integrated thermoluminescence intensities (I) as functions of illumination power densities (P).[26] Below we will demonstrate the I–P relationships under different illumination conditions.

13.3.3.3 Illumination Duration Is Fixed

According to Equation (13.5) and Figure 13.9(a), thermoluminescence intensity of $LaMgGa_{11}O_{19}:Pr^{3+}$ will increase quadratically with the illumination power density. Accordingly, thermoluminescence intensities are recorded after the 450 nm laser illumination for 20 s. To eliminate the effect of optical heating, we illuminate the phosphor at liquid nitrogen temperature ($-196°C$), followed by transferring the phosphor to room temperature for the subsequent measurement. Figure 13.9(b) depicts the integrated thermoluminescence intensity versus the illumination power density (i.e., I–P function) on double-logarithmic scale. As expected, in a valid range of illumination dose (1–10 W s/cm²), the I–P curve can be well fitted by using a quadratic function ($I \propto P^{1.83}$). However, the slope of the I–P curve exhibits a decrease as the illumination dose is greater than 10 W s/cm², as shown in the shadow region in Figure 13.9(b). This result is in accordance with the prediction in Section 13.2.3.2, suggesting that the excitation light–stimulated detrapping becomes competitive with the trapping upon high-power illumination applied in experiment.

13.3.3.4 Illumination Dose Is Fixed

Besides the effect of excitation light–stimulated detrapping, the ambient temperature–stimulated detrapping during the illumination may play an important role in the UCC process. According to the above discussion, a fixed illumination dose implies that a small illumination power density corresponds to long illumination duration. This means that the ambient temperature–stimulated detrapping may become nonnegligible upon illumination with a small power density.

Thermoluminescence curves of $LaMgGa_{11}O_{19}:Pr^{3+}$ are recorded upon illumination at room temperature with a fixed dose of 10 W s/cm². That is, before the thermoluminescence measurements, the phosphor has been illuminated by different power densities with their corresponding exposure durations. Subsequently, we record the thermoluminescence intensities and plot the intensities (I) on double-logarithmic scale against the excitation power densities (P), as presented in Figure 13.9(c). As expected, in a valid range of illumination duration (2–20 s, corresponding to the power density of 5–0.5 W/cm²), the I–P curve is in accordance with Equation (13.6) (i.e., $I \propto P^{0.87}$). However, as the illumination duration is longer than 20 s (i.e., power density is smaller than 0.5 W/cm^2 as the illumination dose has been fixed), the *I-P* curve deviates from the fitting, as shown in the shadow region in Figure 13.9(c). Section 13.2.3.2 has predicted such a deviation, which stems from the contribution of ambient temperature–stimulated detrapping in the UCC process.

13.4 EMISSIONS

13.4.1 Emission Wavelengths

After having presented the trapping/detrapping and excitation properties of persistent phosphors, we now turn our attention to the emission process. To date, the investigations on persistent luminescence are being extensively carried out, covering the emission wavelengths from the ultraviolet light to the near-infrared (NIR).[2–4] In persistent phosphors, two kinds of active centers are involved: traps and emitters. As mentioned above, traps usually store excitation energy and release it gradually to the emitters due to thermal or other physical stimulations, while the emission wavelength of a persistent phosphor is mainly determined by the emitter. Therefore, in the design of persistent phosphors, suitable emitter capable of emitting light with appropriate wavelength is required.

13.4.1.1 Visible-Light Emitters

Since 1996, the world has witnessed intense research and wide applications in persistent luminescence. Up to now, persistent phosphors emitting in the visible spectral region have been extensively studied and some visible persistent phosphors, such as green-emitting $SrAl_2O_4:Eu^{2+},Dy^{3+}$ and blue-emitting $Sr_2MgSi_2O_7:Eu^{2+},Dy^{3+}$, have gained commercial success and are being widely used as night-vision materials in various important fields (e.g., security signs, emergency route signs, traffic signage, dials and displays) because of their sufficiently strong and long (>10 h) persistent luminescence and their ability to be excited by sunlight and room light.[2–4] Here, for clarity, we will introduce the visible persistent phosphors based on the type of emitting ions, with Ce^{3+}, Eu^{2+} and Mn^{2+} as representatives. A list of visible persistent phosphors activated by different emitting ions can be found in **Table 13.1**. Further details can be found in the appropriate references mentioned in the table.

13.4.1.1.1 Ce^{3+}

The emission of Ce^{3+} ion occurs from the lowest crystal-field component of the $5d^1$ configuration to the two levels of the ground state. This gives the Ce^{3+} emission its typical double-band shape. Usually, the Ce^{3+} emission is in the ultraviolet or blue spectral region, but in $Y_3Al_5O_{12}$, it is in the green (crystal-field effect), and in CaS, in the red (covalency effect). Electrons have been considered the charge carries being trapped in the persistent luminescence process. To date, the available persistent phosphors activated by Ce^{3+} include [2–4] silicates (e.g., Lu_2SiO_5), gallates (e.g., $Gd_3Ga_5O_{12}$), phosphates (e.g., YPO_4), sulfides (e.g., CaS). and aluminates (e.g., $CaAl_2O_4$, emission spectrum is presented in Figure 13.10(a)).

13.4.1.1.2 Eu^{2+}

Eu^{2+}-doped phosphors have been extensively studied over the last two decades. Such phosphors usually exhibit broadband luminescence under external excitations. Most of these emissions are attributed to the dipole-allowed $4f^65d{\rightarrow}4f^7$ transitions associated with relatively large transition probabilities. The emissions of Eu^{2+}-doped phosphors cover from violet to infrared spectral regions, depending on the crystal-field environment, covalency and Stokes shift. To date, $SrAl_2O_4:Eu^{2+},Dy^{3+}$ is still the widely used persistent phosphor (its emission spectrum is presented in Figure 13.10(a)). The trapping mechanism of Eu^{2+}-doped phosphors is generally suggested as follows: upon an appropriate excitation, the holes may be trapped by the europium luminescent center, whereas part of the electrons in the conduction band is trapped in the electron traps, from where they escape thermally after some time, yielding the delayed emission.

TABLE 13.1
Visible Emitting Persistent Phosphors

Emitter	Host	Emission Maximum (nm)	Ref.
Ce^{3+}	Lu_2SiO_5	410	[27]
	$Ca_2Al_2SiO_7$	417	[28]
	$BaAl_2O_4$	402,450	[29]
	CaS	508,568	[30]
	$Sr_4Al_{14}O_{25}$	472,511	[31]
	$Y_3Al_5O_{12}$	525	[32]
Pr^{3+}	$CaTiO_3$	612	[33-36]
	Y_2O_2S	606	[37]
	$Ca_{0.8}Mg_{0.2}TiO_3$	612	[38]
	$CaSnO_3$	541	[39]
	$NaNbO_3$	620	[40]
	$La_2Ti_2O_7$	611	[41]
	$Ca_3SnSi_2O_9$	488	[42]
	$Y_3Al_5O_{12}$	490	[43]
	YPO_4	597	[44]
	CaO	620	[45]
	Lu_2O_3	610	[46]
Sm^{3+}	CaS	569	[47]
	$CaSnO_3$	601	[48]
	$Sr_2ZnSi_2O_7$	598	[49]
	SnO_2	607	[50]
	KY_3F_{10}	597	[51]
	CaO	616	[52]
Eu^{3+}	CaO	616	[53]
	Y_2O_2S	614	[54]
	Y_2O_3	611	[55]
	$CaWO_4$	616	[56]
	Ca_2SnO_4	618	[57]
	$CaMoO_4$	616	[58]
	$Sr_2ZnSi_2O_7$	617	[59]
	$Sr_2ZnSi_2O_7$	482	[60]
Eu^{2+}	$SrAl_2O_4$	520	[61]
	$CaAl_2O_4$	440	[62]
	$SrAl_4O_7$	480	[63]
	$SrAl_{12}O_{19}$	400	[64]
	$BaAl_2O_4$	500	[64]
	CaS	650	[65]
	$Sr_4Al_{14}O_{25}$	489	[66]
	$CaGa_2S_4$	555	[67]
	$Sr_2MgSi_2O_7$	470	[68]
	$BaMgAl_{10}O_{17}$	450	[69]
	Ba_2SiO_4	510	[70]
	SrB_2O_4	430	[71]
	Sr_3SiO_5	570	[72]
	$Sr_2Al_2SiO_7$	485	[73]
	$Sr_2P_2O_7$	420	[74]

(Continued)

TABLE 13.1 (*Continued*)
Visible Emitting Persistent Phosphors

Emitter	Host	Emission Maximum (nm)	Ref.
	$Ca_2Si_5N_8$	610	[75]
	$BaMgSiO_4$	537	[76]
	$Ba_5(PO_4)_3Cl$	435	[77]
	$Ca_2ZnSi_2O_7$	580	[78]
	$SrSi_2O_2N_2$	550	[79]
	$BaSi_2O_2N_2$	498	[80]
	$NaAlSiO_4$	550	[81]
	$Ca_3Si_2O_7$	620	[82]
	$Ca_3Mg_3(PO_4)_4$	433	[83]
	$LiSr_4(BO_3)_3$	630	[84]
	$Sr_5(BO_3)_3Cl$	610	[85]
	$Sr_5(PO_4)_3Cl$	467	[86]
	$Ba_3P_4O_{13}$	450	[87]
Tb^{3+}	Lu_2O_3	545	[88]
	SrO	543	[89]
	$CaSnO_3$	545	[90]
	Y_2O_2S	546	[91]
	$CaZnGe_2O_6$	551	[92]
	$Sr_4Al_{14}O_{25}$	542	[93]
	$CaWO_4$	547	[94]
	$CdSiO_3$	544	[95]
	Ca_2SnO_4	545	[96]
	Sr_2SiO_4	546	[97]
Dy^{3+}	$CaAl_2O_4$	477,577,668	[98]
	Sr_2SiO_4	480,575,665	[99]
	$Ca_3MgSi_2O_8$	480,575,667	[100]
	$Ca_3SnSi_2O_9$	484,572,670	[101]
Tm^{3+}	Y_2O_2S	470,495,516,545,588,626	[102]
Mn^{2+}	$ZnGa_2O_4$	504	[103]
	$MgGa_2O_4$	506	[104]
	$MgGeO_3$	670	[105]
	Mg_2SiO_4	650	[106]
	$Y_3Al_5O_{12}$	580	[107]
	$Ca_3(PO_4)_2$	660	[108]
	Mg_2SnO_4	499	[109]
	Zn_2GeO_4	530	[110]
	$CdSiO_3$	587	[111]
	$Zn_3Ga_2Ge_2O_{10}$	520	[112]
Cu^+	ZnS	540	[113]
	BaS	610	[114]
Pb^{2+}	$CdSiO_3$	498	[115]
Bi^{3+}	CaS	448	[116]
	$ZnGa_2O_4$	410,550	[117]
	$CaWO_4$	425	[118]
	Zn_2GeO_4	455	[119]

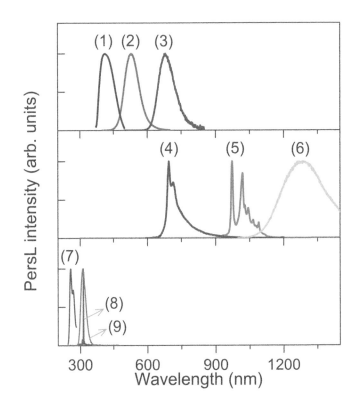

FIGURE 13.10 Normalized emission spectra of persistent luminescence of different phosphors, involving visible persistent phosphors (a): (1) $CaAl_2O_4$:Ce^{3+}, (2) $SrAl_2O_4$:Eu^{2+},Dy^{3+}, (3) $MgGeO_3$:Mn^{2+}; infrared persistent phosphors (b): (4) $Zn_3Ga_2Ge_2O_{10}$:Cr^{3+}, (5) $MgGeO_3$:Yb^{3+}, (6) $Zn_3Ga_2Ge_2O_{10}$:Ni^{2+}; and ultraviolet persistent phosphors (c): (7) $LuBO_3$:Pr^{3+}, (8) $Lu_3Al_2Ga_3O_{12}$:Pr^{3+},Gd^{3+}, (9) $Y_3Al_2Ga_3O_{12}$:Bi^{3+}.

13.4.1.1.3 Mn^{2+}

The Mn^{2+} ion in phosphor has an emission, which consists of a broad band, the maximum of which can vary from green to deep red, depending strongly on the host lattice. The emission originates from the $^4T_1 \rightarrow\ ^6A_1$ transition. Generally, in octahedral crystal field, Mn^{2+} gives an orange to red emission, while the tetrahedrally coordinated Mn^{2+} gives a green emission. The persistent luminescence of Mn^{2+} has been reported [2–4] in silicates (e.g., Zn_2SiO_4), aluminates (e.g., $Y_3Al_5O_{12}$), gallates (e.g., $ZnGa_2O_4$) and germanates (e.g., $MgGeO_3$, emission spectrum is presented in Figure 13.10(a)).

13.4.1.2 Infrared Emitters

Besides the extensively studies on persistent phosphors with visible emissions, in recent years, persistent luminescence in wavelengths beyond the visible spectral region, i.e., in the infrared region, received considerable attention, because the longer wavelength persistent luminescence has promising implications to many advanced applications. An important development in infrared persistent luminescence applications is the possibility of *in vivo* medical imaging using persistent luminescent nanoparticles.[2–4] For instance, the "self-sustained" luminescence of infrared persistent probes would allow the imaging to be conducted in an excitation-free and, hence, autofluorescence-free manner, enabling a high signal-to-noise ratio and exceptional imaging sensitivity. Moreover, for infrared persistent luminescence, it is worth noting that the general evaluation method for visible persistent luminescence by luminance (e.g., in units of cd/m²) is not suitable, since the infrared is outside the visible spectral region. Instead, the units of radiance (e.g., W/sr/m²) are usually applied.

TABLE 13.2
Infrared Persistent Phosphors

Emitter	Host	Emission Maximum (nm)	Ref.
Cr^{3+}	$Gd_3Ga_5O_{12}$	716	[120]
	$La_3Ga_5GeO_{14}$	785	[5, 121]
	$Sr_4Al_{12}O_{25}$	693	[122]
	Ga_2O_3	720	[123]
	$ZnGa_2O_4$	695	[124]
	$Zn_3Ga_2Ge_2O_{10}$	696	[6]
	$Zn_2Ga_3Ge_{0.75}O_8$	697	[125]
	$MgGa_2O_4$	707	[126]
	$LiGa_5O_8$	716	[12]
	$Y_3Al_2Ga_3O_{12}$	700	[127]
Ni^{2+}	$Zn_3Ga_2Ge_2O_{10}$,	1290	[10]
	$LiGa_5O_8$,	1220	[10]
	$La_3Ga_5GeO_{14}$	1430	[10]
Mn^{4+}	$LaAlO_3$	730	[128]
Fe^{3+}	$SrAl_{12}O_{19}$	810	[129]
	$LiGaO_2$	748	[130]
Co^{2+}	$ZnGa_2O_4$	1000	[131]
Pr^{3+}	$MgGeO_3$	1085	[132]
	Ca_2SnO_4	1115	[133]
Nd^{3+}	Sr_2SnO_4	1079	[134]
	$Y_3Al_2Ga_3O_{12}$	1064	[135]
	Ca_2SnO_4	1080	[133]
Ho^{3+}	Ca_2SnO_4	1180	[133]
	$LaAlO_3$,	1200	[136]
Er^{3+}	$SrAl_2O_4$	1530	[137]
	Ca_2SnO_4	1533	[133]
Tm^{3+}	Ca_2SnO_4	1195	[133]
Yb^{3+}	$Gd_2O_2CO_3$	1000	[138]
	$MgGeO_3$	1019	[139]
	$LiGa_5O_8$	1010	[139]
	$Zn_3Ga_2GeO_8$	1007	[139]
	$La_3Ga_5GeO_{14}$	1009	[139]
	Ca_2SnO_4	1000	[133]

The known infrared-emitting ions involve transition metal ions and rare-earth ions. Here, we take Cr^{3+}, Yb^{3+} and Ni^{2+} as samples to introduce the infrared persistent phosphors. A list of reported infrared persistent phosphors doped by different emitting ions can be found in Table 13.2. Further details can be found in the appropriate references mentioned in the table.

13.4.1.2.1 Cr^{3+}

For infrared persistent luminescence, trivalent chromium ion (Cr^{3+}) is a favorable emitting center in phosphors because of its narrow-line emissions (usually near 700 nm) due to the spin-forbidden $^2E \rightarrow {}^4A_2$ transition, or broadband emissions (650–1600 nm) due to the spin-allowed $^4T_2 \rightarrow {}^4A_2$ transition, which strongly depends on the crystal-field environment of the host lattices. As an example, in $Zn_3Ga_2Ge_2O_{10}$:Cr^{3+} phosphor, infrared persistent luminescence after the removal of the excitation source can be obtained, which consists of the $^2E \rightarrow {}^4A_2$ emission peaking at 696 nm, which superimposes on a broad emission band extended from 650 to 1000 nm (Figure 13.10(b)). The broadband emission may be ascribed to the $^4T_2 \rightarrow {}^4A_2$ transitions from some disordered Cr^{3+} ions in the gallogermanate system. The infrared persistent luminescence from Cr^{3+} meets the requirements of many technological applications, such as bioimaging, radiation detection and solar cells. Recently, significant achievements have been made in the bioimaging field using infrared persistent phosphors in the form of nanoparticles as *in vivo* imaging probes.[2–4]

13.4.1.2.2 Yb³⁺

Trivalent ytterbium (Yb^{3+}) has only two manifolds in the 4f shell, i.e., a $^2F_{7/2}$ ground state and a $^2F_{5/2}$ excited state, with an energy difference between the states of approximately 10,000 cm^{-1}. Typical emission for Yb^{3+}-doped phosphors features a fine-structure spectral shape owning to the Stark splitting of the two Yb^{3+} manifolds. This simple and unique energy level structure endows Yb^{3+} with several fascinating optical properties, including no absorption in the visible range, strong absorption at near 980 nm (well-suited for InGaAs diode laser emission) and intense infrared emission ($^2F_{5/2} \rightarrow {}^2F_{7/2}$ transition, emission spectrum of $MgGeO_3$:Yb^{3+} is presented in Figure 13.10(b)). Besides the unique energy level structure, Yb^{3+} also has a unique $4f^{13}$ electronic configuration, which can easily gain one electron to reach more stabilized $4f^{14}$ configuration of the full shell.

13.4.1.2.3 Ni²⁺

For infrared photoluminescence, divalent nickel (Ni^{2+}) is a favorable luminescent center in solids because of its intense broadband emission in the infrared range.[10] In the development of Ni^{2+}-doped phosphors, gallates (e.g., $LiGa_5O_8$) and aluminates (e.g., $MgAl_2O_4$) were frequently used as the hosts. In this sense, we take a phosphor with the composition of $Zn_3Ga_2Ge_2O_{10}$:Ni^{2+} as an example to demonstrate the infrared persistent luminescence performance of Ni^{2+}. After an ultraviolet excitation, the $Zn_3Ga_2Ge_2O_{10}$:Ni^{2+} phosphor exhibits a broadband infrared emission, ranging from ~1050 to ~1600 nm and peaking at ~1290 nm (Figure 13.10(b)). This broadband emission is attributed to the $^3T_2(^3F) \rightarrow {}^3A_2(^3F)$ transition of Ni^{2+}.

13.4.1.3 Ultraviolet Emitters

In stark contrast to the progress in the visible and infrared spectral regions, the research and development at the other end of the spectrum – the shorter wavelength ultraviolet regions are relatively lacking;[22, 140] only a few ultraviolet persistent phosphors were reported in very recent years.

Ultraviolet luminescence holds potential for diverse applications. Further development of the ultraviolet luminescence technology, however, is hindered by the common but very inconvenient photoluminescence form. The further development of ultraviolet persistent luminescence will offer us a unique opportunity to use this special optical phenomenon for some new applications where UV emission is needed but constant external excitation is unavailable. In our view, the achievement of ultraviolet persistent luminescence depends primarily on two factors: (1) appropriate emitting ions that are capable of ultraviolet emissions, such as Ce^{3+}, Pr^{3+}, Gd^{3+}, Pb^{2+} or Bi^{3+} in some suitable phosphor systems; and (2) effective interaction between the activator (ultraviolet emitter or sensitizer) and the energy traps in phosphors.

According to the prerequisites, finding suitable phosphor system containing ultraviolet emitting ion and exploring the trapping performances of the phosphor are essentially required for further developing ultraviolet persistent luminescence. The reported ultraviolet persistent phosphors focus on the Pr^{3+}, Gd^{3+} and Bi^{3+}-doped phosphors, as listed in Table 13.3. Further details can be found in the appropriate references mentioned in the table.

13.4.1.3.1 Pr³⁺

For ultraviolet luminescence, trivalent praseodymium (Pr^{3+}) ion has received considerable attention due to its efficient 4f5d→4f² allowed transition. Persistent luminescence of Pr^{3+}-doped phosphors in the ultraviolet region is predictable, since Pr^{3+} has a tendency to be oxidized in appropriate phosphor

TABLE 13.3
Ultraviolet Persistent Phosphors

Emitter	Host	Emission Maximum (nm)	Ref.
Ce^{3+}	$Sr_2Al_2SiO_7$	400	[141]
Pr^{3+}	Cs_2NaYF_6	250	[142]
	$Ca_2Al_2SiO_7$	268	[140]
	$LaPO_4$	231	[143]
	$Y_3Al_2Ga_3O_{12}$	307	[144]
	$BaLu_2Al_2Ga_2SiO_{12}$	301	[145]
	$LuBO_3$	256	[146]
Tb^{3+}	Cs_2NaYF_6	380	[147]
Pb^{2+}	$Sr_2MgGe_2O_7$	370	[148]
	$Sr_3Y_2Si_6O_{18}$	299	[144]
	$Sr_3Gd_2Si_6O_{18}$	311	[144]
Bi^{3+}	$CdSiO_3$	360	[149]
	$NaLuGeO_4$	400	[150]
	$LiScGeO_4$	361	[151]
	$Y_3Ga_5O_{12}$	316	[11]
	$Y_3Al_5O_{12}$	303	[11]
	YPO_4	240	[152]
	$Y_3Al_2Ga_3O_{12}$	310	[146]
Tm^{3+}	$NaYF_4$	360	[146]
Gd^{3+}	$Lu_3Al_2Ga_3O_{12}$	312	[14]

systems and thus the electron transfer between the praseodymium ion and energy traps may take place. Recently, ultraviolet persistent luminescence has been reported in Pr^{3+}-doped phosphors,[140] the luminescence performances of which have been spectrally characterized in dark environment. For example, in $LuBO_3$:Pr^{3+}, when the excitation ceases, the phosphor gives a long-lasting emission (Figure 13.10(c)). The corresponding emission spectrum is recorded at room temperature.

13.4.1.3.2 Gd^{3+}

Gadolinium ion (Gd^{3+}) features a unique energy level structure, in which the excited levels are at energies higher than 32,000 cm^{-1}. As a consequence, the emission of Gd^{3+} is in the ultraviolet spectral region. Moreover, Gd^{3+} possesses a half-filled $4f^7$ configuration, which is highly stable, so that there are few studies on the persistent luminescence of Gd^{3+}-doped phosphors. Since there is no tendency for Gd^{3+} to be oxidized or reduced in phosphors, a sensitizer is essentially necessary for the achievement of persistent luminescence of Gd^{3+} in phosphors. As a sensitizer for Gd^{3+} persistent luminescence, for instance, Pr^{3+} ion is a good candidate, since electron transfer may take place between the delocalized state of Pr^{3+} and the trap state.[14] As an example, persistent luminescence emission of $Lu_3Al_2Ga_3O_{12}$:Pr^{3+},Gd^{3+} is obtained after the excitation on Pr^{3+}. The emission spectrum is presented in Figure 13.10(c).

13.4.1.3.3 Bi^{3+}

As an effective emitter for the ultraviolet luminescence, trivalent bismuth (Bi^{3+}) ion is also a good candidate due to its inter-configurational $6s6p \rightarrow 6s^2$ transition in some hosts,[11] such as garnets. In the well-established garnet phosphors, Bi^{3+} generally exhibits band-like emissions peaking in the ultraviolet-B (290–320 nm) region (Figure 13.10(c)). To gain insight into the persistent luminescence of Bi^{3+}-doped phosphors, besides the luminescence property of the emitting level itself, the interaction between the Bi^{3+} and the traps is worth exploring.

13.4.2 PERSISTENT ENERGY TRANSFER

The concept of persistent energy transfer is initially applied to $CaAl_2O_4:Ce^{3+},Tb^{3+}$ and $CaAl_2O_4:Ce^{3+},Mn^{2+}$ persistent phosphors,[153–155] in which the long persistence of the donor (i.e., Ce^{3+}) is converted into the long persistence of the acceptor (i.e., Tb^{3+} or Mn^{2+}). A list of persistent phosphors based on persistent energy transfer mechanism can be found in Table 13.4. Further details can be found in the appropriate references mentioned in the table. The investigations on persistent energy transfer have provided much novel insight to design and develop new storage phosphors. Here, we present an interesting example to demonstrate the universality of the persistent energy transfer phenomenon.

We would take ultraviolet persistent phosphors as an example. Ultraviolet light in the wavelength range 309–313 nm is specifically referred to as narrowband ultraviolet-B (NB-UVB), which has exhibited significant advantages in the medical field for the treatment of skin diseases, including psoriasis and vitiligo. The practical NB-UVB luminescence generally originates from phosphors activated by Gd^{3+} ion, which has no absorption in the visible-light region due to its unique energy level structure and no tendency to be oxidized or reduced in phosphor due to its stabilized $4f^7$ electron configuration. To achieve persistent luminescence in Gd^{3+}-doped phosphor, it is essential to introduce sensitizer ion, which should have suitable high-lying state for efficient transfer of the excitation energy to Gd^{3+} ion. As a sensitizer for Gd^{3+} emission, trivalent praseodymium (Pr^{3+}) ion is a good candidate since Pr^{3+}-doped phosphors possess some fascinating optical properties. Up-conversion luminescence from the high-lying $4f5d$ state of Pr^{3+} has been achieved upon excitation of $^3P_0$ or $^1D_2$ level in some phosphors. Moreover, electron transfer may take place between the delocalized state of Pr^{3+} and the trap state, since Pr^{3+} has a tendency to be oxidized in some solids. Accordingly, one can envisage a combination of up-conversion charging and persistent energy transfer. Such a combination enables us to generate Gd^{3+} ultraviolet persistent luminescence after charging by visible-light illumination.

As a consequence, persistent luminescence of Gd^{3+} in the NB-UVB region is expected in appropriate Pr^{3+}–Gd^{3+} co-doped phosphors, in which Pr^{3+} ion may absorb two-step excitation to fill electron traps and then persistently transfer the stored energy to Gd^{3+}.

13.4.3 AMBIENT STIMULATED EMISSION (ASE)

When it mentions persistent luminescence, people would think of glow-in-the-dark and always consider ambient darkness as the prerequisite for such emission. Therefore, persistent luminescence

TABLE 13.4

Persistent Energy Transfer

Emitters	Host	Emission Maximum (nm)	Ref.
$Ce^{3+} \rightarrow Tb^{3+}$	$CaAl_2O_4$	400	[152]
$Ce^{3+} \rightarrow Mn^{2+}$	$CaAl_2O_4$	525	[153]
$Eu^{2+} \rightarrow Mn^{2+}$	$MgSiO_3$	456,660	[156]
$Eu^{2+} \rightarrow Ce^{3+}$	$SrAl_2O_4$	515,760	[156]
$Ce^{3+} \rightarrow Nd^{3+}$	$Y_3Al_2Ga_3O_{12}$	808,1064,1335	[135]
$Eu^{2+} \rightarrow Nd^{3+}$	$SrAl_2O_4$	515,882	[157]
$Ce^{3+} \rightarrow Eu^{2+}$	$Sr_3Al_2O_5Cl_2$	435,620	[77]
$Sm^{3+} \rightarrow Eu^{3+}$	$CaWO_4$	592,616	[158]
$Eu^{2+} \rightarrow Er^{3+}$	$SrAl_2O_4$	525,1530	[137]
$Eu^{2+} \rightarrow Cr^{3+}$	$Sr_4Al_{12}O_{25}$	693	[122]
$Eu^{2+} \rightarrow Mn^{2+}$	$BaMg_2Si_2O_{17}$	620,675	[159,160]
$Pr^{3+} \rightarrow Gd^{3+}$	$Lu_3Al_2Ga_3O_{12}$	312	[14]

is usually considered "ambient temperature–stimulated luminescence". In fact, persistent luminescence may, of course, occur in bright, but such emission is hard to be observed or identified in bright due to the interference of ambient light.

Although the diverse applications of persistent phosphors are benefiting from the glow-in-the-dark feature, it is the requirement of ambient darkness that is hindering the further development of persistent phosphors under different ambient conditions. For persistent phosphors with common emission wavelengths, the weak emission signals are always submerged by the ambient lighting. Accordingly, to make the persistent luminescence detectable under ambient-light illumination, the spectral overlap between the emission of storage phosphor and the ambient light should be avoided. Taking into account the fact that ambient lighting generally falls in the visible spectrum, storage phosphors emitting in the visible-blind region (e.g., ultraviolet or infrared) may be candidates for achieving the detectable glow-in-the-bright emissions.

Taking an ultraviolet emitting $LaMgAl_{11}O_{19}:Gd^{3+}$ phosphor as an example, we introduce a glow-in-the-bright phenomenon by introducing a concept termed ambient stimulated emission, in which the ambient temperature and ambient lighting can simultaneously release the stored energy in storage phosphors and lead to long-lasting emissions beyond the visible spectrum. The inset of Figure 13.11 shows that, after exposure to the X-ray irradiation, Gd^{3+} is dominating the ultraviolet emission, which is recorded at 300K ambient temperature (*AT*) and in 650 lux ambient lighting (*AL*, LED ceiling light). [146] We name such long-lasting luminescence under ambient condition as ambient stimulated emission. Moreover, glow-in-the-dark emission spectrum of the phosphor has also been presented in the inset of Figure 13.11. The glow-in-the-dark emission spectrum is recorded upon the *AT* stimulation in darkness. The two spectra exhibit similar shapes but different intensities at the recorded moment.

To learn the ambient stimulated emission of $LaMgAl_{11}O_{19}:Gd^{3+}$, the time dependence of the emission is investigated by measuring luminescence decay curves after the X-ray irradiation.

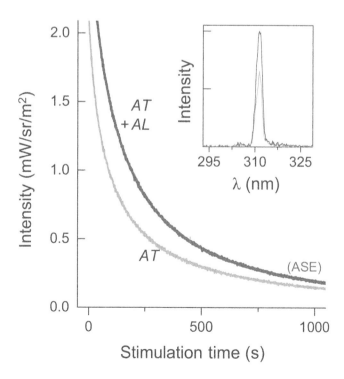

FIGURE 13.11 Time evolutions of the ambient stimulated emission (ASE, at ambient temperature in the ambient lighting) and the glow-in-the-dark emission (at ambient temperature in darkness), respectively. Inset presents the spectra of the ambient stimulated emission and persistent luminescence, respectively.

Figure 13.11 shows that the ambient stimulated emission intensity under continuous *AL* stimulation at 300K (purple curve) is higher than that of glow-in-the-dark emission (pink curve) over the measured timescale. The difference of decay intensities may result from the contribution of *AL* stimulation to the ambient stimulated emission. The glow-in-the-dark decay curve only reflects the thermal emptying of room-temperature traps, while the decay of ambient stimulated emission is due to the release of stored energy caused by both the *AT* thermal stimulation and the *AL* illumination. Moreover, the glow-in-the-bright phenomenon commonly appears in various storage phosphors, involving ultraviolet and infrared phosphors. As an example, we may envisage incorporating a glow-in-the-bright phosphor into marker products that are expected to provide "visible-blind" ultraviolet or infrared taggants for identification, tracking and anti-counterfeiting under general illumination.

13.5 CONCLUSION

Along with the development of the persistent phosphor field, people have realized that the persistent luminescence is a complicated process, which is based on a delicate interplay between emitting ion and traps. Thus, a more comprehensive study of trapping/detrapping mechanisms of persistent luminescence is still needed. Besides the underlying mechanism, novel excitation concept and technique, such as up-conversion charging (UCC) of persistent phosphors, are expected to be developed. Finally, although a lot of research is going on in the visible and infrared spectral regions, and various dopant-host combinations are being developed, only a few studies on ultraviolet emitting persistent phosphors have been reported. Special attention may be paid to the concept of ambient stimulated emission (ASE), in which the ambient temperature and lighting can simultaneously release the stored energy in persistent phosphors and lead to long-lasting emissions beyond the visible spectrum.

REFERENCES

1. Matsuzawa, T., Aoki, Y., Takeuchi, N., and Murayama, Y., *J. Electrochem. Soc.*, 143, 2670, 1996.
2. Smet, P.F., Van den Eeckhout, K., De Clercq, O.Q., and Poelman, D., *Persistent Phosphors in Handbook on the Physics and Chemistry of Rare Earths*. Ed. Bunzli, J.C. and Pecharsky, V.K., Vol. 48. Amsterdam: North-Holland, 2015.
3. Li, Y., Gecevicius, M., and Qiu, J.R., *Chem. Soc. Rev.*, 45, 2090, 2016.
4. Xu, J. and Tanabe, S., *J. Lumin.*, 205, 581, 2019.
5. Jia, D., Lewis, L.A., and Wang, X.J., *Electrochem. Solid-State Lett.* 13, J32, 2010.
6. Pan, Z., Lu, Y., and Liu, F., *Nat. Mater.*, 11, 58, 2012.
7. Wang, X.J., and Jia, D., *Long persistent phosphors in Phosphor Handbook*, 2nd ed. ed Yen, W.M., Shionoya, S., and Yamamoto, H., 793–818. Boca Raton, CRC Press, 2007.
8. Chen, R. and McKeever, S.W.S., *Theory of Thermoluminescence and Related Phenomena*. World Scientific, Singapore, 1997.
9. Bos, A.J.J., *Radiat. Meas.*, 41, S45, 2006.
10. Liu, F., Liang, Y.J., Chen, Y.F., and Pan, Z.W., *Adv. Opt. Mater.*, 4, 562, 2016.
11. Sun, H.X., Gao, Q.Q., Wang, A.Y., Liu, Y.C., Wang, X.J., and Liu, F., *Opt. Mater. Express*, 10, 1296, 2020.
12. Liu, F., Yan, W.Z., Chuang, Y.J., Zhen, Z.P., Xie, J., and Pan, Z.W, *Sci. Rep.*, 3, 1554, 2013.
13. He, Z.Y., Wang, X.J., and Yen, W.M., *J. Lumin.*, 119–120, 309, 2006.
14. Yan, S.Y., Liu, F., Zhang, J.H., Wang, X.J., and Liu, Y.C., *Phys. Rev. Appl.*, 13, 044051, 2020.
15. Jia, D., Wang, X.J., and Yen, W.M., *Phys. Rev. B*, 69, 23511312, 2004.
16. Huang, X., Zhao, X.Y., Yu, Z.C., Liu, Y.C., Wang, A.Y., Wang, X.J., and Liu, F., *Opt. Mater. Express*, 10, 1163, 2020.
17. Blasse, G., de Mello Donega, C., Efryushina, N., Dotsenko, V. and Berezovskaya, I., *Solid State. Commun.*, 92, 687, 1994.
18. Liu, F., Liang, Y.J., and Pan, Z.W., *Phys. Rev. Lett.*, 113, 177401, 2014.
19. Liu, F., Chen, Y.F., Liang, Y.J., and Pan, Z.W., *Opt. Lett.*, 41, 954, 2016.

20. Chen, Y.F., Liu, F., Liang, Y.J., Wang, X.L., Wang, X.J., and Pan, Z.W., *J. Mater. Chem. C*, 6, 8003, 2018.
21. Zheng, H., Wang, X.J., Dejneka, M.J., Yen, W.M., and Meltzer, R.S., *J. Lumin.* 108, 395, 2004.
22. Maiman, T.H., *Phys. Rev. Lett.*, 4, 564, 1960.
23. Gao, Q.Q., Li, C.L., Liu, Y.C., Zhang, J.H., Wang, X.J., and Liu, F., *J. Mater. Chem. C*, 8, 6988, 2020.
24. Pollnau, M., Gamelin, D.R., Lüthi, S.R., and Güdel, H.U., *Phys. Rev. B*, 61, 3337, 2000.
25. Van den Eeckhout, K., Bos, A.J.J., Poelman, D., and Smet, P.F., *Phys. Rev. B*, 87, 045126, 2013.
26. Zhao, X.Y., Li, C.L., Liu, F., and Wang, X.J., *J. Rare Earths*, DOI: 10.1016/j.jre.2021.03.008, 2021.
27. Dorenbos, P., Van Eijk, C.W.E., Bos, A.J.J., and Melcher, C.L., *J. Phys. Condes. Mater.*, 6, 4167, 1994.
28. Kodama, N., Takahashi, T., Yamaga, M., Tanii, Y., Qiu, J., and Hirao, K., *Appl. Phys. Lett.*, 75, 1715, 1999.
29. Jia, D., Wang, X.J., van der Kolk, E., and Yen, W.M., *Opt. Commun.*, 204, 247, 2002.
30. Jia, D., Meltzer, R.S., and Yen, W.M., *J. Lumin.*, 99, 1, 2002.
31. Sharma, S.K., Pitale, S.S., Manzar Malik, M., Dubey, R.N., and Qureshi, M.S., *J. Lumin.*, 129, 140, 2009.
32. Zhang, S., Li, C., Pang, R., Jiang, L., Shi, L., and Su, Q., *J. Rare Earths*, 29, 426, 2011.
33. Pan, Y.X., Su, Q., Xu, H.F., Chen, T.H., Ge, W.K., Yang, C.L., and Wu, M.M., *J. Solid State Chem.*, 174, 69, 2003.
34. Zhang, X., Zhang, J., Ren, X., and Wang, X.J., *J. Solid State Chem.*, 181, 393, 2008.
35. Zhang, X., Zhang, J., Zhang, X., Chen, L., Lu, S., and Wang, X.J., *J. Lumin.*, 122, 958, 2007.
36. Zhang, X., Cao, C., Zhang, C., Chen, L., Zhang, J., and Wang, X.J., *Phys. B: Condensed Matter* 406, 3891, 2011.
37. Lei, B.F., Liu, Y.L., Tang, G.B., Ye, Z.R., and Shi, C.S., *Chem. J. Chin. Univ.*, 24, 208, 2003.
38. Zhang, X.Y., Cheng, G., Mi, X.Y., Xiao, Z.Y., Jiang, W.W., and Hu, J.J., *J. Rare Earths*, 22, 137, 2004.
39. Lei, B., Li, B., Zhang, H., Zhang, L., Cong, Y., and Li, W., *J. Electrochem. Soc.*, 154, H623, 2007.
40. Boutinaud, P., Sarakha, L., and Mahiou, R., *J. Phys. Condes. Mater.*, 21, 025901, 2009.
41. Chu, M.H., Jiang, D.P., Zhao, C.J., and Li, B., *Chin. Phys. Lett.*, 27, 047203, 2010.
42. Xu, X., Wang, Y., Zeng, W., Gong, Y., and Liu, B., *J. Am. Ceram. Soc.*, 94, 3632, 2011.
43. Zhang, S., Li, C., Pang, R., Jiang, L., Shi, L., and Su, Q., *J. Lumin.*, 131, 2730, 2011.
44. Lecointre, A., Bessière, A. and Bos, A.J.J., Dorenbos, P., Viana, B., Jacquart, S., *J. Phys. Chem. C*, 115, 4217, 2011.
45. Jin, Y.H., Hu, Y.H., Chen, L., Wang, X.J., Mou, Z.F., Ju, G.F., and Liang, F., *Mater. Sci. Eng. B*, 178, 1205, 2013.
46. Wiatrowska, A., and Zych, E., *J. Phys. Chem. C*, 117, 11449, 2013.
47. Paulose, P.I., Joseph, J., Rudra Warrier, M.K., Jose, G., and Unnikrishnan, N.V., *J. Lumin.*, 127, 583, 2007.
48. Lei, B., Li, B., Zhang, H., and Li, W., *Opt. Mater.*, 29, 1491, 2007.
49. Zhang, Y.Y., Pang, R., Li., C.Y., Zang., C.Y., and Su, Q., *J. Rare Earths*, 28, 705, 2010.
50. Zhang, J., Ma, X., Qin, Q., Shi, L., Sun, J., Zhou, M., Liu, B., and Wang, Y., *Mater. Chem. Phys.*, 136, 320, 2012.
51. Zhang, J.S., Zhong, H.Y., Sun, J.S., Cheng, L.H., Li, X.P., and Chen, B.J., *Chin. Phys. Lett.*, 29, 017101, 2012.
52. Jin, Y.H., Hu, Y.H., Chen, L., Wang, X.J., Ju, G.F., Mou, Z. F., and Liang, F., *Opt. Commun.*, 311, 266, 2014.
53. Fu, J., *Electrochem. Solid State Lett.*, 3, 350, 2000.
54. Wang, X., Zhang, Z., Tang, Z., and Lin, Y., *Mater. Chem. Phys.*, 80, 1, 2003.
55. Lin, Y., Nan, C.W., Cai, N., Zhou, X., Wang, H., and Chen, D., *J. Alloys Compd.*, 361, 92, 2003.
56. Liu, Z.W., Liu, Y.L., Yuan, D.S., Zhang, J. X., Rong, J.H., and Huang, L.H., *Chin. J. Inorg. Chem.*, 20, 1433, 2004.
57. Lei, B.F., Man, S.Q., Liu, Y. L., and Yue, S., *Chin. J. Inorg. Chem.*, 26, 1259, 2010.
58. Kang, F. W., Hu, Y. H., Wu, H. Y., and Ju, G.F., *Chin. Phys. Lett.*, 28, 107201, 2011.
59. Liu, G., Zhang, Q., Wang, H., and Li, Y., *Mater. Sci. Eng. B*, 177, 316, 2012.
60. Wang, X.J., He, Z.Y., Jia, D., Strek, W., Dariusz, R., Hreniak, D., and Yen, W.M., *Microelectron. J.*, 36, 546, 2005.
61. Palilla, F.C., Levine, A.K., and Tomkus, M.R., *J. Electrochem. Soc.*, 115, 642, 1968.
62. Katsumata, T., Nabae, T., Sasajima, and K., Matsuzawa, T., *J. Cryst. Growth*, 183, 361, 1998.
63. Katsumata, T., Sasajima, K., Nabae, T., Komuro, S., and Morikawa, T., *J. Am. Ceram. Soc.*, 81, 413, 1998.
64. Sakai, R., Katsumata, T., Komuro, S., and Morikawa, T., *J. Lumin.*, 85, 149, 1999.
65. Jia, D., Jia, W., Evans, D.R., Dennis, W.M., Liu, H., Zhu, J., and Yen, W.M., *J. Appl. Phys.*, 88, 3402, 2000.
66. Lin, Y., Tang, Z., and Zhang, Z., *Mater. Lett.*, 51, 14, 2001.

67. Najafov, H., Kato, A., Toyota, H., Iwai, K., Bayramov, A., and Iida, S., *Jpn. J. Appl. Phys.*, 41, 2058, 2002.
68. Lin, Y., Nan, C.W., Zhou, X., Wu, J., Wang, H., Chen, D., and Xu, S., *Mater. Chem. Phys.*, 82, 860, 2003.
69. Jüstel, T., Bechtel, H., Mayr, W., and Wiechert, D.U., *J. Lumin.*, 104, 137, 2003.
70. Yamaga, M., Masui, Y., Sakuta, S., Kodama, N., and Kaminaga, K., *Phys. Rev. B*, 71, 205102, 2005.
71. Zhang, L., Li, C., and Su, Q., *J. Rare Earths*, 24, 196, 2006.
72. Sun, X., Zhang, J., Zhang, X., Luo, Y., and Wang, X.J., *J. Phys. D Appl. Phys.*, 41, 195414, 2008.
73. Ding, Y., Zhang, Y., Wang, Z., Li, W., Mao, D., Han, H., and Chang, C., *J. Lumin.*, 129, 294, 2009.
74. Pang, R., Li, C., Shi, L., and Su, Q., *J. Phys. Chem. Solids*, 70, 303, 2009.
75. Van den Eeckhout, K., Smet, P.F., and Poelman, D., *J. Lumin.*, 129, 1140, 2009.
76. Li, Y., Wang, Y., Gong, Y., Xu, X., and Zhang, F., *Acta Mater.*, 59, 3174, 2011.
77. Ju, G., Hu, Y., Chen, L., and Wang, X., *J. Appl. Phys.*, 111, 113508, 2012.
78. Jiang, L.L., Xiao, S.G., Yang, X.L., Zhang, X.A., Liu, X.H., Zhou, B.Y., and Jin, X.L., *Mater. Sci. Eng. B*, 178, 123, 2013.
79. Qin, J.L., Lei, B.F., Li, J.F., Liu, Y.L., Zhang, H.R., Zheng, M.T., Xiao, Y., and Chao, K.F., *ECS J. Solid State Sci. Technol.*, 2, R60, 2013.
80. Qin, J.L., Zhang, H.R., Lei, B.F., Hu, C.F., Li, J.F., Liu, Y.L., Meng, J.X., Wang, J., Zheng, M.T., and Xiao, Y., *J. Am. Ceram. Soc.*, 96, 3149, 2013.
81. Pang, R., Zhao, R. and Su, Q., *J. Rare Earths*, 32, 792, 2014.
82. Jin, Y.H., Hu, Y.H., Chen, L., Wang, X.J., Mu, Z.F., Ju, G.F., and Wang, T., *Mater. Lett.*, 126, 75, 2014.
83. Ju, G.F., Hu, Y.H., Chen, L., Wang, X.J., and Mu, Z.F., *Opt. Mater.*, 36, 1183, 2014.
84. Li, G., Wang, Y.H., Zeng, W., Han, S.C., Chen, W.B., Li, Y.Y., and Li, H., *Opt. Mater.*, 36, 1808, 2014.
85. Jin, Y.H., Hu, Y.H., Chen, L., and Wang, X.J., *J. Am. Ceram. Soc.*, 97, 2573, 2014.
86. Wu, C.Q., Zhang, J.C., Feng, P.F., Duan, Y.M., Zhang, Z.Y., and Wang, H., *J. Lumin.*, 147, 229, 2014.
87. Guo, H.J., Chen, W.B., Zeng, W., Wang, Y.H., and Li, Y.Y., *ECS Solid State Lett.*, 4, R1, 2015.
88. Zych, E., Trojan-Piegza, J., Hreniak, D., and Strek, W., *J. Appl. Phys.*, 94, 1318, 2003.
89. Kuang, J.Y., Liu, Y.L., Zhang, J.X., Yuan, D.S., Huang, L.H., and Rong, J.H., *Chin. J. Inorg. Chem.*, 21, 1383, 2005.
90. Liu, Z., and Liu, Y., *Mater. Chem. Phys.*, 93, 129, 2005.
91. Liu, C., Che, G., Xu, Z., and Wang, Q., *J. Alloys Compd.*, 474, 250, 2009.
92. Liu, B., Shi, C., and Qi, Z., *J. Phys. Chem. Solids*, 67, 1674, 2006.
93. Zhang, S., Pang, R., Li, C., and Su, Q., *J. Lumin.*, 130, 2223, 2010.
94. Wu, H., Hu, Y., Kang, F., Chen, L., Wang, X., Ju, G., and Mu, Z., *Mater. Res. Bull.*, 46, 2489, 2011.
95. Rodrigues, L.C.V., Brito, H.F., Hölsä, J., Stefani, R., Felinto, M.C.F.C., Lastusaari, M., Laamanen, T., and Nunes, L.A.O., *J. Phys. Chem. C*, 116, 11232, 2012.
96. Jin, Y., Hu, Y., Chen, L., Wang, X., Ju, G., and Mu, Z., *J. Lumin.*, 138, 83, 2013.
97. Wang, Q., Qiu, J.B., Song, Z.G., Zhou, D.C., and Xu, X.H., *Chin. Phys. B*, 23, 064211, 2014.
98. Liu, B., Shi, C., and Qi, Z., *Appl. Phys. Lett.*, 86, 191111, 2005.
99. Kuang, J., and Liu, Y., *Chem. Lett.*, 34, 598, 2005.
100. Chen, Y., Cheng, X., Liu, M., Qi, Z., and Shi, C., *J. Lumin.*, 129, 531, 2009.
101. Wei, R.P., Ju, Z.H., Ma, J.X., Zhang, D., Zang, Z.P., and Liu, W.S., *J. Alloys Compd.*, 486, L17, 2009.
102. Lei, B.F., Liu, Y.L., Tang, G.B., Ye, Z.R., and Shi, C.S., *Chem. J. Chin. Univ.*, 24, 782, 2003.
103. Uheda, K., Maruyama, T., Takizawa, H., and Endo, T., *J. Alloys Compd.*, 262–263, 60, 1997.
104. Matsui, H., Xu, C.N., Akiyama, M., and Watanabe, T., *Jpn. J. Appl. Phys.*, 39, 6582, 2000.
105. Iwasaki, M., Kim, D.N., Tanaka, K., Murata, T., and Morinaga, K., *Sci. Technol. Adv. Mater.*, 2, 137, 2003.
106. Lin, L., Yin, M., Shi, C., and Zhang, W., *J. Alloys Compd.*, 455, 327, 2008.
107. Mu, Z., Hu, Y., Wang, Y., Wu, H., Fu, C., and Kang, F., *J. Lumin.*, 131, 676, 2011.
108. Lecointre, A., Ait Benhamou, R., Bessiére, A., Wallez, G., Elaatmani, M., and Viana, B., *Opt. Mater.*, 34, 376, 2011.
109. Lei, B., Li, B., Wang, X.J., and Li, W., *J. Lumin.* 118, 173, 2006.
110. Sun, Z.X., *Chin. J. Inorg. Chem.*, 28, 1229, 2012.
111. Qu, X.F., Cao, L.X., Liu, W., Su, G., Wang P.P., and Schultz. I., *Mater. Res. Bull.*, 47, 1598, 2012.
112. Xu, X.Q., Ren, J., Chen, G.R., Kong, D.S., Gu, C.J., Chen, C.M., and Kong, L.R., *Opt. Mater. Express*, 3, 1727, 2013.
113. Garlick, G.F.J., and Gibson, A.F., *Proc. Phys. Soc.*, 60, 574, 1948.
114. Lastusaari, M., Laamanen, T., Malkamäki, M., Eskola, K.O., Kotlov, A., Carlson, S., Welter, E., Brito, H.F., Bettinelli, M., Jungner, H., and Hölsä, J., *Eur. J. Mineral.*, 24, 885, 2012.
115. Kuang, J., and Liu, Y., *J. Electrochem. Soc.*, 153, G245, 2006.

116. Garlick, G.F.J., and Mason, D.E., *J. Electrochem. Soc.*, 96, 90, 1949.
117. Zhuang, Y.X., Ueda J., and Tanabe. S., *Opt. Mater. Express*, 2, 1378, 2012.
118. Jin, Y.H., Hu, Y.H., Chen, L., Wang, X.J., Ju, G.F., and Mu, Z.F., *Radiat. Meas.*, 51, 18, 2013.
119. Zhang, S.A., Hu, Y.H., Chen, R., Wang, X.J., and Wang, Z.H., *Opt. Mater.*, 36, 1830, 2014.
120. Blasse, G., Grabmaier, B.C., and Ostertag. M., *J. Alloys Compd.*, 200, 17, 1993.
121. Yan, W., Liu, F., Lu, Y., Wang, X.J., Yin, M., and Pan, Z., *Opt. Express*, 18, 20215, 2010.
122. Zhong, R., Zhang, J., Zhang, X., Lu, S., and Wang, X.J., *Appl. Phys. Lett.*, 88, 201916, 2006.
123. Lu, Y., Liu, F., Gu, Z., and Pan, Z., *J. Lumin.*, 131, 2784, 2011.
124. Bessière, A., Jacquart, S., Priolkar, K., Lecointre, A., Viana, B., and Gourier, D., *Opt. Express*, 19, 10131, 2011.
125. Yang, J., Liu, Y., Yan, D., Zhu, H., Liu, C., Xu, C., Ma, L., Wang, X.J., *Dalton Trans.*, 45, 1364, 2016.
126. Basavaraju, N., Sharma, S., Bessière, A., Viana, B., Gourier, D., and Priolkar, K., *J. Phys. D: Appl. Phys.*, 46, 375401, 2013.
127. Ueda, J., Kuroishi, K., and Tanabe, S., *Appl. Phys. Lett.*, 104, 101904, 2014.
128. Li, S., Zhu, Q., Li, X., Sun, X., and Li, J. G., *J. Alloys Compd.*, 827, 154365, 2020.
129. Kang, R., Dou, X., Lian, H., and Li, Y., *J. Am. Ceram. Soc.*, 103, 258, 2020.
130. Zhou, Z., Yi, X., Xiong, P., Xu, X., Ma, Z., and Peng, M. *J. Mater. Chem. C*, 8, 14100, 2020.
131. Pellerin, M., Castaing, V., Gourier, D., Chanéac, C., and Viana, B., *Proc. SPIE*, 10533, 1053321, 2018.
132. Liang, Y., Liu, F., Chen, Y., Wang, X., Sun, K., and Pan, Z., *Dalton Trans.*, 46(34), 11149, 2017.
133. Liang, Y., Liu, F., Chen, Y., Wang, X., Sun, K., and Pan, Z., *J. Mater. Chem. C*, 5, 6488, 2017.
134. Kamimura, S., Xu, C., Yamada, H., Terasaki, N., and Fujihala, M., *Jpn. J. Appl. Phys.*, 53, 092403, 2014.
135. Xu, J., Tanabe, S., Sontakke, A. D., and Ueda, J., *Appl. Phys. Lett.*, 107, 081903, 2015.
136. Xu, J., Murata, D., So, B., Asami, K., Ueda, J., Heo, J., and Tanabe, S., *J. Mater. Chem. C*, 6, 11374, 2018.
137. Yu, N., Liu, F., Li, X., and Pan, Z., *Appl. Phys. Lett.*, 95, 231110, 2009.
138. Caratto, V., Locardi, F., Costa, G.A., Masini, R., Fasoli, M., Panzeri, L., Martini, M., Bottinelli, E., Gianotti, E., and Miletto, I., *ACS Appl. Mater. Interfaces*, 6, 17346, 2014.
139. Liang, Y., Liu, F., Chen, Y., Wang, X.J., Sun, K., and Pan, Z.W., *Light Sci. Appl.*, 5, e16124, 2016.
140. Wang, X.L., Chen, Y.F., Liu, F., and Pan, Z.W., *Nat. Commun.*, 11, 2040, 2020.
141. Gutiérrez-Martín, F., Fernández-Martinez, F., Díaz, P., Colón, C., and Alonso-Medina, A., *J. Alloys Compd.*, 501, 193, 2010.
142. Yang, Y., Li, Z., Zhang, J., Lu, Y., Guo, S., Zhao, Q., Wang, X., Yong, Z., Li, H., Ma, J., Kuroiwa, Y., Moriyoshi, C., Hu, L., Zhang, L., Zheng, L., and Sun, H., *Light Sci. Appl.*, 7, 88, 2018.
143. Li, H., Liu, Q., Ma, J., Feng, Z., Liu, J., Zhao, Q., Kuroiwa, Y., Moriyoshi, C., Ye, B., Zhang, J., Duan, C., and Sun, H., *Adv. Opt. Mater.*, 8, 1901727, 2020.
144. Wang, X.L., Chen, Y.F., Kner, P.A., and Pan, Z.W., *Dalton Trans.*, 50, 3499, 2021.
145. Yuan, W.H., Tan, T., Wu, H.Y., Pang, R., Zhang, S., Jiang, L.H., Li, D., Wu, Z.J., Li, C.Y., and Zhang, H.J., *J. Mater. Chem. C*, 9, 5206, 2021.
146. Zhao, X.Y., Liu, F., Wang, X.J., and Liu, Y.C., *Phys. Rev. Appl.*, 2021 (in press).
147. Li, Z., Li, H., and Sun, H., *J. Rare Earths*, 38, 124, 2020.
148. Liang, Y., Liu, F., Chen, Y., Sun, K., and Pan, Z., *Dalton Trans.*, 45, 1322, 2016.
149. Yang, Z., Liao, J., Wang, T., Wu, H., Qiu, J., Song, Z. and Zhou, D., *Mater. Express*, 4, 172, 2014.
150. Wang, W., Sun, Z., He, X., Wei, Y., Zou, Z., Zhang, J., Wang, Z., Zhang, Z., and Wang, Y., *J. Mater. Chem. C*, 5, 4310, 2017.
151. Zhou, Z.H., Xiong, P.X., Liu, H.L., and Peng, M.Y, *Inorg. Chem.*, 59, 12920, 2020.
152. Liu, Q., Feng, Z., Li, H., Zhao, Q., Shirahata, N., Kuroiwa, Y., Moriyoshi, C., Duan, C., and Sun, H., *Adv. Opt. Mater.*, 9, 2002065, 2021.
153. Jia, D., Meltzer, R.S., Yen, W.M., Jia, W., and Wang, X.J., *Appl. Phys. Lett.*, 80, 1535, 2002.
154. Wang, X.J., Jia, D., and Yen, W.M., *J. Lumin.*, 102–103, 34, 2003.
155. Jia, D., Wang, X.J., Jia, W., and Yen, W.M., *J. Appl. Phys.*, 93, 148, 2003.
156. Teng, Y., Zhou, J.J., Khisro, S.N., Zhou, S.F., and Qiu, J.R., *Mater. Chem. Phys.*, 147, 772, 2014.
157. Teng, Y., Zhou, J., Ma, Z., Smedskjaer, M.M., and Qiu, J., *J. Electrochem. Soc.*, 158, K17, 2011.
158. Kang, F., Hu, Y., Wu, H., Mu, Z., Ju, G., Fu, C., and Li, N., *J. Lumin.*, 132, 887, 2012.
159. Ye, S., Zhang, J., Zhang, X., Lu, S., Ren, X., and Wang, X.J., *J. Appl. Phys.* 101, 033513, 2007
160. Ye, S., Zhang, J., Zhang, X., and Wang, X.J., *J. Lumin.*, 122, 914, 2007.

14 Color-Tunable Metallophosphors for Organic Light-Emitting Diodes (OLEDs)

Hongyang Zhang and Wai-Yeung Wong

CONTENTS

14.1 METALLOPHOSPHORS BASED ON TRANSITION METALS AND RARE-EARTH METALS

14.1.1 METALLOPHOSPHORS BASED ON TRANSITION METALS

A "transition metal" is described as an element with a partially filled d subshell, including groups 3 to 12 on the periodic table.[1, 2] Among them, some heavy transition metal elements are found to induce the strong spin-orbit coupling (SOC) effect in the organometallic complexes that facilitates the intersystem crossing (ISC) process.[3] As a result, the mixed spin state is formed in which the triplet-excited state is mixed with the singlet-excited state and the spin-forbidden nature of the

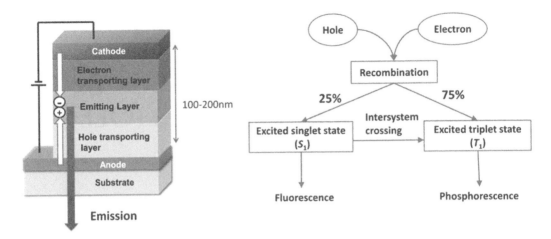

FIGURE 14.1 Device structure of OLED and the luminescence process. Reproduced with permission from Ref [6]. Copyright 2014 National Institute for Materials Science.

radiative relaxation of the triplet-excited state is removed, thus leading to the metallophosphors with highly efficient luminescent performances. As these metallophosphors are capable of using both singlet and triplet excitons, this provides the feasibility of a theoretical maximum internal quantum efficiency (IQE) up to 100% in electroluminescence (EL).[4–6] Figure 14.1 exhibits the typical device structure of organic light-emitting diode (OLED) and the luminescence process.

Typically, complexes of Ir(III), Pt(II), Os(II), Ru(II), Rh(III), Pd(II), Mn(II), Re(I), Cu(I), Ag(I), Au(I) and Au(III) are the most widely studied triplet emitters.[7–16] These phosphorescent heavy transition metal complexes normally possess radiative lifetimes in the range of microseconds (μs) at room temperature, which are longer than the typical lifetimes (ns) of fluorescent organic compounds while advantageously shorter than that (ms-s) of pure organic phosphors, since the relatively short phosphorescent lifetimes can benefit from the faster radiative rate constants (k_r). While the triplet-triplet annihilation (TTA) and/or concentration self-quenching effects will significantly affect the EL efficiencies of fabricated electronic devices, these metallophosphors are usually doped into organic host materials during the fabrication of phosphorescent OLEDs to allow the efficient energy transfer from the host excitons to the metallophosphor dopants (Figure 14.2).[17, 18]

Through the modification of coordinated ligand structures and metal ion centers, the emitting colors, luminescent lifetimes, quantum yields, optoelectronic properties and the associated OLED device performances of metallophosphors can be finely tuned. A mixing of the ligand orbitals and metal orbitals can result in the frontier molecular orbitals of metallophosphors; therefore, both the chelating ligands and the transition metal center contribute to the excited states. In particular, the excited states of metallophosphors can be ascribed to various charge transfers or transitions such as the metal-to-ligand charge transfer (MLCT), ligand-to-metal charge transfer (LMCT), ligand-to-ligand charge transfer (LLCT), intraligand charge transfer (ILCT) and ligand-centered (LC) transitions, which directly determine the photophysical and optoelectronic properties of metallophosphors.[19]

14.1.2 METALLOPHOSPHORS BASED ON RARE-EARTH METALS

According to the definition by the International Union of Pure and Applied Chemistry (IUPAC), rare-earth metals are a set of 17 chemical elements in the periodic table, including the fifteen lanthanides, scandium and yttrium.[20] The metallophosphors based on rare-earth metals mainly contain the inorganic lanthanide compounds and molecular lanthanide complexes.[21, 22] Herein, we

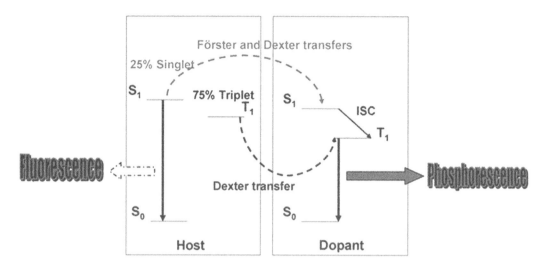

FIGURE 14.2   Schematic energy-level alignment of the singlet-excited state (S_1), triplet-excited state (T_1) and ground state (S_0), as well as the energy transfer and light emission processes in a host-dopant system. Reproduced with permission from Ref. [18]. Copyright The Royal Society of Chemistry and the Centre National de la Recherche Scientifique 2013.

specially focus on the molecular lanthanide complexes. In general, the emissive lanthanide complexes comprise the metal center and chelating ligands, while the ligands are functionalized with some groups serving as a sensitizer or antenna. Figure 14.3 shows two representative models in which the antenna part is attached to the chelating ligands, and both structures can afford efficient energy transfer due to the reduced distance between metal and antenna.[23]

The lanthanide metals possess the intrinsic luminescence that originates from the electronic transitions inside the 4f shell or 5d–4f of the $4f^n5d^m$ ($n = 0$–14, $m = 0/1$) configuration. However, the f–f transitions are formally parity forbidden by the spin and Laporte selection rules. On the other hand, the shielding of the closed $5s^2$ and $5p^6$ shells on the environment of 4f shell results in the narrow emission bands and the long lifetimes of the excited states in the microsecond to millisecond range.[23] The earlier circumstances can cause a relatively weak intrinsic luminescence, and hence, the light-harvesting antenna or sensitizer (commonly the organic fluorophore) is introduced into the molecular lanthanide complex to facilitate the process of energy transfer. Actually, diverse types of excited states and transitions can funnel energy onto the Ln(III) ions, such as the first singlet-excited state S_1, triplet-excited state T_1 and the ILCT of chelating ligands, the LMCT and MLCT of molecular lanthanide complexes, the 4f–5d and intraconfigurational 4f–4f transitions of Ln(III) ions (Figure 14.4).[24, 25]

14.2   BLUE-EMITTING METALLOPHOSPHORS FOR ELECTROLUMINESCENT APPLICATION

14.2.1   BLUE-EMITTING IR(III) METALLOPHOSPHORS

In 1998, Ma et al. reported the first observation of electroluminescence from osmium(II) complexes and demonstrated the emission to be originated from the triplet MLCT-excited states of the metallophosphor.[9] In the same year, Forrest et al. presented the efficient organic light-emitting devices based on a platinum(II) complex, in which a fluorescent dye doped with the metallophosphor was used as the host material to improve the electroluminescent efficiency.[8] Then in the next year, Forrest et al. utilized the green electrophosphorescent iridium(III) complex *fac*-Ir(ppy)$_3$

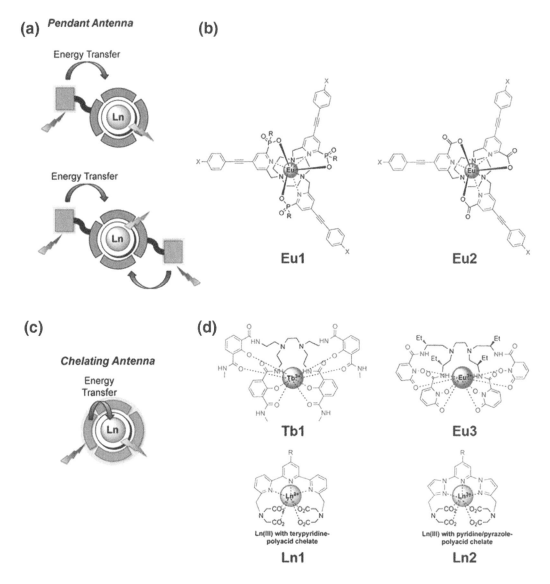

FIGURE 14.3 (a) Schematic of lanthanide chelates that utilize pendant antenna. (b) Examples of reported lanthanide complexes with pendant antenna. (c) Schematic of lanthanide chelates that utilize chelating antenna. (d) Examples of reported lanthanide complexes with chelating antenna. Reproduced with permission from Ref. [23]. Copyright 2013 American Chemical Society.

doped into the 4,4′-N,N'-dicarbazole-biphenyl (CBP) host as the emitter, and the corresponding devices exhibit the external quantum efficiencies (EQEs) as high as 8.0% with a peak luminance up to 100,000 cd/m². [26] Since then, the developments of OLEDs based on metallophosphors grow very fast in the past two decades and are considered the promising candidate material for developing new-generation flat panel displays and energy-saving solid lighting sources, because of their remarkable advantages like high electroluminescent efficiency, full-color/large-area display, flexible panel fabrication, wide-viewing angle and low power consumption.[17]

For OLED applications, metallophosphors that can emit the three primary colors (red, green and blue) are essential for the full-color display. In addition, to realize the white light emission, blue-emitting metallophosphors are truly indispensable. However, due to the high energy of blue

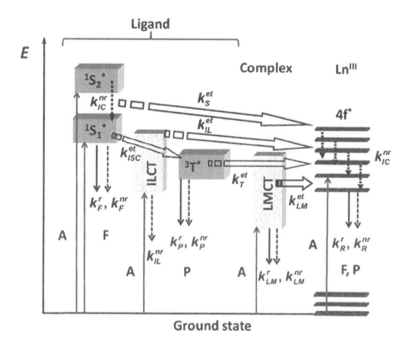

FIGURE 14.4 Schematic representation of energy absorption, migration, emission (plain arrows) and dissipation (dotted arrows) processes in a molecular lanthanide complex. Reproduced with permission from Ref. [24]. Copyright The Royal Society of Chemistry 2010.

light, it requires the emitters with a wider band gap, which is a challenge for the stability of devices. Therefore, the research on developing the blue phosphorescent emitters possessing both the long device operational lifetime and high quantum efficiency become important for the purpose of making OLEDs technology really practical.[18]

Among the numerous blue-emitting metallophosphors, the blue-emitting Ir(III) complexes are one of the most studied and popular triplet emitters. The first reported blue-emitting Ir(III) structure is FIrpic (**Ir1**), and its maximum EL wavelength is 475 nm (Figure 14.5). The doped OLED device affords a high EQE of 5.7% and a power efficiency (PE) of 6.3 lm/W.[27] The peak EQE can be significantly improved to 31.4% by choosing the proper host material and optimizing the device structure.[28]

There are various strategies to tune the emission colors and the EL performances of Ir(III) phosphors. One is to change the auxiliary ligand so as to influence the energy of excited state by modifying the electron density at the metal center. On the basis of FIrpic framework, if the tetrakis(1-pyrazolyl)borate ligand is used instead of the picolinato ligand, the electron-withdrawing group will pull electron density away from the iridium center, thereby stabilizing the metal orbitals and lowering the HOMO energy while the LUMO energy of the complex is largely unaffected.[29] As a result, the HOMO-LUMO gap should increase and the emission of **Ir2** can be blue shifted to 457 nm. The OLEDs based on **Ir2** can give an increase in the maximum EQE from the original 11.6% to the recent 20.8% by optimizing the device structures.[30–32] The pyridyl-tetrazolate ligand is also used to substitute for the picolinato ligand to afford **Ir3** (Figure 14.6), and its electron-withdrawing feature allows the photoluminescence wavelength to show a hypsochromic shift by 10 nm to around 460 nm compared to FIrpic. The optoelectronic devices using **Ir3** as the emitter attain a high EQE of 14.4% and PE of 11.9 lm/W.[33] The Ir(III) complexes with other auxiliary ligands can also be designed to realize efficient blue phosphorescence. For example, **Ir4** can emit the blue-green light at 489 nm with an outstanding current efficiency (CE) of 25.45 cd/A and PE of 23.52 lm/W.[34]

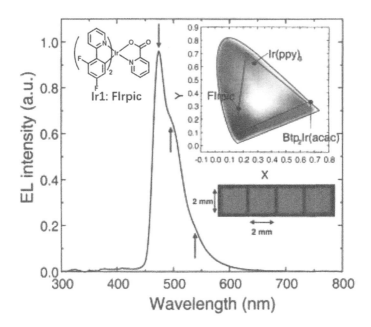

FIGURE 14.5 The structure and electroluminescence spectrum of FIrpic. Reproduced with permission from Ref. [27]. Copyright 2001 American Institute of Physics.

Another strategy to tune the complex structure and property is to change the main cyclometalated ligand. Introduction of electron-withdrawing groups into the phenyl moiety can lower the HOMO level, whereas adding electron-donating groups into the pyridyl moiety can elevate the LUMO level. As a result, the triplet energy of the Ir(III) complex will be enlarged, making the emission colors to be pure-blue or even deep-blue. Lee et al. added the cyano (–CN) group into the phenyl moiety to yield **Ir5**, and the triplet energy can be increased to 2.74 eV, larger than 2.62 eV which is the triplet energy of FIrpic. The electroluminescent peak is blue-shifted to 457 nm with color coordinates of (0.14, 0.19) and a superior quantum efficiency of 22.1% is achieved in the OLED device based on **Ir5**.[35] Using another electron-withdrawing diphenylphosphoryl group to replace the cyano group on the difluorophenyl moiety will likewise improve the emitting color purity of the blue light, and the blue OLED devices based on **Ir6** emitter exhibit the peak EQE of 7.1% and CE of 11.1 cd/A.[36] If the carbon at the 5 position on the phenyl ring is changed with nitrogen, the σ-donation from the phenyl moiety of the cyclometalated ligand to the d orbitals of Ir(III) center would decrease, which can stabilize the HOMO level. Thus, the triplet exciton energy is raised, leading to the blue-shifted

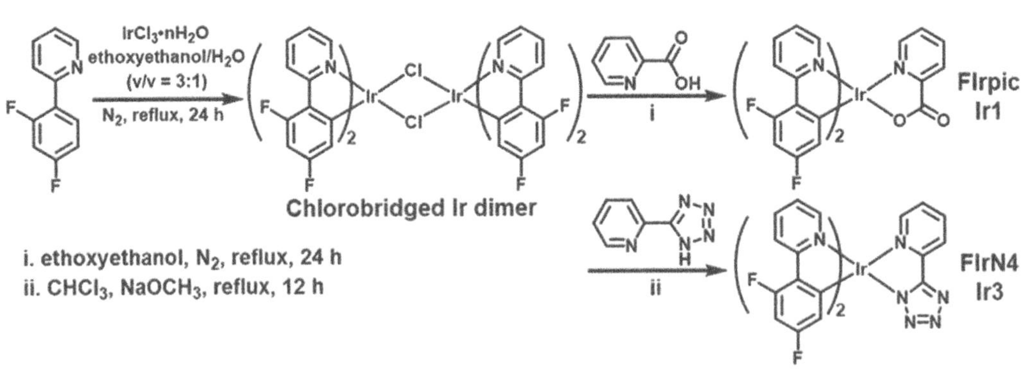

FIGURE 14.6 Synthetic routes and structures of blue-emitting Ir(III) complexes **Ir1** and **Ir3**.

emission of **Ir7** by over 20 nm compared with FIrpic, in which the color coordinates lie at (0.16, 0.20) corresponding to the pure blue region.[37] Concurrently incorporating the electron-withdrawing groups into the phenyl moiety and electron-donating groups into the pyridyl ring can play a synergistic effect on tuning the emission color. For example, Kim et al. put a methyl group on the pyridyl part while added a perfluoro carbonyl substituent into the phenyl part, and the resultant complex **Ir8** can be utilized as the emitter in the phosphorescent OLEDs, showing a peak EQE of 17.1% and Commission Internationale de l'Eclaira (CIE) coordinates of (0.141, 0.158). Notably, the color purity of metallophosphor **Ir8** is further ameliorated. In addition, simultaneously altering the auxiliary ligand can further pull the emission color to a deeper blue range. **Ir9** with the pyridyl-triazolate instead of picolinate (pic) as the chelating ligand can exhibit an emission peak with the shorter wavelength compared to **Ir8**, and the CIE coordinates of (0.147, 0.116) are recorded in its associated OLED device.[38] In addition, employing a stronger field ligand like the ligand containing an N-heterocyclic carbene (NHC) rather than using the pyridyl counterpart can render the complex to have higher energy ligand field (LF) states, enabling it to significantly blue shift the emission to deep blue or even near-ultraviolet region. This class of Ir(III) complexes bearing NHC ligands can display impressive device performances. For example, the OLEDs based on **Ir10** afford an outstanding EQE of 20.6%, CE of 47.6 cd/A and PE of 39.3 lm/W with the sky blue emission [39–41] Likewise, the phenylimidazole derivatives are utilized as the ligands to give the efficient blue phosphorescent iridium complexes, and the resultant **Ir14**-based OLEDs give the maximum EQE of 33.2% with CE up to 73.6 cd/A and PE up to 71.9 lm/W, which is, to our knowledge, the best reported performance for blue OLEDs based on Ir(III) complexes.[42] Figure 14.7 shows some typical examples of blue-emitting Ir(III) phosphors for OLED applications.

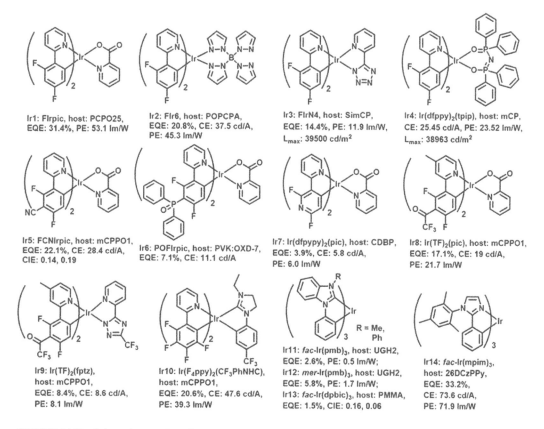

Ir1: FIrpic, host: PCPO25, EQE: 31.4%, PE: 53.1 lm/W

Ir2: FIr6, host: POPCPA, EQE: 20.8%, CE: 37.5 cd/A, PE: 45.3 lm/W

Ir3: FIrN4, host: SimCP, EQE: 14.4%, PE: 11.9 lm/W, L_{max}: 39500 cd/m^2

Ir4: Ir(dfppy)$_2$(tpip), host: mCP, CE: 25.45 cd/A, PE: 23.52 lm/W, L_{max}: 38963 cd/m^2

Ir5: FCNIrpic, host: mCPPO1, EQE: 22.1%, CE: 28.4 cd/A, CIE: 0.14, 0.19

Ir6: POFIrpic, host: PVK:OXD-7, EQE: 7.1%, CE: 11.1 cd/A

Ir7: Ir(dfpypy)$_2$(pic), host: CDBP, EQE: 3.9%, CE: 5.8 cd/A, PE: 6.0 lm/W

Ir8: Ir(TF)$_2$(pic), host: mCPPO1, EQE: 17.1%, CE: 19 cd/A, PE: 21.7 lm/W

Ir9: Ir(TF)$_2$(fptz), host: mCPPO1, EQE: 8.4%, CE: 8.6 cd/A, PE: 8.1 lm/W

Ir10: Ir(F$_4$ppy)$_2$(CF$_3$PhNHC), host: mCPPO1, EQE: 20.6%, CE: 47.6 cd/A, PE: 39.3 lm/W

Ir11: *fac*-Ir(pmb)$_3$, host: UGH2, EQE: 2.6%, PE: 0.5 lm/W; Ir12: *mer*-Ir(pmb)$_3$, host: UGH2, EQE: 5.8%, PE: 1.7 lm/W; Ir13: *fac*-Ir(dpbic)$_3$, host: PMMA, EQE: 1.5%, CIE: 0.16, 0.06

Ir14: *fac*-Ir(mpim)$_3$, host: 26DCzPPy, EQE: 33.2%, CE: 73.6 cd/A, PE: 71.9 lm/W

FIGURE 14.7 Selected examples of blue-emitting Ir(III) phosphors and the performances of their electroluminescent devices.

14.2.2　Blue-Emitting Pt(II) Metallophosphors

Phosphorescent Pt(II) complexes are also an important class of triplet state emitters that can be used as the dopants in the application of blue-emitting OLEDs. However, Pt(II) phosphors are much less explored compared to Ir(III) phosphors, due to their inherent square-planar coordination geometry with strong Pt-Pt interactions. When the emitting layer has a higher concentration of Pt(II) phosphors, excimer will be normally formed so that the emission will remarkably shift to the longer wavelength.[43] Thus, the blue-emitting OLEDs based on Pt(II) metallophosphors usually refer to the doped devices at a low concentration.

According to the coordination sites of ligands, bidentate, tridentate and tetradentate ligands are capable of combining with Pt(II) metal ion to form the blue Pt(II) emitters (Figure 14.8). High-field-strength carbene-based bidentate ligands united with an ancillary ligand of acetylacetonate (acac) derivatives can be used to afford the Pt(II) complexes **Pt1**, **Pt2** and **Pt3** with high emission energy, and EQEs of their doped OLED devices can be as high as 9.8%, but their color purity needs to be further improved.[44–46] One tridentate ligand and one halogen ion can also coordinate to the Pt(II) center to form the metallophosphor. By introducing fluorine on the 1,3-di(2-pyridyl)benzene ligand, it will mainly lower the HOMO energy level of Pt(II) phosphor **Pt4** so that the emission

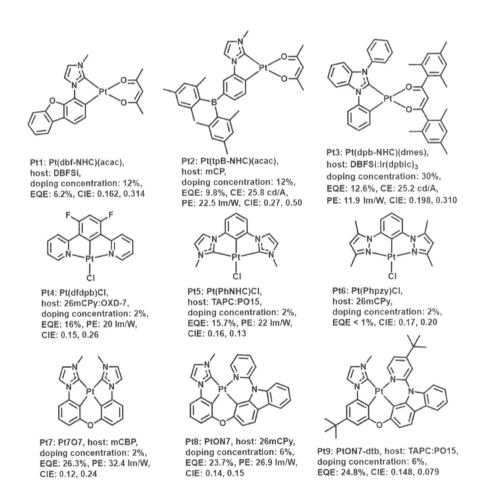

Pt1: Pt(dbf-NHC)(acac),
host: DBFSi,
doping concentration: 12%,
EQE: 6.2%, CIE: 0.162, 0.314

Pt2: Pt(tpB-NHC)(acac),
host: mCP,
doping concentration: 12%,
EQE: 9.8%, CE: 25.8 cd/A,
PE: 22.5 lm/W, CIE: 0.27, 0.50

Pt3: Pt(dpb-NHC)(dmes),
host: DBFSi:Ir(dpbic)₃
doping concentration: 30%,
EQE: 12.6%, CE: 25.2 cd/A,
PE: 11.9 lm/W, CIE: 0.198, 0.310

Pt4: Pt(dfdpb)Cl,
host: 26mCPy:OXD-7,
doping concentration: 2%,
EQE: 16%, PE: 20 lm/W,
CIE: 0.15, 0.26

Pt5: Pt(PhNHC)Cl,
host: TAPC:PO15,
doping concentration: 2%,
EQE: 15.7%, PE: 22 lm/W,
CIE: 0.16, 0.13

Pt6: Pt(Phpzy)Cl,
host: 26mCPy,
doping concentration: 2%,
EQE < 1%, CIE: 0.17, 0.20

Pt7: Pt7O7, host: mCBP,
doping concentration: 2%,
EQE: 26.3%, PE: 32.4 lm/W,
CIE: 0.12, 0.24

Pt8: PtON7, host: 26mCPy,
doping concentration: 6%,
EQE: 23.7%, PE: 26.9 lm/W,
CIE: 0.14, 0.15

Pt9: PtON7-dtb, host: TAPC:PO15,
doping concentration: 6%,
EQE: 24.8%, CIE: 0.148, 0.079

FIGURE 14.8　Selected examples of blue-emitting Pt(II) phosphors and the performances of their electroluminescent devices.

energy would move to a shorter wavelength. The OLED device with **Pt4** as the emitter (dopant concentration of 2%) can result in a peak EQE of 16%, a high PE of 20 lm/W and CIE coordinates of (0.15, 0.26). That belongs to the monomer emission, whereas the excimer emission with longer wavelengths such as the orange/red emission can be observed with the increase of dopant concentration. Therefore, it is possible to realize the white OLED by employing only one doped emitter, which is a distinguishing feature for square-planar blue-emitting Pt(II) complexes. It is noteworthy that as the dopant concentration is raised to 8%, the white OLED based on **Pt4** with a peak EQE of 9.3% and CIE coordinates of (0.33, 0.36) is achieved.[47] Using the N-heterocyclic carbene (NHC) group to substitute for the pyridyl unit in the 1,3-di(2-pyridyl)benzene ligand, the obtained Pt(II) complex **Pt5** can possess a destabilized LUMO energy level. As a result, **Pt5** emits at 448 nm in a dichloromethane solution, and its corresponding electroluminescent device furnishes a maximum EQE of 15.7% and a notable PE of 22 lm/W. By changing the carbene unit with the pyrazolyl unit to afford **Pt6**, its doped PMMA film can emit the deep blue light at 430 nm. However, its low emission efficiency and the unmatched triplet energy with the host material caused the low efficiency of **Pt6** doped OLED device.[48] The complex **Pt7** is synthesized from a symmetrical tetradentate ligand containing two carbene groups, and its doped OLEDs can attain a remarkable EQE up to 26.3%. Moreover, **Pt7** can be employed as the single emitter to furnish the white OLEDs with a maximum EQE of 25.7%, a CRI of 70 and CIE coordinates of (0.37, 0.42).[49] By using a pyridyl-carbazole group to generate the asymmetric metallophosphor **Pt8**, the emission energy can be shifted to a deeper blue range compared to **Pt7** and its doped device gives a peak EQE of 23.7%.[50] Then incorporating the *tert*-butyl groups into **Pt8** can provide **Pt9**. By using a co-host of TAPC and the high-bandgap electron-transporting material PO15 as well as replacing the hole blocking material by PO15 to lessen the potential quenching effects from poor triplet energy alignment with the host or blocking molecules, **Pt9**-based OLED shows an excellent performance with a peak EQE of 24.8% in the pure blue (CIE: 0.148, 0.079) region, remarkably close to the NSTC standard for blue (CIE: 0.14, 0.08).[51]

In addition to the previous Pt(II) phosphors with multidentate ligands, the *trans*-NHC Pt(II) phenylacetylide **Pt10** with four monodentate ligands has been designed and employed as the emitting dopant for phosphorescent OLEDs. A maximum EQE of 8% with the CIE coordinates of (0.20, 0.20) can be achieved on its optimized device (Figure 14.9).[52] All these impressive works suggest that the employment of Pt phosphors is a viable approach for the development of blue OLEDs.

14.2.3　Blue-Emitting Metallophosphors Containing Other Metals

Some other metals including transition metals and rare-earth metals are also capable of combining with organic ligands to form the blue-emitting metallophosphors. Regarding the transition metal-based phosphors (Figure 14.10), both Au(I) and Au(III) complexes have been studied as efficient blue emitters. The carbene-Au(I)-amides are a novel class of linear and two-coordinate metal complexes. They can show a delayed luminescence as the strongly bound carbene ligand acts as the π-acceptor while the amide group is an electron-rich anion acting as an electron donor upon excitation. The HOMO is centered on the amide moiety and the LUMO is mainly localized on the carbene group. As a result, two frontier orbitals are well separated in contribution, followed by a small energy gap between the lowest singlet-and triplet-excited states after charge-transfer excitations. This arrangement enables the energy utilization of both singlet and triplet-excited states and the detailed process of luminescence involved has been subjected to a number of investigations. Thus, the carbene-metal-amides complexes are a new family of efficient emitters with the theoretical IQE up to 100%. For example, **Au1**-based OLED can realize the blue emission with a maximum EQE of 17.3% in a host-free OLED and 20.9% in a doped device.[15] A series of Au(III) complexes containing a tridentate ligand and a monoaryl auxiliary ligand have also been explored for the application of blue OLEDs.

CIE color coordinates

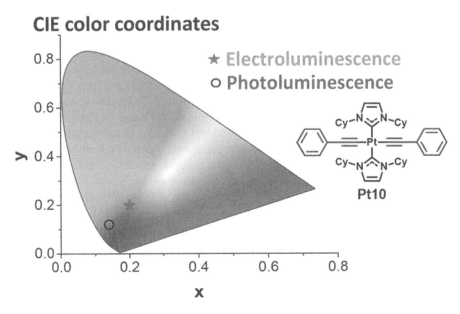

FIGURE 14.9 The chemical structure of *trans*-N-heterocyclic carbene Pt(II) complex **Pt10** and the CIE coordinates of its corresponding electroluminescence in the OLED device and photoluminescence in a doped PMMA film. Reproduced with permission from Ref. [52]. Copyright 2017 American Chemical Society.

Among them, sky-blue-emitting OLED based on **Au2** attains a maximum EQE of 11.3%, PE of 18.9 lm/W and CE of 21.1 cd/A.[53] Cu(I) complexes are another candidate for harvesting the energy of the singlet and triplet-excited states. The photoluminescent quantum efficiency (PLQY) can be up to 90% for **Cu1**, showing a good potential for its application in the OLED.[54] Besides, the homoleptic and heteroleptic Os(II) complexes consisting of bis(tridentate) carbene ligands can also realize the color-tunable phosphorescence with high emission quantum yields. The Os(II) complexes are isoelectric to their Ir(III) analogues showing a lower oxidation state, and their OLED performances are comparable to the devices based on Ir(III) complexes. OLEDs using **Os1** as the dopant in the emitting layer display the blue emission with a maximum EQE of 19.2% and PE of 17 lm/W.[55] Interestingly, the metal assisted delayed fluorescence (MADF)[56] has been observed on a class of blue-emitting Pd(II) complexes, and efficient electroluminescence from the associated OLED device can be realized via the routine triplet emission as well as the delayed singlet emission.[57] It opens up a new avenue for harvesting all the electrogenerated excitons and diversifies the development of OLED emitters.

Lanthanide complexes are also designed and exploited as the efficient blue emitters. Cerium(III) complexes **Ce1-Ce3** in which the lanthanide ion is encapsulated in the cavity formed by two poly-benzimidazole tripodal ligands have been reported (Figure 14.11). **Ce2** is used as the emitting dopant in the electroluminescent devices and its 5% doped device can furnish a CE of 1.5 cd/A and a PE of 0.52 lm/W with CIE (0.18, 0.21). Herein, the surrounding ligand serves as a light-harvesting antenna, in other words, a sensitizer that promotes the transfer of the absorbed energy to the rare-earth metal ion.[58] Recently, the OLEDs with a peak EQE up to 20.8% and a maximum luminance over 100,000 cd/m^2 have been reported based on a dinuclear Ce(III) complex **Ce4** in which the metal is surrounded by the pyrazolyl-borates. The operational lifetime even exceeds a classic phosphorescent OLED based on FIrpic, indicating that the Ce(III) complexes could be a candidate for efficient and stable blue lighting devices.[59]

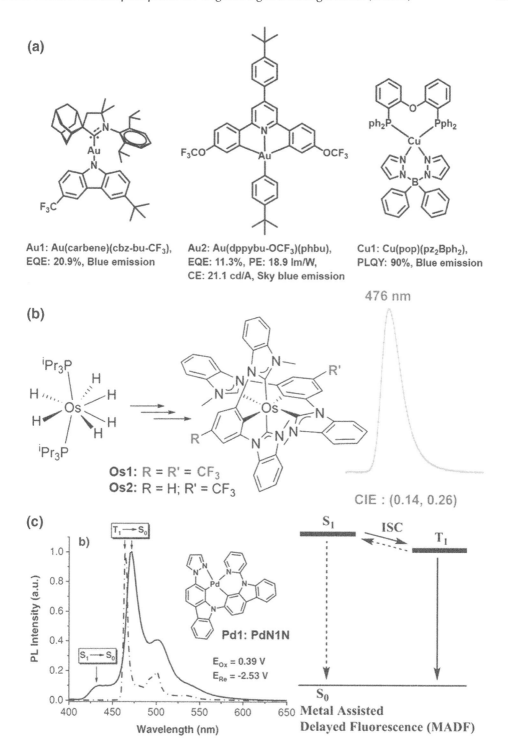

FIGURE 14.10 (a) Selected examples of blue-emitting metallophosphors containing Au(I), Au(III) and Cu(I). (b) The structure and formation of blue-emitting Os(II) complex and its emission in OLED with the CIE coordinates. (c) PL spectra of **Pd1** at room temperature (solid line, dichloromethane) and 77K (dash-dotted line, 2-methyltetrahydrofuran) and illustration of the emission mechanisms for MADF. Reproduced with permission from Refs. [55, 57]. Copyright 2014 American Chemical Society and 2019 WILEY-VCH Verlag GmbH & Co. KGaA, Weinheim.

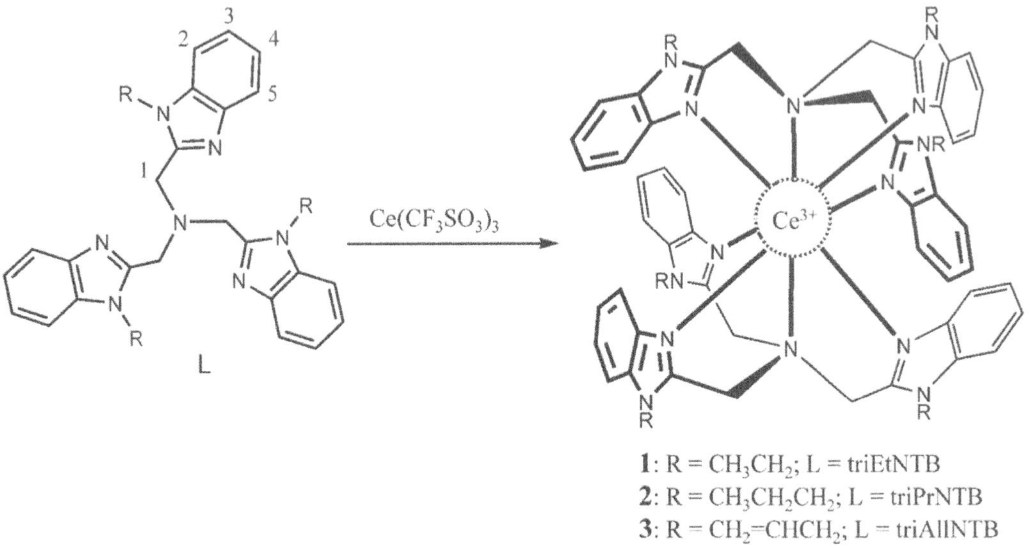

1: R = CH$_3$CH$_2$; L = triEtNTB
2: R = CH$_3$CH$_2$CH$_2$; L = triPrNTB
3: R = CH$_2$=CHCH$_2$; L = triAllNTB

FIGURE 14.11 Structures of the three polybenzimidazole tripodal ligands and the formation of blue-emitting cerium(III) complexes **Ce1-Ce3**. Reproduced with permission from Ref. [58]. Copyright 2007 WILEY-VCH Verlag GmbH & Co. KGaA, Weinheim.

14.3 GREEN-EMITTING METALLOPHOSPHORS FOR ELECTROLUMINESCENT APPLICATION

14.3.1 GREEN-EMITTING IR(III) METALLOPHOSPHORS

The first reported Ir(III) phosphor is Ir(ppy)$_3$ (**Ir15**) with an electroluminescence peak at 510 nm. Its OLED devices show the peak EQE, CE and PE of 8.0%, 28 cd/A and 31 lm/W, respectively.[26] Through the optimization of device structures, Kim and Lee reported that Ir(ppy)$_3$-based OLEDs can reveal the maximum EQE, CE and PE of 30.4%, 93.6 cd/A and 50.0 lm/W by using three bicarbazole derivatives with the bipolar carrier transporting feature as the hosts.[60] The performances of optoelectronic devices are well improved and surpass those of fluorophore-based counterparts. Ir(ppy)$_3$ is termed as a homoleptic organometallic complex. By changing one cyclometalated ligand (ppy) with an auxiliary ligand acetylacetonate (acac), the heteroleptic Ir(ppy)$_2$(acac) (**Ir16**) can be obtained.[61] The initial green OLED based on Ir(ppy)$_2$(acac) gives the EL efficiencies of 12.3%, 38 lm/W and over 50 cd/A.[7] Likewise, the OLED performances are found to be enhanced via the optimized design of devices, exhibiting excellent outputs with the peak EQE and PE of 30.2% and 127.3 lm/W.[62]

Based on the molecular templates of Ir(ppy)$_3$ and Ir(ppy)$_2$(acac), there are two lines of the homoleptic (**Ir17-Ir23**) and heteroleptic (**Ir24-Ir30**) complexes that consist of various group-substituted phenyl-pyridine ligands realizing color-tunable properties plus efficient OLED performances, in which the emissions range from blue-green to green-yellow. Herein, their structures and device indexes are summarized in Figure 14.12.[63, 64] It has been demonstrated that the functionalization of the cyclometalated ligand with main group element units will facilitate carrier injection and charge trapping in the metallophosphor of the emitting layer and further boost the EL efficiencies of OLED devices.

The horizontal orientation of emitting molecules in the OLED is instrumental in promoting the out-coupling efficiency of the device, since the light is emitted mainly in the direction which is vertical to the transition dipole moment (Figure 14.13). Therefore, increasing the horizontal-dipole ratio of metallophosphors in the host material should be taken into consideration for increasing the EL efficiency.[65]

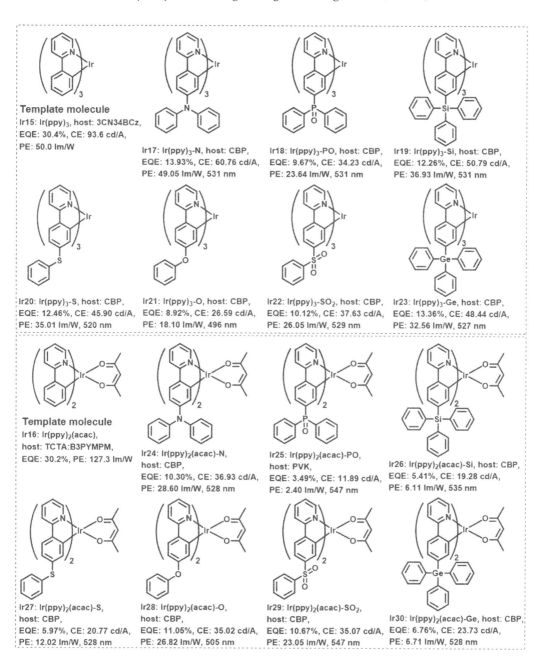

FIGURE 14.12 Upper: structures and OLED performances of green-emitting Ir(III) complex template Ir(ppy)₃ and its derivatized metallophosphors; bottom: structures and OLED performances of green-emitting Ir(III) complex template Ir(ppy)₂(acac) and its derivatized metallophosphors.

On the basis of Ir(ppy)₂(acac) structure, modifications of the coordinated ligands are able to raise the horizontal-dipole ratio. One approach is butylation of the auxiliary ligand. The furnished complex **Ir31** can emit at 524 nm with the PLQY up to 96%. The horizontal-dipole ratio of Ir(III) molecules in the host is also measured as high as 78%. These advantages render an outstanding performance to **Ir31**-based OLED with EQE up to 32.3%.[66] Furthermore, another approach is methylation, propylation or arylation of the main C^N ligands. The resultant **Ir32** in the host

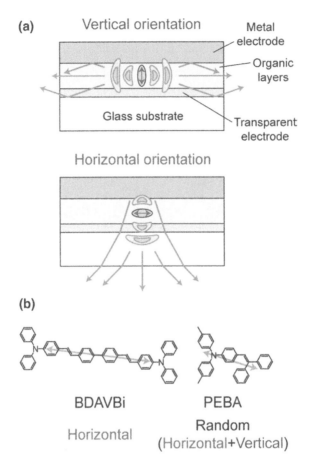

FIGURE 14.13 (a) Schematic diagram of the enhancement of outcoupling efficiency of the OLEDs by horizontal orientation of the transition dipole moment of emitting molecules. (b) Chemical structures used for the comparison of horizontally and randomly oriented emitting molecules. The arrows indicate the direction of the transition dipole moment. Reproduced with permission from Ref. [65]. Copyright The Royal Society of Chemistry 2011.

(TCTA:B3PYMPM) yielded a horizontal-dipole ratio of 83.5%, apart from a nearly unitary PLQY of 98%. Its optimized OLED device affords a peak EQE of 36.0%, which is the top efficiency reported in the green-emitting OLEDs without outcoupling structures.[67] The design and development of new auxiliary and cyclometalated ligands would also reap efficient green-emitting Ir(III) phosphors. Here, we pick up two typical examples **Ir33** and **Ir34** in Figure 14.14. Especially for **Ir34**, the horizontal-dipole ratio of its doped film is recorded as 71% and the OLED device based on this bis-tridentate Ir(III) complex attains an outstanding EQE, PE and CE of 31.4%, 108.7 lm/W and 110.8 cd/A, respectively.[68, 69]

14.3.2 Green-Emitting Pt(II) Metallophosphors

Green-emitting Pt(II) complexes also play a very important part in the application of electrophosphorescent devices. As we have mentioned, the aggregation of planar Pt(II) complexes can cause a red shift of luminescent spectrum. The green emissions of Pt(II) phosphors presented here are normally measured in the dilute solution or doped layer at a low concentration. Likewise, the green OLED devices based on Pt(II) complexes have displayed excellent EL performances, indicating their great potentials for the solid lighting application.

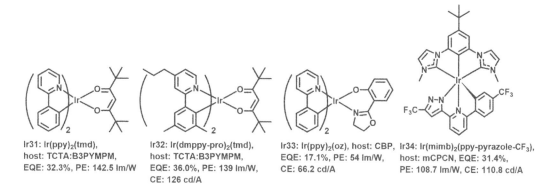

Ir31: Ir(ppy)₂(tmd),
host: TCTA:B3PYMPM,
EQE: 32.3%, PE: 142.5 lm/W

Ir32: Ir(dmppy-pro)₂(tmd),
host: TCTA:B3PYMPM,
EQE: 36.0%, PE: 139 lm/W,
CE: 126 cd/A

Ir33: Ir(ppy)₂(oz), host: CBP,
EQE: 17.1%, PE: 54 lm/W,
CE: 66.2 cd/A

Ir34: Ir(mimb)₂(ppy-pyrazole-CF₃),
host: mCPCN, EQE: 31.4%,
PE: 108.7 lm/W, CE: 110.8 cd/A

FIGURE 14.14   Selected examples of green-emitting Ir(III) phosphors and the performances of their electroluminescent devices.

The synthetic strategy should also focus on the design and functionalization of the chelating ligands. Bidentate ligands including cyclometalated and auxiliary ligands can together coordinate with Pt(II) center to furnish the green-emitting metallophosphors. Specifically, introduction of the electron-withdrawing triarylboron group on ppy ligand can significantly increase the electron transport ability of the complex and further enhance the efficiency of OLED devices. Wang et al. examined the electron transporting ability by fabricating single-carrier (electron-only) devices of **Pt11** and Pt(ppy)(acac) (**Pt12**). The result reveals that device based on the Pt(II) complex containing the triarylboron group can afford a current density 3–4 times larger than that of the device based on the Pt(II) complex without the triarylboron group. As a consequence, the CE and PE can be increased from 14.1 cd/A and 11.7 lm/W for **Pt12** to 34.5 cd/A and 29.8 lm/W for **Pt11** under the same OLED device structure.[70] Besides, if the NHC ligand of N-phenyl-4,5-dimethyl-1,3-thiazol-2-ylidene is used as the cyclometalated ligand and the functionalized acac ligand (dmes) is used as the ancillary ligand, the OLED devices based on **Pt13** can result in an optimum EQE of 12.3%, CE of 37.8 cd/A and PE of 24.0 lm/W.[71] Moreover, the use of (phenyl-oxadiazolyl)phenol as the ancillary ligand can furnish a novel heteroleptic complex **Pt14**. The author claimed that the newly designed ancillary ligand can promote the transport of electrons so that the hole transported from TAPC (hole transporting layer) can be better balanced. Thus, it can suppress triplet-polaron annihilation (TPA) and TTA of the excitons accumulated near the interface of the emitting layer and electron transporting layer. The corresponding device can show a low efficiency roll-off and an improved electroluminescent performance. In detail, the OLED based on **Pt14** gives the peak EQE, CE and PE of 18.0%, 55.6 cd/A and 52.2 lm/W, respectively.[72] One tridentate ligand with a carbazolyl or an alkynyl group can integrate with Pt(II) ion to generate various cyclometalated Pt(II) complexes. For instance, **Pt15** comprises 1,3-bis(N-alkylbenzimidazol-2′-yl)benzene as the tridentate ligand and a carbazole linkage as the ancillary ligand, and its solution-processed OLED shows a maximum EQE, CE and PE of 4.9%, 16.1 cd/A and 2.8 lm/W.[73] **Pt16** with a dendritic carbazole-containing alkynylide as the monodentate ligand leads to a better device performance, i.e., the peak EQE, CE and PE of 10.4%, 37.6 cd/A and 11.4 lm/W.[74] In addition, **Pt17** bearing an asymmetric tridentate ligand in which the adamantane unit can provide an improved solubility and a suppressed aggregation tendency. The OLED devices utilizing it as the phosphorescent dopant exhibit the green emission with a maximum EQE, CE and PE of 5.6%, 15.5 cd/A and 16.4 lm/W.[75] Recently, tetradentate ligands have also drawn an increasing attention due to their rigid structures which are in favor of achieving a higher quantum efficiency. Che et al. developed a family of structurally robust Pt(O^N^C^N) phosphors for high performance OLEDs. The bulky parts at the periphery of the tetradentate ligand would weaken the intermolecular interaction and lessen the formation of excimers, as a result, the impact of TTA and concentration quenching effect can be minimized.

For example, a negligible bathochromic shift of the EL spectrum has been observed in the OLED device of **Pt18** with the spiro linkage. Furthermore, the maximum EQE, CE and PE of 27.6%, 104.2 cd/A and 109.4 lm/W can be achieved in **Pt18**-based OLEDs, and the recorded EQE is the highest reported efficiency among these green-emitting OLEDs based on Pt(II) phosphors so far.[76] **Pt19** with a bridging carbazole unit is measured with the PLQY as high as 0.99, and similarly the bulky substitution on the tetradentate ligand can lead to the suppressed excimer emission and attenuated emission self-quenching. The solution-processed OLEDs utilizing **Pt19** as the emitter attain a peak EQE of 15.55%, CE of 53.63 cd/A and PE of 25.99 lm/W.[77] Figure 14.15 exhibits some typical examples of green-emitting Pt(II) phosphors for OLED applications.

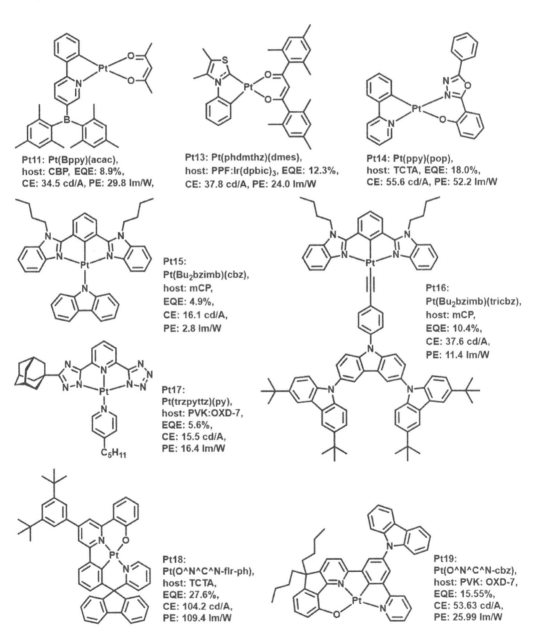

Pt11: Pt(Bppy)(acac), host: CBP, EQE: 8.9%, CE: 34.5 cd/A, PE: 29.8 lm/W,

Pt13: Pt(phdmthz)(dmes), host: PPF:Ir(dpbic)₃, EQE: 12.3%, CE: 37.8 cd/A, PE: 24.0 lm/W

Pt14: Pt(ppy)(pop), host: TCTA, EQE: 18.0%, CE: 55.6 cd/A, PE: 52.2 lm/W

Pt15: Pt(Bu₂bzimb)(cbz), host: mCP, EQE: 4.9%, CE: 16.1 cd/A, PE: 2.8 lm/W

Pt16: Pt(Bu₂bzimb)(tricbz), host: mCP, EQE: 10.4%, CE: 37.6 cd/A, PE: 11.4 lm/W

Pt17: Pt(trzpyttz)(py), host: PVK:OXD-7, EQE: 5.6%, CE: 15.5 cd/A, PE: 16.4 lm/W

Pt18: Pt(O^N^C^N-flr-ph), host: TCTA, EQE: 27.6%, CE: 104.2 cd/A, PE: 109.4 lm/W

Pt19: Pt(O^N^C^N-cbz), host: PVK: OXD-7, EQE: 15.55%, CE: 53.63 cd/A, PE: 25.99 lm/W

FIGURE 14.15 Selected examples of green-emitting Pt(II) phosphors and the performances of their electroluminescent devices.

14.3.3   GREEN-EMITTING METALLOPHOSPHORS CONTAINING OTHER METALS

There are plenty of green-emitting metallophosphors containing other metals. Group 11 metals including Cu(I), Ag(I) and Au(I) can coordinate with the functionalized ligands such as the pyridyl-pyrrolide ligand together with an ancillary phosphine ligand to afford phosphorescent d^{10} metal complexes. It is found that the degree of involvement of a metal d-orbital in the electronic transition is more important than the atomic number of three heavy metals in prompting the spin-orbit coupling. Even though Ag(I) and Au(I) have a larger atomic number in this family, Cu(I) complexes possess a larger metal d-orbital contribution in the lowest lying transition so that they can show the larger rate of ISC and radiative decay rate constant of phosphorescence than those of the Ag(I) and Au(I) complexes. As a result, **Cu2** exhibits an intensive phosphorescence with a relatively high PLQY in both solution and solid state. Moreover, the OLEDs fabricated with **Cu2** as the emitter offer a peak EQE, CE and PE of 6.6%, 20.0 cd/A and 14.9 lm/W (Figure 14.16).[78]

The metal valence state of the previous complexes is +1. The class of carbene-metal-amide also consists of group 11 metal with +1 valence state, utilizing a carbene ligand as the π-acceptor and an amide group as the electron donor. It enables a near-zero singlet-triplet energy gap, so that the energy of triplet-excited states can be harvested. This class of Cu(I), Ag(I) and Au(I) complexes can realize efficient luminescence for OLED application, and their relevant devices can give maximum EQEs of 5.6% (**Cu5**), 13.7% (**Ag3**) and 27.5% (**Au6**), respectively.[13, 14, 79] Besides, a series of three-coordinate Cu(I) complexes can show strong phosphorescence in both solutions and doped films. OLEDs utilizing **Cu6** as the emitter presented a bright green emission with a maximum EQE of 21.3% and CE of 65.3 cd/A.[80] Not only Au(I), but also Au(III) could combine with chelating ligands to afford the gold complexes. When the dopant concentration is optimized at 4% in the CBP host, the device based on **Au7** can attain an optimum EQE of 11.5%, together with a CE of 37.4 cd/A

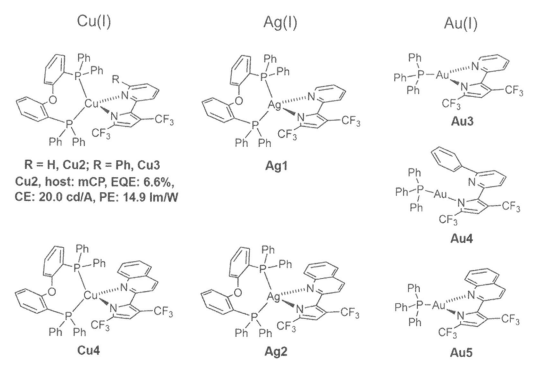

FIGURE 14.16   Structures of group 11 Cu(I), Ag(I), Au(I) complexes and the performance of electroluminescent device based on **Cu2**. Reproduced with permission from Ref. [78]. Copyright 2011 American Chemical Society.

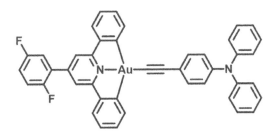

Cu5: Cu(carbene)(amide-cbz), host: PVK, EQE: 5.6%

Ag3: Ag(carbene)(amide-bu), host: mCP, EQE: 13.7%

Au6: Au(carbene)(amide-bu), host: PVK, EQE: 27.5%, CE: 87.1 cd/A, PE: 75.1 lm/W

Cu6: Cu(dtpb)Br, host: mCP, EQE: 21.3%, CE: 65.3 cd/A

Au7: Au(dppy-dfph)(DPA-alkynyl), host: CBP, EQE: 11.5%, CE: 37.4 cd/A, PE: 26.2 lm/W

FIGURE 14.17 Selected examples of green-emitting metallophosphors containing group 11 metals and the performances of their electroluminescent devices.

and a PE of 26.2 lm/W.[81] Figure 14.17 presents some typical examples of green-emitting metallophosphors containing group 11 metals for OLED applications.

Re(I) phosphors are known to be highly emissive; however, the investigations on Re(I) complexes are much less than the other transition metal complexes due to their inferior processability for OLED devices. Nonetheless, **Re1** with a dinuclear structure has been reported in which the emission stems from the ^{3}MLCT. Its processable device using the vacuum-sublimed method affords an emission peak at 514 nm with the maximum CE of 11.0 cd/A and PE of 6.3 lm/W.[82] Mononuclear metallophosphor **Re2** can also realize green phosphorescence; however, its EL device performance could not compete with those of the Ir(III) and Pt(II) ones.[83] Mn(II) complexes constitute another class of materials that are able to harvest the energy of triplet-excited states. Chen et al. reported green phosphorescent solution-processed OLEDs based on the tetrabromide complex **Mn1**. By using the blended host of TCTA and 26DCZPPY, the doped devices offer a peak EQE of 9.6%, CE of 32.0 cd/A and PE of 16.2 lm/W.[84] Huang et al. designed and developed two tetrahedral Mn(II) complexes **Mn2** and **Mn3**. By employing **Mn3** as the emitter and TCTA as the host, the EL efficiencies can be achieved at a high level, i.e., a peak EQE, CE and PE of 9.97%, 32.19 cd/A and 28.09 lm/W, respectively.[85] Selected green-emitting metallophosphors containing Re(I) and Mn(II) cores for OLED applications are shown in Figure 14.18, and the green-emitting metallophosphors are not restricted to contain the abovementioned metals. The diversity and abundance of high-performance materials in this field have demonstrated prospective applications not only in optoelectronic devices, but also in the area of biotechnology, communication and so on.

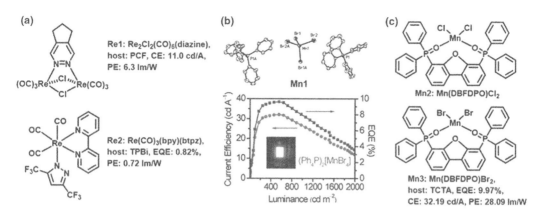

FIGURE 14.18 (a) Green-emitting Re(I) complexes **Re1** and **Re2** and the performances of their electroluminescent devices. (b) Tetrabromide Mn(II) complex **Mn1** and the EQE, CE versus luminance of its electroluminescent device. (c) Tetrahedral Mn(II) complexes **Mn2** and **Mn3** and the performance of the electroluminescent device based on **Mn3**. Reproduced with permission from Ref. [84]. Copyright 2016 WILEY-VCH Verlag GmbH & Co. KGaA, Weinheim.

14.4 YELLOW/ORANGE-EMITTING METALLOPHOSPHORS FOR ELECTROLUMINESCENT APPLICATION

14.4.1 YELLOW/ORANGE-EMITTING IR(III) METALLOPHOSPHORS

The yellow/orange-emitting metallophosphors exert an important role in realizing the full-color display in OLEDs, and they together with the blue-emitting metallophosphors can complement each other to afford the white light emission. Hence, the development of efficient yellow/orange-emitting metallophosphors attracts an increasing attention.

Ir(III) metallophosphors constitute the class of most plentiful species in this family. The emission colors of Ir(III) phosphors are mainly modulated by the energy levels of the cyclometalated ligands, which are usually composed of the aromatic heterocyclic rings containing the elements of H, C, N, S, O and so on (Figure 14.19). For example, the Ir(III) complexes with cyclometalated ligands consisting of C,N-heterocyclic rings can provide the intensely emissive yellow/orange light with excellent OLED performances. Specifically, expanding the π conjugation of 2-phenylpyridine (ppy) ligand can shift the emission of Ir(III) phosphors from green to yellow/orange. When a phenyl ring is fused into the pyridyl moiety of ppy ligand to furnish the two quinolyl C^N ligands, 2-phenylquinoline (pq) and 3-phenylisoquinoline (3-piq), the resultant **Ir35** and **Ir36** emit an orange light at 597 nm in 2-methyltetrahydrofuran solution and a yellow light at 562 nm in dichloromethane solution, respectively. By using them as the emitters, the related OLED devices afford high EQEs of 19.2 and 7.17%, respectively. Notably, **Ir35**-based OLED attains a PE of 16.4 lm/W at a high luminance of 1000 cd/m², and at this luminance, the EQE remains at a high level of 17.1%.[86, 87] If one more phenyl ring is incorporated into the phenyl moiety of ppy ligand to give the cyclometalated ligands consisting of a naphthalene group, which are 2-(2-naphthyl)pyridine (npy) and 2-(1-naphthyl)pyridine (Napy), the resultant **Ir37**- and **Ir38**-based devices can produce a yellow electroluminescence at 551 nm and an orange electroluminescence at 595 nm, respectively. The fabricated OLEDs give the EQEs of 10.5% for **Ir37** and 1.3% for **Ir38**, and the PE and CE of **Ir37** are also remarkable, reaching the maximum efficiency of 21.78 lm/W and 33.64 cd/m², respectively.[88, 89] The cyclometalated C^N ligands containing N,N-heterocyclic rings constitute another class of ligands that can coordinate with iridium center to form the yellow/orange-emitting Ir(III) phosphors with outstanding OLED performances. The phenyl-pyrazine, phenyl-pyrimidine and phenyl-pyridazine derivatives comprise the bulk of the N,N-heterocyclic ligands. For example, 2-methyl-3-phenylpyrazine

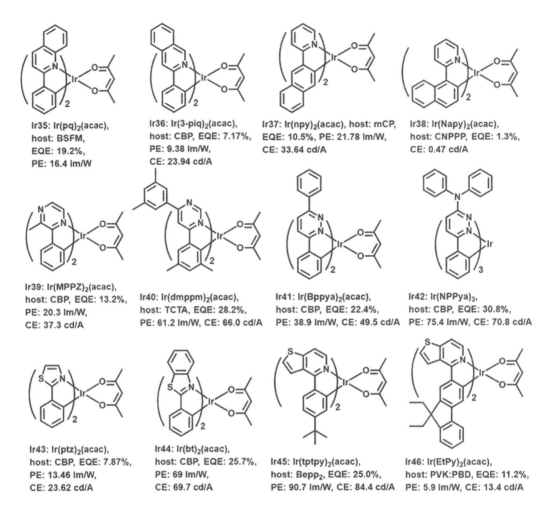

FIGURE 14.19 Selected examples of yellow/orange-emitting Ir(III) phosphors and the performances of their electroluminescent devices.

(MPPZ) together with an acac ancillary ligand can be utilized to combine with iridium to furnish **Ir39**, and the maximum EQE, PE and CE of 13.2%, 20.3 lm/W and 37.3 cd/A with an EL peak at 580 nm are achieved on its fabricated OLEDs.[90] Furthermore, the orange-emitting **Ir40** containing two 4,6-bis(3,5-dimethylphenyl)pyrimidine ligands has a high PLQY of 0.92 and a high horizontal dipole ratio of 0.78 in the doped TCTA film (8%, 40 nm). The monochromic OLED device based on **Ir40** shows excellent performances, i.e., a peak EQE of 28.2%, PE of 61.2 lm/W and CE of 66.0 cd/A, which is one of the best orange phosphorescent devices ever reported. In addition, with the introduction of FIrpic as the efficient blue phosphors, the binary white OLEDs have been realized with EQEs up to 28.6%, which are among the highest efficiency values for the binary white OLEDs.[91] Phenyl-pyridazine derivatives as the cyclometalated ligands have also been employed. **Ir41** bearing 3,6-bis(phenyl)-pyridazine exhibits an intense orange photoluminescence in CH_2Cl_2 solution with the emission peak at 582 nm, and its OLED device can afford a peak EQE of 22.4%, PE of 38.9 lm/W and CE of 49.5 cd/A.[92] By further substituting one phenyl group with a diphenylamino group, the PL peak of the homoleptic **Ir42** is blue-shifted to 558 nm in the CH_2Cl_2 solution, and the performance of its OLED came to the top level of the yellow phosphorescent OLEDs, giving a peak EQE of 30.8%, PE of 75.4 lm/W and CE of 70.8 cd/A.[93] Besides, not only N,N-heterocyclic rings, but also N,S-heterocyclic rings can be integrated with the iridium core to provide the efficient yellow/orange-emitting Ir(III) metallophosphors. **Ir43** bearing 2-phenyl-thiazole as the cyclometalated ligands can produce a yellow electroluminescence with the peak EQE of 7.87%. [94] Furthermore, the incorporation of a phenyl ring on the thiazole moiety is able to significantly improve the OLED performance. The EL efficiencies of the device based on **Ir44** reached the maximum EQE of 25.7% along with PE of 69 lm/W and CE of 69.7 cd/A.[95] 4-(4-*Tert*-butyl-phenyl) thieno[3,2-c]pyridine could interact with iridium center to afford the efficient yellow-emitting **Ir45**, and its fabricated OLEDs furnish a quite high EQE of 25.0%, CE of 84.4 cd/A and PE of 90.7 lm/W. Furthermore, by using the doped emitting layer of FIrpic as the blue luminescent source, a white tandem OLED device is produced, and it could furnish an extremely high EQE of 48.6%, CE of 121.8 cd/A and PE of 67.2 lm/W, which is the state-of-the-art performance.[96] By changing the phenyl moiety with a fluorene derivative, the emission wavelength of the as-prepared **Ir46** exhibits a bathochromic shift. Its solution-processed OLEDs show a maximum EQE, PE and CE of 11.2%, 5.9 lm/W and 13.4 cd/A, respectively.[97] Indeed, there are plenty of Ir(III) complexes comprising diverse molecular structures that have been reported to realize prominent OLED performances on the yellow/orange and even white emissions,[98–102] revealing the great application potentials of efficient Ir(III) phosphors for the full-color lighting and display.

14.4.2 Yellow/Orange-Emitting Pt(II) Metallophosphors

Similar to the green-emitting Pt(II) complexes, we divide the yellow/orange-emitting Pt(II) phosphors into three kinds according to the denticity of ligands, in which the Pt(II) complexes containing bidentate ligands have the most plentiful structures. Introduction of the boron-substituted group into the phenyl moiety rather than pyridine moiety would shift the luminescence of **Pt20** in solution to a longer wavelength of 542 nm compared to that of the green-emitting **Pt11**, and the **Pt20**-based OLEDs hosted by CBP can give the maximum EQE, CE and PE of 9.52%, 30 cd/A and 8.36 lm/W.[103] Replacing the triphenylboron group with the carbazole unit can result in the orange-emitting **Pt21**. Its fabricated OLEDs attain the best performance under 4 wt% doping concentration with the peak EQE, CE and PE of 13.1%, 35.82 cd/A and 25 lm/W.[104] Later, Chou et al. synthesized a series of Pt(II) complexes with the lepidine-based chelating ligands. Among their fabricated OLEDs hosted by CBP, **Pt22**-based OLED shows the most satisfactory EQE, CE and PE of 15.21%, 29.83 cd/A and 11.79 lm/W.[105] On the basis of the green-emitting **Pt11**, incorporation of the arylamine group can dramatically shift the emission of the obtained **Pt23** to the orange-emitting range, and the corresponding OLED affords a good electroluminescent performance with an optimum EQE, CE and PE of 10.6%, 33.2 cd/A and 34.8 lm/W.[106] Adachi et al. reported a linear-shaped Pt(II) phosphor **Pt24** based on Pt(ppy)(acac) (**Pt12**), which is favorable to form a higher horizontal orientation in the emitting layer (EML) of OLED. The **Pt24** with a [(*tert*-butylphenyl)-phenyl]phenylpyridine ligand can afford the yellow light at 550 nm with a PLQY of 0.5 in THF solution, and the maximum EQE of 15.8% is achieved for **Pt24**-based OLEDs. Correspondingly, the light outcoupling efficiency of the device is estimated to be as high as 32%.[107] Besides the abovementioned Pt(II) acetylacetonate complexes, the homoleptic or heteroleptic Pt(II) complexes containing the chelating ligands of two pyridyl-pyrazolates or one neutral imidazolylidene-pyridylidene and one dianionic bipyrazolate, are also able to get impressive EL performances. For example, the homoleptic **Pt25**-based OLEDs show an outstanding EQE, CE and PE of 20.0%, 47.6 cd/A and 50.8 lm/W.[108] The homoleptic **Pt26** presents an emission peak at around 480 nm when it is doped into CBP with the concentration of 5%. When the doping concentration exceeds 25%, it starts to exhibit an absolute emission originated from the excimers and the wavelength peak is red-shifted to around 560 nm. If the doping concentration reaches 17%, a white photoluminescence arising from both the monomer and excimer can be observed.[109] As the square planar conformation of Pt(II) complexes without bulky substituents would induce the stacking Pt–Pt interactions, it can result in the strong intermolecular charge transfer transitions including the metal-metal-to-ligand charge transfer (MMLCT) and lead to the red shift of emission. Therefore, even **Pt26** emits a blue-green light in the doped film at a low concentration, its luminescent color moves to the orange range in the non-doped OLEDs. After optimization, a doping-free device based on **Pt26** demonstrates an exceptionally high EQE of 20.3% and PE of 63.0 lm/W, which is the most efficient Pt(II) emitter-based orange OLED reported to our knowledge.[110] In addition, under a dopant concentration of 8%, the OLED based on heteroleptic **Pt27** with a neutral imidazolylidene-pyridylidene ligand and a dianionic bipyrazolate ligand hosted by 26DCzppy shows an emission peak at 558 nm with a peak EQE of 12.5%, while its non-doped device exhibits an emission peak at 598 nm with a peak EQE of 11.0%.[111] Tridentate ligands together with an alkynyl or a halogen linkage can integrate with Pt(II) ion to furnish complexes such as **Pt28**, **Pt29**, **Pt30** and **Pt31**. However, the EL performances of this kind of yellow/orange-emitting Pt(II) complexes are mediocre,[112–115] probably because the weak-field ligands like bipyridine (bpy) or/and Cl^- leads to increased non-emissive metal-centered (MC) d–d*-excited states.[17] Tetradentate ligands are considered to be rigid so that the associated Pt(II) complexes can normally possess a high PLQY. **Pt32**-based OLED device shows a maximum EQE, CE and PE of 11%, 31 cd/A and 14 lm/W, respectively.[116] The doped OLED containing 8 wt % **Pt33** hosted by TCTA affords a peak EQE, CE and PE of 8.3%, 23 cd/A and 17 lm/W with the emission peak of 564 nm.[117] The cyclometalated complex **Pt34** can harvest a high PLQY up to 0.63 in CH_2Cl_2 solution. By employing it as the emitter hosted by CBP, the corresponding OLED device leads to a maximum EQE of 18.2% with the $CIE_{x,y}$ (0.55, 0.45).[118] Che et al. developed a series of Pt(II) complexes Pt(O^N^C^N) embodying various tetradentate ligands, and the rigidity of ligands render all Pt(II) complexes to possess high quantum efficiencies. The PLQY of the typical complex **Pt35** is up to 0.86 and it displays a yellow emission peak of 553 nm in CH_2Cl_2 solution, and its associated OLED device affords an EL peak of 568 nm at 6% doping concentration with an excellent CE and PE of 74.9 cd/A and 52.1 lm/W.[119] Generally speaking, the performances of yellow/orange-emitting OLEDs based on Pt(II) phosphors are somewhat lagging behind than those of the devices based on Ir(III) phosphors. In spite of this, the features such as the host-free device, the planar-structure inducing the aggregate emission and high PLQYs of Pt(II) complexes composed of the rigid tetradentate ligands are the distinctive advantages on fabricating yellow/orange-emitting OLEDs with the high performance (Figure 14.20).

14.4.3 Yellow/Orange-Emitting Metallophosphors Containing Other Metals

Yellow/orange-emitting metallophosphors containing other metals are not as common as Ir(III) and Pt(III) phosphors. In the latest report, Li et al. incorporated palladium(II), the congener of Pt(II) in the same group, into a rigid molecular architecture to synthesize the tetradentate Pd(II) complex **Pd2**. Its solution shows a blue-green emission originated from the Pd(II) monomer while its aggregate state exhibits a red-shifted emission from the M–M dimer or oligomer. The reason is that as the doping concentration of **Pd2** increases, the Pd–Pd antibonding ($d\sigma^*$) orbital will be raised beyond the ligand π orbitals and the HOMO of its dimer will be predominantly localized on the dual metal centers, which leads to a luminescence dominated by ^{1}MMLCT and ^{3}MMLCT transitions (Figure 14.21). Remarkably, the non-doped OLED based on **Pd2** exhibits a yellow-orange emission with the peak at 588 nm, and an extraordinarily high EQE of 34.8% can be achieved without any outcoupling enhancement. Furthermore, the estimated operational half-lifetime reaches the remarkable 9.59 million hours at 1,000 cd/m², demonstrating the great potential for commercial use in display and lighting.[120] The white OLEDs using single Pd(II) dopant **Pd3** are also realized, affording a peak EQE of 24.2% and PE of 67.9 lm/W.[121]

Being the transition metal elements in the same period of the Periodic Table of Ir and Pt, Os and Au also attract an increasing attention from scientists. Chi et al. reported a class of d^6 Os(II) complexes bearing *cis*-1,2-bis(diphenylphosphino)ethene (dppee) and pyridyl-pyrazole/triazolate(ppz/ptz) in which the isomers show different optoelectronic properties. Taking **Os3** with the *trans*-configuration as an example, it shows an orange emission of 572 nm and a very high quantum efficiency of 0.9. By comparison, **Os4** with the *cis*-configuration emits a yellow light at 547 nm with the low

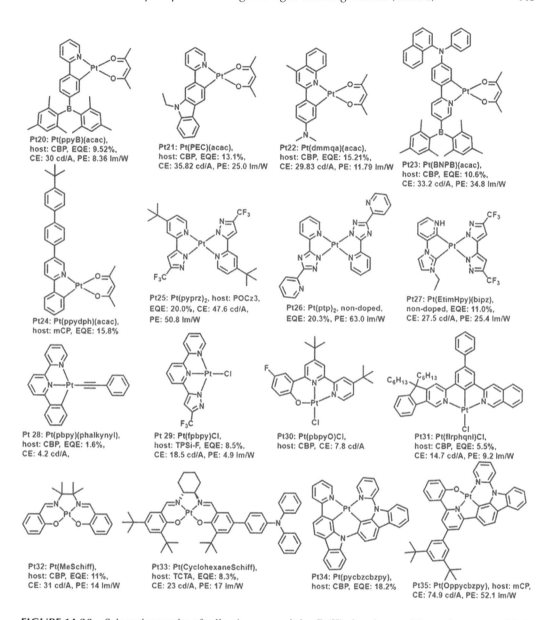

FIGURE 14.20 Selected examples of yellow/orange-emitting Pt(II) phosphors and the performances of their electroluminescent devices.

quantum efficiency of 0.01. The phosphorescent OLEDs based on the *trans*-configured complex **Os3** give a peak EQE, CE and PE of 13.3%, 48.9 cd/A and 16.8 lm/W. Another homologous complex **Os5**-based OLED also furnishes a high EQE of 11.7%, CE of 40.4 cd/A and PE of 10.5 lm/W.[122] Because of the good EL efficiencies of this class of Os(II) phosphors to be utilized as the yellow/orange emitters, it can combine with blue Ir(III) emitters to achieve the pure-white OLEDs with high color rendering. The wide-bandwidth yellow-orange-emitting **Os6** and the blue-emitting Ir(fbppz)$_2$(dfbppy) (**Ir47**) can be employed to fabricate two-component white OLEDs. Outstanding device efficiencies (EQE of 9.6%, CE of 22.5 cd/A and PE of 19.6 lm/W) are achieved with CIE coordinates close to the ideal white light (0.33, 0.33) and a high CRI value of about 80.[123] On the other hand, some researchers have designed and developed the efficient d^{10} Au(I) as well as d^8

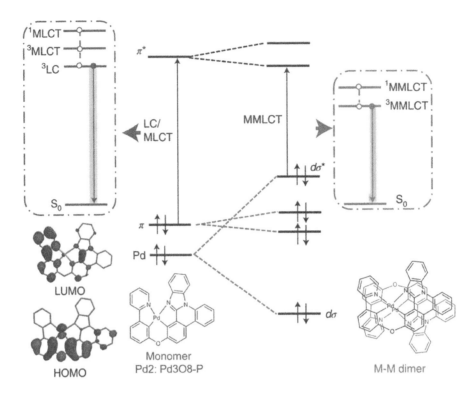

FIGURE 14.21 Schematic diagrams of molecular orbital and energy level for Pd(II) complex **Pd2**, featuring the monomer and dimer emissions. Reproduced with permission from Ref. [120]. Copyright 2020 Springer Nature Limited.

Au(III) phosphors. For example, the carbene-metal-amide as a promising emitter can incorporate Au(I) into the efficient two-coordinate complex **Au8**, and its solution-processed OLED exhibits the excellent EQE, CE and PE of 17.9%, 45.2 cd/A and 33.6 lm/W.[13] Au(III) integrating with a tridentate ligand and a monodentate ligand can also furnish the efficient two-coordinate complex. By enlarging the conjugation of alkynylgold(III) phosphors with bulky substituents or adding the dopant concentration of the alkynylgold(III) phosphors with the planar configuration, the bathochromic shift of the luminescence from the short wavelength range (blue/green) to the longer wavelength range (yellow/orange) can be realized. **Au9** merged with the conjugated fluorene derivatives can display a yellow luminescence. At a dopant concentration of 10 wt%, its corresponding OLED device affords a rational EQE of 9.18%, CE of 30 cd/A, PE of 22.64 lm/W.[16] **Au10** displays a concentration-dependent emission. Using it as the emitter, the OLEDs under the dopant concentration of 10 and 50 wt% can attain the peak EQEs of 7.8 and 6.1%, CEs of 24.0 and 15.6 cd/A, PEs of 14.5 and 13.8 lm/W with the emission maxima at 548 and 580 nm, respectively.[124] Except for these Pd(II), Os(II), Au(I) and Au(III) complexes (Figure 14.22), the research on developing the efficient yellow/orange-emitting phosphors holding other metals for OLED applications is still in its infancy.

14.5 RED-EMITTING METALLOPHOSPHORS FOR ELECTROLUMINESCENT APPLICATION

14.5.1 Red-Emitting Ir(III) Metallophosphors

If Ir(ppy)$_3$ (**Ir15**) or Ir(ppy)$_2$(acac) (**Ir16**) is chosen as the molecular template, in order to shift the emitting color of complexes to the red or even deep red region, normally the employed methods

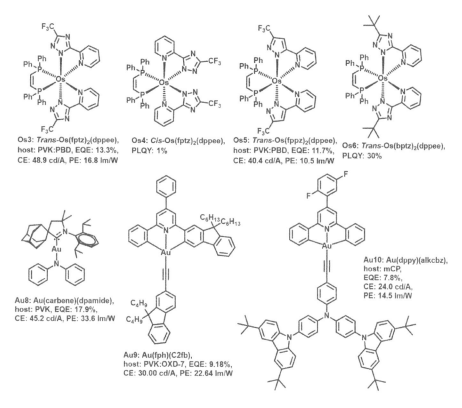

Os3: *Trans*-Os(fptz)₂(dppee),
host: PVK:PBD, EQE: 13.3%,
CE: 48.9 cd/A, PE: 16.8 lm/W

Os4: *Cis*-Os(fptz)₂(dppee),
PLQY: 1%

Os5: *Trans*-Os(fppz)₂(dppee),
host: PVK:PBD, EQE: 11.7%,
CE: 40.4 cd/A, PE: 10.5 lm/W

Os6: *Trans*-Os(bptz)₂(dppee),
PLQY: 30%

Au8: Au(carbene)(dpamide),
host: PVK, EQE: 17.9%,
CE: 45.2 cd/A, PE: 33.6 lm/W

Au9: Au(fph)(C2fb),
host: PVK:OXD-7, EQE: 9.18%,
CE: 30.00 cd/A, PE: 22.64 lm/W

Au10: Au(dppy)(alkcbz),
host: mCP,
EQE: 7.8%,
CE: 24.0 cd/A,
PE: 14.5 lm/W

FIGURE 14.22   Selected examples of yellow/orange-emitting metallophosphors containing other metals and the performances of their electroluminescent devices.

include merging more conjugated groups (like aryl groups) into the phenyl and pyridyl moieties, adding the electron-donating groups (like alkyl groups) on the phenyl moiety or introducing electron-withdrawing groups on the pyridyl moiety (like cyano group). Numerous red Ir(III) emitters with brilliant performance on the OLED devices have been developed and reported, and some typical examples are presented here.

In the early stage, the use of more conjugated benzo[*b*]thienyl instead of phenyl as a part of the cyclometalated ligand can shift the emission of the complex to the red region. **Ir48** can produce the red luminescence at 612 nm in 2-methyltetrahydrofuran solution, while its OLED device hosted by CBP can give the electroluminescence at 617 nm with a mediocre EQE of 6.6% and PE of 2.2 lm/W.[7] 1-(Phenyl)isoquinoline which contains one more benzene ring than 1-(phenyl)pyridine can combine with iridium to afford **Ir49**. The emission peak shows a red shift of about 100 nm compared to that of Ir(ppy)₂(acac) (**Ir16**). By using the device architecture of ITO/HATCN (20 nm)/TAPC (40 nm)/mCP (5 nm)/Ir(piq)₂(acac) in CBP (15 wt%, 30 nm)/TPBi (40 nm)/LiF (1 nm)/Al (100 nm), the **Ir49**-based OLED attains a peak EQE of 16.13% with the emission peak at 626 nm.[125] The homoleptic complex **Ir50** is also used as an emissive dopant in the OLED, and its fabricated device produces a pure red emission with an EQE and PE of 10.3% and 8.0 lm/W.[126] Moreover, researchers have introduced more functional groups into **Ir49** to expand the optoelectronic applications. For example, three methyl groups are introduced into different positions of 1-(phenyl)isoquinoline to afford **Ir51**, and its monochromatic OLED exhibits a deep red emission with a CE and PE of 11 cd/A and 10 lm/W at the luminance of 100 cd/m². Then to demonstrate the importance of the deep red luminescence on the color rendering index (CRI) of the white OLED, the authors designed and fabricated a four-color white OLED using **Ir51** as the source of deep red light. The device exhibits an ultrahigh CRI of 96, and the average of all CRIs at different Munsell hues can be achieved up to 95 at 5 V.[127] Therefore, as one of the three primary-colored lights, the red light plays an irreplaceable role in the high performance of the full-color display and white-light illumination. The electron-rich groups such as the carbazole and triphenylamine derivatives can substitute for the phenyl moiety of 1-(phenyl)isoquinoline ligand for generating more efficient red Ir(III) emitters, because the hole injection and/or transport property of the corresponding metallophosphors can be well improved. The heteroleptic **Ir52** and homoleptic **Ir53** phosphors bearing electron-richer linkages attain superior OLED performances with the maximum EQEs, CEs and PEs of 11.76 and 11.65%, 10.15 and 5.82 cd/A, 5.25 and 3.65 lm/W, respectively.[128, 129] In contrast, under the similar device configuration, the OLED with **Ir49** as the red emitter shows inferior EL efficiencies. [130] When the isoquinoline ring of 1-(phenyl)isoquinoline C^N ligand is functionalized with the strong electron-withdrawing cyano group, **Ir54** can present a significantly stabilized LUMO energy level. Consequently, a significant bathochromic shift of the emission beyond 60 nm can be observed for **Ir54** compared to **Ir49**. The OLEDs based on **Ir54** attain a maximum EQE of 10.62% with an emission peak at 690 nm, which belongs to the deep red region.[125] As we have mentioned in the previous sections, the orientation of transition dipole moments has a great impact on the outcoupling efficiency of OLEDs, and the horizontal orientation of the transition dipole moment of an emitter (parallel to the substrate) rather than the isotropic orientation can enhance the outcoupling efficiency. Kim et al. further demonstrated both the preferred direction of the triplet transition dipole moments of Ir(III) phosphors and the interactive arrangement of molecules in the EML determine the orientation of Ir(III) emitters in OLEDs. They chose NPB:B3PYMPM (molar ratio of 1:1) as the co-host and measured the horizontal dipole ratios of **Ir55** and **Ir59** to be up to 0.8 and 0.82, respectively. The optimized OLED based on **Ir55** affords a very high EQE of 27.1%. Remarkably, the optimized OLED based on **Ir59** furnishes an unprecedentedly high EQE of 35.6% and PE of 66.2 lm/W, the highest reported efficiencies of red-emitting OLEDs using Ir(III) phosphors as the dopants to date.[131] In addition to these outstanding performances of monochromatic OLEDs, white OLED devices using **Ir55** as the red emitter with the PE approaching 100 lm/W have been also reported, which can rival the efficiency of a fluorescent tube.[132] There are also various efficient red Ir(III) phosphors designed from **Ir35**. Effects on the methylation of 2-phenylquinoline and change of the ancillary ligand have been investigated. It shows that the introduction of the electron-donating methyl group to the phenyl moiety would exert a bathochromic effect while the addition of methyl group on the quinoline moiety can result in a hypsochromic shift, and the change of ancillary ligand from acac to tmd leads to an improvement of the device performance. OLEDs based on the homologous derivatives **Ir56**, **Ir57**, **Ir58** and **Ir59** all result in superior EL efficiencies, beyond that of their parent complex **Ir35**.[133] The metallated thienyl ring is considered to have a larger electron donation to the d orbitals of Ir(III) center compared to the phenyl ring; thus, the ^{3}MLCT energy of Ir(III) complex can be reduced. Homoleptic **Ir60** containing the thienyl unit displays a red-shifted emission compared to its phenyl counterpart Ir(dpq)$_3$ (**Ir61**). The high EQE of 21% and CE of 26 cd/A can be achieved on the OLED based on **Ir60** by using a mixed host system comprising TCTA and TPBi, which can provide a better balance of holes and electrons, thereby providing an efficient energy transfer from the co-host materials to the Ir(III) emitters.[134, 135] Furthermore, the devices based on **Ir62** bearing the thienyl group with the use of a newly designed host BIQS can afford a decent performance, i.e., a maximum EQE, CE and PE of 25.9%, 37.3 cd/A and 32.9 lm/W, respectively.[136] Besides, increasing the ligand rigidity and adopting the multidentate coordination mode of Ir(III) complexes are beneficial to achieve a superior luminescent efficiency. OLEDs based on bis-tridentate **Ir63** give a peak EQE, CE and PE of 27.4%, 36.9 cd/A and 36.2 lm/W, respectively.[69] Not only mononuclear Ir(III) complexes, but also dinuclear Ir(III) complexes can realize outstanding OLED performances. Since the process of MLCT is closely associated with the cyclometalated ligand and metal center, introducing one more Ir(III) center will usually strengthen the MLCT and lead to a narrower energy gap between HOMO and LUMO energy levels. The dinuclear **Ir64** can show a red shift of emission by over 70 nm compared to the corresponding mononuclear Ir(ppmd)$_2$(acac) (**Ir65**) in both the PL and EL spectra. Besides, the achieved efficiencies (EQE,

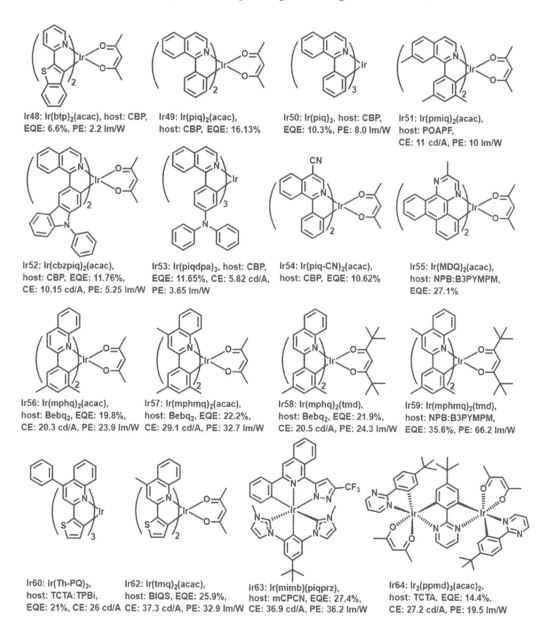

FIGURE 14.23 Selected examples of red-emitting Ir(III) phosphors and the performances of their electroluminescent devices.

CE and PE of 14.4%, 27.2 cd/A and 19.5 lm/W) of the OLED based on the racemic dinuclear **Ir64** are better than those of the OLED based on the mononuclear Ir(III) counterpart.[137] No matter whether the Ir(III) complexes are homoleptic or heteroleptic, bidentate or tridentate, even mononuclear or dinuclear, the development on the red-emitting Ir(III) phosphors has already received enormous and adequate progresses (Figure 14.23).

14.5.2 Red-Emitting Pt(II) Metallophosphors

The first reported Pt(II) phosphor is Pt(OEP) (**Pt36**) by Forrest et al. in 1998, which is also the year when the research on OLEDs using transition metal-based phosphors as the emitters started.[8]

Since then, there is a rapid development of this field. The EQE of OLEDs increased from the primary 4% to the highest 38.8% for red-emitting Pt(II) phosphors. During this process, many theories and techniques have been put forward, employed and developed. For example, to improve the energy transfer in the electrophosphorescent devices, the method of choosing a new host material and adding a new layer to block the exciton diffusion is used, and the EQE of OLEDs based on **Pt36** can be enhanced to 5.6% in which no obvious emissions of other materials are observed in the device.[138] The combination of host and dopant has been demonstrated to be important for ensuring not only the efficiency of devices, but also the stability of emitters. Fukagawa et al. reported highly efficient and stable red OLEDs based on two tetradentate Pt(II) complexes in combination with Bebq$_2$ as the host, and the efficient energy transfer from Bebq$_2$ to the Pt(II) phosphors renders the optimized OLEDs based on **Pt37** a peak EQE of 19.3% and PE of 30.3 lm/W. The operational half-life of its associated phosphorescent device is estimated to be over 10000 h with an initial luminance of 1000 cd/m^2.[139] Furthermore, Li et al. developed a Pt(II) phosphor **Pt38** bearing the rigid tetradentate ligand, and its optimized OLEDs hosted by Bebq$_2$ can attain a maximum EQE of 21.5% with the long operational lifetime as well.[140] The tridentate Pt(II) chlorides/acetylides with the π-extended conjugation structure plus the functional groups can harvest the satisfactory performances of OLED at a low doping concentration. Especially, the maximum EQE, CE and PE of 22.1%, 34.8 cd/A and 18.2 lm/W are realized for the **Pt39**-based OLED.[141] The OLEDs based on bis(pyrrole)-diimine **Pt40** are capable of giving efficient red electroluminescence from the excimer or oligomer with a peak EQE, CE, PE and luminance of 6.5%, 9.0 cd/A, 4.0 lm/W and 11,100 cd/m^2.[142] Besides, the bidentate Pt(II) phosphors represent a good class of red emitters. Bis(8-hydroxyquinolinato) Pt(II) complexes **Pt41** could realize the deep-red to near-infrared (NIR) phosphorescence, and their relevant OLEDs show a maximum EQE of 1.7% and CE of 0.32 cd/A. [143] The research on the crystallinity of the emitting layer has been seldomly reported. Kim and Chi et al. explored the effect of the crystallinity of non-doped Pt(II) emitters on the OLED efficiencies. They found that the crystallinity and the molecular arrangement in the crystal-emitting layer can substantially influence the emitting dipole orientation. The crystal non-doped **Pt42**, **Pt43** and **Pt45** layers with an increased order of structures and packing in the neat layers can provide the preferred horizontal transition dipole ratios of 86%, 90% and 93%, respectively. All of them can furnish remarkable EL efficiencies, especially for the symmetrical **Pt45** that owns a near-unity PLQY. The unprecedentedly high EQE of 38.8%, CE of 62 cd/A and PE of 53.8 lm/W are realized on its resultant OLED device.[144] **Pt44** functionalized with two *tert*-butyl groups which is the analogue of **Pt43** shows a better EL performance compared to that of **Pt43**, i.e., a maximum EQE, CE and PE of 19.0%, 21.0 cd/A and 15.5 lm/W.[145] The homologous emitters **Pt46** and **Pt47** exhibit longer-wavelength emission peaks at 703 and 673 nm in solid-state films, respectively. The OLEDs using these 2-pyrazinyl-pyrazolate Pt(II) complexes attain prominent EQEs beyond 20%. If the light outcoupling hemisphere structure is employed, the EL efficiencies of OLEDs based on Pt(II) materials can be further enhanced.[146] Figure 14.24 presents some typical examples of red-emitting Pt(II) metallophosphors for OLED applications.

14.5.3 RED-EMITTING METALLOPHOSPHORS CONTAINING OTHER METALS

As the congener of Ir(III) phosphors, Rh(III) complexes with cyclometalated ligands to be used as efficient emitters for the OLED application are rarely described, which mainly suffer from the weak luminescence at room temperature due to the presence of non-radiative d–d LF-excited state. Until recently, the limitation on the use of Rh(III) complexes as the triplet emitters in OLED devices has been broken through. By exploiting the strong σ-donating cyclometalated ligand with a lower-lying emissive intraligand (IL)-excited state, three new Rh(III) diphenylquinoxalinato complexes with acetylacetonate homologues as the ancillary ligand exhibit intense luminescence in doped mCP thin films. The OLEDs based on **Rh1** furnish an appreciable EQE of up to 12.2%, together with a maximum CE of 17.5 cd/A and PE of 11.0 lm/W, which is the first example of efficient Rh(III)

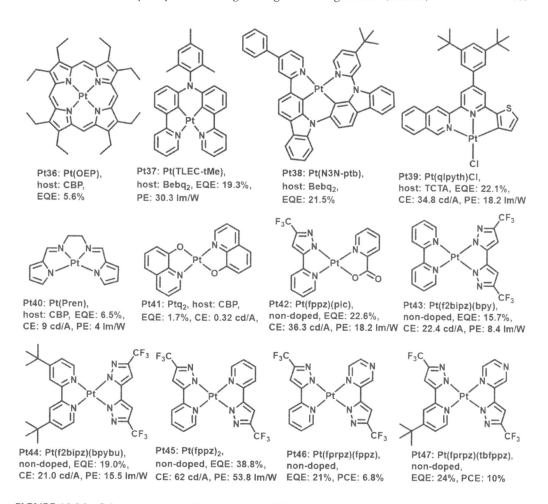

FIGURE 14.24 Selected examples of red-emitting Pt(II) phosphors and the performances of their electroluminescent devices.

phosphors for the OLED application.[11] Being the element in the same period of the Periodic Table of Rh(III), Ru(II) integrated with the isoquinolyl-pyrazole ligands can also afford the red phosphorescent Ru(II) complexes. The OLEDs based on **Ru1** give a fair EQE of 7.03%, CE of 8.02 cd/A and PE of 2.74 lm/W.[147] If Os(II) is used to replace Ru(II), metallophosphors are expected to have an increased MLCT contribution at the excited state manifolds which can raise the radiative rate constant, thereby a better device performance of OLED based on the Os(II) emitter can be achieved. For example, by using a fluorene-based bipolar host material, POAPF, the electrophosphorescent device based on **Os7** gives a superb EQE, CE and PE of 19.9%, 32.8 cd/A and 34.5 lm/W, respectively.[148] Not only the transition metals, but also the rare-earth metals are able to combine with organic ligands to generate the red-emitting metallophosphors (Figure 14.25), namely the molecular lanthanide complexes. The mononuclear **Eu4** possesses the property of visible-light-excited luminescence, yet its related device shows an inferior CE of 0.057 cd/A.[149] **Eu5** with a homo-dinuclear structure displays a bright red emission exhibiting the characteristic $^5D_0 \rightarrow {}^7F_{0-4}$ transitions of Eu(III). Besides, its double-EML OLEDs at the optimum doping concentration of 4.0 wt% afford an impressive EQE of 2.8%, CE of 3.97 cd/A and 3.89 lm/W.[150] Zhang et al. found that the polyfluorination on the alkyl group and the introduction of the larger conjugated naphthyl group into the β-diketone ligand would increase the efficiency of OLEDs based on the corresponding molecular lanthanide complexes. They fabricated the OLEDs based on Ln(phen)

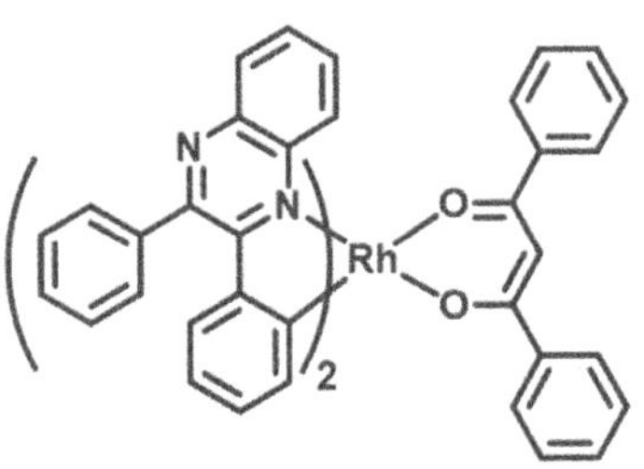

Rh1: Rh(dpqx)₂(dbm), host: mCBP, EQE: 12.2%, CE: 17.5 cd/A, PE: 11.0 lm/W

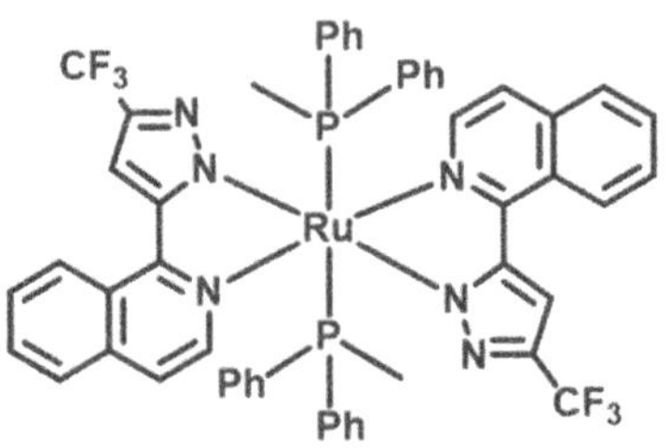

Ru1: Ru(fqlpz)₂(PPh₂Me)₂, host: CBP, EQE: 7.03%, CE: 8.02 cd/A, PE: 2.74 lm/W

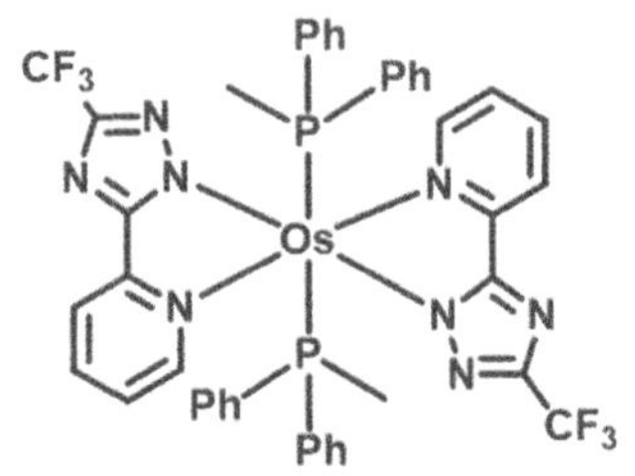

Os7: Os(fptz)₂(PPh₂Me)₂, host: POAPF, EQE: 19.9%, CE: 32.8 cd/A, PE: 34.5 lm/W

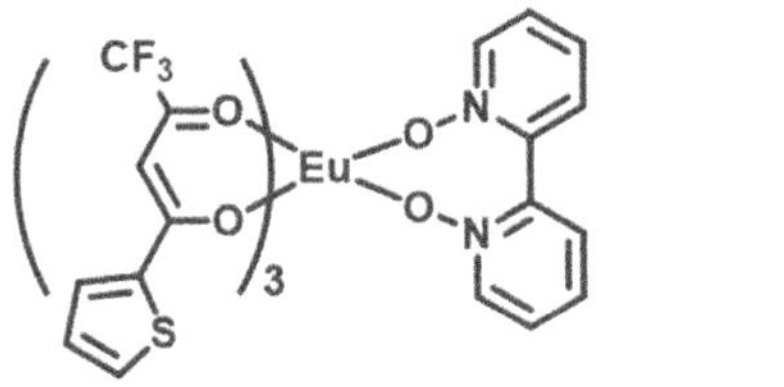

Eu4: Eu(tta)₃(bpyO₂), double-emitting layer, host: TCTA, TPBi, CE: 0.057 cd/A

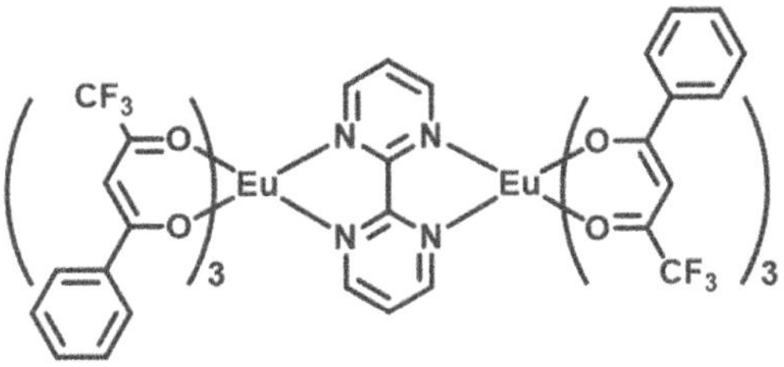

Eu5: Eu₂(btfa)₆bpm, double-emitting layer, host: TCTA, 26DCzPPy, EQE: 2.8%, CE: 3.97 cd/A, PE: 3.89 lm/W

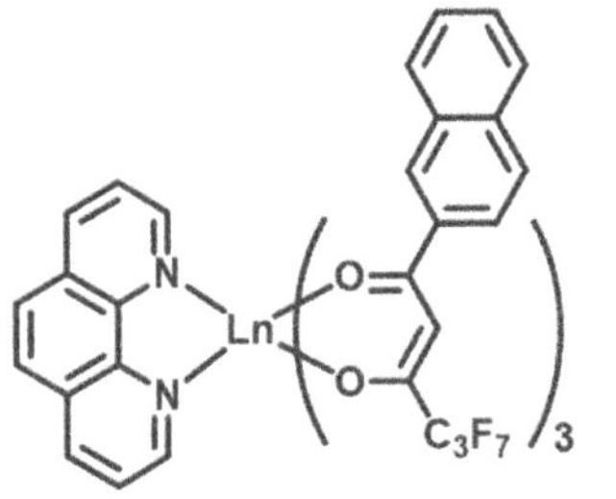

Ln = Eu(III), Sm(III)

Eu6: Eu(phen)(HFNH)₃, host: CBP, CE: 4.14 cd/A, PE: 2.28 lm/W; **Sm1**: Sm(phen)(HFNH)₃, non-doped, CE: 0.18 cd/A

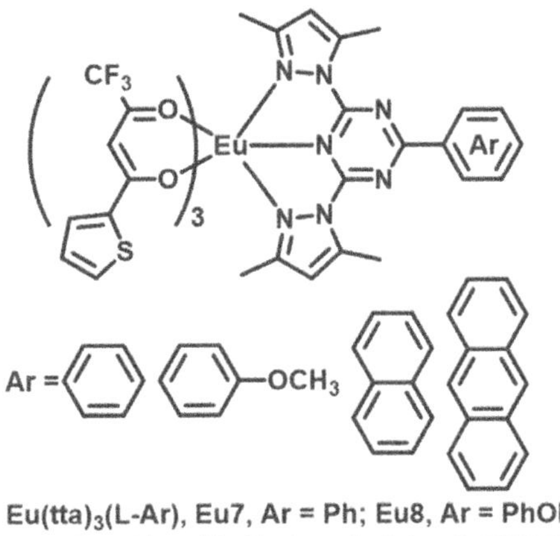

Ar =

Eu(tta)₃(L-Ar), **Eu7**, Ar = Ph; **Eu8**, Ar = PhOMe; **Eu9**, Ar = Nap; **Eu10**, Ar = Ant, host: CBP:PS, CE: 0.47, 0.42, 0.7, 0.35 cd/A, respectively

FIGURE 14.25 Selected examples of red-emitting metallophosphors containing other metals and the performances of their electroluminescent devices.

(HFNH)₃ emitters for Eu(III) and Sm(III) ions. The devices lead to a fair CE of 4.14 cd/A and PE of 2.28 lm/W with a red emission for **Eu6** and an inferior CE of 0.18 cd/A with a reddish-orange emission for **Sm1**.[151] Amini et al. reported Eu(III) trifluoro(thienoyl)acetonates with a series of ligands based on the structure of dipyrazolyltriazin and utilized these Eu(tta)₃(L-Ar) (**Eu7–Eu10**) as the emitting materials in OLEDs. Among them, the best device based on **Eu9** presents a peak luminance of 3156 cd/m² and CE of 0.7 cd/A.[152]

There are also a lot of other lanthanide complexes utilized as the red emitters in OLEDs. Some chemical structures of the homo-dinuclear Eu(III) complexes are shown in Figure 14.26.[150] As mentioned in Section 14.1, the luminescent lanthanide complexes normally comprise various Ln(III) cores and chelating ligands in which the ligands contain many functional groups serving as

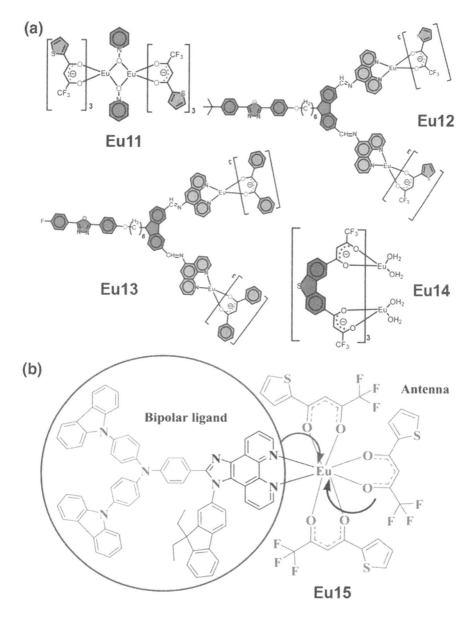

FIGURE 14.26 (a) Selected examples of the homo-dinuclear Eu(III) complexes utilized as the emitter materials for red OLEDs. (b) Chemical structure and energy transfer diagram of the β-diketonate connected Eu(III) complex with the antenna and bipolar ligands. Reproduced with permission from Refs. [150, 153]. Copyright The Royal Society of Chemistry 2020 and The Royal Society of Chemistry and the Centre National de la Recherche Scientifique 2017.

a sensitizer or antenna. Taking **Eu15** as the example, its efficient luminescence is located in the red region with a narrow spectral window. The carbazole unit widens the absorption range and acts as a light-harvesting group. The donor-acceptor structure of the bipolar ligand and good overlap between the bipolar ligand triplet state ($^3\pi\pi^*$) and Eu(III) ion-excited state ($^5D_0$) facilitates the energy transfer from the ligand to Eu(III) ion. Furthermore, effective optical overlap between the absorption spectrum of the acceptor ligand TTA and the emission spectrum of the bipolar ligand Phen-Fl-TPA-CBZ promotes the intramolecular energy transfer. Therefore, efficient energy transfers to the lanthanide center from both the bipolar ligand and TTA ligand can occur, finally enhancing the luminescence of Eu(III).[153] Nevertheless, the efficiencies of OLEDs based on rare-earth metallophosphors still lag behind those of the devices based on the transition metal phosphors currently.

14.6 NEAR-INFRARED-EMITTING METALLOPHOSPHORS FOR ELECTROLUMINESCENT APPLICATION

14.6.1 NEAR-INFRARED-EMITTING IR(III) METALLOPHOSPHORS

The NIR light is regarded as an electromagnetic radiation with a longer wavelength beyond the visible light, normally situated between 700 and 2500 nm.[154, 155] NIR-emitting materials have emerged to show the promising applications in many areas, like the phototherapies,[156, 157] bio-probes/bio-imagings,[158–161] telecommunications,[162] information-secured displays,[163, 164] photovoltaic solar cells [165–167] and OLEDs.[168] In particular, the NIR Ir(III) phosphorescent complexes with their advantages of high quantum efficiency, tunable optoelectronic property, plentiful and variable structures constitute an important class of phosphors in this field.

The most commonly used strategy to extend the emission range of Ir(III) phosphors to the NIR region is to enlarge the degree of π-conjugation of cyclometalated ligands with the purpose of stabilizing the LUMO energy levels of Ir(III) complexes. **Ir66** with one more phenyl ring fused with quinoline compared to the orange-emitting **Ir35** shows a NIR emission at 720 nm. However, the OLEDs by using **Ir66** as the emitter attain an inferior EQE of 1.07%.[169] This can be attributed to two main factors. One is that the extensive conjugation would weaken MLCT contribution at the excited state manifolds of Ir(III) complexes and increase the vibrational decays, leading to the decreased radiative rate constants. Another is the intrinsic "energy gap law" notable in the NIR region in which the coupling between the higher vibrational levels of the S_0 state and the zero vibrational level of the S_1 (or T_1) state can induce an enhanced nonradiative process. Therefore, the design and development of NIR Ir(III) phosphorescent emitters with high EQEs still remains a challenge to date. In spite of this, many efforts have been made. For instance, the incorporation of the N atom into the pyridine ring to furnish the pyrazine or pyridazine ring can significantly red-shift the luminescence. N has a strong electron-withdrawing ability so that it can promote the charge transfer from the metal center to the N-heterocyclic ring and further lower the LUMO level, thereby resulting in a narrow E_g and low-energy emission. **Ir67** with the pyrazine group inside shows a NIR peak at 780 nm that is 60 nm longer than that of **Ir66**. The maximum EQE of its fabricated OLED could be 2.2%, no less than that of **Ir66**.[170] **Ir68** with the pyrazine structure as the coordinated unit and the methoxyethoxy plus hexyl groups for enhancing the solubility is able to be employed to fabricate the solution processed OLEDs with a NIR phosphorescence, and its optimized device affords a maximum EQE of 3.4% with the emission peak at 702 nm.[171] The introduction of a cyano group into the pyridine ring can decrease the electron density on the N-heterocyclic part and strengthen the ^{3}MLCT states of Ir(III) complex. The complex **Ir69** emits strong NIR phosphorescence with a considerable PLQY, and the OLED based on **Ir69** affords a maximum EQE of 9.59% with an EL peak at 706 nm, which is a record-high efficiency at that moment.[125] Additionally, instead of functionalizing the isoquinoline moiety of the cyclometalated ligand, substituting the thienyl moiety by benzo[b]thiophen-2-yl ring can also realize a NIR Ir(III) emitter **Ir70**. It is used as the dopant in a co-host of PVK:OXD7, and its associated OLED device can attain a peak EQE of 3.07% at 714 nm.[172] Homoleptic facial **Ir71** and **Ir72** bearing the extended π-conjugation and the pyridazine structures show the NIR emissions at 765 nm for **Ir71** and at 824 nm for **Ir72** in CH_2Cl_2 solutions. Their PLQYs are up to 17.3% and 5.2%, and the OLEDs based on them give rise to the maximum EQEs of 4.5 and 0.5%, respectively.[173] 2,1,3-Benzothiadiazole is an acceptor unit widely used in the development of optoelectronic materials, and turning its benzene into a pyridine will dramatically lower the LUMO level of Ir(III) complex. Correspondingly, **Ir73**-based OLED with mCP as the host shows an emission at 715 nm with a peak EQE of 4.0%.[174] Moreover, through the replacement of the β-diketonate with the bis-β-diketonate ligand, the Ir(III) coordination can convert the complex from mononuclear to dinuclear framework and a significant red shift of the emission to the NIR range can be observed. The asymmetric dinuclear **Ir74** emits a NIR light at 722 nm while its mononuclear counterpart displays the emission at 698 nm, and the optimized OLED using it as the emitter gives a peak EQE of 3.11% with an EL peak at 720 nm.[175] A series of Ir(III) NIR emitters (**Ir75–Ir79**) which employ the rigid dibenzo[a,c]phenazine (DBPz) unit as the core of cyclometalated ligands have been reported. The rigid and unitary ligands can enhance the PLQYs of Ir(III) complexes so that the OLED devices based on this family of Ir(III) emitters all display attractive EQEs. In particular, by using triphenylamine as the peripheral shell to be anchored in the 3, 6- or 11, 12-positions of DBPz moiety, **Ir75** and **Ir76** can possess a core–shell structure. The optimized OLEDs based on them attain outstanding EQEs of 13.72% and 12.34% with the radiance of 26996 mW/m^2·Sr at 708 nm and 24,647 mW/m^2·Sr at 718 nm, respectively. They should represent the state-of-the-art performance of the device in the Ir(III) phosphor-based NIR-OLEDs.[176–178] Selected typical examples of NIR-emitting Ir(III) phosphors for OLED applications are shown in Figure 14.27.

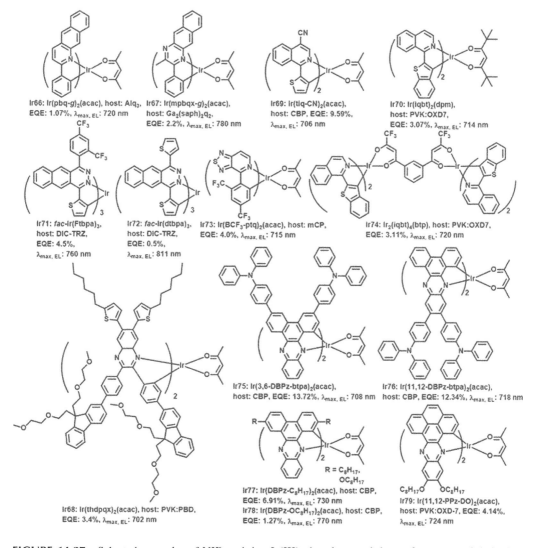

FIGURE 14.27   Selected examples of NIR-emitting Ir(III) phosphors and the performances of their electroluminescent devices.

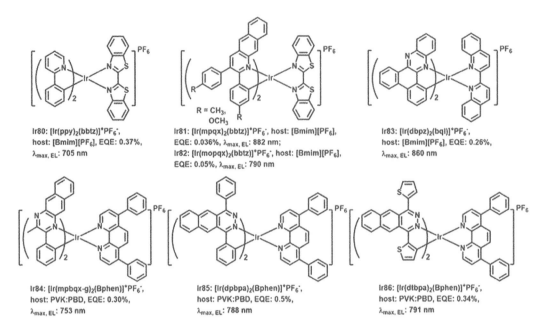

Ir80: [Ir(ppy)₂(bbtz)]⁺PF₆⁻,
host: [Bmim][PF₆], EQE: 0.37%,
λ_max, EL: 705 nm

Ir81: [Ir(mpqx)₂(bbtz)]⁺PF₆⁻, host: [Bmim][PF₆],
EQE: 0.036%, λ_max, EL: 882 nm;
Ir82: [Ir(mopqx)₂(bbtz)]⁺PF₆⁻, host: [Bmim][PF₆],
EQE: 0.05%, λ_max, EL: 790 nm

Ir83: [Ir(dbpz)₂(bql)]⁺PF₆⁻,
host: [Bmim][PF₆], EQE: 0.26%,
λ_max, EL: 860 nm

Ir84: [Ir(mpbqx-g)₂(Bphen)]⁺PF₆⁻,
host: PVK:PBD, EQE: 0.30%,
λ_max, EL: 753 nm

Ir85: [Ir(dpbpa)₂(Bphen)]⁺PF₆⁻,
host: PVK:PBD, EQE: 0.5%,
λ_max, EL: 788 nm

Ir86: [Ir(dtbpa)₂(Bphen)]⁺PF₆⁻,
host: PVK:PBD, EQE: 0.34%,
λ_max, EL: 791 nm

FIGURE 14.28 Selected examples of NIR-emitting cationic Ir(III) phosphors with PF₆ as the anion and the performances of their electroluminescent devices.

In addition to the previous neutral Ir(III) phosphors, cationic Ir(III) complexes (**Ir80–Ir86**) are also capable of realizing NIR emissions (Figure 14.28). Owing to their instability during the vapor deposition process, usually the fabrications of OLEDs or light-emitting electrochemical cells (LEECs) based on the cationic Ir(III) complexes are done by the solution-processed method. Their PLQY, spin–orbit coupling and charge carrier transport are normally lower, not as strong as and not as balanced as the neutral Ir(III) complexes; thus, the EL performance of devices based on the cationic Ir(III) complexes still cannot compare favorably with that of devices based on the neutral Ir(III) complexes.[179–183]

14.6.2 Near-Infrared-Emitting Pt(II) Metallophosphors

To harvest the NIR-emitting Pt(II) phosphors, extension of the π-conjugated structure is the preferable choice. On the basis of red-emitting Pt(OEP) (**Pt36**), a new family of Pt-tetrabenzoporphyrins (**Pt48–Pt51**) have been reported and used as the dopant emitters (4 wt% in Alq₃) in OLEDs, giving rise to devices with the emission maxima ranging from 773 to 792 nm, and the achieved EQEs are also impressive to be as high as 9.2%.[184] By appending the donor-acceptor (D-A) framework of 3,6-substituted salophen chelating ligands on the tetradentate Pt(II) complexes, **Pt52** is obtained and blended into a co-host matrix of PVK:OXD-7. The related polymer OLED devices can produce a NIR emission at 703 nm with the peak EQE of 0.88%.[185] By using the similar strategy, **Pt53** with a simpler D-A structure also shows the NIR emission at 826 nm with the peak EQE of 0.49%.[186] Interestingly, if a chirality is introduced into the NIR Pt(II) emitters, the circularly polarized (CP) light can be generated. The heteroleptic **Pt54** is formed by involving a chiral N^O-Schiff-base ancillary ligand, and its optimized polymer OLEDs attain a maximum EQE of 0.93% with the peak emission wavelength at 732 nm.[187] The excimer emissions on the stacking Pt(II) phosphors can cause the red shift of phosphorescence to the NIR region. The non-doped OLEDs based on Pt(II) chlorides comprising the tridentate 1,3-bis(2-pyridinyl)benzene ligand with various functional groups emit the NIR light induced by excimers. Notably, the EQE of the device based on **Pt55** is recorded as high as 14.5%.[188, 189] This proves that the excimer emission should be a practicable avenue to overcome the energy gap law in achieving efficient NIR OLEDs. Chi and

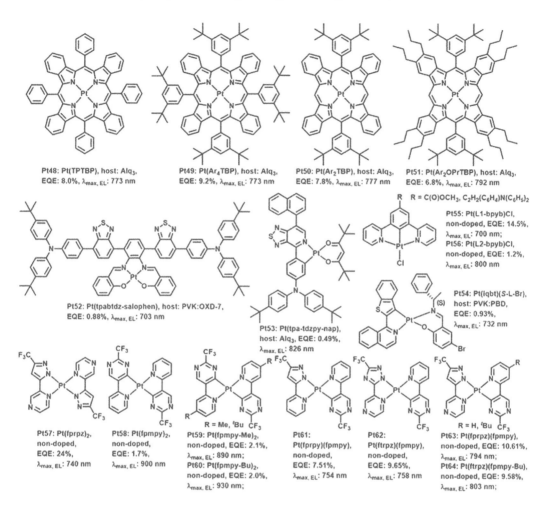

FIGURE 14.29 Selected examples of NIR-emitting Pt(II) phosphors and the performances of their electroluminescent devices.

Chou et al. developed a series of Pt(II) complexes (**Pt57–Pt64**) flanked by various N-heterocyclic ligands including the pyrazolyl-pyrazinate, pyridyl-pyrimidinate, pyrazolyl-pyridinate, triazolyl-pyrimidinate and so forth, which can be representatives of the state-of-the-art Pt(II) emitters utilized in NIR-OLEDs. The highest EQE of 24% is achieved by the OLED based on **Pt57** with the emission at 740 nm. When the EL wavelength maximum exceeds 800 nm, the device based on **Pt64** still attains an outstanding EQE of 9.58%. Moreover, the EQE of **Pt60**-based OLED can reach 2% with no efficiency roll-off, even the emission peak is far at 930 nm. These are the first-class OLED performances among all NIR-emitting materials thus far, and the authors explained that the excimer emission of this family of Pt(II) phosphors is subject to less vibrational quenching, or the exciton delocalization in the Pt-Pt aggregates (dimers/trimers/oligomers) decouples the exciton band from highly vibrational ladders in the S_0 ground state. As a result, such high EQEs of the OLEDs based on this kind of Pt(II) phosphors can be achieved.[146, 190, 191] Figure 14.29 portrays the selected examples of NIR-emitting Pt(II) phosphors for OLED applications.

Likewise, the strategy adopting the dinuclear configuration of Pt(II) complexes is capable of shifting the emission to the region of longer wavelength. Selected examples (**Pt65–Pt70**) are shown in Figure 14.30. All these dinuclear Pt(II) emitters are doped into the host matrix of PVK:OVD-7 to fabricate the OLEDs, except for the Pt-Pt connected dinuclear complexes **Pt65–Pt67**. In the

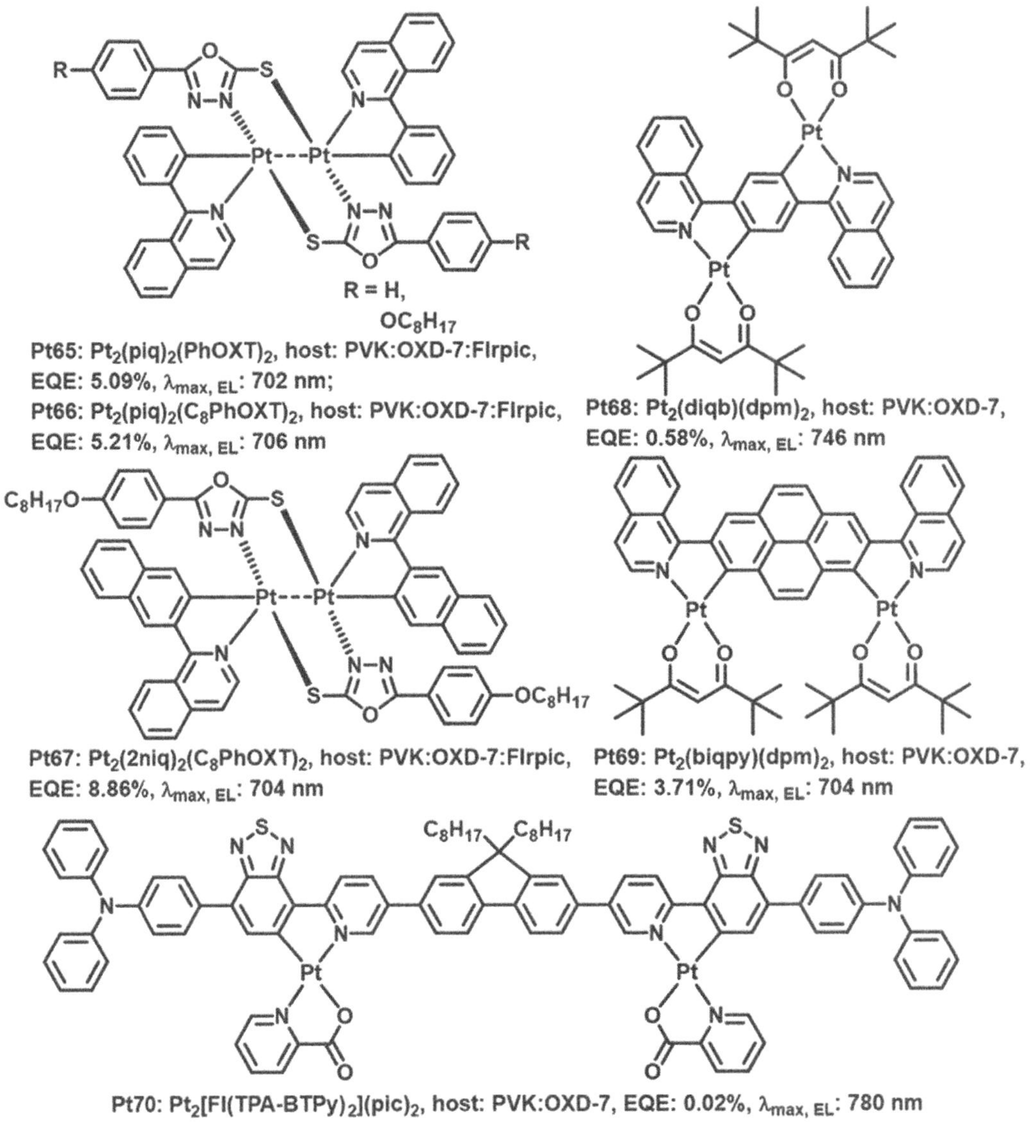

FIGURE 14.30　Selected examples of NIR-emitting Pt(II) phosphors with dinuclear structure and the performances of their electroluminescent devices.

Pt65–Pt67-based OLEDs, the blue-emitting FIrpic (**Ir1**) is additionally introduced into the EML of the devices to play a bridging role between OXD-7 and the dinuclear Pt(II) complexes. The Dexter energy transfer from **Ir1** to **Pt65–Pt67** can be promoted, further improving the OLED performance. As a consequence, **Pt67** doped devices furnish a standout EQE as high as 8.86% with a radiant emittance near 1000 mW/cm^2 under the optimization of devices.[192–195] In addition, high-performance OLEDs with a NIR emission at 716 nm and a maximum EQE of 5.1% are realized by using the Pt(III) complex **Pt71**, which has an octahedral coordination structure and d^7–d^7 electronic configuration.[196]

14.6.3　Near-Infrared-Emitting Metallophosphors Containing Other Metals

Besides Ir(III) and Pt(II) phosphors, other metal-based triplet emitters embodying d^6, d^8, d^{10} transition metals and the lanthanide complexes can realize the luminescence of various colors, including the NIR light.[197–203] Among them, the class of d^6 Os(II) complexes are one of the few efficient emitters applied in the NIR OLEDs, which show good device performances and can compete with those of Ir(III) and Pt(II) emitters. Chi et al. reported a series of Os(II) chelates (**Os8–Os12**) bearing the pyrazinyl/quinolyl azolate ligands, and the stabilized energy of MLCT transitions on the pyrazinyl ligands in the excited states leads to the efficient NIR emission. Especially, the NIR OLED based on **Os11** attains an outstanding device performance with a maximum EQE of 11.5% at the peak emission of 710 nm.[204, 205] There are reports on the d^6 Ru(II) (**Ru2–Ru3**) and Re(I) (**Re3–Re4**) complexes displaying luminescent wavelengths over 700 nm. However, low quantum efficiencies are detected no matter in the solution or device based on these complexes.[12, 206–208] These disadvantages are also present in the NIR-emitting d^8 metal complexes like Pd(II) (**Pd4–Pd7**) and Au(III) (**Au11–Au12**) complexes,[209–211] and the NIR-emitting d^{10} metal complexes like Cu(I) (**Cu7–Cu11**) complexes.[154, 212] The reasons for this are rather complicated. The energy gap law, the weak LF that causes the metal-centered d–d-excited states to be energetically lower than the MLCT-excited states, the strong carrier trapping of phosphors that induces the self-quenching and the inefficient energy transfer from the host to the NIR dopant may all contribute to the low efficiency of luminescence.[146, 211, 213, 214] Selected examples of NIR-emitting metallophosphors containing transition metals for OLED applications are shown in Figure 14.31.

As lanthanide (Ln) metals possess numerous 4f electrons and close energy levels, the Ln metal ions, for example, Nd(III), Er(III), Yb(III), Tm(III) and Ho(III) have been identified with the NIR emission (Figure 14.32).[197] The complexes based on Ln(III) ions demonstrate the relatively

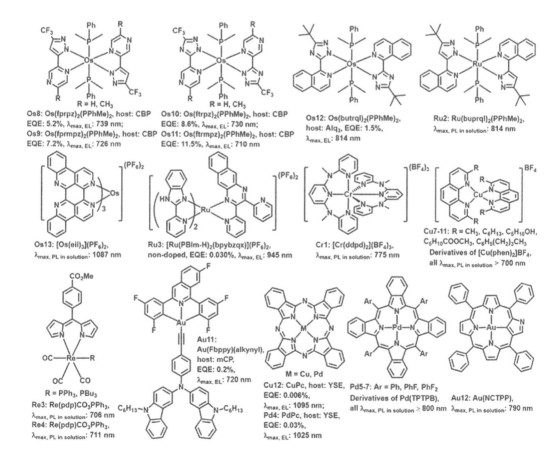

FIGURE 14.31 Selected examples of NIR-emitting metallophosphors containing transition metals and their OLED performances.

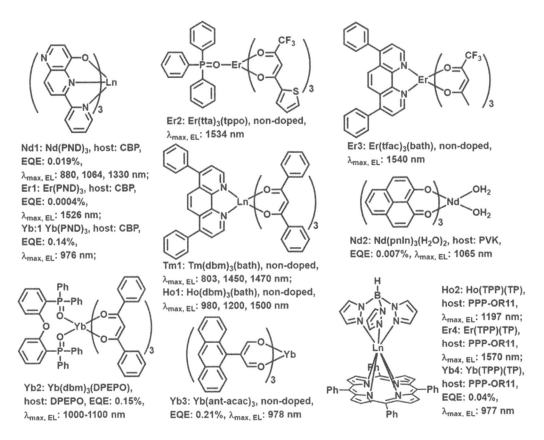

Nd1: Nd(PND)₃, host: CBP, EQE: 0.019%, $\lambda_{max, EL}$: 880, 1064, 1330 nm;
Er1: Er(PND)₃, host: CBP, EQE: 0.0004%, $\lambda_{max, EL}$: 1526 nm;
Yb:1 Yb(PND)₃, host: CBP, EQE: 0.14%, $\lambda_{max, EL}$: 976 nm;

Er2: Er(tta)₃(tppo), non-doped, $\lambda_{max, EL}$: 1534 nm

Tm1: Tm(dbm)₃(bath), non-doped, $\lambda_{max, EL}$: 803, 1450, 1470 nm;
Ho1: Ho(dbm)₃(bath), non-doped, $\lambda_{max, EL}$: 980, 1200, 1500 nm

Er3: Er(tfac)₃(bath), non-doped, $\lambda_{max, EL}$: 1540 nm

Nd2: Nd(pnln)₃(H₂O)₂, host: PVK, EQE: 0.007%, $\lambda_{max, EL}$: 1065 nm

Yb2: Yb(dbm)₃(DPEPO), host: DPEPO, EQE: 0.15%, $\lambda_{max, EL}$: 1000-1100 nm

Yb3: Yb(ant-acac)₃, non-doped, EQE: 0.21%, $\lambda_{max, EL}$: 978 nm

Ho2: Ho(TPP)(TP), host: PPP-OR11, $\lambda_{max, EL}$: 1197 nm;
Er4: Er(TPP)(TP), host: PPP-OR11, $\lambda_{max, EL}$: 1570 nm;
Yb4: Yb(TPP)(TP), host: PPP-OR11, EQE: 0.04%, $\lambda_{max, EL}$: 977 nm

FIGURE 14.32 Selected examples of NIR-emitting metallophosphors containing lanthanide metals and the performances of their electroluminescent devices.

flexible coordination geometries by combination with various ligands. Unlike the transition metal-based complexes in which the organic ligands can significantly influence the emissions, the Ln(III) complexes within organic LFs normally show sharp emission bands with the fixed wavelengths because they are determined by the internal 4f or 5d–4f electronic transitions of Ln(III) ions. The main functions of organic ligands are structure-stabilizing, light-harvesting and energy-transfer-ring. Specifically, Er(III) complexes (**Er1–Er4**) with the $^4I_{3/2} \rightarrow {}^4I_{15/2}$ can show the luminescence at 1.5 μm.[215–217] Nd(III) complexes (**Nd1–Nd2**) exhibit three emission bands at around 0.9, 1.1 and 1.3 μm corresponding to $^4F_{3/2} \rightarrow {}^4I_{9/2}$, $^4F_{3/2} \rightarrow {}^4I_{11/2}$ and $^4F_{3/2} \rightarrow {}^4I_{13/2}$ transitions.[218] $^2F_{5/2} \rightarrow {}^2F_{7/2}$ in Yb(III) complexes (**Yb1–Yb4**) can result in the luminescence at around 1.0 μm.[219–221] The EL maxima at 0.8 and 1.45 μm are observed on the devices based on the Tm(III) complex (**Tm1**), which are assigned to $^3H_4 \rightarrow {}^3H_6$ and $^3H_4 \rightarrow {}^3F_4$ transitions of Tm(III) ion.[222] The NIR emissions at about 1.0, 1.2 and 1.5 μm of the devices based on Ho(III) complexes (**Ho1–Ho2**) originate from the $^5F_5 \rightarrow {}^5I_7$, $^5I_6 \rightarrow {}^5I_8$ and $^5F_5 \rightarrow {}^5I_6$ transitions, respectively.[223] Although the organic ligands have a limited effect on the emission wavelengths of the Ln(III) complexes, they indeed play an important role in transferring energy to Ln(III) ions and charge transporting in the emitting devices, finally influencing the efficiency performance of NIR OLEDs.

14.7 CONCLUDING REMARKS

Since the first report on OLED by Tang and VanSlyke in 1987, the EL efficiencies of OLEDs have increased from the initial EQE of 1.0% and PE of 1.5 lm/W [224] to the ultrahigh EQE of 64.5%

and PE of 283.4 lm/W (recorded with an external outcoupling lens attachment).[225] During this evolution, the organic phosphorescent materials break the restriction of only utilizing the energy of singlet excitons, resulting in the energy harvesting of both singlet and triplet excitons. The metallophosphors, especially the Ir(III) and Pt(II) phosphors, that can achieve color-tunable emissions with the excellent device performances make a crucial contribution to the flourishing development of OLEDs.

On the other hand, the EQE of OLED (η_{EQE}) is given by $\eta_{EQE} = \gamma \eta_{ST} \eta_{PLQY} \eta_{out}$,[107, 226] where γ is the charge balance factor, η_{ST} is the ratio of the singlet with or without triplet excitons, η_{PLQY} is the PLQY efficiency and η_{out} is the outcoupling efficiency. γ can be brought close to 1 via the optimization of the device structure. η_{ST} can be 25% for the fluorophores and 100% for the phosphors. PLQY should be the inherent property of emitters. η_{out} can be determined by many factors, including the refractive index of the device layers, the orientation of the transition dipole moment of the emitters and the outcoupling technique used in the OLED (the use of a half-sphere attachment and/or a patterned surface) [132, 146]. Therefore, the design of highly efficient metallophosphors and the optimization of the OLED devices are the two main directions for attaining superior EL performances, which are of great help to improve the energy conversion efficiency and alleviate the environmental pollution issue (reduction of the carbon footprint).

To realize the commercialization of OLEDs based on metallophosphors, there are still several limitations that need to be overcome. The first is the stability of emitters. The operational lifetime of OLED should meet the requirement of the commercial use. Although some examples of Ir(III) and Pt(II) phosphors have been reported with long operational lifetimes,[99, 118, 139, 140] the stable blue-emitting metallophosphors are very rare due to the high energy of triplet emission. Recently, the blue OLEDs with the use of Ce(III) complex show the operation lifetime 70 times longer than that of a classical OLED based on the Ir(III) complex, which may offer a new direction for developing stable blue OLEDs.[59] The second issue is the color purity of the efficient emission. Getting efficient pure blue- and red-emitting metallophosphors remains a challenge, as the blue phosphorescent complexes may suffer from the mismatch with the energy level of the host material while the red phosphorescent complexes are more subject to the energy gap law. The third limitation is the high cost of metallophosphors based on the transition metals such as Os, Ir, Pt and Au, which belong to the noble metals and they are expensive to use. Thus, the use of the relatively cheaper metals such as Cu and Mn instead of those noble metals to furnish efficient metallophosphors [54, 78, 84, 85] is highly desirable nowadays. In terms of the manufacturing techniques for OLEDs, solution processing rather than vapor deposition processing is becoming the development trend in the future OLED research community, as it is relatively inexpensive, scalable to large areas and available to a wider variety of materials including metallopolymers.[227–229]

Newer and broader applications also emerge in the field of metallophosphors. Examples embrace the use as catalysts and/or photosensitizers in the reduction of CO_2 and hydrogen evolution,[230–232] the use as chemical tools and medicinal agents for health-related applications,[233–235] the phosphorescent soft salts employed as the ion-sensor or bio-probe,[236, 237] the enhancement of power conversion efficiency in polymer solar cells by incorporating triplet-emitting complexes [238, 239] and so on. Overall, the metallophosphors as a kind of versatile materials have already been widely used in many aspects in our daily life and will continuously bring values to our society.

ACKNOWLEDGMENTS

W. Y. Wong thanks the Hong Kong Research Grants Council (PolyU153058/19P), Guangdong-Hong Kong-Macao Joint Laboratory of Optoelectronic and Magnetic Functional Materials (2019B121205002), the Hong Kong Polytechnic University (1-ZE1C), Research Institute for Smart Energy (CDA2) and the Endowed Professorship in Energy from Ms. Clarea Au (847S).

REFERENCES

1. Petrucci RH, Harwood WS, Herring FG (2002) *General Chemistry: Principles and Modern Applications* (8th ed.). Upper Saddle River, NJ: Prentice Hall.
2. Housecroft CE, Sharpe AG (2005). *Inorganic Chemistry* (2nd ed.). Pearson Prentice Hall.
3. Baryshnikov G, Minaev B, Ågren H (2017) Theory and calculation of the phosphorescence phenomenon. *Chem. Rev.* 117: 6500.
4. Yang X, Zhou G, Wong WY (2015) Functionalization of phosphorescent emitters and their host materials by main-group elements for phosphorescent organic light-emitting devices. *Chem. Soc. Rev.* 44: 8484.
5. Wong WY, Ho CL (2009) Functional metallophosphors for effective charge carrier injection/transport: New robust OLED materials with emerging applications. *J. Mater. Chem.* 19: 4457.
6. Suzuri Y, Oshiyama T, Ito H, Hiyama K, Kita H (2014) Phosphorescent cyclometalated complexes for efficient blue organic light-emitting diodes. *Sci. Technol. Adv. Mater.* 15: 54202.
7. Lamansky S, Djurovich P, Murphy D, Abdel-Razzaq F, Lee HE, Adachi C, Burrows PE, Forrest SR, Thompson ME (2001) Highly phosphorescent bis-cyclometalated iridium complexes: synthesis, photophysical characterization, and use in organic light emitting diodes. *J. Am. Chem. Soc.* 123: 4304.
8. Baldo MA, O'Brien DF, You Y, Shoustikov A, Sibley S, Thompson ME, Forrest SR (1998) Highly efficient phosphorescent emission from organic electroluminescent devices. *Nature* 395: 151.
9. Ma Y, Zhang H, Shen J, Che C (1998) Electroluminescence from triplet metal-ligand charge-transfer excited state of transition metal complexes. *Synth. Met.* 94: 245.
10. Chi Y, Chou PT (2007) Contemporary progresses on neutral, highly emissive Os(II) and Ru(II) complexes. *Chem. Soc. Rev.* 36: 1421.
11. Wei F, Lai SL, Zhao S, Ng M, Chan MY, Yam VWW, Wong, KMC (2019) Ligand mediated luminescence enhancement in cyclometalated rhodium(III) complexes and their applications in efficient organic light-emitting devices. *J. Am. Chem. Soc.* 141: 12863.
12. McLean TM, Moody JL, Waterland MR, Telfer SG (2012) Luminescent rhenium(I)-dipyrrinato complexes. *Inorg. Chem.* 51: 446.
13. Di D, Romanov AS, Yang L, Richter JM, Rivett JPH, Jones S, Thomas TH, Jalebi MA, Friend RH, Linnolahti M, Bochmann M, Credgington D (2017) High-performance light-emitting diodes based on carbene-metal-amides. *Science* 356: 159.
14. Romanov AS, Jones STE, Yang L, Conaghan PJ, Di D, Linnolahti M, Credgington D, Bochmann M (2018) Mononuclear silver complexes for efficient solution and vacuum-processed OLEDs. *Adv. Opt. Mater.* 6: 1801347.
15. Conaghan PJ, Matthews CSB, Chotard F, Jones STE, Greenham NC, Bochmann M, Credgington D, Romanov AS (2020) Highly efficient blue organic light-emitting diodes based on carbene-metal-amides. *Nat. Commun.* 11: 1758.
16. Cheng G, Chan KT, To WP, Che CM (2014) Color tunable organic light-emitting devices with external quantum efficiency over 20% based on strongly luminescent gold(III) complexes having long-lived emissive excited states. *Adv. Mater.* 26: 2540.
17. Fan C, Yang C (2014) Yellow/orange emissive heavy-metal complexes as phosphors in monochromatic and white organic light-emitting devices. *Chem. Soc. Rev.* 43: 6439.
18. Ho CL, Wong WY (2013) Small-molecular blue phosphorescent dyes for organic light-emitting devices. *New J. Chem.* 37: 1665.
19. Zhao Q, Li F, Huang C (2010) Phosphorescent chemosensors based on heavy-metal complexes. *Chem. Soc. Rev.* 39: 3007.
20. Connelly NG, Damhus T, Hartshorn RM, Hutton AT (2005) *Nomenclature of Inorganic Chemistry*: IUPAC Recommendations 2005. Cambridge: RSC Publishing.
21. Binnemans K (2009) Lanthanide-based luminescent hybrid materials. *Chem. Rev.* 109: 4283.
22. Boyle TJ, Ottley LAM (2008) Advances in structurally characterized lanthanide alkoxide, aryloxide, and silyloxide compounds. *Chem. Rev.* 108: 1896.
23. Heffern MC, Matosziuk LM, Meade TJ (2014) Lanthanide probes for bioresponsive imaging. *Chem. Rev.* 114: 4496.
24. Eliseeva SV, Bünzli JCG (2010) Lanthanide luminescence for functional materials and bio-sciences. *Chem. Soc. Rev.* 39: 189.
25. Bünzli JCG (2010) Lanthanide luminescence for biomedical analyses and imaging. *Chem. Rev.* 110: 2729.
26. Baldo MA, Lamansky S, Burrows PE, Thompson ME, Forrest SR (1999) Very high-efficiency green organic light-emitting devices based on electrophosphorescence. *Appl. Phys. Lett.* 75: 4.

27. Adachi C, Kwong RC, Djurovich P, Adamovich V, Baldo MA, Thompson ME, Forrest SR (2001) Endothermic energy transfer: a mechanism for generating very efficient high-energy phosphorescent emission in organic materials. *Appl. Phys. Lett.* 79: 2082.

28. Kim M, Lee JY (2014) Engineering the substitution position of diphenylphosphine oxide at carbazole for thermal stability and high external quantum efficiency above 30% in blue phosphorescent organic light-emitting diodes. *Adv. Funct. Mater.* 24: 4164.

29. Li J, Djurovich PI, Alleyne BD, Tsyba I, Ho NN, Bau R, Thompson ME (2004) Synthesis and characterization of cyclometalated Ir(III) complexes with pyrazolyl ancillary ligands. *Polyhedron* 23: 419.

30. Holmes RJ, D'Andrade BW, Forrest SR, Ren X, Li J, Thompson ME (2003) Efficient, deep-blue organic electrophosphorescence by guest charge trapping. *Appl. Phys. Lett.* 83: 3818.

31. Wada A, Yasuda T, Zhang Q, Yang YS, Takasu I, Enomoto S, Adachi C (2013) A host material consisting of a phosphinic amide directly linked donor–acceptor structure for efficient blue phosphorescent organic light-emitting diodes. *J. Mater. Chem. C* 1: 2404.

32. Gong S, Chang YL, Wu K, White R, Lu ZH, Song D, Yang C (2014) High-power-efficiency blue electrophosphorescence enabled by the synergistic combination of phosphine-oxide-based host and electron-transporting materials. *Chem. Mater.* 26: 1463.

33. Yeh SJ, Wu MF, Chen CT, Song YH, Chi Y, Ho MH, Hsu SF, Chen CH (2005) New dopant and host materials for blue-light-emitting phosphorescent organic electroluminescent devices. *Adv. Mater.* 17: 285.

34. Zhu YC, Zhou L, Li HY, Xu QL, Teng MY, Zheng YX, Zuo JL, Zhang HJ, You XZ (2011) Highly efficient green and blue-green phosphorescent OLEDs based on iridium complexes with the tetraphenylimidodiphosphinate ligand. *Adv. Mater.* 23: 4041.

35. Yook KS, Lee JY (2011) Solution processed deep blue phosphorescent organic light-emitting diodes with over 20% external quantum efficiency. *Org. Electron.* 12: 1711.

36. Fan C, Li Y, Yang C, Wu H, Qin J, Cao Y (2012) Phosphoryl/sulfonyl-substituted iridium complexes as blue phosphorescent emitters for single-layer blue and white organic light-emitting diodes by solution process. *Chem. Mater.* 24: 4581.

37. Kang Y, Chang YL, Lu JS, Ko SB, Rao Y, Varlan M, Lu ZH, Wang S (2013) Highly efficient blue phosphorescent and electroluminescent Ir(III) compounds. *J. Mater. Chem. C* 1: 441.

38. Lee S, Kim SO, Shin H, Yun HJ, Yang K, Kwon SK, Kim JJ, Kim YH (2013) Deep-blue phosphorescence from perfluoro carbonyl-substituted iridium complexes. *J. Am. Chem. Soc.* 135: 14321.

39. Chen Z, Suramitr S, Zhu N, Ho CL, Hannongbua S, Chen S, Wong WY (2020) Tetrafluorinated phenylpyridine based heteroleptic iridium(III) complexes for efficient sky blue phosphorescent organic light-emitting diodes. *J. Mater. Chem. C* 8: 2551.

40. Holmes RJ, Forrest SR, Sajoto T, Tamayo A, Djurovich PI, Thompson ME, Brooks J, Tung YJ, D'Andrade BW, Weaver MS, Kwong RC, Brown JJ (2005) Saturated deep blue organic electrophosphorescence using a fluorine-free emitter. *Appl. Phys. Lett.* 87: 243507.

41. Schildknecht C, Ginev G, Kammoun A, Riedl T, Kowalsky W, Johannes HH, Lennartz C, Kahle K, Egen M, Geßner T, Bold M, Nord S, Erk P (2005) Novel deep-blue emitting phosphorescent emitter. *Proc. SPIE* 5937: 59370E.

42. Udagawa K, Sasabe H, Cai C, Kido J (2014) Low-driving-voltage blue phosphorescent organic light-emitting devices with external quantum efficiency of 30%. *Adv. Mater.* 26: 5062.

43. Chen WC, Sukpattanacharoen C, Chan WH, Huang CC, Hsu HF, Shen D, Hung WY, Kungwan N, Escudero D, Lee CS, Chi Y (2020) Modulation of solid-state aggregation of square-planar Pt(II) based emitters: enabling highly efficient deep-red/near infrared electroluminescence. *Adv. Funct. Mater.* 30: 2002494.

44. Unger Y, Meyer D, Molt O, Schildknecht C, Münster I, Wagenblast G, Strassner T (2010) Green–blue emitters: NHC-based cyclometalated [Pt(C^C*)(acac)] complexes. *Angew. Chem. Int. Ed.* 49: 10214.

45. Hudson ZM, Sun C, Helander MG, Chang YL, Lu ZH, Wang S (2012) Highly efficient blue phosphorescence from triarylboron-functionalized platinum(II) complexes of N-heterocyclic carbenes. *J. Am. Chem. Soc.* 134: 13930.

46. Tronnier A, Heinemeyer U, Metz S, Wagenblast G, Muenster I, Strassner T (2015) Heteroleptic platinum(II) NHC complexes with a C^C* cyclometalated ligand – synthesis, structure and photophysics. *J. Mater. Chem. C* 3: 1680.

47. Yang X, Wang Z, Madakuni S, Li J, Jabbour GE (2008) Efficient blue- and white-emitting electrophosphorescent devices based on platinum(II) [1,3-difluoro-4,6-di(2-pyridinyl)benzene] chloride. *Adv. Mater.* 20: 2405.

48. Fleetham T, Wang Z, Li J (2012) Efficient deep blue electrophosphorescent devices based on platinum(II) bis(n-methyl-imidazolyl)benzene chloride. *Org. Electron.* 13: 1430.

49. Li G, Fleetham T, Li J (2014) Efficient and stable white organic light-emitting diodes employing a single emitter. *Adv. Mater.* 26: 2931.

50. Hang XC, Fleetham T, Turner E, Brooks J, Li J (2013) Highly efficient blue-emitting cyclometalated platinum(II) complexes by judicious molecular design. *Angew. Chem. Int. Ed.* 52: 6753.

51. Fleetham T, Li G, Wen L, Li J (2014) Efficient "pure" blue OLEDs employing tetradentate Pt complexes with a narrow spectral bandwidth. *Adv. Mater.* 26: 7116.

52. Bullock JD, Salehi A, Zeman CJ, Abboud KA, So F, Schanze KS (2017) In search of deeper blues: *trans*-N-heterocyclic carbene platinum phenylacetylide as a dopant for phosphorescent OLEDs. *ACS Appl. Mater. Interfaces* 9: 41111.

53. Tang MC, Kwok WK. Lai SL, Cheung WL, Chan MY, Yam VWW (2019) Rational molecular design for realizing high performance sky-blue-emitting gold(III) complexes with monoaryl auxiliary ligands and their applications for both solution-processable and vacuum-deposited organic light-emitting devices. *Chem. Sci.* 10: 594.

54. Czerwieniec R, Yu J, Yersin H (2011) Blue-light emission of Cu(I) complexes and singlet harvesting. *Inorg. Chem.* 50: 8293.

55. Alabau RG, Eguillor B, Esler J, Esteruelas MA, Oliván M, Oñate E, Tsai JY, Xia C (2014) CCC–pincer–NHC osmium complexes: new types of blue-green emissive neutral compounds for organic light-emitting devices (OLEDs). *Organometallics* 33: 5582.

56. Zhu ZQ, Fleetham T, Turner E, Li J (2015) Harvesting all electrogenerated excitons through metal assisted delayed fluorescent materials. *Adv. Mater.* 27: 2533.

57. Zhu ZQ, Park CD, Klimes K, Li J (2019) Highly efficient blue OLEDs based on metal-assisted delayed fluorescence Pd(II) complexes. *Adv. Opt. Mater.* 7: 1801518.

58. Zheng XL, Liu Y, Pan M, Lü XQ, Zhang JY, Zhao CY, Tong YX, Su CY (2007) Bright blue-emitting Ce³⁺ complexes with encapsulating polybenzimidazole tripodal ligands as potential electroluminescent devices. *Angew. Chem. Int. Ed.* 46: 7399.

59. Zhao Z, Wang L, Zhan G, Liu Z, Bian Z, Huang C (2021) Efficient rare earth cerium(III) complex with nanosecond d–f emission for blue organic light-emitting diodes. *Natl. Sci. Rev.* 8: nwaa193.

60. Kim M, Lee JY (2014) Engineering of interconnect position of bicarbazole for high external quantum efficiency in green and blue phosphorescent organic light-emitting diodes. *ACS Appl. Mater. Interfaces* 6: 14874.

61. Lamansky S, Djurovich P, Murphy D, Abdel-Razzaq F, Kwong R, Tsyba I, Bortz M, Mui B, Bau R, Thompson ME (2001) Synthesis and characterization of phosphorescent cyclometalated iridium complexes. *Inorg. Chem.* 40: 1704.

62. Kim SY, Jeong WI, Mayr C, Park YS, Kim KH, Lee JH, Moon CK, Brütting W, Kim JJ (2013) Organic light-emitting diodes with 30% external quantum efficiency based on a horizontally oriented emitter. *Adv. Funct. Mater.* 23: 3896.

63. Zhou G, Wang Q, Ho CL, Wong WY, Ma D, Wang L, Lin Z (2008) Robust tris-cyclometalated iridium(III) phosphors with ligands for effective charge carrier injection/transport: synthesis, redox, photophysical, and electrophosphorescent behavior. *Chem. Asian J.* 3: 1830.

64. Zhou G, Ho CL, Wong WY, Wang Q, Ma D, Wang L, Lin Z, Marder TB, Beeby A (2008) Manipulating charge-transfer character with electron-withdrawing main-group moieties for the color tuning of iridium electrophosphors. *Adv. Funct. Mater.* 18: 499.

65. Yokoyama D (2011) Molecular orientation in small-molecule organic light-emitting diodes. *J. Mater. Chem.* 21: 19187.

66. Kim KH, Moon CK, Lee JH, Kim SY, Kim JJ (2014) Highly efficient organic light-emitting diodes with phosphorescent emitters having high quantum yield and horizontal orientation of transition dipole moments. *Adv. Mater.* 26: 3844.

67. Kim KH, Ahn ES, Huh JS, Kim YH, Kim JJ (2016) Design of heteroleptic Ir complexes with horizontal emitting dipoles for highly efficient organic light-emitting diodes with an external quantum efficiency of 38%. *Chem. Mater.* 28: 7505.

68. Chao K, Shao K, Peng T, Zhu D, Wang Y, Liu Y, Su Z, Bryce MR (2013) New oxazoline- and thiazoline-containing heteroleptic iridium(III) complexes for highly-efficient phosphorescent organic light-emitting devices (PhOLEDs): colour tuning by varying the electroluminescence bandwidth. *J. Mater. Chem. C* 1: 6800.

69. Kuei CY, Tsai WL, Tong B, Jiao M, Lee WK, Chi Y, Wu CC, Liu SH, Lee GH, Chou PT (2016) Bis-tridentate Ir(III) complexes with nearly unitary RGB phosphorescence and organic light-emitting diodes with external quantum efficiency exceeding 31%. *Adv. Mater.* 28: 2795.

70. Hudson ZM, Sun C, Helander MG, Amarne H, Lu ZH, Wang S (2010) Enhancing phosphorescence and electrophosphorescence efficiency of cyclometalated Pt(II) compounds with triarylboron. *Adv. Funct. Mater.* 20: 3426.

71. Leopold H, Heinemeyer U, Wagenblast G, Münster I, Strassner T (2017) Changing the emission properties of phosphorescent C^C*-cyclometalated thiazol-2-ylidene platinum(II) complexes by variation of the β-diketonate ligands. *Chem. Eur. J.* 23: 1118.

72. Lu GZ, Jing YM, Han HB, Fang YL, Zheng YX (2017) Efficient electroluminescence of two heteroleptic platinum complexes with a 2-(5-phenyl-1,3,4-oxadiazol-2-yl)phenol ancillary ligand. *Organometallics* 36: 448.

73. Chan AKW, Ng M, Wong YC, Chan MY, Wong WT, Yam VWW (2017) Synthesis and characterization of luminescent cyclometalated platinum(II) complexes with tunable emissive colors and studies of their application in organic memories and organic light-emitting devices. *J. Am. Chem. Soc.* 139: 10750.

74. Kong FKW, Tang MC, Wong YC, Chan MY, Yam VWW (2016) Design strategy for high-performance dendritic carbazole-containing alkynylplatinum(II) complexes and their application in solution-processable organic light-emitting devices. *J. Am. Chem. Soc.* 138: 6281.

75. Cebrián C, Mauro M, Kourkoulos D, Mercandelli P, Hertel D, Meerholz K, Strassert CA, Cola LD (2013) Luminescent neutral platinum complexes bearing an asymmetric N^N^N ligand for high-performance solution-processed OLEDs. *Adv. Mater.* 25: 437.

76. Cheng G, Kui SCF, Ang WH, Ko MY, Chow PK, Kwong CL, Kwok CC, Ma C, Guan X, Low KH, Suc SJ, Che CM (2014) Structurally robust phosphorescent [Pt(O^N^C^N)] emitters for high performance organic light-emitting devices with power efficiency up to 126 lm W^{-1} and external quantum efficiency over 20%. *Chem. Sci.* 5: 4819.

77. Cheng G, Chow PK, Kui SCF, Kwok CC, Che CM (2013) High-efficiency polymer light-emitting devices with robust phosphorescent platinum(II) emitters containing tetradentate dianionic O^N^C^N ligands. *Adv. Mater.* 25: 6765.

78. Hsu CW, Lin CC, Chung MW, Chi Y, Lee GH, Chou PT, Chang CH, Chen PY (2011) Systematic investigation of the metal-structure-photophysics relationship of emissive d^{10}-complexes of group 11 elements: the prospect of application in organic light emitting devices. *J. Am. Chem. Soc.* 133: 12085.

79. Romanov AS, Yang L, Jones STE, Di D, Morley OJ, Drummond BH, Reponen APM, Linnolahti M, Credgington D, Bochmann M (2019) Dendritic carbene metal carbazole complexes as photoemitters for fully solution-processed OLEDs. *Chem. Mater.* 31: 3613.

80. Hashimoto M, Igawa S, Yashima M, Kawata I, Hoshino M, Osawa M (2011) Highly efficient green organic light-emitting diodes containing luminescent three-coordinate copper(I) complexes. *J. Am. Chem. Soc.* 133: 10348.

81. Au VKM, Wong KMC, Tsang DPK, Chan MY, Zhu N, Yam VWW (2010) High-efficiency green organic light-emitting devices utilizing phosphorescent bis-cyclometalated alkynylgold(III) complexes. *J. Am. Chem. Soc.* 132: 14273.

82. Mauro M, Procopio EQ, Sun Y, Chien CH, Donghi D, Panigati M, Mercandelli P, Mussini P, D'Alfonso G, Cola LD (2009) Highly emitting neutral dinuclear rhenium complexes as phosphorescent dopants for electroluminescent devices. *Adv. Funct. Mater.* 19: 2607.

83. Ranjan S, Lin SY, Hwang KC, Chi Y, Ching WL, Liu CS, Tao YT, Chien CH, Peng SM, Lee GH (2003) Realizing green phosphorescent light-emitting materials from rhenium(I) pyrazolato diimine complexes. *Inorg. Chem.* 42: 1248.

84. Xu LJ, Sun CZ, Xiao H, Wu Y, Chen ZN (2017) Green-light-emitting diodes based on tetrabromide manganese(II) complex through solution process. *Adv. Mater.* 29: 1605739.

85. Qin Y, Tao P, Gao L, She P, Liu S, Li X, Li F, Wang H, Zhao Q, Miao Y, Huang W (2019) Designing highly efficient phosphorescent neutral tetrahedral manganese(II) complexes for organic light-emitting diodes. *Adv. Opt. Mater.* 7: 1801160.

86. Cho YJ, Lee JY (2011) Low driving voltage, high quantum efficiency, high power efficiency, and little efficiency roll-off in red, green, and deep-blue phosphorescent organic light-emitting diodes using a high-triplet-energy hole transport material. *Adv. Mater.* 23: 4568.

87. Li CL, Su YJ, Tao YT, Chou PT, Chien CH, Cheng CC, Liu RS (2005) Yellow and red electrophosphors based on linkage isomers of phenylisoquinolinyliridium complexes: distinct differences in photophysical and electroluminescence properties. *Adv. Funct. Mater.* 15: 387.

88. Lai SL, Tao SL, Chan MY, Lo MF, Ng TW, Lee ST, Zhao WM, Lee CS (2011) Iridium(III) bis[2-(2-naphthyl)pyridine] (acetylacetonate)-based yellow and white phosphorescent organic light-emitting devices. *J. Mater. Chem.* 21: 4983.

89. Zhu W, Zhu M, Ke Y, Su L, Yuan M, Cao Y (2004) Synthesis and red electrophosphorescence of a novel cyclometalated iridium complex in polymer light-emitting diodes. *Thin Solid Films* 446: 128.

90. Ge G, He J, Guo H, Wang F, Zou D (2009) Highly efficient phosphorescent iridium (III) diazine complexes for OLEDs: different photophysical property between iridium (III) pyrazine complex and iridium (III) pyrimidine complex. *J. Organomet. Chem.* 694: 3050.

91. Cui LS, Liu Y, Liu XY, Jiang ZQ, Liao LS (2015) Design and synthesis of pyrimidine-based iridium(III) complexes with horizontal orientation for orange and white phosphorescent OLEDs. *ACS Appl. Mater. Interfaces* 7: 11007.

92. Guo LY, Zhang XL, Zhuo MJ, Liu C, Chen WY, Mi BX, Song J, Li YH, Gao ZQ (2014) Non-interlayer and color stable WOLEDs with mixed host and incorporating a new orange phosphorescent iridium complex. *Org. Electron.* 15: 2964.

93. Guo L Y, Zhang XL, Wang HS, Liu C, Li ZG, Liao ZJ, Mi BX, Zhou XH, Zheng C, Li YH, Gao ZQ (2015) New homoleptic iridium complexes with C^N=N type ligand for high efficiency orange and single emissive-layer white OLEDs. *J. Mater. Chem. C* 3: 5412.

94. Yao C, Jiao B, Yang X, Xu X, Dang J, Zhou G, Wu Z, Lv X, Zeng Y, Wong WY (2013) Tris(cyclometalated) iridium(III) phosphorescent complexes with 2-phenylthiazole-type ligands: synthesis, photophysical, redox and electrophosphorescent behavior. *Eur. J. Inorg. Chem.* 2013: 4754.

95. Fan C, Zhu L, Jiang B, Li Y, Zhao F, Ma D, Qin J, Yang C (2013) High power efficiency yellow phosphorescent OLEDs by using new iridium complexes with halogen-substituted 2-phenylbenzo[*d*]thiazole ligands. *J. Phys. Chem. C* 117: 19134.

96. Kim YJ, Son YH, Kwon JH (2013) Highly efficient yellow phosphorescent organic light-emitting diodes for two-peak tandem white organic light-emitting diode applications. *J. Inf. Disp.* 14: 109.

97. Fan C, Miao J, Jiang B, Yang C, Wu H, Qin J, Cao Y (2013) Highly efficient, solution-processed orange–red phosphorescent OLEDs by using new iridium phosphor with thieno[3,2-c]pyridine derivative as cyclometalating ligand. *Org. Electron.* 14: 3392.

98. Tsuzuki T, Shirasawa N, Suzuki T, Tokito S (2003) Color tunable organic light-emitting diodes using pentafluorophenyl-substituted iridium complexes. *Adv. Mater.* 15: 1455.

99. Chen CH, Hsu LC, Rajamalli P, Chang YW, Wu FI, Liao CY, Chiu MJ, Chou PY, Huang MJ, Chu LK, Cheng CH (2014) Highly efficient orange and deep-red organic light emitting diodes with long operational lifetimes using carbazole–quinoline based bipolar host materials. *J. Mater. Chem. C* 2: 6183.

100. Mei Q, Wang L, Guo Y, Weng J, Yan F, Tian B, Tong B (2012) A highly efficient red electrophosphorescent iridium(III) complex containing phenyl quinazoline ligand in polymer light-emitting diodes. *J. Mater. Chem.* 22: 6878.

101. Zhu M, Zou J, Hu S, Li C, Yang C, Wu H, Qin J, Cao Y (2012) Highly efficient single-layer white polymer light-emitting devices employing triphenylamine-based iridium dendritic complexes as orange emissive component. *J. Mater. Chem.* 22: 361.

102. Gong S, Chen Y, Yang C, Zhong C, Qin J, Ma D (2010) De novo design of silicon-bridged molecule towards a bipolar host: all-phosphor white organic light-emitting devices exhibiting high efficiency and low efficiency roll-off. *Adv. Mater.* 22: 5370.

103. Zhou G, Wang Q, Wang X, Ho CL, Wong WY, Ma D, Wang L, Lin Z (2010) Metallophosphors of platinum with distinct main-group elements: a versatile approach towards color tuning and white-light emission with superior efficiency/color quality/brightness trade-offs. *J. Mater. Chem.* 20: 7472.

104. Yang C, Zhang X, You H, Zhu L, Chen L, Zhu L, Tao Y, Ma D, Shuai Z, Qin J (2007) Tuning the energy level and photophysical and electroluminescent properties of heavy metal complexes by controlling the ligation of the metal with the carbon of the carbazole unit. *Adv. Funct. Mater.* 17: 651.

105. Velusamy M, Chen CH, Wen YS, Lin JT, Lin CC, Lai CH, Chou PT (2010) Cyclometalated platinum(II) complexes of lepidine-based ligands as highly efficient electrophosphors. *Organometallics* 29: 3912.

106. Hudson ZM, Helander MG, Lu ZH, Wang S (2011) Highly efficient orange electrophosphorescence from a trifunctional organoboron–Pt(II) complex. *Chem. Commun.* 47: 755.

107. Taneda M, Yasuda T, Adachi C (2011) Horizontal orientation of a linear-shaped platinum(II) complex in organic light-emitting diodes with a high light out-coupling efficiency. *Appl. Phys. Express* 4: 71602.

108. Huang LM, Tu GM, Chi Y, Hung WY, Song YC, Tseng MR, Chou PT, Lee GH, Wong KT, Cheng SH, Tsai WS (2013) Mechanoluminescent and efficient white OLEDs for Pt(II) phosphors bearing spatially encumbered pyridinyl pyrazolate chelates. *J. Mater. Chem. C* 1: 7582.

109. Li M, Chen WH, Lin MT, Omary MA, Shepherd ND (2009) Near-white and tunable electrophosphorescence from bis[3,5-bis(2-pyridyl)-1,2,4-triazolato]platinum(II)-based organic light emitting diodes. *Org. Electron.* 10: 863.

110. Wang Q, Oswald IWH, Perez MR, Jia H, Gnade BE, Omary MA (2013) Exciton and polaron quenching in doping-free phosphorescent organic light-emitting diodes from a Pt(II)-based fast phosphor. *Adv. Funct. Mater.* 23: 5420.

111. Tseng CH, Fox MA, Liao JL, Ku CH, Sie ZT, Chang CH, Wang JY, Chen ZN, Lee GH, Chi Y (2017) Luminescent Pt(II) complexes featuring imidazolylidene–pyridylidene and dianionic bipyrazolate: from fundamentals to OLED fabrications. *J. Mater. Chem. C* 5: 1420.

112. Lu W, Mi BX, Chan MCW, Hui Z, Che CM, Zhu N, Lee ST (2004) Light-emitting tridentate cyclometalated platinum(II) complexes containing σ-alkynyl auxiliaries: tuning of photo- and electrophosphorescence. *J. Am. Chem. Soc.* 126: 4958.

113. Chen JL, Chang SY, Chi Y, Chen K, Cheng YM, Lin CW, Lee GH, Chou PT, Wu CH, Shih PI, Shu CF (2008) Pt^{II} complexes with 6-(5-trifluoromethyl-pyrazol-3-yl)-2,2′-bipyridine terdentate chelating ligands: synthesis, characterization, and luminescent properties. *Chem. Asian J.* 3: 2112.

114. Kwok CC, Ngai HMY, Chan SC, Sham IHT, Che CM, Zhu N (2005) [(O^N^N)PtX] complexes as a new class of light-emitting materials for electrophosphorescent devices. *Inorg. Chem.* 44: 4442.

115. Yuen MY, Kui SCF, Low KH, Kwok CC, Chui SSY, Ma CW, Zhu N, Che CM (2010) Synthesis, photophysical and electrophosphorescent properties of fluorene-based platinum(II) complexes. *Chem. Eur. J.* 16: 14131.

116. Che CM, Chan SC, Xiang HF, Chan MCW, Liu Y, Wang Y (2004) Tetradentate Schiff base platinum(II) complexes as new class of phosphorescent materials for high-efficiency and white-light electroluminescent devices. *Chem. Commun.* 1484.

117. Zhang J, Zhao F, Zhu X, Wong WK, Ma D, Wong WY (2012) New phosphorescent platinum(II) Schiff base complexes for PHOLED applications. *J. Mater. Chem.* 22: 16448.

118. Zhu ZQ, Klimes K, Holloway S, Li J (2017) Efficient cyclometalated platinum(II) complex with superior operational stability. *Adv. Mater.* 29: 1605002.

119. Lai SL, Tong WY, Kui SCF, Chan MY, Kwok CC, Che CM (2013) High efficiency white organic light-emitting devices incorporating yellow phosphorescent platinum(II) complex and composite blue host. *Adv. Funct. Mater.* 23: 5168.

120. Cao L, Klimes K, Ji Y, Fleetham T, Li J (2021) Efficient and stable organic light-emitting devices employing phosphorescent molecular aggregates. *Nat. Photon.* 15: 230.

121. Fleetham T, Ji Y, Huang L, Fleetham TS, Li J (2017) Efficient and stable single-doped white OLEDs using a palladium-based phosphorescent excimer. *Chem. Sci.* 8: 7983.

122. Cheng YM, Lee GH, Chou PT, Chen LS, Chi Y, Yang CH, Song YH, Chang SY, Shih PI, Shu CF (2008) Rational design of chelating phosphine functionalized Os(II) emitters and fabrication of orange polymer light-emitting diodes using solution process. *Adv. Funct. Mater.* 18: 183.

123. Chang CH, Chen CC, Wu CC, Chang SY, Hung JY, Chi Y (2010) High-color-rendering pure-white phosphorescent organic light-emitting devices employing only two complementary colors. *Org. Electron.* 11: 266.

124. Tang MC, Tsang DPK, Chan MMY, Wong KMC, Yam VWW (2013) Dendritic luminescent gold(III) complexes for highly efficient solution-processable organic light-emitting devices. *Angew. Chem. Int. Ed.* 52: 446.

125. Chen Z, Zhang H, Wen D, Wu W, Zeng Q, Chen S, Wong WY (2020) A simple and efficient approach toward deep-red to near-infrared-emitting iridium(III) complexes for organic light-emitting diodes with external quantum efficiencies of over 10%. *Chem. Sci.* 11: 2342.

126. Tsuboyama A, Iwawaki H, Furugori M, Mukaide T, Kamatani J, Igawa S, Moriyama T, Miura S, Takiguchi T, Okada S, Hoshino M, Ueno K (2003) Homoleptic cyclometalated iridium complexes with highly efficient red phosphorescence and application to organic light-emitting diode. *J. Am. Chem. Soc.* 125: 12971.

127. Li Y, Zhang W, Zhang L, Wen X, Yin Y, Liu S, Xie W, Zhao H, Tao S (2013) Ultra-high general and special color rendering index white organic light-emitting device based on a deep red phosphorescent dye. *Org. Electron.* 14: 3201.

128. Ho CL, Wong WY, Gao ZQ, Chen CH, Cheah KW, Yao B, Xie Z, Wang Q, Ma D, Wang L, Yu XM, Kwok HS, Lin Z (2008) Red-light-emitting iridium complexes with hole-transporting 9-arylcarbazole moieties for electrophosphorescence efficiency/color purity trade-off optimization. *Adv. Funct. Mater.* 18: 319.

129. Zhou G, Wong WY, Yao B, Xie Z, Wang L (2007) Triphenylamine-dendronized pure red iridium phosphors with superior OLED efficiency/color purity trade-offs. *Angew. Chem. Int. Ed.* 46: 1149.

130. Su YJ, Huang HL, Li CL, Chien CH, Tao YT, Chou PT, Datta S, Liu RS (2003) Highly efficient red electrophosphorescent devices based on iridium isoquinoline complexes: remarkable external quantum efficiency over a wide range of current. *Adv. Mater.* 15: 884.

131. Kim KH, Lee S, Moon CK, Kim SY, Park YS, Lee JH, Lee JW, Huh J, You Y, Kim JJ (2014) Phosphorescent dye-based supramolecules for high-efficiency organic light-emitting diodes. *Nat. Commun.* 5: 4769.
132. Reineke S, Lindner F, Schwartz G, Seidler N, Walzer K, Lüssem B, Leo K (2009) White organic light-emitting diodes with fluorescent tube efficiency. *Nature* 459: 234.
133. Kim DH, Cho NS, Oh HY, Yang JH, Jeon WS, Park JS, Suh MC, Kwon JH (2011) Highly efficient red phosphorescent dopants in organic light-emitting devices. *Adv. Mater.* 23: 2721.
134. Giridhar T, Han TH, Cho W, Saravanan C, Lee TW, Jin SH (2014) An easy route to red emitting homoleptic Ir^III complex for highly efficient solution-processed phosphorescent organic light-emitting diodes. *Chem. Eur. J.* 20: 8260.
135. Park YH, Kim YS (2007) Heteroleptic tris-cyclometalated iridium(III) complexes with phenylpridine and diphenylquinoline derivative ligands. *Thin Solid Films* 515: 5084.
136. Fan CH, Sun P, Su TH, Cheng CH (2011) Host and dopant materials for idealized deep-red organic electrophosphorescence devices. *Adv. Mater.* 23: 2981.
137. Yang X, Xu X, Dang JS, Zhou G, Ho CL, Wong WY (2016) From mononuclear to dinuclear iridium(III) complex: effective tuning of the optoelectronic characteristics for organic light-emitting diodes. *Inorg. Chem.* 55: 1720.
138. O'Brien DF, Baldo MA, Thompson ME, Forrest SR (1999) Improved energy transfer in electrophosphorescent devices. *Appl. Phys. Lett.* 74: 442.
139. Fukagawa H, Shimizu T, Hanashima H, Osada Y, Suzuki M, Fujikake H (2012) Highly efficient and stable red phosphorescent organic light-emitting diodes using platinum complexes. *Adv. Mater.* 24: 5099.
140. Fleetham T, Li G, Li J (2015) Efficient red-emitting platinum complex with long operational stability. *ACS Appl. Mater. Interfaces* 7: 16240.
141. Chow PK, Cheng G, Tong GSM, To WP, Kwong WL, Low KH, Kwok CC, Ma C, Che CM (2015) Luminescent pincer platinum(II) complexes with emission quantum yields up to almost unity: photophysics, photoreductive C–C bond formation, and materials applications. *Angew. Chem. Int. Ed.* 54: 2084.
142. Xiang HF, Chan SC, Wu KKY, Che CM, Lai PT (2005) High-efficiency red electrophosphorescence based on neutral bis(pyrrole)-diimine platinum(II) complex. *Chem. Commun.* 1408.
143. Xiang HF, Xu ZX, Roy VAL, Yan BP, Chan SC, Che CM, Lai PT (2008) Deep-red to near-infrared electrophosphorescence based on bis(8-hydroxyquinolato) platinum(II) complexes. *Appl. Phys. Lett.* 92: 163305.
144. Kim KH, Liao JL, Lee SW, Sim B, Moon CK, Lee GH, Kim HJ, Chi Y, Kim JJ (2016) Crystal organic light-emitting diodes with perfectly oriented non-doped Pt-based emitting layer. *Adv. Mater.* 28: 2526.
145. Hsu CW, Zhao Y, Yeh HH, Lu CW, Fan C, Hu Y, Robertson N, Lee GH, Sun XW, Chi Y (2015) Efficient Pt(II) emitters assembled from neutral bipyridine and dianionic bipyrazolate: designs, photophysical characterization and the fabrication of non-doped OLEDs. *J. Mater. Chem. C* 3: 10837.
146. Ly KT, Chen-Cheng RW, Lin HW, Shiau YJ, Liu SH, Chou PT, Tsao CS, Huang YC, Chi Y (2017) Near-infrared organic light-emitting diodes with very high external quantum efficiency and radiance. *Nat. Photon.* 11: 63.
147. Tung YL, Chen LS, Chi Y, Chou PT, Cheng YM, Li EYT, Lee GH, Shu CF, Wu FI, Carty AJ (2006) Orange and red organic light-emitting devices employing neutral Ru(II) emitters: rational design and prospects for color tuning. *Adv. Funct. Mater.* 16: 1615.
148. Chien CH, Hsu FM, Shu CF, Chi Y (2009) Efficient red electrophosphorescence from a fluorene-based bipolar host material. *Org. Electron.* 10: 871.
149. Francis B, Nolasco MM, Brandão P, Ferreira RAS, Carvalho RS, Cremona M, Carlos LD (2020) Efficient visible-light-excitable Eu^3+ complexes for red organic light-emitting diodes. *Eur. J. Inorg. Chem.* 2020: 1260.
150. Ilmi R, Sun W, Dutra JDL, Al-Rasbi NK, Zhou L, Qian PC, Wong WY, Raithby PR, Khan MS (2020) Monochromatic red electroluminescence from a homodinuclear europium(III) complex of a β-diketone tethered by 2,2′-bipyrimidine. *J. Mater. Chem. C* 8: 9816.
151. Yu J, Zhou L, Zhang H, Zheng Y, Li H, Deng R, Peng Z, Li Z (2005) Efficient electroluminescence from new lanthanide (Eu^3+, Sm^3+) complexes. *Inorg. Chem.* 44: 1611.
152. Behzad SK, Amini MM, Ghanbari M, Janghouri M, Anzenbacher P, Ng SW (2017) Synthesis, structure, photoluminescence, and electroluminescence of four europium complexes: fabrication of pure red organic light-emitting diodes from europium complexes. *Eur. J. Inorg. Chem.* 2017: 3644.
153. Rajamouli B, Viswanath CSD, Giri S, Jayasankar CK, Sivakumar V (2017) Carbazole functionalized new bipolar ligand for monochromatic red light-emitting europium(III) complex: combined experimental and theoretical study. *New J. Chem.* 41: 3112.

154. Xiang H, Cheng J, Ma X, Zhou X, Chruma JJ (2013) Near-infrared phosphorescence: materials and applications. *Chem. Soc. Rev.* 42: 6128.
155. Zhang Y, Wang Y, Song J, Qu J, Li B, Zhu W, Wong WY (2018) Near-infrared emitting materials via harvesting triplet excitons: molecular design, properties, and application in organic light emitting diodes. *Adv. Opt. Mater.* 6: 1800466.
156. Li X, Lovell JF, Yoon J, Chen X (2020) Clinical development and potential of photothermal and photodynamic therapies for cancer. *Nat. Rev. Clin. Oncol.* 17: 657.
157. Abdurahman R, Yang CX, Yan XP (2016) Conjugation of a photosensitizer to near infrared light renewable persistent luminescence nanoparticles for photodynamic therapy. *Chem. Commun.* 52: 13303.
158. Maldiney T, Bessière A, Seguin, J, Teston E, Sharma SK, Viana B, Bos AJJ, Dorenbos P, Bessodes M, Gourier D (2014) The in vivo activation of persistent nanophosphors for optical imaging of vascularization, tumours and grafted cells. *Nat. Mater.* 13: 418.
159. Wang J, Ma Q, Wang Y, Shen H, Yuan Q (2017) Recent progress in biomedical applications of persistent luminescence nanoparticles. *Nanoscale* 9: 6204.
160. Wang X, Cui L, Zhou N, Zhu W, Wang R, Qian X, Xu Y (2013) A highly selective and sensitive near-infrared fluorescence probe for arylamine N-acetyltransferase 2 in vitro and in vivo. *Chem. Sci.* 4: 2936.
161. Li X, Li X, Ma H (2020) A near-infrared fluorescent probe reveals decreased mitochondrial polarity during mitophagy. *Chem. Sci.* 11: 1617.
162. Bünzli JCG, Eliseeva SV (2010) Lanthanide NIR luminescence for telecommunications, bioanalyses and solar energy conversion. *J. Rare Earths* 28: 824.
163. Baride A, Meruga JM, Douma C, Langerman D, Crawford G, Kellar JJ, Cross WM, May PS (2015) A NIR-to-NIR upconversion luminescence system for security printing applications. *RSC Adv.* 5: 101338.
164. Kumar P, Singh S, Gupta BK (2016) Future prospects of luminescent nanomaterial based security inks: from synthesis to anti-counterfeiting applications. *Nanoscale* 8: 14297.
165. Koppe M, Egelhaaf HJ, Dennler G, Scharber MC, Brabec CJ, Schilinsky P, Hoth CN (2010) Near IR sensitization of organic bulk heterojunction solar cells: towards optimization of the spectral response of organic solar cells. *Adv. Funct. Mater.* 20: 338.
166. Richards BS (2006) Luminescent layers for enhanced silicon solar cell performance: down-conversion. *Solar En. Mat. Sol. Cells* 90: 1189.
167. van der Ende BM, Aarts L, Meijerink A (2009) Near-infrared quantum cutting for photovoltaics. *Adv. Mater.* 21: 3073.
168. Zampetti A, Minotto A, Cacialli F (2019) Near-infrared (NIR) organic light-emitting diodes (OLEDs): challenges and opportunities. *Adv. Funct. Mater.* 29: 1807623.
169. Qiao J, Duan L, Tang L, He L, Wang L, Qiu Y (2009) High-efficiency orange to near-infrared emissions from bis-cyclometalated iridium complexes with phenyl-benzoquinoline isomers as ligands. *J. Mater. Chem.* 19: 6573.
170. Tao R, Qiao J, Zhang G, Duan L, Chen C, Wang L, Qiu Y (2013) High-efficiency near-infrared organic light-emitting devices based on an iridium complex with negligible efficiency roll-off. *J. Mater. Chem. C* 1: 6446.
171. Cao X, Miao J, Zhu M, Zhong C, Yang C, Wu H, Qin J, Cao Y (2015) Near-infrared polymer light-emitting diodes with high efficiency and low efficiency roll-off by using solution-processed iridium(III) phosphors. *Chem. Mater.* 27: 96.
172. Kesarkar S, Mróz W, Penconi M, Pasini M, Destri S, Cazzaniga M, Ceresoli D, Mussini PR, Baldoli C, Giovanella U, Bossi A (2016) Near-IR emitting iridium(III) complexes with heteroaromatic β-diketonate ancillary ligands for efficient solution-processed OLEDs: structure–property correlations. *Angew. Chem. Int. Ed.* 55: 2714.
173. Xue J, Xin L, Hou J, Duan L, Wang R, Wei Y, Qiao J (2017) Homoleptic facial Ir(III) complexes via facile synthesis for high-efficiency and low-roll-off near-infrared organic light-emitting diodes over 750 nm. *Chem. Mater.* 29: 4775.
174. Chen Z, Wang L, Ho CL, Chen S Suramitr S, Plucksacholatarn A, Zhu N, Hannongbua S, Wong WY (2018) Smart design on the cyclometalated ligands of iridium(III) complexes for facile tuning of phosphorescence color spanning from deep-blue to near-infrared. *Adv. Opt. Mater.* 6: 1800824.
175. Zhou J, Fu G, He Y, Ma L, Li W, Feng W, Lü X (2019) Efficient and low-efficiency-roll-off near-infrared (NIR) polymer light-emitting diodes (PLEDs) based on an asymmetric binuclear iridium(III)-complex. *J. Lumin.* 209: 427.
176. You C, Liu D, Yu J, Tan H, Zhu M, Zhang B, Liu Y, Wang Y, Zhu W (2020) Boosting efficiency of near-infrared emitting iridium(III) phosphors by administrating their π–π conjugation effect of core–shell structure in solution-processed OLEDs. *Adv. Opt. Mater.* 8: 2000154.

177. You C, Liu D, Zhu M, Yu J, Zhang B, Liu Y, Wang Y, Zhu W (2020) $\sigma-\pi$ and p–π conjugation induced NIR-emitting iridium(III) complexes anchored by flexible side chains in a rigid dibenzo[a,c]phenazine moiety and their application in highly efficient solution-processable NIR-emitting devices. *J. Mater. Chem. C* 8: 7079.

178. You C, Liu D, Meng F, Wang Y, Yu J, Wang S, Su S, Zhu W (2019) Iridium(III) phosphors with rigid fused-heterocyclic chelating architectures for efficient deep-red/near-infrared emissions in polymer light-emitting diodes. *J. Mater. Chem. C* 7: 10961.

179. Pal AK, Cordes DB, Slawin AMZ, Momblona C, Pertegás A, Ortí E, Bolink HJ, Zysman-Colman E (2017) Simple design to achieve red-to-near-infrared emissive cationic Ir(III) emitters and their use in light emitting electrochemical cells. *RSC Adv.* 7: 31833.

180. Chen GY, Chang BR, Shih TA, Lin CH, Lo CL, Chen YZ, Liu YX, Li YR, Guo JT, Lu CW, Yang ZP, Su HC (2019) Cationic Ir(III) emitters with near-infrared emission beyond 800 nm and their use in light-emitting electrochemical cells. *Chem. Eur. J.* 25: 5489.

181. Liu YX, Yi RH, Lin CH, Yang ZP, Lu CW, Su HC (2020) Near-infrared light-emitting electrochemical cells based on the excimer emission of a cationic iridium complex. *J. Mater. Chem. C* 8: 14378.

182. Tao R, Qiao J, Zhang G, Duan L, Wang L, Qiu Y (2012) Efficient near-infrared-emitting cationic iridium complexes as dopants for OLEDs with small efficiency roll-off. *J. Phys. Chem. C* 116: 11658.

183. Xin L, Xue J, Lei G, Qiao J (2015) Efficient near-infrared-emitting cationic iridium complexes based on highly conjugated cyclometalated benzo[g]phthalazine derivatives. *RSC Adv.* 5: 42354.

184. Graham KR, Yang Y, Sommer JR, Shelton AH, Schanze KS, Xue J, Reynolds JR (2011) Extended conjugation platinum(II) porphyrins for use in near-infrared emitting organic light emitting diodes. *Chem. Mater.* 23: 5305.

185. Zhang Y, Yin Z, Meng F, Yu J, You C, Yang S, Tan H, Zhu W, Su S (2017) Tetradentate Pt(II) 3,6-substitued salophen complexes: synthesis and tuning emission from deep-red to near infrared by appending donor-acceptor framework. *Org. Electron.* 50: 317.

186. Zhang Y, Chen Z, Wang X, He J, Wu J, Liu H, Song J, Qu J, Chan WTK, Wong WY (2018) Achieving NIR emission for donor-acceptor type platinum(II) complexes by adjusting coordination position with isomeric ligands. *Inorg. Chem.* 57: 14208.

187. Fu G, He Y, Li W, Wang B, Lü X, He H, Wong WY (2019) Efficient polymer light-emitting diodes (PLEDs) based on chiral [Pt(C^N)(N^O)] complexes with near-infrared (NIR) luminescence and circularly polarized (CP) light. *J. Mater. Chem. C* 7: 13743.

188. Cocchi M, Kalinowski J, Virgili D, Williams JAG (2008) Excimer-based red/near-infrared organic light-emitting diodes with very high quantum efficiency. *Appl. Phys. Lett.* 92: 113302.

189. Nisic F, Colombo A, Dragonetti C, Roberto D, Valore A, Malicka JM, Cocchi M, Freeman GR, Williams JAG (2014) Platinum(II) complexes with cyclometallated 5-π-delocalized-donor-1,3-di(2-pyridyl)benzene ligands as efficient phosphors for NIR-OLEDs. *J. Mater. Chem. C* 2: 1791.

190. Wei YC, Wang SF, Hu Y, Liao LS, Chen DG, Chang KH, Wang CW, Liu SH, Chan WH, Liao JL, Hung WY, Wang TH, Chen PT, Hsu HF, Chi Y, Chou PT (2020) Overcoming the energy gap law in near-infrared OLEDs by exciton–vibration decoupling. *Nat. Photon.* 14: 570.

191. Wang SF, Yuan Y, Wei YC, Chan WH, Fu LW, Su BK, Chen IY, Chou KJ, Chen PT, Hsu HF, Ko CL, Hung WY, Lee CS, Chou PT, Chi Y (2020) Highly efficient near-infrared electroluminescence up to 800 nm using platinum(II) phosphors. *Adv. Funct. Mater.* 30: 2002173.

192. Zhang K, Liu Y, Hao Z, Lei G, Cui S, Zhu W, Liu Y (2020) A feasible approach to obtain near-infrared (NIR) emission from binuclear platinum(II) complexes containing centrosymmetric isoquinoline ligand in PLEDs. *Org. Electron.* 87: 105902.

193. Zhang K, Wang T, Wu T, Ding Z, Zhang Q, Zhu W, Liu Y (2021) An effective strategy to obtain near-infrared emission from shoulder to shoulder-type binuclear platinum(II) complexes based on fused pyrene core bridged isoquinoline ligands. *J. Mater. Chem. C* 9: 2282.

194. Zhang YM, Meng F, Tang JH, Wang Y, You C, Tan H, Liu Y, Zhong YW, Su S, Zhu W (2016) Achieving near-infrared emission in platinum(II) complexes by using an extended donor–acceptor-type ligand. *Dalton Trans.* 45: 5071.

195. Xiong W, Meng F, Tan H, Wang Y, Wang P, Zhang Y, Tao Q, Su S, Zhu W (2016) Dinuclear platinum complexes containing aryl-isoquinoline and oxadiazole-thiol with an efficiency of over 8.8%: in-depth investigation of the relationship between their molecular structure and near-infrared electroluminescent properties in PLEDs. *J. Mater. Chem. C* 4: 6007.

196. Wu X, Chen DG, Liu D, Liu SH, Shen SW. Wu CI, Xie G, Zhou J, Huang ZX, Huang CY, Su SJ, Zhu W, Chou PT (2020) Highly emissive dinuclear platinum(III) complexes. *J. Am. Chem. Soc.* 142: 7469.

197. Ibrahim-Ouali M, Dumur F (2019) Recent advances on metal-based near-infrared and infrared emitting OLEDs. *Molecules* 24: 1412.
198. Barbieri A, Bandini E, Monti F, Praveen VK, Armaroli N (2016) The rise of near-infrared emitters: organic dyes, porphyrinoids, and transition metal complexes. *Top. Curr. Chem. (Z)* 374: 47.
199. To WP, Wan Q, Tong GSM, Che CM (2020) Recent advances in metal triplet emitters with d^6, d^8, and d^{10} electronic configurations. *Trends Chem.* 2: 796.
200. Yam VWW, Wong KMC (2011) Luminescent metal complexes of d^6, d^8 and d^{10} transition metal centres. *Chem. Commun.* 47: 11579.
201. Yam VWW, Au VKM, Leung SYL (2015) Light-emitting self-assembled materials based on d^8 and d^{10} transition metal complexes. *Chem. Rev.* 115: 7589.
202. Otto S, Grabolle M, Förster C, Kreitner C, Resch-Genger U, Heinze K (2015) [Cr(ddpd)$_2$]$^{3+}$: a molecular, water-soluble, highly NIR-emissive ruby analogue. *Angew. Chem. Int. Ed.* 54: 11572.
203. Rosenow TC, Walzer K, Leo K (2008) Near-infrared organic light emitting diodes based on heavy metal phthalocyanines. *J. Appl. Phys.* 103: 43105.
204. Yuan Y, Liao JL, Ni SF, Jen AKY, Lee CS, Chi Y (2020) Boosting efficiency of near-infrared organic light-emitting diodes with Os(II)-based pyrazinyl azolate emitters. *Adv. Funct. Mater.* 30: 1906738.
205. Lee TC, Hung JY, Chi Y, Cheng YM, Lee GH, Chou PT, Chen CC, Chang CH, Wu CC (2009) Rational design of charge-neutral, near-infrared-emitting osmium(II) complexes and OLED fabrication. *Adv. Funct. Mater.* 19: 2639.
206. Tung YL, Lee SW, Chi Y, Chen LS, Shu CF, Wu FI, Carty AJ, Chou PT, Peng SM, Lee GH (2005) Organic light-emitting diodes based on charge-neutral Ru^{II} phosphorescent emitters. *Adv. Mater.* 17: 1059.
207. Bergman SD, Gut D, Kol M, Sabatini C, Barbieri A, Barigelletti F (2005) Eilatin complexes of ruthenium and osmium. synthesis, electrochemical behavior, and near-IR luminescence. *Inorg. Chem.* 44: 7943.
208. Xun S, Zhang J, Li X, Ma D, Wang ZY (2008) Synthesis and near-infrared luminescent properties of some ruthenium complexes. *Synth. Met.* 158: 484.
209. Borisov SM, Nuss G, Haas W, Saf R, Schmuck M, Klimant I (2009) New NIR-emitting complexes of platinum(II) and palladium(II) with fluorinated benzoporphyrins. *J. Photochem Photobiol. A: Chem.* 201: 128.
210. Li LK, Tang MC, Cheung WL, Lai SL, Ng M, Chan CKM, Chan MY, Yam VWW (2019) Rational design strategy for the realization of red- to near-infrared-emitting alkynylgold(III) complexes and their applications in solution-processable organic light-emitting devices. *Chem. Mater.* 31: 6706.
211. Toganoh M, Niino T, Furuta H (2008) Luminescent Au(III) organometallic complex of N-confused tetraphenylporphyrin. *Chem. Commun.* 4070.
212. Felder D, Nierengarten JF, Barigelletti F, Ventura B, Armaroli N (2001) Highly luminescent Cu(I)-phenanthroline complexes in rigid matrix and temperature dependence of the photophysical properties. *J. Am. Chem. Soc.* 123: 6291.
213. Büldt LA, Guo X, Vogel R, Prescimone A, Wenger OS (2017) A tris(diisocyanide)chromium(0) complex is a luminescent analog of Fe(2,2′-bipyridine)$_3$$^{2+}$. *J. Am. Chem. Soc.* 139: 985.
214. Nagata R, Nakanotani H, Adachi C (2017) Near-infrared electrophosphorescence up to 1.1 μm using a thermally activated delayed fluorescence molecule as triplet sensitizer. *Adv. Mater.* 29: 1604265.
215. Wei H, Yu G, Zhao Z, Liu Z, Bian Z, Huang C (2013) Constructing lanthanide [Nd(III), Er(III) and Yb(III)] complexes using a tridentate N,N,O-ligand for near-infrared organic light-emitting diodes. *Dalton Trans.* 42: 8951.
216. Ahmed Z, Aderne RE, Kai J, Resende JALC, Padilla-Chavarría HI, Cremona M (2017) Near infrared organic light emitting devices based on a new erbium(III) β-diketonate complex: synthesis and optoelectronic investigations. *RSC Adv.* 7: 18239.
217. Martín-Ramos P, Coya C, Lavín V, Martín IR, Silva MR, Silva PSP, García-Vélez M, Álvarez AL, Martín-Gil J (2014) Active layer solution-processed NIR-OLEDs based on ternary erbium(III) complexes with 1,1,1-trifluoro-2,4-pentanedione and different N,N-donors. *Dalton Trans.* 43: 18087.
218. O'Riordan A, O'Connor E, Moynihan S, Nockemann P, Fias P, Deun RV, Cupertino D, Mackie P, Redmond G (2006) Near infrared electroluminescence from neodymium complex-doped polymer light emitting diodes. *Thin Solid Films* 497: 299.
219. Jinnai K, Kabe R, Adachi C (2017) A near-infrared organic light-emitting diode based on an Yb(III) complex synthesized by vacuum co-deposition. *Chem. Commun.* 53: 5457.
220. Utochnikova VV, Kalyakina AS, Bushmarinov IS, Vashchenko AA, Marciniak L, Kaczmarek AM, Deun RV, Bräse S, Kuzmina NP (2016) Lanthanide 9-anthracenate: solution processable emitters for efficient purely NIR emitting host-free OLEDs. *J. Mater. Chem. C* 4: 9848.

221. Schanze KS, Reynolds JR, Boncella JM, Harrison BS, Foley TJ, Bouguettaya M, Kang TS (2003) Near-infrared organic light emitting diodes. *Synth. Met.* 137: 1013.
222. Zang FX, Hong ZR, Li WL, Li MT, Sun XY (2004) 1.4 μm band electroluminescence from organic light-emitting diodes based on thulium complexes. *Appl. Phys. Lett.* 84: 2679.
223. Zang FX, Li WL, Hong ZR, Wei HZ, Li MT, Sun XY, Lee CS (2004) Observation of 1.5μm photoluminescence and electroluminescence from a holmium organic complex. *Appl. Phys. Lett.* 84: 5115.
224. Tang CW, VanSlyke SA (1987) Organic electroluminescent diodes. *Appl. Phys. Lett.* 51: 913.
225. Lu CY, Jiao M, Lee WK, Chen CY, Tsai WL, Lin CY, Wu CC (2016) Achieving above 60% external quantum efficiency in organic light-emitting devices using ITO-free low-index transparent electrode and emitters with preferential horizontal emitting dipoles. *Adv. Funct. Mater.* 26: 3250.
226. Nowy S, Krummacher BC, Frischeisen J, Reinke NA, Brütting W (2008) Light extraction and optical loss mechanisms in organic light-emitting diodes: influence of the emitter quantum efficiency. *J. Appl. Phys.* 104: 123109.
227. Editorial (2021) Live long and prosper. *Nat. Photon.* 15: 629.
228. Pile DFP (2021) Emitting organically. *Nat. Photon.* 15: 635.
229. Pile DFP (2021) Trailblazing lasing. *Nat. Photon.* 15: 637.
230. Yamazaki Y, Takeda H, Ishitani O (2015) Photocatalytic reduction of CO_2 using metal complexes. *J. Photochem. Photobiol. C* 25: 106.
231. Qiao X, Li Q, Schaugaard RN, Noffke BW, Liu Y, Li D, Liu L, Raghavachari K, Li L (2017) Well-defined nanographene–rhenium complex as an efficient electrocatalyst and photocatalyst for selective CO_2 reduction. *J. Am. Chem. Soc.* 139: 3934.
232. Chen X, McAteer D, McGuinness C, Godwin I, Coleman JN, McDonald AR (2018) RuII photosensitizer-functionalized two-dimensional MoS_2 for light-driven hydrogen evolution. *Chem. Eur. J.* 24: 351.
233. Song H, Kaiser JT, Barton JK (2012) Crystal structure of Δ-[Ru(bpy)$_2$dppz]$_2$+ bound to mismatched DNA reveals side-by-side metalloinsertion and intercalation. *Nat. Chem.* 4: 615.
234. Gabr MT, Pigge FC (2018) Platinum(II) complexes with sterically expansive tetraarylethylene ligands as probes for mismatched DNA. *Inorg. Chem.* 57: 12641.
235. Ho PY, Ho CL, Wong WY (2020) Recent advances of iridium(III) metallophosphors for health-related applications. *Coord. Chem. Rev.* 413: 213267.
236. Ma Y, Liang H, Zeng Y, Yang H, Ho CL, Xu W, Zhao Q, Huang W, Wong WY (2016) Phosphorescent soft salt for ratiometric and lifetime imaging of intracellular pH variations. *Chem. Sci.* 7: 3338.
237. Li J, Ma Y, Liu S, Mao Z, Chi Z, Qian PC, Wong WY (2020) Soft salts based on platinum(II) complexes with high emission quantum efficiencies in the near infrared region for in vivo imaging. *Chem. Commun.* 56: 11681.
238. Qian M, Zhang R, Hao J, Zhang W, Zhang Q, Wang J, Tao Y, Chen S, Fang J, Huang W (2015) Dramatic enhancement of power conversion efficiency in polymer solar cells by conjugating very low ratio of triplet iridium complexes to PTB7. *Adv. Mater.* 27: 3546.
239. Xu X, Feng K, Bi Z, Ma W, Zhang G, Peng Q (2019) Single-junction polymer solar cells with 16.35% efficiency enabled by a platinum(II) complexation strategy. *Adv. Mater.* 31: 1901872.

15 Lanthanide-Based Phosphors for Thermometry

Mochen Jia, Zhiying Wang, and Zuoling Fu

CONTENTS

15.1 INTRODUCTION

Luminescent thermometers have propelled to the forefront in research of thermometry due to their high spatial resolution, remote detection and fast response and are ideally suited for temperature (T) detection of cells, microcircuits and harsh environments (Wang, Wolfbeis, and Meier 2013; Zhu et al. 2016). This kind of noninvasive thermometers can make measurements at submicron scale without physical contact between observers and target analytes, which is not possible with conventional contact thermometers such as liquid-filled glass thermometers, thermocouples and thermistors (Brites et al. 2012; Jaque and Vetrone 2012).

Compared with other luminescent thermometers made up of quantum dots, dyes and polymers, the luminescent thermometers made up of lanthanide-based nanoparticles have attracted more attention by virtue of their rich emission levels, large Stokes/anti-Stokes shifts, exceptional processability, high photochemical stability and low cytotoxicity; thus, it is achievable to design on-demand luminescent thermometers for an enormous variety of applications (Brites, Balabhadra, and Carlos 2019). Especially in the past decade, the lanthanide-based up-conversion probes have shown great potential in biomedical imaging and sensing, because the magic up-conversion process can convert lower-energy near infrared (NIR) irradiation into higher-energy emissions, which refrains from the drawbacks of background autofluorescence, low light penetration depths and severe photodamage to tissues caused by ultraviolet and visible light excitation (Dong et al. 2012; Vetrone et al. 2010). Recently, with the goal of developing more accurate and efficient in vivo diagnostic tools, NIR-to-NIR thermometry (using NIR light for the excitation and emission) is becoming a quickly emerging field (Hemmer et al. 2016).

The temperature readouts of lanthanide-based thermometers are generally reflected by the thermal dependent emission intensity, lifetime, peak shift and excited state absorption, where the

DOI: 10.1201/9781003098676-15

ratiometric technique is required for temperature measurements based on emission intensity or excited state absorption, because this can overcome the signal errors caused by changes of probe concentration and detection efficiency when only one emission or excitation intensity is used as the temperature indicator (McLaurin, Bradshaw, and Gamelin 2013). With the robust, fast and accurate detection capability, the ratiometric luminescent thermometers have been implemented to monitor the local temperature of tissues and microelectronic devices in real time, thereby realizing biomedical pathological diagnosis and microelectronic fault diagnosis (Mi et al. 2019; Zhu et al. 2018). Numerous reports have been devoted to the development of lanthanide-based ratiometric luminescent thermometers, which are the most widely studied and used thermometers.

In lanthanide-based ratiometric thermometry, the ratio signal is generated by one luminescence center or two distinct luminescence centers and is a monotonic function of temperature within a certain temperature range, which allows us to accurately calculate the temperature readouts after obtaining the ratio values based on the collected spectra. This chapter will focus on the strategies and working mechanisms of lanthanide-based ratiometric thermometers, which are divided into thermometers for single-center emission, thermometers for dual-center emission and special infrared-emitting thermometers.

15.2 THERMOMETERS FOR SINGLE-CENTER EMISSION

15.2.1 Overview

Lanthanide-based ratiometric thermometers for single-center emission utilize different thermal dependence of two emission or excitation intensities generated from one lanthanide ion to define thermometric parameters. To date, a series of lanthanide-based ratiometric thermometers for single-center emission have been widely developed, and we simply divide them into three categories according to the physical mechanisms: thermally coupled thermometry, non-thermally coupled thermometry and excited state absorption–based thermometry.

15.2.2 Thermally Coupled Thermometry

For thermally coupled thermometry, the two emission signals originate from two adjacent thermally coupled levels (TCLs) of one lanthanide ion. And the relative population of TCLs follows Boltzmann's law with a quasi-equilibrium state at any temperature. The emission intensity of each transition is proportional to the population of each energy level:

$$I \propto Nh\nu A \qquad (15.1)$$

where N is the population of the energy level. ν and A are the frequency and the spontaneous emission rate of the transition, respectively. h is Planck's constant. According to Boltzmann type population distribution, the luminescence intensity ratio (LIR) of TCLs is given by (Wade, Collins, and Baxter 2003)

$$LIR = \frac{I_2}{I_1} = \frac{N_2 h \nu_2 A_2}{N_1 h \nu_1 A_1} = \frac{g_2 \nu_2 A_2}{g_1 \nu_1 A_1} exp\left(\frac{-\Delta E}{k_B T}\right)$$

$$= B exp\left(\frac{-\Delta E}{k_B T}\right) \qquad (15.2)$$

with

$$B = \frac{g_2 \nu_2 A_2}{g_1 \nu_1 A_1}$$

where I_2 and I_1 denote the integrated intensities from the upper and lower TCLs to a terminal level. Here we calculate the integrated intensity of emission bands as I_i to avoid the errors caused by only collecting the peak intensity. g_i is the degeneracy of the corresponding state. k_B is Boltzmann's constant and ΔE is the energy gap between TCLs. Referring to previous literature (Savchuk et al. 2015), when two emission bands of some TCLs overlap seriously, Equation (15.2) is modified as follows:

$$LIR = AB\exp\left(\frac{-\Delta E}{k_B T}\right) + C \qquad (15.3)$$

where A and C are corrected values. This situation generally occurs in the TCLs composed of two Stark energy levels.

Due to complicated luminescent phenomena of rare earth–ions doping in host lattices, such as phonon-assisted thermal activation and thermal quenching, as well as cross-relaxation and energy transfer between energy levels, the single temperature-dependent luminescent intensity of TCLs usually does not obey Boltzmann's law. However, because the preceding factors have similar effects on the two energy-close levels, their relative population (I_2/I_1) still obeys Boltzmann's law. In most cases, due to the thermal population from the lower energy level, it can be observed that the luminescent intensity from the upper energy level will increase with increasing temperature, or the quenching is relatively slow compared with that from the lower energy level, which is considered the thermally coupled phenomenon. The ΔE of TCLs is usually within 200 to 2000 cm^{-1}, where ΔE over 2000 cm^{-1} will inactivate the thermally coupled phenomenon, while ΔE below 200 cm^{-1} will cause serious overlap of the two emission bands, making it hard to distinguish the two signals. But not all two adjacent energy levels satisfying the energy gap range will generate the thermally coupled phenomenon, and the commonly reported TCLs are summarized in Table 15.1.

Among these TCLs, the $^2H_{11/2}$ and $^4S_{3/2}$ levels of Er^{3+} are the most popular pair due to the suitable energy gap, which have been used to map microscopic temperature in vivo for programming combination cancer therapy (Zhu et al. 2016, 2018). To shed light on the temperature-sensing performance based on TCLs of Er^{3+}, researchers have carried out detailed studies on the two parameters B and ΔE in Equation (15.2). According to the Judd-Ofelt theory, the parameter B can be simplified as follows (Suo et al. 2016):

$$B = \frac{\nu_2^4}{\nu_1^4} \cdot \frac{C_2}{C_1} = \frac{\nu_2^4}{\nu_1^4} \cdot \frac{\Sigma_{\lambda=2,4,6}\Omega_\lambda \left\langle {}^4I_{15/2}\left\|U^{(\lambda)}\right\|^2 H_{11/2}\right\rangle^2}{\Omega_6\left\langle {}^4I_{15/2}\left\|U^{(6)}\right\|^4 S_{3/2}\right\rangle^2} \approx \frac{0.7158\times\Omega_2 + 0.4138\times\Omega_4}{0.2225\times\Omega_6} + \frac{0.0927}{0.2225} \qquad (15.4)$$

TABLE 15.1
The Typical *LIR* Thermometers Based on TCLs of Rare Earth Ions

Rare Earth Ions	Host Material	TCLs	Ref.
Er^{3+}	La_2O_3	$^2H_{11/2}, {}^4S_{3/2}$	Sun et al. (2017)
Tm^{3+}	Y_2O_3	$^1G_{4(2)}, {}^1G_{4(1)}$	Liu et al. (2016)
Tm^{3+}	PbF_2 glass ceramics	$^3F_{2,3}, {}^3H_4$	Xu et al. (2012)
Ho^{3+}	TeO_2 glass	$^5F_4, {}^5S_2$	Singh (2007)
Pr^{3+}	$BaTiO_3$	$^1I_6 + {}^3P_1, {}^3P_0$	Jia et al. (2018)
Nd^{3+}	$CaWO_4$	$^4F_{7/2}, {}^4F_{5/2}$	Xu et al. (2014)
Nd^{3+}	$CaWO_4$	$^4F_{5/2}, {}^4F_{3/2}$	Xu et al. (2014)
Nd^{3+}	$CaWO_4$	$^4F_{7/2}, {}^4F_{3/2}$	Xu et al. (2014)
Sm^{3+}	$Y_4Al_2O_9$	$^4F_{3/2}, {}^4G_{5/2}$	Kaczkan et al. (2014)
Eu^{3+}	$NaEuF_4$	$^5D_1, {}^5D_0$	Tian et al. (2014)
Dy^{3+}	$Y_4Al_2O_9$	$^4I_{15/2}, {}^4F_{9/2}$	Boruc et al. (2012)

where C_i and Ω_λ represent the line strength and Judd-Ofelt intensity parameters of electric-dipole transitions. $\langle \|U^{(\lambda)}\| \rangle$ denotes the reduced matrix element of the unit tensor operator. v_2^4/v_1^4 is approximated as unit one due to the small energy gap between $^2H_{11/2}$ and $^4S_{3/2}$. However, the calculation of Judd-Ofelt theory becomes more complicated for nontransparent crystals and will produce large errors (Hehlen, Brik, and Krämer 2013; León-Luis et al. 2012). For a more convenient comparison, it was found that the B value is positively correlated with the covalence of chemical bond between lanthanide ions and ligands (Liu et al. 2015; Suo et al. 2016). Furthermore, by defining a new parameter α_p (the polarizability of the immediate environment around Er^{3+} ions), the B value was evaluated (Jia et al. 2019):

$$\alpha_p = \sum_i \alpha_b(i) \tag{15.5}$$

where $\alpha_b(i)$ denotes the polarizability of chemical bond volume between Er^{3+} and the ith ligand. Σ represents the summation of all ligands of Er^{3+}. Based on the chemical bond theory of complex crystals (Li and Zhang 2006; Wu and Zhang 1999; Xue and Zhang 1997), the α_p values of Er^{3+}/Yb^{3+} co-doped set of host materials were calculated, and the B value had a linear relationship with the α_p value.

As for the ΔE value, due to the partial overlap of the two emission bands, the selection of the integration interval, etc., the fitted value will have some deviation from the actual value. For different host materials, ΔE values are closely related to local crystal field environments. Usually, the samples with serious energy level splitting allow larger gap between two energy levels centers, that is, lager fitted values of ΔE. The ΔE value was also evaluated by the energy-level splitting factor K_e (Jia et al. 2019), which was defined as follows:

$$K_e = \frac{E_h Z f_i^2}{N} \tag{15.6}$$

where E_h and f_i denote the average homopolar energy and the average ionicity of chemical bond between the lanthanide ion and ligands. Z and N represent the average presented charge and the nearest total coordination number of ligands. And the qualitative relationship between ΔE and K_e could be approximated as $\Delta E = 1000.10 - 180.03 K_e^{-0.5}$.

Next we introduce an example of Er^{3+}/Yb^{3+} co-doped $BaTiO_3$ nanoparticles for thermally coupled thermometry. Under 980 nm irradiation, Yb^{3+} ions, as sensitizers, harvest the 980 nm photons and transfer energy to Er^{3+} ions. Then, the electron of Er^{3+} ions continuously absorbs the energy of two 980 nm photons, pumps to the $^4F_{7/2}$ states from the ground state $^4I_{15/2}$ and generates $^2H_{11/2} \rightarrow ^4I_{15/2}$ (530 nm) and $^4S_{3/2} \rightarrow ^4I_{15/2}$ (554 nm) transitions after multi-phonon relaxation (Figure 15.1(a) and (b)). As the temperature increases, the emission intensity at 530 nm increases first and then decreases, while the emission intensity at 554 nm decreases continuously (Figure 15.1(c)). The integrated intensity from 506 to 538 nm is defined as I_2, and that from 538 to 580 nm is defined as I_1. Using Equation (15.2) to fit the collected data, we obtain a calibration curve shown in Figure 15.1(d), where the solid line is the best fit to the experimental points with $r^2 > 0.997$, and the fitted value of ΔE is 719 cm^{-1}, close to the actual value of 817 cm^{-1}.

Technically, the spectra of each temperature point should be collected several times to get the average value, so as to avoid the accidental error caused by a single collection. We can see that the experimental points in Figure 15.1(d) do not completely match the solid line, that is, there will be errors in the temperature measurements, which is caused by the errors from the detector of the spectrometer, the temperature controller and the experimental sample.

To compare the performance of different *LIR* thermometers, researchers compare their absolute temperature sensitivity S_a, relative temperature sensitivity S_r, temperature uncertainty (or

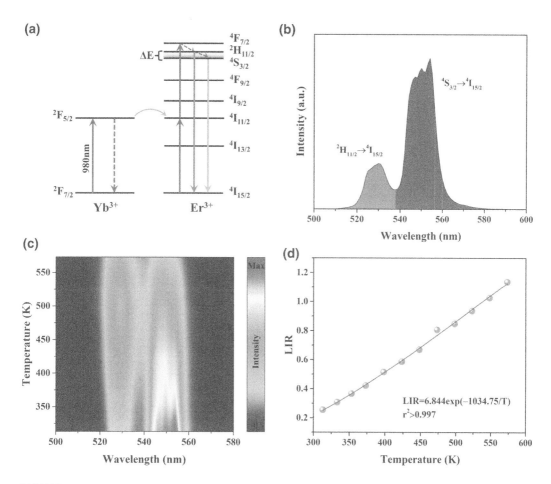

FIGURE 15.1 (a) Scheme of the major energy migration paths of Er^{3+}/Yb^{3+} co-doped $BaTiO_3$ nanoparticles for thermally coupled thermometry. (b) Emission spectrum and integration interval of the transitions of TCLs ($^2H_{11/2}$ and $^4S_{3/2}$). (c) Contour projections of thermal evolution spectra in 313–573 K range. (d) Calibration curve based on Equation (15.2).

temperature resolution) δT and repeatability R. S_a denotes the absolute change of the measured LIR per degree of temperature change:

$$S_a = \left| \frac{\partial LIR}{\partial T} \right| \tag{15.7}$$

S_r denotes the relative change of the measured LIR per degree of temperature change, which is usually expressed as a percentage change per Kelvin temperature (% K^{-1}):

$$S_r = \left| \frac{1}{LIR} \frac{\partial LIR}{\partial T} \right| \tag{15.8}$$

However, for thermometers with different physical mechanisms or the same physical mechanism, it is meaningless to use S_a values to evaluate their performance (Brites, Balabhadra, and Carlos 2019). Actually, S_a does not contribute to δT, and can be easily tuned, such as selecting different spectral integration intervals or doping different concentrations of lanthanide ions for one *LIR* thermometer (Jia et al. 2020a). Therefore, S_r is more comparable between different *LIR* thermometers, and we make a deduction as follows:

δT is the smallest temperature change that can be detected in a given measurement, of which the Taylor's series expansion versus the *LIR* is as:

$$\delta T = \frac{\partial T}{\partial LIR} \delta LIR + \frac{1}{2!} \frac{\partial^2 T}{\partial LIR^2} (\delta LIR)^2 + \cdots + \frac{1}{n!} \frac{\partial^n T}{\partial LIR^n} (\delta LIR)^n \tag{15.9}$$

where δLIR is the uncertainty of the *LIR*. Considering that the first term is dominant, Equation (15.9) can be expressed as:

$$\delta T = \frac{\delta LIR}{S_a} = \frac{1}{S_r} \frac{\delta LIR}{LIR} \tag{15.10}$$

Then we set δLIR as the measurement error of *LIR* (σLIR), according to the error propagation for *LIR*:

$$\frac{\delta LIR}{LIR} = \frac{\sigma LIR}{LIR} \approx \sqrt{\left(\frac{\sigma I_2}{I_2}\right)^2 + \left(\frac{\sigma I_1}{I_1}\right)^2} \tag{15.11}$$

Equation (15.10) can be further expressed as:

$$\delta T = \frac{LIR}{S_a} \sqrt{\left(\frac{\sigma I_2}{I_2}\right)^2 + \left(\frac{\sigma I_1}{I_1}\right)^2} = \frac{1}{S_r} \sqrt{\left(\frac{\sigma I_2}{I_2}\right)^2 + \left(\frac{\sigma I_1}{I_1}\right)^2} \tag{15.12}$$

For *LIR* thermometers, usually the larger S_a is, the larger *LIR* and δLIR are, which cannot determine δT from Equation (15.10) or (15.12), so δT is essentially determined by S_r, $\sigma I_2/I_2$ and $\sigma I_1/I_1$ (Jia et al. 2020a).

Researchers determine the $\delta LIR/LIR$ values caused by detectors based on the signal-to-noise ratio; for example, the $\delta LIR/LIR$ value for the charge-coupled device (CCD) detector is 0.05%. In addition, both the temperature controller and the experimental sample will affect the $\delta LIR/LIR$ values. When we collect the spectrum at a fixed temperature, the temperature controller will produce temperature fluctuations, which depend on the quality of the equipment, usually ±0.1 K. The experimental sample itself will also cause the experimental *LIR* values to deviate from the calibration curve, and the fitting degree and stability of the experimental *LIR* values must be considered. At a certain reference temperature, the experimental sample is repeatedly measured multiple times, and the standard deviation of *LIR* values is defined as the experimental δLIR value. The experimental $\delta LIR/LIR$ value is usually larger than that caused by the detector, because it contains all the influencing factors from the detector, the temperature controller and the experimental sample.

The repeatability of thermometer readings is also a key property, estimated by cycling over a given temperature interval, and some researchers will further quantitatively evaluate the repeatability by:

$$R = 1 - \frac{\max\left(|LIR_C - LIR_i|\right)}{LIR_C} \tag{15.13}$$

where LIR_C is standard value extracted from the calibration curve and LIR_i is the measured value each time.

According to Equations (15.2) and (15.8), we can calculate the S_r of thermally coupled thermometry:

$$S_r = \frac{\Delta E}{k_B T^2} \tag{15.14}$$

The corresponding δT can be written as:

$$\delta T = \frac{k_B T^2}{\Delta E} \frac{\delta LIR}{LIR} \tag{15.15}$$

The main advantage of thermally coupled thermometry is the possibility of predicting the performance of thermometers based on ΔE values before design and synthesis. The S_r of Er^{3+}/Yb^{3+} co-doped $BaTiO_3$ nanoparticles is shown in Figure 15.2(a), where the limited energy difference between TCLs limits the maximum value of S_r to only 1.06% at 313 K. After 50 tests on the sample at 333 K, the experimental δT value is calculated to be ±1.3 K (Figure 15.2(b)). And the repeatability is larger than 97.7% upon five consecutive temperature cycles between 323 and 423 K (Figure 15.2(c)).

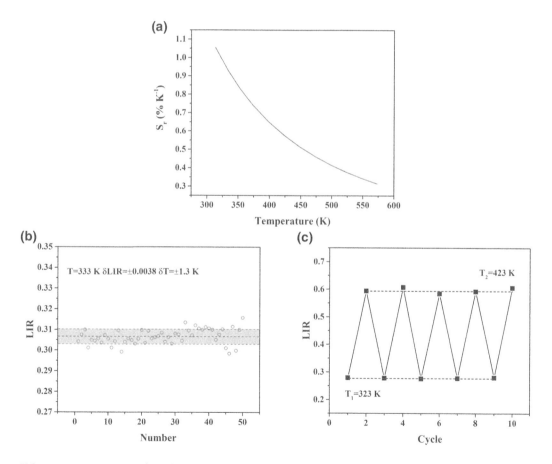

FIGURE 15.2 (a) S_r of Er^{3+}/Yb^{3+} co-doped $BaTiO_3$ nanoparticles. (b) Experimental δT based on 50 measurements at 333 K. The black-dotted line corresponds to the mean LIR and the royal blue dotted lines mark the region within the standard deviation of the measurements. (c) Repeatability in 5 heating-cooling temperature cycles between 323 and 423 K. The royal blue dotted lines indicate the corresponding standard values extracted from the calibration curve.

15.2.3 NON-THERMALLY COUPLED THERMOMETRY

Because the narrow ΔE of TCLs will cause severe spectral overlap and lower S_r values, the researchers began to develop other thermometric mechanisms for ratiometric luminescent thermometers. Here, we classify all other thermometric mechanisms for single-center emission as non-thermally coupled thermometry, which involves two independent emission levels. Although the physical mechanisms of non-thermally coupled thermometry are still controversial, the models are indeed widely used to correlate LIR and T of lanthanide-based luminescent thermometers. Here we try to rationalize the temperature-dependent LIR of non-thermally coupled thermometry. According to Parker's law, the measured luminescent intensity can be written as (Wang, Wolfbeis, and Meier 2013)

$$I = I_e \phi k \varepsilon \mathrm{dc} \tag{15.16}$$

where I_e is the intensity of excitation light, ϕ is the quantum yield of sample, k is the geometrical factor for the setup used, ε is the molar absorbance, d is the penetrated length and c is the concentration of sample. Ideally, the luminescent intensity at different temperatures is only affected by ϕ, which can be expressed as (Wang, Wolfbeis, and Meier 2013)

$$\phi = \frac{W_R}{W_R + W_{NR}} \tag{15.17}$$

where W_R and W_{NR} denote the radiative and non-radiative transition probabilities. W_R is assumed to be temperature independent, and W_{NR} is described by an Arrhenius-type dependence:

$$W_{\mathrm{NR}} \propto \exp\left(\frac{-E}{k_B T}\right) \tag{15.18}$$

where E is the thermal activation energy (the energy gap between the lowest energy level of the excited state and the overlap point of the possible non-radiative decay state).

Solving Equation (15.16), we can get the Mott-Seitz model of emission intensity (Cooke et al. 2004):

$$I(T) = \frac{I_0}{1 + A\exp(-E/k_B T)} \tag{15.19}$$

where $I(T)$ and I_0 are the emission intensities at T and 0 K. A is a pre-exponential constant. And the LIR can be deduced as follows:

$$LIR = \frac{I_2}{I_1} = \frac{I_{2,0}}{I_{1,0}} \frac{1 + A_1\exp(-E_1/k_B T)}{1 + A_2\exp(-E_2/k_B T)} \tag{15.20}$$

where I_i and $I_{i,0}$ denote the integrated intensities of two emissions at T and 0 K. A_i and E_i are the corresponding pre-exponential constants and thermal activation energies of the two emissions in the quenching process. If the two emissions are quenched severely with increasing temperature, i.e., the exponential terms dominate the temperature-dependent emission intensities, we can assume that $A_i\exp(-E_i/k_B T) \gg 1$, or if I_2 is much less affected by temperature than I_1, we can assume that $A_2\exp(-E_2/k_B T) \ll 1$ or as a constant, Equation (15.20) can be approximated as:

$$LIR \approx a + b\exp\left(\frac{c}{T}\right) \tag{15.21}$$

If I_1 is much less affected by temperature than I_2, we can assume that $A_1\exp(-E_1/k_BT) \ll 1$ or as a constant, resulting an S-shaped curve:

$$LIR \approx \frac{a}{1+b\exp(c/T)} \tag{15.22}$$

where a, b and c are fitting parameters. Similarly, due to the rich emission levels and complex luminescence mechanism of rare earth–ions doping in host lattices, the single temperature–dependent emission intensity of non-thermally coupled thermometry does not strictly follow Equation (15.19) in most cases. Different from TCLs, because of the larger energy gap between the two energy levels of non-thermally coupled thermometry, the preceding factors have different effects on the two emission intensities and, therefore, cannot be offset by the ratio. This will not only limit non-thermally coupled thermometry to certain conditions, such as a specific temperature range, a specific rare earth–ion doping or a specific host, but also make the fitting parameters far away from the actual values without physical meaning.

At present, it seems impossible to use only one mechanism to accurately describe the relationship between LIR and T for lanthanide-based non-thermally coupled thermometry. Therefore, based on the fact that a large number of exponential growth models exist in lanthanide-based luminescent thermometers, an empirical exponential growing equation similar to Equation (15.21) is used, where we don't consider its physical meaning:

$$LIR = \frac{I_2}{I_1} = a + b\exp(cT) \tag{15.23}$$

where a, b and c are fitting parameters.

Other researchers have also used the polynomial model (Wang, Jiao, and Fu 2019):

$$LIR = \frac{I_2}{I_1} = a + \frac{b}{T} + \frac{c}{T^2} + \frac{d}{T^3} \tag{15.24}$$

or another form as (Brites et al. 2018; Lu et al. 2017):

$$LIR = \frac{I_2}{I_1} = a + bT + cT^2 + dT^3 \tag{15.25}$$

where a, b, c and d are fitting parameters. These models can be regarded as the Laurent and Taylor expansions of the exponential terms in Equations (15.21) and (15.23), respectively, but these polynomials are more flexible when dealing with complex calibration curves.

In addition, when one of the transitions exhibits a temperature dependence much smaller than the other, LIR and T will show approximately a linear relationship in a short temperature range, and a simple linear model is often used (Kumar et al. 2018):

$$LIR = \frac{I_2}{I_1} = a + bT \tag{15.26}$$

where a and b are fitting parameters.

Recently, a thermometry based on Pr^{3+} ions doping into different kinds of sites has been reported, where Pr^{3+} ions at different sites show different thermal-induced fluorescence quenching behaviors (Wang et al. 2020). An extended exponential model used is as follows:

$$LIR = \frac{I_2}{I_1} = A_0 + A_1\exp\left(\frac{-\Delta E_1}{k_BT}\right) + A_2\exp\left(\frac{-\Delta E_2}{k_BT}\right) \tag{15.27}$$

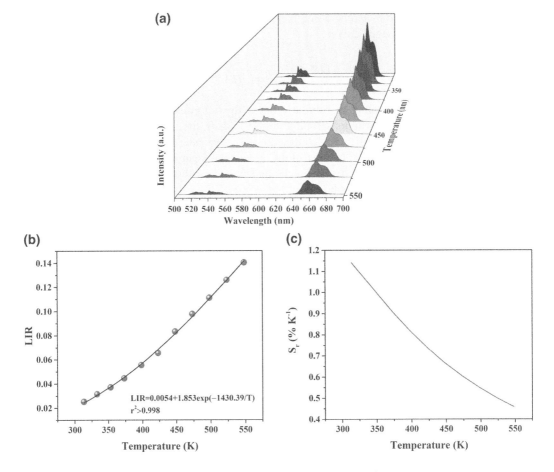

FIGURE 15.3 (a) Temperature evolution emission spectra of Er^{3+}/Yb^{3+} co-doped CaF_2 nanoparticles. (b) Calibration curve of *LIR* (523/656 nm) to temperature based on Equation (15.21), and (c) S_r for non-thermally coupled thermometry.

where ΔE_1 is the energy barrier between 4f5d and 4f transitions of Pr^{3+} ions. ΔE_2 is the energy barrier about the energy transfer between the two different Pr^{3+}-doped sites. A_0, A_1 and A_2 are fitting parameters.

Then, we present an example of Er^{3+}/Yb^{3+} co-doped CaF_2 nanoparticles for non-thermally coupled thermometry, the $^2H_{11/2} \rightarrow ^4I_{15/2}$ (523 nm) and $^4F_{9/2} \rightarrow ^4I_{15/2}$ (656 nm) transitions of Er^{3+} ions are selected as the thermometric signals (Figure 15.3(a)). The energy gap between $^2H_{11/2}$ and $^4F_{9/2}$ is about 4000 cm^{-1}, far exceeding the upper limit ΔE value of TCLs. After collecting the temperature evolution emission spectra, the integrated intensity from 506 to 534 nm is calculated as I_2, and that from 631 to 692 nm is calculated as I_1. Based on Equation (15.21), we get the calibration curve with high fitting degree of $r^2 > 0.998$ (Figure 15.3(b)). The maximum S_r value is 1.14% at 313 K (Figure 15.3(c)).

15.2.4 EXCITED STATE ABSORPTION–BASED THERMOMETRY

The thermometric techniques introduced earlier are all under a single excitation. With the development of lanthanide-based luminescent thermometers, the thermometric technique under multiple excitations has also emerged, which we call excited state absorption–based thermometry in our case. The uniqueness of this thermometry is that the ratio signals are all taken from the intensities of same wavelength or the same emission band, which can minimize the errors originating from different emission wavelengths for temperature detection of biological tissues, but the possible errors caused by different excitation sources should be carefully considered, such as the decay of different excitation power densities resulting in different changes in emission intensity. The current excited state absorption–based thermometry can be divided into two categories:

- Two transition intensities in the excitation spectra are selected as the thermometric parameters.
- The emission intensities of the same transition under two excitation sources are selected as the thermometric parameters.

For the former (Kolesnikov et al. 2019), the ratio of two transition intensities in the excitation spectra by monitoring a given emission shows a high temperature dependence due to the thermal population of the low excited states (Figure 15.4). The thermometric parameters can be fitted by the Boltzmann model of Equation (15.2), where I_2 and I_1 are the integrated intensities of the excitation bands. ΔE denotes the energy gap between the low excited state and the ground state. This thermometry needs the excitation source (such as xenon lamp) to continuously change the excitation wavelength from the ultraviolet to the visible region. The excitation source device is relatively complicate, while the signal acquisition only collects the intensity at given wavelength.

The physical mechanism of the latter is similar to that of the former. The latter requires two excitation sources ($\lambda_{exc,1}$ and $\lambda_{ecx,2}$) and collects the emission intensities of the identical band excited by different excitation sources, where the emission intensity excited by $\lambda_{ecx,2}$ is denoted as I_2 and that excited by $\lambda_{ecx,1}$ is denoted as I_1 (Figure 15.5). Based on Boltzmann-type population distribution, the population of low excited state increases with respect to the ground state at high temperature. Therefore, it is expected that I_2 excited by $\lambda_{ecx,2}$ and I_1 excited by $\lambda_{ecx,1}$ are strongly dependent on temperature and, thus, can be involved in temperature determination (Souza et al. 2016; Trejgis, Bednarkiewicz, and Marciniak 2020). Similarly, the calibration curve uses Boltzmann model, where ΔE denotes the energy gap between the low excited state and the ground state.

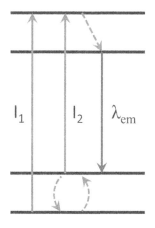 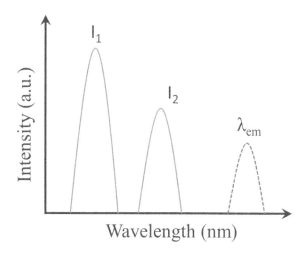

FIGURE 15.4 Scheme of excited state absorption–based thermometry for two excitation intensities as thermometric parameters.

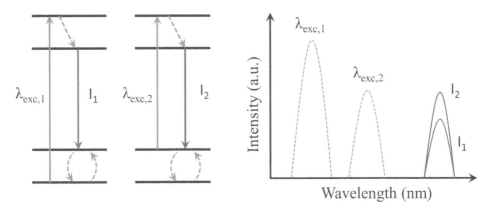

FIGURE 15.5 Scheme of excited state absorption–based thermometry for the emission intensities of the same transition under two excitation sources as thermometric parameters.

15.3 THERMOMETERS FOR DUAL-CENTER EMISSION

15.3.1 OVERVIEW

Lanthanide-based ratiometric thermometers for dual-center emission exploit the different thermal dependence of two emission intensities generated from distinct lanthanide ions, which have interionic energy transfer, indirect interionic energy transfer through excited levels of host or no interaction. A large number of thermometers for dual-center emission with excellent thermometric performance have been reported through reasonable combination of different lanthanide ions and intelligent design of ion distribution. Here we will focus on typical examples of thermometers for dual-center emission, which are divided into two parts, downshifting and up-conversion thermometers.

15.3.2 DOWNSHIFTING THERMOMETERS

Downshifting luminescence is a single-photon process that emits low-energy photons after non-radiative relaxation upon excitation with a high-energy photon (usually in ultraviolet and visible light regions). Some lanthanide ion pairs are used for downshifting thermometry for dual-center emission, such as Tb^{3+}/Eu^{3+}, Tb^{3+}/Pr^{3+} and Eu^{2+}/Eu^{3+}. For the infrared-emitting thermometers composed of downshifting luminescence excited by infrared light, we will introduce them separately in Section 15.4.

Tb³⁺/Eu³⁺-based systems utilize the *LIR* (here expressed as Δ) of $^5D_4 \rightarrow ^7F_5$ (Tb^{3+}) and $^5D_0 \rightarrow ^7F_2$ (Eu^{3+}) transitions as thermometric parameters, which have been widely used to design superior thermometers with higher S_r. As a typical example reported by Wang et al., the S_r of $[(Tb_{0.914}Eu_{0.086})_2 (PDA)_3(H_2O)]\cdot 2H_2O$ (PDA = 1,4-phenylenediacetic acid) nanoparticles, denoted as $Tb_{0.914}Eu_{0.086}$–PDA, is up to 5.96% K^{-1} at 25 K with high reproducibility R (better than 99%) and low temperature uncertainty δT (0.02 K at 25 K) (Wang et al. 2015). The high S_r is a direct result of the two-step concerted temperature-assisted energy transfer from Tb^{3+} to Eu^{3+}, which mainly occurs via the dipole-quadrupole and quadrupole-quadrupole mechanisms. The models of non-thermally coupled thermometry discussed in Section 15.2.3 are also applicable to lanthanide-based ratiometric thermometers for dual-center emission. Due to the quite small $^5D_0$ thermal quenching compared to that of $^5D_4$, the emission intensity of $^5D_0 \rightarrow ^7F_2$ (I_1) is not sensitive to temperature, while that of $^5D_4 \rightarrow ^7F_5$ (I_2) is substantially quenched with rising temperature (Figure 15.6(a) and (b)). Assuming that $A_1\exp(-E_1/k_BT) \ll 1$, the thermometric parameter Δ can be fitted by Equation (15.22) (Figure 15.6(c)), which is called the Mott-Seitz model here. The fitted thermal activation energy for the non-radiative channel of $^5D_4$ is 52.4 ± 2.0 cm^{-1}, which involves inactivation at the ligand level.

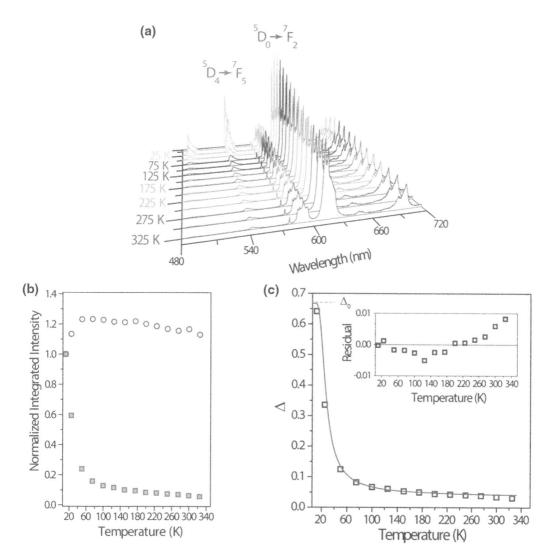

FIGURE 15.6 (a) Temperature-dependent emission spectra (excited at 377 nm, 10–325 K) of the prepared $Tb_{0.914}Eu_{0.086}$–PDA. (b) Normalized integrated intensities of $^5D_0 \rightarrow {}^7F_2$ (red circle) and $^5D_4 \rightarrow {}^7F_5$ (green square) emissions. (c) Calibration curve with the best fit to the experimental points ($r^2 > 0.999$). The inset shows the residuals of the fit. (Adapted from Wang et al. 2015 with permission from Wiley-VCH.)

The pair of Tb^{3+}/Pr^{3+} is used to build ratiometric thermometers by Gao et al. based on different thermal responses of two intervalence charge transfer (IVCT) states (Gao et al. 2016), which possesses universal validity for Tb^{3+}/Pr^{3+} co-doped oxide crystals with d^0 electron configured transition metal ions ($NaGd(MoO_4)_2$, $NaLu(MoO_4)_2$, $NaLu(WO_4)_2$, $LaVO_4$ and $La_2Ti_3O_9$). The ratio signal consists of $^1D_2 \rightarrow {}^3H_4$ (Pr^{3+}) to $^5D_4 \rightarrow {}^7F_5$ (Tb^{3+}) transitions sensitized by IVCT states (such as Tb^{3+}–Mo^{6+} and Pr^{3+}–Mo^{6+}), and the authors believe that the emission intensities depend on the populations of IVCT states, so only the non-radiative channels of IVCT states are considered. The thermometric model used is also similar to that discussed in Equation (15.20), although they attribute its origin to Struck and Fonger theory, where E is called quenching activate energy here, indicating the energy gap from the bottom of IVCT state to the intersection point with the state of lanthanide ions (Figure 15.7). Because of the smaller quenching activation energy for Tb^{3+} emission (or Tb^{3+}–IVCT) compared with that of Pr^{3+} emission (or Pr^{3+}–IVCT), $^5D_4 \rightarrow {}^7F_5$ transition (I_1) tends

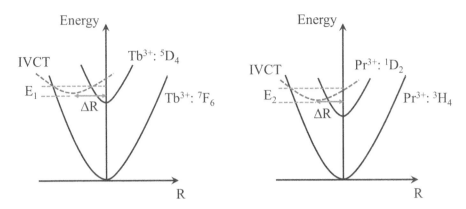

FIGURE 15.7 Configuration coordination diagram of Tb^{3+} and Pr^{3+} with IVCT states. (Adapted from Gao et al. (2016) with permission from Wiley-VCH.)

to be more easily quenched with increasing temperature, while $^1D_2 \rightarrow {}^3H_4$ transition (I_2) shows only a slight decline. Assuming that $A_2\exp(-E_2/k_BT) \ll 1$, the thermometric parameter can be fitted using Equation (15.21). The maximum S_r of Tb^{3+}/Pr^{3+}:$LaVO_4$ reaches as high as 5.30% K^{-1} at 303 K.

Pan et al. reported Eu^{2+}/Eu^{3+} co-doped Sc_2O_3 nanoparticles as luminescent ratiometric nanothermometers with wide temperature range (77–267 K), high repeatability (in excess of 99.94%), and high S_r of 3.06% K^{-1} at 267 K (Pan et al. 2018). Due to the shielding effect of external 5s and 5p shells on 4f orbital electrons, the typical $^5D_0 \rightarrow {}^7F_2$ emission of Eu^{3+} at 612 nm (4f–4f electron transition) is hardly affected by temperature, which serves as an internal reference. Meanwhile, Eu^{3+} is reduced to Eu^{2+} in situ by oleylamine, and the $4f^65d^1 \rightarrow {}^8S_{7/2}$ emission at 403 nm of Eu^{2+} (5d–4f electron transition), which is easily influenced by exterior environment, decreases rapidly with increasing temperature. Consequently, the thermometric parameter (I_{403}/I_{612}) shows a simple linear model with temperature (Equation (15.26)), which can also be described by the Mott-Seitz model (Equation (15.22)).

15.3.3 Up-conversion Thermometers

Up-conversion thermometers for dual-center emission can convert near-infrared excitation into visible emissions, mainly focusing on Er^{3+}, Ho^{3+} and Tm^{3+} ions. With the development of core@shell structure, the typical downshifting luminescent lanthanide ions (such as Tb^{3+} and Eu^{3+}) achieve upconversion luminescence through efficient energy transfer among different lanthanide ions, which can be used to construct up-conversion thermometers.

By doping $Er^{3+}/Ho^{3+}/Tm^{3+}$ in pairs into $YbPO_4$, $NaYb(MoO_4)_2$, $BaTiO_3$, Y_2O_3 and $LaAlO_3$ under 980 nm laser excitation (λ_{ex}), Li et al. have discussed the transition combinations that can be utilized to construct up-conversion thermometers for dual-center emission (Li et al. 2019), which are summarized in Table 15.2 (Er^{3+}/Ho^{3+} are not considered due to the overlap of emissions). The thermometric model adopts Equation (15.21), and it is found that when the two emission levels have a large energy gap (ΔE) without interaction, the thermometric parameter will fit well with Equation (15.21). The maximum S_r of Tm^{3+}/Ho^{3+} co-doped $NaYb(MoO_4)_2$ is as high as 4.23% K^{-1} at 313 K based on $^3F_{2,3} \rightarrow {}^3H_6$ and $^5F_4 \rightarrow {}^5I_8$ transitions.

Xu et al. designed a new type of up-conversion thermometer for dual-center emission using the smart-chemical design of $NaGdF_4$:$Yb/Ho/Ce@NaYF_4$:Yb/Tm core@shell nanostructure (Figure 15.8(a)) (Xu et al. 2016). By doping Ho^{3+} into the core and Tm^{3+} into the shell, the spatial separation of Ho^{3+} and Tm^{3+} inhibits the detrimental energy transfer between them, resulting in intense emission intensities of both activators under 980 nm excitation (Figure 15.8(b)). The thermometric parameter is defined by the red emissions (I_2: both $^5F_5 \rightarrow {}^5I_8$ of Ho^{3+} and $^1G_4 \rightarrow {}^3F_4$ of Tm^{3+}) and the green emission

TABLE 15.2

Summarized the Transitions of Er³⁺, Ho³⁺ and Tm³⁺ Ions as Thermometric Parameters of Up-Conversion Thermometers for Dual-Center Emission

Lanthanide Ions	λ_{ex} (nm)	I_2	I_1	ΔE (cm⁻¹)
Tm³⁺/Er³⁺	980	$^3F_{2,3} \rightarrow {}^3H_6$	$^2H_{11/2} \rightarrow {}^4I_{15/2}$	4746
Tm³⁺/Er³⁺	980	$^3F_{2,3} \rightarrow {}^3H_6$	$^4S_{3/2} \rightarrow {}^4I_{15/2}$	3952
Er³⁺/Tm³⁺	980	$^2H_{11/2} \rightarrow {}^4I_{15/2}$	$^1G_4 \rightarrow {}^3H_6$	2118
Er³⁺/Tm³⁺	980	$^4S_{3/2} \rightarrow {}^4I_{15/2}$	$^1G_4 \rightarrow {}^3H_6$	2912
Tm³⁺/Ho³⁺	980	$^3F_{2,3} \rightarrow {}^3H_6$	$^5F_4 \rightarrow {}^5I_8$	4102
Ho³⁺/Tm³⁺	980	$^5F_4 \rightarrow {}^5I_8$	$^1G_4 \rightarrow {}^3H_6$	2762

Source: Data from Li et al. (2019).

$(I_1: {}^5S_2/^5F_4 \rightarrow {}^5I_8)$. Introducing Ce³⁺ into the core can increase the red to green emission ratio due to the cross relaxation processes between Ce³⁺ and Ho³⁺, thereby tuning the thermometric parameter value. A simple linear model of Equation (15.26) is used to establish the calibration curve.

Another intriguing example was reported by Zheng et al. with NaGdF₄:Yb³⁺/Tm³⁺@NaGdF₄:Tb³⁺/Eu³⁺ core@shell nanostructure (Zheng et al. 2014), which achieves up-conversion emissions of Tb³⁺/Eu³⁺ under 980 nm excitation (Figure 15.9(a)). The Yb³⁺/Tm³⁺ pair is doped into the core to continuously harvest five 980 nm photons and then transfers energy to Gd³⁺ in the core. After energy migration to Gd³⁺ in the shell, Tb³⁺ and Eu³⁺ in the shell show strong up-conversion emissions due to the energy transfer from Gd³⁺ to Tb³⁺ and Eu³⁺. Because the multi-phonon-assisted energy transfer from Tb³⁺ to Eu³⁺ becomes strong with increasing temperature, the emission intensity of 545 nm of Tb³⁺ ($^5D_4 \rightarrow {}^7F_5$) has a linear relationship with temperature, while the emission of 615 nm of Eu³⁺ ($^5D_0 \rightarrow {}^7F_2$) is less affected by temperature changes (Figure 15.9(b)). Therefore, the emission intensity ratio is defined using 545 and 615 nm (denoted as I_{Tb}/I_{Eu}), which is fitted with the linear model of Equation (15.26) (Figure 15.9(c)).

Recently, a noticeable thermal enhancement of Er³⁺, Ho³⁺ and Tm³⁺ up-conversion emissions was observed in Yb³⁺-sensitized nanoparticles with oleic acid (from 9.7 to 57 nm), and Zhou et al.

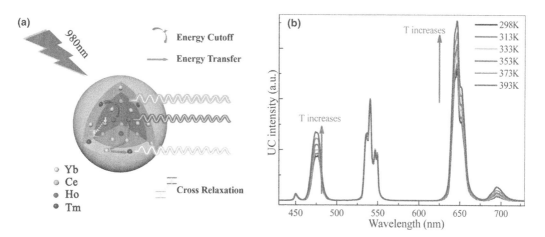

FIGURE 15.8 (a) Schematic diagram of NaGdF₄:Yb/Ho/Ce@NaYF₄:Yb/Tm nanostructure. (b) Temperature evolution emission spectra of the nanoparticles under 980 nm excitation, normalized to the green emission band. (Adapted from Xu et al. (2016) with permission from The Royal Society of Chemistry.)

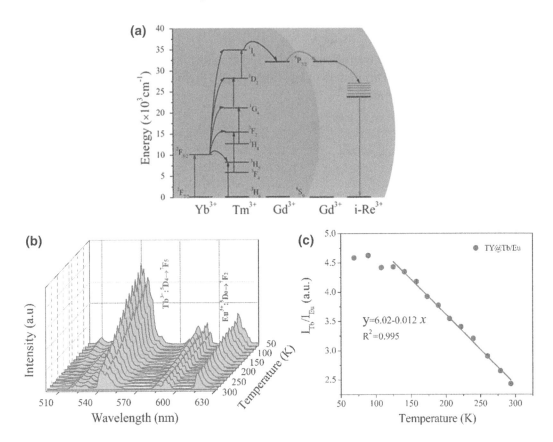

FIGURE 15.9 (a) Simplified energy level diagrams of lanthanide-doped $NaGdF_4@NaGdF_4$ core@shell nanoparticles. (b) Temperature-dependent emission spectra of $NaGdF_4:Yb^{3+}/Tm^{3+}@NaGdF_4:Tb^{3+}/Eu^{3+}$ under 980 nm excitation. (c) Calibration curve with the linear fitting model. (Adapted from Zheng et al. (2014) with permission from The Royal Society of Chemistry.)

proposed that a surface phonon-assisted energy transfer mediated by surface-bound molecules through Yb–O chelating could promote the energy transfer from Yb^{3+} to the activators as the temperature increases (Zhou et al. 2018). Notably, 9.7 nm Yb^{3+}–Tm^{3+} co-doped β-$NaYF_4$ nanoparticles show a 2000-fold enhancement in blue emission at 453 K. Although the physical mechanism that fully explains the thermal enhancement effect is still inconclusive, the fact provides a new idea to design high sensitive luminescent thermometers for dual-center emission. Based on this, Martínez et al. reported the self-calibrated thermometers by combination of large and small up-conversion nanoparticles with oleic acid in a single device (Martínez et al. 2019). As the temperature increases, the emission intensity of large-sized β-$NaYF_4:Yb/Ln$ (>70 nm) nanoparticles exhibits traditional thermal quenching, while that of small-sized (<15 nm) $NaGdF_4:Yb/Ln@NaYF_4$ core@shell nanoparticles (Ln = Tm, Er, Ce/Ho) is thermally enhanced. Therefore, using these high-contrast temperature-dependent characteristics as thermometric parameters will produce higher S_r than the conventional approach. For the thermometer combined by small-sized $NaGd_{0.695}Yb_{0.300}Tm_{0.005}F_4@NaYF_4$ and large-sized $NaY_{0.78}Yb_{0.20}Er_{0.02}F_4$, the thermometric parameter is defined using $\Delta' = I_G/I_S$, where I_G and I_S indicate integrated areas of $^1G_4 \rightarrow {}^3H_6$ (Tm^{3+}) and $^4S_{3/2} \rightarrow {}^4I_{15/2}$ (Er^{3+}) transitions, respectively, and is adjusted to an empirical sigmoidal function:

$$\Delta' = \Delta_2 \frac{\Delta_1 - \Delta_2}{1 + \exp\left((T - T_0)/dT\right)} \quad (15.28)$$

where Δ_1, Δ_2, T_0 and dT are fitting parameters. The corresponding maximum S_r is 5.88% K^{-1} at 339 K, which is more than a six-fold improvement compared to the conventional thermally coupled thermometry based on Er^{3+}.

15.4 INFRARED-EMITTING THERMOMETERS

15.4.1 OVERVIEW

Recently, the luminescent thermometers operating in the so-called biological windows (BWs) have become a hot spot due to the low tissue absorption and scattering of NIR light, which allows them have a remarkable penetration depth to detect the temperature of deep tissues. Usually, three BWs are defined: the first biological window (BW-I) in the range from 650 to 950 nm, the second biological window (BW-II) in the range from 1000 to 1350 nm and the third biological window (BW-III) in the range from 1500 to 1800 nm (Hemmer et al. 2013). According to the operation spectral range of thermometers, here we divide the infrared-emitting thermometers into BW-I and BW-II/III thermometers.

15.4.2 BW-I THERMOMETERS

The vast majority of BW-I thermometers are based on the up-conversion emissions of Er^{3+}, Ho^{3+}, Tm^{3+} and Nd^{3+} ions under NIR excitation. Dong et al. first utilized CaF$_2$:Tm^{3+}/Yb^{3+} nanoparticles to perform the cellular imaging and at the same time demonstrated their ability as BW-I thermometers (Dong et al. 2011). Upon 920 nm excitation, the strong $^3H_4 \rightarrow ^3H_6$ transition of Tm^{3+} ions (around 800 nm) performs successful cell marking without autofluorescence, as shown in Figure 15.10(a),

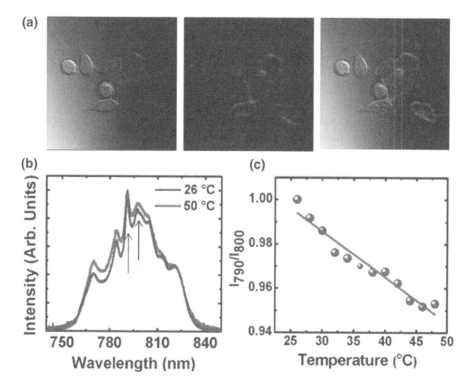

FIGURE 15.10 (a) Optical transmission image of HeLa cells incubated with the CaF$_2$:Tm^{3+}/Yb^{3+} PBS solution, the fluorescence image and the superimposed image (from left to right). (b) Emission spectra at different temperatures and (c) calibration curve with the linear fitting model of CaF$_2$:Tm^{3+}/Yb^{3+} nanoparticles. (Adapted from Dong et al. (2011) with permission from American Chemical Society.)

confirming the potential use of CaF_2:Tm^{3+}/Yb^{3+} nanoparticles in bio-imaging. Meanwhile, due to the thermally coupled effect between the sub-Stark energy levels of $^3H_4$, the intensity ratio of different wavelengths in 800 nm band varies remarkably with temperature, and the most evident change is that composed of 790 and 800 nm (Figure 15.10(b)). The intensity ratio of I_{790}/I_{800} decreases monotonously with temperature and can be fitted with the linear model of Equation (15.26) (Figure 15.10(c)), while the relative thermal-induced change in the ratio is low. In addition, the thermally coupled thermometry based on the sub-Stark energy levels of $^3H_4$ (termed as $^3H_{4(1)}$ and $^3H_{4(2)}$) can also be used to design BW-I thermometers, and the thermometric parameter (I_{796}/I_{804}) for Tm^{3+}/Yb^{3+}/Li^+ co-doped $Bi_{3.84}W_{0.16}O_{6.24}$ thermometers reported by Sun et al. increases with increasing temperature under 980 nm excitation, which is consistent with the model of Equation (15.2) (Sun et al. 2018). But the maximum S_r is only 0.14% K^{-1} at 313 K because of the narrow ΔE between $^3H_{4(1)}$ and $^3H_{4(2)}$ sub-Stark energy levels.

Notably, for SrF_2:Yb/Tm BW-I thermometers reported by Labrador-Páez et al., one peak at 769 nm is attributed to $^1G_4 \rightarrow ^3H_5$ transition (three photon excitation process), whereas the other peak at 795 nm is attributed to $^3H_4 \rightarrow ^3H_6$ transition (two photon excitation process) (Labrador-Páez et al. 2018). Thus, their emission intensities show different dependence on 975 nm excitation power density, which will cause the thermometric parameter (I_{769}/I_{795}) to vary with on-target excitation power density. The thermometers provide a relatively low S_r of 0.24% K^{-1}. Jia et al. design a high sensitive BW-I thermometer based on $NaYb(MoO_4)_2$:Tm^{3+} nanosheets sensitized by Yb^{3+}-MoO_4^{2-} dimers (Jia et al. 2020c). Under the 980 nm laser irradiation, the 691 nm emission ($^3F_{2,3} \rightarrow ^3H_6$ transition) is continuously increased as a result of the thermal enhancing caused by the cooperation of phonon-assisted energy transfer between Yb^{3+} and Tm^{3+} ions and thermally coupled population of $^3F_{2,3}$ state, while the 651 nm emission increases slightly and then decreases seriously because more non-radiative relaxation pathways of high-energy excited state cause the thermal quenching gradually becoming dominant with increasing temperature (Figure 15.11(a) and (b)). The thermometric parameter is defined as the ratio between the high-contrast thermal-dependent emission intensities, which is fitted using the model of Equation (15.23) (Figure 15.11(c)), yielding to a high S_r of 6.5% K^{-1} at 313 K with high repeatability > 98.3% and minimal δT of 0.16 K. After tuning the luminescence process by Yb^{3+}–MoO_4^{2-} dimers, the 651- and 691 nm emissions are both close to two photon excitation process, but their emission intensities still show different dependence on excitation power density. Therefore, the authors calibrate the thermometric parameter based on the excitation power density, and the on-target excitation power density is predicted in the simulated chicken breast tissue. In addition, according to the wavelength-dependent scattering and absorption of tissue, the 651 nm band can be distorted to a much greater extent than the 691 nm band. This will have a serious impact on the reliability of the temperature reading in actual use, depending on the nature of the tissue where the thermometer is located as well as the depth. Based on this, the authors explore the variation of thermometric parameters with penetration depth and successfully correct the temperature readings of thermometer under 2 mm chicken breast tissue.

Mi et al. also utilize the thermal enhancement of Yb^{3+}-sensitized nanoparticles with oleic acid to design an ultrasensitive ratiometric thermometer, the emission signals of which are located in BW-I (Mi et al. 2019). The sandwich-structured nanorods of $NaYF_4$:60%Yb^{3+}, 4%Nd^{3+}/Y/20%Yb^{3+}, 2%Er^{3+} are synthesized, and the 803 nm emission of Nd^{3+} and 654 nm emission of Er^{3+} evolve in opposite thermal enhancing and quenching with increasing temperature respectively, which are selected to define the thermometric parameters fitted by a single exponential function. The maximum S_r is calculated as high as 9.6% K^{-1} at 303 K, and the repeatability R ranges from 97.1 to 99.8% after 10 consecutive temperature cycles, and the δT is lower than 1 K in a temperature range of 300−435 K. However, the oleic acid-modified nanorods are not hydrophilic, which greatly limits their ability to map temperature in tissues. Kaczmarek et al. reported a unique hybrid organic-inorganic material combined with β-$NaGdF_4$:Er^{3+}, Yb^{3+} nanoparticles and periodic mesoporous organosilica (PMO), where β-$NaGdF_4$:Er^{3+}, Yb^{3+} was formed a shell around PMO (Kaczmarek et al. 2020). Interestingly, unlike the traditional green emission of Er^{3+}, they observed the $^2H_{11/2}$, $^4S_{3/2} \rightarrow ^4I_{13/2}$

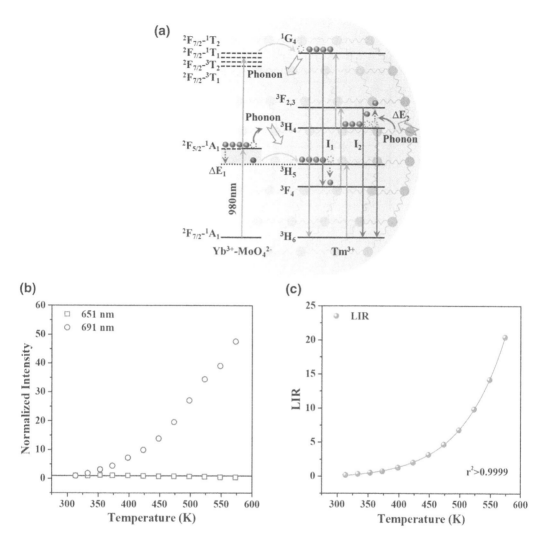

FIGURE 15.11 (a) Simplified energy level diagrams, (b) normalized intensities of 651- and 691 nm emissions with temperature and (c) calibration curve for NaYb(MoO$_4$)$_2$:Tm^{3+} nanosheets. (Adapted from Jia et al. (2020c) with permission from The Royal Society of Chemistry.)

transitions at 790 and 840 nm within BW-I upon laser excitation at 980 nm. Thus, based on the thermally coupled thermometry, the BW-I thermometers give maximum S_r of 0.7% K^{-1} at 293 K, with low temperature uncertainty ($\delta T < 1$ K) and 95–99% repeatability. Moreover, three TCLs of Nd^{3+} ($^4F_{7/2}$, $^4F_{5/2}$ and $^4F_{3/2}$) are often used to design BW-I thermometers, which can generate three pairs of TCLs by combining them in pairs. As an example, for LaPO$_4$:Yb^{3+}/Nd^{3+} BW-I thermometers reported by Suo et al., the thermally coupled thermometry based on $^4F_{7/2}$ and $^4F_{3/2}$ gives a maximum S_r of 3.51% K^{-1} at 280 K (Suo et al. 2020).

It should be noted that water can strongly absorb photons around 980 nm and convert them into heat, which will lead to a temperature rise of the local position irradiated by laser. Thus, the excitation power density should be reduced as much as possible or the excitation source should be shifted out the water absorption range, such as 808 nm. Savchuk et al. reported KLu$_{1-x-y}$Ho$_x$Tm$_y$(WO$_4$)$_2$ BW-I thermometers with tunable sensitivity under 808 nm excitation (Savchuk et al. 2018). The Ho^{3+} emissions may be generated by the energy transfer from Tm^{3+} to Ho^{3+} (Figure 15.12(a)), and the

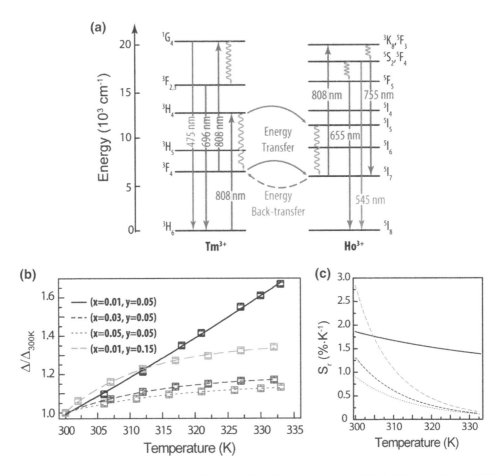

FIGURE 15.12 (a) Simplified energy level diagrams, (b) calibration curves and (c) S_r of $KLu_{1-x-y}Ho_xTm_y(WO_4)_2$ nanoparticles. (Adapted from Savchuk et al. (2018) with permission from The Royal Society of Chemistry.)

thermometric parameter (here expressed as Δ) is defined between the emission intensity centered at 696 nm (Tm^{3+}, $^3F_{2,3} \rightarrow {}^3H_6$) and that centered at 755 nm (Ho^{3+}, $^5S_2/^5F_4 \rightarrow {}^5I_7$). But the authors did not give a complete physical model, and the thermometric parameter was fitted using an empirical exponential growing model of Equation (15.23) (Figure 15.12(b)). The S_r can be tuned by the doping concentration of Tm^{3+} and Ho^{3+}, and the maximum value is 2.8% K^{-1} at 300 K for $KLu_{0.84}Ho_{0.01}Tm_{0.15}(WO_4)_2$ nanoparticles (Figure 15.12(c)). They only considered the relative error of thermometric parameters caused by acquisition setup, took $\delta\Delta/\Delta$ of 0.5% and calculated the minimum δT as 0.2 K.

In addition to the earlier up-conversion BW-I thermometers, Benayas et al. reported a downshifting BW-I thermometers based on Nd^{3+}-doped $Y_3Al_5O_{12}$ nanoparticles (Benayas et al. 2015). Under excitation at 808 nm, two emission lines were observed near 940 nm, which are attributed to $^4F_{3/2}$ (R_2 Stark sublevel, R_1 Stark sublevel) $\rightarrow {}^4I_{9/2}$ (Z_5 Stark sublevel) transitions. The Stark sublevels of $^4F_{3/2}$ follow the thermally coupled phenomenon, and the thermometric parameter is defined the intensity ratio between 938- and 945 nm emissions, which is fitted using Equation (15.2). Due to the narrow ΔE between Stark sublevels, the maximum S_r is only 0.15% K^{-1}. By injecting an aqueous dispersion of the thermometers into chicken tissue, the time evolution of the subtissue temperature is monitored in real time.

15.4.3 BW-II/III THERMOMETERS

Compared with BW-I emissions, BW-II/III emissions are significantly better suited for imaging and temperature sensing of deeper tissues, which provide a higher signal-to-noise ratio by effectively filtering out autofluorescence. The current BW-II/III thermometers rely on the downshifting emissions of Nd^{3+}, Yb^{3+}, Er^{3+}, Ho^{3+} and Tm^{3+} ions under NIR excitation.

The Nd^{3+}/Yb^{3+} pair is commonly used to design BW-II thermometers. Ximendes et al. have carried out detailed research on active-core/active-shell Nd^{3+}- and Yb^{3+}-co-doped LaF_3 nanoparticles, which have been successfully used as subcutaneous thermal probes (Ximendes et al. 2016). Under single beam 790 nm excitation, one single-core and two active-core/active-shell structures show the characteristic emissions of Nd^{3+} ions at 900, 1060 and 1350 nm, and the Yb^{3+} characteristic 1000 nm emission (Figure 15.13(a) and (b)). Due to the $Nd^{3+} \rightarrow Yb^{3+}$ energy transfer and back transfer, the intensity ratio (here expressed as Δ) between 1350 nm of Nd^{3+} and 1000 nm of Yb^{3+} shows a temperature dependence and a linear calibration relation (Figure 15.13(c)). Moreover, compared to single-core structure of LaF_3:Nd/Yb, the core/shell engineering has achieved a four-fold improvement in S_r, and the maximum value is 0.41% K^{-1} at 283 K for LaF_3:Nd@LaF_3:Yb nanoparticles.

In addition, Marciniak et al. designed a double ratiometric optical thermometer based on $NaYF_4$:Yb/Er@$NaYF_4$:Yb/Nd core@shell nanoparticles, one operating in visible green range and the other operating in BW-II (Marciniak et al. 2016). The thermometric parameter is determined by the intensity ratio between 1060 nm emission of Nd^{3+} and 1000 nm emission of Yb^{3+}, yielding a maximum S_r of 2.1% K^{-1} at 370 K, but the thermometer has not been hydrophilized. Similarly,

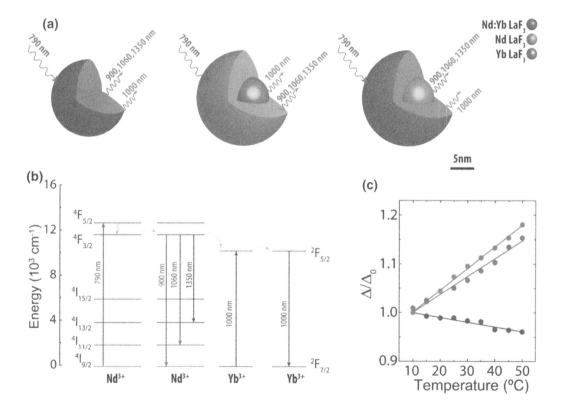

FIGURE 15.13 (a) Schematic diagram of LaF_3:Nd/Yb, LaF_3:Yb@LaF_3:Nd and LaF_3:Nd@LaF_3:Yb nanoparticles. (b) Simplified energy scheme of the Nd^{3+} and Yb^{3+} emitting centers. (c) Calibration curves of LaF_3:Nd/Yb (blue), LaF_3:Yb@LaF_3:Nd (purple) and LaF_3:Nd@LaF_3:Yb (red). (Adapted from Ximendes et al. (2016) with permission from American Chemical Society.)

Cortelletti et al. reported multishell SrF_2:Yb/Tm@Y@Yb/Er/Nd@Nd nanoparticles as optical thermometers in BW-II (Cortelletti et al. 2018). The thermometric parameter is defined using the intensity ratio between 1000 nm of Yb^{3+} and 1060 nm of Nd^{3+}, which decreases almost linearly with temperature. Appropriate doping Er^{3+} ions into the second shell will lead to efficient $Yb^{3+} \rightarrow Er^{3+}$ energy transfer, which is beneficial for the thermometric performance and remarkably increases the S_r value. When the doping concentration of Er^{3+} is 8%, the maximum S_r value is 1.62% K^{-1} at 323 K.

Skripka et al. reported a hydrophilic double rare-earth nanothermometer operating in BW-II/III with core@shell architecture (Skripka et al. 2017). After surface modification of PEG-DOPE phospholipids, $NaGdF_4$:Er/Ho/Yb@$NaGdF_4$:Yb@$NaGdF_4$:Nd/Yb@$NaGdF_4$ nanoparticles can be well dispersed in water (Figure 15.14(a)). Under 806 nm excitation, the Nd^{3+} ions in the second shell absorb the excitation energy and convert part of it into 1340 nm emission in BW-II (Figure 15.14(b)). Meanwhile, part of the energy is transferred throughout the nanoparticles by the network of Yb^{3+} ions, and Er^{3+} and Ho^{3+} ions are sensitized to generate 1550- (BW-III) and 1180 nm (BW-II) emissions, respectively. With elevating temperature, the 1180- (Ho^{3+}) and 1550 nm (Er^{3+}) emissions increase, while the 1340 nm emission (Nd^{3+}) decreases, because of the temperature-dependent phonon-assisted energy transfer and the water assisted non-radiative relaxation. Hence, a double ratiometric approach can be realized by defining LIR (Ho^{3+}/Nd^{3+}, BW-II/II) and LIR (Er^{3+}/Nd^{3+}, BW-III/II) thermometric parameters, which show a linear model in the 20–50°C range. The maximal S_r values are 1.17% and 1.10% °C^{-1} at 20°C (Figure 15.14(c)). The two strategies are reproducible up to 99% and colloidal stability of the thermometers remains unaffected. By taking the $\delta LIR/LIR$ values of

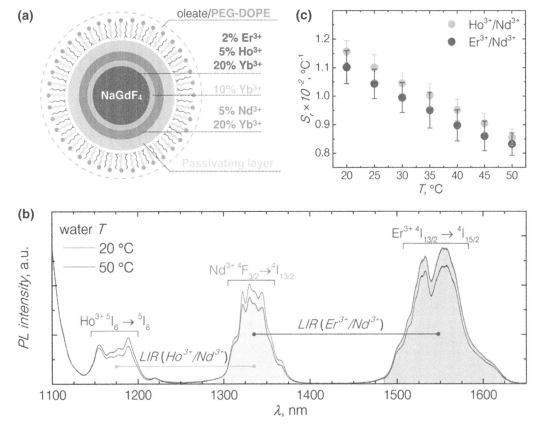

FIGURE 15.14 (a) Schematic diagram of phospholipids-modified $NaGdF_4$:Er/Ho/Yb@$NaGdF_4$:Yb@$NaGdF_4$:Nd/Yb@$NaGdF_4$ nanoparticles. (b) Emission spectra at different temperatures, and (c) S_r of the nanoparticles. (Adapted from Skripka et al. (2017) with permission from The Royal Society of Chemistry.)

LIR (Ho^{3+}/Nd^{3+}) and *LIR* (Er^{3+}/Nd^{3+}) as 1.4% and 0.9% based on the signal-to-noise ratio during spectral acquisition, the minimal δT values are 1.2 and 0.8°C, respectively.

Ximendes et al. also reported a double ratiometric thermal probe operating in BW-II/III based on LaF$_3$:Er^{3+}/Yb^{3+}@LaF$_3$:Yb^{3+}/Tm^{3+} nanoparticles, which successfully recorded a time-resolved 2D subcutaneous thermal video (Ximendes et al. 2017). Under 690 nm excitation, the nanoparticles present the 1230 nm BW-II emission of Tm^{3+}, the 1000 nm BW-II emission of Yb^{3+} and the 1550 nm BW-III emission of Er^{3+}. The double ratiometric parameters are defined as *LIR* (Yb^{3+}/Tm^{3+} BW-II/II) and *LIR* (Yb^{3+}/Er^{3+}, BW-II/III), which are fitted using a linear model in the 20–50°C range. Notably, the maximal S_r for *LIR* (Yb^{3+}/Er^{3+}, BW-II/III) is as large as 5.0% °C^{-1} at 20°C. Jia et al. made a detailed study on Ho^{3+}/Er^{3+} pair doping in a set of host lattices (Gd$_2$O$_3$, Y$_2$O$_3$, Y$_3$Al$_5$O$_{12}$, BaTiO$_3$ and YVO$_4$) as BW-II/III thermometers, where the *LIR* is defined between the 1550 nm emission of Er^{3+} (BW-III) and 1190 nm emission of Ho^{3+} (BW-II) under 980 nm excitation (Jia et al. 2020b). The authors tried to rationalize a physical model through the rate equations of the state populations based on temperature-dependent multi-phonon relaxation (MPR) and phonon assisted energy transfer (PAT) processes, expressed as follows:

$$LIR = \frac{I_{Er}}{I_{Ho}} = \frac{N_{Yb}W_R W_{MPR}A_1}{N_{Yb}W_{PAT}A_2} = a \times \left(\frac{\exp(h\nu/kT)}{\exp(h\nu/kT)-1} \right)^b \tag{15.29}$$

with

$$a = \frac{W_R W_{MPR(0)}A_1}{W_{PAT(0)}A_2}, \, b = \frac{\Delta E_1 - \Delta E_2}{h\nu}$$

where N_{Yb} symbolizes the population of ^{2}F$_{5/2}$ (Yb^{3+}); W_R denotes the rate of resonant energy transfer from ^{2}F$_{5/2}$ (Yb^{3+}) to ^{4}I$_{11/2}$ (Er^{3+}); W_{MPR} and $W_{MPR(0)}$ are the MPR rates of ^{4}I$_{11/2} \rightarrow {}^4$I$_{13/2}$ (Er^{3+}) at T and 0 K; W_{PAT} and $W_{PAT(0)}$ represent the PAT rates of ^{2}F$_{5/2}$ (Yb^{3+})$\rightarrow^5$I$_6$ (Ho^{3+}) at T and 0 K; A_1 and A_2 represent the spontaneous emission rate of corresponding transitions. $h\nu$ is the phonon energy of the host lattice; k is the Boltzmann constant, and ΔE_1 and ΔE_2 are the energy gaps involved, respectively. As a result, the *LIR* of all samples is perfectly fitted using Equation (15.29), and the dominant phonon modes are determined by analyzing the fitting parameters and experimental verification. Moreover, the S_r values can be tuned by the dominant phonon modes, resulting in aqueous Y$_2$O$_3$:Er^{3+}/Ho^{3+}/Yb^{3+} with a maximal S_r of 1.01% °C^{-1} at 65°C.

BW-III thermometers are usually designed using the ^{4}I$_{13/2} \rightarrow {}^4$I$_{15/2}$ downshifting emission of Er^{3+}, based on the thermally coupled thermometry of the Stark sublevels of ^{4}I$_{13/2}$, such as Y$_3$Al$_5$O$_{12}$:Yb^{3+}/Er^{3+}/Ho^{3+}/Cr^{3+} (Zhang et al. 2019) and LiErF$_4$@LiYF$_4$ (Hazra et al. 2020). As mentioned earlier, the dissatisfaction is the low S_r due to the narrow ΔE between Stark sublevels, which limits the δT values. Thus, further improvements and development of new thermometric mechanisms are still needed for BW-III thermometers.

REFERENCES

Benayas, A., B. del Rosal, A. Pérez-Delgado, K. Santacruz-Gómez, D. Jaque, G. A. Hirata, and F. Vetrone. 2015. Nd:YAG near-infrared luminescent nanothermometers. *Advanced Optical Materials* 3: 687–694.

Boruc, Z., M. Kaczkan, B. Fetlinski, S. Turczynski, and M. Malinowski. 2012. Blue emissions in Dy^{3+} doped Y$_4$Al$_2$O$_9$ crystals for temperature sensing. *Optics Letters* 37: 5214–5216.

Brites, C. D. S., P. P. Lima, N. J. O. Silva, A. Millan, V. S. Amaral, F. Palacio, and L. D. Carlos. 2012. Thermometry at the nanoscale. *Nanoscale* 4: 4799–4829.

Brites, C. D. S., K. Fiaczyk, J. F. C. B. Ramalho, M. Sójka, L. D. Carlos, and E. Zych. 2018. Widening the temperature range of luminescent thermometers through the intra- and interconfigurational transitions of Pr^{3+}. *Advanced Optical Materials* 6: 1701318.

Brites, C. D. S., S. Balabhadra, and L. D. Carlos. 2019. Lanthanide-based thermometers: at the cutting-edge of luminescence thermometry. *Advanced Optical Materials* 7: 1801239.

Cooke, D. W., B. L. Bennett, R. E. Muenchausen, J. K. Lee, and M. A. Nastasi. 2004. Intrinsic ultraviolet luminescence from Lu_2O_3, Lu_2SiO_5 and Lu_2SiO_5:Ce^{3+}. *Journal of Luminescence* 106: 125–132.

Cortelletti, P., A. Skripka, C. Facciotti, M. Pedroni, G. Caputo, N. Pinna, M. Quintanilla, A. Benayas, F. Vetrone, and A. Speghini. 2018. Tuning the sensitivity of lanthanide-activated NIR nanothermometers in the biological windows. *Nanoscale* 10: 2568–2576.

Dong, B., B. Cao, Y. He, Z. Liu, Z. Li, and Z. Feng. 2012. Temperature sensing and in vivo imaging by molybdenum sensitized visible upconversion luminescence of rare-earth oxides. *Advanced Materials* 24: 1987–1993.

Dong, N.-N., M. Pedroni, F. Piccinelli, G. Conti, A. Sbarbati, J. E. Ramírez-Hernández, L. M. Maestro, M. C. Iglesias-de la Cruz, F. Sanz-Rodriguez, A. Juarranz, F. Chen, F. Vetrone, J. A. Capobianco, J. G. Solé, M. Bettinelli, D. Jaque, and A. Speghini. 2011. NIR-to-NIR two-photon excited CaF_2: Tm^{3+}, Yb^{3+} nanoparticles: multifunctional nanoprobes for highly penetrating fluorescence bio-imaging. *ACS Nano* 5: 8665–8671.

Gao, Y., F. Huang, H. Lin, J. Zhou, J. Xu, and Y. Wang. 2016. A novel optical thermometry strategy based on diverse thermal response from two intervalence charge transfer states. *Advanced Functional Materials* 26: 3139–3145.

Hazra, C., A. Skripka, S. J. L. Ribeiro, and F. Vetrone. 2020. Erbium single-band nanothermometry in the third biological imaging window: potential and limitations. *Advanced Optical Materials* 8: 2001178.

Hehlen, M. P., M. G. Brik, and K. W. Krämer. 2013. 50th anniversary of the Judd–Ofelt theory: an experimentalist's view of the formalism and its application. *Journal of Luminescence* 136: 221–239.

Hemmer, E., A. Benayas, F. Légaré, and F. Vetrone. 2016. Exploiting the biological windows: current perspectives on fluorescent bioprobes emitting above 1000 nm. *Nanoscale Horizons* 1: 168–184.

Hemmer, E., N. Venkatachalam, H. Hyodo, A. Hattori, Y. Ebina, H. Kishimoto, and K. Soga. 2013. Upconverting and NIR emitting rare earth based nanostructures for NIR-bioimaging. *Nanoscale* 5: 11339–11361.

Jaque, D., and F. Vetrone. 2012. Luminescence nanothermometry. *Nanoscale* 4: 4301–4326.

Jia, M., G. Liu, Z. Sun, Z. Fu, and W. Xu. 2018. Investigation on two forms of temperature-sensing parameters for fluorescence intensity ratio thermometry based on thermal coupled theory. *Inorganic Chemistry* 57: 1213–1219.

Jia, M., Z. Sun, F. Lin, B. Hou, X. Li, M. Zhang, H. Wang, Y. Xu, and Z. Fu. 2019. Prediction of thermal-coupled thermometric performance of Er^{3+}. *The Journal of Physical Chemistry Letters* 10: 5786–5790.

Jia, M., Z. Sun, M. Zhang, H. Xu, and Z. Fu. 2020a. What determines the performance of lanthanidebased ratiometric nanothermometers? *Nanoscale* 12: 20776–20785.

Jia, M., Z. Fu, G. Liu, Z. Sun, P. Li, A. Zhang, F. Lin, B. Hou, and G. Chen. 2020b. NIR-II/III luminescence ratiometric nanothermometry with phonon-tuned sensitivity. *Advanced Optical Materials* 8: 1901173.

Jia, M., Z. Sun, H. Xu, X. Jin, Z. Lv, T. Sheng, and Z. Fu. 2020c. An ultrasensitive luminescent nanothermometer in the first biological window based on phonon-assisted thermal enhancing and thermal quenching. *Journal of Materials Chemistry C* 8: 15603–15608.

Kaczkan, M., Z. Boruc, S. Turczyński, and M. Malinowski. 2014. Effect of temperature on the luminescence of Sm^{3+} ions in YAM crystals. *Journal of Alloys and Compounds* 612: 149–153.

Kaczmarek, A. M., M. Suta, H. Rijckaert, A. Abalymov, I. Van Driessche, A. G. Skirtach, A. Meijerink, and P. Van Der Voort. 2020. Visible and NIR upconverting Er^{3+}–Yb^{3+} luminescent nanorattles and other hybrid PMO-inorganic structures for in vivo nanothermometry. *Advanced Functional Materials* 30: 2003101.

Kolesnikov, I. E., A. A. Kalinichev, M. A. Kurochkin, D. V. Mamonova, E. Y. Kolesnikov, and E. Lähderanta. 2019. Ratiometric optical thermometry based on emission and excitation spectra of YVO_4:Eu^{3+} nanophosphors. *The Journal of Physical Chemistry C* 123: 5136–5143.

Kumar, V., B. Zoellner, P. A. Maggard, and G. Wang. 2018. Effect of doping Ge into Y_2O_3: Ho, Yb on the green-to-red emission ratio and temperature sensing. *Dalton Transactions* 47: 11158–11165.

Labrador-Páez, L., M. Pedroni, A. Speghini, J. García-Solé, P. Haro-González, and D. Jaque. 2018. Reliability of rare-earth-doped infrared luminescent nanothermometers. *Nanoscale* 10: 22319–22328.

León-Luis, S. F., U. R. Rodríguez-Mendoza, P. Haro-González, I. R. Martín, and V. Lavín. 2012. Role of the host matrix on the thermal sensitivity of Er^{3+} luminescence in optical temperature sensors. *Sensors and Actuators B: Chemical* 174: 176–186.

Li, L., and S. Zhang. 2006. Dependence of charge transfer energy on crystal structure and composition in Eu^{3+}-doped compounds. *The Journal of Physical Chemistry B* 110: 21438–21443.

Li, P., M. Jia, G. Liu, A. Zhang, Z. Sun, and Z. Fu. 2019. Investigation on the fluorescence intensity ratio sensing thermometry based on nonthermally coupled levels. *ACS Applied Bio Materials* 2: 1732–1739.

Liu, G., L. Fu, Z. Gao, X. Yang, Z. Fu, Z. Wang, and Y. Yang. 2015. Investigation into the temperature sensing behavior of Yb^{3+} sensitized Er^{3+} doped Y$_2$O$_3$, YAG and LaAlO$_3$ phosphors. *RSC Advances* 5: 51820–51827.

Liu, G., Z. Fu, T. Sheng, Z. Sun, X. Zhang, Y. Wei, L. Ma, X. Wang, and Z. Wu. 2016. Investigation into optical heating and applicability of the thermal sensor bifunctional properties of Yb^{3+} sensitized Tm^{3+} doped Y$_2$O$_3$, YAG and LaAlO$_3$ phosphors. *RSC Advances* 6: 97676–97683.

Lu, H., H. Hao, Y. Gao, D. Li, G. Shi, Y. Song, Y. Wang, and X. Zhang. 2017. Optical sensing of temperature based on non-thermally coupled levels and upconverted white light emission of a Gd$_2$(WO$_4$)$_3$ phosphor co-doped with in Ho(III), Tm(III), and Yb(III). *Microchimica Acta* 184: 641–646.

Marciniak, L., K. Prorok, L. Francés-Soriano, J. Pérez-Prieto, and A. Bednarkiewicz. 2016. A broadening temperature sensitivity range with a core-shell YbEr@YbNd double ratiometric optical nanothermometer. *Nanoscale* 8: 5037–5042.

Martínez, E. D., C. D. S. Brites, L. D. Carlos, A. F. García-Flores, R. R. Urbano, and C. Rettori. 2019. Electrochromic switch devices mixing small- and large-sized upconverting nanocrystals. *Advanced Functional Materials* 29: 1807758.

McLaurin, E. J., L. R. Bradshaw, and D. R. Gamelin. 2013. Dual-emitting nanoscale temperature sensors. *Chemistry of Materials* 25: 1283–1292.

Mi, C., J. Zhou, F. Wang, G. Lin, and D. Jin. 2019. Ultrasensitive ratiometric nanothermometer with large dynamic range and photostability. *Chemistry of Materials* 31: 9480–9487.

Pan, Y., X. Xie, Q. Huang, C. Gao, Y. Wang, L. Wang, B. Yang, H. Su, L. Huang, and W. Huang. 2018. Inherently Eu^{2+}/Eu^{3+} codoped Sc$_2$O$_3$ nanoparticles as high-performance nanothermometers. *Advanced Materials* 30: 1705256.

Savchuk, O. A., J. J. Carvajal, C. D. S. Brites, L. D. Carlos, M. Aguilo, and F. Diaz. 2018. Upconversion thermometry: a new tool to measure the thermal resistance of nanoparticles. *Nanoscale* 10: 6602–6610.

Savchuk, O. A., J. J. Carvajal, M. C. Pujol, E. W. Barrera, J. Massons, M. Aguilo, and F. Diaz. 2015. Ho, Yb: KLu(WO$_4$)$_2$ nanoparticles: a versatile material for multiple thermal sensing purposes by luminescent thermometry. *The Journal of Physical Chemistry C* 119: 18546–18558.

Singh, A. K. 2007. Ho^{3+}: TeO$_2$ glass, a probe for temperature measurements. *Sensors and Actuators A: Physical* 136: 173–177.

Skripka, A., A. Benayas, R. Marin, P. Canton, E. Hemmer, and F. Vetrone. 2017. Double rare-earth nanothermometer in aqueous media: opening the third optical transparency window to temperature sensing. *Nanoscale* 9: 3079–3085.

Souza, A. S., L. A. O. Nunes, I. G. N. Silva, F. A. M. Oliveira, L. L. da Luz, H. F. Brito, M. C. F. C. Felinto, R. A. S. Ferreira, S. A. Júnior, L. D. Carlos, and O. L. Malta. 2016. Highly-sensitive Eu^{3+} ratiometric thermometers based on excited state absorption with predictable calibration. *Nanoscale* 8: 5327–5333.

Sun, Z., G. Liu, Z. Fu, Z. Hao, and J. Zhang. 2018. A novel upconversion luminescent material: Li^{+}- or Mg^{2+}-codoped Bi$_{3.84}$W$_{0.16}$O$_{6.24}$: Tm^{3+}, Yb^{3+} phosphors and their temperature sensing properties. *Dyes and Pigments* 151: 287–295.

Sun, Z., G. Liu, Z. Fu, T. Sheng, Y. Wei, and Z. Wu. 2017. Nanostructured La$_2$O$_3$: Yb^{3+}/Er^{3+}: Temperature sensing, optical heating and bio-imaging application. *Materials Research Bulletin* 92: 39–45.

Suo, H., C. Guo, J. Zheng, B. Zhou, C. Ma, X. Zhao, T. Li, P. Guo, and E. M. Goldys. 2016. Sensitivity modulation of upconverting thermometry through engineering phonon energy of a matrix. *ACS Applied Materials & Interfaces* 8: 30312–30319.

Suo, H., X. Zhao, Z. Zhang, and C. Guo. 2020. Ultra-sensitive optical nano-thermometer LaPO$_4$: Yb^{3+}/Nd^{3+} based on thermo-enhanced NIR-to-NIR emissions. *Chemical Engineering Journal* 389: 124506.

Tian, Y., B. Tian, C. Cui, P. Huang, L. Wang, and B. Chen. 2014. Excellent optical thermometry based on single-color fluorescence in spherical NaEuF$_4$ phosphor. *Optics Letters* 39: 4164–4167.

Trejgis, K., A. Bednarkiewicz, and L. Marciniak. 2020. Engineering excited state absorption based nanothermometry for temperature sensing and imaging. *Nanoscale* 12: 4667–4675.

Vetrone, F., R. Naccache, A. Zamarrón, A. Juarranz de la Fuente, F. Sanz-Rodríguez, L. M. Maestro, E. M. Rodriguez, D. Jaque, J. G. Solé, and J. A. Capobianco. 2010. Temperature sensing using fluorescent nanothermometers. *ACS Nano* 4: 3254–3258.

Wade, S. A., S. F. Collins, and G. W. Baxter. 2003. Fluorescence intensity ratio technique for optical fiber point temperature sensing. *Journal of Applied Physics* 94: 4743–4756.

Wang, S., S. Ma, J. Wu, Z. Ye, and X. Cheng. 2020. A promising temperature sensing strategy based on highly sensitive Pr^{3+}-doped $SrRE_2O_4$ (RE = Sc, Lu and Y) luminescent thermometers. *Chemical Engineering Journal* 393: 124564.

Wang, X. D., O. S. Wolfbeis, and R. J. Meier. 2013. Luminescent probes and sensors for temperature. *Chemical Society Reviews* 42: 7834–7869.

Wang, Z., H. Jiao, and Z. Fu. 2019. Investigation on the up-conversion luminescence and temperature sensing properties based on non-thermally coupled levels of rare earth ions doped $Ba_2In_2O_5$ phosphor. *Journal of Luminescence* 206: 273–277.

Wang, Z., D. Ananias, A. Carné-Sánchez, C. D. S. Brites, I. Imaz, D. Maspoch, J. Rocha, and L. D. Carlos. 2015. Lanthanide–organic framework nanothermometers prepared by spray-drying. *Advanced Functional Materials* 25: 2824–2830.

Wu, Z., and S. Zhang. 1999. Semiempirical method for the evaluation of bond covalency in complex crystals. *The Journal of Physical Chemistry A* 103: 4270–4274.

Ximendes, E. C., W. Q. Santos, U. Rocha, U. K. Kagola, F. Sanz-Rodriguez, N. Fernández, A. da S. Gouveia-Neto, D. Bravo, A. M. Domingo, B. del Rosal, C. D. S. Brites, L. D. Carlos, D. Jaque, and C. Jacinto. 2016. Unveiling in vivo subcutaneous thermal dynamics by infrared luminescent nanothermometers. *Nano Letters* 16: 1695–1703.

Ximendes, E. C., U. Rocha, T. O. Sales, N. Fernández, F. Sanz-Rodríguez, I. R. Martín, C. Jacinto, and D. Jaque. 2017. In vivo subcutaneous thermal video recording by supersensitive infrared nanothermometers. *Advanced Functional Materials* 27: 1702249.

Xu, M., D. Chen, P. Huang, Z. Wan, Y. Zhou, and Z. Ji. 2016. A dual-functional upconversion core@shell nanostructure for white-light-emission and temperature sensing. *Journal of Materials Chemistry C* 4: 6516–6524.

Xu, W., Q. Song, L. Zheng, Z. Zhang, and W. Cao. 2014. Optical temperature sensing based on the near-infrared emissions from Nd^{3+}/Yb^{3+} codoped $CaWO_4$. *Optics Letters* 39, 4635–4638.

Xu, W., X. Gao, L. Zheng, Z. Zhang, and W. Cao. 2012. An optical temperature sensor based on the upconversion luminescence from Tm^{3+}/Yb^{3+} codoped oxyfluoride glass ceramic. *Sensors and Actuators B: Chemical* 173: 250–253.

Xue, D., and S. Zhang. 1997. The origin of nonlinearity in $KTiOPO_4$. *Applied Physics Letters* 70: 943–945.

Zhang, A., Z. Sun, M. Jia, G. Liu, F. Lin, and Z. Fu. 2019. Simultaneous luminescence in I, II and III biological windows realized by using the energy transfer of $Yb^{3+} \rightarrow Er^{3+}/Ho^{3+} \rightarrow Cr^{3+}$. *Chemical Engineering Journal* 365: 400–404.

Zheng, S., W. Chen, D. Tan, J. Zhou, Q. Guo, W. Jiang, C. Xu, X. Liu, and J. Qiu. 2014. Lanthanide-doped $NaGdF_4$ core–shell nanoparticles for non-contact self-referencing temperature sensors. *Nanoscale* 6: 5675–5679.

Zhou, J., S. Wen, J. Liao, C. Clarke, S. A. Tawfik, W. Ren, C. Mi, F. Wang, and D. Jin. 2018. Activation of the surface dark-layer to enhance upconversion in a thermal field. *Nature Photonics* 12: 154–158.

Zhu, X., W. Feng, J. Chang, Y. W. Tan, J. Li, M. Chen, Y. Sun, and F. Li. 2016. Temperature-feedback upconversion nanocomposite for accurate photothermal therapy at facile temperature. *Nature Communications* 7: 10437.

Zhu, X., J. Li, X. Qiu, Y. Liu, W. Feng, and F. Li. 2018. Upconversion nanocomposite for programming combination cancer therapy by precise control of microscopic temperature. *Nature Communications* 9: 2176.

16 Single-Crystal Phosphors

Zhengliang Wang

CONTENTS

16.1 INTRODUCTION

Single-crystal phosphors show excellent optical performances, due to their perfect crystal structures with small defects and grain boundaries.[1] They can find huge applications in lighting, display, laser, and so on.[1–4] Taking their application in white light–emitting diodes (wLEDs) as an example, single-crystal phosphors demonstrate superior performance to their powdery samples (Figure 16.1).[5] First, single-crystal phosphors have much less backscattering and less self-absorption. Second, photons created in single crystals need to pass only one boundary, but they in powdery ones face a lot of grain boundaries that induce much non-radiative loss. At last, single-crystal phosphors have few surface defects, meaning that their performance is less affected by the environment. Hence, crystal phosphors exhibit higher energy–conversion efficiency and higher chemical stability during applications. Besides, crystal phosphors show high thermal conductivity, which is useful for heat dispersion of wLEDs.[6]

The crystal growth is a complex dynamic phase transition process, which is controlled by many parameters such as reaction temperature, press, and solvents.[7,8] In general, the growth process for single-crystal phosphors can be divided into two main steps.[7,8] In the first step, the free ions in the solution aggregate to form the clusters by electrostatic attraction. Then the clusters turn into crystals with the regular shape by oriented attachment.[9] Gebauer et al. confirmed the existence of the

DOI: 10.1201/9781003098676-16

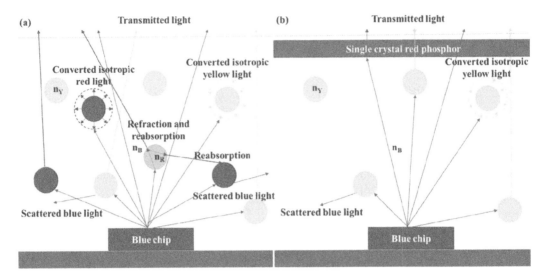

FIGURE 16.1 Luminescence behaviors of (a) powdery and (b) crystal phosphors in wLEDs. (Wang, Z.L., Yang, Z.L., Wang, N, et al. *Adv. Opt. Mater.* 8, 1901512, 2020. With permission)

clusters during the growth process of $CaCO_3$ crystal.[10] The crystal morphologies are influenced not only by the growth environment but also the crystal geometry.[11] For the same crystal geometry, the crystal morphologies can be controlled by the relative growth rates of individual crystal faces.[12] In this chapter, the preparation, structures, luminescent properties, and application of single-crystal phosphors have been discussed.

16.2 PREPARATION OF SINGLE-CRYSTAL PHOSPHORS

According to our investigation, practically crystal phosphors were mainly obtained from the solution or melt salt. To obtain perfect single-crystal phosphors, a lot of methods have been developed by controlling reaction parameters.

16.2.1 Czochralski (CZ) Method

The Czochralski (CZ) method is an important route to obtain transparent crystal phosphors of a large size. A lot of practically crystal phosphors with the high melt temperature, such as $Y_3Al_5O_{12}:Ce^{3+}$ ($YAG:Ce^{3+}$), $Y_3Al_5O_{12}:Nd^{3+}$ ($YAG:Nd^{3+}$), and $Y_3Al_5O_{12}:Yb^{3+}$ ($YAG:Yb^{3+}$) crystals were obtained via this method.[13–18] Taking the preparation process of single-crystal $Y_3Al_5O_{12}:Ce^{3+}$ as an example,[13] a standard process can be described as below. First, stoichiometric raw materials, including Y_2O_3 (99.99%), Al_2O_3 (99.99%), and CeO_2 (99.99%), were mixed thoroughly. Then, the mixed materials were compressed into pieces and calcined at 1200°C for 24 h. The temperature was controlled by a precision temperature controller with a precision of ±0.5°C. The melt temperature kept at 1950 ± 5°C should be favorable for starting the growth. The crystals were grown along the <1 1 1> axis at fixed pulling and rotation rates, 1 mm/h and 10 rpm, respectively. The whole growth process was in the nitrogen atmosphere. The temperature of the reaction system and the rotation rate are important factors to prepare the ideal crystals.

16.2.2 Floating Zone (FZ) Method

The floating zone (FZ) method is often used to prepare inorganic crystals with high melt temperatures.[19–23] $YAG:Ce^{3+}$ crystal was also grown by the FZ method.[19] Halogen lamps were used as the heating source. The feed rods of YAG prepared by the solid-state reaction were heated and melted. Single crystal of YAG:Ce^{3+} was grown under the following conditions: (i) growth rate of 1 mm/h, (ii) feed rotation speed of 15 rpm, (iii) seed rotation speed of 10 rpm, (iv) ambient gas of nitrogen.[19] The FZ technique is very suitable for laboratory-scale studies, but the obtained crystals are limited to their smaller sizes due to the optical heating.[20] Following the similar technology, the green-yellow single crystal of Lu$_3$Al$_5$O$_{12}$:Ce^{3+} (LuAG:Ce^{3+}) was grown in an image furnace.[22]

16.2.3 Liquid-Phase Method

Some ionic crystal phosphors could be obtained from the liquid phase. Different from the CZ and FZ methods with sophisticated facilities and extreme conditions, the conditions for crystal growth in solution are much milder. For this preparation process, some raw materials are dissolved into the appropriate solvent. By vaporing the solvent, free ions in the solution form single crystals. Some Mn^{4+}-activated fluoride crystals were obtained by this method.[4,5,24,25] In a typical process for K$_2$SiF$_6$:Mn^{4+} crystals,[25] the starting materials, including H$_2$SiF$_6$, KF, and K$_2$MnF$_6$, were dissolved into HF solution. The mixed solution filtered through a polytetrafluoroethylene (PTFE) filter was placed in a fume cupboard to keep volatile at room temperature. K$_2$SiF$_6$:Mn^{4+} crystals were formed from the saturated solution after some time. Most of the reported perovskite crystals were also prepared with the liquid growth method.[3,26–30] For example, CsPbI$_3$ single crystals were obtained from dimethylformamide (DMF) solution.[30] First, CsI and PbI$_2$ were dissolved in DMF solution at room temperature. The solution was filtered and placed in a metallic holder. Then the reaction temperature was increased to 110°C. After 3 h, the yellow and needle-like CsPbI$_3$ crystals were observed.

16.3 PRACTICALLY SINGLE-CRYSTAL PHOSPHORS

16.3.1 Single-Crystal YAG Doped with Rare-Earth Ions

Single-crystal Y$_3$Al$_5$O$_{12}$ is an excellent host for phosphors. A series of single-crystal YAG doped with different rare-earth ions (such as Ce^{3+}, Nd^{3+}, Yb^{3+}) have been grown for wLEDs, laser diodes (LDs), and scintillators.[6,13,23,31–44] In this section, the structure, luminescent properties, and application of single-crystal YAG doped with different rare-earth ions are discussed.

16.3.1.1 YAG:Ce^{3+} Crystals

Single-crystal YAG:Ce^{3+} can be obtained by different methods.[13,19] Figure 16.2 gives the typical XRD patterns of the grown YAG:Ce^{3+} single crystals and the pulverized samples.[19] The pulverized samples show similar patterns with that of a conventional powdery YAG:Ce^{3+}. The patterns of the pulverized samples are slightly broadened with more pulverizing. Moreover, the as-grown single crystal shows only the {1 1 0} crystal orientation with the highest intensity and the sharpest width, indicating the anisotropy of the single crystal.

It's well known that YAG:Ce^{3+} belongs to the space group of Ia–3d(230) with the cubic symmetry, which has the primitive structural elements of oxygen tetrahedral, octahedral, and distorted cubes (Figure 16.3). There are two kinds of Al^{3+} sites locating in oxygen tetrahedra (CN = 4) and octahedra (CN = 6), as well as one Y^{3+} site in distorted oxygen cubes. And Y^{3+} ion is coordinated with eight oxygen ions. Due to the same valence and similar ion radii for Y^{3+} (1.019 Å, CN = 8) and Ce^{3+} (1.143 Å, CN = 8), Ce^{3+} ions can easily substitute Y^{3+} sites and form the solid-state compound. The lattice parameters can be calculated by the formula of $\sin^2\theta = \lambda^2/4a^2 (h^2 + k^2 + l^2)$, where λ is the wavelength of X-ray, $(h\,k\,l)$ is Miller index, θ is the diffraction angle, and a is the lattice parameter. The calculated lattice parameter is 12.0178 Å, which is bigger than that of the pure Y$_3$Al$_5$O$_{12}$ host (12.009 Å). This behavior is induced by the substitution of lager Ce^{3+} to Y^{3+}.

The basic properties of single-crystal YAG:Ce^{3+} are summarized in Table 16.1.[6] It can be pointed out that this compound presents a high chemical and physical stability. The high melting temperature

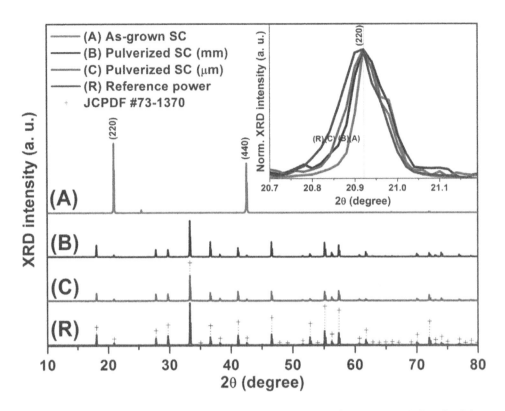

FIGURE 16.2 XRD patterns of the as-grown single crystal (A), the soft (B), and hard (C) pulverizing samples. (Park, K.W., Lim, S.G., Deressa, G., et al. *J. Lumin.* 168, 334, 2015. With permission)

of this oxide, together with its high hardness and the lack of cleavage planes, is indicative of the outstanding resistance of this kind of single crystal even under harsh environments. This is a major advantage in comparison with the oxidation problems for some oxynitride and nitride phosphors.[6]

In the single-crystal YAG:Ce[3+], the ground state of Ce[3+], which has the [Xe]$4f^1$ electronic structure splits into the doublet state of $^2F_{5/2}$ and $^2F_{7/2}$, due to its spin-spin coupling. At room temperature, only the former is occupied by electronics. Because the $4f$ level is shielded in the inner layer, the

TABLE 16.1

Basic Properties of Single-Crystal YAG:Ce[3+] [6]

Crystal structure	Cubic (garnet, space group Ia–3d)
Melting point	~1960°C
Density	4.56 g/cm³
Ce segregation coefficient	0.067
Absorption cross section at 460 nm	3.0×10^{-18} cm²
Absorption cross section at 340 nm	1.0×10^{-18} cm²
Internal quantum efficiency	95–100% in the range RT~300°C
Thermal conductivity	~14 W/(m·K)
Mohs hardness	8.25
Thermal expansion coefficient	8.6×10^{-6} K⁻¹
Specific heat capacity	~590 J/(kg·K)
Refractive index (380–780 nm)	~1.84

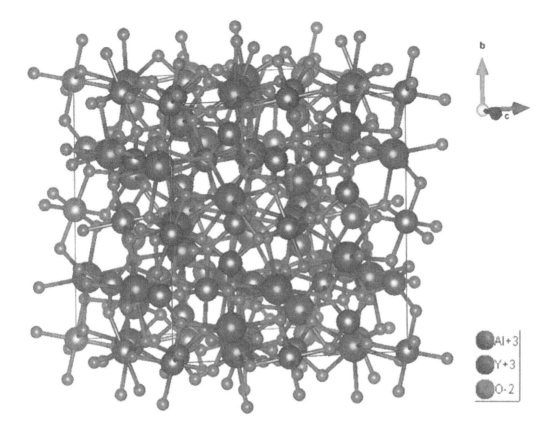

FIGURE 16.3 Crystal structure of YAG: Ce^{3+}.

crystal-field effect is minimal. While, the radial wave function of $5d$ locates in the outside of the $5s^25p^6$ shell, therefore $5d$ states is strongly affected by the crystalline field. In the meantime, the energy of $4f$ and $5d$ electrons in Ce^{3+} is close, which leads to energy level splitting in $5d$ state and forming energy bands in the crystal field.[31,32] Figure 16.4 is the excitation and emission spectra of YAG:Ce^{3+} crystals. The broad excitation band from 350 to 500 nm is due to the $4f$–$5d$ transition of Ce^{3+}. When Ce^{3+} is excited with the blue light at 466 nm, the electrons transmit from the $4f$ ground state to the $5d$ excited state. Most of them immediately transmit from the $5d$ excited state to the $^2F_{7/2}$ and $^2F_{5/2}$ of the $4f$ ground state. The green-yellow emission with a wide spectral band of 500–700 nm provides the theoretical basis that YAG:Ce^{3+} crystals have the immense potential to fabric the wLEDs and LDs. Furthermore, the spectral line strength of YAG:Ce^{3+} crystals is significantly higher than that of YAG:Ce^{3+} powder, and the optical properties of YAG:Ce^{3+} crystals are better than YAG:Ce^{3+} powder.[33]

The thickness of the crystal wafer has an obvious influence on Ce^{3+} emission intensity, which increases with the increasing wafer thickness and reaches a maximum value. The optical transmittance and Ce^{3+} emission intensity of YAG:Ce^{3+} crystal wafer are closely related to wafer thicknesses (Figure 16.5). The absorptions corresponding to the $4f{\rightarrow}5d$ transitions of Ce^{3+} take place at 340 and 465 nm. With the increasing thickness of the crystal wafer, which is induced by the increases of Ce^{3+} ions, the absorption intensity of Ce^{3+} increases continuously.[13] To the contrary, there is a maximum value of the emission intensity along with the increase of YAG:Ce^{3+} crystal wafer thickness. According to the reported investigation,[13] the emission intensity increases continuously, which reaches a maximum at the thickness of 1.5 mm and then declines. Even though, the thicker YAG:Ce^{3+} crystal wafer implies more Ce^{3+} ions in it, the emission intensity with a thickness of

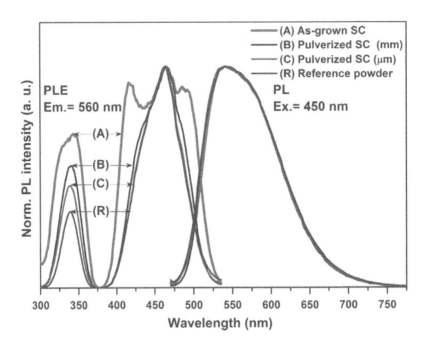

FIGURE 16.4 Excitation and emission spectra of single-crystal YAG:Ce³⁺ and the pulverized samples. (Park, K.W., Lim, S.G., Deressa, G., et al. *J. Lumin.* 168, 334, 2015. With permission)

2 mm is lower than that of YAG:Ce³⁺ crystal wafer with a thickness of 1.5 mm. The reason is that the emitted light by Ce³⁺ ions cannot pass through the crystal wafer entirely.

In contrast to powdery YAG:Ce³⁺, single-crystal YAG:Ce³⁺ belongs to high thermal conductivity and high thermal quenching of luminescence.[34,35] YAG:Ce³⁺ crystals exhibit huge potential on high-brightness wLEDs and LDs.[13,35–41] Figure 16.6 presents the electroluminescence (EL) spectra of the fabricated wLEDs encapsulating YAG:Ce³⁺ crystal wafers with different thicknesses under 350 mA excitation. These spectra are of similar shapes with two broad emission bands in the blue

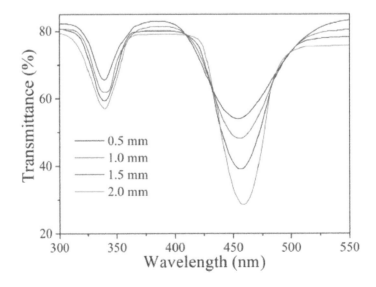

FIGURE 16.5 Optical transmittance of YAG:Ce³⁺ crystal wafer with different thicknesses. (Yang, Y., Wang, X., Liu, B., et al. *J. Lumin.* 157, 204, 2018. With permission)

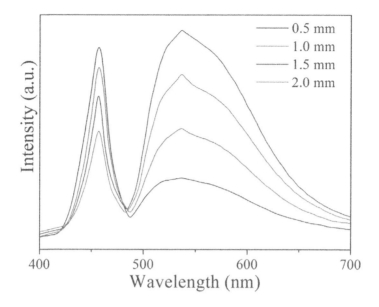

FIGURE 16.6 Electroluminescence spectra of the fabricated wLEDs encapsulating YAG:Ce^{3+} crystal wafers with different thicknesses under 350 mA excitation. (Yang, Y., Wang, X., Liu, B., et al., *J. Lumin.* 204, 157, 2018. With permission)

and yellow regions. With the increase of wafer thickness, the blue emission originating from the LED chip decreases continuously and the yellow from single-crystal YAG:Ce^{3+} increases up to the thickness of 1.5 mm. When the thickness of the crystal wafer exceeds 1.5 mm, both blue and yellow emissions get weaker. The parameters of wLEDs are listed in Table 16.2.[13] The device fabricated with 1.5 mm YAG:Ce^{3+} crystal shows the highest luminous efficacy (LE) of 112.33 lm/W and the lowest color temperature (T_c) of 4458K.

Compared with wLEDs, LDs own a very high output power per wafer area.[38] The LD epitaxial area is much smaller than the wLED area.[39] Moreover, LDs have highly directional light emission, which can be useful for many applications. There are two types of device architectures for the fabrication of LDs, as shown in Figure 16.7.[40] They are named the transmission mode and the reflection mode, respectively. The former is simpler and more compact, but the latter also has an obvious advantage, which enables the phosphors to maintain highly luminescent efficiency in high-power devices. Using single-crystal YAG:Ce^{3+} as a yellow component, one phosphor-converted white LD was fabricated.[40] Under the excitation of a 3.38-W blue laser (~100 W/mm^2), this device exhibits intense white light emission with the luminous flux (LF) of 465 lm, the LE of 145 lm/W, and the T_c of 4980K. So this LD based on single-crystal YAG:Ce^{3+} provides a choice for the design of vehicle headlights, searchlights, projectors, and other high-brightness illumination devices.

TABLE 16.2
CIE Coordinates, T_c Values, and LE of the Fabricated wLEDs[13]

Thickness (mm)	0.5	1.0	1.5	2.0
CIE coordinates (x,y)	0.292, 0.256	0.312, 0.296	0.354, 0.337	0.361, 0.372
T_c (K)	10,572	6836	4458	4530
LE (lm/W)	80.68	107.06	112.33	105.56

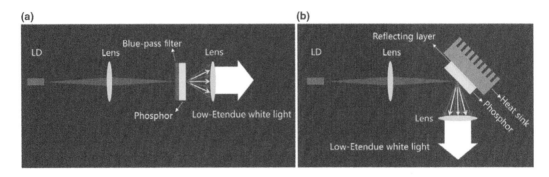

FIGURE 16.7 Schematic structure of the phosphor-converted white LD in (a) transmission mode and (b) reflection mode. (Xu, J., Thorseth, A., Xu, C., et al. *J. Lumin.* 212, 279, 2019. With permission)

16.3.1.2 Rare-Earth Ions–Doped YAG Ce^{3+} Derivants

Single-crystal YAG:Ce^{3+} is more appropriate for cool white sources of high color temperature. Such light sources are satisfactory for a wide variety of high-brightness applications where no particular color characteristics are demanded. However, rendering of green and red colors is necessary due to the requirement of many indoor lighting applications what is white light with lower T_c and high color rendering index ($R_a > 80$). As discussed above, the emission of Ce^{3+} is due to its d–f transition, which is easily influenced by the crystal-field strength of YAG. The emission range of Ce^{3+} could be adjusted by doping other rare-earth ions into YAG crystal. By the compression or expansion of the crystal cell with smaller or larger rare-earth ions, the emission of Ce^{3+} exhibits a blue or redshift. In that case, YAG derivants, especially the mixed garnet phosphors such as $(Y_{1-x}Lu_x)_3Al_5O_{12}$:Ce^{3+} (YLuAG:Ce^{3+}) and $(Y_{1-x}Gd_x)_3Al_5O_{12}$:Ce^{3+} (YGAG:Ce^{3+}), can give a new direction to accomplish such aim.[6,42,43]

For the sake of getting a blueshifted YAG:Ce^{3+} crystal, Lu^{3+} with the ionic radius of 97.7 pm in eight coordination is commonly used to partially or totally substitute Y^{3+} with the ionic radius of 101.9 pm.[6,42] Single-crystal YLuAG:Ce^{3+} can be grown by the CZ technique with different nominal Lu^{3+} concentrations. The emission energy of Ce^{3+} in single-crystal YLuAG:Ce^{3+} linearly increases with the increasing Lu^{3+} concentration.[42] LuAG:Ce^{3+} emission is observed as clear greenish, in contrast to the yellowish one of YAG:Ce^{3+}. The related optical parameters are listed in Table 16.3.[42] The first excitation peak exhibits a blueshift with the Lu^{3+} concentration increase. But, the second excitation band has a little redshift. This is because that the first excited level of Ce^{3+} increases and the second one decreases with the increasing Lu^{3+} content.

Moreover, YLuAG:Ce^{3+} crystal has high internal quantum efficiency (QE) at room temperature and high temperature, indicating that it has a high conversion efficiency and temperature stability.

Single-crystal Lu$_3$Al$_5$O$_{12}$:Ce^{3+} (LuAG:Ce^{3+}) was also prepared by the FZ method.[22] Ce^{3+} occupies the lattice site of Lu^{3+} in LuAG:Ce^{3+} and exhibits similar the luminescent behavior within YAG:Ce^{3+}

TABLE 16.3

Spectroscopic Properties of Single-Crystal YLuAG:Ce^{3+} as a Function of Lu^{3+} Concentration (Ce^{3+} Concentration is Constant with a Value of 0.06 at.%)[42]

Lu (at.%)	λ_{ex2} (nm)	λ_{ex1} (nm)	δ_{ex} (cm^{-1})	λ_{em1} (nm)	λ_{em2} (nm)	δ_{em} (cm^{-1})	ΔS (cm^{-1})
0	338.7	456.6	7623	528.9	574.4	1498	2993
21.0	339.9	454.0	7393	517.7	565.6	1636	2712
29.3	340.3	453.1	7316	516.1	562.9	1609	2694
99.9	344.8	446.0	6580	502.4	545.1	1556	2516

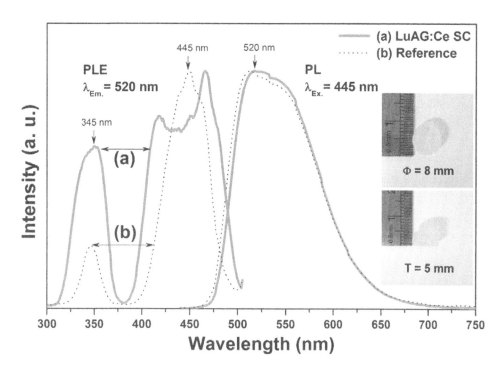

FIGURE 16.8 Excitation and emission of single-crystal LuAG:Ce³⁺ and the polycrystalline powder phosphor as reference. (Kang, T.W., Park, K.W., Ryu, J.H., et al. *J. Lumin.* 191, 35, 2017. With permission)

(Figure 16.8).[22] The broad excitation bands peaking at 460 and 345 nm are due to the $^2F_{5/2}$ level of the $^4f_1$ ground state to the first and second lowest $^5d_1$ states. The broad emission band peaking at 520 nm is due to the *f–d* transition. The grown single crystal shows higher excitation intensity at 345 nm than that of the polycrystal sample, indicating that the single crystal has fewer defects than those of the powdery sample. Single-crystal LuAG:Ce³⁺ also exhibits excellent thermal stability due to fewer intrinsic defects and less phonon generation at high temperatures. When single-crystal LuAG:Ce³⁺ is applied to a 5-W blue LD with a remote-phosphor structure, the LD achieves the extremely stable light emission without an efficiency loss.

In the garnet structure, crystal-field splitting can be increased by expanding the dodecahedral site. For the sake of getting a redshift YAG:Ce³⁺ crystal, Gd³⁺ (105.3 pm) was often used to partially or totally substitute Y³⁺ (101.9 pm). A series of YGAG:Ce³⁺ crystals were grown by the CZ technique with different Gd concentrations.[43] The lattice parameter of YGAG:Ce³⁺ crystals is shown in Figure 16.9.[43] With the increasing content of Gd³⁺, the lattice parameter increases at the rate of 0.001 Å per at% of Y³⁺ substituted at room temperature. The lattice constant also exhibits a rising trend with the increasing temperature. Compared with the emission of YAG:Ce³⁺ crystal, the emission of as-obtained single-crystal YGAG:Ce³⁺ is redshifted with increasing Gd³⁺ content. Meanwhile, YGAG:Ce³⁺ crystal owns high conversion efficiencies, thermal stability, and superior thermal conductivity. Then, a series of wLEDs were obtained using the obtained crystal with different thicknesses. Figure 16.10 illustrates the CIE coordinate and the photographs of these wLEDs.[43] When the thickness of YGAG:Ce³⁺ crystal turns large, the T_c value of wLED gets down to 4098K.

16.3.1.3 Single-Crystals YAG:RE³⁺ (RE = Nd, Yb, Dy)

YAG crystals doped with other rare-earth ions are often used as solid-state laser material and scintillators. For example, single-crystal YAG:Nd³⁺ is one ready laser material used for microchip lasers and supports a huge area of laser applications because of its combined optical, thermal, and mechanical properties.[15] The performance of YAG:Nd³⁺ crystals is better than YAG:Nd³⁺ ceramics.

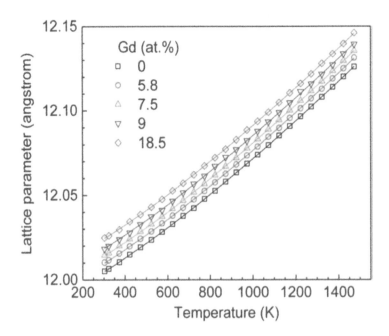

FIGURE 16.9 Lattice parameter of YGAG:Ce^{3+} single crystals as a function of temperature. (Arjoca, S., Inomata, D., Matsushita, Y., et al. *CrystEngComm* 18, 4799, 2016. With permission)

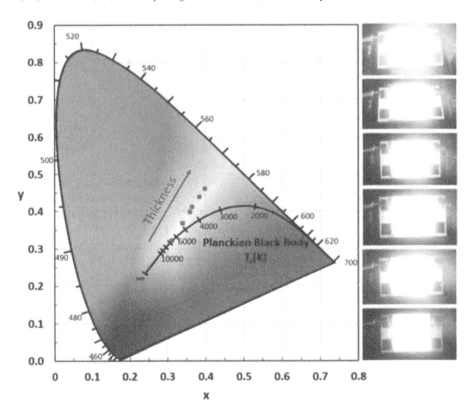

FIGURE 16.10 CIE color coordinate plot and photographs of wLEDs based on the blue LED peaking at 453 nm and the YGAG:Ce^{3+} crystal with different thicknesses. (Arjoca, S., Inomata, D., Matsushita, Y., et al. *CrystEngComm* 18, 4799, 2016. With permission)

YAG:Yb^{3+} crystal also is an excellent laser crystal due to its high thermal conductivity and excellent physical and chemical properties.[16] Single-crystal YAG:Dy^{3+} shows sharp emission lines near 482 and 584 nm, due to the $^4F_{9/2} \rightarrow \,^6H_{15/2}$ and $^4F_{9/2} \rightarrow \,^6H_{13/2}$ transition of Dy^{3+}, respectively. These wavelengths match the high sensitivity region of a Si-avalanche photodiode. So single-crystal YAG:Dy^{3+} may be used as a scintillator.[44]

16.3.2 SINGLE-CRYSTAL YAG:CR^{3+}

Cr^{3+} is an excellent luminescent center for many phosphors.[45–48] Its emission originates from its d–d transition, which is easily influenced by the crystal field of the hosts. Due to the same valence and similar ion radii between Cr^{3+} (0.615 Å, CN=6) and Al^{3+} (0.535 Å, CN=8), Cr^{3+} was easily introduced into YAG. Xu et al. obtained the single-crystal YAG:Cr^{3+} by the CZ method.[45] In the structure of YAG:Cr^{3+}, Cr^{3+} prefers to occupy the Al^{3+}-octahedral center site rather than the Al^{3+}-tetrahedral center site.[49] Figure 16.11 is the excitation and emission spectra of as-grown YAG:Cr^{3+} crystal. The broad excitation band peaking at ~430 nm is ascribed to the $^4A_2$–$^4T_1$ transition of Cr^{3+} in the octahedral crystal field. Red-emitting peak at 687 nm is attributed to the $^2E \rightarrow \,^4A_2$ transition of Cr^{3+} upon 430 nm excitation. Single-crystal YAG:Cr^{3+} also has highly thermal stability and a high internal QE of 92%. At last, one wLED based on YAG:Cr^{3+} and YAG:Ce^{3+} crystals was fabricated. Under the excitation of 20 mA current, the device emits intense white light with the R_a value of 76.4, T_c value of 5236K, and LE value of 118.36 lm/W.

16.3.3 RARE-EARTH IONS–DOPED OTHER CRYSTALS

16.3.3.1 Single-Crystal CaAl$_2$Si$_4$N$_8$:Eu^{2+}

Rare-earth ions–doped nitride/oxynitride phosphors are excellent phosphors, due to their high emission efficiency and high thermal and chemical stabilities. They can find applications on lighting

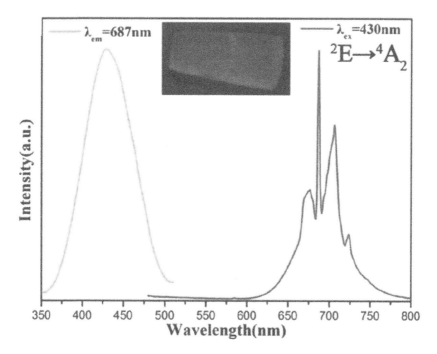

FIGURE 16.11 Excitation and emission spectra of YAG:Cr^{3+}. (Xu, T., Yuan, L., Chen, Y., et al. *Opt. Mater.* 91, 30, 2019. With permission)

and display.[50–53] For example, commercial β-SiALON:Eu^{2+} is used for backlighting as a green component.[51] $CaAlSiN_3$:Eu^{2+} as the unique red-emitting phosphor has been commercialized.[52] Due to the harsh preparation conditions (such as high reaction temperature, atmosphere adjustment), it is not easy to obtain such single-crystal phosphors. The related work was seldom reported. Recently, Hasegawa et al. successfully grew the orange-red emission of $CaAl_2Si_4N_8$:Eu^{2+} single crystal through the vapor phase technique and investigated its crystal structure in detail.[53] The obtained single crystal is formed as a hexagonal column sharp and presents red emission under excitation at 365 nm. It belongs to the trigonal structure with the space group of $P31c$ (no. 159). The lattice parameters are $a = 0.79525(9)$ nm, $c = 0.57712(8)$ nm, $V = 0.31609(8)$ nm^3, and $Z = 2$.[53] In this crystal structure, Si and Al ions occupy two different tetrahedral sites with different crystallographic environments. Each Ca^{2+} occupies the polyhedral center coordinated by seven nitrogen atoms. In consideration of the same valence and similar ion radii for Ca^{2+} (1.06 Å, CN = 7) and Eu^{2+} (1.20 Å, CN = 7), Eu^{2+} replaces the site of Ca^{2+} in $CaAl_2Si_4N_8$:Eu^{2+}.

There are two broad excitation bands in the region from UV to visible light, which is due to the $4f$–$5d$ transition of Eu^{2+}.[53] Single-crystal and powdery $CaAl_2Si_4N_8$:Eu^{2+} phosphors exhibit similar emission spectra with a broad orange-red emission. But the emission wavelength peaking at 590 nm for powder phosphor shows some blueshift, compared with that of the single-crystal phosphor (600 nm). This can be explained by the oxygen content in the phosphor lattice. This work provides a new avenue for growing the high-purity single crystal of nitride phosphor.

16.3.3.2　Single-Crystal $SrAl_2O_4$:Eu^{2+},Dy^{3+}

$SrAl_2O_4$:Eu^{2+},Dy^{3+} is a classical persistent phosphor with intense green emission.[54,55] Single crystal can be obtained by the FZ method.[55] The as-obtained single crystals show the broadband in the near-UV region associated with the electric dipole allowed $4f$–$5d$ transitions and one broad emission band centered at 520 nm. The thermoluminescence spectra of $SrAl_2O_4$:Eu^{2+},Dy^{3+} crystals confirm the presence of very deep traps. Besides, the wavelength dependence of QE for $SrAl_2O_4$:Eu^{2+},Dy^{3+} has been investigated, compared with a Dy^{3+}-free sample. These results shed light on the role of co-dopants in the long-persistent phosphor.

16.3.4　Mn⁴⁺–Doped Fluoride Crystals

Mn^{4+}-activated fluorides are efficient red components for wLEDs, due to their unique luminescent properties.[56–58] In octahedral crystal field, the transition of Mn^{4+} from $^4A_2$ to $^4T_2$ is parity-forbidden, but its ^2E–$^4A_2$ transition is both parity and spin-forbidden.[59–61] In the fluoride system, Mn^{4+} is often experienced by a strong crystal field. Its broad excitation band originated from the $^4A_2$–$^4T_2$ transition is located at the blue-light region, which matches well with the blue emission of GaN chips. Under the phonon assistant, a series of red-emitting peaks can be observed. Hence Mn^{4+}-activated fluoride phosphors are important red components for wLED lighting and backlighting.[56–58]

A lot of fluoride phosphors such as A_2BF_6:Mn^{4+} (A = Na, K, Rb, and Cs; B = Si, Ti, Ge, and Zr) have been investigated.[62–70] All these samples belonged to typical ionic compounds that are composed of alkali-metal cations and $[XF_6]^{2-}$/$[MnF_6]^{2-}$ anions by the electrostatic attraction. The single-crystal samples have been obtained by vaporing the appropriate solvent.[4,5,24,25] Their structures, luminescent properties, and applications are discussed here.

16.3.4.1　Single-Crystals Cs_2XF_6:Mn^{4+} (X = Ge, Si, Ti)

Millimeter-sized single-crystals of Cs_2XF_6:Mn^{4+} (X = Ge, Si, Ti) were grown from the HF solution at room temperature.[5] These crystals emit bright red light under irradiation of blue light (Figure 16.12).[5] Cs_2GeF_6:Mn^{4+} and Cs_2SiF_6:Mn^{4+} crystals are of a similar cubic structure with the space group of Fm–$3m$ (no. 225), while Cs_2TiF_6:Mn^{4+} crystal has the hexagonal structure with the space group of P–$3m1$ (no. 164). The related crystal parameters of Cs_2GeF_6:Mn^{4+} are listed in Table 16.4.[5] In the crystal structure, each Ge^{4+} is coordinated with six F^- and occupies the center of one GeF_6^{2-} octahedron.

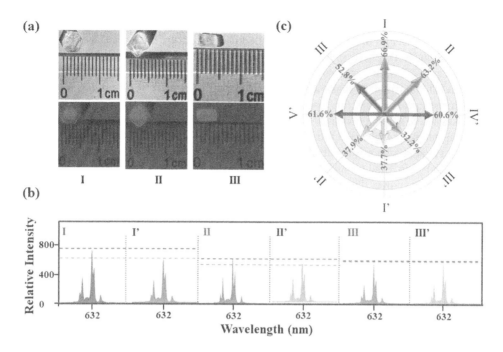

FIGURE 16.12 (a) Photographs, (b) PL spectra, and (c) EQE values of the red-emitting phosphors. (I and I', II and II', as well as III and III': Cs_2GeF_6:Mn⁴⁺, Cs_2SiF_6:Mn⁴⁺, as well as Cs_2TiF_6:Mn⁴⁺ single crystals and their relevantly ground powder samples; IV' and V': commercial K_2SiF_6:Mn⁴⁺ and K_2GeF_6:Mn⁴⁺ powdery samples). (Wang, Z.L., Yang, Z.L., Wang, N., et al., *Adv. Opt. Mater.* 8, 1901512, 2020. With permission)

TABLE 16.4

Crystal Data and Structure Refinement for Cs_2GeF_6:Mn⁴⁺⁵

Identification Code	CGFM
Empirical formula	$Cs_2F_6Ge_{0.96}Mn_{0.04}$
Formula weight	451.82
Temperature	150.00(10)K
Crystal system, space group	Cubic, Fm–$3m$
Unit cell dimensions	$a = 8.9495(4)$ Å alpha = 90°
	$b = 8.9495(4)$ Å beta = 90°
	$c = 8.9495(4)$ Å gamma = 90°
Volume	716.79(11) Å³
Z, Calculated density	4, 4.186 Mg/m³
Absorption coefficient	14.222 mm⁻¹
$F(0\ 0\ 0)$	783.0
2 Theta range for data collection	7.888°–58.622°
Limiting indices	$-11 \leq h \leq 6, -11 \leq k \leq 2, -6 \leq l \leq 10$
Reflections collected/unique	312/72 [R(int) = 0.0264, R_{sigma} = 0.0133]
Refinement method	Full-matrix least-squares on F^2
Data/restraints/parameters	72/0/7
Goodness-of-fit on F^2	1.125
Final R indices [I > 2sigma(I)]	$R_1 = 0.0336, wR_2 = 0.0875$
R indices (all data)	$R_1 = 0.0338, wR_2 = 0.0876$
Largest diff. peak and hole	1.60 and -2.03 e Å³

Considering the same charge and similar volume between GeF_6^{2-} and MnF_6^{2-}, MnF_6^{2-} prefers to occupy the site of GeF_6^{2-} in $Cs_2GeF_6:Mn^{4+}$. This replacement of MnF_6^{2-} can also be observed in $Cs_2SiF_6:Mn^{4+}$ and $Cs_2TiF_6:Mn^{4+}$ crystals and powders.

Mn^{4+} ions in $Cs_2XF_6:Mn^{4+}$ crystals and powders show similar luminescent properties (Figure 16.12).[5] Under excitation at 460 nm, the single-crystal and powdery $Cs_2XF_6:Mn^{4+}$ are of similar emission spectra with the strongest peak at ~630 nm. These sharp peaks are due to the due to the anti-Stokes v_3, v_4, and v_6 and Stokes v_6, v_4, and v_3 phonon sidebands of ^2E to $^4A_2$. The emission intensity of each millimeter-sized crystal sample is higher than that of the corresponding powdery one. Meanwhile, the external QE of each crystal phosphor is significantly higher than that of the powdery sample. The external QE values of $Cs_2GeF_6:Mn^{4+}$ and $Cs_2SiF_6:Mn^{4+}$ crystals are even higher than commercial $K_2SiF_6:Mn^{4+}$ and $K_2GeF_6:Mn^{4+}$.[71] This proves that single-crystal phosphors are more efficient light convertors than powdery samples.

MnF_6^{2-} is not stable and easy to hydrolyze. The fluoride phosphors doped with Mn^{4+} often suffer from poor stability toward moisture.[72–76] Since single-crystal phosphors have few surface defects and surface Mn^{4+}, single-crystal phosphors exhibit excellent stability toward moisture. Figure 16.13 exhibits the comparison of emission intensity for $Cs_2GeF_6:Mn^{4+}$ crystals and powders measured in deionized water at different times.[5] After 7.0 h, the crystal retains its color, but the powder turns black. The integral emission intensity (I_e) of single crystal is 81% of its initial. The I_e value for powdery sample decreases to just 9% of its initial intensity. In weak alkali solution, single-crystal $Cs_2GeF_6:Mn^{4+}$ also remains the higher I_e of 76% of initial intensity than that of the powdery sample (6% of initial intensity). This pattern was also observed between crystal and powdery of $Cs_2SiF_6:Mn^{4+}$ and $Cs_2TiF_6:Mn^{4+}$, respectively. The sudden drop for I_e of the powdery phosphor in deionized water or weak alkali solution is mainly due to the hydrolysis of a large amount of Mn^{4+} ions in the boundaries.

In general, the emission intensity of phosphor decreases with the increasing temperature. This phenomenon is called thermal quenching. Different from rare ions–doped phosphors, Mn^{4+}-activated fluoride phosphors have a higher thermal quenching temperature. Figure 16.14 displays

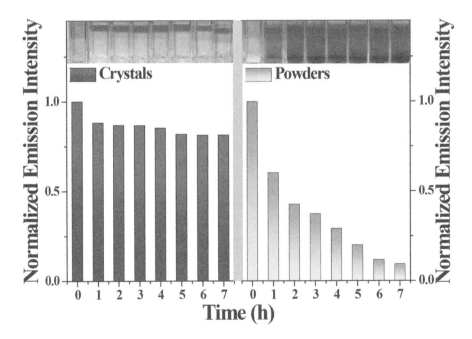

FIGURE 16.13 Emission intensity of $Cs_2GeF_6:Mn^{4+}$ crystals and powders measured in deionized water at different times. (Wang, Z.L., Yang, Z.L., Wang, N., et al., *Adv. Opt. Mater.* 8, 1901512, 2020. With permission)

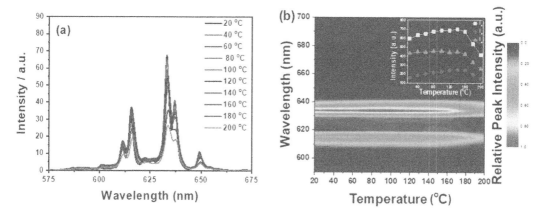

FIGURE 16.14 (a) Emission spectra of $Cs_2GeF_6:Mn^{4+}$ crystal at different temperature under 460 nm light excitation and (b) the temperature-dependent emission intensity of $Cs_2GeF_6:Mn^{4+}$ crystal.

the temperature-dependent emission spectra of single-crystal $Cs_2GeF_6:Mn^{4+}$ upon 460 nm light excitation. All the emission spectra exhibit similar shapes, no obvious shift of emission peaks can be found. With the increasing temperature, the intensities of anti-Stokes (W_a) and Stokes transitions (W_s) show a rising trend, respectively. They reach the top when the temperature is at 140°C. This phenomenon is in accordance with the following equations[69]:

$$W_a(T) = D \cdot \frac{1}{\exp(\hbar\omega/kT) - 1} \tag{16.1}$$

$$W_s(T) = D \cdot \frac{\exp(\hbar\omega/kT)}{\exp(\hbar\omega/kT) - 1} \tag{16.2}$$

where D is the proportional coefficient, $\hbar\omega$ is the energy of the coupled vibronic mode, T is temperature, and k is the Boltzmann constant. So it is easy to deduce that the intensity of anti-Stokes (W_a) and Stokes transitions (W_s) would increase with the rise of temperature. In addition, the intensity of W_a will rise more rapidly than that of W_s with the increasing temperature. The integrated emission intensity (I_e) of $Cs_2GeF_6:Mn^{4+}$ at 140°C is about 1.18 times higher than that at 20°C. Besides, the CIE values according to the emission spectrum at 140°C are calculated to be $x = 0.686$, $y = 0.314$, which is close to that at room temperature. These results indicate that single-crystal $Cs_2GeF_6:Mn^{4+}$ shares excellently thermal quenching resistance and color stability.

Due to their excellent luminescent properties and chemical stability, single-crystal phosphors exhibit huge potentials for the application of wLEDs. The LE of the wLED based on $Cs_2GeF_6:Mn^{4+}$ crystals is as high as 193.0 lm/W. Even under 120 mA, the LE value of the device with single-crystal $Cs_2GeF_6:Mn^{4+}$ remains at about 140.9 lm/W (73% of LE at 20 mA), which is higher than that of the wLED with the powdery one (114.7 lm/W, 64% of LE at 20 mA). This confirms that the red-emitting crystals present a higher blue-light conversion efficiency than their corresponding powders.

To further improve waterproof stability, $Cs_2TiF_6:Mn^{4+}@Cs_2TiF_6$ core-shell structured single crystal was obtained by the epitaxial growing method.[4] The I_e of $Cs_2TiF_6:Mn^{4+}@Cs_2TiF_6$ remained 82.6% of its initial value after 420 min of immersion in deionized water. Using the core-shell structured single crystal with YAG:Ce^{3+} single crystal, one wLED was fabricated with R_a of 90 and T_c of 3155K. Moreover, an LD with high brightness and excellent directionality was obtained by combing $Cs_2TiF_6:Mn^{4+}@Cs_2TiF_6$ and YAG:Ce^{3+} single crystals under blue laser pumping. The LE value of the as-obtained LD increases at first with the rising incident power density and shows an obvious luminance saturation at a power density of 10.462 W/cm^2.[4] Its CIE coordinate and T_c keep good stability under various incident power densities.

16.3.4.2　Single-Crystal K_2GeF_6:Mn^{4+} and K_2SiF_6:Mn^{4+}

Powdery K_2GeF_6:Mn^{4+} and K_2SiF_6:Mn^{4+} have been prepared by many methods, such as the cation exchange method and the coprecipitation approach.[77–82] Due to the highly efficient red emission, these phosphors have more potentials for lighting and backlighting. As discussed above, these powdery phosphors also suffer from instability toward the moisture due to too much Mn^{4+} in the boundaries. Then single-crystals K_2GeF_6:Mn^{4+} and K_2SiF_6:Mn^{4+} were investigated.

Figure 16.15 shows the excitation and emission spectra of K_2GeF_6:Mn^{4+} crystal and commercial K_2GeF_6:Mn^{4+} powder. The two samples exhibit similar spectral shapes. The broad excitation bands in the blue-light region match well with the emission of the GaN chip. A series of red-emitting peaks in these spectra consists of the anti-Stokes and Stokes phonon sidebands of the ^2E-$^4A_2$ transition of Mn^{4+}, with the strongest peak at 630 nm. The zero phonon line (ZPL) of Mn^{4+} emission is difficult to observe in the current coordination environment. Upon the blue-light excitation, the hexagonal plate-like crystals exhibit intense red light with a high color purity of appropriate CIE coordinate values ($x = 0.693$, $y = 0.307$). K_2SiF_6:Mn^{4+} crystals also show similar luminescent properites.[25]

As expected, these crystals exhibit excellent moisture resistance. To investigate the moisture resistance of K_2GeF_6:Mn^{4+} crystal, wLEDs based on KGFM crystal and commercial KGFM powder were aged for 35 days under the high-humidity (85%) and high-temperature (85°C) conditions. The results of two aged wLEDs are listed in Table 16.5. After 35 days under the high-humidity and high-temperature conditions, the wLED with crystal K_2GeF_6:Mn^{4+} still exhibits lower T_c (4090K), higher R_a (91.6), and stable CIE coordinates. The LE (109.3 lm/W) of aged K_2GeF_6:Mn^{4+} crystal-based wLED is almost 74% of its initial value, which is similar to the nonaged wLED with commercial K_2GeF_6:Mn^{4+} (Figure 16.16). However, the LE of wLED with commercial KGFM after aging is approximately 67% of the corresponding initial. Moreover, the emission intensity of K_2SiF_6:Mn^{4+} crystals in deionized water for 12 h still maintains at 97.6% of its initial value.[25]

The as-obtained crystals also exhibit huge potentials for applications in wLEDs and LDs. Two wLEDs are fabricated by combining commercial green phosphor β-SiALON with KGFM crystal

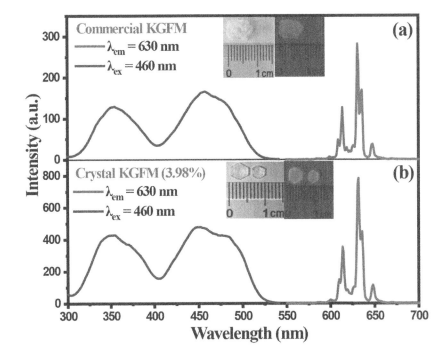

FIGURE 16.15 Excitation and emission spectra of (a) commercial K_2GeF_6:Mn^{4+} powder and (b) K_2GeF_6:Mn^{4+} crystal and inset photographs irradiated by nature and blue light.

TABLE 16.5

Photoelectric Parameters of wLEDs with Single-Crystal K_2GeF_6:Mn^{4+} and Commercial K_2GeF_6:Mn^{4+} Powder after Aging for Different Days

Times (days)	Samples	T_c (K)	R_a	CIE (x,y)	LE (lm/W)
0	K_2GeF_6:Mn^{4+} crystal and YAG:Ce^{3+}	3852	94.7	(0.3799, 0.3562)	148.5
0.5		4053	92.6	(0.3737, 0.3579)	145.9
1		4186	91.5	(0.3690, 0.3566)	140.9
7		4179	90.7	(0.3700, 0.3595)	137.7
14		4079	92.2	(0.3730, 0.3589)	136.8
21		4052	92.7	(0.3738, 0.3584)	136.1
28		4068	92.4	(0.3733, 0.3587)	133.5
35		4090	91.6	(0.3729, 0.3601)	109.3
0	K_2GeF_6:Mn^{4+} powder and YAG:Ce^{3+}	4029	85.5	(0.3806, 0.3813)	110.2
0.5		4062	85.6	(0.3791, 0.3803)	108.3
1		4095	85.3	(0.3776, 0.3789)	101.7
7		4159	85.1	(0.3738, 0.3729)	88.4
14		4165	85.5	(0.3733, 0.3716)	85.9
21		4147	85.2	(0.3739, 0.3717)	83.7
28		4157	84.9	(0.3734, 0.3709)	80.51
35		4149	85.2	(0.3732, 0.3690)	73.4

and commercial K_2GeF_6:Mn^{4+} powder, respectively. They have similar shapes with blue, green, and red emission bands at the same positions (Figure 16.17). Meanwhile, these wLEDs still have high LE values, such as 101.7 lm/W for the wLED based on K_2GeF_6:Mn^{4+} crystal and 98.2 lm/W for the wLED with commercial K_2GeF_6:Mn^{4+} powder. The color gamuts of the two wLEDs are calculated through the commercial red, green, and blue color filters. The wLED with K_2GeF_6:Mn^{4+} crystal has a wide color gamut of 130.5% of National Television Standards Committee (NTSC) value, which is higher than that of wLED with commercial K_2GeF_6:Mn^{4+} (128.2%) (Figure 16.17). As expected, the wLED based on K_2SiF_6:Mn^{4+} crystal has a higher LE of 130.2 lm/W than that of wLED based on the corresponding powdery sample (117.8 lm/W).[25] After lighting for 10 min under a current of

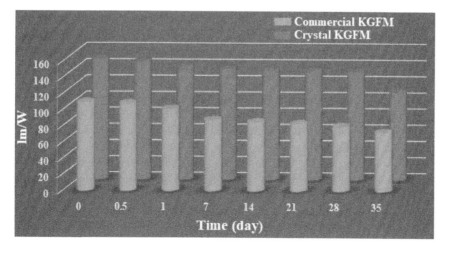

FIGURE 16.16 LE values of warm wLEDs with the K_2GeF_6:Mn^{4+} crystal and commercial K_2GeF_6:Mn^{4+} at different aging periods.

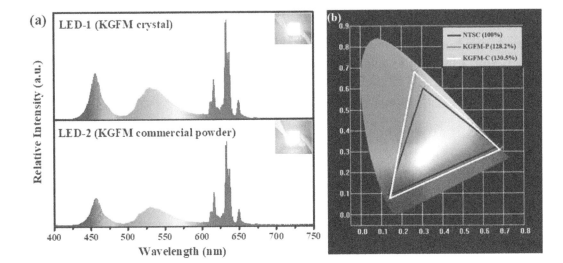

FIGURE 16.17 (a) Luminescent spectra of wLEDs by combining β-SiALON with K_2GeF_6:Mn^{4+} crystal or commercial K_2GeF_6:Mn^{4+} powder, (b) color gamut of the NTSC standard (black triangle) and the wLEDs with K_2GeF_6:Mn^{4+} crystal (white triangle) or commercial K_2GeF_6:Mn^{4+} powder (gray triangle) in the CIE 1931 system.

300 mA, the operating temperature of wLED with the crystal sample is about 106.5°C, while the temperature of wLED with the powdery one is up to 119.6°C. This further proves that the single crystal is beneficial to heat dissipation for wLED. Moreover, the LD based on K_2SiF_6:Mn^{4+} crystal has a high LF of 202.7 lm, low T_c of 3277K, and R_a of 87.9, under a blue-light laser excitation (231.3 W/cm²).

16.3.4.3 Other Fluoride Crystals Doped with Mn^{4+}

Millimeter-sized $(NH_4)_2SiF_6$:Mn^{4+} and $(NH_4)_3SiF_7$:Mn^{4+} single crystals were synthesized via evaporation crystallization and ion exchange of two-step association.[24] $(NH_4)_2SiF_6$:Mn^{4+} and $(NH_4)_3SiF_7$:Mn^{4+} also exhibit intense broad excitation band in the blue-light region (Figure 16.18). As-obtained ammonium fluosilicate crystals exhibit a series of red-emitting peaks. The excitation and emission spectra of crystals have some redshift, compared with those of K_2SiF_6:Mn^{4+} phosphors. This means that Mn^{4+} in $(NH_4)_2SiF_6$:Mn^{4+} and $(NH_4)_3SiF_7$:Mn^{4+} experiences a weaker crystal field. Besides, the decay times of ammonium fluosilicate crystals are shorter than that of K_2SiF_6:Mn^{4+}. A wLED was packaged by using one blue InGaN chip, commercial green phosphor β-SiALON:Eu^{2+}, and $(NH_4)_3SiF_7$:Mn^{4+} (Figure 16.19). Ultrawide color gamut of 122.8% NTSC was obtained from this wLED.

In addition, a set of red-emitting $(NH_4)_2GeF_6$:Mn^{4+} crystals doped with different contents of Mn^{4+} have been prepared by adjusting the ratios between $(NH_4)_2MnF_6$ and GeO_2 (1:20, 1:15, 1:12.5, 1:10, 1:8.5, 1:7.5, 1:5, and 1:2) at room temperature (Figure 16.20). These samples are labeled as NGFM (i), NGFM (ii), NGFM (iii), NGFM (iv), NGFM (v), NGFM (vi), NGFM (vii), and NGFM (viii), respectively. The crystals evolve from the hexagonal structure of the space group *P3m1* to the space group *P6₃mc* with the increasing Mn^{4+} concentration. All crystal samples show similar excitation spectra with two broadbands from 300 to 520 nm (Figure 16.21). The intense excitation bands are due to spin-allowed transitions $^4A_2$-T^4T_1, $^4T_2$ of Mn^{4+}. As Mn^{4+} content increases, the excitation intensity also increases. Under 465 nm light excitation, a series of narrow band-emitting peaks appear in the emission spectra of the crystals. NGFM (i) does not exhibit an obvious ZPL. But the ZPL emission of Mn^{4+} locating at about 622 nm can be found in the other crystal samples. The ZPL intensity strengthens slowly with the increasing content of Mn^{4+}. In the transition process of the

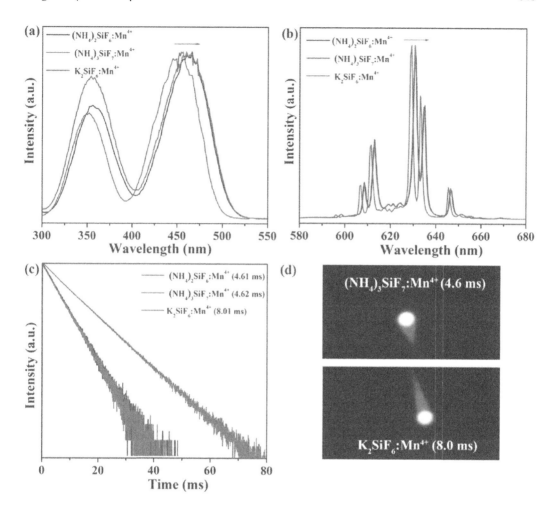

FIGURE 16.18 Normalized (a) excitation, (b) emission spectra, and (c) luminescence decay curves of $(NH_4)_2SiF_6:Mn^{4+}$, $(NH_4)_3SiF_7:Mn^{4+}$, and $K_2SiF_6:Mn^{4+}$. (d) Photographs of the tailing phenomenon of $(NH_4)_3SiF_7:Mn^{4+}$ and $K_2SiF_6:Mn^{4+}$ under 450 nm excitation fast scanning. (Deng, T., Song, E., Zhou, Y., et al., *J. Alloys Comp.* 847, 156550, 2020. With permission)

P3m1 (164) group space to the *P63mc* (186) group space, the D_{3d} symmetry of the *P3m1* group space is slowly distorted to transform the C_{6v} symmetry of the *P63mc* group space. Then, the symmetry around Mn^{4+} slowly decreases as well as the ZPL emission gets stronger. Moreover, the red emission intensities of the samples from NGFM (i) to NGFM (viii) are also improved. The integrated emission intensity of NGFM (viii) is about 15.8 times higher than that of NGFM (i).

At last, $Rb_2XF_6:Mn^{4+}$ crystals (X = Ge, Si, Ti) were also obtained from the HF solution.[83,84] They also provide the similar optical performances with the above fluoride crystals.

16.3.5 Perovskite Crystals

Much attention is paid to halide perovskites with the formula of ABX_3 (A = CH_3NH_3, $CH(NH_2)_2$, Cs; B = Pb, Sn, Ge; and X = Cl, Br, I), due to their inspiring optoelectronic properties.[26–30,85–88] These materials could find applications on solar cells, light-emitting diode, photodetectors, and so on.[85–88] Some single-crystal halide perovskites were grown by the solution method.[89,90] In general, perovskite crystals will form by vaporing the solvent in a precursor solution. The selection

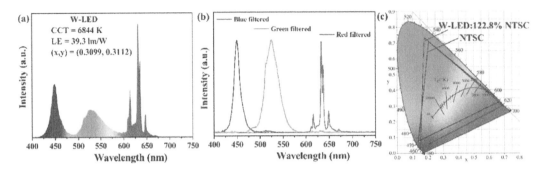

FIGURE 16.19 (a) EL spectra, (b) the corresponding EL spectra through commercial blue, green, and red color filters of fabricated wLED using β-SiALON:Eu^{2+}, $(NH_4)_3SiF_7$:Mn^{4+}, and one blue InGaN chip; (c) color gamut of the NTSC standard (black triangle) and the fabricated wLED (blue triangle). (Deng, T., Song, E., Zhou, Y., et al., *J. Alloys Comp.* 847, 156550, 2020. With permission)

of solvent, the vaporing rate, and reaction temperature are important to obtain perfect crystals. Large perovskite crystals can be synthesized by increasing the temperature of the reaction system.[3] Herein, the optical properties of some perovskite crystals are described briefly.

16.3.5.1 Single-Crystal CH$_3$NH$_3$PbBr$_3$ (MAPbBr$_3$)

The halide perovskite MAPbBr$_3$ is one promising material for miscellaneous optoelectronic applications.[26,91,92] Multi-inch-sized MAPbBr$_3$ single crystals were obtained by a low-temperature-gradient crystallization (LTGC) method.[26] Compared with the crystals grown at the high temperature with twinning defects, the single crystals prepared using the LTGC method show perfect facets without any observable defects (Figure 16.22).[26] The as-prepared MAPbBr$_3$ single crystals exhibit intensive green emission peaking at 547 nm with a narrow half peak width of about 22 nm. They also exhibit superior optoelectronic properties for the next generation of optoelectronic devices.

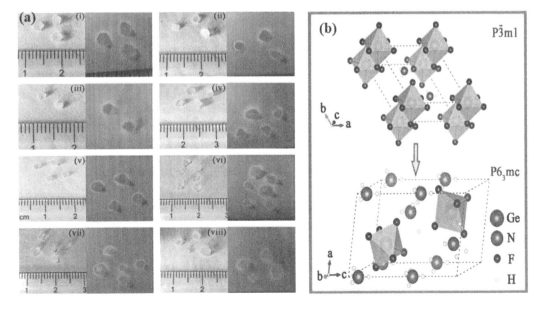

FIGURE 16.20 (a) Photographs of $(NH_4)_2GeF_6$:Mn^{4+} crystals doped with different contents of Mn^{4+} upon nature and blue light, (b) crystal structures of $(NH_4)_2GeF_6$.

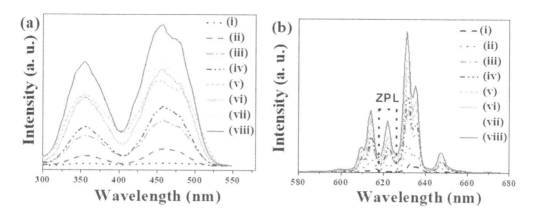

FIGURE 16.21 (a) Excitation and (b) emission spectra of $(NH_4)_2GeF_6:Mn^{4+}$ crystals doped with different contents of Mn^{4+}.

16.3.5.2 Single-Crystal Cs₂InBr₅·H₂O

Red-emitting $Cs_2InBr_5 \cdot H_2O$ single crystals were prepared by a temperature-lowering crystallization method and its crystal structure was confirmed by single-crystal X-ray diffraction.[93] The as-obtained crystal adopted the orthorhombic structure with a space group of *Pnma*. In the structure, each $InBr_5O$ unit is spatially isolated by two Cs^+ ions. Upon irradiation at 365 nm, millimeter-sized $Cs_2InBr_5 \cdot H_2O$ crystals produce intense and broad emission peaking at ~695 nm. The emission intensity increases monotonously with the decreasing temperature from 320 to 78K. Fired at 150°C, $Cs_2InBr_5 \cdot H_2O$ crystal decomposes into a mixture of $Cs_3In_2Br_9$ and CsBr. The dehydrated production yields bright yellow light and quickly turns to red-emitting $Cs_2InBr_5 \cdot H_2O$ after being exposed to air. This process is reversible. So, $Cs_2InBr_5 \cdot H_2O$ single crystal may be used as a water sensor in humidity detection or the detection of traces of water in organic solvents.[93]

16.3.5.3 Single-Crystal Cs₂AgFeCl₆:In³⁺

A set of single-crystal samples of $Cs_2AgFeCl_6:In^{3+}$ were grown by vaporing the solvent.[94] With the doping of In^{3+}, the absorption edges of single crystals display a blueshift, which is derived from its

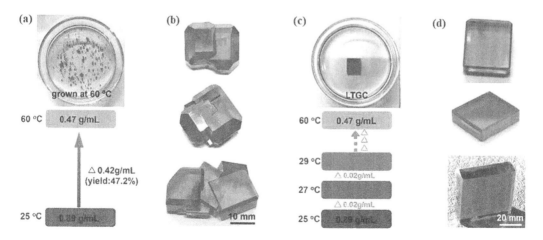

FIGURE 16.22 Photos of MAPbBr₃ single crystals (a and b) grown at 60°C and (c and d) grown using the LTGC method from 25 to 60°C. (Liu, Y.C., Zhang, Y.X., Yang, Z., et al., *Mater. Today* 22, 67, 2019. With permission)

narrowed bandgap with the incorporation of In^{3+} species.[94] Then the energy gap of $Cs_2AgFeCl_6$:In^{3+} turns broad with the increasing In^{3+}. Single-crystal $Cs_2AgInCl_6$ exhibits a broad emission band peaking at 620 nm, which is assigned to the emission via self-trapped exciton recombination for the typical double perovskite single crystal.[95] With In^{3+} content decreasing, the emission peak of $Cs_2AgFeCl_6$:In^{3+} exhibits a slight redshift, and the emission intensity has a downward trend. Moreover, single-crystal $Cs_2AgInCl_6$ displays extreme thermal stability with a decomposable temperature of ~536°C, which benefits the potential applications in photoelectronic and photovoltaic devices.

REFERENCES

1. Li, S.X., Wang, L., Hirosaki, N., and Xie, R.J. *Laser Photonics Rev.* 12, 1800173, 2018.
2. Hu, P., Ding, H., Liu, Y.F., Sun, P., Liu, Z.H., Luo, Z.H., Huang Z.R., Jiang, H.C., and Jiang, J. *Chin. J. Lumin.* 41, 1504, 2020. (In Chinese).
3. Liu, Y.C., Yang, Z., and Liu, S.Z. *Adv. Sci.* 5, 1700471, 2018.
4. Zhou, J.B., Wang, Y.F., Chen, Y.Y., Zhou, Y.Y., Milićević, B., Zhou, L., Yan, J., Shi, J.X., Liu, R.S., and Wu, M.M. *Angew. Chem. Int. Ed.* 59, 202011022, 2020.
5. Wang, Z.L., Yang, Z.L., Wang, N., Zhou, Q., Zhou, J.B., Ma, L., Wang, X.J., Xu, Y.Q., Brik, M.G., Dramićanin, M.D., and Wu, M.M. *Adv. Opt. Mater.* 8, 1901512, 2020.
6. Villora, E.G., Arjoca, S., Inomata, D., and Shimamura, K. *Proc. SPIE* 9768, 976805, 2016.
7. Sun, C.T., Xue, D.F. *Sci. Sin. Chim.* 48, 804, 2018. (In Chinese).
8. Sun, C.T., Xue, D.F. *Sci. Sin. Tech.* 44, 1123, 2014. (In Chinese).
9. Li, D.S., Nielsen, M.H., Lee, J.R.I., Frandsen, C., Banfield, J.F., and DeYoreo, J.J. *Science* 336, 1014, 2012.
10. Gebauer, D., Völkel, A., and Cölfen, H. *Science* 322, 1819, 2008.
11. Prywer, J. *Cryst. Growth Des.* 3, 593, 2003.
12. Krasinski, M.J., and Prywer, J. *Cryst. Growth* 303, 105, 2007.
13. Yang, Y., Wang, X., Liu, B., Zhang, Y., Lv, X., Li, J., Li, S., Wei, L., Zhang, H., and Zhang, C. *J. Lumin.* 204, 157, 2018.
14. Rejman, M., Babin, V., Kucerková, R., and Nikl, M. *J. Lumin.* 187, 20, 2017.
15. Dong, J., Rapaport, A., Bass, M., Szipocs, F., and Ueda, K. *Phys. Stat. Sol.* 202, 2565, 2005.
16. Chen, Y.F., Lim, P.K., Lim, S.J., Yang, Y.J., Hu, L.J., Chiang, H.P., and Tse, W.S. *J. Raman Spectrosc.* 34, 882, 2003.
17. Yang, P.Z., Deng, P.Z., and Xu, J. *J. Cryst. Growth* 218, 87, 2000.
18. Yin, H.B., Deng, P.Z., and Gan, F.X. *J. Appl. Phys.* 83, 3825,1998.
19. Park, K.W., Lim, S.G., Deressa, G., Kim, J.S., Kang, T.W., Choi, H.L., Yu, Y.M., Kim, Y.S., Ryu, J.G., Lee, S.H., and Kim, T.H. *J. Lumin.* 168, 334, 2015.
20. Villora, E.G., Arjoca, S., Shimamura, K., Inomata, D., and Aoki, K. *Proc. SPIE* 8987, 89871u, 2014.
21. Igashira, K., Nakauchi, D., Fujimoto, Y., Kato, T., Kawaguchi, N., and Yanagida, T. *Opt. Mater.* 102, 109810, 2020.
22. Kang, T.W., Park, K.W., Ryu, J.H., Lim, S.G., Yu, Y.M., and Kim, J.S. *J. Lumin.* 191, 35, 2017.
23. He, S., Xia, H., Zhang, J., Zhu, Y., and Chen, B. *Cryst. Res. Technol.* 17, 1700136, 2018.
24. Deng, T., Song, E., Zhou, Y., Chen, J., Liu, W., Deng, H., Zheng, X., and Huang, H. *J. Alloys Comp.* 847, 156550, 2020.
25. Zhou, Y.Y., Yu, C.K., Song, E.H., Wang, Y.J., Ming, H., Xia, Z.G., and Zhang, Q.Y. *Adv. Opt. Mater.* 8, 202000976, 2020.
26. Liu, Y.C., Zhang, Y.X., Yang, Z., Feng, J.S., Xu, Z., Li, Q.X., Hu, M.X., Ye, H.C., Zhang, X., Liu, M., Zhao, K., and Liu, S.Z. *Mater. Today* 22, 67, 2019.
27. Shi, D., Adinolfi, V., Comin, R., Yuan, M., Alarousu, E., Buin, A., Chen, Y., Hoogland, S., Rothenberger, A., Katsiev, K., Losovyj, Y., Zhang, X., Dowben, P.A., Mohammed, O.F., Sargent, E.H., and Bakr, O.M. *Science* 347, 519, 2015.
28. Song, Z., Zhao, J., and Liu, Q.L. *Inorg. Chem. Front.* 6, 2969, 2019.
29. Stranks, S.D., Eperon, G.E., Grancini, G., Menelaou, C., Alcocer, M.J., Leijtens, T., Herz, L.M., Petrozza, A., and Snaith, H.J. *Science* 342, 341, 2013.
30. Saidaminov, M.I., Haque, M.A., Almutlaq, J., Sarmah, S., Miao, X.H., Begum, R., Zhumekenov, A.A., Dursun, I., Cho, N., Murali, B., Mohammed, O.F., Wu, T., and Bakr, O.M. *Adv. Opt. Mater.* 5, 1600704, 2017.

31. Tanner, P.A., Mak, C.S.K., Edelstein, N.M., Murdoch, K.M., Liu, G.K., Huang, J., Seijo, L., and Barandiaran, Z. *J. Am. Chem. Soc.* 125, 13225, 2003.
32. Song, Z., Xia, Z.G., and Liu, Q.L. *J. Phys. Chem. C* 122, 3567, 2018.
33. Pankratov, V., Grigorjeva, L., Chernov, S., Chudoba, T., and Lojkowski, W. *IEEE Trans. Nucl. Sci.* 55, 1509, 2008.
34. Aggarwal, R.L., Ripin, D.J., Ochoa, J.R., and Fan, T.Y. *J. Appl. Phys.* 98, 103514, 2005.
35. Arjoca, S., Víllora, E.G., Inomata, D., Aoki, K., Sugahara Y., and Shimamura, K. *Mater. Res. Express* 2, 055503, 2015.
36. Yang, C., Gu, G.R., Zhao, X.J., Liang, X.J., and Xiang, W.D. *Mater. Lett.* 170, 58, 2016.
37. Balci, M.H., Chen, F., Cunbul, A.B., Svensen, Ø., Akram, M.N., and Chen, X.Y. *Opt. Rev.* 25,166, 2018.
38. Wierer, J.J., Jr., Tsao, J.Y., and Sizov, D.S. *Laser Photonics Rev.* 7, 963, 2013.
39. Cantore, M., Pfaff, N., Farrell, R.M., Speck, J.S., Nakamura, S., and DenBaars, S.P. *Opt. Express* 24, 251040, 2016.
40. Xu, J., Thorseth, A., Xu, C., Krasnoshchoka, A., Rosendal, M., Dam-Hansen, C., Du, B., Gong, Y., and Jensen, O.B. *J. Lumin.* 212, 279, 2019.
41 Xu, J., Yang, Y., Guo, Z., Corell, D.D., Du, B., Liu, B., Ji, H., Dam-Hansen, C., and Jensen, O.B. *Ceram. Int.* 46, 17923, 2020.
42. Arjoca, S., Víllora, E.G., Inomata, D., Aoki, K., Sugahara, Y., and Shimamura, K. *Mater. Res. Express* 1, 025041, 2014.
43. Arjoca, S., Inomata, D., Matsushita Y., and Shimamura, K. *CrystEngComm* 18, 4799, 2016.
44. Seki, M., Kochurikhin, V.V., Kurosawa, S., Suzuki, A., Yamaji, A., Fujimoto, Y., Wakahara, S., Pejchal, J., Yokota, Y., and Yoshikawa, A. *Phys. Stat. Sol.* 9, 2255, 2012.
45. Xu, T., Yuan, L., Chen, Y., Zhao, Y., Ding, L., Liu, J., Xiang, W., and Liang, X. *Opt. Mater.* 91, 30, 2019.
46. Xu, X.X., Shao, Q.Y., Yao, L.Q., Dong, Y., and Jiang, J.Q. *Chem. Eng. J.* 383, 123108, 2019.
47. Zhang, L.L., Wang, D.D., Hao, Z.D., Zhang, X., Pan, G., Wu, H.J., and Zhang, J.H. *Adv. Opt. Mater.* 7, 1900185, 2019.
48. Rajendran, V., Chang, H., and Liu, R.S. *Opt. Mater. X* 1, 100011, 2019.
49 Tang, F., Ye, H.G., Su, Z.C., Bao, Y.T., Guo, W., and Xu, S.J. *ACS. Appl. Mater. Interfaces* 9, 43790, 2017.
50. Xie, R.J., and Hirosaki, N. *Sci. Technol. Adv. Mater.* 8, 588, 2008.
51. Wang, L., Wang, X.J., Kohsei, T., Yoshimura, K., Izumi, M., Hirosaki, N., and Xie, R.J. *Opt. Express* 23, 28707, 2015.
52. Moon, J.W., Min, B.G., Kim, J.S., Jang, M.S., Ok, K.M., Han, K.Y., and Yoo, J.S. *Opt. Mater. Express* 6, 782, 2016.
53. Hasegawa, S., Hasegawa, T., Kim, S.W., Yamanashi, R., Uematsu, K., Toda, K., and Sato, M. *ACS Omega* 4, 9939, 2019.
54. Palilla, E.C., Levine, A.K., and Tomkus, M.R. *J. Electrochem. Soc.* 115, 642, 1968.
55. Delgado, T., Afshani, J., and Hagemann, H. *J. Phys. Chem. C* 123, 8607, 2019.
56. Lin, C.C., Meijerink, A., and Liu, R.S. *J. Phys. Chem. Lett.* 7, 495, 2016.
57. Zhou, Z., Zhou, N., Xia, M., Meiso, Y., and (Bert) Hintzen, H.T. *J. Mater. Chem. C* 4, 9143, 2016.
58. Zhou, Q., Dolgov, L., Srivastava, A.M., Zhou, L., Wang, Z.L., Shi, J.X., Dramicanin, M.D., Brike, M.G., and Wu, M.M. *J. Mater. Chem. C* 6, 2652, 2018.
59. Brik, M.G., and Srivastava, A.M. *J. Lumin.* 133, 69, 2013.
60. Du, M.H. *J. Mater. Chem. C* 2, 2475, 2014.
61. Adachi, S. *J. Lumin.* 197, 119, 2018.
62. Liu, Y., Zhou, Z., Huang, L., Brik, M.G., Si, S.C., Lin, L.T., Xuan, T.T., Liang, H.B., Qiu, J.B., and Wang, J. *J. Mater. Chem. C* 7, 2401, 2019.
63. Song, E.H., Zhou, Y.Y., Yang, X.B., Liao, Z.F., Zhao, W.R., Deng, T.T., Wang, L.Y., Ma, Y.Y., Ye, S., and Zhang, Q.Y. *ACS. Photonics* 4, 2556, 2017.
64. Sijbom, H.F., Verstraete, R., Joos, J.J., Poelman, D., and Smet, P.F. *Opt. Mater. Express* 7, 3332, 2017.
65. Zhou, Q., Tan, H.Y., Zhou, Y.Y., Zhang, Q.H., Wang, Z.L., Yan, J., and Wu, M.M. *J. Mater. Chem. C* 4, 7443, 2016.
66. Pan, Y.X., Chen, Z., Jiang, X.Y., Huang S.M., and Wu, M.M. *J. Am. Ceram. Soc.* 99, 3008, 2016.
67. Wang, Z.L., Zhou, Y.Y., Yang, Z.Y., Liu, Y., Yang, H., Tan, H.Y, Zhang, Q.H., and Zhou, Q. *Opt. Mater.* 49, 235, 2015.
68. Gao, X.L., Song, Y., Liu, G.X., Dong, X.T., Wang J.X., and Yu, W.S. *CrystEngComm* 18, 5842, 2016.
69. Zhu, H.M., Lin, C.C., Luo, W.Q., Shu, S.T., Liu, Z.G., Liu, Y.S., Kong, J.T., Ma, E., Cao, Y.G., Liu, R.S., and Chen, X.Y. *Nat. Commun.* 5, 4312, 2014.

70. Arai, T., and Adachi, S. *J. Appl. Phys.* 110, 063514, 2011.

71. Wu, W.L., Fang, M.H., Zhou, W.L., Lesniweski, T., Mahlik, S., Grinberg, M., Brik, M.G., Sheu, H., Cheng, B.M., Wang, J., and Liu, R.S. *Chem. Mater.* 29, 935, 2017.

72. Zhou, Y.Y., Song, E.H., Deng, T.T., Wang, Y.J., Xia, Z.G., and Zhang, Q.Y. *Adv. Mater. Interfaces* 1, 1802006, 2019.

73. Zhou, Y.Y., Song, E.H., Deng, T.T., and Zhang, Q.Y. *ACS Appl. Mater. Interfaces* 10, 880, 2018.

74. Arunkumar, P., Kim, Y.H., Kim, H.J., Unithrattil, S., and Im, W.B. *ACS Appl. Mater. Interfaces* 9, 7232, 2017.

75. Nguyen, H.D., Lin, C.C., and Liu, R.S. *Angew. Chem. Int. Ed.* 54, 10862, 2015.

76. Wang, B., Lin, H., Xu, J., Chen, H., and Wang, Y. *ACS Appl. Mater. Interfaces* 6, 22905, 2014.

77. Wei, L.L., Lin, C.C., Wang, Y-Y., Fang, M.H., Jiao, H. and Liu, R.S. *ACS Appl. Mater. Interfaces* 7, 10656, 2015.

78. Zhou, W., Fang, M.H., Lian, S., and R.S. Liu, *ACS Appl. Mater. Interfaces* 10, 17508, 2018.

79. Wei, L.L., Lin, C.C., Fang, M.H., Brik, M.G., Hu, S.F., Jiao, H., and Liu, R.S. *J. Mater. Chem. C* 3, 1655, 2015.

80. Oh, J.H., Kang, H., Eo Y.J., Park, H.K., and Do, Y.R. *J. Mater. Chem. C* 3, 607, 2015.

81. Lv, L.F., Jiang, X.Y., Huang, S.M., Chen, X.A., and Pan, Y.X. *J. Mater. Chem. C* 2, 3879, 2014.

82. Kasa, R., and Adachi, S. *J. Electrochem. Soc.* 159, J89, 2012.

83. Sakurai, S., Nakamura, T., and Adachi, S. *ECS J. Solid State Sci. Technol.* 5, 206, 2016.

84. Sakurai, S., Nakamura, T., and Adachi, S. *Jpn. J. Appl. Phys.* 57, 022601, 2018.

85. Green, M.A., Ho-Baillie, A., and Snaith, H.J. *Nat. Photonics* 8, 506, 2014.

86. Li, J.H., Xu, L.M., Wang, T., Song, J.Z., Chen, J.W., Xue, J., Dong, Y.H., Cai, B., Shan, Q.S., Han, B.N., and Zeng, H.B. *Adv. Mater.* 29, 1603885, 2017.

87. Lee, J.W., Choi, Y.J., Yang, J.M., Ham, S., Jeon, S.K., Lee, J.Y., Song, Y.H., Ji, E.K., Yoon, D.H., Seo, S., Shin, H., Han, G.S., Jung, H.S., Kim, D.H., and Park, N.G. *ACS Nano* 11, 3311, 2017.

88. Maeda, K., Eguchi, M., and Oshima, T. *Angew. Chem. Int. Ed.* 53, 13164, 2014.

89. Peng, W., Wang, L.F., Murali, B., Ho, K.T., Bera, A., Cho, N., Kang, C.F., Burlakov, V.M., Pan, J., Sinatra, L., Ma, C., Xu, W., Shi, D., Alarousu, E., Goriely, A., He, J.H., Mohammed, O.F., Wu, T., and Bakr, O.M. *Adv. Mater.* 28, 3383, 2016.

90. Saidaminov, M.I., Abdelhady, A.L., Murali, B., Alarousu, E., Burlakov, V.M., Peng, W., Dursun, I., Wang, L.F., He, Y., Maculan, G., Goriely, A., Wu, T., Mohammed, O.F., and Bakr, O.M. *Nat. Commun.* 6, 7586, 2015.

91. Nguyen, V.-C., Katsuki, H., Sasaki, F., and Yanagi, H. *Jpn. J. Appl. Phys.* 57, 04FL10, 2018.

92. Wei, H., DeSantis D., Wei, W., Deng, Y., Guo, D., Savenije, T.J., Cao, L., and Huang, J. *Nat. Mater.* 16, 826, 2017.

93. Zhou, L., Liao, J.F., Huang, Z.G., Wei, J.H., Wang, X.D., Li, W.G., Chen, H.Y., Kuang, D.B., and Su, C.Y. *Angew. Chem. Int. Ed.* 58, 5277, 2019.

94. Yin, H., Xian, Y.M., Zhang, Y.L., Chen, W.J., Wen, X.M., Rahman, N.U., Long, Y., Jia, B.H., Fan, J.D., and Li, W.Z. *Adv. Funct. Mater.* 30, 2002225, 2020.

95. Slavney, A.H., Hu, T., Lindenberg, A.M., and Karunadasa, H.I. *J. Am. Chem. Soc.* 138, 2138, 2016.

17 Glass Phosphors

Xizhen Zhang and Baojiu Chen

CONTENTS

DOI: 10.1201/9781003098676-17

17.1 INTRODUCTION

Glass materials have already become significant and indispensable in our daily life and have gone through a very long history of more than 4500 years. In fact, we do not know and have no way to know when and where humans first found or produced glasses and used them [1,2]. At present, a picture of glasses in people mind probably is windows, eyeglasses, cups and screens (actually glass is probably not the mainstream material for all of them). The applications of glass are not limited to these fields; in fact, they also cover the information technology, modern medicine, aerospace and so on. The glasses are also important luminescence materials that exhibit wide application prospects in displays, lighting and laser sources. Therefore, in this hand book, we define the luminescent glasses as glass phosphors, and in this chapter, we mainly introduce some corresponding contents related with the glass luminescence materials.

To better understand the glass phosphors, it is necessary to clarify what the glass is. In fact, there is no accurate definition for glass, but the glass can be described in following aspects.

1. In comparison with the ordered arrangements of atoms in crystals, the atoms in glasses are disorderly arranged but not completely randomly dispersed [1]. The atom arrangements in glasses are correlated with the specific compositions of the glasses, for example, in the silicate glasses the $[SiO_4]^{4-}$ tetrahedrons and in borate glasses, the $[BO_3]^{3-}$ groups are existent. Therefore, the glasses are also called as non-crystalline, amorphous and glassy solids.
2. Usually, the glasses are transparent for the visible lights. The transmittable wavelength region for the glasses depends on the glass composition. The glasses are composed of crystalline, non-crystalline and gaseous phases. If the amounts of crystalline and gaseous phases exceed certain values, then the glasses will become opaque. The transparent glass containing nanocrystals or microcrystals is called nanocrystal glass or microcrystal glass. The opaque one which contains microcrystals is called ceramics.
3. Unlike crystals, the glasses are anisotropic. The anisotropically physical properties originate from the glass microstructures.
4. It is also different from crystals that glasses do not exhibit a fixed melting point. As temperature increases, the glasses will soften and then further be melted.
5. Glass exists in a metastable state whose potential energy is higher than its corresponding crystal state (if the crystal state for the same composition is existent). It should be pointed out that in most cases the metastable state for glass is very steady.

17.2 GLASS-GROWING MECHANISM

Figure 17.1 shows the possible processes for a system from liquid to solid. The y axis presents the internal energy of the system, and the x axis presents the temperature. The arrowed solid curves present the relations between the system internal energy and the temperature.

The internal energy of a liquid at high temperature will linearly decrease with decreasing the system temperature. When the temperature of the system reaches the melting point T_m, the internal energy of the system will decrease but the temperature of the system will not change. During temperature-unchanged period, the atoms in the system will be orderly arranged, namely, the system

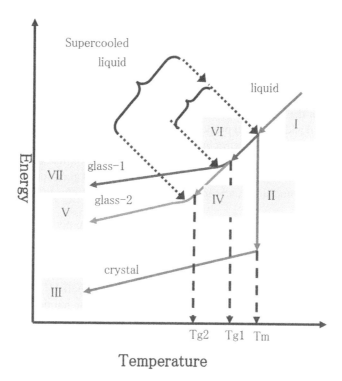

FIGURE 17.1 Glass-growth process.

crystalizes to form the crystal solid. The internal energy for the formed solid is also linearly dependent on the temperature. This process is shown as I → II → III in Figure 17.1.

If the cooling rate defined as temperature decrement per unit time is large, the liquid system will not experience the crystallization process as described earlier. In this sense, the temperature of the liquid system will continue to decrease, and the system keeps liquid state. The liquid whose temperature is lower than the melting point T_m is called supercooled liquid [3]. If the temperature of the supercooled liquid is lower than a temperature T_g (which defined as the glass transition temperature), the system becomes solid, and this solid is just the glass. The internal energy of the glass is linearly dependent on the temperature. This process is shown as I → IV → V and I → VI → VII. The difference between processes I → IV → V and I → VI → VII is the cooling rate. It is obvious that the cooling rate for process I → IV → V is higher than process I → VI → VII. From Figure 17.1, it can also be found that the glass transition temperature T_g is different for different processes with different cooling rates, and the final glass products for different cooling processes are also existent in different metastable states. Typically, the glass transition temperature T_g slightly depends on the cooling rate, the change of cooling rate by an order of magnitude results in the T_g change by only several K [4].

The glass growth depends on many factors including starting materials composition, cooling rate, heating temperature, heating history and melt viscosity. Among these factors, the starting materials composition and cooling rate play a decisive role in comparison with other factors. The starting materials composition is the intrinsic factor, and the cooling rate is the extrinsic factor. For the same growing conditions, the starting materials composition decides whether the glass phase can be obtained or not. Theoretically, the glass can be grown by any composition as long as the cooling rate is high enough. However, it is not possible to achieve the cooling rate as high as we expect via modern techniques. Therefore, the glass composition design is important to prepare glasses.

17.3 GLASS PHOSPHORS AND THEIR CLASSIFICATION

Glasses have many potential and practical applications in various fields, such as handicrafts, household utensils, optical devices and luminescence devices. Luminescence glasses are one kind of luminescence materials, and in this hand book, we define them as glass phosphors. In general, the emissions of glass phosphors originate from the doped luminescence centers or luminescence micro-/nanoparticles in the glass hosts since the self-activated glass phosphors are rarely reported [5,6]. The basic physical and chemical properties and luminescent properties of glass phosphors are dependent on their compositions and structures. Therefore, the classification for the glass phosphors can be carried out based on the glass compositions. For a glass host, its composition contains primary and secondary components. The primary component is called glass former, and the secondary component is called glass modifier. According to the glass compositions, the glasses can be clarified as listed in Table 17.1.

Based on the glass host composition, the glass phosphors can be divided into oxide glass phosphors and non-oxide glass phosphors. The oxide glass phosphors contain oxide formers in their compositions. The non-oxide glass phosphors have no oxide component in their compositions. The oxide glass phosphors include borate glass phosphors [7], phosphate glass phosphors [8], silicate glass phosphors [9], germanate glass phosphors [10], tellurite glass phosphors [11], and aluminate [12] and gallate [13] glass phosphors. The non-oxide glass phosphors include halide and sulfide glass phosphors [14,15]. The halide glass phosphors include fluoride and chloride glass phosphors. Generally, the oxide glass phosphors have better physical and chemical stabilities, better plasticity than the non-oxide glass phosphors. The maximum phonon energies for non-oxide glass oxide phosphors are usually lower than the oxide glass phosphors, thus resulting in higher luminescence quantum efficiencies of luminescence centers in the non-oxide glass phosphors by comparing with the oxide glass phosphors. Moreover, the preparation of oxide glass phosphors is usually easier than the non-oxide glass phosphors. The hosts for the glass phosphors can also be composed of both oxide and non-oxide compounds or more than one glass formers, and these glass phosphors are called composite glass phosphors. Some examples for composite glass phosphors like oxyfluoride [16], borosilicate [17], phosphosilicate [18] and germanotellurite [19] are listed in Table 17.1.

In the luminescence materials, the glasses are excellent hosts that can accommodate various kinds of luminescence centers and complex luminescence groups. Therefore, the glass phosphors can also be classified based on the luminescent origins inside glass hosts. Table 17.2 lists the main glass phosphors with classifications. The luminescence origins include self-activated centers [5,6], ionic centers

TABLE 17.1
Glass Phosphors' Clarification by Compositions

Glass→	Oxide glass→	Borate glass		Ref. [7]
		Phosphate glass		Ref. [8]
		Silicate glass		Ref. [9]
		Germanate glass		Ref. [10]
		Tellurite glass		Ref. [11]
		Aluminate glass		Ref. [12]
		Gallate glass		Ref. [13]
	Non-oxide glass→	Halide glass→	Fluoride glass	Ref. [14]
			Chloride glass	Ref. [14]
		Sulfide glass		Ref. [15]
	Composite glass→	Oxyfluoride glass		Ref. [16]
		Borosilicate glass		Ref. [17]
		Phosphosilicate glass		Ref. [18]
		Germanotellurite glass		Ref. [19]

TABLE 17.2
Glass Phosphors' Classification by Luminescence Origin

Luminescence origin→	Self-activated center			Refs. [5,6]
	Ionic center→	Rare earth ion	Ce^{3+}, Tb^{3+}, Eu^{3+}, Eu^{2+}, Er^{3+}, Nd^{3+}, Yb^{3+}, Pm^{3+}	Refs. [7–10, 15,16]
		Transition metal ion	Cr^{4+}, Mn^{2+}, Cu^+	Refs. [23,24]
		Main group element ion	Bi^{3+}, Sn^{2+}	Ref. [25]
		Color center	F^-, O^{2-}	Ref. [26]
	Complex center		MoO_4^{2-}, VO_4^{3-}, WO_4^{2-}, dye molecule	Refs. [20,21]
	Metal aggregate		Ag_2^+, Ag_3^+, Ag_4^+	Ref. [22]
	Quantum dot		$CsPbBr_3$, $CsPbI_3$, PbS	Refs. [27,28]
	Micro-/nanocrystal		LaF_3:RE^{3+}	Refs. [8, 29]
	Phosphor in glass (PiG)		YAG:Ce^{3+}	Ref. [30]

[10, 15,16], complex centers and dye molecule [20] and metal aggregates as well [21]. The ionic centers include rare earth ions [7–9], transition metal ions [22,23], main group element ions [24] and color centers [25]. The rare earth ions include lanthanide and actinide element ions. The semiconductor quantum dots as luminescence origin can also be introduced into the glass phosphors [26,27]. The emissions of the glass phosphors containing micro- or nanocrystals, in fact, come from the doping centers that locate inside the micro- or nanocrystals, and these glass phosphors are defined as micro- or nanocrystal glass phosphors [8, 28]. PiGs (phosphor in glasses) are newly developed in recent years [29]. The PiGs are produced via some special techniques by using the preprepared phosphors. The metal aggregates include dimer, trimer and tetramer of zero-valence metal atom [30]. In last column of Table 17.2, some representative luminescence origins are listed.

17.4 PREPARATIVE TECHNIQUES OF GLASS PHOSPHORS

We here just introduce some main preparative routes for glass phosphors reported in literatures, and the industrial production technology for the glass phosphors is not involved. The most popular method for producing glass phosphors in modern investigations is called melt-quenching technique, in which the glasses are grown by cooling the melting liquids of the starting materials from high temperature to low temperature with proper cooling rates. The sol-gel technique, in which the very low thermal treatment temperature is its technical characteristic, is often adopted for producing glass phosphors. The containerless processing technique is a recently developed new technique for producing the glass phosphors, and this technique makes the system of low glass-forming ability able to vitrify in bulk form.

17.4.1 MELT-QUENCHING TECHNIQUE

The melt-quenching technique is widely adopted in researching and developing glass phosphors. Most of glass phosphors can be prepared via this melt-quenching route. The common procedure for preparing glass phosphors via melt-quenching route is as follows [31,32].

1. The starting materials including the compounds of glass formers, modifiers and doping luminescence centers are weighed based on the designed stoichiometric ratio. The starting materials are roughly mixed together to obtain the original batch.
2. The batch is loaded into a crucible. The crucible can be of alumina, platinum and graphite. The choice of the crucible is based on the starting material to be calcined. For the

preparation of traditional silicate, borate, phosphate, aluminate, germanate and tellurite glass phosphors, the alumina crucible is suggested to use. For preparation of fluoride glass phosphors, the graphite crucible should be used. If the corrosive component is involved in the starting materials, the platinum crucible should be used.

3. The crucible with starting materials is put into a muffle furnace, and then a calcination procedure following a function of temperature versus calcination time is run. The calcination procedure is usually empirically established or experimentally explored.

4. The melt after calcination is quickly poured into a mold. The mold is made of aluminum, copper or graphite, and its approximate temperature is preset by preheated or precooled process to achieve smaller or larger cooling rate for the melt. After the melt in the mold is cooled, the glass phosphor sample is obtained.

17.4.2 SOL-GEL ROUTE

The sol-gel is also a route for preparing glass phosphors. In comparison with melt-quenching technique, the sol-gel route requires low processing temperature, but a long preparative period. Moreover, the type of glass phosphors that can be prepared via sol-gel technique is limited. The most-reported glass phosphors derived from sol-gel route are phosphate and silicate glass phosphors. In the sol-gel route, the reaction system undergoes the processes from sol-to-gel and gel-to-glass transformation. To prepare the silicate glass phosphors via sol-gel route, the hydrolytic polycondensation of alkoxysilane is adopted. The simplified mechanism of the hydrolysis-polycondensation can be formulized as follows [33].

$$\text{Hydrolysis}: \text{M}(\text{OR})_n + n\text{H}_2\text{O} \rightarrow \text{M}(\text{OH})_n + n\text{ROH}$$

$$\text{Polycondensation}: \text{M}(\text{OH})_n \rightarrow \text{MO}_{n/2} + n/2\text{H}_2\text{O}$$

where $\text{M}(\text{OR})_n$ presents metallic alkoxides, such as $\text{Si}(\text{OR})_4$, $\text{Ti}(\text{OR})_4$ and $\text{Al}(\text{OR})_3$, R usually stands for methyl group and ethyl group. After the abovementioned processes, the gel is obtained and then transformed to glass via thermal treatment.

17.4.3 CONTAINERLESS PROCESSING TECHNIQUE

The containerless processing technique is also a variant melt-quenching route in which the starting materials are melted as in the usual melt-quenching preparation but the cooling process is container-free [34–36]. The melt is suspended in space via counteracting the melt gravity by magnetic, electric fields or air flow. The containerless processing technique can avoid the pollution of the container to the melt and meanwhile improve the cooling rate. Therefore, the containerless processing technique can produce glass phosphors for the systems with low glass-forming ability [37–39].

17.5 ANNEALING TREATMENT FOR GLASS PHOSPHORS

Annealing treatment is a necessary and important process for producing glass phosphors, especially for the glass phosphors derived from melt-quenching route. From the earlier glass phosphor-growing mechanism of melt-quenching route, it is known that it is impossible to achieve identical cooling rate for the different areas of a melt. Therefore, internal stress in the final glass phosphor is inevitable. The internal stress reduces the mechanical strength of the glass phosphor. The main aim of the annealing treatment is to relieve or eliminate the internal stress inside the glass phosphors, and meanwhile it can also improve the uniformity of the glass phosphors.

Annealing temperature and time are two important factors that are usually determined via experiments. The temperature lower than the glass transition temperature is set as the annealing temperature. The annealing temperature for different glass phosphors is different. The higher the annealing temperature, the shorter the annealing time. The annealing treatment is composed of three stages: first stage at which the glass phosphor is heated to the target annealing temperature, second stage at which the constant annealing temperature is kept for certain time and third stage at which the temperature turns down slowly to room temperature. After abovementioned annealing treatment, the glass phosphor is finally obtained.

17.6   DOPING AND FORMING TECHNIQUES OF THE LUMINESCENCE CENTERS

The glass phosphor is composed of the luminescence centers (light-emitting groups) and glass host. The host is a matrix in which the luminescence centers live. The luminescence centers are not always incorporated in the glass host as the glass phosphor is prepared. In some cases, the luminescence centers are introduced after glass host is formed. The luminescence centers in the glass phosphor can be introduced via following techniques.

17.6.1   DIRECT DOPING

The luminescence centers are directly introduced during the preparation of the glass phosphor [17, 20]. The compound containing luminescence centers is one component in the starting materials. Usually, the trivalent rare earth ions, transition metal ions and main group element ions can be doped in the glass phosphors via this technique. Some complex luminescence centers, for instance, MoO_4^{2-} and WO_4^{2-}, can also be incorporated into glass matrix directly.

17.6.2   THERMAL TREATMENT GROWTH

Though the compound contains the element related with luminescence centers, the luminescence centers are not formed during the glass phosphor preparation. The post-thermal-treatment is needed for the growth of the luminescence centers [26]. The glass phosphors containing quantum dots like $CsPbBr_3$ and $CsPbI_3$ are good examples for this doping technique. The compounds Cs_2CO_3, $PbBr_2$, NaBr, PbI_2 and NaI are added into the glass phosphors, but the $CsPbBr_3$ and $CsPbI_3$ quantum dots are not grown during the glass preparation. Thermal treatment helps the growth of these quantum dots. Another example is silver aggregates like Ag_2^+, Ag_3^+ and Ag_4^+ that can also be formed via thermal treatment [30].

17.6.3   REDUCING METHOD

Some elements in high valence are introduced during the glass phosphor preparation, but the corresponding luminescence centers are in their low valence. Therefore, to form the luminescence centers in the glass phosphors, the ions in high valence should be reduced to be in low valence. The reducing reaction can be carried out in the processes of both glass phosphor preparation and post-thermal-treatment. Usually, the reducing atmosphere for changing the ionic valence is used. The reducing atmosphere is pure hydrogen or the mixture of 95% nitrogen and 5% hydrogen. The reducing atmosphere is introduced into the glass phosphor preparation or the post-thermal-treatment.

17.6.4   ION IMPLANTATION

As an ion beam shoots at the surface of a bulk solid, sputtering, scattering and implantation processes may occur. When the ion beam makes the atoms in the bulk solid bump out of the surface, this process is called sputtering; when the ions in the ion beam bounce back from the surface of the solid, or penetrate through the solid, this process is called scattering; when the ions in ion beam sink into the solid, this process is called ion implantation [40,41]. The ion implantation is also a route to dope the luminescence centers into the glass phosphors [42,43]. Usually, luminescence center ions introduced via the ion implantation technique are located at the surface or near-surface of the glass host. The post-thermal-treatment can make the implanted ions re-dispersed into the deep position of the glass host, but the improvement of the distribution depth is limited.

17.6.5 ION EXCHANGE

Ion exchange is a matured technique for producing planar optical waveguides, and meanwhile it is also an effective doping route for luminescence centers into glasses to form glass phosphors. To realize the doping via ion exchange, the glass is immersed in a salt melt containing the luminescence center ions at certain temperature for several hours or days. In the ion exchange process, the luminescence center ions in the salt melt are exchanged for the ions in the glass. The glass after ion exchange is thermally treated at a temperature to make the luminescence center ions further dispersed in the glass. Finally, the glass phosphors can be obtained. Ag^+, Cu^+ and Er^{3+} have been doped into glasses successfully [44–47]. It should be noted that the doping of luminescence center ions via ion exchange are non-uniformly dispersed in the glass host, and the doping concentration gradually decreases with increasing the radial distance from the glass surface. Moreover, the doping concentration ununiformity can hardly be changed by post-thermal-annealing.

17.6.6 MICRO-/NANOCRYSTAL GROWTH

Micro-/nanocrystal glass is one kind of glass phosphors [48–50]. The luminescence micro- or nano-crystals are contained in the glass phosphors, and the emissions usually come from the luminescence center ions, which are involved in the starting materials of the glass phosphors. The micro- or nanocrystals embedded in the glasses grow usually via post-thermal-treatment. In most literatures, the authors have observed the increase of luminescence intensity of the micro-/nanocrystal glass phosphors in comparison with the corresponding glass phosphors without micro-/nanocrystals and concluded that the luminescence center ions entered into the micro-/nanocrystals. However, there are very less evidences to clarify if all luminescence center ions enter into the micro-/nanocrystals or part of the luminescence center ions enter into the micro-/nanocrystals. Therefore, how the micro-/nanocrystals improve the luminescence performance of the glass phosphors leaves an open question to the researchers.

17.6.7 PIG ROUTE

PiG is abbreviation of Phosphor in Glass. It is one type of glass phosphors and developed in recent years [51–54]. The polycrystals phosphor is embedded in the glass matrix of the PiG. The PiGs are usually transparent to the visible light. The transparency of the PiGs depends on both the refractive index difference between the polycrystals and glass matrix and the particle size of the polycrystals. If the refractive index of the polycrystals is nearly close to that of the matrix, the PiG is transparent. If the polycrystals' average size is much smaller than the wavelength of the visible light (500 nm), the PiG is transparent as well [55]. The PiGs are different from the micro-/nanocrystals glass phosphors mainly in their preparative routes. As described above the polycrystals in the micro-/nanocrystals glass phosphors are formed in the glass-forming process or in the post-thermal-treatment process. Differently, the polycrystals in the PiGs are prepared in advance before forming the PiGs. The PiGs are usually prepared in following two approaches. In most cases, the glass host without any dopants is prepared, then the obtained glass is milled to obtain a glass host powder. The glass host powder is mixed with the polycrystal phosphor that is expected to be embedded in the glass host. Finally, the mixture is melted at a proper temperature and then quenched again to form the PiG. Moreover, the PiG can also be formed by melting and quenching the mixture of glass host starting materials (before forming glass) and the polycrystal phosphor. Owing to the special preparation procedure of PiGs, the low glass transition temperature T_g is often required. Therefore, the glasses with low preparation temperatures such as borate, phosphate and tellurite glasses are more preferable and widely adopted as glass hosts for PiGs.

17.7 CHARACTERIZATIONS OF GLASS PHOSPHORS

17.7.1 Differential Thermal Analysis

Differential thermal analysis (DTA) is an important characterization method for glass materials. From the DTA measurement, the endothermic and exothermic dependence of the studied material on the temperature can be obtained. From the DTA curve, the glass transition temperature and the crystallization temperature can be confirmed. Based on the confirmed glass transition temperature and the crystallization temperature, the annealing temperature and the growing temperature for micro-/nanocrystals in glass can be determined. Meanwhile, the thermal stability of the glass can also be quantitatively expressed by the temperature difference between the glass transition temperature and the crystallization temperature.

17.7.2 X-Ray Diffraction

X-ray diffraction (XRD) based on the Bragg diffraction theory is an important tool to identify the solid crystalline matter phase and explore the structure of the crystalline. In the glass science, the XRD is also widely used for confirming the amorphous characteristic and observing the micro-/nanocrystals in the glass host.

17.7.3 Scanning Electron Microscopy/Transmission Electron Microscopy

Scanning electron microscope (SEM) and transmission electron microscopy (TEM) are important techniques for observing the microscopic structure and morphology of the materials. Usually, the SEM/TEM images for the bulk glass are not able to be taken. Therefore, prior to measuring the SEM/TEM images, the bulk glass should be broken and ground to be powdered. With SEM/TEM images the morphology, size and distribution of the crystals embedded in the micro-/nanocrystal glass phosphors and in PiGs can be directly observed.

17.7.4 Raman and Fourier Transform Infrared Spectroscopic Techniques

Raman spectrum is a scattering spectrum that is generated after a fixed wavelength light irradiated at the studied materials. From the Raman spectrum, the stretching vibration and rotation information of the molecules for the studied materials can be obtained. Therefore, the banding relations between the molecules and vibration modes can be clearly seen from the Raman spectrum. For the glass phosphors, the maximum phonon energy can also be found in the Raman spectrum, and the nonradiative transition for the luminescence centers can also be evaluated.

Fourier transform infrared (FTIR) spectrum is measured based on the Michelson interferometer and Fourier transform techniques. In fact, the FTIR is an absorption spectrum that shows the absorption spectrum of the molecular vibration and rotation energy levels. Similar to the Raman spectrum, the same information can also be derived from the FTIR spectrum.

17.7.5 X-Ray Photoelectron Spectroscopy

X-ray photoelectron spectroscopy (XPS) is an effective tool for analyzing the component and valence of the atoms. The XPS spectrum is often used to identify the valence of the doped luminescence centers, such as Mn, Cr, Cu and Ag, and further help scientists to know origin of the emissions.

17.7.6 EXTENDED X-RAY ABSORPTION FINE STRUCTURE SPECTROSCOPY

Extended X-ray absorption fine structure (EXAFS) analysis technique is a very significant tool for understand the molecular composition of amorphous materials. From the EXAFS spectrum, the parameters such as atomic spacing, coordination number and mean square displacement of atoms can be confirmed. From these parameters, the atomic spacing structure of the glass phosphor can be reconstructed. The EXAFS technique can be used to solve some glass material structure problems that are difficult or unable to be solved by other experimental technique.

17.8 SPECIAL SPECTROSCOPIC PROPERTIES OF GLASS PHOSPHORS

The applications of the glass phosphors relay on their optical spectroscopic properties. The optical spectroscopic properties of the glass phosphors mainly depend on the luminescence centers/groups, and glass matrices will also influence or adjust the optical spectroscopic properties of the luminescence centers/groups [56]. Therefore, the matrices are hosts for the luminescence centers/groups, and they play the same roles as the compound matrices for the polycrystalline powdered phosphors. In the optical spectroscopic aspects, the glass phosphors exhibit similar but not the same properties with the powdered polycrystalline phosphors with same dopants. The main differences between the glass phosphors and the polycrystalline phosphors in the optical spectroscopy are described as follows.

17.8.1 ABSORPTION SPECTRUM

The absorption spectrum presents the dependence of absorbance $A(\lambda)$ on wavelength (wavenumber) for the studied sample. The absolute absorption spectrum for the glass phosphor can be measured based on the Lambert-Beer law since the glass phosphor is bulk and transparent [31].

$$A(\lambda) = -\ln\left(\frac{I(\lambda)}{I_0(\lambda)}\right) \tag{17.1}$$

where $I(\lambda)$ is the intensity of the transmitted light from the sample, and $I_0(\lambda)$ presents the intensity for the incident light on the sample. Furthermore, the absorption cross section $\sigma(\lambda)$ can be determined via following equation [32].

$$\sigma_a(\lambda) = \frac{A(\lambda)}{C \cdot L} \tag{17.2}$$

where C is the concentration of the luminescence centers, L is the optical path length of light propagation in the sample. However, for the powdered phosphors, the absolute value of $\sigma_a(\lambda)$ cannot be conformed owing to the existence of intense light scattering and the nondeterministic amount of the loaded sample per unit volume.

17.8.2 INTENSE SELF-ABSORPTION

The self-absorption means that the emitted photon of a luminescent center is captured by another luminescence center of the same type in its propagating path [57]. The luminescence center capturing the photon will release the obtained photon energy via radiative, nonradiative transition or energy transfer. The existence of self-absorption results in the changes of spectral line shape and decreases the luminescence external quantum efficiency. The self-absorption in the glass phosphors is usually more intense than that in powdered phosphors because the propagating paths for the excitation and emission photons in the glass phosphors are longer. In the powdered phosphor case, both the penetration depth of excitation light and the emitting photon propagating path length are short; therefore, the self-absorption effect can be ignored. In fact, it is very difficult to avoid the self-absorption effect in the measurement of emission spectrum for the glass phosphor unless the bulk glass is milled to be particles with very small size. It should be mentioned that the self-absorption for the transition from upper excited state to the lower excited state can be omitted even in the glass since the population of the lower excited state is less. Moreover, the emission cross section for the transition from excited state to the ground state with the actual spectral line shape can be derived from the absorption cross section for the transition from ground state to excited state based on the McCumber theory. This theory is established based on the Einstein relation and presents the relation between the absorption and emission cross sections, which can be mathematically expressed as below [58].

$$\sigma_e^{J \to J'}(v) = \sigma_a^{J' \to J}(v) \frac{Z_l}{Z_u} \exp\left(\frac{\varepsilon - hv}{kT}\right) \tag{17.3}$$

Where $\sigma_a^{J' \to J}$ is the absorption cross section for the transition from ground state J' to excited state J, Z_l and Z_u stand for the degeneracies of the ground state J' and excited state J, ε is free energy, T is absolute temperature and k and h are the Boltzmann constant and Planck's constant. From Equation (17.3), the emission cross section without self-absorption effect can be derived once the absorption cross section is known.

17.8.3   Excitation-Induced Weak Thermal Effect

In the photoluminescent materials, part of the absorbed photo energy converts into photo energies of other wavelengths via upconversion, downconversion and downshift processes, and the rest converts into thermal energy (vibrational energy of lattice). Usually, the photothermal conversion is not preferable for developing novel luminescent materials. However, the photothermal conversion cannot be avoided in any luminescent materials. For the powdered phosphors, the heat energy converted from the excitation light is localized in a very small space area (where the excitation light beam is covered) in the sample since the heat energy can hardly be conducted to the environment through the phosphor particles, thus resulting in very high local temperature at the place where the excitation light beam covers. The high local temperature evokes fluorescence thermal quenching, especially in the case of high-power laser excitation. In comparison with the powdered phosphors, the thermal quenching for the glass phosphors is relatively weak owing to the good thermal conductivity of glass hosts. In the glass phosphors, the generated heat cannot be localized in the spot covered by the excitation light beam but can be quickly conducted to the surrounding environment. The local temperature at the area where the excitation light beam covered is not so high; therefore, the excitation-induced fluorescence thermal quenching is weak.

17.8.4   Spectral Broadening

In the powdered polycrystalline phosphors, the luminescent centers occupy one or more but limited lattice sites. In the glass phosphors, the sites the luminescent centers occupied are more complicated than that in the polycrystalline phosphors. The spectrum of a phosphor is the incoherent superposition of the emissions from the all the luminescent centers occupying different sites in the phosphor. The emission wavelength of the luminescent center occupying a site is usually different from the emission wavelengths of the luminescent centers occupying other sites. Therefore, the spectral lines for the luminescent centers in glass phosphors are usually broadened in comparison with the polycrystalline phosphors owing to more complicated sites in the glass phosphors [59,60].

17.8.5   Highly Adjustable Spectral Properties

The polycrystalline phosphors are formed by the chemical elements in their fixed stoichiometric ratios, and the crystal structures of them are in most cases unchanged when partial elements are replaced by other elements. Therefore, the spectral properties of the luminescent centers in the polycrystalline phosphors can be tuned only in a very limited degree. However, in comparison with polycrystalline phosphors, the compositions for the glass phosphors can be changed in a large range; therefore, the spectral properties such as line widths, transition wavelengths and transition rates can be tuned more greatly [61,62].

17.8.6   RADIATION-INDUCED SPECTROSCOPIC CHANGE

Glasses exist in the metastable state, which is stable but not as stable as crystalline state. Usually, the potential barriers for the glasses are lower than that for the polycrystalline powders. If the barrier energies are close to the energies of electromagnetic, nuclear and thermal radiations, the glass structures and valences of luminescent centers can be changed by the radiations [63–66]. The radiation sources induced the changes can be high-energy ultraviolet light, femtosecond pulsed lasers and nuclear particles. The mechanisms for the changes are photochemical reaction, thermal relaxation and electromagnetic field effect.

17.8.7   MONO-EXPONENTIAL, MULTI-EXPONENTIAL AND NON-EXPONENTIAL FLUORESCENCE DECAYS

Fluorescence decays of the glass phosphors have mono-exponential, multi-exponential and non-exponential forms. If all luminescent centers in a host occupy only one-type site, meanwhile there is no interaction between the luminescent centers, and no interaction between the luminescent centers and any other quenching centers or trap centers, then the fluorescence decay follows mono-exponential function $f(t)$ as below [67].

$$f(t) = Ae^{-\frac{t}{\tau}} \tag{17.4}$$

where A is constant independent from time t, τ is the lifetime for the emitting energy level and is also equal to the average lifetime since the average lifetime $\langle \tau \rangle$ can be calculated via following equation,

$$\langle \tau \rangle = \frac{\int_0^\infty Ae^{-\frac{t}{\tau}} t\, dt}{\int_0^\infty Ae^{-\frac{t}{\tau}}\, dt} = \tau \tag{17.5}$$

If the lifetimes for the luminescent centers located at different sites in the studied material are approximately equal, and the interactions between the luminescent centers, quenching and trap centers can be ignored, then the decay for this luminescent material can be expressed by mono-exponential function as in Equation (17.5). If the lifetimes for the luminescent centers in different sites are deviated obviously from each other, and the interactions mentioned earlier are not existent, then the decay will follow multi-exponential function, which can be expressed as follows [68].

$$f(t) = \sum_{i=1}^{N} A_i e^{-\frac{t}{\tau_i}} \tag{17.6}$$

where N is the number for the different decays from the luminescent centers occupying different sites in the luminescent material. A_i and τ_i have the same meanings as in Equation (17.5). In this sense, the average lifetime for the emitting level can be written as,

$$\langle \tau \rangle = \frac{\int_0^\infty f(t) t\, dt}{\int_0^\infty f(t)\, dt} = \frac{\sum_{i=1}^{N} A_i \tau_i^2}{\sum_{i=1}^{N} A_i \tau_i} \tag{17.7}$$

If the energy transfer from the studied energy level to the other energy level is existent, then the decay of the studied energy level will follow the non-exponential function which has been discovered by Inokuti and Hirayama. In this case, if the interaction is of electric multipoles (including electric dipole – electric dipole, electric dipole – electric quadrupole and electric quadrupole – electric quadrupole interactions), then the decay follows below function [69].

$$f(t) = f(0)\exp\left[-\frac{t}{\tau} - \Gamma\left(1-\frac{3}{s}\right)\frac{c}{c_0}\left(\frac{t}{\tau}\right)^{3/s}\right] \tag{17.8}$$

where c and c_0 are the doping concentration and critical concentration, τ is the lifetime for the studied energy level when the energy transfer is absent, $s = 6$, 8 and 10 is the electric multi-dipole interaction index for the electric dipole – electric dipole, electric dipole – electric quadrupole and electric quadrupole – electric quadrupole interactions, respectively, Γ stands for the Gama function.

If the exchange interaction is existent, then the fluorescent decay will follow the function as below [69].

$$f(t) = f(0)\exp\left[-\frac{t}{\tau} - \Gamma\left(\frac{1}{2}\right)\frac{c}{c_0}\left(\frac{t}{\tau}\right)^{1/2}\left(\frac{1+10.87y+15.5y^2}{1+8.74y}\right)^{3/4}\right] \tag{17.9}$$

In the abovementioned equation, $y = D\tau R_0^{-2}(t/\tau)^{2/3}$, and the other symbols have the same physical meanings as in Equation (17.8). It should be stated that Equations (17.8) and (17.9) present the decays for the donors, and the decays for the accepters will not be changed by the energy transfers. Figure 17.2 shows the examples of the decays in the cases of mono-, multi- and non-exponential functions. In this figure, the τ_0 is fixed to be 10 ms. In the double exponential decay, the other decay constant τ_1 is 50 ms. From Figure 17.2, it can been seen that the addition of long lifetime component results in slow decay, and the presence of energy transfer makes the decay fast. Moreover, it should be pointed out that no matter what type decay can be decomposed by Equation (17.6) very well as long as the N value is large enough. However, the τ_i ($i = 1,\ldots, N$) values are physically meaningless except for the case of multi-exponential decay process.

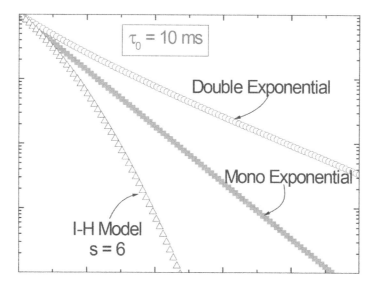

FIGURE 17.2 Plots for mono-, double- and non- (I-H model) exponential decays.

17.8.8 Judd-Ofelt Theory for the Trivalent Rare Earth Ions in Glass Phosphors

The optical transition properties of trivalent rare earth ions in hosts are very important for research and development of novel rare earth-doped luminescent materials. Judd and Ofelt have individually developed a theory for mathematically expressing the optical transitions of trivalent rare earth ions in hosts [70,71]. In this theory, the optical transition intensity was parametrically described. The parameters are widely known as Judd-Ofelt parameters. Once the Judd-Ofelt parameters are known, the radiative transition rates, radiative lifetimes and fluorescence branching ratios can be readily derived via simple algebraic operations. Furthermore, the nonradiative transition rates, energy transfer rates, internal quantum efficiencies and the emission cross sections as well can possibly be derived by assistance of some other fluorescence dynamic measurements.

17.8.8.1 Judd-Ofelt Parameters Derived from Absorption Spectrum

The calculations for the optical transitions of trivalent rare earth ions-doped transparent hosts like glasses or crystals become routine work [72]. Here, we introduce a brief procedure for the routine Judd-Ofelt calculation of trivalent rare earth ions-doped glass phosphors.

Based on the Judd-Ofelt theory, the theoretical oscillator strength of electric dipole component for a transition from state J to state J' can be expressed as [70,71],

$$f_{th}^{ed} = \frac{8\pi^2 mcv}{3h(2J+1)} \frac{(n^2+2)^2}{9n} \sum_{\lambda} \Omega_{\lambda} \langle \Psi J \mid U^{\lambda} \mid \Psi'J' \rangle^2 \tag{17.10}$$

where m and c are the mass of electron and light velocity in the vacuum, h is Planck's constant, π is circular constant, n is refractive index of the host, J is the quantum number of angular momentum for the initial state of the studied transition, $\langle \Psi J | U^{\lambda} | \Psi' J' \rangle$ is the reduced matrix element for the transition from state J to state J', and its value is approximately independent from the host and can be found in the literatures and Ω_{λ} ($\lambda = 2, 4, 6$) are Judd-Ofelt parameters. The theoretical oscillator strength of magnetic component for a transition from state J to state J' can be expressed as [70,71]

$$f_{th}^{md} = \frac{hvn}{6mc(2J+1)} \langle \Psi J | L + 2S | \Psi' J' \rangle^2 \tag{17.11}$$

where $\langle \Psi J | L + 2S | \Psi' J' \rangle$ is the reduced matrix element for the magnetic dipole transition, and its value can be numerically calculated and independent from the host. The total oscillator strength for the studied transition is the sum of the oscillator strengths of electric and magnetic dipole transition components.

$$f_{th} = f_{th}^{ed} + f_{th}^{md} = \frac{8\pi^2 mcv}{3h(2J+1)} \frac{(n^2+2)^2}{9n} \sum_{\lambda} \Omega_{\lambda} \langle \Psi J \mid U^{\lambda} \mid \Psi'J' \rangle^2 + \frac{hvn}{6mc(2J+1)} \langle \Psi J \mid L + 2S \mid \Psi'J' \rangle^2 \tag{17.12}$$

The total oscillator strength of an optical transition for the luminescent centers can also be experimentally confirmed from the absorption spectrum, and its mathematical expression is as follows,

$$f_{ex} = \frac{mc^2}{\pi e^2} \int_{v1}^{v2} \sigma_a(v) dv = -\frac{mc^2}{\pi e^2 CL} \int_{v1}^{v2} \ln\left(\frac{I(v)}{I_0(v)}\right) dv \tag{17.13}$$

where e is the electron charge, $\sigma_a(v)$ is absorption cross section as presented in Equation (17.2) (for the studied transition), and the integral runs over full the spectral rage of the studied transition. Therefore, once the absorption spectrum for the studied glass phosphor is confirmed, the total oscillator strengths for all observed optical transitions can be calculated from Equation (17.13). The experimental oscillator strength derived from Equation (17.13) should be nearly equal to the theoretical oscillator strength derived from Equation (17.12). Therefore, for each absorption transition, we can establish an equation $f_{ex} = f_{th}$ which can be written as,

$$\frac{mc^2}{\pi e^2} \int_{v1}^{v2} \sigma_a(v) dv$$

$$= \frac{8\pi^2 mcv}{3h(2J+1)} \frac{(n^2+2)^2}{9n} \sum_{\lambda} \Omega_{\lambda} \left\langle \Psi J | U^{\lambda} | \Psi' J' \right\rangle^2 + \frac{hvn}{6mc(2J+1)} \left\langle \Psi J | L + 2S | \Psi' J' \right\rangle^2 \quad (17.14)$$

In the abovementioned equation, only the values of three Judd-Ofelt parameters Ω_{λ} are unknown. For all observed absorption transitions, a set of equations can be deduced as

$$f_{th}^i = f_{ex}^i \quad (17.15)$$

If the number of equations in the equation set is equal to three or more than three, then the Judd-Ofelt parameters, Ω_{λ}, can be determined via a least square method.

Furthermore, the radiative transition rate for the transition from J to J' can be calculated by using the derived Judd-Ofelt parameters via following equation [31,32],

$$A_{J-J'} = A_{J-J'}^{ed} + A_{J-J'}^{md} \quad (17.16)$$

In the abovementioned equation,

$$A_{J-J'}^{ed} = \frac{64\pi^4 e^2 v^3 n (n^2+2)^2}{27h(2J+1)} \sum_{\lambda} \Omega_{\lambda} \left\langle \Psi J | U^{\lambda} | \Psi' J' \right\rangle^2 \quad (17.17)$$

$$A_{J-J'}^{md} = \frac{16\pi^4 e^2 v^3 n^3}{3h(2J+1)m^2c^2} \left\langle \Psi J | L + 2S | \Psi' J' \right\rangle^2 \quad (17.18)$$

The total radiative transition rate from J to all lower states can be calculated by following equation,

$$A_J = \sum_{J'} A_{J-J'} \quad (17.19)$$

The radiative lifetime of state J is the reciprocal of the total radiative transition rate A_J and written as

$$\tau_J = \frac{1}{A_J} = \frac{1}{\sum_{J'} A_{J-J'}} \quad (17.20)$$

The fluorescence branching ratio for the transition from J to J' can be written as

$$\beta_{J-J'} = \frac{A_{J-J'}}{\sum_{J'} A_{J-J'}} \quad (17.21)$$

The Judd-Ofelt calculation procedure is widely used for studying the optical transition properties of trivalent rare earth ions-doped transparent hosts including glasses and crystals, and this calculation approach is rooted in the absorption spectra of the studied luminescent materials.

17.8.8.2 Judd-Ofelt Parameters Derived from Fluorescence Decays

For a rare earth ion, its partial energy levels are shown in Figure 17.3. Assuming an electron at the excited state energy level J can radiatively transit to its lower energy level J' with a transition rate $A_{J-J'}$. The total radiative transition rate of energy level J is presented by Equation (17.19), and then by taking Equations (17.17) and (17.18) into Equation (17.19) [72], it is derived that

$$A_J = \sum_{J'}^{m}\left[\frac{64\pi^4 e^2 v^3 n\left(n^2+2\right)^2}{27h(2J+1)}\sum_{\lambda}\Omega_{\lambda}\langle\Psi J\,|\,U^{\lambda}\,|\,\Psi'J'\rangle^2 + \frac{16\pi^4 e^2 v^3 n^3}{3h(2J+1)m^2 c^2}\langle\Psi J\,|\,L+2S\,|\,\Psi'J'\rangle^2\right] \tag{17.22}$$

If the nonradiative transition rate of energy level J can be ignored, and the energy transfer depopulating energy level J can also be ignored or does not exist, then the total radiative transition rate of energy level J is the reciprocal of the fluorescence lifetime of energy level J. Therefore, we get [72],

$$\frac{1}{\tau_J} - \sum_{J'}^{m}\left[\frac{16\pi^4 e^2 v^3 n^3}{3h(2J+1)m^2 c^2}\langle\Psi J|L+2S|\Psi'J'\rangle^2\right]$$
$$= \frac{64\pi^4 e^2 v^3 n\left(n^2+2\right)^2}{27h(2J+1)}\sum_{J'}^{m}\left[\sum_{\lambda}\Omega_{\lambda}\langle\Psi J\,|\,U^{\lambda}\,|\,\Psi'J'\rangle^2\right] \tag{17.23}$$

In Equation (17.23), the fluorescence lifetime can be experimentally measured, and the second item on the left of the equal sign can be directly calculated. The item on the right of the equal sign involves the unknown Judd-Ofelt parameters, but the other physical variables are known. Therefore, an equation can be established for an excited state as long as the fluorescence decay for the excited state is measurable. The Judd-Ofelt parameters can be derived via least square method, if the decays for three or more than three excited states are known. It should be pointed out that the Judd-Ofelt route based on the fluorescence decays is independent from the studied sample shape, morphology and size, and it can be used for trivalent rare earth-doped films, powders and bulks.

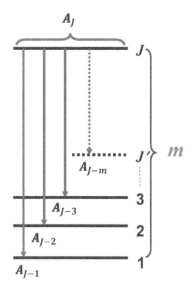

FIGURE 17.3 Partial energy levels and possible optical transitions of a RE ion.

17.9 POTENTIAL AND PRACTICAL APPLICATIONS OF GLASS PHOSPHORS

17.9.1 LASER-WORKING MEDIA

Laser sources are widely used in the fields of daily life, scientific research, healthcare, and aerospace [73]. To generate laser operation, the laser working medium is required. The laser working medium must be a luminescent material with net optical gain for the laser transition [21]. Most glass phosphors listed in Table 17.2 are good candidates for laser working media to achieve all-solid-state laser operations. An obvious advantage of glass phosphors over the luminescent crystals in laser applications is that the glass phosphors can be shaped to bulks, fibers and microspheres. Figure 17.4 shows the simplified laser schemes for the traditional lasers, fiber lasers and microsphere lasers that operate based on the glass phosphors [74–76].

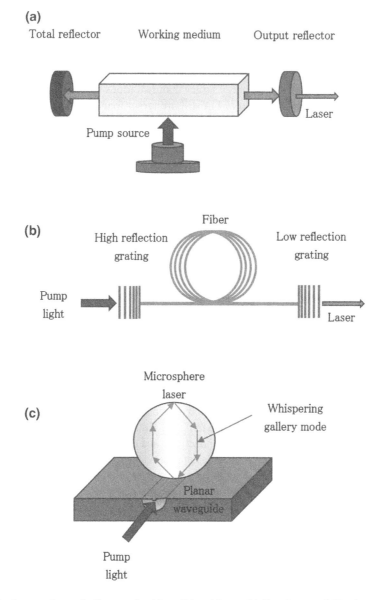

FIGURE 17.4 Laser schematic diagram for (a) traditional laser, (b) fiber laser and (c) microsphere laser.

17.9.2   Optical Waveguide Amplifiers

Optical waveguide is such a device in which the optical signal propagates along the pre-laid path. The waveguide, which is produced on the planar surface of a substrate, is defined as planar waveguide. If the optical signal propagates in an optical fiber, then the fiber is called as fiber waveguide. Optical waveguide amplifier is also an important optical device in integrated optics. In the waveguide amplifier, the optical signal is amplified by the waveguide medium. The glass phosphors are also potential materials for optical waveguide amplifiers. The erbium-doped fiber amplifiers (EDFA) and erbium-doped planar waveguide amplifiers (EDPWA) are known as famous and widely studied waveguide amplifiers [77]. Some other typical waveguide amplifiers operate based on the Nd^{3+}- and Pr^{3+}-doped glass phosphors [78,79].

17.9.3   Light Converters for Solar Cells

Solar cell is a device converting the sunlight photon energy to the electric energy. However, only the photon energy larger than the bandgap energy of the semiconductor material used in the solar cells can be converted to electric energy, and one photon can produce at most one electron-hole pair which can possibly contribute to the electric current. To improve the utilization efficiency of the sunlight photons, the photon with energy larger than two times the bandgap energy can be split to two photons with relatively lower energy but still are larger than the bandgap energy. After this splitting, one sunlight photon can produce two electron-hole pairs, thus resulting in great improvement of the photoelectric conversion efficiency of the solar cell. The photon splitting can be realized via a quantum cutting technique in the rare earth ions-doped glass phosphors, which can be potentially used in solar cells [80,81]. Besides that, two or more sunlight photons with energy smaller than the bandgap energy, could also be bonded together via a frequency upconversion technique to generate one photon with the energy larger than the bandgap energy [82,83]. After this bonding, the sunlight photons with energy lower than bandgap energy can also be effectively used in the photoelectric conversion.

17.9.4   Solid-State Lighting Sources

Solid-state lighting (SSL) is the lighting involving only solid-state modules for electro-optical conversion based on the inorganic and organic semiconductors [84]. The white light-emitting diodes (LEDs) based on the inorganic semiconductor is more attractive owing to their relatively large power output and excellent stability in comparison with the organic light-emitting devices [85]. However, the semiconductor-only lighting devices (composed of multiwavelength emitting semiconductor diodes) still face the problems of high cost and low chromatic stability. Therefore, the combination of single-color LED and photoluminescence phosphor (can be excited by the single-color LED) for generating white light has become mainstream. This combination is called as phosphor-converted white LEDs (pc-WLEDs). Both powdered and glass phosphors can be used in pc-WLEDs. The advantages of the glass phosphors for pc-WLEDs application are good thermal conductivity and easy far-field packaging [86]. Except for the LEDs, the laser diodes (LDs) also exhibit wide application prospects in SSL [87]. The lighting based on LDs is called as laser lighting. The laser lighting is the combination of single-wavelength output laser with the photoluminescence materials including the glass phosphors. Glass phosphors are more preferable in laser lighting thanks to the weak scattering loss.

17.9.5   Displays

Nowadays, the LCDs occupy the main market of the display devices. The backlight sources of the LCD screens are different from the usual lighting sources for daily life. The lighting sources for daily life ask for as wide as possible spectral gamut in visual region to reproduce the natural color

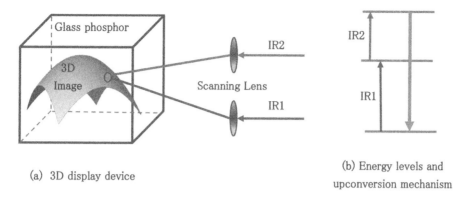

(a) 3D display device

(b) Energy levels and upconversion mechanism

FIGURE 17.5 Schematic diagram for 3D display. (a) Schematic diagram of glass phosphor used as the 3D screen display. (b) Simplified energy level distribution of an upconversion luminescence glass phosphor bulk doped with rare earth ions for the 3D screen display.

of the objects under sunlight. Nevertheless, the LCD backlight requires narrow tricolor (red, green and blue) lights, thus the glass phosphors can be used as the light-emitting materials [88]. Moreover, the glass phosphors can also be used as color filters to generate narrow band tricolor lights when the full-color WLEDs are used in the backlight sources [89].

The volumetric and true three-dimensional (3D) display is still a dream in the display field, and it is expected that the video (containing 3D space information) can be replayed in the true 3D space and can be viewed from any angle in the space [90,91]. Figure 17.5 shows the schematic diagram of a type of 3D display. In this 3D display, a glass phosphor as shown in Figure 17.5(a) is used as the 3D screen in which the 3D images and 3D videos can be displayed. The 3D screen is an upconversion luminescence glass phosphor bulk doped with rare earth ions with the simplified energy level distribution as seen in Figure 17.5(b). When the collimated infrared laser beam IR1 or IR2 individually passes through the 3D screen, we cannot find that any pixels are drawn. When both IR1 and IR2 pass through the 3D screen, and meanwhile they can form a crossover point in the 3D screen, then the crossover point lights up via the upconversion luminescence process as seen in Figure 17.5(b). If the crossover point is moved by following a 3D image in the 3D screen, then the 3D image is displayed in the screen, and the image could be watched from any angle in the 3D space. If the drawn image updates with time and forms a 3D video, then the video can be watched in the 3D space from any angle.

17.9.6 Optical Temperature Sensing

Optical temperature sensing means the temperature of the concerned object can be confirmed via optical spectroscopic measurements [92]. Optical temperature sensing requires that the glass phosphor used for detecting temperature has such a temperature-dependent spectroscopic parameter x whose absolute value can be experimentally measured. For the glass phosphors, the parameters can be the linewidth, peak position and fluorescence lifetime and intensity ratio. For a certain glass phosphor, if any parameter is only dependent on the temperature, then the temperature can be sensed. The dependence of the parameter x on the temperature can be expressed by following equation:

$$x = f(T) \tag{17.24}$$

where x presents the measurable spectroscopic parameter.

If the function presented by Equation (17.24) is explicitly a mathematical expression, then the constants involved in Equation (17.24) can be confirmed by measuring x values at various temperatures and followed by fitting the experimental dependence of x on T to the theoretical Equation (17.24). Once the mathematic presentation of Equation (17.24) is confirmed, the temperature can be sensed by taking the measured x value into Equation (17.24). If the specific mathematical form of Equation (17.24) is not known, then the polynomial expansion can approximately replace it and can be presented as follows.

$$x = \sum_{n=0}^{N} \left(A_n T^n \right) \tag{17.25}$$

In the abovementioned equation, A_n is constant and independent from temperature T, and N is integer whose value is decided by the complexity of dependency of x on T.

The sensitivity S is defined as [93]:

$$S = \frac{dx}{dT} = \frac{df(T)}{dT} \tag{17.26}$$

S reflects the change of parameter x value when the temperature changes by a unit value. It should be clarified that in Equation (17.24), the parameter x should be the only temperature-dependent parameter in the glass phosphor.

17.9.7   RADIATION DETECTION

Radiation is that the electromagnetic waves (microwaves, X-rays and gamma rays) or subatomic particles (alpha-particles, beta-rays and protons) travel through the space. The aim of radiation detection is to discover the types, doses and traces of radiation [94]. The glass phosphors, which can be excited by the radiation rays and emit photons, are good candidate materials for applications in the radiation detection [95]. In this sense, the radiation measurements are carried out by the aid of optical spectroscopic detection [96]. Glass phosphors that can emit photons under radiation excitation are called glass scintillators [97]. It should be stated that the excitation mechanisms of radiations are usually different from the photoluminescence processes since the ionization and photoelectric conversion may happen and the glass host may also participate in the excitation processes [94–100].

17.9.8   LASER-INDUCED ANTI-STOKES FLUORESCENCE COOLING

Laser-induced anti-Stokes fluorescence cooling is an optical refrigeration route, and it was first proposed in 1929, and the first experimental observation was carried out in 1995 [101,102]. The glass phosphors doped with rare earth ions are good choices for achieving optical refrigeration. Figure 17.6 shows the simplified energy-level diagram with the laser-induced anti-Stokes fluorescence process. The ground state E0 and excited state E1 split into three Stark sublevels. The laser with output wavenumber matching the transition wavenumber from the highest sublevel of ground state E1 to the lowest sublevel of the excited state E0 is used to excite the glass phosphor. Under this excitation, the wavenumbers of all emitting photons are larger than the wavenumber of the excitation photon [103–105]. This photoluminescence process results in the decrease of the glass phosphor heat energy, namely, the glass phosphor is cooled. The heat energy of the atom vibrations is brought out by the anti-Stokes fluorescence photons.

17.9.9   OPTICAL FILTERS

Optical filters have a steep absorption edge, high transmittance in pass band and strong absorption (i.e., large optical density) in cutoff band. The optical filters include semiconductor quantum dot

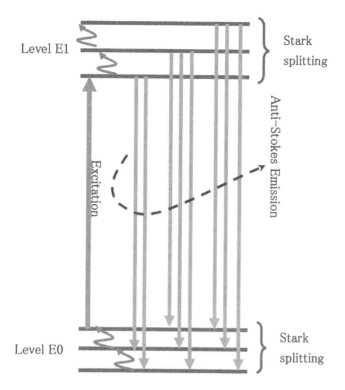

FIGURE 17.6   Energy diagram showing the laser-induced anti-Stokes fluorescence cooling.

glass filters, metal colloids glass filters, and the high/low refractive index dielectric films filters [106,107]. Semiconductor quantum dot glasses are excellent candidates for optical filters. The semiconductor quantum dot glass filters are typically colored by conventional CdS, CdSe, CdTe quantum dots and their alloy compounds [107] and recently developed perovskite $CsPbBr_3$ quantum dots as well [106]. The filtering mechanism for this type of filters is based on the intense optical absorption. The absorption edge is a little larger than the energy gap (Eg) of corresponding bulk semiconductor and it has a redshift with increasing the size of the quantum dots, and thus it is attributed to the quantum confinement effect. For the filters, the transition wavelength defined as the wavelength at half of maximum transmittance is used as work wavelength. Different semiconductor quantum dots have different transition wavelengths; therefore, different filters are produced by introducing different semiconductor quantum dots into the glass host. In addition, the transition wavelength can be adjusted by the alloy compounds of the semiconductor quantum dots, for example, the CdS–CdSe quantum dots.

The metallic colloidal nanoparticle glass filters are typically colored by aggregated nanoparticles of Au, Ag, and Cu in glass matrix [107]. The work mechanism for this type of filters is photons absorption and scattering of the colloidal metallic nanoparticles. Usually, the sizes in ~5–60 nm of the metallic colloidal nanoparticles are effective. The photons absorption is dominant when the nanoparticle size ranges in ~5–40 nm, whereas the photons scattering becomes dominant when the nanoparticle size ranges in ~40–60 nm. The transition wavelength obviously depends on the size of the metallic nanoparticles for this type of filters.

17.9.10   OPTICAL STORAGE

In the present big data age, a large amount of data generates every second, and thus the data storage has attracted much attention [108,109]. The glass phosphors can also be used in the optical storage

[110,111]. Different from electrical storage/memory, the optical storage is storing the data information in the medium via optical spectroscopic techniques. The information can be persistently or temporally stored in the media, and the stored information can be erasable or inerasable, which depends on the various techniques and materials used [108, 112].

REFERENCES

1. Axinte, E. 2011. Glasses as engineering materials: A review. *Mater. Des.* 32:1717–32.
2. Rasmussen, S. C. 2012. *How glass changed the world – The history and chemistry of glass from antiquity to the 13th century.* New York, NY: Springer.
3. Debenedetti, P. G., and Stillinger, F. H. 2001. Supercooled liquids and the glass transition. *Nature* 410:259–67.
4. Ediger, M. D., Angell, C. A., and Nagel, S. R. 1996. Supercooled liquids and glasses, *J. Phys. Chem.* 100:13200–12.
5. Zhu, C. C., Long, Z. W., Wang, Q., Qiu, J. B., Zhou, J. H., Zhou, D. C., Wu, H., and Zhu, R. 2019. Insights into anti-thermal quenching of photoluminescence from $SrCaGa_4O_8$ based on defect state and application in temperature sensing. *J. Lumin.* 208:284–9.
6. Trukhin, A. N., Sharakovski, A., Grube, J., and Griscom, D. L. 2010. Sub-band-gap-excited luminescence of localized states in SiO_2-Si and SiO_2-Al glasses. *J. Non-Cryst. Solids* 356:982–6.
7. Swetha, B. N., and Keshavamurthy, K. 2020. Impact of thermal annealing time on luminescence properties of Eu^{3+} ions in silver nanoparticles embedded lanthanum sodium borate glasses. *Appl. Surf. Sci.* 525:146505.
8. Zheng, H. R., Wang, X. J., Dejneka, M. J., Yen, W. M., and Meltzer, R. S. 2004. Up-converted emission in Pr^{3+}-doped fluoride nanocrystals-based oxyfluoride glass ceramics. *J. Lumin.* 108:395–9.
9. Zhang, X. J., Wang, J., Huang, L., Pan, F. J., Chen, Y., Lei, B. F., Peng, M. Y., and Wu, M. M. 2015. Tunable luminescent properties and concentration-dependent, site-preferable distribution of Eu^{2+} ions in silicate glass for white LEDs applications. *ACS Appl. Mater. Interfaces* 7:10044–54.
10. Kuwik, M., Gorny, A., Zur, L., Ferrar, M., Righini, G. C., Pisarski, W. A., and Pisarska, J. 2019. Influence of the rare earth ions concentration on luminescence properties of barium gallo-germanate glasses for white lights. *J. Lumin.* 211:375–81.
11. Kibrisli, O., Erol, E., Vahedigharehchopogh, N., Yousef, E., Ersundu, M. C., and Ersundu, A. E. 2020. Noninvasive optical temperature sensing behavior of Ho^{3+} and Ho^{3+}/Er^{3+} doped tellurite glasses through up and down-converted emissions. *Sensor. Actuat. B-Chem.* 315:112321.
12. Weber, R., Hampton, S., Nordine, P. C., Key, T., and Scheunemann, R. 2005. Er^{3+} fluorescence in rare-earth aluminate glass. *J. Appl. Phys.* 98:043521.
13. Yoshimoto, K., Ezura, Y., Ueda, M., Masuno, A., and Inoue, H. 2018. 2.7 μm mid-infrared emission in highly erbium-doped lanthanum gallate glasses prepared via an aerodynamic levitation technique. *Adv. Opt. Mater.* 6:1701283.
14. Fan, B., Point, C., Adam, J. L., Zhang, X. H., Fan, X. P., and Ma, H. L. 2011. Near-infrared down-conversion in rare-earth-doped chloro-sulfide glass GeS_2-Ga_2S_3-CsCl:Er,Yb. *J. Appl. Phys.* 110:113107.
15. Weber, J. K. R., Vu, M., Passlick, C., Schweizer, S., Brown, D. E., Johnson, C. E., and Johnson, J. A. 2011. The oxidation state of europium in halide glasses. *J. Phys. Condens. Mat.* 23:495402.
16. Yasukevich, A. S., Rachkovskaya, G. E., Zakharevich, G. B., Trusova, E. E., Komienko, A. A., Dunina, E. B., Kisel, V. E., and Kuleshov, N. V. 2021. Spectral-luminescence properties of oxyfluoride lead-silicate-germanate glass doped with Tm^{3+} ions. *J. Lumin.* 229:117667.
17. Kumar, M., and Rao, A. S. 2020. Concentration-dependent reddish-orange photoluminescence studies of Sm^{3+} ions in borosilicate glasses. *Opt. Mater.* 109:110356.
18. Wang, J. H., Zheng, B. L., and Wang, P. 2020. 3D printed Er^{3+}/Yb^{3+} co-doped phosphosilicate glass based on sol-gel technology. *J. Non-Cryst. Solids* 550:120362.
19. Kang, S. L., Xiao, X. D., Pan, Q. W., Chen, D. D., Qiu, J. R., and Dong, G. P. 2017. Spectroscopic properties in Er^{3+}-doped germanotellurite glasses and glass ceramics for mid-infrared laser materials. *Sci. Rep.* 7:43186.
20. Yu, T. T., Sun, J. S., Hua, R. N., Cheng, L. H., Zhong, H. Y., Li, X. P., Yu, H. Q., and Chen, B. J. 2011. Luminescence of complex ion WO_{12}^{18-} in Dy^{3+} doped nanocrystal Gd_6WO_{12} phosphor. *J. Alloys Compd.* 509:391–5.
21. Zhong, H., Chen, B. J., Fu, S. B., Li, X. P., Zhang, J. S., Xu, S., Zhang, Y. Q., Tong, L. L., Sui, G. Z., and Xia, H. P. 2019. Broadband emission and flat optical gain glass containing Ag aggregates for tunable laser. *J. Am. Ceram. Soc.* 102:1150–6.

22. Chen, D. Q., Zhou, Y., and Zhong, J. S. 2016. A review on Mn^{4+} activators in solids for warm white light-emitting diodes. *RSC Adv.* 6:86285–96.

23. Meng, X. G., and Tanaka, K. 2007. Intense greenish emission from d^0 transition metal ion Ti^{4+} in oxide glass. *Appl. Phys. Lett.* 90:051917.

24. Peng, M. Y., Da, N., Krolikowski, S., Stiegelschmitt, A., and Wondraczek, L. 2009. Luminescence from Bi^{2+}-activated alkali earth borophosphates for white LEDs. *Opt. Express.* 17:21169–78.

25. Wang, X., Zhang, G. D., Zhang, Y. J., Xie, X. P., Cheng, G. H., and Li, W. N. 2020. Photochemical response triggered by ultrashort laser Gaussian-Bessel beams in photo-thermo-refractive glass. *Opt. Express* 28:31093.

26. Guo, L. Z., Zhang, X. Z., Zhang, Y. H., Yu, T., Cheng, C. H., Cheng, Y., Li, X. P., Zhang, J. S., Xu, S., Cao, Y. Z., and Chen, B. J. 2021. Color-adjustable $CsPbBr_{3-x}I_x$ quantum dots glasses for wide color gamut display. *J. Non-Cryst. Solids* 551:120432.

27. Zhang, X. Z., Guo, L. Z., Zhang, Y. H., Cheng, C. H., Cheng, Y., Li, X. P., Zhang, J. S., Xu, S., Cao, Y. Z., Sun, J. S., Cheng, L. H., and Chen, B. J. 2020. Improved photoluminescence quantum yield of $CsPbBr_3$ quantum dots glass ceramics. *J. Am. Ceram. Soc.* 103:5028–35.

28. Wang, X. J., Huang, S. H., Reeves, R., Wells, W., Dejneka, M. J., Meltzer, R. S., and Yen, W. M. 2001. Studies of the spectroscopic properties of Pr^{3+} doped LaF_3 nanocrystals/glass. *J. Lumin.* 94:229–33.

29. Lin, H., Hu, T., Cheng, Y., Chen, M. X., and Wang, Y. S. 2018. Glass ceramic phosphors: Towards long-lifetime high-power white light-emitting-diode applications – A review. *Laser Photonics Rev.* 12:1700344.

30. Nikl, M., Solovieva, N., Apperson, K., Birch, D. J. S., and Voloshinovskii. A. Scintillators based on aromatic dye molecules doped in a sol-gel glass host. *Appl. Phys. Lett.* 86:101914.

31. Zheng, Y. F., Chen, B. J., Zhong, H. Y., Sun, J. S., Cheng, L. H., Li, X. P., Zhang, J. S., Tian, Y., Lu, W. L., Wan, J., Yu, T. T., Huang, L. B., Yu, H. Q., and Lin, H. 2011. Optical transition, excitation state absorption, and energy transfer study of Er^{3+}, Nd^{3+} single-doped, and Er^{3+}/Nd^{3+} codoped tellurite glasses for mid-infrared laser applications. *J. Am. Ceram. Soc.* 94:1766–72.

32. Wang, B., Cheng, L. H., Zhong, H. Y., Sun, J. S., Tian, Y., Zhang, X. Q., and Chen, B. J. 2009. Excited state absorption cross sections of $^4I_{13/2}$ of Er^{3+} in ZBLAN. *Opt. Mater.* 31:1658–62.

33. Colomban, P. 1996. Raman studies of inorganic gels and of their sol-to-gel, gel-to-glass and glass-to-ceramics transformation. *J. Raman Spectros.* 27:747–58.

34. Yu, J., Kohara, S., Itoh, K., Nozawa, S., Miyoshi, S., Arai, Y., Masuno, A., Taniguchi, H., Itoh, M., Takata, M., Fukunaga, T., Koshihara, S., Juroiwa, Y., and Yoda, S. 2009. Comprehensive structural study of glassy and metastable crystalline $BaTi_2O_5$. *Chem. Mater.* 21:259–63.

35. Pan, X., Yu, J., Liu, Y., Yoda, S., Yu, H., Zhang, M., Ai, F., Jin, F., and Jin, W. 2011. Thermal, mechanical, and upconversion properties of Er^{3+}/Yb^{3+} co-doped titanate glass prepared by levitation method. *J. Alloys Compd.* 509:7504–7.

36. Yoshimoto, K., Masuno, A., Inoue, H., and Watanabe, Y. 2012. Transparent and high refractive index La_2O_3-WO_3 glass prepared using containerless processing. *J. Am. Ceram. Soc.* 95:3501–4.

37. Suzuki, F., Sato, F., Oshita, H., Yao, S., Nakatsuka, Y., and Tanaka, K. 2018. Large faraday effect of borate glasses with high Tb^{3+} content prepared by containerless processing. *Opt. Mater.* 76:174–7.

38. Li, Q., Xiang, M., Chen, Z., Wang, X., Zhao, C., Qiu, J., Yu, J., and Chang, J. 2016. Er^{3+}/Yb^{3+} co-doped bioactive glasses with up-conversion luminescence prepared by containerless processing. *Ceram. Int.* 42:13168–75.

39. Mukherjee, S., Zhou, Z., Johnson, W. L., and Rhim, W. K. 2004. Thermophysical properties of Ni-Ni and Ni-Nb-Sn bulk metallic glass-forming melts by containerless electrostatic levitation procession. *J. Non-Cryst. Solids* 337:21–8.

40. He, C., Chen, L., Zhang, D. W., Hong, J. H., Jin, G. Y., Zhang, J., Boeker, J., Liu, R. J., Jin, H., Lv, Y. M., and Chen, J. FinFET doping with PSG/BSG glass mimic doping by ultra low energy ion implantation. 16th International Workshop on Junction Technology. Shanghai, P. R. China, 2016, P64–7.

41. Zhu, Q. F., Wang, Y., Shen, X. L., Guo, H. T., and Liu, C. X. 2018. Optical ridge waveguides in magneto-optical glasses fabricated by combination of silicon ion implantation and femtosecond laser ablation. *IEEE Photonics J.* 10:2400507.

42. Maurizio, C., Cesca, T., Trapananti, A., Kalinic, B., Scian, C., Mazzoldi, P., Battaglin, G., and Mattei, G. 2014. Effect of ultrasmall Au-Ag aggregates formed by ion implantation in Er-implanted silica on the 1.54 μm Er^{3+} luminescence. *Nucl. Instrum. Meth. B* 326:11–4.

43. Chao, L. C., Lee, B. K., Chi, C. J., Cheng, J., Chyr, I., and Steckl, A. J. 1999. Rare earth focused ion beam implantation utilizing Er and Pr liquid alloy ion sources. *J. Vac. Sci. Technol. B* 17:2791–4.

44. Sgibnev, Y., Asamoah, B., Nikonorov, N., and Honkanen, S. 2020. Tunable photoluminescence of silver molecular clusters formed in Na^+-Ag^+ ion-exchanged antimony-doped photo-thermo-refractive glass matrix. *J. Lumin.* 226:117411.

45. Mardegan, M., and Cattaruzaa, E. 2016. Cu-doped photovoltaic glasses by ion exchange for sunlight down-shifting. *Opt. Mater.* 61:105–10.
46. Sgibnev, Y. M., Nikonorov, N. V., and Ignatiev, A. I. 2016. Luminescence of silver clusters in ion-exchanged cerium-doped photo-thermo-refractive glasses. *J. Lumin.* 176:292–7.
47. Salavcova, L., Spirkova, J., Martin, M., Mackova, A., Oswald, J., Langrova, A., Vacik, J. 2007. Localised doping of Li-silicate glasses by Er^{3+} ion exchange to fabricate thin optical layers. *Opt. Mater.* 29:753–9.
48. Rajesh, D., and Camargo, A. S. S. Nd^{3+} doped new oxyfluoro tellurite glasses and glass ceramics containing $NaYF_4$ nano crystals – 1.06 μm emission analysis. *J. Lumin.* 207:469–76.
49. Cao, J. K., Hu, F. F., Chen, L. P., Guo, H., Duan, C. K., and Yin, M. 2017. Optical thermometry based on up-conversion luminescence behavior of Er^{3+} doped KYb_2F_7 nano-crystals in bulk glass ceramics. *J. Alloys Compd.* 693:326–31.
50. Yu, H., Li, S., Qi, Y., Lu, W., Yu, X., Xu, X., and Qiu, J. 2018. Optical thermometry based on up-conversion emission behavior of Ba_2LaF_7 nano-crystals embedded in glass matrix. *J. Lumin.* 194:433–9.
51. Chung, W. J., and Nam, Y. H. 2019. Review – A review on phosphor in glass as a high power LED color converter. *ECS J. Solid. State SC* 9:016010.
52. Zhang, Y. J., Zhang, Z. L., Liu, X. D., Shao, G. Z., Shen, L. L., Liu, J. M., Xiang, W. D., and Liang, X. J. 2020. A high quantum efficiency $CaAlSiN_3:Eu^{2+}$ phosphor-in-glass with excellent optical performance for white light-emitting diodes and blue laser diodes. *Chem. Eng. J.* 401:125983.
53. Shih, H. K., Liu, C. N., Cheng, W. C., and Cheng, W. H. 2020. High color rendering index of 94 in white LEDs employing novel $CaAlSiN_3:Eu^{2+}$ and $Lu_3Al_5O_{12}:Ce^{3+}$ co-doped phosphor-in-glass. *Opt. Express* 28:28218–25.
54. Dou, B. L., Hua, Y. J., Lei, R. S., Deng, D. G., Huang, F. F., and Xu, S. Q. 2020. Ln^{3+} doped phosphor-in-glass: A new choice of color filter for wide-color gamut white light-emitting diodes. *J. Colloid Interface Sci.* 563:139–44.
55. Beall, G. H., and Pinckney, L. R. 1999. Nanophase glass-ceramics. *J. Am. Ceram. Soc.* 82:5–16.
56. Yu, C. F., Chen, B. J., Zhang, X. Z., Li, X. P., Zhang, J. S., Xu, S., Yu, H. Q., Sun, J. S., Cao, Y. Z., and Xia, H. P. 2020. Influence of Er^{3+} concentration and Ln^{3+} on the Judd-Ofelt parameters in LnOCl (Ln = Y, La, Gd) phosphors. *Phys. Chem. Chem. Phys.* 22:7844–52.
57. Mattarelli, M., Montagna, M., Zampedri, L., Chiasera, A., Ferrari, M., Righini, G. C., Fortes, L. M., Goncalves, M. C., Santos, L. F., and Almeida, R. M. 2005. Self-absorption and radiation trapping in Er^{3+}-doped TeO_2-based glasses. *Europhys. Lett.* 71:394–9.
58. Sui, G., Chen, B., Zhang, X., Li, X., Zhang, J., Xu, S., Sun, J., Cao, Y., Wang, X., Zhang, Y., Zhang, Y., and Zhang, X. 2020. Radiative transition properties of Yb^{3+} in Er^{3+}/Yb^{3+} co-doped $NaYF_4$ phosphor. *J. Alloys Compd.* 834:155242.
59. Jha, A., Shen, S., and Naftaly, M. 2000. Structural origin of spectral broadening of 1.5-μm emission in Er^{3+} doped tellurite glasses. *Phys. Rev. B* 62:6215–27.
60. Jaba, N., Mansour, H. B., and Champagnon, B. 2009. The origin of spectral broadening of 1.53 μm emission in Er^{3+} doped zinc tellurite glass. *Opt. Mater.* 31:1242–7.
61. Xiang, X. Q., Lin, H., Xu, J., Cheng, Y., Wang, C. Y., Zhang, L. Q., and Wang, Y. S. 2019. $CsPb(Br,I)_3$ embedded glass: Fabrication, tunable luminescence, improved stability and wide-color gamut LCD application. *Chem. Eng. J.* 378:122255.
62. Zhang, Y. H., Chen, B. J., Zhang, X. Z., Li, X. P., Zhang, J. S., Xu, S., Wang, X., Zhang, Y. Q., Wang, L., Li, D. S., and Lin, H. 2020. Full color white light, temperature self-monitor, and thermochromatic effect of Cu^+ and Tm^{3+} codoped germanate glasses. *J. Am. Ceram. Soc.* 104:350–60.
63. Shao, C., Guo, M., Zhang, Y., Zhou, L., Guzik, M., Boulon, G., Yu, C., Chen, D., and Hu, L. 2020. 193 nm eximer laser-induced color centers in $Yb^{3+}/Al^{3+}/P^{5+}$ doped silica glasses. *J. Non-Cryst. Solids* 544:1920198.
64. Hu, Y., Zhang, W., Ye, Y., Zhao, Z., and Liu, C. 2020. Femtosecond-laser-induced precipitation of $CsPbBr_3$ perovskite nanocrystals in glasses for solar spectral conversion. *ACS Appl. Nano Mater.* 3:850–7.
65. Heidepriem, H. E., and Ehrt, D. 2002. Ultraviolet laser and X-ray induced valence changes and defect formation in europium and terbium doped glasses. *Phys. Chem. Glasses* 43C:38–47.
66. Shakhgildyan, G. Y., Ziyatdinova, M. Z., Vetchinnikov, M. P., Lotarev, S. V., Savinkov, V. I., Presnyakova, N. N., Lopatina, E. V., Vilkovisky, G. A., and Sigaev, V. N. 2020. Thermally-induced precipitation of gold nanoparticles in phosphate glass: Effect on the optical properties of Er^{3+} ions. *J. Non-Cryst. Solids* 550:120408.
67. Zhang, Y., Chen, B., Xu, S., Li, X., Zhang, J., Sun, J., Zhang, X., Xia, H., and Hua, R. 2018. A universal approach for calculating the Judd-Ofelt parameters of RE^{3+} in powdered phosphors and its application for the beta-$NaYF_4:Er^{3+}/Yb^{3+}$ phosphor derived from auto-combustion-assisted fluoridation. *Phys. Chem. Chem. Phys.* 20:15876.

68. Zhang,Y., Chen, B., Xu, S., Li, X., Zhang, J., Sun, J., Zhang, X., Xia, H., and Hua, R. 2019. Reply to the 'Comment on "A universal approach for calculating the Judd-Ofelt parameters of RE^{3+} in powdered phosphors and its application for the beta-$NaYF_4$:Er^{3+}/Yb^{3+} phosphor derived from auto-combustion-assisted fluoridation" by Zhang, D., Xu, Q., and Zhang', Y. *Phys. Chem. Chem. Phys.* 21:10840.

69. Inokuti, M., and Hirayama, F. 1965. Influence of energy transfer by the exchange mechanism on donor luminescence. *J. Chem. Phys.* 43:1978–89.

70. Judd, B. R. 1962. Optical absorption intensities of rare-earth ions. *Phys. Rev.* 127:750–61.

71. Ofelt, G. S. 1962. Intensities of crystal spectra of rare-earth ions. *J. Chem. Phys.* 37:511–20.

72. Luo, M., Chen, B., Li, X., Zhang, J., Xu, S., Zhang, X., Cao, Y., Sun, J., Zhang, Y., Wang, X., Zhang, Y., Gao, D., and Wang, L. 2020. Fluorescence decay route of optical transition calculation for trivalent rare earth ions and its application for Er^{3+}-doped $NaYF_4$ phosphor. *Phys. Chem. Chem. Phys.* 22:25177–83.

73. Jha, A., Richards, B., Jose, G., Fernandez, T. T., Joshi, P., Jiang, X., and Lousteau J. 2012. Rare-earth ion doped TeO_2 and GeO_2 glasses as laser materials. *Prog. Mater. Sci.* 57:1426–91.

74. Malyarevich, A. M., Yumashev, K. V., and Lipovskii, A. A. 2008. Semiconductor-doped glass saturable absorbers for near-infrared solid-state lasers. *Appl. Phys. Rev.* 103:081301.

75. Chiasera, A., Dumeige, Y., Feron, P., Ferrari, M., Jestin, Y., Conti, G. N., Pelli, S., Soria, S., and Righini, G. C. 2010. Spherical whispering-gallery-mode microresonators. *Laser Photon. Rev.* 4:457–82.

76. Pollnau, M., and Jackson, S. D. 2001. Erbium 3-μm fiber lasers. *IEEE J. Sel. Top. Quant.* 7:30–40.

77. Wilkinson, J. S., and Hempstead, M. 1997. Advanced materials for fiber and waveguide amplifiers. *Curr. Opin. Solid. St. M* 2:194–9.

78. Doddoji, R., Vazquez, G. V., Trejo-Luna, R., and Gelija, D. 2019. Spectroscopic and waveguide properties of Nd^{3+}-doped oxyfluorosilicate glasses. *App. Phys. B: Laser. Opt.* 125:117.

79. Liu, H. L., Luo, S. Y., Xu, B., Xu, H. Y., Cai, Z. P., Hong, M. H., and Wu, P. F. 2017. Femtosecond-laser micromachined Pr:YLF depressed cladding waveguide: Raman, fluorescence, and laser performance. *Opt. Mater. Express* 7:3990–7.

80. Ye, S., Zhu, B., Chen, J. X., Luo, J., and Qiu, J. R. 2008. Infrared quantum cutting in Tb^{3+}, Yb^{3+} codoped transparent glass ceramics containing CaF_2 nanocrystals. *Appl. Phys. Lett.* 92:141112.

81. Chen, D. Q., Yu, Y. L., Wang, Y. S., Huang, P., and Weng, F. Y. 2009. Cooperative energy transfer up-conversion and quantum cutting down-conversion in Yb^{3+}:TbF_3 nanocrystals embedded glass ceramics. *J. Phys. Chem. C* 113:6406–10.

82. Castro, T., Manzani, D., and Riveiro, S. J. L. 2018. Up-conversion mechanisms in Er^{3+}-doped fluoro-indate glasses under 1550 nm excitation for enhancing photocurrent of crystalline silicon solar cell. *J. Lumin.* 200:260–4.

83. Balaji, S., Ghosh, D., Biswas, K., Allu, A. R., Gupta, G., and Annapurna, K. 2017. Insights into Er^{3+}-Yb^{3+} energy transfer dynamics upon infrared similar to 1550 nm excitation in a low phonon fluoro-tellurite glass system. *J. Lumin.* 187:441–8.

84. Dandrade, B. W., and Forrest, S. R. 2004. White organic light-emitting devices for solid-state lighting. *Adv. Mater.* 16:1585–95.

85. Schubert, E. F., and Kim, J. K., 2005. Solid-state light sources getting smart. *Science* 308:1274–8.

86. He, M., Jia, J., Zhao, J., Qiao, X., Du, J., and Fan, X. 2021. Glass-ceramic phosphors for solid state lighting: A review. *Ceram. Int.* 47:2963–80.

87. Li, S., Wang, L., Hirosaki, N., and Xie, R. J. 2018. Color conversion materials for high-brightness laser-driven solid-state lighting. *Laser Photonics Rev.* 12:1800173.

88. Wang, G., Yang, X., Li, D. Y., Sun, X. Y., Fang, Y. X., Zhang, Y., and Su, S. C. 2020. High color gamut quantum dots LED and its application research in backlight display. *Spectrosc. Spect. Anal.* 40:1113–9 (in Chinese).

89. Zhang, L., Li, H., Xiang, X., Cheng, Y., Hua, C., Wang, C., Lin, S., Xu, J., and Wang, Y. 2019. Nanostructured NaF_3 glass ceramic: An efficient bandpass color filter for wide-color-gamut white LED. *J. Eur. Ceram. Soc.* 39:2155–60.

90. Refai, H. H. 2009. Static volumetric three-dimensional display. *J. Disp. Technol.* 5:391–7.

91. Downing, E., Hesselink, L., Ralston, J., and Macfarlane, R. 1996. A three-color, solid-state, three-dimensional display. *Science* 273:1185–9.

92. Tong, L., Li, X., Zhang, J., Xu, S., Sun, J., Zheng, H., Zhang, Y., Zhang, X., Hua, R., Xia, H., and Chen, B. 2017. $NaYF_4$:Sm^{3+}/Yb^{3+}@$NaYF_4$:Er^{3+}/Yb^{3+} core-shell structured nanocalorifier with optical temperature probe. *Opt. Express* 25:16047–58.

93. Tong, L., Li, X., Zhang, J., Xu, S., Sun, J., Cheng, L., Zheng, H., Zhang, Y., Zhang, X., Hua, R., Xia, H., and Chen, B. 2017. Microwave-assisted hydrothermal synthesis, temperature quenching and laser-induced heating effect of hexagonal microplate β-$NaYF_4$:Er^{3+}/Yb^{3+} microcrystals under 1550 nm laser irradiation. *Sensor. Actuat. B-Chem.* 246:175–80.

94. Schuyt, J. J., and Williams, G. V. M. 2021. Dual electrical and optical detection of ionizing radiation: Radiation-induced currents and radioluminescence in $NaMgF_3$: Sm. *Mater. Res. Bull.* 135:111122.

95. Greskovich, C., and Duclos, S. 1997. Ceramic scintillators. *Annu. Rev. Mater. Sci.* 27:69–88.

96. Chen, Y. P., and Luo, D. L. 2012. Development of containing 6 Li glass scintillators for neutron detection. *J. Inorg. Mater.* 27:1121–8.

97. Wantana, N., Kaewnuam, E., Kim, H. J., Kang, S. C., Ruangtaweep, Y., Kothan, S., Kaewkhao, J. 2020. X-ray/proton and photoluminescence behaviors of Sm^{3+} doped high-density tungsten gadolinium borate scintillating glass. *J. Alloys Compd.* 849:156574.

98. Kawano, N., Kawaguchi, N., Okada, G., Fujimoto, Y., and Yanagida, T. 2020. Radiation response properties of Dy-doped B_2O_3-Al_2O_3-SrO glasses. *Optik* 224:165613.

99. Teng, L. M., Zhang, W. N., Chen, W. P., Cao, J. K., Sun, X. Y., and Guo, H. 2020. Highly efficient luminescence in bulk transparent Sr_2GdF_7:Tb^{3+} glass ceramic for potential X-ray detection. *Ceram. Int.* 46:10718–22.

100. Saha, S., Kim, H. J., Aryal, P., Tyagi, M., Barman, R., Kaewkhao, J., Kothan, S., and Kaewjaeng, S. 2020. Synthesis and characterization of borate glasses for thermal neutron scintillation and imaging. *Radiat. Meas.* 134:106319.

101. Pringsheim, P. 1929. Two comments on the difference between luminescence and temperature radiation. *Z. Phys. A* 57:739–46 (in German).

102. Epstein, R. I., Buchwald, M. I., Edwards, B. C., Gosnell, T. R., and Mungan, C. E. 1995. Observation of laser-induced fluorescent cooling of a solid. *Nature* 377:500–3.

103. Fernandez, J., Garcia-Adeva, A. J., and Balda, R. 2006. Anti-Stokes laser cooling in bulk erbium-doped materials. *Phys. Rev. Lett.* 97:033001.

104. Sui, G. Z., Li, X. P., Cheng, L. H., Sun, J. S., Zhang, J. S., Zhang, X. Q., Xia, H. P., Hua, R. N., and Chen, B. J. 2015. Theoretical evaluation on laser cooling of ZBLAN:Er^{3+} glass with in situ optical temperature sensing. *Sensor. Actuat. B-Chem.* 220:362–8.

105. Ha, S. T., Shen, C., Zhang, J., and Xiong, Q. H. 2016. Laser cooling of organic-inorganic lead halide perovskites. *Nat. Photon.* 10:115–22.

106. Zhang, X., Lin, M., Guo, L., Zhang, Y., Cheng, C., Sun, J., Cheng, Y., Cao, Y., Xu, S., Li, X., Zhang, J., and Chen, B. 2021. Long-wavelength pass filter using green $CsPbBr_3$ quantum dots glass. *Opt. Laser Technol.* 138:106857.

107. Wang, C. Y., and Tao, Y. 2006. *Design and adjustment for glass composition.* Beijing: Chemical Industry Press. Chapter 11 (in Chinese).

108. Yu, J. B., Luo, M. T., Lv, Z. Y., Huang, S. M., Hsu, H. H., Kuo, C. C., Han, S. T., and Zhou, Y. 2020. Recent advances in optical and optoelectronic data storage based on luminescent nanomaterials. *Nanoscale* 12:23391–423.

109. van de Nes, A. S., Braat, J. J. M., and Pereira, S. F. 2006. High-density optical data storage. *Rep. Prog. Phys.* 69:2323–63.

110. Fjjita, K., Tanaka, K., Hirao, K., and Soga, N. 1998. High-temperature persistent spectral hole burning of Eu^{3+} ions in silicate glasses: New room-temperature hole-burning materials. *J. Opt. Soc. Am. B* 15:2700–5.

111. Ueda, J., Hashimoto, A., and Tanabe, S. 2019. Orange persistent luminescence and photodarkening related to paramagnetic defects of nanodoped CaO-Ga_2O_3-GeO_2 glass. *J. Phys. Chem. C* 123:29946–53.

112. Yu, J., Luo, M., Lv, Z., Huang, S., Hsu, H., Kuo, C. C., Han, S. T., and Zhou, Y. 2020. Recent advance on optical and optoelectronic data storage based on luminescent nanomaterials. *Nanoscale* 12:23391–423.

18 X-Ray Phosphors and Scintillators

Mingwei Wang and Yuxue Liu

CONTENTS

18.1 INTRODUCTION

X-ray phosphors and scintillators are applied in devices for high-energy physics (HEP), astrophysics, industrial nondestructive inspection and medical diagnosis.[1–4] Generally, the term "X-ray phosphors" was used when the applications required screens fabricated by phosphors in the form of powders[5] and the term "scintillators" was adopted when the related device was constructed by luminescent materials in the form of single crystal, ceramics or glass.[6–8] Because X-ray phosphors or scintillators can be defined as luminescent materials, which can convert energy of absorbed X-rays or ionizing radiation into low-energy ultraviolet (UV) or visible emission, the two terms can be used in an interchangeable way.

At present, with rapid developments of preparation and detection methods and medical applications involving in X-ray phosphors and scintillators, more multidisciplinary studies have been concentrated on the efforts to acquire high-quality X-ray phosphors and scintillators and to understand the fundamental mechanisms of different scintillation processes.

DOI: 10.1201/9781003098676-18

18.2 X-RAY PHOSPHORS

Because X-ray phosphors can be defined as materials, which absorb X-rays and convert the high-energy ionizing radiation into low-energy UV or visible emission,[9] it is convenient and effective to detect ionizing radiation, and they have exhibited significant applications in high-energy physics, safety monitoring, mineral exploration, oil well exploration and medical imaging. In this section, we introduce some important X-ray phosphors and their optical properties and potential applications.

18.2.1 X-Ray Phosphors for Medical Imaging

18.2.1.1 Phosphors for X-Ray Fluorescent Screens

X-ray fluorescent screens are mainly used in radiological group health examination and in radiological fluoroscopy for luggage check at airports.[10] In X-ray fluorescent imaging technique, the intensity of X-rays passing through the object changes with the point-to-point absorption/reflection of the object to X-rays, which excite the fluorescent screen and produce fluorescence of different intensity.[11] In around 1930, $(Zn, Cd)S:Ag^+$ phosphors with bright yellow-green fluorescence under X-ray excitation were used in the X-ray fluorescent screens.[12] Their structure was shown in Figure 18.1.[13] $(Zn, Cd)S:Ag^+$ phosphors were dispersed in polymer adhesive glue and deposited on one side of plastic base. After drying, a transparent protective film with a thickness of about 20–30 μm was coated on the phosphors layer with the thickness of about 100–300 μm.

Usually, the molar ratio of ZnS to CdS determines the emission wavelength. When their molar ratio is 7:3, the emission peak is situated at 540 nm, matching the spectral sensitivity of human eye.[14] Because the brightness is a very important factor, X-ray phosphors with average grain size of 20–40 μm are required. In particular, terbium-doped gadolinium oxysulfide ($Gd_2O_2S:Tb^{3+}$) phosphors are preferred to be used rather than $(Zn, Cd)S:Ag^+$ for X-ray fluorescent screens. $Gd_2O_2S:Tb^{3+}$ phosphors were synthesized through homogenous precipitation method via two steps.[15] First, $Gd(NO_3)_3$ and $Tb(NO_3)_3$ aqueous solution with a certain molar ratio were weighed and mixed, then they were diluted to deionized water and heated for 30 min at 90°C. After that, urea solution was added to the mixed solution, and they were stirred for 30 min at 90°C. The obtained solution was aged overnight. The precipitates appeared on the bottom of the beaker. Then, they were separated by centrifugation, washed and dried at 100°C for 24 h. Mixture of oxalates was then thermally decomposed to obtain the yellowish $Gd_2O_3:Tb^{3+}$ powders. Second, the mixture of sulfur and $Gd_2O_3:Tb^{3+}$ powders was placed into a quartz tube and the sulfurization reaction was performed at 900°C in argon atmosphere. After the reaction was kept for 1 h, the sulfurization was stopped and only the argon gas flow was supplied until the sample was cooled to room temperature.

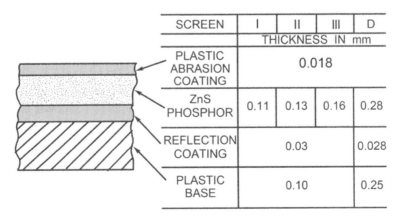

SCREEN	I	II	III	D
	THICKNESS IN mm			
PLASTIC ABRASION COATING	0.018			
ZnS PHOSPHOR	0.11	0.13	0.16	0.28
REFLECTION COATING	0.03			0.028
PLASTIC BASE	0.10			0.25

FIGURE 18.1 Structure of an X-ray fluorescent screen. (Reproduced from Herglotz, H. K., *Rev. Sci. Instrum.* 39, 1658–1659, 1968. With the permission of AIP Publishing.)

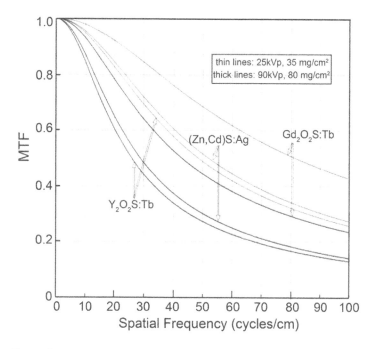

FIGURE 18.2 Comparison of the sharpness of $Gd_2O_2S:Tb^{3+}$ and $(Zn, Cd)S:Ag^+$ fluorescent screens, using a 90 kVp W spectrum with 80 mg/cm screen thickness for general radiographic conditions and a 25 kVp Mo spectrum with 35 mg/cm screen. (Kandarakis, I., Cavouras, D., Panayiotakis, G. S. and Nomicos, C. D., *Phys. Med. Biol.* 42, 1351–1373, 1997. With permission.)

$Gd_2O_2S:Tb^{3+}$ phosphors exhibit narrow emission bands in the range of 400–700 nm originating from the transitions from the $^5D_3/^5D_4$ state to the ground state ($^7F_{j\ (=6, 5, 4, 3, 2, 1, 0)}$) of Tb^{3+}. In particular, the strong green emission band (~546 nm) originates from the $^5D_4 \rightarrow {}^7F_5$ transition of Tb^{3+}. The $Gd_2O_2S:Tb^{3+}$ phosphor-based fluorescent screens are superior to that of $(Zn, Cd)S:Ag^+$ phosphor screens involving in the following parameters, speed,[16] which is the product of X-ray absorption coefficient and emission efficiency, and sharpness,[17] which is characterized by modulation transfer function (MTF). On the basis of the fact that MTF is the capacity of the detector to transfer the modulation of the input signal at a given spatial frequency to its output, the comparative sharpnesses of fluorescent screens fabricated by $Gd_2O_2S:Tb^{3+}$ and $(Zn, Cd)S:Ag^+$ phosphors are given in Figure 18.2.[17]

The phosphors for X-ray fluorescent screens require the following characteristics.[18,19]

1. High emission efficiency, which can minimize the patient's radiation exposure dose and can obtain a radiograph of high quality image.
2. High X-ray absorption, which can make luminescent materials absorb more X-ray energy and turn them into visible light.
3. Emission spectrum matched to the spectral sensitivity of the radiographic film.
4. High intensifying factor (radiographic speed), S, which can be defined as given in Eq. (18.1).

$$S = \int E_\lambda \cdot F_\lambda d\lambda \tag{18.1}$$

where E_λ is the emission spectrum of the fluorescent screen and F_λ is the spectral sensitivity of the radiographic film.

5. Short afterglow time, which can decrease double-exposure with the preceding image and can lower the contrast by raising the base density of the radiographic film.
6. Durability, which makes the phosphors resistant to X-ray radiation and moisture.
7. Narrow particle distribution of the phosphors, which can make the phosphor particles in the luminescent layer on the screen evenly distributed and densely arranged.

18.2.1.2 Phosphors for X-Ray Intensifying Screens

X-ray intensifying screens were used to improve X-ray response speed of radiographic film, thereby it led to the shortened irradiation time of the detected object and reduced dose of X-ray radiation received by human body.[20] Their structure is similar to that of X-ray fluorescent screen as shown in Figure 18.3. The radiographic film is sandwiched between two intensifying screens, each of which is coated with a layer of X-ray luminescent material on a supporter base, afterward a protective surface layer is applied to the surface. A radiographic film is coated with silver halide photosensitive emulsion layers on both sides. The phosphors for X-ray intensifying screen require the same characteristics like X-ray fluorescent screens. The typical phosphors include $CaWO_4$, $(Zn, Cd)S:Ag^+$, $BaFCl:Eu^{2+}$, $LaOBr:Tb^{3+}$ and $Gd_2O_2S:Tb^{3+}$.[21]

In 1896, $CaWO_4$ was the first compound to be used as an X-ray luminescent material.[12] It contains heavy atomic tungsten with a density of 6.1 g/cm^3. Its emission spectrum includes a broadband blue emission peak at 420 nm, which is very sensitive to the spectral sensitivity of blue-sensitive radiographic film. Because of its stable chemical properties, radiation resistance, low price and easy availability, it has been used up to now. $CaWO_4$ were synthesized by wet precipitation and high-temperature solid-phase method. The details can be found in 2nd edition of phosphor handbook.[19]

The emission spectra of $BaFX:Eu^{2+}$ (X = Cl, Br and I) exhibit broad bands at 390 nm, which is very similar to the spectral sensitivity of blue-sensitive radiographic film.[22] Their intensifying factors are four times that of $CaWO_4$.[21] Because their emission wavelengths are relatively shorter and their crossover effect is smaller, their image sharpnesses are better. However, the intensifying screen is not dense due to the large particle size distribution of the phosphors, it will lead to the increased scattering of X-ray fluorescence in the luminescent layer and decreased resolution. Although the density of $BaFCl:Eu^{2+}$ is 4.7, it still has a higher X-ray absorption efficiency in the wavelength region of medical X-rays due to the X-ray K absorption edge of Ba ion located at 37.4 keV, as shown in Figure 18.4(a).[21]

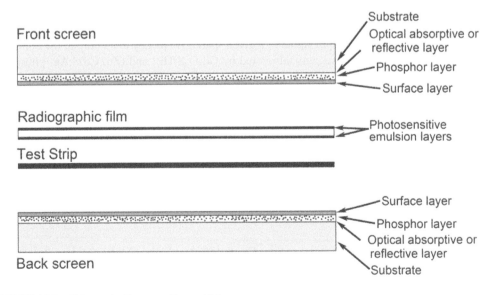

FIGURE 18.3 Schematic diagram of intensifying screens.

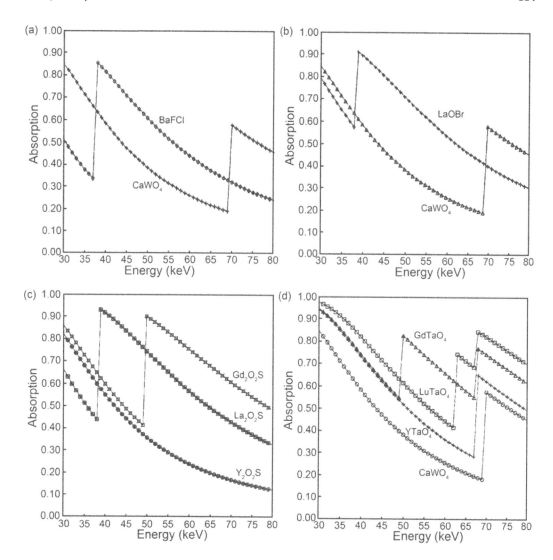

FIGURE 18.4 X-ray absorption coefficients of different phosphors ((a) BaFCl and CaWO$_4$ (b) LaOBr and CaWO$_4$ (c) Y$_2$O$_2$S, La$_2$O$_2$S and Gd$_2$O$_2$S (d) GdTaO$_4$, LuTaO$_4$, YTaO$_4$ and CaWO$_4$) for X-ray intensifying screens with 200 μm. (Brixner, L. H., *Mater. Chem. Phys.* 16, 253–281, 1987. With permission.)

LaOBr:Tb^{3+} and LaOBr:Tm^{3+} phosphors with green and blue emissions have already be used in commercial X-ray intensifying screens.[23] Rabatin et al. found that LaOBr:Tm^{3+} exhibits blue emission with a shorter wavelength compared to Tb^{3+}-doped LaOBr, so it is more effective in reducing the ambiguity caused by cross-effect. The comparative X-ray absorption spectra of LaOBr and CaWO$_4$ are shown in Figure 18.4(b).[21] Moreover, LaOBr:Tm^{3+} is the fastest X-ray intensifying screen, which is four times as fast as CaWO$_4$. However, due to the shortcomings of halide phosphors easily degraded by water, various moisture-proof treatments were required.

Rare earth–doped LaOBr phosphors were prepared by a solid state sintering method. For example, La$_2$O$_3$, NH$_4$Br, Tb$_4$O$_7$ and a small amount of KBr were evenly mixed, and the mixture was put into the crucible and heated at 450°C for 2 h. Then it was heated to 1000°C for 30 min. After cooling, LaOBr:Tb^{3+} powders were soaked in diluted hydrochloric acid, washed with water and washed with ethanol and then dried in vacuum.[21]

In addition, R_2O_2S (R = Y, La, Gd) are chemically very stable and insoluble in water; thus, Y_2O_2S, La_2O_2S and Gd_2O_2S were used as good matrix for luminescent materials. Because their densities were 4.9, 5.73 and 7.34 g/cm^3, respectively, they are highly efficient at absorbing X-rays as shown in Figure 18.4(c).[21] Meanwhile, Eu^{3+}, Tb^{3+} and other rare-earth ions are easy to be introduced into them. For example, Gd_2O_2S:Tb^{3+} is a typical X-ray luminescent material. Especially, these phosphors easily can be synthesized by coprecipitation and high-temperature solid-phase method, the details can be found in 2nd edition of phosphor handbook.[19]

Because $RTaO_4$ (R = Y, La, Gd) hosts contain Ta, which has an atomic number of 73, and their K absorption edges are around 60 keV, they also have higher X-ray absorption coefficient compared to $CaWO_4$ as shown in Figure 18.4(d).[21] For example, for Nb-doped $YTaO_4$ phosphor, its emission spectrum exhibits a broad band at 420 nm originating from the charge transfer state transition of NbO_4^{3-} ions. Its X-ray luminescence efficiency reach to 8.9%, suggesting it is a good X-ray luminescence material. These $RTaO_4$ (R = Y, La, Gd) materials can be prepared by solid-solid sintering reaction through mixing R_2O_3 and Ta_2O_5 powders at high temperature. Through doping different lanthanide elements into $RTaO_4$ (R = Y, La, Gd, Lu), the emissions from UV light to red light can be generated.[24]

18.2.1.3   X-Ray Phosphors for Photostimulable Storage Screens

X-ray photostimulable storage phosphors were usually used in the medical applications of computer X-ray imaging technology.[25] The first radiographic imaging by means of photostimulated luminescence (PSL) was reported by Berg et al. in 1947.[26] As BaFBr:Eu^{2+} was used to make an imaging plate, the first generation of computational X-ray imaging system was constructed in 1983.[22] After stopping high-energy irradiation, the radiographic image is stored in an imaging plate, served as a surface detector. Through point-by-point scanning the imaging plate using a He–Ne laser beam in an image reader, PSL emissions from the scanned pixels are acquired by a photomultiplier tube (PMT) and optical signals are converted into electric signals as a function of time.[21] By processing these signals through a computer, the digital information is reassembled into an optical image that can be displayed on a computer.[22]

For a storage phosphor after stopping high-energy particle irradiation, part of electron/hole pairs do not recombine to transfer their energy to UV or visible emission. Alternatively, their energy is stored in traps. When the storage phosphor is exposed to a red or near-infrared light or heat, the trapped electrons and/or holes in traps are released. Thus, the radiative recombination becomes possible and PSL is observed.[27]

In addition to the application of medical diagnosis, the computational X-ray imaging system can also be applied to the determination of X-ray single crystal structure.[28] Because image plate is a surface detector faster than a point detector and the radiation damage to the crystal is less, it is conducive to the determination of organic crystal structure, proteins, genes and other biological samples.

Many types of X-ray photostimulable storage phosphors have been reported as follows: BaFX:Eu^{2+} (X = Cl,Br,I),[29–31] CsBr:Eu^{2+},[32] $SrAl_2O_4$:Eu^{2+},Dy^{3+},[33] 12CaO·7Al$_2$O$_3$(C12A7):Tb^{3+},[5] Lu_2O_3:Tb^{3+},Pr^{3+},[34] $CaTiO_3$:Pr^{3+},[35] $LiTaO_3$:Bi^{3+},[36] YPO_4:Pr^{3+},Ln^{3+} (Ln = Nd, Er, Ho, Dy).[37] Typical photostimulable phosphors and their specific performances are given in Table 18.1.

The good photostimulable storage phosphors for X-ray imaging have to meet the following requirements:[27]

1. X-ray absorption efficiency and high emission efficiency are high, i.e., the compound contains heavy atoms.
2. Specific defects exist in the host, which can be used as traps for carriers.
3. The stimulation wavelength should be far away from excitation wavelength to avoid the interference of photoluminescence to PSL signals.
4. The decay time of luminescence should be shorter than 1 μs to avoid the overlap of luminescence of adjacent scanning points when the laser beam is on.

TABLE 18.1
Typical X-Ray Photostimulable Storage Phosphors and Their Performances

Phosphors	Peak Wavelength of Stimulation Spectrum (nm)	Peak Wavelength of Emission Spectrum (nm)	Decay Time of PSL	References
$BaFBr:Eu^{2+}$	600	390	0.8 μs	[22]
$BaFCl:Eu^{2+}$	550	385	7.4 μs	[22]
$BaFI:Eu^{2+}$	610, 660	410	0.6 μs	[22]
$CsI:Na^+$	720	338	0.7 μs	[38]
$12CaO \cdot 7Al_2O_3:Tb^{3+}$	N/A	541	2.26 μs	[5]
$RbBr:Tl^+$	680	360	0.3 μs	[39]
$LiTaO_3:Bi^{3+}$	N/A	430	1.46, 5.78 ns	[36]
$Lu_2O_3:Tb^{3+}$	N/A	545	N/A	[40]
$Lu_2O_3:Pr^{3+}$	N/A	631	N/A	[41]
$SrAl_2O_4:Dy^{3+}, Eu^{2+}$	532, 1064	520	0.108 μs	[33]
$LiLuSiO_4:Ce, Tm$	N/A	401	N/A	[42]
$Y_2SiO_5:Ce^{3+}$	<500, 620	410	0.035 μs	[43]
$CsBr:Eu^{2+}$	680	440	0.7 μs	[32, 44]
$ZnGa_2O_4:Mn^{2+}$	N/A	505	51 ms	[45]
$BaBr_2:Eu^{2+}$	580, 760	400	0.5 μs	[46]
$BaCl_2:Ce^{3+}$	570	~349, ~373	N/A	[47]
$RbI:Tl^+$	730	420	N/A	[48]
$LiGa_5O_8:Cr^{3+}$	N/A	716	N/A	[44]
$CaS:Eu^{2+},Sm^{3+}$	Broad NIR region	650	0.31 μs	[48]
$Ba_3(PO_4)_2:Eu^{2+}$	475 nm but has a tail extending to 800 nm	410	N/A	[49]

5. Luminescent intensity has a broad linear relationship with the change of X-ray radiation dose.
6. The phosphor should be stable under some working conditions, i.e., its performances cannot degrade when it is exposed to humidity and daylight.

The conventional photostimulable storage phosphors, $BaFX:Eu^{2+}$ (X = Cl, Br, I), are layered materials with a tetragonal PbFCl or Matlockite structure. The emission and stimulation spectra (excitation spectra of PSL) of $BaFCl:Eu^{2+}$, $BaFBr:Eu^{2+}$ and $BaFI:Eu^{2+}$ phosphors are shown in Figure 18.5.[50] Their emission peaks around 400 nm are due to the 5d $\rightarrow$ 4f allowed transition of Eu^{2+} ions. In particular, it is found that $BaFCl:Eu^{2+}$ exhibited the shortest emission and stimulation wavelengths among these phosphors. For $BaFCl_{1-x}Br_x:Eu^{2+}$, strong photostimulable luminescence can be observed under red or near infrared (NIR) laser excitation. Furthermore, the decay time of $BaFCl:Eu^{2+}$ is 8 μs.[51] With increasing Br^- doping concentration, the decay time gradually becomes short. When $x = 1$, the decay time of $BaFBr:Eu^{2+}$ (0.75 μs) is the shortest.[52]

In 1984, Takahashi et al. determined the energy level diagram of $BaFBr:Eu^{2+}$ phosphors through their optical properties, electrical characteristics and electron spin resonance tests.[53] Subsequently, they gave the X-ray storage and photoluminescence mechanisms of $BaFBr:Eu^{2+}$ as shown in Figure 18.6.[54] They believed that, as X-rays irradiate $BaFBr:Eu^{2+}$, Eu^{2+} ions are ionized and converted to Eu^{3+} ions either directly or by trapping holes. For the electrons excited to the conduction band, part of them are trapped by halogen vacancies (V_F^- and V_{Br}^-) in the crystal to form F centers. At the same time, part of holes in the valence band are captured by Eu^{2+}, and Eu^{2+} are converted to Eu^{3+}, thereby X-ray energy is temporarily stored in the crystal. When $BaFBr:Eu^{2+}$ was stimulated by He–Ne laser (wavelength of 632 nm) irradiation to acquire PSL, trapped electrons are released from

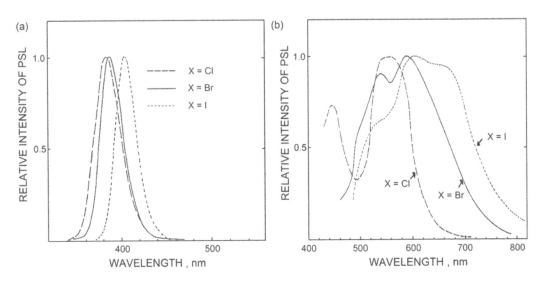

FIGURE 18.5 (a) Emission and (b) stimulation spectra of BaFCl:Eu^{2+}, BaFBr:Eu^{2+} and BaFI:Eu^{2+}. (Kato, H., Miyahara, J. and Takano, M., *Neurosurg. Rev.* 8, 53–62, 1985. With permission.)

F centers to the conduction band and then they combine with Eu^{3+} luminescence centers to form the excited Eu^{2+}. After the electrons in the excited state of Eu^{2+} relax to its 4f^{6}5d level, the 4f^{6}5d–4f^7 transition happens and blue fluorescence at 390 nm can be observed.

In addition, von Seggern et al. proposed that, for X-ray irradiated BaFBr:Eu^{2+}, the electrons are captured by F$^+$ center to form F center (V_{Br}^- + e$^-$), and the holes are combined with Eu^{2+} to become Eu^{3+}.[55] After NIR laser excitation, the trapped electrons are captured by Eu^{3+} ions through a tunneling process, and then PSL is found. During this time, the electrons are not excited to the conduction band. Afterwards, Wang et al. proposed a multi-tunneling model based on von Seggern's tunneling model.[56]

For photostimulated storage screen fabricated by BaFBr:Eu^{2+} phosphors synthesized by high-temperature sintering in an inert atmosphere, the image resolution does not exceed 5 lp/mm, and its sharpness is poor.[57] Therefore, the particle morphology of luminescent materials and the production process of image plate should be improved. Many preparation methods have been reported,

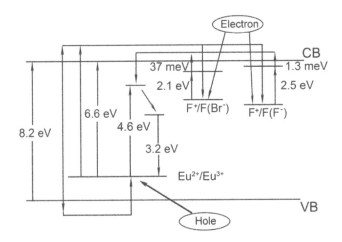

FIGURE 18.6 The energy level diagram and the PSL process in BaFBr:Eu^{2+}. (Iwabuchi, Y., Mori, N., Takahashi, K., Matsuda, T. and Shionoya, S., *Jpn. J. Appl. Phys.* 33, 178–185, 1994. With permission.)

such as nonstoichiometric sintering, coprecipitation of water-based solutions and one-step synthesis method by annealing $BaCO_3$ and ammonium halides with trace amounts of EuF_3 in a reducing atmosphere.[58,59] Wang et al. prepared nanocrystalline $BaFBr:Eu^{2+}$ phosphors with an average size of about 200 nm using one-step synthesis method.[29] This in turn may render higher sensitivities and higher resolution in computed radiography, due to the smaller particle size, in comparison with the ~5 μm particles that are typically used in commercial imaging plates.

In addition to improving the fabrication of conventional materials, it is also important to develop new materials for photostimulable storage screens. For example, $12CaO \cdot 7Al_2O_3$ (abbreviated as C12A7) is one kind of new materials. Its unit cell consists of 12 cages forming the positively charged framework and two free O^{2-} anions occupying randomly 2 of the 12 cages to maintain the charge neutrality.[60] The equivalent charge of each cage is +1/3, which can be regarded as F^+ center that can capture electrons.[61] In 2012, McLeod et al. found the existed cage conduction band (CCB) in the forbidden band in C12A7.[62] Then, Li et al. reported that the special cage existed in C12A7 not only can be used as an electron trap but also can modulate different optical and electrical properties by loading different anion groups in the cages. They also studied the effects of different encaged anions on the optical storage performances of $C12A7:0.5\%Tb^{3+}$ phosphors, discussed the related luminescence mechanism and realized X-ray image with higher spatial resolution (15 lp/mm) as shown in Figure 18.7.[63] They found that its X-ray absorption capacity was increased and its X-ray dynamic range was extended after Sr doping.[64]

Due to the special structure of cubic Lu_2O_3, which contains oxygen vacancies, it has attracted widespread attention. Especially, because noncentrosymmetric C_2 and centrosymmetric S_6 sites exist, there are two different sites for Eu^{3+} replacing Lu^{3+} in the host.[65] Despite the population of C_2 sites triples the abundance of S_6 ones in the host, it was found that Eu^{3+} entering the host at high concentrations has a tendency to preferentially occupy C_2 sites.[66] In addition, Zych et al. reported that $Lu_2O_3:Pr^{3+}$, Hf^{4+} phosphors prepared at 1700°C in a reducing atmosphere showed a long-term energy storage for 9 months.[41]

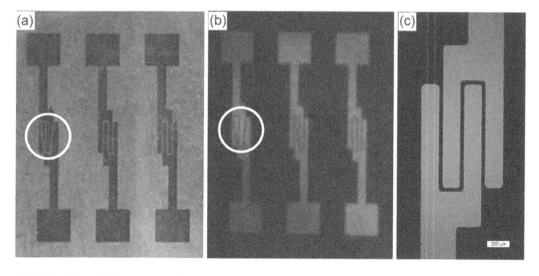

FIGURE 18.7 (a) X-ray images (upon 808 nm stimulation) with carved four-crossed-finger-fork pattern formed on the surfaces of the $C12A7:0.5\%Tb^{3+}$ pellets treated under different conditions with the delay times of 2 and 8 days, respectively, after stopping X-ray irradiation. (b) The optimized X-ray image with a 30-min delay time after stopping X-ray irradiation for 30 min. (c) The steel mask plate with a pattern of carved four-crossed-finger-fork. The width of each finger in the middle of the pattern is 160 mm and the width of the concave part of each fork is 220 mm. (Li, S., Liu, Y., Liu, C., Yan, D., Zhu, H., Yang, J., Zhang, M., Xu, C., Ma, L. and Wang, X., *Mater. Des.* 134, 1–9, 2017. With permission.)

18.2.2 X-Ray Phosphors for Thermoluminescent Dosimetry

When X- or γ-ray irradiated luminescent materials were heated, thermoluminescence (TL) can be observed. For the detection of high-energy particle rays, the determination of their doses is one of the important fields for TL applications. Usually, various types of high-energy particles have different ionization capabilities and penetrating abilities.[67] Thus, there are two types of radiation dosimeters designed by using luminescence techniques: ray luminescence dosimeter and thermoluminescent dosimeter (TLD). The former is a direct measurement of dose magnitude by monitoring radioluminescence intensity of scintillators upon X- or γ-ray irradiation. The latter is to measure dose scale by recording the TL intensity of X-ray phosphors after stopping high-energy particles irradiation. Due to the small size and simplicity of TL dosimeters, it is widely used in various fields such as personal dosimetry, environmental radiation monitoring, nuclear medicine, high-altitude aerospace and geology.[68,69]

Many standard commercial dosimeters are now available, such as LiF:Mg,Cu,P (TLD-700H), Al_2O_3:C (TLD-500), $CaSO_4$:Dy (TLD-900) and CaF_2:Dy (TLD-200) phosphors.[69–71] Each of these phosphors merely was used in a narrow dose range. Therefore, it is essential to develop new X-ray phosphors whose TL intensity exhibits a linear response over a wide dose range. For TL phosphors, they also need to meet other requirements such as small energy dependence (sufficient sensitivity), repeatability, high stability and low fading besides wide dose range.[67,72] Among them, small energy dependence of X-ray phosphors is a key characteristic factor and plays an important role in designing dosimeters. Usually, the larger the effective atomic number of phosphors, the larger the dependence of its sensitivity on the radiation energy. For TL dosimeter designed by phosphors with high-energy dependence, X-ray phosphors are usually added with a "shield" to reduce or even eliminate their strong dependence on energy. Typical photon energy dependences of LiF:Mg, Ti (TLD-100) and (LiF:Mg,Cu,Na,Si) TL detectors with small energy response are shown in Figure 18.8.[73]

For TL phosphors like TLD-100, a single-peaked glow curve and its peak position located near 200°C (~473K) are preferred as shown in Figure 18.9.[74] When the glow curve is consisted of

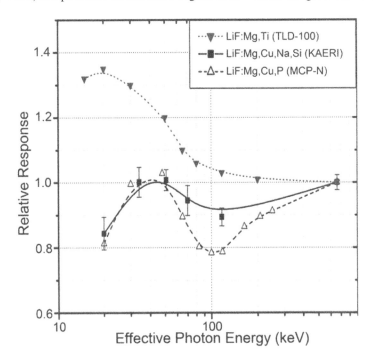

FIGURE 18.8 Photon energy response of LiF:Mg,Cu,Na,Si TL detector, LiF:Mg,Ti (TLD-100) and LiF:Mg,Cu,P (MCP-N, Poland) in bare condition. Responses are normalized to 662 keV photons from a ^{137}Cs source. (Jung, H., Lee, K. J. and Kim, J. L., *Appl. Radiat. Isot.* 59, 87–93, 2003. With permission.)

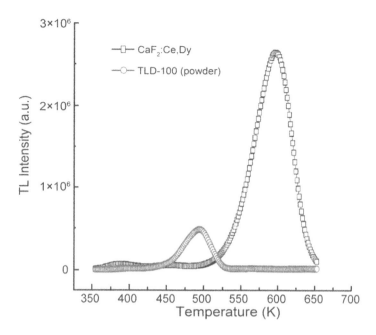

FIGURE 18.9 TL intensity of CaF_2:Ce,Dy compared with TL intensity of TLD-100, dose 1 Gy. (González, P. R., Mendoza-Anaya, D., Mendoza, L. and Escobar-Alarcón, L., *J. Lumin.* 195, 321–325, 2018. With permission.)

multipeaks on the low-temperature side, an undesirable fading effect occurs. While, for the glow curve, if overlapped multipeaks on the high-temperature side exist, longer time is required to dissipate the energy remaining after stopping radiation.[19]

For LiF:Mg,Cu,P,B and LiF:Mg,Cu,P phosphors, their linear ranges are in the region of 0.1–10 Gy.[75] For Dy-doped $CaNa_2(SO)_4$ phosphors synthesized by a coprecipitation method, the TL intensity as a function of γ-ray dose ranging from 10 Gy to 30 kGy exhibited a liner response in a wide range as shown in Figure 18.10.[76] While the dose response becomes nonlinear above 10^4 Gy. On the basis of the fact that the intensity of TL increases linearly with increasing irradiation dose, the irradiation dose is detected. For a dosimeter, it is preferred to have a larger dose range. However, many TLD materials only exhibit a linear relationship within a certain dose range. Above certain doses, this relationship exhibits superlinearity or sublinearity. The reason for this superlinearity can be explained by the fact that a new glow peak appears on the high-temperature side, i.e., trapping process of multi-traps exists.[77]

The phenomenon of TL fading in the dark is related to the shape of the glow curve in the low-temperature region (shallow traps). That is to say, for phosphors with shallow traps, fading is caused by the gradually thermal activation of trapped electrons at room temperature.[78] This phenomenon complicates the correction process in the measurement of radiation. Furthermore, fading also happens through exposure to visible light. This is because the charge carriers excited by the radiation can be released not only by heat, but also by optical energy.[45]

For different TL phosphors, their TL mechanisms might be different. Usually, they are related to thermal activation and tunneling processes.[79] Here, we introduce the thermal activation mechanism in the case of TL originating from a first order kinetic process (trapped electrons participate in luminescence once they are thermally released). The luminescent intensity as a function of temperature, $I(T)$, can be expressed to be Eq. (18.2).[80]

$$I(T) = n_0 s \exp\left(\frac{-\varepsilon}{kT}\right) \exp\left[\left(-\frac{s}{\beta}\right)\int_{T_0}^{T} \exp\left(\frac{-\varepsilon}{kT'}\right)dT'\right] \tag{18.2}$$

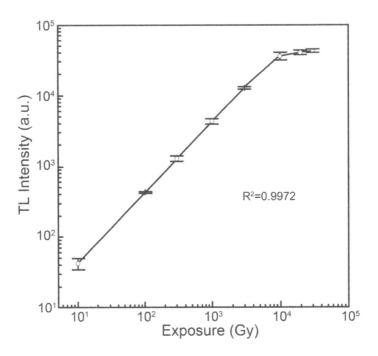

FIGURE 18.10 TL response curve for $CaNa_2(SO_4)_2$:Dy nanophosphor. (Bhadane, M. S., Dahiwale, S. S., Sature, K. R., Patil, B. J., Bhoraskar, V. N. and Dhole, S. D., *Radiat. Meas.* 96, 1–7, 2017. With permission.)

where n_0 is the number of trapped electrons at the temperature, T, when the excitation is applied, $s \cdot \exp(-\varepsilon/kT)$ is the probability of releasing from the captured electrons in the trap (s is frequency factor and ε is trap depth), T_0 is the temperature during excitation and $\beta = dT/dt$ is the rate of temperature increasing.

Finally, commonly used TLD phosphors and their related parameters are given in Table 18.2.

18.2.3 PHOSPHORS FOR X-RAY-INDUCED PHOTODYNAMIC THERAPY

At recently, more researches are focused on X-ray phosphors for photodynamic therapy (PDT).[97–100] For example, in X-ray-induced PDT (X-PDT), after the nanosensitizer systems have accumulated at the tumor site, X-rays will be introduced to stimulate the scintillating nanophosphors (ScNPs) to generate emission light. The emission light can be harvested to activate the nearby photosensitizer (PS) ant it leads to $^1O_2$ generation (Figure 18.11(a)),[101] which is very toxic to the tumor cells or induces tissue ischemia.[102] There are two types of reactions (as shown in Figure 18.11(b)) that occur after PS absorbs energy. Electrons from the ground state (S_0) are pumped to singlet-excited states (S_1). Some of the absorbed energy will be released via intersystem crossing and the promoted electrons will move to a triplet-excited state (T_1). In type 1 reaction, the excited PS transfers an electron to surrounding biomolecules, producing reactive oxygen species, such as hydroxyl radicals, hydrogen peroxide.[103] Alternatively, in type II pathway, an excited PS transfers energy directly to molecular oxygen, forming $^1O_2$, which is extremely reactive due to the pairing of two electrons into one of the antibonding orbitals.[104]

Conventional PDT is activated by visible light that has poor tissue penetration depths, limiting its applicability in the clinic. In 2006, the idea of using scintillator nanoparticles to stimulate PSs was first proposed by Chen et al.[105] Unlike visible or NIR light, X-rays afford superior tissue penetration and can potentially expand the scope of PDT in clinical applications. The demonstrations of X-PDT for *in vivo* tumor therapy using $SrAl_2O_4$:Eu^{2+} nanoscintillators are shown in Figure 18.12.[106]

TABLE 18.2
TLD Materials and Their Related Parameters

Materials	Dosimetry Peak/°C	Z_{eff}	Sensitivity	Dosage Range	Application	Fading	References
LiF (TLD-100)	195	8.2	1.0	0.1 mGy–1 kGy	Medical personnel and patient dosimetry	Negligible	[81]
LiF:Mg,Cu,P (TLD-100H)	195	8.2	15	0.1 μGy–10 kGy	Environmental radiation monitoring	Negligible	[82]
CaF$_2$:Tm (TLD-300)	129, 153 and 194	16.3	N/A	1 Gy–1 kGy	Mixed field dosimetry	Fading of 5% for the third peak and 15% for the second peak of this phosphor have been reported after storage time of 30 days	[83]
LiF:Mg,Cu,P (TLD-600H)	195	8.2	15	1 μGy–10 Gy	Dose measurement in fast neutron, γ rays radiation fields	Negligible	[84]
LiF:Mg,Ti (TLD-700)	195	8.2	1.0	Below ~100 μGy–3 kGy	Estimation of high doses of γ rays	5%/year	[85]
LiF:Mg,Cu,P (TLD-700H)	195	7.4	10	Below ~100 μGy–3 kGy	Clinical dosimetry	5%/year	[86]
Ca$_2$F:Dy (TLD-200)	180	16.3	~12.5	0.1 μGy–10 kGy	Estimation of high doses of γ rays	5%/50 days	[87]
Al$_2$O$_3$:C (TLD-500)	209	10.2	2.9	0.5 μGy–1 Gy	Environmental radiation monitoring	5%/year	[88]
Al$_2$O$_3$:Cr	201	10.2	N/A	100 Gy–20 kGy	Dosimetry of food and seed irradiations	2% in 30 days	[89]
CaSO$_4$:Dy (TLD-900)	220	15.5	20	1 μGy–1 kGy	Thermal neutron dosimetry	8% in 6 months	[90]
CaF$_2$:Mn (TLD-400)	260	16.3	1.7–13	1 μGy–3 kGy	Environmental radiation monitoring	15% in 2 weeks	[91]
NaLi$_2$PO$_4$:Ce	181, 220 and 297	10.8	8	0.1 Gy–1 kGy	Medical and environmental dosimetry, dosimetry of food irradiations	15% in 2 months	[92]
CaB$_4$O$_7$:Cu, Mn	124, 256	12.5	N/A	0.05 Gy–3 kGy	Personal and environmental dosimetry	12% and 23% of its original value after 1 and 2 month storage	[93]
Li$_2$B$_4$O$_7$:Mn (TLD-800)	220	15.5	~12.5	100 μGy–1 kGy	Neutron dosimetry	8%/half a year	[94]
MgB$_4$O$_7$:Mn, Tb	202, 377	8.23	N/A	0.1 Gy–5.0 kGy	Radiation dosimetry	~10% in a month after storing in dark at RT	[95]
CaNa$_2$(SO$_4$)$_2$	164	N/A	N/A	10 Gy–10 kGy	High radiation dosimetry	N/A	[76]
K$_2$Ca$_2$(SO$_4$)$_3$:Eu	160	N/A	N/A	2 Gy–1 kGy	Estimation of high doses of γ rays	N/A	[96]

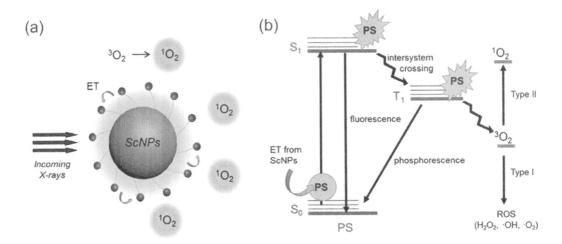

FIGURE 18.11   Scintillating nanoparticles act as an X-ray transducer to generate $^1O_2$ through the energy transfer process (a) and diagram of the PDT mechanism that occurs when energy is transferred from ScNPs to activate the PS. (b) The PS's electrons from the ground state (S_0) will absorb energy and move to singlet-excited states (S_1). Some of the absorbed energy will be released via intersystem crossing, and the promoted electron will move to a triplet-excited state (T_1). This triplet state has a relatively long half-life, allowing energy to be transferred to nearby oxygen molecules. This generates $^1O_2$ in most cases via the type II pathway, which can damage the cells in the surrounding area. (Reprinted with permission from Kamkaew, A., Chen, F., Zhan, Y., Majewski, R. L. and Cai, W., *ACS Nano* 10, 3918–3935, 2016. Copyright 2016 American Chemical Society.)

For lanthanides with high density and high atomic numbers, they are the most widely used elements in X-ray phosphors.[107] In particular, $ALnF_4$ (A = Alkaline, Ln = Lanthanide) nanoparticles are very suitable for X-PDT owing to their band gap in the range of 9–10 eV. For example, Zhong et al. synthesized $NaCeF_4$:Gd,Tb ScNPs by a solvothermal method, as described in Figure 18.13, and have realized multimodal imaging-guided X-PDT of deep tumors.[108] Wang et al. found that $LiLuF_4$:Ce@SiO_2@Ag_3PO_4@Pt(IV) nanoparticles could be used to enhance the curative effects of X-PDT.[109]

To enhance cancer cell killing efficiency and to reduce the radiation dose, nanoscintillators must meet the following requirements:[105]

1. Nanoscintillators must produce a strong radioluminescence upon X-ray excitation.
2. The emission spectrum of nanoscintillators must match the absorption spectrum of PS. For example, Ma et.al. found that, for TBrRh123-ZnS:Cu,Co system, due to the overlap of their emission and absorption spectra, as shown in Figure 18.14,[110] the efficient energy transfer from ZnS:Cu,Co nanoscintillators to TBrRh123 was realized.
3. Nanoscintillators absorb ionizing radiation more strongly than the surrounding tissue. Thus, it leads to the reduced radiation damage to the tissue for a given incident radiation dose.
4. Nanoscintillators must be easily attached to or linked with PSs.
5. Nanoscintillators must be non-toxic, water-soluble and stable in biological environments.

In addition, Liu et al. found that afterglow emission induced by X-ray irradiation also can realize X-PDT in the absence of external irradiation.[111] In the above case, for ZnS:Cu,Co nanoscintillators after stopping X-ray irradiation, the afterglow lasts a few hundred seconds as shown in Figure 18.15(a).[111] Meanwhile, it is found that the afterglow intensity can quickly increase with increasing irradiation time and almost reaches a saturated value as the irradiation time exceeds

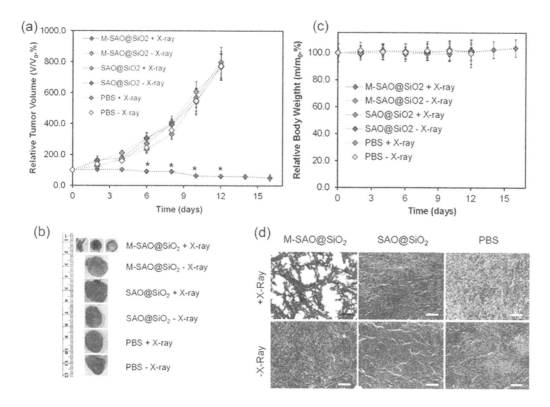

FIGURE 18.12 X-PDT for in vivo tumor therapy. (a) Tumor growth curves (V/V₀%, $n = 5$). Significant tumor suppression and shrinkage was observed with animals injected with M-SAO@SiO₂ nanoparticles and irradiated by X-ray. In all the control groups, tumors grew rapidly and in a comparable pace. By day 14, all the animals in the control groups had either died or been euthanized for meeting at least one humane end point. The error bars represent ± S.E.M. *$P < 0.05$. (b) Photographs of representative tumors taken from Groups 1–6. (c) Body weight curves. No significant decrease of body weight was observed with X-PDT-treated animals. The error bars represent ± S.E.M. (d) H&E staining on tumor tissues taken from Groups 1–6. Compared to all the controls, where densely packed neoplastic cells were observed throughout the mass, tumors treated by X-PDT manifested drastically impacted tumor architectures and significantly reduced cell density. Scale bars, 100 μm. (Reprinted with permission from Chen, H., Wang, G. D., Chuang, Y. J., Zhen, Z., Chen, X., Biddinger, P., Hao, Z., Liu, F., Shen, B., Pan, Z. and Xie, J., *Nano Lett.* 15, 2249–2256, 2015. Copyright 2015 American Chemical Society.)

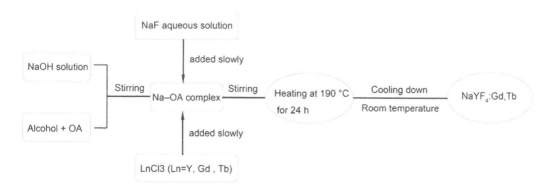

FIGURE 18.13 Production process of the NaYF₄:Gd, Tb phosphor.

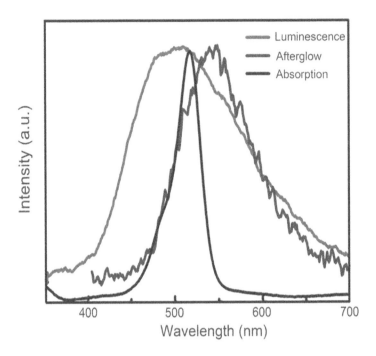

FIGURE 18.14 X-ray excited luminescence (red) and afterglow (blue) spectra of ZnS:Cu,Co nanoparticles. The absorption spectrum of TBrRh123 is displayed in black. The spectra are normalized in comparison. (Reproduced from Ma, L., Zou, X., Bui, B., Chen, W., Song, K. H. and Solberg, T., *Appl. Phys. Lett.* 105, 013702, 2014. With the permission of AIP Publishing.)

1 min as shown in Figure 18.15(b). It is striking that, for $Zn_2Ga_{3-x-y}Cr_xNd_yGe_{0.75}O_8$ nanoparticles, the X-ray excitation strategy accompanied by NIR-I afterglow emission exhibited high tissue penetration capability and NIR-I afterglow can last several hours.[112] Although the X-PDT technology is just starting to begin, many kinds of nanoscintillators for X-PDT exist. Many nanoscintillators for X-PDT and their related parameters are given in Table 18.3.

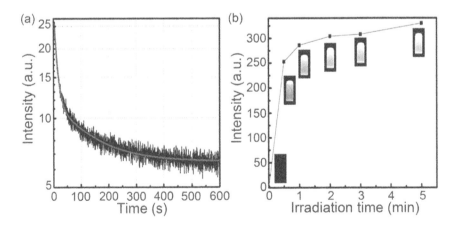

FIGURE 18.15 Afterglow decay of ZnS:Cu,Co nanoparticles excited by X-ray irradiation with fitting in red curve. Residuals are shown above the decay curve. (b) Afterglow intensities of ZnS:Cu,Co nanoparticles as a function of X-ray irradiation time. The inset photos are afterglow images taken by a low noise CCD camera after the X-ray irradiation is removed. (Reproduced from Ma, L., Zou, X., Bui, B., Chen, W., Song, K. H. and Solberg, T., *Appl. Phys. Lett.* 105, 013702, 2014. With the permission of AIP Publishing.)

TABLE 18.3

Nanoscintillators Used for X-PDT and Their Related Parameters

X-Ray Phosphors	Photosensitizer (Absorption)	Peak Wavelength (nm)	Particle Size (nm)	X-Ray Energetics	Preparation Method	References
$Tb_2O_3@SiO_2$	5-(4-Carboxyphenyl)-10,15,20-triphenyl porphyrin (520 nm)	540, narrow peaks	10	44 keV, 11 Gy	Chemical synthesis	[97]
$ZnS:Cu,Co$	TBrRh123 (518 nm)	510, a broad peak	4	120 keV, 2 Gy	Wet chemistry	[113]
$SrAl_2O_4:Eu^{2+}$	Merocyanine 540 (540 nm)	520, a broad peak	407	50 keV, 1–10 Gy	Carbothermal reaction using a vapor-phase deposition method	[106]
$LaF_3:Tb$	Rose Bengal (560 nm)	540, narrow peaks	40	75 keV	Hydrothermal method	[114]
$LiYF_4:Ce$	ZnO (290 nm)	305, 325 broad peak	35	220 keV, 8 Gy	Improved thermal decomposition method	[115]
CeF_3	Verteporfin (370, 420 nm)	340	9	6 MeV, 30 MeV, 1–6 Gy	Coprecipitation method	[98]
$LiGa_5O_8:Cr$	2,3-Naphthalocyanine (775 nm)	~720	100	50 keV, 5 Gy	Sol-gel method	[99]
Au	Verteprofin (365, 690 nm)	Not determined	12	6 MeV, 6 Gy	Chemical synthesis	[116]
$NaLuF_4:Gd,Eu@NaLuF_4:Gd@NaLuF_4:Gd,Tb$	Rose Bengal (560 nm)	543, narrow peaks	25	160 keV, 5 Gy	Thermal decomposition method	[100]
CdS	Tretrakis(o-aminophenyl) porphyrin (400 nm)	420	N/A	N/A	N/A	[105]
TiO_2–Au	Triphenylphosphine	525	20	N/A	Sol-gel method	[117]
SiC/SiO_x core/shell nanowires	Tetracarboxyphenyl porphyrin derivative (550 nm)	545	Diameter 40 nm	6 MeV, 0.4–2 Gy	Low-cost carbothermal method	[118]
GdEuC12 micelle	Hyp	N/A	4.6	400 mA	Chemical synthesis	[119]
$LaF_3:Ce^{3+}$	Protoporphyrin IX (409 nm)	520	2 μm	90 keV, 3 Gy	Wet chemistry	[120]

18.3 SCINTILLATORS

Luminescent bulk materials that emit UV or visible light under an action of high-energy particles become scintillators.[121] According to their structures, they can be divided into single crystals, glass and ceramic scintillators. Their applications are both widespread and at the forefront of new technologies, especially the application characteristics of high-energy physics experiments and medicine imaging. The following section will mainly focus on introducing inorganic crystal scintillators and their optical properties and applications.

18.3.1 INORGANIC CRYSTAL SCINTILLATORS

Inorganic crystal scintillators are currently widely used in many detection systems addressing different fields, such as nuclear, high-energy and astrophysics experiments, homeland security, industrial applications, medical imaging and non-destructive inspection.[122–124] More than one century after the first use of inorganic crystal scintillators, their researches are still very active as detector technologies are progressing. Because the functionalities and performances of ionizing radiation systems are changing, the related inorganic scintillator materials have been greatly improved and extended as shown in Figure 18.16.[12,122]

There are two kinds of emission mechanisms in scintillators: luminescence without activators (intrinsic luminescence) and luminescence with activators (extrinsic luminescence). The former represents free-exciton luminescence, selftrapped exciton (STE) luminescence, Auger-free luminescence and self-activation luminescence. The latter refers to defect-based luminescence originating from doped cations in the host.[125]

The scintillation properties required for practical detectors depend on the application. A scintillator is coupled with a photodetector, such as a PMT, a Si-photodiode (Si-PD) and complementary metal-oxide semiconductor (CMOS), which has a function to convert scintillation photons to electrons via a photoelectric conversion.[126–128] Although specific requirements for scintillation properties used in different fields are different, most of the applications using scintillators were based on light yield, emission wavelength, energy resolution, scintillation decay time, afterglow and high cross-section with the radiation type of interest.[4] Characteristics and required parameters of typical inorganic crystal scintillators are given in Table 18.4.

Generally, there are two methods for growing large-sized crystals: Czochralski method and Bridgman method. For example, oxides and binary compound halides crystals were grown by

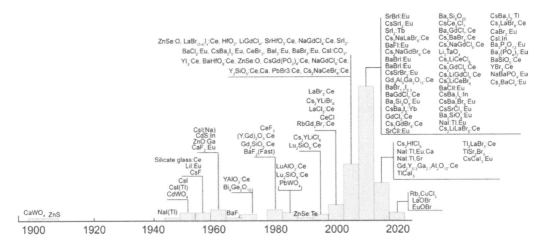

FIGURE 18.16 Development history of major inorganic scintillator materials. (See Table 18.4 for references.)

TABLE 18.4

Typical Scintillators and Their Required Parameters

Scintillators	Density (g/cm³)	Z_{eff}	Light Yield (Photons/MeV)	Decay Time (ns)	Emission Maximum (nm)	Radiation Length (cm)	Energy Resolution (%, at 662 keV)	Hygroscopicity	References
NaI:Tl⁺	3.67	50.8	40,000	230	415	2.59	6.7	Yes	[129]
CsI:Tl⁺	4.53	54.1	48,000	1050	550	1.86	6.6	Slightly	[130]
LiI:Eu²⁺	4.08	52.3	12,000	1400	470–485	2.18	~3.5	Yes	[131]
CaF₂:Eu²⁺	3.19	17.1	19,000	940	420	6.72	N/A	No	[132]
BaF₂	4.89	52.7	~3333	0.6/620	220/310	2.03	10	Yes	[133]
CeF₃	6.16	53.3	2000	30	375	1.66	N/A	No	[134]
LaBr₃:Ce³⁺	5.3	46.9	61,000	30 (90%)	370	1.88	< 3	Yes	[135]
LaCl₃:Ce³⁺	3.86	N/A	46,000	25 (65%)	330	N/A	3.3	Yes	[136]
LuF₃:Ce³⁺	8.3	50.2	8000	28	310	N/A	N/A	Yes	[137]
LuI₃:Ce³⁺	5.7	60.5	98,000	33	475	N/A	4.6	Yes	[138]
PrCl₃:Ce³⁺	4.0	51.5	21,000	17	340	N/A	8.4	Yes	[139]
CeCl₃	3.9	50.4	28,000	25	360	N/A	N/A	Yes	[136]
ZnS:Ag	4.09	27.4	60,000 (α)	72 (α)	450	3.45	N/A	No	[140]
LiF-ZnS:Ag	2.36	N/A	160,000/nth	~1000	450	~6.0	N/A	No	[141]
Tl₂LaBr₅:Ce³⁺	11.8	67	43,000	25	375, 415	N/A	6.3	Yes	[142]
CaWO₄	6.1	75.6	15,800	6800	420	N/A	6.3	No	[143]
CdWO₄	7.9	64.2	16,000	5000	470	1.1	6.5	No	[144]
ZnWO₄	7.87	64.9	11,000	5000	475	1.16	9.6	No	[145]
PbWO₄	8.28	75.6	2000–3000	<3/<20	420	0.89	N/A	No	[146]
Lu₂O₃:Eu	9.4	N/A	30,000ᵇ	>10⁶	611	N/A	N/A	No	[147]
Ba₃(PO₄)₂	5.25	38.8	27,000	459	420	N/A	N/A	No	[148]
Gd₂O₂S:Pr,Ce	7.3	61.1	40,000	3000	511	N/A	N/A	No	[149]
Bi₄Ge₃O₁₂	7.1	75.2	9000	300	480	1.12	9.0	No	[150]
Bi₄Si₃O₁₂	6.8	77.3	12,000	100	480	1.15	N/A	No	[151]
BaSiO₃:Ce	N/A	52.1	1500	43–47	400	N/A	N/A	No	[152]
Lu₂SiO₅:Ce³⁺	7.4	66.4	27,000	40	420	1.14	10	No	[153]
Lu₂Si₂O₇:Ce³⁺	6.2	46.4	26,000	38	378	N/A	9.5	No	[154]

(Continued)

TABLE 18.4 (Continued)
Typical Scintillators and Their Required Parameters

Scintillators	Density (g/cm³)	Z_{eff}	Light Yield (Photons/MeV)	Decay Time (ns)	Emission Maximum (nm)	Radiation Length (cm)	Energy Resolution (%, at 662 keV)	Hygroscopicity	References
$Gd_2SiO_5:Ce^{3+}$	6.71	59.5	9000	30 60/600	430	1.38	7.0	No	[155]
$Lu_{2x}Gd_{2-2x}SiO_5:Ce^{3+}$	7.3	~63	30,000–39,000	30–40	410–430	N/A	~8	No	[156]
$Lu_{0.25}Y_{0.75}Al_5O_{12}:Pr^{3+}$	6.2	44.1	33,000	N/A	N/A	N/A	4.4	No	[157]
$YAlO_3:Ce^{3+}$	5.4	25.6	17,000	26	370	N/A	5.7	No	[158]
$LuAlO_3:Ce^{3+}$	8.3	64.9	11,400	17	365	1.08	23	No	[159]
$LuBO_3:Ce^{3+}$	6.8	66	50,000	21	375, 410	1.28	N/A	No	[160]
$Lu_3Al_5O_{12}:Ce^{3+}$	6.7	44.3	12,500	70	510	1.45	N/A	No	[161]
$K_2YF_5:Pr^{3+}$	3.1	23.9	6900	20	240	N/A	N/A	Yes	[162]
$Cs_2LiLaCl_6:Ce^{3+}$	3.3	41.4	35,000	1/40	400	N/A	7.1	Yes	[163]
$Cs_2LiYCl_6:Ce^{3+}$	3.3	38.1	21,000	1/35	376	N/A	6.0	Yes	[164]
$Cs_2LiYCl_6:Pr^{3+}$	3.3	38.1	10,000	1/35	315	N/A	15.0	Yes	[165]
$CaI_2:Eu$	3.96	48	110,000	790	470	N/A	8.0	Yes	[166]
$RbGd_2Br_7:Ce^{3+}$	4.8	50.6	56,000	43	420	N/A	3.8	Yes	[167]
$K_2LaI_5:Ce^{3+}$	4.4	52.4	57,000	24	401	N/A	4.2	Yes	[168]
$Cs_2LiLaBr_6:Ce^{3+}$	3.3	44.1	60,000	55	410	N/A	2.9	Yes	[163]
$K_2BaI_4:Eu$	4.12	49.4	63,000	1500	448	N/A	2.9	Yes	[169]
$Cs_2NaGdCl_6$	3.52	N/A	27,000	37	374, 402	N/A	4.0	Yes	[170]
Cs_3GdCl_6	3.56	N/A	24,500	36	381, 406	N/A	4.5	Yes	[170]
$Cs_2NaGdBr_6$	4.19	N/A	48,000	35	388, 417	N/A	3.3	Yes	[170]
Cs_3GdBr_6	4.14	N/A	47,000	37	396, 424	N/A	4.0	Yes	[170]
$CsBa_2I_5$	4.8	N/A	80,000–97,000	48 (1%) 383 (6%) 1500 (68%) 9900 (25%)	435	N/A	3.8	Yes	[171]
$SrI_2:Eu^{2+}$	4.6	49.4	85,000	1200	422	N/A	3.7	Yes	[172]
$KCaI_3$	3.81	N/A	$72,000 \pm 3000$	1060	466	N/A	3	Yes	[173]
$KSr_2I_5:Eu$	4.39	N/A	94,000	990	445	N/A	2.4	Yes	[174]

Czochralski technique[175] and Bridgman technique[176], respectively. Due to the different equipment used in the two methods, the performances of scintillators will differ significantly. For multicomponent halides-based crystal scintillators, the modified Bridgman-Stockbarger technique was developed.[142] However, this technique is not suitable for industrial production. In addition, the micro-pulling down[177] and optical float zone[178] techniques have been used increasingly to prepare small-sized crystals. In particular, micro-pulling down has the capability to produce shaped crystals or fibers with well controlled dimensions. Meanwhile the optical float zone technique has the advantage of not using a crucible and thus being free from typical issues of compatibility of scintillator, crucible and atmosphere.[179]

Usually, the luminous efficiency of oxide crystals is generally higher than that of fluorides, and their efficiency values are determined by their energy band structure (E_g, ΔE_v). The band gap (E_g) of oxides is smaller (mostly 4–7 eV) and the valence band width (ΔE_v) is relatively larger (10 eV) than those of fluorides (whose E_g is mostly 6–14 eV and ΔE_v is 6 eV). If E_g increases, the energy loss related to non-radiative recombination increases and the relative efficiency decreases under ionizing radiation. If ΔE_v increases, luminescent centers might be excited more effectively.

The following section will introduce some applications of inorganic crystal scintillators.

18.3.1.1 Scintillators for Nuclear and High Energy Physics Detectors

In HEP and nuclear physics experiments, total absorption electromagnetic calorimeters made of inorganic crystals are known for their superb energy resolution and detection efficiency for photon and electron measurements.[180] That is to say that, an inorganic crystal calorimeter is preferred for those experiments where precision measurements of photons and electrons are crucial for their physics missions. Thus, the following requirements become important: (1) high-density materials with high stopping power to high-energy particles and short radiation lengths X_0 (large absorption coefficients). (2) Decay constants as short as or shorter than tens of nanoseconds. (3) Radiation hardness against radiation environments. For example, in many large accelerators, crystal ball NaI:Tl calorimeter, L3 BGO calorimeter, BELLE CsI:Tl calorimeter, kTeV-undoped CsI calorimeter, compact muon solenoid (CMS) PWO calorimeter and HDME LYSO calorimeter were used.[122] Among all existing crystal calorimeters, the CMS PWO crystal calorimeter, consisting of 75,848 crystals of 11 m³, is the largest and it has the shortest radiation length of $X_0 = 0.92$ cm.[122] Because of its superb energy resolution and detection efficiency, the CMS PWO calorimeter has played an important role for the discovery of Higgs boson by CMS experiment.[181]

Because so many electromagnetic calorimeters are used in high-energy particle physics detectors, many efforts were made to develop ultrafast and radiation-hard inorganic scintillators for future HEP experiments at the energy and intensity frontiers. For example, Chen et al. found that yttrium doping in BaF_2 crystals can suppress the slow scintillation component in BaF_2 while maintaining the sub-ns fast component.[182] In addition, dramatic improvements in the transmittance at short wavelengths, radiation hardness and suppression of slow components were obtained by doping trivalent rare-earth ions such as La^{3+}, Gd^{3+} and Y^{3+} in PWO crystal.[183] In particular, Cerium-doped silicate-based heavy crystal scintillators, such as GSO, LSO and LYSO, have been explored for HEP experiments due to their high stopping power and fast bright scintillation.[184] For $LaBr_3$:Ce crystals, good scintillation performances such as light yield, decay time and energy resolution were achieved.[135] It can be found that the scintillation decay time of $LaBr_3$:Ce related to the 5d → 4f transition of Ce^{3+} is about 15 ns and its light yield is about 70,000 ph/MeV.

For HEP experiments at future lepton colliders, inorganic scintillators also have been proposed to build a homogeneous hadron calorimeter to achieve unprecedented jet mass resolution by dual readouts of both Cherenkov and scintillation light.[185,186] For this application, development of cost-effective crystal detectors is a crucial issue because of the huge crystal volume and small radiation hardness required in the lepton collider environment.[187]

18.3.1.2 Scintillators for Neutron Detection

The increasing use of spallation neutrons for fundamental investigations of the structure of matters, the need for thermal neutron detectors is rapidly expanding.[179] Probes of thermal neutrons instead of X- and γ-rays can give information on light atoms like H, C, N and O in materials. Because a neutron is an uncharged neutral particle, it is impossible to directly detect it. Thus, neutron detection mainly utilizes neutrons to react with specific isotopes in the detection materials (such as He-3, ^{6}Li, ^{10}B, ^{155}Gd or ^{157}Gd), releasing α particles or γ-rays to indirectly detect neutrons.[188,189] Among them, ^{3}He has been the standard for neutron detection and it is widely believed that its performance cannot be surpassed. However, ^{3}He gas is expensive and in limited supply. As an alternative to ^{3}He gas, many efforts have been made to develop scintillators containing ^{6}Li and ^{10}B, which have high interaction probabilities against thermal neutrons.[190] For scintillators to detect intense thermal neutrons with a good position resolution, three requirements (large light output, fast decay and suppression of background γ-rays due to inefficiency or discriminating capability) must be met.[191]

To realize the above purpose, commercial ^{6}Li-enriched elpasolite and LiCaAlF$_6$ scintillators have been developed.[192,193] The former has a chemical composition of Cs$_2$AmRe(Cl,Br)$_6$, where Am means alkali metal elements and Li is chosen to detect neutrons. For neutron detector using elpasolite-based scintillators, such as Ce-doped Cs$_2$LiYCl$_6$ (CLYC), the light output of more than 70,000 photons/neutron, which is second only to that of ^{6}LiF–ZnS:Ag (160,000 photons/neutron), was reached.[194] However, for Eu^{2+}-doped LiCaAlF$_6$ scintillators, their non-hygroscopicity is better than that of Ce-doped Cs$_2$LiYCl$_6$ in practical applications.[193,195] In addition, for Na, Eu-codoped LiCaAlF$_6$ (LiCAF:Eu, Na) crystalline scintillators, the improved scintillation properties was realized under thermal neutron exposure.[196] For example, the light yield (monitored at 370 nm originating from the 5d–4f transition of Eu^{2+}) of Eu 2%, Na 2%-codoped LiCAF reached 40,000 photons/neutron, which was about 30% higher than that of Eu:LiCAF.

Up to now, the brightest ^{10}B-based bulk crystalline scintillators has been developed, which is Ce-doped CaB$_2$O$_4$, exhibiting 2200 photons/neutron under neutron excitation.[197] In the future, more efforts are made to develop new scintillators with high light yield for practical detector applications.

18.3.1.3 Scintillators for Industrial Applications

In industrial applications, inorganic crystal scintillators are widely used in well logging.[198] This technique is to characterize the properties of subsurface minerals through lowering the radiation detector into a drilled borehole.[125,199] That is to say that underground natural resources were searched for by drilling boreholes and remotely operating a γ-ray detector, called a sonde, to measure the natural radioactivity of geologic formations. In some cases, the sonde also carries a γ-ray source or a neutron source to irradiate geologic formation. γ-rays coming back to the sonde were detected with a scintillator.[19] To achieve higher count rate capability and detection efficiency, scintillators used for well logging are required to possess the following performances, such as large volumes, high atomic number, short decay time, large light output and small temperature-dependent light yield.[200]

Conventional NaI:Tl$^+$ as faster and higher Z_{eff} scintillators has been most widely used in well logging.[201] Recently, new high-performance scintillators, such as lanthanum halides and cerium-doped silicate-based heavy crystal scintillators, have been developed.[125,199] For example, LaBr$_3$:Ce scintillators exhibited small temperature-dependent light yield, which minimally changes at up to 300°C.[202] It makes LaBr$_3$:Ce scintillators a potential candidate for NaI:Tl$^+$ scintillators. However, because of the presence of the naturally radioactive La^{138} (lanthanum) isotope in lanthanum halides, substantial self-radiation background appeared during detecting γ-ray signal.[200] Moreover, rare-earth halides are highly hygroscopic and require special atmosphere-tight sealing. Fortunately, Ce-doped inorganic complex oxide crystals, such as Y-Lu aluminum garnets, perovskites and orthosilicates, are non-hygroscopic, hard, chemically and mechanically stable. Although (Gd, Y)$_2$SiO$_5$ (GYSO) has higher temperature-dependent light yield than that of LaBr$_3$:Ce, GYSO-based detectors, it is attractive for counting measurements due to their high γ-ray efficiency.[203]

Besides, expensive $Lu_2SiO_5:Ce^{3+}$ crystal exhibited the highest light yield up to ~30,000 ph/MeV.[204] Meanwhile, $Gd_2SiO_5:Ce^{3+}$ (GSO:Ce^{3+}) crystal has relatively high volumetric detection efficiency, rapid decay and a temperature dependence of light output similar to that of NaI:Tl⁺.[155] For Ce-doped $Gd_2Si_2O_7$ (GPS:Ce) crystal, the light output is 2.5 times as that of GSO:Ce^{3+} crystals and the decay time is 46 ns.[205] Although GPS:Ce crystal is a high-performance scintillator, it is difficult to obtain large-sized uniform crystal due to the appearance of impurities (Gd_2O_3 or/and SiO_2) in GPS crystal.

Currently, in well-logging industry, there still remains the issues on developing new inorganic crystal scintillators used in the detection technologies capable of operation to temperatures above 175°C in the tough downhole environment.[200]

18.3.1.4 Scintillators for γ-Ray Camera

Gamma camera for medical imaging uses γ-rays as a light source to detect their intensity distribution. Usually, γ-ray camera, referred as Anger γ-camera, takes pictures of γ-rays emitted from isotopes [⁹⁹ᵐTc(141 keV), ¹²³I(159 keV), ²⁰¹Tl(~75 keV)] introduced into human bodies.[206] It is mainly composed of collimator, scintillator, PMT and the related electronic system. The γ-rays passing through a collimator are detected with a single scintillator plate which is read out with several tens of PMTs.[207] At present, some γ-cameras have used pixelated detectors, where the field of view is covered by an array of individual scintillators with a face size of 2–3 mm.[207,208]

For γ-camera, scintillators with high atomic numbers and high density are required to realize high detection efficiency. For example, scintillators, such as NaI:Tl, CsI:Tl, CsI:Na and LaBr₃:Ce crystals, have been used in γ-camera.[209]

18.3.1.5 Scintillators for Positron Emission Tomography

Positron emission tomography (PET) is a very powerful medical diagnostic technique and applied in the fields of neurology, oncology, biology and cardiology.[210] PET imaging is realized by means of two 511-keV quanta, which are emitted approximately collinearly when a positron, emitted by a radiopharmaceutical, such as ¹⁸F, ¹¹C, ¹³N and ¹⁵O, introduced into a patient's body, annihilates with a nearby electron in tissue.[211] The incoming direction of the two annihilation photons detected in coincidence is given by the straight line connecting the two hit crystals, also called line of response. In conventional PET, because a patient being scanned is usually surrounded by rings of scintillator detectors, many annihilations give many lines of response and in principle the three-dimensional (3D) reconstruction of the lines of response represents the radiopharmaceutical distribution, i.e., the image.[212] In time-of-flight (TOF) PET, the approximate location of annihilation is determined along the line of annihilation by using the measured difference in arrival times as shown in Figure 18.17.[128,208] This TOF information spatially constrains the location of the event and it leads to the improved signal-to-noise ratio in the reconstructed image.[213]

To realize high-quality imaging, scintillators for PET are required to possess good performances, such as fast decay, large light output, high stopping power (high probability that a 511-keV γ-ray will be totally absorbed by the detector), high spatial resolution (ability to determine the interaction location of γ-rays in the detector to a small spatial volume), good energy resolution (to reject scattered events) and sub-100 ps time resolution.[214]

For PET application, NaI:Tl⁺ crystal scintillator was first used and then replaced by BGO.[128] Since 2006, Ce-doped LYSO scintillators has been used in the first commercially available TOF-PET scanner due to their excellent performances such as high light output, fast decay time and good stopping power.[215] Furthermore, for LaBr₃:Ce crystal scintillators with a lower atomic number, it is found that their scintillation properties are better than those of Ce-doped LSO or LYSO crystals. For example, an array of $4 \times 4 \times 30$ mm³ LaBr₃ (5% Ce) readout with large-size PMTs could achieve a timing resolution of ~313 ps and an average energy resolution of ~5.1% full width at half maxima at 511 keV.[216] In particular, it is found that the energy resolution of LaBr₃:Ce^{3+} can reach to 2.0% at 662 keV using ¹³⁷Cs source after Sr^{2+} doping, as shown in Figure 18.18.[217]

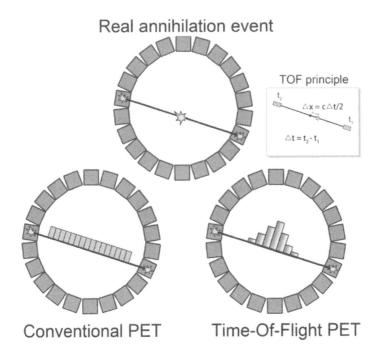

FIGURE 18.17 Compared to conventional PET, the estimated time-of-flight difference (Δt) between the arrival times of photons on both detectors in TOF-PET allows localization (with a certain probability) of the point of annihilation on the line of response. In TOF-PET, the distance to the origin of scanner (Δx) is proportional to the TOF difference via the relation: $\Delta t : \Delta x = c\Delta t/2$, where c is the speed of light. t_1 is the arrival time on the first detector, and t_2 is the arrival time on the second detector. (Vandenberghe, S., Mikhaylova, E., D'Hoe, E., Mollet, P. and Karp, J. S., *EJNMMI Phys.* 3, 3, 2016. With permission.)

Recently, Ce-doped LuI_3 and Ce or Pr-doped $Gd_3Al_2Ga_3O_{12}$ (GAGG) crystal scintillators have shown promising characteristics in the application of TOF-PET. For example, $LuI_3:Ce^{3+}$ shows a record-breaking light yield of 98,000 photons/MeV at 662 keV compared to $LaCl_3:Ce^{3+}$ and $LaBr_3:Ce^{3+}$.[138] For Ce- or Pr-doped GAGG scintillator, its light yield is about 60,000 ph/MeV. It is found that, despite Ca and Mg codoping resulted in the decreased light yield, its timing coincidence resolution was significantly improved.[218–220] For Ce-doped LuI_3 and Ce or Pr-doped GAGG crystal scintillators, there still remains a challenge to fabricate large-sized crystals.[221–224]

18.3.1.6 Scintillators for Computed Tomography

Computer tomography (CT) scanning has become a widely used technique to realize transversal images of human body.[225] Its schematic diagram for CT imaging is shown in Figure 18.19.[226] An X-ray fan beam with a 1D-position-sensitive detector on a fixed construction rotates around the patient body in the center with a typical frequency of 0.5–1 Hz and a large number of attenuation profiles are recorded during one 360° rotation. Via applying simultaneously the slow gradual movement of the patient through the CT machine (so-called multi-slice spiral scan mode), even volume 3D reconstruction images, i.e., the cross-sectional images of the body, were reconstructed.[4]

Scintillators for CT are required to possess good optical performances, such as large absorption coefficient, large light output, fast decay (<1 ms), weak afterglow (<0.01% at 3 ms after termination of irradiation) and minimization of both radiation damage and gain hysteresis (<2% at ~5 Gy). In addition, the emission spectrum of scintillators should have a good overlap with the spectral sensitivity of the photodetectors. Recently, in the application of CT, Si-based photodiode arrays in the spectral range of 400–900 nm are usually used.[227] For $CsI:Tl^+$ scintillator with 550 nm emission, its light output is almost as large as that of $NaI:Tl^+$, but its afterglow (0.7–1.5% after 20 ms) is not weak

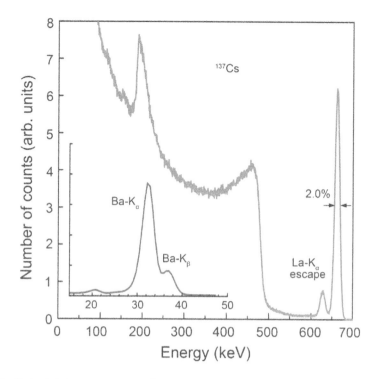

FIGURE 18.18 Pulse height spectrum of a ^{137}Cs source measured with a Sr^{2+}-codoped $LaBr_3$:5%Ce crystal and a Hamamatsu R6231-100 super bialkali PMT. The inset shows the 10–50 keV region on a reduced vertical scale. (Reproduced from Alekhin, M. S., de Haas, J. T. M., Khodyuk, I. V., Krämer, K. W., Menge, P. R., Ouspenski, V. and Dorenbos, P., *Appl. Phys. Lett.* 102, 161915, 2013. With the permission of AIP Publishing.)

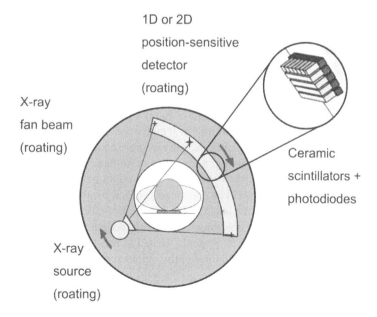

FIGURE 18.19 Principle of CT imaging. (van Eijk, C. W. E., *Nucl. Instrum. Meth. A.* 509, 17–25, 2003. With permission.)

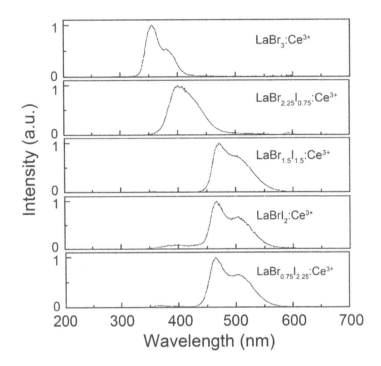

FIGURE 18.20 Radioluminescence spectra of $LaBr_xI_y:Ce^{3+}$ ($x = 3$, $y = 0$; $x = 2.25$, $y = 0.75$; $x = 1.5$, $y = 1.5$; $x = 1$, $y = 2$ and $x = 0.75$, $y = 2.25$) at RT. (Reproduced from Birowosuto, M. D., Dorenbos, P., Krämer, K. W. and Güdel, H. U., *J. Appl. Phys.* 103, 103517, 2008. With the permission of AIP Publishing.)

enough. For CsI:Tl,Yb scintillator, Yb doping leads to an increase of light yield reaching 90,000 ph/ MeV and a decrease of afterglow intensity down to 0.035% at 80 ms.[228] For Pr-doped Gd_2O_2S (GSO) and Eu^{3+}-doped $(Lu, Gd)_2O_3$ scintillators, their yellow-green or red emissions originating from the 4f–4f transitions of rare-earth ions are well-matched to the spectral sensitivity of Si-PD.[125] In particular, $LaBr_3:Ce^{3+}$ scintillator with UV emission has attracted wide attention because of their high light yield (60,000–70,000 photons/MeV).[229] After I^- doping in $LaBr_3:Ce^{3+}$ scintillator, its emission wavelength falls into the visible region as shown in Figure 18.20 and is matched to the sensitivity of Si-PD detector.[230]

18.3.1.7 Scintillators for X-Ray Image Intensifier

In X-ray medical examination system, the design of X-ray image intensifier is to instantly convert an X-ray image into a corresponding optical image. It consists of a vacuum glass tube, image input screen and photocathode, high-voltage electrode, positive pole and image output screen as shown in Figure 18.21.[212] X-rays penetrating through a specimen or human body excite the input phosphor screen and they are converted into visible light. Then visible light irradiates photocathode located on the back of the input phosphor screen. The generated photoelectrons are accelerated through the focusing electrodes and converged on the anode. These converged electrons subsequently excite the output phosphor screen to realize a visible image. The intermediate image is acquired by means of advanced charge-coupled device (CCD) systems. Usually, in X-ray image intensifier, the input phosphor screen is made of CsI:Na scintillator and the output screen is made of ZnS:Cu, (Zn, Cd) S:Cu or (Zn, Cd)S:Ag phosphors.[231]

The scintillators used in the input screen are required to possess the following performaces:[232]

1. High X-ray absorption efficiency, i.e., inorganic scintillator is made of elements having large atomic numbers (high density).

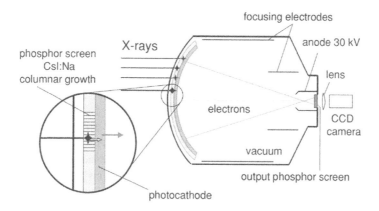

FIGURE 18.21 Schematic of an image intensifier system. (van Eijk, C. W. E., *Phys. Med. Biol.* 47, R85–R106, 2002. With permission.)

2. High emission efficiency upon X-ray excitation.
3. The emission wavelength matches to the spectral response of the photocathode.

18.3.1.8 Scintillators for Flat Detectors

Flat detectors, first introduced in the late 1990s, have become the gold standard for digital radiography (DR), full-field mammography, fluoroscopic and endovascular imaging as well as imaging in hybrid surgical environments.[233] Usually, flat detectors can be divided into two categories, i.e., direct conversion flat detector and indirect conversion flat detector.[234]

Direct conversion flat panel detector is mainly composed of an amorphous selenium (a-Se) layer and a silicon thin film transistor (TFT) array. A-Se can be deposited directly onto an active matrix of a-Si where each pixel consists of a charge collecting electrode and readout TFT. The operating principle can be simply described as the following processes. First, the photoelectrons are generated during the X-ray absorption process. Second, the directly created electron-hole pairs are collected by applying an electric field across the detector. Finally, under the trigger of the control circuit, the digital image is acquired by A/D conversion.[126]

Indirect conversion flat detector is consisted of a scintillator layer, an amorphous silicon photodiode circuitry layer and a silicon TFT array, as shown in Figure 18.22.[126] When X-ray photons reach the scintillator, visible light proportional to the incident energy is emitted and then recorded

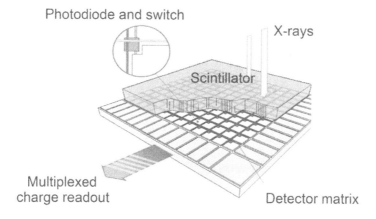

FIGURE 18.22 Schematic view of an indirect conversion flat detector based on columnar CsI and an active matrix of a-Si photodiodes. (Spahn, M., *Nucl. Instrum. Meth. A.* 731, 57–63, 2013. With permission.)

by an array of photodiodes and converted into electrical charges. These charges are then read out by a TFT array similar to that of direct conversion DR systems.[4] Because the indirect conversion flat detector has replaced image intensifiers in angiography and enabled DR with instant image display and high-dose efficiency, it has become the main type of flat detectors in practical applications.[235–238] In flat imaging panels, scintillators with high atomic numbers have been selected to be directly deposited over the large area matrix of a:Si–H photodiodes. For example, thallium-doped cesium iodide (CsI:Tl) with high atomic numbers (55 and 53 for Cs and I) exhibited good X-ray absorption properties. Usually, CsI:Tl scintillator with needlelike structure was grown by physical vapor deposition. It is found that an individual CsI:Tl crystal shows 5–10 μm in diameter and up to 600 μm or more in length. This needlelike structure facilitates the photodiode to collect green light emitted by the CsI:Tl scintillator upon X-ray irradiation.[239] Meanwhile the emission wavelength of green light is well matched to that corresponding to the maximum photo-efficiency of a-Si:H photodiode. In addition, the indirect conversion flat detectors using Tb-doped gadolinium sulfide (Gd$_2$O$_2$S:Tb) scintillator have shown some advantages, such as fast imaging and low cost, but it possesses lower gray-scale dynamic range.[240]

Recently, new scintillators, such as perovskite-type materials, have been developed to fabricate the indirect conversion flat detector due to their excellent performances. For example, CsPbBr$_3$ nanocrystals exhibit a strong emission peak at ~550 nm, as shown in Figure 18.23(a),[241] and a light yield of ~33,000 photons/MeV.[242] Furthermore, for CsPbBr$_3$ nanocrystals, it is found that their

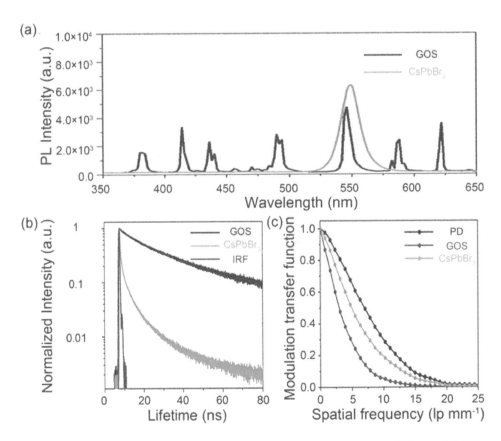

FIGURE 18.23 (a) PL spectra and (b) Dynamic PL decay curves monitored at ~550 nm of CsPbBr$_3$ and GOS:Tb scintillators upon 285 nm UV-light irradiation. (c) MTF of pristine Si-PD, CsPbBr$_3$ scintillators and conventional GOS:Tb scintillators. (Heo, J. H., Shin, D. H., Park, J. K., Kim, D. H., Lee, S. J. and Im, S. H., *Adv. Mater.* 30, 1801743, 2018. With permission.)

average lifetime (~2.87 ns) is shorter than that of Tb-doped GOS, as shown in Figure 18.23(b),[241] indicating that they have faster response time than that of Tb-doped GOS. Meanwhile the spatial resolution of CsPbBr$_3$ nanocrystals on Si-PDs is 9.8 lp/mm higher than that (6.2 lp/mm) of Tb-doped GOS scintillator on Si-PDs as shown in Figure 18.23(c).[241]

In addition, new flat detectors using CMOS technology have recently been used and have shown the improved radiation resistance and increased spatial resolution due to smaller pixel size.[243]

18.3.2 CERAMIC SCINTILLATORS

Transparent ceramics are often believed to be as an alternative to single crystals when a specific geometrical form is not easily achieved with single crystals and in cases where they are produced at a lower cost.[122] Usually, to fabricate transparent ceramic scintillators, there are mainly three kinds of preparation methods, such as high-temperature sintering,[244] hot-pressing[245] and spark plasma sintering (SPS).[246] Before sintering, these methods are required to obtain powders with uniform size distribution.[247]

For example, Ce-doped Lu$_3$Al$_5$O$_{12}$ (LuAG) ceramic scintillators were fabricated by SPS method at 1700°C using nanopowders.[246] For the optimized ceramic scintillators upon X-ray excitation, its radioluminescence intensity at 514 nm is significantly higher than that of BGO standard scintillator and is even slightly higher than that of Ce-doped LuAG single crystal as shown in Figure 18.24.[246] Many attempts suggest that Ce-doped LuAG ceramic scintillator might serve as a candidate for fast and robust flat detector in a severe radiation environment.[248]

At present, there still remain unsolved issues on the difficulty in achieving good transparency of ceramic scintillators with non-cubic structure and the fabrication of transparent hygroscopic-halide-based ceramic scintillators.[249]

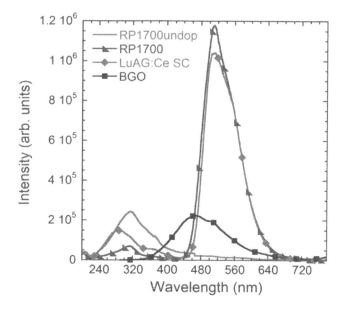

FIGURE 18.24 X-ray-excited radioluminescence spectra for the undoped LuAG and LuAG:Ce (RP1700undop, RP1700) samples sintered at 1700°C from nanopowder with a rapid pre-heating, compared with that of the BGO standard scintillator and LuAG:Ce single crystal. (Pejchal, J., Babin, V., Beitlerova, A., Kucerkova, R., Panek, D., Barta, J., Cuba, V., Yamaji, A., Kurosawa, S., Mihokova, E., Ito, A., Goto, T., Nikl, M. and Yoshikawa, A., *Opt. Mater.* 53, 54–63, 2016. With permission.)

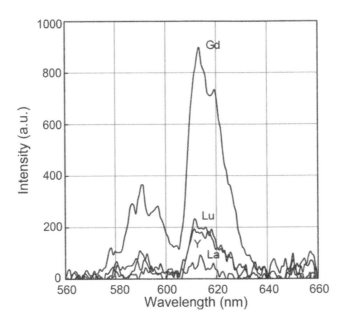

FIGURE 18.25 Emission spectra for $35SiO_2-15B_2O_3-20AlF_3-30Ln_2O_3-0.2Eu_2O_3$ (Ln = Y, La, Gd, Lu) glasses under X-ray excitation. (Fu, J., Kobayashi, M., Sugimoto, S. and Parker, J. M., *Mater. Res. Bull.* 43, 1502–1508, 2008. With permission.)

18.3.3 GLASS SCINTILLATORS

Glass scintillators are usually used to detect X-rays and neutrons due to their unique advantages, such as easy synthesis, low cost, high optical homogeneity, large volume production and various shaping.[250,251] The conventional technique to prepare glass scintillators is the melt quenching method.[71] For example, Eu^{3+}-doped borosilicate glasses were prepared by this method. The addition of Ln_2O_3 (Ln = Y, La, Gd and Lu) in the glasses can increase the density of the doped glass scintillators from 4.0 to 6.1 g/cm³ in a high atom number order (Y, La, Gd and Lu). It is found that the glass scintillator containing Gd exhibits a higher light yield than the other glasses as shown in Figure 18.25.[252] In recent years, glass scintillators containing Gd also have been succeed in various germinate, phosphate and silicate glasses and the improved performances were obtained.[250,253] Recently, the need for scintillating glass fibers for tracking detectors has become a driving force for new exploration.[254]

18.3.4 ORGANIC SCINTILLATORS

Organic scintillators are widely used for the detection of conventional radiation dose, especially for the intensity detection of α, β, γ rays, neutrons and protons due to their low price, light mass, high efficiency and fast timing resolution.[255,256] Organic scintillators are conjugated hydrocarbon compounds, which feature an extended conjugated system of π-electrons in the double carbon bonds of the molecule.[229] They can exist both as solid crystals, as plastics or as liquid solutions. Liquid scintillators and plastic scintillators are in principle quite similar to one another, as they generally consist of an aromatic organic matrix, and one or more dissolved fluorescent dyes which enable light emission.[117,257] Some conventional organic scintillators and the related properties are given in Table 18.5. Among them, it can be found that anthracene has the highest light yield (16,000 photons/MeV).[258,259]

Because the optical performances of organic scintillators, such as light yield, energy resolution and absorption efficiency, are usually poor than those of inorganic scintillators. At present, more

TABLE 18.5
The Related Properties of the Conventional Organic Scintillators

Scintillator	ρ (g/cm³)	Z_{eff}	Light yield (Photons/MeV)	λ (nm)	τ (ns)	Type	References
Anthracene	1.25	5.24	16,000	447	30	Organic crystal	[258, 259]
Stilbene	1.16	5.14	8000	410	4.5	Organic crystal	[259, 261]
2,5-Diphenyloxazole	1.06	5.52	8800	405	7	Organic crystal	[262]
Naphthalene	0.96	5.18	2000	348	80	Organic crystal	[263]
p-Terphenyl	N/A	N/A	~9200	440	5	Liquid	[259]

efforts are made to synthesize organic-inorganic hybrid scintillators, such as $(C_6H_5C_2H_4NH_3)_2PbBr_4$ crystal, which has light yield of 14,000 photons/MeV and very fast decay time (11 ns). [260]

18.4 OUTLOOK

For X-ray phosphors and scintillators, both of them require high X-ray absorption efficiency and high light yield. With the rapid developments of preparation and detection methods and medical applications, more multidisciplinary studies have been concentrated on the efforts to acquire high-quality X-ray phosphors and scintillators used in different applications. Recently, halide perovskite nanocrystal scintillators has become a focus of interest due to their high light quantum yield (150,000–300,000 photons/MeV) at low temperature. However, there still remains a challenge to acquire halide perovskite scintillators with such high light yield at room temperature. In the future, the synthesis of nanocrystals[264] and nanostructure fabrications[265] may hold the keys for developing better X-ray phosphors and scintillators.

Nowadays, much attention has been focused on medical diagnosis and therapy. Especially, more efforts were made to develop new nanophosphors for X-PDT. It is believed that a further area of future exploration of X-PDT is in vasculature targeting.[266] In the future, multifunctional nanoplatforms constructed by nanophosphors for X-PDT and other nanomedicines may be required in the aforementioned field.

REFERENCES

1. Mao, R., Zhang, L. and Zhu, R.-Y. 2008. Optical and scintillation properties of inorganic scintillators in high energy physics, *IEEE Transactions on Nuclear Science* 55, 2425–2431.
2. Yoshida, E., Hirano, Y., Tashima, H., Inadama, N., Nishikido, F., Moriya, T., Omura, T., Watanabe, M., Murayama, H. and Yamaya, T. 2013. The X'tal cube PET detector with a monolithic crystal processed by the 3D sub-surface laser engraving technique: Performance comparison with glued crystal elements, *Nuclear Instruments and Methods in Physics Research, Section A: Accelerators, Spectrometers, Detectors and Associated Equipment* 723, 83–88.
3. Yanagida, T., Fujimoto, Y., Kurosawa, S., Kamada, K., Takahashi, H., Fukazawa, Y., Nikl, M. and Chani, V. 2013. Temperature dependence of scintillation properties of bright oxide scintillators for well-logging, *Japanese Journal of Applied Physics* 52(7 PART 1), 076401.
4. Nikl, M. 2006. Scintillation detectors for X-rays, *Measurement Science and Technology* 17, R37–R54.
5. Li, S., Liu, Y., Liu, C., Yan, D., Zhu, H., Xu, C., Ma, L. and Wang, X. 2017. Improvement of X-ray storage properties of C12A7:Tb³⁺ photo-stimulable phosphors through controlling encaged anions, *Journal of Alloys and Compounds* 696, 828–835.
6. Cheon, J. K., Kim, S., Rooh, G., So, J. H., Kim, H. J. and Park, H. 2011. Scintillation characteristics of $Cs_2LiCeBr_6$ crystal, *Nuclear Instruments and Methods in Physics Research, Section A: Accelerators, Spectrometers, Detectors and Associated Equipment* 652, 205–208.
7. Seeley, Z., Cherepy, N. and Payne, S. 2013. Two-step sintering of $Gd_{0.3}Lu_{1.6}Eu_{0.1}O_3$ transparent ceramic scintillator, *Optical Materials Express* 3, 908–912.

8. Struebing, C., Lee, G., Wagner, B. and Kang, Z. 2016. Synthesis and luminescence properties of Tb doped LaBGeO$_5$ and GdBGeO$_5$ glass scintillators, *Journal of Alloys and Compounds* 686, 9–14.

9. Blasse, G. and Grabmaier, B. C. 1994. *Luminescent Materials*, Springer: Berlin, Germany, p 232.

10. Eilbert, R. F. 2009. X-ray technologies. In *Aspects of Explosives Detection*, Elsevier, pp 89–130.

11. Debnath, R. 2016. *Advances in Functional Luminescent Materials and Phosphors*, John Wiley & Sons, ebook, pp 425–472.

12. Derenzo, S. E., Weber, M. J., Bourret-Courchesne, E. and Klintenberg, M. K. 2003. The quest for the ideal inorganic scintillator, *Nuclear Instruments and Methods in Physics Research Section A: Accelerators, Spectrometers, Detectors and Associated Equipment* 505, 111–117.

13. Herglotz, H. K. 1968. Fluorescent screens for monochromatic X-ray diffraction patterns on polaroid film, *Review of Scientific Instruments* 39, 1658–1659.

14. Lehmann, W. 1963. Emission spectra of (Zn,Cd)S phosphors, *Journal of the Electrochemical Society* 110, 754.

15. Hassani, M., Sarabadani, P. and Aghda, A. H. 2019. Synthesis and characterization of Gd$_2$O$_2$S:Tb^{3+} phosphor powder for X-ray imaging detectors, *Journal of Nanostructures* 9, 616–622.

16. Alves, R. V. and Buchanan, R. A. 1973. Properties of Y$_2$O$_2$S:Tb, X-ray intensifying screens, *IEEE Transactions on Nuclear Science* 20, 415–419.

17. Kandarakis, I., Cavouras, D., Panayiotakis, G. S. and Nomicos, C. D. 1997. Evaluating X-ray detectors for radiographic applications: A comparison of ZnSCdS:Ag with Gd$_2$O$_2$S:Tb and Y$_2$O$_2$S:Tb screens, *Physics in Medicine and Biology* 42, 1351–1373.

18. Gurvich, A. M., Il'ina, M. A. and Katomina, R. V. 1987. Basic parameters and test methods for medical X-ray screens, *Biomedical Engineering* 21, 215–219.

19. Yen, W. M., Shionoya, S. and Yamamoto, H. 2006. *Phosphor Handbook*, CRC Press, Taylor & Francis Group, second edition.

20. Ardran, G. M., Crooks, H. E. and James, V. 1969. Testing X-ray cassettes for film-intensifying screen contact, *Radiography* 35, 143–145.

21. Brixner, L. H. 1987. New X-ray phosphors, *Materials Chemistry and Physics* 16, 253–281.

22. Sonoda, M., Takano, M., Miyahara, J. and Kato, H. 1983. Computed radiography utilizing scanning laser stimulated luminescence, *Radiology* 148, 833–838.

23. Rabatin, J. G. 1982. Luminescence of rare earth activated lutetium oxyhalide phosphors, *Journal of the Electrochemical Society* 129, 1552–1555.

24. Hartenbach, I., Lissner, F., Nikelski, T., Meier, S. F., Müller-Bunz, H. and Schleid, T. 2005. About lanthanide oxotantalates with the formula MTaO$_4$ (M = La–Nd, Sm–Lu), *Zeitschrift für Anorganische und Allgemeine Chemie* 631, 2377–2382.

25. von Seggern, H. 1992. X-ray imaging with photostimulable phosphors, *Nuclear Instruments and Methods in Physics Research Section A: Accelerators, Spectrometers, Detectors and Associated Equipment* 322, 467–471.

26. Berg, O. E. and Kaiser, H. F. 1947. The X-ray storage properties of the infra-red storage phosphor and application to radiography, *Journal of Applied Physics* 18, 343–347.

27. Leblans, P., Vandenbroucke, D. and Willems, P. 2011. Storage phosphors for medical imaging, *Materials* 4, 1034–1086.

28. Amemiya, Y., Wakabayashi, K., Tanaka, H., Ueno, Y. and Miyahara, J. 1987. Laser-stimulated luminescence used to measure X-ray diffraction of a contracting striated muscle, *Science* 237, 164–168.

29. Wang, X. and Riesen, H. 2015. Mechanochemical synthesis of an efficient nanocrystalline BaFBr:Eu^{2+} X-ray storage phosphor, *RSC Advances* 5, 85506–85510.

30. Chen, W., Kristianpoller, N., Shmilevich, A., Weiss, D., Chen, R. and Su, M. 2005. X-ray storage luminescence of BaFCl:Eu^{2+} single crystals, *Journal of Physical Chemistry B* 109, 11505–11511.

31. Liang, Q., Yang, X. and Li, Z. 2011.Preparation and optical spectroscopy of BaFCl:Eu^{2+}A photoluminescent X-ray storage phosphor, In International Conference on Digital Printing Technologies, pp 528–531.

32. Nanto, H., Takei, Y., Nishimura, A., Nakano, Y., Shouji, T., Yanagita, T. and Kasai, S. 2006. Novel X-ray image sensor using CsBr:Eu phosphor for computed radiography, In Progress in Biomedical Optics and Imaging – Proceedings of SPIE.

33. Aitasalo, T., Dereń, P., Hölsä, J., Jungner, H., Krupa, J.-C., Lastusaari, M., Legendziewicz, J., Niittykoski, J. and Stręk, W. 2003. Persistent luminescence phenomena in materials doped with rare earth ions, *Journal of Solid State Chemistry* 171, 114–122.

34. Kulesza, D., Bolek, P., Bos, A. J. J. and Zych, E. 2016. Lu$_2$O$_3$-based storage phosphors. An (in)harmonious family, *Coordination Chemistry Reviews* 325, 29–40.

35. Zhang, X., Cao, C., Zhang, C., Xie, S., Xu, G., Zhang, J. and Wang, X.-J. 2010. Photoluminescence and energy storage traps in $CaTiO_3$:Pr^{3+}, *Materials Research Bulletin* 45, 1832–1836.

36. Hu, R., Zhang, Y., Zhao, Y., Wang, X., Li, G. and Wang, C. 2020. UV–Vis-NIR broadband-photostimulated luminescence of $LiTaO_3$:Bi^{3+} long-persistent phosphor and the optical storage properties, *Chemical Engineering Journal* 392, 124807.

37. Lecointre, A., Bessière, A., Bos, A. J. J., Dorenbos, P., Viana, B. and Jacquart, S. 2011. Designing a red persistent luminescence phosphor: The example of YPO_4:Pr^{3+},Ln^{3+} (Ln = Nd, Er, Ho, Dy), *Journal of Physical Chemistry C* 115, 4217–4227.

38. Kano, T., Takahashi, T., Okajima, K., Umetani, K., Ataka, S., Yokouchi, H. and Suizuki, R. 1986. Laser-stimulable transparent CsI:Na film for a high quality X-ray imaging sensor, *Applied Physics Letters* 1986, 1117–1118.

39. von Seggern, H., Meijerink, A., Voigt, T. and Winnacker, A. 1989. Photostimulation mechanisms of X-ray-irradiated RbBr:Tl, *Journal of Applied Physics* 66, 4418–4424.

40. Kulesza, D. and Zych, E. 2013. Managing the properties of Lu_2O_3:Tb, Hf storage phosphor by means of fabrication conditions, *Journal of Physical Chemistry C* 117, 26921–26928.

41. Wiatrowska, A. and Zych, E. 2013. Traps formation and characterization in long-term energy storing Lu_2O_3:Pr, Hf luminescent ceramics, *Journal of Physical Chemistry C* 117, 11449–11458.

42. Dobrowolska, A., Bos, A. J. J. and Dorenbos, P. 2019. High charge carrier storage capacity in lithium lutetium silicate doped with cerium and thulium, *Physica Status Solidi – Rapid Research Letters* 13(3), 1800502.

43. Meijerink, A. and Blasse, G. 1991. Photostimulated luminescence and thermally stimulated luminescence of some new X-ray storage phosphors, *Journal of Physics D: Applied Physics* 24, 626–632.

44. Liu, F., Yan, W., Chuang, Y.-J., Zhen, Z., Xie, J. and Pan, Z. 2013. Photostimulated near-infrared persistent luminescence as a new optical read-out from Cr^{3+}-doped $LiGa_5O_8$, *Scientific Reports* 3, 1554.

45. Luchechko, A., Zhydachevskyy, Y., Ubizskii, S., Kravets, O., Popov, A. I., Rogulis, U., Elsts, E., Bulur, E. and Suchocki, A. 2019. Afterglow, TL and OSL properties of Mn^{2+}-doped $ZnGa_2O_4$ phosphor, *Scientific Reports* 9(1),9544.

46. Iwase, N., Tadaki, S., Hidaka, S. and Koshino, N. 1994. Photostimulated luminescence of $BaBr_2$:Eu, *Journal of Luminescence* 60–61, 618–619.

47. Selling, J., Corradi, G., Secu, M. and Schweizer, S. 2005. Comparison of the luminescence properties of the X-ray storage phosphors $BaCl_2$:Ce^{3+} and $BaBr_2$:Ce^{3+}, *Journal of Physics Condensed Matter* 17, 8069–8078.

48. Amitani, K., Kano, A., Tsuchino, H. and Shimada, F. 1986. X-ray imaging system utilizing new photostimulable phosphor, In Proceedings of SPIE's Conference and Exhibition: Electric Imaging, 26th Fall Symposium, 180–183.

49. Schipper, W. J., Hamelink, J. J., Langeveld, E. M. and Blasse, G. 1993. Trapping of electrons by H^+ in the X-ray storage phosphor $Ba_3(PO_4)_2$:Eu^{2+}, La^{3+}, *Journal of Physics D: Applied Physics* 26, 1487–1492.

50. Kato, H., Miyahara, J. and Takano, M. 1985. New computed radiography using scanning laser stimulated luminescence, *Neurosurgical Review* 8, 53–62.

51. Zhao, W., Mi, Y., Su, M., Song, Z. and Xia, Z. 1996. Response time of photostimulated luminescence of $BaFX$:Eu^{2+} and BaFCl:Pr^{3+} under near-infrared light stimulation, *Journal of the Electrochemical Society* 143, 2346–2348.

52. Dong, Y. and Su, M.-Z. 1995. Luminescence and electro-conductance of BaFBr:Eu^{2+} crystals during X-irradiation and photostimulation, *Journal of Luminescence* 65, 263–268.

53. Takahashi, K., Kohda, K., Miyahara, J., Kanemitsu, Y., Amitani, K. and Shionoya, S. 1984. Mechanism of photostimulated luminescence in BaFX:Eu^{2+} (X=Cl,Br) phosphors, *Journal of Luminescence* 31–32, 266–268.

54. Iwabuchi, Y., Mori, N., Takahashi, K., Matsuda, T. and Shionoya, S. 1994. Mechanism of photostimulated luminescence process in BaFBr:Eu^{2+} phosphors, *Japanese Journal of Applied Physics* 33, 178–185.

55. von Seggern, H., Voigt, T., Knüpfer, W. and Lange, G. 1988. Physical model of photostimulated luminescence of X-ray irradiated BaFBr:Eu^{2+}, *Journal of Applied Physics* 64, 1405–1412.

56. Li, J., Huang, F., Wang, L., Yu, S. Q., Torimoto, Y., Sadakata, M. and Li, Q. X. 2005. High density hydroxyl anions in a microporous crystal: $[Ca_{24}Al_{28}O_{64}]^{4+}$·$4(OH^-)$, *Chemistry of Materials* 17, 2771–2774.

57. Rowlands, J. A. 2002. The physics of computed radiography, *Physics in Medicine and Biology* 47, R123–R166.

58. Kobayashi, H., Tanaka, S., Iwase, N. and Yoshiyama, H. 1988. Photostimulated luminescence in BaFBr:Eu^{2+} evaporated thin films, *Journal of Luminescence* 40–41, 819–820.

59. Hesse, S., Zimmermann, J., von Seggern, H., Meng, X., Fasel, C. and Riedel, R. 2009. Synthesis and functionality of the storage phosphor BaFBr:Eu^{2+}, *Journal of Applied Physics* 105(6), 063505.

60. Huang, J., Valenzano, L. and Sant, G. 2015. Framework and channel modifications in mayenite (12CaO·7Al_2O_3) nanocages by cationic doping, *Chemistry of Materials* 27, 4731–4741.

61. Hayashi, K., Matsuishi, S., Kamiya, T., Hirano, M. and Hosono, H. 2002. Light-induced conversion of an insulating refractory oxide into a persistent electronic conductor, *Nature* 419, 462–465.

62. McLeod, J. A., Buling, A., Kurmaev, E. Z., Sushko, P. V., Neumann, M., Finkelstein, L. D., Kim, S.-W., Hosono, H. and Moewes, A. 2012. Spectroscopic characterization of a multiband complex oxide: Insulating and conducting cement 12CaO·7Al_2O_3, *Physical Review B – Condensed Matter and Materials Physics* 85(4), 045204.

63. Li, S., Liu, Y., Liu, C., Yan, D., Zhu, H., Yang, J., Zhang, M., Xu, C., Ma, L. and Wang, X. 2017. Design, fabrication and characterization of nanocaged 12CaO·7Al_2O_3:Tb^{3+} photostimulable phosphor for high-quality X-ray imaging, *Materials and Design* 134, 1–9.

64. Li, S., Liu, C., Zhu, H., Yan, D., Xu, C. and Liu, Y. 2017. The improved storage properties of C12A7:Tb^{3+} X-ray phosphors via Sr^{2+} doping, *Materials Research Bulletin* 94, 140–146.

65. Trojan-Piegza, J. and Zych, E. 2010. Afterglow luminescence of Lu_2O_3:Eu ceramics synthesized at different atmospheres, *The Journal of Physical Chemistry C* 114, 4215–4220.

66. Concas, G., Spano, G., Zych, E. and Trojan-Piegza, J. 2005. Nano- and microcrystalline Lu_2O_3:Eu phosphors: Variations in occupancy of C_2 and S_6 sites by Eu^{3+} ions, *Journal of Physics Condensed Matter* 17, 2597–2604.

67. Murthy, K. V. R. 2014. Thermoluminescence and its applications: A review, *Defect and Diffusion Forum* 347, 35–73.

68. Sahare, P. D., Singh, M. and Kumar, P. 2014. TL characteristics of Ce^{3+}-doped $NaLi_2PO_4$ TLD phosphor, *Journal of Radioanalytical and Nuclear Chemistry* 302, 517–525.

69. Hochman, M. B. M. and Ypma, P. J. M. 1984. Thermoluminescence as a tool in uranium exploration, *Journal of Geochemical Exploration* 22, 315–331.

70. Fox, P. J., Akber, R. A. and Prescott, J. R. 1988. Spectral characteristics of six phosphors used in thermoluminescence dosimetry, *Journal of Physics D: Applied Physics* 21, 189–193.

71. Duragkar, A., Muley, A., Pawar, N. R., Chopra, V., Dhoble, N. S., Chimankar, O. P. and Dhoble, S. J. 2019. Versatility of thermoluminescence materials and radiation dosimetry – A review, *Luminescence* 34, 656–665.

72. Holt, J. G., Edelstein, G. R. and Clark, T. E. 1975. Energy dependence of the response of lithium fluoride TLD rods in high energy electron fields, *Physics in Medicine and Biology* 20, 559–570.

73. Jung, H., Lee, K. J. and Kim, J.-L. 2003. A personal thermoluminescence dosimeter using LiF:Mg,Cu,Na,Si detectors for photon fields, *Applied Radiation and Isotopes* 59, 87–93.

74. González, P. R., Mendoza-Anaya, D., Mendoza, L. and Escobar-Alarcón, L. 2018. Luminescence and dosimetric properties of CaF_2:Ce, Dy phosphor, *Journal of Luminescence* 195, 321–325.

75. Preto, P. D., Vidyavathy, B. and Dhabekar, B. S. 2017. Preparation, thermoluminescence, photoluminescence and dosimetric characteristics of LiF:Mg,Cu,P,B phosphor, *Indian Journal of Physics* 91, 25–31.

76. Bhadane, M. S., Dahiwale, S. S., Sature, K. R., Patil, B. J., Bhoraskar, V. N. and Dhole, S. D. 2017. TL studies of a sensitive $CaNa_2(SO_4)_2$:Dy nanophoshor for gamma dosimetry, *Radiation Measurements* 96, 1–7.

77. Nikiforov, S. V. and Kortov, V. S. 2014. Simulation of sublinear dose dependence of thermoluminescence with the inclusion of the competitive interaction of trapping centers, *Physics of the Solid State* 56, 2064–2068.

78. França, L. V. S. and Baffa, O. 2020. Boosted UV emission on the optically and thermally stimulated luminescence of CaB_6O_{10}:Gd,Ag phosphors excited by X-rays, *Applied Materials Today* 21, 100829.

79. Yang, J., Zhao, Y., Meng, Y., Zhu, H., Yan, D., Liu, C., Xu, C., Zhang, H., Xu, L., Li, Y. and Liu, Y. 2020. Irradiation-free photodynamic therapy in vivo induced by enhanced deep red afterglow within NIR-I bio-window, *Chemical Engineering Journal* 387, 124067.

80. Randall, J. T. and Wilkins, M. H. F. 1945. Phosphorescence and electron traps. I. The study of trap distributions, *Proceedings of the Royal Society of London* A184, 347–364.

81. Massillon-Jl, G., Gamboa-Debuen, I. and Brandan, M. E. 2006. Onset of supralinear response in TLD-100 exposed to ^{60}Co gamma-rays, *Journal of Physics D: Applied Physics* 39, 262–268.

82. Pereira, J. S., Pereira, M. F., Rangel, S., Caldeira, M., Carvalhal, G., Santos, L. M., Cardoso, J. V. and Alves, J. G. 2019. Type testing of LiF:Mg,Cu,P (TLD-100H) whole-body dosemeters for the assessment of Hp(10) and Hp(0.07), *Radiation Protection Dosimetry* 184, 216–223.

83. Zahedifar, M. and Sadeghi, E. 2012. Synthesis and dosimetric properties of the novel thermolumines-cent CaF_2:Tm nanoparticles, *Radiation Physics and Chemistry* 81, 1856–1861.

84. Romanyukha, A., Minniti, R., Moscovitch, M., Thompson, A. K., Trompier, F., Colle, R., Sucheta, A., Voss, S. P. and Benevides, L. A. 2011. Effect of neutron irradiation on dosimetric properties of TLD-600H (^{6}LiF:Mg,Cu,P), *Radiation Measurements* 46, 1426–1431.

85. Berger, T. and Hajek, M. 2008. On the linearity of the high-temperature emission from ^{7}LiF:Mg,Ti (TLD-700), *Radiation Measurements* 43, 1467–1473.

86. Voss, S. P., Sucheta, A., Romanyukha, A., Moscovitch, M., Kennemur, L. K. and Benevides, L. A. 2011. Effect of TLD-700H (LiF: Mg, Cu, P) sensitivity loss at multiple read-irradiation cycles on TLD reader calibration, *Radiation Measurements* 46, 1590–1594.

87. Binder, W. and Cameron, J. R. 1969. Dosimetric properties of CaF_2: Dy, *Health Physics* 17, 613–618.

88. Benevides, L., Romanyukha, A., Hull, F., Duffy, M., Voss, S. and Moscovitch, M. 2010. Light induced fading in the OSL response of Al_2O_3:C, *Radiation Measurements* 45, 523–526.

89. Shinsho, K., Maruyama, D., Yanagisawa, S., Koba, Y., Kakuta, M., Matsumoto, K., Ushiba, H. and Andoh, T. 2018. Thermoluminescence properties for X-ray of Cr-doped Al_2O_3 ceramics, *Sensors and Materials* 30, 1591–1598.

90. Kafadar, V. E. and Majeed, K. F. 2014. The effect of heating rate on the dose dependence and thermolu-minescence characteristics of $CaSO_4$:Dy (TLD-900), *Thermochimica Acta* 590, 266–269.

91. Danilkin, M., Lust, A., Kerikmäe, M., Seeman, V., Mändar, H. and Must, M. 2006. CaF_2:Mn extreme dosimeter: Effects of Mn concentration on thermoluminescence mechanisms and properties, *Radiation Measurements* 41, 677–681.

92. Sahare, P. D., Ali, N., Rawat, N. S., Bahl, S. and Kumar, P. 2016. Dosimetry characteristics of $NaLi_2PO_4$:Ce^{3+} OSLD phosphor, *Journal of Luminescence* 174, 22–28.

93. Erfani Haghiri, M., Saion, E., Soltani, N., Wan Abdullah, W. S., Navasery, M., Ebrahim Saraee, K. R. and Deyhimi, N. 2014. Thermoluminescent dosimetry properties of double doped calcium tetraborate (CaB_4O_7:Cu-Mn) nano-phosphor exposed to gamma radiation, *Journal of Alloys and Compounds* 582, 392–397.

94. Rahimi, M., Zahedifar, M. and Sadeghi, E. 2018. Synthesis, optical properties and thermoluminescence dosimetry features of manganese doped $Li_2B_4O_7$ nanoparticles, *Radiation Protection Dosimetry* 181, 360–367.

95. Sahare, P. D., Singh, M. and Kumar, P. 2015. Synthesis and TL characteristics of MgB_4O_7:Mn, Tb phos-phor, *Journal of Luminescence* 160, 158–164.

96. Mandlik, N., Sahare, P. D., Kulkarni, M. S., Bhatt, B. C., Bhoraskar, V. N. and Dhole, S. D. 2014. Study of TL and optically stimulated luminescence of $K_2Ca_2(SO_4)_3$:Cu nanophosphor for radiation dosimetry, *Journal of Luminescence* 146, 128–132.

97. Bulin, A.-L., Truillet, C., Chouikrat, R., Lux, F., Frochot, C., Amans, D., Ledoux, G., Tillement, O., Perriat, P., Barberi-Heyob, M. and Dujardin, C. 2013. X-ray-induced singlet oxygen activation with nanoscintillator-coupled porphyrins, *Journal of Physical Chemistry C* 117, 21583–21589.

98. Clement, S., Deng, W., Camilleri, E., Wilson, B. C. and Goldys, E. M. 2016. X-ray induced singlet oxy-gen generation by nanoparticle-photosensitizer conjugates for photodynamic therapy: Determination of singlet oxygen quantum yield, *Scientific Reports* 6, 19954.

99. Chen, H., Sun, X., Wang, G. D., Nagata, K., Hao, Z., Wang, A., Li, Z., Xie, J. and Shen, B. 2017. $LiGa_5O_8$:Cr-based theranostic nanoparticles for imaging-guided X-ray induced photodynamic therapy of deep-seated tumors, *Materials Horizons* 4, 1092–1101.

100. Hsu, C.-C., Lin, S.-L. and Chang, C. A. 2018. Lanthanide-doped core-shell-shell nanocomposite for dual photodynamic therapy and luminescence imaging by a single X-ray excitation source, *ACS Applied Materials and Interfaces* 10, 7859–7870.

101. Kamkaew, A., Chen, F., Zhan, Y., Majewski, R. L. and Cai, W. 2016. Scintillating nanoparticles as energy mediators for enhanced photodynamic therapy, *ACS Nano* 10, 3918–3935.

102. Huang, Z., Xu, H., Meyers, A. D., Musani, A. I., Wang, L., Tagg, R., Barqawi, A. B. and Chen, Y. K. 2008. Photodynamic therapy for treatment of solid tumors – Potential and technical challenges, *Technology in Cancer Research and Treatment* 7, 309–320.

103. Maisch, T., Baier, J., Franz, B., Maier, M., Landthaler, M., Szeimies, R.-M. and Bäumler, W. 2007. The role of singlet oxygen and oxygen concentration in photodynamic inactivation of bacteria, *Proceedings of the National Academy of Sciences of the United States of America* 104, 7223–7228.

104. Marchetti, B. and Karsili, T. N. V. 2016. An exploration of the reactivity of singlet oxygen with biomo-lecular constituents, *Chemical Communications* 52, 10996–10999.

105. Chen, W. and Zhang, J. 2006. Using nanoparticles to enable simultaneous radiation and photodynamic therapies for cancer treatment, *Journal of Nanoscience and Nanotechnology* 6, 1159–1166.

106. Chen, H., Wang, G. D., Chuang, Y.-J., Zhen, Z., Chen, X., Biddinger, P., Hao, Z., Liu, F., Shen, B., Pan, Z. and Xie, J. 2015. Nanoscintillator-mediated X-ray inducible photodynamic therapy for in vivo cancer treatment, *Nano Letters* 15, 2249–2256.

107. Sudheendra, L., Das, G. K., Li, C., Stark, D., Cena, J., Cherry, S. and Kennedy, I. M. 2014. NaGdF$_4$:Eu^{3+} nanoparticles for enhanced X-ray excited optical imaging, *Chemistry of Materials* 26, 1881–1888.

108. Zhong, X., Wang, X., Zhan, G., Tang, Y., Yao, Y., Dong, Z., Hou, L., Zhao, H., Zeng, S., Hu, J., Cheng, L. and Yang, X. 2019. NaCeF$_4$:Gd, Tb scintillator as an X-ray responsive photosensitizer for multimodal imaging-guided synchronous radio/radiodynamic therapy, *Nano Letters* 19, 8234–8244.

109. Wang, H., Lv, B., Tang, Z., Zhang, M., Ge, W., Liu, Y., He, X., Zhao, K., Zheng, X., He, M. and Bu, W. 2018. Scintillator-based nanohybrids with sacrificial electron prodrug for enhanced X-ray-induced photodynamic therapy, *Nano Letters* 18, 5768–5774.

110. Ma, L., Zou, X., Bui, B., Chen, W., Song, K. H. and Solberg, T. 2014. X-ray excited ZnS:Cu, Co afterglow nanoparticles for photodynamic activation, *Applied Physics Letters* 105, 013702.

111. Liu, X., Chen, Y., Li, H., Huang, N., Jin, Q., Ren, K. and Ji, J. 2013. Enhanced retention and cellular uptake of nanoparticles in tumors by controlling their aggregation behavior, *ACS Nano* 7, 6244–6257.

112. Jiang, R., Yang, J., Meng, Y., Yan, D., Liu, C., Xu, C. and Liu, Y. 2020. X-ray/red-light excited ZGGO:Cr,Nd nanoprobes for NIR-I/II afterglow imaging, *Dalton Transactions* 49, 6074–6083.

113. Ma, L., Zou, X., Bui, B., Chen, W., Song, K. H. and Solberg, T. 2014. X-ray excited ZnS:Cu, Co afterglow nanoparticles for photodynamic activation, *Applied Physics Letters* 105(1), 013702.

114. Elmenoufy, A. H., Tang, Y., Hu, J., Xu, H. and Yang, X. 2015. A novel deep photodynamic therapy modality combined with CT imaging established via X-ray stimulated silica-modified lanthanide scintillating nanoparticles, *Chemical Communications* 51, 12247–12250.

115. Zhang, C., Zhao, K., Bu, W., Ni, D., Liu, Y., Feng, J. and Shi, J. 2015. Marriage of scintillator and semiconductor for synchronous radiotherapy and deep photodynamic therapy with diminished oxygen dependence, *Angewandte Chemie – International Edition* 54, 1770–1774.

116. Clement, S., Chen, W., Anwer, A. G. and Goldys, E. M. 2017. Verteporfin conjugated to gold nanoparticles for fluorescent cellular bioimaging and X-ray mediated photodynamic therapy, *Microchimica Acta* 184, 1765–1771.

117. Hajagos, T. J., Liu, C., Cherepy, N. J. and Pei, Q. 2018. High-Z sensitized Plastic scintillators: A review, *Advanced Materials* 30(27), 1706956.

118. Fabbri, F., Rossi, F., Attolini, G., Salviati, G., Dierre, B., Sekiguchi, T. and Fukata, N. 2012. Luminescence properties of SiC/SiO$_2$ core-shell nanowires with different radial structure, *Materials Letters* 71, 137–140.

119. Kaščáková, S., Giuliani, A., Lacerda, S., Pallier, A., Mercère, P., Tóth, É. and Réfrégiers, M. 2015. X-ray-induced radiophotodynamic therapy (RPDT) using lanthanide micelles: Beyond depth limitations, *Nano Research* 8, 2373–2379.

120. Zou, X., Yao, M., Ma, L., Hossu, M., Han, X., Juzenas, P. and Chen, W. 2014. X-ray-induced nanoparticle-based photodynamic therapy of cancer, *Nanomedicine* 9, 2339–2351.

121. Yanagida, T. 2017. Recent progress of transparent ceramic scintillators, *Advances in Science and Technology* 98, 44–53.

122. Dujardin, C., Auffray, E., Bourret-Courchesne, E., Dorenbos, P., Lecoq, P., Nikl, M., Vasil'Ev, A. N., Yoshikawa, A. and Zhu, R.-Y. 2018. Needs, trends, and advances in inorganic scintillators, *IEEE Transactions on Nuclear Science* 65, 1977–1997.

123. Nikl, M. and Yoshikawa, A. 2015. Recent R&D trends in inorganic single-crystal scintillator materials for radiation detection, *Advanced Optical Materials* 3, 463–481.

124. Lu, L., Sun, M., Lu, Q., Wu, T. and Huang, B. 2021. High energy X-ray radiation sensitive scintillating materials for medical imaging, cancer diagnosis and therapy, *Nano Energy* 79, 105437.

125. Seferis, I., Michail, C., Valais, I., Zeler, J., Liaparinos, P., Fountos, G., Kalyvas, N., David, S., Stromatia, F., Zych, E., Kandarakis, I. and Panayiotakis, G. 2014. Light emission efficiency and imaging performance of Lu$_2$O$_3$:Eu nanophosphor under X-ray radiography conditions: Comparison with Gd$_2$O$_2$S:Eu, *Journal of Luminescence* 151, 229–234.

126. Spahn, M. 2013. X-ray detectors in medical imaging, *Nuclear Instruments and Methods in Physics Research, Section A: Accelerators, Spectrometers, Detectors and Associated Equipment* 731, 57–63.

127. Anger, H. O. 1958. Scintillation camera, *Review of Scientific Instruments* 29, 27–33.

128. Vandenberghe, S., Mikhaylova, E., D'Hoe, E., Mollet, P. and Karp, J. S. 2016. Recent developments in time-of-flight PET, *EJNMMI Physics* 3, 3.

129. van Eijk, C. W. E., Dorenbos, P., van Loef, E. V. D., Krämer, K. and Güdel, H. U. 2001. Energy resolution of some new inorganic-scintillator gamma-ray detectors, *Radiation Measurements* 33, 521–525.

130. Sakai, E. 1987. Recent measurements on scintillator-photodetector systems, *IEEE Transactions on Nuclear Science* 34, 418–422.

131. Bhattacharya, P., Wart, M., Miller, S., Brecher, C. and Nagarkar, V. V. 2019. Codoped lithium sodium iodide with Tl^+ and Eu^{2+} activators for neutron detector, *IEEE Transactions on Nuclear Science* 66, 2136–2139.

132. Holl, I., Lorenz, E. and Mageras, G. 1988. A measurement of the light yield of common inorganic scintillators, *IEEE Transactions on Nuclear Science* 35, 105–109.

133. Laval, M., Moszyński, M., Allemand, R., Cormoreche, E., Guinet, P., Odru, R. and Vacher, J. 1983. Barium fluoride – Inorganic scintillator for subnanosecond timing, *Nuclear Instruments and Methods In Physics Research* 206, 169–176.

134. Moses, W. W. and Derenzo, S. E. 1990. The scintillation properties of cerium-doped lanthanum fluoride, *Nuclear Instruments & Methods in Physics Research. Section A* 299, 51–56.

135. van Loef, E. V. D., Dorenbos, P., van Eijk, C. W. E., Krämer, K. and Güdel, H. U. 2001. High-energy-resolution scintillator: Ce^{3+} activated $LaBr_3$, *Applied Physics Letters* 79, 1573–1575.

136. van Loef, E. V. D., Dorenbos, P., van Eijk, C. W. E., Krämer, K. and Güdel, H. U. 2000. High-energy-resolution scintillator: Ce^{3+} activated $LaCl_3$, *Applied Physics Letters* 77, 1467–1468.

137. Birowosuto, M. D. and Dorenbos, P. 2009. Novel γ- and X-ray scintillator research: On the emission wavelength, light yield and time response of Ce^{3+} doped halide scintillators, *Physica Status Solidi (A) Applications and Materials Science* 206, 9–20.

138. Birowosuto, M. D., Dorenbos, P., van Eijk, C. W. E., Krämer, K. W. and Güdel, H. U. 2006. High-light-output scintillator for photodiode readout: $LuI_3:Ce^{3+}$, *Journal of Applied Physics* 99(12), 123520.

139. Birowosuto, M. D., Dorenbos, P., van Eijk, C. W. E., Krämer, K. W. and Güdel, H. U. 2007. Thermal quenching of Ce^{3+} emission in PrX_3 ($X \leq Cl$, Br) by intervalence charge transfer, *Journal of Physics Condensed Matter* 19(25), 256209.

140. Raue, R., Shiiki, M., Matsukiyo, H., Toyama, H. and Yamamoto, H. 1994. Saturation of ZnS:Ag, Al under cathode-ray excitation, *Journal of Applied Physics* 75, 481–488.

141. van Eijk, C. W. E. 2012. Inorganic scintillators for thermal neutron detection, *IEEE Transactions on Nuclear Science* 59, 2242–2247.

142. Kim, H. J., Rooh, G., Khan, A. and Kim, S. 2017. New $Tl_2LaBr_5:Ce^{3+}$ crystal scintillator for γ-rays detection, *Nuclear Instruments and Methods in Physics Research, Section A: Accelerators, Spectrometers, Detectors and Associated Equipment* 849, 72–75.

143. Moszyński, M., Balcerzyk, M., Czarnacki, W., Nassalski, A., Szczęśniak, T., Kraus, H., Mikhailik, V. B. and Solskii, I. M. 2005. Characterization of $CaWO_4$ scintillator at room and liquid nitrogen temperatures, *Nuclear Instruments and Methods in Physics Research, Section A: Accelerators, Spectrometers, Detectors and Associated Equipment* 553, 578–591.

144. Kinloch, D. R., Novak, W., Raby, P. and Toepke, I. 1994. New developments in cadmium tungstate, *IEEE Transactions on Nuclear Science* 41, 752–754.

145. Danevich, F. A., Henry, S., Kraus, H., McGowan, R., Mikhailik, V. B., Shkulkova, O. G. and Telfer, J. 2008. Scintillation properties of pure and Ca-doped $ZnWO_4$ crystals, *Physica Status Solidi (A) Applications and Materials Science* 205, 335–339.

146. Derenzo, S. E., Moses, W. W., Cahoon, J. L., Perera, R. C. C. and Litton, J. E. 1990. Prospects for new inorganic scintillators, *IEEE Transactions on Nuclear Science* 37, 203–208.

147. Miller, S. R., Nagarkar, V. V., Tipnis, S. V., Shestakova, I., Brecher, C., Lempicki, A. and Lingertat, H. 2003. $Lu_2O_3:Eu$ scintillator screen for X-ray imaging, In Proceedings of SPIE – The International Society for Optical Engineering, pp 167–172.

148. Derenzo, S., Bizarri, G., Borade, R., Bourret-Courchesne, E., Boutchko, R., Canning, A., Chaudhry, A., Eagleman, Y., Gundiah, G., Hanrahan, S., Janecek, M. and Weber, M. 2011. New scintillators discovered by high-throughput screening, *Nuclear Instruments and Methods in Physics Research, Section A: Accelerators, Spectrometers, Detectors and Associated Equipment* 652, 247–250.

149. Greskovich, C. and Duclos, S. 1997. Ceramic scintillators, *Annual Review of Materials Science* 27, 69–88.

150. Kapusta, M., Pawelke, J. and Moszyński, M. 1998. Comparison of YAP and BGO for high-resolution PET detectors, *Nuclear Instruments and Methods in Physics Research Section A: Accelerators, Spectrometers, Detectors and Associated Equipment* 404, 413–417.

151. Ishii, M., Harada, K., Hirose, Y., Senguttuvan, N., Kobayashi, M., Yamaga, I., Ueno, H., Miwa, K., Shiji, F., Yiting, F., Nikl, M. and Feng, X. Q. 2002. Development of BSO ($Bi_4Si_3O_{12}$) crystal for radiation detector, *Optical Materials* 19, 201–212.

152. Nakamura, F., Kantuptim, P., Nakauchi, D., Kato, T., Kawaguchi, N. and Yanagida, T. 2020. Scintillation properties of BaSiO$_3$:Ce crystals by the floating zone method, *Materials Research Bulletin* 131, 110961.

153. Melcher, C. L. and Schweitzer, J. S. 1992. Cerium-doped lutetium oxyorthosilicate: A fast, efficient new scintillator, *IEEE Transactions on Nuclear Science* 39, 502–505.

154. Pauwels, D., Le Masson, N., Viana, B., Kahn-Harari, A., van Loef, E. V. D., Dorenbos, P. and van Eijk, C. W. E. 2000. A novel inorganic scintillator: Lu$_2$Si$_2$O$_7$:Ce^{3+} (LPS), *IEEE Transactions on Nuclear Science* 47, 1787–1790.

155. Dorenbos, P., de Haas, J. T. D. and van Eijk, C. W. V. 1995. Non-proportionality in the scintillation response and the energy resolution obtainable with scintillation crystals, *IEEE Transactions on Nuclear Science* 42, 2190–2202.

156. Shimizu, S., Kurashige, K., Usui, T., Shimura, N., Sumiya, K., Senguttuvan, N., Gunji, A., Kamada, M. and Ishibashi, H. 2006. Scintillation properties of Lu$_{0.4}$Gd$_{1.6}$SiO$_5$:Ce (LGSO) crystal, *IEEE Transactions on Nuclear Science* 53, 14–17.

157. Drozdowski, W., Brylew, K., Wojtowicz, A. J., Kisielewski, J., Świrkowicz, M., Łukasiewicz, T., de Haas, J. T. M. and Dorenbos, P. 2014. 33000 photons per MeV from mixed (Lu$_{0.75}$Y$_{0.25}$)$_3$Al$_5$O$_{12}$:Pr scintillator crystals, *Optical Materials Express* 4, 1207–1212.

158. Moszyński, M., Kapusta, M., Wolski, D., Klamra, W. and Cederwall, B. 1998. Properties of the YAP:Ce scintillator, *Nuclear Instruments and Methods in Physics Research Section A: Accelerators, Spectrometers, Detectors and Associated Equipment* 404, 157–165.

159. Moszyński, M., Wolski, D., Ludziejewski, T., Kapusta, M., Lempicki, A., Brecher, C., Wiśniewski, D. and Wojtowicz, A. J. 1997. Properties of the new LuAP:Ce scintillator, *Nuclear Instruments and Methods in Physics Research Section A: Accelerators, Spectrometers, Detectors and Associated Equipment* 385, 123–131.

160. Mansuy, C., Nedelec, J. M., Dujardin, C. and Mahiou, R. 2004. Scintillation of sol-gel derived lutetium orthoborate doped with Ce^{3+} ions, *Journal of Sol-Gel Science and Technology* 32, 253–258.

161. Mareš, J. A., Jacquier, B., Pédrini, C. and Boulon, G. 1988. Fluorescence decays and lifetimes of Nd^{3+}, Ce^{3+} and Cr^{3+} in YAG, *Czechoslovak Journal of Physics B* 38, 802–816.

162. Dorenbos, P., Visser, R., van Eijk, C. W. E., Khaidukov, N. M. and Korzhik, M. V. 1993. Scintillation properties of some Ce^{3+} and Pr^{3+} doped inorganic crystals, *IEEE Transactions on Nuclear Science* 40, 388–394.

163. Glodo, J., van Loef, E., Hawrami, R., Higgins, W. M., Churilov, A., Shirwadkar, U. and Shah, K. S. 2011. Selected properties of Cs$_2$LiYCl$_6$, Cs$_2$LiLaCl$_6$, and Cs$_2$LiLaBr$_6$ scintillators, *IEEE Transactions on Nuclear Science* 58, 333–338.

164. Bessiere, A., Dorenbos, P., van Eijk, C. W. E., Krämer, K. W. and Güdel, H. U. 2004. New thermal neutron scintillators: Cs$_2$LiYCl$_6$:Ce^{3+} and Cs$_2$LiYBr$_6$:Ce^{3+}, *IEEE Transactions on Nuclear Science* 51, 2970–2972.

165. van Loef, E. V. D., Glodo, J., Higgins, W. M. and Shah, K. S. 2005. Optical and scintillation properties of Cs$_2$LiYCl$_6$:Ce^{3+} and Cs$_2$LiYCl$_6$:Pr^{3+} crystals, *IEEE Transactions on Nuclear Science* 52, 1819–1822.

166. Cherepy, N. J., Payne, S. A., Asztalos, S. J., Hull, G., Kuntz, J. D., Niedermayr, T., Pimputkar, S., Roberts, J. J., Sanner, R. D., Tillotson, T. M., van Loef, E., Wilson, C. M., Shah, K. S., Roy, U. N., Hawrami, R., Burger, A., Boatner, L. A., Choong, W.-S. and Moses, W. W. 2009. Scintillators with potential to supersede lanthanum bromide, *IEEE Transactions on Nuclear Science* 56, 873–880.

167. Dorenbos, P., Van't Spijker, J. C., Frijns, O. W. V., van Eijk, C. W. E., Krämer, K., Güdel, H. U. and Ellens, A. 1997. Scintillation properties of RbGd$_2$Br$_7$:Ce^{3+} crystals; fast, efficient, and high density scintillators, *Nuclear Instruments and Methods in Physics Research, Section B: Beam Interactions with Materials and Atoms* 132, 728–731.

168. van Loef, E. V. D., Dorenbos, P., van Eijk, C. W. E., Krämer, K. W. and Güdel, H. U. 2005. Scintillation properties of K$_2$LaX$_5$:Ce^{3+} (X=Cl, Br, I), *Nuclear Instruments and Methods in Physics Research Section A: Accelerators, Spectrometers, Detectors and Associated Equipment* 537, 232–236.

169. Stand, L., Zhuravleva, M., Chakoumakos, B., Johnson, J., Lindsey, A. and Melcher, C. L. 2016. Scintillation properties of Eu^{2+}-doped KBa$_2$I$_5$ and K$_2$BaI$_4$, *Journal of Luminescence* 169, 301–307.

170. Samulon, E. C., Gundiah, G., Gascón, M., Khodyuk, I. V., Derenzo, S. E., Bizarri, G. A. and Bourret-Courchesne, E. D. 2014. Luminescence and scintillation properties of Ce^{3+}-activated Cs$_2$NaGdCl$_6$, Cs$_3$GdCl$_6$, Cs$_2$NaGdBr$_6$ and Cs$_3$GdBr$_6$, *Journal of Luminescence* 153, 64–72.

171. Cherepy, N. J., Sturm, B. W., Drury, O. B., Hurst, T. A., Sheets, S. A., Ahle, L. E., Saw, C. K., Pearson, M. A., Payne, S. A., Burger, A., Boatner, L. A., Ramey, J. O., van Loef, E. V., Glodo, J., Hawrami, R., Higgins, W. M., Shah, K. S. and Moses, W. W. 2009. SrI$_2$ scintillator for gamma ray spectroscopy, In *Proceedings of SPIE – The International Society for Optical Engineering*.

172. Cherepy, N. J., Hull, G., Drobshoff, A. D., Payne, S. A., van Loef, E., Wilson, C. M., Shah, K. S., Roy, U. N., Burger, A., Boatner, L. A., Choong, W.-S. and Moses, W. W. 2008. Strontium and barium iodide high light yield scintillators, *Applied Physics Letters* 92(8), 083508.

173. Lindsey, A. C., Zhuravleva, M., Stand, L., Wu, Y. and Melcher, C. L. 2015. Crystal growth and characterization of europium doped KCaI$_3$, a high light yield scintillator, *Optical Materials* 48, 1–6.

174. Stand, L., Zhuravleva, M., Lindsey, A. and Melcher, C. L. 2015. Growth and characterization of potassium strontium iodide: A new high light yield scintillator with 2.4% energy resolution, *Nuclear Instruments and Methods in Physics Research, Section A: Accelerators, Spectrometers, Detectors and Associated Equipment* 780, 40–44.

175. Yoshikawa, A., Kamada, K., Kurosawa, S., Shoji, Y., Yokota, Y., Chani, V. I. and Nikl, M. 2016. Crystal growth and scintillation properties of multi-component oxide single crystals: Ce:GGAG and Ce:La-GPS, *Journal of Luminescence* 169, 387–393.

176. Gektin, A., Vasyukov, S., Galenin, E., Taranyuk, V., Nazarenko, N. and Romanchuk, V. 2016. Strontium iodide: Technology aspects of raw material choice and crystal growth, *Functional Materials* 23, 473–477.

177. El Hassouni, A., Lebbou, K., Goutaudier, C., Boulon, G., Yoshikawa, A. and Fukuda, T. 2003. SBN single crystal fibers grown by micro-pulling down technique, *Optical Materials* 24, 419–424.

178. Shonai, T., Higuchi, M. and Kodaira, K. 2001. High-speed float zone growth of heavily Nd-doped YVO$_4$ single crystals, *Journal of Crystal Growth* 233, 477–482.

179. Melcher, C. L. 2005. Perspectives on the future development of new scintillators, *Nuclear Instruments and Methods in Physics Research Section A: Accelerators, Spectrometers, Detectors and Associated Equipment* 537, 6–14.

180. Zhu, R.-Y. 2018. The next generation of crystal detectors, *Radiation Detection Technology and Methods* 2(1), 2.

181. Ter-Pogossian, M. M., Phelps, M. E., Hoffman, E. J. and Mullani, N. A. 1975. A positron emission transaxial tomograph for nuclear imaging (PETT), *Radiology* 114, 89–98.

182. Chen, J., Yang, F., Zhang, L., Zhu, R.-Y., Du, Y., Wang, S., Sun, S. and Li, X. 2018. Slow scintillation suppression in yttrium doped BaF$_2$ Crystals, *IEEE Transactions on Nuclear Science* 65, 2147–2151.

183. Annenkov, A. N., Auffray, E., Chipaux, R., Drobychev, G. Y., Fedorov, A. A., Géléoc, M., Golubev, N. A., Korzhik, M. V., Lecoq, P., Lednev, A. A., Ligun, A. B., Missevitch, O. V., Pavlenko, V. B., Peigneux, J.-P. and Singovski, A. V. 1998. Systematic study of the short-term instability of PBWO$_4$ scintillator parameters under irradiation, *Radiation Measurements* 29, 27–38.

184. Mao, R., Zhang, L. and Zhu, R.-Y. 2012. Crystals for the HHCAL detector concept, *IEEE Transactions on Nuclear Science* 59, 2229–2236.

185. Driutti, A., Para, A., Pauletta, G., Briones, N. R. and Wenzel, H. 2011. Towards jet reconstruction in a realistic dual readout total absorption calorimeter, *Journal of Physics: Conference Series* 293, 012034.

186. Pauwels, K., Dujardin, C., Gundacker, S., Lebbou, K., Lecoq, P., Lucchini, M., Moretti, F., Petrosyan, A. G., Xu, X. and Auffray, E. 2013. Single crystalline LuAG fibers for homogeneous dual-readout calorimeters, *Journal of Instrumentation* 8, P09019–P09019.

187. Yang, F., Yuan, H., Zhang, L. and Zhu, R.-Y. 2015. BSO crystals for the HHCAL detector concept, *Journal of Physics: Conference Series* 587, 012064.

188. Kouzes, R. T., Ely, J. H., Lintereur, A. T., MacE, E. K., Stephens, D. L. and Woodring, M. L. 2011. Neutron detection gamma ray sensitivity criteria, *Nuclear Instruments and Methods in Physics Research, Section A: Accelerators, Spectrometers, Detectors and Associated Equipment* 654, 412–416.

189. Runkle, R. C., Bernstein, A. and Vanier, P. E. 2010. Securing special nuclear material: Recent advances in neutron detection and their role in nonproliferation, *Journal of Applied Physics* 108(11), 111101.

190. Yanagida, T. 2018. Inorganic scintillating materials and scintillation detectors, *Proceedings of the Japan Academy Series B: Physical and Biological Sciences* 94, 75–97.

191. Milbrath, B. D., Peurrung, A. J., Bliss, M. and Weber, W. J. 2008. Radiation detector materials: An overview, *Journal of Materials Research* 23, 2561–2581.

192. Kawaguchi, N., Yanagida, T., Novoselov, A., Kim, K. J., Fukuda, K., Yoshikawa, A., Miyake, M. and Baba, M. 2008. Neutron responses of Eu^{2+} activated LiCaAlF$_6$ scintillator, In IEEE Nuclear Science Symposium Conference Record, pp 1174–1176.

193. Yanagida, T., Yoshikawa, A., Yokota, Y., Maeo, S., Kawaguchi, N., Ishizu, S., Fukuda, K. and Suyama, T. 2009. Crystal growth, optical properties, and α-ray responses of Ce-doped LiCaAlF$_6$ for different Ce concentration, *Optical Materials* 32, 311–314.

194. Glodo, J., Higgins, W. M., van Loef, E. V. D. and Shah, K. S. 2008. Scintillation properties of 1 inch Cs$_2$LiYCl$_6$:Ce crystals, *IEEE Transactions on Nuclear Science* 55, 1206–1209.

195. Yanagida, T., Kawaguchi, N., Fujimoto, Y., Fukuda, K., Yokota, Y., Yamazaki, A., Watanabe, K., Pejchal, J., Uritani, A., Iguchi, T. and Yoshikawa, A. 2011. Basic study of Europium doped LiCaAlF$_6$ scintillator and its capability for thermal neutron imaging application, *Optical Materials* 33, 1243–1247.

196. Yanagida, T., Yamaji, A., Kawaguchi, N., Fujimoto, Y., Fukuda, K., Kurosawa, S., Yamazaki, A., Watanabe, K., Futami, Y., Yokota, Y., Uritani, A., Iguchi, T., Yoshikawa, A. and Nikl, M. 2011. Europium and sodium codoped LiCaAlF$_6$ scintillator for neutron detection, *Applied Physics Express* 4(10), 106401.

197. Fujimoto, Y., Yanagida, T., Kawaguchi, N., Kurosawa, S., Fukuda, K., Totsuka, D., Watanabe, K., Yamazaki, A., Yokota, Y. and Yoshikawa, A. 2012. Characterizations of Ce^{3+}-doped CaB$_2$O$_4$ crystalline scintillator, *Crystal Growth & Design* 12, 142–146.

198. Melcher, C. L. 1989. Scintillators for well logging applications, *Nuclear Instruments & Methods in Physics Research. Section B* 40–41, 1214–1218.

199. Odom, R. C., Tiller, D. E. and Wilson, R. D. 2005. Experiments on closely spaced detector candidates for carbon/oxygen logging, *Petrophysics* 46, 188–198.

200. Nikitin, A. and Bliven, S. 2010. Needs of well logging industry in new nuclear detectors, In IEEE Nuclear Science Symposium conference record, pp 1214–1219.

201. Yanagida, T., Fujimoto, Y., Kurosawa, S., Kamada, K., Takahashi, H., Fukazawa, Y., Nikl, M. and Chani, V. 2013. Temperature dependence of scintillation properties of bright oxide scintillators for well-logging, *Japanese Journal of Applied Physics* 52, 076401.

202. Iltis, A., Mayhugh, M. R., Menge, P., Rozsa, C. M., Selles, O. and Solovyev, V. 2006. Lanthanum halide scintillators: Properties and applications, *Nuclear Instruments and Methods in Physics Research, Section A: Accelerators, Spectrometers, Detectors and Associated Equipment* 563, 359–363.

203. Guo, W., Jacobson, L., Truax, J., Dorffer, D. and Kwong, S. 2010. A new three-detector 1-11/16-inch pulsed neutron tool for unconventional reservoirs, In SPWLA 51st Annual Logging Symposium, 2010.

204. Wang, Y., Rhodes, W. H., Baldoni, G., van Loef, E., Glodo, J., Brecher, C., Nguyen, L. and Shah, K. S. 2009. Lu$_2$SiO$_5$:Ce optical ceramic scintillator, In Proceedings of SPIE – The International Society for Optical Engineering.

205. Kawamura, S., Kaneko, J. H., Higuchi, M., Fujita, F., Homma, A., Haruna, J., Saeki, S., Kurashige, K., Ishibashi, H. and Furusaka, M. 2007. Investigation of Ce-doped Gd$_2$Si$_2$O$_7$ as a scintillator material, *Nuclear Instruments and Methods in Physics Research, Section A: Accelerators, Spectrometers, Detectors and Associated Equipment* 583, 356–359.

206. Fujii, T., Tanaka, M., Yazaki, Y., Kitabayashi, H., Koizumi, T., Hongo, M., Sekiguchi, M., Itoh, A., Gomi, T. and Yano, K. 1997. [Myocardial scintigraphic studies with 123I-MIBG, 201Tl and 99mTc-PYP in patients with cardiac amyloidosis], *Kaku Igaku* 34, 1033–1039.

207. Hutton, B. F. 2014. The origins of SPECT and SPECT/CT, *European Journal of Nuclear Medicine and Molecular Imaging* 41, S3–S16.

208. Vandenberghe, S., Mikhaylova, E., D'Hoe, E., Mollet, P. and Karp, J. S. 2016. Recent developments in time-of-flight PET, *EJNMMI Physics* 3, 3.

209. Madsen, M. T. 2007. Recent advances in SPECT imaging, *The Journal of Nuclear Medicine* 48, 661–673.

210. Cherry, S. R., Jones, T., Karp, J. S., Qi, J., Moses, W. W. and Badawi, R. D. 2018. Total-body PET: Maximizing sensitivity to create new opportunities for clinical research and patient care, *Journal of Nuclear Medicine* 59, 3–12.

211. Perrin, D. M. and Ting, R. 2012. *Radiolabeled Compounds and Compositions, Their Precursors and Methods for Their Production*, The University of British Columbia: Vancouver, British Columbia, Canada; United States.

212. van Eijk, C. W. E. 2002. Inorganic scintillators in medical imaging, *Physics in Medicine and Biology* 47, R85–R106.

213. Conti, M. 2009. State of the art and challenges of time-of-flight PET, *Physica Medica* 25, 1–11.

214. Lewellen, T. K. 2008. Recent developments in PET detector technology, *Physics in Medicine and Biology* 53, R287–R317.

215. Surti, S., Kuhn, A., Werner, M. E., Perkins, A. E., Kolthammer, J. and Karp, J. S. 2007. Performance of Philips Gemini TF PET/CT scanner with special consideration for its time-of-flight imaging capabilities, *Journal of Nuclear Medicine* 48, 471–480.

216. Kuhn, A., Surti, S., Karp, J. S., Raby, R. S., Shah, K. S., Perkins, A. E. and Muehllehner, G. 2004. Design of a lanthanum bromide detector for time-of-flight PET, *IEEE Transactions on Nuclear Science* 51, 2550–2557.

217. Alekhin, M. S., de Haas, J. T. M., Khodyuk, I. V., Krämer, K. W., Menge, P. R., Ouspenski, V. and Dorenbos, P. 2013. Improvement of γ-ray energy resolution of $LaBr_3$:Ce^{3+} scintillation detectors by Sr^{2+} and Ca^{2+} co-doping, *Applied Physics Letters* 102(16), 161915.

218. Nikl, M., Kamada, K., Babin, V., Pejchal, J., Pilarova, K., Mihokova, E., Beitlerova, A., Bartosiewicz, K., Kurosawa, S. and Yoshikawa, A. 2014. Defect engineering in Ce-doped aluminum garnet single crystal scintillators, *Crystal Growth and Design* 14, 4827–4833.

219. Nagura, A., Kamada, K., Nikl, M., Kurosawa, S., Pejchal, J., Yokota, Y., Ohashi, Y. and Yoshikawa, A. 2015. Improvement of scintillation properties on Ce doped $Y_3Al_5O_{12}$ scintillator by divalent cations co-doping, *Japanese Journal of Applied Physics* 54(4), 04DH17.

220. Petrosyan, A. G., Ovanesyan, K. L., Derdzyan, M. V., Ghambaryan, I., Patton, G., Moretti, F., Auffray, E., Lecoq, P., Lucchini, M., Pauwels, K. and Dujardin, C. 2015. A study of radiation effects on LuAG:Ce(Pr) co-activated with Ca, *Journal of Crystal Growth* 430, 46–51.

221. Shah, K. S., Glodo, J., Klugerman, M., Higgins, W., Gupta, T., Wong, P., Moses, W. W., Derenzo, S. E., Weber, M. J. and Dorenbos, P. 2004. LuI_3:Ce – A new scintillator for gamma ray spectroscopy, *IEEE Transactions on Nuclear Science* 51, 2302–2305.

222. Glodo, J., Shah, K. S., Klugerman, M., Wong, P., Higgins, B. and Dorenbos, P. 2005. Scintillation properties of LuI_3:Ce, *Nuclear Instruments and Methods in Physics Research, Section A: Accelerators, Spectrometers, Detectors and Associated Equipment* 537, 279–281.

223. Nikl, M., Ogino, H., Krasnikov, A., Beitlerova, A., Yoshikawa, A. and Fukuda, T. 2005. Photo- and radioluminescence of Pr-doped $Lu_3Al_5O_{12}$ single crystal, *Physica Status Solidi (A)* 202, R4–R6.

224. Ogino, H., Yoshikawa, A., Nikl, M., Pejchal, J. and Fukuda, T. 2007. Growth and luminescence properties of Pr-doped $Lu_3(Ga,Al)_5O_{12}$ single crystals, *Japanese Journal of Applied Physics* 46, 3514–3517.

225. Ammon, J., Frik, W., Karstens, J. H., Rübben, H. and Schoffers, J. 1979. [Total body computer tomography or the urogenital system (author's transl)], *Urologe A* 18, 1–13.

226. van Eijk, C. W. E. 2003. Inorganic scintillators in medical imaging detectors, *Nuclear Instruments and Methods in Physics Research Section A: Accelerators, Spectrometers, Detectors and Associated Equipment* 509, 17–25.

227. Ronda, C., Wieczorek, H., Khanin, V. and Rodnyi, P. 2016. Review-scintillators for medical imaging: A tutorial overview, *ECS Journal of Solid State Science and Technology* 5, R3121–R3125.

228. Wu, Y., Ren, G., Nikl, M., Chen, X., Ding, D., Li, H., Pan, S. and Yang, F. 2014. CsI:Tl^+,Yb^{2+}: Ultra-high light yield scintillator with reduced afterglow, *CrystEngComm* 16, 3312–3317.

229. Maddalena, F., Tjahjana, L., Xie, A., Zeng, S., Wang, H., Coquet, P., Drozdowski, W., Dujardin, C., Dang, C. and Birowosuto, M. D. 2019. Inorganic, organic, and perovskite halides with nanotechnology for high-light yield X- and γ-ray scintillators, *Crystals* 9(2), 88.

230. Birowosuto, M. D., Dorenbos, P., Krämer, K. W. and Güdel, H. U. 2008. Ce^{3+} activated $LaBr_{3-x}I_x$: High-light-yield and fast-response mixed halide scintillators, *Journal of Applied Physics* 103, 103517.

231. Pitchford, G. 2001. Radiotherapy physics: In practice (second edition), *Physics in Medicine and Biology* 46, 899–899.

232. Hiroshi, K., Atsuya, Y. and Keiichi, S. 1993. High-MTF X-ray image intensifier, In Proc.SPIE.

233. Webb, S. 1990. *The Physics of Medical Imaging*, Adam Hilger: Bristol.

234. Körner, M., Weber, C. H., Wirth, S., Pfeifer, K.-J., Reiser, M. F. and Treitl, M. 2007. Advances in digital radiography: Physical principles and system overview, *Radiographics* 27, 675–686.

235. Antonuk, L. E., El-Mohri, Y., Siewerdsen, J. H., Yorkston, J., Huang, W., Scarpine, V. E. and Street, R. A. 1997. Empirical investigation of the signal performance of a high-resolution, indirect detection, active matrix flat-panel imager (AMFPI) for fluoroscopic and radiographic operation, *Medical Physics* 24, 51–70.

236. Chabbal, J., Chaussat, C., Ducourant, T., Fritsch, L., Michailos, J., Spinnler, V., Vieux, G., Arques, M., Hahm, G., Hoheisel, M., Horbaschek, H., Schulz, R. and Spahn, M. 1996. Amorphous silicon X-ray image sensor, In Proceedings of SPIE – The International Society for Optical Engineering, pp 499–510.

237. Granfors, P. R., Albagli, D., Tkaczyk, J. E., Aufrichtig, R., Netel, H., Brunst, G., Boudry, J. and Luo, D. 2001. Performance of a flat panel cardiac detector, In Proceedings of SPIE – The International Society for Optical Engineering, pp 77–86.

238. Yamazaki, T., Tamura, T., Nokita, M., Okada, S., Hayashida, S. and Ogawa, Y. 2004. Performance of a novel 43 cm × 43 cm flat-panel detector with CsI:Tl scintillator, In Proceedings of SPIE – The International Society for Optical Engineering 1 ed., pp 379–385.

239. Miyata, E., Miki, M., Tawa, N. and Miyaguchi, K. 2005. X-ray responsivities of direct-scintillator-deposited charge-coupled device, *Japanese Journal of Applied Physics* 44, 1476–1484.

240. Yorkston, J. 2007. Recent developments in digital radiography detectors, *Nuclear Instruments and Methods in Physics Research Section A: Accelerators, Spectrometers, Detectors and Associated Equipment* 580, 974–985.

241. Heo, J. H., Shin, D. H., Park, J. K., Kim, D. H., Lee, S. J. and Im, S. H. 2018. High-performance next-generation perovskite nanocrystal scintillator for nondestructive X-ray imaging, *Advanced Materials* 30(40), 1801743.

242. Wang, L., Fu, K., Sun, R., Lian, H., Hu, X. and Zhang, Y. 2019. Ultra-stable $CsPbBr_3$ perovskite nanosheets for X-ray imaging screen, *Nano-Micro Letters* 11(1), 52.

243. Bamji, C., Mehta, S., Elkhatib, T. A. T. 2011. *CMOS Three-Dimensional Image Sensor Detectors with Assured Non Collection of Late Arriving Charge, More Rapid Collection of Other Charge, and with Improved Modulation Contrast*. CANESTA, INC: Sunnyvale, CA, United States.

244. Wang, S. F., Zhang, J., Luo, D. W., Gu, F., Tang, D. Y., Dong, Z. L., Tan, G. E. B., Que, W. X., Zhang, T. S., Li, S. and Kong, L. B. 2013. Transparent ceramics: Processing, materials and applications, *Progress in Solid State Chemistry* 41, 20–54.

245. Podowitz, S. R., Gaumé, R. and Feigelson, R. S. 2010. Effect of europium concentration on densification of transparent $Eu:Y_2O_3$ scintillator ceramics using hot pressing, *Journal of the American Ceramic Society* 93, 82–88.

246. Pejchal, J., Babin, V., Beitlerova, A., Kucerkova, R., Panek, D., Barta, J., Cuba, V., Yamaji, A., Kurosawa, S., Mihokova, E., Ito, A., Goto, T., Nikl, M. and Yoshikawa, A. 2016. Luminescence and scintillation properties of $Lu_3Al_5O_{12}$ nanoceramics sintered by SPS method, *Optical Materials* 53, 54–63.

247. Cherepy, N. J., Seeley, Z. M., Payne, S. A., Beck, P. R., Swanberg, E. L., Hunter, S., Ahle, L., Fisher, S. E., Melcher, C., Wei, H., Stefanik, T., Chung, Y. S. and Kindem, J. 2014. High energy resolution transparent ceramic garnet scintillators, In Proceedings of SPIE – The International Society for Optical Engineering.

248. Zhu, R.-Y. 2019. Ultrafast and radiation hard inorganic scintillators for future HEP experiments, *Journal of Physics: Conference Series* 1162, 012022.

249. Podowitz, S. R., Gaumé, R. M., Hong, W. T., Laouar, A. and Feigelson, R. S. 2010. Fabrication and properties of translucent SrI_2 and $Eu:SrI_2$ scintillator ceramics, *IEEE Transactions on Nuclear Science* 57, 3827–3835.

250. Sun, X.-Y., Yu, X.-G., Wang, W.-F., Li, Y.-N., Zhang, Z.-J. and Zhao, J.-T. 2013. Luminescent properties of Tb^{3+}-activated B_2O_3–GeO_2–Gd_2O_3 scintillating glasses, *Journal of Non-Crystalline Solids* 379, 127–130.

251. Thomas, S., Rasool, Sk. N., Rathaiah, M., Venkatramu, V., Joseph, C. and Unnikrishnan, N. V. 2013. Spectroscopic and dielectric studies of Sm^{3+} ions in lithium zinc borate glasses, *Journal of Non-Crystalline Solids* 376, 106–116.

252. Fu, J., Kobayashi, M., Sugimoto, S. and Parker, J. M. 2008. Eu^{3+}-activated heavy scintillating glasses, *Materials Research Bulletin* 43, 1502–1508.

253. Sun, X.-Y., Jiang, D.-G., Wang, W.-F., Cao, C.-Y., Li, Y.-N., Zhen, G.-T., Wang, H., Yang, X.-X., Chen, H.-H., Zhang, Z.-J. and Zhao, J.-T. 2013. Luminescence properties of B_2O_3-GeO_2-Gd_2O_3 scintillating glass doped with rare-earth and transition-metal ions, *Nuclear Instruments and Methods in Physics Research, Section A: Accelerators, Spectrometers, Detectors and Associated Equipment* 716, 90–95.

254. Rielage, K., Arisaka, K., Atac, M., Binns, W. R., Christl, M. J., Dowkontt, P., Epstein, J. W., Hink, P. L., Israel, M. H., Leopold, D., Pendleton, G. N. and Wallace, D. B. 2001. Characterization of a multianode photomultiplier tube for use with scintillating fibers, *Nuclear Instruments and Methods in Physics Research Section A: Accelerators, Spectrometers, Detectors and Associated Equipment* 463, 149–160.

255. Bass, C. D., Beise, E. J., Breuer, H., Heimbach, C. R., Langford, T. J. and Nico, J. S. 2013. Characterization of a ^6Li-loaded liquid organic scintillator for fast neutron spectrometry and thermal neutron detection, *Applied Radiation and Isotopes* 77, 130–138.

256. Birks, J. B. 1951. Scintillations from organic crystals: Specific fluorescence and relative response to different radiations, *Proceedings of the Physical Society. Section A* 64, 874–877.

257. Hajagos, T. J. 2017. Plastic scintillators for pulse shape discrimination of particle types in radiation detection.

258. Porter, F. T., Freedman, M. S., Wagner Jr, F. and Sherman, I. S. 1966. Response of NaI, anthracene and plastic scintillators to electrons and the problems of detecting low energy electrons with scintillation counters, *Nuclear Instruments and Methods* 39, 35–44.

259. Yanagida, T., Watanabe, K. and Fujimoto, Y. 2015. Comparative study of neutron and gamma-ray pulse shape discrimination of anthracene, stilbene, and p-terphenyl, *Nuclear Instruments and Methods in Physics Research, Section A: Accelerators, Spectrometers, Detectors and Associated Equipment* 784, 111–114.

260. Kawano, N., Koshimizu, M., Okada, G., Fujimoto, Y., Kawaguchi, N., Yanagida, T. and Asai, K. 2017. Scintillating organic-inorganic layered perovskite-type compounds and the gamma-ray detection capabilities, *Scientific Reports* 7(1),14754.

261. Zaitseva, N., Glenn, A., Carman, L., Paul Martinez, H., Hatarik, R., Klapper, H. and Payne, S. 2015. Scintillation properties of solution-grown trans-stilbene single crystals, *Nuclear Instruments and Methods in Physics Research, Section A: Accelerators, Spectrometers, Detectors and Associated Equipment* 789, 8–15.

262. Adrova, N. A., Koton, M. M. and Florinsky, F. S. 1957. Preparation of 2,5-diphenyloxazole and its. scintillation efficiency in plastics, *Bulletin of the Academy of Sciences of the USSR Division of Chemical Science* 6, 394–395.

263. Smeltzer, J. C. 1950. Energy dependence of the naphthalene scintillation detector, *Review of Scientific Instruments* 21, 669.

264. Chen, Q., Wu, J., Ou, X., Huang, B., Almutlaq, J., Zhumekenov, A. A., Guan, X., Han, S., Liang, L., Yi, Z., Li, J., Xie, X., Wang, Y., Li, Y., Fan, D., Teh, D. B. L., All, A. H., Mohammed, O. F., Bakr, O. M., Wu, T., Bettinelli, M., Yang, H., Huang, W. and Liu, X. 2018. All-inorganic perovskite nanocrystal scintillators, *Nature* 561, 88–93.

265. Knapitsch, A. and Lecoq, P. 2015. Review on photonic crystal coatings for scintillators, *International Journal of Modern Physics A* 29(30), 1430070.

266. Cline, B., Delahunty, I. and Xie, J. 2019. Nanoparticles to mediate X-ray-induced photodynamic therapy and Cherenkov radiation photodynamic therapy, *Wiley Interdisciplinary Reviews: Nanomedicine and Nanobiotechnology* 11(2), e1541.

19 Phosphors-Converting LEDs for Agriculture

Bingfu Lei and Weibin Chen

CONTENTS

19.1 INTRODUCTION

Photosynthesis is the oldest and most important biochemical reaction on the earth, which means that some organisms such as plants and algae convert the collected energy into chemical energy.[1,2] Photosynthesis decomposes atmospheric water (H_2O) and reduces carbon dioxide (CO_2), producing oxygen (O_2) and biomass in the form of carbohydrates, such as glucose ($C_6H_{12}O_6$). About 95% of the dry matter in plants is organic matter synthesized by photosynthesis. The productivity of crops mainly depends on photosynthetic efficiency and economic factors. The incident energy from solar irradiance is collected by specific photoreceptors chlorophylls in plants or algae. In the absence of significant adverse conditions, the energy efficiency of field crops converting from sunlight can reach up to 5%, but the general availability is only about 1.1%. Therefore, improving photosynthesis efficiency is an important way to increase crop yields. Artificial photosynthesis[3–5] tends to imitate the natural photosynthesis process of plants or artificial leaves[6–8], algae, and bacteria[9,10] by using more efficient artificial light sources to split water into hydrogen or solidify carbon dioxide into carbohydrates.

The general definition of photosynthesis by the incident energy of photosynthetically active radiation (PAR) is that part of the solar spectrum between 400 to 700 nm[11], although this wavelength range is still controversial due to the newly discovered redshifted chlorophyll photoreceptors[12] and the role of yellow-green and ultraviolet light[13–15] in plant growth process. PAR is essential for photosynthesis because it provides energy for plants to use in photosynthesis reactions and partly converts

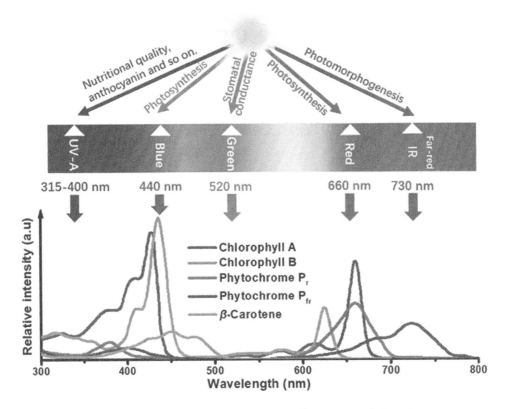

FIGURE 19.1 The absorption spectra of chlorophyll A and B, β-carotene, and phytochromes.

physical solar energy into biomass that carries biological energy.[16] PAR based on flux density is defined as photosynthetic photon flux density (PPFD).

Photosynthetic pigments are closely related to light harvesting and energy transfer during photosynthesis. The pigments that can absorb light energy for photosynthesis include chlorophyll A and B, β-carotene, and phytochromes, as shown in Figure 19.1, Chlorophylls have maximum sensitivities in the blue and red regions, around 380–480 and 600–700 nm, respectively.[17] β-carotene mainly absorbs blue light and is thought as an auxiliary photoreceptor of chlorophyll, while phytochrome P_f and P_{fr} have strong absorptions in the red region at 660 nm and in the far-red (700- to 800 nm) region at 730 nm, respectively.[18]

Cryptochromes are blue/ultraviolet (UV)-A light-sensing photoreceptors involved in regulating various growth and developmental responses in plants. UV-A light receptors[19] can perceive and respond to blue light (400–500 nm) throughout the green plant world. There are photoreceptors sensitive to blue- and UV-A light in the cryptochrome signal transduction system. The response includes the production of anthocyanins and carotenoids in plants and fungi and the influence of the behavioral rhythms of fruit flies and mammals. UV-B has a certain effect on the growth and development of plants.[15]

The phytochrome photosystem is composed of two interconvertible phytochromes P_r and P_{fr}, as shown in Figure 19.2. The photomorphic reaction controlled by the phytochrome is usually related to the red (R)/far-red (FR) ratio (R/FR).[12,20] In particular, long-wavelength light (at 640–660 nm) closely matches the absorption spectra of chlorophyll A and B for the maximum photosynthesis efficiency[21], while the quantum yield decreases rapidly at the wavelength band above ~680 nm. That is to say, Emerson's effect.[22] It can be seen from the Emerson's effect that only slight changes in the peak wavelength of the 660 nm red light source in a scale from 10–20 nm will have a great impact on the maximum absorbance and photosynthetic efficiency of chlorophyll.

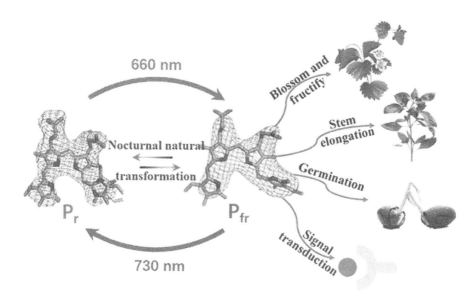

FIGURE 19.2 The mutual transformation and functional schematic diagram of P_r and P_{fr}.

Although the photosynthetic efficiency declines rapidly under long-wavelength illumination above 680 nm, when the wavelength reaches longer, the initial decline is replaced by an unexpected increase. From the detectable O_2 evolution, it appears that this increased benefit continues until 780 nm. The quantum yield of O_2 evolution at 745 nm reaches almost 20% at 650 nm. Similar to photosystem I, extremely long-wavelength chlorophyll may exist in the integrated photosystem II antenna system.[23] Specifically, the effect of different red light (660 ± 30 nm)/far red light (730 ± 30 nm) radiance ratio (R/FR) on photosynthetic characteristics and chlorophyll fluorescence parameters are important factors in achieving high photosynthetic efficiency.[24,25] The effect of yellow-green (500–600 nm) light on plant growth has not been well demonstrated. It was found that yellow (580–600 nm) light appears to inhibit lettuce growth by suppressing the formation of chlorophyll or chloroplasts, and green light can also restore blue-stimulated stomatal opening. Overall, the current research results support the opinion that the green light sensory system and the red and blue-light receptors regulate development and growth together.[26] Therefore, the simultaneous combination of red and blue light is essential for effective photosynthesis, but it cannot alone provide a final solution for the optimal growth of certain specific plants.[27] To achieve the maximum output of photosynthesis, it is also necessary to evenly distribute and absorb light quanta among the photosystems.

However, many artificial lighting sources, such as commonly used incandescent lamps and high-pressure sodium lamps, not only have spectral mismatch problems, but also possess high energy consumption or environmental threats. At the same time, the use of energy-saving and environmentally-friendly light-emitting diodes (LEDs) in solid-state lighting is a very effective means that can meet specific standards in gardening throughout the year. LED plays a role in various horticultural lighting, including for tissue culture, future greenhouse or supplementary lighting plants, and plant photoperiod and vertical farming. In principle, LED lighting sources have several unique advantages over traditional horticultural lighting systems, including the ability to control spectral composition, very high light radiation levels with low heat output, and continuous effective light output for many years without replacement. LED is the first light source capable of truly controlling the composition of the spectrum, allowing the wavelength to match the photoreceptor of the plant for providing better production and affecting the morphology and composition of the plant.

LED solid-state lighting has been widely used in horticultural lighting. These semiconductor lighting devices usually use single red (AlGaInP) and blue (InGaN) LED chips to build PAR for indoor plants. In addition to saving energy, lighting formulas in gardening also include lighting aspects (spectrum, intensity), required uniformity, location and duration, climate conditions and other environmental aspects, and expected effects. The study of the spectral effects of light on plant development has been carried out for a long time,[1,19] but people do not fully understand it due to the complexity of the mechanism that regulates the response of plants to light and the differences between different plant species.

For horticultural lighting, using single red (AlGaInP) and blue (InGaN) LED chips without phosphor conversion (PC) has some disadvantages. The current problems with the combination of red- and blue-LEDs mainly include the different working lifetime of LEDs at different emission wavelengths and the lower quantum efficiency of red-LEDs compared to the blue-LEDs.[28] In addition, the fatal point is that production of red-LEDs consumes a large amount of rare metal sources and produces toxic exhaust gas. At present, the most efficient LED light-emitting wavelength is 380–460 nm, which also provides performance-improving opportunities for most PC materials.

Compared with expensive red-LEDs with narrow band emission in wavelength, the red emission of rare earth (RE)-doped phosphors with wider wavelength can well cover the absorption spectrum of plants.[29,30] A wide range of excitation wavelengths of RE ion-doped phosphors can also easily avoid the dependence of the emission wavelength of red (AlGaInP) and blue (InGaN) LEDs on efficiency and temperature. A PC device excited by blue- or near ultraviolet-LEDs (NUV-LEDs) overcomes the inconsistency of the forward bias driving current between the blue- and red-LEDs. Therefore, blue- or NUV-LEDs excited luminescent phosphors were developed for horticultural lighting. In principle, phosphors with emission match the absorption spectrum of photosynthesis can be used in horticultural lighting applications. These phosphors mainly include inorganic hosts doped with REs and transition metals, such as silicates, nitrides, sulfides, and phosphates,[31–34] but this chapter will introduce the recent use of horticultural lighting based on new blue-(~450 nm), red- (~660 nm), and far-red (~730 nm) emitting phosphors.

In addition, the intensity of light is a significant factor affecting plant growth.[35,36] High-intensity illumination is beneficial for secondary metabolic processes of plants, such as self-repairing and active oxygen quenching. This has forced horticultural light source development towards high energy density, high output power, and high stability. The well-known limitation of pc-LEDs: as the input power density increases, the efficiency will drop, i.e., thermal quenching. This requires relatively low input power density during operation, so it is difficult to achieve high radiant intensity levels for artificial crop cultivation.[37] Therefore, the laser diode (LD) and high-power LEDs are expected to become the next high-intensity lighting technology, which poses a challenge for PC devices. In this chapter, we will pay close attention to the luminescent glass-ceramic materials used for high-power horticultural LED applications.

19.2 BLUE PHOSPHORS

At present, the research on blue phosphors has an increasing trend, because blue phosphors are excited mainly by near-ultraviolet light, which can provide higher excitation energy for the phosphors, making LEDs more efficient. Such phosphors are mainly based on aluminates, silicates, phosphates, and nitrides. Eu^{2+}, Ce^{3+} with the $4f^6 5d \rightarrow 4f^7$ transition can emit blue light in these hosts, so there are many studies on blue phosphors using Eu^{2+}, Ce^{3+} as the activating ions.[38–43]

19.2.1 ALUMINATE BLUE PHOSPHOR

RE activated aluminate is an important class of luminescent substrate. Such systems are not only transparent to photons in the visible region, with excellent heat resistance, high temperature resistance, radiation resistance, chemical stability, and high luminous efficiency, but also becomes favorable conditions multifunctional material. For now, most of them are commercial blue phosphors. For example, $BaMgAl_{10}O_{17}:Eu^{2+}$(BAM) is an excellent blue phosphor,[44,45] which has high quantum efficiency (about 80%), good chromaticity, and strong absorption band under vacuum ultraviolet excitation.

Ekambaram et al.[46] used Eu^{2+} as the activating ion and $BaMgAl_{10}O_{17}$, $BaMg_2Al_{16}O_{27}$, $BaO\cdot6Al_2O_3$ ($x = 0.64–1.8$), and $LaMgAl_{11}O_{19}$ as host, respectively. The phosphor prepared by the combustion method is excited at 254 nm and emits blue light with a wavelength of 435–462 nm. At the same time, the study found that when $xBaO\cdot6Al_2O_3$ ($x = 0.64–1.8$) is used as the host, the content of Ba increases and the emission spectrum of the phosphor is redshifted. Because this type of phosphor is prepared by the combustion method, it has the advantages of simple, fast, and low-cost preparation.

Commercial BAM blue phosphors are usually prepared by high-temperature solid-phase method. This method suffer from high synthesis temperature, uneven raw material mixing, abnormal growth of crystal grains to form large clumps, and new impurity phases.[47] Therefore, in recent years, researchers have committed to finding new preparation methods such as sol-gel method and spray pyrolysis method; they optimized the preparation on the basis of high-temperature solid-phase method or lower temperature by adding flux and doping a small number of other elements to improve its luminescent performance and thermal stability.

Zhu et al.[48] prepared the $BaMgAl_{10}O_{17}:Eu^{2+}$ blue phosphor by the sol-gel method. Morphology of the samples presented as uniform granular, with average particle diameter of 5–10 μm. The excitation spectrum of the sample has several broad bands between 230 and 430 nm, corresponding to the $4f^7 \rightarrow 4f^65d^1$ transition of Eu^{2+}. The emission spectrum consists of a single broad emission band between 420 and 520 nm, which corresponds to the $4f^65d^1 \rightarrow 4f^7$ transition of Eu^{2+}, as shown in Figure 19.3. The sample synthesized at 1500°C shows the strongest emission intensity. Compared with the traditional high-temperature solid-phase method, the sol-gel method makes the Eu^{2+} ions uniformly distributed, reduces the possibility of concentration quenching, and improves the luminescence performance of the phosphor.

It is well known that single-layer graphene has a high light transmittance up to 97.7% for ultraviolet-visible light. Yin et al. calcined BAM samples with a mixture of nitrogen and C_2H_2 at 700°C for 30 min to obtain carbon-coated BAM. The as-prepared coating layers may be made up of a carbon compound which is similar to graphene. In the excitation spectrum of the sample that monitored at 458 nm, a broadband between 250 and 400 nm composed peaks at 310 nm can be obtained. The emission spectrum obtained under 310 nm excitation appears as a broadband between 425 and 500 nm with center at 458 nm. The SEM image of the sample shows that as the number of carbon layers increases, the rough surface of the BAM becomes smoother, and the emission of

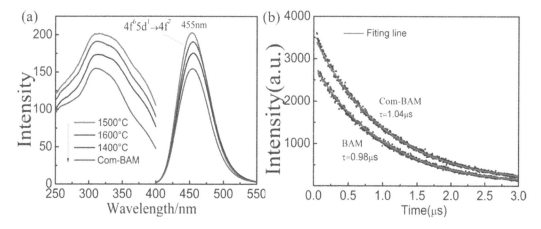

FIGURE 19.3 (a) Excitation and emission spectra of BAM prepared by sol-gel method, calcination temperature of 1400, 1500, and 1600°C. (b) The decay curves of the AS-BAM obtained at 1500°C and the Com-BAM. (Zhu, Q.Q., Xu, X., and Hao, L.Y., *Int. J. Appl. Ceram. Technol.* 12(4), 760–764, 2015. With permission.)

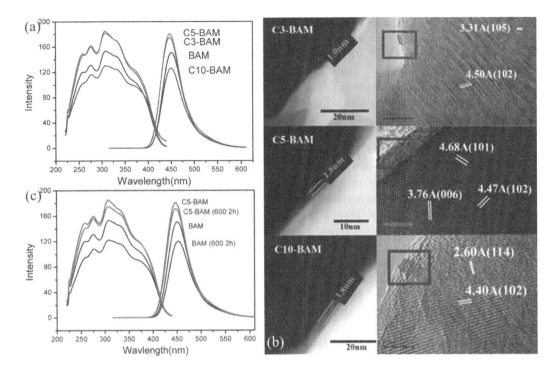

FIGURE 19.4 (a) Excitation and emission spectra of carbon-coated BAM and uncoated BAM phosphors. The excitation and monitoring wavelengths are 310 and 458 nm, respectively. (b) DF-STEM and HRTEM images of carbon-coated BAM. (c) Excitation and emission spectra of carbon-coated and uncoated BAM phosphors before and after heat treatment at 600°C for 2 h in air. (Yin L.J. et al. *J. Phys. Chem. C.* 120, 2355–2361, 2016. With permission.)

carbon-coated BAM is stronger than that of non-carbon-coated BAM under the same excitation conditions. Carbon-coated BAM phosphors have better oxidation resistance at high temperatures than uncoated BAM phosphors (as shown in Figure 19.4). These results all indicate that the carbon coating method on phosphor particles has great prospects for improving lighting.[49]

19.2.2 SILICATE BLUE PHOSPHOR

Silicate has huge and abundant reserves on the earth and has a wide range of distribution. The synthesis method of silicate phosphor is simple, its chemical and physical properties are very stable, and it has a wide-spectrum excitation band. Its excitation spectrum range can match the emission spectrum of NUV-LED chip, and it is also a good matrix for RE fluorescent materials. The excitation and/or emission spectrum of silicate phosphor is easily modulated by microstructure, and the light conversion efficiency is high, and the crystal has good light transmittance.[21–23] Therefore, it has received extensive attention and in-depth research in the field of LED solid-state lighting.

For example, $BaSc_2Si_3O_{10}:Eu^{2+}$ phosphor exhibits an emission of 380–550 nm under excitation at 330 nm, and the main emission peak is 445 nm.[50] $Ca_{1.65}Sr_{0.35}SiO_4:Ce^{3+}$ has a strong absorption between 250 and 400 nm, and its strongest emission peak is 440 nm. This phosphor can adjust the spectrum by replacing Si with Al/Ga/B.[51] $K_2ZrSi_3O_9:Eu^{2+}$ has a strong absorption at 300–450 nm, with a peak emission spectrum of 465 nm and a full-width at half-maximum (FWHM) of 57 nm. The spectrum can be adjusted by replacing the atom pairs.[52]

Typically, the β phase existing in the Sr_2SiO_4 will transform into the α phase. With the doping of Mg^{2+} ions into the silicate host, the β phase will disappear, the intensity of the excitation and emission spectra will be improved, and the decay time will be prolonged. The excitation spectrum

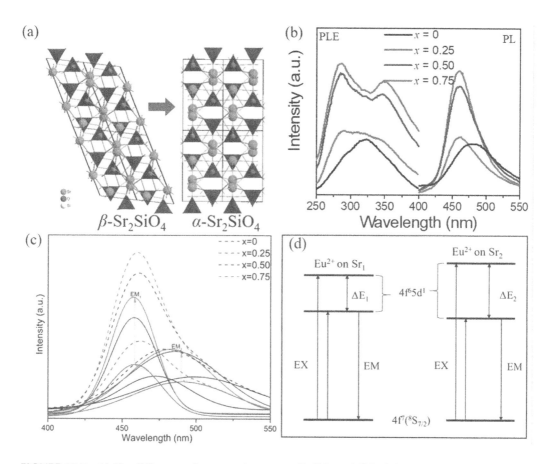

FIGURE 19.5 (a) The difference of structure between α-Sr_2SiO_4 and β-Sr_2SiO_4; (b) PLE and PL spectra of $Sr_{(1.992-x)}Mg_xSiO_4$:0.008Eu^{2+} ($x = 0$, 0.25, 0.50, 0.75); (c) the multi-peak fitting results of the PL emission spectra of samples by the Gaussian method; (d) the Schematic energy-level transition diagram of Eu^{2+} occupied different site. (Yang L.X. et al. *Opt. Mater.* 75, 887–892, 2018. With permission.)

of $Sr_{(1.992-x)}Mg_xSiO_4$:0.008Eu^{2+} monitored at 460 nm shows that the broadband range is 250–400 nm, and the excitation peak in the range of 280–375 nm is high at the same time, with peaks at 280 and 305 nm. This can be attributed to the 4f–5d transition of Eu^{2+}. Excite the sample with 350 nm to obtain the emission spectrum, the peak range is 400–550 nm, corresponding to 4f^{6}5d^1–4f^7 ($^8S_{7/2}$) transition of Eu^{2+}, as shown in the Figure 19.5.[53]

The advantage of silicate-based phosphors is that there are many variants of anionic groups, and various kinds of activated ions can be introduced into the matrix lattice, which is strongly affected by the crystal field, and the band will be relatively split, and the strongest peak position will be more obviously redshift. Eu^{2+}-doped silicate phosphor has a wide excitation spectrum bandwidth and a wide range of continuous and adjustable emission spectrum in the visible region, high light conversion efficiency, good aging resistance, and excellent chemical stability.[54] The disadvantage of silicate-based phosphors is that the anionic groups in the silicate-based phosphors are prone to produce variants under high temperature and low temperature conditions, and it is difficult to control the phase transition.[55,56] Alkaline-earth silicate is relatively alkaline and is easily attacked by electrophilic groups (such as CO^{2-}, H^+) during the reaction, which will increase the electron cloud density of oxygen atoms on the $[SiO_4]^{4-}$ group. Under humid circumstances, the surface of the phosphor is easy to deliquesce, reducing the luminescent efficiency of the phosphor. Therefore, the synthesis conditions must be strictly controlled and appropriate post-processing methods must be adopted.

19.2.3 PHOSPHATE BLUE PHOSPHOR

Phosphate materials are used as the host of luminescent materials. Its usage is second only to aluminate. It has good chemical stability, low synthesis temperature, cheap raw materials, strong absorption of $[PO_4]^{3-}$ in the ultraviolet and NUV regions, and a special light damage threshold.[57,58] Phosphate compounds can provide a large amount of crystal field environment for the activation center. Further, rare earth ions-doped phosphate phosphor has excellent thermal stability and charge stability,[59,60] which is a hot topic in materials science.

Orthophosphate ($ABPO_4$, A = alkali element; B = alkaline earth element) has a rigid three-dimensional network of strictly $[PO_4]$ tetrahedra, which is easy for stabilization and reduction of RE ions, so it is suitable for high-density and high-energy excitation quantum excitation environment. There are many kinds of orthophosphates, with various crystal lattice types. $ABPO_4$ space group of compounds are the following six kinds: $P31c$, $P6_3$, $P-3m1$, $Pnma$, $Pna2_1$, and $P2_1/c$, shown in Figure 19.6. The abundant space group structure and replaceable A and B ions provide different doping environments for RE ions, realizing colorful light emission.

Usually Eu^{2+}- or Ce^{3+}-doped $ABPO_4$ has a strong broadband excitation peak in the wavelength range of 250–420 nm. The peak position of the excitation peak and emission peak, also the intensity of luminescence, varies with the changes of A and B. Therefore, the phosphor of $ABPO_4$ type has certain application potential in the field of near-ultraviolet excited LEDs.

In recent years, a large number of Eu^{2+}-doped orthophosphate blue phosphors have been reported successively, such as $LiSrPO_4$, $NaBaPO_4$, and $RbSrPO_4$. Table 19.1 lists the spectral peak position, thermal stability, and space group of typical Eu^{2+}-doped $ABPO_4$ phosphors. As can be seen, almost all $ABPO_4$ type phosphors have good thermal stability.

Aiming at improving the thermal stability of $ABPO_4$, Qiao et al.[71] designed and synthesized the $P-3m1$ phosphate blue phosphor $K_2BaCa(PO_4)_2:Eu^{2+}$ (KBCP). This phosphor has a high quantum efficiency and excellent thermal stability. KBCP: 3% Eu^{2+} at 200°C remains no quenching. Under the excitation of 350 nm near-ultraviolet light, KBCP:Eu^{2+} emits blue light with a peak at 460 nm,

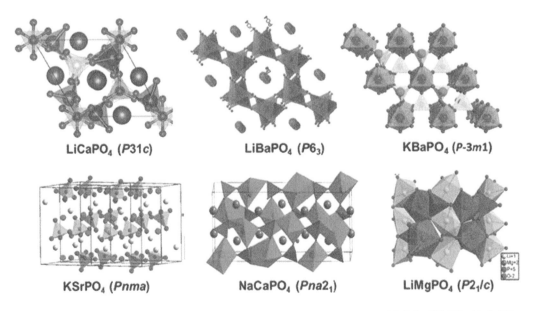

FIGURE 19.6 Lattice structures and space group types of $LiCaPO_4$, $LiBaPO_4$, $KBaPO_4$, $KSrPO_4$, $NaCaPO_4$, and $LiMgPO_4$ compounds. (Provided by Qiao J.W.)

TABLE 19.1

The Emission Peak, Thermal Stability Parameters, and Space Group of ABPO$_4$ Phosphor

ABPO$_4$:Eu^{2+}	Emission Peak (nm)	PL Intensity at 150°C (Compared with 25°C)	Space Group	Reference
LiSrPO$_4$:Eu^{2+}	445	68%	$P6_3$	[61]
LiMgPO$_4$:Eu^{2+}	450	–	$P6_1/c$	[62]
NaSrPO$_4$:Eu^{2+}	450	91%	$Pna2_1$	[63, 64]
NaBaPO$_4$:Eu^{2+}	435	83%	P-$3m1$	[65, 66]
NaMgPO$_4$:Eu^{2+}	437	–	–	[67]
Na$_2$CaBaPO$_4$:Eu^{2+}	451	–	P-$3m1$	[68]
RbSrPO$_4$:Eu^{2+}	450	80%	$P2_1/c$	[69]
RbBaPO$_4$:Eu^{2+}	430	88%	$Pmna$	[70]

which is attributed to the 5d → 4f energy level transition of Eu^{2+} ions. The KBCP lattice contains three kinds of cationic polyhedrons with different positions (as shown in Figure 19.7). After Eu^{2+} ion doping, it will selectively occupy the different valence K2 and K3 sites, and at the same time introduce cation defect V$_K$. Density functional theory (DFT) calculation and thermal luminescence spectra analysis show that there is an appropriate depth of oxygen defect V$_O$ in the phosphor. Under the electrostatic action of the V$_K$ defect, the electrons captured by V$_O$ are transferred to the 5d level electrons through the conduction band, which effectively compensates for the electron energy loss caused by heat ionization.

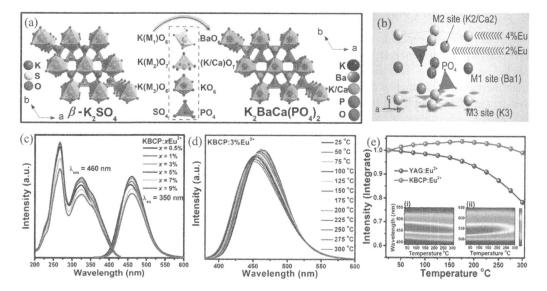

FIGURE 19.7 (a) Schematic diagram of the construction process from β-K$_2$SO$_4$ prototype to KBCP new phase based on co-substitution of polyhedrons; (b) KBCP:8%Eu^{2+} crystal structure and Eu^{2+} occupancy rate in different cation sites; (c) the excitation and emission spectra of KBCP:xEu^{2+}(x = 0.5%–9%) phosphors; (d) Within the range of 25–300°C, the emission spectrum of KBCP:3%Eu^{2+} sample; (e) Comparison of thermal quenching performance between KBCP:3%Eu^{2+} and commercial YAG: Ce^{3+} phosphor. (Qiao J.W. et al. *J. Am. Chem. Soc.* 140, 9730–9736, 2018. With permission.)

19.2.4 Borate Blue Phosphor

Borate phosphors have attracted much attention due to their high luminescence efficiency and good color purity. At the same time, borate phosphors have good ultraviolet transmission, high damage threshold, and good thermal stability. In the past reports, many borates with excellent photoluminescence properties have been discovered and used in various applications.

Xiao et al.[72] prepared a $Ba_2Lu_5B_5O_{17}$:Ce^{3+} blue phosphor by high temperature solid phase method. The excitation spectrum of the sample shows a broadband of 250–375 nm, and peaks at 260, 312, and 348 nm, corresponding to the 5d–4f transition of Ce^{3+}. The emission spectrum of the sample under 348 nm excitation shows an asymmetric blue emission band with a peak at 443 nm, with a FWHM of about 100 nm. When the concentration of Ce^{3+} is 1%, the emission intensity reaches the highest. Under the excitation of 348 nm, the light-emitting internal quantum efficiency of the $Ba_2Lu_5B_5O_{17}$:Ce^{3+} blue phosphor can be as high as 92%, which is equivalent to the commercial blue phosphor BAM.

Martin Hermus et al.[73] determined the crystal structure of barium yttrium borate ($Ba_2Y_5B_5O_{17}$) through ab initio global optimization algorithm, DFT calculation, and combined with Rietveld's refinement of synchrotron X-ray powder diffraction data. The structure is synthesized by a high-temperature solid-state route and consists of Y and Ba central polyhedrons sharing edges and corners and BO_3 triangular planes. Ba and Y occupy four separate positions in the crystal, two of which are completely occupied by Y, and two are statistically a mixture of Y and Ba. Substituting Ce^{3+} into the Y^{3+} structure can produce blue photoluminescence under UV light excitation (composite $\lambda_{em} =$ 443 nm). The compound has high emission efficiency, the external quantum yield is 70%, and it is stable with temperature changes, and the quenching temperature is 400K. However, the excitation peak of most borate phosphors is at ~310 nm. Unfortunately, it cannot be excited by high-efficiency 375–405 nm NUV-LED chips.

19.2.5 Other Blue Phosphors

Other blue phosphors include halides, nitrides, and oxides. In the past few years, these phosphors have played an important role, especially in development of solid-state lighting and bioluminescent labeled optical display systems. Halogenated phosphate represented by $Ca_2(PO_4)Cl$:Eu^{2+} has photoluminescence performance comparable to BMA.[74] Its excitation spectrum is even slightly wider than that of BMA, and it can be more efficiently excited by 375–405 nm NUV-LED chips. Similar halophosphate phosphors also include $Sr_5(PO_4)_3Br$:Eu^{2+} and $(Sr,Ba)_5(PO_4)_3Cl$:Eu^{2+}[75].

Nitride phosphors are mainly concentrated in the yellow-red light region, and the research on blue light-emitting nitrides are still relatively rare. Recently, Tang et al. studied $BaSi_3Al_3O_4N_5$:Eu^{2+} (BSAON) blue phosphor, which has similar luminescent intensity to commercial $BaMgAl_{10}O_{17}$:Eu^{2+} (BAM) blue phosphor under vacuum ultraviolet and NUV excitation. In addition, the phosphor also has excellent thermal and chemical stability, high quenching temperature and high quantum efficiency.[76]

Oxide system phosphors have a certain structural rigidity, so their physical and chemical properties are stable and their moisture resistance is strong. Due to its good chemical durability, excellent light output and environmental protection, oxide-based phosphors have attracted more and more attention in recent years, for example, tungstate, carbonate, niobate, and lutetium. Here, we will focus on introducing a $SrLu_2O_4$:Ce^{3+} phosphor with great potential in the blue-light source of plant lighting.

The crystal structure of $SrLu_2O_4$ belongs to the orthorhombic crystal system, the space group is *Pnma*, and its strongest excitation peak is 405 nm, which is in perfect agreement with the near-ultraviolet chip. The main emission peak is 460 nm and the FWHM is about 90 nm, which is nearly twice that of BAM.[77] Moreover, the thermal stability of the phosphor is very good, and the luminous intensity is 86% under room temperature at 150°C (as shown in Figure 19.8).

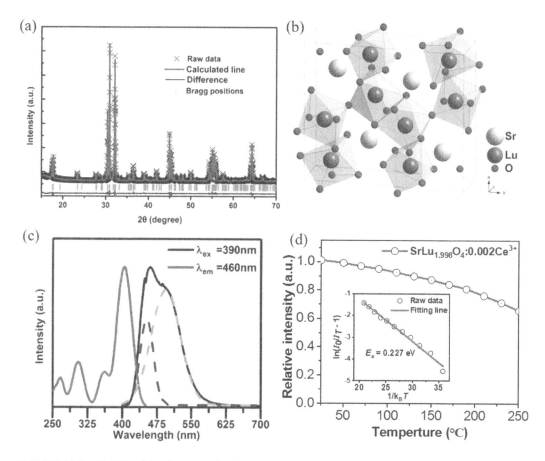

FIGURE 19.8 (a) Rietveld refinements for $SrLu_2O_4$. (b) Crystal structure schematic diagram of $SrLu_2O_4$. (c) PLE and PL spectra of $SrLu_{1.998}O_4$: $0.002Ce^{3+}$ with the emission band fitted with two Gaussian curves (dashed). (d) Temperature dependence of the integrated emission intensities in $SrLu_{1.998}O_4$: $0.002Ce^{3+}$ excited at 405 nm. (Zhang S. et al. *Sci. Rep.* 8, 10463, 2018. With permission.)

19.3 RED PHOSPHOR

Most red phosphor uses Eu^{2+}, Mn^{4+} ion doping or Mn^{2+} and Ce^{3+}/Eu^{2+} ions co-doping phosphors were obtained.

19.3.1 Eu^{2+}-DOPED RED PHOSPHOR

Compared with blue phosphors, Eu^{2+}-doped red phosphors are still less studied. For the development and performance improvement of new Eu^{2+}-doped red phosphors, it is of great significance to fully understand the intrinsic relationship between Eu^{2+} ion site occupation and luminescence performance in the matrix structure of luminescent materials, and to establish effective phosphor design principles. RE Eu^{2+} has a typical 4f–5d transition, and Eu^{2+} doped phosphors have broadband emission characteristics. In different crystal field environments, there is a big difference between the Eu^{2+} emission wavelength which can be tuned from ultraviolet to near infrared spectral regions.

Recently, Eu^{2+}-doped red phosphors have also been reported successively, and their emission wavelength can be absorbed by the photosensitive pigment P_R. For example, Zhang et al.[78] successfully synthesized $SrLiAl_3N_4$:Eu^{2+} red phosphor, which can be excited by a 450 nm blue LED chip and emitted 650 nm deep red light; Lee and his group[79] successfully prepared Sr^{2+}-doped

$BaZnS_3:Eu^{2+}$ red phosphor, Sr^{2+} doping makes the emission peak of $BaZnS_3:Eu^{2+}$ blueshift from 680 to 660 nm, which is consistent with the absorption wavelength of the photosensitive pigment P_R; $Ba_3GdNa(PO_4)_3F:Eu^{2+}$ phosphor reported by Chen et al. has a blue emission band (436 nm) and a red emission band (630–680 nm). This phosphor can not only provide blue light absorbed by lutein and carotene but also meet the requirements of the photosensitive pigment P_R. But the bad fact is that Eu^{2+}-doped red oxide phosphors are still less studied.

For the nitride-based red phosphor, Eu^{2+} ions are condensed in the tetrahedral grid of MN_4 (M = Si, Al, Li), and polarized N^{3-} ions are surrounded by Eu^{2+} ions, which has strong electro-negativity.[80,81] The 5d energy level has a large center of mass shift, which leads to red light emission in the nitride.[82] For $CaS:Eu^{2+}$, the strong Eu^{2+}–S^{2-} covalent bond and the low coordination number (CN) CaS_6 polyhedron lead to large center of mass displacement and crystal field splitting, thereby realizing red light emission[83]. Therefore, increasing the center of mass displacement of the 5d energy level is an effective way to reduce the 5d–4f energy level difference.

However, compared with Eu^{2+}–N^{3-} and Eu^{2+}–S^{2-}, the weak Eu^{2+}–O^{2-} covalent bond makes it difficult to achieve a large center of mass displacement in the oxide phosphor. Another way to obtain red oxide phosphors is to find an oxide matrix with strong crystal field strength, and the crystal field strength is closely related to the CN of the polyhedron occupied by RE ions. Generally, the CNs of transition metals and main group elements is 2–6, while the CNs of RE ions are ≥ 6.

Qiao et al.[84–86] selected $A_3LnSi_2O_7$ (A = K, Rb, Ln = Y, Gd, Lu) compounds with abundant cation sites as the research object and investigated the relationship among matrix modulation and RE ion site occupation and luminescence properties. The design concept of Eu^{2+} ions occupying low-coordination and small-radius polyhedrons to achieve large redshift emission is proposed, and a series of new red and near-infrared phosphors have been developed on this basis.

The $Rb_3YSi_2O_7$ unit cell contains three different cationic polyhedrons, YO_6, $Rb1O_9$, and $Rb2O_6$. After Eu^{2+} ion doping, it will selectively occupy the low CN of YO_6 and $Rb2O_6$ polyhedrons, causing large 5d energy level crystal field splits. As a result, the $Rb_3YSi_2O_7$: Eu^{2+} phosphor can achieve a broadband red emission of 550–750 nm under excitation at 450 nm.[85]

Based on the design concept of Eu^{2+} ions occupying low-coordination and small-radius polyhedrons to achieve large redshift emission, through $Rb^+ \rightarrow K^+$, $Y^{3+} \rightarrow Lu^{3+}$ substitution, Xia et al. successfully designed the first near-infrared phosphor $K_3LuSi_2O_7:Eu^{2+}$ that can be excited by 460 nm blue light and has an emission peak at 740 nm. The structure and spectrum analysis showed that Eu^{2+} ions selectively occupy the low-coordinated LuO_6 and $K2O_6$ polyhedrons, resulting in abnormal near-infrared luminescence.[86]

Not only that, Xia et al. utilized a regression model to predict the emission wavelengths of Eu^{2+}-doped phosphors by revealing the relationships between the crystal structure and luminescence property (as shown in Figure 19.9). The emission wavelengths of $[Rb_{(1-x)}K_{(x)}]_3LuSi_2O_7:Eu^{2+}$ ($0 \leq x \leq 1$) phosphors, as examples for the data-driven photoluminescence tuning, are successfully predicted on the basis of the existing data of only eight systems, also consistent with the experimental results. These phosphors can be excited by blue light and exhibit broadband red and near-infrared emissions ranging from 619 to 737 nm.[87]

These findings in Eu^{2+}-doped silicate phosphors indicate that data-driven computations through the regression mode would have bright application in discovering novel phosphors with a targeted emission wavelength.

19.3.2 Mn⁴⁺-Doped Red Phosphor

The transition metal Mn^{4+} has the advantages of abundant resources and low price. Mn^{4+} with $3d^3$ electronic configuration can exhibit strong absorption from NUV to blue light and can emit wavelengths in the region of 630–780 nm in different crystal field environments. Generally, Mn^{4+} can stably exist in the octahedral coordination unit cell, but the wavelength of the excited light will be strongly affected by the Mn^{4+} coordination bond.

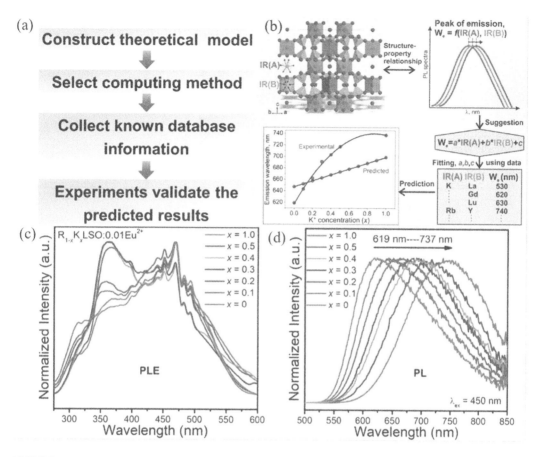

FIGURE 19.9 (a) Full data-driven steps by regression analysis. (b) Crystal structure model of $A_3BSi_2O_7$ and the relationship between structure and luminescence property; and assumed formula and the related database for prediction. Normalized PLE (a) and PL (b) spectra of as-prepared $Rb_{1-x}K_xLuSi_2O_7$:$0.01Eu^{2+}$ ($0 \le x \le 1$). (Lai S.Q. et al. *J. Phys. Chem. C.* 11, 5680–5685, 2020. With permission.)

The red emission of Mn^{4+} mainly comes from Mn^{4+}-doped oxides, for example, Ren et al.[88] synthesized Ba_2LaSbO_6:Mn^{4+} double perovskite-type red phosphor. Under the excitation of NUV light at 365 nm, the emission peak of the phosphor is 678 nm. Liu et al.[89] reported that emission wavelength range of powder of $CaAl_4O_7$:Mn^{4+} red phosphor is 620–680 nm, which can meet the demand of P_R for red light.

Mn^{4+}-doped fluoride is also used to make red phosphor. Shao et al.[90] produced a red magnesium fluorogermanate ($Mg_{28}Ge_{7.5}O_{38}F_{10}$:$Mn^{4+}$) phosphor activated by Mn^{4+}. The excitation wavelength of this sample ranges from NUV to blue light and can emit red light (620–670 nm). Among them, due to the asymmetric coupling of O^{2-}–Mn^{4+} and the Stokes effect, there are two main peaks at 653 and 660 nm in the emission band. The $Mg_{3.5}Ge_{1.25}O_6$:Mn^{4+} phosphor studied by Lee et al.[91] has two strong peaks at 632 and 660 nm. Interestingly, by adding MgF_2, the intensity of the phosphor will be greatly improved.

19.3.3 Mn^{2+}- and Ce^{3+}/Eu^{2+}-co-Doped Red Phosphor

The electronic configuration of the transition metal Mn^{2+} is $3d^5$. Generally, the emission wavelength range of Mn^{2+} is 500–700 nm, and the red emission can be achieved by adjusting the crystal field environment of Mn^{2+}. However, the d-d transition of Mn^{2+} belongs to the forbidden transition so that

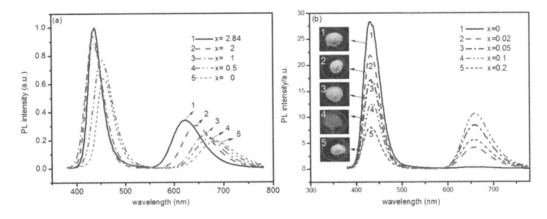

FIGURE 19.10   (a) PL spectra of $Ba_xSr_{2.84-x}Mg_{0.9}Si_2O_8$:0.06$Eu^{2+}$, 0.1$Mn^{2+}$ (x = 2.84, 2, 1, 0.5, 0), (b) emission spectra and photographs of $Ba_{1.3}Sr_{1.64}Mg_{1-x}Si_2O_8$:0.06$Eu^{2+}$, $x$$Mn^{2+}$ phosphors upon 355 nm light irradiation (x = 0, 0.02, 0.05, 0.1, 0.2). (Lu, Q.F., Li, J. and Wang, D.J. *Curr. Appl. Phys.* 13, 1506–1511, 2013. With permission.)

the excitation efficiency is low. Therefore, other sensitizer ions can be used to transfer energy to achieve high-efficiency luminescence of Mn^{2+}. In general, Ce^{3+} or Eu^{2+} can be used as a sensitizer ion, and its 5d excited state electrons transit to the ground state ($^2F_{5/2}$, $^2F_{7/2}$) to emit near-ultraviolet or blue light.[92] Ce^{3+}/Eu^{2+} not only has a sensitizing effect, but also Ce^{3+}/Eu^{2+} can meet the blue-light demand of plants.

The $NaSr_4(BO_3)_3$:Ce^{3+}, Mn^{2+} phosphor synthesized by Zhang et al.[93] has two emission bands. Among them, the blue-light emission at 440 nm is due to the Ce^{3+}:5d^1 $\rightarrow$ 4f^1 transition, which is similar to the demand for blue light by plants; the red emission at 660 nm is due to the Mn^{2+}: $^4T_1$ (^4G)$\rightarrow$$^6A_1$ (^6S) transition, which provides red light for the P_R absorption. Lu et al.[94] prepared a single-phase $(Ba,Sr)_3MgSi_2O_8$:Eu^{2+}, Mn^{2+} phosphor. The study found that the best ratio is $Ba_{1.3}Sr_{1.64}Mg_{0.9}Si_2O_8$:0.06$Eu^{2+}$, 0.1$Mn^{2+}$, and the phosphor has red emission peak at 660 nm and a strong blue emission peak at 430 nm that is consistent with the plant absorption spectrum. The study also found that the Ba-Sr ratio affects the crystal field distribution of the activator in the host lattice and promotes the energy transfer from Eu^{2+} to Mn^{2+}, which can make the peak position redshift, even to the far-red light region, as shown in Figure 19.10.

19.4   FAR-RED PHOSPHOR

Compared with the more mature blue and red phosphors, far-red phosphors are the hottest research in recent years, especially the research based on Mn^{4+}, Cr^{3+}-doped oxides.

19.4.1 Mn^{4+}-Doped Far-Red Phosphor

Mn^{4+} is a red activator with a wide range of sources. Its outer electronic structure is $3d^3$ and its energy level transitions are relatively abundant. Zhou et al.[95] used the sol-gel method to prepare Mn^{4+}-activated $La(Mg,Ti)_{1/2}O_3$ phosphor. As shown in Figure 19.11, the sample can emit far-infrared light with a peak center of 708 nm under the excitation of ultraviolet light (345 nm) or blue light (487 nm), and its optimal Mn^{4+} doping mole fraction is 0.8%. Except for the lower internal quantum efficiency (27.2%), the phosphor can basically meet the needs of plants for far-infrared light absorption.

Huang et al.[96,97] studied the tungstate as the host for Mn^{4+}-doping phosphors $Ca_3La_2W_2O_{12}$:Mn^{4+} and $NaLaMgWO_6$:Mn^{4+}. The host material of the latter belongs to the double perovskite structure. As a low-cost, easy-to-manufacture, high chemical stability, and superior optical performance material, tungstate is often used as a phosphor-doping host. Since Mn^{4+} has a $3d^3$ electronic

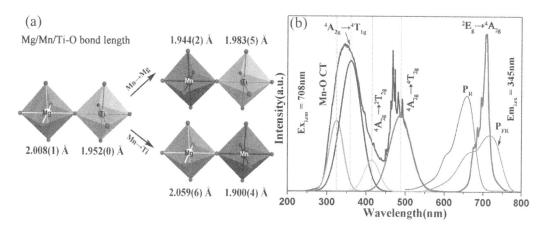

FIGURE 19.11 (a) Structure of $[MgO_6]$–$[TiO_6]$ in cubic $La(Mg,Ti)_{1/2}O_3$ and the change of the Mn/Ti/Mg–O bond length as Mn atoms enter the Mg or Ti site. (b) Room temperature PLE (λ_{em} = 708 nm) and PL (λ_{ex} = 345 nm) spectra of $La(Mg,Ti)_{1/2}O_3$: 0.8% Mn^{4+} and the absorption spectra of phytochrome P_r and P_{fr}. (Zhou Z.W. et al. *ACS Appl. Mater. Inter.* 9, 6177–6185, 2017. With permission.)

configuration, all three types of phosphors can be excited by NUV light and blue light. The difference is that $Ca_3La_2W_2O_{12}$:Mn^{4+} phosphor has an optimal doping mole fraction of Mn^{4+} of 0.8%, and an internal quantum efficiency of 47.9%. The double perovskite structure of $NaLaMgWO_6$:Mn^{4+} phosphor increases the internal quantum efficiency to 60%. In addition to quantum efficiency, the temperature characteristics of phosphors are also an important parameter for measuring phosphors as devices. The luminescent intensity of the phosphor with a double perovskite structure $NaLaMgWO_6$:Mn^{4+} at a temperature of 423K drops to 57% compared to room temperature, while the phosphor $Ca_3La_2W_2O_{12}$:Mn^{4+} is only 29%.

Titanate as a double perovskite structure can also be used as a host to be doped. Yang et al.[98] prepared La_2ZnTiO_6:Mn^{4+} phosphor by high-temperature solid-state reaction. The sample can emit far red light with a peak center wavelength of 708 nm at excitation wavelengths of 342 and 504 nm. It is worth noting that Ti^{4+} and Mn^{4+} in titanate have similar ionic radii and the same oxides, so there is no need to compensate for the charge. Mn^{4+} is easier to replace Ti^{4+}, which is also the focus of researchers.

Zhou et al.[100] also reported the preparation of a new red phosphor $Ca_{14}Ga_{10-m}Al_mZn_6O_{35}$ (CGAZO) by co-doping (Dy^{3+} and Mn^{4+}). The phosphor has five emission peaks, the highest peak is at 715 nm. Experimental studies have shown that using Dy^{3+} as a sensitizer and Mn^{4+} as an activator greatly improves the intensity of the emitted light of the sample. The author also studied the absorption spectra of samples with different Al^{3+} mole fractions. The results show that after Al^{3+} replaces Ga^{3+}, the absorption of the sample in the ultraviolet region will be enhanced, while the blue light will be weakened. It is worth mentioning that the sample has good quantum efficiency and thermal stability, its internal quantum efficiency reaches 91.7%, and the luminous intensity at 150°C is 75.96% at room temperature. To verify the effect of these phosphors on plant growth, a series of LEDs were prepared by combining 460 nm blue chips and CGAZO: Dy^{3+}, Mn^{4+} phosphors. Figure 19.14 shows the appearance of the LED device under different Al^{3+} mole fractions, and the electroluminescence (EL) spectrum of a blue-red LED prepared under a 150 mA drive circuit. Compared with traditional commercial LEDs, the CGAZO: Dy^{3+}, Mn^{4+}-LED has stronger far-infrared light, so it is particularly suitable for plant lighting. Figure 19.12 shows the growth of tobacco plants under two LED illuminations. Obviously, under same growth time and different light treatments, there are significant differences in the height of tobacco plants. Treatment with plant growth LED lights containing CGAZO: Dy^{3+}, Mn^{4+} phosphors can promote the growth of tobacco seedlings and shortening germination time, and reduces the germination rate during the cultivation period.

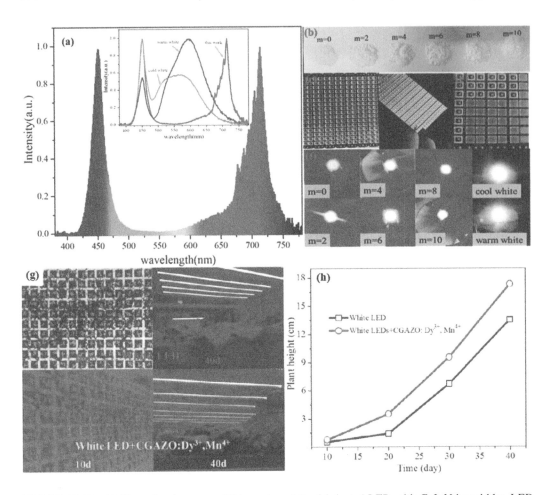

FIGURE 19.12 (a) Electroluminescence (EL) spectra of the fabricated LED with GaInN-based blue LED chip and CGAZO: Dy^{3+}, Mn^{4+} phosphor. The inset shows the EL spectra of the commercial LED with cold white light and warm white light and driven with a current of 150 mA. (b) Photographs of phosphors with different Al contents under daylight lamp, (c) GaInN-based blue LED chip, (d) fluid dispensing, (e) the as-packaged LED with blue chip and CGAZO: Dy^{3+}, Mn^{4+} phosphors, (f) the appearance of the commercial LED and the as-fabricated LED devices with different Al concentrations in operation. (g) Photographs of indoor tobacco plant cultivation at 10 and 40 d under different light treatment; (h) the relationships between average plant height (the number of tobacco plants is 100) and growth time. Growth time recording started after the 7 days seedling stage. (Zhou Z. et al. *J. Mater. Chem. C* 5, 8201–8210, 2017. With permission.)

Zhong et al.[101] proposed the chemical unit engineering of Ge^{4+} + M^+ (M = Li, Na, K) instead of Ga^{3+} + Ca^{2+} in the phosphor $Ca_{14}Ga_{10}Zn_6O_{35}$:Mn^{4+} and prepared four types of phosphors $Ca_{14}Ga_{9.85}Zn_6O_{35}$:$0.15Mn^{4+}$ and $Ca_{13.6}Li_{0.4}Ga_{9.45}Ge_{0.4}Zn_6O_{35}$:$0.15Mn^{4+}$, $Ca_{13.4}Na_{0.6}Ga_{9.25}Ge_{0.6}Zn_6O_{35}$:$0.15Mn^{4+}$ and $Ca_{13.6}K_{0.4}Ga_{9.45}Ge_{0.4}Zn_6O_{35}$:$0.15Mn^{4+}$. Experimental results show that the luminous intensity and internal quantum efficiency of the phosphor have been significantly improved. Figure 19.13 shows that by doping Ge^{4+} + $Li^+/Na^+/K^+$, the intensity of the emitted light becomes 169.4%, 195%, and 198.9% as before. The internal quantum efficiencies are respectively 49.8%, 47.0%, and 50.9%. In further experimental comparison, it is also found that the doping of Ge^{4+} + Li^+ will cause a redshift of the emission peak of the phosphor (~3.6 nm). What is gratifying is that the Mn^{4+} mole fraction is fixed, but the quantum efficiency of the sample is improved, which shows that more Mn^{4+} enters the crystal through chemical unit engineering.

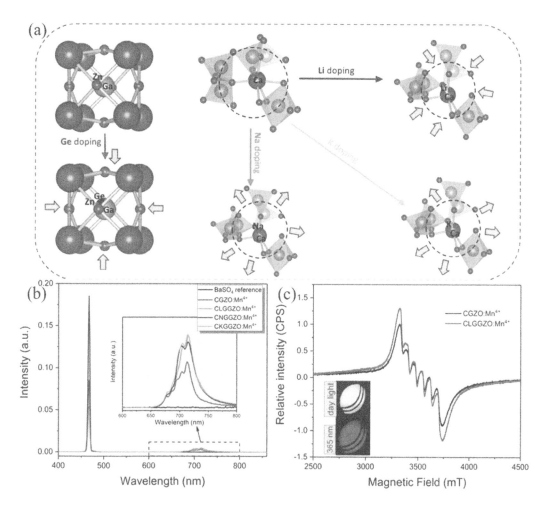

FIGURE 19.13 (a) Schematic illustration of the effect of $Ge^{4+}-M^+$ (M = Li, Na, K) doping on local crystal structures around Ga and Ca. (b) The measurements of QE of CGZO:Mn^{4+} and CMGGZO:Mn^{4+} (M = Li, Na, K) phosphors, the inset is the enlarged pattern range from 600 to 800 nm; (c) EPR spectra of CGZO:Mn^{4+} and CLGGZO:Mn^{4+} phosphor. (Zhong Y. et al. *Chem. Eng. J.* 374, 381–391, 2019. With permission.)

19.4.2 CR^{3+}-DOPED FAR-RED PHOSPHOR

It is well known that transition metal ions have unfilled d orbitals, the electronic structure is $3d^n$ (0 < n < 10). Resembled Mn^{4+}, the spectral properties of Cr^{3+} are strongly affected by the crystal field. As the Cr^{3+} ion has a $[Ar]3d^3$ electron configuration, the three valence electrons are not shielded by outer shells and this gives rise to a strong interaction with the crystal field and lattice vibrations due to the spatial extension of the d electron wave functions in the crystals. The strong crystal field interaction for Cr^{3+} ion gives the possibility to tune photoluminescence properties of Cr-phosphors.

Malysa et al.[102] reported the optical properties of $X_3Sc_2Ga_3O_{12}$ (X = Lu, Y, Gd, La) garnets doped with Cr^{3+} showing efficient $^4T_2 \rightarrow {}^4A_2$ broadband NIR emission between 600 and 1000 nm. The chromium-doped garnets were investigated in view of their capability for blue to NIR conversion in high-power phosphor-converted LEDs (pcLEDs). The pcLED based on $Y_3Sc_2Ga_3O_{12}$: Cr^{3+} is a good light source for P_{fr}, but unfortunately its quantum efficiency isn't strong.

Yu et al.[103] achieved broadband near infrared emission with an efficiency of 75% by doping 60% Cr^{3+} in Na_3AlF_6 material, with a peak value of 720 nm and a FWHM of 95 nm (as shown in

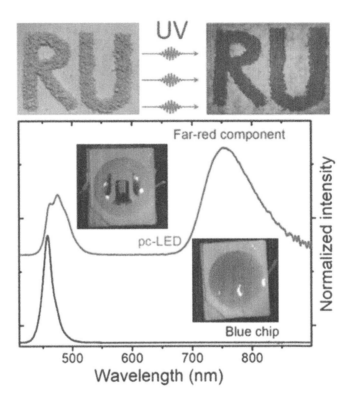

FIGURE 19.14   Optical images of Na_3AlF_6:60% Cr^{3+} phosphors shined by natural white light (left top) and UV light (right top), and the EL spectrum of a bare blue-emitting InGaN LED device and the InGaN LED device encapsulated with a silicone Na_3AlF_6: 60% Cr^{3+} phosphor slurry. (Yu D.C. et al. *ACS Appl. Electron. Mater.* 1, 2325–2333, 2019. With permission.)

Figure 19.14). However, the excitation location was 420 nm, which did not match the wavelength of the most efficient blue LED chip (445–470 nm).

At present, there is still a lack of research on Cr^{3+}-doped phosphors that can be used for plant lighting sources, especially, the Cr^{3+}-doped phosphors with high luminescent efficiency and high thermal stability will be the pursuit of researchers.

19.5   GLASS CERAMIC FOR HIGH-POWER HORTICULTURAL PC-LED

The development of new RE luminescent materials provides ideas for the realization of LED plant growth lamps and plant lighting systems with wide emission spectrum and high energy conversion efficiency. However, so far, plant lighting LED lamps based on the 'LED chip + red phosphor in silicone (PiS)' model cannot meet the high illuminance requirements for plant growth. High-efficiency and high-power LEDs are an inevitable means to meet the requirements of high light quantum numbers for plant growth. High-power mean chip input current and junction temperature rising, high-power LED chip junction temperature up to 150–200°C, result in a high junction temperature, which will reduce the life of the plant LED lighting device. At the same time, the red-to-blue ratio will also change with the use of time. Compared with traditional plant lighting technology, the 'LED + luminescent glass ceramic' solution has excellent thermal conductivity, thermal stability, and light stability at high junction temperatures and has, therefore, become a preface for high-power LED research. To achieve high-power LED output, adjust its performance to match plant lighting application requirements and improve its heat transfer performance; our research group has made some meaningful attempts:

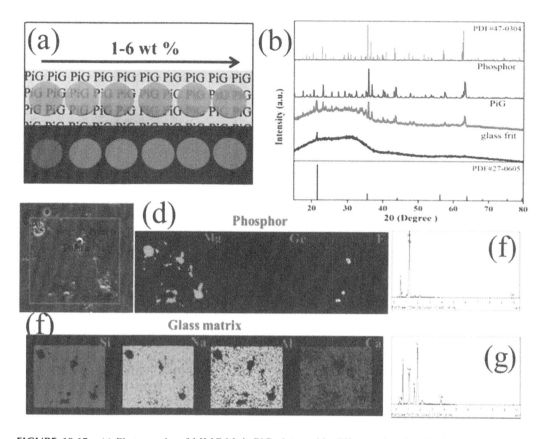

FIGURE 19.15 (a) Photographs of MMG:Mn^{4+}-PiG plates with different phosphor-doping concentrations irradiated by the fluorescent light and 365 nm UV light, respectively. (b) XRD patterns of MMG:Mn^{4+} phosphor, precursor glass frit, and 6 wt% MMG:Mn^{4+}-PiG and standard data of Mg$_{3.5}$Ge$_{1.25}$O$_6$ (PDF#47-0304) as well as SiO$_2$ (PDF#27-0605). (c) SEM image of 6 wt% MMG:Mn^{4+}-PiG. (d and f) SEM mappings inside the red frame of (c) of the elementals correspond to MMG:Mn^{4+} phosphor and glass matrix. (e and g) EDS analysis of point a and point b of (c). (Deng J.K. et al. *J. Mater. Chem. C* 6, 1738–1745, 2018. With permission.)

In response to the requirements of high-power LED output on materials, after exploratory research and development, we reported a highly thermally stable Mn^{4+} ion-activated red glass ceramic suitable for high-power plant LED lighting applications (as shown in Figure 19.15)[104] and studied its luminescent properties in detail, including thermal quenching property, physical and chemical and optical stability, and weather resistance. Its thermal conductivity is as high as 1.671 W·m^{-1}·K^{-1}, which is eight times than that of resin packaging materials such as epoxy or silicone. The luminescent glass ceramic has good absorption and extern quantum efficiency (~27.5%). Planting experiments were carried out with 36 pieces of luminescent glass ceramics prepared and 420 nm high-power chips encapsulated plant illuminators. Compared with ordinary red and blue chips assembly LED plant growth lights, the bioaccumulation of milk Chinese cabbage increased by 48.9%, and other major indicators such as vitamin C, soluble protein, soluble sugar, and total chlorophyll content increased to varying degrees.

The photosynthesis needs broadband spectrum in the blue region according to the physiological requirements of plants. But the pc-LED based on excitation of blue-light chip results in a narrow blue-light emission spectral range inevitably. Focus on this problem, we reported that a dual emitting phosphor-in-glass(PiG) plate was prepared toward high power and dual broadband emitter, in which BaMgAl$_{10}$O$_{17}$:Eu^{2+} and CaAlSiN$_3$:Eu^{2+} provide blue and red emission (as shown in Figure 19.16), respectively.[105] The splicing PiG plate showed several merits, including a high quantum efficiency

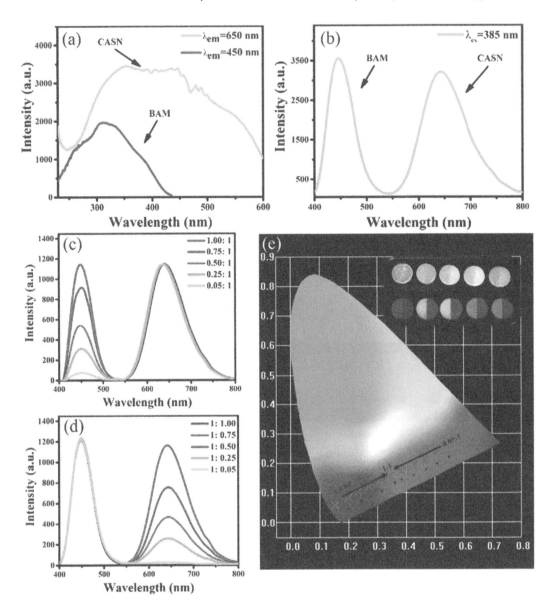

FIGURE 19.16 (a) PLE spectra of the BAM phosphor and CASN phosphor; (b) PL spectrum of Dual-PiGP; (c) PL spectra of Dual-PiGP under red emission unchanged tuning blue emission; (d) PL spectrum of Dual-PiGP under blue emission unchanged tuning red emission; (e) CIE color coordinates of Dual-PiGP with different red-blue ratio; and insets of (e) show the photo under daylight and 365 nm UV light. (Li M.C. et al. *J. Mater. Chem. C* 7, 3617–3622, 2019. With permission.)

of 93.90%, a superior thermal stability (80.1%@150°C of the peak intensity at 25°C), and dual broad spectrum with the FWHM of 50 and 106 nm at 446 and 645 nm, respectively.

In addition, we also prepared a one-piece glass-ceramics with red and blue dual-emission composed of a glass matrix with embedded $Ba_{1.3}Sr_{1.7}MgSi_2O_8$:$Eu^{2+}$, Mn^{2+} (BSMS) phosphors.[106] The obtained samples show an external quantum efficiency of 45.3%, outstanding thermal stability, and a specific emission spectrum that highly matches the absorption of chlorophyll and β-carotene. We revealed the concentric distribution of thermal energy on the surface of the BSMS-PiG and BSMS-PiS, as shown in Figure 19.17, BSMS-PiG shows a broader radial temperature gradient zone than BSMS-PiS.

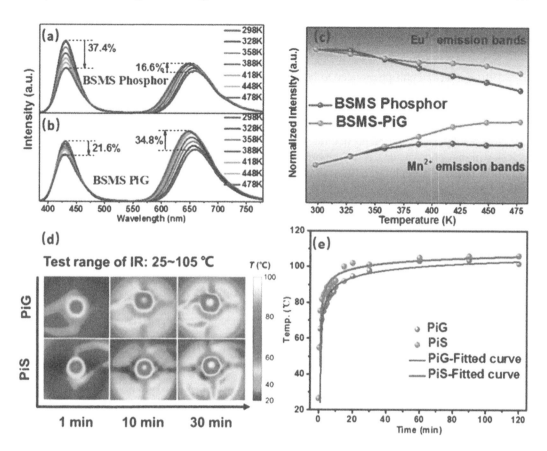

FIGURE 19.17 (a) Temperature-dependent PL spectra of the BSMS phosphor and (b) 7 wt% BSMS-PiG recorded from 298 to 478K. (c) The variation of the corresponding normalized PL intensities of the BSMS phosphor and 7 wt% BSMS-PiG. (d) Thermal images of BSMS-converted LED chips and BSMS-PiS-converted LED chips with 1 A current after different operating times, and (e) temperature as a function of the time of operation. (Chen W.B. et al. *J. Mater. Chem. C* 8, 3996–4002, 2020. With permission.)

This phenomenon is because the BSMS-PiG has a higher thermal conductivity (~1.242 W·m⁻¹·K⁻¹) than that of BSMS-PiS (~0.16 W·m⁻¹·K⁻¹) and contributes to the spreading of the heat. This can weaken the thermal quenching of the embedded BSMS phosphor. But unfortunately, when the LED chip is powered on continuously for a period of time, the surface temperature difference between the two is not large (only ~3%). Although the PiG technology has greatly enhanced the radiation resistance of the conversion devices, the core problem is the development of heat-resistant quenching phosphors.

19.6 SUMMARY AND PERSPECTIVES

This chapter briefly reviewed the mechanism of phosphors applied in agriculture and then summarized the recent advances of luminescent materials with potential applications in plant lighting. Moreover, the development trend of glass ceramic technology in high-power LED horticultural light sources was discussed. In the future of practical agricultural application of the phosphor, challenges exist and a thorough study is needed, including but not limited to the following:

1. The wavelength range of PAR in lighting sources is recognized to mainly located at red and blue band for photosynthetic reaction, but the role of green-yellow band needs to be further evaluated.

2. In the early stage of plant growth, the blue light with broad wavelengths is needed to satisfy root development. Blue phosphors (with small Stokes shifts) that can be efficiently excited by near-UV LED chips are still lacking.

3. Efficient far-red phosphors need to be further developed and suitable luminescent materials should have a main wavelength of 730 nm and a half-width at half-maximum of 80 nm to cover the absorption spectrum of P_{fr}.

4. LD technology featured in its high brightness is evolving into the next-generation lighting technology, applicable to horticulture lighting. Blue LD-excited luminescent glass ceramics in phosphor-in-glass is an important option. The core problem is the development of heat-resistant quenching red/far-red phosphors.

REFERENCES

1. Ceriani, M.F., Darlington, T.K., Staknis, D., Más, P., Petti, A.A., J., C., Weitz and Kay, S.A. 1999. Light-dependent sequestration of TIMELESS by cryptochrome. *Science* 285: 553–556.
2. Ahmad, M., Jarillo, J., Smirnova, O., Anthory, R. and Cashmore. 1998. Cryptochrome blue-light photoreceptors of Arabidopsis implicated in phototropism. *Nature* 392: 720–723.
3. Scholes, G.D., Fleming, G.R., Olaya-Castro, A. and van Grondelle, R. 2011. Lessons from nature about solar light harvesting. *Nat Chem* 3: 763–774.
4. Dugar, D. and Stephanopoulos, G. 2011. Relative potential of biosynthetic pathways for biofuels and bio-based products. *Nat Biotechnol* 29: 1074–1078.
5. Sheehan, J. 2009. Engineering direct conversion of CO_2 to biofuel. *Nat Biotechnol* 27: 1128–1129.
6. Liu, C., Hwang, Y.J., Jeong, H.E. and Yang, P. 2011. Light-induced charge transport within a single asymmetric nanowire. *Nano Lett* 11: 3755–3758.
7. Li, C., Wang, F. and Yu, J.C. 2011. Semiconductor/biomolecular composites for solar energy applications. *Energy Environ Sci* 4: 100–113.
8. Zhou, H., Li, X., Fan, T., Osterloh, F.E., Ding, J., Sabio, E.M., Zhang, D. and Guo, Q. 2010. Artificial inorganic leafs for efficient photochemical hydrogen production inspired by natural photosynthesis. *Adv Mater* 22: 951–956.
9. Chisti, Y. and Yan, J. 2011. Energy from algae: Current status and future trends. *Appl Energy* 88: 3277–3279.
10. Torkamani, S., Wani, S.N., Tang, Y.J. and Sureshkumar, R. 2010. Plasmon-enhanced microalgal growth in miniphotobioreactors. *Appl Phys Lett* 97: 043703.
11. Pinho, P., Jokinen, K. and Halonen, L. 2012. Horticultural lighting – present and future challenges. *Lighting Res Technol* 44: 427–437.
12. Fankhauser, C. 2001. The phytochromes, a family of red/far-red absorbing photoreceptors. *J Biol Chem* 276: 11453–11456.
13. Folta, K.M. and Maruhnich, S.A. 2007. Green light: a signal to slow down or stop. *J Exp Bot* 58: 3099–9111.
14. Kim, H.-H., Goins, G.D., Wheeler, R.M. and Sager, J.C. 2004. Green-light supplementation for enhanced lettuce growth under red-and blue-light-emitting diodes. *Hortic Sci* 39: 1617–1622.
15. Waring, J., Underwood, G.J. and Baker, N.R. 2006. Impact of elevated UV-B radiation on photosynthetic electron transport, primary productivity and carbon allocation in estuarine epipelic diatoms. *Plant Cell Environ* 29: 521–534.
16. Ge, S., Smith, R.G., Jacovides, C.P., Kramer, M.G. and Carruthers, R.I. 2010. Dynamics of photosynthetic photon flux density (PPFD) and estimates in coastal northern California. *Theor Appl Climatol* 105: 107–118.
17. Li, M., Zhang, H., Zhang, X., Deng, J., Liu, Y., Xia, Z. and Lei, B. 2018. Cr^{3+} doped $ZnGa_2O_4$ far-red emission phosphor-in-glass: Toward high-power and color-stable plant growth LEDs with responds to all of phytochrome. *Mater Res Bull* 108: 226–233.
18. Fang, M.H., De Guzman, G.N.A., Bao, Z., Majewska, N., Mahlik, S., Grinberg, M., Leniec, G., Kaczmarek, S.M., Yang, C.W., Lu, K.-M., Sheu, H.S., Hu, S.F. and Liu, R.S. 2020. Ultra-high-efficiency near-infrared $Ga_2O_3:Cr^{3+}$ phosphor and controlling of phytochrome. *J Mater Chem C* 8: 11013–11017.
19. Anthory, R., Cashmore, Jose, A., Jarillo, Wu, Y.-J. and Liu, D. 1999. Cryptochromes: blue light receptors for plants and animals. *Science* 284: 760–765.

20. Pettai, H., Oja, V., Freiberg, A. and Laisk, A. 2005. Photosynthetic activity of far-red light in green plants. *Biochim Biophys Acta* 1708: 311–321.

21. Joachim, W. and Alfred, R.H. 1987. State transitions in the green alga *Scenedesmus obliquus* probed by time-resolved chlorophyll fluorescence spectroscopy and global data analysis. *Biophys J* 52: 717–728.

22. Robert, E., Ruth, C. and Carl, C. 1957. Some factors influencing the long-wave limit of photosynthesis. *PNAS* 43: 133–143.

23. Pettai, H., Oja, V., Freiberg, A. and Laisk, A. 2005. The long-wavelength limit of plant photosynthesis. *FEBS Lett* 579: 4017–4019.

24. Lee, M.J., Park, S.Y. and Oh, M.-M. 2015. Growth and cell division of lettuce plants under various ratios of red to far-red light-emitting diodes. *Horticult Environ Biotechnol* 56: 186–194.

25. Finlayson, S.A., Hays, D.B. and Morgan, P.W. 2007. phyB-1 sorghum maintains responsiveness to simulated shade, irradiance and red light: far-red light. *Plant Cell Environ* 30: 952–962.

26. Oh, J.H., Kang, H., Park, H.K. and Do, Y.R. 2015. Optimization of the theoretical photosynthesis performance and vision-friendly quality of multi-package purplish white LED lighting. *RSC Adv* 5: 21745–21754.

27. Akira, Y. and Kazuhiro, F. 2012. Plant lighting system with five wavelength-band light-emitting diodes providing photon flux density and mixing ratio control. *Plant Methods* 8: 46.

28. Horng, R.-H., Han, P. and Wuu, D.-S. 2008. Phosphor-free white light from InGaN blue and green light-emitting diode chips covered with semiconductor-conversion AlGaInP epilayer. *IEEE Photonics Technol Lett* 20: 1139–1141.

29. Cao, R., Shi, Z., Quan, G., Chen, T., Guo, S., Hu, Z. and Liu, P. 2017. Preparation and luminescence properties of Li_2MgZrO_4:Mn^{4+} red phosphor for plant growth. *J Luminescence* 188: 577–581.

30. Shi, L., Han, Y.-j., Zhao, Y., Li, M., Geng, X.-y., Zhang, Z.-w. and Wang, L.-j. 2019. Synthesis and photoluminescence properties of novel Sr_3LiSbO_6:Mn^{4+} red phosphor for indoor plant growth. *Opt Mater* 89: 609–614.

31. Chen, J., Guo, C., Yang, Z., Li, T., Zhao, J. and McKittrick, J. 2016. Li_2SrSiO_4:Ce^{3+}, Pr^{3+} phosphor with blue, red, and near-infrared emissions used for plant growth LED. *J Am Ceram Soc* 99: 218–225.

32. Yang, X., Zhang, Y., Zhang, X., Chen, J., Huang, H., Wang, D., Chai, X., Xie, G., Molokeev, M.S., Zhang, H., Liu, Y. and Lei, B. 2019. Facile synthesis of the desired red phosphor $Li_2Ca_2Mg_2Si_2N_6$:Eu^{2+} for high CRI white LEDs and plant growth LED device. *J Am Ceram Soc* 103: 1773–1781.

33. Jang, M.-K., Cho, Y.-S. and Huh, Y.-D. 2020. Photoluminescence properties of Eu^{2+} activator ions in the SrS–Ga_2S_3 system. *J Alloys Compd* 828: 154424.

34. Zhou, Z., Zhang, N., Chen, J., Zhou, X., Molokeev, M.S. and Guo, C. 2018. The Vis-NIR multicolor emitting phosphor $Ba_4Gd_3Na_3(PO_4)_6F_2$: Eu^{2+}, Pr^{3+} for LED towards plant growth. *J Ind Eng Chem* 65: 411–417.

35. Yang, Y.J., Zhang, S.B., Wang, J.H. and Huang, W. 2019. Photosynthetic regulation under fluctuating light in field-grown *Cerasus cerasoides*: a comparison of young and mature leaves. *Biochim Biophys Acta Bioenerg* 1860: 148073.

36. Yang, Y.J., Zhang, S.B. and Huang, W. 2019. Photosynthetic regulation under fluctuating light in young and mature leaves of the CAM plant *Bryophyllum pinnatum*. *Biochim Biophys Acta Bioenerg* 1860: 469–477.

37. Wierer, J.J., Tsao, J.Y. and Sizov, D.S. 2013. Comparison between blue lasers and light-emitting diodes for future solid-state lighting. *Laser Photonics Rev* 7: 963–993.

38. Zhang, X., Zhang, J., Wu, X., Yu, L., Liu, Y., Xu, X. and Lian, S. 2020. Discovery of blue-emitting Eu^{2+}-activated sodium aluminate phosphor with high thermal stability via phase segregation. *Chem Eng J* 388: 124289.

39. Li, C., Wang, X.M., Chi, F.F., Yang, Z.-P. and Jiao, H. 2019. A narrow-band blue emitting phosphor $Ca_8Mg_7Si_9N_{22}$:Eu^{2+} for pc-LEDs. *J Mater Chem C* 7: 3730–3734.

40. Böhnisch, D., Rosenboom, J., García-Fuente, A., Urland, W., Jüstel, T. and Baur, F. 2019. On a blue emitting phosphor $Na_3RbMg_7(PO_4)_6$:Eu^{2+} showing ultra high thermal stability. *J Mater Chem C* 7: 6012–6021.

41. Zhang, X., Zhang, J., Ma, W., Liao, S., Zhang, X., Wang, Z., Yu, L. and Lian, S. 2019. From nonluminescence to bright blue emission: boron-induced highly efficient Ce^{3+}-doped hydroxyapatite phosphor. *Inorg Chem* 58: 13481–13491.

42. Duke, A.C., Hariyani, S. and Brgoch, J. 2018. $Ba_3Y_2B_6O_{15}$:Ce^{3+}—a high symmetry, narrow-emitting blue phosphor for wide-gamut white lighting. *Chem Mater* 30: 2668–2675.

43. Li, P., Wang, Z., Yang, Z. and Guo, Q. 2014. $Ba_2B_2O_5$:Ce^{3+}: a novel blue emitting phosphor for white LEDs. *Mater Res Bull* 60: 679–681.

44. Wang, Y.F., Wang, Y.F., Zhu, Q.-Q., Hao, L.-Y., Xu, X., Xie, R.-J., Agathopoulos, S. and Srivastava, A. 2013. Luminescence and structural properties of high stable Si-N-doped $BaMgAl_{10}O_{17}:Eu^{2+}$ phosphors synthesized by a mechanochemical activation route. *J Am Ceram Soc* 96: 2562–2569.

45. Bui, H.V., Nguyen, T., Nguyen, M.C., Tran, T.A., Tien, H.L., Tong, H.T., Nguyen, T.K.L. and Pham, T.H. 2015. Structural and photoluminescent properties of nanosized $BaMgAl_{10}O_{17}:Eu^{2+}$ blue-emitting phosphors prepared by sol-gel method. *Adv Nat Sci: Nanosci Nanotechnol* 6: 035013.

46. Ekambaram, S. and Patil, K.C. 1997. Synthesis and properties of Eu^{2+} activated blue phosphors. *J Alloys Compd* 248: 7–12.

47. Zhou, Q., Wang, M. and Wang, Y. 2019. Research progress on rare earth ion doped blue phosphors. *Guangzhou Chem Ind* 47: 20–29.

48. Zhu, Q.Q., Xu, X. and Hao, L.Y. 2015. Synthesis of high-performance $BaMgAl_{10}O_{17}:Eu^{2+}$ phosphor through a facile aqua-suspension method. *Int J Appl Ceram Technol* 12: 760–764.

49. Yin, L.-J., Dong, J., Wang, Y., Zhang, B., Zhou, Z.-Y., Jian, X., Wu, M., Xu, X., van Ommen, J.R. and Hintzen, H.T. 2016. Enhanced optical performance of $BaMgAl_{10}O_{17}:Eu^{2+}$ phosphor by a novel method of carbon coating. *J Phys Chem C* 120: 2355–2361.

50. Wang, Q., Zhu, G., Xin, S., Ding, X., Xu, J., Wang, Y. and Wang, Y. 2015. A blue-emitting Sc silicate phosphor for ultraviolet excited light-emitting diodes. *Phys Chem Chem Phys* 17: 27292–27299.

51. Li, K., Shang, M., Lian, H. and Lin, J. 2015. Photoluminescence properties of efficient blue-emitting phosphor alpha-$Ca_{1.65}Sr_{0.35}SiO_4:Ce^{3+}$: color tuning via the substitutions of Si by Al/Ga/B. *Inorg Chem* 54: 7992–8002.

52. Ding, X., Zhu, G., Geng, W., Mikami, M. and Wang, Y. 2015. Novel blue and green phosphors obtained from $K_2ZrSi_3O_9:Eu^{2+}$ compounds with different charge compensation ions for LEDs under near-UV excitation. *J Mater Chem C* 3: 6676–6685.

53. Yang, L., Wang, J.S., Zhu, D.C., Pu, Y., Zhao, C. and Han, T. 2018. The effect of doping Mg^{2+} on structure and properties of $Sr_{(1.992-x)}MgxSiO_4$: $0.008Eu^{2+}$ blue phosphor synthesized by co-precipitation method. *Opt Mater* 75: 887–892.

54. Kim, D., Kim, S.C., Bae, J.S., Kim, S., Kim, S.J. and Park, J.C. 2016. Eu^{2+}-activated alkaline-earth halophosphates, $M_5(PO_4)_3X:Eu^{2+}$ (M = Ca, Sr, Ba; X = F, Cl, Br) for NUV-LEDs: site-selective crystal field effect. *Inorg Chem* 55: 8359–8370.

55. Park, J.K., Lim, M.A., Choi, K.J. and Kim, C.H. 2005. Luminescence characteristics of yellow emitting $Ba_3SiO_5:Eu^{2+}$ phosphor. *J Mater Sci* 40: 2069–2071.

56. Lim, M.A., Park, J.K., Kim, C.H. and Park, H.D. 2003. Luminescence characteristics of green light emitting $Ba_2SiO_4:Eu^{2+}$ phosphor. *J Mater Sci Lett* 22: 1351–1353.

57. Poort, S.H.M., W, J. and Blasse, G. 1997. Optical properties of Eu^{2+}-activated orthosilicates and ortho-phosphates. *J Alloys Compd* 260: 93–97.

58. Silva, E.N., Ayala, A.P., Guedes, I., Paschoal, C.W.A., Moreira, R.L., Loong, C.-K., and Boatner, L.A. 2006. Vibrational spectra of monazite-type rare-earth orthophosphates. *Opt Mater* 29: 224–230.

59. Tang, Y.S., Hu, S.F., Lin, C.C., Bagkar, N.C. and Liu, R.-S. 2007. Thermally stable luminescence of $KSrPO_4:Eu^{2+}$ phosphor for white light UV light-emitting diodes. *Appl Phys Lett* 90: 151108.

60. Wu, Z.C., Shi, J.X., Wang, J., Gong, M.L. and Su, Q. 2006. A novel blue-emitting phosphor $LiSrPO_4:Eu^{2+}$ for white LEDs. *J Solid State Chem* 179: 2356–2360.

61. Lin, C.C., Shen, C.C. and Liu, R.S. 2013. Spiral-type heteropolyhedral coordination network based on single-crystal $LiSrPO_4$: implications for luminescent materials. *Chemistry* 19: 15358–15365.

62. Zhang, S., Huang, Y., Shi, L. and Seo, H.J. 2010. The luminescence characterization and structure of Eu^{2+} doped $LiMgPO_4$. *J Phys Condens Matter* 22: 235402.

63. Bandi, V.R., Jeong, J., Shin, H.-J., Jang, K., Lee, H.-S., Yi, S.-S. and Jeong, J.H. 2011. Thermally stable blue-emitting $NaSrPO_4:Eu^{2+}$ phosphor for near UV white LEDs. *Opt Commun* 284: 4504–4507.

64. Yim, D.K., Song, H.J., Cho, I.-S., Kim, J.S. and Hong, K.S. 2011. A novel blue-emitting $NaSrPO_4:Eu^{2+}$ phosphor for near UV based white light-emitting-diodes. *Mater Lett* 65: 1666–1668.

65. Tang, W. and Chen, D. 2009. Photoluminescent properties of $ABaPO_4:Eu$ (A=Na, K) phosphors prepared by the combustion-assisted synthesis method. *J Am Ceram Soc* 92: 1059–1061.

66. Zhang, S., Huang, Y., Nakai, Y., Tsuboi, T. and Seo, H.J. 2011. The luminescence characterization and thermal stability of Eu^{2+} ions-doped $NaBaPO_4$ phosphor. *J Am Ceram Soc* 94: 2987–2992.

67. Tang, W. and Zheng, Y. 2010. Synthesis and luminescence properties of a novel blue emitting phosphor $NaMgPO_4:Eu^{2+}$. *Luminescence* 25: 364–366.

68. Cao, Y., Ding, J., Ding, X., Wang, X. and Wang, Y. 2017. Tunable white light of multi-cation-site $Na_2BaCa(PO_4)_2:Eu,Mn$ phosphor: synthesis, structure and PL/CL properties. *J Mater Chem C* 5: 1184–1194.

69. Yim, D.K., Song, H.J., Roh, H.-S., Kim, S.-J., Han, B.S., Jin, Y.-H., Hong, H.-S., Kim, D.-W. and Hong, K.S. 2013. Luminescent properties of RbSrPO$_4$:Eu^{2+} phosphors for near-UV-based white-light-emitting diodes. *Eur J Inorg Chem* 2013: 4662–4666.

70. Song, H.J., Yim, D.K., Roh, H.-S., Cho, I.S., Kim, S.-J., Jin, Y.-H., Shim, H.-W., Kim, D.-W. and Hong, K.S. 2013. RbBaPO$_4$:Eu^{2+}: a new alternative blue-emitting phosphor for UV-based white light-emitting diodes. *J Mater Chem C* 1: 500–505.

71. Qiao, J., Ning, L., Molokeev, M.S., Chuang, Y.-C., Liu, Q. and Xia, Z. 2018. Eu^{2+} site preferences in the mixed cation K$_2$BaCa(PO$_4$)$_2$ and thermally stable luminescence. *J Am Chem Soc* 140: 9730–9736.

72. Xiao, Y., Hao, Z., Zhang, L., Wu, H., Pan, G.-H., Zhang, X. and Zhang, J. 2018. An efficient blue phosphor Ba$_2$Lu$_5$B$_5$O$_{17}$:Ce^{3+} stabilized by La$_2$O$_3$: photoluminescence properties and potential use in white LEDs. *Dyes Pigments* 154: 121–127.

73. Hermus, M., Phan, P.-C. and Brgoch, J. 2016. Ab initio structure determination and photoluminescent properties of an efficient, thermally stable blue phosphor, Ba$_2$Y$_5$B$_5$O$_{17}$:Ce^{3+}. *Chem Mater* 28: 1121–1127.

74. Yu, R., Guo, C., Li, T. and Xu, Y. 2013. Preparation and luminescence of blue-emitting phosphor Ca$_2$PO$_4$Cl:Eu^{2+} for n-UV white LEDs. *Curr Appl Phys* 13: 880–884.

75. Deressa, G., Park, K.W., Jeong, H.S., Lim, S.G., Kim, H.J., Jeong, Y.S. and Kim, J.S. 2015. Spectral broadening of blue (Sr, Ba)$_5$(PO$_4$)$_3$Cl:Eu^{2+} phosphors by changing Ba^{2+}/Sr^{2+} composition ratio for high color rendering index. *J Luminescence* 161: 347–351.

76. Tang, J.-Y., Xie, W.-J., Huang, K., Hao, L.-Y., Xu, X. and Xie, R.-J. 2011. A high stable blue BaSi$_3$Al$_3$O$_4$N$_5$:Eu^{2+} phosphor for white LEDs and display applications. *Electrochem Solid-State Lett* 14: J45–J47.

77. Zhang, S., Hao, Z., Zhang, L., Pan, G.-H., Wu, H., Zhang, X., Luo, Y., Zhang, L., Zhao, H. and Zhang, J. 2018. Efficient blue-emitting phosphor SrLu$_2$O$_4$:Ce^{3+} with high thermal stability for near ultraviolet (~400 nm) LED-chip based white LEDs. *Sci Rep* 8: 10463.

78. Zhang, Y., Zhang, X., Zhang, H., Zhuang, J., Hu, C., Liu, Y., Wu, Z., Ma, L., Wang, X. and Lei, B. 2019. Improving moisture stability of SrLiAl$_3$N$_4$:Eu^{2+} through phosphor-in-glass approach to realize its application in plant growing LED device. *J Colloid Interface Sci* 545: 195–199.

79. Lee, C.-W., Petrykin, V. and Kakihana, M. 2009. Synthesis and effect of Sr substitution on fluorescence of new Ba$_{2-x}$Sr$_x$ZnS$_3$: Eu^{2+} red phosphor: Considerable enhancement of emission intensity. *J Cryst Growth* 311: 647–650.

80. Poesl, C. and Schnick, W. 2017. Crystal structure and nontypical deep-red luminescence of Ca$_3$Mg[Li$_2$Si$_2$N$_6$]:Eu^{2+}. *Chem Mater* 29: 3778–3784.

81. Maak, C., Strobel, P., Weiler, V., Schmidt, P.J. and Schnick, W. 2018. Unprecedented deep-red Ce^{3+} luminescence of the nitridolithosilicates Li$_{38.7}$RE$_{3.3}$Ca$_{5.7}$[Li$_2$Si$_{30}$N$_{59}$]O$_2$F (RE = La, Ce, Y). *Chem Mater* 30: 5500–5506.

82. Tang, Z., Zhang, Q., Cao, Y., Li, Y. and Wang, Y. 2020. Eu^{2+}-doped ultra-broadband VIS-NIR emitting phosphor. *Chem Eng J* 388: 124231.

83. Avci, N., Cimieri, I., Smet, P.F. and Poelman, D. 2011. Stability improvement of moisture sensitive CaS:Eu^{2+} micro-particles by coating with sol–gel alumina. *Opt Mater* 33: 1032–1035.

84. Qiao, J., Ning, L., Molokeev, M.S., Chuang, Y.-C., Zhang, Q., Poeppelmeier, K.R. and Xia, Z. 2019. Site-selective occupancy of Eu^{2+} toward blue-light-excited red emission in a Rb$_3$YSi$_2$O$_7$:Eu phosphor. *Angew Chem Int Ed Engl* 58: 11521–11526.

85. Qiao, J., Amachraa, M., Molokeev, M., Chuang, Y.-C., Ong, S.P., Zhang, Q. and Xia, Z. 2019. Engineering of K$_3$YSi$_2$O$_7$ to tune photoluminescence with selected activators and site occupancy. *Chem Mater* 31: 7770–7778.

86. Qiao, J., Zhou, G., Zhou, Y., Zhang, Q. and Xia, Z. 2019. Divalent europium-doped near-infrared-emitting phosphor for light-emitting diodes. *Nat Commun* 10: 5267.

87. Lai, S., Zhao, M., Qiao, J., Molokeev, M.S. and Xia, Z. 2020. Data-driven photoluminescence tuning in Eu^{2+}-doped phosphors. *J Phys Chem Lett* 11: 5680–5685.

88. Ren, Y., cao, R., Chen, T., Su, L., Cheng, X., Chen, T., Guo, S. and Yu, X. 2019. Photoluminescence properties of Ba$_2$LaSbO$_6$:Mn^{4+} deep-red-emitting phosphor for plant growth LEDs. *J Luminescence* 209: 1–7.

89. Liu, S.X., Xiong, F.B., Lin, H.F., Meng, X.G., Lian, S.Y. and Zhu, W.Z. 2018. A deep red-light-emitting phosphor Mn^{4+}:CaAl$_4$O$_7$ for warm white LEDs. *Optik* 170: 178–184.

90. Shao, Q., Lin, H., Hu, J., Dong, Y. and Jiang, J. 2013. Temperature-dependent photoluminescence properties of deep-red emitting Mn^{4+}-activated magnesium fluorogermanate phosphors. *J Alloys Compd* 552: 370–375.

91. Lee, S.J., Jung, J., Park, J.Y., Jang, H.M., Kim, Y.-R. and Park, J.K. 2013. Application of red light-emitting diodes using $Mg_{3.5}Ge_{1.25}O_6$:Mn^{4+} phosphor. *Mater Lett* 111: 108–111.

92. Li, M., Zhang, J., Han, J., Qiu, Z., Zhou, W., Yu, L., Li, Z. and Lian, S. 2017. Changing Ce^{3+} content and codoping Mn^{2+} induced tunable emission and energy transfer in $Ca_{2.5}Sr_{0.5}Al_2O_6$:Ce^{3+},Mn^{2+}. *Inorg Chem* 56: 241–251.

93. Zhang, X., Qiao, X. and Seo, H.J. 2011. Luminescence properties of novel Ce^{3+}, Mn^{2+} doped $NaSr_4(BO_3)_3$ phosphors. *Curr Appl Phys* 11: 442–446.

94. Lu, Q.F., Li, J. and Wang, D.J. 2013. Single-phased silicate-hosted phosphor with 660 nm-featured band emission for biological light-emitting diodes. *Curr Appl Phys* 13: 1506–1511.

95. Zhou, Z., Zheng, J., Shi, R., Zhang, N., Chen, J., Zhang, R., Suo, H., Goldys, E.M. and Guo, C. 2017. Ab initio site occupancy and far-red emission of Mn^{4+} in cubic-phase $La(Mg,Ti)_{1/2}O_3$ for plant cultivation. *ACS Appl Mater Interfaces* 9: 6177–6185.

96. Huang, X. and Guo, H. 2018. Finding a novel highly efficient Mn^{4+}-activated $Ca_3La_2W_2O_{12}$ far-red emitting phosphor with excellent responsiveness to phytochrome PFR: towards indoor plant cultivation application. *Dyes Pigments* 152: 36–42.

97. Huang, X., Liang, J., Li, B., Sun, L. and Lin, J. 2018. High-efficiency and thermally stable far-red-emitting $NaLaMgWO_6$:Mn^{4+} phosphors for indoor plant growth light-emitting diodes. *Opt Lett* 43: 3305–3308.

98. Yang, Z., Yang, L., Ji, C., Xu, D., Zhang, C., Bu, H., Tan, X., Yun, X. and Sun, J. 2019. Studies on luminescence properties of double perovskite deep red phosphor La_2ZnTiO_6:Mn^{4+} for indoor plant growth LED applications. *J Alloys Compd* 802: 628–635.

99. Lu, W., Lv, W., Zhao, Q., Jiao, M., Shao, B. and You, H. 2014. A novel efficient Mn^{4+} activated $Ca_{14}Al_{10}Zn_6O_{35}$ phosphor: application in red-emitting and white LEDs. *Inorg Chem* 53: 11985–11990.

100. Zhou, Z., Xia, M., Zhong, Y., Gai, S., Huang, S., Tian, Y., Lu, X. and Zhou, N. 2017. Dy^{3+}@Mn^{4+} co-doped $Ca_{14}Ga_{10-m}Al_mZn_6O_{35}$ far-red emitting phosphors with high brightness and improved luminescence and energy transfer properties for plant growth LED lights. *J Mater Chem C* 5: 8201–8210.

101. Zhong, Y., Gai, S., Xia, M., Gu, S., Zhang, Y., Wu, X., Wang, J., Zhou, N. and Zhou, Z. 2019. Enhancing quantum efficiency and tuning photoluminescence properties in far-red-emitting phosphor $Ca_{14}Ga_{10}Zn_6O_{35}$:Mn^{4+} based on chemical unit engineering. *Chem Eng J* 374: 381–391.

102. Malysa, B., Meijerink, A. and Jüstel, T. 2018. Temperature dependent Cr^{3+} photoluminescence in garnets of the type $X_3Sc_2Ga_3O_{12}$ (X = Lu, Y, Gd, La). *J Luminescence* 202: 523–531.

103. Yu, D., Zhou, Y., Ma, C., Melman, J.H., Baroudi, K.M., LaCapra, M. and Riman, R.E. 2019. Non-rare-earth Na_3AlF_6:Cr^{3+} phosphors for far-red light-emitting diodes. *ACS Appl Electron Mater* 1: 2325–2333.

104. Deng, J., Zhang, H., Zhang, X., Zheng, Y., Yuan, J., Liu, H., Liu, Y., Lei, B. and Qiu, J. 2018. Ultrastable red-emitting phosphor-in-glass for superior high-power artificial plant growth LEDs. *J Mater Chem C* 6: 1738–1745.

105. Li, M., Zhang, X., Zhang, H., Chen, W., Ma, L., Wang, X., Liu, Y. and Lei, B. 2019. Highly efficient and dual broad emitting light convertor: an option for next-generation plant growth LEDs. *J Mater Chem C* 7: 3617–3622.

106. Chen, W., Zhang, X., Zhou, J., Zhang, H., Zhuang, J., Xia, Z., Liu, Y., Molokeev, M.S., Xie, G. and Lei, B. 2020. Glass-ceramics with thermally stable blue-red emission for high-power horticultural LED applications. *J Mater Chem C* 8: 3996–4002.

20 AC-Driven LED Phosphors

Hang Lin and Yuansheng Wang

CONTENTS

20.1 INTRODUCTION: BACKGROUND

Currently, the mainstream product in commercial LED market is direct current (DC)-LED driven by DC. Since DC-LEDs generate light only when they are forward biased, it is required to transform the input alternating current (AC) from city power into DC for directionally driving the electron-hole recombination in p-n junction of LED (Figure 20.1(a)). Unfortunately, during the AC-DC conversion, several tough technical issues arise[1-6]: (1) there is 15–30% unnecessary electric power consumption that decreases the system efficiency; (2) the generated massive heat accelerates yellowing and aging of the organic encapsulant; (3) the limited lifespan (less than 20,000 h) of electrolytic capacitor series in AC-DC converter restricts the actual service life of LED (the electro-component is broken first, when the chip die is still workable); and (4) the increased module volume makes the appearance design of LED lamp harder. To avoid these problems resulted from operating DC-LED with AC, a scheme of AC-LED (sometimes, also described as high-voltage LED, HV-LED) was proposed (Figure 20.1(b)), showing tremendous merits of not only lower price, but also higher energy utilization efficiency, more compacted volume, and longer service life. However, AC-LED cannot emit immediately until the voltage across the circuit is higher than its turn-on voltage (V_F), which indicates that a time gap of 5–20 ms (correlating with the AC frequency and the AC circuit design) is unavoidable in every AC cycle.[7] Such lighting flicker degrades the light quality and is harmful to human health.[8] How to restrain it is one key issue to develop the AC-LED technology.

As a special kind of storage material, the persistent luminescence (PersL) phosphor recently has been demonstrated applicable to compensate the lighting flicker of AC-LED.[5-7] PersL, a phenomenon that the light can still last for durations after removal of the excitation source, originates from the continuous recombination of electrons/holes which are captured/immobilized in the traps upon

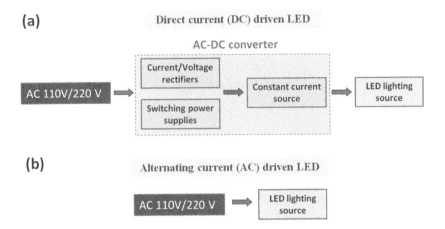

(a) Direct current (DC) driven LED

AC-DC converter

AC 110V/220 V → [Current/Voltage rectifiers / Switching power supplies] → [Constant current source] → [LED lighting source]

(b) Alternating current (AC) driven LED

AC 110V/220 V → [LED lighting source]

FIGURE 20.1 Schematic illustration of the application scheme for (a) DC-LED and (b) AC-LED.

charging and then are gradually released from the traps under environmental stimuli (such as, heat, light, and stress).[9–12] PersL behaviors, including PersL lifetime and PersL brightness, are known to be closely related with the nature of traps (species, concentration, and depth) in host.[13–16] In Figure 20.2, it schematically illustrates the working mechanism of PersL phosphor for AC-LED. When the AC input voltage (V_{in}) is higher than V_F, the PersL phosphor converts the blue/UV light emitted from chip die into the white light. During this period, the defect centers in host can store the excited charge carriers. When $V_{in}<V_F$, the supply current drops to zero, but AC-LED does not immediately go out, thanks to the compensation effect of PersL. Therefore, the lighting flicker effect can be greatly suppressed.

In the following sections, we will give a brief introduction to the construction of AC-LED, the working principle, and the flickering effect. Emphasis is made on the designing principle of AC-LED phosphor and its up-to-date progress. Finally, some challenging issues are proposed, which should be solved for further developing the PersL phosphor-converted AC-LED technology.

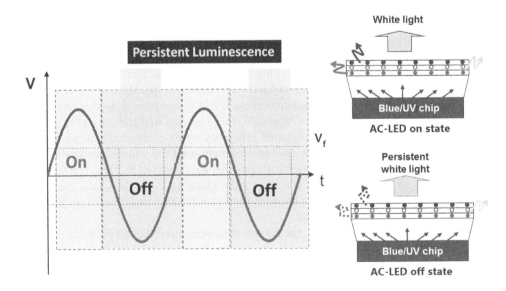

FIGURE 20.2 Schematic illustration of using persistent phosphors to compensate AC-LED flicker.

20.2 ABOUT AC-LED

AC-LED is an emerging solid-state lighting source with great superiorities over the DC-LED counterparts. It has a very short history but is developing very fast. In 2000, the first AC-LED module was proposed for decorative purpose (*e.g.*, the light strings of a Christmas tree).[17] Up to date, AC-LED products have penetrated into various fields of general lightings. Seoul Semiconductor Co., the world biggest AC-LED manufacturer, launched a series of AC-LED products with maximal high luminous efficacy (LE) of 168 lm/W, together with CCT of 5000K, CRI of 80, junction temperature of 85°C, and power of 40 W.[18]

20.2.1 AC-LED CONFIGURATION

The simplest AC-LED configuration is the LED light string employing a plurality of LED lamp beads wired in series form (Figure 20.3(a)).[17] In this case, AC-LED is on state only in the positive period of AC cycle, and so, the severe lighting flicker with a flash rate of 50–60 Hz occurs. There have been several circuit topology designs proposed to make AC-LED also workable in the negative AC cycle; and correspondingly, the flash rate is increased to 100–120 Hz, which is insensitive to human eyes for the persistence of vision. Among them, the antiparallel circuit with two series blocks reversed (Figure 20.3(b)) is often used, enabling two series blocks flash alternatively (one brightens in the positive AC cycle and the other one brightens in the negative AC cycle).[19,20] Equivalent to the antiparallel circuit, the Wheatstone bridge (WB) circuit (Figure 20.3(c))[4] and the ladder circuit (Figure 20.3(d))[3] also drive LED module during the opposite phases of AC power. The only difference is the public circuit segment on which the loaded LEDs work all the time. In these two cases, the light output efficiency of AC-LED is improved and the cost is reduced, due to the increased chip area utilization ratio.[1]

In the conventional design of AC-LED, there should be amounts of LED lamp beads to disperse the high forward AC voltage. One typical example is the huge lighting array with 697 packaged white LEDs driven by 100-V AC reported by Tamura.[21] However, such design brings about tough issues of large volume, high cost, and unreliability, arousing great concerns. To address these

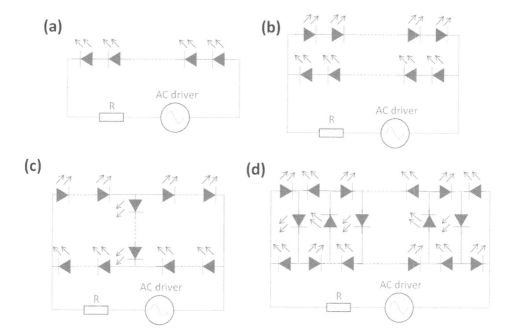

FIGURE 20.3 Diagrams of (a) series, (b) antiparallel, (c) WB, and (d) ladder circuit topologies for AC-LED.

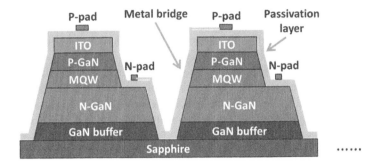

FIGURE 20.4 Schematic illustration of the monolithic AC-LED series array structure.

issues, in 2002, Ao *et al.* proposed and demonstrated a brand-new chip design, *i.e.*, the monolithic blue LED series arrays operated under high AC voltage,[22] with the aids of sophisticated semiconductor manufacturing technique. For example, through etching the light emitting materials into the insulating substrate, the luminous region of a single chip can be separated into a number of mutually insulated micrometer-sized luminous units, enabling the monolithic integration (Figure 20.4). The same year, III-N Technology Co. in the United States announced a similar integrated "single chip" AC-LED for standard 110/220-V AC voltage operation.[23] Industrial Technology Research Institute (ITRI) in Taiwan took great efforts to further develop the "single chip" technology. ITRI's monolithic integration circuit has evolved from antiparallel (first generation), to WB (second generation), and to Schottky barrier diodes (SBDs) modified WB (third generation),[1,24] as presented in Figure 20.5.

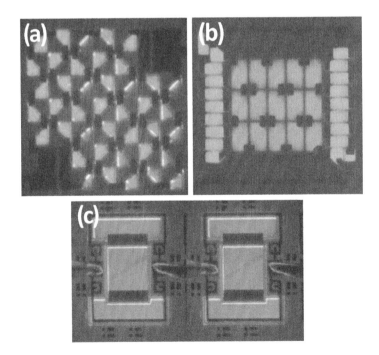

FIGURE 20.5 Physical photographs of (a) antiparallel-type (first generation), (b) WB-type (second generation), and (c) SBD-type (third generation) AC-LEDs developed by ITRI. (Yeh, W. Y., Yen, H. H., and Chan, Y. J. *Proc. SPIE* 7939, 793910, 2010. With permission)

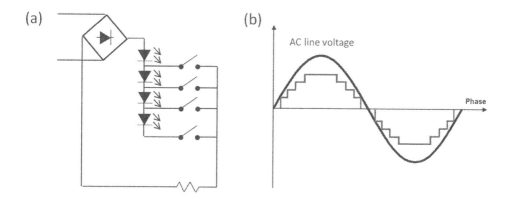

FIGURE 20.6 (a) Circuit design and (b) current waveform on state using a segmented linear constant current driver for AC-LED.

For the SBDs-rectified AC-LED, its designing idea is to integrate four GaN SBDs with micro-LEDs on a single chip, whereupon the low breakdown voltage and fast failure problems of rectifying micro-LEDs in WB-AC-LED will be well solved and the chip area utilization ratio can be greatly increased.[1]

20.2.2 AC-LED DRIVING SCHEME

An appropriate AC driver is required to control the electric current state of AC-LED for upgrading power factor and degrading total harmonic distortion. Seoul Semiconductor Co. employed a segmented linear constant current driver to control the current waveform. In this driving scheme, the mains electricity rectified by one bridge circuit flows through one LED string, on which LEDs are lit up/go out depending on the magnitude of AC voltage (a signal voltage is compared with a predetermined reference voltage); and correspondingly, the output current presents a stairs-type rise/fall (Figure 20.6).[25] There have been some other AC driver designs. For example, Jung *et al.* presented a new pulse width control dimmer using two active switches, solving the problem of chopping AC sine wave with noticeably deformed output waveform.[26] Park *et al.* proposed a linear current regulator controlled by simple digital logics and comparators to limit the current within a predefined range.[27] Hwu *et al.* designed a bridgeless driver by using the field programmable gate array with dimmable control.[28] Yu *et al.* demonstrated an AC driver with current glitch eliminated, showing a high power factor of 0.998 and a low total harmonic distortion of 5.1%.[29]

In addition, what is worthy to be mentioned is the AC-LED light engine, featuring with the integration of drivers and other auxiliary electronic control units on board. Its designing concept is to promote the standardization of lighting products, so as to realize the interchangeability of LED light source products from different manufacturers. Notably, the common chip-on-board (COB) LED module cannot be called as the light engine, since the driver is not on board. Seoul Semiconductor Co. is at the cutting edge of AC-LED light engine technology. In 2017, the breakthrough in size reduction of the MICRO DRIVER Series LED driver benefits to achieving high-performance AC-LED light engine with high compactness, improved efficiency, and low cost.[30]

20.3 LIGHTING FLICKER

For LEDs, inherently it does not show flicker; nevertheless, when coupled with power electronics, it can exhibit more pronounced flicker than that of the conventional electric lighting source, due to fast response to the current change (as short as magnitude of nanoseconds).[31] The drawing of AC power with two opposite phases, the interaction between drivers and dimmer electronics, or the disturbances in electricity network result in unintentional light flicker, while sometimes the intentional light modulation should be carried out to control intensity and color.[32]

20.3.1 Photometric Flicker

Illuminating Engineering Society (IES) defines the flicker as "variations of luminance in time"[33] or "a rapid and repeated change over time in the brightness of light".[34] Commission Internationale de L'Eclairage (CIE) defines the flicker as "impression of unsteadiness of visual perception induced by a light stimulus whose luminance or spectral distribution fluctuates with time".[35] Considering the past definition without consideration of interaction with movements, the latest version of CIE standard introduces a new term "temporal light artifact" (TLA). TLA is defined as "change in visual perception, induced by a light stimulus the luminance or spectral distribution of which fluctuates with time, for a human observer in a specified environment". The concept of TLA covers three specific cases of observing the temporal light modulation: (1) a static observer in a static environment (flicker); (2) a static observer in a non-static environment (stroboscopic effect); and (3) a non-static observer in a static environment (phantom array effect).

Light flicker can be categorized into "visible flicker" and "invisible flicker", of which the former one indicates the luminous modulation is sensed and perceived, while the latter one indicates the luminous modulation is sensed but not perceived.[31] In common situations, 60 Hz is a boundary frequency determining whether the light flicker is visible or invisible. Of course, it varies (60–100 Hz) between individuals and depends on the amplitude, brightness, and visual size of lighting source. This boundary frequency at which a flickering light source fuses into an apparently constant source is called as the critical fusion frequency (CFF).[36] Above CFF, the light flicker is appreciated only in terms of its effects on spatial perception, such as the stroboscopic effect or phantom array effect.[35]

There are two acknowledged metrics to characterize the light flicker: percent flicker (PF) and flicker index (FI) (Figure 20.7).[37] PF, also named as peak-to-peak contrast or Michelson contrast, is defined by Equation (20.1), characterizing the fluctuation depth of light waveform. FI, defined by

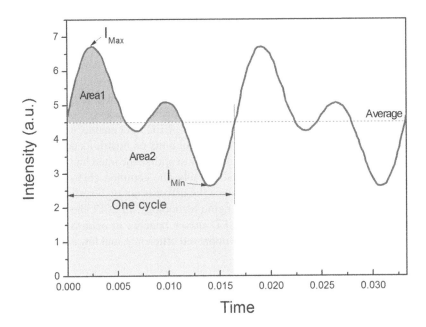

FIGURE 20.7 Determining percent flicker and flicker index in a periodic waveform. (Adapted from IEEE Std 1789–2015)

Equation (20.2), considers not only amplitude but also shape and duty cycle of the periodic light waveform; therefore, it is generally regarded as a more reliable metric than PF.

$$\text{Percent flicker} = 100\% \times \frac{\left(I_{\text{Max}} - I_{\text{Min}}\right)}{\left(I_{\text{Max}} + I_{\text{Min}}\right)} \tag{20.1}$$

$$\text{Flicker index} = \frac{\text{Area 1}}{\left(\text{Area 1} + \text{Area 2}\right)} \tag{20.2}$$

Poplawski *et al.* compared three typical light waveforms to reveal the difference between PF and FI for photometric flicker.[38] As presented in Figure 20.8, in spite of having an identical PF of 100% at a frequency of 120 Hz, the calculated FI values for triangle (0.250), sine (0.318), and square (0.500) waveforms are quite distinct from each other. It appears that the simple periodic waveforms that transition faster from their low levels to their high levels result in higher FI values. One useful conclusion can be drawn is that, at low modulation amplitude (<40%), PF is reliable and preferred to use for its simplicity to calculate, while at high modulation amplitude (>40%), FI should be used since it takes accounts of the effect of the other waveform characteristics.[38]

The stroboscopic effect refers to the interaction of moving objects with temporally modulated light, making the object appear to move discretely rather than continuously (variation in motion perception).[39] Unlike the abovementioned flicker, it can be perceived under certain conditions even at frequency as high as 500–800 Hz.[39,40] Bullough *et al.* designed an experiment attempting to quantitatively describe the stroboscopic effect, where the participants were asked to wave a white plastic rod back and forth underneath the luminaire and to report whether they could perceive the stroboscopic effect (*e.g.*, observing the striated images of the rod). Based on the experimental

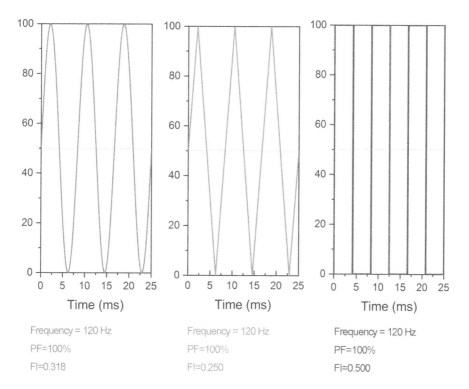

Frequency = 120 Hz
PF=100%
FI=0.318

Frequency = 120 Hz
PF=100%
FI=0.250

Frequency = 120 Hz
PF=100%
FI=0.500

FIGURE 20.8 Waveform properties and flicker metrics for three periodic waveforms. (Adapted from Poplawski, M. E. 2013)

results, he proposed an empirical model to predict the detection percentages (d) for stroboscopic effects from flicker[41]:

$$d = \left[\frac{(25p + 140)}{(f + 25p + 140)} \right] \times 100\% \tag{20.3}$$

where p is PF and f is frequency. Perz *et al.* designed another experiment of observing a white dot on a black rotating turntable that revolved at a certain fixed speed.[39] A new measure called the stroboscopic visibility measure (SVM) is demonstrated effective to characterize the stroboscopic effect. The SVM method, using the relative energy of the Fourier components normalized with the threshold modulation depth of simple sine waveforms, is defined as follows[39]:

$$\text{SVM} = \sqrt[n]{\sum_{m=1}^{\infty} \left(\frac{C_m}{T_m} \right)^n} \quad \begin{cases} <1 & \text{not visible} \\ =1 & \text{just visible} \\ >1 & \text{visible} \end{cases} \tag{20.4}$$

where C_m is the amplitude of the mth Fourier component and T_m is the visibility threshold for the stroboscopic effect for a sine wave at the frequency of the mth Fourier component. According to Equations (20.3) and (20.4), the photometric stroboscopic effect is highly frequency-dependent, which is quite different from PF and FI. The magnitude of the stroboscopic effect also depends on the rate of object motion and the viewing conditions.[33]

The "phantom array effect" usually occurs when one saccade in the dark across a point source of light blinking rapidly on and off.[42] During eye saccades, a visual illusion of variation in shape or spatial positions of objects can be seen. Roberts performed an experiment of utilizing the phantom array effect to discriminate modulated light from steady light.[43] It was demonstrated that perception of the phantom array effect could occur at high temporal frequency up to ~3 kHz and its magnitude is correlative with the spatial frequency, contrast, and extent (number of cycles) of the intrasaccadic stimulation. Up to date, there still has been no mathematical model to predict the phantom array effect.

20.3.2 Biological Effects of Light Flicker and Risk Assessment

In IEEE PAR1789,[34] the detailed descriptions of possible biological effects induced by light flicker have been put forward, including photosensitive epilepsy, dizziness, migraines, headaches, general malaise, impaired ocular motor control, and impaired visual performance. The visible flicker with frequency below CFF results in serious consequences to human health. The invisible flicker above CFF still induces potential health risks. Notably, the risk increases monotonically along with brightness, contrast, and exposing time, and it also depends on the distance between viewer and lighting source. When considering the stroboscopic effect or phantom array effect, the light flicker even beyond 200 Hz could bring about disastrous consequence: for example, in the industrial spaces with moving machinery, the illusions of slow or stoppage of machine brought by the stroboscopic effect can be of grave danger.[44]

To protect against the described health risks, IEEE has summarized the recommended operating area as a function of frequency and Modulation (Mod.%) (Figure 20.9). Three practices were proposed[34]:

Practice 1 – To limit adverse biological effect of lighting flicker, Mod.% should be less than 0.025 × frequency or 0.08 × frequency when the frequency is below 90 Hz or between 90 and 1250 Hz; and no restriction on Mod.% is made when the frequency is above 1250 Hz.

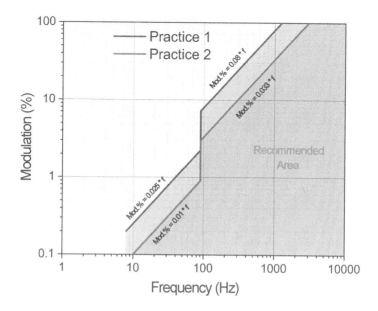

FIGURE 20.9 Recommended practices to protect against the described health risks. (Adapted from IEEE Std 1789–2015)

Practice 2 with more restrictions – To ensure no observable effect level, Mod.% should be reduced by 2.5 times below the limited biological effect level given in Recommended Practice 1.
Practice 3 – For any lighting source, under all operating scenarios, Mod.% should be less than 5% when the frequency is below 90 Hz.

20.3.3 CHROMATIC FLICKER

Not only flicker in luminance but also flicker in color (chromatic flicker) can be visually perceived.[45] The chromatic flicker in AC-LED is produced due to the necessity of using two or three multicolor phosphors to improve color quality of the lighting source. It is very difficult to balance the decay times of these phosphors, and so, a color point shift may occur during the AC cycles. In comparison with the flicker in blue/red region, the flicker in green/yellow region is more sensitive to human eyes. Fortunately, in the temporal domain, one cannot distinguish the chromatic flicker faster than 25 Hz,[46] since the chromatic channels of human vision shows poor temporal sensitivity.[32,47]

20.4 ABOUT AC-LED PHOSPHORS

The idea of using slow-decay phosphor to reduce flicker sensitivity of lighting source can date back to 1960s when the researchers attempted to achieve flickerless regeneration rate for cathode ray tube (CRT) displays.[48,49] Thereafter, this scheme was demonstrated also applicable to visual display unit (VDU) screens,[50] fluorescent lamps,[51] and AC-LEDs.[52]

There are two kinds of slow-decay phosphors for AC-LEDs. The first one is the phosphor with intrinsic millisecond-ranged fluorescent decay, such as, Mn^{2+} activated phosphor.[53] Despite its effectiveness, the optical saturation takes place at high excitation fluxes, resulting in the reduction of absorption/emission efficiency and the occurrence of color variation.[54,55] The second one is the PersL phosphor with PersL decay in the millisecond-ranged time window, which shows great promises. Unfortunately, the material system of such phosphors is still rather limited. Researchers are always in pursuit of long PersL time, while little attentions are paid to short afterglow in the millisecond range.

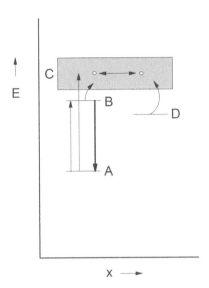

FIGURE 20.10 Schematical illustration of the PersL mechanism.

20.4.1 THEORETICAL FUNDAMENTALS

Figure 20.10 schematically illustrates a luminescence model that is generally accepted.[56–60] The abscissa denotes the position of the electron in an insulating solid. The ordinate represents the energy of an electron. Upon excitation, electrons of active ions can be prompted from ground level *A* to excited level *B*, or from ground level *A* to conduction band (CB) *C*, dependent on the photon energy. The electrons in *B* can also be further promoted to *C* via photoionization, when the energy gap between *B* and *C* is small enough. The electrons in *C* are free from the luminescence center and may move through the solid until they are captured by a trap level *D*. The fluorescence generates when the electrons return from *B* to *A*. While the generation of PersL is much more complicated, requiring a thermal or optical stimulation to release the immobilized electrons from *D*, and then return to *B* via *C*. During this process, the retrapping of electrons in shallow traps may take place. The PersL behavior is closely related with the trap properties. Generally speaking, PersL time is determined by the depth of traps, and PersL brightness relies on the concentration of traps, especially for those shallow ones.

Garlick and Wilkins studied the PersL decay of various types during the first few milliseconds in sulfide (*e.g.*, ZnS:Cu and ZnS:Ag) and clarified the millisecond afterglow is determined by the time that electrons spend in traps, rather than that electrons spend in moving through the phosphor.[56] Importantly, from the theoretical point of view, PersL decay behaviors in the timescale of milliseconds were predicted, laying fundamentals to study the AC-LED phosphors. In this section, we will summarize the main theoretical points of this work to get insights into this unique PersL decay behavior, which is different from that of the familiar one with long duration up to seconds, minutes, or hours.

- Excitation intensity dependent PersL decay behavior
  The probability *p* of an electron escaping from a trap can be expressed as

$$p = se^{-E/kT} \tag{20.5}$$

where *E* is the trap depth, *k* the Boltzmann constant, *T* the absolute temperature, and *s* a constant ~10^{-8} second^{-1}. According to this equation, the escaping rate of electrons is slow when the trap depth is deep. The deep traps could be perceived as filled up after prolonged excitation; in contrast, the shallow traps are only partly filled, because the quantity of electrons escaped from traps is appreciable. An equilibrium is expected to be established when the rate of supplied electrons to traps is equal to the rate of escaped electrons from traps. In the case of long afterglow, PersL decay behavior is influenced by the excitation intensity insignificantly once the deep traps are filled to capacity. Unlike that, in the case of short afterglow, the afterglow will decay more rapidly as the excitation intensity increases, since more of the shallower traps will be filled, leading to the average a shorter time spent in traps.[56]

- Excitation time dependent PersL decay behavior

Let's consider the situations of prolonged excitation versus pulsed excitation. Assuming the number of traps of all depths in the phosphor is equal and the cross section for electron capture of the traps does not vary much with their depth, one can make the following inferences: after prolonged excitation, all traps are filled to an equilibrium extent with the shallower traps being filled less than the deeper traps; whereas, upon the pulsed excitation with the same intensity, such equilibrium is not established, in which case, the fraction of numbers of electrons in shallower traps is larger, and, therefore, the initial decay should be more rapid.[56]

- Temperature dependent PersL decay behavior

The temperature influences PersL decay in a more complicated manner; however, one can predict it based on the shape of thermoluminescence (TL) glow curve. The latent logic is as follows: the number of traps (N_E) in different depths (E) can be derived from TL glow curve, considering N_E is proportional to the height of glow curve and E is proportional to the glow temperature (T_G). In the TL measurement, when the phosphor is warmed at a constant rate, electrons escaping from traps of depth E cause a peak-of-light emission that reaches a maximum at a temperature T_G given by

$$t_1 = \frac{1}{S} \exp\left(-\frac{E}{kT_G} \right) \tag{20.6}$$

To connect the decay curve with the glow curve, Equations (20.5) and (20.6) are combined together to produce:

$$\frac{T_G}{T} = \frac{\log St}{\log St_1} \tag{20.7}$$

Corresponding temperatures on the two curves will be in constant ratio given by Equation (20.7), and, hence, if plotted on the same logarithmic scale of temperature, one curve will be displaced relative to the other one a constant distance along the temperature scale. Dependent on the working temperature relative to the temperature range of TL glow curve, the decay behavior should be different from case to case. To figure out the inherent physical processes, please refer to the original.[56]

20.4.2 Design Principles

To design PersL phosphor for high-performance AC-LED, three key factors should be considered: quantum efficiency (QE), optical storage capacity (OSC), and persistent lifetime. QE reflects the ability of converting excited/absorbed photons into emissive photons. OSC is a reflection of the number of effective trapping centers in PersL phosphor. These two factors determine the brightness of AC-LED phosphors on state. The persistent lifetime is majorly influenced by the depth of trapping centers. To compensate dead time of AC-LED in the millisecond range, the trap depth should be shallow.

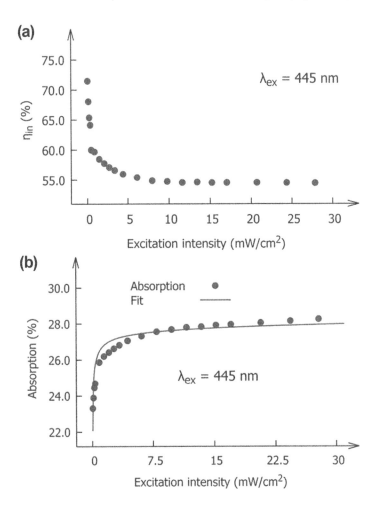

FIGURE 20.11 Intensity dependence of the (a) internal quantum efficiency and (b) absorption of $SrAl_2O_4{:}Eu_{2+},Dy_{3+}$ under excitation with blue light. (Van der Heggen, D., Joos, J. J., and Smet, P. F., *ACS Photon.* 5, 4529, 2018. With permission.)

Van der Heggen *et al.* took an in-depth study on the PersL properties of $SrAl_2O_4{:}Eu^{2+}$, Dy^{3+} applicable to AC-LED and found QE and OSC in such kind of phosphors are greatly limited by a detrapping process called as the optically stimulated luminescence (OSL).[61] OSL is a two-photon process: the first photon ionizes the activators and simultaneously the charge will be trapped; then the second photon stimulates the detrapping process to yield PersL. Unlike the common sense that QE is independent on the excitation intensity, QE decreases with the increase of excitation intensity in $SrAl_2O_4{:}Eu^{2+}$, Dy^{3+} (Figure 20.11(a)). Similar abnormal phenomenon also takes place by examining the excitation intensity-dependent absorption of $SrAl_2O_4{:}Eu^{2+}$, Dy^{3+}, showing a first increasing and then levering off (Figure 20.11(b)). The OSL mechanism explains these phenomena well: on one hand, QE of OSL should not go beyond 50% in theory, and so, its occurrence will remarkably reduce the QE in phosphor; on the other hand, OSL turns the initially nonabsorbing defect centers into optically active centers, the absorption cross section of which is much larger than that of luminescent ion (*e.g.*, Eu^{2+}), and, therefore, the rearrangement of charge carriers induced by OSL improves the absorption cross section in phosphor. The lesson of this work is to design AC-LED phosphor with defects that show a limited optical response to the excitation light.

Tan *et al.* proposed a mathematical model to predict the luminescent performance of AC-LED with time gap compensated by slow-decay phosphor.[62] In this model, three key factors are considered

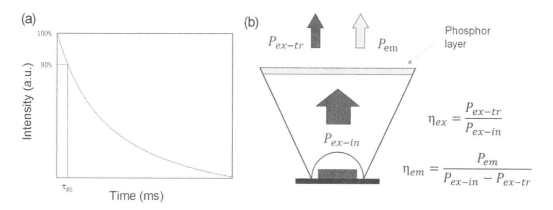

FIGURE 20.12 Definitions of (a) τ_{80} and (b) η_{ex} and η_{em}. (Tan, J. C., and Narendran, N., *J. Lumin.* 167, 21, 2015. With permission.)

(Figure 20.12): (1) τ_{80}, defined as the time for the phosphor decaying to 80% of initial intensity after ceasing the excitation; (2) η_{ex}, defined as extraction efficiency of excitation light; and (3) η_{em}, defined as extraction efficiency of emission light. YAG:Ce^{3+} with high QE is used to mimic the slow-decay phosphor with τ_{80} varying from 0.12 to 0.28 ms. The simulation and experiment show the self-consistent results that when τ_{80} increases, or η_{ex} decreases, or η_{em} increases: FI decreases; LE increases; and CIE x and y increase. The explanation for these phenomena is the increase in time-average intensity of emission light. In the other experiment (Figure 20.13), when fixing η_{ex}, LE, and CIE (x, y), it is found the FI decreases as τ_{80} increases, and through extrapolation to reach FI=0.1, τ_{80} is determined as 0.4 ms; meanwhile, it is found that, to maintain white light, η_{em} should decrease to compensate the effect of increasing τ_{80}. The authors also performed a human factors study to confirm the potential of using slow-decay phosphors to improve acceptability of light output of AC-LEDs. Finally, a set of recommendations for luminescent properties of slow-decay phosphor were proposed to construct the white AC-LED with minimal flicker:

- $\tau_{80} = 0.4$ ms
- $\eta_{ex} = 22\%$
- $\eta_{em} = 60\%$
- $\lambda_p = 555$ nm
- FWHM = 115 nm

Bandgap engineering is an effective approach to tune the electrical and optical properties of inorganic materials through solidification of foreign elements into the host lattice to form solid solution. Recently, in the pioneering works of Tanabe and Ueda,[63–65] this approach was demonstrated effective to manipulate the luminescent behaviors of PersL phosphors by affecting both the charging and detrapping processes. As can be seen from Figure 20.14, through bandgap engineering, on one hand, the CB in host is lowered down, which allows for electrons' jumping over the barrier between excited state of activators and CB via photoionization; on the other hand, the trap depth gets closer to CB. The former effect leads to more efficient electrons' charging via CB, instead of quantum tunneling, whereupon the blue-light-excited PersL is possible when choosing host with strong crystal field strength for Ce^{3+} and Eu^{2+} doping. The latter effect facilitates the electrons' releasing from traps, which enables the greatly shortened PersL lifetime. Notably, the resultant blue-excited PersL and greatly shortened PersL lifetime are favorable for AC-LED application. Following this designing principle, we developed three kinds of PersL phosphor for AC-LED, including the $Gd_3Al_{5-x}Ga_xO_{12}:Ce^{3+}$,[5] $Mg_3Y_2(Si_{1-x}Ge_x)_3O_{12}:Ce^{3+}$,[66] and $Lu_2CaMg_2(Si_{1-x}Ge_x)_3O_{12}:Ce^{3+}$.[67]

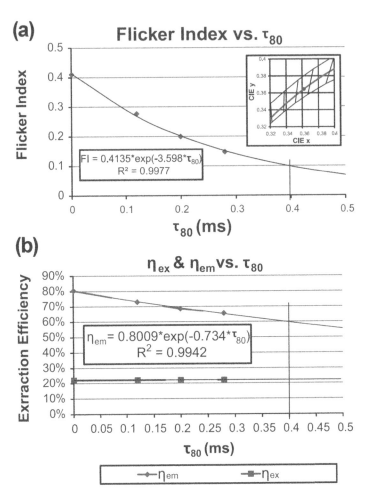

FIGURE 20.13 (a) Flicker index and (b) ηex and ηem versus $\tau 80$, with chromaticity fixed on the black body locus as in the inset of part (a). (Tan, J. C., and Narendran, N., *J. Lumin.* 167, 21, 2015. With permission.)

20.4.3 MATERIAL SYSTEMS

There have been surges of works on PersL materials with long afterglow for emergency lighting, identification markers, luminous paints, and biomedical applications. In contrast, the material system of PersL phosphor for AC-LEDs is rather limited (some of which are summarized in Table 20.1).

- $M_{1-x-y}Si_2O_{2-w}N_{2+2w/3}:Eu_x,R_y$

  In 2012, Liu *et al.* announced in US patent 2012/0013243 a kind of AC-LED phosphor, $M_{1-x-y}Si_2O_{2-w}N_{2+2w/3}:Eu_x,R_y,$[7] wherein M is at least one alkaline earth element (*i.e.*, Ca, Sr, Ba), R is a transition metal ion (*e.g.*, Mn^{2+}) or a lanthanide ion (*e.g.*, Ce^{3+}, Dy^{3+}), $0<x\leq1$, $0<y<1$, and $0\leq w<4$. With adjusting the N/O ratio, the emissive color can be altered from blue to yellow. For a specific phosphor with composition of $SrSi_2O_2N_2:Eu$, Mn, its decay behavior was compared with that of $SrSi_2O_2N_2:Eu$, showing half-life of the former (6.2 ms) much longer than that of the latter (0.0008 ms), as exhibited in Figure 20.15. The greatly prolonged decay was ascribed to the long lifetime spent in traps induced by Mn^{2+} doping.[52] It also proposed an AC-LED device design using this kind of phosphor as color converter.[7]

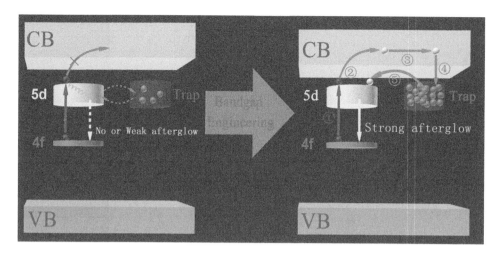

FIGURE 20.14 Schematic illustration of applying bandgap engineering concept to achieve stronger PersL brightness and shorter PersL time. (Lin, H., Wang, B., Huang, Q. M., Huang, F., Xu, J., Chen, H., Lin, Z. B., Wang, J. M., Hu, T., and Wang, Y. S. *J. Mater. Chem. C.* 4, 10329, 2016. With permission.)

TABLE 20.1
Overview of State-of-the-Art AC-LED Phosphors and Their Properties

Material	Type	Excitation Wavelength (nm)	Lifetime	Percent Flicker (%)	Flicker Index	References
$Sr_5(PO_4)_3Cl:Eu^{2+}Mn^{2+}$	Fluorescence	420	4 ms (9%) + 13.4 ms (91%)	30	0.08	[45]
$SrSi_2O_2N_2:Eu^{2+},Mn^{2+}$	PersL	460	<5 ms	–	–	[7, 52]
$SrAl_2O_4:Eu^{2+},Ce^{3+},Li^+$	PersL	420	99.57 ms	–	–	[68]
$SrAl_2O_4:Eu^{2+},R^{3+}$ (R=Y, Dy)	PersL	450	Seconds order	–	–	[69]
$SrAl_2O_4:Eu^{2+}, Dy^{3+}$	PersL	365	428.83 ms	–	–	[70]
$SrAl_{2-x}Si_xO_{4-x}N_x:Eu^{2+}, Dy^{3+}$	PersL	365	765.34 ms	–	–	[71]
(Sr, Ca, Ba) $Al_2O_4:Eu^{2+}, Dy^{3+}$	PersL	370	219.91–795.11 ms	–	–	[72]
$Gd_3Al_2Ga_3O_{12}:Ce^{3+}$	PersL	460	115 ms	69	–	[5]
$Mg_3Y_2(Ge_{1-x}Si_x)_3O_{12}:Ce^{3+}$	PersL	460	74–360 ms	71	–	[66]
$Lu_2CaMg_2(Si_{1-x}Ge_x)_3O_{12}:Ce^{3+}$	PersL	460	180 ms to >1 second	64.1	–	[67]
$Gd_3Al_2Ga_3O_{12}:Ce^{3+},Cr^{3+}$ ceramics	PersL	442	τ_1=2.31 ms τ_2=353.9 ms	59.8	–	[73]
$Gd_3Al_2Ga_3O_{12}:Ce^{3+}$	PersL	450	$\tau_{80\%}$=15 ms	–	–	[74, 75]
$Gd_3Ga_2(Al_{3-y}Si_y)$ $(O_{12-y}N_y):Ce^{3+}$	PersL	460	17.5–25 seconds	–	–	[76]
$SrGa_{12}O_{19}:Dy^{3+}$	PersL	260	Seconds order	–	–	[77]
$Ca_4Ti_3O_{10}:Pr^{3+}$	PersL	365	25 ms	–	–	[78]
$CaS:Yb^{2+}, Cl^-$	PersL	496	–	–	–	[79]

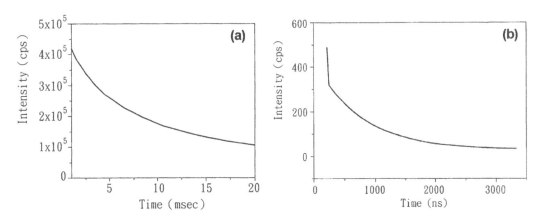

FIGURE 20.15 Decay curves of (a) $SrSi_2O_2N_2$:Eu, Mn and (b) $SrSi_2O_2N_2$:Eu upon blue light excitation. (*Source:* US Patent 20120013243, 2012)

The millisecond half-life of phosphor was believed to be able to compensate for the dead time generated during AC voltage conversion, and so, to reduce the lighting flicker; however, no direct flickering measurements were performed.

- $(Ca,Sr,Ba)Al_2O_4$:Eu^{2+}, Ln^{3+}

  $(Ca,Sr,Ba)Al_2O_4$:Eu^{2+}, Ln^{3+} is one of the few PersL phosphors capable of being charged by blue light, despite UV light being the most effective charging wavelength. Upon co-doping with auxiliary ion Ln^{3+}, the PersL properties can be greatly improved for the introduction of appropriate trap states. These phosphors are known for the extremely bright afterglow and the ultra-long duration even overnight. As a matter of fact, some of these phosphors yield bright afterglow in the millisecond range, which can greatly reduce the lighting flicker of AC-LED.

  Aizawa and Katsumata *et al.* screened the PersL decay behaviors in the millisecond time window for a series of $SrAl_2O_4$:Eu^{2+}, Ln^{3+} (Ln = Y, La, Ce, Pr, Nd, Sm, Gd, Tb, Dy, Ho, Er, Tm, Yb, and Lu) PersL phosphors,[80] as presented in Figure 20.16. Equation (20.8), including two or three exponential terms, was adopted for curve fitting to estimate the persistent lifetime:

$$I(t) = I_1 \exp(-t / \tau_1) + I_2 \exp(-t / \tau_2) + I_3 \exp(-t / \tau_3)$$ (20.8)

where $I(t)$ is the PL intensity at time t, I_1, I_2, and I_3 are the PL intensity of the faster, intermediate, and slower decay components at $t = 0$; τ_1, τ_2, and τ_3 are the lifetimes of faster, intermediate, and slower PL decays, respectively. The calculated lifetimes are summarized in Table 20.2. As can be seen, lifetimes of faster decay components in all samples show almost the same value (τ_1: 0.1–0.2 ms), which was considered originating from the Eu^{2+}: $5d$–$4f$ transition. For the Nd-, Ho-, Dy-, Yb-doped samples, the intermediate and slower decay components lead to the prolonged lifetimes at the millisecond timescale. A systematical trap analysis was performed on the TL glow curves of $SrAl_2O_4$:Eu^{2+}, Ln^{3+}.[81] As exhibited in Figure 20.17, the relationship between trap depth (E_t) and trap density (n_0) for various auxiliary ions follows along a linear equation of $n_0 = 5.5 \times 10^5 \exp(-13.3E_t)$. Evidently, the trap depth at around $E_t = 0.5$ eV with trapped carrier density more than $n_0 = 10^2$ generates stronger afterglow.

Chen *et al.* first proposed and demonstrated this kind of material applicable to AC-LED. In their work,[68] the luminescence intensity of $SrAl_2O_4$:Eu^{2+} was found significantly enhanced by co-doping with Ce^{3+} and Li^+, due to the suppression of impurity phases. $SrAl_2O_4$:Eu^{2+}, Ce^{3+}, Li^+ yields luminescent decay with lifetime of 36.5 ms and able to compensate the dead time of AC-LED. When examining the PersL properties of $SrAl_2O_4$: Eu^{2+}, Dy^{3+} and $SrAl_2O_4$: Eu^{2+}, Y^{3+},[69] it was found

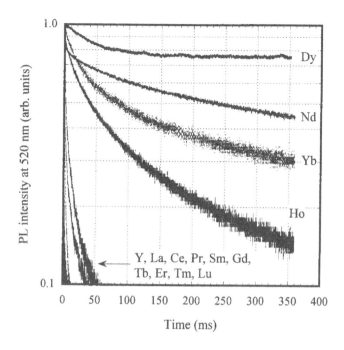

FIGURE 20.16 Decay curves of $SrAl_2O_4:Eu^{2+}$, Ln^{3+} (Ln = Y, La, Ce, Pr, Nd, Sm, Gd, Tb, Dy, Ho, Er, Tm, Yb, and Lu). Decay curves are measured using blue LED modulated by 1 Hz at 273K. (Aizawa, H., Katsumata, T., Takahashi, J., Matsunaga, K., Komuro, S., Morikawa, T., and Toba, E. *Rev. Sci. Instrum.* 74, 1344, 2003. With permission.)

TABLE 20.2

Calculated PersL Lifetime in the Millisecond Time Window for a Series of $SrAl_2O_4:Eu^{2+}$, Ln^{3+} (Ln = Y, La, Ce, Pr, Nd, Sm, Gd, Tb, Dy, Ho, Er, Tm, Yb, and Lu) Phosphors[80]

Auxiliary Activator Elements	Faster Decay, τ_1 (ms)	Intermediate Decay, τ_2 (ms)	Slower Decay, τ_3 (ms)
Eu (nondoped)	0.122	–	–
Y	0.143	–	–
La	0.158	–	–
Ce	0.180	–	–
Pr	0.157	–	–
Nd	0.130	7.03	787.73
Sm	0.112	–	–
Gd	0.229	–	–
Tb	0.141	–	–
Dy	0.121	3.45	1140.40
Ho	–	10.95	266.57
Er	0.157	–	–
Tm	0.110	–	–
Yb	0.112	32.29	535.11
Lu	0.126	–	–

Source: Data from Aizawa (2003).

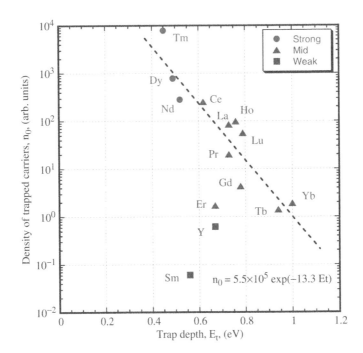

FIGURE 20.17   The relationship between trap depth (E_t) and trap density (n_0) for $SrAl_2O_4$:Eu^{2+}, Ln^{3+} doped with various auxiliary Ln^{3+} ions. (Katsumata, T., Toyomane, S., Sakai, R., Komuro, S., and Morikawa, T. *J. Am. Ceram. Soc.* 89, 932, 2006. With permission)

that the doped Dy^{3+} helps to prolong the PersL time, while the doped Y^{3+} is efficient to enhance the PersL intensity. Then, a prototype device was constructed by coating the phosphor-in-silicone encapsulant on a blue chip driven by AC, where the green PersL is clearly observed after cutting off the AC power (Figure 20.18(a)). The effect of PersL on compensating the dark duration of AC-LED was also demonstrated by using a high-speed microscope for the first time (Figure 20.18(b)).

Zhang's group attempted to optimize PersL properties of $SrAl_2O_4$:Eu^{2+}, Dy^{3+} through using different flux or via composition substitution, aiming to improve PersL intensity and shorten PersL time for better catering to the requirements of AC-LED.[70–72] For example, it was found that the addition of $SrCl_2$ flux, instead of the commonly used H_3BO_3 flux, results in the maximum improvement of integrated emission intensity of phosphor reaching up to 13.56%; meanwhile, the average lifetime is shortened to 428.83 ms.[70] In the other work, when gradually substituting (Al, O)$^+$ with (Si, N)$^+$, the PersL time of $SrAl_{2-x}Si_xO_{4-x}N_x$:Eu^{2+}, Dy^{3+} is found shortened from 845.86 to 765.34 ms.[71] In the work on the (Sr, Ca, Ba) Al_2O_4:Eu^{2+}, Dy^{3+} PersL phosphor,[72] the lifetime largely decreases from 1088.5 to 521.6 ms in $Sr_{0.90-x}Ca_xAl_2O_4$:0.05 Eu^{2+}, 0.05 Dy^{3+} as x increases, and the lifetime is greatly shortened from 842.04 to 219.91 ms in $Sr_{0.90-y}Ba_yAl_2O_4$:0.05 Eu^{2+}, 0.05 Dy^{3+} as y increases.

Van der Heggen *et al.* made a comparison between the $SrAl_2O_4$:Eu^{2+}, Dy^{3+} PersL phosphor and the YAG:Ce^{3+} photoluminescent phosphor when they were pumped by AC at different timescale.[61] As shown in Figure 20.19(a), YAG:Ce^{3+} perfectly follows the emission profile of blue LED, while there is an obvious charging process for $SrAl_2O_4$:Eu^{2+}, Dy^{3+} when the blue LED is switched on, followed by a slow-decay process when the blue LED is switched off. In the case of reducing the pulse width to 5 ms (Figure 20.19(b)), nothing changes for YAG:Ce^{3+} due to the intrinsic short decay time of 65 ns, resulting in an obvious lighting flicker with PF of 100%; in contrast, a fairly constant emission retains for $SrAl_2O_4$:Eu^{2+}, Dy^{3+}, effectively reducing the lighting flicker.

- Ce^{3+}-activated garnet-structured phosphors

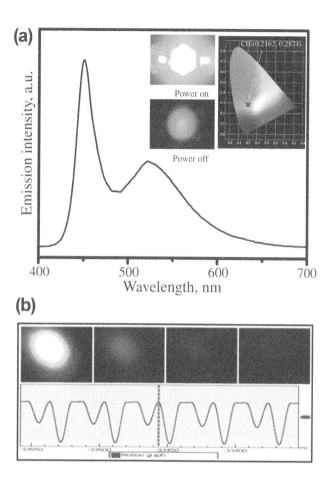

FIGURE 20.18 (a) Electroluminescent spectrum of AC-LED device packaged by using the $Sr_{0.97}Al_2O_4$:0.02 Eu^{2+}, Y^{3+} PersL phosphor. (b) Luminescent variation of white AC-LED using $Sr_{0.97}Al_2O_4$:0.02 Eu^{2+}, Y^{3+} PersL phosphor as color converter. (Chen, L., Zhang, Y., Xue, S. C., Deng, X. R., Luo, A. Q., Liu, F. Y., and Jiang, Y. *Funct. Mater. Lett.* 6, 1350047, 2013. With permission)

The first reported AC-LED phosphor is the Ce^{3+}-activated garnet phosphor, $aY_2O_3 \cdot bAl_2O_3 \cdot c$ SiO_2: mCe·nB·xNa·yP ($1 \leq a \leq 2$, $2 \leq b \leq 3$, $0.001 \leq c \leq 1$, $0.0001 \leq m \leq 0.6$, $0.0001 \leq n \leq 0.5$, $0.0001 \leq x \leq 0.2$, $0.0001 \leq y \leq 0.5$), developed by Zhang *et al.* in 2010.[82] This PersL phosphor is excitable by UV or blue light with wavelength of 200–500 nm and yields yellow emission with wavelength of 530–570 nm. During the excitation, the PersL phosphor can store energy and then emit yellow afterglow at room temperature or yellow TL when heated. This characteristic enables its application in AC-LED, since energy is continuously supplied from defect centers in the off-time. A high-speed CCD capable of taking 300 pictures per second demonstrated the effectiveness of this phosphor for AC-LED. During the dead time of 3.33, 6.66, and 9.99 ms after ceasing the excitation, the luminescence intensity (1527, 1510, and 1505, respectively; in arbitrary unit) remains almost constant, while only the background noise signal was detected in the reference sample by using the common YAG:Ce^{3+} as color converter.

In 2014, our group paid attention on one kind of Ga-doped garnet-structured PersL phosphors $(RE_3Al_{t-x}Ga_xO_{12}:Ce^{3+}$, RE = Y, Gd, Lu, $x = 0$–5). Ga plays a key role in engineering the host bandgap, allowing electrons' delocalization to CB under blue-light excitation and thus facilitating electrons' storage at the traps to achieve bright afterglow luminescence (Figure 20.20).[83] Such blue-light-activated PersL phosphors are expected to fulfill the requirements of AC-LED application.

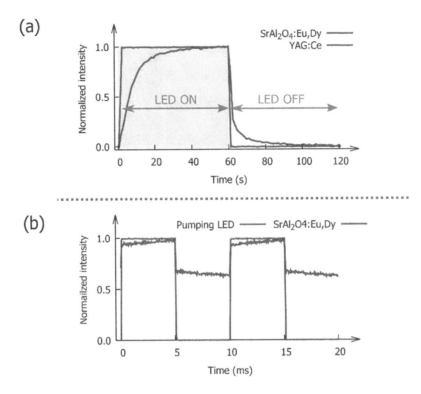

FIGURE 20.19 A comparison between the SrAl$_2$O$_4$:Eu^{2+}, Dy^{3+} PersL phosphor and the YAG:Ce^{3+} photoluminescent phosphor when they are pumped by AC at timescales of (a) seconds and (b) milliseconds. (Van der Heggen, D., Joos, J. J., and Smet, P. F., *ACS Photon.* 5, 4529, 2018. With permission)

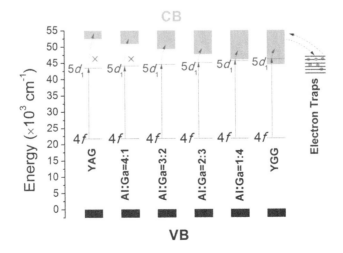

FIGURE 20.20 Schematic illustration of modulating PersL properties via bandgap engineering in Ce^{3+}:Y$_3$Al$_{5-x}$Ga$_x$O$_{12}$ ($x = 0-5$) phosphors. (Wang, B., Lin, H., Yu, Y. L., Chen, D. Q., Zhang, R., Xu, J., and Wang, Y. S. *J. Am. Ceram. Soc.* 97, 2539, 2014. With permission)

On the study of $Gd_3Al_{5-x}Ga_xO_{12}$: Ce^{3+},[5] we identified that the best PersL properties are attained when the ratio of Al:Ga equals to 2:3. The optimized $Gd_3Al_2Ga_3O_{12}$:Ce^{3+} phosphor powders were further introduced into a low-melting glass matrix, forming PersL phosphor-in-glass (PiG), to improve its heat/moisture resistivity. The steady-state spectroscopy demonstrates this PersL PiG is efficiently excited by blue light and emits yellowish green light. The PersL excitation (PersLE) spectrum of PiG shows that the intense PersL is realized by Ce^{3+} excitation in the blue region (Figure 20.21(a)). The measured afterglow decay curve (Figure 20.21(b)) comprises a fast decay component and a slow one (even lasting for 1 hour). In the fast decay process, the afterglow luminescence is bright, whereas in the slow decay process it becomes very weak. By carefully measuring the sample in the initial 1-second decay at a step of 10 ms, as presented in the inset of Figure 20.21(b), the curve could be well fitted with an average lifetime of 115 ms, indicating its potential to compensate the AC time gap. In a proof-of-concept experiment, we fabricated a remote-type AC-LED

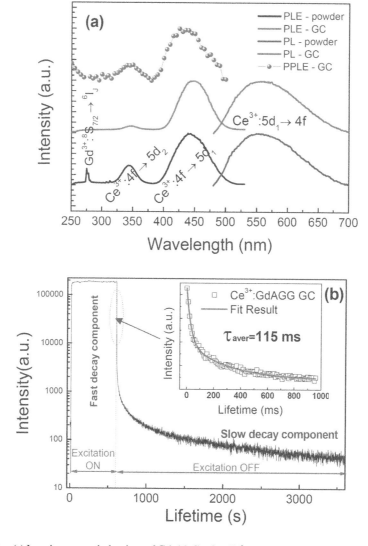

FIGURE 20.21 (a) Luminescence behaviors of $Gd_3Al_2Ga_3O_{12}$:Ce^{3+} powder and corresponding PiG. (b) PersL decay curve of $Gd_3Al_2Ga_3O_{12}$:Ce^{3+} PiG; inset in part (b) shows the initial 1-second decay at a step of 10 ms after the stoppage of excitation source. (Lin, H., Wang, B., Xu, J., Zhang, R., Chen, H., Yu, Y. L., and Wang, Y. S. *ACS Appl. Mater. Interfaces* 6, 21264, 2014. With permission)

based on the PiG containing the commercial YAG:Ce^{3+} and the homemade Gd$_3$Al$_2$Ga$_3$O$_{12}$:Ce^{3+} and connected it to an AC bridge circuit. The corresponding AC-LED emits bright white light when the power is on and yellowish green PersL when the power is off (Figure 20.22(a) and (b)). Although the persistent color is different from the white color in operation, such chromatic flicker is not sensitive to human eyes (Section 20.3.3). To evaluate the ability of the PersL PiG in reducing light flicker effect, the luminescence intensity variation of LED driven in AC periodic cycles is measured (Figure 20.22(c)). As expected, 100% PF is found in the YAG:Ce^{3+} PiG, while only the 69% one is found in the YAG:Ce^{3+}+Gd$_3$Al$_2$Ga$_3$O$_{12}$:Ce^{3+} PersL PiG.

In our following work,[66] a new kind of Mg$_3$Y$_2$(Ge$_{1-x}$Si$_x$)$_3$O$_{12}$:Ce^{3+} ($x = 0$–1) inverse-garnet PersL phosphors were developed, whose afterglow can be efficiently activated by the blue light and the

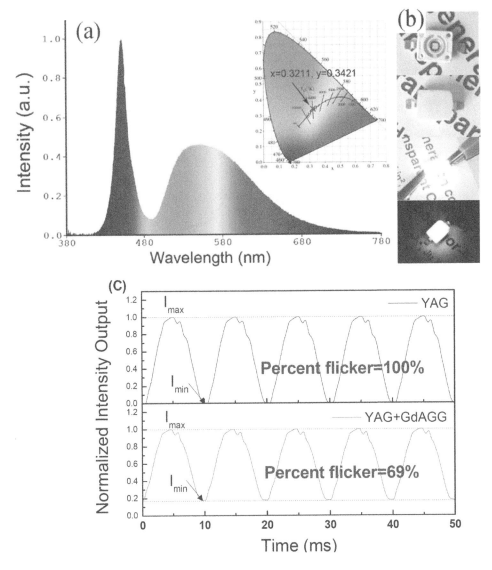

FIGURE 20.22 (a) Electroluminescent spectrum of the remote-type AC-LED encapsulated by Gd$_3$Al$_2$Ga$_3$O$_{12}$:Ce^{3+} PiG. (b) Photographs of the used blue chip, the assembled LED covered with PiG, and its luminescence behavior when power is on/off. (c) Luminescence intensity variations of LED driven in periodic AC cycles. (Lin, H., Wang, B., Xu, J., Zhang, R., Chen, H., Yu, Y. L., and Wang, Y. S. *ACS Appl. Mater. Interfaces* 6, 21264, 2014. With permission)

PersL decays in the millisecond range. It was experimentally demonstrated that Si doping tailors the host bandgap, so that both the electron charging and detrapping in the PersL process are optimized. To explore the origin of the millisecond afterglow, a series of TL analyses are performed (Figure 20.23). With the aid of "initial rise" analyses, the shallowest depth of traps is determined as ~0.1 eV, which is considered to be the origin of the millisecond afterglow. According to the inconsistent variation trend between the bandage energy and the energy of trap level, one can deduce that the trap is also variable upon changing Si/Ge ratio, and so the trap type should be related with Si/Ge, possibly ascribing to the $V_{Ge/Si}$–Ce^{3+}–V_O defect clusters. A more detailed analysis on trap properties is performed by measuring the excitation time- or excitation temperature (T_{exc})-dependent TL curves, from which, we identify three types of traps continuously distributed in the host (Figure 20.24). Finally, an AC-LED prototype device is fabricated, which exhibits warm white emission with a reduced PF of 71.7%.

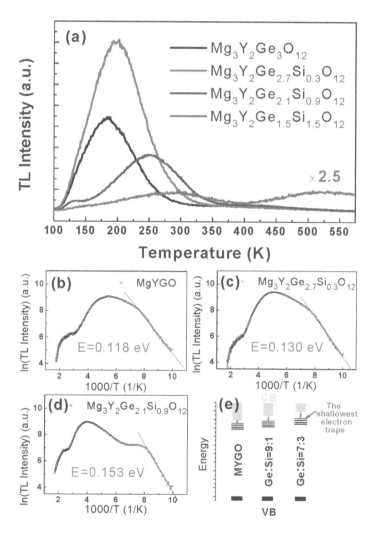

FIGURE 20.23 (a) Ge/Si ratio-dependent TL curves measured at a heating rate of 1K/s after 445 nm irradiation for 5 minutes. (b)–(d) Initial rise analyses of corresponding TL curves revealing the shallowest occupied traps. (e) Schematic illustration of the variation in the energy band and the trap depth. (Lin, H., Xu, J., Huang, Q. M., Wang, B., Chen, H., Lin, Z. B., Wang, Y. S. *ACS Appl. Mater. Interfaces* 7, 21835, 2015. With permission)

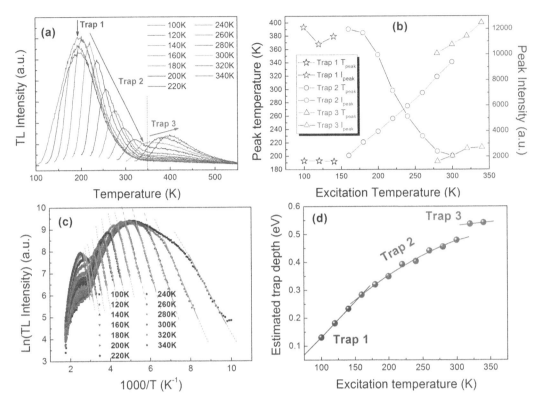

FIGURE 20.24 (a) Excitation temperature-dependent TL curves by preheating/cooling the $Mg_3Y_2Ge_{2.7}Si_{0.3}O_{12}$ sample at different T_{exc} (excitation wavelength: 445 nm, heating rate: 1K/s); (b) dependence of T_{peak} and I_{peak} on the T_{exc}. (c) Initial rise analyses on all the TL curves as a function of T_{exc}. (d) Estimated trap depth as a function of T_{exc}. (Lin, H., Xu, J., Huang, Q. M., Wang, B., Chen, H., Lin, Z. B., Wang, Y. S. *ACS Appl. Mater. Interfaces* 7, 21835, 2015. With permission)

Focusing on the $Lu_2CaMg_2(Si_{1-x}Ge_x)_3O_{12}:Ce^{3+}$ solid solution phosphors,[67] we demonstrate that the bandgap engineering design is efficient in manipulating the electrons' charging and detrapping processes to realize blue-light-excited PersL with adjustable afterglow in the millisecond range. A series of TL measurements accompanied by the "initial rising method" analyses were performed to take a closer look at the PersL mechanism. In Figure 20.25(a)–(c), one can see the variation of equilibrium state between charging and detrapping. The inflection point shifts toward the low temperature side when decreasing the ratio of Si/Ge, indicating the electron detrapping tends to take control; therefore, the persistent lifetime gradually shortens. By plotting the estimated trap depth versus T_{exc} (Figure 20.25(d)–(f)), the trap depth distribution can be roughly determined. The good linear fit in $Lu_2CaMg_2Ge_3O_{12}:Ce^{3+}$ sample indicates that the detrapping rate is constant, and, so, it is considered to have the uniform trap depth distribution. There are two different fit functions for the Si:Ge = 1:1 and Si:Ge = 1:2 samples, probably attributing to two different kinds of traps. The shallowest trap depth is lower than 0.1 eV, which might be responsible for the millisecond afterglow.

Some other research groups also have paid attention to the garnet-structured PersL phosphor applicable to AC-LED.

In the work by Tanabe's group,[73] $Gd_3Al_2Ga_3O_{12}:Ce^{3+}$, Cr^{3+} (GA2G3G:Ce,Cr) phosphor plate was fabricated. The doping of auxiliary ion Cr^{3+} greatly improves the PersL behavior by providing appropriate electron traps. In Figure 20.26(a), one can see that, upon the blue-light irradiation for 10 ms, the photoluminescence intensity of GA2G3G:Ce,Cr is much weaker than that of GA2G3G:Ce; however, after ceasing the excitation, the PersL of GA2G3G:Ce,Cr becomes much stronger. The PersL

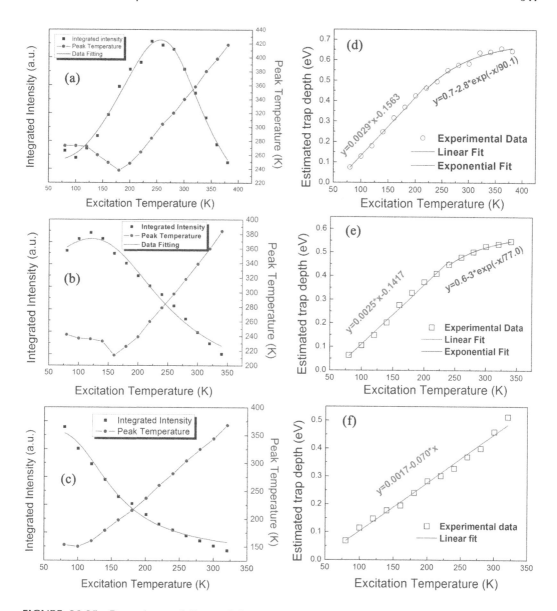

FIGURE 20.25 Dependence of T_{peak} and I_{int} on T_{exc}, obtained from the TL spectra of (a) Si:Ge=1:1, (b) Si:Ge=1:2, and (c) LCMGO samples. Estimated trap depths in the (d) Si:Ge=1:1, (e) Si:Ge=1:2, and (f) LCMGO samples as a function of the T_{exc}. (Lin, H., Wang, B., Huang, Q. M., Huang, F., Xu, J., Chen, H., Lin, Z. B., Wang, J. M., Hu, T., and Wang, Y. S. *J. Mater. Chem. C*. 4, 10329, 2016. With permission)

decay curves for these samples can be decomposed into two components, *i.e.*, $\tau_1 = 2.19$ ms and $\tau_2 = 7.70$ ms for GA2G3G:Ce; and $\tau_1 = 2.31$ ms and $\tau_2 = 353.9$ ms for GA2G3G:Ce,Cr. The first term is ascribed to be the rapid PersL decay and the second term is inferred to contribute the suppression of flicker. An experimental setup of using blue laser diode to stimulate PersL phosphor was established to demonstrate the effectiveness for AC flicker suppression, in which blue light is modulated by a pulse generator with the modulation frequency of 50 Hz and the duty cycle of 5:5. As presented in Figure 20.26(b), the maximal PersL intensity of GA2G3G:Ce,Cr is about 7.5 times stronger than that of GA2G3G:Ce; and correspondingly, the flicker suppression is more prominent in the former sample (PF = 59.8%) than that in the latter one (PF = 96.6%).

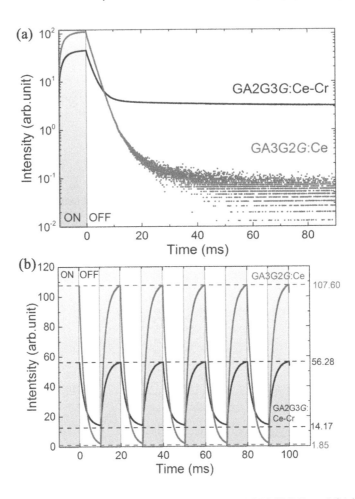

FIGURE 20.26 (a) PersL decay curves in the millisecond range of GA3G2G:Ce and GA2G3G:Ce, Cr samples. (b) The time evolution of luminescence intensity of a blue LD with GA3G2G:Ce and GA2G3G:Ce, Cr samples in AC periodic cycles. (Asami, K., Ueda, J., and Tanabe, S. *J. Sci. Technol. Lighting*, 41, 89, 2018. With permission)

Liu's group and Wang's group took efforts to improve the QE of $Gd_3Al_2Ga_3O_{12}$:Ce^{3+} PersL phosphor by using a two-step solid-state reaction method.[74,75] In the first step, the raw materials were sintered in ample O_2 atmosphere to suppress the evaporation of Ga_2O_3. In the second step, the obtained $Gd_3Al_2Ga_3O_{12}$:Ce^{3+} powders were treated in a reducing CO atmosphere to enhance Ce^{3+} state. Thereby, the QE of $Gd_3Al_2Ga_3O_{12}$:Ce^{3+} can reach up to 81.9%. Together with the PersL time τ_{80} of 15 ms, the synthesized $Gd_3Al_2Ga_3O_{12}$:Ce^{3+} fulfills the requirements of AC-LED phosphor proposed by Tan *et al.*[62]

Based on a cation-anion co-substitution scheme, *i.e.*, [Si^{4+}+N^{3-}] for [Al^{3+}+O^{2-}], Wang's group developed the $Gd_3Ga_2(Al_{3-y}Si_y)(O_{12-y}N_y)$:$Ce^{3+}$ solid solution PersL phosphors for AC-warm LEDs.[76] The induced stronger $5d$ orbital nephelauxetic effect and bigger $5d$ energy level splitting result in the longer wavelength of excitation/emission from Ce^{3+} and the broadening of emission spectra. $Gd_3Ga_2(Al_{3-y}Si_y)(O_{12-y}N_y)$:$Ce^{3+}$ exhibits sufficient red component and constant PersL decay within 10 ms, suitable for the blue-light-excited AC-warm LEDs.

- Other inorganic phosphors

In patent CN 10207644,[84] a series of AC-LED phosphors were announced by Zhang *et al.*, including $CaS:Eu^{2+}$; $CaS:Bi^{2+}$, Tm^{3+}; $ZnS:Tb^{3+}$; $CaSrS_2:Eu^{2+}$, Dy^{3+}; $SrGa_2S_4:Dy^{3+}$; $Ga_2O_3:Eu^{3+}$; $(Y,Gd)BO_3:Eu^{3+}$; $Zn_2SiO_4:Mn^{2+}$; $YBO_3:Tb^{3+}$; $Y(V,P)O_4:Eu^{3+}$; $SrAl_2O_4:Eu^{2+}$; $SrAl_2O_4:Eu^{2+}$, B^{3+}; $SrAl_2O_4:Eu^{2+}$, Dy^{3+}, B^{3+}; $Sr_4Al_{14}O_{25}:Eu^{2+}$; $Sr_4Al_{14}O_{25}:Eu^{2+}$, Dy^{3+}, B^{3+}; $BaAl_2O_4:Eu^{2+}$; $CaAl_2O_4:Eu^{2+}$; $Sr_3SiO_5:Eu^{2+}$, Dy^{3+}; $BaMgAl_{10}O_{17}:Eu^{2+}$, Mn^{2+}; $Tb(acac)_2(AA)phen$; $Y_2O_2S:Eu^{3+}$; $Y_2SiO_5:Tb^{3+}$; $SrGa_2S_4:Ce^{3+}$; $Y_3(Al,Ga)_5O_{12}:Tb^{3+}$; $Ca_2Zn_4Ti_{15}O_{36}:Pr^{3+}$; $CaTiO_3:Pr^{3+}$; $Zn_2P_2O_7:Tm^{3+}$; $Ca_2P_2O_7:Eu^{2+}$, Y^{3+}; $Sr_2P_2O_7:Eu^{2+}$, Y^{3+}; $Lu_2O_3:Tb^{3+}$; $Sr_2Al_6O_{11}:Eu^{2+}$; $Mg_2SnO_4:Mn^{2+}$; $CaAl_2O_4:Ce^{3+}$, Tb^{3+}; $Sr_4Al_{14}O_{25}:Tb^{3+}$; $Ca_{10}(PO_4)_6(F,Cl)_2:Sb^{3+}$, Mn^2; $Sr_2MgSi_2O_7:Eu^{2+}$; $Sr_2CaSi_2O_7:Eu^{2+}$; $Zn_3(PO4):Mn^{2+}$, Ga^{3+}; $CaO:Eu^{3+}$; $Y_2O_2S:Mg^{2+}$, Ti^{3+}; $Y_2O_2S:Sm^{3+}$; $SrMg_2(PO_4)_2:Eu^{2+}$, Gd^{3+}; $BaMg_2(PO_4)_2:Eu^{2+}$, Gd^{3+}; $Zn_2SiO_4:Mn^{2+}$, As^{5+}; $CdSiO_3:Dy^{3+}$; and $MgSiO_3:Eu^{2+}$, Mn^{2+}, whose luminescence decaying to $1/e$ of the maximal intensity are in the millisecond range. These slow-decay phosphors are fluorescent phosphors or PersL phosphors. Eight preferred embodiments by coupling some of these phosphors with UV/blue LED chip were provided to show the effect of suppression of AC flicker, as listed in Table 20.3. A high-speed scientific camera was used to record light fluctuation of the corresponding LEDs driven by AC. According to Table 20.4, the brightness of AC-LEDs encapsulated with slow-decay phosphors are relatively stable during AC cycles; in contrast, the Reference 1 sample based on YAG:Ce with lifetime of 100 ns shows remarkable light fluctuation. The Reference 2 sample based on $SrAl_2O_4:Eu$, Dy and $Y_2O_2S:Eu$, Mg, Ti with lifetime higher than 1 second also shows small light fluctuation, unfortunately, its brightness is relatively weaker, probably ascribing to the insufficient release of energy stored during excitation.

Except for the slow-decay inorganic phosphors mentioned earlier, some other phosphors have been developed for AC-LED. For example, using $Sr_5(PO_4)_3Cl:Eu^{2+}$, Mn^{2+} as phosphor color converter, Korte *et al.* demonstrated a low PF of 30% and a low FI of 0.08 in the corresponding AC-LEDs.[45] Lai *et al.* developed a single-phased white light PersL phosphor, $SrGa_{12}O_{19}:Dy^{3+}$, whose emission intensity remained when excited by a flickering light source with a chopping speed or off-time of a few seconds.[77] Li *et al.* developed a red-emitting PersL phosphor $Ca_4Ti_3O_{10}:Pr^{3+}$ with a decay time of 25 ms for AC-LED.[78] Li *et al.* developed a red PersL phosphor $CaS:Yb^{2+}$, Cl^- potentially applicable for AC-LED, the attenuation of the afterglow luminescence of which is little in the first second.[79]

- Organic phosphorescent materials

TABLE 20.3

Eight Preferred Examples by Coupling Some AC-LED Phosphors with UV/Blue LED Chip to Suppress AC Flicker[84]

Examples	LED Chip (Wavelength)	Luminescent Materials	Lifetime of Light Emitting Materials (ms)
1	Ultraviolet (254 nm)	45 wt% $Zn_2P_2O_7:Tm^{3+}$+55 wt% $Zn_3(PO_4)_2:Mn^{2+}$, Ga^{3+}	10
2	Ultraviolet (254 nm)	$CaAl_2O_4:Dy^{3+}$	25
3	Ultraviolet (310 nm)	15 wt% $Sr_2P_2O_7:Eu^{2+}$, Y^{3+}+30 wt% $Sr_4Al_{14}O_{25}:Eu^{2+}$, Dy^{3+}, B^{3+}+15 wt% $Ca_4O(PO_4)_2:Eu^{2+}$+40 wt% $Zn_3(PO_4)_2:Mn^{2+}$, Ga^{3+}	30
4	Ultraviolet (365 nm)	10 wt% $Sr_2P_2O_7:Eu^{2+}$, Y^{3+}+30 wt% $Sr_4Al_{14}O_{25}:Eu^{2+}$+60 wt% $Y_2O_2S:Eu^{3+}$	14
5	Violet (400 nm)	50 wt% $SrMg_2(PO_4)_2:Eu^{2+}$, Gd^{3+}+50 wt% $Ca_4O(PO_4)_2:Eu^{2+}$	4
6	Violet (400 nm)	40 wt% $Sr_4Al_{14}O_{25}:Eu^{2+}$+60 wt% $Y_2O_2S:Eu^{3+}$	1
7	Blue (450 nm)	30 wt% $SrAl_2O_4:Eu^{2+}$, B^{3+}+70 wt% $CaS:Eu^{3+}$	100
8	Blue (460 nm)	60 wt% $Y_2O_2S:Mg^{2+}$, Ti^{3+}+40 wt% $SrAl_2O_4:Eu^{2+}$	48

Source: Data from Zhang, H. J. 2012 in CN Patent 102074644.

TABLE 20.4

Luminous Brightness of the Photos Shot Within 20 ms by a High-Speed Scientific Camera for the Constructed AC-LED Using AC-LED Phosphors (Examples 1–8) or Common Phosphors (References 1 and 2) as Color Converters[84]

Brightness (a.u.)	Time					
	3.33 ms	6.66 ms	9.99 ms	13.32 ms	16.65 ms	19.98 ms
Reference 1	2265	3466	0	2153	3570	0
Reference 2	746	998	670	702	965	712
Example 1	2931	3025	1455	3187	3443	1665
Example 2	3140	3373	1654	2884	3437	1877
Example 3	3200	3423	1506	3135	3362	1656
Example 4	2910	3190	1652	2723	3245	1850
Example 5	2250	2734	1468	2114	2800	1420
Example 6	2109	2636	1150	2213	2858	1163
Example 7	2017	2420	1569	2115	2654	1510
Example 8	1879	2000	1270	1746	2123	1303

Source: Data from Zhang, H. J. 2012 in CN Patent 102074644.

Phosphorescence from organic molecules due to the triplet-to-singlet radiative transition is "forbidden" in nature, exhibiting slow radiative decay from long-lived excited states with lifetime typically in a range of microseconds to even seconds.[9] Principally, this kind of material can be used for AC-LED, especially for the AC-driven organic light-emitting devices (AC-OLEDs). In comparison with DC-OLEDs, AC-OLEDs show advantages of[85] (1) the free-of-charge accumulation due to frequent reversal of the applied electric field, which may reduce triplet-exciton annihilation occurring at high current densities; (2) the effectively prevented electrochemical reactions between the organic layer and the electrodes by the employed inert dielectric layer, which improves the resistivity to oxygen and moisture in device; and (3) free of the complicated back-end electronics, which reduces power loss and allows for the design of plug-and-play and flexible light sources.

However, the phosphorescence in organic materials is hardly observed at room temperature, unless the rare heavy metal ions, such as Ir^{3+}, Pt^{2+}, and Ru^{3+}, are incorporated to facilitate the intersystem crossing.[86] To achieve cost-effective optoelectronics application, the metal-free organic room-temperature phosphorescent (RTP) materials are required. The designing of such RTP materials was achieved by suppressing non-radiative deactivation process of triplet excitons through locking and rigidifying the molecular conformations/avoiding contact with quenching species or slowing the radiative deactivations of triplet excited state by adjusting the extent of spin-orbit coupling and vibrational coupling.[87]

Recently, Li *et al.* reported the first case of using metal-free RTP materials for AC-LED.[88] A facile molecular design on *N*-phenyl-carbazole (Cz) that incorporates Br atom or methoxy group into *o*-BrCz was found to significantly enhance the intersystem crossing rate constant and improve the phosphorescence quantum yield. Among the compounds (namely, *o*-BrCz, *m*-BrCz, *p*-BrCz, DBrCz, and MeBrCz) with different substitutes and/or positions of heavy bromide atom or methoxyl group, the DBrCz and MeBrCz show the best luminescent performance with quantum yields up to 24.53 and 27.81% in solid powders. The phosphorescent lifetimes of DBrCz and MeBrCz are as long as 68.69 and 148.51 ms and able to fill the time gap of AC-LED (Figure 20.27). In a demonstration experiment, the RTP materials-coated LED yields white light when the power is turned on and yellow phosphorescence light when the power is turned off. The PF of DBrCz- and MeBrCz-coated AC-LEDs are 78 and 83%, much lower than that (~100%) of the reference AC-LED using weak *o*-BrCz RTP material as encapsulant (Figure 20.28).

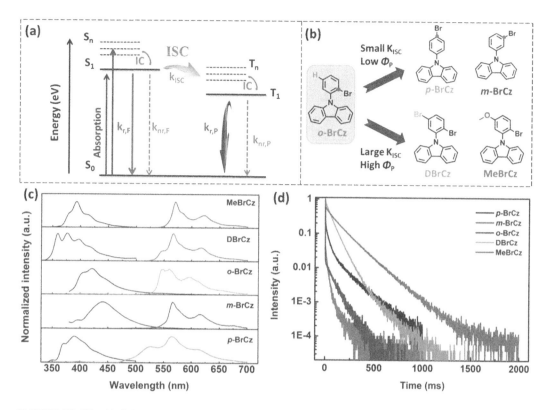

FIGURE 20.27 (a) Schematical illustration of phosphorescence mechanism. (b) Structural conformation of *p*-BrCz, *m*-BrCz, *o*-BrCz, DBrCz, and MeBrCz. (c) Normalized fluorescence (left) and phosphorescence spectra (right) of these compounds in solid powders. (d) Normalized phosphorescence decay curves of these compounds. (Li, B. W., Gong, Y. B., Wang, L., Lin, H., Li, Q. Q., Guo, F. Y., Li, Z., Peng, Q., Shuai, Z. G., Zhao, L. C., and Zhang, Y. *J. Phys. Chem. Lett.* 10, 7141, 2019. With permission)

20.5 CHALLENGES AND OUTLOOKS

20.5.1 Challenges

The study on AC-LED phosphor with short-decay afterglow is still at its primitive stage on both the fundamental and application aspects. The involved trapping and detrapping processes of charge carriers are rather complicated and difficult to be analyzed, since the shallow traps have a quick response to external stimuli, even in a slight level. Despite there has been some material systems with short decay in the millisecond range, these phosphors exhibit low QE and/or poor thermal quenching behaviors, which hinders their practical applications.

20.5.2 Outlooks

Looking forward, future work includes but not limits to the following aspects:

1. The improved knowledge on understanding the physical mechanism behind PersL phenomenon and the new analysis methods to quantitatively/semi-quantitatively characterize the charging/detrapping processes would greatly benefit the design of novel AC-LED phosphors.
2. When considering AC-LED in operation, the AC current change, the AC frequency modulation, and the variation in environmental temperature would substantially change the

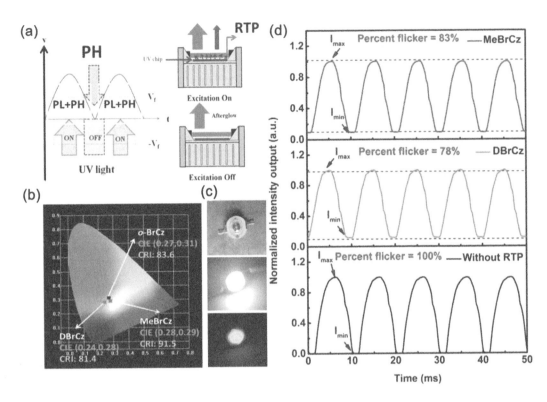

FIGURE 20.28 (a) Scheme of using RTP materials to construct AC-LED with a bridge circuit design. (b) CIE coordinates of DBrCz- and MeBrCz-coated LEDs driven by 350 mA. (c) The physical photographs of 365 nm LED (top), the MeBrCz-coated LED on state (middle) and off state (bottom). (d) Luminescence intensity variations of DBrCz- and MeBrCz-coated LED devices driven in AC periodic cycles. (Li, B. W., Gong, Y. B., Wang, L., Lin, H., Li, Q. Q., Guo, F. Y., Li, Z., Peng, Q., Shuai, Z. G., Zhao, L. C., and Zhang, Y. *J. Phys. Chem. Lett.* 10, 7141, 2019. With permission)

PersL decay behaviors of AC-LED phosphors and thus influence their compensation effect in the off-time. These effects are totally overlooked in the up-to-date work and should be paid great attentions in future.

3. Most of the developed AC-LED phosphors are far away from practical applications. Exploring new AC-LED phosphors with desirable luminescence characteristics will never stop the step. Except for the relatively matured blue-light-excited AC-LED phosphors, there is still plenty of room for the study of UV-light-excited AC-LED phosphors. Especially, the study on the organic phosphorescent materials for AC-OLED has only just started.

4. It is possible to combine the AC circuit modulation technology and the AC-LED phosphor technology to achieve high performance AC-LED with no lighting flicker.

REFERENCES

1. Yeh, W. Y., Yen, H. H., and Chan, Y. J. 2010. The development of monolithic alternating current light-emitting diode. *Proc. SPIE* 7939:793910.
2. Ao, J. P. 2011. Monolithic integration of GaN-based LEDs. *J. Phys.: Conf. Ser.* 276:012001.
3. Onushkin, G. A., Lee, Y. J., Yang, J. J., Kim, H. K., Son, J. K., Park, G. H., and Park, Y. J. 2009. Efficient alternating current operated white light-emitting diode chip. *IEEE Photon. Technol. Lett.* 21:33–35.
4. Yen, H. H., Yeh, W. Y., and Kuo, H. C. 2007. GaN alternating current light-emitting device. *Phys. Stat. Sol. A* 204:2077–2081.

5. Lin, H., Wang, B., Xu, J., Zhang, R., Chen, H., Yu, Y. L., and Wang, Y. S. 2014. Phosphor-in-glass for high-powered remote-type white AC-LED. *ACS Appl. Mater. Interfaces* 6:21264–21269.

6. Su, Q., Li, C. Y., and Wang, J. 2004. Some interesting phenomena in the study of rare earth long lasting phosphors. *Opt. Mater.* 36:1894–1900.

7. Yeh, C. W., Li, Y., Wang, J., and Liu, R. S. 2012. Appropriate green phosphor of $SrSi_2O_2N_2$: Eu^{2+}, Mn^{2+} for AC LEDs. *Opt. Express* 20:18031–18034.

8. Korte, S., Enseling, D., and Justel, T. 2018. Measurement approach for monitoring time-dependent intensity variations of commercial light sources. *ECS J. Solid State Sci. Technol.* 7:R3148–R3157.

9. Xu, J., and Setsuhisa, T. 2019. Persistent luminescence instead of phosphorescence: history, mechanism, and perspective. *J. Lumin.* 205:581–620.

10. Van den Eeckhout, K., Smet, P. F., and Poelman, D. 2010. Persistent luminescence in Eu^{2+}-doped compounds: a review. *Materials* 3:2536–2566.

11. Li, Y., Geceviciusa, M., and Qiu, J. R. 2016. Long persistent phosphors-from fundamentals to applications. *Chem. Soc. Rev.* 45:2090–2136.

12. Smet, P. F., Botterman, J., Van den Eeckhout, K., Korthout, K., and Poelman, D. 2014. Persistent luminescence in nitride and oxynitride phosphors: a review. *Opt. Mater.* 36:1913–1919.

13. Jia, D. D., Wang, X. J., and Yen, W. M. 2004. Delocalization, thermal ionization, and energy transfer in singly doped and codoped $CaAl_4O_7$ and Y_2O_3. *Phys. Rev. B* 69:235113.

14. Bos. A. J. 2007. Theory of thermoluminescence. *Radiat. Meas.* 41:S45–S56.

15. Matsuzawa, T., Aoki, Y., Takeuchi, N., and Murayama, Y. 1996. New long phosphorescent phosphor with high brightness, $SrAl_2O_4$:Eu^{2+}, Dy^{3+}. *J. Electrochem. Soc.* 143:2670–2673.

16. Dorenbos, P. 2005. Mechanism of persistent luminescence in Eu^{2+} and Dy^{3+} codoped aluminate and silicate compounds. *J. Electrochem. Soc.* 152:H107–H110.

17. Allen, M. R. 2000. W.O. Patent 200013469.

18. Seoul Semiconductor Co. 2017. Seoul Semiconductor Presents New Acrich COB Product Line-up at Japan Lighting Fair 2017. www.seoulsemicon.com/en/company/press_view/284.

19. Clauberg, B. 2005. US Patent 6853150.

20. Michael, M. 2011. E.P. Patent 1731003.

21. Tamura, T., Setomoto, T., and Taguchi, T. 2000. Illumination characteristics of lighting array using 10 candela-class white LEDs under AC 100 V operation. *J. Lumin.* 87–89:1180–1182.

22. Ao, J. P., Sato, H., Mizobuchi, T., Kawano, S., Muramoto, Y., Lee, Y. B., Sato, D., Ohno, Y., and Sakai, S. 2002. Monolithic blue LED series arrays for high-voltage AC operation. *Phys. Stat. Sol. A* 194:376–379.

23. Jiang, H. X., Lin, J. Y., and Jin, S. X. 2004. US Patent 2004080941.

24. Yen, H. H., Chi, J. Y., Yeh, W. Y., Lee, T. C., Lin, M. T., and Huang, S. P. 2007. US Patent 0131942.

25. Yamamoto, K., and Kok Keong, R. L. 2008. US Patent 2008/0094000 A1.

26. Jung, H. M., Kim, J. H., Lee, B. K., and Yoo, D. W. A new PWM dimmer using two active switches for AC LED lamp. International Power Electronics Conference (IPEC), 1547–1551, June 2010.

27. Park, C., and Rim, C. T. Filter-free AC direct LED driver with unity power factor and low input current THD using binary segmented switched LED strings and linear current regulator. Applied Power Electronics Conference and Exposition (APEC), 870–874, 2013.

28. Hwu, K. I., and Tu, W. C. 2013. Controllable and dimmable AC LED driver based on FPGA to achieve high PF and Low THD. *IEEE Trans. Ind. Inform.* 9:1330–1342.

29. Yu, D. J., Wu, S. Y., Yu, Q., Chen, W. B., Feng, C. Y., and Liu, Y. 2015. A high power factor AC LED driver with current glitch eliminated. *Analog. Integr. Circ. Sig. Process* 83:209–216.

30. Seoul Semiconductor Co. 2017. Seoul Semiconductor Introduces an Acrich based Compact LED Driver. www.seoulsemicon.com/en/company/press_view/370.

31. IEEE Std. 1789–2015. IEEE Recommended Practices for Modulating Current in High-Brightness LEDs for Mitigating Health Risks to Viewers.

32. CIE TN 006:2016. Visual Aspects of Time-Modulated Lighting Systems – Definitions and Measurement Models.

33. DiLaura, D. L., Houser, K. W., Mistrick, R. G., and Steffy, G. R. 2011. *The Lighting Handbook*. New York, NY: Illuminating Engineering Society.

34. Wilkins, A., Veitch, J., and Lehman, B. LED lighting flicker and potential health concerns: IEEE standard PAR1789 update. Energy Conversion Congress and Exposition (ECCE), Atlanta, GA, 171–178, 2010.

35. CIE 2011. CIE S 017/E:2011. ILV: International Lighting Vocabulary.

36. Kelly, D. H. 1972. Flicker. In *Handbook of Sensory Physiology*, eds. Jameson, D., and Hurvich, L. M. 273–302. Berlin-Heidelberg; New York, NY: Springer-Verlag.

37. Rea, M. S. 2000. *IESNA Lighting Handbook*. New York, NY: Illuminating Engineering.

38. Poplawski, M. E., and Miller, N. M. Flicker in solid-state lighting: measurement techniques, and proposed reporting and application criteria. Proceedings of the CIE Centenary Conference, Paris, France, 188–202, 2013.

39. Perz, M., Vogels, I. M. L. C., Sekulovski, D., Wang, L., Tu, Y., and Heynderickx, I. E. J. 2015. Modeling the visibility of the stroboscopic effect occurring in temporally modulated light systems. *Light. Res. Technol.* 47:281–300.

40. Hershberger, W. A., Jordan, J. S., and Lucas, D. R. 1998. Visualizing the perisaccadic shift of spatiotopic coordinates. *Percept. Psycho.* 60:82–88.

41. Bullough, J. D., Sweater Hickcox, K., Klein, T. R., Lok, A., and Narendran, N. 2012. Detection and acceptability of stroboscopic effects from flicker. *Light. Res. Technol.* 44:477–483.

42. Hershberger, W. A., and Jordan, J. S. 1998. The phantom array: a perisaccadic illusion of visual direction. *Psychol. Rec.* 48:21–32.

43. Roberts, J. E., and Wilkins, A. J. 2013. Flicker can be perceived during saccades at frequencies in excess of 1 kHz. *Light. Res. Technol.* 45:124–132.

44. BI EN 12464-1:2011. Light and Lighting – Lighting in Work Places – Part 1: Indoor Work Places.

45. Korte, S., Enseling, D., and Justel, T. 2018. Measurement approach for monitoring time-dependent intensity variations of commercial light sources. *ECS J. Solid State Sci. Technol.* 7:R3148–R3157.

46. Jiang, Y., Zhou, K., and He, S. 2007. Human visual cortex responds to invisible chromatic flicker. *Nat. Neurosci.* 10:657–662.

47. Ciganek, L., and Ingvar, D. H. 1969. Colour specific features of visual cortical responses in man evoked by monochromatic flashes. *Acta Physiol. Scand.* 76:82–92.

48. Dill, A. B., and Gould, J. D. 1970. Flickerless regeneration rates for CRT displays as a function of scan order and phosphor persistence. *Hum. Factors* 12:465–471.

49. Gould, J. D. 1968. Visual factors in the design of computer-controlled CRT displays. *Hum. Factors* 10:359–376.

50. Bauer, D. 1987. Use of slow phosphors to eliminate flicker in VDUs with bright background. *Displays* 8:29–32.

51. Wilkins, A. J., and Clark, C. 1990. Modulation of light from fluorescent lamps. *Light. Res. Technol.* 22:103–109.

52 Liu, R. S., Yeh, C. W., Hsu, H. W., Li, W. H., Chang, J. C., Lan, Y. B. 2012. US Patent 20120013243.

53. Korte, S., Lindfeld, E., and Justel, T. 2018. Flicker reduction of AC LEDs by Mn^{2+} doped apatite phosphor. *ECS J. Solid State Sci. Technol.* 7:R21–R26.

54. Setlur, A. A., Shiang, J. J., and Happek, U. 2008. Eu^{2+}-Mn^{2+} phosphor saturation in 5 Mm light emitting diode lamps. *Appl. Phys. Lett.* 9:081104.

55. Bicanic, K. T., Li, X., Sabatini, R. P., Hossain, N., Wang, C.-F., Fan, F., Liang, H., Hoogland, S., and Sargent, E. H. 2016. Design of phosphor white light systems for high-power applications. *ACS Photon.* 3:2243–2248.

56. Garlick, G. F. J., and Wilkins, M. H. F. 1945. Short period phosphorescence and electron traps. *Proc. R. Soc. Lond. Ser. A* 184:408–433.

57. Randall, J. T., and Wilkins, M. H. F. 1945. The phosphorescence of various solids. *Proc. R. Soc. Lond. Ser. A* 184:347–364.

58. Raukas, M., Basun, S. A., Van Schaik, W., Yen, W. M., and Happek, U. 1996. Luminescence efficiency of cerium doped insulators: The role of electron transfer processes. *Appl. Phys. Lett.* 69:3300–3302.

59. Dorenbos, P. 2005. Thermal quenching of Eu^{2+} 5d–4f luminescence in inorganic compounds. *J. Phys.: Condens. Matter.* 17:8103–8111.

60. Van den Eeckhout, K., Smet, P. F., and Poelman, D. 2010. Persistent luminescence in Eu^{2+}-doped compounds: a review. *Materials* 3:2536–2566.

61. Van der Heggen, D., Joos, J. J., and Smet, P. F. 2018. On the importance of evaluating the intensity dependency of the quantum efficiency: impact on LEDs and persistent phosphors. *ACS Photon.* 5:4529–4537.

62. Tan, J. C., and Narendran, N. 2015. Defining phosphor luminescence property requirements for white AC LED flicker reduction. *J. Lumin.* 167:21–26.

63. Ueda, J., Tanabe, S., and Nakanishi, T. 2011. Analysis of Ce^{3+} luminescence quenching in solid solutions between $Y_3Al_5O_{12}$ and $Y_3Ga_5O_{12}$ by temperature dependence of photoconductivity measurement. *J. Appl. Phys.* 110:053102.

64. Ueda, J., Kuroishi, K., and Tanabe, S. 2014. Bright persistent ceramic phosphors of Ce^{3+}-Cr^{3+}-codoped garnet able to store by blue light. *Appl. Phys. Lett.* 104:101904.

65. Ueda, J., Dorenbos, P., Bos, A. J. J., Kuroishi, K., and Tanabe, S. 2015. Control of electron transfer between Ce^{3+} and Cr^{3+} in the $Y_3Al_{5-x}Ga_xO_{12}$ host via conduction band engineering. *J. Mater. Chem. C* 3:5642–5651.

66. Lin, H., Xu, J., Huang, Q. M., Wang, B., Chen, H., Lin, Z. B., and Wang, Y. S. 2015. Bandgap tailoring via Si doping in inverse-garnet $Mg_3Y_2Ge_3O_{12}$:Ce^{3+} persistent phosphor potentially applicable in AC-LED. *ACS Appl. Mater. Interfaces* 7:21835–21843.

67. Lin, H., Wang, B., Huang, Q. M., Huang, F., Xu, J., Chen, H., Lin, Z. B., Wang, J. M., Hu, T., and Wang, Y. S. 2016. $Lu_2CaMg_2(Si_{1-x}Ge_x)_3O_{12}$:$Ce^{3+}$ solid-solution phosphors: bandgap engineering for blue-light activated afterglow applicable to AC-LED. *J. Mater. Chem. C* 4:10329–10338

68. Chen, L., Zhang, Y., Liu, F. Y., Luo, A. Q., Chen, Z. X., Jiang, Y., Chen, S. F., and Liu, R. S. 2012. A new green phosphor of $SrAl_2O_4$:Eu^{2+},Ce^{3+},Li^+ for alternating current driven light-emitting diodes. *Mater. Res. Bull.* 47:4071–4075.

69. Chen, L., Zhang, Y., Xue, S. C., Deng, X. R., Luo, A. Q., Liu, F. Y., and Jiang, Y. 2013. The green phosphor $SrAl_2O_4$:Eu^{2+}, R^{3+} (R=Y, Dy) and its application in alternating current light-emitting diodes. *Funct. Mater. Lett.* 6:1350047.

70. Li, B. W., Zhang, J. H., Zhang, M., Long, Y. B., and He, X. 2015. Effects of $SrCl_2$ as a flux on the structural and luminescent properties of $SrAl_2O_4$:Eu^{2+}, Dy^{3+} phosphors for AC-LEDs. *J. Alloys Compd.* 651:497–502.

71. Li, B. W., Xie, Q. D., Qin, H. H., Zhang, M., He, X., Long, Y. B., and Xing, L. S. 2016. Optical properties of $SrAl_{2-x}Si_xO_{4-x}N_x$: Eu^{2+}, Dy^{3+} phosphors for AC-LEDs. *J. Alloys Compd.* 679:436–441.

72. Xie, Q. D., Li, B. W., He, X., Zhang, M., Chen, Y., and Zeng, Q. G. 2017. Correlation of structure, tunable colors, and lifetimes of (Sr, Ca, Ba)Al_2O_4: Eu^{2+}, Dy^{3+} phosphors. *Materials* 10:1198.

73. Asami, K., Ueda, J., and Tanabe, S. 2018. Flicker suppression of AC driven white LED by yellow persistent phosphor of Ce^{3+}–Cr^{3+} co-doped garnet. *J. Sci. Technol. Light.* 41:89–92.

74. Liu, Y. F., Liu, P., Wang, L., Cui, C. E., Jiang, H. C., and Jiang, J. 2017. Enhanced optical performance by two-step solid-state reactions to synthesize the yellow persistent $Gd_3Al_2Ga_3O_{12}$:Ce^{3+} phosphor for AC-LED. *Chem. Commun.* 77:10636–10639.

75. Liu, P., Liu, Y. F., Cui, C. E., Wang, L., Qiao, J. W., Huang, P., Shi, Q. F., Tian, Y., Jiang, H. C., and Jiang, J. 2018. Enhanced luminescence and afterglow by heat-treatment in reducing atmosphere to synthesize the $Gd_3Al_2Ga_3O_{12}$: Ce^{3+} persistent phosphor for AC-LEDs. *J. Alloys Compd.* 731:389–396.

76. Mao, A. J., Zhao, Z. Y., Wang, J. T., Yang, C. X., Ren, J. J., and Wang, Y. H. 2019. Crystal structure and photo-luminescence of $Gd_3Ga_2(Al_{3-y}Si_y)(O_{12-y}N_y)$:$Ce^{3+}$ phosphors for AC-warm LEDs. *Chem. Eng. J.* 368:924–932.

77. Lai, J. N., Long, Z. W., Qiu, J. B., Zhou, D. C., Zhou, J. H., Zhu, C. C., Hu, S. H., Zhang, K., and Wang, Q. 2019. Warm white light emitting from single composition $SrGa_{12}O_{19}$:Dy^{3+} phosphors for AC-LED. *J. Am. Ceram. Soc.* 103:335–345.

78. Zhuang, Y. F., Li, T. Y., Yuan, P., Li, Y. Q., Yang, Y. M., and Yang, Z. P. 2019. The novel red persistent phosphor CaS:Yb^{2+}, Cl^- potentially applicable in AC LED. *Appl. Phys. A* 125:141.

79. Li, X. S., and Zhao, L. T. 2020. UV or blue light excited red persistent perovskite phosphor with millisecond lifetime for use in AC-LEDs. *Luminescence* 35:138–143.

80. Aizawa, H., Katsumata, T., Takahashi, J., Matsunaga, K., Komuro, S., Morikawa, T., and Toba, E. 2003. Long afterglow phosphorescent sensor materials for fiber-optic thermometer. *Rev. Sci. Instrum.* 74:1344–1349.

81. Katsumata, T., Toyomane, S., Sakai, R., Komuro, S., and Morikawa, T. 2006. Trap levels in Eu-doped $SrAl_2O_4$ phosphor crystals co-doped with rare-earth elements. *J. Am. Ceram. Soc.* 89:932–936.

82. Zhang, H. J., Zhang, M., Li, C. Y., Zhao, K., and Zhang, H. 2010. CN Patent 101705095.

83. Wang, B., Lin, H., Yu, Y. L., Chen, D. Q., Zhang, R., Xu, J., and Wang, Y. S. 2014. Ce^{3+}/Pr^{3+}: YAGG: a long persistent phosphor activated by blue-light. *J. Am. Ceram. Soc.* 97:2539–2545.

84. Zhang, H. J., Zhang, M., Li, C. Y., Zhao, K., Li, D. M., and Zhang, L. 2012. CN Patent 102074644.

85. Pan, Y. F., Xia, Y. D., Zhang, H. J., Qiu, J., Zheng, Y. T., Chen, Y. H., and Huang, W. 2017. Recent advances in alternating current-driven organic light-emitting devices. *Adv. Mater.* 29:1701441.

86. Hirata, S. 2017. Recent advances in materials with room temperature phosphorescence: photophysics for triplet exciton stabilization. *Adv. Opt. Mater.* 5:1700116.

87. Xu, S., Chen, R. F., Zheng, C., and Huang, W. 2016. Excited state modulation for organic afterglow: materials and applications. *Adv. Mater.* 28:9920–9940.

88. Li, B. W., Gong, Y. B., Wang, L., Lin, H., Li, Q. Q., Guo, F. Y., Li, Z., Peng, Q., Shuai, Z. G., Zhao, L. C., and Zhang, Y. 2019. High efficient organic room-temperature phosphorescent luminophores through tuning triplet states and spin-orbit coupling with incorporation of a secondary group. *J. Phys. Chem. Lett.* 10:7141–7147.

21 Phosphors for Solar Cells

Donglei Zhou and Hongwei Song

CONTENTS

21.1 INTRODUCTION

Sunlight, which is free and abundant in most parts of the world, can be captured by solar cells (SCs) and transformed into electricity.[1,2] It has been reported that the radiation power of the sunlight that reaches the earth's surface in one day is 8659 times of the current total power of human civilization. The energy irradiated to the earth in a day by the sun is about 1.49×10^{22} Joules.[3,4] As a result, the use of the solar energy is expected to have the potential to meet a large portion of future energy consumption requirements and help solve the human energy crisis. SCs, which can convert the sunlight into electricity, have been developed for several decades.[5,6] The crystalline silicon SC (SSC) is considered as the first-generation SC with the power conversion efficiency (PCE) as high as ~26%, which has been widely used in many parts of the world.[7–10] The first monocrystalline SSC with an efficiency of 6% came out in Bell Labs in 1954.[11] So far, the world record for the conversion efficiency of monocrystalline silicon cells is 26.7%, and that of polycrystalline silicon is 23.3% (Figure 21.1).[12] The second-generation SC is the inorganic compound thin-film SCs, including the amorphous silicon, CIGS, CdTe and GaAs SCs.[11] The first amorphous silicon thin-film SC was fabricated in RCA laboratory in the United States in 1976.[13] The conversion efficiency of amorphous silicon thin-film SCs is the lowest among thin-film SCs, the world record is only 14.0%,[14] The conversion efficiency of polycrystalline silicon thin-film SCs is 21.2%[15] and the highest PCE of CIGS thin-film SCs is 22.6%.[16] The PCE of CdTe thin-film SCs also reached 23.4%.[17] The third-generation SC combines the inorganic semiconductor thin-film SC, dye-sensitized SC (DSSC), organic SC and perovskite SC (PSC).[17–19] In particular, the PSCs are attracting extensive worldwide attentions as its

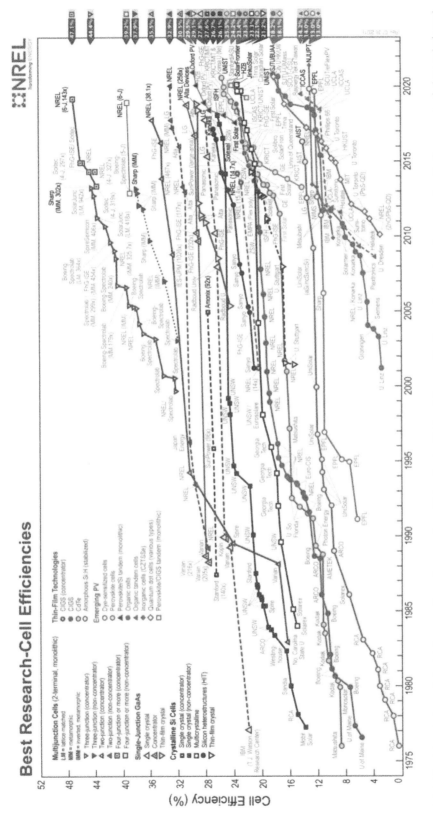

FIGURE 21.1 A chart of the highest confirmed conversion efficiencies by National Renewable Energy Laboratory for research cells for a range of photovoltaic technologies, plotted from 1976 to the present. (www.nrel.gov/pv/cell-efficiency.html)

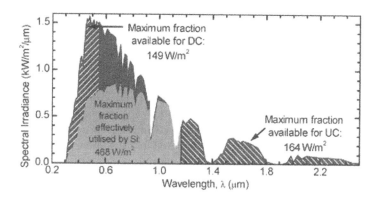

FIGURE 21.2 AM 1.5G spectrum showing the fraction (highlighted in green) absorbed by a typical silicon-based photovoltaic (PV) cell and the spectral regions that can be utilized through quantum cutting and up-conversion processes (highlighted in purple and red, respectively). (Adapted with permission from Ref. [21]. Copyright 2007, Elsevier B.V.)

photoelectric conversion efficiency has been rapidly increased from the initial 3.9 to 25.5% up to April, 2021.[20] However, the current annual solar energy usage is well below 1% of total energy consumption, while fossil fuels account for over 90% of the energy consumption. SCs with high PCE, stable performance and low cost must be developed.

A key problem limiting the conversion efficiency of photovoltaic cells is their limited spectral response to a full solar spectrum. The spectral distribution of sunlight at Air Mass 1.5 global (AM 1.5G) is composed of photons with wide wavelengths ranging from ultraviolet (UV) to infrared (280–2500 nm, 0.5–4.4 eV), but current photovoltaic cells only utilize a relatively small fraction of the solar photons (Figure 21.2).[21] Each single-junction semiconductor SC only responds to a narrow range of solar photons with energy matching the characteristic bandgap (E_g) of the material, which induces the mismatch between the incident solar spectrum and the spectral absorption properties of the material. Only photons with energy (E_{ph}) higher than the bandgap can be absorbed, and electron-hole (e-h) pair with energy greater than E_g can be created, but the excess energy cannot be effectively used and dissipated as heat rapidly by thermalization to the edges of the conduction band (CB) and valence band (VB). Photons with energy smaller than the bandgap cannot be absorbed and do not make any contribution to the carrier generation. The third mechanism of losses is recombination of e-h pairs close to or at the surface. This kind of losses mainly affects photons with high energy or short wavelength, because these photons absorbed in the surface region are likely be affected by imperfect collection and, hence, result in reduced spectral response. These fundamental spectral losses amount to the loss of approximately 50% of the incident solar energy. For instance, the spectral response of the crystalline SSC with a bandgap energy of 1.1 eV covers the range of the UV, visible and near-infrared (NIR) (300–1100 nm), and the maximum theoretical efficiency of a crystalline-SSC can approach up to 31% under AM 1.5G illumination, as defined by the Shockley-Queisser limit.[22]

21.2 SPECTRAL CONVERSION

Spectral conversion is to convert the solar light that cannot or cannot be efficiently absorbed to the wavelengths that can be utilized by the SC, which can modify the incident solar spectrum to obtain a better match with the wavelength-dependent conversion efficiency of the SC.[23,24] The advantage of the spectral conversion is that it can be applied on the existing SCs and does not need to change the structure of the SCs as the spectral converters and the SCs can be optimized separately.

The common methods using spectral converters to enhance the efficiency of SCs include spectral converters on the front surface of the SCs, spectral converters inside the SCs and spectral converters with a reflector on the back surface of the SCs.[1,25,26] There are three different types of luminescent materials including up-conversion, downshifting and quantum cutting, which are currently used to enhance the efficiency of the SCs.[27–29]

Up-conversion is a process to absorb two or more photons with low energy (sub-bandgap) and give one high-energy photon. The limiting efficiency of a SC in combination with an up-conversion material has been discussed in a detailed balance model by Trupke et al.[30] A maximum value of 47.6% was calculated in the case for a SC in combination with an upconverter for nonconcentrated light and with no restriction of the solid angle into which luminescence can be emitted. Regards also predicted that efficiency as high as 35% can be achieved for crystalline silicon with an upconverter calculated for the standard 1000-W/m^2 AM 1.5 solar spectrum.

Downshifting is a process to absorb one photon with high energy and give one photon with high energy. Downshifting can lead to an efficiency increase by shifting photons to a spectral region where the SC has a higher response, which can basically improve the blue response of the SC and was predicted to result in up to 10% relative efficiency increase. Until now, two main classes of PV devices based on downshifting have been investigated: a luminescent solar concentrator (LSC) and a planar downshifting layer. A variety of downshifting materials have been explored, including lanthanide-doped phosphors and glasses, semiconductor quantum dots (QDs) and organic lanthanide complexes.

Quantum cutting can transform one photon with high energy into two photons with low energy, which is the inverse of up-conversion process. This process could minimize the energy loss caused by thermalization of hot charge carriers after the absorption of high-energy photons. The theoretical quantum efficiency of inner quantum cutting is 200%, but the actual quantum efficiency is lower due to concentration quenching and parasitic absorption processes. The quantum cutting is predicted to be able to raise the efficiency above the Shockley-Queisser limit. The quantum cutting materials are mainly focused on the lanthanide-doped phosphors, such as Yb^{3+}, Er^{3+}, Tm^{3+}, Pr^{3+}, Ho^{3+}, Ce^{3+} and three combinations.

This chapter focuses on the development of variety of phosphors and their performance on photovoltaic SCs. We will begin with a basic concept on SCs and phosphors, followed by introducing the operating principles of phosphors as spectral converters to enhance the efficiency of SCs. Finally, an overview of the variety of phosphors and applications on SCs is presented.

21.3 SOLAR CELLS

SCs typically consist of light-absorbing semiconductor materials that can absorb the sunlight and create the photocurrent. The semiconductor materials absorb the solar energy, pump the electrons to the higher energy states and form the photocurrent. The semiconductor materials are doped with positive charge (p-type) or the negative charge (n-type). If two differently doped semiconductor layers are combined, a space charge region is formed on the boundary of the layers called p-n junction. Sunlight falls on the semiconductor p-n junction to form a new hole-electron pair. Under the action of the built-in electric field in the p-n junction, the light-generated holes flow to the p-area, and the photo-generated electrons flow to the n-area, and a current is generated after the circuit is turned on. This is the working principle of photovoltaic effect SCs.

The parameters of a SC consist of the short-circuit current (J_{sc}), open-circuit voltage (V_{oc}), fill factor (FF) and the PCE.

Short-circuit current (J_{sc}): When the positive and negative electrodes of the SC are short-circuited and voltage=0, the current at this time is the short-circuit current (J_{sc}) of the cell. The unit of the short-circuit current is ampere (A), and the short-circuit current changes with the change of light intensity.

Open-circuit voltage (V_{oc}): When the positive and negative poles of the SC are not connected to the load and current=0, the voltage between the positive and negative poles of the SC is the open-circuit voltage (V_{oc}), and the unit of the open-circuit voltage is volt (V). The open-circuit voltage of a monolithic SC does not change with the increase or decrease of the cell area.

Peak current (I_m): The peak current is also called the maximum working current or the best working current. The peak current refers to the working current when the SC outputs the maximum power, and the unit of the peak current is ampere (a).

Peak voltage (U_m): The peak voltage is also called the maximum working voltage or the best working voltage. The peak voltage refers to the working voltage when the SC outputs the maximum power, and the unit of the peak voltage is V. The peak voltage does not change with the increase or decrease of the cell area.

Peak power (P_m): Peak power is also called maximum output power or optimal output power. The peak power refers to the maximum output power of the SC under normal working or test conditions, that is, the product of the peak current and the peak voltage:

$$P_m = I_m \times U_m \tag{21.1}$$

The unit of peak power is watts (W). The peak power of the SC depends on the solar irradiance, the solar spectral distribution and the operating temperature of the cell. Therefore, the measurement of the SC should be carried out under standard conditions. The measurement standard is the European Commission's 101 standard, and the conditions are: irradiation degree 1 kW/m², spectrum AM 1.5 and test temperature 25°C.

FF: The FF is also called the curve factor, which refers to the ratio of the maximum output power of the SC to the product of the open-circuit voltage and the short-circuit current. The calculation formula is:

$$FF = \frac{P_m}{\left(I_{sc} \times V_{oc}\right)} \tag{21.2}$$

FF is an important parameter for evaluating the output characteristics of SCs. The higher its value, the more rectangular the output characteristics of SCs and the higher the photoelectric conversion efficiency of the cells.

Figure 21.3 presents the common structure of monocrystalline SSC, heterojunction with intrinsic thin-layer SSC, DSSC, organic SC and PSC.

The common silicon-based SC contains: (1) a front metal grid; (2) an antireflection coating (ARC) (usually silicon nitride); (3) a diffused p-n junction, which is formed by diffusing a phosphorus dopant into a p-type wafer doped with boron; (4) a p-type absorber layer with surface texturing on both sides; (5) a back-surface field (BSF) fabricated by using an aluminum paste, followed by annealing to form a p⁺ region at the back of the cell and also to lower the contact resistance; and (6) a back metal contact.

The structure of the DSSC consists of a titanium dioxide layer (semiconductor)-coated photo-anode electrode, a counter electrode (CE) used as a cathode, a sensitizer and an electrolyte as shown in Figure 21.3(c).[34]

Organic SCs (Figure 21.3(d))[35] involve majority charge carriers since holes are located in the highly occupied molecular orbital of donor phase and electrons are present in the lowest unoccupied molecular orbital of acceptor phase and their movements result in photovoltaic current. Organic SCs made up from blends of conjugated polymers or conjugated organic compounds (donor) and fullerenes (acceptor).

The common structure of PSC usually contains: (1) fluorine-doped tin oxide (FTO) or indium tin oxide (ITO) glass substrate; (2) electron transfer layer, including TiO_2, SnO_2 and PEDOT: PSS; (3) perovskite active layer, including $CH_3NH_3PbI_3$, $FAPbI_3$ and $CsPbI_3$; (4) hole transport layer, including Spiro-OMeTAD, [6,6]-phenyl-C61-butyric acid methylester (PCBM) and MoO_3; and (5) metal electrode.

Figure 21.4 shows the typical spectral response curves of c-Si SC, CIGSe SC and $Cs_{0.05}(MA_{0.17}FA_{0.83})_{0.95}Pb(I_{0.83}Br_{0.17})_3$ PSC. An effective light converter should be able to convert sunlight outside the spectral response into the spectral response range via up-conversion, downshifting or quantum

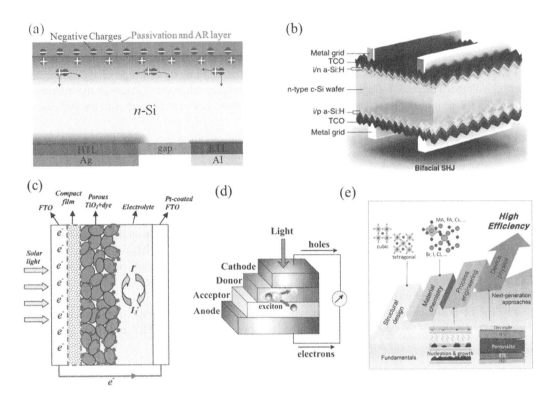

FIGURE 21.3 Schematic structure of (a) interdigitated back contact monocrystalline silicon solar cell[31] (adapted with permission from Ref. [32]. Copyright 2019, Elsevier B.V.), (b) heterojunction with intrinsic thin-layer silicon solar cell[32] (adapted with permission from Ref. [33]. Copyright 2020, Elsevier B.V.), (c) dye-sensitized solar cell (adapted with permission from Ref. [34]. Copyright 2017, Elsevier B.V.), (d) organic solar cell (adapted with permission from Ref. [35]. Copyright 2017, Elsevier B.V.) and (e) perovskite solar cell[33] (adapted with permission from Ref. [33]. Copyright 2017, American Chemical Society).

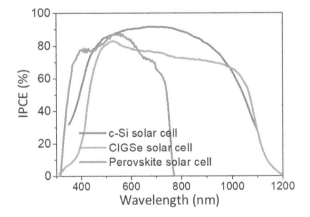

FIGURE 21.4 Spectral response of c-Si solar cell, CIGSe solar cell and $Cs_{0.05}(MA_{0.17}FA_{0.83})_{0.95}Pb(I_{0.83}Br_{0.17})_3$ perovskite solar cell.

cutting process without affecting the light response in the spectral response range. The c-Si SCs work efficiently in the 300- to 1100 nm spectral region, but they show very low spectral response to the UV light and short-wavelength light. The ideal downshifting and quantum cutting materials should satisfy the following requirements: (1) the excitation range shorter than 400 nm; (2) the emission wavelength in the range of 400–1100 nm; (3) low excitation power threshold (<100 W/m^2); and (4) high light conversion efficiency and high transparency to the light at 400–1100 nm. For the short-wavelength sunlight (wavelength >1100 nm), the up-conversion materials are necessary. The ideal upconverters should be able to absorb the light above 1100 nm and convert them to emission around 1000 nm or in the visible region. For the PSCs with a bandgap of 1.55 eV, the light absorption region is limited to a maximum wavelength of ~800 nm. More than half of the sunlight is lost because it cannot be absorbed by the SC. Therefore, the light converters are necessary for most of the current SCs.

21.4 UP-CONVERSION MATERIALS FOR PHOTOVOLTAIC APPLICATIONS

Up-conversion process can convert two or more low-energy photons to a high-energy photon, which is a nonlinear anti-Stokes optical process. The up-conversion process was first discovered in 1959 when it was called "quantum counteraction",[36] and then discovered independently by Auzel in the 1960s.[37] Since that time, a large number of research work on up-conversion has been carried out, which focused on the fundamentally theoretical study or practically applied investigation. The up-conversion processes are only possible in trivalent lanthanide ions with metastable and long-lived intermediate levels acting as storage reservoirs for the pump energy. As shown in Figure 21.5,[38] the Er^{3+}, Tm^{3+} and Yb^{3+} ions, with ladder-like energy levels for photon absorption and subsequent energy transfer steps, were generally selected as activators to give rise to efficient visible emissions under low excitation power densities. To enhance the up-conversion efficiency, the Yb^{3+} ion is usually used as a sensitizer. The absorption cross section of $^2F_{5/2}$–$^2F_{7/2}$ transition (Yb^{3+} ion) is larger than the transitions of other lanthanide ions.

The typical up-conversion processes are shown in Figure 21.6.[1] In contrast to common nonlinear processes including two photon absorption and second harmonic generation, these up-conversion

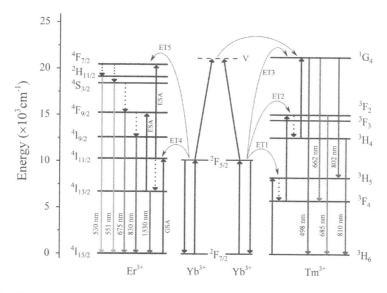

FIGURE 21.5 The energy-level diagram of Yb^{3+}, Er^{3+} and Tm^{3+} ions and involved energy transfer mechanisms of photoluminescence emissions under 980 nm excitation. (Adapted with permission from Ref. [38]. Copyright 2018, Elsevier B.V.)

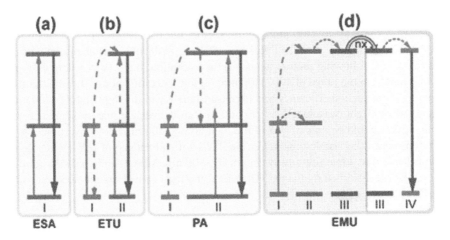

FIGURE 21.6 Proposed typical up-conversion processes. (a) Excited state absorption (ESA); (b) energy transfer up-conversion (ETU); (c) photon avalanche (PA); (d) energy migration-mediated up-conversion (EMU). (Adapted with permission from Ref. [1]. Copyright 2013, Royal Society of Chemistry)

processes combine the benefit of high quantum efficiency without the need for intense coherent excitation sources, with the inherent advantages of large anti-Stokes shift. The excited state absorption (ESA) is a process in a single ion, which involves sequential absorption of two (or more) photons by an excited ion via a real intermediary energy level and leads to the population of that ion to a higher excited state. For the energy transfer up-conversion (ETU), two ions participate in which two pump photons excite two neighboring ions to a metastable energy level through ground-state absorption (GSA). The excited ions then exchange energy nonradiatively, promoting one excited ion to an upper emitting state and demoting the other ion to the ground state. Photon avalanche (PA) is an unconventional mechanism as it could lead to strong upconverted emission without any resonant GSA when the pump power is above a certain threshold value. The energy migration-mediated up-conversion (EMU) usually occurs in a core-shell structure nanoparticle involving the use of four types of lanthanide ions. The energy can migrate from one ion to another because of the similar energy gap. Through EMU, efficient tunable up-conversion emission can be realized in various ions without long-lived intermediary energy states. For photovoltaic applications, the majority up-conversion mechanisms are based on the combination of ESA and EMU processes.

The use of up-conversion materials to enhance photovoltaic performance was first reported by Gibart's group in 1996 using the Er^{3+}–Yb^{3+}-doped vitroceramic in bifacial GaAs SCs.[39] From then on, many researchers have focused on enhancing the performance of photovoltaic SCs using up-conversion materials, as shown in Table 21.1. The Yb^{3+} codoped with $Er^{3+}/Tm^{3+}/Ho^{3+}$ ions can absorb the light around 980 nm and emit at visible, which can be used to enhance the performance of photovoltaic devices with wide bandgap semiconductors, such as GaAs SC, amorphous Si SCs and PSCs. The phosphors singly doped with $Er^{3+}/Tm^{3+}/Ho^{3+}$ ions have the ability to convert long-wavelength NIR light into short-wavelength NIR and visible emissions, which is applicable in narrow bandgap c-Si SCs (E_g=1.12 eV; 1100 nm).

21.4.1 UP-CONVERSION PHOSPHORS FOR C-SI SOLAR CELLS

Er ions. Er^{3+}-doped up-conversion phosphors are the most promising up-conversion materials for c-Si SCs due to the GSA of Er^{3+} in the range of 1480–1580 nm ($^4I_{15/2}$–$^4I_{13/2}$ transition). The GSA centered at about 1540 nm results in up-conversion via an ETU mechanism, giving rise to four emission bands: $^4I_{11/2}$–$^4I_{15/2}$, 980 nm; $^4I_{9/2}$–$^4I_{15/2}$, 810 nm; $^4F_{9/2}$–$4I_{15/2}$, 660 nm; and $^4S_{3/2}$–$^4I1_{5/2}$, 550 nm (Figure 21.7).

TABLE 21.1
Selected Up-conversion Phosphors Used for Photovoltaic Application and Their Performance

Dopant Ions	Phosphor	Excitation (nm)	Solar Cell Type	Enhancement	Ref.
23% Er^{3+}	β-NaYF$_4$	1523	c-Si	1.33–1.80% of EQE	[39]
20% Er^{3+}	β-NaYF$_4$	1523	c-Si	3.0% of PCE	[40]
25% Er^{3+}	β-NaYF$_4$	1450–1600	c-Si	–	[41]
20% Er^{3+}	β-NaYF$_4$	1480–1580	c-Si	3.4% of EQE	[42]
Er^{3+}	β-NaYF$_4$	1523	c-Si	2.5% of EQE	[43]
20% Er^{3+}	β-NaYF$_4$	1523	c-Si	0.34% of EQE	[44]
30% Er^{3+}	BaY$_2$F$_8$	1520	c-Si	–	[45]
Ho^{3+}	Oxyfluoride glass ceramics	1170	c-Si	–	[46]
Ho^{3+}–Yb^{3+}	–	980	c-Si	1.2 mA/cm^2 of J_{sc}	[47]
Yb^{3+}–Er^{3+}–Gd^{3+}	NaYF$_4$	980	a-Si	0.14% of EQE	[35]
20% Yb^{3+}/1% Er^{3+} 25% Yb^{3+}/1% Ho^{3+} 60% Yb^{3+}/0.5% Tm^{3+}	β-NaYF$_4$	980	a-Si: H	0.055% of PCE	[48]
Yb^{3+} (18%), Er^{3+} (2%)	β-NaYF$_4$	980	a-Si: H	0.03% of EQE	[49]
18% Yb, 2% Er	β-NaYF$_4$	980	a-Si: H	–	[50]
10% Er^{3+}	β-NaYF$_4$	980 and 1560	a-Si: H	0.54 μA of photocurrent	[51]
Yb^{3+}, Er^{3+}	β-NaYF$_4$	980	DSSC	Relative 17% of PCE	[52]
Yb^{3+}, Er^{3+}	β-NaYF$_4$@SiO$_2$@TiO$_2$	980	DSSC	Relative 26.39% of PCE	[53]
Yb^{3+}, Er^{3+}	β-NaYF$_4$@SiO$_2$@TiO$_2$	980	DSSC	Relative 29.41% of PCE	[54]
Yb^{3+}, Er^{3+}	β-NaYF$_4$@SiO$_2$@Au	980	DSSC	–	[56]
Yb^{3+}, Er^{3+}	β-NaYF$_4$	980	DSSC	–	[57]
Yb^{3+}, Er^{3+}	β-NaYF$_4$	980	DSSC	Relative 16% of PCE	[58]
Yb^{3+}, Er^{3+}	β-NaYF$_4$	980	DSSC	Relative 5.45% of PCE	[59]
Yb^{3+}, Er^{3+}	β-NaYF$_4$	980	DSSC	Relative 12.5% of PCE	[60]
Yb^{3+}, Er^{3+}	β-NaYF$_4$	980	DSSC	Relative 10% of PCE	[61]
Yb^{3+}, Er^{3+}	NaCsWO$_3$@NaYF$_4$@NaYF$_4$ coupled with LSPR	980	PSC	PCE: 17.99–18.89%	[64]
Yb^{3+}, Er^{3+}	NaLuF$_4$@NaLuF$_4$	980	PSC	PCE: 15.51–15.86%	[65]
Yb^{3+}, Er^{3+}	β-NaYF$_4$	980	PSC	PCE: 18.60–19.70%	[63]
Er^{3+}	$mCu_{2-x}S$@SiO$_2$@Er$_2$O$_3$	980	PSC	Relative 10% of PCE	[128]
Yb^{3+}, Er^{3+}	β-NaYF$_4$@SiO$_2$	980	PSC	Relative 21.0% of PCE	[129]
Ho^{3+}	NaYbF$_4$	–	PSC	Relative 28.8% of PCE	[130]
Yb^{3+}, Er^{3+}	TiO$_2$	980	PSC	Relative 20.8% of PCE	[131]
Yb^{3+}, Er^{3+}	NaGdF$_4$	976	Organic solar cell	PCE: 2.81–3.61%	[66]
Yb^{3+}, Er^{3+}	NaYF$_4$	980	Organic solar cell	–	[67]

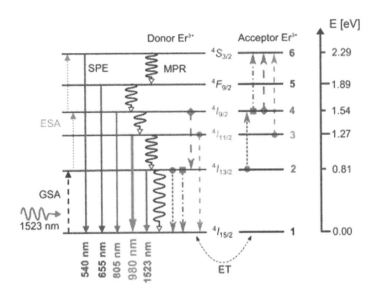

FIGURE 21.7 Up-conversion processes via the ETU mechanism between two Er^{3+} ions under 1523 nm light excitation. (Adapted with permission from Ref. [42]. Copyright 2014, Elsevier B.V.)

The most used up-conversion material is the rare earth (RE)-doped fluoride, such as $NaYF_4$:Yb, Er and YF_3:Yb, Er phosphors. Georgios E. Arnaoutakis et al.[40] in 2015 reported the application of β-$NaYF_4$: (23%) Er phosphor as upconverters in c-Si SC. To concentrate the sunlight, the dielectric-filled compound parabolic concentrators were used as integrated optics on the rear side of a planar bifacial SSC together with upconverters. An increase of external quantum efficiency (EQE) was obtained from 1.33 to 1.80% under an excitation of 1523 nm at an irradiance of 0.024 W/cm². The enhancement can be attributed to the concentration of the excitation on the up-conversion phosphor and the efficient collection of up-conversion luminescence.

Structure of c-Si SCs with up-conversion phosphors. The common structure of the SC with upconverters is shown Figure 21.8.[41] The β-$NaY_{0.8}Er_{0.2}F_4$ phosphor was adopted as the upconverter, which can be excited by 1523 nm light and exhibits efficient up-conversion luminescence at 980 nm. A system geometry, LSC, was assumed, which allowed for additional geometric concentration onto the upconverter that covered only a fraction of the LSC surface. The result of the calculation indicated that efficiency of an optimized SC can be increased by 3.0% relative using an upconverter. The SCs combined with the up-conversion layer β-$NaY_{0.8}Er_{0.2}F_4$ embedded in the polymer perfluorocyclobutyl showed an EQE of 1.69% measured under 1508 nm monochromatic excitation with an irradiance power of 1091 W/m².

Relationship with solar power. The efficiency of upconverter and performance on SCs are closely associated with the excitation power. The power density of solar energy is too low to achieve the efficient up-conversion luminescence. The solar concentrator was adopted to focus the solar energy to get a high enhancement of SC performance. The upconverter material β-$NaYF_4$: 25% Er^{3+} was applied in the bifacial SSC.[42] An external up-conversion quantum yield of 2.0% was obtained in the range from 1450 to 1600 nm under low solar concentration of 50 suns. This corresponded to a potential increase of the short-circuit current density of 3.89 mA/cm² (Figure 21.9). Further increasing the solar concentration to 77 suns, the external short-circuit current density from up-conversion reached 4.03 mA/cm².

Relationship with Er^{3+} ions concentrations. The performance of Er^{3+} ions-doped upconverters on c-Si SCs had an intensive relationship with concentrations of Er^{3+} ions. As shown in Figure 21.10,[43] the monocrystalline BaY_2F_8 doping with 10, 20 and 30% Er^{3+} were synthesized, which were grown in a Czochralski furnace and subsequently attached to an adapted planar bifacial SSC. The results showed that 30% Er^{3+}-doped BaY_2F_8 sample showed the best optical performance,

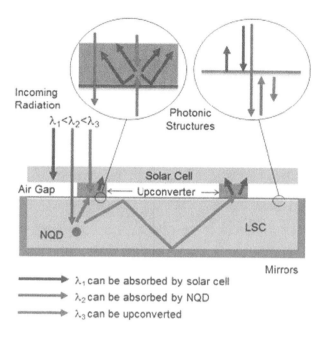

FIGURE 21.8 Concept of an advanced upconverter system. It consists of a bifacial solar cell that can utilize higher energy photons (λ_1) and an upconverter that can convert low-energy photons (λ_3) into higher energy photons (λ_1). (Adapted with permission from Ref. [41]. Copyright 2014, Elsevier B.V.)

which displayed a high peak absorption coefficient of 45.1 cm^{-1} at 1493 nm and a high external (internal) up-conversion quantum yield of 9.5±0.7% under 4740±250 W/m^2 monochromatic irradiation at 1520 nm. The SSC displayed a 33.4 mA/cm^2 short-circuit current density under one-sun AM 1.5G standard measurement conditions. After coupled with the 30% Er^{3+}-doped BaY$_2$F$_8$ monocrystalline sample, an additional short-circuit current density arising from up-conversion luminescence of 17.2±3.0 mA/m^2 at an illumination with a solar concentration of 94±17 suns. This result equivalented to a record relative enhancement of the short-circuit current of 0.55±0.14% at one sun.

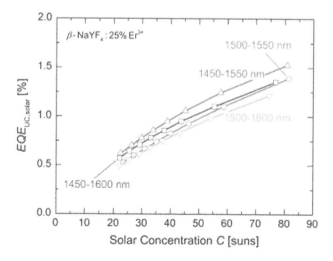

FIGURE 21.9 EQE that can be expected under concentrated radiation with the solar spectrum in the respective spectral range EQE$_{UC,solar}$ is determined by applying the spectral mismatch correction method to the EQE$_{UC}$ data. (Adapted with permission from Ref. [42]. Copyright 2014, Elsevier B.V.)

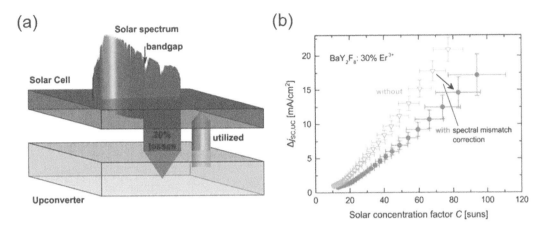

FIGURE 21.10   (a) The sunlight with lower energy than the bandgap energy can be converted to photons with higher energy through up-conversion process. (b) Internal and external up-conversion quantum yield of 30% Er^{3+}-doped BaY_2F_8 sample under illumination of 1520 nm. (Adapted with permission from Ref. [43]. Copyright 2015, Elsevier B.V.)

Ho^{3+} ion. The Er^{3+}-doped fluorides for enhancing the c-Si SCs was strongly limited owing to the small solar spectral range around 1540 nm that is used. The Ho^{3+} ion has a relatively wide absorption band in the 1150–1225 nm spectral range due to $^5I_8$–$^5I_6$ transition. Fernando Lahoz presented Ho^{3+}-doped oxyfluoride glass ceramics in 2008,[44] which can be excited by the 1170 nm light and give out the up-conversion luminescence via ETU at 650 and 910 nm. The author stated that this was a promising upconverter for Si SCs. Moreover, owing to the transparency of the glass ceramics around 1540 nm, this upconverter could complement the already existing Er^{3+}-doped upconverter phosphors. They proposed a double-layer structure, where a first Ho^{3+}-doped GC layer would be placed directly at the rear of a bifacial SC, followed by a second Er^{3+}-doped upconverter phosphor and a mirror. Both layers can contribute to the efficiency of the Si SC in an additive way. In 2011, Lahoz et al.[45] displayed the Ho^{3+}–Yb^{3+}-codoped upconverter with an enhanced up-conversion emission due to the energy transfer from Ho^{3+} to Yb^{3+} ions. They observed an increment of photocurrent of about 1.2 mA/cm^2.

Both Er^{3+} and Ho^{3+} ions. Chen et al. separately incorporated the Er^{3+} and Ho^{3+} into the core and shell layer of a nanoparticle to form a core-shell structure.[46–49] The core-shell design can extend the NIR wavelength range excitable for up-conversion emission and furthermore reduce the luminescence quenching induced by the cross relaxation through precise modulation to the activator concentrations and distance between activators and donors. As a proof-of-concept experiment, the authors fabricated monodisperse $NaGdF_4$:Er^{3+}@$NaGdF_4$:Ho^{3+}@$NaGdF_4$ core-shell-shell nanoparticles. The shell layer can enhance the luminescence through passivation to the surface defects and achieve the upconversion luminescence with both Er^{3+} and Ho^{3+} ions.

21.4.2   UP-CONVERSION FOR AMORPHOUS SILICON SOLAR CELLS

The bandgap of the amorphous Si SCs is about 1.75 eV, which can absorb NIR light shorter than 708 nm. To extend the spectral response range, up-conversion materials with the ability to convert the sub-bandgap NIR light ($\lambda>700$ nm) to visible emission are needed. Particularly, the upconversion phosphors, such as $NaYF_4$: Yb/Er, have visible emission bands located in the spectral region, which can be an appropriate candidate to combine with the amorphous Si SCs.

To achieve best performance of upconverters on SSCs, various structures of SCs were studied. Meanwhile, noble metal nanoparticles with plasmonic resonance effect were also adopted to enhance the upconversion luminescence (UCL) through magnifying the electric field surrounding the up-conversion nanoparticles. Li et al.[36] reported the up-conversion NaYF$_4$: Yb/Er/Gd nanorods in combination with gold nanostructures. The up-conversion luminescence was enhanced by the localized surface plasmonic resonance of gold nanostructures. After having applied the upconverter on the front of the a-Si SC, the device showed 16–72 folds of enhancement of the photocurrent under irradiation of 980 nm light. The SC using NaYF$_4$:Yb/Er/Gd nanorods with gold nanoparticles obtained an external photocurrent of 1.16 mA and an EQE of 0.14% under 980 nm laser (power: 1100 mW).

Besides the up-conversion process, the light scattering effect of up-conversion particles also played an important role in enhancing the performance of SCs. The upconverters can serve as light scattering centers to enhance the light harvest as the efficiency of UCL is usually quite low. Figure 21.11 shows the typical structures of applying upconverters on SSCs,[50] which presented three up-conversion phosphors: β-NaYF$_4$:Yb^{3+} 20%/Er^{3+} 1%, b-NaYF$_4$:Yb^{3+} 25%/Ho^{3+} 1% and b-NaYF$_4$:Yb^{3+} 60%/Tm^{3+} 0.5%. Two different coupling ways of upconverter on a-Si: H SC were demonstrated: by dissolving the powder in the polydimethylsiloxane (PDMS) and by compressing it into solid slice with a tablet-pressing machine (Figure 21.11(a) and (b)). The performance of a-Si: H SC coupled with upconverter was studied under both 980 nm laser and AM 1.5G illumination. The short-current was increased from 18.29 to 18.52 mA/cm^2 and the PCE was enhanced by about 0.055%. The β-NaYF$_4$ doped with Yb^{3+}, Er^{3+}/Tm^{3+}/Ho^{3+} phosphors has the ability to absorb the light around 980 nm and emit at the visible region (Figure 21.11(c)), which contributed to the enhancement of the performance of a-Si: H SC. In addition, the NaYF$_4$ nanoparticles can act as effective scattering centers in the range of 400 nm to the visible region, which resulted in the extension of light path length within the electrodes and, therefore, the increase of light harvesting.

The combination of up-conversion phosphors with organic polymer can fabricate a high-quality film, which does not influence the mobility of the device. In the year of 2009, X. D. Zhang et al.[51] fabricated an up-conversion layer using solidified mixture of PDMS and NaYF$_4$:18% Yb, 2% Er upconversion phosphors. This up-conversion layer was integrated with the a-Si: H SC with the structure of glass/SnO$_2$/p/i/n/ZnO/upconverter/Al/glass. Compared with the SC without upconverter, the short-circuit current density was enhanced from 15.99 to 17 mA/cm^2.

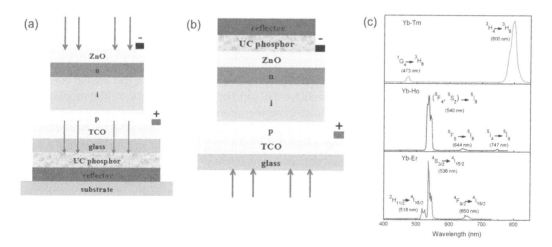

FIGURE 21.11 Schematic structure of p-i-n type a-Si: H solar cells with upconverters under illumination of two light sources (a) 980 nm infrared laser (300 mW), (b) AM 1.5G, 25°C. (c) Up-conversion emission spectra of samples β-NaYF$_4$:Yb^{3+} 20%/Er^{3+} 1%, β-NaYF$_4$:Yb^{3+} 25%/Ho^{3+} 1%, β-NaYF$_4$:Yb^{3+} 60%/Tm^{3+} 0.5% (980 nm, 40 mW). (Adapted with permission from Ref. [50]. Copyright 2016, Elsevier B.V.)

The absorption cross section of up-conversion phosphors is small and only light with specific wavelengths can be absorbed due to the $4f–4f$ transition properties of RE ions. Thus, the enhancement of SC performance is quite little only using an excitation light with one wavelength. Yongsheng Chen et al.[52] applied the β-NaYF$_4$:Er^{3+}(10%) microprisms to the back of a thin-film hydrogenated a-Si: H SC to expand the spectral response to the sub-bandgap NIR light. 0.3 and 0.01 μA of photocurrent were achieved under irradiation of 980- and 1560 nm diode laser with 80 mW/cm^2, respectively. Under irradiation of both 980- and 1560 nm light, an external photocurrent of 0.54 μA was obtained. The enhancement was mainly attributed to the red emission of Er^{3+} ions excited by the 980- and 1560 nm light.

21.4.3 UP-CONVERSION PHOSPHORS FOR OTHER SOLAR CELLS

DSSCs. DSSCs are the promising next-generation photovoltaic cells that can be used to create low-cost, flexible solar panels. Differing from the conventional Si-based SCs, DSSCs primarily consist of photosensitive dyes and other substances such as an electrolyte solution and metal oxide nanoparticles. The efficiency of DSSCs are currently limited by the limited absorption range to the solar spectrum. The most commonly used dyes in current DSSCs are ruthenium-based dyes, including N$_3$ ([Ru(dcbpyH$_2$)$_2$-(NCS)$_2$], dcbpyH$_2$=2,20-bipyridyl-4,40-dicarboxylic acid), N719 ([(C$_4$H$_9$)$_4$N]$_2$[Ru(dcbpyH)$_2$(NCS)$_2$]) and N749 ([(C$_4$H$_9$)$_4$N]$_3$-[Ru(tcterpy)(NCS)$_3$]$_3$H$_2$O, tcterpy=4,40,400-tricarboxy-2,20:60,200-terpyridine). As a result of their large optical bandgap of 1.8 eV, these dyes have an absorption threshold below ~700 nm. To achieve higher efficiencies for DSSCs, light absorption must be extended into the NIR spectral region without sacrificing their performance in the visible region.

The most commonly used upconverter was Yb^{3+}- and Er^{3+}-doped up-conversion nanoparticles that can absorb the light around 980 nm and emit at the visible region. The integration of upconverters in DSSCs usually adopted the structure as shown in Figure 21.12. Based on the structure of typical DSSC, the upconverter was coated or mixed with a dye-sensitized TiO$_2$ film as the photoanode. The UCL can be generated pumped by the infrared light. The UCL can be absorbed by reabsorption process. The excited electrons located at the green emission level (^{2}H$_{11/2}$–^{4}I$_{15/2}$, ^{4}S$_{3/2}$–^{4}I$_{15/2}$) of hexagonal NaYF$_4$:Yb^{3+}, Er^{3+} can also be quenched via a nonrelaxation path due to the matched energy level. A work by Jie Chang et al.[53] adopted the TiO$_2$/NaYF$_4$:Yb^{3+}, Er^{3+} nano-heterostructures in DSSCs. After introducing the nano-heterostructures, the overall efficiency of the DSSC device is 17% higher than the reference cells. The nano-heterostructure enabled the efficient electron injection from upconversion nanoparticle (UCNP) to the CB of TiO$_2$.

As mentioned earlier, light scattering effect plays an important role on enhancing the efficiency of SCs. Therefore, the photon up-conversion and light scattering effect both contribute to the enhancement of efficiency. Figure 21.12(a) presented a work[54] applying the unique core/double-shell-structured β-NaYF$_4$:Er^{3+}, Yb^{3+}@SiO$_2$@TiO$_2$ mesoporous microspheres and employed in DSSCs internally. The results indicated that these mesoporous microspheres can act as effective scattering centers to enhance the light harvesting. The coating of amorphous SiO$_2$ can effectively passivate the surface defects, which can avoid the electron trapping. The optimal properties with J_{sc} of 14.95 mA/cm^2 and η of 9.10% was obtained, which was enhanced by 26.39% compared with the reference cell.

Noble metal nanoparticles with plasmonic effect also make a noticeable contributions to the efficiency. The core-shell-structured β-NaYF$_4$:Er^{3+}, Yb^{3+}@SiO$_2$@Au nanocomposites with Au nanoparticles attached on the surface was presented by Peng Zhao et al.[55] This core-shell structure was applied to DSSCs as a multifunctional layer on top of TiO$_2$ layer as shown in Figure 21.12(b). The introduction of Au nanoparticles can enhance the UCL through amplifying the emission light field by surface plasmon resonance effect.

As shown in Figure 21.12(c),[56] the DSSCs were fabricated by mixing the up-conversion phosphor NaYF$_4$:Yb^{3+}, Er^{3+} nanocrystalline with P25 and TiO$_2$ hollow microspheres. The metal platinum

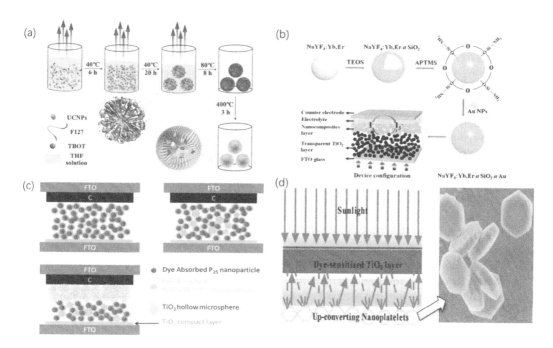

FIGURE 21.12 (a) Schematic for the experimental process for core/double-shell-structured β-NaYF$_4$:Yb^{3+}, Er^{3+}@SiO$_2$@TiO$_2$ hexagonal sub-microprisms and the structure of DSSC device (adapted with permission from Ref. [54]. Copyright 2016, Elsevier B.V.). (b) Schematic for NaYF$_4$:Yb^{3+}, Er^{3+}@SiO$_2$ applied in DSSCs (adapted with permission from Ref. [55]. Copyright 2014, Royal Society of Chemistry). (c) Schematic for DSSC structure based on P25/NaYF$_4$:Yb^{3+}, Er^{3+} with TiO$_2$ hollow microspheres (adapted with permission from Ref. [56]. Copyright 2019, Elsevier B.V.). (d) Schematic for DSSC structure based on one internal TiO$_2$ transparent layer with an external rear layer β-NaYF$_4$:Yb^{3+}, Er^{3+} nanoplatelets (adapted with permission from Ref. [57]. Copyright 2011, American Chemical Society).

of DSSCs was replaced by the low-cost carbon CE. The DSSCs fabricated with NaYF$_4$:Yb^{3+}, Er^{3+} achieved a PCE of 5.27% under one sun illumination, which displayed an increase of 16% compared with the benchmark P25 cell (4.55%). The increase of PCE was mainly attributed to the increase of J_{sc}, which can be attributed to the enhanced light harvesting at NIR region of up-conversion phosphors and the light scattering to visible light of TiO$_2$ hollow microsphere.

Besides the layer structure, the core-shell structure of upconverter and TiO$_2$ was also commonly adopted in DSSCs. The nano- or sub-micrometer-sized up-conversion nanoparticles included the surface defects and ligands with high-energy vibrational modes, which can lead to quenching effect by multi-phonon relaxation process. The core-shell structure can effectively passivate the defects and trapping centers on the surface and avoid the quenching effect.

As shown in Figure 21.12(d), β-NaYF$_4$:Er^{3+}, Yb^{3+} hexagonal nanoplatelets were synthesized using a hydrothermal route and applied to DSSC as a bifunctional layer.[57] The β-NaYF$_4$:Er^{3+}, Yb^{3+} bifunctional layer is placed on the external side of the CE, which can simplify DSSC fabrication and not affect the thickness of internal transparent TiO$_2$ film. About 10% enhancement of photocurrent and overall efficiency of the DSSCs by the introduction of the nanoplatelet-based external layer was achieved.

PSCs. PSC was first reported in 2009 with 3.8% PCE.[58] After that, extensive studies were focused on PSCs due to their excellent photovoltaic performance including excellent light absorption, high carrier mobility, low exciton binding energy (50 meV) and a simple preparation process. Up to now, the PCE of PSCs has reached 25.5%. However, the excellent light response of perovskite materials (FAPbI$_3$, MAPbI$_3$, CsPbI$_3$, etc.) only covers the visible light range (300–800 nm), which only takes up a small portion of total sunlight energy (~49%). To make full use of infrared light, a great effort has been devoted to extending the response of PSCs to the NIR region. UC materials can convert the NIR photon to UV and visible photons, which has been considered as the promising strategy to expand the light response of PSCs. The UC materials can act as an interlayer inside the PSCs or mixing with electronic transport layer, improving the light conversion efficiency by optical management. In addition to light up-conversion process, the UC materials can also serve as the light scattering centers and enhance the light absorption of PSCs.

Upconverters on the surface of PSCs. As shown in Figure 21.13(a),[47] the $NaYF_4$:Yb^{3+}, Er^{3+}/ $NaYF_4$:Yb^{3+}, Tm^{3+}/Ag composite layers were introduced on the backlight side of the PSCs device by the pulsed laser deposition, on the basis of a novel NiO/Ag/NiO transparent electrode. This design enables multifunctional effects, harvesting NIR light and converting to visible, plasmonic scattering, reflection and luminescent enhancement. In addition, a typical UV-to-visible $Eu(TTA)_2$(Phen) MAA downshifting layer was introduced on the incident light side of PSCs. Benefiting from this design, the UV and NIR light can converted to the response range of PSCs (Figure 21.13(b)). Overall, the up-conversion and downshifting effects give rise to the highest PCE performance of 19.5% and noticeable current density (J_{sc}) value of 27.1 mA/cm^2 among the reported transparent PSCs.

Upconverters inside the PSCs. The β-$NaYF_4$:Yb, Er up-conversion nanocrystals (NCs) have been embedded in situ into a $CH_3NH_3PbI_3$ layer to fabricate NIR-enabled planar PSCs.[59] The CH_3NH_3I-capped up-conversion NCs generated from the ligand exchange were mixed with the

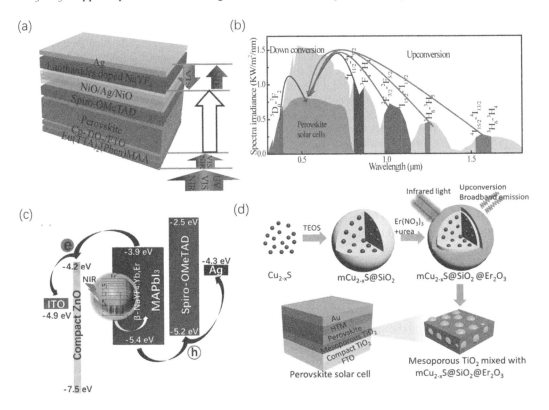

FIGURE 21.13 (a) Schematic configuration of a complete PSCs with $NaYF_4$ upconverters (adapted with permission from Ref. [47]. Copyright 2018, Elsevier). (b) AM 1.5G spectra shows the fraction absorbed by PSCs and the spectral regions that can be utilized through the up-conversion and downshifting processes. (c) Schematic of energy transfer within an up-conversion nanocrystals-embedded planar PSC device (adapted with permission from Ref. [59]. Copyright 2017, Royal Society of Chemistry). (d) Fabrication process of mCSE nanocomposites and their applications in PSCs (adapted with permission from Ref. [48]. Copyright 2017, Royal Society of Chemistry).

perovskite precursor and served as nucleation sites for the NCs-mediated heteroepitaxial growth of perovskite; moreover, the in situ embedding of NCs into the perovskite layer was realized during a spin coating process. As shown in Figure 21.13(c),[59] the incident photons with energy lower than the bandgap of perovskite, which could not be harvested by perovskite, were absorbed by embedded up-conversion NCs and then transferred to perovskite absorbable high-energy photons to contribute to the additional photocurrents of the perovskite active layer. Another work (Figure 21.13(d)) demonstrated the efficient photon energy UC in semiconductor plasmonic $mCu_{2-x}S@SiO_2@Er_2O_3$ (mCSE) nanocomposites,[48] where the broadband semiconductor plasmon (800–1600 nm) of $mCu_{2-x}S$ serves as an antenna to sensitize UC of Er_2O_3 nanoparticles. The excitation range was expanded, ranging from 800 to 1600 nm. The highly efficient mCSE nanocomposites were utilized to improve the PCE of PSCs. The expansion of the NIR response (800–1000 nm) and considerable improvement of PCE were obtained, with an optimum PCE of 17.8%. The mCSE composites in PSCs enhanced the photocurrent via electron transfer from oxygen defects to the CB of TiO_2 under irradiation of sunlight. Under irradiation of 15 suns, the electron transfer and reabsorption of UCL both contributed to the enhancement of PCE.

Another work[60] reported the localized surface plasmon resonance (LSPR)-enhanced UCNPs applied in the hole transport layer of PSCs. The $NaCsWO_3@NaYF_4@NaYF_4$:Yb, Er nanoparticles were synthesized with $NaCsWO_3$ as LSPR center. The UCL was enhanced about 124 folds with appropriate concentration of $NaCsWO_3$. After integrating the core-shell nanoparticles into hole transport layer, the PCE of PSCs reached 18.89%, which was enhanced 17.99% relatively compared with the reference cell of 16.01%.

As shown in Figure 21.14,[61] the core-shell-structured $NaLuF_4$:Yb,Er@$NaLuF_4$ UCNPs were introduced into the hole transport layer in the $CsPbI_3$-based PSCs. The solar spectrum is a wide wavelength ranging from 300 to 1100 nm, covering the UV, visible and NIR regions. It can be seen that the EQE of $CsPbI_3$ is sensitive to UV and visible light. The UCNPs in the hole transport layer can convert the light to the UV and visible wavelengths, which can broaden the NIR response of PSCs. Compared to the reference cell, the short-circuit current density J_{sc} reached up to 19.17 mA/cm^2 (18.81 mA/cm^2) and the PCE reached to 15.86% (15.51%). However, the NIR response was not

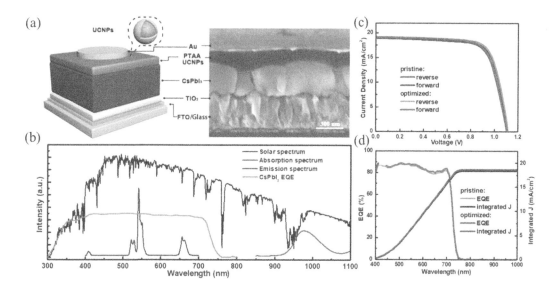

FIGURE 21.14 (a) Schematic structure and cross-sectional SEM image of the solar cell device. (b) Solar spectrum, absorption and emission spectra of UCNPs and EQE curves of $CsPbI_3$ PSC. (c) J-V curves under both the reverse and forward scan directions. (d) EQE curves and the integrated photocurrents. (Adapted with permission from Ref. [61]. Copyright 2019, American Chemical Society)

observed from the incident photon-to-current conversion efficiency (IPCE) results. Due to the low photoluminescence quantum yield (PLQY) of UCL, the up-conversion effect of adjusting the light absorption edge from visible to NIR range for extending the spectral absorption contributed quite little. The UCNPs were primarily served as scattering centers that can extend the sunlight optical path through combining with scattering and reflecting sunlight.

As discussed earlier, PSCs integrated with UCNPs have achieved noticeable improvement of PCEs. For the UCNPs located outside the PSCs, the improvement of performance can be attributed to photon conversion and light scattering effect. For the UCNPs integrated inside the PSCs, UCNPs can not only broaden the NIR response of PSCs due to the LSPR-enhanced UCL but also increase the visible light reabsorption of PSCs through the scattering and reflection effect. Moreover, UCNPs can modify the perovskite film by filling hole and gaps at the grain boundary and passivate the surface defects. In conclusion, the photon up-conversion is not the main reason that leads to the improvement of PSC performance due to the low absorption cross section, limited absorption range and low photon conversion efficiency of UCNPs, whereas the light scattering and passivation effects play a nonnegligible role on performance of PSCs.

Organic SCs. Organic SCs own excellent properties, such as light-weight and flexible, leading to inexpensive, large-scale production for solar energy conversion. One of the major energy losses in organic SCs is the sub-bandgap transmission due to the mismatch between the absorption properties of polymer materials and the solar spectrum. The current organic SCs usually adopt the organic molecules with large bandgap, which leads that it can only harvest the visible light. The organic SCs with best performance are usually fabricated using heterojunctions comprising poly(3-hexylthiophene) (P3HT) and the fullerene derivative PCBM. To extend the light response range of organic SCs to the NIR range, UCNPs are the appropriate candidate for photon conversion. For example, Xiao Jin et al. demonstrated the feasibility of up-conversion in organic SC.[62] They incorporated the $NaGdF_4$:Yb/Er nanoparticles into the photocatalytic titania layer and the PCE was enhanced from 2.81 to 3.61%. The response at visible region was also enhanced due to the inhibited charge recombination. The results showed that Yb concentration plays an important role in up-conversion, because it can inhibit the back-energy transfer process and absorb considerable low-energy photons. The $NaGdF_4$:Yb/Er nanoparticles can be excited at 976 nm and emit the green and red light.

In another work, lanthanide-doped $NaYF_4$:Yb/Er UCNPs were laminated on top of the organic device, as shown in Figure 21.15.[63] The sun light illuminates the SC from the ITO electrode side, the sub-bandgap photons can pass through the SCs and propagate into the UC converter, then which can be converted to visible photons via up-conversion process. Afterward, the converted visible photons can be absorbed by the SC and produce photocurrents.

Discussion. Before the conversion efficiency of photovoltaic devices can be further improved through use of up-conversion materials, it is necessary to put some effort into addressing a number

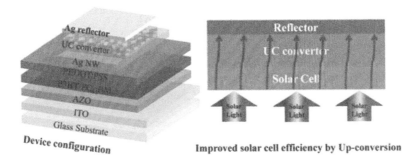

FIGURE 21.15 Schematic diagram shows the configuration of semitransparent organic solar cells with lanthanide-doped UC converter ($NaYF_4$:Yb/Er [20/2 mol%]) and the work principle. (Adapted with permission from Ref. [63]. Copyright 2015, Elsevier B.V.)

of challenges. First, the up-conversion phosphors can only absorb a small fraction of solar spectrum due to the weak and narrowband absorption. It is essential to develop up-conversion phosphors with a broadband absorption. Second, the high threshold of excitation power limits the performance of up-conversion phosphors on SCs. The power density of solar energy is relatively low to pump upconversion phosphors. In many cases, the concentrated solar power is needed, which leads to a complicated manufacture process. Third, the up-conversion quantum yield is quite low currently, lower than 10%. Although about 20% relative enhancement was observed (Table 21.1), the light scattering effect also contributed a lot to the enhancement of PCE. Therefore, there are many challenges to overcome before the industrial application of upconversion-sensitized SC.

21.5 DOWNSHIFTING PHOSPHORS FOR PHOTOVOLTAIC APPLICATIONS

Downshifting process can absorb one photon with high-energy and give one photon with high-energy, which can be utilized in many photovoltaic devices with poor UV spectral response. Downshifting can lead to an efficiency increase by shifting photons to a spectral region where the SC has a higher response, which can basically improve the blue response of the SC, and was predicted to result in up to 10% relative efficiency increase. Until now, two main classes of PV devices based on downshifting have been investigated: a LSC and a planar downshifting layer. A variety of downshifting materials have been explored, including lanthanide-doped phosphors and glasses, semiconductor QDs and organic-lanthanide complexes.

The ideal luminescent downshifting materials should own the following properties: (1) broadband absorption, particularly in the low spectral response region of the SC; (2) high absorption coefficient and high luminescence quantum efficiency; (3) high transmittance that does not affect the absorption to the light at other wavelengths; (4) large Stokes shift to minimize the self-absorption energy losses due to the spectral overlap between the absorption and emission bands; (5) good long-term stability.

21.5.1 LANTHANIDE-DOPED PHOSPHORS AND GLASSES

Lanthanide ions own abundant $4f–5d$ energy levels, which can absorb the UV light and emit at visible and NIR regions. Many SCs, such as SSCs, DSSCs, organic SCs, display poor spectral response at a specific region. The lanthanide-doped phosphors can absorb the light in the poor response region of SCs, which can increase the PCE of SC through photon energy conversion.

The Tm^{3+}-doped fluoride ZLAG glasses were applied to enhance the performance of c-Si SCs (Figure 21.16(a)).[64] The Tm^{3+}-doped fluoride ZLAG glass is coated on the surface of a c-Si SC. The increase of PCE can be attributed to the luminescence at 650 and 800 nm. In another work by Jihuai Wu et al.,[65] the RE compound YF_3:Eu^{3+} was mixed in TiO_2 film of DSSC. The YF_3:Eu^{3+} transfers UV light to visible light through photon energy downshifting, which increases the incident harvest and the photocurrent of DSSC. The introduction of Eu^{3+} ions can elevate the Fermi level of TiO_2 film and heighten the photovoltage of SCs. The PCE of DSSC reached 7.74%, which was enhanced by 32% compared to the reference cells.

The enhanced luminescent downshifting emissions were observed in Eu-doped phosphors under the effects of localized surface plasmon resonance from silver nanoparticles and applied to enhance the performance of SSCs (Figure 21.16(b)).[66] The plasmonic effect of silver nanoparticles can magnify the local electric field around the luminescent centers. Compared to the SC with only an anti-reflection layer, the introduction of silver nanoparticles with Eu-doped phosphors enhanced the luminescent intensity by 1.1243 folds at 514 nm. The EQE response was increased from 57.28 to 62.05%. The PCE was enhanced from 12.04 to 12.89% with 7.1% relative enhancement.

In another work,[67] the Eu-doped SiO_2 phosphors were applied to enhance the performance of GaAs SCs. The antireflective ITO layer was used to reduce the reflective loss. The relative enhancement of PCE was as high as 21.07%, whereas the cells coated a SiO_2 layer doped with Eu^{3+} ions. The red phosphor $CaAlSiN_3$:Eu^{2+} was reported as the luminescent downshifting material in DSSCs to

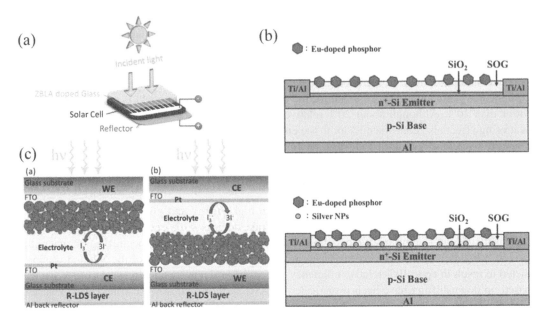

FIGURE 21.16 (a) Schematic view of J-V characteristics measurement configuration (adapted with permission from Ref. [64]. Copyright 2016, Elsevier B.V.). (b) Schematic diagrams of a sample with Ag NPs and with a silicate glass layer containing Eu-doped phosphors (adapted with permission from Ref. [66]. Copyright 2020, Elsevier B.V.). (c) Scheme showing the working principle of a DSSC with a reflective layer of luminescence downshifting materials (adapted with permission from Ref. [68]. Copyright 2013, American Chemical Society).

enhance the light-harvesting efficiency at the wavelength region 400–550 nm.[68] The downshifting layer was coated on the CE side and working electrode (WE) size, respectively. An Al back reflector was used to reduce the light loss (Figure 21.16(c)). Relative enhancements greater than 200% in IPCE near 500 nm and 40–54% in J_{sc} were achieved as shown in Figure 21.17.[69]

21.5.2　Semiconductor Quantum Dots

Semiconductor QDs are a class of photon energy downshifting materials that have been widely applied in SCs. The semiconductor QDs, including CdS, CdSe and CdTe, have shown excellent features such as broadband absorbance, high quantum yield and tunable emission properties.

The CdS QD-embedded silica films were applied as luminescent downshifting layer to improve the performance of crystalline SSCs.[70] The CdS QD-embedded silica film was placed on the front side of SSC. The short-circuit current of SSC with CdS QDs was improved by 4.0% compared to the reference cells. Another work by Jea-Gun Park synthesized the $Cd_{0.5}Zn_{0.5}S$–ZnS core (4.2 nm in diameter)-shell (1.2 nm in thickness) QDs (Figure 21.18(a)–(c)),[71] which can absorb the UV light (250–450 nm) and emit blue light at 442 nm. The luminescent quantum yield of the core-shell QDs was ~80%. The QDs were coated on the SiN_x film textured p-type SSC (Figure 21.18(a)) and got 30% enhancement of EQE at 300–450 nm (Figure 21.18(d)). The PCE was enhanced ~1.08% with relative enhancement of ~6.04% compared with the reference SCs (Figure 21.18(e)).

21.5.3　Dye

The organic dyes can absorb the incident solar light below the cutoff point of a SC. As shown in Figure 21.19(a),[72] the N719 dye can absorb the solar light below 550 nm. The N719 dye was mixed with TiO_2 and was integrated in DSSCs. The long persistent phosphors (LPP) can enable

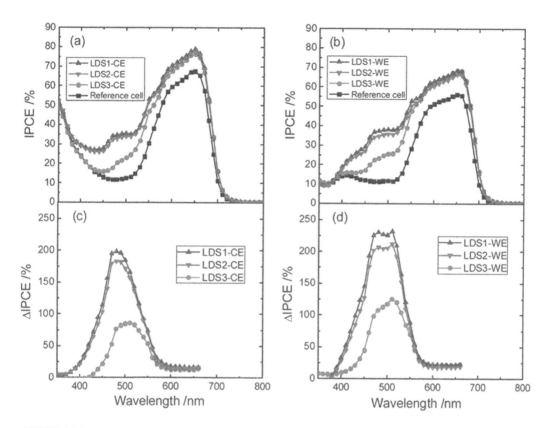

FIGURE 21.17 (a) IPCE spectra of devices with and without the downshifting layer coated on either (a) the counter electrode side or (b) the working electrode side, and the percentage enhancement of IPCE for devices with an R-LDS layer coated on either (c) the counter electrode side or (d) the working electrode side. (Adapted with permission from Ref. [69]. Copyright 2017, Elsevier B.V.)

the device to work in the dark weather. Single dyes usually display narrow absorption bands. Several dyes can be employed together to extend their absorption range. L. Danos et al. reported the performance of a two-dye-doped luminescent downshifting layer that uses excitation energy transfer between the dyes.[73] By using two dyes, the light absorption covered from 300 to 515 nm. The relative increase of PCE was observed were up to 10% for the CdTe SCs under AM 1.5G sunlight (Figure 21.19(b)).

Discussion. The downshifting phosphors were usually deposited on the front face or inside of SCs, which have brought noticeable improvement on SC performance. For the case of downshifting phosphors inside the a SC, light scattering effect and tuning on carrier mobility also contributed a lot to the performance improvement besides the fluorescence downshifting process. The improvement would be quite lower if only the contribution of photon conversion is considered. For the case of downshifting phosphors on the front face of a SC, the fluorescence downshifting process should be the main contributor for the performance improvement. The increase in PCE when the fluorescence conversion layer is placed on the front face is the real increase caused by fluorescence conversion. For high-efficiency SCs, such as c-Si and PSCs, theoretical and experimental studies have shown that the PCE can be enhanced 5–10% relatively caused by the downshifting process. Moreover, the existence of downshifting phosphors can also reduce the damage to photovoltaic materials and devices benefiting from the light conversion of UV to visible, especially for the new-emerging SCs, which is helpful for the long-term stability of devices.

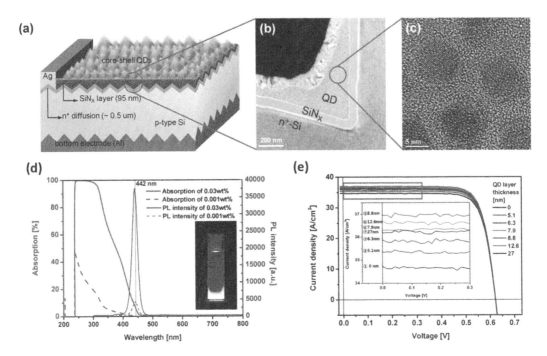

FIGURE 21.18 (a) Schematic structure of p-type silicon solar cell coated with the $Cd_{0.5}Zn_{0.5}S$–ZnS core-shell QD layer, TEM (b) and high-resolution images (c) of the $Cd_{0.5}Zn_{0.5}S$–ZnS core-shell QD layer coated on textured pyramid. (d) Absorption, photoluminescence and emitting light photography dependency on QD concentration in solution. (e) J-V curves of p-type silicon solar-cells coated with $Cd_{0.5}Zn_{0.5}S$–ZnS core-shell QDs. (Adapted with permission from Ref. [71]. Copyright 2014, Royal Society of Chemistry)

21.6 QUANTUM CUTTING PHOSPHORS FOR SOLAR CELLS

Using the luminescence phenomenon of the spectra conversion material can make the incident solar spectra better match the absorption characteristics of the SC and further improve the PCE of the SCs, which can exceed the Shockley-Queisser limit in theoretically.[1] To realize this kind of spectra conversion processes, the material with large absorption cross section, high luminescent efficiency

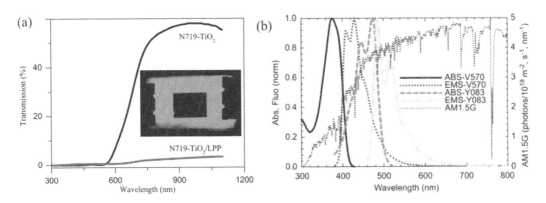

FIGURE 21.19 (a) Optical transmission curves of $N719$-TiO_2 and $N719$-TiO_2/LPP photoanodes. The inset shows the fluorescent image of the $N719$-TiO_2/LPP photoanode (adapted with permission from Ref. [72]. Copyright 2017, Royal Society of Chemistry). (b) Normalized absorption and fluorescence spectra of the individual dyes Violet–V570 and Yellow–Y083 downshifting layers spin coated on glass substrates plotted together with the AM 1.5G photon flux (adapted with permission from Ref. [76]. Copyright 2012, Elsevier B.V.)

and stable performance is crucial. QDs, metal luminescence, RE photoluminescent materials, silicon NCs and organic luminescent materials have all been used to achieve spectral conversion and improve the efficiency of SCs. RE ions have rich intermediate energy levels and chemical stability, the manufacturing process is mature, so they have been considered to be an ideal spectra conversion material suitable for application on SCs. To obtain a larger absorption cross section, a variety of RE ions may be doped into the host material at the same time to obtain the desired up-conversion or down-conversion emission. Compared with the down-conversion photoluminescence process, the up-conversion photoluminescence process has the weak intensity of photoluminescence and low conversion efficiency. Therefore, the current research on spectra conversion in SCs is mainly focused on down-conversion, especially for quantum cutting phosphor materials.

21.6.1 CONCEPT OF QUANTUM CUTTING

Due to the energy of high-energy UV photons being more than twice that of low-energy visible photons or NIR photons, the luminescent material can absorb one high-energy photon and produce two low-energy photons, which makes the emission of photons theoretically achievable. The process of the material absorbing a high-energy photon and then emitting two or more visible photons is quantum cutting, and the theoretical PLQY of this process is more than 100% and always happens in RE ions.[58] Quantum cutting was first proposed by Dexter in 1957[74] and was experimentally proved until 1974 from quantum cutting of YF_3:Pr^{3+} independently by Sommerdijk et al.[75] and Piper et al.[76] The quantum cutting process of RE ions is shown in Figure 21.20, where I and II represent two different ions, and the horizontal line represents energy level.[77] Quantum cutting can be divided into single RE ion cascade emission and energy transfer between RE ion pairs:

1. A schematic diagram of a single RE ion cascade emission model is given (Figure 21.20(a)). A single RE I absorbs a high-energy photon and is excited to a high-energy excited state and then generates a two-photon cascade emission through a two-step radiation. This model is often happened in some single-doped Pr^{3+}, Gd^{3+} and Tm^{3+} luminescent materials.[76,78–80]
2. Quantum cutting between RE ion pairs (Figure 21.18(b)–(d)). Quantum cutting of additional emission through cross relaxation between RE ion pairs. The model given in

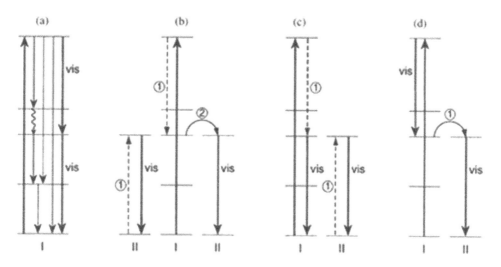

FIGURE 21.20 (a) Quantum cutting on a single ion I by the sequential emission of two visible photons. (b) The possibility of quantum cutting by a two-step energy transfer. (c and d) The remaining two possibilities involve only one energy transfer step from ion I to ion II. (Adapted with permission from Ref. [77]. Copyright 2010, Elsevier B.V.)

Figure 21.20(b) shows the quantum cutting emission generated by cross relaxation between RE ion pairs. First, the RE ions I is excited to a high energy level to transfer partial energy to the neighboring RE ions II through cross relaxation, so that it is excited to an excited state, and the ion I returns to an intermediate excited state. Further, RE ion I transfers the remaining energy to the neighboring ion II. This process often occurs in some Gd^{3+}, Eu^{3+} ion pairs.[81–83] Figure 21.20(c) shows another the quantum cutting process produced by the two-step energy transfer. Ion I transfers part of energy to neighbor ion II through cross relaxation, causing them to be excited into the excited state. The ion I located at the intermediate excited state returns to the ground state by emitting a low-energy photon, and the ion II transitions from the excited state to the ground state and also emits a low-energy photon, resulting in two-photon emission. The typical representative of this process is the Gd^{3+}–Tb^{3+} ion pair.[84–86] Difference from the model in Figure 21.20(b) and (c), Figure 21.20(d) presents that the ion I transitions from the high-energy excited state to the intermediate excited state through radiation and produces a low-energy photon in the process. The process thereafter is similar to that described earlier. Since the possibility of quantum cutting was proposed theoretically, the cascade emission process of Pr^{3+} has been observed in a series of fluorides such as YF_3.[75,76,87]

Solar energy has the advantages of being clean, cheap and efficient, attracting worldwide attention and becoming the most popular clean energy and research hotspot.[88] However, the low spectral response of SCs to UV and blue wavelengths (300–450 nm) of sunlight limits their practical PCE.[89] To solve this problem, various strategies have been proposed, such as tandem PSCs and SSCs and insert the photoluminescent layers into SCs.[10,90,91]

21.6.2 MECHANISM OF QUANTUM CUTTING APPLIED TO SOLAR CELLS

A representative mechanism is shown in Figure 21.21 (left).[92] A photoluminescence conversion layer converts one high-energy photon into two or more low-energy photons, which is located on the surface of the conventional single junction SCs. The photoluminescence conversion layer is electronically isolated from the SCs, absorbs two or more low-energy photons emitted by the conversion layer and leads to the generation of e-h pairs in the SCs, meaning that each incident high-energy photon will produce more than one e-h pair in the SCs. In different structures, the photoluminescence conversion layer can be located on the front surface or the back surface of the SC, in which

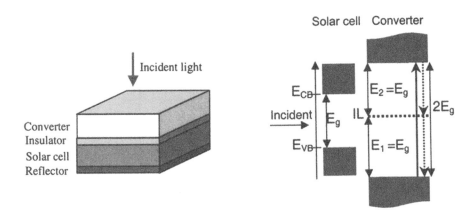

FIGURE 21.21 Mechanism of quantum cutting applied to solar cells.[92] (Adapted with permission from Ref. [92]. Copyright 2002, American Institute of Physics)

the conversion layer located on the back surface needs a double-sided SC to avoid the absorption of high-energy photons and subsequent e-h pairs heating inside the SC, and the SC material must be transparent. These conditions exclude all semiconductor materials used in SCs but can be achieved by dye molecules in DSSCs.

In principle, every three-level system can achieve down-conversion of incident photons into two or more photons with lower energy. For the conversion of solar energy, spectral down-conversion is more meaningful, on the right side of Figure 21.21, a photoluminescence conversion layer that simulates a semiconductor with an intermediate energy level E_1 above the edge of the VB is shown. The material of down-conversion layer with a bandgap E_g and contains impurity levels (ILs) with energy E_1 above the VB edge. The radiation transition occurs between the VB and the CB (solid arrow) or between one of the bands and the IL (dashed arrow). In an ideal situation, IL is located at the center of the bandgap of the conversion layer, $E_1 = E_2 = E_g$.

21.6.3 Quantum Cutting Phosphors Applied on Solar Cells

Among the various photoluminescent materials, quantum cutting of RE ions are considered an effective way to realize enhanced conversion efficiency enhancement of SCs.[1,93,94] Moreover, phosphors doped with RE ions have great applications in different types of SCs.

1. Semiconductor-based SCs.
   Semiconductor-based SCs have been demonstrated to have great commercial application value, the most commonly used materials include mono and polycrystalline silicon, amorphous silicon and III-Vs, II-VIs and I-III-VI2s compounds.[95,96] Since the work reported by Trupke et al.,[92] which discussed the use of a down-converting material in combination with a SC to improve the efficiency of conventional SCs, different quantum cutting materials have been explored as down-conversion layers for SCs. To obtain the strong photoluminescence of RE ions, the absorption of RE ions needs to match with the bandgap of the host materials. A heterostructure down-converter composed of Y_2O_3:[(Tb^{3+}–Yb^{3+}), Li^+] quantum cutting phosphor and ZnSe was proposed,[97] the ZnSe phase was used to absorb the incident light, and transfer the energy to Tb^{3+}–Yb^{3+} quantum cutting couple ion pairs. This work indicates that phosphor/semiconductor heterostructure based on RE ions is effective to realize the broadband down-conversion, which can be used to boost the energy efficiency of SSCs. Among the different semiconductor systems used in SCs, crystalline silicon c-Si is the most prevalent. SSCs have been extensively researched and are a mature technology that has been commercialized. However, the low spectral response at UV and blue wavelengths (300–450 nm) limit the photoelectric conversion efficiency (PCE) for c-Si SCs.[89] A successful example about a Pr^{3+}–Yb^{3+}-codoped ZBLA (family of glasses with a composition of ZrF_4–BaF_2–LaF_3–AlF_3–NaF) as the quantum cutting down-conversion layer applied to a commercial SSC.[98] The possible energy transfer processes involved in the down-conversion mechanism are described in Figure 21.22(a), under the presence of Yb^{3+}, after the excitation of the levels $^3P_0$, $^3P_1$ or $^3P_2$ of Pr^{3+}, a depopulation of the excited state can occur through two sequential resonant energy transfer steps between Pr^{3+} and Yb^{3+} with $^1G_4$ acting as the intermediate level by two routes. Tm^{3+}, Yb^{3+}-codoped tellurite glass is a promising composite material that converts one UV (240–400 nm) and blue (450–490 nm) photon into two NIR (920–1100 nm) photons by two-step energy transfer, which matches with the optimal spectral response of SSCs.[99] Furthermore, the quantum cutting of Yb^{3+}, Er^{3+} ion pair in CaF_2 NCs was applied to commercial SSCs.[100] Two-step energy transfer process between the Er^{3+} and Yb^{3+} ions (Figure 21.22(b)) makes the quantum cutting layer have potential to improve the PCE of c-SSCs. Based on the earlier discussion, the NIR quantum cutting effect opens a way to improve the energy efficiency of semiconductor-based SCs.

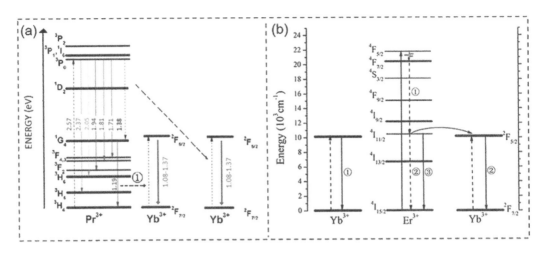

FIGURE 21.22 (a) Schematic energy-level diagram of Pr^{3+} and Yb^{3+} ions explaining the energy transfer process between the dopants route 1 and route 2 (adapted with permission from Ref. [98] Copyright 2015, Elsevier). (b) Schematic energy-level diagram of $Er^{3+}-Yb^{3+}$ codoped GCs, showing the two-step energy transfer mechanism for the NIR emission under 460 nm excitation (adapted with permission from Ref. [100]. Copyright 2016, Royal Society of Chemistry).

2. DSSCs.

DSSCs consist of two conductive glass electrodes and an electrolyte. Electrodes are usually coated with FTO or ITO. One electrode is the anode, also called the WE, which is mainly screen printed with titanium dioxide or zinc oxide and is sensitized with dye to absorb photons in the visible region. The other electrode is the cathode, or called CE, covered with a thin platinum film (which acts as a catalyst). An electrolyte is between the two electrodes, usually an organic solvent with a redox system. DSSCs are a good substitute for traditional semiconductor-based SCs because of their low cost and relatively easy manufacturing process.[101]

In DSSCs, RE ions-doped oxide hosts (such as TiO_2) are the most hopeful quantum cutting material for conversion layer. Recent years, RE ion-doped materials have been studied as photoanodes on TiO_2 films.[102–104] The RE ions-doped phosphors for DSSCs applications was summarized, oxide SiO_2, ZnO, $SrTiO_3$ and Nb_2O_5 doped with RE-based phosphors as a down-conversion layer were explored, which showed a great potential for DSSCs.[105]

3. Other SCs.

Most scientific research about SCs with additional quantum cutting materials as down-conversion layers are related to semiconductor-based SCs and DSSCs. Recently, some third-generation SCs are also explored.

One solution to take advantage of the photovoltaic quantum cutting effect is to build the LSCs, which is usually fabricated using highly transparent glass plates or transparent polymer sheets with photoluminescent species. The photoluminescence of the particles is randomly emitted and guided to the SC tank for total reflection.[106] LSCs have many advantages, such as the heat generated can be well dissipated to the collector plate area, and the collection of direct light and diffuse light also improves the performance of PV devices. Luo and coworkers synthesized Yb^{3+}-doped perovskite NCs and introduced the concept of quantum cutting LSCs for the first time.[107] The as-prepared NCs has a PLQY up to 164±7% and almost zero self-absorption loss of PL photons (Figure 21.23(a)), making the LSC internal optical efficiency (η_{int}), a new upper limit of 150%, which is almost independent of the size of the LSC. And the LSC was fabricated by coating a mixture of NCs and poly(methyl methacrylate) as a thin film. The top inset of Figure 21.23(b) is a photograph of LSC

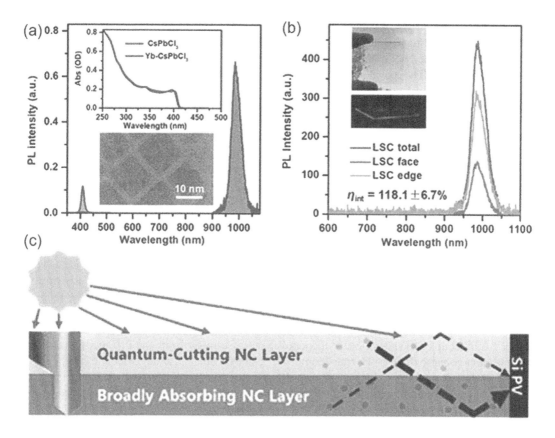

FIGURE 21.23 (a) PL spectra (solid lines with shading) of undoped (purple) and Yb^{3+}-doped (dark red) $CsPbCl_3$ NCs excited with a 365 nm light source. The top inset shows their absorption spectra. The bottom inset is a TEM image of doped NCs. (b) Total (dark red), face (light red) and edge (orange) emissions measured for a 5 cm × 5 cm QC-LSC using Yb^{3+}-doped $CsPbCl_3$ NCs. The η_{int} of this LSC is measured to be 118.1±6.7%. The top inset is the picture of the LSC under sunlight; the bottom inset shows the edge emission from the LSC under UV illumination and with a 570 nm longpass filter taken using a cell phone (MI 8 SE) camera (adapted with permission from Ref. [107], Copyright 2019, American Chemical Society). (c) Schematic of the proposed monolithic bilayer LSC. The top layer contains quantum cutting NCs (e.g., $Yb^{3+}:CsPb(Cl_{1-x}Br_x)_3$ NCs) and the bottom layer contains broadly absorbing NCs (e.g., $CuInS_2$ NCs) (adapted with permission from Ref. [108]. Copyright 2019, Royal Society of Chemistry).

under sunlight, showing a completely transparent appearance, which is suitable for building-integrated solar windows. This work demonstrates the quantum cutting effect has potential applications in the field of LSC. Moreover, Gamelin et al. combined the experimental and computational studies about Yb^{3+}-doped $CsPb(Cl_{1-x}Br_x)_3$ NCs on LSCs.[108] The proposed device structure was illustrated in Figure 21.23(c); in this process, high-energy light is absorbed by the NCs, generate lower energy light through a quantum cutting process, and the lower energy light is transmitted by the top layer and absorbed in the bottom layer. The work showed that Yb^{3+}-doped $CsPb(Cl_{1-x}Br_x)_3$ NCs have high potential to serve as unique LSC luminophores due to their large effective Stokes shifts and extraordinarily high PLQYs.

21.6.4 QUANTUM CUTTING PHOSPHORS BASED ON Yb^{3+} IONS

The ideal NIR quantum cutting materials for Si SCs should be able to convert the UV-green (300–500 nm) part of the solar spectra to NIR photons. Among the RE ions with quantum cutting emissions, Yb^{3+} ions are most suitable candidates for Si SCs. The cooperative quantum cutting has been extensively investigated in $Tb^{3+}-Yb^{3+}-$, $Tm^{3+}-Yb^{3+}-$, and $Pr^{3+}-Yb^{3+}$-codoped systems.

$Tb^{3+}-Yb^{3+}$ couple. NIR quantum cutting for $Tb^{3+}-Yb^{3+}$-codoped systems was first reported in 2005 on $Yb_xY_{1-x}PO_4$:Tb^{3+} powder phosphors.[109] Subsequently, Tb^{3+}, Yb^{3+}-codoped materials have been extensively studied, but the issue of the underlying energy transfer mechanism from Tb^{3+} to Yb^{3+} ions is still debatable.

Some researchers believed that a nonlinear second-order down-conversion process should be responsible for the cooperative quantum cutting.[110] The slope of Yb^{3+} luminescence intensity versus the excitation power was found to be nearly 0.5, indicating the emission of two lower energy photons upon the absorption of one photon. On the other hand, two different groups found the slope of luminescence intensity curves to be near 1 instead of 0.5, indicating a one-photon process for the Yb^{3+} NIR emission.[111,112]

$Tm^{3+}-Yb^{3+}$ couple. The NIR quantum cutting for the $Tm^{3+}-Yb^{3+}$ couple was reported in many host materials, including $GdAl_3(BO_3)_4$,[113] YPO_4,[114] glasses,[115,116] glass ceramics[117] and Y_2O_3-based transparent ceramics.[118] In the $Tm^{3+}-Yb^{3+}$ couple system, NIR quantum cutting occurs upon excitation at the $^1G_4$ (Tm^{3+}) level. The energy of the $^1G_4-^3H_6$ transition in Tm^{3+} is about twice as large as the $^2F_{5/2}-^2F_{7/2}$ transition of Yb^{3+}. Thus, the excited state $^1G_4$ (Tm^{3+}) can simultaneously transfer the energy to two neighboring Yb^{3+} ions, subsequently resulting in NIR emission in the range of 950–1100 nm.

$Pr^{3+}-Yb^{3+}$ couple. Two energy transfer mechanisms have been proposed for quantum cutting in the $Pr^{3+}-Yb^{3+}$ couple: (1) second-order cooperative energy transfer from one Pr^{3+} ion to two different Yb^{3+} ions because the $^3P_0-^3H_4$ transition of Pr^{3+} (~20,700 cm^{-1}) is about twice the energy of the $^2F_{5/2}-^2F_{7/2}$ transition of Yb^{3+}[113,119,120]; (2) first-order resonant energy transfer via two sequential steps involving cross relaxation between Pr^{3+} ($^3P_0-^1G_4$) and Yb^{3+} ($^2F_{7/2}-^2F_{5/2}$), followed by a second energy transfer process from Pr^{3+} ($^1G_4-^3H_4$) to Yb^{3+} ($^2F_{7/2}-^2F_{5/2}$).

$Er^{3+}-Yb^{3+}$ couple. The combination of Er^{3+} and Yb^{3+} is well known for up-conversion research, it also gives rise to efficient visible to NIR quantum cutting.[121] The Er^{3+} ion has excited states at ~20,000 cm^{-1} ($^4F_{7/2}$) and 10,000 cm^{-1} ($^4I_{11/2}$) that allow for a two-step energy transfer process, raising two neighboring Yb^{3+} ions to the $^2F_{5/2}$ excited state around 10,000 cm^{-1}.

$Nd^{3+}-Yb^{3+}$ couple. In the Nd^{3+}, Yb^{3+}-codoped system, NIR quantum cutting process could be realized from the $^2G_{9/2}$ (Nd^{3+}) level (~21,000 cm^{-1}) via the cross relaxation between Nd^{3+} ($^2G_{9/2}-^4F_{3/2}$) and Yb^{3+} ($^2F_{7/2}-^2F_{5/2}$), followed by a second energy transfer step from the $^4F_{3/2}$ (Nd^{3+}) level (~11,547 cm^{-1}) to the $^2F_{5/2}$ (Yb^{3+}) level (~10,000 cm^{-1}), which could lead to emission of two NIR photons from Yb^{3+} ions.[122,123]

$Ho^{3+}-Yb^{3+}$ couple. The $^5F_3-^5I_8$ transition of Ho^{3+} is approximately twice the energy required for the $^2F_{5/2}-^2F_{7/2}$ transition of Yb^{3+}. However, the second-order cooperative quantum cutting process of Ho^{3+} ($^5F_3-^5I_8$) and Yb^{3+} ($^2F_{5/2}-^2F_{7/2}$) is unlikely because that the $^5F_3$ (Ho^{3+}) excited state can decay rapidly to the next $^5S_2$, $^5F_4$ states.[124]

$Dy^{3+}-Yb^{3+}$ couple. NIR quantum cutting through the $Dy^{3+}-Yb^{3+}$ couple has been investigated to a lesser extent. The first work about this ion pair was demonstrated in 2011.[125] Under the excitation of 430 nm blue light, NIR quantum cutting could be achieved in Dy^{3+}, Yb^{3+}-codoped zeolites through a two-step energy transfer process from the $^4F_{9/2}$ (Dy^{3+}) level to two neighboring Yb^{3+} ions.

21.6.5 Yb^{3+} Ions Doped All Inorganic Perovskite Quantum Cutting Phosphors

Despite the quantum cutting materials based on $RE^{3+}-Yb^{3+}$ couples displaying effective photon down-conversion ability, they are still far from practical application due to the low PLQY. A main factor for the low conversion efficiency is the small absorption cross section (typically on the order of 10^{21} cm^{-2}) of the RE ions. Therefore, the quantum cutting efficiency is estimated in most cases; the actual application in SSCs is not ideal. An appropriate host material should meet the following conditions: (1) a large absorption cross section; (2) low defect density; (3) low phonon energy.

TABLE 21.2

Synthesis Method and PLQY of CsPbX$_3$:Yb^{3+} Quantum Cutting NCs

Perovskite Material	Synthesis Method	Applications	PLQY (%)	Ref.
CsPbCl$_3$:Yb^{3+} NCs	Hot-injection	–	143	[126]
CsPbCl$_{1.5}$Br$_{1.5}$:Yb^{3+}, Ce^{3+} NCs	Hot-injection	Silicon solar cells	146	[127]
CsPbClBr$_2$:Yb^{3+}–Pr^{3+}–Ce^{3+} NCs	Hot-injection	CIGS solar cells/silicon solar cells	173	[133]
CsPbCl$_3$:Yb^{3+} NCs	Hot-injection	–	170	[134]
CsPb(Cl$_{1-x}$Br$_x$)$_3$:Yb^{3+} film	Two-step solution-deposition method	–	193	[131]
CsPb(Cl$_{1-x}$Br$_x$)$_3$:Yb^{3+} film	Scalable single-source vapor deposition	–	183	[132]
CsPbX$_3$:Yb^{3+} NCs (X=Cl, Br, I) NCs	Postsynthesis strategy	–	–	[130]

In 2017, Song's group observed the luminescence of RE ions in the RE^{3+}-doped CsPbX$_3$ perovskite.[126] The energy transfer from CsPbX$_3$ perovskite host to RE^{3+} solves the issue about the low photon conversion efficiency due to the large absorption cross section of perovskite host. Subsequently, the same group reported a meaningful work about applying Yb^{3+}-doped CsPbX$_3$ perovskite on Si SCs, which enhance the IPCE of Si SCs from 18.8 to 21.5%.[127] This work demonstrates the excellent optical properties of quantum cutting of Yb^{3+}, which has a great application prospect. Furthermore, other research groups also carried out studies on this issue, and this important finding has attracted extensive interest.[128,129]

Synthesis methods. The quantum cutting of CsPbX$_3$:Yb^{3+} NCs has been achieved in NCs and films; the synthesis methods are summarized in Table 21.2. Hot-injection method is the most commonly used method to synthesize NCs.[126] Since Song' group reported the synthesis of RE ions-doped perovskites by this method,[126] many researchers modified this method to incorporate Yb^{3+} more effectively. For example, Milstein et al. used metal acetates instead of metal halides due to the higher solubility of metal acetates in the long-chain organic solvents. However, the PLQY of the samples synthesized by this method is lower than those synthesized by other hot-injection method, but there is no systematic study has been conducted on the exact relation between the choice of the precursors and PLQY yet. Moreover, the successful incorporating Yb^{3+} into perovskite can be realized via ion exchange.[130]

Furthermore, Yb^{3+}-doped perovskite film has also been explored. The film is usually prepared by two-step spin coating.[131] In addition, single-source vapor deposition (SSVD) method was also developed to deposit the RE ions-doped perovskite film.[132]

Applications on the Si SCs. Quantum cutting material of Yb^{3+}-doped perovskite as a down-conversion layer has wide application prospect on Si SCs. This layer converts high-energy photons into multiple low-energy photons that can be harvested by the SCs. The photoelectric response of this designed device structure between 350 and 450 nm has been greatly improved, which corresponds to the absorption range of perovskite NCs. Thus, the down-conversion layer has two advantages for the SCs, (1) the photon multiplication by quantum cutting; (2) the down-conversion of the high-energy photons to an energy range where the EQE of the SC is high.[127,133]

The first work about the application of Yb^{3+}-doped perovskite on SSCs was reported by Song' group.[135] Figure 21.24(a) shows the UV–vis absorption and PL spectra of Yb^{3+}-doped CsPbCl$_x$Br$_{1-x}$ QDs, although the doping of Yb^{3+} suppressed the exciton emission intensity, the total emission intensity increased by about two times compared with undoped QDs. The optimal PLQY up to 146% for CsPbCl$_{1.5}$Br$_{1.5}$ codoped with Yb (7.1%) and Ce (2%), in which Ce^{3+} provides an intermediate energy level. In the presence of Ce^{3+}, electrons on the conduction will relax to the 5d state of Ce^{3+} and then transfer to the Yb^{3+} ions. Furthermore, the doped perovskite QDs were explored as a down-converter to enhance the performance of the SSCs due to their excellent quantum cutting properties and the exact spectral matching with SSCs (Figure 21.24(b)). The QDs were self-assembled

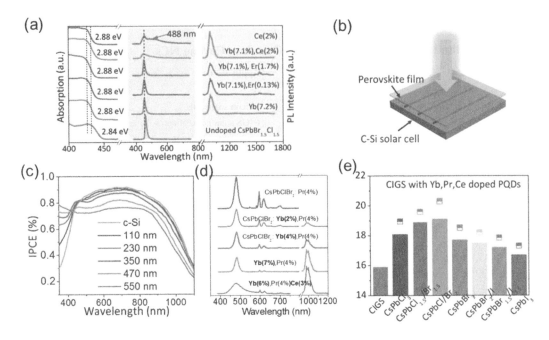

FIGURE 21.24 (a) Absorption spectra (left), visible emission spectra (middle) and near-infrared emission spectra (right, excited by 365 nm light) of $CsPbCl_{1.5}Br_{1.5}$ perovskite QDs codoped with different RE ions. (b) Schematic diagram of perovskite film to enhance the PCE of crystalline-silicon solar cells. (c) IPCE curves of SSCs coated with different thickness of perovskite films (adapted with permission from Ref. [135]. Copyright 2017, Wiley). (d) Comparison of emission spectra by varying the proportions of doping ions in $CsPbClBr_2$ QDs. (e) PCEs of CIGS solar cells coated with 230 nm $CsPbCl_xBr_yI_{3-x-y}$:Yb^{3+}, Pr^{3+}, Ce^{3+} perovskite films and the theoretical maximum PCEs induced by the perovskite films (adapted with permission from Ref. [133]. Copyright 2019, American Chemical Society).

on the surface of the commercial single-crystal SSCs via a liquid-phase deposition method. The IPCE curve (Figure 21.24(c)) can show that due to the effective down-conversion, the coating of perovskite QDs has greatly improved the photoelectric responses at 350–450 nm, and the extraordinary PCE enhancement from 18.1 to 21.5%. This work has opened a new door for promoting the optical properties of perovskite QDs and expanding the application of quantum cutting to SCs. Followed by this work, the same group reported another meaningful work.[133] To further improve the quantum cutting efficiency of Yb^{3+}, the codoped and tridoped methods were explored to improve the quantum cutting emission of QDs (Figure 21.24(d)). It was found that Yb^{3+}–Pr^{3+} as well as Yb^{3+}–Ce^{3+} pairs could effectively sensitize the emission of Yb^{3+} owing to Pr^{3+} and Ce^{3+} ions offering intermediate energy states close to the exciton transition energy of the QDs, a 173% PLQY were obtained in the Yb^{3+}–Pr^{3+}–Ce^{3+}-tridoped $CsPbClBr_2$. And the tridoped QDs were designed as the down-converter for $CuIn_{1-x}Ga_xSe_2$ (CIGS) as well as the SSCs, which leads to an enhancement of the PCE of as high as ~20% (Figure 21.24(e)). This enhancement of performance is the highest compared with SCs sensitized by other phosphors (Table 21.3). This work demonstrated efficient lanthanide-tridoped PQDs and provided a universal and effective way to enhance the performance of a class of photovoltaic SCs with a response in the NIR region, which will be of great significance in the development of impurity-doped perovskite QDs and the photovoltaic industry area. Prof. Daniel Gamelin, a chemist at the University of Washington in Seattle, highlighted this strategy to enhance solar harvest and says "for solar energy conversion, this combination of materials is almost exactly what you want". He predicted that topping a high-end silicon cell with the ytterbium perovskite should enable it to convert 32.2% of the energy it absorbs as sunlight into electricity, up from 27%—a 19.2% boost.

TABLE 21.3

Performance Enhancement of Solar Cells Induced by Luminescent Phosphors

Sample	Solar Cell	PCE (%)	Enhanced PCE (%)	Relative Enhanced Proportion (%)	Ref.
ZnSe quantum dots	Si nanowire solar cells	5.5	5.8	5.5	[136]
ZnS nanoparticles	Si nanotips solar cell	6.57	7.2	9.6	[137]
n-ZnO nanowire	Si solar cell	8.97	9.51	6.0	[138]
Eu-doped phosphor	Si solar cell	12.56	13.86	10.4	[139]
YBO$_3$:Ce, Yb	Si solar cell	13.550	14.071	3.8	[140]
Silicon quantum dot	Si solar cell	17.2	17.5	1.7	[141]
Graphene quantum dot	Silicon heterojunction solar cells	14.77	16.55	12	[10]
Silicon quantum dot	Si solar cell	18	18.8	4.4	[142]
Yb/Ce codoped perovskite quantum dots	Si solar cell	18.1	21.5	18.8	[127]
Yb/Pr/Ce codoped perovskite quantum dots	CIGSe solar cell	15.9	19.1	20.1	[133]

Discussion. In a very short and limited time, the research team has made amazing progress in the study of quantum cutting. Especially, the application of quantum cutting to boost the efficiency of SCs seems around the corner. In the near future, it will be proved whether quantum cutting can increase the efficiency of SCs on a large scale in the industrial area.

Yb^{3+}-doped $CsPbX_3$ NCs (X = Cl, Br) and thin films show very high PLQYs with a current record of 193%. The production of these high PLQYs is attributed to the quantum cutting of Yb^{3+}, during which one high-energy excitation photon is converted into two low-energy photons. The excited emission energy of Yb^{3+} $^2F_{5/2}$–$^2F_{7/2}$ transitions (1.26 eV) is suitable for SSCs. This perovskite material opens a new opportunity for application in down-conversion, in which a quantum cutting layer is deposited on the top of SCs, shaping the solar spectra to be more suitable for the bandgap of the SC.

However, when the $CsPbX_3$ NCs is exposed to ambient atmospheric conditions, fast decrease shown in PLQY results from the ionic character of the material and the metastable structure. These instabilities of oxygen water and UV-light have a negative effect on lead halide perovskites, especially for mixed halide perovskites. If the instability problem can be solved, such as encapsulation into glass, the quantum cutting of Yb^{3+}-doped perovskite will be further away from industrialization.

21.6.6 PERSPECTIVES AND CHALLENGES

Although the quantum cutting applied to SCs have been widely studied, the single doped and codoped phosphor materials are still far from practical application due to the low excitation efficiency. The main limiting factor for conversion efficiency is the small absorption cross section (usually on the order of 10^{-21} cm^2) of RE ions resulted from 4f–4f parity-forbidden transitions.[108] Recently, many researchers paid attention to of the dipole-allowed 4f–5d transitions of RE ions (such as Ce^{3+}, Eu^{2+} and Yb^{2+}), which have much higher absorption cross sections of up to 10^{-18} cm^2. And these RE ions codoped with Yb^{3+} could generate the effective NIR-quantum cutting and are more suitable as the down-conversion layer for SCs.

Moreover, the selection of host material is also essential for quantum cutting efficiency, which will further affect the PCE of SCs. In the past, most researchers chose glass as the host material. The first work about quantum cutting in glass host was reported by Park and coworkers,[143] they prepared Na_2O–CaO–SiO_2 glasses doped with Tm^{3+} and Yb^{3+}, and the efficiency of SCs with doped glasses shows a real efficiency enhancement. Theoretically speaking, the NIR quantum cutting luminescent material of RE^{3+} can be used in SCs and improves their conversion efficiency. But realizing the perfect combination of high-efficiency NIR quantum cutting materials and SCs is also an urgent problem in this field. With the development, the emergence of halide perovskite provides a good host for RE^{3+} ions, such as all-inorganic lead halide perovskite, Sn-based halide perovskite and double perovskite etc., which can transfer energy to RE^{3+} via Förster resonance. The excellent and controllable optical and electric properties of halide perovskite make them suitable for SCs. Since Song' group reported the work about applying Yb^{3+}-doped $CsPb(ClBr)_3$ NCs on SSCs,[127] more and more researchers have studied the role of quantum cutting of RE^{3+}-doped halide perovskite on SCs and have given them high praise. A report in Science in 2019 highlighted the Yb^{3+}-doped perovskite quantum cutting phosphors. "This is one of the most exciting results I've seen in a long time", says Michael McGehee, a perovskite expert at Stanford University in Palo Alto, California. The boost in efficiency is very significant for SCs, which makes Yb^{3+}-doped perovskite phosphors the most promising applicable materials in the future solar industry.

21.7 CONCLUSION AND OUTLOOK

In this chapter, fundamentals for luminescent materials as spectral converters are presented in the context of enhancing SC efficiency. The main challenge of designing SCs is to minimize energy losses due to the spectral mismatch between the SC and incident solar spectrum. To make full use of solar energy, upconversion, downshifting and quantum cutting phosphors have shown great promise as spectral converters to reduce the spectral mismatch and enhance the performance of SCs.

Up-conversion phosphors can convert the sub-bandgap light into light with higher energy, which can be harvested by SCs. However, the development of upconverter on SCs in the future still encounters many challenges, such as the narrow absorption band, the high excitation power threshold and the low up-conversion quantum yield. Several efforts have been paid to overcome these problems, such as introducing plasmonic materials to enhance the up-conversion efficiency and concentrating the solar energy. However, these approaches will lead to a complicated and expensive manufacture process of SC.

Downshifting phosphors enables SCs the ability to harvest the short-wavelength light. Lanthanide-doped phosphors and glasses, semiconductor QDs, dye and organolanthanide complexes can be used in LSCs and planar downshifting layers for enhancing PCE of SCs. However, the fabrication of a highly efficient LSC remains complex, and the optimal configuration may involve a combination or design layout of different luminescent materials.

The quantum cutting phosphors had been considered the effective converters and offered the ability to utilize the high-energy above-bandgap sunlight, which provided potential for efficiency enhancement of narrow bandgap SCs. The quantum cutting phosphors can achieve a quantum yield close to 200%, which has been studied for several decades. Although phosphors with nearly 200% theoretical quantum yield have been reported, the actual quantum yield measured in the experiment was far from 200% theoretical quantum yield. Nowadays, this predicament has been broken since the discovery of quantum cutting phosphors Yb^{3+}-doped perovskite QDs, which achieved a measured quantum yield of 180%. The Yb^{3+}-doped perovskite QD is a nearly-perfect quantum cutting phosphors with large absorption cross section to UV light and efficient emission at 1000 nm, which brought about 20% PCE enhancement for SSCs. The Yb^{3+}-doped perovskite QD is one of the most promising materials for the practical applications in solar industry area. Several approaches should be investigated toward industrial manufacture, such as the developing easy manufacture technique, film preparation and long-time stability.

REFERENCES

1. Huang, X.; Han, S.; Huang, W.; Liu, X., Enhancing solar cell efficiency: the search for luminescent materials as spectral converters. *Chem. Soc. Rev.*, *42*, 173–201, 2013.
2. Scholes, G. D.; Fleming, G. R.; Olaya-Castro, A.; van Grondelle, R., Lessons from nature about solar light harvesting. *Nat. Chem.*, *3*, 763–774, 2011.
3. Lewis, N. S., Toward cost-effective solar energy use. *Science*, *315*, 798–801, 2007.

4. de la Mora, M. B.; Amelines-Sarria, O.; Monroy, B. M.; Hernández-Pérez, C. D.; Lugo, J. E., Materials for downconversion in solar cells: perspectives and challenges. *Sol. Energy Mater. Sol. Cells*, *165*, 59–71, 2017.

5. Cotter, J. E.; Guo, J. H.; Cousins, P. J.; Abbott, M. D.; Chen, F. W.; Fisher, K. C., P-type versus n-type silicon wafers: prospects for high-efficiency commercial silicon solar cells. *IEEE Trans. Electron Devices*, *53*, 1893–1901, 2006.

6. Green, M. A., The passivated emitter and rear cell (PERC): from conception to mass production. *Energy Mater. Sol. Cells*, *143*, 190–197, 2015.

7. Bahabry, R. R.; Kutbee, A. T.; Khan, S. M.; Sepulveda, A. C.; Wicaksono, I.; Nour, M.; Wehbe, N.; Almislem, A. S.; Ghoneim, M. T.; Torres Sevilla, G. A.; Syed, A.; Shaikh, S. F.; Hussain, M. M., Solar cells: corrugation architecture enabled ultraflexible wafer-scale high-efficiency monocrystalline silicon solar cell. *Adv. Energy Mater.*, *8*, 1870055, 2018.

8. Liu, W.; Zhang, L.; Yang, X.; Shi, J.; Yan, L.; Xu, L.; Wu, Z.; Chen, R.; Peng, J.; Kang, J.; Wang, K.; Meng, F.; De Wolf, S.; Liu, Z., Damp-heat-stable, high-efficiency, industrial-size silicon heterojunction solar cells. *Joule*, *4*, 913–927, 2020.

9. Xue, M.; Nazif, K. N.; Lyu, Z.; Jiang, J.; Lu, C.-Y.; Lee, N.; Zang, K.; Chen, Y.; Zheng, T.; Kamins, T. I.; Brongersma, M. L.; Saraswat, K. C.; Harris, J. S., Free-standing 2.7 μm thick ultrathin crystalline silicon solar cell with efficiency above 12.0%. *Nano Energy*, *70*, 104466, 2020.

10. Tsai, M.-L.; Tu, W.-C.; Tang, L.; Wei, T.-C.; Wei, W.-R.; Lau, S. P.; Chen, L.-J.; He, J.-H., Efficiency enhancement of silicon heterojunction solar cells via photon management using graphene quantum dot as downconverters. *Nano Lett.*, *16*, 309–313, 2016.

11. Repins, I.; Contreras, M. A.; Egaas, B.; DeHart, C.; Scharf, J.; Perkins, C. L.; To, B.; Noufi, R., 19·9%-efficient ZnO/CdS/CuInGaSe$_2$ solar cell with 81·2% fill factor. *Prog. Photovolt: Res. Appl.*, *16*, 235–239, 2008.

12. Masuko, K.; Shigematsu, M.; Hashiguchi, T.; Fujishima, D.; Kai, M.; Yoshimura, N.; Yamaguchi, T.; Ichihashi, Y.; Mishima, T.; Matsubara, N.; Yamanishi, T.; Takahama, T.; Taguchi, M.; Maruyama, E.; Okamoto, S., Achievement of more than 25% conversion efficiency with crystalline silicon heterojunction solar cell. *IEEE J. Photovolt.*, *4*, 1433–1435, 2014.

13. Temple-Boyer, P.; Scheid, E.; Faugere, G.; Rousset, B., Residual stress in silicon films deposited by LPCVD from disilane. *Thin Solid Films*, *310*, 234–237, 1997.

14. Pagliaro, M.; Ciriminna, R.; Palmisano, G., Flexible solar cells. *ChemSusChem*, *1*, 880–891, 2008.

15. Tennyson, E. M.; Frantz, J. A.; Howard, J. M.; Gunnarsson, W. B.; Myers, J. D.; Bekele, R. Y.; Sanghera, J. S.; Na, S.-M.; Leite, M. S., Photovoltage tomography in polycrystalline solar cells. *ACS Energy Lett.*, *1*, 899–905, 2016.

16. Saifullah, M.; Ahn, S.; Gwak, J.; Ahn, S.; Kim, K.; Cho, J.; Park, J. H.; Eo, Y. J.; Cho, A.; Yoo, J.-S.; Yun, J. H., Development of semitransparent CIGS thin-film solar cells modified with a sulfurized-AgGa layer for building applications. *J. Mater. Chem. A*, *4*, 10542–10551, 2016.

17. Major, J. D.; Al Turkestani, M.; Bowen, L.; Brossard, M.; Li, C.; Lagoudakis, P.; Pennycook, S. J.; Phillips, L. J.; Treharne, R. E.; Durose, K., In-depth analysis of chloride treatments for thin-film CdTe solar cells. *Nat. Commun.*, *7*, 13231, 2016.

18. Han, G.; Zhang, S.; Boix, P. P.; Wong, L. H.; Sun, L.; Lien, S.-Y., Towards high efficiency thin film solar cells. *Prog. Mater. Sci.*, *87*, 246–291, 2017.

19. Cole, J. M.; Pepe, G.; Al Bahri, O. K.; Cooper, C. B., Cosensitization in dye-sensitized solar cells. *Chem. Rev.*, *119*, 7279–7327, 2019.

20. Yoo, J. J.; Seo, G.; Chua, M. R.; Park, T. G.; Lu, Y. L.; Rotermund, F.; Kim, Y. K.; Moon, C. S.; Jeon, N. J.; Correa-Baena, J. P.; Bulovic, V.; Shin, S. S.; Bawendi, M. G.; Seo, J., Efficient perovskite solar cells via improved carrier management. *Nature*, *590*, 587–593, 2021.

21. Richards, B. S., Enhancing the performance of silicon solar cells via the application of passive luminescence conversion layers. *Sol. Energy Mater. Sol. Cells*, *90*, 2329–2337, 2006.

22. Shockley, W.; Queisser, H. J., Detailed balance limit of efficiency of p-n junction solar cells. *J. Appl. Phys.*, *32*, 510–519, 1961.

23. Strümpel, C.; McCann, M.; Beaucarne, G.; Arkhipov, V.; Slaoui, A.; Švrček, V.; del Cañizo, C.; Tobias, I., Modifying the solar spectrum to enhance silicon solar cell efficiency—an overview of available materials. *Sol. Energy Mater. Sol. Cells*, *91*, 238–249, 2007.

24. Conibeer, G., Third-generation photovoltaics. *Mater. Today*, *10*, 42–50, 2007.

25. Eliseeva, S. V.; Bünzli, J.-C. G., Lanthanide luminescence for functional materials and bio-sciences. *Chem. Soc. Rev.*, *39*, 189–227, 2010.

26. Chen, D.; Wang, Y.; Hong, M., Lanthanide nanomaterials with photon management characteristics for photovoltaic application. *Nano Energy*, *1*, 73–90, 2012.

27. Ferro, S. M.; Wobben, M.; Ehrler, B., Rare-earth quantum cutting in metal halide perovskites—a review. *Mater. Horiz.* 2021. DOI:10.1039/D0MH01470B.

28. Ho, W.-J.; Lin, W.-C.; Liu, J.-J.; Syu, H.-J.; Lin, C.-F., Enhancing the performance of textured silicon solar cells by combining up-conversion with plasmonic scattering. *Energies*, *12*, 4119, 2019.

29. Tao, R.; Fang, W.; Li, F.; Sun, Z.; Xu, L., Lanthanide-containing polyoxometalate as luminescent down-conversion material for improved printable perovskite solar cells. *J. Alloys Compd.*, *823*, 153738, 2020.

30. Trupke, T.; Shalav, A.; Richards, B. S.; Würfel, P.; Green, M. A., Efficiency enhancement of solar cells by luminescent up-conversion of sunlight. *Sol. Energy Mater. Sol. Cells*, *90*, 3327–3338, 2006.

31. Yang, Z.; Lin, H.; Chee, K. W. A.; Gao, P.; Ye, J., The role of front-surface charges in interdigitated back contact silicon heterojunction solar cells. *Nano Energy*, *61*, 221–227, 2019.

32. Liu, Y.; Li, Y.; Wu, Y.; Yang, G.; Mazzarella, L.; Procel-Moya, P.; Tamboli, A. C.; Weber, K.; Boccard, M.; Isabella, O.; Yang, X.; Sun, B., High-efficiency silicon heterojunction solar cells: materials, devices and applications. *Mater. Sci. Eng. R Rep.*, *142*, 100579, 2020.

33. Kim, J. Y.; Lee, J.-W.; Jung, H. S.; Shin, H.; Park, N.-G., High-efficiency perovskite solar cells. *Chem. Rev.*, *120*, 7867–7918, 2020.

34. Richhariya, G.; Kumar, A.; Tekasakul, P.; Gupta, B., Natural dyes for dye sensitized solar cell: a review. *Renew. Sust. Energ. Rev.*, *69*, 705–718, 2017.

35. Kumavat, P. P.; Sonar, P.; Dalal, D. S., An overview on basics of organic and dye sensitized solar cells, their mechanism and recent improvements. *Renew. Sust. Energ. Rev.*, *78*, 1262–1287, 2017.

36. Li, Z. Q.; Li, X. D.; Liu, Q. Q.; Chen, X. H.; Sun, Z.; Liu, C.; Ye, X. J.; Huang, S. M., Core/shell structured $NaYF_4$:Yb^{3+}/Er^{3+}/Gd^{+3} nanorods with Au nanoparticles or shells for flexible amorphous silicon solar cells. *Nanotechnology*, *23*, 025402, 2011.

37. Auzel, F., Upconversion and anti-stokes processes with f and d ions in solids. *Chem. Rev.*, *104*, 139–174, 2004.

38. Kshetri, Y. K.; Hoon, J. S.; Kim, T.-H.; Sekino, T.; Lee, S. W., Yb^{3+}, Er^{3+} and Tm^{3+} doped α-Sialon as upconversion phosphor. *J. Lumin.*, *204*, 485–492, 2018.

39. Gibart, P.; Auzel, F.; Guillaume, J.-C.; Zahraman, K., Below band-gap IR response of substrate-free GaAs solar cells using two-photon up-conversion. *Jpn. J. Appl. Phys.*, *35*, 4401–4402, 1996.

40. Arnaoutakis, G. E.; Marques-Hueso, J.; Ivaturi, A.; Fischer, S.; Goldschmidt, J. C.; Krämer, K. W.; Richards, B. S., Enhanced energy conversion of up-conversion solar cells by the integration of compound parabolic concentrating optics. *Sol. Energy Mater. Sol. Cells*, *140*, 217–223, 2015.

41. Rüdiger, M.; Fischer, S.; Frank, J.; Ivaturi, A.; Richards, B. S.; Krämer, K. W.; Hermle, M.; Goldschmidt, J. C., Bifacial n-type silicon solar cells for upconversion applications. *Sol. Energy Mater. Sol. Cells*, *128*, 57–68, 2014.

42. Fischer, S.; Fröhlich, B.; Steinkemper, H.; Krämer, K. W.; Goldschmidt, J. C., Absolute upconversion quantum yield of β-$NaYF_4$ doped with Er^{3+} and external quantum efficiency of upconverter solar cell devices under broad-band excitation considering spectral mismatch corrections. *Sol. Energy Mater. Sol. Cells*, *122*, 197–207, 2014.

43. Fischer, S.; Favilla, E.; Tonelli, M.; Goldschmidt, J. C., Record efficient upconverter solar cell devices with optimized bifacial silicon solar cells and monocrystalline BaY2F8:30% Er^{3+} upconverter. *Sol. Energy Mater. Sol. Cells*, *136*, 127–134, 2015.

44. Lahoz, F., Ho^{3+}-doped nanophase glass ceramics for efficiency enhancement in silicon solar cells. *Opt. Lett.*, *33*, 2982–2984, 2008.

45. Lahoz, F.; Pérez-Rodríguez, C.; Hernández, S. E.; Martín, I. R.; Lavín, V.; Rodríguez-Mendoza, U. R., Upconversion mechanisms in rare-earth doped glasses to improve the efficiency of silicon solar cells. *Sol. Energy Mater. Sol. Cells*, *95*, 1671–1677, 2011.

46. Lian, H. Z.; Hou, Z. Y.; Shang, M. M.; Geng, D. L.; Zhang, Y.; Lin, J., Rare earth ions doped phosphors for improving efficiencies of solar cells. *Energy*, *57*, 270–283, 2013.

47. Li, H.; Chen, C.; Jin, J.; Bi, W.; Zhang, B.; Chen, X.; Xu, L.; Liu, D.; Dai, Q.; Song, H., Near-infrared and ultraviolet to visible photon conversion for full spectrum response perovskite solar cells. *Nano Energy*, *50*, 699–709, 2018.

48. Zhou, D.; Liu, D.; Jin, J.; Chen, X.; Xu, W.; Yin, Z.; Pan, G.; Li, D.; Song, H., Semiconductor plasmon-sensitized broadband upconversion and its enhancement effect on the power conversion efficiency of perovskite solar cells. *J. Mater. Chem. A*, *5*, 16559–16567, 2017.

49. Chen, D.; Lei, L.; Yang, A.; Wang, Z.; Wang, Y., Ultra-broadband near-infrared excitable upconversion core/shell nanocrystals. *Chem. Commun.*, *48*, 5898–5900, 2012.

50. Qu, B.; Jiao, Y.; He, S.; Zhu, Y.; Liu, P.; Sun, J.; Lu, J.; Zhang, X., Improved performance of a-Si:H solar cell by using up-conversion phosphors. *J. Alloys Compd.*, *658*, 848–853, 2016.

51. Zhang, X. D.; Jin, X.; Wang, D. F.; Xiong, S. Z.; Geng, X. H.; Zhao, Y., Synthesis of NaYF4: Yb, Er nanocrystals and its application in silicon thin film solar cells. *Phys. Status Solidi C*, *7*, 1128–1131, 2010.

52. Chen, Y.; He, W.; Jiao, Y.; Wang, H.; Hao, X.; Lu, J.; Yang, S.-E., β-NaYF$_4$:Er^{3+} (10%) microprisms for the enhancement of a-Si:H solar cell near-infrared responses. *J. Lumin.*, *132*, 2247–2250, 2012.

53. Chang, J.; Ning, Y.; Wu, S.; Niu, W.; Zhang, S., Effectively utilizing NIR light using direct electron injection from up-conversion nanoparticles to the TiO$_2$ photoanode in dye-sensitized solar cells. *Adv. Funct. Mater.*, *23*, 5910–5915, 2013.

54. Wang, Y.; Zhu, Y.; Yang, X.; Shen, J.; Li, X.; Qian, S.; Li, C., Performance optimization in dye-sensitized solar cells with β-NaYF$_4$:Yb^{3+},Er^{3+}@SiO$_2$@TiO$_2$ mesoporous microspheres as multi-functional photoanodes. *Electrochim. Acta*, *211*, 92–100, 2016.

55. Zhao, P.; Zhu, Y.; Yang, X.; Jiang, X.; Shen, J.; Li, C., Plasmon-enhanced efficient dye-sensitized solar cells using core–shell-structured β-NaYF$_4$:Yb,Er@SiO$_2$@Au nanocomposites. *J. Mater. Chem. A*, *2*, 16523–16530, 2014.

56. Cai, W.; Zhang, Z.; Jin, Y.; Lv, Y.; Wang, L.; Chen, K.; Zhou, X., Application of TiO$_2$ hollow microspheres incorporated with up-conversion NaYF$_4$:Yb^{3+}, Er^{3+} nanoparticles and commercial available carbon counter electrodes in dye-sensitized solar cells. *Sol. Energy*, *188*, 441–449, 2019.

57. Shan, G.-B.; Assaaoudi, H.; Demopoulos, G. P., Enhanced performance of dye-sensitized solar cells by utilization of an external, bifunctional layer consisting of uniform β-NaYF4:Er^{3+}/Yb^{3+} nanoplatelets. *ACS Appl. Mater. Interfaces*, *3*, 3239–3243, 2011.

58. Kojima, A.; Teshima, K.; Shirai, Y.; Miyasaka, T., Organometal halide perovskites as visible-light sensitizers for photovoltaic cells. *J. Am. Ceram. Soc.*, *131* (17), 6050–6051, 2009.

59. Meng, F.-L.; Wu, J.-J.; Zhao, E.-F.; Zheng, Y.-Z.; Huang, M.-L.; Dai, L.-M.; Tao, X.; Chen, J.-F., High-efficiency near-infrared enabled planar perovskite solar cells by embedding upconversion nanocrystals. *Nanoscale*, *9* (46), 18535–18545, 2017.

60. Xu, F.; Sun, Y.; Gao, H.; Jin, S.; Zhang, Z.; Zhang, H.; Pan, G.; Kang, M.; Ma, X.; Mao, Y., High-performance perovskite solar cells based on NaCsWO$_3$@NaYF$_4$@NaYF4:Yb,Er upconversion nanoparticles. *ACS Appl. Mater. Interfaces*, *13*, 2674–2684, 2021.

61. Liang, L.; Liu, M.; Jin, Z.; Wang, Q.; Wang, H.; Bian, H.; Shi, F.; Liu, S., Optical management with nanoparticles for a light conversion efficiency enhancement in inorganic γ-CsPbI$_3$ solar cells. *Nano Lett.*, *19*, 1796–1804, 2019.

62. Jin, X.; Li, H.; Chen, Z.; Zhang, Q.; Li, F.; Sun, W.; Li, D.; Li, Q., Sodium gadolinium fluoride nanophosphor-based solar cells: toward subbandgap light harvesting and efficient charge transfer. *IEEE. J. Photovolt.*, *7*, 199–205, 2017.

63. Chen, W.; Hou, Y.; Osvet, A.; Guo, F.; Kubis, P.; Batentschuk, M.; Winter, B.; Spiecker, E.; Forberich, K.; Brabec, C. J., Sub-bandgap photon harvesting for organic solar cells via integrating up-conversion nanophosphors. *Org. Electron.*, *19*, 113–119, 2015.

64. Maalej, O.; Merigeon, J.; Boulard, B.; Girtan, M., Visible to near-infrared down-shifting in Tm^{3+} doped fluoride glasses for solar cells efficiency enhancement. *Optic. Mater.*, *60*, 235–239, 2016.

65. Wu, J.; Wang, J.; Lin, J.; Xiao, Y.; Yue, G.; Huang, M.; Lan, Z.; Huang, Y.; Fan, L.; Yin, S.; Sato, T., Dual functions of YF3:Eu^{3+} for improving photovoltaic performance of dye-sensitized solar cells. *Sci. Rep.*, *3*, 2058, 2013.

66. Ho, W.-J.; Chen, J.-C.; Liu, J.-J.; Ho, C.-H., Enhancing luminescent down-shifting of Eu-doped phosphors by incorporating plasmonic silver nanoparticles for silicon solar cells. *Appl. Surf. Sci.*, *532*, 147434, 2020.

67. Ho, W.-J.; Bai, W.-B.; Liu, J.-J.; Shiao, H.-P., Efficiency enhancement of single-junction GaAs solar cells coated with europium-doped silicate-phosphor luminescent-down-shifting layer. *Thin Solid Films*, *660*, 651–656, 2018.

68. Hosseini, Z.; Huang, W.-K.; Tsai, C.-M.; Chen, T.-M.; Taghavinia, N.; Diau, E. W.-G., Enhanced light harvesting with a reflective luminescent down-shifting layer for dye-sensitized solar cells. *ACS Appl. Mater. Interfaces*, *5*, 5397–5402, 2013.

69. Rajeswari, R.; Susmitha, K.; Jayasankar, C. K.; Raghavender, M.; Giribabu, L., Enhanced light harvesting with novel photon upconverted Y$_2$CaZnO$_5$:Er^{3+}/Yb^{3+} nanophosphors for dye sensitized solar cells. *Sol. Energy*, *157*, 956–965, 2017.

70. Cheng, Z.; Su, F.; Pan, L.; Cao, M.; Sun, Z., CdS quantum dot-embedded silica film as luminescent down-shifting layer for crystalline Si solar cells. *J. Alloys Compd.*, *494*, L7–L10, 2010.

71. Baek, S.-W.; Shim, J.-H.; Park, J.-G., The energy-down-shift effect of Cd$_{0.5}$Zn$_{0.5}$S–ZnS core–shell quantum dots on power-conversion-efficiency enhancement in silicon solar cells. *Phys. Chem. Chem. Phys.*, *16*, 18205–18210, 2014.

72. Tang, Q.; Wang, M.; Wang, Z.; Sun, W.; Shang, R., A long persistence phosphor tailored quasi-solid-state dye-sensitized solar cell that generates electricity in sunny and dark weathers. *Chem. Commun.*, *53*, 4815–4817, 2017.

73. Danos, L.; Parel, T.; Markvart, T.; Barrioz, V.; Brooks, W. S. M.; Irvine, S. J. C., Increased efficiencies on CdTe solar cells via luminescence down-shifting with excitation energy transfer between dyes. *Sol. Energy Mater. Sol. Cell*, *98*, 486–490, 2012.

74. Dexter, D. L., Possibility of luminescent quantum yields greater than unity. *Phys. Rev.*, *108*, 630–633, 1957.

75. Sommerdijk, J. L.; Bril, A.; de Jager, A. W., Two photon luminescence with ultraviolet excitation of trivalent praseodymium. *J. Lumin.*, *8*, 341–343,1974.

76. Piper, W. W.; DeLuca, J. A.; Ham, F. S., Cascade fluorescent decay in Pr³⁺-doped fluorides: achievement of a quantum yield greater than unity for emission of visible light. *J. Lumin.*, *8*, 344–348, 1974.

77. Zhang, Q. Y.; Huang, X. Y., Recent progress in quantum cutting phosphors. *Prog. Mater. Sci.*, *55*, 353–427, 2010.

78. Wegh, R. T.; Donker, H.; Meijerink, A.; Lamminmäki, R. J.; Hölsä, J., Vacuum-ultraviolet spectroscopy and quantum cutting for Gd³⁺ in LiYF₄. *Phys. Rev. B*, *56*, 13841–13848, 1997.

79. Rodnyi, P. A.; Mikhrin, S. B.; Dorenbos, P.; van der Kolk, E.; van Eijk, C. W. E.; Vink, A. P.; Avanesov, A. G., The observation of photon cascade emission in Pr³⁺-doped compounds under X-ray excitation. *Opt. Commun.*, *204*, 237–245, 2002.

80. Yang, Z.; Lin, J. H.; Su, M. Z.; Tao, Y.; Wang, W., Photon cascade luminescence of Gd³⁺ in GdBaB₉O₁₆. *J. Alloys Compd.*, *308*, 94–97, 2000.

81. Wegh, R. T.; Donker, H.; Oskam, K. D.; Meijerink, A., Visible quantum cutting in Eu³⁺-doped gadolinium fluorides via downconversion. *J. Lumin.*, *82*, 93–104, 1999.

82. You, F.; Wang, Y.; Lin, J.; Tao, Y., Hydrothermal synthesis and luminescence properties of NaGdF₄:Eu. *J. Alloys Compd.*, *343*, 151–155, 2002.

83. Liu, B.; Chen, Y.; Shi, C.; Tang, H.; Tao, Y., Visible quantum cutting in BaF₂:Gd,Eu via downconversion. *J. Lumin.*, *101*, 155–159, 2003.

84. Hou, D.; Liang, H.; Xie, M.; Ding, X.; Zhong, J.; Su, Q.; Tao, Y.; Huang, Y.; Gao, Z., Bright green-emitting, energy transfer and quantum cutting of Ba₃Ln(PO₄)₃: Tb³⁺ (Ln = La, Gd) under VUV-UV excitation. *Opt. Express*, *19*, 11071–11083, 2011.

85. Han, B.; Liang, H.; Huang, Y.; Tao, Y.; Su, Q., Vacuum ultraviolet–visible spectroscopic properties of Tb³⁺ in Li(Y, Gd)(PO₃)₄: tunable emission, quantum cutting, and energy transfer. *J. Phys. Chem. C*, *114*, 6770–6777, 2010.

86. Xie, M.; Tao, Y.; Huang, Y.; Liang, H.; Su, Q., The quantum cutting of Tb³⁺ in Ca₆Ln₂Na₂(PO₄)₆F₂ (Ln = Gd, La) under VUV–UV excitation: with and without Gd³⁺. *Inorg. Chem.*, *49*, 11317–11324, 2010.

87. Sommerdijk, J. L.; Bril, A.; de Jager, A. W., Luminescence of Pr³⁺-activated fluorides. *J. Lumin.*, *9*, 288–296, 1974.

88. Gong, J.; Li, C.; Wasielewski, M. R., Advances in solar energy conversion. *Chem. Soc. Rev.*, *48*, 1862–1864, 2019.

89. Tiedje, T.; Yablonovitch, E.; Cody, G. D.; Brooks, B. G., Limiting efficiency of silicon solar cells. *IEEE Trans. Electron Devices*, *31*, 711–716, 1984.

90. Zhang, Y.; Ye, L.; Hou, J., Precise characterization of performance metrics of organic solar cells. *Small Methods*, *1*, 1700159, 2017.

91. McMeekin, D. P.; Sadoughi, G.; Rehman, W.; Eperon, G. E.; Saliba, M.; Hörantner, M. T.; Haghighirad, A.; Sakai, N.; Korte, L.; Rech, B.; Johnston, M. B.; Herz, L. M.; Snaith, H. J., A mixed-cation lead mixed-halide perovskite absorber for tandem solar cells. *Science*, *351*, 151–155, 2016.

92. Trupke, T.; Green, M. A.; Würfel, P., Improving solar cell efficiencies by down-conversion of high-energy photons. *J. Appl. Phys.*, *92*, 1668–1674, 2002.

93. Li, Q.; Jin, X.; Yang, Y.; Wang, H.; Xu, H.; Cheng, Y.; Wei, T.; Qin, Y.; Luo, X.; Sun, W.; Luo, S., Nd₂(S, Se, Te)₃ colloidal quantum dots: synthesis, energy level alignment, charge transfer dynamics, and their applications to solar cells. *Adv. Funct. Mater.*, *26*, 254–266, 2016.

94. van der Ende, B. M.; Aarts, L.; Meijerink, A., Lanthanide ions as spectral converters for solar cells. *Phys. Chem. Chem. Phys.*, *11*, 11081–11095, 2009.

95. Avrutin, V.; Izyumskaya, N.; Morkoç, H., Semiconductor solar cells: recent progress in terrestrial applications. *Superlattices Microstruct.*, *49*, 337–364, 2011.

96. Tanabe, K., A review of ultrahigh efficiency III-V semiconductor compound solar cells: multijunction tandem, lower dimensional, photonic up/down conversion and plasmonic nanometallic structures. *Energies*, *2*, 504–530, 2009.

97. Wu, X.; Meng, F.; Zhang, Z.; Yu, Y.; Liu, X.; Meng, J., Broadband down-conversion for silicon solar cell by ZnSe/phosphor heterostructure. *Opt. Express*, 22, A735–A741, 2014.

98. Merigeon, J.; Maalej, O.; Boulard, B.; Stanculescu, A.; Leontie, L.; Mardare, D.; Girtan, M., Studies on Pr^{3+}–Yb^{3+} codoped ZBLA as rare earth down convertor glasses for solar cells encapsulation. *Opt. Mater.*, 48, 243–246, 2015.

99. Zhou, X.; Shen, J.; Wang, Y.; Feng, Z.; Wang, R.; Li, L.; Jiang, S.; Luo, X., An efficient dual-mode solar spectral modification for c-Si solar cells in Tm^{3+}/Yb^{3+} codoped tellurite glasses. *J. Am. Ceram. Soc.*, 99, 2300–2305, 2016.

100. Tai, Y.; Wang, H.; Wang, H.; Bai, J., Near-infrared down-conversion in Er^{3+}–Yb^{3+} co-doped transparent nanostructured glass ceramics for crystalline silicon solar cells. *RSC Adv.*, 6, 4085–4089, 2016.

101. Grätzel, M., Dye-sensitized solar cells. *J. Photochem. Photobiol. C*, 4, 145–153, 2003.

102. Kim, H. J.; Song, J. S.; Kim, S. S., Efficiency enhancement of solar cell by down-conversion effect of Eu^{3+} doped $LiGdF_4$. *J. Korean Phys. Soc.*, 4, 609–613, 2004.

103. Dai, W. B.; Lei, Y. F.; Li, P.; Xu, L. F., Enhancement of photovoltaic performance of TiO_2-based dye-sensitized solar cells by doping $Ca_3La_{3(1-x)}Eu_{3x}(BO_3)_5$. *J. Mater. Chem. A*, 3, 4875–4883, 2015.

104. Hafez, H.; Saif, M.; Abdel-Mottaleb, M. S. A., Down-converting lanthanide doped TiO_2 photoelectrodes for efficiency enhancement of dye-sensitized solar cells. *J. Power Sources*, 196, 5792–5796, 2011.

105. Yao, N.; Huang, J.; Fu, K.; Deng, X.; Ding, M.; Xu, X., Rare earth ion doped phosphors for dye-sensitized solar cells applications. *RSC Adv.*, 6, 17546–17559, 2016.

106. Shea, J. J., Handbook on the physics and chemistry of rare earths, vol. 56. *IEEE. Electr. Insul. M*, 36, 57–57, 2020.

107. Luo, X.; Ding, T.; Liu, X.; Liu, Y.; Wu, K., Quantum-cutting luminescent solar concentrators using ytterbium-doped perovskite nanocrystals. *Nano. Lett.*, 19, 338–341, 2019.

108. Cohen, T. A.; Milstein, T. J.; Kroupa, D. M.; MacKenzie, J. D.; Luscombe, C. K.; Gamelin, D. R., Quantum-cutting Yb^{3+}-doped perovskite nanocrystals for monolithic bilayer luminescent solar concentrators. *J. Mater. Chem. A*, 7, 9279–9288, 2019.

109. Vergeer, P.; Vlugt, T. J. H.; Kox, M. H. F.; den Hertog, M. I.; van der Eerden, J. P. J. M.; Meijerink, A., Quantum cutting by cooperative energy transfer in $Yb_xY_{1-x}PO_4$:Tb_{3+}. *Phys. Rev. B*, 71, 014119, 2005.

110. Stręk, W.; Bednarkiewicz, A.; Dereń, P. J., Power dependence of luminescence of Tb^{3+}-doped $KYb(WO_4)_2$ crystal. *J. Lumin.*, 92, 229–235, 2001.

111. Terra, I. A. A.; Borrero-González, L. J.; Figueredo, T. R.; Almeida, J. M. P.; Hernandes, A. C.; Nunes, L. A. O.; Malta, O. L., Down-conversion process in Tb^{3+}–Yb^{3+} co-doped Calibo glasses. *J. Lumin.*, 132, 1678–1682, 2012.

112. Ye, S.; Katayama, Y.; Tanabe, S., Down conversion luminescence of Tb^{3+}–Yb^{3+} codoped SrF_2 precipitated glass ceramics. *J. Non-Cryst. Solids*, 357, 2268–2271, 2011.

113. Zhang, Q. Y.; Yang, G. F.; Jiang, Z. H., Cooperative downconversion in $GdAl_3(BO_3)_4$:RE^{3+},Yb^{3+} (RE=Pr, Tb, and Tm). *Appl. Phys. Lett.*, 91, 051903, 2007.

114. Xie, L.; Wang, Y.; Zhang, H., Near-infrared quantum cutting in YPO_4:Yb^{3+}, Tm^{3+} via cooperative energy transfer. *Appl. Phys. Lett.*, 94, 061905, 2009.

115. Zhang, Q.; Zhu, B.; Zhuang, Y.; Chen, G.; Liu, X.; Zhang, G.; Qiu, J.; Chen, D., Quantum cutting in Tm^{3+}/Yb^{3+}-codoped lanthanum aluminum germanate glasses. *J. Am. Ceram. Soc.*, 93, 654–657, 2010.

116. Lakshminarayana, G.; Qiu, J., Near-infrared quantum cutting in RE^{3+}/Yb^{3+} (RE=Pr, Tb, and Tm): GeO_2–B_2O_3–ZnO–LaF_3 glasses via downconversion. *J. Alloys Compd.*, 48, 582–589, 2009.

117. Ye, S.; Zhu, B.; Luo, J.; Chen, J.; Lakshminarayana, G.; Qiu, J., Enhanced cooperative quantum cutting in Tm^{3+}-Yb^{3+} codoped glass ceramics containing LaF_3 nanocrystals. *Opt. Express*, 16, 8989–8994, 2008.

118. Lin, H.; Zhou, S.; Hou, X.; Li, W.; Li, Y.; Teng, H.; Jia, T., Down-conversion from blue to near infrared in Tm^{3+}–Yb^{3+} codoped Y_2O_3 transparent ceramics. *IEEE Photonics Technol. Lett.*, 22, 866–868, 2010.

119. Chen, X. P.; Huang, X. Y.; Zhang, Q. Y., Concentration-dependent near-infrared quantum cutting in $NaYF_4$:Pr^{3+}, Yb^{3+} phosphor. *J. Appl. Phys.*, 106, 063518, 2009.

120. Lin, H.; Yan, X.; Wang, X., Synthesis and blue to near-infrared quantum cutting of Pr^{3+}/Yb^{3+} co-doped Li_2TeO_4 phosphors. *Mater. Sci. Eng. B*, 176, 1537–1540, 2011.

121. Eilers, J. J.; Biner, D.; van Wijngaarden, J. T.; Krämer, K.; Güdel, H. U.; Meijerink, A., Efficient visible to infrared quantum cutting through downconversion with the Er^{3+}–Yb^{3+} couple in $Cs_3Y_2Br_9$. *Appl. Phys. Lett.*, 96, 151106, 2010.

122. Meijer, J.-M.; Aarts, L.; van der Ende, B. M.; Vlugt, T. J. H.; Meijerink, A., Downconversion for solar cells in YF_3:Nd^{3+},Yb^{3+}. *Phys. Rev. B*, 81, 035107, 2010.

123. Chen, D.; Yu, Y.; Lin, H.; Huang, P.; Shan, Z.; Wang, Y., Ultraviolet-blue to near-infrared downconversion of Nd^{3+}-Yb^{3+} couple. *Opt. Lett.*, 35, 220–222, 2010.

124. Lin, H.; Chen, D.; Yu, Y.; Yang, A.; Wang, Y., Near-infrared quantum cutting in Ho^{3+}/Yb^{3+} codoped nanostructured glass ceramic. *Opt. Lett.*, *36*, 876–878, 2011.

125. Bai, Z.; Fujii, M.; Hasegawa, T.; Imakita, K.; Mizuhata, M.; Hayashi, S., Efficient ultraviolet-blue to near-infrared downconversion in Bi-Dy-Yb-doped zeolites. *J. Phys. D Appl. Phys.*, *44*, 455301, 2011.

126. Pan, G.; Bai, X.; Yang, D.; Chen, X.; Jing, P.; Qu, S.; Zhang, L.; Zhou, D.; Zhu, J.; Xu, W.; Dong, B.; Song, H., Doping lanthanide into perovskite nanocrystals: highly improved and expanded optical properties. *Nano Lett.*, *17*, 8005–8011, 2017.

127. Milstein, T.; Kluherz, K.; Kroupa, D.; Erickson, C.; De Yoreo, J.; Gamelin, D., Anion exchange and the quantum-cutting energy threshold in ytterbium-doped CsPb(Cl$_{1-x}$Br$_x$)$_3$ perovskite nanocrystals. *Nano Lett.*, *19*, 1931–1937, 2019.

128. Crane, M. J.; Kroupa, D. M.; Gamelin, D. R., Detailed-balance analysis of Yb^{3+}:CsPb(Cl$_{1-x}$Br$_x$)$_3$ quantum-cutting layers for high-efficiency photovoltaics under real-world conditions. *Energy Environ. Sci.*, *12*, 2486–2495, 2019.

129. Erickson, C. S.; Crane, M. J.; Milstein, T. J.; Gamelin, D. R., Photoluminescence saturation in quantum-cutting Yb^{3+}-doped CsPb(Cl$_{1-x}$Br$_x$)$_3$ perovskite nanocrystals: implications for solar downconversion. *J. Phys. Chem. C*, *123*, 12474–12484, 2019.

130. Mir, W. J.; Mahor, Y.; Lohar, A.; Jagadeeswararao, M.; Das, S.; Mahamuni, S.; Nag, A., Postsynthesis doping of Mn and Yb into CsPbX$_3$ (X = Cl, Br, or I) perovskite nanocrystals for downconversion emission. *Chem. Mater.*, *30*, 8170–8178, 2018.

131. Kroupa, D. M.; Roh, J. Y.; Milstein, T. J.; Creutz, S. E.; Gamelin, D. R., Quantum-cutting ytterbium-doped CsPb(Cl$_{1-x}$Br$_x$)$_3$ perovskite thin films with photoluminescence quantum yields over 190%. *ACS Energy Lett.*, *3*, 2390–2395, 2018.

132. Crane, M. J.; Kroupa, D. M.; Roh, J. Y.; Anderson, R. T.; Smith, M. D.; Gamelin, D. R., Single-source vapor deposition of quantum-cutting Yb^{3+}:CsPb(Cl$_{1-x}$Br$_x$)$_3$ and other complex metal-halide perovskites. *ACS Appl. Energy Mater.*, *2*, 4560–4565, 2019.

133. Zhou, D.; Sun, R.; Xu, W.; Ding, N.; Li, D.; Chen, X.; Pan, G.; Bai, X.; Song, H., Impact of host composition, codoping, or tridoping on quantum-cutting emission of ytterbium in halide perovskite quantum dots and solar cell applications. *Nano Lett.*, *19*, 6904–6913, 2019.

134. Milstein, T. J.; Kroupa, D. M.; Gamelin, D. R., Picosecond quantum cutting generates photoluminescence quantum yields over 100% in ytterbium-doped CsPbCl$_3$ nanocrystals. *Nano Lett.*, *18*, 3792–3799, 2018.

135. Zhou, D. L.; Liu, D. L.; Pan, G. C.; Chen, X.; Li, D. Y.; Xu, W.; Bai, X.; Song, H. W., Cerium and ytterbium codoped halide perovskite quantum dots: a novel and efficient downconverter for improving the performance of silicon solar cells. *Adv. Mater.*, *29*, 1704149, 2017.

136. Jung, J.-Y.; Zhou, K.; Bang, J. H.; Lee, J.-H., Improved photovoltaic performance of Si nanowire solar cells integrated with ZnSe quantum dots. *J. Phys. Chem. C*, *116*, 12409–12414, 2012.

137. Huang, C.-Y.; Wang, D.-Y.; Wang, C.-H.; Chen, Y.-T.; Wang, Y.-T.; Jiang, Y.-T.; Yang, Y.-J.; Chen, C.-C.; Chen, Y.-F., Efficient light harvesting by photon downconversion and light trapping in hybrid ZnS nanoparticles/Si nanotips solar cells. *ACS Nano*, *4*, 5849–5854, 2010.

138. Zhu, L.; Wang, L.; Pan, C.; Chen, L.; Xue, F.; Chen, B.; Yang, L.; Su, L.; Wang, Z. L., Enhancing the efficiency of silicon-based solar cells by the piezo-phototronic effect. *ACS Nano*, *11*, 1894–1900, 2017.

139. Ho, W.-J.; Deng, Y.-J.; Liu, J.-J.; Feng, S.-K.; Lin, J.-C., Photovoltaic performance characterization of textured silicon solar cells using luminescent down-shifting Eu-doped phosphor particles of various dimensions. *Materials*, *10*, 21, 2017.

140. Hao, Y.; Wang, Y.; Hu, X.; Liu, X.; Liu, E.; Fan, J.; Miao, H.; Sun, Q., YBO$_3$: Ce^{3+}, Yb^{3+} based near-infrared quantum cutting phosphors: synthesis and application to solar cells. *Ceram. Int.*, *42*, 9396–9401, 2016.

141. Pi, X.; Zhang, L.; Yang, D., Enhancing the efficiency of multicrystalline silicon solar cells by the inkjet printing of silicon-quantum-dot ink. *J. Phys. Chem. C*, *116*, 21240–21243, 2012.

142. Sgrignuoli, F.; Ingenhoven, P.; Pucker, G.; Mihailetchi, V. D.; Froner, E.; Jestin, Y.; Moser, E.; Sànchez, G.; Pavesi, L., Purcell effect and luminescent downshifting in silicon nanocrystals coated back-contact solar cells. *Sol. Energy Mater. Sol. Cells*, *132*, 267–274, 2015.

143. Park, W. J.; Oh, S. J.; Kim, J. K.; Heo, J.; Wagner, T.; Strizik, L., Down-conversion in Tm^{3+}/Yb^{3+} doped glasses for multicrystalline silicon photo-voltaic module efficiency enhancement. *J. Non-Cryst. Solids*, *383*, 181–183, 2014.

Index

Printed and bound by CPI Group (UK) Ltd, Croydon, CR0 4YY

17/10/2024

01775701-0005